TRANSGENIC ANIMALS

TRANSGENIC ANIMALS
GENERATION AND USE

Edited by

Louis Marie Houdebine

Unité de Différenciation Cellulaire
Institut National de la Recherche Agronomique
France

harwood academic publishers
Australia • Canada • China • France • Germany • India
Japan • Luxembourg • Malaysia • The Netherlands
Russia • Singapore • Switzerland • Thailand • United Kingdom

Copyright © 1997 OPA (Overseas Publishers Association) Amsterdam B.V. Published in The Netherlands by Harwood Academic Publishers.

All rights reserved.

No part of this book may be reproduced or utilized in any form or by any means, electronic or mechanical, including photocopying and recording, or by any information storage or retrieval system, without permission in writing from the publisher. Printed in Singapore.

Amsteldijk 166
1st Floor
1079 LH Amsterdam
The Netherlands

British Cataloguing in Publication Data

Transgenic animals : generation and use
  1. Transgenic animals   2. Animal genetic engineering   3. Genetic transformation
  I. Houdebine, Louis Marie
575.1'0724

ISBN 90-5702-069-6 (soft cover)

CONTENTS

Preface — xiii

Contributors — xv

PART I OVERVIEW

1. **Overview** — 1
   Louis Marie Houdebine

PART II THE TECHNIQUES FOR GENE TRANSFER

SECTION A   GENE TRANSFER INTO MAMMAL EMBRYOS

2. **Transgenic Rats** — 7
   J.J. Mullins, G. Brooker and L.J. Mullins

3. **The Generation of Transgenic Rabbits** — 11
   C. Viglietta, M. Massoud and L.M. Houdebine

4. **Techniques and Problems in Producing Transgenic Pigs** — 15
   V.G. Pursel

5. **Gene Transfer into Goat Embryos** — 19
   W.G. Gavin

6. **The Generation of Transgenic Sheep by Pronuclear Microinjection** — 23
   Y. Gibson and A. Colman

7. **Gene Microinjection into Bovine Pronuclei** — 27
   M. Gagné, F. Pothier and M.-A. Sirard

8. **DNA Delivery by Cytoplasmic Injection** — 37
   W.H. Velander

9. **The Direct Gene Transfer through Mammal Spermatozoa** — 41
   K. Schellander and G. Brem

SECTION B   *IN VITRO* MAMMAL EMBRYO MANIPULATION

10. ***In Vitro* Generation of One Cell Embryos in Sheep and Goat** — 45
    N. Crozet

11. ***In Vitro* Generation of Bovine Embryos** — 51
    F.A.M. de Loos and F.R. Pieper

12. The Use of Cryopreserved Fertilized Ova to Generate Transgenic Mammals 55
    N. Tada

SECTION C GENE TRANSFER INTO BIRD EMBRYOS

13. The Complete *In Vitro* Development of Quail Embryo 61
    T. Ono

14. The Microinjection of DNA into Early Chicken Embryo 69
    M. Naito

15. Production of Chimeric Chickens as Intermediates for Gene Transfer 75
    R.J. Etches, M.E. Clark, A.M. Verrinder Gibbins and M.B. Cochran

16. The Use of Retroviral Vectors for Gene Transfer into Bird Embryo 83
    C.M. Ronfort, C. Legras and G. Verdier

17. Transgenic Chickens by Liposome-Sperm-Mediated Gene Transfer 95
    E.J. Squires and D. Drake

18. Ballistic Transfection of Avian Primordial Germ Cells *In Ovo* 101
    K. Simkiss and G. Luke

SECTION D GENE TRANSFER INTO FISH AND LOWER VERTEBRATES

19. Transgenic Salmonids 105
    R.H. Devlin

20. Transferring Foreign Genes into Zebrafish Eggs by Microinjection 119
    P. Collas, H. Husebye and P. Alestrøm

21. Gene Injection into *Xenopus* Embryos 123
    S.-G. Gong and L.D. Etkin

22. The Techniques using Electroporation to Generate Transgenic Fish 129
    K. Inoue and S. Yamashita

SECTION E GENE TRANSFER INTO INVERTEBRATES

23. Gene Transfer in *Drosophila Melanogaster* 133
    K. Kaiser

24. Genetic Transformation in *Caenorhabditis Elegans* 137
    D. Thierry-Mieg, K. Naert and C. Bonnerot

25. Gene Transfer Technology in Marine Invertebrates 151
    E. Miahle, V. Boulo, J.-P. Cadoret, V. Cedeño, C. Rousseau, E. Motte, S. Gendreau
    and E. Bachère

SECTION F THE USE OF ES CELLS FOR GENE TRANSFER AND GENE TARGETING

26. **What are ES Cells?** 157
    J.-E. Fléchon

27. **Formation of Mouse Chimeric Embryos from ES Cells** 167
    A. Nagy

28. **Bovine Pluripotent Stem Cells for Transgenic Vectors** 173
    N. Strelchenko

29. **Towards the Use of Primordial Germ Cells for Germline Modification** 179
    P.J. Donovan, J.L. Resnick, L. Cheng and L.F. Lock

SECTION G THE DIRECT GENE TRANSFER INTO SOMATIC CELLS OF ANIMALS

30. **The Adenovirus Vectors for Gene Delivery** 189
    J.F. Dedieu, E. Vigne, C. Orsini, M. Perricaudet and P. Yeh

31. **The Development of Adeno-Associated Virus as a Vector for *In Vivo* Gene Therapy** 197
    R.J. Samulski

32. **Advances Toward Synthetic Retro-Vectoring** 205
    C.P. Hodgson, M.A. Zink, F. Solaiman and G. Xu

33. **The Biolistic** 209
    P. Couble

PART III THE FATE OF MICROINJECTED DNA INTO MAMMAL EMBRYOS

SECTION A THE INTEGRATION AND MODIFICATION OF FOREIGN DNA

34. **The Fate of Microinjected Genes in Preimplantation Embryos** 215
    R.J. Wall and T.G. Burdon

35. **Chromosomal Insertion of Foreign DNA** 219
    J.O. Bishop

36. **Detection of Transgenes in the Preimplantation Stage Embryo** 225
    G.T. O'Neill

37. **Genetic Mosaicism in the Generation of Transgenic Mice** 233
    Y. Echelard

38. **Transgene Methylation in Transgenic Animals** 237
    K.J. Snibson and T.E. Adams

39. **Influence of Genomic Imprinting on Transgene Expression** 245
    J.R. Chaillet

SECTION B THE VECTORS FOR TRANSGENESIS

40. The Use of Episomal Vectors for Transgenesis — 251
    J. Attal, M.G. Stinnakre, M.C. Théron, M. Terqui and L.M. Houdebine

41. Insulation of Transgenes from Chromosomal Position Effects — 257
    A.E. Sippel, H. Saueressig, M.C. Hubler, N. Faust and C. Bonifer

42. Transgene Rescue — 267
    A.J. Clark

43. The Use of P1 Bacteriophage Clones to Generate Transgenic Animals — 273
    S.P.A. McCormick, C.M. Allan, J.M. Taylor and S.G. Young

44. Bacterial Artificial Chromosomes and Animal Transgenesis — 283
    K. Dewar, B.W. Birren and H. Abderrahim

45. The Use of Yeast Artificial Chromosomes for Transgenesis — 289
    T. Umland, L. Montoliu and G. Schütz

46. Gene Transfer Using Microdissected Chromosome Fragments — 299
    J. Richa and C.W. Lo

47. Prospects for Directing Temporal and Spatial Gene Expression in Transgenic Animals — 303
    L. Hennighausen and P.A. Furth

48. Mouse Genetic Manipulation via Homologous Recombination — 307
    S. Viville

49. Improvements in Transgene Expression by Gene Targeting — 323
    A.J. Kind and A. Colman

50. Nuclear Antisense RNA as a Tool to Inhibit Gene Expression — 327
    G.G. Carmichael

51. Controlled Genetic Ablation by Immunotoxin-Mediated Cell Targeting — 331
    K. Kobayashi, I. Pastan and T. Nagatsu

52. Reporter Genes and Detection of Mutational Activity in Mice — 337
    P.J. Stambrook and J.A. Tischfield

PART IV THE USE OF TRANSGENIC ANIMALS

SECTION A THE STUDY OF GENE FUNCTIONS

53. Gene Activity in the Preimplantation Mouse Embryo — 345
    S. Forlani and J.-F. Nicolas

54. Insertional Mutagenesis in Transgenic Mice — 361
    U. Rüther

55.	**Enhancer-Trap Studies of the *Drosophila* Brain** J.D. Armstrong and K. Kaiser	365
56.	**Transgenic Strategies for the Study of Mouse Development: An Overview** C. Babinet	371
57.	**Gene Transfer in Laboratory Fish: Model Organisms for the Analysis of Gene Function** C. Winkler and M. Schartl	387
58.	**The Preparation of Human Antibodies from Mice Harbouring Human Immunoglobin Loci** M. Brüggemann	397

SECTION B THE GENERATION OF ANIMAL MODELS FOR BIOMEDICAL AND PHARMACEUTICAL STUDIES

59.	**Standardization of Transgenic Lines: from Founder to an Established Animal Model** D. Carvallo, G. Canard and D. Tucker	403
60.	**The Use of Mouse Knockouts to Study Tumor Suppressor Genes** C.J. Kemp	411
61.	**The Generation of New Immortalized Cell Lines through Transgenesis** M. Theisen and A. Pavirani	421
62.	**The Use of Transgenic Mammals for AIDS Studies** T. Sorg and M. Mehtali	427
63.	**The Role of Mouse Models in the Development of New Therapies for Cystic Fibrosis** G. McLachlan and D.J. Porteous	435
64.	**Transgenic Animals in Atherosclerosis Research** M.W. Miller and E.M. Rubin	445
65.	**Genetically Engineered Mice in Obesity Research** B.B. Lowell	449
66.	**The Generation and Use of Transgenic Animals for Xenotransplantation** J.L. Platt and J.S. Logan	455

SECTION C THE PREPARATION OF RECOMBINANT PROTEINS FROM TRANSGENIC ANIMALS

67.	**The Preparation of Recombinant Proteins from Mouse and Rabbit Milk for Biomedical and Pharmaceutical Studies** M.G. Stinnakre, M. Massoud, C. Viglietta and L.M. Houdebine	461

68. **Production of Complex Human Pharmaceuticals in the Milk of Transgenic Goats Using the Goats Beta Casein Promoter** — 465
P. DiTullio, K.M. Ebert, J. Pollock, T. Edmunds and H.M. Meade

69. **Purification of Recombinant Proteins from Sheep's Milk** — 469
G. Wright and A. Colman

SECTION D THE MODIFICATION OF MILK COMPOSITION

70. **The Modification of Milk Protein Composition through Transgenesis: Progress and Problems** — 473
J.-C. Mercier and J.-L. Vilotte

SECTION E THE PROTECTION OF ANIMALS AGAINST DISEASES

71. **The Use of Host-Derived Antiviral Genes** — 483
E. Meier and H. Arnheiter

72. **The Use of Antisense RNA and Ribozyme Sequences to Prevent Infectious Diseases in Transgenic Animals** — 489
L. Han and T.A. Wagner

73. **Antibody Encoding Transgenes – Their Potential Use in Congenital and Intracellular Immunisation of Farm Animals** — 495
M. Müller, U.H. Weidle and G. Brem

74. **Intracellular Antibodies (Intrabodies): Potential Applications in Transgenic Animal Research and Engineered Resistance to Pathogens** — 501
S.D. Jones and W.A. Marasco

SECTION F TOWARDS THE INTRODUCTION OF NEW GENETIC TRAITS IN DOMESTIC ANIMALS BY TRANSGENESIS

75. **Improved Wool Production from Insulin-Like Growth Factor 1 Targeted to the Wool Follicle in Transgenic Sheep** — 507
D.W. Bullock, S. Damak, N.P. Jay, H.-Y. Su and G.K. Barrell

76. **The Transfer of Isolated Genes Having Known Functions** — 511
K.A. Ward

77. **Recent Progress in Mammalian Genomics and its Application for the Selection of Candidate Transgenes in Livestock Species** — 519
M. Georges

PART V THE NEW PROBLEMS GENERATED BY TRANSGENIC ANIMALS

78. **To Save or Not to Save: The Role of Repositories in a Period of Rapidly Expanding Development of Genetically Engineered Strains of Mice** — 525
J.J. Sharp and L.E. Mobraaten

79.	**TBASE: The Relationalized Database of Transgenic Animals and Targeted Mutations** A.V. Anagnostopoulos	533
80.	**The Patenting of Transgenic Animals** J. Warcoin	553
81.	**The Biosafety Problems of Transgenic Animals** L.M. Houdebine	559
82.	**Food Safety Evaluation of Transgenic Animals** M.A. Miller and J.C. Matheson III	563
83.	**From a Moral Point of View: Ethical Problems of Animal Transgenesis** E. Schroten	569
Index		575

PREFACE

Fifteen years ago, when transgenesis suddenly appeared as a distinct possibility for experimenters, nobody imagined that it would so quickly become a popular technique and an inevitable tool, not only for molecular geneticists but for all biologists. Techniques for gene transfer and transgene expression have made considerable progress. Simple DNA microinjection is still very commonly used, but gene targeting through homologous recombination, which is inducible even now, is increasingly preferred by experimenters. Transgene expression is now better controlled and even specifically inducible by exogenous factors. All these techniques have quite significantly transformed the experimental approaches taken by biologists. Molecular biologists are rediscovering the complexity of living organisms and physiologists are aided in their study of biological functions at a molecular level.

Despite these considerable advances, transgenesis still faces several problems. Gene transfer remains limited to a relatively small number of species. It is still relatively difficult and expensive in large animals, moreover there is no possibility of gene targeting. Progress will probably result from a better knowledge of reproduction. Embryonic cells and possibly spermatic cells should become the ideal gene vehicles for many species in the future.

The applications of animal transgenesis are more limited than expected and are still essentially restricted to the medical and pharmaceutical fields. This is in clear contrast to applications in plants. Several transgenic plants are already on the market, whereas transgenic animals will not be available until the next century.

Basic and applied research with transgenic animals has raised new problems, biosafety being one. These problems can find satisfactory solutions as long as experiments are carried out in confined areas, but become much more complex when the dissemination of transgenic organisms is envisaged.

The physiological, social and philosophical problems related to animal transgenesis are far from negligible. Some people consider gene transfer a violation of nature and therefore fundamentally unacceptable. Consumer acceptance of new products obtained by gene transfer will probably meet resistance unless accompanied by educative measures. Transgenesis offers a new possibility in animal exploitation. Yet limits on any exploitation of animals (including transgenics) cannot be defined easily.

The present book attempts to address all these problems and offers the reader the possibility of taking into consideration most of the aspects of his science.

This is certainly an ambitious goal, especially in a field where reality changes so rapidly and the editor is very grateful to the authors of the different chapters for their contributions, especially as they had a relatively short period of time to write their manuscripts. The technical problems of animal transgenesis will probably be solved progressively, however definite answers cannot be provided to most of the other questions. Understanding the rules which govern living organisms is a very exciting venture, and biotechnologies, including transgenesis, undoubtedly open up new possibilities for mankind. It remains that "Maîtriser notre maîtrise de la nature" [To control our control of nature], as suggested by Michel Serres, is more than ever a priority.

CONTRIBUTORS

Hadi Abderrahim
Centre d' Estudes du Polymorphisme Humain,
Paris, France

Timothy E. Adams
Center for Animal Biotechnology, School of Veterinary Science, The University of Melbourne, Parkville Victoria 3052, Australia

Peter Alestrøm
Department of Biochemistry, Norwegian College of Veterinary Medicine, P.O. Box 8146 Dep.
0033 Oslo, Norway

Charles M. Allan
Gladstone Institute of Cardiovascular Disease and Cardiovascular Research Institute, University of California, P.O. Box 419100
San Francisco, CA 94141-9100, USA

Anna V. Anagnostopoulos
Division of Biomedical Information Sciences, The Johns Hopkins University School of Medicine, 2024 E. Monument Street
Baltimore, MD 21205-2236, USA

J. Douglas Armstrong
Division of Molecular Genetics, IBLS, Ponte Corvo Building, Church Street, Univeristy of Glasgow,
Glasgow G11 5JS, UK

Heinz Arnheiter
Laboratory of Developmental Neurogenetics, National Institute of Neurological Disorders and Stroke, National Institutes of Health
Bethesda, MD 20892, USA

Joé Attal
INRA, Unité de Différenciation Cellulaire, Domaine de Vilvert
78352 Jouy-en-Josas Cedex, France

Charles Babinet
Unité de Biologie du Développement, URACNRS 1960 Institut Pasteur, 25, rue du Dr. Roux
75724 Paris Cedex 15, France

Evelyne Bachère
Unité de Recherche "Défense et Résistance chez les Invertébrès Marins" (DRIM), Institut Français de Recherche et d' Exploitation de la Mer (IFREMER) – Centre National de la Recherche Scientifique (CNRS), Université de Montpellier II
34000 Montpellier, France

G.K. Barrell
Animal and Veterinary Sciences Group, Lincoln University, P.O. Box 84
Canterbury, New Zealand

Bruce W. Birren
Whitehead Institute, MIT Center for Genome Research, 1 Kendall Square, Building 300
Cambridge, MA 02139, USA

John O. Bishop
Centre for Genome Research, Department of Genetics University of Edinburgh, King's Buildings, West Mains Road
Edinburgh EH9 3JQ, UK

Constanze Bonifer
Institut für Biologie III der Albert-Ludwigs-Universität Freiburg, Schänzlestrasse 1
D-79104 Freiburg, Germany

Claire Bonnerot
CRBM, CNRS, BP5051,
34033 Montpellier, France

Viviane Boulo
Unité de Recherche "Défense et Resistance chez les Invertébrès Marins" (DRIM), Institut Français de Recherche et d'Exploitation de la Mer (IFREMER) – Centre National de la Recherche Scientifique (CNRS), Université de Montpellier II
34000 Montpellier, France

Gottfried Brem
Institut für Tierzucht und Genetik, Veterinärmedizinische Universität Wien,

Linke Bahngasse 11
A-1030 Wien, Austria

Gillian Brooker
Centre for Genome Research, The University of Edinburgh, King's Buildings, West Mains Road Edinburgh EH9 3JQ, UK

Marianne Brüggemann
Development and Differentiation Laboratory, The Babraham Institute, Babraham Hall, Babraham Cambridge, CB2 4AT, UK

D.W. Bullock
Centre for Molecular Biology and Animal and Veterinary Sciences Group, Lincoln University, P.O. Box 84 Canterbury, New Zealand

Thomas G. Burdon
USDA-ARS, Building 200, Room 16, BARC-East Beltsville, MD 20705 USA

Jean-Paul Cadoret
Unité de Recherche "Défense et Résistance chez les Invertébrès Marins" (DRIM), Institut Français de Recherche et d'Exploitation de la Mer (IFREMER)–Centre National de la Recherche Scientifique (CNRS), Université de Montpellier II 34000 Montpellier, France

Georges Canard
Transgenic Alliance, Domaine des Oncines 69210 Saint-Germain sur L'Arbresle, France

Gordon G. Carmichael
Department of Microbiology, University of Connecticut Health Center Farmington, CT 06030, USA

Dorothée Carvallo
Château de Villendry, 37510 Villendry, France

Virna Cedeño
Centro Nacional de Acuacultura I Investigaciones Marinas (CENAIM), Escuela Politecnica (ESPOL) Campus Proserpina, P.O. Box 09 01 4519 Guayaquil, Ecuador

J. Richard Chaillet
Department of Biological Sciences, University of Pittsburgh Pittsburgh, PA 15260, USA

Linzhao Cheng
Systemix Inc., 3155 Porter Drive Palo Alto, CA 94304, USA

A.J. Clark
Division of Molecular Biology, Roslin Institute, Roslin Midlothian, EH25 9PS, UK

M.E. Clark
Ontario Agricultural College, Department of Animal and Poultry Science, University of Guelph, Guelph Ontario, NIG 2W1, Canada

M.B. Cochran
Ontario Agricultural College, Department of Animal and Poultry Science, University of Guelph, Guelph Ontario, NIG 2W1, Canada

Philippe Collas
Department of Biochemistry, Norwegian College of Veterinary Medicine, P.O. Box 8146 Dep. 0033 Oslo, Norway

Alan Colman
Pharmaceutical Proteins Limited, Roslin Edinburgh EH25 9PP, UK

Pierre Couble
Centre de Génétique Moléculaire et Cellulaire, CNRS UMR 106-INRA LA 810, Université Claude Bernard Lyon 1, 43, Boulevard du 11 Novembre 1918 69622 Villeurbanne Cedex, France

Nicole Crozet
INRA, Unité Biologie de la Fécondation, Station de Recherches de Physiologie Animale 78352 Jouy-en-Josas Cedex, France

S. Damak
Centre for Molecular Biology, Lincoln University, P.O. Box 84 Canterbury, New Zealand

J.F. Dedieu
CNRS URA 1301/Rhône-Poulenc Rorer Gencell, Laboratoire de Génétique des Virus Oncogènes, Institut Gustave Roussy, PRII, 39, rue Camile Desmoulins 94805 Villejuif Cedex, France

Robert H. Devlin
Fisheries and Oceans Canada, West Vancouver Laboratory, 4160 Marine Drive, West Vancouver, BC V7V 1N6, Canada

Contributors

K. Dewar
Department of Biology, University of Pennsylvania,
Philadelphia, PA 19104, USA

Paul DiTullio
Genzyme Transgenics Corp., 1 Mountain Road
Framingham, MA 01701-9322, USA

Peter J. Donovan
Cell Biology and Differentiation Group, ABL-Basic
Research Program, NCI Frederick Cancer Research
and Development Center, Building 539, P.O. Box B
Frederick, MD 21702-1201, USA

D. Drake
Ontario Agricultural College, Department of Animal
and Poultry Science, University of Guelph, Guelph
Ontario, N1G 2W1, Canada

K.M. Ebert
School of Medicine, Dental Medicine and Veterinary
Medicine, Tufts Univeristy, 200 Westboro Rd,
N. Grafton, MA 01536, USA

Yann Echelard
Genzyme Transgenics Corp., 1 Mountain Road
Framingham, MA 01701-9322, USA

T. Edmunds
Genzyme Transgenics Corp and Genzyme Corp.,
1 Mountain Rd, Framingham,
MA 01701, USA

Robert J. Etches
Ontario Agricultural College, Department of Animal
and Poultry Science, University of Guelph, Guelph
Ontario, N1G 2W1, Canada

Laurence D. Etkin
Department of Molecular Genetics, University of Texas,
M.D. Anderson Cancer Center, Box 45,
6723 Bertner Avenue
Houston, TX 77030, USA

Nicole Faust
Institut für Biologie III der Albert-Ludwigs-Universität
Freiburg, Schänzlestrasse 1
D-79104 Freiburg, Germany

J.-E. Fléchon
INRA, Laboratoire de Biologie Cellulaire et Moléculaire,
Domaine de Vilvert
78352 Jouy-en-Josas, Cedex, France

Sylvie Forlani
Unité de Biologie Moléculaire du Développement, URA
1947 du CNRS Institut Pasteur,
25, rue du Dr. Roux
75724 Paris Cedex 15, France

Priscilla A. Furth
Division of Infectious Diseases, Department of
Medicine, University of Marylund Medical School
and Baltimore Veterans Affairs Medical Center
Baltimore, MD 21201, USA

Marc Gagné
Immunovaltée, 2750 Einstein Street, Office 110
Sainte-Foy (Québec), G1P 4R1 Canada

William G. Gavin
Genzyme Transgenics Corp., 1 Mountain Road
Framingham, MA 01701-9322, USA

Sylvain Gendreau
Unité de Recherche "Défense et Résistance chez les
Invertébres Marins" (DRIM), Institut Français de
Recherche et d'Exploitation de la Mer
(IFREMER)–Centre National de la Recherche
Scientifique (CNRS), Université de Montpellier II
34000 Montpellier, France

Michel Georges
Department of Genetics, Faculty of Veterinary Medicine
University of Liège, 20, Boulevard de Colonster
400-Liège, Belgium

Yvonne Gibson
Pharmaceutical Proteins Limited, Roslin
Edinburgh EH25 9PP, UK

Siew-Ging Gong
Department of Orthodontics and Pediatric Dentistry
University of Michigan, School of Dentistry,
Ann Arbor, MI 48109, USA

Lei Han
Department of Laboratory Medicine and Pathology
Medical School Box 609, Room 6-159, Jackson Hall,
UMHC, 321 Caurch Street S.E.
Minneapolis, MN 55455, USA

Lothar Hennighausen
Laboratory of Biochemistry and Metabolism, National
Institute of Diabetes, Digestive and Kidney Diseases
National Institutes of Health, Building 10,
Room 9N113
Bethesda, MD 20892-1812, USA

Clague P. Hodgson
Division of Cancer Biology, Creighton Cancer Center, Creighton University School of Medicine, 2500 California Plaza
Omaha, NE 68178, USA

Louis Marie Houdebine
Unite de Différenciation Cellulaire, INRA
78352 Jouy-en-Josas Cedex, France

Matthias C. Huber
Institut für Biologie III der Albert-Ludwigs-Universität Freiburg, Schänzlestrasse 1
D-79104 Freiburg, Germany

Harald Husebye
Department of Biochemistry, Norwegian College of Veterinary Medicine, P.O. Box 8146 Dep.
0033 Oslo, Norway

Koji Inoue
Central Research Laboratory, Nippon Suisan Kaisha Ltd., 559-6 Kitano-machi, Hachioji
Tokyo 192, Japan

N.P. Jay
Animal and Veterinary Sciences Group, Lincoln University, P.O. Box 84
Canterbury, New Zealand

Susan Dana Jones
IntraImmune Therapies Inc., PO Box 15599,
Boston, MA 02215, USA

Kim Kaiser
Division of Molecular Genetics, IBLS, Ponte Corvo Building, Church Street, University of Glasgow,
Glasgow G11 5JS, UK

Christopher J. Kemp
Program in Cancer Biology, Division of Public Health Science, Fred Hutchinson Cancer Research Center, 1124 Columbia Street, C1015
Seattle, WA 98104, USA

Alexander J. Kind
Pharmaceutical Proteins Limited, Roslin
Edinburgh EH25 9PP, UK

Kazuto Kobayashi
Institute for Comprehensive Medical Science, Fujita Health University, Toyoake
Aichi 470-11, Japan

Catherine Legras
Centre de Génétique Moléculaire et Cellulaire, CNRS UMR 106 – INRA LA 810, Université Claude Bernard Lyon 1, 43, Boulevard du 11 Novembre 1918
69622 Villeurbanne Cedex, France

Cecilia W. Lo
School of Arts and Sciences, Department of Biology Goddard Laboratories, University of Pennsylvania
Philadelphia, PA 19104-6017, USA

Leslie F. Lock
Cell Biology and Differentiation Group, ABL-Basic Research Program, NCI Frederick Cancer Research and Development Center, Building 539, P.O. Box B Frederick,
MD 21702-1201, USA

John S. Logan
Nextran, Princeton
New Jersey, USA

Frans A.M. de Loos
Pharming B.V., Niels Bohrweg 11-13
2333 CA Leiden, The Netherlands

Bradford B. Lowell
Division of Endocrinology, Department of Medicine, Beth Israel Hospital and Harvard Medical School, 330 Brooklyn Avenue
Boston, MA 02215, USA

G. Luke
School of Animal and Microbial Sciences, The University of Reading, Whiteknights, P.O. Box 228
Reading RG6 6AJ, UK

Wayne A. Marasco
Division of Human Retrovirology, Dana Farber Cancer Institute, Harvard Medical School, 44 Binney Street
Boston, MA 02115, USA

Micheline Massoud
INRA, Laboratoire de Physiologie Animale, Domaine de Vilvert
78352 Jouy-en-Josas Cedex, France

John C. Matheson III
Center for Veterinary Medicine, U.S. Food and Drug Administration, 7500 Standish Place
Rockville, MD 20855, USA

Sally P.A. McCormick
Gladstone Institute of Cardiovascular Disease and
  Cardiovascular Research Institute, University of
  California P.O. Box 419100
San Francisco, CA 94141-9100, USA

Gerry McLachlan
MRC Human Genetics Unit, Western General Hospital
  Crewe Road
Edinburgh EH4 2XU, UK

H.M. Meade
School of Medicine, Dental Medicine and
  Veterinary Medicine, Tufts University,
  200 Westboro Rd, N. Grafton,
MA 01536, USA

Majid Mehtali
Department of Gene Therapy, TRANSGENE,
  11, rue de Molsheim
67000 Strasbourg Cedex, France

Ellen Meier
Blümle Life Sciences Building,
  233 South 10th Street, Room 408,
Philadelphia PA 19107, USA

Jean-Claude Mercier
INRA CRJ, Laboratoire de Génétique Biochimique et de
  Cytogénétique
78352 Jouy-en-Josas Cedex, France

Eric Mialhe
Centro Nacional de Acuacultura I Investigaciones
  Marinas (CENAIM), Escuela Politecnica (ESPOL),
  Campus Proserpina, P.O. Box 09 01 4519
Guayaquil, Ecuador

Margaret Ann Miller
Center for Veterinary Medicine, U.S. Food and Drug
  Administration, 7500 Standish Place
Rockville, MD 20855, USA

Miles W. Miller
Human Genome Center, Life Sciences Division,
  Berkeley National Laboratory, University of
  California, 1 Cyclotron Road (M/S 74-157)
Berkeley, CA 94720, USA

Larry E. Mobraaten
Induced Mutant Resource, Jackson Laboratory,
  600 Main Street, Bar Harbor
Maine 04609-1500, USA

Lluis Montoliu
Department of Biochemistry and Molecular Biology,
  Faculty of Veterinary Medicine, University of
  Barcelona, 08193 Bellaterra,
Barcelona, Spain

Emeric Motte
Centro Nacional de Acuacultura I Investigaciones
  Marinas (CENAIM), Escuela Politecnica (ESPOL)
  Campus Proserpina, P.O. Box 09 01 4519
Guayaquil, Ecuador

Mathias Müller
Institut für Tierzucht und Genetik, Veterinärmedizinische
  Universität Wien, Linke Bahngasse 11
A-1030 Wien, Austria

John J. Mullins
Center for Genome Research, The University of
  Edinburgh, King's Buildings, West Mains Road
Edinburgh EH9 3JQ, UK

Linda J. Mullins
Center for Genome Research, The University of
  Edinburgh, King's Buildings, West Mains Road
Edinburgh EH9 3JQ, UK

Karine Naert
CRBM, CNRS, BP5051,
34033 Montpellier, France

Toshiharu Nagatsu
Institute for Comprehensive Medical Science,
  Fujita Health University, Toyoake
Aichi 470-11, Japan

Andras Nagy
Samuel Lunenfeld Research Institute, Mount Sinai
  Hospital, 600 University Avenue
Toronto, Ontario M5G 1X5, Canada

M. Naito
National Institute of Animal Industry, Tsukuba
  Norindanchi, P.O. Box 5
Ibaraki 305, Japan

Jean-François Nicolas
Unité de Biologie Moléculaire du Développement, URA
  1947 du CNRS Institut Pasteur, 25, rue du
  Dr. Roux
75724 Paris Cedex 15, France

G.T. O'Neill
BBSRC/MRC Neuropathogenesis Unit, Institute of

*Animal Health, Ogston Building, King's Buildings
    West Mains Road
Edinburgh EH9 3JF, UK*

Tamao Ono
*Laboratory of Developmental Biology, Faculty of
    Agriculture, Shinshu University
Ina 399-45, Japan*

C. Orsini
*CNRS URA 1301/Rhône-Poulenc Rorer Gencell,
    Laboratoire de Génétique des Virus Oncogènes,
    Institut Gustave Roussy, PRII, 39, rue Camille
    Desmoulins,
94805 Villejuif Cedex, France*

Ira Pastan
*Laboratory of Molecular Biology, Division of Cancer
    Biology, National Cancer Institute, National
    Institutes of Health, Bethesda, MD 20892, USA*

Andrea Pavirani
*Department of Molecular and Cellular Biology,
    TRANSGENE, 11, rue de Molsheim
67082 Strasbourg Cedex, France*

M. Perricaudet
*CNRS URA 1301/Rhône-Poulenc Rorer Gencell,
    Laboratoire de Génétique des Virus Oncogènes,
    Institut Gustave Roussy,PR.II, 39, rue Camille
    Desmoulins, 94805 Villejuif Cedex, France*

Frank R. Pieper
*Gene Pharming Europe B.V., Leiden University
    P.O. Box 9502
2300 RA Leiden, The Netherlands*

Jeffrey L. Platt
*Departments of Surgery, Pediatrics and Immunology
    Duke University Medical Center, Box 2605
Durham, NC 27710, USA*

J. Pollock
*Genzyme Transgenics Corp and Genzyme Corp.,
1 Mountain Rd, Framingham, MA 01701, USA*

David J. Porteous
*MRC Human Genetics Unit, Western General Hospital,
    Crewe Road
Edinburgh EH4 2XU, UK*

François Pothier
*Centre de Recherche en Biologie de la Reproduction,
    Université Laval, Sainte-Foy (Québec)
G1K 7P4, Canada*

Vernon G. Pursel
*U.S. Department of Agriculture, Agricultural Research
    Service, Gene Evaluation and Mapping Laboratory,
    BARC-East, Building 200, Room 2
Beltsville, MD 20705, USA*

James L. Resnick
*Department of Molecular Genetics and Microbiology
    University of Florida, 1600 Southwest Archer Road
Gainesville, FL 32610, USA*

Jean Richa
*Department of Genetics, University of Pennsylvania,
    415 Curie Boulevard
Philadelphia, PA 19104-6145, USA*

Corinne Marie Ronfort
*Centre de Génétique Moléculaire et Cellulaire, CNRS
    UMR 106 – INRA LA 810, Université Claude
    Bernard Lyon 1, 43, Boulevard du 11 Novembre 1918
69622 Villeurbanne Cedex, France*

Christophe Rousseau
*Centro Nacional de Acuacultura I Investigaciones
    Marinas (CENAIM), Escuela Politecnica (ESPOL),
    Campus Proserpina, P.O. Box 09 01 4519
Guayaquil, Ecuador*

Edward M. Rubin
*Human Genome Center, Life Sciences Division,
    Berkeley National Laboratory, University of
    California, 1 Cyclotron Road (M/S 74-157)
Berkeley, CA 94720, USA*

Ulrich Rüther
*Institut für Molekularbiologie, Medizinische
    Hochschule, Hannover OE 5250
30625 Hannover, Germany*

R. Jude Samulski
*Gene Therapy Center, University of North Carolina
    at Chapel Hill, 7119 Thurston-Bowles Building,
    CB 7352
Chapel Hill, NC 27599-7352, USA*

Harald Saueressig
*Molecular Neurobiology Laboratory, The Salk Institute
    10010 N Torrey Pines Road
La Jolla, CA 92037, USA*

Manfred Schartl
*Physiological Chemistry I, Biocenter, University of
    Würzburg, Am Hubland
97074 Würzburg, Germany*

Karl Schellander
Institut für Tierzucht und Genetik,
 Veterinärmedizinische Universität Wien,
 Linke Bahngasse 11
A-1030 Wien, Austria

Egbert Schroten
Universitair Centrum voor Bio-Ethiek en
 Gezondheidsrecht, Universiteit Utrecht,
 Heidelberglaan 2
3584 CS Utrecht, The Netherlands

Günther Schütz
Division of Molecular Biology of the Cell I, German
 Cancer Research Center, Im Neuenheimer Feld 280
D-69120 Heidelberg, Germany

John J. Sharp
Induced Mutant Resource, Jackson Laboratory,
 600 Main Street, Bar Harbor
Maine 04609-1500, USA

Ken Simkiss
School of Animal and Microbial Sciences,
 The University of Reading, Whiteknights,
 P.O. Box 228
Reading RG6 6AJ, UK

Albrecht E. Sippel
Institut für Biologie III der Albert-Ludwigs-Universität
 Freiburg, Schänzlestrasse 1
D-79104 Freiburg, Germany

Marc-André Sirard
Centre de Recherche en Biologie de la Reproduction
 Université Laval, Sainte-Foy (Québec)
G1K 7P4, Canada

Kenneth J. Snibson
Center for Animal Biotechnology, School of Veterinary
 Science, The University of Melbourne, Parkville
Victoria 3052, Australia

Tania Sorg
Department of Gene Therapy, TRANSGENE, 11, rue de
 Molsheim
67000 Strasbourg Cedex, France

E.J. Squires
Ontario Agricultural College, Department of Animal
 and Poultry Science, University of Guelph, Guelph
Ontario, N1G 2W1, Canada

Peter J. Stambrook
Department of Cell Biology, Neurobiology and Anatomy
University of Cincinnati, College of Medicine,
 P.O. Box 670521
Cincinnati, OH 45267-0521, USA

Marie Georges Stinnakre
INRA, Laboratoire de Génétique Biochimique, Domaine
 de Vilvert
78352 Jouy-en-Josas Cedex, France

N. Strelchenko
Animal Breeding Service Global Inc., DeForest
WI 53532-0459, USA

H.-Y. Su
Centre for Molecular Biology and Animal and
 Veterinary Sciences Group, Lincoln University
P.O. Box 84
Canterbury, New Zealand

N. Tada
Hoechst Japan Limited, Pharma Research and
 Development Division, 3-2 Minamidai 1 Chome,
 Kawagoe, Saitama 350-11, Japan

John M. Taylor
Department of Physiology, Gladstone Institute of
 Cardiovascular Disease and Cardiovascular Research
 Institute, University of California, P.O. Box 419100
San Francisco, CA 94141-9100, USA

Michel Terqui
INRA, Laboratoire de Physiologie de la Reproduction
 des Mammifères Domestiques, Domaine de Vilvert
78352 Jouy-en-Josas Cedex, France

Manfred Theisen
Department of Molecular and Cellular Biology,
 TRANSGENE, 11, rue de Molsheim
67082 Strasbourg Cedex, France

Marie Claire Théron
INRA, Unité de Differenciation Cellulaire, Domain
 de Vilvert
78352 Jouy-en-Josas Cedex, France

Danielle Thierry-Mieg
CRBM, CNRS, BP5051, 34033
Montpellier, France

Jay A. Tischfield
Department of Medical Genetics, Indiana University
 School of Medicine, 975 West Walnut Street,
 Room 1B 130
Indianapolis, IN 46202-5251, USA

David Tucker
Zeneca Pharmaceuticals, Mereside, Alderley Park,
Macclesfield, Cheshire SK10 4TG, UK

Thorsten Umland
Division of Molecular Biology of the Cell I, German
   Cancer Research Center, Im Neuenheimer Feld 280
D-69120 Heidelberg, Germany

William H. Velander
Department of Chemical Engineering,
   Virginia Polytechnic Institute and State University
Blacksburg, VA 24061-0315, USA

Gérard Verdier
Centre de Génétique Moléculaire et Cellulaire, CNRS
   UMR 106-INRA LA 810, Université Claude
   Bernard Lyon 1, 43, Boulevard du
   11 Novembre 1918
69622 Villeurbanne Cedex, France

A.M. Verrinder Gibbins
Ontario Agricultural College, Department of Animal
   and Poultry Science, University of Guelph, Guelph
Ontario, NIG 2W1, Canada

Céline Viglietta
INRA, Unité de Différenciation Cellulaire, Domaine
   de Vilvert
78352 Jouy-en-Josas Cedex, France

E. Vigne
CNRS URA 1301/Rhône-Poulenc Rorer Gencell,
   Laboratoire de Génétique des Virus Oncogènes,
   Institut Gustave Roussy, PRII, 39, rue Camille
   Desmoulins
94805 Villejuif Cedex, France

Jean-Luc Vilotte
INRA CRJ, Laboratoire de Génétique Biochimique et de
   Cytogénétique
78352 Jouy-en-Josas Cedex, France

Stéphane Viville
Institut de Génétique et de Biologie Moléculaire et
   Cellulaire, CNRS-INSERM-Université Louis
   Pasteur, Collège de France, Parc d'Innovation,
   1, rue Laurent Fries BP 163
67404 Illkirch-C.U. de Strasbourg, France

Thomas A. Wagner
Department of Laboratory Medicine and Pathology
Medical School Box 609, Room 6-159, Jackson Hall,
   UMHC, 321 Caurch Street S.E.
Minneapolis, MN 55455, USA

Robert J. Wall
USDA-ARS, Building 200, Room 16, BARC-East
Beltsville, MD 20705, USA

Jacques Warcoin
Cabinet Regimbeau, 26 Avenue Kléber
75116 Paris Cedex, France

Kevin A. Ward
Division of Animal Production, CSIRO, Institute of
   Animal Production and Processing, Locked bag 1,
   Delivery Centre,
Blacktown, NSW 2148, Australia

Ulrich H. Weidle
Institut für Tierzucht und Genetik,
   Veterinärmedizinische Universität Wien Linke
   Bahngasse 11
1030 Wien, Austria

Christoph Winkler
Physiological Chemistry I, Biocenter, University of
   Würzburg, Am Hubland
97074 Würzburg, Germany

Gordon Wright
Pharmaceutical Proteins Limited, Roslin
Edinburgh EH25 9PP, UK

Shinaya Yamashita
Central Research Laboratory, Nippon Suisan Kaisha
   Ltd., 559-6 Kitano-machi, Hachioji
Tokyo 192, Japan

P. Yeh
CNRS URA 1301/Rhône-Poulenc Rorer Gencell,
   Laboratoire de Génétique des Virus Oncogènes,
   Institut Gustave Roussy, PRII, 39, rue Camille
   Desmoulins
94805 Villejuif Cedex, France

Stephen G. Young
Gladstone Institute of Cardiovascular Disease,
   Cardiovascular Research Institute and
   Department of Medicine, University of California,
   P.O. Box 419100 San Fransisco,
CA 94141-9100, USA

PART I

1. Overview

Louis Marie Houdebine
Unité de Différenciation Cellulaire, Institut National de la Recherche Agronomique, 78352 Jouy-en-josas Cedex, France

About fifteen years after the pioneer experiments which showed that a stable gene transfer into animals, an expression and a phenotypic effect of the transgenes were possible (Gordon *et al.*, 1980 and Palmiter *et al.*, 1982), it is interesting to consider the state of art in animal transgenesis. This technique has already provided an invaluable amount of information for understanding the mechanisms which govern the life of animals. However, the situation is somewhat contrasted. The applications of transgenesis in agronomy for animal production remain very limited, in comparison to the numerous successes in plant production. This is obviously due to the technical problems which persist in generating transgenic farm animals at a reasonable cost. The real success of animal transgenesis lies undoubtedly in the field of basic research. Many laboratories throughout the world routinely use transgenic mice, Drosophila and Caenorhabditis elegans, to study gene and biological functions. Relevant biological models are also created each year by gene transfer to study human diseases and new pharmaceuticals. Transgenic farm animals are ready to be used as the source of recombinant proteins for pharmaceutical use. Recently, very encouraging experiments have shown that some human genes transferred into pigs can protect their grafted heart from hyperacute rejection caused by the activation of the primate complement. The introduction of new genetic traits in domestic animals by transgenesis is still in its infancy. The success is limited not only by the techniques for generating transgenic animals but also by the lack of valid identified genes.

The mapping of domestic animal genomes will certainly provide researchers with genes of interest.

One may imagine that the first success in this field will be the protection of animals against diseases and the modification of milk composition. The improvement of the essential biological functions for breeding (growth, prolificity, meat quality, etc...) seems much less accessible due to the great complexity of the biological mechanisms involved. The generation of transgenic animals has raised totally new problems such as the conservation and the patenting of the animals, the biosafety of the gene transfer experiments for researchers and for the environment, the quality of the new products originating from the animals and also the ethics of animal genome manipulation.

In the present book, all these points are considered. For this purpose, many authors have been solicited, in order to have the point of view of many different people. The authors were asked to write relatively short chapters providing the most relevant information in their field. For these reasons, the chapters have no standard structure and are of variable length.

The Generation of Transgenic Animals

The first group of chapters describes the methods used to generate different transgenic mammals such as rat (J.J. Mullins), rabbit (C. Viglietta *et al.*), pig (V.G. Pursel), goat (W.G. Gavin), sheep (Y. Gibson) and cow (M. Gagné *et al.*, and A.M. Deloos) are described. The possibility of enhancing the integration rate of the foreign gene by adding cow repeated DNA sequences to the vector is considered by M. Gagné *et al.*. The efficiency of polycation-DNA complex injected into embryo cytoplasm rather than pronuclei is evaluated by W.H. Velander.

A critical evaluation of gene transfer into cow through spermatozoa is given by K. Schellander et al.

In practice, the success of transgenesis, specially for farm animals, is highly dependent on the techniques of reproduction. The possibility of using one-cell embryos generated after *in vitro* oocyte maturation and *in vitro* fertilization in sheep, goat and cow rather than by conventional superovulation is depicted by N. Crozet and A.M. Deloos. The use of one-cell embryos kept frozen before microinjection is shown by N. Tada. The cloning of embryos harbouring interesting transgenes by nuclear transfer into the cytoplasm of oocytes is expected to accelerate diffusion of the new genetic traits (N. Strelchenko).

Gene transfer into bird embryos is still not a fully resolved problem. Several complementary approaches are studied: the direct gene microinjection and *in vitro* development of the embryos (T. Ono and N. Naito), the use of retroviral vectors (C. Ronfort et al.) of biolistics (K. Simkiss) and liposomes (E.J. Squires) to transfer a foreign genes into embryos or spermatozoa. The possibility of culturing chicken embryonic cells and generating chimaerae offers quite interesting additional possibilities (R.J. Etches).

Gene transfer into lower vertebrates is possible after microinjection into early embryo cytoplasm (D.H. Delvin and P. Collas et al.). In fish (K. Inoue), as in chicken (E.J. Squires et al.) and mammals (K. Schellander et al.), gene transfer through spermatozoa appears possible but with a rather low yield and often with rearrangement or inactivation of the transgenes.

In invertebrates, direct microinjection into early embryos is also the technique routinely used. In Drosophila, the use of a transposon as a vector greatly facilitates the integration of the foreign DNA (Kaiser et al.). Gene transfer is carried out in the nematode C. elegans (D. Thierry-Mieg et al.) in marine invertebrates (E. Mialhe) and in arthropods (Presnail and Hoy, 1992).

The Use of Cellular Vectors for Transgenesis

The use of cellular vectors to carry the foreign DNA into the embryo is an attractive possibility. Cultured spermatozoa precursors, ES cells and EG cells can theoretically be used for this purpose. Spermatozoa fully matured or not can be injected into oocytes and give rise to fertilization and embryo development with an acceptable yield. Alternatively, cultured spermatogonial cells can be reintroduced into a recipient testis and give rise to the birth of offspring by normal fertilization (Brinster et al., 1994). This pioneer experiment is very encouraging. Longer periods of spermatogonial cell culture are necessary however to allow gene integration and homologous recombination. The number of cultured cells to be reintroduced into the testis to colonize this organ to a significant degree is probably so high for farm animals that this approach may turn out to be applicable only to laboratory animals. ES and EG cells can be fused to developing embryos to generate chimaerae. The success remains limited to mouse. Attempts to establish stable ES or EG cell lines from other species have failed so far. This may be due to the lack of basic knowledge on what are really totipotent cells. This point is discussed by J.E. Fléchon whereas the possible use of totipotent cells is described by N. Strelchenko, P.J. Donovan and A. Nagy. The possibility to generate lambs after transfer of nuclei from cultured sheep embryonic cells into cytoplasm of oocyte have opened new avenues for transgenesis (Campbell et al., 1996).

The Direct Gene Transfer into Somatic Cells

The direct gene transfer into somatic cells is the technical basis for gene therapy. It can be used in some cases as a relatively simple substitute to transgenesis to study gene function *in vivo*. It is also an attractive tool for vaccination using injections of pure DNA (Krishnan et al., 1995) for *in vivo* transfection to study gene activity (P. Couble) and for the evaluation of mammary cells to perform post-translational modifications of recombinant proteins (Archer et al., 1994). Retroviral vectors (C. Ronfort et al.), adenoviral vectors (J.F. Dedieu et al. and Nakayima et al., 1995) adeno-associated viral vectors (R.J. Salmulski) or retrosposons (C.P. Hogson) can be used for this purpose. Interestingly, viral vectors can be used in adults to express transiently and locally the Cre recombinase to induce a local gene recombination in transgenic animals harbouring the integrated lox recombination sequence (see Chapter 48 by S. Viville). Somewhat unexpectedly, DNA injected intravenously into pregnant mice is spontaneously transferred to fetuses (T. Subamoto et al., 1995). If this observation is confirmed, this simple technique might be useful to study gene expression and action in mouse foetus without prior transgenesis.

The Fate of Transgenes

The fate of the foreign DNA microinjected into embryos is not well known. In mammals, tandem arrays of foreign genes are integrated whereas in lower vertebrates a random association of individual copies of the foreign gene occurs (R.H. Delvin). The mechanism of persistence and integration has been studied by R.J. Wall et al., J.O. Bishop, Chen *et al.*, 1995 and Pawlik *et al.*, 1995. Transgenic founder animals are more frequently mosaic than originally imagined in mammals (Y. Echelard) and also in fish (R.H. Devlin) in the first generation. Animals are no more mosaic in offspring as was expected.

Detection of transgenes in preimplantation embryos would be very helpful for farm animals to reduce the number of recipient females. A systematic study carried out by G.T. O'Neil revealed that no reliable method is presently available. The use of the Vargula luciferase gene as a reporter co-transgene is a potentially interesting tool (Thompson *et al.*, 1995). The tyrosinase gene which has a phenotype effect on pigmentation may be used to study the presence of a coinjected gene of interest. This test can be used reliably but only in newborns and adults (Umland *et al.* and Methot *et al.*, 1995).

The transgenes may be methylated (K.J. Snibson) and subjected to genomic imprinting (J.R. Chaillet) which lead to their inactivation. These phenomena are interesting models to study the mechanisms of gene expression and they may limit the long term use of transgenic animals.

The Vectors for Transgenesis

Transgenesis has pointed out somewhat unexpectedly several aspects of the control mechanism of gene expression: introns are required for transgenes and generally not for genes transfected into cultured cells (Palmiter *et al.*, 1991 and Attal *et al.*, 1995). The role of intron seems quite complex. Interestingly, they seem to participate in some cases in nucleosome phasing (Lin *et al.*, 1995).

Episomal vectors would be of great help to generate transgenic animals. Attempts have had limited success so far (J. Attal *et al.*). This kind of vectors have the potential advantage of generating transgenic animals with a high yield and of being independent of host chromatin. Only a few transgenes are independent of their site of integration in chromatin and have a copy number-dependent expression. A few gene insulators have been partially identified and used successfully when added to the gene constructs (A. Sippel *et al.*).

The coinjection of a gene known to be functional as a transgene with the gene construct of interest may significantly enhance the expression of the latter (A.J. Clark). Relatively long DNA fragments can be faithfully reconstituted from overlapping fragments with good yield before integration into the animal chromatin (Pieper *et al.*, 1992). Long genomic DNA fragments cloned in P_1 phages (Sternberg, 1992 and S.P.A. McCormick *et al.*) in BAC vectors (Birren) or in YAC vectors (T. Umland *et al.*) are expected to bring most, if not all of the regulatory elements leading to a more predictable and confident expression of transgenes. Interestingly, the YAC vectors allow a spontaneous homologous recombination with the host genome with a relatively high frequency (T. Umland *et al.*). A combination of homologous recombination with YAC vectors in yeast to introduce new sequences in the vector and a standard microinjection into pronuclei can potentially lead to gene replacement. If this fact is confirmed, the use of ES cells for gene targeting may no longer be compulsory.

Contiguous YAC and P_1 vectors covering up to 2 Mb of a chromosome can also be used to generate transpolygenic mice which constitute a partial *in vivo* library. These long DNA fragments may be stably integrated without any rearrangement and the genes they contain are expressed. This elegant approach has been used to identify and to study the gene responsible for Down's syndrome located in chromosome 21 in human and in chromosome 16 in mouse (Smith *et al.*, 1995). No more than 8 lines of mice harbouring a DNA fragment in a YAC or a P_1 were sufficient to cover the 2 Mb.

Fragments of chromosomes can be microinjected and integrated into an animal genome (Richa *et al.*) bringing simultaneously several loci to the host.

The control of the transgene by an external stimulus not acting on host genes is highly desirable. Several systems and one using tetracycline as an inducer are available and efficient in transgenic animals (L. Hennighausen *et al.*).

Gene targeting by homologous recombination has become an essential tool for gene knock-out or replacement (S. Viville). A foreign gene replacing an endogenous gene in a given locus may be expressed quite effectively (A.J. Kind *et al.*).

Gene inactivation may be obtained by methods other than knock-out. Antisense RNA and specially those having no poly A end and thus concentrated in the nuclear compartment seem to work efficiently (G.G. Carmichael). Alternatively, ribozymes may reinforce the antisense action of an RNA. These antisense oligonucleotides inhibit efficiently specific genes such as rev from the HIV genome. These vectors can potentially be used as an alternative to more conventional antisense vectors.

Secreted or nonsecreted (intrabodies) antibodies may selectively inhibit the product of a specific gene (S.D. Jones *et al.*). The overexpression of trans-dominant negative mutant proteins may also greatly attenuate the action of a protein, in a specific manner (Chang *et al.*, 1994).

Genetic ablation which leads to the specific (B.B. Lowell) or even controlled destruction of a given cell type *in vivo* (K. Kobayashi *et al.*) is also a means to inhibit a given biological function.

Reporter genes are often added to vectors to follow gene expression in the different cell types of transgenic animals. Various reporter genes can be used, according to the problem to be solved (P.J. Stambrook).

The Study of Gene Functions

In many cases, transgenes are efficient tools for studying the mechanisms which govern specific gene expression and the role of genes in the control of biological functions. This is, for example, the case in early embryo (S. Forlani *et al.*) or in the development in mammals (C. Babinet) in Drosophila (K. Kaiser) and in C. elegans (D. Thierry-Mieg). Random or controlled insertion mutagenesis may lead to the identification of genes controlling development or other biological functions (V. Rütter).

Gene rearrangement during lymphocyte differentiation and the preparation of human antibodies can be carried out using transgenic mice harbouring an immunoglobulin gene locus (M. Brüggemann).

Studying genes in other quite different areas such as drug transporters (Postan *et al.*, 1991), transcription mechanism (Scott Strutters *et al.*, 1991), prions (Baldwin *et al.*, 1995) and animal behaviour (Yagi *et al.*, 1993) can be done using transgenesis.

The Models for Biomedical Studies

Numerous models have been created using transgenesis in mouse, rat, rabbit, and pig to study human diseases. There are however few relevant models. Animals which really mimic a human disease are rare. This may be due to the fact that the animal species is too different from humans or that the transferred gene is not properly expressed or insufficient when used alone. An animal model is satisfactory and utilizable only if the effect of the transgene is predictable. Genetic background sometimes has an important effect on the transgene effect (D. Carvallo).

Transgenic animals have been generated to study cancer (C.J. Kemp) AIDS (T. Sorg *et al.*), cystic fibrosis (G. McLachlan *et al.*), atherosclerosis (M.W. Miller *et al.*) obesity (B.B. Lowell) and many other diseases. Transfer of oncogenes into animals can also lead to the isolation of immortalized cellular clones from tumours (A. Pavirani).

In some cases, a transgenic animal may not only be a useful model to study a human disease but also a tool to define vectors for gene therapy (G. McLahan *et al.*).

Xenotransplantation of pig organs to humans is expected to become a reality when appropriate genes inhibiting hyperacute rejection and other related phenomena are transferred to the animals (J.L. Platt *et al.*).

The Preparation of Recombinant Proteins

The use of transgenic animals as living fermentors is becoming a reality (Houdebine, 1994). Several milligrams of recombinant proteins per ml milk have been obtained from mouse, rabbit (M.G. Stinnakre *et al.*), sheep (G. Wright *et al.*), goat (P. DiTullio) and cow (A.M. Loos) milk. The proteins are not in all cases fully glycosylated or post-transcriptionally matured. The engineering of the mammary cells by transferring a gene like that coding for furin proved to be possible and to allow the secretion of fully cleaved human protein C in the milk of double transgenic pigs (Drews *et al.*, 1995).

The Modification of Milk Composition

Milk represents about 30% of the proteins of human food in developed countries. Ruminant milk is not fully adapted to human consumption or to the dairy industry. Modifications of milk composition are possible and may lead to the preparation of milk having different properties (J.C. Mercier).

Milk may also be the vehicle for various factors having bacteriostatic actions such as lactoferrin, lysozyme and various antibodies. It may also contain antigenes active by the oral route. A relatively large variety of milk with specific properties is expected to be available in the next century.

The Protection of Animals against Diseases

Several cellular and molecular mechanisms can be used to block the infection of a cell or living organism. The success obtained with transgenic plants leads us to be optimistic for animals. Overexpression of capsid or envelope viral proteins may hamper the formation of viral particles (Salter and Crittenden, 1989). Antisense RNA and ribozymes may block selectively the expression of viral genes (G.G. Carmichael and Han *et al.*). Nucleic acid sequences used as decoys may theoretically in some cases trap limiting viral proteins such as RNA polymerase and thus strongly attenuate viral genome replication or transcription.

Secreted (M. Müller *et al.*) or nonsecreted antibodies (S.D. Jones *et al.*) may also inhibit propagation of pathogens in animals.

Natural resistance genes involving the activation of the host defense system or any other mechanisms can be potentially used (E. Meier *et al.*).

The Introduction of New Genetic Traits in Herds

A very limited number of specific genes have been transferred into domestic animals to improve production. The GH gene remains one interesting possibility namely in fish and pig if the transgenes are expressed at a moderate level or in a controlled manner (R.H. Devlin).

Enhanced wool growth has been observed in sheep harbouring the IGF-1 gene driven by the keratin gene promoter (D.W. Bullock *et al.*).

Several other interesting projects are under study to improve animal production (K. Ward). Genome mapping will progressively provide researchers with interesting genes to be transferred into birds via transgenesis (M. Georges).

The New Problems Generated by Transgenic Animals

Transgenic animals are often unique as gene transfer has occurred in a random fashion. In practice, transgenic animals are generally studied for one of their properties. There is generally a considerable wastage of potentially interesting experimental animals and of relevant information. Due to the high cost of animal breeding, most of the experimental animals are rapidly eliminated at the end of the projects. A repository for transgenic animals is available in the U.S. (J.J. Sharp) and is being developed in the E.U. (in Rome). The transgenics can be kept as living animals, as frozen semen or as frozen embryos.

A specific database (TBASE) potentially containing all the relevant information on conventional transgenic and knock-out animals (still living or not) is now available (A.V. Anagnostopoulos). The use of this database has become inevitable due to the overwhelming and increasing number of experimental data.

Patenting transgenic animals may be a convenient way to protect researchers and companies. This legal protocol is still a matter of controversy in the E.U. (J. Warcoin).

A transgenic animal is fundamentally a more or less unknown living organism since the transgene effect may be very complex and quite unexpected. Guidelines for the maintenance of transgenic animals in facilities have been defined (L.M. Houdebine). No general rules have yet been proposed to control the dissemination of transgenic animals in the environment. No candidate is presently available. As is the case for transgenic plants, these problems will be solved on a case by case basis in due time.

New products originating from transgenic animals (pharmaceuticals, organs and food) are progressively going to be available for humans. New rules for the use of these products have been defined (M.A. Miller *et al.*; Report from The Food and Drug Administration). They will be adapted on a case by case basis taking advantage of experience.

A certain number of people are afraid of the so-called gene manipulations which seem particularly diabolical when applied to animals and potentially to humans. Even if, obviously, many of these fears are due to misinformation of public opinion, genuine and new ethical problems have emerged with transgenesis. E. Schroten has given his personal view on this point.

References

Archer, J.S., Kennan, W.S., Gould, M.N. and Bremel, R.D. (1994) Human growth hormone (hGH) secretion in milk of goats after direct transfer of the hGH gene into the mammary gland by using replication-defective retrovirus vectors. *Proc. Natl. Acad. Sci. USA* **91**, 6840–6844

Attal, J., Cajero-Juarez, Marco and Houdebine, L.M. (1995) A simple method of DNA extraction from whole tissues and blood using glass powder for detection of transgenic animals by PCR. *Transgenic Res.* **4**, 149–150

Baldwin, M.A., Cohen, F.E. and Prusiner, S.B. (1995) Prion protein isoforms, a convergence of biological and structural investigations. *J. Biol. Chem.* **270**, 19197–19200

Brinster, R.L. and Zimmermann, J.W. (1994) Spermatogenesis following male germ-cell transplantation. *Proc. Natl. Acad. Sci. USA* **91**, 11298–11302

Campbell, K.H.S., McWhir, J., Ritchie, W.A. and Wilmut, I. (1996) Sheep cloned by nuclear transfer from a cultured cell line. *Nature* **380**, 64–66

Chang, P.Y., Benecke, H., Le Marchand-Brustel, Y., Lawitts, J. and Moller, D.E. (1994) Expression of a dominant-negative mutant human insulin receptor in the muscle of transgenic mice. *J. Biol. Chem.* **269**, 16034–16040

Chen, C.M., Choo, K.B. and Cheng, W.T.K. (1995) Frequent deletions and sequence aberrations at the transgene junctions of transgenic mice carrying the papillomavirus regulatory and the SV40 TAg gene sequences. *Transgenic Research* **4**, 52–59

Drews, R., Paleyanda, R.K., Lee, T.K., Chang, R.R., Rehemtulla, A., Kaufman R.J., Drohan, W.N. and Lubon, H. (1995) Proteolytic maturation of protein C upon engineering the mouse mammary gland to express furin. *Proc. Natl. Acad. Sci. USA* **92**, 10462–10466

Gordon, J.W., Scangos, G.A., Plotkin, D.J., Barbosa, J.A. and Ruddle, F.H. (1980) Genetic transformation of mouse embryos by microinjection of purified DNA. *Proc. Natl. Acad. Sci. USA* **77**, 7380–7384

Houdebine, L.M. (1994) Production of pharmaceutical proteins from transgenic animals. *J. Biotechnol.* **34**, 269–287

Krishnan, S., Haensler, J. and Meulien, P. (1995) Paving the way towards DNA vaccines. *Nature Medicine* **1**, 521–587

Liu, K., Sandgren, E.P., Palmiter, R.D. and Stein, A. (1995) Rat growth hormone gene introns stimulate nucleosome alignment *in vitro* and in transgenic mice. *Proc. Natl. Acad. Sci. USA* **92**, 7724–7728

Methot, D., Reudelhuber, T.L. and Silversides, D.W. (1995) Evaluation of tyrosinase minigene co-injection as a marker for genetic manipulations in transgenic mice. *Nucleic Acids Res.* **23**, 4551–4556

Nakajima, M., Hutchinson, H.G., Fujinaga, M., Hayashida, W., Morishita, R., Zhang, L., Horiuchi, M., Pratt, R.E. and Dzau, V.J. (1995) The angiotensin II type 2 (AT2) receptor antogonizes the growth effects of the AT1 receptor: gain-of-function study using gene transfer. *Proc. Natl. Acad. Sci. USA* **92**, 10663–10667

Palmiter, R.D., Brinster, R.L., Hammer, R.E., Trumbauer, M.E., Rosenfeld, M.G., Birnberg, N.C. and Evans, R.M. (1982) Dramatic growth of mice that develop from eggs microinjected with metallothionein-growth hormone fusion genes. *Nature* **300**, 611–615

Palmiter, R.D., Sandgren, E.P., Avarbock, M.R., Allen, D.D. and Brinster, R.L. (1991) Heterologous introns can enhance expression of transgenes in mice. *Proc. Natl. Acad. Sci. USA* **88**, 478–482

Pastan, I., Willingham, M.C. and Gottesman, M. (1991) Molecular manipulations of the multidrug transporter: a new role for transgenic mice. *FASEB J.* **5**, 2523–2528

Pawlik, K.M., Sun, C.W., Higgins, N.P. and Townes, T.M. (1995) End joining of genomic DNA and transgene DNA in fertilized mouse eggs. *Gene* **165**, 173–181

Pieper, F.R., De Wit, I.C.M., Pronk, A.C.J., Kooiman, P.M., Strijker, R., Krimpenfort, P.J.A., Nuyens, J.H. and De Boer, H.A. (1992) Efficient generation of functional transgenes by homologous recombination in murine zygotes. *Nucleic Acids Res.* **20**, 1259–1264

Presnail, J.K. and Hoy, M.A. (1992) Stable genetic transformation of a beneficial arthropod, Metaseiulus occidetalis (Acari: Phytoseiidae), by a microinjection technique. *Proc. Natl. Acad. Sci. USA* **89**, 7732–7736

Smith, D.J., Zhu, Y., Zhang, J., Cheng, J.F. and Rubin, E.M. (1995) Construction of a panel of transgenic mice containing a contiguous 2-Mb set of YAC/P_1 clones from human chromosome 21q22.2. *Genomics* **27**, 425–434

Sternberg, N.T. (1992) Cloning high molecular weight DNA fragments by the bacteriophage P_1 system. *TIG* **8**, 11–16

Stevens, A.M. and Yu-Lee L.-Y. (1994) Multiple prolactin-responsive elements mediate G1 and S phase expression of the interferon regulatory factor-1 gene. *Mol. Endocrinol.* **8**, 345–355

Struthers, R.S., Vale, W.W., Arias, C., Sawchenko, P.E. and Montminy, M.R. (1991) Somatotroph hypoplasia and dwarfism in transgenic mice expressing a non-phosphorylatable CREB mutant. *Nature* **350**, 622–624

Thompson, E.M., Adenot, P., Tsuji, F.I. and Renard, J.P. (1995) Real time imaging of transcriptional activity in live mouse preimplantation embryos using a secreted luciferase. *PNAS USA, Developmental Biology* **92**, 1317–1321

Yagi, T., Aizawa, S., Tokunaga, T., Shigetani, Y., Takeda, N. and Ikawa, Y. (1993) A role for Fyn tyrosine kinase in the suckling behaviour of neonatal mice. *Nature* **366**, 742–745

Points to consider in the manufacture and testing of therapeutic products for human use derived from transgenic animals. Prepared by: U.S. Food and Drug Administration Center for Biologics Evaluation and Research. 1995

PART II, SECTION A

2. Transgenic Rats

John J. Mullins*, Gillian Brooker and Linda J. Mullins

*Centre for Genome Research, The University of Edinburgh,
West Mains Road, Edinburgh, EH9 3JQ, UK*

Introduction

To date the vast majority of transgenic experiments have been performed in the mouse, reflecting the cost effectiveness of using a small species with high fecundity and rapid generation time. However, despite the inherent constraints of extending transgenic analysis to the rat, for certain applications this species is more appropriate simply because of its size. The need for multiple analyses to be performed on a single sample of, for example, blood, may mean that insufficient material can be obtained from a mouse. Additionally, the rat is more amenable to chronic studies in which sequential sampling is essential. For long-term monitoring which requires surgical implantation, e.g. blood pressure measurement via telemetry, the rat is, once again, the species of choice because of its size. In this section we will briefly outline aspects of general husbandry which must be considered when using the rat, and will highlight some technical details specific to the generation of transgenics in this species.

Animal Husbandry

Approximately 4–5 times more animal space is required to run a transgenic rat programme of equivalent size to that typically required for the mouse. Should animal house space be limiting, economic considerations may justify the use of a commercial breeder for the production and supply of donor animals. However, if this route is chosen, the quality and exact age of the animals supplied must be guaranteed by the supplier.

The gestation time of the rat is typically 20–24 days, compared to 19–21 days in the mouse, and the age of sexual maturation is significantly longer for the rat than for the mouse (8–12 weeks in the rat as opposed to 5–8 weeks in the mouse). Therefore the derivation of a transgenic rat line takes considerably longer than for an equivalent strain of transgenic mouse. Similar variation in litter size is observed in the rat compared to the mouse and litter size is strain-dependent.

One problem which may be encountered if the recipient females are in any way stressed, is cannibalism, where the mother will eat some or all of her pups. Should this become a serious problem, then where possible, one should consider transferring the embryos to a recipient female from a robust strain such as Sprague Dawley, irrespective of the genetic origin of the zygotes. The fostering of pups at birth (or within 1–2 days of age) to an experienced mother may also overcome difficulties with poor mothering, but the success rate depends very largely on the skill of the animal care staff and the quality of the animal husbandry.

Other Considerations

It is essential that the rat colony should be free of any infection, before a transgenic programme is undertaken. In addition, the rat seems to be particularly sensitive to environmental stress such as adverse humidity, or over-crowding. For the most valuable animals, in particular the transgenic founders, it is advisable to use a reliable identification system, such as microchip-tagging. The finance and time invested in the generation of transgenic animals is significant, and problems in identification may result in repeat testing, costly delays, and

*Author for Correspondence.

possibly even the loss of founder animals through mistaken identity.

Technical Aspects of Transgenic Rat Production

Choice of Strain

The choice of strain used for superovulation has a major bearing on the number of recoverable fertilized eggs. Generally speaking, outbred strains such as Sprague Dawley, or F_1 hybrid animals between two inbred strains yield higher numbers of eggs than inbred strains. Additionally, the inbred strains generally exhibit poorer breeding performance, producing reduced litter sizes and being less capable mothers. In the final analysis, the choice of strain will reflect the constraints of the experimental design which may necessitate the use of a particular strain despite other technical drawbacks. However, it must be remembered that the choice of strain has far-reaching implications due to the effect of the genetic background on the expression of the transgene or its resulting phenotype. The selection of suitable controls deserves careful consideration. For example, if one chooses to use an outbred Sprague Dawley strain, then by definition the derivation of a transgenic line will create a new 'substrain' and care must then be taken in choosing suitable control animals for a given experiment. Unless the transgenic animals have been produced on a pure inbred background or careful backcrossing has been performed, the optimal controls would be non-transgenic litter mates. It is also worth pointing out that even Sprague Dawley rats from different sources may exhibit different phenotypes when expressing the identical transgene (see below).

Superovulation

In our experience superovulation in the rat is best achieved by the procedure of Armstrong and Opavsky (1988). An 'Alzet' osmotic minipump (Alza Corporation, USA) is implanted, subcutaneously, in 30–34 day-old virgin female rats. The age at which superovulation is initiated is critical and should be optimised for each strain. Females should be as old as possible but must be used prior to the natural (strain-dependent) onset of vaginal opening. The miniosmotic implant releases follicle-stimulating hormone (FSH) [porcine 'Folltropin', Vetrepharm, Ontario, Canada; or ovine 'Ovagen', Genus, Newcastle-Upon-Tyne, England] in a constant controlled manner. Forty eight hours after the initiation of FSH infusion an injection of human chorionic gonadotrophin (Sigma, England) is given and the females mated 12 hours later with a proven stud male. Depending on their size the osmotic minipumps can be used more than once if they are carefully cleaned and refilled. This makes them more cost-effective, but the success rate is usually lower. On the morning following mating fertilised eggs are flushed from the oviduct under a dissecting microscope. Unlike the mouse, the mated female rat often does not have a visible plug which may be deep-seated, and can sometimes only be observed using a probe. Using the superovulation protocol, between 30 and 110 fertilized eggs can be recovered from a single female, depending on the strain of rat. The percentage of eggs which will be healthy (30% to 90%), and the time of appearance of pronucleii, can also vary enormously between strains. Therefore, parameters such as the light/dark regime, timings of drug administration, mating, egg collection and microinjection should be optimised for any new strain of rat being considered as a transgene host.

Culturing of Embryos

Once the fertilised eggs have been recovered from a superovulated female, surrounding cumulus cells are removed with hyaluronidase (Calbiochem, USA). Most commercial sources supply a relatively low specific activity enzyme, and for the efficient removal of cumulus cells a specific activity in the range of 500 units/mg is preferable. The cells are rinsed in M2 culture medium and then maintained in M16 culture medium, exactly as for mouse microinjection protocols (Hogan et al., 1986) until the two pronucleii become visible (usually within 1–4 hours). At this time, DNA is microinjected into the male pronucleus of the zygote. Due to the increased elasticity of the rat oocyte membrane compared to that of the mouse, it is usually more time-consuming to perform microinjection in the rat. However having adapted to using rat zygotes, some people prefer using them to working with the mouse. We have commonly observed that the tip of the microinjection needle must be pushed through the pronucleus and then drawn back into it in order to achieve penetration of the membrane. Following microinjection, culturing zygotes to the two-cell stage allows the selection of healthy zygotes for reimplantion, however the majority of damaged zygotes are easily identified within 2–3 hours of injection and we find it preferable to reimplant zygotes on the same day as the microinjection.

Ovarian Blood Supply

Injected eggs are surgically implanted into the infundibulum of the recipient pseudopregnant female. Due to the highly vascularised nature of the bursa membrane in the rat, there may be excessive bleeding when it is ruptured, completely obscuring the entrance to the infundibulum. This can be prevented either by cauterisation, or by the topical application of adrenalin. Typically 15 injected zygotes are introduced into each oviduct, and the embryos are allowed to develop to term. The use of sterile surgical technique is essential to achieve a high pregnancy success rate and the treatment of the recipient females with an antibiotic at the time of surgery may further improve the results.

Other Considerations

Screening of G_0 animals is performed using standard procedures such as Southern blot hybridisation or Polymerase Chain Reaction (PCR) screening using DNA isolated from a tail biopsy or a blood sample. We have noted a higher degree of mosaicism and multiple insertion sites in transgenic rat founders than has been our experience with mice. Although care must be taken to identify such animals, the transgene insertions can be segregated through breeding and individual sublines analysed.

Transgenic Rat Models of Disease

Certain fields of research such as neurobiology, reproductive and cardiovascular biology lend themselves to the use of larger rodents, simply because of the database of existing research, the availability of specific models, or the inability to miniaturise monitoring systems to permit data acquisition in the mouse. The number of laboratories which routinely generate transgenic rats is still small. However, a growing number of transgenic rat models have been developed in the last few years and they have also been used for the expression of transgene-encoded proteins in milk, and for xenotransplantation experiments. Of particular interest are the models where expression of HLA-B27 and human β2-microglobulin led to the development of arthritis (Hammer *et al.*, 1990), and the expression of mouse *Ren*-2 renin caused hypertension (Mullins *et al.*, 1990). In both cases, the respective gene, expressed in a number of transgenic mouse lines, failed to elicit any equivalent phenotype demonstrating that the choice of species can have a considerable bearing on the outcome of the experiment. The hypertension model provides an example of genetic background effects and the importance of strain selection. When the *Ren*-2 transgene was bred onto Sprague Dawley strains from two different sources, a difference in the phenotype became apparent with the progeny of one cross having a significantly higher incidence of malignant hypertension than those of the other (Whitworth *et al.*, 1995).

Future Developments

With increased understanding of locus control regions and matrix attachment sites, transgene construction will be continually refined, to produce better animal models. Attempts are being made to deliver corrective transgenes to specific anatomical sites such as the eye, as a means to correct an inborn defect. The size of the rat makes it more amenable than the mouse to such surgical procedures and therefore potentially valuable in the development of gene therapy methodologies. One noticeable gap in the experimental possibilities for the rat is the lack of Embryonic Stem cells (ES cells). The application of embryonic stem (ES) cell technology in this species will have a dramatic effect on the use of the rat in transgenic experiments and to this end a number of groups are attempting to isolate rat ES cells. Iannaconne *et al.* (1995) have reported the isolation of pluripotent rat ES cells which can contribute to chimaeras, however to date no germline transmission has been demonstrated. The availability of such tools must remain a major goal for germline manipulation studies in the rat.

References

Armstrong, D.T. and Opavsky, M.A. (1988) Superovulation of immature rats by continuous infusion of follicle-stimulating hormone. *Biology of Reproduction* **39**, 511–518

Hammer, R.E., Maika, S.D., Richardson, J.A., Tang, J-P. and Taurog, J.D. (1990) Spontaneous inflammatory disease in transgenic rats expressing HLA-B27-associated human disorders. *Cell* **63**, 1099–1112

Hogan, B., Costantini, F. and Lacy, E. (1986) Manipulating the Mouse Embryo – A Laboratory Manual. Cold Spring Harbor Laboratories

Iannaccone, P.M., Taborn, G.U., Garton, R.L., Caplice, M.D. and Brenin, D.R. (1994) Pluripotent embryonic stem cells from the rat are capable of producing chimeras. *Dev. Biol.* **163**, 288–292

Mullins, J.J., Peters, J. and Ganten, D. (1990) Fulminant hypertension in transgenic rats harbouring the mouse Ren-2 Gene. *Nature* **344**, 541–544

Whitworth, C., Fleming, S., Kotelevtsev, Y., Manson, L., Brooker, G., Cumming, A. and Mullins, J.J. (1995) A genetic model of malignant phase hypertension in rats. *Kidney Int.* **47**, 529–535

PART II, SECTION A

3. The Generation of Transgenic Rabbits

Céline Viglietta[1], Micheline Massoud[2] and Louis Marie Houdebine[1]

[1]Unité de Differenciation Cellulaire, [2]Laboratoire de Physiologie Animale,
Institut National de Recherche Agronomique, 78352 Jouy-en-Josas Cedex, France

The first transgenic rabbits were obtained more than 10 years ago (Hammer et al., 1985). No particular difficulty appeared to generate these animals. Rabbits are sexually mature after 5–6 months. The average number of offspring per pregnancy is about 8, and up to 40 embryos can be obtained per female after superovulation. Pronuclei are as visible as in the mouse. Embryo transfer into recipient rabbit females was a routinely used technique long before gene transfer was possible. Yet, the number of transgenic rabbits still remains very limited. The rabbit offers several advantages over the mouse. It is a much larger animal and a certain number of manipulations can be more easily performed in this species than in the mouse. Due to its size, the rabbit can also be a laboratory and an industrial producer of recombinant proteins (Stinnakre et al., 1996). Human α-antitrypsin was produced in the blood of transgenic rabbits at the concentration of 1 mg/ml (Massoud et al., 1991). Foreign antibodies were also obtained in the blood (Weidle et al., 1991) and milk (Castro et al., 1996) of transgenic rabbits. Human interleukin 2 (Bühler et al., 1990) tissue plasminogen activator (Riego et al., 1993), chymosin (Brem et al., 1995), human erythropoietin (Castro et al., 1995), human IGF-1 (Brem et al., 1994) and human superoxide dismutase (Hansson et al., unpublished data) were secreted in milk from transgenic rabbits in amounts ranging form a few ng/ml to several mg/ml.

The relatively large size of the rabbit is also a drawback. Due to the required space, somewhat costly facilities must be used to breed this animal. The fact that rabbit is sexually mature after several months still increases the cost of the experiments.

The rabbit also shows a specific interest. It is a good model for some particular studies. This is clearly the case for the study of lipid metabolism which is almost similar in rabbit and human but quite different in mouse. For this reason, transgenic rabbits expressing human hepatic lipase gene (Fan et al., 1994), human apolipoprotein A1 (Duverger et al., 1995), human apolipoprotein B100 (Fan et al., 1996 and Duverger et al., unpublished data), lecithin-cholesterol acyl transferase (Hoeg et al., 1995 and Brousseau et al., 1995), human apolipoprotein E (Fan et al., 1995) human 15-lipoxygenase (Shen et al., 1995) and other human genes related to lipid metabolism (Duverger et al., unpublished data) have been obtained. They offer satisfactory models for the study of atherosclerosis.

Similarly, rabbit cells show an exceptional capacity to replicate the human immunodeficiency virus. Transgenic rabbits expressing in their T lymphocytes the human CD4 gene, which is one essential element of the virus receptor, have been generated (Dunn et al., 1995 and Snyder et al., 1995). These animals can be infected more efficiently by the virus than control animals (Dunn et al., 1995).

Obtaining monoclonal antibodies from rabbits would be of interest. This implies that immortalized rabbit lymphocyte lines are available to generate hybridomas. Transgenesis using c-myc (Knight et al., 1988 and Sethupathi et al., 1994) and IL-6 genes (A. Pavirani et al., unpublished data) was performed but with no successful generation of utilizable cell lines. More recently, an association of c-myc and v-abl transgenes led to the production of stable hybridomas and of rabbit monoclonal antibodies (Spieker-Polet et al., 1995).

Transgenic rabbits expressing rabbit papilloma and EJ-ras genes were obtained several years ago (Peng et al., 1993). These rabbits appeared to be a potentially better model than previous ones to study a specific variety of skin cancer.

Rabbits are also bred for human consumption and are appreciated by hunters. Improving existing rabbit lines and specially enhancing their resistance to viral diseases such as streptohaemmoragy or myxomatosis may be justified. Attempts to protect transgenic rabbits against infection by adenovirus 5 have been done (Ernst et al., 1991).

The Experimental Conditions to Generate Transgenic Rabbits

In the rabbit, ovulation does not occur spontaneously, it is provoked by mating. A female which accepts the male becomes therefore normally pregnant. A better receptivity towards the male can be obtained by keeping the females before mating for at least one week in long days (18 hours of light) after a longer period in short days (8 hours of light).

Superovulation enhances quite significantly the number of embryos and up to 40 embryos per female can be obtained in this routine. Superovulation can be obtained after different treatments with gonadotropins (Brem et al., 1985; Knight et al., 1988; Voss et al., 1990; Ernst et al., 1991 and Fan et al., 1994). We routinely use the following protocol. Each female is treated intramuscularly for 3 days by porcine FSH. On the first day, 250 µg are injected on morning and evening. On the second day, 625 µg are injected also on morning and evening. On the third day, in the morning, 250 µg are injected. At 3 p.m. the female is mated and 330 µg of porcine LH are then injected intramuscularly. The next day at 9.30 a.m. the embryos are collected by flushing. By using the relatively long delay between mating and the sacrifice of animals, the embryos need not to be treated by hyaluronidase and microdissected to remove surrounding cells which are spontaneously released. Embryos are maintained in medium B2 Ménézo at 39°C. Microinjections are carried out immediately. Embryos are then reimplanted in both uterine horns (10 embryos on each side) of recipient females prepared by mating with a vasectomized male. The recipient females must preferably be mated the day before reimplantation of embryos.

The identification of transgenes can be carried out from pieces of tail or ears. These tissue samples can be collected about 5 days after birth. This can be done sooner or later if needed. Ear may be preferred to tail. The different ways of collecting pieces of ear may also allow individual identification of animals. PCR analysis after a DNA extraction with Geneclean (Attal et al., 1995) is efficient with this material.

To the best of our knowledge, essentially New Zealand rabbits have been used successfully to generate transgenics. About 15% of manipulated embryos give rise to newborn animals. About 1.5% of the reimplanted embryos become transgenic offsprings. This yield is systematically lower than in the mouse. It is not known if other lines of rabbit would improve or reduce the efficiency of microinjection. Changing the time of microinjections after fertilization does not seem to improve the survival of embryos significantly, the proportion of transgenic newborns or the transmission of the transgenes to progeny. Microinjections are generally less easily performed later than 20 hours after mating. Optimizing breeding conditions can lead to a better yield of surviving transgenic rabbits (Wang et al., 1996 and Ramos et al., 1996).

Transgenes are transmitted from rabbit founders to offspring usually with good yield. Some of the animals are obviously mosaic, as judged by the transmission frequency of the transgenes. No systematic study seems to have been done with this species. An approximative estimation suggests that the degree of mosaicism might be essentially similar in mouse and rabbit.

ES cells from rabbit would be very useful for many purposes. Although several publications reported that ES cells from rabbit were obtained (Graves and Moreadith, 1993), it seems that the establishment of rabbit ES cells remains to be done.

References

Attal, J., Cajero-Juarez, M. and Houdebine, L.M. (1995) A simple method of DNA extraction from whole tissues and blood using glass powder for detection of transgenic animals by PCR. *Transgenic Res.* **4**, 149–150

Brem, G., Brenig, B., Goodman, H.M., Selden, R.C., Graf, F., Kruff, B., Springman, K., Hondele, J., Meyer, J., Winnacker, E.L. and Kräublich, H. (1985) Production of transgenic mice, rabbits and pigs by microinjection into pronuclei. *Zuchthygiene* **20**, 251–252

Brem, G., Hartl, P., Besenfelder, U., Wolf, E., Zinovieva, N. and Pfaller, R. (1994) Expression of synthetic cDNA sequences encoding human insulin-like growth factor-1 (IGF-1) in the mammary gland of transgenic rabbits. *Gene* **149**, 351–355

Brem, G., Besenfelder, U., Zinovieva, N., Seregi, J., Solti, L. and Hartl, P. (1995) Mammary gland specific expression of chymosin constructs in transgenic rabbits. *Theriogenology* **43**, 175

Brousseau, M.E., Zech, L.A., Berard, A., Valsman, B.L., Meyn, S.M., Powell, D., Santamarina-Fojo, S., Brewer Jr., H.B. and Hoeg, J.M. (1995) LCAT overexpression in transgenic rabbits delays ApoA-1 catabolism in a gene dose-dependent manner and enhances LDL remodelling. *Circulation* **92**, 8

Bühler, T.A., Bruyère, T., Went, D.F., Stranzinger, G. and Bürki, K. (1990) Rabbit β-casein promoter directs secretion of human interleukin-2 into the milk of transgenic rabbits. *Biotech.* **8**, 140–143

Castro, F.O., Aguirre, A., Fuentes, P., Ramos, B., Rodriguez, A. and De La Fuente, J. (1995) Secretion of human erythropoetin by mammary gland explants from lactating transgenic rabbits. *Theriogenology* **43**, 184

Castro, F.O., Limonta, J., Gavilondo, J. and De La Fuente, J. (1996) Expression of humanized chimaeric antibodies in transgenic mouse milk and in CHO cells: implications for biotechnology. *Theriogenology* **45**, 340

Dunn, C.S., Methali, M., Houdebine, L.M., Gut, J.P., Kirn, A. and Aubertin, A.M. (1995) Human immuno-deficiency virus type 1 infection of human CD4-transgenic rabbits. *J. Gen. Virol.* **76**, 1327–1336

Duverger, N., Emmanuel, F., Viglietta, C., Attenot, F., Viry, I., Houdebine, L.M. and Denèfle, P. (1995) Overexpression of human apolipoprotein A-I in normal and Watanabe rabbits. *Atherosclerosis* 115

Duverger, N., Viglietta, C., Berthou, L., Emmanuel, F., Tailleux, A., Nihoul, L., Laine, B., Fievet, C., Castro, G., Frichart, J.C., Houdebine, L.M. and Denèfle, P. Transgenic rabbits expressing human apolipoprotein A-I in the liver. Arteriosclerosis. Thrombosis and Vascular Biology (in press)

Ernst, L.K., Zakcharchenko, V.I., Suraeva, N.M., Ponomareva, T.I., Miroshnichenko, O.I., Prokof'ev, M.I. and Tikchonenko, T.I. (1991) Transgenic rabbits with antisense RNA gene targeted at adenovirus H5. *Theriogenology* **35**, 1257–1271

Fan, J., Wang, J., Bensadoun, A., Lauer, S.J., Dang, Q., Mahley, R.W. and Taylor, J.M. (1994) Overexpression of hepatic lipase in transgenic rabbits leads to a marked reduction of plasma high density lipoproteins and intermediate density lipoproteins. *Proc. Natl. Acad. Sci. USA* **91**, 8724–8728

Fan, J., McCormick, S.P.A., Krauss, R.M., Taylor, S., Quan R., Taylor, J.M. and Young, G. Overespression of human apolipoprotein B100 transgenic rabbits in increased levels of low density of lipoproteins and decreased levels of high density lipoproteins. *Arterioscler. Thromb. Vasc. Biol.* (in press)

Graves, K.H. and Moreadith, R.W. (1993) Derivation and characterization of putative pluripotential embryonic stem cells from preimplantation rabbit embryos. *Mol. Reprod. Dev.* **36**, 424–433

Hammer, R.E., Pursel, V.G., Rexroad Jr., C.E., Wall, R.J., Bolt, D.J., Ebert, K.M., Palmiter, R.D. and Brinster, R.L. (1985) Production of transgenic rabbits, sheep and pigs by microinjection. *Nature* **315**, 680–683

Hoeg, M., Santamarina-Fojo, B., Vaisman, B., Hoyt Jr., R.F., Feldman, S., Cornhill, F., Herderick, E., Talley, G., Wood, D., Marcovina, S. and Brewer Jr., H.B. (1995) Lecithin-cholesterol acyl transferase transgenic rabbits: hyperalphalipoproteinemia and diet induced atherosclerosis. *Circulation* **92**, 8

Knight, K.L., Spieker-Polet, H., Kazdin, D.S. and Or, V.T. (1988) Transgenic rabbits with lymphocytic leukemia induced by the c-myc oncogene fused with the immunoglobulin heavy chain enhancer. *Proc. Natl. Acad. Sci. USA* **85**, 3130–3134

Massoud, M., Bischoff, R., Clesse, D., Dalemans, W., Pointu, H., Attal, J., Schultz, H., Stinnakre, M.G., Pavirani, A. and Houdebine, L.M. (1991) Expression of active recombinant human a1-antitrypsin in transgenic rabbits. *J. Biotech.* **18**, 193–204

Peng, X., Olson, R.O., Christian, C.B., Lang, C.M. and Kreider, J.W. (1993) Papillomas, and carcinomas in transgenic rabbits carrying EJ-*ras* DNA and cottontail rabbit papillomavirus DNA. *J. Virol.* **67**, 1698–1701

Ramos, B., Pichardo, D., Aguilar, A., Puentes, P. and Castro, F.O. (1996) The use of multiparous does as recipients of microinjected embryos improve survival of the litters at weaning. *Theriogenology* **45**, 350

Riego, E., Limonta, J., Aguilar, A., Perez, A., De Armas, R., Solano, R., Ramos, B., Castro, F.O. and De La Fuente, J. (1993) Production of transgenic mice and rabbits that carry and express the human tissue plasminogen activator cDNA under the control of a bovine alpha S1 casein promoter. *Theriogenology* **39**, 11731–185

Sethupathi, P., Spieker-Polet, H., Polet, H., Yam, P., Tunyaplin, C. and Knight, K.L. (1994) Lymphoid and non-lymphoid tumors in E_k-myc transgenic rabbits. *Leukemia* **8**, 2144–2155

Shen, J., Kühn, H., Petho-Schramm, A. and Chan, L. (1995) Transgenic rabbits with the integrated human 15-lipoxygenase gene driven by a lysozyme promoter: macrophage-specific expression and variable positional specificity of the transgenic enzyme. *FASEB J.* **9**, 1623–1631

Snyder, B.W., Vitale, J., Milos, P., Gosselin, J., Gillespie, F., Ebert, K., Hague, B.F., Kindt, T.J., Wadsworth, S. and Leibowitz, P. (1995) Developmental and tissue-specific expression of human CD4 in transgenic rabbits. *Mol. Reprod. Dev.* **40**, 419–428

Spieker-Polet, H., Sethupathi, P., Yam, P. and Knight, K.L. (1995) Rabbit monoclonal antibodies: generating a fusion partner to produce rabbit-rabbit hybridomas. *Proc. Natl. Acad. Sci. USA* **92**, 9348–9352

Stinnakre, M.G., Massoud, M., Viglietta, C. and Houdebine, L.M. (1996) The preparation of recombinant proteins from mouse and rabbit milk for biomedical and pharmaceutical studies (this book)

Voss, A.K., Sandmöller, A., Suske, G., Strojek, R.M., Beato, M. and Hahn, J. (1990) A comparison of mouse and rabbit embryos for the production of transgenic animals by pronuclear microinjection. *Theriogenology* **34**, 813–824

Wang, B., Page, R.L. and Yang, X. (1996) Improved transgenic efficiency in rabbits attributed to successful microinjection, embryo transfer and animal husbandry procedures. *Theriogenology* **45**, 342

Weidle, U.H., Lenz, H. and Brem, G. (1991) Genes encoding a mouse monoclonal antibody are expressed in transgenic mice, rabbits and pigs. *Gene* **98**, 185–191

PART II, SECTION A

4. Techniques and Problems in Producing Transgenic Pigs

Vernon G. Pursel

US Department of Agriculture, Agricultural Research Service, Beltsville Agricultural Research Center,
Gene Evaluation and Mapping Laboratory, Beltsville, Maryland 20705, USA

Why Create Transgenic Pigs?

Transgenic pigs have been produced for a variety of purposes that fall in the following categories: (1) to improve their productivity, (2) to enhance their resistance to diseases, (3) for biomedical products, and (4) for models of human diseases.

In comparison to the other large farm species, the pig offers advantages of producing a litter, thus fewer recipients are required, a gestation period of only 114 days, and a generation interval of only one year. These characteristics in combination also provide for ease of subsequent propagation and expansion of transgenic lines. Although pigs are available in three sizes (domestic, mini, and micro) all transgenic pigs thus far produced have been from full-sized domestic swine. The reasons for this are: domestics are in greater supply, they cost less per animal, and they usually have larger litters than mini- or micro-swine. In addition, genetic research with mini- and micro-swine is also hindered by suppliers' purchasing agreements that usually prohibit their propagation.

Improve Productivity

Transfer of genes for improved animal productivity traits, such as feed conversion, rate of gain, reduction of fat, and improved quality of meat may in the future have a dramatic impact on the swine industry. These productivity traits are controlled by numerous genes, most of which have not been identified. Much of the early transgenic pig research was devoted to transfer of growth hormone genes. Marketing of pigs with elevated growth hormone offers considerable potential value to both the producer and the consumer. Even if only the improved efficiency in feed utilization is considered, the potential savings amounts to more than US $1,000 million annually in lower feed costs if applied across the market hog population in the USA alone. In addition, the producer could market 10% to 15% more pigs with the same fixed assets, the consumer could benefit from having pork available with less fat at a lower cost, and the environment could benefit from 15% less waste matter being generated.

Enhance Health and Disease Resistance

Increasing disease resistance by transgenesis could have a major impact on the welfare of swine and the economy of swine production. Müller and Brem (1994) suggest possibilities for reducing disease susceptibility include transfer of genes: (1) known to modulate non-specific host defense mechanisms, such as cytokines, major histocompatibility complex (MHC) proteins, and T-cell receptors; (2) for specific disease resistance, such as the Mx1 gene (Staeheli et al., 1993); (3) encoding an immunoglobulin specific for a common pathogen (Weidle et al., 1991 and Lo et al., 1991); (4) for a viral protein that is able to interfere with viral replication or attachment to target cells; and (5) for antisense RNA as antipathogenic agents.

Biomedical Products

Transgenic pigs may play an important part in supplying rare biomedical products for use in human medicine. Human hemoglobin (Swanson et al., 1992 and Sharma et al., 1994) and human protein C, an essential blood clotting factor (Velander et al.,

1992), are currently being produced in swine for potential medical applications. At present, extensive research is underway to insert genes into swine so that in the future pigs may supply organs for xenotransplantation into humans. Initially, genes for inhibitors of complement are being transferred to prevent hyperacute rejection of transplanted organs (Cozzi *et al.*, 1994). Other rejection events may be prevented in the future by transfer of additional genes.

Human Disease Models

Although transgenic animal models are being used extensively in biomedical research, these studies have been primarily confined to mouse. An exception is recent production of transgenic pigs with a defective rhodopsin gene as a model for investigation of retinitis pigmentosa (Petters, 1994). Similarities of size and retinal physiology and anatomy of the pig and human eye makes pig a far better model than mouse. While mouse may be an effective transgenic model for certain biomedical problems, it has proven to be ineffective for others (e.g., cardiovascular diseases and sickle cell anemia). In the future we can expect much more use of swine for transgenic research involving the digestive and cardiovascular systems, which are quite similar for human and swine.

Methods Used to Produce Transgenic Pigs

Microinjection is the primary method used to transfer genes into swine. Successful gene transfer is dependent on surgical recovery of one-cell or two-cell zygotes for microinjection and subsequent surgical transfer into recipients that have either ovulated at the same time as the donor or several hours after the donor.

Control of Estrus and Ovulation

The method used to control the estrous cycle of swine depends on their reproductive status. Prepuberal females are injected with pregnant mares serum gonadotropin (PMSG) to stimulate follicular development and induce estrus. Cycling females are fed altrenogest (an oral progestogen) for 14–21 days (Davis *et al.*, 1979). The CL of pregnant females can be regressed by prostaglandin F2α injections. The day after altrenogest withdrawal or prostaglandin treatment, PMSG is injected to stimulate follicular development. Ovulation is induced by injection with human chorionic gonadotropin (hCG) 72 h after the PMSG injection in prepuberal and pregnant females and 78–82 h after the PMSG injection in cycling females. Donor females are either bred with a boar or artificially inseminated twice between hCG injection and ovulation (40 h after hCG).

Ovum Recovery, Microinjection and Transfer

Ovum are usually recovered during surgical laparotomy performed between 51 and 66 h after the hCG injection. A cannula is inserted into the infundibulum and culture medium is flushed through the oviduct by inserting a blunt 20-gauge needle into the tip of the uterine horn and through the uterotubal junction. Ova are collected in a sterile petri dish and located using a steriomicroscope. Since lipid granules in the cytoplasm interfere with nuclear visualization, pig ova are centrifuged at $15,000 \times g$ for 8–10 min to stratify the cytoplasm so that pronuclei are visible with use of DIC microscopy (Wall *et al.*, 1985). Pig ova have prominent nucleoli that aid in locating a pronucleus for microinjection.

Injected ova are usually cultured a minimum duration before 20 or more are transferred into one oviduct of each recipient female. Ova in a small volume of medium are aspirated into the lumen of a flexible sterile tube, which is then inserted through the infundibulum, passed into the lower half of the oviduct, and the ova are expelled as the tube is withdrawn. Pig ova can also be transferred into the oviducts of donor females immediately after ova have been recovered without compromising the pregnancy rate or subsequent litter size (Brem *et al.*, 1989).

Transgenic Efficiencies

The entire process of producing transgenic swine is quite inefficient. Typically, about 70% of treated cycling donors exhibit estrus, ovulate and produce fertile ova. Nuclei or pronuclei are visible in only about 70% of recovered ova. Between 50% and 80% of mature recipients become pregnant and farrow about 6 pigs per litter. In contrast, only 25% to 50% of prepuberal recipients farrow with a litter size averaging 4 (Brem *et al.*, 1989). The percentage of gene-injected embryos that develop into transgenic pigs varies from 0.3% to 4.0% with a mean of 0.9% (Pursel and Rexroad, 1993). In two-thirds of completed experiments, 50% or more of transgenic pigs have expressed their transgene. In addition, breeding studies with transgenic pigs indicate mosaicism is a definite problem, with about 20% of founders failing to transmit the transgene to progeny. Another 20% to 30% transmit the transgene to less than 50% of their progeny, presumably due to mosaicism in the germ cells (Pursel *et al.*, 1990).

References

Brem, G., Springmann, K., Meier, E., Krausslich, H., Brenig, B., Muller, M. and Winnaker, E.-L. (1989)

Factors in the success of transgenic pig programs. In: R.B. Church (ed.), *Transgenic Models in Medicine and Agriculture*, Vol. 116, Wiley-Liss, New York, pp. 61–72

Cozzi, E., Langford, G.A., Richards, A., Elsome, K., Lancaster, R., Chen, P., Yannoutsos, N. and White, D.J.G. (1994) Expression of human decay accelerating factor in transgenic pigs. *Transplant. Proc.* **26**, 1402–1403

Davis, D.L., Knight, J.W., Killian, D.B. and Day, B.N. (1979) Control of estrus in gilts with a progestogen. *J. Anim. Sci.* **49**, 1506–1509

Lo, D., Pursel, V., Linton, P.J., Sandgren, E., Behringer, R., Rexroad, C., Palmiter, R.D. and Brinster, R.L. (1991) Expression of mouse IgA by transgenic mice, pigs and sheep. *European J. Immunol.* **21**, 1001–1006

Müller, M. and Brem, G. (1994) Transgenic strategies to increase disease resistance in livestock. *Reprod. Fertil. Dev.* **6**, 605–613

Petters, R.M. (1994) Transgenic livestock as genetic models of human disease. *Reprod. Fertil. Dev.* **6**, 643–645

Pursel, V.G., Hammer, R.E. Bolt, D.J., Palmiter, R.D. and Brinster, R.L. (1990) Genetic engineering of swine: Integration, expression and germline transmission of growth-related genes. *J. Reprod. Fertil. Suppl.* **41**, 77–87

Pursel, V.G. and Rexroad, C.E., Jr. (1993) Status of research with transgenic farm animals. *J. Anim. Sci.* **71** (Suppl. 3), 10–19

Sharma, A., Martin, M.J., Okabe, J.F., Truglio, R.A., Dhanjal, N.K., Logan, J.S. and Kumar, R. (1994) An isologous porcine promotor permits high level expression of human hemoglobin in transgenic swine. *Bio/Technology* **12**, 55–59

Staeheli, P., Pitossi, F. and Pavlovic, J. (1993) Mx proteins: GTPases with antiviral activity. *Trends Cell Biol.* **3**, 268–272

Swanson, M.E., Martin, M.J., O'Connell, J.K., Hoover, K., Lago, W., Huntress, V., Parsons, C.T., Pinkert, C.A., Pilder, S. and Logan, J.S. (1992) Production of functional human hemoglobin in transgenic swine. *Bio/Technology* **10**, 557–559

Velander, W.H., Johnson, J.L., Page, R.L., Russell, C.G., Subramanian, A., Wilkins, T.D., Gwazdauskas, F.C., Pittius, C. and Drohan, W.N. (1992) High level expression in the milk of transgenic swine using the cDNA encoding human Protein C. *Proc. Natl. Acad. Sci. USA* **89**, 12003–12008

Wall, R.J., Pursel, V.G., Hammer, R.E. and Brinster, R.L. (1985) Development of porcine ova that were centrifuged to permit visualization of pronuclei and nuclei. *Biol. Reprod.* **32**, 645–651

Weidle, U.H., Lenz, H. and Brem, G. (1991) Genes encoding a mouse monoclonal antibody are expressed in transgenic mice, rabbits and pigs. *Gene* **98**, 185–191

PART II, SECTION A

5. Gene Transfer into Goat Embryos

William G. Gavin DVM

Genzyme Transgenics Corporation, One Mountain Road, Framingham, MA 01701-9322, USA

Animal Selection

Although there are potentially a number of applications for transgenesis in the caprine species, the following discussion will only focus on altering milk production. With this in mind, animal selection criteria will focus on selecting for and maximization of milk production.

The breeds of goats that are best suited for milk production include those of European origin. The list of breeds includes, but are not limited to: the Alpine, the Saanen and the Toggenburg. When selecting these animals, type and conformation are critical to a production setting. The discussion on genetic selection is beyond the scope of this text and one is directed to other information sources [1,2]. Once a population of animals is selected, subpopulations of donor and recipient animals needs to be identified. The optimal population of animals to be used as donors are virgin does 6 months and older. The optimal population of animals to be used as recipients are multiparous does, which would put their ages at approximately $1-1\frac{1}{2}$ years old. The importance of having multiparous does as recipients is to eliminate the potential for birthing difficulties that may be seen with primiparous animals.

Superovulation

There are a number of different superovulation protocols that may be used in the goat and most are derived from the bovine field. The protocol that follows produces optimal embryos for pronuclear injection [6]. The timing of estrus in the donors is synchronized on Day 0 with the placement of a 6 mg subcutaneous progesterone ear implant. Prostaglandin ($PGF_{2\alpha}$), at a dose of 5–10 mg, is administered after the first seven days to shut down the endogenous synthesis of progesterone. Starting on Day 13, after insertion of the implant, a total of 18 mg of follicle-stimulating hormone is given intramuscularly over three days in twice-daily injections. These injections follow a decreasing dose format starting with 4 mg the first day, ending with 2 mg on the third day of FSH-P treatment, and the implant is removed on Day 14. Twenty-four hours following implant removal, the donor animals are mated several times to fertile males over a two-day period. Minor changes in hormone dosages are made if it is used outside of the normal breeding season. The FSH-P can be increased to a total of 24 mg with equivalent results [2]. The dose format for this regime starts at 5 mg on the first day of FSH-P treatment and ends with 3 mg on the third day. With the above protocol, recoveries should average 5.4 fertilized injectable 1-cell embryos per doe (in-house data).

Embryo Collection

Surgery for embryo collection occurs on the second day following breeding (or 72 hours following implant removal). Superovulated does are removed from food and water 36 hours prior to surgery. Does are administered 0.8 mg/kg valium, IV, followed immediately by 5.0 mg/kg ketamine, IV. The animal is unconscious within 20–30 seconds. Following intubation, Halothane is administered at 2.5% with a 2 L/min oxygen flow rate. The animal is lifted onto a gurney, surgically prepared, and the reproductive tract is exteriorized through a midline laparotomy incision. Corpora lutea and unruptured follicles are counted to evaluate superovulation results and to predict the number of embryos that should be collected by oviductal flushing. A polyethylene cannula is placed in the ostium. The cannula is held in place with either a single temporary ligature of 3–0 Prolene or by simply applying pressure with two fingers to secure and seal the ostium around the cannula. The later approach reduces tissue handling and thereby decreases adhesion formation seen on subsequent procedures. A 20 gauge needle is inserted into the uterus approximately 0.5 cm from the uterotubal junction. Ten to twenty ml of sterile phosphate buffered saline (PBS) is flushed through the cannulated oviduct and collected in a Petri dish. This procedure is repeated on the opposite side and then the reproductive tract returned to the abdomen. Before closure, 10–20 ml of a sterile saline glycerol with heparin solution is used to bathe the reproductive tract where it was physically handled to decrease post-surgical adhesion formation. The linea alba is closed with simple interrupted sutures of 0 Polydioxone (PDS), the subcuticular fat with a simple continuous using 2.0 PDS, and the skin with a continuous subcuticular pattern of 2.0 PDS followed by skin staples. The animals are returned to a stall for post-surgical observation and recovery.

Embryo Evaluation and Microinjection

The caprine oocytes/embryos are collected from the PBS oviductal flushings on a stereomicroscope and evaluated. Separation of fertilized, unfertilized and abnormal embryos is performed. Embryos are washed and cultured short term in Ham's F12 medium containing 10% fetal bovine serum at 37°C in a humidified gas chamber containing 5% CO_2 in air prior to microinjection. Fertilized one-cell goat embryos are placed in a microdrop of medium under oil on a glass depression slide [3,5]. Zygotes having two visible pronuclei are immobilized on a flame-polished holding micropipette on a Nikon Diaphot inverted microscope using Normarski optics. Location and visualization of the pronuclei is achieved by repeated rotation of the embryo on the holding micropipette. A pronucleus is microinjected with a DNA construct in injection buffer (Tris-EDTA) using a fine glass microneedle [6]. On occasion, only one pronuclei may be visible for injection, however there are times when both pronuclei are visible and aligned so that both may be injected. After microinjection, the embryos are placed into culture for a short period of time to evaluate post injection survival. If the embryo is fertilized but pronuclei are not visible, the embryos are placed back into culture until pronuclei are visible. Once post injection survival has been evaluated, the embryos are maintained in culture until the recipient animals are prepared for embryo transfer.

Embryo Transfer

Estrus synchronization in recipient animals is induced by placement of a 6 mg progesterone ear implant. On Day 13 after insertion of the implant, the animals are given a single non-superovulatory injection (400 I.U.) of pregnant mare serum gonadotropin. Recipient females are mated to vasectomized males to ensure estrus synchrony. An optimal recipient will have between 1–5 total corpora lutea. Animals having more than eight total ovulation points on the ovaries are deemed unsuitable as recipients and are not used. This protocol needs minor dose changes if used outside of the normal breeding season. Increases to a total of 600–750 I.U. of PMSG have been reported elsewhere with equivalent results [2]. The surgical procedure is identical to that outlined for embryo collection, except that the oviduct is not cannulated. The embryos are transferred in a minimal volume of Ham's F12 containing 10% FBS into the oviductal lumen, via the fimbria, using a glass micropipette [6]. Incision closure and post-operative care are the same as for donor animals.

Gestation and Parturition

Recipient does are first evaluated by ultrasonography approximately Day 35 from the first day of estrus to detect pregnancies. Pregnancy determination can be performed as early as approximately Day 21, but embryological/fetal developmental competence may still be lost at this early stage. A confirmatory ultrasound at Day 55 will provide the most reliable indication of a viable pregnancy with fetal number and viability evaluated. Pregnant does are monitored daily throughout pregnancy and appropriate prekidding procedures performed [7]. Parturition may occur naturally or be induced, based on facilities and labor. It is this author's experience that, to avoid complications during parturition, allowing the doe to give birth naturally is favorable. If necessary, the exogenous induction of labor can be performed on approximately day 148 by administering 40 mg of $PGF_{2\alpha}$ split into two doses, one in the morning and one 8 hours later. Parturition will occur approximately 36 hours following the first injection [2].

Identification of Transgenesis

The protocol for identification of transgenic kids is described elsewhere in this volume. At present, frequencies of transgenesis in the caprine species range from 6–10% of live offspring or 1–3% of injected and transferred embryos.

References

1. American Dairy Goat Association (ADGA), P.O. Box 865, 209 West Main Street, Spindale, North Carolina, 28160 (USA). Phone: (704) 286-3801
2. Ebert, K.M., Selgrath, J.P., DiTullio, P., Denman, J., Smith, T.E., Memon, M.A., Schindler, J.E., Monastersky, G.M., Vitale, J.A. and K. Gordon (1991) Transgenic production of a variant of human tissue-type plasminogen activator in goat milk: generation of transgenic goats and analysis of expression. *Biotechnology* **9**, 835–838
3. Hammer, R.E., Pursel, V.G., Rexroad, C.E., Jr., Wall, R.J., Bolt, D.J., Ebert, K.M., Palmiter, R.D. and R.L. Brinster (1985) Production of transgenic rabbits, sheep and pigs by microinjection. *Nature* **315**, 680–683
4. International Goat Association (IGA), 216 Wachusett Street, Rutland, MA 01543-2099 (USA). Phone: (508) 886-2221
5. Rexroad, C.E., Jr., Hammer, R.E., Bolt, D.J., Mayo, K.E., Frohman, L.A., Palmiter, R.D. and R.L. Brinster. (1989) Production of transgenic sheep with growth-regulating genes. *Molecular Reproduction and Development* **1**, 164–169
6. Selgrath, J.P., Memon, M.A., Smith, T.E. and K.M. Ebert (1990) Collection and Transfer of microinjectable embryos from dairy goats. *Theriogenology* **34**, 1195–1205
7. Smith, M.C. and D.M. Sherman (1994) *Goat Medicine*, Lea and Febiger, Pennsylvania (USA), pp. 565–577

PART II, SECTION A

6. The Generation of Transgenic Sheep by Pronuclear Microinjection

Yvonne Gibson and Alan Colman

PPL Therapeutics Ltd, Roslin, Edinburgh, Scotland, UK

Transgenic animals are produced by the introduction of foreign [usually] DNA into the genome of the chosen host species. This can be accomplished by one of several routes, microinjection (Gordon and Ruddle, 1981), retroviral infection (Jaenish, 1988) or site directed integration using embryonic stem (ES) cells (reviewed by Bradley *et al.*, 1992). As ES cells are currently only available for the mouse, the preferred route for generation of transgenic livestock is currently the first option, microinjection. In this article we will limit ourselves to the case of the generation of transgenic sheep.

In general, donor female animals are superovulated, artificially inseminated and fertilised eggs are then collected. Following this, foreign DNA is injected into the eggs which are subsequently returned to recipient mothers where they are carried to term. Upon parturition, a small percentage of the progeny will be found to be transgenic.

Superovulation

Donor ewes are treated with intravaginal progesterone-impregnated sponges [Intervet] left *in situ* for 11–15 days and removed two days prior to artificial insemination [AI]. Superovulation is induced by a regime of ovine follicle stimulating hormone [P.M.S.G., Intervet] administered in eight intramuscular injections of ~0.125 units per injection twice daily beginning five days before AI. To synchronise ovulation, the donor animals are injected intramuscularly with synthetic releasing hormone analogue one day prior to AI.

Fertilisation

Freshly collected semen is diluted with equal parts of sterile phosphate buffered saline. Fertilisation is achieved by artificial insemination by intrauterine laparoscopy under sedation and local anesthesia. 0.2 ml of the diluted semen is injected per uterine horn. Sedation is provided by intravenous injection of acetyl promazine at a dose rate of 0.05 ml per 10 kg together with a Temgesic injection at a dose rate of 0.16 ml per 10 kg bodyweight. Immediately pre- or post-AI donors are given an intramuscular injection of Amoxypen [Vet Drug] at the manufacturers recommended dose rate.

Embryo Recovery

Fertilised eggs are recovered one day post-AI under general anaesthesia induced by an intravenous injection of 5% thiopentone sodium at a dose rate of 3 ml per 10 kg bodyweight. Anaesthesia is maintained by inhalation of 1–2% Halothane/O_2/N_2O. To recover the fertilised eggs, a laparotomy incision is made, and the uterus exteriorised. The eggs are recovered by retrograde flushing of the oviducts with Ovum Culture Media [Advanced Protein Products, Brierly Hill, West Midlands, UK]. After flushing, the uterus is returned to the abdomen, and the incision is closed. Donors are allowed to recover post-operatively following an intramuscular injection of Amoxypen [Vet Drug] at the manufacturer's recommended does rate.

Microinjection

For injection into fertilised eggs, the DNA expression cassettes are removed from their respective vectors by digestion with appropriate restriction enzymes. The expression units are recovered by conventional methods, such as electro-elution followed by phenol extraction and ethanol precipitation, sucrose density gradient centrifugation, or combinations of these approaches.

DNA is injected into eggs essentially as described by Hogan and colleagues (Hogan et al., 1994). In a typical injection, eggs in a dish of ovum culture medium are located using a stereo zoom microscope ($\times 50$ or $\times 63$ magnification preferred). The eggs are secured and transferred to the centre of a glass slide on an injection rig using, for example a mouth pipette complete with capillary tube. Viewing the eggs at lower (e.g. $\times 4$) magnification is used at this stage. Using the holding pipette of the injection rig, the eggs are positioned centrally on the slide. Individual eggs are sequentially secured to the holding pipette for injection. For each injection process, the holding pipette/egg is positioned in the centre of the viewing field. The injection needle is then positioned directly below the egg. Preferably using $\times 40$ Normarski objectives, both manipulator heights are adjusted to focus both the egg and the needle. The pronuclei are located by rotating the egg and adjusting the holding pipette assembly as necessary. Once the pronucleus has been located, the height of the manipulator is altered to focus the pronuclear membrane. The injection needle is positioned below the egg such that the needle tip is in a position below the centre of the pronucleus. The position of the needle is then altered using the injection manipulator assembly to bring the needle and the pronucleus into the same focal plane. The needle is moved via the joy stick on the injection manipulator assembly, to a position to the right of the egg. With a short, continuous jabbing movement, the pronuclear membrane is pierced to leave the needle tip inside the pronucleus. Pressure is applied to the injection needle via the glass syringe until the pronucleus swells to approximately twice its volume. At this point, the needle is slowly removed. Reverting to lower (e.g. $\times 4$) magnification, the injected egg is moved to a different area of the slide, and the process is repeated with another egg.

All fertilised eggs surviving pronuclear microinjection can be returned to synchronised recipient ewes on the same day or cultured in vitro at 38.5 C in an atmosphere of 5% CO_2, 5% O_2, 90% N_2 and about $\sim 100\%$ humidity in a bicarbonate buffered synthetic oviduct medium supplemented with 20% v/v vasectomised ram serum. The serum may be heat inactivated at 56 C for 30 minutes and stored frozen at 20 C prior to use. The fertilised eggs are cultured for a suitable period of time to allow early embryo mortality (caused by the manipulation techniques) to occur. These dead or arrested embryos are discarded. Embryos having developed to 5 or 6 cell divisions are transferred to synchronised recipients ewes.

Recipients

Each recipient ewe is treated with in intravaginal progesterone impregnated sponge [Intervet] inserted for 11–15 days which is removed 2 days prior to embryo transfer. This is followed by an intramuscular injection of follicle stimulating hormone substitute [~ 400 iu, P.M.S.G., Intervet]. Recipients are tested for oestrus with a vasectomised ram. Embryo return is carried out under general anaesthesia as described above. The uterus is exteriorised via a laparotomy incision. Embryos are returned to one or both uterine horns only in ewes with at least one suitable corpea lutea. After replacement of the uterus, the abdomen is closed and the recipient allowed to recover.

Progeny Screening

During embryogenesis, the injected DNA integrates in a random fashion in the genomes of a small number of the developing embryos. Potential transgenic offspring are screened for the presence of this heterologous DNA via blood samples and tissue biopsies.

A 10 ml blood sample is taken from the jugular vein into an EDTA vacutainer. Tissue samples are taken by tail biopsy generated by normal animal husbandry practices. DNA is extracted from sheep blood by first separating white blood cells. A 10 ml sample of blood is diluted in 20 ml of Hank's Buffered Saline [HBS]. Ten ml of the diluted blood is layered over 5 ml of histopaque [Sigma] in each of two 15 ml screw capped tubes. The tubes are centrifuged at 3000 rpm (2000 $\times$ g max.), low brake for 15 minutes at room temperature. White cell interfaces are removed to a clean 15 ml tube and diluted to 15 ml in HBS. The diluted cells are spun at 3000 rpm for 10 minutes at room temperature, and the cell pellet is recovered and resuspended in 2–5 ml of 0.3 M NaAc, 50 mM KCL, 1.5 mM $MgCl_2$, 10 mM Tris-HCL [pH 8.5], 0.5% NP40, 0.5% Tween 20. Proteinase K is added to a concentration of 200 µg/ml and the mixture is incubated overnight at 55 C. Tail samples are digested similarly. DNA is extracted from either source using an equal volume of phenol/chloroform ($\times 3$) and chloroform/isoamly alcohol ($\times 1$). The DNA is then precipitated by adding 2 volumes of ethanol, and the tube inverted to mix. The DNA is spooled out using a clean glass rod with a sealed end. The spool is washed in 70% ethanol, the DNA is allowed to partially dry, prior to being redissolved in TE (10 mM Tris-HCL, 1 mM EDTA, pH 7.4).

DNA samples from tail and blood are analyzed by Southern blotting. Animals exhibiting the expected band pattern following this are termed "Founder Transgenics" or "G_0s". These are bred to produce families or "lines" of transgenic animals. A line would be deemed "satisfactory" and suitable for flock development if it is genetically stable and members exhibit high and stable expression levels of the desired protein in the milk through sequential lactations of individual ewes and from lactations of subsequent generations. These principles have been demonstrated and discussed (Carver et al., 1993).

References

Bradley, A., Hasty, P., Davis, A. and Ramirez-Solis, R. (1992) Modifying the mouse: design and desire. *Bio/Technology* **10**, 534–539

Carver, A.S., Dalrymple, M.A., Wright, G., Cottom, D.S., Reeves, D.B., Gibson, Y.H., Keenan, J.K., Barrass, D., Scott, A.R., Colman, A. and Garner, I. (1993) Transgenic livestock as bioreactors: stable expression of human alpha-1-antitrypsin by a flock of sheep. *Bio/Technology* **11**, 1263–1270

Gordon, J.W. and Ruddle F.H. (1981) Integration and stable germline transmission of genes injected into mouse pronuclei. *Science* **214**, 1244–1246

Hogan, B., Constantini, F. and Lacy, E. (1994) *Manipulating the mouse embryo: a laboratory manual*. Cold Spring Harbor Laboratory Press, New York

Jaenish, R. (1988) Transgenic Animals. *Science* **240**, 1468–1474

PART II, SECTION A

7. Gene Microinjection into Bovine Pronuclei

Marc Gagné[1], François Pothier[2] and Marc-André Sirard[2]

[1]*Immunova ltée, 2750 Einstein street, Office 110, Sainte-Foy (Québec) G1P 4R1, Canada*
[2]*Department of Animal Sciences, Laval University, Sainte-Foy (Québec) G1K 7P4, Canada*

Introduction

Transgenic technology has been developed and perfected in the laboratory mouse. Since the early 1980s (Gordon *et al.*, 1980 and Brinster *et al.*, 1981), hundreds of different genes have been introduced into various mouse strains. These studies have contributed to an understanding of gene regulation, tumor development, immunological specificity, molecular genetics of development, and other fundamental biological processes of interest. Transgenic mice have also played a role in examining the feasibility of therapeutic human drug production by domesticated animals and in the creation of transgenic animal strains that act as biomedical models to various human diseases, and even in the near future as a supply of replacement organs.

In domestic farm animals however, the efficiency of transgene integration tends to be significantly lower than in mice. Until now, the percentage of gene-injected embryos that develop into transgenic animals varied from 0.3% to 4.0% for pigs, 0.1% to 4.4% for sheep and 0.7% to 3.2% for cattle. To improve the rate of transgenesis in the higher species, several methods have been attempted to improve the integration of transgenes. Moreover, all these methods allow the efficient introduction of foreign DNA sequences into germ line without the considerable skills and expensive equipement required for pronuclear injection. Microprojectile cell bombardment technique (Zelenin *et al.*, 1993), retroviral insertion (Kim *et al.*, 1993), sperm mediated gene transfer (Gagné *et al.*, 1991) and the currently popular stable transfection of embryonic stem cells (ES) (Cherny *et al.*, 1994), are all delivery, vector or transport systems to genetically modify mammals (Figure 7.1.). Each technique has its advantages, in some case numerous, in regard to pronuclear microinjection. Unfortunately, neither method has demonstrated its ability to produce transgenic cattle.

Currently, the only reliable method for the introduction of new genetic material in livestock species has been gene transfer by microinjection. Several disadvantages are associated to this approach (Wilmut and Whitelaw, 1994 and Clark *et al.*, 1994). It is characterized by a low efficiency, with less than 1% of injected zygotes surviving to become transgenic offspring; the random integration can induce severe genomic mutations, including reciprocally if regulatory elements (enhancers or silencers) are present in the neighbouring integration site region. These mutations may influence the expression of the transgenes in such a way that it is often impossible to anticipate its transcriptional rate.

Despite these several constraints on the commercial adoption of this technology, microinjection is the preferred method for the study of transgenesis. Several important parameters that influence the frequency of integration have been described in the mouse (Brinster *et al.*, 1985) but have not been systematically investigated in other species. On this point, the significant difference between transgene integration in cattle and mice following pronuclear microinjection constitutes in itself, a clue that trangenesis processing does not follow similar patterns in different species. Therefore, it is conceivable that some molecular events are carried out differently in bovine zygotes.

In the present chapter, we describe attempt to improve the efficiency of gene incorporation in the bovine, by injecting oocytes and zygotes at different

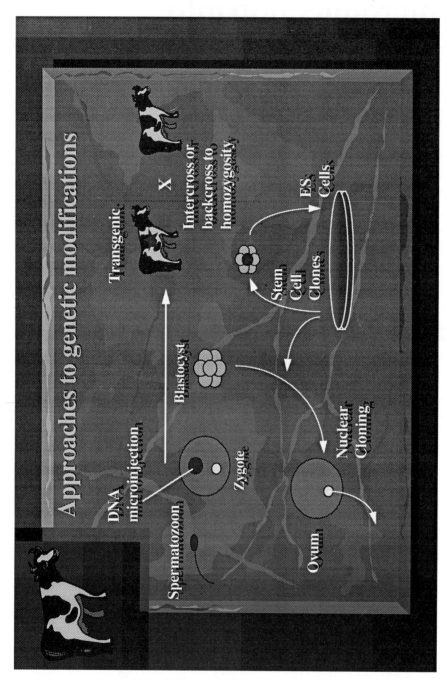

Figure 7.1. Approaches to genetic modifications in bovine.

stages of the cell cycle, by following the fate of the foreign DNA in treated embryos with regard to the time of injection of the zygotic replication phase, by exploiting the principle of restriction endonuclease-mediated integration demonstrated in other organisms, and finally by flanking the transgenes with highly repeated endogenous genomic regions.

Cellular Site of Microinjection

Injection of cloned DNA into ova and zygotes has been used to examine the expression of a reporter gene under the control of different promoters. Such studies have determined that a number of 5' flanking elements controlling gene expression are active either at specific stages in oocytes and embryos or during the maturation of oocytes and early and late development of embryos (Bonnerot et al., 1991). Although these observations are quite essential to understanding development and cell differentiation, they do not indicate the pathways followed by the injected DNA or how the cells processes these foreign molecules. Transgenesis in mammals depends on many critical parameters, such as the site of microinjection and the concentration and topological form of the DNA injected. In studies on the mouse, Brinster and colleagues (1985) showed the following: the pronuclear-delivered DNA allowed genomic integration at a rate at least 10-fold higher than that of cytoplasmic injection; that foreign DNA should be injected in linear form instead of in circular or supercoiled forms; and that it should be injected in low amounts because of its toxic effect on the survival of the embryos, even if the integration efficiency improves as the DNA 001concentration increases. The presence of microinjected DNA was detected at a high rate in the mouse at the morula (44%) and blastocyst (26%) stages (Burdon and wall, 1992). Although integration in rodents resulting in transgenesis is lower than that, it is still relatively high, ranging from 6 to 10%. The fate of injected DNA itself has also been studied. In lower species such as Xenopus and sea urchin, DNA injected into the cytoplasm of maturing and fertilized oocytes can result in an end-to-end ligation (Goedecke et al., 1992), episomal replication (McMahon et al., 1985; Harland and Laskey, 1980 and Etkin et al., 1984), and degradation (Maryon and Carroll, 1989 and Harland et al ., 1983).

It appears that some specific characteristics make bovine embryos more refractory of foreign DNA integration than the other species mentioned. Two hypothetical mechanisms, acting separately or synchronously, could be implicated in this type of resistance. First, the integration of an injected construct in bovine genome may occur in a manner that impairs the genetic physiology of the embryo, leading to transgenic cell death. Second, the embryo could protect the integrity of its genome by having an efficient mechanism of degradation of foreign molecules, preventing injected DNA to have enough time to reach the chromatin.

Powell and colleagues (1992) have shown that injected DNA undergoes ligation, and they observed a certain level of degradation in maturing bovine oocytes and zygotes. The susceptibility of DNA constructs to endogenous enzymatic digestion has been reported in the nucleus of the somatic cells with respect to the stage of the cell cycle and gene transcription activity (Brown, 1984; and Groudine and Weintraub, 1982). Higher sensitivity to digestion appears to be related to genes that are expressed because of the formation of short hypersensitive sequences during the uncoiling of DNA.

As observed in the mouse (Brinster et al., 1985), the number of bovine embryos surviving during the first cleavage stages of development decreases linearly with the concentration of plasmid injected (Gagné et al., 1994). The rate of the first cleavage observed (24–32 h post fertilization) decreased as DNA concentration insreased, indicating a certain level of toxicity of the foreign DNA. A correlation between the amount of injected DNA and expression was observed (Bonnerot et al., 1987 and Gagné et al., 1994). Despite the importance of molecular factors in the expression of the foreign gene in host cells, variables such as the time and site of injection must also be considered.

DNA microinjection into cytoplasm of fertilized mouse oocytes allowed integration in ten percent (0.6–0.8%) of pronuclear injections (Brinster et al., 1985). Wyllie et al. (1977) observed that at an earlier stage of development, linear DNA injected into the cytoplasm did not penetrate the germinal vesicle, even 48 h following injection, and was degraded when injected into the nucleus and cytoplasm of unmatured Xenopus oocytes. However, Etkin (1984) estimated that following cytoplasmic microinjection into newly fertilized Xenopus oocytes, integration occurs in about 60–70% of surviving eggs injected with linear DNA molecules. This type of internalization of foreign DNA has been explained by a fusion of micronucleus-enclosed plasmids with the nucleus (Forbes et al., 1983). In bovine the results are comparable with those obtained in nonactivated pig, sheep and bovine oocytes at metaphase II (Powell et al., 1992), where the expression of an active injected gene construct into nuclei was detected, but not following injection into cytoplasm.

It has been proposed that injected DNA into bovine oocytes does not benefit from mechanisms such as complexing to proteins or special termini structure (Powell et al., 1992), methylation (Razin and Riggs, 1980) or release of specific nuclease inhibition factors (Wyllie, 1977), which all potentially contribute to protect genomic chromatin against degradation. In addition to its own toxic effect when injected into bovine oocytes, injected DNA fragment is more detrimental to the development if it is expressed during the earlier cleavage stages that when it is quiescent. Cytoplasmic injection does not allow transcription of pAGS-lacZ either in artificially activated nor fertilized bovine oocytes, suggesting that foreign DNA is not internalized actively nor passively into forming nuclei or already formed pronuclei. Maturing and fertilized bovine oocytes seem to have efficient machinery for degradation of DNA microinjected either into the nucleus or the cytoplasm (Gagné et al., 1994).

Influence of the Zygotic S-Phase

Few studies describe the molecular events involved in chromosomal integration of foreign DNA (Hamada et al., 1993; and Bishop and Smith, 1989). The state and spatial distribution of chromatin and other nuclear components in interphase nuclei are critical aspects in cell biology. It is during interphase when DNA is accessible for replication, and most molecular activities take place. In somatic cells, the rate of recombination of injected DNA constructs has been found to be directly related to nuclear state; it occurs 10–15 times more frequently in cells microinjected at the S phase than in those microinjected in cells at either G phase (Wong and Capecchi, 1987). In zygotes, during the process of pronuclei formation, the dynamic and structural aspects of chromatin change. When the sperm fuses with the oocyte, it activates the oocyte, which then directs the remodeling of gamete chromatin into male and female pronuclei capable of DNA synthesis (reviewed in Longo, 1985; Zirkin et al., 1989 and Perreault, 1990). Sensitivity to DNA damage is related to the stage of nuclear events occurring mainly before and during the period of DNA synthesis preceeding pronuclei breakdown (Bandriff and Gordon, 1992). Experiments using X-ray irradiation revealed that development of mouse zygotes was less affected by treatment at later stages of DNA replication (Jacquet et al., 1983 and Domon 1982). Also, the fact that foreign DNA becomes integrated in a minority injected embryos, usually at a single site, suggests that the frequency of integration is determined by rare molecular events. The suggestion has been made that the low frequency of chromosomal breaks constrains the number of integration events (Brinster et al., 1985). Thus, mechanisms of DNA repair could be important not only to eliminate highly lethal DNA double-strand breaks from duplex DNA, but also to provide the fundamental mechanism for the rejoining of DNA elements in recombination processes (Roth and Wilson, 1988).

Considering these observations, it is conceivable that the introduction of a microneedle, rather than the deposit of foreign DNA into a pronucleus would induce some breaks into the chromatin which might affect the normal evolution of the embryos. The observation that the frequency of DNA integration is also increased by chromatin breaks of transfected cells (Perez et al., 1985) confirms this fact. It has been observed that a balance exists between susceptibility to damage and capacity for repair depending on the degree of chromatin condensation (Barone et al., 1989).

Few studies have been devoted to the effect of pronuclear microinjection on the development of bovine embryos (Wall and Hawk, 1988; Hawk et al., 1989; Gagné et al., 1990 and Thomas et al., 1993). Recently, Thomas et al. (1993) found no differences between bovine zygotes injected at 18 and 22 h post-injection (HPI) in regard to the rate of embryonic development, whereas the study of Kubish and colleagues (1994) showed an important difference in development after microinjection at 18 and 23 HPI. The accessibility to pronuclei being obviously an important aspect for gene transfer in domestic animal species, few studies, however, have considered the recombination process of foreign DNA in transgenesis at the molecular level instead of at the subcellular level (Bishop and Smith, 1989).

At the early, mid and late period of the S phase, the microinjection procedure had a clearly negative effect on in vitro development (Gagné et al., 1995). However, it was observed in this study that the stage of lowest sensitivity in the first cell cycle in bovine zygotes occurs in the middle of the pronuclear stages, when genomic DNA is actively replicating. It seems that microinjection provokes two effects: an immediate effect expressed by the inability of the egg to undergo cleavage, and a delayed one, leading to developmental arrest mainly after the third division.

Peura and Jänne (1994) demonstrated that breaking through both cytoplasmic and pronuclear membranes does not decrease the development of bovine embryos, which suggests that the poor survival of the embryos is due mostly to chromatin damage. The higher development of embryos injected at 18 HPI suggests that mechanisms required to repair DNA are qualitatively or quantitatively more competent, which is reflected by a greater capacity to develop. This confirms several systems where DNA is repaired more rapidly in actively replicating chromatin that in mitotic chromatin (Lehmann et al., 1992).

These results are similar to experiments where the lowest sensitivity of mouse zygotes to irradiation was observed at the mid-period compared with the beginning of the S phase (Domon, 1982). A gradual increase in sensitivity to irradiation treatment from the maximum rate of DNA synthesis to the first cleavage also appears from the studies of Dufrain and Casarett (1975). Other experiments of Gagné et al. (1995a) indicate that resistance of the embryos to microinjection is synchronized with the cytoplasm during the interphase instead of during the stage of the pronuclear S phase and might depend on timed-activated enzymes. This study suggests also that the ability to resist microinjection is based on a time-specific activation of the factors already existing at a precise time and involved in the repair processes, confirming previous observation that protein synthesis is not required for DNA repair (Saxena et al., 1990).

Microinjection should not be performed at the end or after the first S phase following fertilization. If intergration occurs in using the molecular systems of replication and repair, injected genes will be integrated in the second cell cycle, therefore, the possibility of producing mosaic animals would be considerably increased. Nevertheless, an important question remains to be answered. Does the higher sensitivity observed early during the pronuclear S phase indicate a greater accessibility of the chromatin to the incorporation of an injected gene? Or does the higher resistance observed at the middle of this phase reflect the activity level of the repair mechanisms, which would truly participate in the molecular transgenesis processes?

Restriction Enzyme-mediated Integration

A variety of experiments have shown that double-strand DNA breaks can stimulate recombination in mammalian cells between two DNA molecules simultaneously introduced into cells (Anderson et al., 1986; Brenner et al., 1985; Brouillette et al., 1987 and Farzaneh et al., 1988). Both single-strand annealing (SSA) and double-strand break repair (DSBR) have been proposed as models to account for this stimulation. It is supposed that double-strand breaks are substrates for the action of an exonuclease that exposes homologous complementary single-strand DNA ends on the same or on different DNA molecules. Annealing of the complementary strands is then followed by a repair process which generates the recombination. The DSBR model has also been used to explain recombination between transfected and chromosomal DNAs (Jasin et al., 1985 and Smithies et al., 1985). In this model, double-strand DNA ends are converted to single strands by exonuclease action and then paired with homologous DNA duplexes. Several experiments have demonstrated that both breakage (Ashram et al., 1984; Calos et al., 1983 and Lebkowski et al., 1984) and efficient end joining (Roth et al., 1985 and Wake et al., 1984) occur in transfected DNA in a variety of animal cells.

Integration of linear DNA fragments into genomic restriction sites has been demonstrated in lower organisms (Shaulsky et al., 1995 and Adachi et al., 1994), and intentional breaks made in DNA molecules that are transfected into somatic cells stimulate recombination events at the break sites in diverse species, including prokaryotes, fungi, and cultured cells from mammals (Lin et al., 1984; Haber et al., 1988 and Szostak et al., 1983). Restriction enzymes (RE), which produce blunt- or cohesive-end double-strand breaks at specific DNA sequences, could be excellent tools to stimulate recombination of homologous ends of transfected and genomic DNA molecules. Several studies on the use of RE have reported on the mechanisms of aberration induction and for possibly producing very corresponding genomic cleavage (Bryant, 1984; Natarajan and Obe, 1984; and Obe and Natarajan, 1985).

In mammalian embryos it has been suggested that the frequency of chromosomal breaks constrains the number of integration events (Brinster et al., 1985). In bovine embryos the possibility of raising the ligation of ends of injected DNA by creating double-strand breaks induced by RE co-injected with foreign DNA molecules, and which creates cohensive 3' protruding ends was examined (Gagné et al., 1996a). It was observed in this study that cytoplasmic injection of foreign linearized DNA molecules alone or in the presence of a restriction enzyme at low concentrations did not alter the maturation and fertilization of bovine oocytes. Injection soon after insemination greatly impaired the development of embryos, and did not allow detection of the injected genes by polymerase chain reaction (PCR) at the morula stage. The pronuclear injection of high concentration of restriction endonuclease is harmful to the development of embryos, which suggests the cleavage action of the enzyme. The two models for explaining the rejoining of cohesive ends, namely direct ligation, or partial single strand annealing followed by cleaning up remaining free overlaps, have not allowed to improve the genomic insertion of foreign DNA after co-injection with restriction endonuclease into maturing and inseminated bovine oocytes. Since double-strand breaks result from the presence of restriction endonuclease in the cells (Bryant, 1984; Obe and Winkel, 1985; and Bishop and Smith, 1989) and stimulates the recombination of transfecting DNA molecules (Anderson et al., 1986; Brenner et al., 1985 and Brouillette et al., 1987), further studies could determine which enzyme, type of DNA ends, concentration and time of injection before or after insemination might contribute to improved transgenesis in bovine embryos.

Flanking the Transgene with Repeated Genomic Sequences

As genes introduced by microinjecting directly into fertilized eggs are integrated into the genome at a low rate and at random chromosomal sites, alternative methods using transfection of embryonic stem cells (ES) have been developed. Evans and Kaufman (1981) reported that it is possible to establish cell lines derived from the inner cell mass of mouse blastocysts in culture. They also demonstrated that the cells are capable of resuming their full developmental potential. Therefore, once driven to select a desired phenotype, transfected ES cells are then introduced into the blastocyst where they can contribute to the formation of the developing mouse (Thomas et al., 1986; Thomas and Capecchi, 1987 and Doetschman et al., 1987).

The ES approach provides the potential for production of transgenic offspring with site-specific insertion. Smithies and colleagues (1985) developed a technique for selecting cells after a homologous recombination (HR) event between an endogenous gene and a DNA construct carrying a copy of the gene. This work has been extended for replacement of a targeted gene (Kuehn et al., 1987; Doetschman et al., 1987 and Capecchi, 1980, 1989) or to cause gene deletion as well as introduction of foreign DNA molecules (Thomas and Capecchi, 1987). Using this principle, mice expressing a site-specific transgene were produced and germ-line transmission occurred (Thomas et al., 1989). Although this development could provide the potential for immediate specific gene deletion or replacement in livestock production, it would require generations of selection.

Homologous recombination (HR) between transferred and chromosomal DNA, also called "gene targeting", has been used in several experiments to modify genes of cultured mouse ES cells (Smithies et al., 1985; Thomas and Capecchi, 1990; Mansour et al., 1988; Charron et al., 1990; Jasin et al., 1990; and Koller and Smithies, 1989). The recombination efficiency has been highly variable. In these studies, one HR event was found for every 10^2–10^5 nonhomologous integration events. In Xenopus eggs, recombination was performed by three types of processes; simple ligation, illegitimate end joining, and homologous recombination (Lehman et al., 1993). The gene targeting approach was also tested for mouse transgenesis by microinjecting the DNA construct directly into the fertilized oocytes (Brinster et al., 1989). This experiment demonstrated the feasibility of targeting foreign DNA to correct a mutant gene that is inactive in fertilized mouse eggs. MHC class II E alpha gene deletion was corrected by HR in 1 of 500 transgenic mice that incorporated the injected DNA. It has been clearly demonstrated that the homologous recombination machinery is very efficient in murine zygotes (Pieper et al., 1992).

Therefore, the annealing pathway of HR appears to be functional in a number of higher eukaryotic systems, including mammalian embryos. In a study of Gagné et al. (1996b), the possibility of improving HR transgene incorporation as reported by Brinster et al. (1989) was attempted in flanking the 3' only or both ends of the foreign marker DNA with halves of a bovine Satellite (BS) DNA (Plucienniczak et al., 1982). This D-light centromeric sequence was estimated as existing in at least a thousand copies in the genome. A high frequency of recombination during the S phase was observed in centromeric regions of Drosophila genomes (Grell, 1978). Gene targeting in the bovine satellite DNA sequences was performed by microinjection in interphasic bovine zygotes (Gagné et al., 1996b). This approach offers many advantages:

(1) to improve transgenetic efficiency in bovine species;
(2) to prevent mosaisism resulting from transgenes introduced by microinjection into eggs that are not usually represented in the germ line of the first generation;
(3) to drive the integration of foreign DNA so as to control the influence of the regions around the integration site and to prevent the alteration of a vital gene by insertion;
(4) to select embryos showing homologous integration by the PCR technique before transfer into recipient cows.

Microinjection of foreign DNA with homologous sequences had unexpected effects on the development of embryos. It was observed that even injected at very low concentration, the presence of genomic repeated transfecting homologous DNA impaired the development of the embryos. In this study, the microinjection of a construct without flanking homologous nucleic regions was significantly higher. To our knowledge, the harmful effect of the homologous region of the transfecting DNA either in cultured cells or in embryos on the subsequent divisions has not previously been reported. It has been observed that homologous recombination peaks in early to mid-S phase (Wong and Capecchi, 1987). We can suppose also that the thousands of BS sequences used as genomic targets synchronously replicate in the pronuclei. Consequently, homologous recombination of injected constructs at this time is potentially high.

In this context, if the HR insertion occurs at too many sites it would interfere with the processing enzymes involved in the replication or repair mechanisms that have been found to follow the replication activity level (Lehman et al., 1992). For unknown reasons, we did not observe differences in the degree of influence on the development between the uni- and bilateral homology of injected DNA ends.

Conventional techniques like restriction enzyme digestion and Southern blot are generally used for the molecular analysis of events such as HR. However, it was previously reported (Gagné et al., 1991) that there is not enough genomic DNA in a single bovine embryo to perform these analyses. In addition, it was not possible to use the expression of a foreign gene as a recombination marker because we have also observed (Gagné et al., 1994) that the transcription of the E coli lacZ gene driven by the 5′ flanking region of the chicken β-actin gene could be supported by embryonic cells without being integrated in eukaryokic cells (Zimmer and Gruss, 1989, Joyner et al., 1989; and Kim and Smithies, 1988).

Current recombination models (Meselson and Radding, 1975 and Szostak et al., 1983) require homology between two DNA sequences on both of a recombination site in order to initiate a recombination reaction. In the study of Gagné et al. (1996b), since one of injected constructs shared a homologous region with only one side of the chromosomal BS sequences, these seemed to be inadequate substrates for insertion of the foreign DNA, but the 1.5 kb fragment identifying targeted recombinants after PCR analysis was detected in approximately 27% of the embryos tested. However, it is still difficult for the construct with one homologous side to detect whether the PCR HR-product was due to an uncleaved contaminating plasmid. PCR detection of nonhomologous DNA injected in bovine embryos has been achieved in 5% of the embryos (Gagné et al., 1994). Furthermore, in this later study a rapid and constant decreasing rate of detection of injected constructs throughout development was observed, suggesting a rapid and constant degradation or loss of foreign DNA. Thus, the significant higher rate of PCR response in amplifying the 1.5 kb fragment observed in this study suggests that the PCR response would be due to a homologous recombination. The observation that the frequency of DNA integration is increased by addition of homologous sequence on one side in mammalian cells are also demonstrated in other experiments (Adair et al., 1989; Aratani et al., 1992 and Berinstein et al., 1992).

In the case of the two-sided homologous constuct, 38% of embryos retained (beyond the 12-cell stage) enabled amplification of the 750 bp fragment used to identify the HR event. For this construct, a sample of each PCR HR-positive reaction mixture was further analyzed using PCR-primers allowing the detection of a possible contamination plasmid that could have been responsible of the positive response. However, in 10 of the 12 samples, the plasmid was undoubtedly absent from the first PCR reaction mixture, meaning that amplification of the 750 bp fragment could be truly due to an HR integration of the injected DNA. These results confirm observations where PCR HR-detection was consistent with Southern blot analysis (Aratani et al., 1991). For the two samples showing the band specific to the plasmid used to clone the injected marker fragment, it is not clear that it was due to a free or an integrated complete construct, since in the study by Aratani and coworkers (1992), it was observed that closed, circular vectors were probably cut by nucleases within cells and then integrated into the genome. These preliminary data suggest that homologous recombination by using a highly repeated genomic sequence can make transgenesis more sucessful in bovine embryos.

Conclusion

In conclusion, in all experiments using different systems for foreign gene transport into germ cells, DNA incorporation into the transformed cells was demonstrated. Unfortunately, it does not always mean that, except for the PCR method in the case of gene targeted transgene, the transgenes are well integrated into the host genome, and methods to detect early transgenic embryos, particularly in farm animals, are not still performed. It seems that the molecular level is the critical aspect to consider not only in attempts to produce transgenic domestic animals. It is likely that when the mechanisms of genomic integration are clarified, it will be possible to specifically modify any type of cells, and efficiently transform any animal.

References

Adachi, H., Hasebe, T., Yoshinaga, K., Ohta, T. and Sutoh, K. (1994) Isolation of dictyostelium-discoideum cytokinesis mutants by restriction enzyme-mediated integration of the blasticidin-s tesistance marker. *Biochem. Biophy. Res. Comm.* **3**, 1808–1814

Adair, G.M., Nairn, R.S., Seidman, J.H., Brotherman, K.A., MacKinnon, K.C. and Scheere, J.B. (1989) Targeted homologous recombination at the endogenous adenine phosphoribosyltransferase locus in Chinese hamster cells. *Proc. Natl. Acad. Sci. USA* **86**, 4572–4578

Anderson, R.A. and Eliason, S.L. (1986) Recombination of homologous DNA fragments transfected into mammalian cells occurs predominantly by terminal pairing. *Mol. Cell. Biol.* **6**, 3246–3252

Aratani, Y., Okazaki, R. and Koyama, H. (1992) End extension repair of introduced targeting vectors mediated by homologous recombination in mammalian cells. *Nucl. Acids* **20**, 4795–4801

Ashram, C.R. and Davidson, R.L. (1984) High spontaneous mutation frequency in shuttle vector sequences recovered from mammalian cellular DNA. *Mol. Cell. Biol.* **4**, 2266–2272

Bandriff, B.F. and Bordon, L.A. (1992) Spatial distribution of sperm-derived chromatin in zygotes determined by fluorescence *in situ* hybridization. *Mutat. Res.* **296**, 33–42

Barone, F., Belli, M., Pazzaglis, S., Sapora, O. and Tabocchini, M.A. (1989) Radiation damage and chromatin structure. *Ann. 1st Super. Sanita.* **25**, 59–67

Berinstein, N., Pennell, N., Ottaway, C.A. and Shulman, M.J. (1992) Gene replacement with one-sided homologous recombination. *Mol. Cell. Biol.* **12**, 360–367

Bishop, J.O. and Smit, P. (1989) Mechanism of chromosomal integration of microinjected DNA. *Mol. Biol. Med.* **6**, 283–298

Bonnerot, C., Rocancourt, D., Briand, P., Grimber, G. and Nicolas, J.F. (1987) A β-galactosidase hybrid protein targeted to nuclei as a marker for developmental studies. *Proc. Natl. Acad. Sci. USA* **84**, 6795–6799

Bonnerot, C., Vernet, M., Grimber, G., Briand, P. and Nicolas, J.F. (1991) Transcriptional selectivity in early mouse embryos: a qualitative study. *Nucl. Acids* **19**, 7251–7257

Brenner, D.A., Smigocki, A.C. and Camerini-Otero, R.D. (1985) Effect of insertions, deletions, and double-strand breaks on homologous recombination in mouse cells. *Mol. Cell. Biol.* **5**, 684–691

Brinster, R.L., Chen, H.Y., Trumbauer, M., Senear, A.W., Warren, R. and Palmiter, R.D. (1981) Somatic expression of herpes thymidine kinase in mice following injection of a fusion gene into eggs. *Cell* **27**, 223–231

Brinster, R.L., Braun, R.E., Lo, D., Avarbock, M.R., Oram, F. and Palmiter, R.D. (1989) Targeted correction of major histocompatibility class IIE alpha gene by DNA microinjected into mouse cells. *Proc. Natl. Acad. Sci. USA* **86**, 7087–7090

Brinster, R.L., Chen H.Y., Trumbauer, M.E., Yagle, M.K. and palmiter, R.D. (1985) Factors affecting the efficiency of introducing foreign DNA into mice by microinjecting eggs. *Proc. Natl. Acad. Sci. USA* **82**, 4438–4442

Brouillette, S. and Chartrand, P. (1987) Intermolecular recombination assay for mammalian cells that produces recombinants carrying both homologous and nonhomologous junctions. *Mol. Cell. Biol.* **7**, 2248–2255

Brown, D.D. (1984) The role of stable complexes that repress and active eucaryotic genes. *Cell* **37**, 359–365

Bryant, P.E. (1984) Enzymatic restriction of mammalian cell DNA using Pvu II and Bam HI: evidence for the double-strand break original of chromosomal aberrations. *Int. J. Radiat. Biol.* **46**, 57–65

Burdon, T.G. and Wall, R.J. (1992) Fate of microinjected genes in preimplantation mouse embryos. *Mol. Reprod. Dev.* **33**, 436–442

Calos, M.P., Lebkowski, J.S. and Botchan, M.R. (1983) High mutation frequency in DNA transfected into mammalian cells. *Proc. Natl. Acad. Sci. USA* **80**, 3015–3019

Capecchi, M.R. (1980) High efficiency transformation by direct microinjection of DNA into cultured mammalian cells. *Cell* **22**, 479–488

Capecchi, M.R. (1989) The new mouse genetics: altering the genome by gene targeting. *TIG* **5**, 70–76

Chakrabarti, A. and Seidman, M.M. (1986) Intramolecular recombination between transfected repeated sequences in mammalian cells is nonconservative. *Mol. Cell. Biol.* **6**, 2520–2526

Charron, J., Malynn, B.A., Robertson, E.J., Goff, S.P. and Alt, F.W. (1990) High frequency disruption of the c-myc gene in embryonic stem and pre-B cell lines by homologous recombination. *Mol. Cell. Biol.* **10**, 1799–1804

Cherny, R.A., Stokes, T.M., Merei, J., Lom, L., Brandon, M.L. and Williams, R.L. (1994) Strategies for the isolation and characterization of bovine embryonic stem cells. *Reprod. Fertil. Dev.* **6**, 569–575

Clark, A.J., Bissinger, P., Bullock, D.W., Damak, S., Wallace, R., Whitelaw, C.B.A. and Yull, F. (1994) Chromosomal position effects and the modulation of transgene expression. *Reprod. Fertil. Dev.* **6**, 589–598

Doetschman, T., Gregg, R.G., Maeda, N., Hooper, M.L., Melton, W., Thompson, S. and Smithies, O. (1987) Targetted correction of a mutant HPRT gene in mouse embryonic stem cells. *Nature Lond.* **330**, 576–578

Domon, M. (1982) Radiosensitivity variation during the cell cycle in pronuclear mouse embryos *in vitro*. *Cell Tissue Kinet.* **15**, 89–98

DuFrain, R.J. and Casarett, A.P. (1975) Response of the pronuclear mouse embryo to X-irradiation *in vitro*. *Rad. Res.* **63**, 494–500

Etkin, L.D., Pearman, B., Roberts, M. and Bektesh, S.L. (1984) Replication, integration and expression of exogenous DNA injected into fertilized eggs of Xenopus leavis. *Differentiation* **26**, 194–202

Evans, M.F. and Kaufman, M.H. (1981) Establishment in culture of pluripotential cells from mouse embryos. *Nature* **292**, 154–156

Farzaneh, F., Panayotou, G.N., Bowler, L.D., Hardas, B.D., Broom, T., Walthher, C. and Shall, S. (1988) ADP-ribosylation is involved in the integration of foreign DNA into the mammalian cell genome. *Nucl. Acids* **16**, 1119–1126

Forbes, D.J., Kirschner, M.W. and Newport, J.W. (1983) Spontaneous formation of nucleus like structures around bacteriophage DNA microinjected into Xenopus eggs. *Cell* **34**, 13–23

Gagné, M., Pothier, F. and Sirard, M.-A. (1995) Effect of microinjection time during the post-fertilization S-phase on bovine embryonic development. *Mol. Reprod. Dev.* **41**, 185–195

Gagné, M., Pothier, F. and Sirard, M.-A. (1990) Developmental potential of early bovine zygoes submitted to centrifugation and microinjection following *in vitro* maturation of oocytes. *Theriogenology* **34**, 417–425

Gagné, M., Pothier, F. and Sirard, M.-A. (1994) Comparison of cytoplasmic and nuclear microinjection of non transcribed and silent plasmids: expression and degradation after activation and fertilization of bovine oocytes. *Transgenics* **1**, 293–304

Gagné, M., Pothier, F. and Sirard, M.-A. (1996b) Recombination of foreign DNA with homologous terminating ends following microinjection into bovine embryos. Nature Biotechnology (submitted)

Gagné, M., Pothier, F. and Sirard, M.-A. (1996a) Study of genomic recombination of foreign DNA co-injected with restriction endonuclease into maturing and inseminated bovine oocytes. *Transgenics* 2, 79–88

Gagné, M., Potheir, F. and Sirard, M.-A. (1991) Electroporation of bovine spermatozoa to carry foreign DNA in oocytes. *Mol. Repro. Dev.* 29, 6–16

Goedecke, W., Vielmetter, W. and Pfeiffer, P. (1992) Activation of a system for the joining of nonhomologous DNA ends during Xenopus egg maturation. *Mol. Reprod. Dev.* 12, 811–816

Gordon, J.W. and Ruddle, F.H. (1981) Integration and stable germ line transmission of genes injected into mouse pronuclei. *Science* 214, 1244–1246

Grell, R.F. (1978) High frequency recombination in centromeric and histone regions of Drosophila genosmes. *Nature Lond.* 272, 78–80

Groudine, M. and Weintraub, H. (1982) Propagation of globin DNAase I-hypersensitive sites in absence of factor required for induction: A possible mechanism for determination. *Cell* 30, 131–139

Haber, J.A., Borts, R.H., Connoly, B., Lichten, M., Rudin, N. and White, C.I. (1988) Physical monitoring of meiotic and mitotic recombination in yeast. *Prog. Nucleic Acid. Mol. Biol.* 35, 209–259

Hamada, T., Saski, H., Seki, R. and Sakaki, Y. (1993) Mechanism of chromosomal integration of transgenes in microinjected mouse eggs: sequence analysis of genome-transgene and transgene-transgene junctions at two loci. *Gene* 128, 197–202

Harland, R.M., Weintraub, H. and McKnight, S.L. (1983) Transcription of DNA injected into *Xenopus* oocytes is influenced by template topology. *Nature Lond.* 302, 38–43

Harland, R. and Laskey, R. (1980) Regulated replication of DNA microinjected into eggs of Xenopus leavis. *Cell* 21, 761–771

Hawk, H.W., Wall, R.J. and Conley, H.H. (1989) Survival of DNA injected cow embryos temporarily cultured in rabbit oviducts. *Theriogenology* 32, 243–253

Jacquet, P., Kervyn, G. and De Clercq, G. (1983) Studies *in vitro* on mouse-egg radiosensitivity from fertilization up to the first cleavage. *Mutat. Res.* 110, 351–365

Jasin, M., de Villier, J., Weber, F. and Schaffner, W. (1985) High frequency of homologous recombination in mammalian cells between endogenous and introduced SV40 genomes. *Cell* 43, 695–703

Jasin, M., Elledge, S.J., Davis, R.W. and Berg, P. (1990) Gene targeting at the human CD4 locus by epitope addition. *Gene Dev.* 4, 157–166

Joyner, A.L., Skarnes, W.C. and Rossant, J. (1989) Production of a mutation in mouse En-2 gene by homologous recombination in embryonic stem cells. *Nature Lond.* 338, 153–6

Kim, H.S. and Smithies, O. (1988) Recombinant fragment assay for gene targeting based on the polymerase chain reaction. *Nucl. Acids* 16, 8887–8903

Kim, T., Leibfried-Rutledge, M.L. and First, N.L. (1993) Gene transfer in bovine blastocysts using replication-defective retroviral vectors packaged with Gibbon ape leukemia virus envelopes. *Mol. Reprod. Dev.* 35, 105–113

Kispa, A. and Loomis, W.F. (1994) REMI-RELP mapping in the dictyostelium genome, *Genetics* 138, 665–674

Koller, B.H. and Smithies, O. (1989) Inactivating the β2 microglobulin locus in mouse embryonic stem cells by homologous recombination. *Proc. Natl. Acad. Sci. USA* 86, 8932–8935

Kubish, H.M., Hernandez-Kedezma, J.J., Sikes, J.D. and Roberts, R.M. (1994) Kinetics of pronuclear formation of IVM/IVF-derived bovine and their development rates following DNA microinjection. *Theriogenology* 41, 230

Kuesh, M.R., Bradley, A., Robertson, E.J. and Evans, M.J. (1987) A potential enimal model for Lesch-Nyhan syndrome through introduction of HPRT mutations into mice. *Nature Lond.* 326, 295–298

Lebkowski, J.S., DuBridge, R., Antell, E.A., Greisen, K.S. and Calos, M.P. (1984) Transfected DNA is mutated in monkey, mouse, and human cells. *Mol. Cell. Biol.* 4, 1951–1960

Lehman, C.W., Clemens, M., Worthylake, D.K., Trautman, J.K. and Corroll, D. (1993) Homologous and illegitimate recombination in developing Xenopus oocytes and eggs. *Mol. Cell. Biol.* 13, 6797–6806

Lehmann, A.R., Hoeijmakers, J.H.J., van Zeeland, A.A., Backendorf, C.M.P., Bridges, B.A., Collins, A., Fuchs, R.P.D., Margison, G.P., Montesano, R., Moustacchi, E., Natarajan, A.T., Radman, M., Sarasin, A., Seeberg, E., Smith, C.A., Stefanini, M., Thompson, L.H., van der Schans, G.P., Weber, C.A. and Zdzienicka, M.Z. (1992) Workshop on DNA repair. *Mut. Res.* 273, 1–28

Lin, F.L., Sperle, K. and Sternberg, N. (1984) Model for homologous recombination during transfer of DNA into mouse L cells: role for DNA ends in the recombination process. *Mol. Cell. Biol.* 4, 1020–1034

Longo, F. (1985) Pronuclear events during fertilization. In: C.B. Metz, A. Monroy (eds.), *Biology of Fertilization*, Vol. 3. Orlando, Florida, Academic Press, pp. 251–293

Mansour, S.L., Thomas, K.R. and Capecchi, M.R. (1988) Disruption of the proto-oncogene int-2 in mouse embryo-derived stem cells: a general strategy for targeting mutations to non-selectable genes. *Nature Lond.* 336, 348–352

Maryon, E. and Carroll, D. (1989) Degradation of linear DNA by a strand-specific exonuclease activity in Xenopus leavis oocytes. *Mol. Cell. Biol.* 9, 4862–4871

McMahon, A.P., Flyzanis, C.N., Hough-Evans, B.R., Katula, K.S., Britten, R.J. and Davidson, E.H. (1985) Introduction of cloned DNA into sea urchin egg cytoplasm: replication and persistence during embryogenesis. *Dev. Biol.* 108, 420–430

Meselson, M.S. and Radding, C.M. (1975) A general model for genetic recombination. *Proc. Natl. Acad. Sci. USA* 78, 6354–6358

Natarajan, A.T. and Obe, G. (1984) Molecular mechanisms involved in the production of chromosomal aberratoions, III. Restriction endonuclease. *Chromosoma* 90, 120–127

Obe, G. and Natarajan, A.T. (1985) Chrosomal aberrations induced by the restriction endonuclease Alu I in

Chinese hamster ovry cells: influence of duration of treatment and potentation by cytosine arabinoside. *Mutat. Res.* **152**, 205–210

Obe, G. and Winkel, E.U. (1985) The chromosome-breaking activity of the restriction endonucleas Alu I in CHO cells is independent of the S-phase of the cell cycle. *Mutat. Res.* **152**, 25–29

Perez, C.F., Botchan, M.R. and Tobias, C.A. (1985) DNA-mediated gene transfer efficiency is enhanced by ionizing and ultraviolet irradiation of rodent cells *in vitro*. *Rad. Res.* **104**, 200–213

Perreault, S.D. (1990) Regulation of sperm nuclear reactivation during fertilization. In: B.D. Bavister, J. Cummings, E.R.S. Roldan (eds.), *Fertilization in Mammals*. Norwell, Massachusetts: Serono Symposia USA, pp. 285–296

Pieper, F.R., de Wir, I.C.M., Pronk, A.C.J., Kooiman, P., Strijker, R., Krimpenfort, P.J.A., Nuyens, J.H. and de Boer, H.A. (1992) Efficient generation of functional transgenes by homologous recombination in murine zygotes. *Nucl. Acids* **20**, 1259–1264

Peura, T.T. and Jänne, J. (1994) Effects of fluid injection or membrane piercing on the development of *in vitro* produced bovine zygotes. *Theriogenology* **41**, 273

Plucienniczak, A., Skowronski, J. and Jaworski, J. (1982) Nucleotide sequence of bovine 1.715 satellite DNA and its relation to other bovine satellite sequences. *J. Mol. Biol.* **158**, 293–304

Powell, D.J., Galli, C. and Moor, R.M. (1992) The fate of DNA injected into mammalian oocytes and zygotes at different stages of the cell cycle. *J. Reprod. Fert.* **95**, 211–220

Razin, A. and Riggs, A.D. (1980) DNA methylation and gene function. *Science* **210**, 604–610

Roth, D.B. and Wilson, J.H. (1988) Illegitimate recombination in mammalian cells. In: R. Kucherlapati, G. Smith (eds.), *Genetic Recombination*. Washington, DC. American Society Microbiology, pp. 621–653

Roth, D.B. and Wilson, J.H. (1985) Relative rates of homologous and nonhomologous recombination in transfected DNA. *Proc. Natl. Acad. Sci. USA* **82**, 3355–3359

Saxena, J.K., Hays, J.B. and Ackerman, E.J. (1990) Excision repair of UV-damaged plasmid DNA in Xenopus oocytes is mediated by DNA polymerase alpha (and/or delta). *Nucl. Acids* **18**, 7425–7431

Shaulsky, G., Kuspa, A. and Loomis W.F. (1995) A multidrug-resistance transporter serine-protease gene is required for prestalk specalization in dictyostelium. *Genes and Development* **9**, 1111–1122

Smithies, O., Gregg, R.G., Boggs, S.S., Koralewski, M.A. and Kucherlapati, R.S. (1985) Insertion of DNA sequences into the human chromosomal β-globin locus via homologous recombination. *Nature Lond.* **317**, 230–234

Szostak, J.W., Orr-Weaver, T.L., Rothstein, R.J. and Stahl, F.W. (1983) The double-strand break repair model for recombination. *Cell* **33**, 25–35

Thomas, K.R. and Capecchi, M.R. (1987) Site-directed mutagenesis by gene targeting in mouse embryo-derived stem cells. *Cell* **51**, 503–512

Thomas, K.R., and Capecchi, M.R. (1990) Targeted disruption of the mouse Int-1 proto-oncogene results in severe abnormalities in mid-brain and cerebellar development. *Nature Lond.* **346**, 847–850

Thomas, K.R., Folger, K.R. and Capecchi, M.R. (1986) High frequency targeting of genes to specific sites in the mamalian genome. *Cell* **44**, 419–428

Thomas, W.K., Schnieke, A. and Seidel, G.E. Jr. (1993) Methods for producing transgenic bovine embryos from *in vitro* matured and fertilized oocytes. *Theriogenology* **40**, 679–688

Thomas, S., Clarke, A.R., Pow, A.M., Hooper, M.L. and Melton, D.W. (1989) Germline transmission and expression of a corrected HPRT gene, produced by gene targeting in embryonic stem cells. *Cell* **56**, 313–321

Wake, C.T., Gudewicz, T., Porter, T., White, A. and Wilson, J.H. (1984) How damaged is the biologically active subpopulation of transfected DNA? *Mol. Cell. Biol.* **4**, 387–398

Wall, R.J. and Hawk, H.W. (1988) Development of centrifuged cow zygotes cultured in rabbit oviducts. *J. Reprod. Fertil.* **82**, 673–680

Wilmut, I. and Whitelaw, C.B.A. (1994) Strategies for production of pharmaceutical proteins in milk. *Reprod. Fertil. Dev.* **6**, 625–630

Wong, E. and Capecchi, M.R. (1987) Homologous recombination between coinjected DNA sequences peaks in early to mid-S phase. *Mol. Cell. Biol.* **7**, 2294–2295

Wyllie, A.H., Gurdon, J.B., Price, J. (1980) Nuclear localisation of an oocyte component required for the stability of injected DNA. *Nature Lond.* **268**, 150–152

Zelenin, A.V., Alimov, A.A., Zelenina, I.A., Semenova, M.L., Rodova, M.A., Chernov, B.K. and Kolesnikov, V.A. (1993) Transfer of foreign DNA into the cells of developing mouse embryos by microprojectile bombardment. *FEBS* **315**, 29–32

Zimmer, A. and Gruss, P. (1989) Production of chimaeric mice containing embryonic stem (ES) cells carrying a homoeobox Hox 1.1 allele mutated by homologous recombination. *Nature Lond.* **338**, 150–153

Zirkin, B.R., Perreault, S.D. and Naish, S.J. (1989) Formation and function of the male pronucleus during mammalian fertilizaton. In: H. Schatten, G. Schatten (eds.), *Molecular Biology of Fertilization*. Orlando, Florida: Academic Press pp. 91–114

PART II, SECTION A

8. DNA Delivery by Cytoplasmic Injection

William H. Velander

Department of Chemical Engineering, Virginia Polytechnic Institute and University, Blacksburg, Virginia 24062, USA

Microinjection of DNA is currently the most efficient delivery method for making transgenic animals, and in particular for making transgenic livestock. Many transgenic livestock have been made using pronuclear microinjection (Hammer *et al.*, 1985; Wall *et al.*, 1991; Ebert *et al.*, 1991; Wright *et al.*, 1991 and Velander *et al.*, 1992a). Transgenic frequencies obtained from pronuclear microinjection typically range from 5% to 30% of the total animals born. Alternatively, transgenic animals have been made using direct retroviral infection of embryos (Jaenisch and Mintz, 1974) or embryonic stem cells (Evans and Kaufman, 1981). However, the combination of culturing stem cells, subsequent transfection with retroviral DNA, screening for viable transfectants, and reimplantation of these selected cells into embryos is labor intensive. In addition, transgenic animal production by retroviral infection is limited to small (5–6 kb) foreign DNA constructs (Jaenisch, 1976), results in a high frequency of germline mosaicism (Anderson, 1992), and has been most practised in the making of transgenic chickens (Salter *et al.*, 1986). Stem cell lines are available in only a few animal species such as mice and for livestock have been reported only recently in pig (Gerfen *et al.*, 1995). Unlike retroviral infection, microinjection can be used to introduce multiple, independent constructs to create multigenic transgenic animals (Greenberg *et al.*, 1991; Clark *et al.*, 1992 and Drews *et al.*, 1995). Furthermore, DNA constructs as large as 33 kb have been delivered by pronuclear microinjection as multiple overlapping fragments which subsequently integrate after homologous recombination (Pieper *et al.*, 1992). Recently, transgenic mice have been made by microinjection of a 248 kb yeast artificial chromosome which contained a 82 kb DNA construct (Peterson *et al.*, 1993). In summary and relative to other methods of DNA delivery, microinjection has advantages of a less labor intensive process that accommodates a wide range of construct sizes and can simultaneously deliver multiple, independent gene constructs. Thus, there is considerable impetus to further develop microinjection methods as a means of DNA delivery and the production of transgenic animals.

The production of transgenic animals by cytoplasmic injection of DNA potentially offers several advantages to pronuclear microinjection. First, pronuclei need not be targeted and thus the injection process is more quickly and easily done. Second, centrifugation of zygotes to stratify lipid so as to permit visualization of pronuclei would not be necessary for species such as pigs (Wall *et al.*, 1985) and cows (Biery *et al.*, 1988). This may be more advantageous in sheep since pronuclei are difficult to visualize even after centrifugation (Hammer *et al.*, 1985). In addition, replication of the genome may occur before DNA has been injected into the pronucleus resulting in undesired somatic and germline mosaicism (Wilkie *et al.*, 1986). Thus, injecting during the pre-pronuclear phase could widen the time frame over which integration may more optimally occur in the zygote. However, in the case of cytoplasmic injection, the foreign DNA may be more subject to enzymatic degradation than for pronuclear injection. Degradation of DNA is suggested by past studies in which no transgenic animals were produced by cytoplasmic injection of purified construct DNA alone (Brinster *et al.*, 1985). These results suggest a need to stabilize injected DNA against degradation so as to enable transport to the host genome and while also providing a conformational state which is compatible with integration (Page et al., 1995).

A relatively efficient pathway for transgenesis exists in murine zygotes for foreign DNA introduced by cytoplasmic microinjection of mixtures containing poly-L-lysine and construct DNA during pronuclear stage (Page et al., 1995). These mixtures contained large polylysine/DNA aggregates that tended to clog injection pipettes which were drawn to less than about 1 ϕ micron in tip diameter. The highest integration frequencies were obtained from cytoplasmic injections at DNA concentrations of 50 µg/ml at lysine to phosphate (L:P) ratios of about 0.5:1 or greater. A transgenic frequency of about 13% of pups was obtained from cytoplasmic injection of a polylysine/DNA mixture at a L:P ratio of 1:1 and a DNA concentration of 50 µg/ml. No transgenic pups were obtained from microinjection of DNA alone at 50 µg/ml into the cytoplasm. The transgenic frequency for the pronuclear microinjection of DNA without polylysine at about 5 µg/ml was about 22%. In general, we have found that optimal transgenesis occurs in mice from pronuclear injection at about 3–10 µg/ml of non-agarose-purified DNA fragments (unpublished results and Velander et al., 1992a) in tris-EDTA buffer (Brinster et al., 1985). Our studies suggest that the presence of polylysine may affect the governing pathways of DNA transport from the cytoplasm into the pronucleus, enzymatic degradation and integration.

Pronuclear injections of DNA alone at 15 µg/ml or higher resulted in low pregnancy rates and also low embryo viability in vitro. While the pipette tip diameters used in cytoplasmic injection were about two-fold larger than used for pronuclear injections, the viability of the pronuclear injected zygotes at 5 µg/ml DNA alone was similar to those cytoplasmically injected with DNA at 50 µg/ml with and without polylysine; about 16%, 19%, and 22% of transferred embryos resulted in pups from pronuclear injection of 5 µg/ml DNA alone, and cytoplasmic injections of 50 µg/ml DNA complexed with polylysine at an L:P of 1:1 and DNA alone, respectively. About 30 to 40 microinjected embryos were transferred into each recipient for each of the injection treatment groups. While the above studies did not optimize the conditions at which cytoplasmic injection of DNA and polylysine could produce transgenic animals, the overall efficiency of transgenesis based upon injected embryos transferred for cytoplasmic injection (about 2.4%) was similar to pronuclear injection (about 3.6%). The application of polylysine/DNA delivery by cytoplasmic injection will likely require further optimization of DNA concentration, injection needle bore diameter due to the large size of polylysine/DNA aggregates, and time of injection before transgenic livestock can be efficiently obtained.

The transgenic mice produced from cytoplasmic injection of polylysine/DNA mixtures had genotypic and phenotypic characteristics similar to those produced by pronuclear microinjection. The incidence of germline mosaicism was about 50% for transgenic mice produced both by cytoplasmic and pronuclear microinjection. The transmission of the transgene by founder males indicates that transgenesis was not due to integration into mitochondrial DNA. Furthermore, Southern analysis showed similar restriction banding patterns for transgenic mice produced by pronuclear and cytoplasmic injection. The level of expression of the mammary gland specific transgene into the milk was similar in transgenic mice produced by pronuclear (Velander et al., 1992b) and cytoplasmic injection (Page et al., 1995). This indicates that insertions into similarly active chromosomal domains had occurred for both DNA injected into the cytoplasm and pronu-cleus. In summary, the process of DNA integration likely occurred at an early cell stage resulting in a similar genotype and phenotype from DNA introduced by either pronuclear or cytoplasmic injection.

Our experiments with transgenesis from cytoplasmic injection of polylysine/DNA mixtures show that the process is dependent upon the concentration of DNA and the L:P ratio. Past studies have shown that the presence of polylysine can greatly alter the aqueous solution behavior of DNA fragments ranging from 0.4 to 100 kbp (Leng and Felsenfeld, 1966, Shapiro et al., 1969 and Laemmli, 1975). In the absence of polylysine, linearized DNA fragments in low ionic strength buffer will be monomeric and typically have lengths which are proportional to molecular weight. Both light scattering and electron microscopy show that polylysine causes DNA to condense into multimeric aggregates of about 170 nm in mean diameter for DNA sizes up to 100 kbp (Shapiro et al., 1969). These aggregates contain individual DNA rod-like fragments which have a nearly constant axial dimension with an end to end wound structure that increases in cross-sectional diameter in proportion to molecular weight. Thus, the polylysine/DNA aggregate contains DNA monomers having a conformation which are markedly different than that of aqueous DNA alone. However, the polylysine/DNA aggregate has properties which can make DNA delivery difficult. As is commonly done for pronuclear injections, our initial studies used lower ionic strength mixtures which promotes stronger electrostatic attractions between DNA host and potential guest molecules like polylysine. Using dynamic light scattering, we have detected aggregates having a hydrodynamic diameter in excess of 200 nm with polylysine/DNA mixtures having L:P ratios greater than 2:1. Because the inner diameter of a typical microinjection needle is about 1000 nm, the formation of large aggregates may have prevented DNA from being efficiently injected into some zygotes. For example, the frequency of detection of the injected construct by PCR was much lower in zygotes microinjected with the 2:1 L:P ratio mixture than for a L:P of about 1:1. Aggregate formation as detected by light scattering was considerably less for an L:P ratio of 1:1 than for a L:P of 2:1. In summary, optimal conditions for transgenesis from the microinjection of polylysine/DNA may include higher ionic strengths which limit the formation of larger aggregates that may not be efficiently delivered by microinjection.

As is the case for pronuclear delivery of DNA alone, the mechanism by which polylysine enables DNA integration is not known. In either case, DNA integration appears to be a random process possibly dependent on a breakage-and-repair mechanism caused by the injection(Brinster et al., 1985). However and in contrast to pronuclear injection of purified DNA, cytoplasm injection of pure DNA seems to have no efficient pathway for transgenesis. Furthermore, electroporation of linearized DNA which had been injected into the perivitelline space of murine embryos resulted in a low transgenic efficiency and only germline mosaicism (Nemec et al., 1989). Another alternative to the delivery of DNA by direct injection is through extracellular introduction of liposome encapsulated DNA. While liposome encapsulated DNA has resulted in genetic transformation of mammalian cells cultured in vitro (Felgner et al., 1987 and Behr et al., 1989), microinjection into the blastocoel cavity (Reed et al., 1988) and perivitelline space (Loskutoff et al., 1986) of preimplantation embryos has not produced in transgenic animals. Similarly, transfection of mammalian cells occurs by receptor-mediated endocytosis of DNA complexed with a surface receptor ligand-polylysine conjugate (Wu and Wu, 1987; 1988a,1988b). Here the polylysine was used to complex the surface receptor protein to the DNA by electrostatic interaction (Li et al., 1973). In contrast, our studies showed transfection of murine embryos with polylysine/DNA complexes in the absence of a receptor protein. Because both the tertiary and quaternary conformations of DNA fragments are greatly altered by polylysine alone, the polylysine/DNA complex may be a more favorable substrate for achieving integration in embryos from injection into the cytoplasm than purified DNA alone.

References

Anderson, G.B. (1992) Isolation and use of embryonic stem cells from livestock species. *Animal Biotech.* **3**, 165–175

Behr, J.-P., Demeneix, B., Loeffer, J.-P. and Perez-Mutul, J. (1989) Efficient gene transfer into mammalian primary endocrine cells with lipopolyamine-coated DNA. *Proc. Natl. Acad. Sci. USA* **86**, 6982–6986

Biery, K.A., Bondioli, K.R. and De Mayo, F.J. (1988) Gene transfer by pronuclear injection in the Bovine. *Theriogenology* **29**, 224

Brinster, R.L., Chen, H.Y., Trumbauer, M.E., Yagle, M.K. and Palmiter, R.D. (1985) Factors affecting the efficiency of introducing foreign DNA into mice by microinjection. *Proc. Natl. Acad. Sci. USA* **82**, 4438–4442

Clark, A.J., Cowper, A., Wallace, R., Wright, G. and Simons, J.P. (1992) Rescuing transgene expression by co-integration. *Biotechnology* **10**, 1450–1454

Drews, R., Paleyanda, R.K., Lee, T.K., Chang, R.R., Rehemtulla, A., Kaufman, R.J., Drohan, W.N. and Lubon, H. (1995) Proteolytic maturation of protein C upon engineering the mouse mammary gland to express furin. *Proc. Natl. Acad. Sci. USA* **92**, 10462–10466

Ebert, K.M., Selgrath, J. P., DiTullio, P., Denman, J., Smith, T.E., Memon, M.A., Schindler, J.E., Monastersky, G.M., Vitale, J.A. and Gordon, K. (1991) Transgenic production of a variant of human tissue-type plasminogen activator in goat milk: generation of transgenic goats and analysis of expression. *Bio/Technology* **9**, 835–838

Evans, M.J. and Kaufman, M.H. (1981) Establishment in culture of pluripotential cells from mouse embryos. *Nature* **292**, 154–156

Felgner, P.L., Gadek, T.R., Holm, M., Roman, R., Chan, H.W., Wenz, M., Northrop, J.P., Ringold, G.M. and Danielsen, M. (1987) Lipofectin: a highly efficie, lipid-mediated DNA-transfection procedure. *Proc. Natl. Acad. Sci. USA* **84**, 7413–7417

Gerfen, R.W. and Wheeler, M.B. (1995) Isolation of embryonic cell-lines from porcine blastocysts. *Animal Biotechnology* **6**(1), 1–14

Greenberg, N.M., Anderson, J.W., Hsueh, A.J.W., Nishimori, K., Reeves, J.J., DeAvila, D.M., Ward, D.N. and Rosen, J.M. (1991) Expression of biologically active heterodimeric bovine follicle-stimulating hormone in milk of transgenic mice. *Proc. Natl. Acad. Sci. USA* **88**, 8327–8331

Hammer, R.E., Pursel, V.G., Rexroad, C.E. Jr., Wall, R.J., Bolt, D.J., Ebert, K.M., Palmiter, R.D. and Brinster, R.L. (1985) Production of transgenic rabbits, sheep and pigs by microinjection. *Nature* (London) **315**, 680–683

Jaenisch, R. (1976) Germ line integration and mendelian transmission of the exogenous moloney leukemia virus. *Proc. Natl. Acad. Sci. USA* **73**, 1260–1264

Jaenisch, R. and Mintz, B. (1974) Simian virus 40 DNA sequences in healthy adult mice derived from preimplantation blastocysts injected with viral DNA. *Proc. Natl. Acad. Sci. USA* **71**, 1250–1254

Laemmli, U.K. (1975) Characterization of DNA condensates induced by poly(ethylene oxide) and polylysine. *Proc. Natl. Acad. Sci. USA* **72**, 4288–4292

Leng, M. and Felsenfeld, G. (1966) The preferential interactions of polylysine and polyarginine with specific base sequences in DNA. *Biochem. J.* **56**, 1325–1332

Li, H.J., Chang, C. and Weiskopf, M. (1973) Helix-coil transition in nucleoprotein-chromatin structure. *Biochem. J.* **12**, 1763–1771

Loskutoff, N.M., Roessner, C.A. and Kraemer, D.C. (1986) Preliminary studies on liposome-mediated gene transfer: effects on survivability of murine zygotes. *Theriogenology* **25**, 169

Nemec, L.A., Skow, L.C., Goy, J.M. and Kraemer, D.C. (1989) Introduction of DNA into Murine Embryos by Electroporation. *Theriogenology* **31**, 233

Page, R.L., Butler, S.P., Subramanian, A., Gwazdauskas, F.C., Johnson, J.L. and Velander, W.H. (1995) Transgenesis in mice by cytoplasmic injection of polylysine/DNA mixtures. *Transgenic Research* **4**, 353–360

Peterson, K.R., Clegg, C.H., Huxley, C., Josephson, B.M., Haugen, H.S., Furukawa, T. and Stamatoyannopoulos, G. (1993) Transgenic mice containing 248-kb yeast artificial chromosome carrying the human β-globin locus display proper developmental control of human globin genes. *Proc. Natl. Acad. Sci. USA* **90**, 7593–7597

Pieper, F.R., de Wit, Ineke C.M., Pronk, Arjan C.J., Kooiman, P.M., Strijker, R., Krimpenfort, P.J.A., Nuyens, J.H. and de Boer, H.A. (1992) Efficient generation of functional transgenes by homologous recombination in murine zygotes. *Nucleic Acids Research* **20**, 1259–1264

Reed, M.L., Roessner, C.A., Womack, J.E., Dorn, C.C. and Kraemer, D.C. (1988) Microinjection of liposome-encapsulated DNA into murine and bovine blastocysts. *Theriogenology* **29**, 293

Salter, D.W., Smith, E.J., Hughes, S.H., Wright, S.E., Fadly, A.M., Witter, R.L. and Crittenden, L.B. (1986) Gene insertion into the chicken germ line by retroviruses. *Poultry Sci.* **65**, 1445–1458

Schapiro, J.T., Leng, M. and Felsenfeld, G. (1969) Deoxyribonucleic Acid-Polylysine Complexes. Structure and Nucleotide Specificity. *Biochem. J.* **8**, 3219–3232

Velander, W.H., Johnson, J.L., Page, R.L., Russell, C.G., Subramanian, A., Wilkins, T.D., Gwazdauskas, F.G., Pittius, C. and Drohan, W.N. (1992a) High level expression of a heterologous protein in the milk of transgenic swine using the cDNA encoding human protein C. *Proc. Natl. Acad. Sci. USA* **89**, 12003–12007

Velander, W.H., Page, R.L., Morcol, T., Russell, C.G., Canseco, R., Young, J.M., Drohan, W.N., Gwazdauskas, F.G., Wilkins, T.D. and Johnson, J.L. (1992b) Production of biologically active protein C in the milk of transgenic mice. *Annals New York Acad. Sci.* **665**, 391–403

Wall, R.J., Pursel, V.G., Hammer, R.E. and Brinster, R.L. (1985) Development of porcine ova that were centrifuged to permit visualization of pronuclei and nuclei. *Biol. Reprod.* **32**, 645–651

Wall, R.J., Pursel, V.G., Shamay, A.V., McKnight, R.A., Pittius, C.W. and Hennighausen, L. (1991) High-level synthesis of a heterologous milk protein in the mammary glands of transgenic swine. *Proc. Natl. Acad. Sci. USA* **88**, 1696–1700

Wilkie, T.M., Brinster, R.L. and Palmiter, R.D. (1986) Germline and somatic mosaicism in transgenic mice. *Developmental Biology* **118**, 9–18

Wright, G., Carver, A., Cottom, D., Reeves, D., Scott, A., Simons, P., Wilmut, I., Garner, I. and Colman, A. (1991) High level expression of active human alpha-1-antitrypsin in the milk of transgenic sheep. *Bio/Technology* **9**, 830–834

Wu, G.Y. and Wu, C.H. (1987) Receptor-mediated *in vitro* gene transformation by a soluble DNA carrier system. *J. Biol. Chem.* **262**, 4429–4432

Wu, G.Y. and Wu, C.H. (1988a) Evidence for targeted gene delivery to hep G hepaoma cells *in vitro*. *Biochem. J.* **27**, 887–892

Wu, G.Y. and Wu, C.H. (1988b) Receptor-mediated gene delivery and expression *in vivo*. *J. Biol. Chem.* **263**, 14621–14624

PART II, SECTION A

9. The Direct Gene Transfer Through Mammal Spermatozoa

K. Schellander* and G. Berm

Institute of Animal Breeding and Genetics, Veterinary University of Vienna, A-1030 Vienna, Linke Bahnagasse 11, Austria

Introduction

In 1971 Brackett et al., reported the first evidence of the uptake of a heterologous genome by mammalian spermatozoa. In this key report ejaculated rabbit sperm were exposed to SV40 DNA in a Krebs-Ringer-Phosphate buffer. With [^{3}H]-thymidine labeled DNA they showed that 15.9% of the input radioactivity was recovered from the spermatozoan DNA. After autoradiography 30–35% of the spermatozoa contained radioactive material mostly localized in the postacrosomal area. Following uterine insemination of rabbit with spermatozoa exposed to SV40, the ability of those sperm to fertilize eggs was demonstrated. When the fertilized ova were cocultivated with CV1-cells, infectious virus was recovered. Polar bodies, zonae pellucidae (from fertilized and unfertilized eggs) did not show any cytopathic effects on CV1-cells.

Although the gene transfer technology in mammals developed to an enormously useful tool (Palmiter and Brinster 1986) to address a broad range of questions in biology, Brackett's discovery did not attract wide attention for application to produce transgenic animals.

In 1989 Lavitrano et al., attracted widespread attention from the scientific as well as public community. A simple and efficient technique to produce transgenic mice was described. In this technique DNA was mixed with spermatozoa before their use *in vitro* fertilization. In about 30% of the obtained offspring the foreign DNA was integrated and germ-line transmitted. Moreover, the gene-construct (pSV-CAT) was expressed in tissues of adult F1 individuals.

This finding prompted many scientists to repeat the experiment and/or to investigate possible mechanisms or modified techniques in sperm-mediated gene transfer.

Methods Developed for DNA Treatment of Sperm

The basic principle of sperm-mediated gene transfer is a coincubation of the DNA fragments in the sperm suspension. DNA (in supercoiled, plasmid or linearized form) is simply added to the sperm between 30 min (Lavitrano et al., 1989) and 120 min (Brackett et al., 1971) prior to the use of this suspension for fertilization resp. insemination. In both protocols a concentration of 1 mg DNA/10^6 sperms is added. In Brackett's technique DMSO was additionally added to the fertilization medium. The incubation conditions in mice during sperm-DNA coincubation are similiar to sperm preincubation necessary for normal IVF (Hogan et al., 1986).

To increase the efficiency of DNA-uptake two modifications of this technique have been described. We have used (Bachiller et al., 1990, 1991) liposomes to transfect DNA into the sperm cells. These cationic lipids interact with the negatively charged nucleic acid molecules forming complexes in which the nucleic acid is coated by the lipids (Felgner et al., 1987). The positive outer surface of the complex can then associate with the negatively charged cell membrane, allowing the internalization of the nucleic acid. In this technique, the DNA is mixed with cationic lipids just before addition of this complex to the sperm suspension.

* Author for Correspondence.

Gagné et al. (1991) used electroporation to increase DNA uptake of bovine sperm cells. Association of DNA (1mg/10^7 sperm plasmid-DNA) with the spermatozoa was increased severalfold using electric pulses between 500–1000 V and capacitance of 1 or 25 µFarads. Similar findings were reported also by Horan et al. (1992a) in pig, where electroporation increased the DNA-binding of pig spermatozoa.

In the three methods described the fertilization capacity of DNA-treated sperm was tested *in vitro* fertilization experiments using standard protocols. The fertilization rate of the oocyte was depressed in a varying degree, depending on the concentration of DNA and liposomes and the electric conditions used during electroporation. When DNA-treated sperm were used for artificial insemination in pig (Gandolfi et al., 1989) and cattle (Schellander et al., 1995) moderate reduction of the number of born offspring was observed.

Interaction of DNA with Sperm Cells

Sperm cells of mice (Lavitrano et al., 1989; Hochi et al., 1990; Gavora et al., 1991; Bachiller et al., 1991; Francolini et al., 1993 and Camaioni et al., 1992), rabbit (Brackett et al., 1971), sheep, goat, buffalo (Castro et al., 1990), pig (Gandolfi et al., 1989; Castro et al., 1990; Horan et al., 1992a,b and Camaioni et al., 1992) and cattle (Castro et al., 1990; Gagné et al., 1991 and Camaioni et al., 1992) associate with DNA when they are coincubated with DNA molecules under conditions allowing the sperm to remain vital. This association was termed DNA-binding ability (Lavitrano et al., 1989) and is defined as the proportion of sperm with attached exogenous DNA following extensive washing after DNA treatment. This binding ability ranges from 30–35% in rabbits (Brackett et al., 1971), 47–52% in mice, 70–78% in pig and 39–47% in cattle (Camaioni et al., 1992).

The binding activity is influenced by the vitality of sperms, since DNA does not associate with dead sperms (Castro et al., 1990). However, sperms immobilized by low temperatures (0 C) do not lose their binding activity (Lavitrano et al., 1989). Epididymal sperms and those from ejaculates, which are thoroughly washed free from seminal plasma bind DNA much more effectively than sperm from diluted ejaculates (Camaioni et al., 1992). This indicates that seminal plasma contains factors blocking the sperm permeability. These factors could either interfere direct with the DNA binding site on the sperm surface or bind directly to the DNA (Lauria and Gandolfi 1993). Seminal plasma contains negatively charged polyanions, such as glycoaminoglycans. Lavitrano et al. (1992a) have shown that heparin totally prevented DNA-binding of sperms.

Therefore, since the charge density plays an important role in binding, longer DNA fragments (7 kb) associate more efficiently than shorter ones (Lavitrano et al., 1992a). This may indicate that binding is not only a passive process. Since DNA and polyanions bind to the same region of the sperm, it has been indicated that specific binding sites are present on certain areas of the sperm surface. Lavitrano et al., (1992a,b) have shown the presence of a protein in the sperm heads which specifically binds DNA molecules.

Atkinson et al. (1991) described four discrete patterns of the topographical association of DNA with cattle sperm. In the vast majority (84.8%) DNA was bound to the posterior head region. Less than 1% was located on the outer acrosomal cap, around 2% at the equatorial segment and in 12.7% DNA binding was confined in the extreme posterior head region. When samples were DNase treated, in 19% of the sperm the DNA remained associated with the sperm head. Interestingly, the topographical association pattern was altered: 87.5% exhibited labeling in the extreme posterior head region and 11.7% in the posterior head region. This localization was confirmed also by others in rabbits (Brackett et al., 1971), mice, cattle, pigs and man (Camaioni et al., 1992). While Lavitrano et al. (1992a) have evidence for a DNA-specific binding protein present in sperm head extracts, Wu et al. (1990) show that binding is mediated by MHC class 1 molecules localized in the postacrosomal region of the mouse sperm head.

Binding of DNA to sperm heads is not a peculiarity of an *in vitro* system. Sato et al. (1994) injected the pSV2-CAT plasmid directly via a needle into the mouse testis, expecting either more easy transfection of actively proliferating precursor cells of spermatozoa or that testicular spermatozoa during their different maturation steps are more susceptible to transformation than epididymal or ejaculated sperm. The injected DNA was present in the testis tissue and DNA extract from ejaculated sperm. However, none of 43 fertilized 1-cell eggs produced from DNA-treated animals showed a presence of the exogenous DNA. This indicates, that no stable integration of the testes-injected DNA in the sperm or their precursors were obtained by this technique.

The mechanisms of DNA uptake by sperm are not known. At least one part of the associated DNA resists washing procedures and intensive DNase treatment (Brackett et al., 1971; Bachiller et al., 1991 and Atkinson et al., 1991). Using laser confocal microscopy, we have shown evidence that the exogenous DNA is present in or near the nucleus of the sperm head. However, no information is available whether the DNA is intact or not. Since the DNA of mammalian sperm heads is highly compacted and functionally organized (Ward and Coffey 1991), it remains to be elucidated how external DNA can be "placed" and transported in the sperm nucleus. On the other hand, Camaioni et al. (1992) suggested that DNA bound to the sperm surface may be also an important mechanism of the sperm-mediated DNA transport into the egg, because transport of sperm surface molecules into the egg cytoplasm has been demonstrated (Gundersen et al., 1986).

Success of Sperm-Mediated DNA Transfer in Mammals

The fate of foreign DNA associated with sperm following *in vitro* fertilization has been determined to some extent in several species. Transgenic offspring with stable germline integration produced with sperm cells as vectors for the introduction of foreign DNA, have been produced so far by one research group in 1989 (Lavitrano et al., 1989). The foreign DNA was integrated into the genome in approximately 30% of the mice. Furthermore, the transgene, pSV2CAT, was inherited by the offspring via paternal and maternal transmission. The CAT gene was expressed in F1 individuals, mainly in tails and muscle. However, the integration pattern of the foreign DNA in the transgenic animals showed, that the introduced DNA head had undergone some rearrangement. It was a peculiarity, that in the five independent mice shown, the same rearrangement has arisen – a fact never observed in independent mice obtained after pronuclear microinjection. No one was able to repeat the experiment of Lavitrano et al. (1989), although intensive research was carried out in independent labs (Brinster et al., 1989). Erikson (1990) suggested as explanation for the phenomenon that intermediate factors (embryonic retroposon activity) are needed between the foreign DNA on sperm and the nucleus.

Evidence for the presence of exogenous DNA introduced with DNA-treated sperm into oocytes have been demonstrated by PCR analyses of morulae and blastocysts in mice (Hochi et al., 1990) and cattle (Gagné et al., 1991 and Schellander et al., 1995). Although in 46 of 50 embryonic stages (from 1-cell stage up to blastocyst stage) exogenous DNA was detected by PCR, none of 130 mice resulted from such embryos were transgenic. This indicates that the detected DNA was either not stably integrated into the genome or integrated in a truncated form, since the cytoplasm of the germ or egg could by a site of intensive digestion of exogenous DNA. This could also explain why we found in our study on DNA-Southern blot analysis in about 1% of 458 analyzed mice pups, obtained after IVF with DNA-treated sperm, faint signals of unexpected fragment length. The intensity of the signal as well as the length of the fragments suggested less than one copy of a truncated construct per cell (Bachiller et al., 1991).

Gandolfi et al. (1989) have reported on the production of transgenic piglets using surgical insemination with pSV2CAT-DNA treated spermatozoa. They found in the genome of 12 individuals out of 102 the pSV2CAT-DNA, however the transgene was not transmitted to the F1 progeny. Recently, we have used DNA-treated, ejaculated sperm for artificial insemination in cattle (Schellander et al., 1995). *In vitro* association studies, using radio-labeled DNA and DNase treatment, showed that about 10% of the DNA remained associated with the sperm cells. When DNA-treated sperm were used for *in vitro* production of blastocysts, in 13% of the PCR-analyzed embryos the exogenous DNA was detected. For insemination 40×10^6 sperm, obtained with an artificial vagina (washed twice in TALP, incubated for 1 h at 37 C with 40 mg linearized DNA), were applied via routine artificial insemination procedures into synchronized heifers. From 300 inseminated animals 86 calves or fetuses were obtained. DNA of one animal showed a probe specific signal after Southern analysis. The Southern signal was not of the expected size, suggesting a rearrangement during the process of fertilization or integration.

Conclusions

There is ample evidence, that foreign DNA associates with specific regions of the sperm heads of several mammalian species. In cattle and mice uptake and internalization of the associated DNA into the sperm head have been shown.

The fate of the sperm associated DNA after fertilization is not clear: while sperm-transported DNA was found in post-fertilization stages, no clear-cut data are available on the production of germ-line transgenic mammals by this technique. Only very limited additional information on the sperm-DNA uptake and transport into eggs has emerged in the last years.

A reliable technique of sperm-mediated gene transfer in mammals remains to be developed.

References

Atkinson, P.W., Hines, F.R., Beaton. S., Matthaer, K.I., Reed, K.C. and Bradley, M.P. (1991) Association of exogenous DNA with cattle and insect spermatozoa *in vitro*. Mol. Reprod. Dev. **29**, 1–5

Bachiller, D., Dotti, C., Schellander, K. and Ruther, U. (1990) Investigation of the use of sperm as a vehicle for the introduction of exogenous DNA into oocytes. *Cold Spring Harb. Symp. Quant. Biol.* 178

Bachiller, D., Schellander, K., Peli, J. and Ruther, U. (1991) Liposome-mediated DNA uptake by sperm cells. *Mol. Reprod. Dev.* **30**, 194–200

Brackett, B.G., Baranska, W., Sawicki, W. and Koprowski, H. (1971) Uptake of heterologous genome by mammalian spermatozoa and its transfer to ova through fertilization. *Proc. Natl. Acad. Sci. USA* **68**, 353–357

Brinster, R.L., Sandgren, E.P., Behringer, R.R. and Palmiter, R.D. (1989) No simple solution for making transgenic mice. *Cell* **59**, 239–244

Camaioni, A., Russo, M.A., Odorisio, T., Gandolfi, F., Fazio, V.M. and Siracusa, G. (1992) Uptake of exogenous DNA by mammalian spermatozoa: specific localization of DNA on sperm heads. *J. Reprod. Fertil.* **96**, 203–212

Castro, F.O., Hernandez, O., Uliver, C., Solano, R., Milanos, C., Aguilar, A., Perez, A., De Armas, R., Herrera, I. and De la Fuente, J. (1990) Introduction of foreign DNA into the spermatozoa of farm animals. *Theriogenology* **34**, 1099–1110

Erickson, R.P. (1990) Are intermediate vectors needed between foreign DNA on sperm and the nucleus? *TIG* **6**, 31

Felgner, P.L., Gadek, T.R., Holm, M., Roman, R., Chan, H.W., Wenz, M., Northrop, J.P., Ringold, G.M. and Danielsen, M. (1987) Lipofection: A highly efficient, lipid-mediated DNA-transfection procedure. *Proc. Natl. Acad. Sci. USA* **84**, 7413–7417

Francolini, M., Lavitrano, M., Lamia, C.L., French, D., Frati, L., Cotelli, F. and Spadafora, C. (1993) Evidence for nuclear internalization of exogenous DNA into mammalian sperm cells. *Mol. Reprod. Dev.* **34**, 133–139

Gandolfi, F., Lavitrano, M., Camaioni, A., Spadafora, C., Siracusa, C. and Lauria, A. (1989) The use of sperm-mediated gene transfer for the generation of transgenic pigs. *J. Reprod. Fert. Abstr. Ser.* **4**, 10

Gagné, M.P., Pothier, F. and Sirard, M.A. (1991) Electroporation of bovine spermatozoa to carry foreign DNA in oocytes. *Mol. Reprod. Dev.* **29**, 6–15

Gavora, J.S., Benkel, B., Hasada, H., Cantwell, W.J., Fiser, P., Teather, R.M., Nagai, J. and Sabour, M.P. (1991) An attempt at sperm-mediated gene transfer in mice and chickens. *Can. J. Anim. Sci.* **71**, 287–291

Gunderson, G.G., Medill, L. and Shapiro, B.M. (1984) Sperm surface proteins are incorporated into the egg membrane and cytoplasm after fertilization. *Dev. Biol.* **113**, 207–217

Hochi, S., Ninomiya, T., Mizuno, A., Honma, M. and Yuki, A. (1990) Fate of exogenous DNA carried into mouse eggs by spermatozoa. *Anim. Biotechnol.* **1**, 25–30

Hogan, B.L., Constatini, F. and Lacy, E. (1986) *Manipulating the Mouse Embryo*, Cold Spring Harbour Laboratory, New York.

Horan, R. Powell, R., Bird, J.M (1992a) Effects of electropermeabilization on the association of foreign DNA with pig sperm. *Arch. Androl.* **28**, 105–114

Horan, R., Powell, R., Gannon, F. and Houghton, J.A. (1992b) The fate of foreign DNA associated with pig sperm following the *in virto* fertilization of zona-free hamster ova and zona-intact pic ova. *Arch. Androl.* **29**, 199–206

Lauria, A. and Gandolfi, F. (1993) Recent advances in sperm cell mediated gene transfer. *Mol. Reprod. Dev.* **36**, 255–257

Lavitrano, M., Camaioni, A., Fazio, V.M., Dolci, S., Farace, M.G. and Spadafora C. (1989) Sperm cells as vectors for introducing foreign DNA into eggs: Genetic transformation of mice. *Cell* **57**, 717–723

Lavitrano, M., French, D., Zani, M., Frati, L. and Spadafora, C. (1992a) The interaction between exogenous DNA and sperm cells. *Mol. Reprod. Dev.* **31**, 161–169

Lavitrano, M., Zani, M., Lulli, V., French, D., Sperandino, S., Maione, B., Francolini, M., Frati, L. and Spadafora, C. (1992b) Mechanism of sperm-DNA interaction: Spermatozoa as vectors for introducing exogenous DNA into eggs. In: A. Lauria and F. Gandolfi (eds.), *Embryonic Development and Manipulation: Trends in Research and Application*, Portland Press, London, pp. 165–174

Palmiter, R.D., Brinster, R.L. (1986) Germ-line transformation of mice. *Annu. Rev. Genet.* **20**, 465–499

Sato, M., Iwase, R., Kasai, K. and Tada, N. (1994) Direct injection of foreign DNA into mouse testis as a possible alternative of sperm-mediated gene transfer. *Anim. Biotechnol.* **5**, 19–31

Schellander K., Peli J., Schmoll F. and Brem, G. (1995) Artificial insemination in cattle with DNA-treated sperm. *Animal Biotech.* **6**, 41–50

Ward, W.S., Coffey, D.S. (1991) DNA packaging and organization in mammalian spermatozoa: Comparison with somatic cells. *Biol. Reprod.* **44**, 569–574

Wu, G.M., Nosek, K., Mori, E. and Mori, T. (1990) Binding of foreign DNA to mouse sperm mediated by ist MHC class II structure. *Amer. J. Reprod. Immunol.* **24**, 120–126

PART II, SECTION B

10. *In Vitro* Generation of One Cell Embryos in Sheep and Goat

Crozet Nicole

INRA, Unité Biologie de la Fécondation,
Station de Physiologie Animale, 78352 Jouy-en-Josas Cedex, France

Introduction

Production of transgenic offsprings requires large numbers of one-cell embryos at the pronuclear stage, for gene injection. Such zygotes can be obtained *in vivo* following superovulation and insemination. However, the procedure is expensive and leads to variable yields of embryos depending on the technique used, the individual response of the donor, semen quality and other factors (Brebion *et al.*, 1992). Furthermore, spreading of ovulation and fertilization over several hours, which is common in these species, result in eggs at different stages at the time of recovery. Efficient *in vitro* techniques as a low-cost source of embryos at specific stages are therefore essential for the development of new biotechnologies such as transgenesis. Recent advances have contributed to improve the *in vitro* production of zygotes from domestic species. To generate one-cell embryos, different steps must be achieved *in vitro*: the maturation of ovarian oocytes (IVM), the capacitation of spermatozoa and fertilization events (IVF) (Figure 10.1). By using *in vitro* methods for IVM and IVF, it is now possible to routinely produce sheep and goat one-cell zygotes. Ovaries collected at the slaughterhouse can provide an abundant and economic source of immature oocytes for IVM. Procedures for sperm capacitation and IVF conditions will ensure synchronized sperm – egg penetration and consequently, the production of synchronous populations of one-cell zygotes at the stage required.

Maturation of Oocytes *In Vitro*

Development to term has been achieved after transfer of sheep and goat oocytes matured and fertilized *in vitro* to recipient females (sheep: Cheng *et al.*, 1986 and Crozet *et al.*, 1987; goat: Crozet *et al.*, 1993; Keskintepe *et al.*, 1994 and Cognié *et al.*, 1995). However, for producing *in vitro* oocytes with full developmental capacity, it is necessary to select oocytes competent for supporting meiotic maturation and embryonic development, and to use culture conditions that ensure full oocyte maturation.

Mammalian oocytes become competent to resume meiosis and to complete nuclear maturation until metaphase of the second meiotic division (MII) towards the end of their growth phase. Since the acquisition of meiotic competence occurs progressively during follicular growth, size of donor follicles must be considered when collecting oocytes for IVM. In goats, only 8% of oocytes from antral follicles smaller than 2 mm in diameter reached MII *in vitro*, whereas 96% of oocytes from follicles larger than 3 mm were able to progress to MII under the same conditions (Table 10.1). Most

Table 10.1. Meiotic stages reached *in vitro* after 27 hours of culture by goat oocytes from different sized follicles.

Follicular size	*No. oocytes*	*% of oocytes in*			
		GV	GVBD	MI	MII
< 0.5 mm	45	82	7		
0.5–0.8 mm	50	10	86	4	
1–1.8 mm	50	2	30	60	8
2–3 mm	52			44	56
> 3 mm	194			4	96

from Desmedt *et al.*, 1994.
GV: germinal vesicle; GVBD: GV breakdown; M I: metaphase of the 1st meiotic division; M II: metaphase of the 2nd meiotic divisoin.

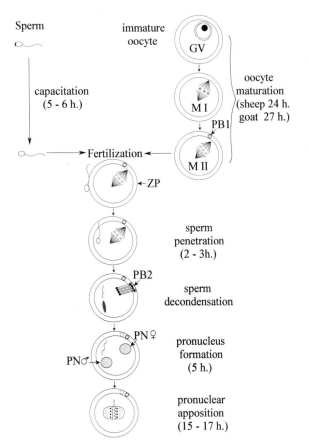

Figure 10.1. Generation of sheep and goat one-cell zygotes *in vitro*. Oocyte maturation: after the germinal vesicle (GV) has broken down, the chromosomes align on the metaphase plate of the 1st meiotic division (MI); they migrate to the spindle poles and the 1st division occurs, leading to the extrusion of the 1st polar body (PB1) containing one set of maternal chromosomes; the second set remaining in the ooplasm align on the metaphase plate of the 2nd meiotic division (MII).
Fertilization: after binding to the zona pellucida (ZP), the capacitated sperm penetrates and activates the mature oocyte, which completes the 2nd meiotic division and extrudes the 2nd polar body (PB2). The male and female pronucleus (♂ PN, ♀ PN) form around the decondensed male chromatin and maternal chromosomes; they enlarge thereafter, and migrate towards the centre of the egg.

of the oocytes originating from medium antral follicles (2–3 mm) resumed meiosis, but only half of them progressed to MII. Although the situation is less clear in the sheep, only half of the oocytes originating from follicles 1–1.5 mm in diameter completed nuclear maturation *in vitro*, whereas 90% of oocytes from follicles larger than 1.6 mm reached MII (Cognié, personal communication).

It is also important to stress that developmental competence of the oocyte develops progressively during follicular growth. In goat oocytes, developmental capacity is acquired after the acquisition of meiotic competence. Following IVM, IVF and embryo co-culture with oviduct epithelial cells (Table 10.2), 46% of oocytes originating from follicles 2–3 mm cleaved but the majority did not progress to the morula stage; oocytes from follicles 3–5 mm yielded, 35% morulae and 12% blastocysts whereas significantly more blastocysts (26%) were obtained from oocytes derived from follicles larger than 5 mm. However, ovulated oocytes fertilized and cultured *in vitro* under the same conditions yielded 41% blastocysts, indicating that not only the size of the follicle from which it originates, but also the conditions of maturation (*in vivo* or *in vitro*) influence the developmental potential of the oocyte. In sheep, oocytes originating from follicles 1–2 mm yielded 13% morulae and 13% blastocysts, whereas 29% and 40% blastocysts were obtained from oocytes isolated from follicles 2–3.5 mm and larger than 3.5 mm, respectively (Table 10.2). This emphasizes the necessity to select donor follicles according to their size, and not only to discard those judged atretic on morphological criteria. It is worth noting that only a small proportion of oocytes isolated from antral follicles can be used for IVM, to ensure full development. For instance, an average of 8 oocytes from follicles larger than 5 mm can be obtained per FSH-primed goat.

The age of the donor female may also influence the quality of the oocyte. The capacity of fertilized oocytes from prepubertal ewes for supporting development *in vitro* is lower than those of adult ewes (Wright *et al.*, 1976). Although similar rates of nuclear maturation were achieved *in vitro* for oocytes from prepubertal and adult goats following IVF, the incidence of fertilization abnormalities was higher for oocytes from the prepubertal group (Martino *et al.*, 1995).

Full developmental competence is also dependent on the conditions for oocyte maturation. It is well known that fully grown mammalian oocytes released from their follicular environment are able to undergo spontaneous nuclear maturation and to reach MII *in vitro*. However such *in vitro* matured oocytes are not necessarily competent for supporting normal fertilization and embryonic development (Thibault *et al.*, 1970 and Moor *et al.*, 1977). Developmental incompetence may result from incomplete cytoplasmic maturation under inadequate culture conditions. Follicular components play an essential role in promoting full oocyte maturation. The oocyte is coupled to adjacent cumulus cells by junctional complexes which mediate the passage of nutrients and signal molecules from the follicle to the maturing oocyte. In order to improve the cytoplasmic maturation and developmental competence of sheep

Table 10.2. *In vitro* developmental competence of goat and sheep oocytes from follicles of different size cultured for 8 days after fertilization.

	Source of oocytes	No. oocytes inseminated	Cleavage rate %	% Morulae	% Blastocysts
Goat*	F: 2–3 mm	68	46	10	6
	F: 3.1–5 mm	93	55	35	12
	F: > 5 mm	143	69	29	26
	Ovulated	128	75	20	41
Sheep†	F: 1–2 mm	60	65	13	13
	F: 2–3.5 mm	55	82	13	29
	F: > 3.5 mm	18	56		40

*from Crozet *et al.*, 1995; † from Cognié *et al.*, unpublished data. F: Follicle.

oocytes matured outside the follicle, a system which mimicked the intrafollicular conditions was developed by Staigmiller *et al.* (1984). Intact cumulus-oocytes complexes were co-cultured, in non-static conditions, with granulosa cells, in 2 ml of TCM 199 medium supplemented with FSH, LH, oestradiol 17-β and fetal calf serum. This system, very efficient for sheep IVM, has been successfully extended to goat oocytes (De Smedt *et al.*, 1992). Other culture methods, originally proposed for cattle oocytes, have been more recently used for sheep and goat IVM: granulosa cells were omitted, but entire cumulus-oocyte complexes were cultured in microdrops; the medium, usually TCM 199, was supplemented with gonadotropins and serum from various sources, or follicular fluid. The developmental potential of oocytes matured under these different conditions was attested by their ability to reach the morula-blastocyst stage *in vitro*: morula-blastocyst rates ranging from 23% to 42% were obtained. It should be noted, however, that comparative experiments using the same source of oocytes are required to evaluate the real efficiency of each IVM method in terms of cytoplasmic maturation.

The success of the maturation process may be influenced by other factors such as the temperature of incubation. Body temperature (38.5–39 C) is optimal for full maturation of sheep and goat oocytes. Maintaining isolated sheep follicles at room temperature for a few hours prior to IVM, does not hamper the maturation process of enclosed oocytes, whereas handling goat follicle-enclosed oocytes in these conditions may induce further abnormalities during meiotic maturation. Moreover, the time required to complete nuclear maturation *in vitro* is slightly different in sheep and goat oocytes, 24 hours and 27 hours respectively.

Sperm Capacitation and Fertilization *In Vitro*

Ejaculated spermatozoa must undergo physiological changes, known as capacitation, to acquire fertilizing ability. Once they are capacitated, spermatozoa can undergo the acrosome reaction, a preliminary event necessary for oocyte penetration. The capacitation process involves the removal of sperm-coating materials and is accompanied by an efflux of membrane cholesterol. It has been demonstrated that sterol-binding proteins from serum are efficient in removing cholesterol from human sperm plasma membrane (Langlais *et al.*, 1988). Addition of 20% heat-inactivated sheep serum to the culture medium promoted ^{3}H-cholesterol efflux from labelled ram spermatozoa, in a time-dependent manner. Moreover, exposure of ram sperm to sheep serum dramatically increased the proportion of spermatozoa undergoing the acrosome reaction at the zona surface (Huneau *et al.*, 1994). High fertilization rates (around 85%) were achieved, both in sheep and goats, using culture media supplemented with sheep serum (Crozet *et al.*, 1987 and De Smedt *et al.*, 1992). By raising the Ca^{++} concentration in the fertilization medium, it was also possible to reduce the variations existing amongst males and even between ejaculates of a given male, and to increase the fertilization rate achievable *in vitro* by individual ejaculates from various rams (Huneau *et al.*, 1989). These culture conditions, initially proposed for fresh sperm capacitation are also efficient for frozen semen. The duration of capacitation is 5–6 hours and 2–3 hours for sperm from fresh and frozen ejaculates, respectively. The method is currently being used in several laboratories working on ovine or caprine IVF. Heparin has been shown to be a potent activator of bull sperm capacitation. It has been widely used for bovine IVF and was recently applied to goat IVF (Younis et al., 1991). In the goat, heparin was shown to increase sperm–egg penetration, when added to IVF medium containing 10% sheep serum and to significantly increase the blastocyst rate (31.3% versus 20.5% in the control). However, the quality of the embryos produced when heparin is added to the IVF medium is questionable since 5/20 blastocysts transferred to recipient goats gave rise to viable youngs compared to 11/18 in the control group (Cognié et al., 1995). It is interesting to note that in this experiment the developmental rate (61%) of the blastocysts from the control group (IVM in M199 supplemented with 10% follicular fluid and IVF in medium containing 10% sheep serum), was close to the developmental rate obtained with *in vivo* produced embryos.

The efficiency of an *in vitro* capacitation system can be evaluated by the time required for sperm–egg penetration after gamete mixing. It should be noted that early sperm–egg penetration prevents *in vitro* oocyte ageing which may be detrimental for further development. When capacitation and fertilization occurs in the presence of 20% sheep serum, half of sheep and goat oocytes are penetrated within 3 hours of *in vitro* insemination. Decondensing sperm nuclei were found in sheep oocytes as early as 2 hours post-insemination, and 3 hours later both the male and the female pronuclei were formed; they enlarged thereafter and migrated to the centre of the egg (Crozet, 1988) (Figure 10.1). At 17 hours post-insemination, fertilized sheep and goat oocytes contain fully developed pronuclei in close apposition. Data on the timing of sperm–egg penetration and fertilization events during sheep and goat IVF, using media supplemented with heparin, are lacking. For applications such as gene injection into pronuclei, it would be also interesting to know the timing of DNA replication in sheep and goat eggs fertilized *in vitro*.

The incidence of polyspermy varies between IVF experiments, but is relatively high (10–20%) compared to natural conditions. This probably results from the large numbers of spermatozoa in proximity to the oocyte during IVF. Unfortunately, culture conditions which would maintain sperm motility at low concentrations comparable to those *in vivo* are lacking. However, even after discarding polyspermic eggs, an average of 60% normally *in vitro* fertilized sheep and goat eggs can be routinely obtained.

The incubation temperature is a crucial factor for successful fertilization. As for IVM, a body temperature of 38.5–39 C is required for IVF in ruminants.

Conclusion

Nowadays, sperm capacitation and fertilization *in vitro* are not limiting steps to generate one-cell sheep and goat zygotes. The *in vitro* production of viable zygotes with full developmental potential is mainly hampered by the quality of oocytes produced through IVM. As discussed above, developmental incompetence may be related to the origin of the oocytes or to suboptimal IVM conditions. Strict selection of the donor follicle may markedly improve the quality of matured oocytes, but only a small number of oocytes per ovary can be used. It is obvious that further studies are needed in order to improve the developmental competence of the oocytes produced *in vitro*. Given the present state of the techniques, when the source of ovaries is abundant there is no problem in obtaining sufficient oocytes which, after selection, will be matured and fertilzed *in vitro* to produce one-cell embryos for further use. When the source of immature oocytes is limited, an alternative could be to use ovulated oocytes from superovulated females and to fertilize them *in vitro*; this applies mainly to goats, in which an average of 15 ovulated oocytes per female can be recovered after FSH treatment. Using this procedure, the main advantage of IVF, namely the obtention of populations of eggs at the stage required, will be conserved.

References

Brebion, P., Baril, G., Cognié, Y. and Vallet, J.C. (1992) Embryo transfer in sheep and goats. *Ann. Zootech.* **41**, 331–340

Cheng, W.T.K., Moor, R.M. and Polge, C. (1986) *In vitro* fertilization of pig and sheep oocytes matured *in vivo* and *in vitro*. *Theriogenology* **25**, 146–149

Cognié, Y., Poulin, N., Pignon, P., Sulon, J., Beckers, J.F. and Guérin, Y. (1995) Does heparin affect developmental ability of IVF goat oocytes? *11th Meeting of the European Embryo Transfer Association*, Hannover, Germany

Crozet, N. (1988) Fine structure of sheep fertilization *in vitro*. *Gamete Res.* **19**, 291–303

Crozet, N., Huneau, D., De Smedt, V., Théron, M.C., Szöllösi, D., Torrés, S. and Sévellec, C. (1987) *In vitro* fertilization with normal development in the sheep. *Gamete Res.* **16**, 159–170

Crozet, N., De Smedt, V., Ahmed-Ali, M. and Sévellec, C. (1993) Normal development following *in vitro* oocyte maturation and fertilization in the goat. *Theriogenology* **39**, 206

Crozet, N., Ahmed-Ali, M. and Dubos, M.P. (1995) Developmental competence of goat oocytes from follicles of different size categories following maturation, fertilization and culture *in vitro*. *J. Reprod. Fert.* **103**, 293–298

De Smedt, V., Crozet, N., Ahmed-Ali, M., Martino, A. and Cognié, Y. (1992) *In vitro* maturation and fertilization of goat oocytes. *Theriogenology* **37**, 1049–1060

De Smedt, V., Crozet, N. and Gall, L. (1994) Morphological and functional changes accompanying the acquisition of meiotic competence in ovarian goat oocytes. *J. Exper. Zool.* **269**, 128–139

Huneau, D., Crozet, N. and Ahmed-Ali, M. (1994) Estrous sheep serum as a potent agent for ovine IVF: effect on cholesterol efflux from spermatozoa and the acrosome reaction. *Theriogenology* **42**, 1017–1028

Huneau, D. and Crozet, N. (1989) *In vitro* fertilization in the sheep: effect of elevated calcium concentration at insemination. *Gamete Res.* **23**, 119–125

Keskintepe, L., Darwish, G.M., Kenimer, A.T. and Brackett, B.G. (1994) Term development of caprine embryos derived from immature oocytes *in vitro*. *Theriogenology* **47**, 527–535

Langlais, J., Kan, F.W.K., Granger, L., Raymond, L., Bleau, G. and Roberts, K.D. (1988) Identification of sterol acceptors that stimulate cholesterol efflux from human spermatozoa during *in vitro* capacitation. *Gamete Res.* **20**, 185–201

Martino, A., Mogas, T., Palomo, M.J. and Paramio, M.T. (1995) *In vitro* maturation and fertilization of prepubertal goat oocytes. *Theriogenology* **43**, 473–485

Moor, R.M. and Trounson, A.O. (1977) Hormonal and follicular factors affecting maturation of sheep oocytes *in vitro* and subsequent developmental capacity. *J. Reprod. Fert.* **49**, 101–109

Staigmiller, R.B. and Moor, R.M. (1984) Effect of follicle cells on the maturation and developmental competence of ovine oocytes matured outside the follicle. *Gamete Res.* **9**, 221–229

Thibault, C. and Gérard, M. (1970) Facteur cytoplasmique nécessaire à la formation du pronucleus mâle dans l'ovocyte de lapine. *C. R. Acad. Sci. Paris* **270**, 2025–2026

Wright, R.W., Anderson, Jr. G.B., Cupps, P.T., Drost, M. and Bradford, G.E. (1976) *In vitro* culture of embryos from adult and prepuberal ewes. *J. Animal Sci.* **42**, 912–917

Younis, A.I., Zuelke, K.A., Harper, K.M., Oliveira, M.A.I. and Brackett, B.G. (1991) *In vitro* fertilization of goat oocytes. *Biol. Reprod.* **44**, 1177–1182

PART II SECTION B

11. *In Vitro* Generation of Bovine Embryos

F.A.M. de Loos and F.R. Pieper

Pharming BV, Niels Bohrweg 11-13, 2333 CA Leiden, The Netherlands

Introduction

Generation of Transgenic Cattle

Pronuclear microinjection of bovine zygotes is currently the method of choice for the production of transgenic cattle. As the efficiency of gene integration in cattle is among the lowest reported for the production of transgenic livestock, the number of oocytes required to generate several transgenic calves is relatively large (Krimpenfort *et al.*, 1991 and for review see Wall, 1996). Therefore it is essential to have a system in place that allows efficient generation of large numbers of microinjectable zygotes and, ultimately, transferable microinjected embryos.

In principle, two major systems for the generation of bovine embryos can be discriminated. One is based on the use of zygotes, recovered from the oviduct of donor cows via surgery or slaughter after superovulation and artificial insemination (superovulation or *in vivo* protocol). The other system is based on the *in vitro* generation of zygotes, in which case oocytes are collected from slaughterhouse ovaries and are matured and fertilized *in vitro*.

In vivo versus *in vitro* generation of zygotes

The capacity of bovine zygotes generated *in vivo* to develop to the morula/blastocyst stage is ca. 2-fold higher than that of embryos generated *in vitro* (Van Soom *et al.*, 1992 and McCaffrey *et al.*, 1992). In addition, the superovulation protocol allows control over the origin of the oocytes via selection of donor animals with regard to health status and genetic background. On the other hand, however, this system provides limited numbers of zygotes, which have to be recovered via surgery or slaughter.

The *in vitro* system for the generation of bovine embryos, characterized by *in vitro* maturation (IVM) and fertilization (IVF) steps, can provide an almost unlimited supply of inexpensive oocytes, mainly depending on the number of ovaries obtainable from the slaughterhouse. Moreover, the accessibility and synchronous development of the embryos at the one-cell stage greatly facilitate microinjection procedures. The low developmental capacity of zygotes generated *in vitro* is mostly due to suboptimal oocyte maturation (De Loos *et al.*, 1992).

Injected zygotes from both systems can be cultured *in vitro* (IVC) or transferred to an intermediate host. Originally, IVC conditions entailed co-culture of embryos with other cell types (e.g. bovine oviduct epithelial cells) and complex culture media, but more defined culture conditions have been established (for review see Thompson, 1996). IVC of zygotes recovered from donors selected for health status should preferably be carried out under completely defined conditions, to avoid the risk of infection via serum or co-culture with primary cells. The use of intermediate hosts (such as sheep and rabbits) to support embryo development from the zygote to blastocyst stage is relatively expensive, while the final recovery of transferable embryos does not exceed that of the IVC system (Ellington *et al.*, 1990b and Jura *et al.*, 1994). The quality of these embryos, however, with respect to cell number and development (Ectors *et al.*, 1993), and most likely resistance to freeze/thawing is higher than that of IVC embryos.

Some of the advantages of the superovulation and the *in vitro* embryo generation protocols (i.e., donor selection and a full IVM/IVF/IVC system, respectively) can be combined by the use of transvaginal ultrasound-guided oocyte recovery (TVR) from live donors. The technique of *in vivo* oocyte collection was developed in The Netherlands (Pieterse et al., 1989) under the name Ovum Pick-Up (OPU) and has developed rapidly as it can be used in cattle breeding programs (Kruip et al., 1994). Although some laboratories have used TVR oocytes on a small scale (e.g. Krisher et al., 1995), it is unclear whether this technique can be used in a large scale production system for the generation of transgenic cattle. Issues that need to be resolved include the logistics of large scale TVR, developmental capacity and injectability of the oocytes and the number of oocytes collected per donor.

These aspects were investigated in an experiment in which 50 Holstein-Friesian heifers were used as donors for TVR.

Material and Methods

Trans-vaginal Recovery of Oocytes

Donor animals, Holstein-Friesian heifers, were screened for health status before they were transported to the farm.

Donor animals were restrained in a shute and sedated. An Aloka SSD-1200 scanner equipped with a trans-vaginal 7.5 MHz mechanical sector transducer was used to visualize the ovaries. All visible follicles were punctured and follicular fluid was collected in a conical tube. Follicular fluid was filtered and oocyte cumulus complexes were searched directly after recovery.

Oocytes were matured in 0.5 ml M199 supplemented with 10% FCS, LH (Sigma, cat. no. L9773) and FSH (Sigma, cat. no. F-8001) (1 µg/ml) and ITS (Sigma, cat. no. I1884) for 24–28 hours in 4 well plates in groups of 35–50 oocytes. All cultures were performed at 39°C in 5% CO_2 in humidified air. Frozen/thawed sperm was centrifuged through a Percoll gradient and used at a concentration of 0.5×10^6 sperm cells/ml. Fertilization was performed in 4 well plates in 0.5 ml TALP medium supplemented with heparin (2 µg/ml) and PHE (Penicillamine 20 µM, Hypotaurine 10 µM and Epiniphrine 1 µM). After removal of the cumulus cells and centrifugation of the zygotes, to visualize the pronuclei, a 43 Kb linear DNA fragment was injected in one of the pronuclei. Embryos were cultured in 0.5 ml M199+10% FCS and ITS on a bovine oviduct epithelial cell monolayer. Morulae and blastocysts were transferred at day 7 to synchronized heifers, one embryo per heifer.

Results

In 26 collection days a total of 50 different heifers was subjected to TVR of oocytes. In total 156

Table 11.1. Embryo development of transvaginal recovered (TVR) oocytes and slaughterhouse (SLH) oocytes in an *in vitro* culture system.

	Oocytes n	Cleaved n (%)	Day 8 embryos n (%)
TVR	319	186 (58.3)	76 (23.8)
SLH	985	689 (69.9)	246 (24.9)

Table 11.2. Efficiencies of the steps involved in the process from collecting oocytes via transvaginal recovery to pregnancy. Percentages indicate the proportion of embryos or cells that successfully complete each step.

	Number	Percentage
Oocytes collected	1192	—
Oocytes matured/fertilized	957	80.3
Zygotes injected	726	75.9
Cleavage	447	61.6
Morula/blastocyst	42	9.4
Transferred	40	95.2
Pregnant	14	35.0

puncture sessions were performed. One puncture session is defined as a collection from one animal at one time point at which all visible follicles on both ovaries are punctured. The number of oocytes collected per session varied from 0 to 22. A mean number of 6 oocytes per collection were of sufficient quality for IVC.

The developmental rates of not injected TVR- and slaughterhouse derived oocytes appeared to be identical (Table 11.1).

The efficiencies of the steps in DNA microinjection and *in vitro* development are given in Table 11.2. Subjecting TVR oocytes to DNA microinjection reduced their development from 40.8% (23.8% on total oocytes in culture) to 9.4% (compare Tables 11.1 and 11.2). For slaughterhouse oocytes this drop in development after DNA microinjection was somewhat higher, although no parallel studies were performed to confirm this observation.

Discussion

The *in vivo* collection of bovine zygotes for DNA microinjection potentially is useful as it allows selection of donor animals. The feasibility of producing transgenic cattle via this system has been demonstrated (Seidel, 1993 and Bowen et al., 1994).

However, the limited numbers of injectable zygotes that can be recovered by superovulation and the costs associated with this system constitute major hurdles in a program aimed at producing transgenic animals. Recovery of oocytes from live donor animals via TVR may well provide an alternative. While cows can be superovulated every two months they yield a mean of approximately 8-18 zygotes per session (Ellington *et al.*, 1990a), TVR can be performed twice weekly (Gibbons *et al.*, 1994; our unpublished data), yielding 5-7 zygotes per session (40-56 zygotes per donor in a two month period). Thus, TVR greatly reduces the number of donor animals required, and allows much more efficient use of selected donor cows.

Conflicting data have been reported concerning *in vitro* development of TVR-derived embryos. Superior (Van der Schans *et al.*, 1992), equal (Gibbons *et al.*, 1994) or relatively poor *in vitro* development (Den Daas *et al.*, 1994) in comparison to slaughterhouse-derived oocytes have been reported. In our hands, TVR- and slaughterhouse-derived oocytes cultured in parallel displayed identical developmental capacities.

In Table 11.3 the characteristics of the three zygote generation systems are compiled. Any embryo generation system used for the production of transgenic cattle must yield high numbers of injectable oocytes. In this respect, oocyte recovery from slaughterhouse-derived oocytes will remain superior. However, TVR yields can be further improved by selection of donor cows. The total number of oocytes obtained via TVR is mainly limited by the number of oocytes recovered per donor but also by the number of animals that can be collected from. Large individual differences between donors with respect to number of oocytes retrieved and subsequent embryo development have been reported (Hasler *et al.*, 1995; our unpublished data). This offers the possibility of increasing embryo yields via donor selection, especially via elimination of poor donors.

Thus, TVR not only allows for selection of donor animals for genetic background and health, but also for relatively rapid selection of donor cows.

We conclude that TVR can be applied for the production of transgenic cattle, and provides advantages over both *in vitro* and *in vivo* embryo generation.

Acknowledgement

We thank M. Salaheddine and A. Garcia for the collection of oocytes via transvaginal recovery. For the technical assistance we greatly acknowledge the support of Th. van Beneden, C. van Dillen, S. Hengst, A. Rademakers, T. Schoenmaker and C. Samuel.

References

Bowen, R., Reed, M., Schnieke, A., Seidel Jr, G., Stacey, A., Thomas, W. and Kajikawa, O. (1994) Transgenic cattle resulting from biopsied embryos: expression of c-ski in a transgenic calf. *Biology of Reproduction* **50**, 664-668

De Loos, F., Maurik, P., Van Beneden, T. and Kruip, T. (1992) Structural aspects of bovine oocyte maturation *in vitro*. *Molecular Reproduction and Fertility* **31**, 208-214

Den Daas, N. and Merton, S. (1994) *In vitro* embryo production, its use. *10th scientific meeting AETE.*, Lyon, September 1994

Ectors, F., Thonon, F., Delval, A., Fontes, R., Touati, K., Beckers, J. and Ectors, F. (1993) Comparison between culture of bovine embryos *in vitro* versus development in rabbit oviducts and *in vivo*. *Livestock Production Science* **36**, 29-34

Ellington, J., Farrell, P., Simkin, M. and Foote, R. (1990a) Method for obtaining bovine zygotes produced *in vivo*. *Am. J. Vet. Res.* **51**, 1708-1710

Ellington, J., Farrell, P., Simkin, M., Foote, R., Goldman, E. and McGrath, E. (1990b) Development and survival after transfer of cow embryos cultured from 1-2 cells to morulae or blastocysts in rabbit oviducts or in a simple medium with bovine oviduct epithelial cells. *J. Reprod. Fert.* **89**, 293-299

Gibbons, J.R., Beal, W.E., Krisher, R.L., Faber, E.G., Pearson, R.E. and Gwazdauskas, F.C. (1994) Effects of once versus twice weekly transvaginal aspiration on bovine oocyte recovery and embryo development. *Theriogenology* **41**, 206

Hasler, J., Henderson, W., Hurtgen, P., Jin, Z., McCauley, A., Mower, S., Neely, B., Shuey, L., Stokes, J. and Trimmer, S. (1995) Production, freezing and transfer of bovine IVF embryos and subsequent calving results. *Theriogenology* **43**, 141-152

Jura, J., Kopchick, J., Chen, W., Wagner, T., Modlinski, J., Reed, M., Knapp, J. and Smorag, Z. (1994) *In vitro* and *in vivo* development of bovine embryos from zygotes

Table 11.3. Characteristics of Three Zygote Generation Systems. *In vivo* generation; superovulation of the donor and surgical recovery of the zygotes. *In vitro* SLH; *in vitro* protocol starting with slaughterhouse derived oocytes and *In vitro* TVR; *in vitro* protocol starting with oocytes collected from live donors by means of transvaginal recovery.

	In vivo	*In vitro* SLH	*In vitro* TVR
Costs	high	low	intermediate
Labor intensive	high	low	intermediate
Yield of zygotes	low	unlimited	intermediate
Donor selection	possible	not possible	possible

and 2-cell embryos microinjected with exogenous DNA. *Theriogenology* **41**, 1259–1266

Krimpenfort, P., Rademakers, A., Eyestone, W., Schans, A. van der, Broek, S. van de, Kooiman, P., Kootwijk, E., Platenburg, G., Pieper, F., Strijker, R. and Boer, H. de (1991) Generation of transgenic dairy cattle using "*in vitro*" embryo production. *Biotechnology* **9**, 844–847

Krisher, R., Gibbons, J. and Gwazdauskas, F. (1995). Nuclear transfer in the bovine using microinjected donor embryos: Assessment of development and deoxyribonucleic acid detection frequency. *Journal of Dairy Science* **78**, 1282–1288

Kruip, T., Boni, R., Wurth, Y., Roelofsen, M. and Pieterse, M. (1994) Potential use of ovum pick-up for embryo production and breeding in cattle. *Theriogenology* **42**, 675–684

McCaffrey, C., Lu, K. and Sreenan, J. (1992) Development of *in vivo* and *in vitro* fertilised (IVF) cattle ova during *in vitro* culture. *12th International Congress on Animal Reproduction*. The Hague, August 1992

Pieterse, M., Vos, P., Kruip, T., Wurth, Y., Beneden T. van, Willemse, A. and Taverne, M. (1991) Transvaginal ultrasound guided follicular aspiration of bovine oocytes. *Theriogenology* **35**, 19–24

Seidel, G. (1993) Resource requirements for transgenic livestock research. *Journal of Animal Science* **71**, 26–33

Thompson, J. (1996) Defining the requirements for the bovine embryo culture. *Theriogenology* **45**, 27–40

Van der Schans, A., van Rens, B.T.T.M., Van der Westerlaken, L.A.J. and de Wit, A.C.C. (1992) Bovine embryo production by repeated transvaginal oocyte collection and *in vitro* fertilization. *12th International Congress on Animal Reproduction*, The Hague, August 1992

Van Soom, A. and De Kruif, A. (1992) A comparative study of *in vivo* and *in vitro* derived bovine embryos. *12th International Congress on Animal Reproduction*. The Hague, August 1992

Wall, R. (1996) Transgenic Livestock: progress and prospects for the future. *Theriogenology* **45**, 57–68

PART II, SECTION B

12. The Use of Cryopreserved Fertilized Ova to Generate Transgenic Mammals

N. Tada

Hoechst Japan Limited, Pharma Research and Development Division, 3-2 Minamidai 1 Chome, Kawagoe, Saitama, 350-11 Japan

Cryopreservation of mammalian embryos has been greatly improved since successful freezing was first reported for mice (Whittingham *et al.*, 1972 and Wilmut, 1972). The original technique (the so-called slow-cooling method) involves slow cooling (0.5°C/min) to −60°C or below before transfer into liquid nitrogen (LN_2). This technique is now widely used, but it requires a very expensive programmed freezer for accurate control of cooling and warming. A new technique, vitrification, was first developed for cryopreservation of mouse 8-cell embryos by Rall and Fahy (1985). Vitrification (solidification of a solution without crystallization owing to elevated viscosity during the process of cooling) can be performed rapidly and simply without a programmed freezer. Therefore, in the past four or five years, many investigators have attempted to improve and simplify vitrification of mammalian embryos using various concentrations of permeating and nonpermeating cryoprotectants (Scheffen *et al.*, 1986; Kono and Tsunoda, 1988; Nakagata, 1989; Kasai *et al.*, 1990; Nagashima *et al.*, 1991; Schiewe *et al.*, 1991; Ishimori *et al.*, 1992 and Tada *et al.*, 1993). On the other hand, during the past 13 years, many transgenic mice have been produced for study of the biological functions of isolated genes and for developing human disease-models (Brinster and Palmiter, 1986; Gordon, 1989; Hanahan, 1989 and Wilmut, *et al.*, 1991). The procedure itself is long and arduous: (1) many pronucleate stage eggs must be collected at once, and (2) precise timing is required to ensure that the injected eggs are transferred into recipient females in synchrony with the donors of the eggs. For this to be achieved, large colonies of mice must always be maintained, a laborious and expensive undertaking. To overcome these problems, it would be preferable to utilize cryopreserved pronucleate stage mouse eggs for microinjection, because (1) maintenance of a large colony of mice would not be required, and (2) some routine work (e.g., injection of gonadotropins to induce superovulation and collection of eggs from oviducts) for each microinjection could be omitted. It is also difficult for investigators always to obtain a constant number of eggs (more than 100) because the number of fertilized eggs collected often varies from day to day for unknown reasons, despite breeding of mice in a strictly controlled environment. It is also difficult always to obtain the needed number of pseudopregnant females (recipients); sometimes there are more than needed, but sometimes too few recipients are available. Furthermore, *in vitro* fertilized ova have proven to be useful as recipients of exogenous DNA. This is advantageous in that many stage-matched fertilized eggs (e.g., more than 100) can be obtained at once. In contrast, acquisition of *in vivo* fertilized eggs is rather difficult, because it requires a number of cages for male mice for mating, and the obtained eggs are often in a mixture of different one-cell stages (i.e., early and late one-cell stages).

Leibo *et al.* (1991) were the first to report successful production of several transgenic mice by microinjection of foreign DNA into pronucleate stage eggs that had been frozen by a slow-cooling procedure. They were able to produce several transgenic mice each with five kinds of transgenes. In their study, a total of 506 eggs were frozen and thawed, and 90% of them appeared morphologically normal. The *in vivo* developmental rate of eggs that had been frozen-thawed and injected with DNA was slightly less than that for unfrozen and

DNA-injected eggs (13% vs. 22%). We also attempted to produce several transgenic mice derived from vitrified-injected eggs (Tada et al., 1995b). In our studies, 11% of vitrified eggs that had been DNA-injected developed into full-term foetuses. This discrepancy in findings may be insignificant, since similar low *in vivo* developmental rates have been recorded even for fresh eggs injected with DNA (31% for FVB/N strain mice (Taketo et al., 1991); and 19% for C57BL × SJL hybrid mice (Brinster et al., 1985). With regard to the overall efficiency of transgenic production in our experiments, 4.6% (8/175) transgenic mice were obtained from transferred cryopreserved eggs. These rates are very similar to those previously reported: 4.1% (28/683) from cryopreserved eggs (Leibo et al., 1991), and 5.0–5.9% from fresh eggs (Brinster et al., 1985 and Taketo et al., 1991). Given these and our own findings, it is reasonable to conclude that transgenesis is not affected by cryopreservation of fertilized eggs. Findings of the above studies and our own are shown in Table 12.1. Interestingly, it has been reported that in only one study was substantial enhancement of transgene integration observed when cryopreserved zygotes were used as donors (Leibo et al., 1991). To explain this discrepancy, they noted that the frozen-thawed zygotes had been incubated for 1 hour prior to injection and that this short period of incubation might have affected the efficiency of transgenic production. However, in our experiments, no increase in transgenic efficiency was observed despite 1–2 hours incubation before microinjection (Tada et al., 1995b). The variables which affect gene integration are myriad and poorly understood. The skill of the individual performing the manipulations, the quality of DNA preparation, the age of the culture medium, differences in mouse strains, the number of hours after mating that eggs are retrieved, and the time after retrieval that the eggs are injected, the speed with which a group of eggs is injected (which is inversely proportional to the time embryos spend on the microscope stage): all of these factors impact on the procedure. The above discrepancy thus remains unexplained.

The use of cryopreserved fertilized eggs, as for the generation of transgenic mice, also enables efficient production of other transgenic large domestic animals. For example, transgenic cows can be produced from cryopreserved eggs derived from fertilization of ovarian oocytes isolated from superovulated females by frozen-thawed spermatozoa (Figure 12.1). Furthermore, it will be possible to amplify transgenic cattle embryos using a nuclear transfer technique (Willadsen, 1986; Prather et al., 1987; Westhusin et al., 1991 and Barnes et al., 1993) from *in vitro* selected embryos that harbor foreign DNA (Figure 12.1). We have already reported successful *in vitro* selection of mouse embryos harboring the neomycin resistant gene (*neo*) in the presence of G-418 (Tada et al., 1995a). We found that (i) 48% of embryos survived after microinjection and subsequent cultivation in a medium containing 200 µg/ml of G-418, which completely inhibited the development of normal mouse embryos, and that (ii) all of the surviving embryos tested possessed the foreign DNA (Tada et al., 1995a). This method of *in vitro* selection of preimplantation embryos carrying exogenous DNA before transfer will enable a large number of recipients to be saved. The optimal concentrations of G-418 for killing normal bovine embryos must be determined, since we already have found that bovine embryos are capable of surviving even in the presence of 200 µg/ml of G-418 (unpublished data). Furthermore, mammalian pronucleate stage eggs appears

Table 12.1. Efficiency of Production of Transgenic mice from Fresh and Cryopreserved eggs

Treatment	No. of eggs surviving after injection	No. of newborns (%)	No. of transgenics (%)	Overall (%)	Authors
Fresh	170	33/170 (19)	10/33 (30)	10/170 (5.9)	Brinster et al. (1985)
Fresh	301	93/301 (31)	15/93 (16)	15/301 (5.0)	Taketo et al. (1991)
Cryopreserved	683	88/683 (13)	28/88 (32)	28/683 (4.1)	Leibo et al. (1991)
Cryopreserved	175	20/175 (11)	8/20 (40)	8/175 (4.6)	Tada et al. (1995b)

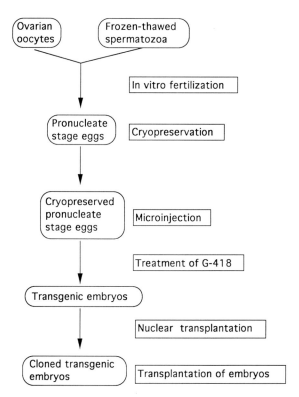

Figure 12.1. Procedure for efficient production of transgenic cows.

to be more sensitive to cooling processes than embryos at any more advanced stage of development in mice (Scheffen et al., 1986; Hsu et al., 1986; Kono et al., 1987; Li and Trounson, 1991; and Tada et al., 1993b), cows (Massip et al., 1986), rabbits (Smorag et al., 1989) and sheep (Szell et al., 1990). In these reports, the cryosurvival rates of mouse pronuclate stage eggs (1-cell stage) were relatively low when vitrification techniques (Hsu et al., 1986 and Kono et al., 1987) and the standard slow freezing method (Whittingham et al., 1972) were used. In general, mammalian pronuclate stage eggs are more sensitive to the cryopreservation process than embryos at any other more advanced stage of development. This may be because the susceptibilities of embryos to the freezing/thawing process differ among species, and/or this may reflect physiological changes in the permeability characteristics of embryos due to embryonic development, as pointed out by Schneider (1986). The post-thawing viability of pronuclate stage eggs must thus be determined for other animals including cows.

The procedures used for cryopreservation and subsequent microinjection for mice described below.

1. Intraperitoneal injection of 5 IU pregnant mare's serum gonadotropin (PMSG) and 5 IU human chorionic gonadotropin (hCG) 46–48 hours later into egg-donor female mice (induction of superovulation).
2. Removal of oviducts 14–16 hours after hCG injection.
3. Collection of unfertilized eggs from oviducts and placement in a drop (400 µl) of Krebs-Ringer-based TYH medium (Toyoda et al., 1971) under paraffin oil equilibrated with 5% CO_2 in air at 37°C.
4. Addition of 10 µl sperm suspension from the cauda epididymides of male mice into the drop containing unfertilized eggs and incubation of the sperm–egg mixture for 6 hours at 37°C.
5. Determination which fertilized eggs possess the 2nd polar body and male pronucleus 6 hours after insemination.
6. Transfer of only the fertilized eggs to a drop (50 µl) of phosphate-buffer based PB1 medium (Whittingham, 1971a) under paraffin oil equilibrated with 5% CO_2 in air at 37°C and incubation prior to vitrification.
7. Transfer of fertilized eggs into 30 µl of vitrification solution consisting of 2.75 M dimethylsulfoxide, 2.75M propylene glycol and 1.0 M sucrose (hereafter designated DPS) in PB1 (Tada et al., 1993a) in a cryotube (No. 366656, Nunc) and soon direct plunging of the cryotube into liquid nitrogen.
8. Storage of the cryotube before warming.
9. Warming of the cryotube by immersion in a 40°C waterbath, then immediate supplementation with 300 µl of PB1 containing 0.3 M sucrose, and then keep it for 1 minute at 40°C.
10. Placement of the egg-containing solution in the cryotube onto a plastic culture dish.
11. Washing of the recovered eggs by transferring them through 3 drops (100 µl) of M16 medium (Whittingham, 1971b) covered with paraffin oil, and incubation for 1–2 hours at 37°C prior to microinjection of DNA constructs.
12. Microinjection of DNA into male pronuclei of the vitrified-warmed fertilized eggs using the procedures described by Hogan et al. (1986).

With this protocol, 80–88% of vitrified eggs appeared morphologically normal after warming (Tada et al., 1995b). The rate of in vitro development of vitrified, microinjected eggs to the two-cell stage appeared to be decreased compared with that of the unvitrified, microinjected eggs. However, 17–19% of the transferred eggs developed into live young, with no difference in efficiency of development into live young between the vitrified and unvitrified eggs after injection and subsequent transfer. Overall, the percentages of live young identified as transgenic were approximately the same for the vitrified and unvitrified groups.

This review describes the successful use of cryopreserved pronucleate stage eggs to generate transgenic animals. This system will enable efficient and convenient production of transgenic animals particularly in conditions in which space for animal breeding and manpower are markedly limited.

References

Barnes, F., Endebrock, M., Looney, C., Powell, R., Westhusin, M.E. and Bondioli, K.R. (1993) Embryo cloning in cattle: the use of *in vitro* matured oocytes. *J. Reprod. Fert.* **97**, 317–320

Brinster, R.L., Chen, H.Y., Trumbauer, M.E., Yagle, M.K. and Palmiter, R.D. (1985) Factors affecting the efficiency of introducing foreign DNA into mice by microinjecting eggs. *Proc. Natl. Acad. Sci. USA* **82**, 4438–4442

Brinster, R.L. and Palmiter, R.D. (1986) Introduction of genes into the germ line of animals. *Harvey Lect.* **80**, 1–38

Gordon, J. W. (1989) Transgenic animals. *Intl. Rev. Cytol.* **115**, 171–229

Hanahan, D. (1989) Transgenic mice as probes into complex systems. *Science* **8**, 1265–1275

Hogan, B., Costantini, F. and Lacy, E. (1986) *Manipulating the Mouse Embryo*. Cold Spring Harbor Press, New York.

Hsu, T.T., Yamakawa, H., Yamanoi, J. and Ogawa, S. (1986) Survival and transfer test of mouse early embryos frozen by vitrification method. *Jpn. J. Anim. Reprod.* **32**, 106–109

Ishimori, H., Takahashi, Y. and Kanagawa, H. (1992) Viability of vitrified mouse embryos using various cryoprotectant mixtures. *Theriogenology* **37**, 481–487

Kasai, M., Komi, J.H., Takakamo, A., Tsudera, H., Sakurai, T. and Machida, T. (1990) A simple method for mouse embryo cryopreservation in a low toxicity vitrification solution, without appreciable loss of viability. *J. Reprod. Fert.* **89**, 91–97

Kono, T. and Tsunoda, Y. (1987) Frozen storage of mouse embryos by vitrification. *Jpn. J. Anim. Reprod.* **33**, 77–81 (In Japanese)

Kono, T. and Tsunoda, Y. (1988) Ovicidal effects of vitrification solution and the vitrification-warming cycle and establishment of the proportion of toxic effects of nuclei and cytoplasm of mouse zygotes. *Cryobiology* **25**, 197–202

Krimpenfort, P., Rademakers, A., Eyestone, W., van der Schans, A., van der Broek, S., Kooiman, P., Kootwijk, E., Platenburg, G., Pieper, F., Strijker, P. and de Boer, H. (1991) Generation of transgenic dairy cattle using '*in vitro*' embryo production. *Biotechnol.* **9**, 844–848

Leibo, S.P., DeMayo, F.J. and O'Malley, B. (1991) Production of transgenic mice from cryopreserved fertilized ova. *Mol. Reprod. Dev.* **30**, 313–319

Li, R. and Trounson, A. (1991) Rapid freezing of the mouse blastocysts: Effects of cryoprotectants and of time and temperature of exposure to cryoprotectant before direct plunging into liquid nitrogen. *Reprod. Fert. Dev.* **3**, 175–183

Massip, A., van der Zwalmen, P., Scheffen, B. and Ectors, F. (1986) Pregnancies following transfer of cattle embryos preserved by vitrification. *Cryo-Letters* **7**, 270–273

Nagashima, H., Kobayashi, Y., Yamakawa, H. and Ogawa, S. (1991) Cryopreservation of mouse half-morulae and chimeric embryos by vitrification. *Mol. Reprod. Dev.* **30**, 220–225

Nakagata, N. (1989) Survival of two-cell mouse embryos derived from fertilization *in vitro* after ultrarapid freezing and thawing. *Jpn. J. Fert. Steril.* **34**, 470–473 (In Japanese)

Prather, R.S., Barnes, F.L., Sims, M.M., Robl, J.M., Eyestone, W.H. and First, N.L. (1987) Nuclear transplantation in the bovine embryos: Assessment of donor nuclei and recipient oocyte. *Biol. Reprod.* **37**, 859–866

Rall, W.F. and Fahy, G.M. (1985) Ice-free cryopreservation of mouse embryos at −196°C by vitrification. *Nature* **313**, 573–574

Scheffen, B., van der Zwalmen, P. and Massip, A. (1986) A simple and efficient procedure for preservation of mouse embryos by vitrification. *Cryo-Letters* **7**, 260–269

Schiewe, M.C., Rall, W.F., Stuart, L.D. and Wildt, D.E. (1991) Analysis of cryoprotectant, cooling rate and *in situ* dilution using conventional freezing or vitrification for cryopreserving sheep embryos. *Theriogenology* **36**, 279–293

Schneider, U. (1986) Cryobiological principles of embryo freezing. *J. in vitro Fert. Embryo Transfer* **3**, 3–9

Smorag, Z., Gajda, B., Wieczorek, B. and Jura, J. (1989) Stage-dependent viability of vitrified rabbit embryos. *Theriogenology* **31**, 1227–1231

Szell, A., Zhang, J. and Hudson, R. (1990) Rapid cryopreservation of sheep embryos by direct transfer into liquid nitrogen vapour at −180°C. *Reprod. Fert. Dev.* **2**, 613–618

Tada, N., Sato, M., Amann, E. and Ogawa, S. (1993a) A simple and rapid method for cryopreservation of mouse 2-cell embryos by vitrification: beneficial effect of sucrose and raffinose on their cryosurvival rate. *Theriogenology* **40**, 333–344

Tada, N., Sato, M., Amann, E. and Ogawa, S. (1993b) Stage-dependent viability of mouse preimplantation embryos vitrified with sugar-containing solutions. *J. Reprod. Dev.* **39**, 139–144

Tada, N., Sato, M., Hayashi, K., Kasai, K. and Ogawa, S. (1995a) *In vitro* selection of transgenic mouse embryos in the presence of G-418. *Transgenics* **1**, 535–540

Tada, N., Sato, M., Kasai, K. and Ogawa, S. (1995b) Production of transgenic mice by microinjection of DNA into vitrified pronucleate stage eggs. *Transgenic Res.* **4**, 208–213

Taketo, M., Schroeder, A.C., Mobraaten, L.E., Gunning, K.B., Hanten, G., Fox, R.R., Roderick, T.H., Stewart, C.L., Lilly, F., Hansen, C.T. and Overbeek, P.A. (1991) FVB/N: An inbred mouse strain preferable for transgenic analyses. *Proc. Natl. Acad. Sci. USA* **88**, 2065–2069

Toyoda, Y., Yokoyama, M. and Hoshi, T. (1971) Studies on the fertilization of eggs by fresh epididymal sperm. *Jpn. J. Anim. Reprod.* **16**, 147–151 (In Japanese)

Westhusin, M.E., Pryor, J.H. and Bondioli, K.R. (1991) Nuclear transfer in the bovine embryo: a comparison of 5-day, 6-day, frozen-thawed, and nuclear transfer donor embryos. *Mol. Reprod. Dev.* **28**, 119–123

Whittingham, D.G. (1971a) Survival of mouse embryos after freezing and thawing. *Nature* **233**, 125–126

Whittingham, D.G. (1971b) Culture of mouse ova. *J. Reprod. Fert.* (Suppl) **14**, 7–21

Whittingham, D.G., Leibo, S.P. and Mazur, P. (1972) Survival of mouse embryos frozen to −196°C and −269°C. *Science* **178**, 411–414

Willadsen, S.M. (1986) Nuclear transplantation in sheep embryos. *Nature* **320**, 63–65

Wilmut, I. (1972) The effect of cooling rate, warming rate, cryoprotective agent and stage of development on survival of mouse embryos. *Life Science* **11**, 1071–1079

Wilmut, I., Hooper, M.L. and Simmons, J.P. (1991) Genetic manipulation of mammals and its application in reproductive biology. *J. Reprod. Fert.* **92**, 245–279

PART II, SECTION C

13. The Complete *In Vitro* Development of Quail Embryo

Tamao Ono

Faculty of Agriculture, Shinshu University, Ina 399-45, Japan

Introduction

A number of attempts have been made to produce transgenic birds. For example, DNA has been injected into a fertilized single-cell ovum (Sang and Perry, 1989; Perry et al., 1991; Naito et al., 1991, 1994a; Ono et al., 1994a and Love et al., 1994), dispersed blastoderm cells have been injected into a host embryo at the blastoderm stage and resulted in germ line chimeras (Petitte et al., 1990; Carsience et al., 1993 and Ono et al., 1995a), cells with gene lipofection have been injected into blastoderm stage embryos (Brazolot et al., 1989 and Ono et al., 1994b) and DNA has been transfected with a retroviral vector into blastoderm cells (Souza et al., 1984; Shuman and Shoffner, 1986; Salter et al., 1986, 1987; Hippenmeyer et al., 1988; Bosselman et al., 1989; Crittenden et al., 1989; and Salter and Crittenden, 1989). When the production of transgenic offspring is the target, DNA transfection into primordial germ cells (PGCs) with a retroviral vector (Allioli et al., 1994), ballistic DNA transfection into the germinal crescent (Li et al., 1995) and transfusion of liposome-mediated DNA into circulating PGCs (Watanabe et al., 1994 and Ono et al., 1995b) are the alternative strategies. Germ line chimeras have been produced by transferring foreign PGCs (Tajima et al., 1993 and Naito et al., 1994b), this procedure will serve as a powerful tool for the genetic improvement of birds and yield useful products. Moreover, there is a possibility that endangered species can be rescued by transferring their PGCs into the embryos of a more prolific species.

Successful hatching of cultured avian embryos was made by Ono and Wakasugi (1984) for quail cultured from 2.5-day embryos and by Rowlett and Simkiss (1987) for chickens from 3-day embryos. To improve transgenesis, a method which allows the culturing of the earlier developmental stages and which results in hatchlings is required. Perry (1988) has reported such a complete culture method for chickens, which has proved to be applicable for gene transfer studies.

Chickens have been widely used for the study of transgenesis in birds and the methods for transgenesis mentioned above were initially developed using chick embryos. However, Japanese quail may serve as better experimental animal for the transgenesis due to its smaller body size, faster growth rate and shorter generation interval. In this chapter, emphasis will be placed on the protocol for the manipulation of quail embryos such as the introduction of foreign genes and cells, and the subsequent *in vitro* culturing of the manipulated embryos resulting in hatchlings.

Microinjection of DNA into the Naked Ovum at the Single-Cell Stage and the Subsequent Culturing of the Embryo

Collection of the Ovum

The ovum is obtained from the infundibulum or the upper magnum 1–2 h after the preceding oviposition. When the ovum is already covered with a thick albumen capsule, the capsule is removed with forceps in Ca^{2+} – and Mg^{2+} – free Dulbecco's phosphate-buffered saline (PBS(−)) at 41.5°C.

Preparation of Micropipettes

Glass capillary tubings (G-1, Narishige, Tokyo) are siliconized (5% dimethyldichlorosilane/chloroform),

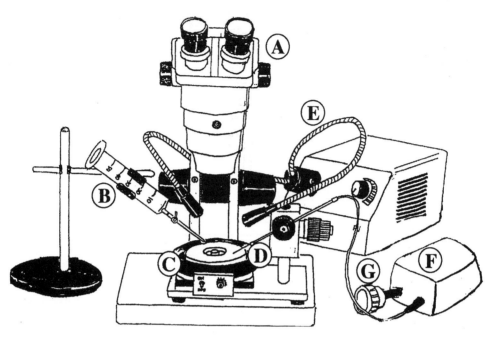

Figure 13.1. A schematic drawing for the preparation of a micropipette. The protocol is described in the text. A, dissecting microscope with ocular micrometer; B, 50-ml syringe with a two-way cock and needle; C, micropipette grinder; D, micropipette holder with micropipette; E, fiber light system; F, air pump; G, membrane holder with filter.

rinsed with distilled water and dried. They are then drawn out with a micropipette puller (PB-7, Narishige) making two micropipettes available from one piece of tubing. The tips are beveled at a 30° angle to an outer diameter of 10–15 μm with the aid of a pipette grinder (EG-4, Narishige) irrigated with dripping distilled water while the resulting glass dust is blown with an air pump (SPP-6GA, Techno Takatsuki, Osaka) connected to a micropipette holder (HI-4A, Narishige; Figure 13.1). The size of the tip is monitored under a dissecting microscope using an ocular micrometer. The micropipettes thus made are sterilized at 160°C for 2 h.

Gene Introduction

The micropipette is attached to a manipulator and connected to a delivery injector (M-152 and IM-6, Narishige) with Teflon tubing (1 and 2 mm inner and outer diameters, respectively) filled with distilled water (Figure 13.2) The naked ovum is placed on the mechanical stage of remodeled microscope, the arm above the stage is cut off (BHT, Olympus), and to this a manipulator (M-152, Narishige) and a micropipette holder are attached. Lighting (LGW, Olympus) is set horizontally and a dissecting microscope (SZ60, Olympus) connected to a video camera (MTV-H, Olympus) is set aslant and held in place with a universal base (VS-4, Olympus). The ovum on the stage is moved vertically with the microscope's focusing knob and the micropipette is inserted vertically into the center of the ovum while being monitored (SSM-121, Sony) with the aid of an image-processor system (Image Sigma-III, Nippon Avionics, Tokyo).

Culture

When trying to introduce a foreign gene into the nucleus of the avian ovum we find, however, that it is difficult to recognize the nucleus through the capsular thick albumen. This difficulty may be one of the reasons for the low rate of success in transgenesis in birds. If a naked ovum can be used for culture, the introduction of genes is achieved into the center of the ovum that is more readily visible. The injected naked ovum is then cultured through three consecutive steps, i.e., systems Q1a, Q2 and Q3 (Figure 13.3, Ono et al., 1994a, 1996)

System Q1a. In system Q1a, thick albumen of hen's egg is used as the culture medium. The egg is cracked open into a 90 mm polystyrene culture dish and the thick albumen collected by sucking it off with a 10 ml syringe. The pH of the albumen is adjusted to 7.2–7.4 by bubbling CO_2 through it, any resulting foam is then removed from albumen using spatula and the albumen is warmed at 41.5°C under 20% CO_2 until use. In normal quail development

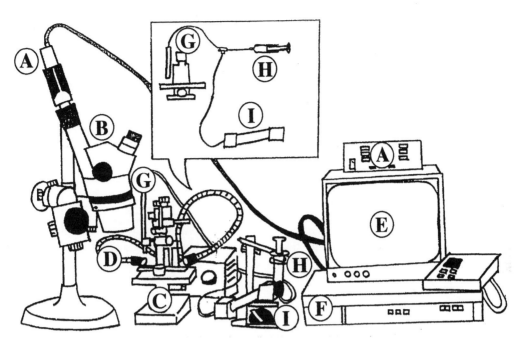

Figure 13.2. A schematic drawing of the micromanipulation system for the avian ovum or embryo. Details are described in the text. A, video camera; B, dissecting microscope with universal base; C, remodeled microscope base with mechanical stage; D, fiber light system; E, monitor TV; F, image-processor system; G, manipulator and micropipette holder; H, 3 ml syringe; I, delivery injector.

the pH of albumen is 7.0–7.3 while the embryo remains in the oviduct (Ono *et al.*, 1996). The ovum is placed in 20 ml polypropylene cup (55-079-03, Iuchi, Osaka; 32 mm tall, 30 and 35 mm lower and upper diameters, respectively) and the cup is filled with the prepared thick albumen. The open surface of the cup is tightly sealed by placing another cup inside it making sure that any air pockets are eliminated. The ovum is submerged just below the surface and incubated without agitation for 24 h at 41.5°C and saturated humidity under 20% CO_2. A cell culture incubator is used for system Q1a culture.

System Q2. In system Q2, the shell of a quail egg is cut horizontally at a level of 19 mm in diameter at its narrow end using a diamond disk (Z411, Minitor, Osaka) attached to a flexible shaft (No. 710, Sunflag, Osaka) and an electric drill (No. 508, Sunflag). The emptied shell is used as a bed for embryo culture. Thin albumen of hen's egg is used as the culture medium. The egg is cracked open into a 90 mm polystyrene culture dish and the egg yolk with its capsular thick albumen scooped out using spatula. The pH of albumen is 7.3–7.4 at the time of oviposition but it rises rapidly to 7.7 after 2 h and to 8.8 during 24 h of storage at 12°C (Ono *et al.*, 1996). In the normal quail embryo the pH of albumen is about 7.1 after 1 day of incubation at 37.5°C (Ono *et al.*, 1996).

Thus, non-pH-adjusted albumen can be used for system Q2. The embryo with yolk from system Q1a is transferred into the new bed shell and the shell is filled with thin albumen of hen's egg. The open surface of the bed shell is sealed tightly with a piece of cling film (Saran Wrap) and a pair of polyvinyl chloride rings (made from water pipe; 7 mm tall, 20 and 26 mm inner and outer diameters, respectively, and with four stainless steel screw projections attached to the outside), which is secured by elastic bands (13 mm diameter; No. 7, O'Band, Osaka) hooked around the screws. The culture-set thus made is placed in an incubator with the long axis of the shell held horizontally, and the embryo is then cultured for 48–53 h at 37.5°C and 70% relative humidity in an atmosphere of 100% air while being rocked round the long axis at a 90° angle at 30 min intervals. A laboratory egg incubator with an automatic turner (P-008-B special model for embryo culture, Showa Furanki Inst., Saitama) is used for the culture systems Q2 and Q3.

System Q3. In system Q3, the narrow end half of a small hen's egg shell cut circularly with the diamond disk and is then used as a bed for embryo culture. The embryo from the system Q2 is transferred to the new bed shell, and the shell is sealed with cling film using thin albumen of hen's egg

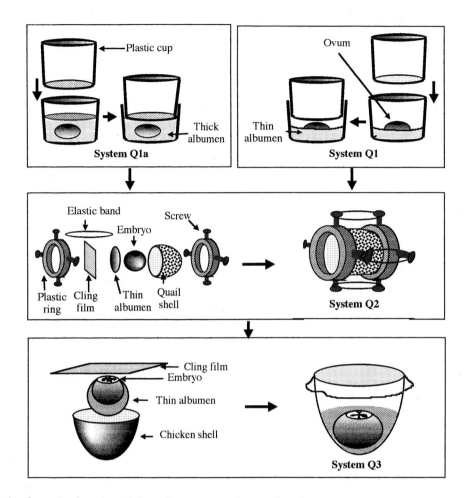

Figure 13.3. A schematic drawing of the culture system for quail embryos. Details of the culture systems and culture conditions are described in the text.

as adhesive. The culture conditions are similar to those in the system Q2 except that the cling film surface is directed upward and rocking is done round the short axis of the shell at a 30° angle. One or two days before the expected hatching time the cling film is perforated to facilitate embryonic respiration. Just before hatching the cling film is replaced with 60-mm polystyrene dish and the culture-set is placed in a hatcher and not rocked.

Microinjection of DNA into the Single-cell Stage Ovum with an Albumen Capsule and the Subsequent Culturing of the Embryo

Collection of the Ovum and Gene Introduction

When trying to introduce a foreign gene into the cytoplasm, the single-cell stage ovum covered with a thick albumen capsule is used. The ovum is obtained from the magnum 2.75 hrs after the preceding oviposition. The protocol for gene introduction is similar to that described above, however, the albumen capsule represents an obstacle to the observation of the ovum.

Culture

The injected ovum is cultured through three consecutive steps, i.e., systems Q1, Q2 and Q3 (Figure 13.3). In system Q1, thin albumen of hen's egg is used as the culture medium. In the original protocol (Ono *et al.*, 1994a) its pH was not adjusted. However, it is recommended that the pH of albumen is adjusted to 7.2–7.4 by bubbling CO_2 through it, any resulting foam is removed from the albumen using a spatula and the albumen warmed to 41.5°C until use. The ovum is placed in 20 ml polypropylene cup and the prepared thin albumen added up to the ovum's equatorial level. The open surface of the cup is tightly sealed by placing another cup inside it. The ovum is incubated without agitation for 24 h at 41.5°C and 100% air of saturated humidity.

After removal of the thick albumen capsule, the embryo from system Q1 is transferred into system Q2. The subsequent procedures are the same as described above.

Introduction of Cells and Genes to the Blastoderm Stage Embryo and the Subsequent Culturing of the Embryo

Cell Transfer for Chimera

Preparation of Donor Cells. Newly laid eggs of quail, chicken, etc. at EG&K stage X (Eyal-Giladi and Kochav, 1976) are used as the cell donor. The protocol for chick embryos has been reported by Petitte et al. (1990) and some modifications are made here. The egg is cracked open and albumen capsule is removed from yolk using forceps. The vitelline membrane adjacent to the blastoderm is lifted using forceps and cut around the blastoderm. The blastoderm is placed in Leibovitz's L-15 culture medium containing 50 units of penicillin and 50 µg of streprtomycin per ml. The blastoderm is gently removed from the vitelline membrane using fine forceps (No. 5, Inox), transferred into PBS(−) using a pasteur pipette and the adherent yolk washed off. The blastoderm cells are incubated for 10 min at 37°C in 1% trypsin (0152-13-1, type 1:250, DIFCO) in PBS(−). Foetal calf serum (FCS) is added to the cell suspension (1 to 100 v/v) and the cells are dispersed throughly by repeated pipetting and spun down in PBS(−) twice at 80G for 4 min. Finally, cells collected from 20 blastoderms are suspended in 400 µl L-15.

Injection of the Cells. The content of the recipient egg is placed in a 10 ml polypropylene cup (55-079-02, Iuchi; 24 mm tall, 24 and 28 mm lower and upper diameters, respectively) with the blastodisc upward. The micropipette, beveled at a 35° angle to an outer diameter of 40–130 µm and filled with the cell suspension is vertically inserted into the subgerminal cavity of the embryo through the surrounding albumen and vitelline membrane.

Gene Introduction. The protocol for the chick embryos (Brazolot et al., 1991) is modified as follows. DNA (1 µg) and cationic liposomes (4 µl; Lipofectin, BRL) are mixed together in 10 µl distilled water, and incubated for 15 min at room temperature. This is then added to a 90 µl blastoderm cell suspension (see above), the mixture is incubated at 37°C for 2 h and FCS is then added (1 to 200 v/v) to stop lipofection. The cells are spun down in L-15 twice at 80G for 4 min. The subsequent injection protocol is the same as described above.

Culture

The embryo at EG & K stage X thus manipulated is cultured through two consecutive steps, i.e., systems Q2 and Q3 (Figure 13.3). It is not necessary to remove the albumen capsule.

Introduction of PGCs and Genes to the Vascularized Embryo and the Subsequent Culturing of the Embryo

Transfer of PGCs

One to One Transfer. A micropipette, beveled at a 35° angle to an outer diameter of 80–100 µm, is used for the removal of blood containing PGCs and transfusion. Embryos at H&H stages 12–16 (Hamburger and Hamilton, 1951) are used. To reduce the number of endogenous PGCs and to prevent high blood pressure after the transfusion blood is removed from the marginal vein of the receipient embryo using suction tubing with a mouth-piece (05-2000-00, Drummond). The perforation in the recipient's vein is sealed with fine protein foam produced from 10% FCS in L-15. The foam is pre-loaded into the micropipette, blood containing PGCs is taken from the donor embryo, transfused into the recipient embryo and perforation sealed with the foam (Yasuda et al., 1992).

Transfer of Pooled PGCs. The protocol for chick PGCs (Yasuda et al., 1992) is modified for quail cells. About 100 µl blood collected from donor embryos is suspended in 1 ml L-15. The cells are spun down at 200G for 5 min, the pellets are dispersed in 1 ml of 16% Ficoll (type 400-DL, F9378, Sigma) /L-15, overlaid with 100 µl of 6.3% Ficoll/PBS(−) and spun down at 800G for 30 min. The PGCs-rich fraction is located in the surface of 16% Ficoll layer. The fraction is collected, it is then diluted in 1 ml L-15 and spun down twice at 200G for 5 min. Finally, pellets are resuspended in 100–200 µl L-15. The subsequent injection protocol is the same as described above.

Liposome-mediated DNA Transfer into PGCs *In Vivo*

The protocol for chick embryos (Watanabe et al., 1994) is modified as follows. DNA (15.5 µg), cationic liposomes (10 µl) and distilled water (14.5 µl) are mixed together, and incubated for 15 min at room temperature. The mixture is transfused into the marginal vein of the embryo at H&H stages 12–16 using a micropipette, beveled at a 20° angle to an outer diameter of 30 µm.

Culture

The embryo at H&H stages 12–16 is cultured using system Q3. In a one to one transfer, the donor embryo is also cultured *in vitro* up to day 8 of development and its sex can be determined by morphological and histological observations of the gonads.

References

Allioli, N., Thomas, J.L., Chebloune, Y., Nigon, V.M., Verdier, G. and Legras, C. (1994) Use of retroviral vectors to introduce the express and β-galactosidase marker gene in cultured chicken primordial germ cells. *Dev. Biol.* **165**, 30–37

Bosselman, R.A., Hsu, R.-Y., Boggs, T., Hu, S., Bruszewski, J., Ou, S., Kozar, L., Martin, F., Green, C., Jacobsen, F., Nicolson, M., Schults, J.A., Semon, K.M., Rishell, W. and Stewart, R.G. (1989) Germ line transmission of exogenous genes in the chicken. *Science* **243**, 533–535

Brazolot, C.L., Petitte, J.N., Etches, R.J. and Verrinder Gibbins, A.M. (1991) Efficient translocation of chicken cells by Lipofectin, and introduction of transfected blastodermal cells into the embryo. *Mol. Reprod. Devel.* **30**, 304–312

Carsience, R.S., Clark, M.E., Verrinder Gibbins, A.M. and Etches, R.J. (1993) Germline chimeric chickens from dispersed donor blastodermal cells and compromised recipient embryos. *Development* **117**, 669–675

Crittenden, L.B., Salter, D.W. and Federspiel, M.J. (1989) Segregation, viral phenotype, and proviral structure of 23 avian leukosis virus inserts in the germ line of chickens. *Theor. Appl. Genet.* **77**, 505–515

Eyal-Giladi, H. and Kochav, S. (1976) From cleavage to primitive streak formation: a complementary normal table and a new look at the first stages of development of the chick. *Dev. Biol.* **49**, 321–337

Hamburger, V. and Hamilton, H.L. (1951). A series of normal stages in the development of the chick embryo. *J. Morphol.* **88**, 49–92

Hippenmeyer, P.J., Krivi, G.G. and Highkin, M.K. (1988) Transfer and expression of the bacterial NPT-II gene in chick embryos using a Schmidt-Ruppin retrovirus vector. *Nucleic Acids Res.* **16**, 7619–7631

Li, Y., Behnam, J. and Simkiss, K. (1995) Ballistic transfection of avian primordial germ cells *in ovo*. *Transgenic Res.* **4**, 26–29

Love, J., Gribbin, C., Mather, C. and Sang, H. (1994) Transgenic birds by DNA microinjection. *Bio/Technology* **12**, 60–63

Naito, M., Agata, K., Otsuka, K., Kino, K., Ohta, M., Hirose, Perry, M.M. and Eguchi, G. (1991) Embryonic expression of β-actin-lacZ hybrid gene injected into the fertilized ovum of the domestic fowl. *Int. J. Dev. Biol.* **35**, 69–75

Naito, M., Sasaki, E., Otaki, M. and Sakurai, M. (1994a) Introduction of exogeneous DNA into somatic and germ cells of chickens by microinjection into the germinal disc of fertilized ova. *Mol. Reprod. Dev.* **37**, 167–171

Naito, M., Tajima, A., Yasuda, Y. and Kuwana, T. (1994b) Production of germline chimeric chickens, with high transmission rate of donor-derived gametes, produced by transfer of primordial germ cells. *Mol. Reprod. Dev.* **39**, 153–161

Ono, T., Murakami, T., Mochii, M., Agata, K., Kino, K., Otsuka, K., Ohta, M., Mizutani, M., Yoshida M. and Eguchi, G. (1994a) A complete culture system for avian transgenesis, supporting quail embryos from the single-cell stage to hatching. *Dev. Biol.* **161**, 126–130

Ono, T., Muto, S., Mizutani, M., Agata, K., Mochii, M., Kino, K., Otsuka, K., Ohta, M., Yoshida, M. and Eguchi, G. (1994b) Production of quail chimera by transfer of early blastodermal cells and its use for transgenesis. *Jpn. Poult. Sci.* **31**, 119–129

Ono, T., Muto, S., Matsumoto, T. and Yoshida, M. (1995a) Production of quail chimeras by transfer of early blastoderm cells: plumage chimeras and a germline chimera without plumage mixture. *Jpn. Poult. Sci.* **32** 252–256

Ono, T., Muto, S., Matsumoto, T., Mochii, M. and Eguchi, G. (1995b) Gene transfer into circulating primordial germ cells of quail embryos. *Exp. Anim.* **44**, 275–278

Ono, T., Murakami, T., Tanabe, Y., Mizutani, M., Mochii, M. and Eguchi, G. (1996c) Culture of naked quail (*Cotrurnix coturnix Japonica*) ova *in vitro* for avian transgenesis: Culture from the single-cell stage to hatching with pH-adjusted chicken albumen. *Comp. Biochem. Physiol.* **113A**, 287–292

Ono, T. and Wakasugi, N. (1984) Mineral content of quail embryos cultured in mineral-rich and mineral-free conditions. *Poultry Sci.* **63**, 159–166

Perry, M.M. (1988) A complete culture system for the chick embryo. *Nature* **331**, 70–72

Perry, M.M., Morrice, D., Hettle, S. and Sang, H. (1991) Expression of exogenous DNA During the early development of the chick embryo. *Roux's Arch. Dev. Biol.* **200**, 312–319

Petitte, J.N., Clark, M.E., Liu, A.M. Verrinder Gibbins, A.M. and Etches, R.J. (1990) Production of somatic and germline chimeras in the chickens by transfer of early blastodermal cells. *Development* **108**, 185–189

Rowlett, K. and Simkiss, K. (1987) Explanted embryo culture: *in vitro* and *in ovo* techniques for domestic fowl. *Brit. Poult. Sci.* **28**, 91–101

Salter, D.W. and Crittenden, L.B. (1989) Artificial insemination of a dominant gene for resistance to avian leukosis virus into the chicken. *Theor. Appl. Genet.* **77**, 457–461

Salter, D.W., Smith, E.J., Hughes, S.H., Wright, S.E. and Crittenden, L.B. (1987) Transgenic chicken: insertion of retroviral genes into the chicken germ line. *Virology* **157**, 236–240

Salter, D.W., Smith, E.J., Hughes, S.H., Wright, S.E., Fradly, A.M., Witter, R.L. and Crittenden, L.B. (1986) Gene insertion into the chicken germ line by retroviruses. *Poultry Sci.* **65**, 1445–1458

Sang, H. and Perry, M.M. (1989) Episomal replication of cloned DNA injected into the fertilized ovum of the hen, *Gallus domesticus*. *Mol. Rep. Dev.* **1**, 98–106

Shuman, R.M. and shoffner, R.N. (1986) Gene transfer by avian retroviruses. *Poultry Sci.* **65**, 1437–1444

Souza, L.M., Boone, T.C., Murdock, D., Langley, K., Wypych, J., Fenton, D., Johnson, S., Lai, P.H., Everett,

R., Hsu, R.-Y. and Bosselman, R. (1984) Application of recombinant DNA technologies to study on chicken growth hormone, *J. Exp. Zool.* **232**, 465–473

Tajima, A., Naito, M., Yasuda, Y. and Kuwana, T. (1993) Production of germ line chimera by transfer of primordial germ cells in the domestic chicken (*Gallus domesticus*). *Theriogenology* **40**, 509–519

Yasuda, Y., Tajima, A., Fujimoto, T. and Kuwania, T. (1992) A method to obtain avian germ-line chimeras using isolated primordial germ cells. *Reprod. Fert.* **96**, 521–528

Watanabe, M., Naito, M., Sasaki, E., Sakurai, M., Kuwana, T. and Oishi, T. (1994) Liposome-mediated DNA transfer into chicken primordial germ cells *in vivo*. *Mol. Reprod. Dev.* **38**, 268–274

PART II, SECTION C

14. The Microinjection of DNA into Early Chicken Embryo

M. Naito

National Institute of Animal Industry,
Tsukuba Norindanchi, P.O. Box 5, Ibaraki 305, Japan

Summary

Manipulation of early chicken embryo became possible by developing the *ex vivo* embryo culture technique from single-cell stage to hatching. Microinjection of DNA into the germinal disc of fertilized ova was performed and the manipulated embryos were cultured *in vitro* and then in recipient eggshells. Expression of the injected DNA was observed in the embryonic and extra-embryonic tissues, but the expression pattern was mosaic. After hatching, the injected DNA was detected in the somatic and germ cells of chickens, although the efficiency of introduction of injected DNA into chickens was low. It was observed that in one of these chickens the injected DNA was integrated into the host chromosome and transmitted stably to the next generation. These results indicate that microinjection of exogenous DNA into fertilized ova is a feasible method for introducing DNA into chickens.

Introduction

The gene transfer technique is quite useful for studying avian embryology and genetic improvement of chickens (Naito, 1993; Perry and Sang, 1993; Simkiss, 1993; Sang, 1994 and Bulfield, 1995). Manipulation of the early avian embryo was hampered by the distinctive reproductive system of the laying hen. The avian embryo contains a large amount of yolk, and the early stage embryo develops in the hen's oviduct parallel to the egg formation process. After oviposition, the embryo develops inside the eggshell until hatching. Recently, a chicken embryo culture technique from single-cell stage to hatching has been devised, and this technique opens the way to manipulate the early avian embryo. DNA microinjection into early chicken embryo has become possible by this *ex vivo* embryo culture. Here, introduction of exogenous DNA into somatic and germ cells of chickens by microinjection into the fertilized ova is described.

Development of Early Chicken Embryo and Embryo Culture

Fertilization in chickens takes place in the infundibulum of the oviduct within 15 minutes after ovulation (Figure 14.1). Thick albumen is then formed

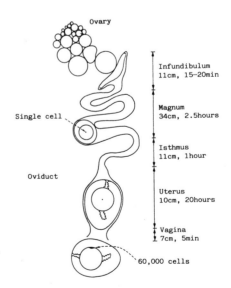

Figure 14.1. Egg formation and early embryonic development.

around the yolk in the magnum. The first cleavage division of the fertilized ovum starts about 4.5 hours after ovulation, and the timing coincides roughly with the entry of the ovum into the isthmus, where the shell membranes are formed. While the egg stays in the isthmus, the embryo develops to the 4–8 cell stage. When the egg enters into the uterus, the albumen absorbs the so-called "plumping fluid" and the embryo develops to the 256 cell stage within 4 hours after entering into the uterus, then eggshell is formed around the shell membrane. At the time of oviposition, the embryo consists of about 60,000 cells, and a growth ring composed of the area opaca and area pellucida is formed (Eyal-Giladi and Kochav, 1976 and Kochav et al., 1980). After laying, embryonic development continues inside the eggshell.

The developmental stage appropriate for DNA microinjection into fertilized ovum is at about 2.5 hours after ovulation (about 2.75 hours after the preceding egg is laid, Naito et al., 1990b) when the male and female pronuclei are enlarged and centrally placed in apposition (Perry, 1987). The fertilized ovum at the single-cell stage obtained from the posterior portion of the magnum can be cultured ex vivo until hatching, involving three stages; fertilized ovum to blastoderm formation (system I) for 1 day, embryogenesis (system II) for 3 days, and embryonic growth (system III) for 18 days (Perry, 1988). The culture method is subsequently modified (Naito et al., 1990a) and the whole procedure is as follows (Figure 14.2). The egg obtained from the magnum is incubated stationary with culture medium in a glass jar for 24 hours at 41°C. The embryo (yolk) is then transferred to a small recipient eggshell after removing the thick albumen and culture medium. The shell is filled with thin albumen and sealed with cling film and secured by plastic rings and elastic bands. The reconstituted egg is incubated for 3 days at 38°C with rocking through an angle of 90°. Then, the contents of the egg is transferred to a large recipient eggshell, sealed with cling film and incubated for 18 days until hatching at 38°C with rocking through an angle of 30° (Figure 14.3). The rate of hatching was 7–15% (Perry, 1988 and Perry and Mather, 1991), and it was 34.4% using the modified method (Naito et al., 1990a). This ex vivo culture method has also been used for the growth in culture for hatching early uterine embryos obtained by induced premature oviposition (Naito and Perry, 1989).

Microinjection of DNA

For microinjection of DNA, micropipettes are made by pulling 1.0 mm siliconized microcapillary tubing. The tips are beveled down (25–30°) to an outside diameter of 10–15 μm. The micropipette is held by a micromanipulator and connected to a microinjector with a teflon tube. The pressure in the micropipette can be monitored by connecting a pressure-lock gas syringe to the mid point of the tube. The tube needs to be filled with distilled water and the micropipette containing DNA also needs to be filled with liquid paraffin in order to remove the air from the tube and micropipette (Naito et al., 1991).

In chickens, male and female pronuclei are masked by the opaque cytoplasm, and it is difficult to

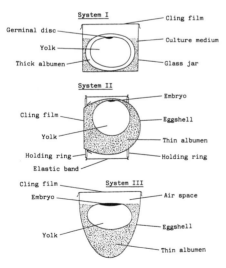

Figure 14.2. Chicken embryo culture method from single-cell stage to hatching.

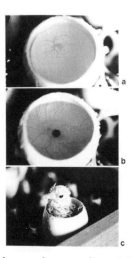

Figure 14.3. Chicken embryos cultured from single-cell stage. (a) 4-day incubated embryo; (b) 8-day incubated embryo; (c) chick at hatch.

Figure 14.4. DNA microinjection into the germinal disc of fertilized ovum at the single cell stage.

Expression of Exogenous DNA in the Early Chicken Embryo

In order to study the expression of injected DNA, a plasmid DNA containing *E.coli* β-galactosidase (*lacZ*) gene under the control of RSV enhancer and chicken β-actin gene promoter/enhancer (MiwZ) was constructed (Suemori *et al.*, 1990 and Naito *et al.*, 1991). The circular form plasmid DNA was injected into the germinal disc of fertilized ova and the injected ova were cultured *ex vivo* for 4 days (Naito *et al.*, 1991). Expression of the *lacZ* gene was detected by X-gal staining. Analysis of the *lacZ* gene expression after one day of incubation indicates that the injected DNA does not diffuse throughout the blastoderm. Viability of the injected embryos on day 4 of incubation was 42.3% (55/130). Expression of the *lacZ* gene was observed in 63.6% (35/55) of the embryos that survived, and 40.0% (22/55) of them were expressed in the embryonic tissues and 23.6% (13/55) of them were expressed in the extra-embryonic tissues only. Embryos in Figure 14.5 show the expression of *lacZ* gene in the embryonic and extra-embryonic tissues, but the expression pattern is mosaic. The incidence of expression was higher in the neural tube than in other organs and tissues.

Sang and Perry (1989) reported that linear DNA molecules injected into the germinal disc are ligated rapidly after injection to form random concatamers of head-to-tail, head-to-head or tail-to-tail, and the injected DNA was gradually lost during the embryonic development and no evidence was obtained about the incorporation of the injected DNA into the host chromosomes. Perry *et al.* (1991) investigated the expression of DNA injected into the cytoplasm of fertilized chick ova, which was in circular form and contained *lacZ* gene under the control of cytomegalovirus immediate early promoter. The *lacZ* gene expression first appeared at the mid-cleavage stage at 12 hours of incubation. At

distinguish the male pronucleus that will fuse with the female pronucleus due to the presence of supernumerary male pronuclei. It is, therefore, impossible to inject DNA into one of the pronuclei which contributes to the zygote nucleus, so DNA is injected into the cytoplasm near the pronuclei. The micropipette is inserted into the central area of the germinal disc through the albumen capsule and vitellin membrane (Figure 14.4). A continuous flow system is preferably employed for DNA microinjection in order to inject DNA into the central part of the germinal layer close to the upper surface because embryonic tissues and primordial germ cells may originate from the central zone of the area pellucida (Ginsburg and Eyal-Giladi, 1987).

Since high doses of DNA may be very toxic to chick embryos, the concentration of DNA for injection should be 0.1 mg/ml or less (Naito *et al.*, 1991 and Perry *et al.*, 1991). Otherwise, development of most embryos is arrested at early blastoderm stages. The injection volume of DNA is 3 nl (Perry *et al.*, 1991) or 10–50 nl (Naito *et al.*, 1991).

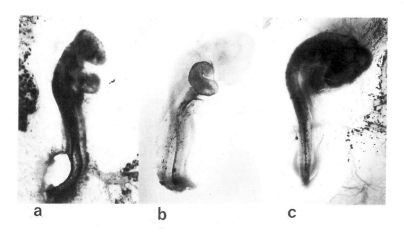

Figure 14.5. Expression of *lacZ* gene injected into the germinal disc of fertilized ovum in 4-day incubated embryos (a) *lacZ* gene expression throughout the embryo. (b) *lacZ* gene expression mainly in the heart and neural tube of the embryo. (c) mosaic expression of *lacZ* gene in the embryo. (a, b: from Naito *et al.*, 1991).

later stages, positive cells were scattered in the vicinity of the primitive streak of most embryos, but after gastrulation they were present in the embryonic tissues of only 7% of surviving embryos. These results show that DNA injected into the fertilized ova persist episomally and are gradually lost during embryonic development. However, the injected DNA is assumed to be incorporated at an early stage in at least one case (Figure 14.5a), suggesting that by the method employed here (Sang and Perry, 1989; Naito et al., 1991 and Perry et al., 1991) the injected DNA could be incorporated into the host chromosomes even though it has low efficiency.

Introduction of Exogenous DNA into Chickens

Based on the analysis of the expression of injected DNA in the early chicken embryo, the possibility of introduction of exogenous DNA into hatched chickens by microinjection of DNA into the fertilized ova were pursued. The plasmid DNA containing lacZ gene under the control of chicken β-actin gene promoter (pAcZ) was injected in linearized form into the germinal disc of fertilized ova (Naito et al., 1994). The injected embryos were cultured ex vivo until hatching. The rate of hatching was 11.8% (31/263), and 19 males and 6 females were matured. DNA from blood and semen samples of the 25 matured chickens were analyzed for the presence of the injected DNA by Southern hybridization. The injected DNA was detected in the blood DNA of one male and in the sperm DNA in another male.

Restriction digestion analysis of the injected DNA suggests that it was not rearranged and was organized as head-to-tail multimers. The copy numbers of the DNA were 0.07–0.02 in the blood DNA of one male per diploid genome, and 0.02–0.015 in the sperm DNA of another male, indicating that the exogenous DNA was present in limited populations of the blood and sperm cells.

Love et al. (1994) also succeeded in introducing exogenous DNA into chickens by a similar procedure. They produced 7 healthy chickens, 5.5% of the total number of injected ova, reached maturity. One of these, a cockerel was identified as mosaic with approximately one copy of the injected DNA per 10 genome equivalent. The cockerel transmitted the exogenous DNA to 3.4% (14/412) of his G_1 offspring, and G_2 transgenic offspring were produced (81.5%, 106/130) according to normal Mendelian inheritance of the transgene by mating the G_1 transgenic birds, suggesting that stable transmission of exogenous DNA was obtained.

Thus, the cytoplasmic injection of DNA into the germinal disc of fertilized ovum results in generating transgenic chickens. The overall efficiency of obtaining chickens positive for the injected DNA was about 0.8% (2/263, Naito et al., 1994; 1/127, Love et al., 1994). It was about 3.5 times higher than that when DNA was injected in the cytoplasm of mouse eggs (0.22%, 2/890, Brinster et al., 1995). If DNA can be injected into one of the pronuclei which will contribute to the zygote nucleus, the efficiency of introducing exogenous DNA into chickens or into host chromosomes would be increased by 10 times or more.

Conclusions

The results indicate that microinjection of exogenous DNA into fertilized ova is a feasible method for introducing DNA into chickens. Direct gene transfer is quite useful for improving economically important traits in chickens, accelerates the genetic improvement and makes possible the utilization of genes from different animal species. Also, gene transfer technique will contribute to understanding the various biological phenomena in chickens. Having a new role for chickens, the gene transfer technique can be used in generating transgenic chickens which produce pharmaceutical materials into eggs.

Acknowledgements

Our study shown here was carried out by collaborating with the following scientists: M.M. Perry, K. Nirasawa, T. Oishi, K. Agata, K. Otsuka, K. Kino, M. Ohta, K. Hirose, G. Eguchi, E. Sasaki, M. Ohtaki and M. Sakurai. The study was supported by Special Funds from the Science and Technology Agency and the Ministry of Agriculture, Forestry and Fisheries of the Japanese Government.

References

Brinster, R.L., Chen, H.Y., Trumbauer, M.E., Yagle, M.K. and Palmiter, R.D. (1985) Factors affecting the efficiency of introducing foreign DNA into mice by microinjecting eggs. *Proc. Natl. Acad. Sci. USA* **82**, 4438–4442

Bulfield, G. (1995) Biotechnology and the poultry industry. *Proc. 11th AVIAGEN Symp.*, Krakow, Poland, May–June, pp. 1–5

Eyal-Giladi, H. and Kochav, S. (1976) From cleavage to primitive streak formation: a complementary normal table and a new look at the first stages of development of the chick. I. General morphology. *Dev. Biol.* **49**, 321–337

Ginsburg, M. and Eyal-Giladi, H. (1987) Primordial germ cells of the young chick blastoderm originate from the central zone of the area pellucida irrespective of the embryo-forming process. *Development* **101**, 209–219

Kochav, S., Ginsburg, M. and Eyal-Giladi, H. (1980) From cleavage to primitive streak formation: a complementary normal table and a new look at the first stages of development of the chick. II. Microscopic anatomy and cell population dynamics. *Dev. Biol.* **79**, 296–308

Love, J., Gribbin, C., Mather, C. and Sang, H. (1994) Transgenic birds by DNA microinjection. *Bio/Technol.* **12**, 60–63

Naito, M. (1993) Egg genetics: embryo manipulation and genetic improvement of egg production. *Proc. 10th AVIAGEN Symp.*, Nitra, Slovakia, June, pp. 41–47

Naito, M., Agata, K., Otsuka, K., Kino, K., Ohta, M., Hirose, K., Perry, M.M. and Eguchi, G. (1991) Embryonic expression of β-actin-*lacZ* hybrid gene injected into the fertilized ovum of the domestic fowl. *Int. J. Dev. Biol.* **35**, 69–75

Naito, M., Nirasawa, K. and Oishi, T. (1990a) Development in culture of the chick embryo from fertilized ovum to hatching. *J. Exp. Zool.* **254**, 322–326

Naito, M., Nirasawa, K. and Oishi, T. (1990b) Duration of egg formation in hens selected for increased rate of lay under 23 h and 24 h light-dark cycles. *Brit. Poult. Sci.* **31**, 371–375

Naito, M. and Perry, M.M. (1989) Development in culture of the chick embryo from cleavage to hatch. *Brit. Poult. Sci.* **30**, 251–256

Naito, M., Sasaki, E., Ohtaki, M. and Sakurai, M. (1994) Introduction of exogenous DNA into somatic and germ cells of chickens by microinjection into the germinal disc of fertilized ova. *Mol. Reprod. Dev.* **37**, 167–171

Perry, M.M. (1987) Nuclear events from fertilisation to the early cleavage stages in the domestic fowl (*Gallus domesticus*) *J. Anat.* **150**, 99–109

Perry, M.M. (1988) A complete culture system for the chick embryo. *Nature* **331**, 70–72

Perry, M.M. and Mather, C.M. (1991) Satisfying the needs of the chick embryo in culture, with emphasis on the first week of development. In: Tullett, S.G., (ed.), *Avian Incubation*, Butterworth-Heinemann, London, pp. 91–105

Perry, M., Morrice, D., Hettle, S. and Sang, H. (1991) Expression of exogenous DNA during the early development of the chick embryo. *Roux's Arch. Dev. Biol.* **200**, 312–319

Perry, M.M. and Sang, H. (1993) Transgenesis in chickens. *Transgenic Res.* **2**, 125–133

Simkiss, K. (1993) Surrogate eggs, chimaeric embryos and transgenic birds. *Comp. Biochem. Physiol.* **104A**, 411–417

Sang, H. (1994) Transgenic chickens – methods and potential applications. *Trends Biotechnol.* **12**, 415–420

Sang, H. and Perry, M.M. (1989) Episomal replication of cloned DNA injected into the fertilized ovum of the hen, *Gallus domesticus*. *Mol. Reprod. Dev.* **1**, 98–106

Suemori, H., Kadokawa, Y., Goto, K., Araki, I., Kondoh, H. and Nakatsuji, N. (1990) A mouse embryonic stem cell line showing pluripotency of differentiation in early embryos and ubiquitous β-galactosidase expression. *Cell Differ. Dev.* **29**, 181–186

PART II, SECTION C

15. Production of Chimeric Chickens as Intermediates for Gene Transfer

R.J. Etches,* M.E. Clark, A.M. Verrinder Gibbins and M.B. Cochran

Department of Animal and Poultry Science, University of Guelph, Guelph, Ontario, N1G 2W1, Canada

Introduction

The production of transgenic animals requires genetic modification of cells that retain the ability to be incorporated into the germline and produce sperm and eggs. Under ideal conditions, genetic modification would be introduced into the newly fertilized zygote and all cells within the embryo, including sperm and eggs, would possess the modified genotype. In the case of chickens, however, microinjection of DNA into the newly fertilized zygote is fraught with technical difficulties (see chapter by Naito). To obviate these difficulties, an alternate strategy for the production of transgenic chickens *via* chimeric intermediates has been developed using either blastodermal cells or primordial germ cells. Theoretically, genetic modifications can be introduced into these cells *in vitro*, and the genetically modified cells can be returned to the germline of a recipient embryo. Offspring of chimeras derived from the genetically modified cells that contributed to the germline can be identified after hatching and used as founder animals to establish stocks of transgenic chickens that express the genetic modification. This chapter will review the achievements in embryo manipulation that have given this strategy enormous potential and will describe the techniques that have yet to be developed in order to fulfil the potential for germline chimeras as intermediates in the production of transgenic chickens.

Development of the Germline in Birds

Fertilization in birds occurs in the infundibulum of the oviduct within 15 min after ovulation (Perry, 1987). During the next 4–5 h, the newly fertilized egg descends into the magnum and isthmus of the oviduct where the egg white proteins and shell membranes are secreted around the ovum. The first cleavage division occurs when the egg enters the shell gland. As the shell is formed during the next 19–22 h, the embryo develops into a blastoderm containing 50,000 to 60,000 cells (Eyal-Giladi, Kochav, 1976; Kochav *et al.*, 1980 and Watt *et al.*, 1993). Development of the embryo between the beginning of shell formation and oviposition has been divided into ten stages (Eyal-Giladi and Kochav, 1976 and Kochav *et al.*, 1980). In the absence of apoptosis, 17 cell divisions would be required to produce an embryo containing 65,000 cells and the average length of the cell cycle would be approximately 70 min. These estimates of the number of cells divisions and the length of the cell cycle, however, are probably conservative because cells are shed from the undersurface of the embryo between stages VII and X (Eyal-Giladi and Kochav, 1976 and Kochav *et al.*, 1980).

Primordial germ cells can be distinguished in the germinal crescent of stage 2–15 (Hamburger and Hamilton, 1951) embryos, in the blood of embryos between stages 15 and 25 (H&H), and in the gonadal ridge between stage 16 (H&H) onwards by their substantial deposits of glycogen (Meyer, 1964; Fujimoto *et al.*, 1976 and Fargeix *et al.*, 1981) and their content of alkaline phosphatase (Swartz, 1982). At various stages during the migration of the primordial germ cells from the germinal crescent to the

*Author for Correspondence.

gonad, they exhibit α and β galactose, N-acetyllactosamine, and N-N' diacetylchitobiose residues (Didier *et al.*, 1990 and Yoshinaga *et al.*, 1992). Primordial germ cells isolated from the germinal crescent, from the vascular system, or from the undifferentiated gonad of quail display epitopes that are recognized by the QH-1 (Pardanaud *et al.*, 1987) and B4 (Ginsburg *et al.*, 1989) antibodies in quail and by the EMA-1 (Hahnel and Eddy, 1986 and Urven *et al.*, 1988), FC10.2 (Loveless *et al.*, 1990), 2C9 (Maeda *et al.*, 1994), 7B7 and 1B3 (Halfter *et al.*, 1996) antibodies in chickens.

The events that occur as primordial germ cells become committed to the germline in the early embryo are unresolved. The EMA-1 epitope is expressed on some cells in stage XI chicken embryo (Urven *et al.*, 1988 and Karagenc *et al.*, 1995) indicating that primordial germ cells may be present at this time. Evidence supporting the presence of morphologically unrecognizable primordial germ cells in stage X (E-G&K) chicken embryos can be inferred from the presence of committed primordial germ cells in cultures derived from the central disc, but not the area opaca, of stage X blastoderms (Ginsburg and Eyal-Giladi, 1987). The presence of cells that are destined for the germline can also be inferred from the observation that somatic and germline chimeras are produced more frequently from cells taken from the central disc rather than the area opaca of stage X (E-G&K) embryos (Petitte *et al.*, 1993). In the quail, QH-1 positive cells are present in embryos at the time of oviposition (Pardanaud *et al.*, 1987) and all of the cells in embryos up to and including stage X (E-G&K) are B4 positive (Ginsburg *et al.*, 1989). Although all of these data provide evidence indicating that the precursors of primordial germ cells are present in the central disc of avian embryos before or at the time of oviposition, it is not yet clear if these cells are committed to the germline or if they retain the ability to enter both the somatic and germline lineages. Without this information, it is difficult to choose between the merits of using blastodermal cells, whose commitment is as yet undetermined, and committed primordial germ cells as donor cells to form chimeric intermediates to produce transgenic chickens. Incorporation of genetically modified cells into the germline would be maximized if the entire population of blastodermal cells derived from the central disc of stage X (E-G&K) embryos remains uncommitted. If some cells within stage X (E-G&K) embryos are committed to the germline, however, it may be advantageous to access this sub-population. In practice, the choice between using blastodermal cells and primordial germ cells to access the germline is determined by the ease with which cells can be harvested and returned to recipient embryos and the ensuing rate of germline trans-

mission. Incorporation of strategies such as gene targetting mediated by homologous recombination, which requires the ability to culture cells for extended periods, into transgenic technology for chickens will depend on the relative ease of culturing blastodermal cells and primordial germ cells while retaining their ability to contribute to the germline.

Chimeras made with Blastodermal Cells

Blastodermal cells have been harvested from Stage X (Eyal-Giladi and Kochav, 1976) embryos and yielded germline chimeras (Carsience *et al.*, 1993; Thoraval *et al.*, 1994 and Kagami *et al.*, 1995). Briefly, embryos were recovered from eggs within 2 hours following oviposition using sterile filter paper rings (Lucas and Jamroz, 1961), and placed into Dulbecco's phosphate buffered saline containing 5.6 mM glucose at pH 7.2 (PBS-G). Excess yolk was carefully washed off, stage X (E-G&K) embryos were selected and the area pellucida was cut out with a hair loop. The embryos were then incubated with calcium and magnesium free phosphate-buffered saline (CMF-PBS) supplemented with 2% chick serum for 10 min at room temperature. The CMF-PBS was replaced with 0.05% trypsin (w/v) in 0.02% EDTA (w/v) and incubated for not more than 5–10 min or until a population of individual cells was obtained. The cells were then washed with DMEM and 10% FBS (v/v), resuspended into a small volume of medium and injected into recipient stage X (E-G&K) White Leghorn embryos as described below.

Recipient embryos were prepared from eggs collected within 2 h of oviposition and exposed to approximately 600 rads of irradiation from a ^{60}Co source in a Gammacell 200 (AECL, Chalk River) to retard development for approximately 24 h. A hole was made in the side of the shell above the recipient embryo and approximately 500 donor cells were injected into the subgerminal cavity of the recipient. The egg was sealed with two layers of shell membrane and a layer of Opsite and, in early experiments, was incubated at 37.5 C and 50% relative humidity for 18 days and for three days at 36.5 C and 80% relative humidity until the chick hatched. During the first 24 h of development, donor-derived cells continue to proliferate rapidly whereas development of the irradiated recipient is severely impeded. This protocol has typically yielded chimeras with substantial contributions to the germline and in some cases, the entire germline is donor-derived (Table 15.1). If the donor embryo is from a coloured breed of chicken and the recipient has white feathers, the extent of somatic chimerism can be judged visually at hatch. In most cases, there is no evidence of germline chimerism in the absence of

Table 15.1. The proportion of donor-derived plumage and the rate of germline transmission of somatic and putative chimeras hatched after injecting approximately 500 cells into irradiated stage X (Eyal-Giladi and Kochav, 1976) recipient embryos.

Chimera	Proportion of ♂donor-derived plumage	Number of offspring derived from the		Rate of germline transmission (%)
		donor	recepient	
♂1919[1]	75	0	558	0
♂1928[1]	70	25	1232	2
♂1967[1]	10	135	1134	11
♂821[1]	75	604	546	53
♂1701[1]	75	40	201	17
♂926[1]	30	0	568	0
♂933[1]	90	201	327	38
♂959[1]	90	130	512	20
♂973[1]	95	0	137	0
♂6311[2]	+	0	151	0
♂6308[2]	+++	0	12	0
♀6354[2]	+++	0	14	0
♀6351[2]	+++	3	104	3
♀6349[2]	+	31	101	31
♀6346[2]	+	8	87	9
♂065[3]	20	398	539	43
♂075[3]	30	377	708	35
♂048[3]	99	9	271	3
♂042[3]	99	869	0	100
♂068[3]	45	62	1001	6
♂082[3]	20	0	449	0
♀050[3]	99	1	43	2
♀021[3]	85	3	54	5
♀028[3]	3	15	45	25
♀046[3]	60	0	96	0
♀032[3]	95	0	113	0
♀034[3]	35	0	103	0

[1] Kagami et al. (1995); [2] Thoraval et al. (1994); [3] Carsience et al. (1993); + small patches of donor-derived plumage; +++ more than 75% donor-derived plumage.

somatic chimerism estimated by feather pigmentation. However, there is no correlation between the extent of somatic chimerism estimated from feather pigmentation and the rate of germline transmission (Figure 15.1).

A significant practical problem of the protocol described above has been poor hatchability of embryos from injected eggs. Typically, between 5% and 15% of injected embryos develop into live chicks while the remainder die between the seventh day of incubation and hatch (Carsience et al., 1993 and Kagami et al., 1995). Embryonic mortality has been attributed to puncture of the inner shell membrane rather than the injection *per se* and has been associated with deformation of the chorioallantois around the hole through which the injection was made (unpublished data). Deformation of the chorioallantois during development can be avoided by transferring potentially chimeric embryos from their original shell to a surrogate shell prepared as described by Rowlett and Simkiss (1985, 1987). Briefly, surrogate shells were prepared from eggs that were 25 g heavier than the egg containing the recipient embryo. The blunt end of the shell was removed from the egg and the contents were discarded. The shell containing the injected embryo was broken open and the embryo was placed into a shallow bowl lined with Saran Wrap. The Saran Wrap was formed into a pouch and the embryo was lifted in the pouch from the bowl into the surrogate shell. The Saran Wrap was then removed leaving the embryo in the surrogate shell and the open end of the surrogate shell was covered with Saran Wrap which was sealed to the shell with albumen. The eggs were returned to an incubator which rotated through 60 per hour until the 19th day of incubation when they were transferred to a hatcher. The Saran Wrap was punctured when the embryos began to breathe and was removed just before the chicks left the surrogate shell. The hatchability of injected and sham-injected embryos transferred to surrogate shells was 32% and 26%, respectively (Figure 15.2). Ninety percent of the male chicks and 63% of the female chicks that developed in surrogate shells were somatically chimeric (Table 15.2). Four of the nine somatically chimeric males and 3 of the 5 somatically chimeric females produced donor-derived offspring (Table 15.2).

Chimeras made with Primordial Germ Cells

Primordial germ cells isolated from the germinal crescent and blood have been successfully transferred into recipient embryos and contributed to the

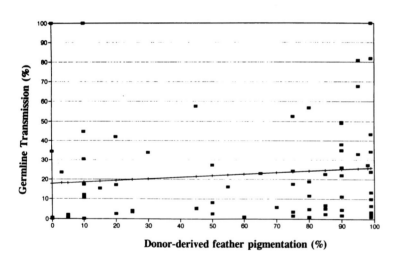

Figure 15.1. The relationship between donor-derived feather pigmentation and the production of donor-derived gametes in chimeras made by injecting blastodermal cells from Barred Plymouth Rocks into irradiated White Leghorn recipients. The correlation coefficient between somatic and germline chimerism was 0.12, which was not statistically significant. Note that three germline chimeras have been observed that produced 0.2, 35 and 100% donor-derived offspring and exhibited no evidence of somatic chimerism. From Etches and Verrinder Gibbins (1996).

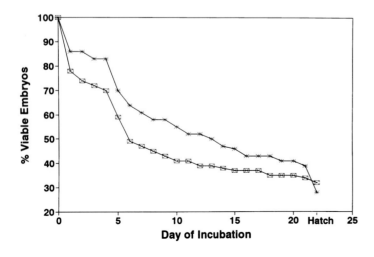

Figure 15.2. The proportion of viable embryos throughout incubation following transfer into a surrogate shell (stars), and transfer to a surrogate shell following injection of approximately 500 donor cells from stage X (E-G&K) embryos (boxes). The embryos were transferred to a surrogate shell on the fourth day following injection. Forty-eight embryos were transferred to surrogate shells after sham injection and 71 embryos were transferred after injection of donor cells. There was no significant difference ($X^2 = 0.326$; $P \geq 0.05$) between the proportion of viable embryos which emerged from the surrogate shells.

Table 15.2. The proportion of donor-derived plumage and the rate of germline transmission of somatic and putative chimeras hatched after injecting approximately 500 cells into irradiated stage X (Eyal-Giladi and Kochav, 1976) recipient embryos and subsequent transfer to surrogate shells.

Chimera	Proportion of donor-derived plumage	Number of offspring derived from the donor	Number of offspring derived from the recepient	Rate of germline transmission (%)
♂1766	25	0	570	0
♂1768	50	0	492	0
♂1791	50	199	44	81
♂1793	35	1	459	0.22
♂1768	99	86	113	43
♂1797	99	0	373	0
♂1770	80	267	204	57
♂1671	99	0	460	0
♂1694	50	0	576	0
♂1795	0	0	438	0
♀1757	40	0	97	0
♀1778	15	0	108	0
♀1673	90	13	46	22
♀1698	95	46	22	68
♀1696	90	3	63	5
♀1750	0	0	114	0
♀1783	0	0	111	0
♀1700	0	0	87	0

germline of chimeras (Wentworth et al., 1989; Vick et al., 1993a,b and Naito et al., 1994). Briefly, primordial germ cells have been harvested from the germinal crescent of stage 11 (H&H) embryos by enzymatic digestion and injected into the vasculature of busulphan-treated stage 15 (H&H) embryos to reduce the population of endogenous germ cells (Vick et al., 1993a). Approximately 40% of the busulphan-treated embryos that were injected with primordial germ cells hatched (Vick et al., 1993a) and approximately 50% of the putative chimeras produced donor-derived chicks (Table 15.3). The mean rate of germline transmission was approximately 35% (Vick et al., 1993a). When primordial germ cells from the germinal crescent have been infected with retroviral vectors, DNA has been transferred to the chicken genome and incorporated into the germline to produce transgenic chickens (Vick et al., 1993b) and turkeys (Wentworth et al., 1995).

When primordial germ cells were harvested from blood of stage 13–15 embryos, the preparation of donor cells was separated from erythrocytes on

Table 15.3. The number of donor-derived and recipient-derived offspring among the first 20 chicks produced by six breeding groups made up of six male and 12 female chimeras. The chimeras were made by injecting primordial germ cells isolated from the germinal crescent of stage 11 (H&H) embryos into the vasculature of stage 15 (H&H) embryos. From Vick et al. (1993a).

Breeding Group	Number of offspring derived from the donor	Number of offspring derived from the recepient	Rate of germline transmission (%)
1	20	0	0
2	20	0	0
3	20	0	0
4	17	3	3
5	15	5	25
6	11	9	45

a Ficoll gradient (Naito et al., 1994). Approximately 200 donor primordial germ cells were injected into the dorsal aorta of stage 15 recipient embryos and the resulting chimeras were incubated in surrogate shells (Naito et al., 1994). The endogenous population of primordial germ cells in recipient embryos was depleted by removing 4–10 µL of blood from the dorsal aorta of the recipient embryo a few hours before the donor primordial germ cells were injected (Naito et al., 1994). Between 18 and 28% of recipient embryos injected with donor primordial germ cells hatched and most of these chimeras produced donor-derived offspring (Table 15.4).

The Potential Efficacy of Producing Transgenic Chickens using Blastodermal Cells and Primordial Germ Cells

The production of transgenic chickens using blastodermal cells is facilitated because the presence of somatic contributions from the donor provides an index of the success of the injection at the time of hatch. In the absence of somatic chimerism, the likelihood that the donor cells contribute to the germline is very low (Carsience et al., 1993). In contrast, the contribution of donor cells to the recipient following injections of primordial germ cells cannot be evaluated until the birds mature 6 months after hatching. Of chicks that exhibit somatic chimerism at hatch following injection of blastodermal cells, approximately 40% also have donor-derived contributions to the germline (Tables 15.1 and 15.2). In comparison, the yield of germline chimerism when primordial germ cells were harvested from the germinal crescent and injected into busulphan-treated recipients was very similar (Table 15.3). By contrast, the yield of germline chimeras when donor primordial germ cells were harvested from blood and injected into recipients in which endogenous primordial

Table 15.4. The rate of germline transmission when primordial germ cells isolated from blood from stage 13–15 (H&H) Barred Plymouth Rocks (BPR) or White Leghorns (WL) were injected into the dorsal aorta of stage 14–15 WL or BPR, respectively. Data from Naito et al. (1994).

Chimera	Donor-Recipient	Number of offspring derived from the		Rate of germline transmission (%)
		donor	recepient	
♂W-8224*	WL-BPR	57	13	81
♂W-8226*	WL-BPR	631	189	77
♂W-8229*	WL-BPR	57	11	84
♀W-8231*	WL-BPR	229	10	96
♂W-8237*	BPR-WL	50	205	20
♂W-8238*	BPR-WL	55	125	31
♂W-8240*	BPR-WL	80	296	21
♂W-8245*	BPR-WL	298	346	46
♂W-8246*	BPR-WL	47	436	10
♂W-8250*	BPR-WL	49	355	12
♀W-8236*	BPR-WL	0	268	0
♀W-8241*	BPR-WL	4	195	2
♀W-8244*	BPR-WL	32	252	32
♀W-8248*	BPR-WL	26	256	26
♀W-8251*	BPR-WL	5	235	1
♂W-8233	BPR-WL	86	502	15
♂W-8234	BPR-WL	79	429	16
♂W-8239	BPR-WL	60	248	20
♂W-8249	BPR-WL	17	290	6
♀W-8243	BPR-WL	162	98	62

* Indicates that 4–10 µL of blood was removed from the recipient prior to injection of donor primordial germ cells.

germ cells were depleted was greater than 90% (Table 15.4).

One of the advantages of using blastodermal cells to produce germline chimeras is their relative abundance. Large numbers of blastodermal cells can be isolated from unincubated eggs and prepared as dissociated cells within one or two hours. By contrast, isolation of primordial germ cells from blood requires considerable technical skill and the yield limits the number of cells available for genetic modification. Since the efficiency of incorporation of foreign DNA into cells is generally low, it may be difficult to obtain enough genetically modified primordial germ cells to inject into recipient embryos. While the difficulty of obtaining large numbers of genetically modified cells pertains to both blastodermal and primordial germ cells, it may be easier to obtain sufficient numbers of cells from a large population of blastodermal cells than a limited number of primordial germ cells. A potential technique for isolating transfected blastodermal cells has been developed using fluorescence-activated cell sorting. Cells transfected with a lacZ gene which encodes β-galactosidase have been sorted using the fluorescent substrate C_{12}-FDG and retained their ability to contribute to the germline (Speksnijder, 1995). By co-transfecting cells with lacZ and a gene of interest, it may be possible to produce a population of cells that is destined for the germline and in which the gene of interest is present at a high frequency. The relative efficacy of producing transgenic chickens using blastodermal cells or primordial germ cells will depend upon the time required to produce germline chimeras from them. For example, the yield of germline chimeras from chicks that hatch following injection of primordial germ cells exceeds the yield following injection of blastodermal cells by at least twofold (cf Tables 15.1. and 15.4.). If more than twice as many chimeras can be produced in an equal amount of time using blastodermal cells, however, it may be more effective to produce a larger number of blastodermal-cell chimeras and subsequently identify those with contributions to the germline.

The production of transgenic chickens via chimeric intermediates would be facilitated if the ability to proliferate blastodermal cells or primordial germ cells in culture while retaining their ability to colonize the germline were available. For example, if primordial germ cells could be induced to proliferate in vitro, the difficulty of obtaining these cells from donor embryos would be obviated. The ability to maintain either blastodermal cells or primordial germ cells in selective media would provide the opportunity to eliminate cells in which foreign genes had not incorporated and is an essential component of strategies for the production of "knock out" mutants or other gene targeting events by homologous recombination (Cappechi, 1989). To date, germline chimeras have been made with blastodermal cells cultured for 48 h (Etches et al., 1993 and Etches et al., 1996a) and somatic chimeras have been made with blastodermal cells cultured for up to 7 days (Pain et al., 1995). Homologous recombination has been accomplished in blastodermal cells (Liu, 1995) although the ability of these cells to colonize recipient embryos has not yet been tested. Nevertheless, it seems likely that techniques to manipulate blastodermal cells in culture will become available to facilitate the production of transgenic chickens through chimeric intermediates.

References

Cappechi, M.R. (1989) Altering the genome by homologous recombination. *Science* **244**, 1288–1292

Carsience, R.S., Clark, M.E., Verrinder Gibbins, A.M. and Etches, R.J. (1993) Germline chimeric chickens from dispersed donor blastodermal cells and compromised recipient embryos. *Development* **117**, 669–675

Didier, E., Didier, P., Fargeix, N., Guillot, J. and Thiery, J.-P. (1990) Expression and distribution of carbohydrate sequences in chick germ cells: a comparative study with lectins and the NC-1/HNK-1 monoclonal antibody. *Int. J. Dev. Biol.* **34**, 421–431

Etches, R.J., Carsience, R.S., Fraser, R.M., Clark, M.E. Toner, A. and Verrinder Gibbins, A.M. (1993) Avian chimeras and their use in manipulation of the avian genome. *Poultry Science* **72**, 882–889

Etches, R.J., Clark, M.E., Toner, A., Liu, G. and Verrinder Gibbins, A.M. (1996a) Contributions to somatic and germline lineages of chicken blastodermal cells maintained in culture. Manuscript in preparation

Etches, R.J. and Verrinder Gibbins, A.M. (1996b) Strategies for the production of transgenic chickens. In Expression and detection of recombinant genes. Edited by R.S. Tuan, Humana Press, Totowa, NJ (in press)

Eyal-Giladi, H. and Kochav, S. (1976) From cleavage to primitive streak formation: a complementary normal table and a new look at the first stage of the development of the chick I. General morphology. *Dev. Biol.* **49**, 321–337

Fargeix, N., Didier, E. and Didier, P. (1981) Early sequential development in avian gonads. An ultrastructural study using selective glycogen labelling in the germ cells. *Reprod. Nutr. Devel.* **21**, 479–496

Fujimoto, T., Ninomiya, T. and Ukeshima, A. (1976) The origin, migration and morphology of the primordial germ cells in the chick embryo. *Anat. Rec.* **185**, 139–154

Ginsburg, M. and Eyal-Giladi H. (1987) Primordial germ cells of the young chick blastoderm originate from the central zone of the area pellucida irrespective of the embryo forming process. *Development* **101**, 209–219

Ginsburg, M., Hochman, J. and Eyal-Giladi H. (1989) Immunohistochemical analysis of the segregation process of the quail germline lineage. *Int. J. Dev. Biol.* **33**, 389–395

Hahnel, A.C. and Eddy, E.M. (1986) Cell surface markers of mouse primordial germ cells defined by two monoclonal antibodies. *Gamete Research* **15**, 25–34

Halfter, W., Schurer, B., Hasselhorn, H.-M., Christ, B., Gimbel, E. and Epperlein, H.H. (1996) An ovomucin-like protein on the surface of migrating primordial germ cells of the chick and rat. *Development*, in press

Hamburger, V. and Hamilton, H.L. (1951) A series of normal stages of development of the chick embryo. *J. Morph.* **88**, 49–92

Kagami, H., Clark, M.E., Verrinder Gibbins, A.M. and Etches, R.J. (1995) Sexual differentiation of chimeric chickens containing ZZ and ZW cells in the germline. *Mol. Reprod. Dev.* **42**, 379–388

Karagenc, L., Ginsburg, M., Eyal-Giladi, H. and Petitte, J.N. (1995) Immunohistochemical analysis of germline segregation in preprimitive streak chick embryos using stage-specific embryonic antigen-1 (SSEA-1). *Poultry Science* **74**, Supplement 1, 26

Kochav, S., Ginsburg, M. and Eyal-Giladi, H. (1980) From cleavage to primitive streak formation: a complementary normal table and a new look at the first stage of the development of the chick I. General morphology. *Dev. Biol.* **79**, 296–307

Liu, G. (1995) Targetted modification in the genome of chicken blastodermal cells. Ph.D. Thesis, University of Guelph

Loveless, W., Bellairs, R., Thorpe, S.J., Page, M. and Feizi, T. (1990) Developmental patterning of the carbohydrate antigen FC10.2 during early embryogenesisin the chick. *Development* **108**, 97–106

Lucas, A.M. and Jamroz, C. (1961) Atlas of Avian Hematology, U.S. Department of Agriculture, Washington, D.C. p. 225

Maeda, S., Ohsako, S., Kurohmaru, M., Hayashi, Y. and Nishida, T. (1994) Analysis for the stage specific antigen of the primordial germ cells in the chick embryo. *J. Vet. Med. Sci.* **56**, 315–320

Meyer, D.B. (1964) The migration of primordial germ cells in the chick embryo. *Dev. Biol.* **10**, 154–190

Naito, M., Tajima., A., Yasuda, Y. and Kuwana, T. (1994) Production of germline chimeric chickens with high transmission rate of donor-derived gametes produced by transfer of primordial germ cells. *Mol. Reprod. Devol.* **39**, 153–161

Pain, B., Cochran, M., Clark, M.E., Sakurai, M., Samarut, J. and Etches, R.J. (1995) Identification and *in vitro* maintenance of putative pluripotential chicken embryonic stem cells. Proceedings of the 35th National Institute for Basic Biology Conference, Okazaki, Japan

Pardanaud, L., Buck, C. and Dieterelen-Lievre, F. (1987) Early germ cell segregation and distribution in the quail blastodisc. *Cell Differentiation* **22**, 47–60

Perry, M.M. (1987) Nuclear events from fertilization to the early cleavage stages in the domestic fowl (*Gallus domesticus*). *J. Anat.* **150**, 99–109

Petitte, J.N., Clark, M.E., Liu, G., Verrinder Gibbins, A.M. and Etches, R.J. (1990) Production of somatic and germline chimeras in the chicken by transfer of early blastodermal cells. *Development* **108**, 85–189

Petitte, J.N., Brazolot, C.L., Clark, M.E., Liu, G., Verrinder Gibbins, A.M. and Etches, R.J. (1993) Accessing the chicken genome using germline chimeras. In *Manipulation of the Avian Genome*. Edited by R.J. Etches and A.M. Gibbins. CRC Press, Boca Raton

Rowlett, K. and Simkiss, K. (1985) The surrogate egg. *New Scientist* **107**, 42–44

Rowlett, K. and Simkiss, K. (1987) Explanted embryo culture: *In vitro* and *in ovo* techniques for domestic fowl. *British Poultry Science* **28**, 91–101

Speksnijder, G.J. (1995) M.Sc. Thesis, University of Guelph, in preparation

Swartz, W.J. (1982) Acid and alkaline phosphatase activity in migrating primordial germ cells of the early chick embryo. *Anat. Rec.* **202**, 379–385

Thoraval, P., Lasserre, F., Coudert, F. and Dambrine, G. (1994) Production of germline chimeras obtained from Brown and White Leghorns by transfer of early blastodermal cells. *Poultry Sci.* **73**, 1897–1905

Urven, L.E., Erickson, C.A., Abbott, UK and McCarrey, J. (1988) Analyses of germline development in the chick embryo using an antimouse EC cell antibody. *Development* **103**, 299–304

Vick, L., Luke, G. and Simkiss, K. (1993a) Germ-line chimeras can produce both strains of fowl with high efficiency after partial sterilization. *J. Reprod. Fert.* **98**, 637–641

Vick, L., Li, Y. and Simkiss, K. (1993b) Transgenic birds from transformed primordial germ cells. *Proc. Roy. Soc., Lond. B.* **251**, 179–182

Watt, J.M., Pettite, J.M. and Etches, R.J. (1993) Early development of the chick embryo. *J. Morph.* **214**, 1–18

Wentworth, B.C., Tsai, H., Hallett, J.H., Gonzales, D.S. and Rajcic-Spasojevic, G. (1989) Manipulation of the avian primordial germ cells and gonadal differentiation. *Poultry Science* **68**, 999–1010

Wentworth, B.C., Tsai, H. and Wentworth, A.L. (1995) Primordial germ cells for genetic modification of poultry. *Proceedings of the Beltsville Symposium XX: Biotechnology's role in the genetic improvement of farm animals.* United States Department of Agriculture, Beltsville

Yoshinaga, K. Fujimoto, T., Nakamura, M. and Terakura, H. (1992) Selective lectin binding sites of primordial germ cells in chick and quail embryos. *Anat. Rec.* **233**, 625–632

PART II, SECTION C

16. The Use of Retroviral Vectors for Gene Transfer into Bird Embryo

Corinne Ronfort, Catherine Legras and Gérard Verdier

Centre de Génétique Moléculaire et Cellulaire, CNRS UMR 5534-INRA UA 810, Université Claude Bernard, 43 boulevard du 69622 Villeurbanne, France

Due to the particular physiology of the development of birds, introduction of recombinant DNA into the avian germ line involves a technology different from that used for insertion of genes into mammalian embryos. The particularities and techniques performed to overcome gene transfer difficulties in birds are reported in other chapters of this section. At the moment, despite a high number of cells in the embryo at oviposition, direct inoculation of retroviral vectors into early embryos seems one of the most efficient methods to obtain transgenic lines of birds. By this way, introduction of foreign genes into chickens has been successfully performed by using either avian replication – competent or avian replication – defective vectors.

Here, we review the basic principles of retroviral vector design and production. Prospects and limits of retroviral vectors technology are also discussed and the main data obtained for gene transfer into bird embryos are briefly reported.

Retrovirus Biology and Retroviral Vector Design

The retroviral particle contains two positive-strand RNA genomes. An unique feature of the retroviral life cycle is the conversion of viral RNA into DNA which ultimately becomes integrated into the host cell's genome. The integration results in the establishment of a provirus whose molecular configuration is precisely defined and invariant (Varmus, 1988). The retrovirus genome can be divided functionally into *trans*- and *cis*-acting sequences (Figure 16.1A). *Trans* functions are represented by *gag*, *pol* and *env* products. *Gag* encodes the structural components of the capsid, *pol* encodes the RNA-dependent DNA polymerase (reverse-transcriptase) and a nuclease (integrase), and *env* encodes the surface envelope glycoproteins. *Cis* sequences are found at the 5' and 3' ends of the genome (Figure 16.1A). The Long Terminal Repeat (LTR) contains transcription and integration signals. Initiation of reverse transcription depends on *cis*-acting sequences located in the *pbs* sites and the packaging of diploid RNA genomes into the viral particles depends upon a *cis*-acting sequence called Ψ sequence.

The first type of avian retroviral vectors is based on replication-competent (RC) viruses that retains *trans* and *cis* sequences, whereas the second type is based on replication-defective viruses in which the most part of retroviral genes were removed and replaced by foreign DNA to create recombinant vectors (Figure 16.1B). Thus, packaging of defective retroviral vectors into virions requires that the viral products deleted in the viral genome be supplied in some other way. The use of these replication-defective vectors is possible thanks to the generation of packaging cell lines that provide the *gag/pol* and *env* proteins (*trans* functions) needed to assemble infectious particles (Figure 16.2).

Replication-Competent (RC) Vectors (Figure 16.1B1)

The basic replication-competent vector (RCAS) was derived from the Schmidt-Ruppin A strain of Rous Sarcoma Virus (SR-RSV-A) in which the *src* gene and its flanking repeat sequences were removed and replaced by a fragment containing the *src* splice acceptor site and an unique *cla*I site suitable for insertion of the transgene (Hughes and Kosik, 1984;

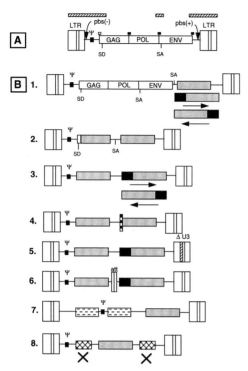

Figure 16.1. A. Structure of a provirus. Abbreviations: LTR: Long Terminal Repeat; Ψ: signal for packaging of the viral RNA; pbs(−): primer binding site for the minus-strand initiation; pbs(+): primer binding site for the plus-strand initiation; SD: splice donor site; SA: splice acceptor site; GAG, POL and ENV: viral protein encoding sequences, □: initiation codon for translation, ■: stop codon of translation. Shaded boxes above the provirus indicate the location of *cis*-acting sequences.
B. Vectors. (1) Replication-competent vectors containing the transgene placed under transcriptional control of either the retroviral promoter (LTR) or an internal promoter in the two possible orientations. (2–8) Replication-defective vectors corresponding to double expression vectors (DE): splice vectors (2), vectors with an internal promoter (VIP vectors) (3), or IRES vectors containing an internal ribosome entry site (4). In SIN vectors, U3 region of the 3′ LTR has been partially deleted (5) whereas an additional integration sequence *att* has been introduced into att vectors (6). Vectors with repetitive sequences contain either direct repeat sequences (E-vectors) or inverted complementary sequences (IC vectors) (7) while gene-targeted vectors muct lead to integration using homologous recombination process (crosses) instead of retroviral random integration (8). □: viral sequences; ▨: genes to be transferred; ■: non retroviral promoter; ▩: IRES sequences; ▨: repetitive sequences; ▨: sequences having homologies with genomic DNA sequences.

and Hughes *et al.*, 1987). This splice acceptor site permits expression of the transgene from the promoter located within the viral LTR, *via* a spliced subgenomic mRNA. RCAS vectors were improved: (1) by introducing an eucaryotic leader and an ATG

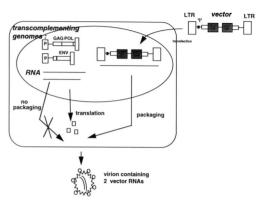

Figure 16.2. Basic principles of retroviral vectors production by a packaging cell line. Packaging cell line contain the transcomplementing genomes that encode the *gag*, *pol* and *env* viral proteins but are deleted of the Ψ sequence. Since the Ψ sequence is required for efficient packaging of RNA into virions, the resulting packaging cell lines produce "empty virions", i.e. virions that do not contain viral RNA. When a Ψ+ retroviral vector is introduced into the packaging cells, its RNA is selectively packaged into virions. P: non retroviral promoter. White boxes: viral genes. Shaded boxes: genes to be transferred.

initiation codon to promote the translation of the transgene (Hughes *et al.*, 1990); (2) by substituting the *gag* and *pol* genes with the corresponding region of the Bryan high-titer strain of RSV (BH-RSV) to enhance replication (Petropoulos and Hughes, 1991; and Federspiel and Hughes, 1994). Moreover, to reduce the oncogenic potential of RCAS vectors, the LTRs were replaced by the endogenous RAV-0 LTRs (Greenhouse *et al.*, 1988). Resulting vectors were named RCOS.

Other RC vectors, derived from the above-mentioned ones, correspond to the RCAN vectors that lack the splice acceptor site. In these vectors, the expression of the transgene was allowed from an internal promoter. The cassettes containing the trangene placed under transcriptional control of an internal promoter are inserted in place of v-*src*, in the same orientation to the 5′ LTR, or in reverse orientation. RCON vectors are similar to RCAN except that they contain the LTR from RAV-0 (Greenhouse *et al.*, 1988). Various ubiquitous promoters, available for expression in a wide range of cell types, have been introduced into RCAN or RCON vectors, as the chicken β-*actin* promoter (Petropoulos and Hughes, 1991 and Petropoulos *et al.*, 1992), the tyrosine kinase (*tk*) promoter (Hughes *et al.*, 1990 and Akiyama *et al.*, 1994) or the mouse metallothionein (mMT) promoter (Petropoulos and Hughes, 1991). The expression of the transgenes from these internal promoters was increased when the cassette was inserted in the forward orientation (Petropoulos and Hughes, 1991). Two tissue-specific promoters have also been used, the chicken squeletal muscle α-actin (α_{sk}-*actin*) promoter whose expression is restricted to striated muscle (Petropoulos *et al.*, 1992) and the rat phosphoenolpyruvate carboxykinase (PEPCK) promoter that is restricted to liver (Cook *et al.*, 1993).

Avian Leukosis and Sarcoma Viruses (ALSV) retroviruses have been divided into five envelope subgroups on the basis of distinct receptor recognition (Wesis, 1984). To produce RC vectors pseudotyped in different subgroups, the *env* gene of these different vectors was removed and replaced by the *env* gene from either the B or D subgroup of RSV, to generate RCAS(B), RCAN(B), RCOS(B) RCON(B), or RCAS(D), RCAN(D), RCOS(D), RCON(D) vectors, respectively (Hughes *et al.*, 1987, 1990).

The major advantages of these RC vectors are (1) the high titer of viral vector stocks (10^6 particles per ml for RCAS and RCAN vectors and 10^5 particles per ml for RCOS and RCON vectors); and (2) the efficient spread of these RC vectors to most cell types in culture or to embryo. Conversely, the major disadvantages of these RC vectors are: (1) the limited size of transgene that can be introduced (about 2 to 2.5 kb); and (2) the disease associated with chronic viral infection.

Replication-defective Retroviral Vectors

There are two major origins of avian defective retroviral vectors. The first one is derived from the ALSV and the second from the ReticuloEndotheliosis Viruses (REV).

ALSV-based retroviral vectors were derived: (1) from the SR-A strain of RSV by substituting viral genes by sequences to be transferred (Gray *et al.*, 1988 and Galileo *et al.*, 1990); and (2) from natural defective oncogenic viruses as either the Avian Erythroblastosis Virus-ES4 (AEV-ES4) (Benchaibi *et al.*, 1989; Stoker *et al.* 1990) or the UR2 (Garber *et al.*, 1991), in which the viral oncogenes were deleted and replaced by the transgene. A major advantage of ALSV-based retroviral vectors is that viral stocks can be concentrated 100 folds by ultracentrifugation and stored at -80 C without decrease of viral titers.

REV-based retroviral vectors were mainly derived from the Spleen Necrosis Virus (SNV). Similarly to previous approaches, the structural *gag*, *pol* and *env* genes were replaced by the transgene (Shimotohono and Temin, 1981; Bandyopadhyay and Temin, 1984; and Hippenmeyer and Krivi, 1991).

While the ALSV-based vectors are strictly avitropic and then should offer great safety for investigations, one should be cautious in the use of REV retroviruses because their host range includes rat, dog and primates (Koo *et al.*, 1991).

Description of Classic Double Expression Vectors (DE Vectors). To express two genes, three main vector designs have been developed: vectors utilizing splice sequences, vectors using an internal promoter, and finally, vectors utilizing a ribosome entry site.

Vectors with Splice Sequences (*Splice Vectors*) (Figure 16.1B2) In these splice vectors, the promoter located in the 5' LTR is used to express both genes. One gene is inserted in 5' position and expressed from the genomic viral RNA, while the second gene, inserted in 3' position, is expressed from the subgenomic RNA. A set of ALSV splice vectors, called NL vectors, has been designed in our laboratory. They carry the *neo* selectable gene, conferring resistance to G418, and the *lacZ* reporter gene encoding for the β-Galactosidase (Cosset *et al.*, 1991a). These NL vectors were improved by exchanging the *cis*-acting sequences from AEV with the corresponding sequences of the Rous Associated Virus-2 (RAV-2) to enhance the production of vector particles, the expression of transgene (Cosset *et al.*, 1991a) and the stability of proviral vector expression (Molina *et al.*, 1995). Titers were ranging from 10^5 to 10^6 particles per ml (Cosset *et al.*, 1991a, 1993).

Vectors with Internal Promoters (*VIP-vectors*) (Figure 16.1B3) Retroviral LTR are strong promoters able to express in a great number of tissues. Insertion of an internal promoter into the vector has the potential advantage of cell-specific expression of the transgene. Several double expression vectors have been engineered in which the first gene was expressed from the LTR and the second gene was expressed from an inserted internal promoter.

SNV recombinant retroviruses in which the chicken *tk* promoter (Emerman and Temin, 1984a, 1986a,b), the mouse MTI promoter or the SV40 early promoter (Emerman and Temin, 1986a) were inserted together with two selectable genes, have been constructed. Cell clones infected with such retroviruses usually do not fully express both genes. Indeed, the 3' gene is usually suppressed when expression of the 5' gene is selected, and *vice versa*. Large deletions or other rearrangements in the integrative proviruses do not account for the suppression of the second selectable gene (Emerman and Temin, 1984a). The amount of suppression is rather correlated with a decrease in the amount of steady-state RNA transcribed from the suppressed gene's promoter (Emerman and Temin, 1986a). The amount of suppression was neither dependent on the relative strength of the promoters (Emerman and Temin, 1986a; and Hippenmeyer and Kiwi, 1991) nor dependent on the distance between the promoters (Emerman and Temin, 1986b). This suppression is *cis*-acting, epigenetic and reversible when the virus is recovered (Emerman and Temin, 1984a). The mechanism of suppression is unknown.

To circumvent this problem, Bandyopadhyay and Temin (1984) have inserted the *tk* promoter either in the same orientation as the 5' LTR or in the opposite orientation. In the first orientation, virus production was efficient, whereas *tk* expression was inefficient. When the promoter was in opposite orientation, *tk* expression was efficient probably as a result of the absence of interference by the opposite promoter but virus production was inefficient.

In some cases, deletion of internal structures were observed (Bandyopadhyay and Temin, 1984; and Emerman and Temin, 1984b).

Such transcriptional interference process has also been described *in vivo*. The REV LTR was actively transcribed in a variety of tissues from transgenic chickens, while the HSV-1 internal *tk* gene promoter was not expressed (Briskin *et al.*, 1991).

The same transcriptional interference process was encountered with ALSV-derived vectors in which deletion of one unit was frequently observed when the second unit was selected for expression (Drynda and Verdier, unpublished data). However, a functional ALSV double expression vector with an internal promoter have been described once (Flamant *et al.*, 1993).

Vectors with Internal Entry Site (IRES-vectors) (Figure 16.1B4) Picornavirus have been shown to contain a sequence element in their 5' region that acts as an internal ribosome entry site (IRES), allowing cap-independent initiation of translation: ribosomes bind internally at the initiating AUG without scanning the 5' non translated region of the transcript (Jang *et al.*, 1988).

A double expression SNV vector containing the IRES sequence of the Encephalomyocarditis virus (EMCV), instead of a splice acceptor site, was constructed. Both genes can therefore be transcriptionally expressed from the LTR with one of the gene being translated from an IRES sequence. In infected cells, a single bicistronic RNA derived from an intact wild type provirus was able to express both genes efficiently. Such virus was found at titer as high as 2×10^5 particles per ml (Koo *et al.*, 1992).

Ghattas *et al.* (1991) have compared splicing vectors, vectors with internal promoters and IRES vectors derived from ALSV viruses. IRES vectors were found to be more effective than the two other vectors to express two exogenous genes.

Design of Retroviral Vectors with Higher Safety. The above-reported classic double expression vectors present several disadvantages. One of them is that the retroviral LTR enhancer can influence internal promoters, making regulation of the passenger gene difficult. Another problem is that the promoter and enhancer sequences into LTR can activate the transcription of downstream cellular genes after integration into the genome of target cells. The last problem is that the proviral vector retains all *cis*-acting sequences after integration into target cells. Thus, either in the event of infection of these target cells with a replication-competent virus or in the event of complementation of vector provirus with resident endogenous retroviral structures, the possible spread of retroviral vector could occur. To address these problems, various strategies have led to develop more safety retroviral vectors.

Self-Inactivating Vectors (SIN vectors) (Figure 16.1B5) SIN retroviral vectors contain a deletion of the promoter and/or the enhancer of the 3' LTR. This deletion does not impair vector RNA transcription in the producer cells. Since the 3' U3 region serves as template for both U3 during reverse transcription, proviruses that lack U3 from both LTR are formed after one round of replication. Having lost its promoter, the retrovirus cannot further replicate or influence nearby promoters. Passenger genes of the vector can therefore be expressed from internal promoters.

REV SIN vector constructions have involved deletion of all the U3 sequences, except for 10 bp required for integration, from the 3' LTR (Dougherty and Temin, 1987). This deletion resulted in inability to propagate virus efficiently. Indeed, 3' end processing did not occurred at the end of R despite the sequence essential for virus polyadenylation was still present. By adding a SV40 poly(A) site to the vector at the end of the U5, Dougherty and Temin (1987) obtained higher-titer virus stock. However, titer was still lower by a factor of 5 than the non deleted vector. REV SIN vectors were also found to be genetically unstable since reconstitution of the deleted U3 in several SIN vectors was observed at high frequency (Olson *et al.*, 1992). The sequences that reconstituted the U3 region of the vector LTR were derived from sequences present in the producer cell. To overcome this problem, almost completely U3-free retroviral vectors were developed. The promoter and enhancer of the 5' LTR was replaced with those of the Cytomegalovirus (CMV) immediate-early promoter (Olson *et al.*, 1994) while the U3 of 3' LTR was deleted. Such vectors were found to be genetically stable; they replicate with efficiencies similar to those of vector possessing two wild type LTR.

ALSV SIN vectors have been developed once. In these vectors, the U3 LTR was deleted for 39 nucleotides removing the CCAAT box. This vector enables the production of a rather low titer that could be explained by the observation that SIN RNA are slightly under-represented in viral particles, probably as a result of deletions in the downstream U3 LTR. Therefore, RNA packaging could be a limiting step in the replication of these vectors (Flamant et al., 1993).

Vector with Additional Integration Sequences (att vectors) (Figure 16.1B6). Duplication of LTR produces an *att* sequence at the junction of the LTR doublet. This sequence is recognized by the viral integrase which results in the integration of the viral DNA into the host cell DNA (Varmus, 1988). Chebloune et al. (1990) and Panganiban and Talbot (1993) have constructed ALSV- and REV-based vectors respectively carrying an additional integration sequence located internally in the vector. Data have shown that such an internal sequence could be functional. The ALSV-derived vectors were based upon double expression vectors, expressing one selectable marker gene from the LTR promoter, and a second selectable gene from an internal SV40 promoter. Upon infection, such vectors were integrated: (1) either by the natural *att* sequence allowing expression of the LTR-driven gene; (2) or by the additional *att* sequence and the 3' LTR together, allowing expression of the internal promoter-driven gene. These vectors were produced with titers as high as 10^4 particles per ml (Chebloune et al., 1990; and Drynda and Verdier, unpublished data).

Vectors with Repetitive Sequences (Figure 16.1B7): Julias et al. (1995) have developed novel self-inactivating retroviral vectors, called E^- vectors, that use the template-switching properties of reverse transcriptase. One of these REV-based retroviral vectors contains large direct repeats, represented by the *neo* gene, flanking the Ψ packaging signal. Selection of infected cells for resistance to G418 led to the deletion of Ψ sequence in the resulting provirus.

A similar type of vectors, named IC vectors, able to delete one part of their genome during replication have been designed in our group. They contain inverted complementary sequences susceptible to form an hairpin-like structure. During reverse transcription, this hairpin structure may promote template jumping by reverse transcriptase and the deletion of inverted repeats. If the inverted complementary sequences are inserted flanking an essential *cis*-acting sequence such as the Ψ signal, we hope to obtain the deletion of both the hairpin structure and the Ψ signal (Faraut and Legras unpublished data).

Gene-targeted Vectors (Figure 16.1B8) The optimal vector would permit control of transcription of the foreign genes, allowing their expression in the target cells, but inhibiting it in non target cells, and permitting regulation by developmental timing or hormonal stimuli. The most basic approach would be to customize existing tissue-specific or hormone-specific transcription controls. However, cellular enhancers of transcription and locus controls are often ill-defined and may reside far from the desirable promoters. Any transgene targeted to its native site should consequently be expressed appropriately. Other potentials of gene-targeted vectors include the generation of animals homozygous for disruptions or site-directed mutations of specific genes. Homologous recombination has been successfully used to target the integration of DNA transferred to cells by physical means (Capecchi, 1989) but the frequencies were rather low. A combination of the efficient gene delivery system provided by retroviral vectors with homologous recombination should improve the frequency of gene targeting. The feasibility of such an approach has already been demonstrated (Ellis and Bernstein, 1989). To tentatively develop gene-targeted retroviral vectors, we have used a replacement strategy that leads to the integration of vector sequences located inside the homologous region while sequences at the borders of the vector should be lost. Moreover, the retroviral vectors have been deleted for the *att* sequences, resulting in inhibition of integration by the natural viral mechanism. With such vectors, a great part of retroviral sequences (especially *cis*-acting sequences) should be eliminated after integration by homologous recombination (Ronfort unpublished data).

Packaging Cell Lines

Packaging cells are designed to synthetize all retroviral proteins necessary for the production of infectious particles. Introduction of a retroviral vector containing *cis*-acting retroviral signals into packaging cells results in production of viruses which can infect target cells (Figure 16.2). Two types of packaging cell lines have been engineered (Figure 16.3): one type is used for the production of ALSV-based vectors; the other for the production of REV-based vectors.

Design of Safe Packaging Cell Lines

The packaging cells should produce vector particles but not replication-competent (RC) viruses because of viremia, associated diseases and pathologies. Thus, the transcomplementing genomes (i.e., genomes encoding *the gag/pol* and *env* products) must be deleted for *cis* functions. The first avian transcomplementing genomes were designed by deleting the

Ψ signal for the packaging of viral RNA (Stoker and Bissell, 1988 and Watanabe and Temin, 1983; Figure 16.3a, 3d). However, although deletion of the Ψ signal reduces the amount of virus RNA that is packaged into virions, the inhibition is far from complete. Evidence that deletion of the Ψ signal is not sufficient to prevent packaging comes from studies with a Ψ-deleted ALSV genome that was packaged into particles at rates corresponding to approximately 1 to 2% of a wild-type RNA (Girod et al., 1995). Moreover, in the REV-based C3 line, this deletion does not completely prevent spread of a RC virus (Hu et al., 1987). In the Q2bn packaging line (Stoker and Bissell, 1988), the pbs(−) sequence has also been deleted. Therefore, the resulting RNA cannot be reverse-transcribed in case of misincorporation into viral particles. However, only one recombinant event between the overlapping sequences of the packaging-competent vector and the packaging-defective transcomplementing genome is sufficient to restore the Ψ signal resulting in replication-competent virus production (Figure 16.3a). While there is no release of replication-competent viruses by this packaging cell line, such a virus could arise when the packaging line was transfected with the replication-defective vector.

The pHF-g clone (Savatier et al., 1989, Figure 16.3b) was constructed from a transcomplementing genome in which not only the packaging signal was deleted but also the 3' region (i.e. LTR and pbs(+)), the last one being replaced by an heterologous polyadenylation site. Thus, even if the resulting RNA is packaged, it cannot be reverse-transcribed. Moreover, sites required for integration of the resulting DNA are missing. Thus, two recombination events should be necessary to generate replication-competent viruses. It has been shown that this packaging cell line did not produce replication-competent viruses. However, vector titers obtained from this packaging line were always 10 to 20 times lower than those obtained with other lines (see below).

In the *Isolde* and *SEnta* packaging cell lines (Cosset et al., 1990; 1992, Figure 16.3c), to further reduce the probability of generating replication-competent viruses *via* intermolecular recombinations, the viral genes were expressed from two separate plasmids, one expressing the *gag/pol* genes and the other expressing the *env* gene. In both plasmids, the packaging sequence was deleted and the 3' LTR was replaced by the polyadenylation sequence from heterologous viruses. These alterations were shown to fully abolish the generation of replication-competent viruses.

Finally, the most effective improvement involves more extensive deletion in the two separate transcriptional units. This new design of transcomplementing genomes requires the expression of *gag/pol* and *env* genes by using heterologous promoters, as exemplified by the REV-derived DSN and DAN cell lines (Dougherty et al., 1989; Figure 16.3e).

Packaging of Unexpectable Viral Genomes. Another safety issue concerns contaminant particles that carry either the *gag/pol* or the *env* genes; such particles are difficult to detect because they are not replication-competent particles. The first particles of that kind are those bearing the transcomplementing genomes. Their transfer to target cells along with the vector was illustrated by the *Isolde* packaging cell line. As each of these genomes carry a selectable gene (Figure 16.3c), titers of contaminant particles could be estimated between 0 and 6 infectious particles per ml of supernatants (Girod et al., 1995). It was hypothesized that transfer of these genomes would involve: (1) misincorporation of these packaging-defective RNA into virions; and (2) recombination events during the reverse transcription steps following copackaging of a transcomplementing RNA and a vector RNA (Girod et al., 1996). The second type of contaminant virus corresponds to particles having retroviral endogenous RNA. Most chicken cell lines contain endogenous elements which are structurally related to their exogenous counterparts (Smith 1986, Ronfort et al., 1991). In a chicken-derived packaging cell, an ALSV-based vector stock was found to be contaminated with 1% of particles bearing endogenous RNA (Ronfort et al., 1995a, 1995b). Thus, to prevent production of these two types of particles, it is recommended: (1) to reduce homologous overlaps between the vector and the transcomplementing genomes in order to reduce risks of recombination and therefore to reduce the transfer of the transcomplementing genomes; (2) to use cells that do not contain endogenous proviruses, such as the permanent quail cells (Moscovici et al., 1977) or cells in which these retroviral endogenous elements have reduced homologous overlaps with the vector sequences.

Titers According to the Packaging Cell Lines

Another consideration in the choice of a packaging cell line is the titer at which the retroviral vector is generated. For the REV-based lines, virus titers produced by the D17-C3 packaging line were comprising between 10^2 to 10^3 particles per ml (Watanabe and Temin, 1983). The C3A2 subclone (Dornburg and Temin, 1988) was found to produce vector titers of 10^4 to 10^5 particles per ml. A 10^7 particles per ml titer was reported once with this C3A2 line (Dornburg and Temin, 1988). The DSN line (Dougherty et al., 1989) was reproducibly found to produce vectors titers of 10^4 to 10^5 particles per ml.

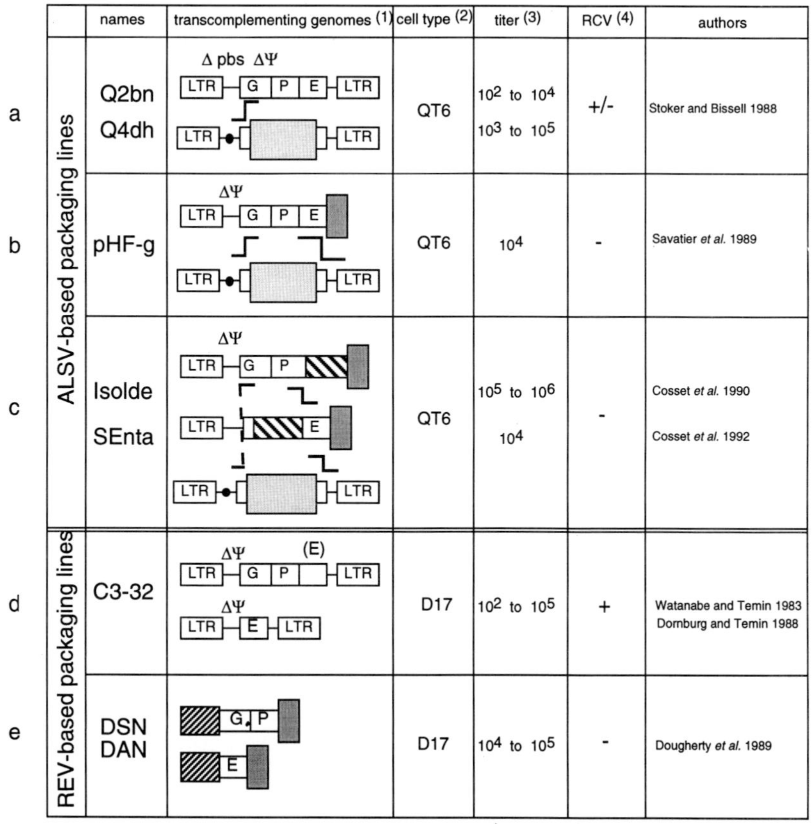

		names	transcomplementing genomes (1)	cell type (2)	titer (3)	RCV (4)	authors
a	ALSV-based packaging lines	Q2bn Q4dh	Δpbs ΔΨ [LTR]-[G|P|E]-[LTR] / [LTR]-•-[▨]-[LTR]	QT6	10^2 to 10^4 / 10^3 to 10^5	+/−	Stoker and Bissell 1988
b		pHF-g	ΔΨ [LTR]-[G|P|E]-▨ / [LTR]-•-[▨]-[LTR]	QT6	10^4	−	Savatier et al. 1989
c		Isolde SEnta	ΔΨ [LTR]-[G|P|▨]-▨ / [LTR]-[▨|E]-▨ / [LTR]-•-[▨]-[LTR]	QT6	10^5 to 10^6 / 10^4	−	Cosset et al. 1990 Cosset et al. 1992
d	REV-based packaging lines	C3-32	ΔΨ (E) [LTR]-[G|P]-[LTR] / ΔΨ [LTR]-[E]-[LTR]	D17	10^2 to 10^5	+	Watanabe and Temin 1983 Dornburg and Temin 1988
e		DSN DAN	▨[G|P]▨ / ▨[E]▨	D17	10^4 to 10^5	−	Dougherty et al. 1989

Figure 16.3. Currently available avian packaging cell lines.
(1) Structure of the transcomplementing genomes that provide all the proteins required for viral particle assembly. Potential points for homologous recombination between the transcomplementing genomes and the vector genome leading to replication-competent virus are shown by ⌐; they are only indicated for the ALSV-based packaging cell lines. G, P and E: *gag, pol* and *env* encoding sequences. (E): Presence of an *env* gene that is not functional. Δpbs and ΔΨ indicate deletion of the pbs(−) sequence and of the Ψ packaging signal respectively. • Ψ+. : gene intended to be transferred.■: heterologous signal of polyadenylation; ▨ selectable gene; ▨: non retroviral promoter.
(2) Quail QT6 cells or canine D17 cells.
(3) infectious particles per ml.
(4) RCV = replication competent virus, : absent, +: present, +/ : present at low level.

For the ASLV-based packaging cell lines, titers of 10^2 to 10^5 have been obtained with the Q2bn and Q4dh lines (Stoker and Bissell, 1988). By using the *Isolde* packaging cell line, titers of 10^5 to 5×10^5 infectious particles per ml can be reproducibly obtained (Cosset et al., 1991a). Moreover, a 50 to 100-fold concentration can be expected by ultracentrifugation leading to titers as high as 5×10^6 to 3×10^7 particles per ml (Thomas et al., 1992).

Introduction of a vector by an infection of the packaging cells instead of a transfection was shown to facilitate the creation of high-titer vector-producing cells. Indeed, infection lead to specific integration of the vector in active regions of the chromatin while transfection frequently displays some rearrangements of the constructs. However, since packaging lines that produce one given envelope specificity are resistant to infection with viruses coated with the same envelope, infection of the packaging cell line is only possible when using pseudotypes of the vector, i.e. vector viruses coated with envelopes from other origin than those produced by the packaging cells. The *Isolde* cell line produce vectors as particles of subgroup A (Cosset et al., 1990). ALSV-based packaging cell lines that provide B, C and E ALSV envelopes have also been generated (Cosset et al., 1992). The availability of non-A subgroup pseudotypes of retroviral vectors offers the possibility of infecting *Isolde* cells. This technique could give rise, on average, to 10 times higher titers (Cosset et al., 1993).

In order to generate an extremely high-titer producer cell line, the ping-pong strategy described with the murin system (Bestwick et al., 1988) has been adapted by cocultivating two avian vector-producer cell lines having two different host-ranges (A and E). In such a coculture, one cell type can be infected by virus packaged in the other partner and *vice versa*. This should allow an increase in the copy number of the proviral genome and then permit an increase in viral titers. Such experiments have been successfull with Murine Leukemia Virus (MuLV)-based vectors (Bestwick *et al.*, 1988; and Kozak and Kabat, 1990). Viral titers up to 10^{10} infectious particles per ml have been reported (Bodine *et al.*, 1990). However, this method failed to amplify avian vectors in ping-pong protocols by using ALSV-based packaging cells (Cosset *et al.*, 1993). Data suggest that empty particles or envelope protein released by packaging cells inhibit the amplification of the vector. Moreover, ping-ponging caused emergence of replication-competent viruses.

Other Developments in the Design of Packaging Cell Lines

An important improvement could come with the use of Vesicular Stomatitis Virus G (VSV-G) pseudotyping procedure described for the MuLV-based vectors (Burns *et al.*, 1993 and Yee *et al.*, 1994). Attempts to concentrate retroviral vectors by physical methods such as ultracentrifugation and filtration may result in loss of infectious virus, due to the instability of the *env* protein. VSV is a member of the rhabdovirus family; it can be concentrated by ultracentrifugation without loss of infectivity. Thus the method consists of constructing pseudotypes in which the genome of one virus (i.e. retrovirus) is coated by the envelope protein of a second virus (VSV). The great advantage of the method lies in the ability to concentrate the generated virus by ultracentrifugation to titers ranging from 10^8 to 10^9 particles per ml. As VSV-G interacts with a phospholipid component of the cell membrane to mediate viral entry, VSV-G pseudotype vectors have been shown to infect not only mammalian cells but also many other cell types.

Another possibility to increase the viral titer could be the use of the bovine papillomavirus (BPV) gene for amplification of the transcomplementing plasmid. To amplify the packaging plasmids, this sequence was inserted in the transcomplementing constructs used to generate a murine packaging cell line (Takahara *et al.*, 1992). High-titer vector viruses were produced: 5×10^5 to 1×10^6 particles per ml instead of 25×10^4 particles per ml reported for the parental transcomplementing plasmids that lacked the BPV gene for amplification.

Another improvement of packaging cell lines would be to control the cell type in which the virus enters. The retroviral component that mediates interaction with receptor and entry of the virus into the cell is the envelope glycoproteins. As described above, ALSV retroviruses have been divided into 5 envelope subgroups on the basis of distinct receptor recognition. The availability of ALSV-based packaging cell lines that provide different subgroups of envelope (subgroup A, B, C and E (Cosset *et al.*, 1990; 1992)) is of particular interest since it can extend the tropism of vectors to cells that were previously not susceptible to infection with a particular subgroup. Therefore, ALSV-based vectors can now be used not only for infection of most chicken strains but also for infection of other avian species. Moreover, it is possible to restrict the tropism of vector to specific cell types that bear the corresponding receptor while the neighbour cells do not (Jaffredo *et al.*, 1993).

A complementary approach is the control of the tissue specificity for gene transfer. It might be possible to construct packaging cell lines providing viral envelopes modified for their binding domain with the usual receptor, so that the modified product can recognize specific cell surface molecules other than the currently used receptor. ALSV subgroup A mutant envelopes in which a small RGD-containing peptide was inserted, which is a ligand for cell surface receptors of the integrin superfamilly, have been constructed. The viral particles bearing this peptide were shown to specifically use integrins as receptors to initiate infection of mammalian cells (Valsesia-Wittman *et al.*, 1994). However, because of the low efficiency of this envelope, today, these data should be considered as a general principle for reorientation of viral tropism by minimal modifications of the receptor binding domain of the retroviral envelope protein.

Gene Transfer into Bird Embryos by Using Retroviral Vectors

Retroviral vectors have been mainly used in somatic transgenesis, for gene expression studies during development by using reporter genes (Fekete and Cepko, 1993a; Garber *et al.*, 1991; Thomas *et al.*, 1992 and Jaffredo *et al.*, 1993), for cell lineage (Gray *et al.*, 1988 and Mikawa *et al.*, 1991), for gene expression in chickens (Souza *et al.*, 1984; Hippenmeyer *et al.*, 1988) or inhibition of gene expression in specific cell types by using antisense sequences (Galileo *et al.*, 1992), and for viral protection against viruses (Hunt *et al.*, 1988; Chebloune *et al.*, 1991 and Cosset *et al.*, 1991b).

For germinal transgenesis, the first germ-line insertions have beeen obtained by infecting unincubated embryos with replication-competent recombinant ALSV carrying the coding sequences of RAV1 (Salter *et al.*, 1986; Salter and Crittenden, 1989 and

Crittenden et al., 1989). Viremic G_0 males transmitted provirus sequences to G_1 progeny at frequencies ranging from 1 to 11%. RC vectors expressing the neomycin phosphotransferase (NPT-II) gene or the bovine Growth hormone (GH) gene (Kopchick et al., 1991 and Chen et al., 1990) have been also used to inoculate young embryos. The frequency of transmission varied from 0 to 40%. However, practical applications of retroviral vector mediated germinal transgenesis require replication-defective vectors. Unincubated embryos were microinjected with REV-based vectors carrying the tk and Growth Hormone (GH) genes. About 8% of G_0 males were found mosaic in their semen and all analyzed mosaic males transmitted vector sequences to G_1 progeny at a frequency that varied from 2 to 8% (Bosselman et al., 1989; 1990). Lee and Shuman (1990) have also reported successful gene transfer into quails by using a REV-based vectors, but a low efficiency of germinal transmission was obtained (0.06% of G_1 transgenic quails). Another success, in chicken, has been obtained by using an ALSV-based retroviral vectors, carrying the neo selectable gene and the reporter lacZ gene. Microinjection of this ALSV-based retroviral vectors beneath unincubated embryo blastoderms results in infection of germ line stem cells. One male was found mosaic for its semen and vector sequences of this male were transmitted to G_1 progeny at the frequency of 2.7% (Dambrine et al., 1993 and Thoraval et al., 1995).

More recently, collection of embryonic stem cells or primordial germ cells (PGCs) as tools for gene transfer offers several advantages and different techniques can then be used to introduce foreign DNA into the bird genome (Vick et al., 1993; Allioli et al., 1994, 1996; and see other chapters of this section). Nevertheless, retroviral infection remains an efficient way to transfer foreign DNA into such isolated embryonic cells prior to their introduction into a recipient embryo.

Conclusion

Defective avian retroviral vector technology presents two main interests. The first concerns somatic or germinal gene transfers in early bird embryos. Whatever were the goal of these studies, the authors concentrated their efforts to obtain an efficient system for gene transfer: vectors and packaging cell lines. A second interest of avian retroviral vector system concerns their use as models to determine the safety of retroviral mediated gene transfer. In fact, retroviral vectors are presently used to gene transfer for research, for industrial and clinical purposes and mainly for human gene therapy. They offer many peculiar advantages as gene transfer vehicles including efficient delivery and integration of a foreign gene into the host genome. However, several safety problems had to be taken into consideration before these vectors can be utilizable tools. The major problem associated with the use of retroviral vectors is the generation of infectious viruses that can be indefinitely transmitted. Available avian retroviral vectors as well as packaging cell lines can help to define safety parameters of retroviral mediated gene transfer. Conversely, the avian retroviral systems can be taken as a basis to develop constructions of novel retroviral vectors which will be more efficient with optimal targeted infection, integration and expression.

Acknowledgements

We thank all members of our group for data obtained and helpful discussions regarding the related literature.

References

Allioli, N., Thomas, J.-L., Chebloune, Y., Nigon, V.M., Verdier, G. and Legras, C. (1994) Use of retroviral vectors to introduce and express the beta-galactosidase marker gene in cultured chicken primordial germ cells. Dev. Biol. 165, 30–37

Allioli, N., Verdier, G. and Legras, C. (1995) Use of gonadal primordial germ cells (PGCs) as tool for gene transfer in chicken. Chap. 42: Strategies in avian transgenesis. Expression and detection of recombinant genes. Methods in Molecular Biology (in press)

Akiyama, T., Whitaker, B., Federspiel, M., Hughes, S.H., Yamamoto, H., Takeuchi, T. and Brumbaugh, J. (1994) Tissue-specific expression of mouse tyrosinase gene in cultured chicken cells. Exp. Cell Res. 214, 154–162

Bandyopadhyay, P.K. and Temin, H.M. (1984) Expression of complete chicken thymidine kinase gene inserted in a retrovirus vector. Mol. Cell. Biol. 4, 749–754

Benchaïbi, M., Mallet, F., Thoraval, P., Savatier, P., Xiao, J.H., Verdier, G., Samarut, J. and Nigon V.M. (1989) Avian retroviral vectors derived from avian defective leukemia virus: role of the translational context of the inserted gene on efficiency of the vectors. Virology 169, 15–26

Bestwick, R.K., Kozak, S.L. and Kabat, D. (1988) Overcoming interference to retroviral superinfection results in amplified expression and transmission of cloned genes. Proc. Natl. Acad. Sci. USA 85, 5404–5408

Bodine, D.M., McDonagh, K.T., Brandt, S.J., Ney, P.A., Agricola, B., Byrne, E. and Nienhuis, A.W. (1990) Development of a high-titer retrovirus producer cell line capable of gene transfer into rhesus monkey hematopoietic stem cells. Proc. Natl. Acad. Sci. USA 87, 3738–3742

Bosselman, R.A., Hsu, R.-Y., Boggs, T., Hu, S., Bruszewski, J., Ou, S., Kozar, L., Martin, F., Green, C., Jacobsen, F., Nicolson, M., Schultz, J.A., Semon, K.M., Rishell, W.

and Stewart, R.G. (1989) Germline transmission of exogenous genes in the chicken. *Science* **243**, 533–535

Bosselman, R.A., Hsu, R.-Y., Briskin, M.J., Boggs, T., Hu, S., Nicolson, M., Souza, L.M., Schultz, J.A., Rishell, W. and Stewart, R.G. (1990) Transmission of exogenous genes into the chicken. *J. Reprod. Fert. Suppl.* **41**, 183–195

Briskin, M.J., Hsu, R.Y., Boggs, T., Schultz, J.A., Rishell W. and Bosselman, R.A. (1991) Heritable retroviral transgenes are highly expressed in chickens. *Proc. Natl. Acad. Sci. USA* **88**, 1736–1740

Burns, J.C., Friedmann, T., Driever, W., Burrascano, M. and Yee, J.K. (1993) Vesicular stomatitis virus G glycoprotein pseudotype retroviral vectors: concentration to very high titer and efficient gene transfer into mammalian and nonmammalian cells. *Proc. Natl. Acad. Sci. USA* **90**, 8033–8037

Capecchi M.R. (1989) Altering the genome by homologous recombination. *Science* **244**, 1288–1292

Chebloune, Y., Cosset, F.L., Legras, C., Drynda, A., Faure, C., Nigon, V.M. and Verdier, G. (1990) Construction of AEV-based suicide vectors expressing transferred gene under control of a non retroviral promoter. *Proc. 4th World Congress on Genetics Applied to Livestock Production* Edinburgh, UK, July 1990

Chebloune, Y., Rulka, J., Cosset, F.L., Valsesia, S., Ronfort, C., Legras, C., Drynda, A., Kuzmak, J., Nigon, V.M. and Verdier, G. (1991) Immune response and resistance to Rous sarcoma virus challenge of chickens immunized with cell-associated glycoproteins provided with a recombinant avian leukosis virus. *J. Virol.* **65**, 5374–5380

Chen, H.Y., Garber, E.A., Mills, E., Smith, J., Kopchick, J.J., DiLella, A.G. and Smith, R.G. (1990) Vectors, promoters, and expression of genes in chick embryos. *J. Reprod. Fert. Suppl.* **41**, 173–182

Cook, R.F., Cook, S.J., Savon, S., McGrane, M., Hartitz, M., Hanson, R.W. and Hodgson, C.P. (1993) Liver-specific expression of a phosphoenolpyruvate carboxykinase-neo gene in genetically modified chickens. *Poultry Sci.* **72**, 554–567

Cosset, F.L., Legras, C., Chebloune, Y., Savatier, P., Thoraval, P., Thomas, J.L., Samarut, J., Nigon, V.M. and Verdier, G. (1990) A new avian leukosis virus-based packaging cell line that uses two separate transcomplementing helper genomes. *J. Virol.* **64**, 1070–1078

Cosset, F.L., Legras, C., Thomas, J.L., Molina, R.M., Chebloune, Y., Faure, C., Nigon, V.M. and Verdier, G. (1991a) Improvement of avian leukosis virus (ALV)-based retrovirus vectors by using different *cis*-acting sequences from ALVs. *J. Virol.* **65**, 3388–3394

Cosset, F.L., Bouquet, J.F., Drynda, A., Chebloune, Y., Rey-Senelongue, A., Kohen, G., Nigon, V.M., Desmettre, P. and Verdier, G. (1991b) Newcastle disease virus (NDV) vaccine based on immunization with avian cells expressing the NDV hemagglutinin-neuraminidase glycoprotein. *Virology* **185**, 862–866

Cosset, F.L., Ronfort, C., Molina, R.M., Flamant, F., Drunda, A., Benchaï bi, M., Valsesia, S., Nigon, V.M. and Verdier, G. (1992) Packaging cells for avian leukosis virus-based vectors with various host ranges. *J. Virol.* **66**, 5671–5676

Cosset, F.L., Girod, A., Flamant, F., Drynda, A., Ronfort, C., Valsesia, S., Molina, R.M., Faure, C., Nirgon, V.M. and Verdier, G. (1993) Use of helper cells with two host ranges to generate high-titer retroviral vectors. *Virology* **193**, 385–395

Crittenden, L.N., Salter, D.W. and Federspiel, M.J. (1989) Segregation, viral phenotype, and proviral structure of 23 avian leukosis virus inserts in the germ line of chickens. *Theor. Appl. Genet.* **77**, 505–515

Dambrine, G., Afanassieff, M., Cosset, F.L., Lasserre, F., Verdier, G. and Thoraval, P. (1993) Use of avian retroviral vectors for gene transfer in chickens. *The 31st NIBB Conference Japan-France Collaborative workshop on gene manipulation in Aves.* National Institute for Basic Biology, Okasaki 444, Japan, March 1993

Dornburg, R. and Temin, H.M. (1988) Retroviral vector system for the study of cDNA gene formation. *Mol. Cell Biol.* **8**, 2328–2334

Dougherty, J.P. and Temin, H.M. (1987) A promoterless retroviral vector indicates that there are sequences in U3 required for 3' RNA processing. *Proc. Natl. Acad. Sci.* **84**, 1197–1201

Dougherty, J.P., Wisniewski, R., Yang, S., Rhode, B.W. and Temin, H.M. (1989) New retrovirus helper cells with almost no nucleotide sequence homology to retrovirus vectors. *J. Virol.* **63**, 3209–3212

Ellis, J. and Bernstein, A. (1989) Gene targeting with retroviral vectors: recombination by gene conversion into regions of nonhomology. *Mol. Cell. Biol.* **9**, 1621–1627

Emerman, M. and Temin, H.M. (1984a) Genes with promoters in retrovirus vectors can be independently suppressed by an epigenetic mechanism. *Cell* **39**, 459–467

Emerman, M. and Temin, H.M. (1984b) High-frequency deletion in recovered retrovirus vectors containing exogenous DNA with promoters. *J. Virol.* **50**, 42–49

Emerman, M. and Temin, H.M. (1986a) Quantitative analysis of gene suppression in integrated retrovirus vectors. *Mol. Cell. Biol.* **6**, 792–800

Emerman, M. and Temin, H.M. (1986b) Comparison of promoter suppression in avian and murine retrovirus vectors. *Nucleic Acids Res.* **14**, 9381–9396

Federspiel, M.J. and Hughes, S.H. (1994) Effects of the gag region on genome stability: avian retroviral vectors that contain sequences from the Bryan strain of Rous sarcoma virus. *Virology*, **203**, 211–220

Feteke, D.M. and Cepko, C.L. (1993) Replication-competent retroviral vectors encoding alkaline phosphatase reveal spatial restriction of viral gene expression/transduction in the chick embryo. *Mol. Cell. Biol.* **13**, 2604–2613

Flamant, F., Aubert, D., Legrand, C., Cosset, F.L. and Samarut, J. (1993) Importance of 3' non-coding sequences for efficient retrovirus-mediated gene transfer in avian cells revealed by self-inactivating vectors. *J. Gen. Virol.* **74**, 39–46

Galileo, D.S., Gray, G.E., Owens, G.C., Majors, J. and Sanes, J.R. (1990) Neurons and glia arise from a commun progenitor in chicken optic tectum: demonstration with two retroviruses and cell type-specific antibodies. *Proc. Natl. Acad. Sci.* **87**, 458–462

Garber, E.A., Rosenblum, C.I., Chute, H.T., Scheidel, L.M., Chen, H. (1991) Avian retroviral expression of luciferace. *J. Virol.* **185**, 652–660

Ghattas, I.R., Sanes, J.R., Majors, J.E. (1991) The encephalomyocarditis virus internal ribosome entry site allows efficient coexpression of the two genes from a recombinant provirus in cultured cells and in embryos. *Mol. Cell. Biol.* **11**, 5848–5859

Girod, A., Cosset, F.L., Verdier, G. and Ronfort, C. (1995) Analysis of ALV-based packaging cell lines for production of contaminant defective viruses. *Virology* **209**, 671–675

Girod, A., Drynda, A., Cosset, F.L., Verdier, G. and Ronfort, C. (1996) Homologous and nonhomologous retroviral recombinations are both involved in the transfer by infectious particles of defective ALV-derived transcomplementing genomes, *J. Virol.* **70**, 5651–5657

Gray, G.E., Glover, J.C., Majors, J. and Sanes, J.R. (1988) Radial arrangement of clonally related cells in the chicken optic tectum: lineage analysis with a recombinant retrovirus. *Proc. Natl. Acad. Sci.* **85**, 7356–7360

Greenhouse, J.J., Petropoulos, C.J., Crittenden, L.B. and Hughes, S.H. (1988) Helper-independent retrovirus vectors with Rous-Associated virus type O long terminal repeats. *J. Virol.* **62**, 4809–4812

Hippenmeyer, P.J. and Krivi, G.G. (1991) Gene expression from heterologous promoters in a replication-defective avian retrovirus vector in quail cells. *Poultry Sci.* **70**, 982–992

Hippenmeyer, P.J., Krivi, G.G. and Highkin, M.K. (1988) Transfer and expression of the bacterial NPT-II gene in chick embryos using a Schmidt-Ruppin retrovirus vector. *Nucleic Acids Res.* **16**, 7619–7632

Hu, S., Bruszewski, J., Nicolson, M., Tseng, J., Hsu, R.Y. and Bosselman, R. (1987) Generation of competent virus in the REV helper cell line C3. *Virology* **159**, 446–449

Hughes, S. and Kosik, E. (1984) Mutagenesis of the region between the env and src of the SR-A strain of Rous Sarcoma Virus for the purpose of constructing helper-independent vectors. *Virol.* **136**, 89–99

Hughes, S.H., Greenhouse, J.J., Petropoulos, C.J. and Sutrave, P. (1987) Adaptor plasmids simplify the insertion of foreign DNA into helper-independent retroviral vectors. *J. Virol.* **61**, 3004–3012

Hughes, S.H., Petropoulos, C.J., Federspiel, M.J., Sutrave, P., Forry-Schaudies, S. and Bradac, J.A. (1990) Vectors and genes for improvement of animal strains. *J. Reprod. Fert. Suppl.* **41**, 39–49

Hunt, L.A., Brown, D.W., Robinson, H.L., Naeve, C.W. and Webster, R.G. (1988) Retrovirus-expressed hemagglutinin protects against lethal Influenza Virus infections. *J. Virol.* **62**, 3014–3019

Jaffredo, T., Molina, R.M., Al Moustafa, A.E., Gautier, R., Cosset, F.L., Verdier, G. and Dieterlen-Lievre, F. (1993) Patterns of integration and expression of retroviral, non-replicative vectors in avian embryos: embryo developmental stage and virus subgroup envelope modulate tissue-tropism. *Cell Adhesion and Communication* **1**, 119–132

Jang, S.K., Krausslich, H.G., Nicklin, M.J.H., Duke, G.M., Palmemberg, A.C. and Wimmer, E. (1988) A segment of the 5' nontranslated region of encephalomyocarditis virus RNA directs internal entry of ribosomes during *in vitro* translation. *J. Virol.* **62**, 2636–2643

Julias, J.G., Hash, D. and Pathak, V.K. (1995) E-vectors: development of novel self-inactivating and self-activating retroviral vectors for safer gene therapy. *J. Virol.* **69**, 6839–6846

Koo, H.M., Brown, A.M.C., Ron, Y. and Dougherty, J.P. (1991) Spleen necrosis virus, an avian retrovirus, can infect primate cells. *J. Virol.* **65**, 4769–4776

Koo, H.M., Brown, A.M., Kaufman, R.J., Prorock, C.M., Ron, T.Y. and Dougherty, J.P. (1992) A Spleen Necrosis virus-based retroviral vector which expresses two genes from a dicistonic mRNA. *J. Virol.* **186**, 669–675

Kopchick, J.J., Mills, E., Rosenblum, C., Taylor, J., Macken, F., Leung, F., Smith, J. and Chen, H. (1991) Methods for the introduction of recombinant DNA into chickens embryos. *Biotechnology* **16**, 275–293

Kozak, S.L. and Kabat, D. (1990) Ping-pong amplification of a retroviral vector achieves high-level gene expression: human growth hormone production. *J. Virol.* **64**, 3500–3508

Lee, M.R. and Shuman, R.M. (1990) Transgenic quail produced by retrovirus vector infection transmit and express a foreign marker gene. *Proc. 4th World Congress on Genetics Applied to Livestock Production*, Edinburgh, UK, July 1990

Mikawa, T., Fischman, D.A., Dougherty, J.P. and Brown, A.M.C. (1991) *In vivo* analysis of a new *lacZ* retrovirus vector suitable for cell lineage marking in avian and other species. *Exp. Cell. Res.* **195**, 516–523

Molina, R.M., Chebloune, Y., Verdier, G. and Legras, C. (1995) Influence of expression and *cis*-acting sequences from avian leukosis viruses (ALVs) on stability of (ALV)-based retrovirus vectors. *C.R. Acad. Sci. Paris* **318**, 541–551

Moscovici, C., Moscovici, M.G., Jimenez, H., Lai, M.M.C., Hayman, M.J. and Vogt, P.K. (1977) Continuous tissue culture cell lines derived from chemically induced tumors of japanese quail. *Cell* **11**, 95–103

Olson, P., Temin H.M. and Dornburg R. (1992) Unusually high frequency of reconstitution of long terminal repeats in U3-minus retrovirus vectors by DNA recombination or gene conversion. *J. Virol.* **66**, 1336–1343

Olson, P., Nelson, S. and Dornburg, R. (1994) Improved self-inactivating retroviral vectors derived from spleen necrosis virus. *J. Virol.* **68**, 7060–7066

Panganiban, A.T. and Talbot, K.J. (1993) Efficient insertion from an internal long terminal repeat (LTR)-LTR sequence on a Reticuloendotheliosis virus vector is imprecise and cell specific. *J. Virol.* **67**, 1564–1571

Petropoulos, C.J. and Hughes, S.H. (1991) Replication-competent retrovirus vectors for the transfer and expression of gene cassettes in avian cells. *J. Virol.* **65**, 3728–3737

Petropoulos, C.J., Payne, W., Salter, D.W. and Hughes, S.H. (1992) Appropriate *in vivo* expression of muscle-specific promoter by using avian retroviral vectors for gene transfer. *J. Virol.* **66**, 3391–3397

Ronfort, C., Afanassieff, M., Chebloune, Y., Dambrine, G., Nigon, V.M. and Verdier, G. (1991) Identification and structure analysis of endogenous proviral sequences in a Brown Leghorn chicken strain. *Poultry Sci.* **70**, 2161–2175

Ronfort, C., Girod, A., Cosset, F.L., Legras, C., Nigon, V.M., Chebloune, Y. and Verdier, G. (1995a) Defective retroviral endogenous RNA is efficiently transmitted by infectious particles produced on an avian retroviral vector packaging cell line. *Virology* **207**, 271–275

Ronfort, C., Chebloune, Y., Cosset, F.L., Faure, C., Nigon, V.M. and Verdier, G. (1995b) Structure and expression of endogenous retroviral sequences in the permanent LMH chicken cell line. *Poultry Sci.* **74**, 127–135

Salter, D.W., Smith, E.J., Hughes, S., Wright, S.E., Fadly, A.M., Writter, R.L. and Crittenden, L.B. (1986) Gene insertion into the chicken germ line by retroviruses. *Poultry Sci.* **65**, 1445–1458

Salter, D.W., Smith, E.J., Hughes, S.H., Wright, S.E. and Crittenden, L.B. (1987) Transgenic chickens: insertion of retroviral genes into the chicken germ line. *Virology* **157**, 236–240

Salter, D.W. and Crittender, L.B. (1989) Artificial insertion of a dominant gene for resistance to avian leukosis virus into the germ line of the chicken. *Theor. Apple. Genet.* **77**, 457–461

Savatier, P., Bagnis, C., Thoraval, P., Poncet, D., Belakebi, M., Mallet, F., Legras, C., Cosset, F.L., Thomas, J.L., Chebloune, Y., Faure, C., Verdier, G., Samarut, J. and Nigon, V. (1989) Generation of a helper cell line for packaging avian leukosis virus-based vectors. *J. Virol.* **63**, 513–522

Shimotohno, K. and Temin, H.M. (1981) Formation of infectious progeny virus after insertion of Herpes Simplex thymidine kinase gene into DNA of an avian retrovirus. *Cell* **26**, 67–77

Shuman, R.M. and Shoffner, R.N. (1986) Symposium: molecular approaches to poultry breeding. Gene transfer by avian retroviruses. *Poultry Sci.* **65**, 1437–1444

Smith, E.J. (1986) Endogenous avian leukemia virus. In De Boer, (ed), *Avian Leukosis*, Martinus Nijhoff Publishing, Boston, MA

Souza L.M., Boone, T.C., Murdock, D., Langley, K., Wypych, J., Fenton, D., Johnson, S., Lai, P.H., Everett, R., Hsu, R.-Y. and Bosselman, R. (1984) Application of recombinant DNA technologies to studies on chicken growth hormone. *J. Exp. Zoology* **232**, 465–473

Stoker, A.W. and Bissell, M.J. (1988) Development of avian sarcoma and leukosis virus-based vector-packaging cell lines. *J. Virol.* **62**, 1008–1015

Stoker, A.W., Hatier, C. and Bissel, M.J. (1990) The embryonic environment strongly attenuates v-src oncogenesis in mesenchymal and epithelial tissues, but not in endothelia. *J. Cell Biol.* **111**, 217–228

Takahara, Y., Hamada, K. and Housman, D.E. (1992) A new retrovirus packaging cell for gene transfer constructed from amplified long terminal repeat-free chimeric proviral genes. *J. Virol.* **66**, 3725–3732

Thomas, J.L., Afanassieff, M., Cosset, F.L., Molina, R.M., Ronfort, C., Drynda, A., Legras, C., Chebloune, Y., Nigon, V.M. and Verdier, G. (1992) *In situ* expression of helper-free avian leukosis virus (ALV)-based retrovirus vectors in early chick embryos. *Int. J. Dev. Biol.* **36**, 215–227

Thoraval, P., Afanassieff, M., Cosset, F.L., Lasserre, F., Verdier, G., Coudert, F. and Dambrine, G. (1995). Germ line transmission of exogenous genes in chickens using helper free ecotropic avian leukosis virus-based vectors. *Transgenic Research* **4**, 369–376

Valsesia-Wittman, S., Drynda, A., Deleage, G., Aumilley, M., Heard, J.M., Danos, O., Verdier, G. and Cosset, F.L. (1994) Modifications in the binding domain of avian retrovirus envelope protein to redirect the host range of retroviral vectors. *J. Virol.* **68**, 4609–4619

Varmus, H. (1988) Retroviruses. *Science* **240**, 1427–1435

Vick L., Li, Y. and Simkiss, K. (1993) Transgenic birds from transformed primordial germ cells. *Proc. R. Soc. Lond.* **251**, 179–182

Watanabe, S. and Temin, H.M. (1983) Construction of a helper cell line for avian reticuloendotheliosis virus cloning vectors. *Mol. Cell. Biol.* **3**, 2241–2249

Weiss R.A. (1984) Experimental biology and assay. In R.A. Weiss, N.Teich, H.E. Varmus, and J. Coffin (eds), *RNA tumors virus, 2nd ed.*, Cold Spring Harbor Laboratory, Cold Spring Harbor, NY

Yee, J.K., Miyanohara, A., LaPorte, P., Bouic, K., Burns, J.C. and Friedmann, T. (1994) A general method for the generation of high-titer, pantropic retroviral vectors: highly efficient infection of primary hepatocytes. *Proc. Natl. Acad. Sci. USA* **91**, 9564–9568

PART II, SECTION C

17. Transgenic Chickens by Liposome-Sperm-Mediated Gene Transfer

E.J. Squires and D. Drake

Department of Animal and Poultry Science, University of Guelph, Guelph, Ontario, N1G 2W1, Canada

Introduction

A number of alternatives exist for the generation of transgenic animals, such as the direct microinjection of foreign DNA into the pronucleus (Murphy and Hanson, 1987) and the use of retroviruses as vectors (Brown and Scott, 1987). However, neither of these two methods are entirely suitable for the generation of transgenic birds. The use of sperm cells as vectors to carry foreign DNA to the egg is potentially a simple method for producing transgenic animals in all species (Lauria and Gandolfi, 1993). Attempts to use sperm cells for this purpose have been largely ineffective (Brinster, 1989 and Gavora, 1991), probably due to the need for efficient loading of DNA into the sperm cells.

Liposomes have been used for the transfer of DNA to cells both in vitro (Schaefer-Ridder et al., 1982) and to somatic cells in liver in vivo (Nicolau et al., 1987). Synthetic cationic lipids, such as lipofectin (Felgner et al., 1987) and DOTAP, have been useful for gene transfer in tissue culture and into intact chick embryos (Watanabe et al., 1994). Liposomes have been used to study the effect of different phospholipids on the motility and the acrosome reaction of bull sperm (Graham et al., 1986, 1987) and liposomes prepared from some phospholipids have been shown not to reduce sperm cell fertility. The use of liposomes along with sperm cells has been reported (Bachiller et al., 1991; Rottman et al., 1992 and Nakanishi and Iritani, 1993) but no stable transgenics were generated. Alternatively, DNA can be transferred to sperm cells by electroporation, but this has resulted in decreased fertility (Nakanishi and Iritani, 1993).

We have systematically investigated various liposome preparations for efficacy of encapsulation of DNA and transfer of DNA to sperm cells while still maintaining their fertility (Squires and Drake, 1993). We have then used the optimum conditions we have found to generate transgenic chickens as determined by PCR analysis of genomic DNA (Squires and Drake, 1994). Similar experiments are necessary if this method is to be used for gene transfer in other species.

DNA Incorporation into Liposomes

Liposomes were prepared by the reverse phase evaporation procedure (Szoka and Papahadjopoulos, 1978), since this method results in the highest encapsulation efficiency, using vortexing instead of sonicating to minimize shearing of the DNA (Nicolau et al., 1987). Sufficient DNA was included in the preparations to provide 1×10^{-6} ng of DNA per sperm cell (at 100% incorporation of the DNA into the liposomes). Lipid was dissolved in chloroform/ether 1:1 (v/v) and BPSE (Beltsville Poultry Semen Extender; Sexton, 1978) containing the DNA was added such that the ratio of organic solvent to aqueous was 3:1 (v/v). This mixture was then vortexed vigorously for 4 min, the organic solvent was removed using a rotary evaporator and the lipid was held in a water bath at 40–45°C for 30 min. The liposomes were then suspended in BPSE and centrifuged at $100,000 \times g$ for 30 min at 10°C. The liposome pellet was then suspended in the required volume of BPSE before use.

We determined the percentage of DNA that was incorporated into the liposomes using ^{32}P labelled DNA and measuring the amount of radioactivity that was recovered in the liposome pellet. The efficiency of trapping of the DNA into liposomes containing 100 μmol/ml lipid increased as the fatty acid carbon chain length increased from C12 (dilauroylphosphatidylcholine, DLPC) to C16 (dipalmitoylphosphatidylcholine, DPPC), with the highest efficiencies obtained with egg yolk phosphatidylcholine (EYPC). The efficiency of trapping of the DNA into the liposomes tended to increase as the concentration of lipids in the liposome preparation was increased, particularly with liposomes made from EYPC. However, liposomes produced from levels of lipid higher than 40 μmol of total lipid per ml of BPSE inhibited the motility of sperm cells. All subsequent liposome preparations were therefore prepared from 40 μmol of total lipid per ml of BPSE and diluted before use.

Including cholesterol into liposomes made from DPPC or dimyristoylphosphatidylcholine (DMPC) had little effect on the trapping efficiency of DNA. However, including 10 mol% phosphatidylserine (PS) increased the trapping efficiency approximately two fold and including stearylamine (SA) into DPPC, DLPC or EYPC liposomes increased the trapping efficiency of DNA as much as 8 fold. In DPPC liposomes, the trapping efficiency increased with increasing concentration of SA with maximal efficiency at 5 mol% SA, while in DLPC and EYPC liposomes the maximal efficiency was obtained with 0.5 mol% SA. The inclusion of dilauroyl-lysophosphatidylcholine (LPC) into the lipid mixture increased the trapping efficiency in DPPC liposomes 2 fold and in EYPC liposomes almost 5 fold but had little effect on DLPC liposomes. This effect was similar over the range of 1 mol% to 20 mol% LPC in DPPC and DLPC liposomes. Combining LPC and SA together in the liposomes resulted in approximately the same trapping efficiency as with SA alone in DLPC and EYPC liposomes but lowered the trapping efficiency of DPPC liposomes containing SA alone by one half.

Transfer of DNA to Sperm

Semen was collected from mature white leghorn roosters by lumbar massage into one volume of BPSE. Sperm cell numbers were counted after dilution of a sample with formaldehyde in 1% saline and the remaining semen was then diluted to an appropriate concentration with BPSE. The live sperm cells were examined microscopically to ensure a high degree of motility.

Liposomes (0.1 ml) containing ^{32}P labelled DNA were mixed with 0.1 ml of sperm cells and incubated at 40°C for 10 min. In other experiments, DNA was mixed with lipofectin reagent and kept at room temperature for 15 min before mixing with sperm cells. The mixture was then layered over 0.8 ml of 0.25 M sucrose, and centrifuged at 5000 × g for 5 min to separate the sperm cells from the liposomes. The sperm cell pellet was then taken up in 0.2 ml of BPSE and washed twice more by centrifugation through sucrose. The transfer of DNA to the sperm cells was estimated by measuring the amount of the ^{32}P labelled DNA in the sperm cell pellet after each washing step. The amount of DNA associated with the sperm cell pellet decreased dramatically in the first washing through sucrose but remained constant after three washes.

The transfer efficiency to sperm was greater with liposomes containing 10 μmol/ml lipid rather than 100 μmol/ml lipid when pure phospholipids were used. Cholesterol has been shown to have a stabilizing effect on liposomes (New, 1990), but we found that cholesterol had no effect on either the encapsulation of DNA into liposomes or the transfer efficiency of DNA to sperm. Phosphatidylserine increased the encapsulation efficiency of the DNA into the liposomes but lowered the efficiency of DNA transfer to sperm from DPPC, DMPC and DLPC liposomes. In experiments using 25×10^7 sperm, including 1 mol% LPC in EYPC and DPPC liposomes had little effect on transfer efficiency, while at 20 mol% LPC, transfer efficiency from DPPC liposomes was increased 2 fold compared to liposomes without LPC. Including SA in the lipid mixtures increased the transfer of DNA from DPPC, DMPC and EYPC liposomes. The addition of LPC along with SA in DPPC and EYPC liposomes increased the transfer efficiency of DNA to sperm cells compared to liposomes with LPC alone. Lysophospholipids as well as SA have been implicated in enhancing the fusion of liposomes with membranes (Martin and MacDonald, 1976). Stearylamine introduces a positive charge on the liposomes that is important in sequestering the DNA into the liposomes and increasing the interaction between the liposomes and the sperm cells, while LPC increases the fusion of the liposomes with the sperm cells.

The transfer efficiency of DNA from liposomes made from only DPPC or EYPC was decreased somewhat by increasing the number of sperm from 5×10^7 to 25×10^7, while with lipofectin the transfer efficiency was 7 times higher at the lower levels of sperm.

Liposome Composition and Sperm Fertility

Sperm cells were incubated with the liposome-DNA for 10 min at 40°C and the mixture of sperm cells and liposomes was used directly for artificial insemination of white leghorn hens. Inseminations were made with a volume of 0.2 ml. In some experiments, intramagnal catheters were placed

(Lakshmanan et al., 1990) and used for insemination of the hens. The eggs were collected daily and held at 13°C. Eggs were set at weekly intervals and incubated at 37°C and 54% relative humidity. After 7 days, the number of fertile eggs was estimated by candling the eggs.

Untreated chicken semen collected into BPSE remained fully fertile for at least one hour after collection. The percentage of fertile eggs produced decreased from about 90% to about 50% when hens were inseminated with 5×10^7 sperm rather than 12.7×10^7 sperm that had been incubated with DNA without liposomes. Sperm cells were rendered totally infertile after incubation with liposomes made with DLPC, while liposomes made from DMPC or EYPC reduced the fertility of sperm cells to about 20% of the control values. Liposomes containing 100 µmol/ml DPPC also reduced the fertility of sperm cells, but this effect was not seen at lower levels of DPPC. Exposing sperm cells to DPPC liposomes that contained 1 mol% SA lowered fertility of the sperm cells, while the presence of 0.5 mol% SA had no effect on fertility of the sperm cells. Including 10 mol% LPC only in DPPC liposomes had little effect on the fertility of sperm cells but LPC tended to alleviate the inhibitory effect of SA on fertility of the sperm cells when present in DPPC liposomes along with SA. This result was somewhat suprising since lysophospholipids have been implicated in the acrosome reaction (Meizel, 1984), and sperm cells treated with LPC might be expected to undergo the acrosome reaction and thus have a very limited time in which to remain fertile. Exposing sperm cells to 0.006 µmol/ml lipofectin reagent reduced their fertility to about 80% of control values. Increasing the concentration of lipofectin reagent to 0.06 µmol/ml reduced the fertility of sperm to about 67%, while exposure to 0.6 µmol/ml lipofectin rendered the sperm totally infertile.

Thus, the most effective method for the liposome mediated transfer of DNA to sperm is using liposomes composed of DPPC, LPC and SA since this results in high trapping efficiency of the DNA into the liposomes, high transfer of the DNA to the sperm and high fertility rates of the sperm. Lipofectin reagent diluted to 0.06 µmol/ml (50 µg/ml) is also an effective method for transferring DNA to sperm, but lower numbers of sperm cells (5×10^7) must be used to obtain high levels of transfer of DNA to the sperm and this results in lowered rates of fertility than obtained using DPPC liposomes and 25×10^7 sperm.

Detection of Transgenics

We have used a VLDLCAT construct made by linking the promoter sequences of the VLDL gene from chicken to bacterial chloramphenicol acetyl transferase (CAT) in pGEM (supplied by R. Deeley, Queen's University, Kingston, Ontario, Canada). A 7.6 kb fragment containing the VLDLCAT construct was removed from the plasmid by digestion with BamHI. DNA was encapsulated into liposomes prepared from 10 µmol/ml DPPC, 5% SA and 10% LPC or mixed with lipofectin reagent at 0.006 or 0.06 µmol/ml and transferred to sperm as outlined above.

Chicks produced from these experiments were bled at hatch and genomic DNA was prepared from the blood by phenol-chloroform extraction. The presence of the transgene in genomic DNA was determined using 30 amplification cycles of the Polymerase Chain Reaction (PCR) with primers that amplify a 586 bp fragment across the VLDL-CAT junction. This fragment contained diagnostic PvuII and HindIII restriction sites that were used to confirm that the correct sequences were amplified. As a positive control for amplification, primers for the endogenous VLDL gene were used. After amplification, the PCR reaction mixture was run on a 2% agarose gel, blotted to nitrocellulose and probed with CAT sequences to identify the VLDLCAT fragments. A bird was considered positive only after positive results were obtained from 3-4 successive blood samples.

PCR screening of transgenic birds generated with DPPC, SA, LPC liposomes and sperm introduced via an intramagnal catheter indicated that 13% of the chicks produced from the liposome-treated sperm were positive for the transgene. In the second generation, produced by mating with control birds, 19% of the offspring of one bird and 37% of the offspring of a second bird were positive for the transgene. In the third generation, produced by mating the second generation with control birds, from 7-24% of the birds were positive for the transgene. In a second experiment using 0.06 µmol/ml lipofectin and birds with intramagnal catheters, 20% of the first generation, 21% of the second generation and from 18-31% of the third generation were positive for the transgene. In a third experiment, lipofectin reagent at 0.006 µmol/ml was used to transfer DNA to the sperm cells, which were used to inseminate both noncatheterized hens and hens that had an intramagnal catheter. In both groups of hens, 24% of the chicks produced in the first generation were positive for the transgene. In the second generation, 20% of the offspring were positive and in the third generation, from 3-35% of the birds carried the transgene.

CAT activity was measured in extracts of liver from birds that were treated with estrogen (1 mg/kg intramuscularly). The "direct diffusion" method of Neuman et al. (1987) and Eastman (1987) was used.

An ELISA procedure obtained from Boehringer Mannheim was used to measure levels of CAT protein in liver extracts. However, attempts to measure either CAT activity or CAT protein were not conclusive, with levels in transgenic birds similar to controls.

These results demonstrate the generation of stable transgenic chickens by liposome-sperm-mediated gene transfer. A high efficiency of transgenics was produced in all experiments with an average efficiency of 20% in the first generation. There was also no difference in the efficiency of production of transgenic birds when sperm was introduced via an intramagnal catheter or intravaginally, suggesting that sorting out of liposome-treated sperm does not occur in the lower part of the reproductive tract of the hen. We also noted that the positive amplification control VLDL primers were much more efficient at amplifying the endogenous VLDL gene than the VLDLCAT primers were in amplifying the transgene. Thus, the values for the efficiency of production of transgenics may be slightly underestimated, since some DNA samples may not have properly amplified in the PCR reaction.

The percentage of positive offspring in the second generation was expected to be 50%, but averaged 24%. This could be due to mosaicism in the first generation, as has been demonstrated for transgenics produced by other methods (Bosselman et al., 1989; Overbeek et al., 1991 and Love et al., 1994). The percentage of positive transgenics in the third generation was also expected to be 50%, but our results averaged 19% (range of 3–35%). It appears that the transgene may have been lost from some genetic lines.

We were unable to detect expression of the transgene in a significant number of animals, although some birds did express small amounts of activity. This lack of expression may be due to degradation of the transgene, since our PCR protocol amplifies only a portion of the VLDLCAT gene. This hypothesis is supported by results obtained by Bachiller et al. (1991) who reported the presence of partial fragments of the foreign transgene in some animals. Alternatively, the position of insertion in the genome or other unknown factors related to this particular construct may have prevented expression. Further work is needed to resolve these issues before this method can be used to reliably generate transgenic animals.

Summary

The effect of different types of liposomes on the efficiency of trapping of DNA into the liposomes, transfer of DNA from the liposomes to chicken sperm cells and the fertility of the sperm cells was determined. Increasing the concentration of lipid in the liposome preparations increased the trapping efficiency of DNA into liposomes but lowered the transfer of DNA to sperm. Including stearylamine (SA) in the liposomes increased the incorporation of DNA into the liposomes and the DNA transfer to sperm cells, while including lauroyllysophosphatidylcholine (LPC) along with SA resulted in the highest transfer efficiency from liposomes to sperm. The transfer of DNA from liposomes to sperm cells was lowered by increasing the number of sperm cells, while decreasing the number of sperm cells lowered the fertility. The sperm cells remained fertile after exposure to low levels of DPPC or lipofectin reagent or to high levels of SA and LPC. The best conditions for liposome-mediated gene transfer to chicken sperm cells are thus using either lipofectin reagent at 0.006 to 0.06 µmol/ml and 5×10^7 sperm or with DPPC liposomes comprised of 10 µmol/ml total lipid including 5 mol% SA and 20 mol% LPC with 2.5×10^8 sperm cells. We obtained from 15–25% of the first generation offspring positive for the foreign gene as determined by Polymerase Chain Reaction analysis of genomic DNA. The foreign gene was also present in a high percentage of the second and third generation offspring produced by test-mating with control birds. However, we were unable to detect significant amounts of CAT protein or activity from these birds. It is possible that these transgenic birds carry only partial copies of the foreign gene.

References

Bachiller, D., Schellander, K., Peli, J. and Ruther, U. (1991) Liposome-mediated DNA uptake by sperm cells. *Mol. Reprod. Dev.* 30, 194–200

Brinster, R.L., Sandgren, E.P., Behringer, R.R. and Palmiter, R.D. (1989) No simple solution for making transgenic mice. *Cell* 59, 239–241

Brown, A.M.C. and Scott, M.R.D. (1987) Chapter 9: Retroviral vectors. In D.M. Glover (ed.), *DNA Cloning A Practical Approach*, Vol. III, IRL Press Ltd., Oxford, pp. 189–212

Eastman, A. (1987) An improvement to the novel rapid assay for chloramphenicol acetyltransferase gene expression. *Biotechniques* 5(8), 730–732

Erickson, R.P. (1990) Are intermediate vectors needed between foreign DNA on sperm and the nucleus? *TIG* 6, 31

Felgner, P.L., Gadek, T.R., Holm, M., Roman, R., Chan, H.W., Wenz, M., Northrop, J.P., Ringold, G.M. and Danielsen, M. (1987) Lipofection: A highly efficient, lipid-mediated DNA-transfection procedure. *Proc. Natl. Acad. Sci. USA* 84, 7413–7417

Gavora, J.S., Benkel, B., Sasada, H., Cantwell, W.J., Fiser, P., Teather, R.M., Nagai, J. and Sabour, M.P. (1991) An attempt at sperm-mediated gene transfer in mice and chickens. *Can. J. Anim. Sci.* **71**, 287–291

Graham, J.K., Foote, R.H. and Parrish, J.J. (1986) Effect of dilauroylphosphatidylcholine on the acrosome reaction and subsequent penetration of bull spermatozoa into zona-free hamster eggs. *Biol. Reprod.* **35**, 413–424

Graham, J.K., Foote, R.H. and Hough, S.R. (1987) Penetration of zona-free hamster eggs by liposome-treated sperm from the bull, ram, stallion, and boar. *Biol. Reprod.* **37**, 181–188

Lakshmanan, N., Duby, R.T. and Smyth Jr., J.R. (1990) Intramagnal catheterization: A novel method for intramagnal insemination of chickens. *Poultry Sci.* **69**, Suppl. 1, 77

Lauria, A., Gandolfi, F. (1993) Recent advances in sperm cell mediated gene transfer. *Mol. Reprod. Dev.* **36**, 255–257

Love, J., Gribbin, C., Mather, C. and Sang, H. (1994) Transgenic birds by DNA microinjection. *Biotechnology* **12**, 60–63

Martin, F.J. and Macdonald, R.C. (1976) Lipid vesicle-cell interactions: II. Induction of cell fusion. *J. Cell Biol.* **70**, 506–514

Meizel, S. (1984) The importance of hydrolytic enzymes to an exocytotic event, the mammalian sperm acrosome reaction. *Bio. Rev.* **59**, 125–157

Murphy, D. and Hanson, J. (1987) Chapter 10: The production of transgenic mice by the microinjection of cloned DNA into fertilized one-cell eggs. In D.M. Glover (ed.), *DNA Cloning A Practical Approach*, Vol. III, IRL Press Ltd., Oxford, pp. 213–248

Nakanishi, A. and Iritani, A. (1993) Gene transfer in the chicken by sperm-mediated methods. *Mol. Reprod. Dev.* **36**, 258–261

Neumann, J.R., Morency, C.A. and Russian, K.O. (1987) A novel rapid assay for chloramphenicol acetyltransferase gene expression. *Biotechniques* **5**(5), 444–447

New, R.R.C. (1990) Chapter 2: Preparation of liposomes. In R.R.C. New (ed.), *Liposomes A Practical Approach* IRL Press Ltd., Oxford, pp. 33–104

Nicolau, C., Legrand, A. and Grosse, E. (1987) Liposomes as carriers for *in vivo* gene transfer and expression. *Methods in Enzymol.* **149**, 157–176

Overbeek, P.A., Aguilar-Cordova, E., Hanten, G., Schaffner, D., Patel, P., Lebovitz, R.M. and Lieberman, M.W. (1991) Coinjection strategy for visual identification of transgenic mice. *Trans. Res.* **1**, 31–37

Rottmann, O.J., Antes, R., Höfer, P. and Maierhofer, G. (1992) Liposome mediated gene transfer via spermatozoa into avian egg cells. *J. Anim. Breed. Genet.* **109**, 64–70

Schaefer-Ridder, M., Wang, Y. and Hofschneider, P.H. (1982) Liposomes as gene carriers: Efficient transformation of mouse L cells by thymidine kinase gene. *Science* **215**, 166–168

Sexton, T.J. and Fewlass, T.A. (1978) A new poultry semen extender 2. Effect of the diluent components on the fertilizing capacity of chicken semen stored at 5°C. *Poult. Sci.* **57**, 277–284

Squires, E.J. and Drake, D. (1993) Liposome-mediated DNA transfer to chicken sperm cells. *Anim. Biotech.* **4**(1), 71–88

Squires, E.J. and Drake, D. (1994) Transgenic chickens by liposome-sperm-mediated gene transfer. *Proc. 5th World Congress Genet. Appl. Livestock Prod.* Vol. 21, Guelph, Ontario, 7 August 1994, pp. 350–353

Szoka Jr., F. and Papahadjopoulos, D. (1978) Procedure for preparation of liposomes with large internal aqueous space and high capture by reverse-phase evaporation. *Proc. Natl. Acad. Sci. USA* **75**(9), 4194–4198

Watanabe, M., Naito, M., Sasaki, E., Sakuri, M., Kuwana, T. and Oishi, T. (1994) Liposome-mediated DNA transfer into chicken primordial germ cells *in vivo*. *Mol. Reprod. Dev.* **38**, 268–274

PART II, SECTION C

18. Ballistic Transfection of Avian Primordial Germ Cells *In Ovo*

K. Simkiss and G. Luke

School of Animal and Microbial Sciences, The University of Reading, Reading, RG6 6AJ, UK

The introduction of foreign molecules into plant cells by firing high velocity microprojectiles through the cell walls was developed by Sanford *et al.*, (1987). In the decade since the concept was first advanced the system has been described as ballistic ("biolistic") transformation, microprojectile bombardment, a particle acceleration method and a "gene gun". The technique has undergone considerable scrutiny and refinement and the applications have spread from plant cells to animal cells and tissues, prokaryotes and even organelles (Sanford *et al.*, 1993).

The method has two essential components. The first is to accelerate microscopic particles to supersonic speeds so as to enable them to pass through any restraining systems in either the medium (air) or the tissue (various membranes). The second is to coat these particles with macromolecules (usually nucleic acids although the method can also be used with proteins and other substances) and to release these into the cell contents (Sanford, 1988).

The propulsive systems that are used have almost always relied on shock waves in gases. In the original technique a gunpowder charge was used (Klein *et al.*, 1987) but this has been modified to include air guns (Dard, 1993), electric discharges through water (Christou *et al.*, 1990), and compressed gases including air, nitrogen and helium. In order to accelerate the microprojectiles in the μm size range they are initially carried on a macroprojectile (a plunger) (Kikkert, 1993) or disc (Sanford, 1988), that releases them when it hits a baffle placed near the target material. As the macroprojectile hits this stop the microprojectiles detach and penetrate the tissue. In order to quantify this process and obtain reproducible results the final flight of the isolated microprojectiles occurs in a partial vacuum.

The microprojectiles that act as "bullets" in this system must be in the μm size range if they are to lodge inside cells without destroying them. The velocity and density of these particles determines their penetration and most experiments have used either gold or tungsten materials. Particles of gold can either be purchased in the 1–2 μm range or produced by precipitation in the laboratory. The particles are usually round and of relatively uniform size (Sanford *et al.*, 1993). Tungsten particles are considerably cheaper; they cover a wider size range but tend to be less regular in size and shape. There are some indications that tungsten can be toxic to some cells (Kikkert, 1993).

The difficulty of coating the particles with nucleic acids is a major source of variation and usually involves precipitating the DNA onto the surface of the microprojectile by adding $CaCl_2$ and spermidine. In order to keep the particles suspended this is normally done in a vortexed tube to which freshly prepared DNA ($1\,\mu g \cdot \mu l^{-1}$) $CaCl_2$ ($2.5\,mol \cdot l^{-1}$) and spermidine ($100\,mmol \cdot l^{-1}$) are added sequentially. There are a large number of preferred recipes including buffer components (to control pH), glycerol (to keep particles in suspension) and ethanol (to aid precipitation) which are added at various stages (Sanford *et al.*, 1993).

The particles are then loaded into the accelerating system and discharged into the cell preparation. The intention in a typical experiment is to carry the DNA into the cell nucleus where the DNA may be freed and able to integrate into the host DNA during a DNA break/repair activity (Bishop and Smith, 1989). This occurs in about 5% of cases with transient expression from non-integrated DNA being more common.

Variants on the Ballistic Process

Virtually all aspects of the original ballistic concept have been modified in the ten years since it was advocated. This is very surprising as the original system was based on suitably quantified aspects such as gas pressure, distance projectile, travel before 50% loss of velocity, effect of vacuum and depth of penetration (Klein et al., 1992). The impression that is now being produced in that it is remarkably easy to administer nucleic acids into the cytosol of animal cells where the cellulose wall is not present (Acsadi et al., 1991). Among the more remarkable of these ballistic techniques are the following modifications.

(a) *Absence of macroprojectile system* (Sautter et al., 1991 and Sautter, 1993). In this technique microprojectiles are introduced directly into the gas stream from an air gun by using Bernoulli's force to draw them out of a delivery tube.

(b) *Absence of a microprojectile component* (Furth et al., 1992). This approach is based on an air delivery vaccine gun that drives liquid directly into the tissues. Thus no particles are necessary and the solutions of DNA in this system can be used to transfect cells mm or even cm beneath the skin surface.

(c) *Absence of a partial vacuum over the target* (Williams et al., 1991). This is a variant of the original ballistic instrument which was modified to be hand held so that it could be placed directly over the living tissue. When used in this way it was found that microprojectiles were capable of penetrating at least 500 μm into liver without the need for any reduction in air pressure to facilitate the particle flight path.

The Use of Ballistics on the Avian Embryo

The avian embryo has a number of features that make it particularly attractive for the use of ballistic transfection of the germ-line. These are:

(a) the embryo is accessible without recourse to surgery on the parent,
(b) the embryo lies immediately beneath the shell
(c) the primordial germ cells are easily identified in the germinal crescent region at developmental stage 12 (Hamburger and Hamilton, 1951),
(d) the germinal crescent is extraembryonic,
(e) there are few cells beneath the crescent region so that little damage will be done to embryonic tissues by the ballistic process.

In our experiments we have successfully introduced plasmid DNA into the germ line of the bird by using a dental needle-less syringe. This is a simple spring system which was loaded with tungsten particles with a mean diameter of 1.5 μm but covering a considerable size range. In use roughly 0.5 cm^2 of the shell of a 2d incubated egg was removed and the egg tilted so that the embryo was covered by the shell. DNA coated particles were then fired into the exposed germinal crescent, the hole sealed and the egg returned to the incubator. In the hands of an experienced operator, several embryos can be transfected per minute. Because a variety of different sized particles are used and because the material beneath the germinal crescent is mainly yolk proteins, it is not crucial to determine either the velocity of the projectiles nor their depth of penetration.

Hatchlings from these experiments contain the foreign DNA in their sperm and transmit it to their offspring (Li et al., 1995). It is not clear whether this DNA has integrated into the genome.

References

Acsadi, G., Dickson, G., Love, D.R., Jani, A., Walsh, F.S., Gurusinghe, A., Wolff, J.A. and Davies, K.E. (1991) Human dystrophin expression in mdx mice after intramuscular injection of DNA constructs. *Nature* **353**, 815–818

Hamburger, V. and Hamilton, H.L. (1951) A series of normal stages in the development of the chick embryo. *J. Morphol.* **88**, 49–92

Bishop, J.O. and Smith, P. (1989) Mechanism of chromosomal integration of microinjected DNA. *Mol. Biol. Med.* **6**, 283–298

Christou, P., McCabe, D.E., Martinell, B.J. and Swain, W.F. (1990) Soybean genetic engineering – commercial production of transgenic plants. *TIB Tech.* **8**, 145–151

Furth, P.A., Shamay, A., Wall, R.J. and Henninghausen, L. (1992) Gene transfer into somatic tissues by jet injection. *Anal. Biochem.* **205**, 365–368

Kikkert, J.R. (1993) The Biolistic PDS-1000/He device. *Plant Cell, Tissue and Organ Culture* **33**, 221–226

Klein, T.M., Wolf, E.D., Wu, E., Sanford, J.C. (1987) High velocity microprojectiles for delivering nucleic acids into living cells. *Nature* **327**, 70–73

Klein, T.M., Arentzen, R., Lewis, P.A. and Fitzpatrick-McElliott, S. (1992) Transformation of microbes, plants and animals by particle bombardment. *Biotechnology* **10**, 286–291

Li, Y., Behnam, J. and Simkiss, K. (1995) Ballistic transfection of avian primordial germ cell *in ovo*. *Transgenic Research* **4**, 26–29

Oard, J. (1993) Development of an airgun device for particle bombardment. *Plant Cell, Tissue and Organ Culture* **33**, 247–250

Sanford, J.C. (1988) The biolistic process. *Trends in Biotech.* **6**, 299–302

Sanford, J.C., Klein, T.M., Wolf, E.D. and Allen, W. (1987) Delivery of substances into cells and tissues using a particle bombardment process. *Particulate Science & Tech.* **5**, 27–37

Sanford, J.C., Smith, F.D. and Russell, J.A. (1993) Optimizing the biolistic process for different biological applications. *Methods in Enzymology* **217**, 483–509

Sautter, C. (1993) Development of a microtargeting device for particle bombardment of plant meristems. *Plant Cell, Tissue and Organ Culture* **33**, 251–257

Sautter, C., Waldner, H., Neuhaus-Url, G., Neuhas, G. and Potrykus, I. (1991) Micro-targeting: high efficiency gene transfer using a novel approach for the acceleration of microprojectiles. *Biotech.* **9**, 1080–1085

Vain, P., Keen, N., Murillo, J., Rathus, C., Nemes, C. and Finer, J.J. (1993) Development of the particle inflow gun. *Plant Cell, Tissue and Organ Culture* **33**, 237–246

Williams, R.S., Johnston, S.A., Riedy, M., Devit, M.J., McElligot, S.G. and Sanford, J.C. (1991) Introduction of foreign genes into tissues of living mice by DNA-coated microprojectiles. *Proc. Nat. Acad. Sci.* **88**, 2726–2730

PART II, SECTION D

19. Transgenic Salmonids

Robert H. Devlin
Fisheries and Oceans Canada, West Vancouver Laboratory,
4160 Marine Drive, West Vancouver, B.C., Canada V7V 1N6

Introduction

The development of technologies for introducing new genetic information into fish has been realized over the past decade, allowing both investigations of gene function and providing altered phenotypes for enhancing aquaculture production of food fish. Although evidence for DNA transfer in fish has been observed since the 1970s (Vielkind et al., 1971), it was not until the mid 1980s that transgenesis with purified gene constructs was achieved in several fish species from different families (goldfish, Zhu et al., 1985; rainbow trout, Chourrout et al., 1986 and Maclean et al., 1987; loach, Zhu et al., 1986; catfish, Dunham et al., 1987 and Atlantic salmon, Fletcher et al., 1988). A great deal of information has been generated on transgenesis in non-salmonid fish species (see excellent reviews by Maclean and Penman 1990; Chen and Powers 1990; Houdebine and Chourrout 1991; Fletcher and Davies 1991; Hackett 1993; Pandian and Marian 1994; and Gong and Hew 1995) which will be mentioned here only when pertinent to the production of transgenic salmonids.

In salmonid aquaculture, production efficiency is very important for the economic viability of the operation. Perhaps one of the most important factors influencing production is growth rate and its association with feed conversion efficiencies. It has been well documented that salmonids possess a remarkable ability to respond to exogenously supplied growth hormone (see McLean and Donaldson, 1993), and thus, it is natural that more than half of the research on transgenic salmonids has been conducted with gene constructs designed to influence growth (see Table 19.1). Other factors important to aquaculture that are subject to transgenic manipulation include (1) controlling sexual maturation to prevent carcass deterioration near the end of the life cycle, (2) reduction in viability or induction of sterility to ensure reproductive containment of transgenic fish in the event of escapement into the natural environment, (3) control of sex differentiation (in some salmonids, males can undergo precocious maturation and, although they experience rapid growth rates, they mature at a smaller size), (4) improvement in survival (enhancing disease resistance via the acquired and non-specific immune systems), and (5) modifying the biochemical characteristics of the flesh to enhance nutritional and/or organoleptic qualities.

The objective of the present review is to provide a brief summary of transgenic research in salmonid fishes, with emphasis both on methodology and on potential implementation of the technology in aquaculture. Further, the question of why dramatic growth enhancement has been achieved in transgenic salmonids relative to the more modest or absent effects seen in other species will be explored.

DNA Transfer and Retention

Microinjection

Currently, DNA transfer into the germ line of salmonids has only been achieved by microinjection into the cytoplasm of fertilized eggs in early development, and this is the most common method used to date. Compared to microinjection in other species (mammals and other fish), the procedure with salmonid eggs is somewhat simpler owing to the vast numbers of gametes that can be obtained and

Table 19.1. Summary of gene transfer research in salmonids by species.

Species	Promoter/ enhancer	Coding region	Transfer method	Retention (%) Early[a]	Retention (%) Late[b]	Integration	Expression	Phenotype	Transmission (%)	References
Rainbow trout (*Oncorhynchus mykiss*)	SV40	hGH cDNA/ α-globin intron	m[c]	40.0 (c)[d] 59.4 (1)[e]						Chourrout et al., 1986
	mMT	rGH	m	3.7		SB[f]	+Protein			Maclean et al., 1987
	Drosophila hsp 70	CAT	m	25.0						Gibbs et al., 1988
	SV40	hGH cDNA/ α-globin intron	m		74.4	SB	−Protein	no growth stim.	F$_1$: 8.2–28.6	Guyomard et al., 1989
	MHC-I H2-K	hGH	m		46.2		−Protein	no growth stim.	F$_1$: 16	
	mMT	hGH	m		37.5		−Protein	no growth stim.		
	mMT	rGH	m				−Protein	no growth stim.		
	SV40/MMTV	rGH	m				−Protein			
	mMT	hGH	m	75.0		SB	+mRNA +Protein			Rokkones et al., 1989
	mMT	rGH	m	1.0 (c)		SB				Penman et al., 1990
	mMT	bGH	inj.[g]	0–18.0 (1) 4.0						Chandler et al., 1990
	carp α-globin	carp α-globin	m	40.0	39.0	SB	+Protein			Yoshizaki et al., 1991
	RSV/chicken β-actin	lacZ	m	4/5				Blue stain		Inoue et al., 1991
	mMT	rGH	m	SB		SB	−Protein			Penman et al., 1991 Maclean et al., 1992
	RSV	bGH cDNA/ α-globin intron	inc.[h]	0.0						Chourrout and Perrot, 1992
	CMV	CAT	inc.	0.0						
	RSV/chicken β-actin	CAT	m	60.0		SB	+Protein			Yoshizaki et al., 1992
	SV40	bGH cDNA/ α-globin intron	m			SB, PFGE[i] In situ[j]	−Protein		F$_2$: ∼50% ∼75%	Tewari et al., 1992
	CMV	CAT	m	+			+Protein			Inove et al., 1993
	RSV/chicken β-actin	trout GHI cDNA	m	8/10			+mRNA	1.12 × growth		
	mIgκ	CAT	m		67.2	SB	+Protein	tissue-specific expression		Michard-Vanhée et al., 1994
	mIgκ/SV40	CAT	m		43.5		+Protein			
	CMVIE	CAT	m	2.3 (c)			+Protein			Iyengar and Maclean, 1995
	carp β-actin	CAT	m	9.1 (1)			+Protein (prehatch) +Protein			
	opAFP	chinook GH1 cDNA	m		13.7			3.2 × growth		Devlin et al., 1995a
	sockeye MT-B	sockeye GH1	m		∼10.0			∼10 × growth		Devlin et al. (unpub.)
Cutthroat trout (*O. clarki*)	opAFP	chinook GH1 cDNA	m					∼10 × growth		Devlin et al., 1995a

Table 19.1. (Continued)

Species	Construct					Result		Reference
Atlantic salmon (*Salmo salar*)	flounder AFP	flounder AFP	m		6.7	SB	+Protein	Fletcher et al., 1988
	mMT	lacZ	m	+				McEvoy et al., 1988
	mMT	hGH	m		20.0	SB		Rokkones et al., 1989
	Atlantic GH1	Atlantic GH1	m		~5.0			Lorens et al., 1990
	flounder AFP	flounder AFP	m		3.1	SB	+Protein	Davies et al., 1990
								F_1: 15–64 Shears et al., 1991
								F_2: ~50.0
	opAFP	chinook GH1 cDNA	m		1.9		−Protein	3–5× growth	Du et al., 1992a
	sockeye MT-B	sockeye GH1	m		~2.0			~5× growth	Devlin et al., (unpub.)
Chinook salmon (*O.tshawytscha*)	RSV	lacZ	e^k	5–10					Sin et al., 1993
	RSV	lacZ	e	+(sperm)					Symonds et al., 1994a
	RSV	lacZ	e	0–44.0 (fry)		−			Symonds et al., 1994b
	opAFP	chinook GH1 cDNA	m		1.9		−Protein	6.2× growth, acromegaly	Devlin et al., 1995a
Coho salmon (*O. kisutch*)	opAFP	chinook GH1 cDNA	m		3.9–4.7	SB	+Protein	3.6–10.7× growth, acromegaly	Devlin et al., 1995a and unpub.
	opAFP	chinook GH1 cDNA	m				+Protein	embryonic growth, acromegaly, reduced viability, colour change	Devlin et al., 1995b and unpub.
								F_1: 0–18.9
	sockeye MT-B	sockeye GH1	m		6.2		+Protein	11× growth stim., acromegaly, reduced viability, colour change	Devlin et al., 1994
	sockeye MT-B	sockeye GH1	m			SB	+Protein		Devlin et al. (unpub.)
								F_1: 2.8–12.2
	sockeye H3	sockeye GH1	m		5.0			6× growth	Devlin et al. (unpub.)
	sockeye MT-B	chinook IGF-1	m		0.0				Devlin and Yesaki (unpub.)
	sockeye H3	chinook IGF-1	m		0.0				

a. < six months old; b. > six months old; c. microinjection; d. circular DNA; e. linear DNA; f. determined by Southern blot; g. 33 gauge needle injection; h. Incubation; i. Pulse-field gel electrophoresis; j. Cytogenetic localization by *in situ* hybridization; k. electroporation. Species abreviations in construct designations: m = mouse, h = human, r = rat, b = bovine.

because they are relatively resistant to the injection procedure (viabilities higher than 80% can be routinely obtained).

Although the large size, thick chorion and yolky structure of the salmonid egg precludes visualization of the male or female pronuclei, the animal pole region of the egg is easily identifiable. In unfertilized eggs, the animal pole cytoplasm is observed as a clearing amongst yolk droplets – this condition persists as long as the eggs are maintained in physiological saline solution. In some species (e.g. Atlantic salmon and some domestic rainbow trout) the opening of the mycropyle is clearly discernible as a small depression in the chorion's surface (a crater) and can aid in locating a suitable injection site, although in other species (some wild strains of coho and chinook salmon) the micropyle is more difficult to rapidly locate. In these latter species, localization of the micropyle can be enhanced by incubating fertilized eggs in $0.8 \times$ fish saline solution (Ginsburg, 1963) which initiates formation of the mycropylar plug seen as a small opaque spot. After fertilization and transfer to fresh water, the chorion continues to harden for approximately two hours, and the cytoplasm begins coalescing into a small mound prior to the first cleavage division which occurs some 8–10 hours post fertilization at 10°C. Injection into male or female pronuclei (which remain separate until after the first cell cycle) is not necessary for gene transfer, probably because considerable movement of cytoplasm occurs which brings injected DNA into the vicinity of nuclear membrane dissolution and reformation during division. However, since DNA may be excluded from the pronuclei until late in the first cell cycle (i.e. after S phase), mosaic incorporation of DNA injected into the cytoplasm is expected to be common (see below).

Several variations for microinjection of salmonid eggs have been developed including (1) drilling small holes in the chorion to allow passage of the microinjection needle (Chourrout et al., 1986), (2) injection through the micropyle (Fletcher et al., 1988), and (3) injection through chorions kept soft by treatment with glutathione (Yoshizaki et al., 1991) or by fertilizing and retaining eggs in physiological saline (Devlin et al., 1994a). Penman et al., (1990) have found that injection into the germinal disc region provided a higher frequency of gene retention than mycropylar injection, although, in general, both methods appear to be approximately equally effective in generating transgenic salmonids (see Table 19.1). Microinjection procedures vary, but volumes injected range from 1 to 20 nL with DNA concentrations between approximately 4.0 and 500 ng/μL containing 10^6 to 10^8 gene copies (Fletcher and Davies, 1991). While micromanipulator controlled injection needles of 5–10 μm tip diameter are most commonly used, successful gene transfer has also been reported for injections performed by hand (Gibbs et al., 1988 and personal communication), or by using large (33 gauge) needles and injecting into the perivitelline space (Chandler et al., 1990). Frequencies of gene retention can vary dramatically depending on the time elapsed since microinjection: if gene transfer is tested at the alevin stage, values of approximately 50% can be obtained, whereas testing conducted with fish of six to twelve months of age generally yield frequencies of 1–5% (Table 19.1). It is important to note that exceptions are also found where high frequencies of gene retention can be found in old fish and vice versa, indicating that many variables still require refinement in the injection procedure and/or the methods and criteria for transgene identification. Although high frequencies of gene retention may appear desirable, this may prove detrimental during subsequent strain development if founder fish contain many independently segregating transgene insertions that complicate the establishment of true-breeding strains.

In most cases, linear DNA is retained more effectively than the circular form in early development (Chourrout et al., 1986; Penman et al., 1990 and Iyengar et al., 1995), and the frequency of germline transformation appears to be very low for circular DNA. It is thus desirable to inject linear DNA molecules, preferably purified away from vector sequences to obviate problems associated with gene silencing by non-fish sequences (incorrect chromosomal environment) as well as for reasons of public acceptance of engineered foods where the use of homologous DNA is desirable.

Other Methods

Several other methods of gene transfer have been evaluated in salmonids, primarily to circumvent the technical demands and time requirements needed for microinjection of large numbers of zygotes. Although incubation of sperm in DNA solutions does not appear to result in gene transfer (Chourrout and Perrot, 1992), electroporation has been shown to enhance DNA-sperm association (Symonds et al., 1994a) and some evidence for transfer to embryos has been obtained (Sin et al., 1993 and Symonds et al., 1994b). Treatment of sperm with DNA-liposome complexes has been successful in a non-salmonid fish (see Szelei et al., (unpublished) in Hackett, 1993) although we have not been successful using this approach for salmonids (J. Squires and R. Devlin, unpublished observations). Delivery of foreign genes by particle bombardment of Rainbow trout eggs has been investigated (Zelenin et al.,

1991), and, although not yet demonstrated for salmonids, successful transformation of fish has also been achieved by retroviral vectors (Lin et al., 1994). Disney et al. (1988) have also described a method for creating transgenic salmonids via the transmission of large chromosomal fragments between species. Ultimately, development of embryonic stem cells that can be used for precise manipulation of host genes will be desirable for salmonids. Progress has been made in developing ES cells for medaka (Wakamatsu et al., 1994) and zebrafish (Sun et al., 1995), and in salmonids, evidence for transplantation of blastomeres into developing embryos has been obtained (Nilsson and Cloud, 1992).

Integration

Considerable evidence has been obtained that suggests foreign DNA can be retained in salmonids developing from injected eggs (see above), and in some cases Southern blot data have suggested that the DNA may have integrated into the host genome. The DNA can be organized into tandem arrays of random concatemers ligated end to end in all possible combinations (Guyomard et al., 1989). The cited evidence for integration may include (1) uncut DNA migrating as a high-MW complex suggesting that the DNA is covalently attached to genomic DNA, and (2) fragments other than the expected sizes are observed, presumably derived from the juxtaposition of constructs with the host genome. While this evidence is consistent with integration, other phenomenon could generate similar data (e.g. the DNA may be arranged as large extrachromosomal concatemers with some copies possessing rearrangements). However, compelling evidence for integration has been recently obtained, showing that transgenes in rainbow trout were contained in large fragments observable by pulse-field electrophoresis, and that the DNA could be located cytogenetically in the chromosomes by *in situ* hybridization (Tewari et al., 1992). Southern blot analysis has shown that the integrated DNA is arranged as randomly-ligated concatemers, and interestingly, Tewari et al. (1992) showed that this arrangement can adopt complex secondary structures (hairpin fragments comigrating with end fragments) that can produce Southern blot data which may falsely suggest the presence of extrachromosomal DNA. Multiple integration events can occur in founder transgenic fish (Guyomard et al., 1989; Penman et al., 1991; Tewari et al., 1992 and Devlin, unpublished), with copy numbers ranging up to 2000 per cell (Tewari et al., 1992), but insufficient data is currently available to conclude that random integration is occurring throughout the genome. In zebrafish, Ivics et al. (1993) have shown that co-injection of a viral integrase protein can increase the rate of transgenesis, however the use of repeat sequences to facilitate homologous recombination were unsuccessful (He et al., 1992).

Germline Transmission

The integration of DNA in a mosaic fashion in founder transgenic animals usually results in transmission to F_1 progeny at frequencies lower than expected for Mendelian segregation (50% for each independent insert in the same cell), presumably because germline tissue is a mixture of transgenic and non-transgenic cells. Thus, from five studies of transmission to F_1 progeny in salmonids consisting of 25 crosses from founder transgenic parents (see Table 19.1), inheritance averages 15.6% (SE +/− 3.12). Assuming equal contribution of transgenic and nontransgenic blastomeres to subsequent gonadal tissue, and that single insertions are much more common than multiple events, this frequency suggests that integration occurs most commonly at the 2 to 4-cell stage of development. In two investigations where transmission has been observed to F_2 progeny (Shears et al., 1991 and Tewari et al., 1992), Mendelian frequencies have been detected, strongly suggesting that F_1 individuals have inherited a stably-integrated transgene in all cells.

Constructs and Vectors, and Evidence for Expression

Expression in transgenic fish is variable between different founder animals due to gene copy number, site of integration, and the exact nature of the construct employed. Gene constructs used to date in salmonids fall into three classes: (1) those designed to enhance growth (GH and IGF-I genes), (2) antifreeze genes to enhance freeze resistance, and (3) reporter genes. Further, the constructs can be subdivided into those with viral and/or mammalian components, those comprised of fish DNA, and those derived entirely from salmonid gene sequences (see Table 19.1). Hackett (1993) has summarized early studies examining the activity of different promoters in fish tissue-culture cells, and more recent studies have followed (Bearzotti et al., 1992; Sharps et al., 1992; Winkler et al., 1992; Moav et al., 1992; Bétancourt et al., 1993; Chan and Devlin 1993 and Sekkali et al., 1994). To simplify construct synthesis, several studies specifically report the production of vectors designed for the production of transgenic fish (Du et al., 1992b; Moav et al., 1992; Cavari et al., 1993; Hong et al., 1993; Devlin et al., 1994b and Takagi et al., 1994). Recently, fish vectors have been produced that contain border elements to provide a protected chromosomal domain to provide more uniform and predictable expression (Caldovic and Hackett, 1995).

Non-fish Promoters

Most investigation has been performed with mammalian or viral promoters. Several studies have shown that the mouse metallothionein (mMT), SV40, CMVIE, and RSV promoters are capable of directing transcription in early salmonid embryos (McEvoy et al., 1988; Rokkones et al., 1989; Inoue et al., 1991; Yoshizaki et al., 1992; Tewari et al., 1992 and Michard-Vanhée et al., 1994). Although most studies have involved measurement of mRNA or protein early in development when transcription from transient DNA may predominate, one study has demonstrated expression from the CMV promoter retained late in development and presumably integrated into the genome (Michard-Vanhée et al., 1994). Several studies have been unable to detect expression from the mMT or SV40 promoters in transgenic salmonids (Guyomard et al., 1989 and Penman et al., 1991), however, in these cases, the results are complicated by the use of mammalian GH genes (see below). Expression from the mMT, RSV, and SV40 promoters driving a reporter gene known to be detectable in fish cells has not been examined from integrated transgenes, thus making evaluation of the utility of these regulatory sequences somewhat uncertain. Other non-fish promoters shown to be active in salmonids are from the Drosophila hs70 gene (Gibbs et al., 1988), the chicken β-actin gene (with the RSV enhancer: Inoue et al., 1991; 1993, and Yoshizaki et al., 1992), and the mouse immunoglobulin-κ gene (Michard-Vanhée et al., 1994), the latter showing tissue-specific expression in white blood cells.

Fish Promoters

Several fish promoters have been tested in transgenic salmon, and were found to be active in all cases where expression was examined. The flounder antifreeze gene (including its promoter) has been transferred to Atlantic salmon (Fletcher et al., 1988 and Shears et al., 1991) and has been shown to produce moderate levels (in the μg/mL range) of antifreeze protein in transgenic fish (Davies et al., 1990), although not at a level sufficient to confer freeze tolerance (but see Wang et al., 1995 for studies with goldfish). The promoter from the antifreeze gene of ocean pout (*Macrozoarces americanius*) has been used to drive the expression of chinook salmon GH cDNA in Atlantic, coho, and chinook salmon, and in rainbow and cutthroat trout (Du et al., 1992a; and Devlin et al., 1995a and b). Although evidence for elevated GH levels was not observed in transgenic Atlantic salmon with this construct (Du et al., 1992a), transgenic coho salmon possess at least 20-fold more GH than controls (R.H. Devlin and P. Swanson, unpublished). The carp β-actin promoter has been shown to be active in rainbow trout (Iyengar and Maclean 1995) in early development, and Yoshizaki et al. (1991) have transferred the carp α-globin gene into rainbow trout but expression was not measured in this study.

Salmonid Promoters

The metallothionein-B (MT-B) and histone H3 promoters from sockeye salmon have been shown to be active in salmonid tissue-culture cell lines (Chan and Devlin, 1993), and both result in growth enhancement in transgenic salmonids when fused to the sockeye salmon type-1 GH gene (Devlin et al., 1994a; unpublished). The MT-B promoter construct (OnMTGH1) elevated plasma GH levels some 40 fold in transgenic coho over controls (Devlin et al., 1994, see Table 19.2). The H3 promoter is weaker than the MT-B promoter when transfected into tissue-culture cells (Chan and Devlin 1993), and stimulates growth to a lesser degree in transgenic coho (unpublished).

Table 19.2. Growth performance and serum GH levels of transgenic and control coho salmon at twelve months postfertilization. Transgenic individuals contain the OnMTGH1 construct (Devlin et al., 1994a).

Group	n	Weight (g)	Length (cm)	Condition factor[a]	Weight SGR[b]	Serum GH (ng/mL)
Uninjected controls	792	13.5 ± 0.1	10.4 ± 0.03	1.16 ± 0.01	1.23 ± 0.01	ND[c]
Transgenic	74	158.5 ± 13.7	21.6 ± 0.8	1.24 ± 0.01	1.89 ± 0.04	11.0 ± 3.53
Non-transgenic[d]	999	18.6 ± 0.8	11.3 ± 0.06	1.15 ± 0.01	1.30 ± 0.01	$0.26^e \pm 0.03$

a. Condition factor = Weight/Length^3 × 100; b. SGR = Specific Growth Rate (% body weight gain per day = $\Delta \ln W / \Delta t$ (days) × 100; c. Not Determined; d. This group may contain individuals transgenic in tissues other than adipose fin; e. The levels of GH in 5 out of 9 individuals in this group were at the minimum detection limit of the radio-immune assay (0.2 ng/mL).

Phenotypic Effects

While the goal of most transgenic salmonid research is to alter phenotypic expression of important culture characteristics, this has proven more difficult than was originally suspected based on the dramatic results obtained with growth enhancement in transgenic mice (Palmiter et al., 1982). With the exception of phenotypic expression of lac-Z reporter gene activity and small growth effect observed in potentially transgenic trout microinjected with a trout GH gene construct (Inoue et al., 1991; 1993), early work in salmonids did not result in phenotypic effects (see Table 19.1) despite making remarkable progress in developing techniques for gene insertion and structural analysis of transgenes.

Mammalian GH Gene Constructs. For growth enhancement studies, initial work in salmonids was largely conducted with gene constructs derived from mammalian and viral promoters, and mammalian GH genes (Table 19.1). It has been well established that mammalian GH is capable of stimulating growth in salmonids (see McLean and Donaldson, 1993), and consequently, it seems reasonable that if expression could be achieved in transgenic salmon, growth enhancement would result. To date, however, gene constructs derived from non-fish sequences have not stimulated growth in founder transgenics nor F_1 and F_2 progeny (see Table 19.1). While the promoters used in these constructs appear to be capable of driving expression in fish tissue-culture cells (see above), transiently in embryos (see references in Table 19.1) and in some integrated transgenes (Tewari et al., 1992 and Michard-Vanhée et al., 1994), it has not been possible to detect GH protein derived from integrated mammalian GH genes in salmonids (Guyomard et al., 1989; Penman et al., 1991; Tewari et al., 1992; but see Rokkones et al., 1989 for transient production of hGH).

There are several possible reasons why mammalian constructs are not capable of producing functional GH protein in transgenic salmonids. Rokkones et al. (1989) and Friedenreich and Schartl (1990) showed that transcripts derived from transient copies of mammalian GH constructs were variable in size, implying that incorrect transcription initiation/termination or imprecise splicing was occurring. Further, expression is sensitive to the presence or absence of intron sequences (Friedenreich and Schartl 1990; Bearzotti et al., 1992 and Bétancourt et al., 1993) such that mammalian or viral introns may inhibit production of functional protein. This inhibitory effect on expression in fish cells may arise either from transcriptional inhibition or from incorrect RNA splicing, and the effects may be specific to each promoter/intron combination (see Bétancourt et al., 1993). All non-fish GH constructs tested in transgenic salmonids have contained mammalian introns, from either the natural mammalian GH gene, or, if a cDNA was employed, from another mammalian gene (see Table 19.1). In the case of constructs containing cDNAs, the intron was placed downstream of the coding region, suggesting that expression may have been disrupted by transcriptional inhibition or by affecting mRNA stability due to incorrect splicing.

It is possible that some promoters are transcriptionally active transiently but not in integrated transgenes, however, in at least one case, Tewari et al. (1992) have demonstrated expression from integrated CMV/CAT constructs. It is also possible that transgenic fish are only recovered when mammalian GH constructs insert in transcriptionally silent chromosomal regions, if, for example, all active inserts resulted in dominant lethality due to abnormalities arising from overexpression (see abnormalities below). We suspect that dominant lethality may account for our inability to recover transgenic salmon overexpressing IGF-I (see Table 19.1), since it is known that elevated levels of this protein can affect viability by inducing hypoglycemia (Skyrud et al., 1989 and McCormick et al., 1992), perhaps due to cross binding to the insulin receptor. However, when transgenic fish are generated with GH gene constructs (see below), a range of phenotypic expression is observed in most cases (Du et al., 1992a and Devlin et al., 1994, 1995a), and at least some mammalian GH transgenes would be expected to produce GH levels compatible with survival.

It is also possible that the codon differences observed between fish and mammalian genes could contribute to reduced efficiency of translation of mRNAs derived from mammalian genes (particularly if correctly spliced mRNAs are generated only rarely), however, it should be noted that even bacterial coding regions have been shown to produce functional protein in salmonids (see Table 19.1) and other fish species (Hackett, 1993).

Fish GH Gene Constructs. Gene constructs have been recently developed which result in growth enhancement of transgenic salmonids, and in both cases they are composed entirely of fish promoters and coding regions. Du et al. (1992a) were the first to show that dramatic growth enhancement of transgenic salmonids was possible using a construct consisting of the chinook salmon GH1 cDNA under the control of the ocean pout antifreeze protein promoter and termination regions. This construct (opAFPGHc) resulted in a 3 to 5-fold increase in weight of Atlantic salmon, and has subsequently been shown to enhance the growth rate of coho and chinook Pacific salmon as well as rainbow and cutthroat trout (Devlin et al., 1995a). Another gene construct (OnMTGH1), consisting entirely of salmonid gene sequences (Devlin et al., 1994), has been used to accelerate growth of coho, chinook, and Atlantic salmon, and rainbow trout up to 11 fold (Table 19.2 and unpublished observations). This construct, which is identical in concept to the construct used to growth accelerate transgenic mice (Palmiter et al., 1982), contains the metallothionein-B promoter from sockeye salmon fused to the GH1 gene from the same species (Figure 19.1). Some indication that additional copies of the intact salmon GH1 gene (promoter and coding region) are capable of increasing growth of Atlantic salmon has also been reported (Lorens et al., 1990).

Both constructs (opAFPGHc and OnMTGH1) contain promoters that have been shown to be active in salmonid tissue-culture cells (Gong et al., 1991 and Chan and Devlin et al., 1993). Evidence for elevated levels of salmonid GH in vivo is observed in Pacific but not Atlantic salmon (see above), perhaps reflecting differences in the growth physiology between salmonid species, culture conditions, or the GH assay employed. Since both constructs are capable of stimulating comparable levels of growth, it appears that the presence of salmonid introns are not essential for expression, nor do they result in complete inactivity when used with a heterologous salmonid or fish promoter. Thus, it appears that salmonid transcriptional and RNA processing machinery is better able to handle constructs derived from fish than mammalian sources (see Bétancourt et al., 1993). A comparison of the activity of integrated constructs with promoters, genes and cDNAs derived from mammalian or viral, fish, and salmonid sources is required to precisely elucidate the variables influencing expression in transgenic salmon.

Overexpression of GH in transgenic salmonids can affect phenotypes other than overall body growth. For example, skin pigmentation appears to be lightened in salmon containing either the opAFPGHc or OnMTGH1 constructs, perhaps by influencing hormonal control of melanocyte development and/or condensation. This phenotype can be used as a reliable visual marker for identifying GH transgenic individuals, particularly prior to first feeding (Devlin et al., 1995b).

The largest transgenic salmonids can display morphological abnormalities (Devlin et al., 1995a, b, and unpublished) that superficially resemble acromegaly (Figure 19.1), a syndrome observed in mammals with elevated GH synthesis. Not all fish display the abnormal phenotype (smaller transgenic fish tend to be completely normal), and other species (e.g. rainbow trout) are not affected as severely. In affected fish, abnormalities in the cranium, operculum, and lower jaw, which superficially appear to be caused by overgrowth of cartilage, can result in serious morphological disruptions that may ultimately affect viability (Devlin et al., 1995a). These abnormalities can be observed in older founder transgenic coho salmon and in transgenic F_1 progeny prior to first feeding (Devlin et al., 1995b); presumably the latter are more severely affected because higher levels of GH are found in these non-mosaic progeny. A gene construct identical to OnMTGH1 but containing the histone H3 promoter from sockeye salmon stimulates growth to a lesser degree in transgenic coho salmon and also induces a much lower incidence of acromegaly (Table 19.1 and unpublished). It is unknown whether these abnormalities are caused directly by elevated levels of GH, or are mediated by other downstream endocrine factors such as insulin-like growth factor I or II. However, transgenic fish may provide excellent model systems to allow investigation of the mechanisms of hormone action, receptor regulation, and signal transduction, and how these normally interplay to maintain endocrine homeostasis in normal fish.

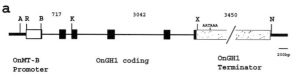

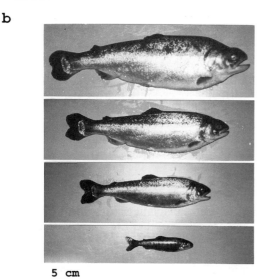

Figure 19.1. (a) Structure of OnMTGH1 gene construct. (b) Growth-enhanced transgenic fish (top three fish) shown relative to non-transgenic sibling (bottom) at ten months of age. Note cranial abnormalities in largest transgenic fish.

Growth-accelerated transgenic salmonids undergo precocious smoltification (the physiological adaptation which allows survival in sea-water environments) up to two years before the natural transformation, displaying both enhanced silver colouration and osmoregulatory ability. GH levels naturally rise during smoltification (Sweeting et al., 1985) and treatment of fish with somatotropins can accelerate this process (Shrimpton et al., 1995). This effect may have considerable commercial value since one limiting factor in the production of salmonids is the juvenile rearing phase which is conducted in cold fresh water (resulting in low growth rates) or in heated water (incurring considerable expense), prior to transfer to warmer sea-water conditions.

Comparison to Growth in Other Transgenic Species. The growth effects observed in transgenic salmonids appears to be greater than that observed in other transgenic animals containing GH gene constructs (see Table 19.3). For mammals, limited stimulation arises presumably because growth senescence is reached shortly after reaching maturity, whereas for salmonids, growth continues throughout their entire lives and effects can compound throughout this time. For other fish species, growth stimulation ranges to as much as 4.6 fold, although in some cases where high growth is observed, insufficiently controlled data precludes a statistical evaluation. Several factors may be contributing to these variable growth responses, including (1) the source of construct (see above) and its structure (e.g. exclusion of signal peptide), (2) variation in copy number and site of integration, (3) GH may not be limiting growth to the same degree as in salmonids, and (4) the biology of the fish dramatically differ (i.e. growth rates of cold-water salmonids are seasonally very low whereas some warmer-water species display more rapid uniform growth which may be already near maximal). A comparison of the activity of salmonid gene constructs in other species should help clarify some of these issues.

Commercial Implementation

From the research conducted to date, it is clear that transgenic salmonids can be created with modified phenotypes for application in aquaculture. However, as discussed above, some problems have been encountered (e.g. acromegaly) that will need to be resolved. Transgenic fish for aquaculture will require both the use of appropriate gene constructs as well as genetic selection among founders and within strains (see Dunham et al., 1992) for the

Table 19.3. Growth enhancement in transgenic animals using GH genes.

Species	Promoter	Coding region	Growth effect	References
Mouse	mMT	rGH gene	~2	Palmiter et al., 1982
Pig	mMT	bGH gene	1.11	Pursel et al., 1989
Salmonids	opAFP,	chinook cDNA	3–10.7	see Table 19.1
	sockeye MT-B or H3	sockeye GH1	6–11	
Loach	mMT	hGH gene	2	Zhu et al., 1986
	mMT	hGH gene	+	Enikolopov et al., 1989
	opAFP	chinook GH1 cDNA	2.5	Tsai et al., 1995
Common carp	mMT	hGH gene	1.09	Zhu et al., 1992
	RSV	rtGH1 cDNA	1.21–1.4	Zhang et al., 1990
	RSV	rtGH1 cDNA	0.73–1.4	Chen et al., 1993
	mMT	hGH gene	+	Wu et al., 1994
Crucian carp	mMT	hGH gene	1.78	Zhu et al., 1992
Catfish	RSV	coho GH cDNA	1.26	Dunham et al., 1992
Pike	RSV	bGH cDNA	0–1.2	Gross et al., 1992
Medaka	chicken β-actin	hGH gene	1.2–1.53	Lu et al., 1992
	mMT	hGH gene	~1.46	
Zebrafish	RSV	coho GH cDNA	1.7	Zhao et al., 1993
	mMT	rGH	+/−	Pandian et al., 1991
Tilapia	CMV	tilapia GH cDNA	1.82	Martinez et al., 1996

development of true-breeding stocks with predictable and uniform phenotypes. For the implementation of transgenic salmonids into commercial aquaculture, careful evaluation of new strains will be required to show that other pleiotropic effects do not cause unexpected effects on performance. For example, growth-stimulated fish may appear to have considerable advantage over nontransgenic strains, but many other factors such as flesh quality, feed conversion efficiencies and disease resistance will need careful examination before a final conclusion can be drawn regarding the utility of a particular strain. Whether germline transgenic fish will ultimately be used in aquaculture depends on the benefits relative to other competing technologies (i.e. for growth enhancement, other alternatives include application of slow-release formulations of GH protein, or somatic cell transformation of muscle (Rhaman and Maclean, 1992 and Chan *et al.*, 1994). Concern over potential ecological impacts of escaped transgenic salmonids must be considered (Kapascinski and Hallerman, 1991) and effective methods of physical and biological (sterilization) containment need refinement (Devlin and Donaldson, 1992) to ensure that the risk of reproductive interaction between escaped transgenic and wild fish stocks is minimized.

References

Bearzotti, M., Perrot, E., Michard-Vanhée, C., Jolivet, G., Attal, J., Théron, M.C., Puissant, C., Dreano, M., Kopchick, J.J., Powell, R., Gannon, F., Houdebine, L.M. and Chourrout, D. (1992) Gene expression following transfection of fish cells. *J. Biotechnol.* **26**, 315–325

Bétancourt, O.H., Attal, J., Théron, M.C., Puissant, C., Houdebine, L.M. and Bearzotti, M. (1993) Efficiency of introns from various origins in fish cells. *Mol. Mar. Biol. Biotech.* **2**, 181–188

Caldovic, L. and Hackett, P.B. (1995) Development of position-independent expression vectors and their transfer into transgenic fish. *Mol. Mar. Biol. Biotech.* **4**, 51–61

Cavari, B., Funkenstein, B., Chen, T.T., Gonzalez-Villasenor, L.I. and Schartl, M. (1993) Effect of growth hormone on the growth rate of the gilthead seabream (*Sparus aurata*), and use of different constructs for the production of transgenic fish. *Aquaculture* **111**, 189–197

Chan, W.K. and Devlin, R.H. (1993) Polymerase chain reaction amplification and functional characterization of sockeye salmon histone H3, metallothionein-B, and protamine promoters. *Mol. Mar. Biol. Biotech.* **2**, 308–318

Chan, W.K., Tan, J.H., Lim, L.S. and Devlin, R.H. (1994). In Chou, L.M., Munro, A.D., Lam, T.J., Chen, T.W., Cheong, L.K.K., Ding J.K., Hooi, K.K., Khoo, H.W., Phang, V.P.E., Shim K.F. and Tan, C.M., (eds). Third Asian Fisheries Forum, Asian Fisheries Society, Manila, Philippines

Chandler, D.P., Welt, M. and Leung, F.C. (1990) Development of a rapid and efficient microinjection technique for gene insertion into fertilized salmonoid eggs. *Tech. Rep. Batelle Pac. Northw. Labs.* Richland, WA (USA), 22

Chen, T. and Powers, D.A. (1990) Transgenic Fish. *Trends in Biotechnology* **8**, 209–215

Chen, T.T., Kight, K, Lin, C.M., Powers, D.A., Hayat, M., Chatakondi, N., Ramboux, A.C., Duncan, P.L. and Dunham, R.A. (1993) Expression and inheritance of RSVLTR–rtGH1 complementary DNA in the transgenic common carp, (*Cyprinus carpio*). *Mar. Mol. Biol. Biotech.* **2**, 88–95

Chourrout, D., Guyomard, R. and Houdebine, L.M. (1986) High efficiency gene transfer in rainbow trout (*Salmo gairdneri* Rich) by microinjection into egg cytoplasm. *Aquaculture* **51**, 143–150

Chourrout, D. and Perrot, E. (1992) No transgenic rainbow trout produced with sperm incubated with linear DNA. *Mol. Mar. Biol. Biotech.* **1**, 282–285

Davies, P.L., Hew, C.L., Shears, M.A. and Fletcher, G.L. (1990) Antifreeze protein expression in transgenic salmon. In R.B. Church (ed.), *Transgenic models in Medicine and Agriculture*, Wiley-Liss, New York, NY, pp. 141–161

Devlin, R.H. and Donaldson, E.M. (1992) Containment of genetically altered fish with emphasis on salmonids. In C.L. Hew and G.L. Fletcher, (eds.), *Transgenic Fish* World Scientific, Singapore, pp. 229–265

Devlin, R.H., Yesaki, T.Y., Biagi, C.A., Donaldson, E.M., Swanson, P. and Chan, W.-K. (1994a) Extraordinary salmon growth. *Nature* **371**, 209–210

Devlin, R.H., Yesaki, T.Y., Biagi, C., Donaldson, E.M. and Chan, W.-K. (1994b) Production and breeding of transgenic salmon. In C. Smith, J.S. Gavora, B. Benkel, J. Chesnais, W. Fairfull, J.P. Gibson, B.W. Kennedy and E.B. Burnside, (eds.), *Proceedings of the 5th World Congress on Genetics Applied to Livestock Production.* Guelph, Ontario, August 7–12, 1994, Vol. 19, pp. 372–378

Devlin, R.H., Yesaki, T.Y., Donaldson, E.M., Du, S.-J. and Hew, C.L. (1995a) Production of germline transgenic Pacific salmonids with dramatically increased growth performance. *Can. J. Fish. Aquat. Sci.* **52**, 1376–1384

Devlin, R.H., Yesaki, T.Y., Donaldson, E.M. and C.L. Hew. (1995b) Transmission and phenotypic effects of an antifreeze/GH gene construct in coho salmon (*Oncorhynchus kisutch*). *Aquaculture* (in press)

Disney, J.E., Johnson, K.R., Banks, D.K. and Thorgaard, G.H. (1988) Maintenance of foreign gene expression in adult transgenic rainbow trout and their offspring. *J. Exp. Zool.* **248**, 335–344

Du, S.J., Gong, Z., Fletcher, G.L., Shears, M.A., King, M.J., Idler, D.R. and Hew, C.L. (1992a) Growth enhancement in transgenic Atlantic salmon by the use of an "all fish" chimeric growth hormone gene construct. *Bio/Technology* **10**, 176–180

Du, S.J., Gong, Z., Hew, C.L., Tan, C.H. and Fletcher, G.L. (1992b) Development of an all-fish gene cassette for gene transfer in aquaculture. *Mol. Mar. Biol. Biotech.* **1**, 290–300

Dunham, R.A., Eash, J., Askins, J. and Townes, T.M. (1987) Transfer of the metallothionein-human growth hormone fusion gene into channel catfish. *Trans. Am. Fish. Soc.* **116**, 87–91

Dunham, R.A., Ramboux, A.C., Duncan, P.L., Hayat, M., Chen, T.T., Lin, C.M., Kight, K., Gonzalez-Villasenor, I. and Powers, D.A. (1992) Transfer, expression and inheritance of salmonid growth hormone in channel catfish, Ictalurus punctatus, and effects on performance traits. *Mol. Mar. Biol. Biotech.* **1**, 380–389

Enikolopov, G.N., Benyumov, A.O., Barmintsev, A., Zelenina, L.A., Sleptsova, L.A., Doronin, Y.K., Golichenkov, V.A., Grashchuk, M.A., Georgiev, G.P., Rubtsov, P.M., Skryabin, K.G. and Baev, A.A. (1989) Advanced growth of transgenic fish containing human somatotropin gene. *Doklady Akademii Nauk SSSR* **301**, 724–727

Fletcher, G.L., Shears, M.A., King, M.J., Davies, P.L. and Hew, C.L. (1988) Evidence for antifreeze protein gene transfer in Atlantic salmon. *Can. J. Fish. Aquat. Sci.* **45**, 352–357

Fletcher, G. and Davies, P.L. (1991) Transgenic fish for aquaculture. *Genetic Engineering* **13**, 331–369

Friedenreich, H. and Schartl, M. (1990) Transient expression directed by homologous and heterologous promoter and enhancer sequences in fish cells. *Nucleic Acids Research* **18**, 3299–3305

Gibbs, P.D.L., Gray, A. and Thorgaard, G.H. (1988) Transfer of reporter genes into zebrafish and rainbout trout. *Aquaculture International Congress* Vancouver, C., Canada September 6–9, 1988 p. 56

Ginsburg, A.S. (1963) Sperm-egg association and its relationship to the activation of the egg in salmonid fishes. *J. Embryol. Exp. Morphol.* **2**, 13–33

Gong, Z., Hew, C.L. and Vielkind, J.R. (1991) Functional analysis and temporal expression of promoter regions from fish antifreeze protein genes in transgenic Japanese medaka embryos. *Mol. Mar. Biol. Biotech.* **1**, 64–72

Gong, Z. and Hew, C.L. (1995) Transgenic fish in aquaculture and developmental biology. In R.A. Pederson and G.P. Schatten (eds.), *Current Topics in Developmental Biology* Academic Press, San Diego, CA, Vol. 30, pp. 175–214

Gross, M.L., Schneider, J.F., Moav, B. Moav, N., Alvarez, C., Myster, S.H., Liu, Z., Hallerman, E.M., Hackett, P.B., Guise, K.S., Faras, A.J. and Kapuscinski, A.R. (1992) Molecular analysis and growth evaluation of northern pike (*Esox lucius*) microinjected with growth hormone genes. *Aquaculture* **103**, 253–273

Guyomard, R., Chourrout, D., Leroux, C., Houdebine, L.M. and Pourrain, F. (1989) Integration and germ line transmission of foreign genes microinjected into fertilized trout eggs. *Biochimie* **71**, 857–863

Hackett, P.B., (1993) The molecular biology of transgenic fish. In P.W. Hochachka and T.P. Mommsen (eds.), *Biochemistry and Molecular Biology of Fishes Molecular Biology Frontiers*. Elsevier, Amsterdam, Vol. 2, pp. 207–240

He, L., Zhu, Z., Faras, A.J., Guise, K.S., Hackett, P.B. and Kapuscinski, A.R. (1992) Characterization of AluI repeats of zebrafish (*Brachydanio rerio*). *Mol. Mar. Biol. Biotech.* **1**, 125–135

Hong, Y., Winkler, C., Brem, G. and Schartl, M. (1993) Development of a heavy metal-inducible fish-specific expression vector for gene transfer *in vitro* and *in vivo*. *Aquaculture* **111**, 215–226

Houdebine, L.M. and Chourrout, D. (1991) Transgenesis in fish. *Experientia* **47**, 891–897

Inoue, K., Yamashita, S., Akita, N., Mitsuboshi, T., Nagahisa, E., Shiba, T. and Fujita, T. (1991). Histochemical Detection of Foreign Gene Expression in Rainbow Trout. *Nippon Suisan Gakkaishi* **57**, 1511–1517

Inove, K., Yamada, S. and Yamashita, S. (1993) Introduction, expression, and growth-enhancing effect of rainbow trout growth hormone cDNA fused to an avian chimeric promoter in rainbow trout fry. *J. Mar. Biotechnol.* **1**, 131–134

Ivics, Z., Izsv k, Z. and Hackett, P.B. (1993) Enhanced incorporation of transgenic DNA into zebrafish chromosomes by a retroviral integration protein. *Mol. Mar. Biol. Biotech.* **2**, 162–173

Iyengar, A. and Maclean, N. (1995) Transgene concatemerization and expression in rainbow trout (*Oncorhynchus mykiss*). *Mol. Mar. Biol. Biotech.* **4**, 248–254

Kapuscinski, A.R. and Hallerman, E.N. (1991) Implications of introduction of transgenic fish into natural ecosystems. *Can. J. Fish. Aquat. Sci.* **48**, 99–107

Lin, S., Gaiano, N., Culp, P., Burns, J.C., Friedmann, T., Yee, J.-K. and Hopkins, N. (1994) Integration and germline transmission of a pseudotyped retroviral vector in zebrafish. *Science* **265**, 666–669

Lorens, J.B., Male, R., Nerland, A., Totland, G., Telle, W., Olsen, L. and Lossius, I. (1990) Atlantic Salmon carrying multiple copies of a salmon growth hormone gene: Microinjection of a sGH I gene construct in fertilized eggs from salmo salar. Norwegion Biochemical Society 26th Winter Meeting Gol, Norway 11–13 January 1990. Abstract p. 126

Lu, J.K., Chen, T.T., Chrisman, C.L., Andrisani, O.M. and Dixon, J.E. (1992) Integration, expression, and germline transmission of foreign growth hormone genes in medaka (Oryzias Iatipes). *Mol. Mar. Biol. Biotech.* **1**, 366–375

Maclean, N., Penman, D. and Zhu, Z. (1987) Introduction of novel genes into fish. *Bio/Technology* **5**, 257–261

Maclean. N. and Penman, D. (1990) The application of gene manipulation to aquaculture. *Aquaculture* **85**, 1–20

Maclean, N., Iyengar, A., Rahman, A., Sulaiman, Z. and Penman, D. (1992) Transgene transmission and expression in rainbow trout and tilapia. *Mol. Mar. Biol. Biotech.* **1**, 355–365

Martinez, R., Estrada, M.P., Berlanga, J., Guillen, I., Hernandez, O., Cabrera, E., Pimentel, R., Morales, R., Herrera, F., Morales, A., Pina, J., Abad, Z., Sanchez, V., Melamed, P., Lleonart, R., de la Fuente, J. (1996) Growth enhancement in transgenic tilapia by ectopic expression of tilapia growth hormone. *Mol. Mar. Biol. Biotech.* **5**, 62–70

McCormick, S.D., Kelly, K.M., Young, G., Nishioka, R.S. and Bern, H.A. (1992) Stimulation of coho salmon growth by insulin-like growth factor 1. *Gen. Comp. Endocrinol* **86**, 398–407

McEvoy, T., Stack, M., Keane, B., Barry, T., Sreenan, J. and Gannon, F. (1988) The expression of a foreign gene in salmon embryos. *Aquaculture* **68**, 27–37

McLean, E. and Donaldson, E.M. (1993) The role of somatotropin in growth in poikilotherms. In M.P. Schreibman, C.G. Scanes and P.K.T. Pang (eds.), *The endocrinology of growth development and metabolism in vertebrates*. Academic Press, New York, NY, pp. 43–71

Michard-Vanhée, C., Chourrout, D., Strömberg, S., Thuvander, A. and Pilström, L. (1994) Lymphocyte expression in transgenic trout by mouse immunoglobulin promoter/enhancer. *Immunogenetics* **40**, 1–8

Moav, B., Liu, Z., Groll, Y. and Hackett, P.B. (1992) Selection of promoters for gene transfer into fish. *Mol. Mar. Biol. Biotech.* **1**, 338–345

Nilsson, E.E., and Cloud, J.G. (1992) Rainbow trout chimeras produced by injection of blastomeres into recipient blastulae. *Proc. Natl. Acad. Sci.* **89**, 9425–9428

Palmiter R. D., Brinster, R.L., Hammer, R.E., Trumbauer, M.E., Rosenfeld, M.G., Birnberg, N.C. and Evans, R.M. (1982) Dramatic growth of mice that develop from eggs microinjected with metallothionein-growth hormone fusion genes. *Nature* **300**, 611–615

Pandian, T.J., Kavumpurath, S., Mathavan, S. and Dharmalingam, K. (1991) Microinjection of rat growth-hormone gene into zebrafish egg and production of transgenic zebrafish. *Current Science* **60**, 596–600

Pandian, T.J. and Marian, L.A. (1994) Problems and prospects of transgenic fish production. *Current Science* **66**, 635–649

Penman, D.J., Beeching, A.J., Penn, S. and Maclean, N. (1990) Factors affecting survival and integration following microinjection of novel DNA into rainbow trout eggs. *Aquaculture* **85**, 35–50

Penman, D.J., Beeching, A.J., Penn, S., Rhaman, A., Sulaiman, Z. and Maclean, N. (1991) Patterns of transgene inheritance in rainbow trout (*Oncorhynchus mykiss*). *Mol. Reprod. Dev.* **30**, 201–206

Pursel, V.G., Pinkert, C.A., Miller, K.F., Bolt, D.J., Campbell, R.G., Palmiter, R.D., Brinster, R.L. and Hammer, R.E. (1989) Genetic Engineering of Livestock, *Science* **244**, 1281–1288

Rahman, A. and Maclean, N. (1992) Fish transgene expression by direct injection into fish muscle. *Mol. Mar. Biol. Biotechnol.* **1**, 296–289

Rokkones, E., Alestrom, P., Skjervold, H. and Gautvik, K.M. (1989) Microinjection and expression of a mouse metallothionein human growth hormone fusion gene in fertilized eggs. *J. Comp. Physiol.* B **158**, 751–758

Sekkali, B., Belayew, A. Martial, J.A., Hellemans, B.A., Ollevier, F. and Volckaert, F.A. (1994) A comparative study of reporter gene activities in fish cells and embryos. *Mol. Mar. Biol. Biotech.* **3**, 30–34

Sharps, A., Nishiyama, K, Collodi, P. and Barnes, D. (1992) Comparison of activities of mammalian viral promoters directing gene expression *in vitro* in zebrafish and other fish cell lines. *Mol. Mar. Biol. Biotech.* **1**, 426–431

Shears, M.A., Fletcher, G.L., Hew, C.L., Gauthier, S. and Davies, P.L. (1991) Transfer, expression and stable inheritance of antifreeze protein genes in Atlantic salmon (*Salmo salar*). *Mar. Mol. Biol. Biotech.* **1**, 58–63

Shrimpton, J.M., Devlin, R.H., McLean, E., Byatt, J.C., Donaldson, E.M. and Randall, D.J. (1995) Increases in gill corticosteroid receptor abundance and saltwater tolerance in juvenile coho salmon (*Oncorhynchus kisutch*) treated with growth hormone and placental lactogen. *Gen. Comp. Endocrinol.* **98**, 1–15

Sin, F.Y.T., Bartley, A.L., Walker, S.P., Sin, I.L., Skymonds, J.E., Hawke, L. and Hopkins, C.L. (1993) Gene transfer in chinook salmon (*Oncorhynchus tshawytscha*) by electroporating sperm in the presence of pRSV-lacZ DNA. *Aquaculture* **117**, 57–69

Skyrud, T., Andersen, O., Alestrom, P. and Gautvik, K.M. (1989) Effects of recombinant human growth hormone and insulin-like growth factor I on body growth and blood metabolites in brook trout (*Salvelinus fontinalis*). *Gen. Comp. Endocrinol.* **75**, 247–255

Sun, L., Bradford, C.S., Ghosh, C., Collodi, P. and Barnes, D.W. (1995) ES-like cell cultures derived from early zebrafish embryos. *Mol. Mar. Biol. Biotech.* **4**, 193–199

Sweeting, R.M., Wagner, G.F. and McKeown, B.A. (1985) Changes in plasma glucose, amino acid nitrogen and growth hormone during smoltification and seawater adaptation in coho salmon, (*Oncorhynchus kisutch*). *Aquaculture* **45**, 185–197

Symonds, J.E., Walker, S.P. and Sin, F.Y.T. (1994a) Electroporation of salmon sperm with plasmid DNA: evidence of enhanced sperm/DNA association. *Aquaculture* **119**, 313–327

Symonds, J.E., Walker, S.P., Sin, F.Y.T. and Sin, I. (1994b) Development of a mass gene transfer method in chinook salmon: optimization of gene transfer by electroporated sperm. *Mol. Mar. Biol. Biotechnol.* **3**, 104–111

Takagi, S., Sasado, T., Tamiya, G., Ozato, K., Wakamatsu, Y., Takeshita, A. and Kimura, M. (1994) An efficient expression vector for transgenic medaka construction. *Mol. Mar. Biol. Biotechnol.* **3**, 192–199

Tewari, R., Michard-Vanhée, C., Perrot, E. and Chourrout, D. (1992) Mendelian transmission, structure and expresssion of transgenes following their injection into the cytoplasm of trout eggs. *Transgenic Research* **1**, 250–260

Tsai, H.J., Tseng, F.S. and Liao, I.C. (1995) Electroporation of sperm to introduce foreign DNA into the genome of loach (*Misgurnus anguillicaudatus*). *Can. J. Fish. Aquat. Sci.* **52**, 776–787

Vielkind, J., Vielkind, U., von Grotthus, E. and Anders, F. (1971) Uptake of bacterial H^3-DNA into fish embryos. *Experientia* **27**, 347–348

Wakamatsu, Y., Ozato, K. and Sasado, T. (1994) Establishment of a pluripotent cell line derived from a medaka (*Oryzias latipes*) blastula embryo. *Mol. Mar. Biol. Biotech.* **3**, 185–191

Wang, R. and Zhang, P., Gong, Z. and Hew, C.L. (1995) Expression of the antifreeze protein gene in transgenic goldfish (*Carassius auratus*) and its implication in cold adaption. *Mol. Mar. Biol. Biotech.* **4**, 20–26

Winkler, C., Hong, Y., Wittbrodt, J. and Schartl, M. (1992) Analysis of heterologous and homologous promoters and enhancers *in vitro* and *in vivo* by gene transfer into Japanese Medaka (*Oryzias latipes*) and Xipophorus. *Mol. Mar. Biol. Biotech.* **1**, 326–337

Wu, T., Yang, H., Dong, Z., Xia, D., Shi, Y., Ji, X., Shen, Y., Sun, W. (1994) The integration and expression of human growth gene in blunt snout bream and common carp. *J. Fish. China Shuichan Xuebao* **18**, 284–289

Yoshizaki, G., Oshiro, T. and Takashima, F. (1991) Introduction of carp α-Globin Gene into rainbow trout. *Nippon Suisan Gakkaishi* **57**, 819–824

Yoshizaki, G., Kobayashi, S., Oshiro, T. and Takashima, F. (1992) Introduction and expression of CAT gene in rainbow trout. *Nippon Suisan Gakkaishi* **58**, 1659–1665

Zelenin, A.V., Alimov, A.A., Barmintzev, V.A., Beniumov, A.O., Zelenina, I.A., Krasnov, A.M. and Kolesnikov, V.A. (1991) The delivery of foreign genes into fertilized fish eggs using high-velocity microprojectiles, *FEBS Letters* **287**, 118–120

Zhang P., Hayat, M., Joyce, M., Gonzalez-Villasenor, C., Lin, L.I., C.M., Dunham, R.A., Chen, T.T., and Powers, D.A. (1990) Gene transfer, expression and inheritance of pRSV-rainbow trout-GHcDNA in the carp, Cyprinus carpio (Linnaeus). *Molec. Reprod. Dev.* **25**, 3–13

Zhao, X., Zhang, P.J. and Wong, T.K. (1993) Application of Baekonization: a new approach to produce transgenic fish. *Mol. Mar. Biol. Biotech.* **2**, 63–69

Zhu, Z., Li, G., He, L. and Chen, S. (1985) Novel gene transfer into the fertilized eggs of goldfish (*Carassius auratus* L. 1758). *Z. Angew. Ichthyol.* **1**, 31–34

Zhu, Z., Xu, K., Li, G., Xie, Y. and He, L. (1986) Biological effects of human growth hormone gene injected into the fertilized eggs of loach Misgurnus anguillicaudatus (Cantor). *Kexue Tongbao* **31**, 988–990

Zhu, Z. (1992) Generation of Fast Growing Transgenic Fish: Methods and Mechanisms. In C.L Hew and G.L. Fletcher, (eds.), *Transgenic Fish*, World Scientific Publishing, Singapore, pp. 92–119

PART II, SECTION D

20. Transferring Foreign Genes into Zebrafish Eggs by Microinjection

Philippe Collas, Harald Husebye and Peter Alestrøm

Department of Biochemistry, Norwegian College of Veterinary Medicine, PO Box 8146 Dep., 0033 Oslo, Norway

The rapid development of transgenic fish technology has created a demand for efficient methods of foreign DNA transfer into fertilized fish eggs. Zebrafish have been increasingly used for transgenics studies mainly because they exhibit a short generation interval and lay eggs frequently. Several methods have been developed for gene transfer into fish eggs, but to date microinjection remains the most efficient. We describe here a microinjection procedure routinely used in our laboratory for DNA transfer into zebrafish eggs.

Introduction

The importance of transgenic fish technology in the world is rapidly increasing. The potential economic benefits of transgenic fish technology to aquaculture are no doubt enormous, with the introduction of novel desirable traits and their transfer to broodstocks (Flechter and Davies, 1991). Transgenic fish may also be used as bioreactors to produce commercially and medically valuable compounds. Transgenic fish are also widely used as models in developmental biology, disease resistance and growth control studies.

Small tropical fish with short generation intervals, such as the zebrafish (*Danio rerio*), have proven to be valuable tools for reproductive and developmental molecular biology and experimental medicine. The zebrafish presents advantages over mammalian systems, in that large quantities of eggs (fertilized or not) can be collected year round, and that following micromanipulation, eggs can be easily raised in a simple *in vitro* environment. Zebrafish embryo genetic manipulations have resulted in the development of homozygous mutants (Kimmel, 1989) and transgenic lines (Stuart et al., 1988, 1990; Pandian et al., 1991; Alestrøm et al., 1992, 1994; Bayer and Campos-Ortega, 1992; Powers et al., 1992; Westerfield et al., 1992; Ivics et al., 1993; Zhao et al., 1993 and Lin et al., 1994).

To date, egg microinjection remains the most successful method of gene transfer into zebrafish genomes (Stuart et al., 1988, 1990 and Pandian et al., 1991). A limitation of this method, however, is that eggs must be handled individually. Alternative strategies for batchwise gene transfer have been developed which include electroporation (Powers et al., 1992; Muller et al., 1993 and Zhao et al., 1993), sperm-mediated gene transfer (Khoo et al., 1992 and Muller et al., 1992), and particle bombardment (Zelenin et al., 1991 and Kavumpurath et al., 1993). Additionally, zebrafish embryonic stem cell lines with potential use for transgenesis are being developed (Wakamatsu et al., 1994 and Sun et al., 1995). Another limitation of microinjection is that injection of DNA directly into the nucleus is not possible in fish, unlike mammals. Microinjection into the cytoplasm requires the injection of high gene copy numbers and results in slow kinetics of DNA transfer into the nucleus and chromosomal integration. This leads to a high incidence of mosaic animals. Nevertheless, microinjection remains the most commonly used gene transfer method in fish, and ambitions to improve this procedure include injection of DNA-peptide complexes (Ivics et al., 1993 and Collas et al., manuscript submitted), with the aim of improving nuclear uptake and integration frequency. This paper describes the gene microinjection procedure into zebrafish eggs used in our laboratory.

Methods

Breeding and Collection of Fertilized Eggs

Adult zebrafish are maintained in dechlorinated water as described by Westerfield (1993) and Brand *et al.* (1995). Fish are fed daily with commercial flakes, low-fat and high-protein 0.5–0.8 mm pellets and 48-hr brine shrimp larvae. In the evening prior to breeding and egg colletion, ten females and five males are placed in a 5 l spawning tank containing a 2 mm mesh 3 cm from the bottom to allow egg recovery. Newly fertilized eggs are collected within 10 min after onset of light and at 20 min intervals during the following hour.

Dechorionation

Eggs are washed extensively in declorinated water at room temperature (RT), rinsed with 10% Hank's balanced salt solution (HBSS; 0.137 M NaCl, 5.4 mM KCl, 0.25 mM Na_2HPO_4, 0.44 mM KH_2PO_4, 1.3 mM $CaCl_2$, 1 mM $MgSO_4$, 4.2 mM $NaHCO_3$) and transferred into a 15 ml petri dish. Intact one-cell stage eggs are placed into 3 ml of 10% HBSS. Pronase E (from porcine pancreas) is added to a final concentration of 0.5 mg/ml (Stuart *et al.*, 1990). Eggs are incubated in pronase at RT for 5 min with gentle agitation during which the chorion turns slightly brown and becomes pliable under gentle pressure with tweezers. The dish containing the eggs is then emptied into 300 ml of dechlorinated water at RT. Eggs are washed five times by allowing them to settle to the bottom of a beaker, discarding the water and adding another 300 ml of water. During each wash, as the eggs settle, the softened chorions come off the eggs. After the last wash, most of the water is removed and the eggs transferred carefully into 10% HBSS with a pasteur pipette. Intact eggs are placed into a new dish of 10% HBSS and should be immediately injected or placed in culture.

Microinjection Solution

The microinjection solution consists of 50 ng/μl plasmid DNA in 0.25 M KCl and 0.2% phenol red. This solution is is-osmotic with the egg interior and the phenol red allows visualization of the injected solution inside the egg. Typically, 200 μl of solution are made and filtered through a 0.22 μm cellulose acetate centrifuge filter into a 1.5 ml conical tube at 4,000 g for 8 min at 4°C. The solution can be stored at 4°C for several weeks. Before use, we recommend to spin the solution at 10,000 g for 5 min at 4°C to pellet any precipitate or debris in order to avoid possible clogging of the injection pipette.

Pipette Manufacture

Glass capillary tubes (1.2 mm OD, 0.94 mm ID; Sutter Instruments, Co.; Novato, CO., USA) are pulled with a horizontal pipette puller (Narishige; New York, NY, USA). Pipettes are stored dry protected from dust and can be kept at RT for months.

Pipette Loading

The microinjection pipette is held vertically tip down and 1 μl of injection solution (DNA) back-loaded. The solution is drawn towards the pipette tip by capilllarity within 3 min. Loaded pipettes are held vertically, tip down, in a humidified container prior to use to avoid dessication.

Microinjection

Dechorionated embryos at the one- to two-cell stage are placed on a depression slide in a drop of 300 μl of 10% HBSS at RT, on the stage of a stereo dissecting microscope. Eggs are oriented by gently shaking the slide so that the blastodiscs are clearly visible (Figure 20.1A, arrows). Eggs are injected using a coarse micromanipulator and a Narishige pico-injector. The pipette loaded with DNA solution is inserted horizontally into the manipulator's arm, the arm tilted forward to an angle of 80° and the pipette carefully plunged into the drop containing the eggs. Gently tapping the pipette tip once on the bottom of the slide breaks off the tip to produce an opening of approximately 3–5 μm. A narrower opening would make injection of the solution difficult, whereas a larger tip would be detrimental to the egg. Approximately 250 pl of solution are injected into each egg. The injection volume can easily be adjusted with settings on the pico-injector. The volume can be determined in preliminary experiments by injecting known concentrations of ^{32}P-γ-ATP or equivalent radiolabelled compounds into eggs and measuring radioactivity. Microinjection is performed by carefully penetrating the egg plasma membrane with the pipette, delivering a single pulse with the pico-injector and carefully withdrawing the tip. The drop of injected solution should be visible inside the egg (Figure 20.1B, arrow). By moving the pipette only vertically and sliding the slide containing the eggs on the microscope stage, up to 60 eggs can be injected per minute. Injected eggs are removed with a pasteur pipette, washed once in a 4 ml dish of 10% HBSS and placed in culture.

Embryo Culture

Embryos are raised for the first four days in 10% HBSS at 28.5°C in regular atmosphere. Dead or deformed embryos should be removed from the dish 4 hr after injection and daily thereafter. The young fish are then transferred to 28°C dechlorinated water and raised as described in

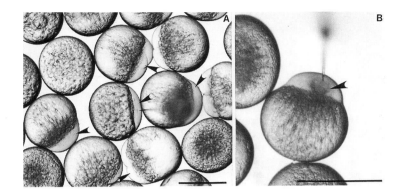

Figure 20.1. Microinjection into a fertilized zebrafish egg. (A) Newly fertilized (one-cell) and early two-cell stage zebrafish eggs oriented on the microinjection slide so that blastodiscs become clearly visible (arrows). (B) Droplet of DNA solution containing phenol red visible inside the egg at the site of injection in the blastodisc (arrow). Bar, 500 μm.

Westerfield (1993). Embryo viability is assessed at 4, 24, 48, 72 and 96 h after injection, according to morphological criteria described by Westerfiled (1993).

Results and Discussion

The gene microinjection procedure described results in. high embryo survival rates and transient gene expression frequencies. Both embryo survival and gene expression, however, are affected by the number of plasmid copies injected in each egg. Injection of 10^2 to 10^6 copies of a plasmid bearing a CMV promoter and a luciferase reporter gene (pCMVL, Gibbs et al., 1994) results in ca. 70% live embryos at 48 h (in our hands not different from the survival rate of control injected embryos), but 10^7 or more copies appear detrimental (Collas et al., submitted). Transient luciferase expression on photographic film at 24 h is undetectable after injection of 10^2 to 10^4 pCMVL copies whereas 10^5 and 10^6 copies routinely promote expression in 80–100% of embryos. Expression frequency is reduced with $> 10^6$ plasmid copies presumably due to the detrimental effect of such elevated plasmid copy numbers (Collas et al., submitted).

Although transgenes may be transiently expressed at a very high efficiency, rates of stable integration and germ line transmission of the transgene remain in the order of a few percent. Thus a mandatory step in the establishing transgenic fish lines is analysis of transgenesis. Transient expression can be assessed using reporter genes such as luciferase (Sato et al., 1992 and Gibbs et al., 1994), β-galactosidase (Westerfield et al., 1992; Lin et al., 1994 and Takagi et al., 1994) or green fluorescent protein (Chalfie et al., 1994). Assessment of transgene integration into the fish genome may be done by Southern blotting and inverse polymerase chain reaction (PCR) analyses of injected individuals, and transmission through the germ line by PCR analysis of F1 offspring. Integration and germ-line transmission efficiencies are currently investigated in our laboratory.

An advantage of microinjection is the possibility of controlling the site of injection in the egg. To our knowledge, no studies have reported the influence of the site of microinjection on the efficiency of transgene expression and integration events. In species where dechorionation is difficult or impractical, such as the rainbow trout, microinjections are usually performed through the micropyle. In dechorionated zebrafish eggs several injections sites are available. Eggs have been commonly injected directly in the blastodisc or in the yolk region adjacent to the blastodisc (Husebye, unpublished). However, we have recent evidence that eggs can be injected at a random location in the yolk without affecting transient reporter (luciferase) gene expression at 24 h (Collas, unpublished). In case of random injection sites, however, the kinetics of DNA transfer from the yolk (or cytoplasm) to the nucleus is greatly affected and, as one might expect, DNA transfer occurs more rapidly when eggs are injected in the blastodisc as opposed to the vegetal pole (Collas, unpublished). The effects of microinjection site on stable transgene integration and degree of mosaicism remain to be investigated.

References

Alestrøm, P., Kisen, G., Klungland, H. and Andersen, Ø. (1992) Fish GnRH gene and molecular approaches for control of sexual maturation – Development of a transgenic fish model. *Mol. Mar. Biol. Biotech.* **1**, 376–379

Alestrøm, P., Husebye, H., Kavumpurath, S. and Kisen, G. (1994) Zebrafish, a vertebrate model for transgene expression and biological function. *Anim. Biotech.* **5**, 147–154

Bayer, T.A. and Campos-Ortega, J.A. (1992) A transgene containing LacZ is expressed in primary sensory neurons in zebrafish. *Development*, **115** 421–426

Brand, M., Beuchle, D., Endres, F., Haffter, P., Hammerschmidt, M., Mullins, M., Schulte-Merker, S., Nüsslein-Volhard, C., Lück, R., Jürgen, K. and Schwarz, S. (1995) Keeping and raising zebrafish (*Danio rerio*) in Tübingen. *The Zebrafish Science Monitor* 3(5), 2–7

Chalfie, M., Tu, Y., Euskirchen, G., Ward, W.W. and Prasher, D.C. (1994) Green fluorescent protein as a marker for gene expression. *Science* 263, 802–805

Flechter, L.G. and Davies, P.L. (1991) Transgenic fish for aquaculture. In J.K. Setlow, (ed.), *Transgenic Fish for Aquaculture in Genetic Engeneering*, Plenum Press, New York, pp. 331–370

Gibbs, P.D.L., Peek, A. and Thorgaard, G. (1994) An *in vivo* screen for the luciferase transgene in zebrafish. *Mol. Mar. Biol. Biotech.* 3, 307–316

Ivics, Z., Iszvack, Z. and Hacket, P. (1993) Enhanced incorporation of transgenic DNA into zebrafish by a retroviral integration protein. *Mol. Mar. Biol. Biotech.* 2, 162–173

Kavumpurath, S., Andersen, Ø., Kisen, G. and Alestrøm, P. (1993) Gene transfer methods and luciferase gene expression in zebrafish, *Brachydanio rerio* (Hamilton). *Israeli J. Aquacult.* 45, 154–163

Khoo, H.W., Ang, L.H., Lim, H.B. and Wong, K.Y. (1992) Sperm cells as vectors for introducing foreign DNA into zebrafish. *Aquaculture* 107, 1–19

Kimmel, C. (1989) Genetics and early development in zebrafish. *Trends Genet.* 5, 283–288

Lin, S., Yang, S. and Hopkins, N. (1994) LacZ expression in germline transgenic zebrafish can be detected in living embryos. *Dev. Biol.* 161, 77–83

Muller, F., Ivics, Z., Erdelyi, F., Papp, T., Varadi, L., Horvath, L., Maclean, N. and Orban, L. (1992) Introducing foreign genes into fish eggs with electroporated sperm as a carrier. *Mol. Mar. Biol. Biotech.* 1, 276–281

Muller, F., Lele, Z., Varadi, L., Menczel, L. and Orban, L. (1993) Efficient transient expression system based on square pulse electroporation and *in vivo* luciferase assay of fertilized fish eggs. *FEBS Lett.* 324, 27–32

Pandian, T.J., Kavumpurath, S., Mathavan, S. and Dharmalingam, K. (1991) Microinjection of growth hormone gene and production of transgenic zebrafish. *Current Science* 60, 596–600

Powers, D.A., Hereford, L., Cole, T., Chen, T.T., Lin, C.M., Kight, K., Creech, K. and Dunham, R. (1992) Electroporation: a method for transferring genes into the gametes of zebrafish (*Brachydanio rerio*), channel catfish (*Ictalurus punctatus*), and common carp (*Cyprinus carpio*). *Mol. Mar. Biol. Biotech.*, 1, 301–308

Sato, A., Komura, J.I., Masahito, P. and Matsukuma, S. (1992) Firefly luciferase gene transmission and expression in transgenic medaka (*Oryzias latipes*). *Mol. Mar. Biol. Biotech.* 1, 318–325

Stuart, G.W., McMurray, J.V. and Westerfield, M. (1988) Replication, integration and stable germ-line transmission of foreign sequences injected into early zebrafish embryos. *Development* 103, 403–412

Stuart, G.W., Vielkind, J.R., McMurray, J.V. and Westerfield, M. (1990) Stable lines of transgenic zebrafish exhibit reproducible patterns of transgene expression. *Development* 109, 577–584

Sun, L., Bradford, C.S., Ghosh, C., Collodi, P. and Barnes, D (1995) ES-like cell cultures derived from early zebrafish embryos. *Mol. Mar. Biol. Biotech.* 4, 193–199

Takagi, S., Sasado, T., Tamiya G., Ozato, K., Wakamatsu, Y., Takeshita, A. and Kimura, M. (1994) An efficient expression vector for transgenic medaka construction. *Mol. Mar. Biol. Biotech.* 3, 192–199

Wakamatsu, Y., Ozato, K. and Sasado, T. (1994) Establishment of a pluripotent cell line derived from a medaka (*Oryzias latipes*) blastula embryo. *Mol. Mar. Biol. Biotech.* 3, 185–191

Westerfield, M., Wegner, J., Jegalian, B.G., DeRobertis, E.M. and Puschel, A.W. (1992) Specific activation of mammalian *Hox* promoters in mosaic transgenic zebrafish. *Genes. Devel.* 6, 591–598

Westerfield, M. (1993) *The Zebrafish Book*, University of Oregon Press, Eugene, OR

Zelenin, A.V., Alimov, A.A., Barmintzev, V.A., Beniumov, A.O., Zelenina, I.A., Krasnov, A.N. and Kolesnikov, V.A. (1991) The delivery of foreign genes into fertilized fish eggs using high-velocity microprojectiles. *FEBS Lett.* 287, 118–120

Zhao, X., Zhang, P.J. and Wong, T.K. (1993) Application of Beakonization: a new approach to produce transgenic fish. *Mol. Mar. Biol. Biotech.* 2, 63–69

21. Gene Injection into *Xenopus* Embryos

Siew-Ging Gong[1] and Laurence D. Etkin[2],*

[1] *Department of Orthodontics and Pediatric Dentistry, University of Michigan, School of Dentistry, Ann Arbor, Michigan 48109, USA*
[2] *Department of Molecular Genetics, University of Texas M.D. Anderson Cancer Center, Houston, Texas 77030, USA*

Potential Uses of Transgenic Frogs

One of the major problems in the use of amphibians as an experimental system is the fact that they are not amenable to genetic analysis. This is due to their relatively long generation time which makes it extremely difficult to analyze the expression of genes in other than the founder generation. Despite this problem it is clear that injecting genes into embryos has been useful and resulted in the acquisition of an important body of information regarding DNA replication and the regulation of gene expression (Gurdon and Melton, 1981; Etkin, 1982; Etkin and DiBerardino, 1983; for technical aspects see Vize *et al.*, 1991 and Sargent and Mathers, 1991). In addition, laboratories are developing procedures that have the potential to allow the utilization of *Xenopus laevis* as a transgenic system. In this brief chapter we will analyze some of the work that has been done and discuss the potential of *Xenopus* as a system to analyze the function of exogenous genes.

Fate and Replication of Injected DNA

The fate of injected DNAs during early embryonic development of *Xenopus* has been well studied. Injected DNAs form chromatin structures, replicate and are expressed (Bendig and Williams, 1983; Etkin *et al.*, 1984; 1987 and Etkin and Balcells, 1985); however, the behavior of the injected DNAs is dependent upon its being injected as supercoiled plasmid or linear form (Harland and Laskey, 1980; Mechali and Kearsey, 1984; Marini *et al.*, 1988 and Hines and Benbow, 1982). Within the first few cell cycles following injection the exogenous circular or linear DNA is incorporated in nucleus-like structures that resemble authentic nuclei in that they contain a well formed nuclear membrane and possess other features typical of nuclei (Forbes *et al.*, 1983; Shiokawa *et al.*, 1986 and Etkin and Pearman, 1987).

Injected supercoiled plasmids replicate reaching their maximal abundance at the late gastrula-neurula stage and gradually diminish in amount becoming almost undetectable by the late tailbud stages. Linear DNAs, on the other hand, concatenate into high molecular weight complexes, replicate, and persist much longer during development and are often detectable in the different organs of adult frogs (Bendig, 1981; Rusconi and Schaffner, 1981; Etkin and Pearman, 1987; Etkin *et al.*, 1984; Marini *et al.*, 1988 and Fu *et al.*, 1989).

One of the major problems encountered in using the *Xenopus* system for the generation of transgenic animals is asymmetric distribution of injected DNA resulting in mosaic embryos. This interferes with the ability to assess the function of genes since the products of the exogenous DNA sequences are also expressed in a random pattern throughout the embryo. This property is not unique to the amphibian but is typical of all founder generations in all organisms in which there are no specific vectors directing integration into the genome in the pronuclei. One possible means to solve this problem in *Xenopus* may be to inject the DNA sequences into the germinal vesicles of oocytes after which the injected oocytes would be matured and fertilized using several available procedures (see Heasman *et al.*, 1991). Using this approach it may be feasible

---

*Author for Correspondence.

to integrate the exogenous sequences into the maternal pronucleus.

Expression of Injected Genes

In the amphibian embryo, transcription is not detectable until the mid-blastula stage of development. At this time, there is an increase in transcription from the embryonic genome (Newport and Kirschner, 1982 and Etkin, 1988). Exogenous DNA injected into the fertilized egg also follows this same pattern of expression in that there is no detectable transcription until the blastula stage of development. Injected genes coding for rabbit β-globin (Rusconi and Schaffner, 1981), sea urchin histones (Bendig, 1981), *Drosophila* alcohol dehydrogenase (Etkin *et al.*, 1984), yeast tRNA (Newport and Kirschner, 1982), ribosomal RNA (Busby and Reeder, 1983), adult *Xenopus* α- and β-globin (Bendig and Williams, 1983, 1984) and chloramphenicol acetyl transferase (CAT) (Etkin and Balcells, 1985) are all first expressed at the mid-late blastula stage of development. The early embryo therefore, may be viewed as a transient expression system in which detectable transcription on injected templates peaks at the blastula-gastrula stage of development and then declines in parallel with the loss of the extrachromosomal plasmid molecules.

Use of the *Xenopus* System to Study Promoter Elements

A number of cloned *Xenopus* genes have been shown to be regulated in the correct temporal manner when introduced into embryos (Table 21.1). These include the promoters of *Xenopus* rRNA genes (Busby and Reeder, 1983), GS17 (Krieg and Melton, 1985 and Krieg and Melton, 1987), cardiac actin gene (Wilson *et al.*, 1986 and Mohun *et al.*, 1986) and keratin (Jonas *et al.*, 1989). Furthermore, Krieg and Melton (1987) were able to define a 74-base enhancer responsible for activating transcription of the GS17 gene at the MBT. Gong *et al.* (1995) demonstrated that, temporally, the CAT/xnf7 (*Xenopus* nuclear factor 7) fusion gene was expressed following the MBT which is in accordance with the timing of the endogenous xnf7 gene.

The study of tissue-specific regulation of injected genes using the *Xenopus* system has presented a greater challenge. Currently, there are only a few examples where there was correct tissue-specific expression of injected genes (Table 21.1). Examples include those of the α-actin (Mohun *et al.*, 1984), expressed only in muscle, XK81A1 (Jonas *et al.*, 1989) a keratin expressed only in epidermis, and the XMyoDa (Leibham *et al.*, 1994), expressed in a

Table 21.1. Correct temporal and spatial expression of injected genes in embryos.

Correct Temporal Regulation	
Xenopus rRNA	Busby and Reeder, 1983
GS17	Krieg and Melton, 1985, 1987
Cardiac Actin	Wilson *et al.*, 1986, Mohun *et al.*, 1986
Cytoskeletal actin	Brennan, 1990
XK81A1 (Keratin)	Jonas *et al.*, 1989
hsp 70	Krone and Heikkila, 1989
Xnf7	Gong *et al.*, 1995
Xenopus Hepatocyte Growth Factor	Nakmamura *et al.*, 1996
Correct Spatial Expression	
α-actin	Mohun *et al.*, 1984
XK81A1 (Keratin gene)	Jonas *et al.*, 1989
XmyoDa	Leibham *et al.*, 1994
Xnf7	Gong *et al.*, 1995
Xenopus Hepatocyte Growth Factor	Nakamura *et al.*, 1996

somite-specific manner. Using the *Xenopus* system, Jonas *et al.* (1989) found a sequence in the 5′ flanking region of the XK81A1 keratin gene responsible for its tissue-specificity in epidermal cells, while activation of the α-actin gene in muscle requires the presence of the CArG box, part of the "serum response element" a short distance upstream from the initiation site. In the examples mentioned above, expression of the injected genes was mosaic, which is commonly seen due to the unequal distribution of the injected DNA in the embryo.

The *Xenopus* system also was used to dissect the cis-acting sequences involved in the temporal and spatial patterns of expression of the xnf7 gene (Gong *et al.*, 1995). The xnf7 gene is expressed maternally and again during the neurula stage of development in the dorsal region of the embryo (Reddy *et al.*, 1991 and Gong *et al.*, 1995). To study the promoter elements involved in the transcription of the xnf7 gene in *Xenopus* oocytes and embryos, a chimeric gene was constructed using various regions of the 5′ promoter elements of xnf 7 fused to a reporter gene encoding chloramphenicol acetyl transferase (CAT). These constructs were injected into the GVs of oocytes to analyze the promoter elements involved in the basal levels of transcription and into fertilized eggs to analyze the elements involved in the developmental and spatial regulation of expression of the gene.

The results of this analysis demonstrated that the elements controlling the basal level of transcription were contained within the highly G + C rich region 121 basepairs upstream of the transcriptional initiation site. The binding sites of the *trans*-acting factors involved in basal regulation of the gene were mapped by DNase footprinting. This analysis showed binding of factors to the E2F and Sp1 binding sites, located between −68 and −37 and another Sp1 site located at −92 to −84. The fact that 2 Sp1 and an E2F sites were bound by factors in the oocyte extract was not surprising in view of the fact that these binding sites have been implicated in the basal regulation of many other TATA-less promoters.

From the analysis of the CAT activity in embryos microinjected with the CAT/xnf7 fusion genes, the elements involved in the spatial and temporal control of the gene were found to reside within 421 bp upstream of the transcriptional initiation site. Finer mapping using deletion mutants of the promoter showed that the elements responsible for conferring tissue specificity were actually between −421 and −121. Two interesting putative *trans*-acting factor binding sites are present in this region; one an AP1 binding site and the other a Pu box. Further work is needed to define how these two sites function in the tissue-specific expression of xnf7; however, it is clear that this strategy is extremely useful in dissecting the intricate workings of eukaryotic promoters.

Possibilities of Transgenic *Xenopus*

One of the important questions is whether or not the *Xenopus* embryo can be used as a transgenic system. To date the only successful reported attempt demonstrated that an injected exogenous DNA sequence can be transmitted through the germ-line to the next generation; however, in no case was expression of the gene detected (Etkin and Pearman, 1987). In addition, it was not clear to what degree the exogenous DNA was integrated into the genome since it may have been carried through the germ-line as an episomal form. Perhaps expression of the gene was inhibited by hypermethylation which may be reversed if passed through the germ-line for another generation.

One of the major problems was the low degree of integration or passage through the germ-line since only about 5% of the parental generation passed the transgene to the next generation. A possible solution to this problem would be the generation of new vectors that would enhance the degree of integration into the germ-line of the parental generation.

Etkin and Roberts (1983) attempted a novel approach to overcome the problems of mosaicism and to enhance the efficiency of germ-line transfer of exogenous genes. They utilized the procedure of nuclear transplantation to transplant nuclei from embryos that were injected with exogenous genes. The idea was to make clones of embryos that would have the gene of interest integrated into the genome. This idea was elaborated upon by Kroll and Gerhart (1994) and Chan and Gurdon (1996). Their approach was to transfect *Xenopus* tissue culture cells with an exogenous gene afterwhich the nuclei would be transplanted to fertilized eggs. Thus, one would produce clones of frogs derived from the nucleus of the tissue culture cell. The major obstacle with this approach is that the tissue culture cells used were aneuploid and their nuclei could not support normal development. The use of euploid cells would help to overcome this problem.

Future Directions

The development of a transgenic system for *Xenopus* would be extremely useful for attacking problems in the areas of the molecular aspects of gene regulation as well as fundamental embryological problems. One of the areas that requires intensive study is that of the development of vectors to enhance efficient and site directed integration into the genome. Another area would be to develop approaches that would enable one to select embryos or cells in which the gene was integrated and was being expressed. This might be accomplished using sensitive cell sorting devices to select cells expressing a specific cell surface antigen or fluorescent molecule such as the green fluorescent protein. Also, it would be possible to develop lines of transgenic frogs that express specific gene products under the direction of inducible promoters so that one could produce a product at a specific time and within specific groups of cells or tissues. These animals could be important resources for the pharmaceutical industry in that one could produce oocytes overexpressing specific proteins useful in drug assay tests. In addition genetically marked or altered cells could be used for embryological experiments such as tissue grafting or to produce chimeric embryos or as genetic markers in nuclear transplantation experiments.

References

Bendig, M.M. and Williams, J.G. (1983) Replication and expression of *Xenopus laevis* globin genes injected into fertilized *Xenopus* eggs. *PNAS USA* **80**, 6197–6201

Bendig, M.M. (1981) Persistence and expression of histone genes injected into *Xenopus laevis* eggs in early development. *Nature* (London) **292**, 65

Bendig, M.M. and Williams, J.G. (1984) Differential expression of the *Xenopus laevis* tadpole and adult β-globin genes when injected into fertilized *Xenopus laevis* eggs. *Mol. and Cell. Biol.* **4**, 5670

Brennan, S.M. (1990) Transcription of endogenous and injected cytoskeletal actin genes during early embryonic development in *Xenopus laevis*. *Differentiation* **44**, 111–121

Busby, S.J. and Reeder, R.H. (1983) Spacer sequences regulate transcription of ribosomal gene plasmids injected into *Xenopus* embryos. *Cell* **34**, 989–996

Chan, A. and Gurdon, J.B. (1996) Nuclear Transplantation from stably transfected cultured cells of *Xenopus*. *Int. J. Dev. Biol.* **40**, 441–451

Etkin, L.D. (1982). Analysis of the mechanisms involved in gene regulation and cell differentiation by microinjection of purified genes and somatic cell nuclei into amphibian oocytes and eggs. *Differentiation* **21**, 149

Etkin, L.D. and Pearman, B. (1987) Distribution, expression, and germ line transmission of exogenous DNA sequences following microinjection in *Xenopus laevis* eggs. *Development* **99**, 15–23

Etkin, L.D. and Roberts, M. (1983). Transmission of integrated sea urchin histone genes by nuclear transplantation in *Xenopus laevis*. *Science* **221**, 67

Etkin, L.D. (1988) Regulation of the mid-blastula transition in amphibians in *Developmental Biology A comprehensive synthesis* **5**, 209–226

Etkin, L.D. and DiBerardino, M.A. (1983) Expression of nuclei and purified genes microinjected into oocytes and eggs. *Eukaryotic Genes, their structure, activity and regulation* Chapter 9, eds., Maclean, Gregory and Flavell. Butterworth and Co

Etkin, L.D. Pearman, B. Roberts, M. and Bektesh, S. (1984) Replication, integration and expression of exogenous DNA injected into fertilized eggs of *Xenopus laevis*. *Differentiation* **26**, 194–202

Etkin, L.D. and Balcells, S. (1985) Transformed *Xenopus* embryos as a transient expression system to analyze gene expression at the mid blastula transition. *Developmental Biology* **108**, 173–178

Etkin, L.D. Pearman, B. and Ansah-Yiadom, R. (1987) Replication of injected DNA templates in *Xenopus* embryos. *Expt. Cell Res.* **169**, 468–477

Forbes, D.J., Kirschner, M. and Newport, J. (1983) Spontaneous formation of nucleus-like structures around bacteriophage DNA microinjected in *Xenopus* eggs. *Cell* **34**, 13–23

Fu, Y., Hosokawa, K. and Shiokawa, K. (1989) Expression of circular and linerized bacterial chloramphenicol acetyl transferase genes with or without viral promoters after injection into fertilized eggs, unfertilized eggs, and oocytes of *Xenopus laevis*. **198**, 148–156

Gong, S.G., Reddy, B.A. and Etkin, L.D. (1995) Two forms of *Xenopus* nuclear factor 7 have overlapping spatial but different temporal patterns of expression during development. *Mech. of Develop.* **52**, 305–318

Gurdon, J.B. and Melton, D.A. (1981) Gene transfer in amphibian eggs and oocytes. *Annual Review in Genetics* **15**, 189–218

Harland, R. and Laskey, R. (1980) Regulated Replication of DNA microinjected into eggs of *X. laevis*. *Cell* **21**, 761–771

Heasman, J., Holwill, S. and Wylie, C.C. (1991) Fertilization of cultured *Xenopus* oocytes and use in studies of maternally inherited molecules in *Methods in Cell Biology*, Vol. 36, (B. Kay and H.B. Peng, eds.) pp. 214–231

Hines, P. and Benbow, R. (1982) Initiation of replication at specific origins in DNA molecules microinjected into unfertilized eggs of the frog *Xenopus laevis*. *Cell* **30**, 459–468

Jonas, E.A., Snape, A.M. and Sargent, T.D. (1989) Transcriptional regulation of a *Xenopus* embryonic epidermal keratin gene. *Development* **106**, 399–405

Krieg, P.A. and Melton, D.A. (1985) Developmental regulation of a gastrula-specific gene injected into fertilized *Xenopus* eggs. *EMBO* **4**(13A), 3463–3471

Krieg, P.A. and Melton, D.A. (1987) An enhancer responsible for activating transcription at the midblastula transition in *Xenopus* development. *PNAS USA* **84**, 2331–2335

Kroll, K.L. and Gerhart, J.C. (1994) Transgenic *X. laevis* embryos from eggs transplanted with nuclei of transfected cultured cells. *Sci.* **266**, 650–653

Krone, P.H. and Heikkila, J.J. (1989) Expression of microinjected HSP70/CAT and HSP30/CAT chimeric genes in developing *Xenopus laevis* embryos. *Dev.* **106**, 271–281

Leibham, D. Wong, M.W., Cheng, T.C., Schroeder, S., Weil, P.A., Olson, E.N. and Perry, M. (1994). Binding of TFIID and MEF2 to the TATA element activates trans-cription of the *Xenopus MyoDa* promoter. *Molecular and Cellular Biology* **14**(1), 686–699

Marini, N., Etkin, L.D. and Benbow, R. (1988) Persistence and Replication of plasmid DNA microinjected into early embryos of *Xenopus laevis*. *Dev. Biol.* **127**, 421–434

Mechali, M. and Kearsey, S. (1984) Lack of specific sequence requirement for DNA replication in *Xenopus* eggs compared with high sequence specificity in yeast. *Cell* **38**, 55–64

Mohun, T.J., Brennan, S., Dathan, N., Fairman, S. and Gurdon, J.B. (1984). Cell type-specific activation of actin genes in the early amphibian embryo. *Nature* **311**, 716–721

Mohun, T.J., Garret, N. and Gurdon, J.B. (1986) Upstream sequences required for tissue-specific activation of the cardiac actin gene in *Xenopus laevis* embryos. *EMBO* **5**, 3185–3193

Nakamura, H., Tashiro, K. and Shiokawa, H. (1996) Isolation of *Xenopus* HGF gene promoter and its functional analysis in embryos and animal caps. *Roux's Arch. Dev Biol.*, in press

Newport, J. and Kirschner, M. (1982) A major developmental transition in early *Xenopus* embryo, I. Characterization and timing of cellular changes at the midblastula stage. *Cell* **30**, 675–686

Reddy, B.A., Kloc, M. and Etkin, L. (1991) The cloning and characterization of a maternally expressed novel zinc finger phosphoprotein (xnf7) in *Xenopus laevis*. *Dev. Biol.* **148**, 107–116

Rusconi, S. and Schaffner, W. (1981) Transformation of frog embryos with a rabbit beta-globin gene. *PNAS USA* **78**(8), 5051–5055

Sargent, T. and Mathers, P. (1991) Analysis of class II gene regulation. In Methods in Cell Biology, Vol. 36 (B. Kay and H.B. Peng, eds.) pp. 347–365

Shiokawa, K., Sameshima, M., Tashiro, K., Miura, T., Nakakura, N. and Yamana, K. (1986) Formation of nucleus-like structures in the cytoplasm of lambda injected fertilized eggs and its partitioning into blastomeres during early embryogenesis of *Xenopus laevis*. *Develop. Biol.* **116**, 539–542

Vize, P., Melton, D., Hemmati-Brivanlou, A. and Harland, R. (1991) Assays for gene functioning in developing *Xenopus* embryos. In *Methods in Cell Biology*, Vol. 36 (B. Kay and B. Peng, eds.), pp. 367–387

Wilson, C., Cross, G.S. and Woodland, H.R. (1986) Tissue-specific expression of actin genes injected into *Xenopus* embryos. *Cell* **47**, 589–599

PART II, SECTION D

22. The Techniques using Electroporation to Generate Transgenic Fish

Koji Inoue and Shinya Yamashita*

Central Research Laboratory, Nippon Suisan Kaisha, Ltd., Kitano-machi, Hachioji, Tokyo 192, Japan

Introduction of Foreign Genes by Electroporation

Transgenic fish have been generated mainly by microinjection of foreign DNA into the cytoplasm of fertilized eggs or into the pronucleus (germinal vesicle) (Ozato et al., 1989; Chen and Powers, 1990; and Fletcher and Davies, 1991). Although microinjection is an efficient and well-established method, it is a complicated and time consuming operation that requires a great deal of skill. Thus, only limited numbers of eggs can be treated in one experiment. Since fish usually spawn an enormous number of eggs in one spawning, development of a more simple and easy method is expected to enhance the efficiency of experiments. Application of electroporation, which has frequently used to transform cultured cells (Nickoloff, 1995), to fish eggs and sperm has been attempted to develop a mass foreign gene transfer method in fish. Successful foreign gene transfer into fish by electroporation has been reported from several laboratories in recent years.

Since microinjection into a small target is quite difficult, it has been applied to oocytes and fertilized eggs, which are relatively large. The target of electroporation is, on the other hand, not limited by the size of the target. Small target is rather easy to be treated in electroporation. Thus, electroporation has been applied to both fertilized eggs and sperm.

Electroporation of Fertilized Eggs

Foreign gene transfer into fertilized eggs can be achieved by simply immersing eggs in an appropriate buffer containing foreign DNA and applying electric pulses with suitable parameters. For example, we successfully introduced foreign plasmid into fertilized eggs of medaka (*Oryzias latipes*) as follows (see also Inoue et al., 1990, 1995; Inoue, 1992 and Ozato et al., 1992).

1. Collect clusters of fertilized eggs of medaka from the abdomen of the female about 20 min after spawning. Medaka spawn eggs everyday at the beginning of the light period under the controlled photoperiod.
2. Separate eggs by cutting off the attachment filament.
3. Mix eggs and mannitol buffer (0.25 M mannitol, 0.1 mM $CaCl_2$, 0.1 mM $MgCl_2$, 0.2 mM Tris-HCL, pH 7.5) containing 100 μg/ml linearized foreign plasmid.
4. Apply 5 electric pulses of 750 V/cm, 50 μs.
5. Rinse eggs with distilled water and incubate at 26 C until hatching.

Using the simple protocol above, 4–7% of survived larvae were found to have the foreign DNA (Inoue et al., 1990 and Inoue, 1992).

One of the most important factors affecting the survival rate and transformation efficiency is the pulse conditions. Specific conditions should be determined for each apparatus, fish species, buffer, etc. Several conditions previously reported were shown in Table 22.1.

The state of chorion is also an important factor. Inoue et al. (1990), Buono and Linser (1991, 1992) and Murakami et al. (1994) successfully introduced a transgene into fertilized eggs of medaka and zebrafish (*Brachydanio rerio*) without removing the

---

*Author for Correspondence.

Table 22.1. Foreign gene transfer into fish by electroporation of fertilized eggs.

Species	Pulse strength (V/cm)	Pulse length (m/sec)	Pulse number	Electroporator	Buffer	%Survival	% Transformation	Reference
Medaka	750	0.05	5	GTE-1 (Shimadzu)	ME	25	4	Inoue et al. (1990)
	250–500	0.32–0.40	3	ECM600 (BXT)	YS	70	20	Lu et al. (1992)
	750	0.05	5	GTE-1 (Shimadzu)	ME	100	26	Murakami et al. (1994)
Zebrafish	125	7–10	3	Gene Pulser (Bio-Rad)	PBS (Ca-free)	68	65	Buono and Linser (1991)
African catfish	100	0.2	16	Homemade system	DW	60	24–47.5	Müller et al. (1993)

MB, Mannitol Buffer; YS, Yamamoto's Solution; DW, Dechlorionated Water, PBS, Phosphate-Bufferd Saline.

chorion. On the other hand, Müller et al. (1993) reported that foreign gene transfer into dechorionated eggs of African catfish (*Clarias gariepinus*) was successful but that into intact or partially dechorionated eggs was not.

The developmental stage and orientation of eggs may be also important. In general, eggs before first cleavage should be used to reduce the possibility of mosaic incorporation. In most experiments, fertilized eggs at 1–4 cell stages are used for mass electroporation. Yamaha et al. (1988) introduced exotic reagents, propidium iodide, fluorescent dextran lysine and horseradish peroxidase instead of DNA by electroporation into the cytoplasm of dechorionated eggs of goldfish (*Carassius auratus*) just after fertilization. It was shown by fluorescent microphotographs that introduced reagents were distributed widely in the cytoplasm surrounding the surface of eggs just after electroporation and they move to the animal pole during blastdisk formation. This result suggests that the orientation of eggs is not so important if eggs just after fertilization were used. Murakami et al. (1994) introduced a plasmid containing a luciferase gene into one-cell stage eggs of medaka just before cleavage using a localized electric field applied by thin film electrodes formed on a glass plate. In this system, the orientation of eggs seems very important. Introduction of foreign DNA into unfertilized eggs may be difficult because unfertilized eggs may be activated by application of electric pulses and lose ability to be fertilized (our unpublished observation).

Concerning the form of foreign DNA, linearized plasmid is usually used. In some experiments, however, it has been shown that circular plasmid can be also transferred (Müller et al., 1993).

Recently, an alternative method of electroporation, termed "Beakonization" was proposed (Zhao et al., 1993). It was reported that high voltage but low current electric pulses can be applied to eggs/DNA mixture from the electrodes placed apart from the mixture in this method. Several results of this method were summarized in Table 22.2.

The transmission of transgenes to offspring from the founder fish generated by electroporation of fertilized eggs has already been achieved (Inoue et al., 1990 and Zhao et al., 1993). It seems possible to generate stable transgenic lines by electroporation of fertilized eggs. Expression of transgenes has been also achieved in some experiments (Buono and Linser, 1991, 1992; Yamaha, 1992 and Müller et al., 1993).

Sperm-Mediated Foreign Gene Transfer

Since the report by Lavitrano et al. (1989) demonstrating the generation of transgenic mice by merely incubating sperm cells in foreign DNA solution, similar attempts have been made in fish. Most of them, however, have been unsuccessful (for example, Chourrout and Perrot, 1992; and Müller et al., 1992). The only exception is the report by Khoo et al. (1992) that demonstrated successful introduction of foreign DNA into zebrafish by only incubating sperm in a DNA solution. The introduced sequence was existed in the adult fish and transferred to F_1 and F_2 generations but was still supposed to exist extrachromosomally.

It became possible, however, to introduce foreign genes into the sperm by applying electric pulses to sperm/DNA mixture as summarized in Table 22.3. Müller et al. (1992) demonstrated successful foreign

Table 22.2. Foreign gene transfer into fish by beakonization.

Species	Pulse strength (kV)	Pulse time (μsec)	Burst time (sec)	Pulse number	Cycles	Buffer	% Transformation	Reference
Zebrafish	2.5–10	160	0.8	2^6–2^{11}	4	TE/0.008 mM	35–75	Powers et al. (1992)
Common carp						NaCl	0–100	
Channel catfish							0–100	
Black Porgy (Acanthopagrus schlegeli)	3	160	0.4	2^{11}	8	ASW	15	Tsai and Tseng (1994)

ASW, Artificial Seawater.

gene transfer into common carp (*Cyprinus carpio*), tilapia (*Oreochromis niloticus*) and African catfish by electroporation of sperm in a solution containing a linear or circular foreign plasmid. Symonds *et al.* (1994a) also indicated that application of electric pulses to sperm/DNA mixture enhances association between sperm and DNA. They successfully introduced linear or circular pRSV-lacZ into chinook salmon (*Oncorhynchus tshawytscha*) by inseminating eggs with electroporated sperm (Sin *et al.*, 1993 and Symonds *et al.*, 1994b). However, germline transmission of transgenes introduced by electroporation of sperm has never demonstrated in these reports.

While Müller *et al.* (1992) indicated the expression of HSVtk/neo construct by an enzymatic assay, expression of foreign genes introduced by electroporation of sperm still seems difficult to be achieved. One problem in electroporation of sperm may be the amount of foreign DNA carried by sperm. In some cases, existence of transgenes is detectable by polymerase chain reaction (PCR) but undetectable by Southern blotting (Symonds *et al.*, 1994b). It seems that such individuals are mosaic for transgene and the proportion of transgene-positive cells in the whole body may be very low. In addition, foreign DNA seems to remain extrachromosomal in many cases. These may be reasons for undetectable expression.

Conclusion

The three essential steps, introduction, expression and germ-line transmission of transgenes have been achieved by electroporation of fertilized eggs as mentioned above. These results suggest that transgenic lines expressing foreign genes can be generated by electroporation as by microinjection. This method will become a practical method for foreign gene transfer into fish.

Although the gene transfer by electroporation of sperm is not completely established yet and it is

Table 22.3. Foreign gene transfer into fish by electroporation of sperm.

Species	Pulse strength (V/cm)	Pulse length (m/sec)	Pulse number	Electroporator	Buffer	% Transformation	References
Chinook salmon	625	18.6	1	Cell-Porator (BRL)	HBS	10	Sin et al. (1993)
	1000	27.4	2	Cell-Porator (BRL)	HBS	44–85	Symonds et al. (1994)
Common carp Tilapia African catfish	750–2250	5–15		Homemade system	2.8% potassium citrate or 1.8% sodium citrate	2.6–3.5	Müller et al. (1992)

HBS, HEPES-Buffered Saline.

still unknown whether stable lines can be generated by this technique, it still seems a potential method. Sperm is far smaller in size than eggs. If sperm-mediated gene transfer system were established, more individuals can be treated at a time than using eggs, regardless of sizes of eggs. Electroporation of sperm may become an excellent system for molecular biological study as well as for breeding of economically important species.

References

Buono, R.J. and Linser, P.J. (1991) Transgenic zebrafish by electroporation. *Bio-Rad US/EG Bulletin* **1354**, 1–3

Buono, R. J. and Linser, P.J. (1992) Transient expression of RSVCAT in transgenic zebrafish made by electroporation. *Mar. Mol. Biol. Biotechnol.* **1**, 271–275

Chen, T.T. and Powers, D.A. (1990) Transgenic fish. *Trends Biotechnol.* **8**, 209–215

Chourrout, D. and Perrot, E. (1992) No transgenic rainbow trout produced with sperm incubated with linear DNA. *Mol. Mar. Biol. Biotechnol.* **1**, 282–285

Fletcher, G.L. and Davies, P.L. (1991) Transgenic fish for aquaculture. In J.K. Setlow (ed.) *Genetic engineering Vol. 13.* pp. 331–371

Inoue, K. (1992) Expression of reporter genes introduced by microinjection and electroporation in fish embryos and fry. *Mol. Mar. Biol. Biotechnol.* **1**, 266–270

Inoue, K., Yamashita, S., Hata, J., Kabeno, S., Asada, S., Nagahisa, E. and Fujita, T. (1990) Electroporation as a new technique for producing transgenic fish. *Cell. Differ. Dev.* **29**, 123–128

Inoue, K., Hata, J. and Yamashita, S. (1995) Transformation of fish cells and embryos. In J.A. Nickoloff (ed.) *Methods in molecular biology*, Vol. 48, *Animal cell electroporation and electrofusion protocols*, Humana Press Inc., Totowa, New Jersey, pp. 245–251

Khoo, H.-W., Ang, L.-H., Lim, H.-B. and Wong, K.-Y. (1992) Sperm cells as vectors for introducing foreign DNA into zebrafish. *Aquaculture* **107**, 1–19

Lavitrano, M., Camaioni, A., Fazio, V.M., Dolci, S., Farace, M.G. and Spadafora, C. (1989) Sperm cells as vectors for introducing foreign DNA into eggs: genetic transformation of mice. *Cell* **57**, 717–723

Lu J.-K., Chrisman, C.L., Andrisani, O.M., Dixson, J.E. and Chen, T.T. (1992) Integration, expression, and germ-line transmission of foreign growth hormone genes in medaka (*Oryzias latipes*). *Mol. Mar. Biol. Biotechnol.* **1**, 366–375

Müller, F., Ivics, Z., Erdélyi, F., Papp, T., Varadi, L., Horváth, L., Maclean, N. and Orbán, L. (1992) Introducing foreign genes into fish eggs with electroporated sperm as a carrier. *Mol. Mar. Biol. Biotechnol.* **1**, 276–281

Müller, F., Lele, Z., Váladi, L., Menczel, L. and Orbán, L. (1993) Efficient transient expression system based on square pulse electroporation and in vivo luciferase assay of fertilized fish eggs. *FEBS Letters* **324**, 27–32

Murakami, Y., Motohashi, K., Yano, K., Ikebukuro, K., Yokoyama, K., Tamiya, E. and Karube, I. (1994) Micromachined electroporation system for transgenic fish. *J. Biotechnol.* **34**, 35–42

Nickoloff, J.A. (1995) *Animal cell electroporation and electrofusion protocols. Methods in molecular biology*, Vol. 48, Human Press Inc., Totowa, NJ

Ozato, K., Inoue, K. and Wakamatsu, Y. (1989) Transgenic fish: biological and technical problems. *Zool. Sci.* **6**, 445–457

Ozato, K., Inoue, K. and Wakamatsu, Y. (1992) Gene transfer and expression in medaka embryos. In C.L. Hew and G.L. Fletcher (eds.), *Transgenic fish*, World Scientific Publishing, Singapore, pp. 27–43

Powers, D.A., Hereford, L., Cole, T., Creech, K. Chen, T.T., Lin, C.M., Kight, K. and Dunham, R. (1992) Electroporation: a method for transferring genes into the gametes of zebrafish (*Brachydanio rerio*), channel catfish (*Ictalurus punctatus*), and common carp (*Cyprinus carpio*). *Mar. Mol. Biol. Biotechnol.* **1**, 301–308

Sin, F.Y.T., Bartley, A.L., Walker, S.P., Sin, I.L., Symonds, J.E., Hawke, L. and Hopkins, C.L. (1993) Gene transfer in chinook salmon (*Oncorhynchus tshawytscha*) by electroporating sperm in the presence of pRSL-lacZ DNA. *Aquaculture* **117**, 57–69

Symonds, J.E., Walker, S.P. and Sin, F.Y.T. (1994a) Electroporation of salmon sperm with plasmid DNA: evidence of enhanced sperm/DNA association. *Aquaculture* **119**, 313–327

Symonds, J.E., Walker, S.P., Sin, F.Y.T. and Sin I. (1994b) Development of a mass gene transfer method in chinook salmon: optimization of gene transfer by electroporated sperm. *Mol. Mar. Biol. Biotechnol.* **3**, 104–111

Tsai, H.-J. and Tseng, F.-S. (1994) Electroporation of a foreign gene into black porgy *Acanthopagrus schlegeli* embryos. *Fisheries Science* **60**, 787–788

Xie, Y., Liu, D., Zou, J., Li, G. and Zhu, Z. (1993) Gene transfer via electroporation in fish. *Aquaculture* **111**, 207–213

Yamaha, E. (1992) Development of a method for introducing genes into fish cells. In Japan Fisheries Resource Conservation Association (ed.) *Suisanbaiteku donyu kiban seibi jigyou hokokusho*, Japan Fisheries Resource Conservation Association, Toyomi, Chuo-Ku-, Tokyo, Japan, pp. 121–134 (in Japanese)

Yamaha, E., Matsuoka, M. and Yamazaki, F. (1988) Introduction of exotic reagents into denuded eggs of goldfish and crucian carp by electroporation. *Nippon Suisan Gakkaishi* **54**, 2043

Zhao, X., Zhang, P.J. and Wong, T.K. (1993) Application of beakonization: a new approach to produce transgenic fish. *Mol. Mar. Biol. Biotechnol.* **2**, 63–69

23. Gene Transfer in *Drosophila Melanogaster*

Kim Kaiser

Division of Molecular Genetics, IBLS, University of Glasgow, Glasgow G11 6NU, UK

The *Drosophila* P Element

Germ-line transformation of *Drosophila* relies upon the availability of suitable transposon vectors. Of all *Drosophila* transposons, the most heavily exploited has been the P element, a Class II transposable element (for reviews see Engels, 1989; Kaiser et al., 1995). Biological effects of P element transposition attracted the attention of researchers well before the causative agent was identified by molecular biologists. Crosses that we now know to mobilise P elements were observed to cause a diversity of effects including high rates of mutation, chromosomal rearrangement, male recombination (normally absent in *D. melanogaster*), and aberrant gonadal development with consequent sterility. The syndrome is known as hybrid dysgenesis.

Full-length P elements (2.9 kb) have four long open reading frames (Figure 23.1a), encoding an 87 kD transposase (O'Hare and Rubin, 1983). The third intron is not removed in somatic cells, thus restricting transposase activity to the germ-line (Rio, 1991). A P element derivative from which the third intron has been removed (Δ2,3) produces high levels of functional transpose in both the germ-line and soma (Laski et al., 1986). P element sequences required in *cis* for transposition are contained within 138 bp at the 5' end and 150 bp at the 3' end (Mullins et al., 1989). This includes a 31 bp terminal inverted repeat.

Germ Line Transformation

More than a decade has now passed since the P element was first engineered as a transformation vector, and used for the generation of transgenic flies (Spradling and Rubin, 1982; and Rubin and Spradling, 1982). Germ-line transformation of *Drosophila* is achieved by injecting plasmid DNA containing a suitably manipulated P element into embryos undergoing the transition between syncitial and cellular blastoderm (Figure 23.1b). The earliest nuclei to cellularise are a group of germ-line progenitors that migrate to the posterior pole. P element DNA injected into this region can become internalised during cellularisation, and in the presence of P element transposase will transpose to the genome.

An autonomous P element provides its own transposase. P elements engineered as vectors dispense with this ability, but retain sequences required in *cis* for transposition. It is therefore necessary to provide transposase from another source. Different options are the co-injection of a helper element that produces transposase but that cannot itself transpose (see Figure 23.1b; Karess and Rubin, 1984); co-injection of purified transposase (Kaufman and Rio, 1991); injection of the construct into embryos that express transposase endogenously, often from an immobile variant of Δ2,3 (see Laski et al., 1986 and Robertson et al., 1988). The frequency with which transformants are recovered appears inversely related to transposon length, though transformation with cosmid sized pieces of DNA can be achieved (Haenlin et al., 1985).

Transposition is not a frequent event on a per molecule basis, but can nevertheless provide acceptable transformation efficiencies. It is usually the case that only a single transposon becomes integrated. Transient expression has been observed following introduction of DNA into embryos by electroporation (Kamdar et al., 1992). Though not yet demonstrated, germ-line transformation may also be possible by such means.

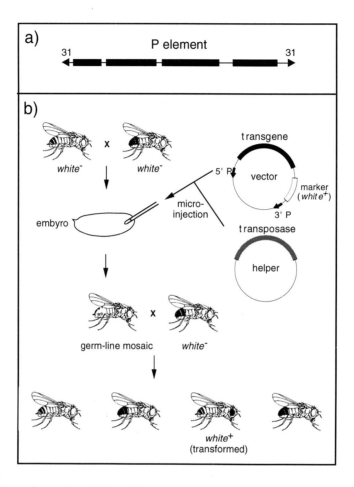

Figure 23.1. P-element transformation of *Drosophila*.
(a) The *Drosophila* P element. Black boxes represent long open reading frames that encode P element transposase; arrowed elements indicate terminal 31 bp inverted repeats.
(b) *Drosophila* germ-line transformation. A P element vector, propagated as a component of a bacterial replicon, contains the transgene one wishes to introduce into the *Drosophila* genome, together with a marker gene (*white*⁺) that enables transformants to be recognised by virtue of red eye colour in a white eyed background. DNA injected at the posterior pole prior to cellularisation can become incorporated into germ-line precursors, and occasional transposition will occur from the injected plasmid to the *Drosophila* genome. Adults that develop from injected embryos are genetic mosaics with respect to the presence of the transposon in their germ-line. In the next generation, however, there will be a few true-breeding individuals that are recognisable by virtue of carrying the marker (here shown rescuing an eye pigmentation defect). Where transposase has been provided by the injected embryos themselves, a cross that separates the transformation construct from the transposase source may be required to ensure that the former is stably maintained at the site of insertion.

A transformed phenotype will not usually be recognised in flies developing from injected eggs. It is rather the progeny of injected individuals that provide the first evidence of successful transformation, since their somatic as well as their germ-line cells contain the P element construct. In order that successful transformation can be easily recognised, P element vectors have been engineered to carry a variety of markers. These include genes that rescue visible phenotypic defects, e.g. loss of eye or body pigmentation, or abnormal eye morphology. Alternative markers, such as Adh and neomycin resistance, confer the ability to survive on selective media (see Ashburner, 1989).

Expression of Transgenes

A wealth of information concerning gene function and control can be obtained by the generation of transgenic individuals. In the case of a newly cloned gene, rescue of a mutant phenotype by transformation provides confirmation that a wild-type version of the mutant allele has indeed been cloned, and identifies the extent of flanking sequences required for normal expression. Transformation with manipulated versions of the gene can then be used to address specific aspects of gene function and control in the context of the whole organism. Transformation constructs that include a heat shock promoter allow timing of gene expression to be manipulated, and thus allow one to address questions concerning development.

It is important to recognise that there can be pronounced effects of genomic flanking sequences on the expression of genes contained within transformation constructs. It is thus advisable to generate several transformed lines, each containing an independent insertion. These can be generated either as primary transformants, or via remobilisation of a construct by a cross that provides Δ2,3. It should be noted that P element transposition is non-random with respect to insertion site. Moreover, sequences contained within a P element construct can have a pronounced effect on insertion specificity (Kassis et al., 1992).

Enhancer-trap Elements

As mentioned above, it is often desirable to express transgenes under the control of cell-type specific regulatory elements. A strategy known as 'enhancer-trapping' can provide a rich source of such elements. An enhancer-trap element is a modified P element, close to one end of which lies a 'reporter' gene. Due to the lack of a transcriptional enhancer, the reporter has a negligible level of intrinsic expression. In order for it to be expressed at significant levels, the transposon must insert close to an endogenous Drosophila enhancer. Reporter activity in a line with only one insertion thus reflects the temporal and spatial expression characteristics of a flanking gene. The reporter of first generation enhancer-trap elements was the lacZ gene of E. coli (O'Kane and Gehring, 1987).

There is now a second generation enhancer-trap element, P{GAL4}, that allows more direct targeting of a desired gene product to the marked cells (Brand and Perrimon, 1993; Kaiser, 1993; Brand and Dormand, 1995). The reporter of P{GAL4} is a yeast transcription factor that is functional in Drosophila, and that can be used to direct expression of other transgenes placed under the control of a GAL4-dependent promoter (UAS_G). A cross between a fly with a new GAL4 insertion and a fly containing UAS_G-lacZ for example, causes β-galactosidase to be expressed in a pattern that reflects GAL4 distribution. Any number of other transgenes can then substitute for lacZ. A particularly attractive feature of this system is that any UAS_G-transgene construct can be used in conjunction with any P{GAL4} line. Examples of the use of the P{GAL4} system are discussed elsewhere in this volume (Armstrong and Kaiser, 1996).

Other Drosophila Transposon Vectors

Among other transposable elements that have been successfully re-introduced into the Drosophila germline are *hobo* and *mariner*. Functional *hobo* elements have a structure similar to that of P elements, and have been used as transformation vectors and enhancer-trap elements (Blackman et al., 1989). The *mariner* element has been introduced into D. melanogaster by P element transformation, but can also transform this species unaided (Garza et al., 1991 and Lidholm et al., 1993). *Mariner*-like elements have also been found in a wide range of other species (Robertson, 1993). This wide host distribution suggests that *mariner* could be developed as an integrative system for transformation of a wide range of animal species or phyla.

References

Ashburner, M. (1989) *Drosophila, A laboratory handbook*, Cold Spring Harbor Laboratory Press, Cold Spring Harbor, NY

Armstrong, J.D. and Kaiser, K. (1996) Enhancer-trap studies of the *Drosphila* brain. This volume

Blackman, R.K., Koehler, M.M.D., Grimaila, R. and Gelbart, W.M. (1989) Identification of a fully functional *hobo* transposable element and its use in germ-line transformation of *Drosophila*. *EMBO J.* **8**, 211–217

Brand, A.H. and Dormand, E.L. (1995) The GAL4 system as a tool for unravelling the mysteries of the *Drosophila* nervous system. *Current Opinion in Neurobiology* **5**, 572–578

Brand, A.H. and N. Perrimon (1993) Targeted gene expression as a means of altering cell fates and generating dominant phenotypes. *Development* **118**, 401–415

Engels, W.R. (1989) P elements in *Drosophila melanogaster*. In D.E. Berg and M.M. Howe, (eds.), *Mobile DNA*. American Society for Microbiology, pp. 437–484

Garza, D, Medhora, M., Koga, A. and Hartl, D.L. (1991) Introduction of the transposable element *mariner* into the germline of *Drosophila melanogaster*. *Genetics* **128**, 303–310

Haenlin, M., Steller, H., Pirrotta, V. and Mohier, E. (1985) A 43 kilobase cosmid P transposon rescues the fs(1)K10 morphogenetic locus and three adjacent *Drosophila* developmental mutants. *Cell* **40**, 827

Kaiser, K. (1993) Second generation enhancer traps. *Current Biology* **3**, 560–562

Kaiser, K., Sentry, J. and Finnegan, D. (1995) Eukaryotic transposable elements as tools to study gene structure and function. In D.J. Sherratt (ed.) *Mobile Genetic Elements* IRL Press, pp. 69–100

Kamdar, P., von Allmen, G. and Finnerty, V. (1992) Transient expression of DNA in *Drosophila* via electroporation. *Nucl. Acids Res.* **20**, 3526

Karess, R.E and Rubin, G.M. (1984) Analysis of P transposable element functions in *Drosophila*. *Cell* **38**, 135

Kassis, J.A., Noll, E., Vansickle, E.P., Odenwald, W.F. and Perrimon, N. (1992) Altering the insertional specificity of a *Drosophila* transposable element. *Proc. Natl. Acad. Sci. USA*. **89**, 1919–1923

Kaufman, P.D. and Rio, D.C. (1991) Germline transformation of *Drosophila melanogaster* by purified P element transposase. *Nucl. Acid. Res.* **19**, 6336

Laski, F.A., Rio, D.C. and Rubin, G.M. (1986) Tissue-specificity of *Drosophila P element* transposition is regulated at the level of mRNA splicing. *Cell* **44**, 7

Lidholm, D.D., Lohe, A.R. and Hartl, D.L. (1993) The transposable element *mariner* mediates germline transformation in *Drosophila melanogaster*. *Genetics* **134**, 859–868

Mullins, M.C., Rio, D.C. and Rubin, G.M. (1989) Cis-acting DNA sequence requirements for P element transposition. *Genes and Development* **3**, 729

O'Hare, K. and Rubin, G.M. (1983) Structures of P transposable elements and their sites of insertion and excision in the *Drosophila melanogaster* genome. Cell **34**, 25–35.

O'Kane, C.J. and Gehring, W.J. (1987) Detection *in situ* of genomic regulatory elements in *Drosophila*. *Proc. Natl. Acad. Sci. USA* **84**, 9123–9127

Rio, D.C. (1991) Regulation of *Drosophila P element* transposition. *TIG* **7**, 282

Robertson, H.M. (1993) The *mariner* transposable element is widespread in insects. *Nature* **362**, 241

Robertson, H.M., Preston, C.R., Phillis, R.W., Johnson-Schlitz, D.M., Benz, W.K. and Engels, W.R. (1988) A stable genomic source of P element transposase in *Drosophila melanogaster*. *Genetics* **118**, 461–470

Rubin, G.M. and Spradling, A.C. (1982) Genetic transformation of *Drosophila* with transposable element vectors. *Science* **218**, 348–353

Spradling, A.C. and Rubin, G.M. (1982) Transposition of cloned P elements into *Drosophila* germ line chromosomes. *Science* **218**, 341–347

PART II, SECTION E

24. Genetic Transformation in *Caenorhabditis Elegans*

Danielle Thierry-Mieg, Karine Naert and Claire Bonnerot
CRBM, CNRS, BP5051, 34033 Montpellier, France

The free living nematode *Caenorhabditis elegans* was chosen by Sydney Brenner in 1965 (Brenner, 1974) as a model animal to study development and the nervous system. The phylum includes a large number of parasitic species of medical or agricultural interest, but the choice was driven by many other good reasons that are detailed in the book "The Nematode *C. elegans*" (ed. Wood et al., 1988). This book presents a portrait of the worm before transformation was commonly used.

C. elegans is an especially simple organism; nematodes are among the very first triploblastic metazoans, having arisen 500 millions years ago. The adult *C. elegans* hermaphrodite has 959 somatic nuclei, and is essentially invariant in cell number or organization in different individuals. Because the animal is transparent, one can follow its development at the cellular level in living specimens using Nomarski microscopy. This property has permitted the description of the entire cell lineage, that is the timing of all divisions from the zygote to the adult and the positions and genealogical relationships of all cells in hermaphrodites and males (reviewed in Sulston, 1988). The complete anatomy has been resolved at electron microscopy level (reviewed in White, 1988; Sulston and White, 1988).

C. elegans is grown easily in the laboratory, on agar plates with *Escherichia coli* as a food source. Moreover, the whole organism can be revived after being stored frozen for years in liquid nitrogen. *C. elegans* is particularly well suited for genetic analysis: its life cycle is short (two and a half days in optimal conditions) and its two sexes, self-fertilizing hermaphrodite and males, allow for both clonal propagation and crossing (see Sulston and Hodgkin, 1988). An important consequence of this mode of reproduction is that our reference strain is truly homozygote and identical in all worm laboratories. Approximately 1700 genes defined by mutation and 600 rearrangements have been mapped to date.

Finally, *C. elegans* has a compact genome of 100 megabases, with about one putative gene per 5 kb. Physical mapping is complete, with an average of eight fold coverage by cosmids and seven by YACs. Half of the genome is now sequenced, and the other half should be completed by 1998 (Waterston and Sulston, 1995). An expression map of the genome is being built by Kohara et al. (pers. comm.), who already tag-sequenced 15 000 cDNA, representing about 4000 of the 13 000 predicted genes, and is now investigating their location in the genome and their expression pattern by *in situ* hybridization (Tabara et al., 1996). All in all, a picture pointing to a good conservation of mechanisms and molecules from nematodes to men emerges.

A special feature of the *C. elegans* community, 2000 people in 150 labs, is the belief that cooperativity and openness are the most efficient motors for research. Information is exchanged long before publication in the Worm Breeders Gazette, and a devoted database, *acedb* (Durbin and Thierry-Mieg, pers. comm.), in which data about people working on the worm, bibliography, genes, clones and sequences are collated, is available by anonymous ftp at *ftp://ncbi.nlm.nih.gov/repository/acedb*.

Genetic transformation consists in introducing, maintaining and expressing an extra piece of DNA in the whole organism. It provides a powerful functional test of gene activity *in vivo*. Its impact on prokaryotic, yeast and Drosophila genetics prompted the development, in the eighties, of a method to transform the worm. Microinjection in *C. elegans* germline was first carried out by Kimble *et al.* (1982) who rescued a nonsense mutation by injection of suppressor tRNA in the syncytial gonad. Stinchcomb *et al.* (1985) then showed that injected DNA could be maintained in transgenic lines as extrachromosomal arrays. Fire (1986) used a marker to select integrated transformants. Mello *et al.* (1991) introduced a more versatile marker and optimized conditions yielding high frequencies of transformed lines. DNA transformation is now commonly used in all the worm community for rescue experiments, mutation characterization, expression studies, gene interactions studies, mosaic analysis and inter-specific complementation tests.

Transformation

Microinjection Technique

The Setting. Each *C. elegans* adult hermaphrodite contains two reflexed gonad arms (Figure 24.1). The distal part of each gonad is a syncytium where a common core of cytoplasm is surrounded by germ line nuclei. The distal nuclei are mitotic, while more proximal are progressing through meiotic prophase. Near the bend in the gonad, plasma membranes begin to fully encompass individual oocyte nuclei, which arrest in diakinesis of meiosis I.

In contrast to many other organisms, transformation is not performed in the zygote, but rather in the germ line of the adult. The DNA is injected either in maturing oocytes (Fire, 1986), or in the syncytial gonad (Kimble, 1982; Stinchcomb *et al.*, 1985 and Mello *et al.*, 1991). This last method is easily performed (Figure 24.1) and is now the most widely used. Detailed protocols can be found in a recent review by Mello and Fire (1995).

Briefly, worms are immobilized by gentle desiccation on a dried 2% agarose pad on a microscope slide under a mineral oil layer. An injection needle containing DNA in TE (10 mM Tris, 0.1 mM EDTA) is introduced in each syncytial gonad, which blows like a balloon. Although most labs follow the procedure described in Mello and Fire (1995), using compressed nitrogen to generate the pressure and an inverted microscope, a simple static air pressure system such as a 50 ml plastic syringe connected by teflon tubing to the microinjection needle can be used, and injections can be performed on a regular upright microscope equipped with a good 20x Nomarski objective with extra long working

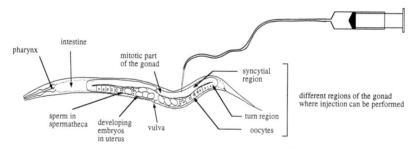

Figure 24.1. Schematic representation of injection of an adult *C. elegans* hermaphrodite.
The hermaphrodite's head is to the left (see the pharynx) and its tail to the right. The two reflexed arms of the gonad join at the vulva. The germ cells in the distal area of the gonad undergo mitosis, then meiotic prophase starts in a syncytium, with nuclei surrounding a common core of cytoplasm. Progressing around the turn of the gonad, cell membranes enclose the nuclei to form immature oocytes. Mature oocytes arrested in prophase of the first meiotic division pass through the spermatheca where they are fertilized by sperm made earlier by the hermaphrodite, during the fourth larval stage. The fertilized eggs start their development in the uterus before they are laid through the vulva. About three hundred progeny are laid per animal in the course of three days. For the injection, a dehydrated agarose pad is made by depositing on a microscope slide one melted drop of 2% agarose in water, flattened by dropping another slide or a coverslip of a size chosen according to the preferred gluing strength: – the smaller and lighter the coverslip, the thicker, more absorbent and stickier the pad is, the quicker one must inject – . The pad is dehydrated by cooking one hour at 100°C. In our lab, to perform the injection, ten to twenty young or middle-aged adults are caught on a platinum wire loaded with bacteria, then transferred under paraffin oil onto the pad. One patiently lets them go and comfortably lay on the pad, which sticks them immobilized. The DNA solution is carefully injected in the syncytial gonad. The pressure of about two bars, obtained by compressing the air in the syringe to about half of its original volume, is maintained constant during the experiment, since the needle will not flow in the oil, but will as soon as it penetrates the animal. The worms are then rehydrated in recovery buffer or directly in M9 buffer, as described by Mello and Fire (1995).

distance (about 1 cm). Movement of the standard microscope stage then drives the needle into the worm, so the needle holder can be a simple and cheap micromanipulator.

Nature and Quality of the Injected DNA. A wild variety of sources of nucleic acids has been successfully used for microinjection into worms. These include RNA, genomic DNA, PCR and restriction fragments or cloned sequences. There are no constraints on the constructs or the vectors. Plasmids, phages, cosmids or yeast artificial chromosomes (YAC) containing inserts from many species, including nematodes, Drosophila, yeast and man, have been successfully introduced and maintained in the worm. The injected sample must however be sufficiently pure to avoid toxicity.

For plasmid and cosmid preparations, a simple alkaline lysis followed by a CTAB (hexadecyl-trimethyl-ammonium bromide, Sigma H6269) precipitation gives DNA suitable for injection (see the protocol in Mello and Fire, 1995). Purification over a CsCl gradient is not needed. Usually, the whole plasmid or cosmid DNA, including the vector, is injected directly in its supercoiled form, at concentrations of 5 to 200 µg/ml. Cosmids from the wild type reference strain, covering above 80% of the chromosomes, are available from the Cambridge collection (see Coulson et al., 1988). Since worm introns are small, most genes fit in a cosmid or two. If no cosmid seems to cover the area of interest, a YAC can be used. YAC DNA is separated from yeast chromosomes by pulsed field electrophoresis and then recovered from the gel slice by standard techniques. A method in which yeast is used to recombine the YAC with subclones from a local cosmid to selectively produce shorter YAC derivatives has been developed by Miller et al. (1993).

Fate of the Injected DNA

Recombination and Ligation. If the DNA concentration is sufficient (>100 µg/ml), the sequences newly injected in the syncytial gonad undergo efficient recombination reactions, including ligation of blunt ends of linearized DNA molecules and homologous recombination between injected circular or linear molecules within shared regions of homology. This leads to the formation of extrachromosomal arrays, one or more megabases long, containing hundreds of copies of the exogenous molecules (Stinchcomb et al., 1985 and Mello et al., 1991).

Recombination among the injected molecules during array formation is so efficient that a gene injected in two pieces (two circular plasmids or PCR products) overlapping by only 600 basepairs will rescue lines as well as a single plasmid containing the entire gene (Aroian et al., 1990 and Mello et al., 1991). The ability of the worm to reconstitute a gene or a molecule from its overlapping fragments with the highest fidelity is a feature with large impact: it relaxes the problems of getting very large inserts, it facilitates the precise functional mapping of mutations (see section **Molecular Identification of a Genetic Locus**), it also allows efficient cotransformation by co-injection of various types of DNA, provided they share some homology. For example, two plasmids coinjected at similar concentrations and sharing only the antibiotic resistance area will cosegregate in 90% of the lines.

Such high levels of recombination among injected molecules have also been observed in mammalian cells, Drosophila, Xenopus, sea urchin and fish, but the special feature of C. elegans is that these arrays can be maintained as supernumerary chromosomes in a propagating strain.

Transient Expression. Transient expression of the injected DNA can occur immediately after injection, or more generally in the progeny of the injected animals (Fire, 1986). It probably reflects the presence of newly formed arrays of exogenous DNA in some of the cells of the animal, which is thought to be mosaic. Altogether, 10% to 40% of the embryos derived from a germ cell that received the DNA will display transient expression. A single injection in the syncytial gonad affects about one hundred nuclei and yields 15 to 50 transiently expressing progeny per animal (Mello et al., 1991). In contrast, 1 or 2 such progeny will arise from a worm injected in its 10 to 20 maturing oocytes.

Extrachromosomal Lines. The extrachromosomal array can segregate by chance to the germ line in a few percent of the progeny. This leads to the establishment of a few transgenic lines per injected animal. Separate progeny from a single injected animal inherit independent assemblies, often visible as extra dots in transgenic lines on DAPI stained oocytes (4,6-diamidino-2-phenylindole is a fluorescent DNA dye, used at 1µg/ml) (Stinchcomb et al., 1985 and Mello et al., 1991). The extrachromosome then replicates and imperfectly segregates in mitosis and meiosis, behaving genetically like a free duplication.

The loss of the extrachromosomal array in mitosis (see the pioneer work of Stinchcomb et al., 1985) can be exploited for mosaic analysis (see section **Mosaic Analysis**). For meiosis, each line is characterized by an average level of transmission, ranging from 5% to 95%, so that the stock usually has to be maintained by manual selection. In fact, a closer look shows that large variations among individuals within a stock persist even after 9 generations (C.B. unpublished data, Figure 24.2). A plausible cause

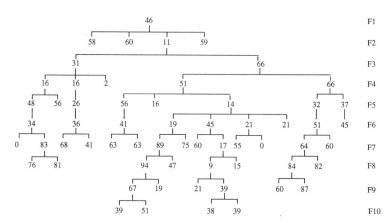

Figure 24.2. Transmission of an injected roller transgene.
In this experiment, plasmid pRF4 encoding *rol6d* (*su1006*) (Mello et al., 1991) was injected at 100 μg/ml in the syncytial gonad. The progeny of one of the F1 Rollers produced by the injected hermaphrodite was followed by randomly picking individual Rollers through 9 generations (F2 to F10). The percentage of Roller animals in each individual progeny is indicated. The average number of progeny per animal was 237 ± 59. Notice the high variability in the transmission rate.

is that a variable number of the minichromosome, i.e. more than two, are segregating in the line. Indeed, DAPI staining shows that some oocytes contain more than one extra dot.

Despite the widespread use of worm genetics, we still know surprisingly little about the mechanisms of chromosome segregation in mitosis as well as meiosis, or about the genes involved. Although invaluable, the maintenance of the extrachromosome is not understood. It may even not be necessary to inject nematode sequences to get a stable array (Stinchcomb et al., 1985). The stability may relate to the diffuse nature of the kinetochores in *C. elegans* (Albertson and Thomson, 1982; 1993) or to the presence in the mini-chromosome of nematode chromosomal fragments playing the roles of telomeres and/or centromeres (see Mc Kim et al., 1988; 1993 and Villeneuve, 1994).

Integrated Lines. Extrachromosomal arrays can be forced to integrate in a chromosome by standard gamma or X-ray irradiation (see protocol in Mello and Fire, 1995) or, conveniently, by UV irradiation (300 J/m^2 at 254nm in a UV cross linker, available in most molecular biology laboratories, Mitani, 1995). The major advantage of forcing integration is that the array will then segregate in a Mendelian fashion, eliminating the need to manually select the transformed animals. Of course, the transformant will still contain a large number of copies of the injected DNA. Also one should outcross the stock to remove the eventual mutations induced by the irradiation.

Alternatively, if a small number of integrated copies is critical, it is worth mentioning that, during microinjection, spontaneous events of integration of a few copies, typically one to ten, in a single chromosomal site may occur anywhere in the genome. These events are very rare, but similar rates of about one integrated line per 20 to 100 injected nematodes are obtained after syncytial gonad or maturing oocytes injections (Fire, 1986; and Mello et al., 1991). To get this frequency of low copy integration, Mello coinjects 0.7 to 1 mg/ml of a 50-mer oligonucleotide in syncytial injections, while Fire aims at the oocyte's nucleus. In both cases, a plausible interpretation is that integration requires the reactivation of chromosomal recombination or repair mechanisms, which are no longer active in the areas injected, but can be stimulated by physical damage, such as the injection needle hitting the chromosomes during nuclear injections or the oligos triggering mutations and reparation in the syncytial gonad.

Markers Used for Cotransformation

It is often advisable, though not required in particular for rescue experiments (see for instance Way and Chalfie, 1988) to coinject a marker or a carrier with the DNA of interest. Two types of benefits are expected: first, use of a visible or selectable dominant marker makes the screening and further stock keeping easier, especially for low transmission lines. Second, use of a carrier DNA as neutral as possible can prevent the adverse consequences of dosage effects, by keeping a low copy number of the gene in an otherwise large extrachromosomal array, composed mainly of the carrier.

The most common marker today is probably the semi-dominant *rol-6d* allele *rol-6*(*su1006*). Plasmid pRF4, containing a 4 kb EcoRI genomic fragment from the mutant, was first used by Mello et al. (1991). It encodes a mutant collagen that makes the worm roll along its longitudinal axis towards the right and consequently move in circles, in contrast to the straight sinusoidal movement of the wild type (Kramer et al., 1990). As a marker, rol-6d has the advantage of not killing the worms or making them sterile even when present in hundreds of copies. Just a few copies can be recognized in a wild type background. Injection in rol-6 loss-of-function mutants, phenotypically wild type, may ease the selection of both transiently rescued and transmitting transformants, raising the apparent rates of success (Mello et al., 1991). One limitation of rol-6d as a marker lies in the difficulty to recognize the roller phenotype in animals that do not move, for example in some uncoordinated mutants. The problem may be circumvented by injecting heterozygote mutants. Also, some wild type genes encoding other cuticular proteins are needed for the realization of the roller phenotype of su1006, so that for example the loss of function allele sc103 of the collagen encoding gene sqt-1 suppresses rol-6d (Kramer and Johnson, 1993). The suppression of the roller phenotype by the sqt-1 allele should also ease cell identification in transgenics by restoring a normal morphology. As a carrier, no general statement can be made, but a number of cloned genes appear expressed and regulated properly in arrays with rol-6d. However rol-6 is not neutral; its influence in some arrays may lead to a bias against late larval expression.

Wild type cloned genes selected by their ability to rescue the corresponding recessive visible mutants are also used as cotransformation markers. These genes include dpy-20 (clone pMH86 Clark et al., 1995), lin-15 (clone pJM23, Huang et al., 1994), unc-4 (clone pNC4-21, Miller and Niemeyer, 1995) or the antisense unc-22 construct (Fire et al., 1991) In theory, rescue of lethal or sterile alleles (e.g., fem-1, Spence et al., 1990, or pha-1ts, Granato et al. 1994) should provide a positive self-selecting screen for transformed lines. In this vein, the first marker used for transformation was a sup-7d semidominant allele which encodes an amber suppressor tRNA, selected by its suppression of a sterile allele (Fire, 1986). But, as expected, this marker has general effects on translation. Consequently, it is viable or subviable below 5 or 6 copies per worm but lethal above. This lethality prevents sup-7d being used as a carrier for extrachromosomal lines, but makes it a useful marker to select for low copy number integration (Fire, 1990b).

Caveats and Hints

Poisoning and Dosage Effects. If the DNA preparation is insufficiently pure or if the presence of a large amount of an exogenous sequence is deleterious to the worm, toxicity can result, leading to too few or dead progeny and to a lack of transformants. The poison effect may be due to the overrepresentation of a piece of DNA that, among other possibilities, swamps a transcription factor, or to the overexpression of a gene encoded by the injected DNA. One should then test a series of dilutions of the DNA of interest, down to 1 µg/ml, with a concentration of the carrier (i.e. a cotransformation marker or just any DNA, such as Bluescript) kept around 100 µg/ml. Alternatively, an integrated low copy transformant could be sought.

Expression of the gene can also lead to more subtle effects, such as reduced fertility in the transformed lines. It is worth trying to rescue at different temperatures, since these effects are often temperature sensitive.

Position Effects. Some position effects have been described in integrated and in extrachromosomal lines (e.g., Mello et al., 1991 and Okkema et al., 1993). They probably reflect the action of enhancers brought close to the gene of interest. In the case of integrated lines, a variety of specific position effects depending on the integration site are expected. In an extrachromosome, composed mostly of a marker with characterized biases, the effects might be more constant. The ideal solution would be to use, as carrier for extrachromosomal lines, neutral pieces of DNA. But they may or may not exist.

Induction of Mutations. In worms, like in Drosophila and probably in all organisms, injection of DNA is mutagenic. Of course, integration of the exogenous DNA in the chromosome may lead to gene disruption and homozygous lethality. This happens in about one-third of the integrative transformations (see for instance Fire and Waterston, 1989; and MacMorris et al., 1992) in line with the coding potential of the whole genome ($\sim 30\%$ ORF). But in addition, a number of mutations, including suppressors, are induced upon microinjection of DNA in the gonad, resulting in pseudo-transformants that do not contain the exogenous DNA (e.g., Fire, 1986; Fire and Waterston, 1989; and Fire et al., 1991, DTM, unpublished observation). In our opinion, this is too frequent to be overlooked, and it remains essential to confirm the presence of the injected DNA in the transformants, by Southern blot or atleast by PCR.

Applications

Molecular Identification of a Genetic Locus

Mutant Rescue. Genetic transformation experiments have been widely used in the worm to clone mutationally defined genes. The gene of interest is first mapped precisely by classical genetics, using visible markers (Brenner, 1974) or molecular polymorphic markers (Williams et al., 1992), the genetic and physical maps are compared, the position of the gene is narrowed down to a few cosmids. The relevant cosmids or YACs from the Cambridge wild type collection are injected, in pools or alone, with or without a cotransformation marker, into mutant animals preferably containing a loss-of-function allele, and the progeny are screened for rescue. For genes with lethal, sterile or unhealthy phenotypes, the DNA is introduced into heterozygote rather than homozygote mutants. Alternatively, the cosmid and a marker, usually *rol-6d*, can be injected into a wild type strain and the mutation tested subsequently by genetic crosses. A collection of such transformants, containing various cosmids from the sequencing project and the *rol-6d* plasmid, is being built by Ann Rose and David Baillie's labs and made available to the community (*http://darwin.mbb.sfu.ca/imbb/dbaillie/cosmid.html*, this project is funded by CGAT). The outcome of any injection experiment should be carefully examined, in search of effects reflecting overexpression of a wild type gene and leading to a visible hypermorphic phenotype; the concentrations should be adapted consequently. Han and Sternberg (1990) discovered in this way that *lin-34* is an allele of *let-60* (see also Wu and Han, 1994). Once rescue has been obtained, restriction digests (e.g., Kim and Horvitz, 1990; beware of ligation), overlapping clones, and ultimately subclones may be tested to define the smallest possible rescuing fragment.

Confirming Gene Identification. Transformation rescue is quite a convincing argument that a gene is cloned, but it does not constitute a definitive proof. Indeed, the rescue could be due to a secondary effect linked to the overexpression of a wild type suppressor gene encoded in the clone or to the induction of suppressor mutations. Confirming indications can be obtained by introducing various frameshift, deletions or stop mutations in the clone and testing them for elimination of the rescuing activity (e.g., Hengartner and Horvitz, 1994). This is especially suited if the gene is part of a polycistronic unit, which could include 16% of the nematode genes (Zorio et al., 1994). In addition, transformation with fusion genes containing pieces of cDNA can help define which of various alternatively spliced products is responsible for the rescue (e.g., Perry et al., 1993 and Rhind et al., 1995). Sequencing of a few alleles will also prove that the appropriate gene has been identified. But if the gene is large or if sequencing needs be minimal, an elegant method using microinjection again has been developed by B. Williams and P. Hoppe (pers. comm.) to narrow down to a few hundred basepairs the localization of the relevant mutation. The method takes advantage of the technique of long PCR, which allows routine amplification of DNA fragments up to 30 kb (Barnes, 1994), and of what was learnt about homologous recombination during array formation from Mello et al. (1991). The useful observation is that, if a gene is injected in two or more pieces overlapping by only 295 basepairs, a double stranded break located very close to the region of homology will trigger recombination to reconstitute faithfully a complete mosaic gene, yielding a number of transformed lines comparable to that obtained after injection of the full linear plasmid.

The principle of the experiment is depicted in Figure 24.3. Long PCR fragments are copied from the wild type clone and complementary overlapping short fragments are copied from the mutant genomic DNA of a gene of interest. DNA can be made from a single embryo if the mutant cannot be maintained as homozygote (Williams et al., 1992) After microinjection of a mixture of these PCR fragments, the regions of the mutant able to rescue a null allele when associated to the complementary wild type long PCR fragment are functionally wild type, while those not rescuing are suspect of containing a mutation.

Description of a Pattern of Gene Expression

Once a gene is cloned, one may want to determine its expression pattern during development and to analyze the sequences required for its proper expression. Transformation using reporter genes in transcriptional or translational fusions is an appropriate tool, and a growing collection of expression vectors is made available to the community by Andy Fire (e-mail: *fire@mail1.ciwemb.edu*, documentation at ftp address *ciw1.ciwemb.edu*) (Fire et al., 1990a; and Fire, Xu, *and coll.* pers. comm.).

Reporter Genes. The most commonly used reporter to date is the *E. Coli LacZ* gene, encoding β-galactosidase. Its activity can be detected with a sensitive histochemical reaction (Fire, 1992) or by using specific antibodies (e.g., Herman et al., 1995). A nuclear localization signal can be inserted to concentrate the staining and ease cell identification; a synthetic secretion signal or a mitochondrial matrix localization signal may also be useful (Fire et al., 1990a). The reaction can be performed on fixed animals or simply after permeabilization of the cuticle; this allows to describe the pattern of expression in the mother without killing its progeny retained in the uterus (Fire, 1992; see for instance Stone and Shaw, 1993). *LacZ* has been used in translational fusions, and Miller and Niemeyer (1995) succeeded in generating a fusion protein that still rescued the *unc-4* mutant while the cells in which it was expressed could be stained.

Alternative reporters, using peptide tagging with a defined epitope which may not interfere with the rescuing activity, such as the nine aminoacids from influenza hemagglutinin recognized by monoclonal antibody 12CA5 (Field *et al.*, 1988; and Bloom and Horvitz, pers. comm.) or any other recognizable modification can also be used (e.g., MacMorris *et al.*, 1992).

Lately, a new reporter gene encoding an intrinsically fluorescent protein, the green fluorescent protein (GFP), from the jellyfish *Aequoria victoria* was described by Chalfie *et al.* (1994). This reporter allows the direct visualization of gene activity in live and developing animals (see for instance Troemel *et al.*, 1995), opening a realm of exciting possibilities for the future. The detection of the original GFP was less sensitive than that of β-gal, but it is steadily improving (Fire, Xu *and coll.*, pers. comm.), for example by the introduction of mutations in the site of fluorochrome formation (following Heim *et al.*, 1994), by the addition of introns, or by targeting the molecule to mitochondria or nuclei. GFP is a small protein, so that it may behave well as a mere dye, tagging a selected protein in the expressing live cells eventually without disrupting gene function.

Patterns of Gene Expression. For the plain description of the pattern of expression of a wild type gene, direct *in situ* hybridization to transcripts, as described by Seydoux and Fire (1994), Tabara *et al.* (1996) or Birchall *et al.* (using FISH, 1995) or immunohistolocalization seem more direct and may soon be as sensitive as the reporter gene approach.

However, transformation using reporters in which pieces or variants of promoters or enhancers can be inserted is the only means to perform

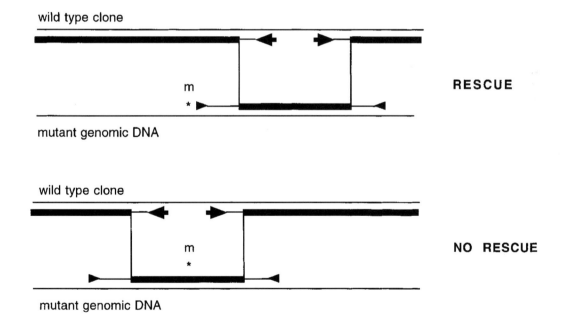

Figure 24.3. Precise mapping of a mutation by rescue experiments (B. Williams and P. Hoppe, pers. comm.).
You need a wild type clone or DNA fragment able to rescue a loss-of-function allele of the locus and the total genomic DNA isolated from the allele m* to be mapped.
Pairs of long oligos (one example depicted, large arrows) are used to generate by long PCR a scan of the rescuing clone. A window is left open, the size of which determines the precision of the mapping. Matching pairs of oligos (small arrows), 300 to 500 bp away from the long ones and in the reverse orientation, are used to generate, by regular PCR, fragments copied on the mutant genomic DNA. Mixtures of corresponding long wild type and short mutant fragments are injected together with a comarker in mutant (or wild type) animals.
The worms will efficiently recombine the molecules and reconstitute faithfully a mosaic gene, shown in bold line. Rescue is assessed directly in the progeny (or after appropriately crossing comarker-bearing transformants). Rescue proves that the fragment from the mutant is functionally wild type, whereas lack of rescue identifies the fragment containing the mutation.

functional analysis of the expression pattern. Indeed, many groups studied the promoter and regulating elements of their preferred genes (at least thirty papers; see for instance Way and Chalfie, 1989; MacMorris et al., 1992; Hill and Sternberg, 1992; and Okkema et al., 1993) while Ian Hope and his collaborators (e.g., Hope, 1991 and Lynch et al., 1995) tested random and sequence-based genomic fragments.

Some interesting patterns of expression have been described, and a few general remarks can be made. Noticeably, among all promoters studied to-date, no clone expressed in all cells at all time was found: a constitutive ubiquitous promoter may not exist, or else it may not be expressed fully in reporter constructs. As in other species, even promoters that are inducible by heat or metals show a specific pattern of expression (e.g., Stringham et al., 1992). Another observation is that elements necessary for proper regulation are scattered throughout the gene. As expected, most are in the 5' area, but many lie in the 3'UTR (see for instance Ahringer and Kimble, 1991; Wightman et al., 1993; Goodwin et al., 1993; Okkema et al., 1993; Evans et al., 1994 and Jones et al., 1995) and some are in introns (Okkema et al., 1993; and Wu and Han, 1994) and possibly in exons as well.

Yet another puzzling observation is the absence of expression in the germ line, at all stages of development, among about 100 promoters tested to date. Since we expect ten to fifty per cent of the genes expressed there, lack of expression probably reflects an artefact. It is especially suspect when presumptive promoters of maternal effect genes, such as *glp-1* or *skn-1*, known to be expressed in the germ line (Priess et al., 1987; Austin and Kimble, 1987; 1989; Bowerman et al., 1993 and Crittenden et al., 1994), are used. Two explanations come to mind: there may be a difference in the mechanism of somatic versus germinal transcriptional or translational control, for instance the regulating elements are not where we expect them. Alternatively, the structure of the chromatin may be different in the germ line, resulting in germ line inactivation of the genes in the array, for instance through heterochromatization of the large number of identical copies (investigated by Kelly, W., Xu, S. and Fire, A. pers. comm.). Both explanations may apply, since maternal effect genes are reputedly more difficult to rescue by transformation than zygotic genes. A way around the expression problem was devised by Evans et al. (1994), who directly injected in the gonad RNA transcribed *in vitro* from various constructs, in order to define the role of the 3'UTR of the *glp-1* gene. Similarly, injection of antisense in the syncytial gonad may help to identify a maternal effect gene (e.g. Guo and Kemphues, 1995).

Identification of Factors Controlling the Pattern of Expression. Once transformed lines containing fusions with reporter genes have been generated, they can be used to identify mutations that alter the pattern of expression. Existing mutations can be tested by crossing them into transformed lines (e.g., Way and Chalfie, 1989; Hill and Sternberg, 1992; and Cowing and Kenyon, 1992), and new mutations can be identified after mutagenesis of the transformants (Jia et al., 1995). For this purpose, the use of *gfp* reporters, visible in the live animal, should be especially useful.

Problems and Limitations in Expression Studies. The identification of gene expression patterns using reporter fusions is not always straightforward (see for example Mello et al., 1992; Okkema et al., 1993 and Krause et al., 1994). Lynch et al. (1995) found that only 18 of 34 putative promoters in C. elegans could drive detectable β-galactosidase expression. A number of reasons could account for these results. Some genes may be ill-predicted by *genefinder*, some may belong to polycistronic transcripts, so that their 5' regulatory sequences could be far from the coding region (Zorio et al., 1994, but gene ZK637.5, studied by Lynch et al. is a counterexample). It is also possible that inherent limitations of the method, such as the lack of germline expression discussed above, contribute to give this incomplete picture.

A limitation, found with transient expression, and to a lesser extent, in established lines, is that weak or minimal promoters often lead to a low and mosaic expression of β-galactosidase. This cannot be completely accounted for by the mosaicism of the transformants, since it happens in integrated lines as well (e.g., Mello et al., 1992 and Krause et al. 1994). In part the variability and weakness in expression can be overcome by examining many transformed animals in several transformed lines. But Fire and collaborators recently found that the addition of 1 to 12 synthetic introns scattered throughout the transcript highly increases the expression driven by the weak promoters, both transiently and in established lines, possibly by facilitating RNA export from the nucleus (Seydoux, Xu, Ahnn and Fire, pers. comm.). The addition of introns also leads to an increased background in promoterless constructs, which seems relieved by addition of a 5'-decoy sequence. We should know soon if this new set of vectors solved that problem.

In addition, ectopic expression is sometimes seen, especially with short promoter segments. With the standard vectors we use, expression occurs more prominently in the gut and the pharynx, probably because of weak promoter or enhancer signals in the vector. Unfortunately, no background free vector was yet found (Fire, Xu *and coll.*, pers. comm.).

To confirm that the expression pattern of the reporter construct reflects the expression of the endogenous gene, other approaches such as *in situ* hybridization, or immunohistolocalization should be used.

Dissecting the Gene-Function Relationship

The Tools. A growing collection of vectors for ectopic expression is also made available by Fire (see Table 1 in Mello and Fire, 1995 by Fire, Xu *and coll.*, pers. comm. and *ftp://ciw1.ciwemb.edu*). These have been used to express or misexpress a wide variety of proteins (see for example Salser and Kenyon, 1992, or Perry *et al.*, 1993). Interestingly, transformation has an autocatalytic effect on knowledge, the results generated on some genes can be used to design finer tools to study other genes, allowing to foresee more and more means to force, induce, control or target the expression of a gene or of its pieces. Specific transcription patterns can be imposed by using appropriate promoters. For instance, the function of a molecule may be ascertained by expressing it in an ectopic group of cells. A good example is the expression of the presumptive guidance receptor *unc-5* in mechanosensory cells, which suffices to redirect the nerve processes toward the *unc-6* signal molecule (Hamelin *et al.*, 1993). Translation of a gene can also be controlled by fusing translational signals from 3'UTR regulative regions (see section **Description of a Pattern of Gene Expression**). Protein domains leading to a defined cellular localization (e.g., Bloom, Hartwieg and Horvitz, pers. comm., Huang *et al.*, 1995) may allow to target a gene product to this specific place. Even death can be targeted to the cells expressing a gene by fusing the promoter of interest to a cell autonomous dominant lethal mutant, such as *mec-4d* (Driscoll and Chalfie, 1991, e.g. Maricq *et al.*, 1995).

Inducing Mutations In Vitro. Once a mutant has been rescued by a cloned fragment, it can be used as a test tube to assess the phenotype of any variant made *in vitro*, by directed mutagenesis of the clone (e.g., Stern *et al.*, 1993; and Hoppe and Waterston, pers. comm.). If the recipient contains a complete loss-of-function mutation, the transformed worms will express only the *in vitro* engineered gene. In this case, structure–function relationships can be directly assessed.

If no loss-of-function allele is available as recipient, as is the case for genes for which only molecular data is available, the *in vitro* mutated gene will be coexpressed with the endogenous wild type gene, therefore only dominant gain-of-function mutants are expected to give an observable phenotype. Dominant mutations could include hypermorphs (excess function), obtained for instance by overexpressing the gene under an inducible promoter or simply by dosage effects, neomorphs (new function which does not compete with the wild type), resulting perhaps from ectopic expression of the wild type gene, or neomorphs or antimorphs (competing poisons) produced from mutations known to interfere with gene function from studies of similar molecules, in the nematode or in other species (e.g., Roehl and Kimble, 1993) or from the expression of the antisense gene (Fire *et al.*, 1991).

In all cases, a search for suppressors of the mutant phenotype may shed light on other gene products involved in the realization of that phenotype. This may be achieved by classical genetics, or by transformation with adequately expressed (or overexpressed) genes of the pathway (e.g. Rhind *et al.*, 1994).

Dissecting Molecules. A number of questions about the function of individual protein domains can be answered by expressing fusion genes. A few examples illustrating the versatility of the method are mentioned below.

Specific domains of a multifunctional molecule can be selectively removed to test their function and assess separability (e.g. Klein and Meyer, 1993). On the other hand, if dominant phenotypes are associated to hypermorphic alleles of a gene, individual regions of the gene can be overexpressed, for instance under a heat shock promoter. In the case of *glp-1*, this allowed to define the critical area, and to show that it is functionally conserved among genes of the same family, since *lin-12* gain-of-function phenotypes were also apparent in the transformed animals (e.g. Roehl and Kimble, 1993). If the intracellular localization of a protein depends on a specific protein domain, the effects on this localization of other mutations can be tested using transformants in which the domain is fused to a reporter protein (e.g. Huang *et al.*, 1995). Mosaic molecules can also be used to study the interchangeability of promoters or coding regions of proteins of the same family (Hoppe and Waterston, pers. comm., Gao and Kimble, 1995).

Studies Across Species. Finally, the same vectors can be used to express heterologous proteins. This provides a powerful interspecific complementation test (e.g., Vaux *et al.*, 1992; Stern *et al.*, 1993; Wu and Han, 1994; and Hunter and Kenyon, 1995). If complementation occurs, it may even enable to test in the worm candidate mutations from other organisms.

Mosaic Analysis

The fact that a gene is expressed in a given cell does not imply that this expression is biologically significant to the organism. Only mosaic analysis allows to define the cells in which gene expression is both necessary and sufficient for the animal to display a wild type phenotype. In a sense, mosaic analysis is complementary to the description of pattern expression in very much the same way as the study of mutants is to the description of the wild type, they both ask about significance. The principle, reviewed by Herman (1995), consists in analyzing the phenotypes of a large set of randomly mosaic individuals in which some cells are genotypically mutant and other genotypically wild type. This allows one to determine the focus of action of the gene and to conclude if the mutation acts cell autonomously, i.e. in the cell in which it is produced. Non autonomy may happen if the gene product is secreted or involved in cell–cell communication (e.g. Hunter and Wood, 1992).

Genetic mosaics are usually generated in *C. elegans* by the spontaneous somatic loss of a free duplication carrying the wild-type allele in an animal otherwise homozygous mutant (see references in Herman, 1995). The mutant clone is then identified and the event located precisely on the lineage tree by looking at an independent cell autonomous marker. One such marker is *ncl-1*, which results in enlarged nucleoli easily observed by Nomarski microscopy in nearly all cells, except perhaps in the germ line and the intestine. Nucleoli are marginally enlarged also in embryos derived from a $ncl-1^+$ mother, because of maternal rescue (Hedgecock and Herman, 1995). Other cell autonomous markers whose focus of action have already been described can also be used. Such markers, with phenotypes that are easy to score, include *ace-1*, *unc-93*, *unc-3*, *unc-36*, *unc-30*, *unc-5*, *dpy-17*, *dpy-4*, *dpy-1*. These markers are quick tools for the preliminary coarse focus assignment, while *ncl-1* can be used for fine mapping.

Because the arrays obtained in transformation experiments behave as free duplications, and are lost in mitosis at reasonable frequencies (of the order of one per cent per mitosis, dependent on the strain), transgenic lines can be used for mosaic analysis (Leung-Hagesteijn et al., 1992). In addition, the $ncl-1^+$ gene was recently cloned (cosmid C33C3, Waring, pers. comm.), as were most of the other genes usable as markers (e.g. clone R1p16 for unc-36, from Lobel), so mosaic analysis can be done in a multiply mutant line cotransformed with the cloned markers and the wild type gene of interest (e.g., Lackner et al., 1994; Simske and Kim, 1995 and Herman et al., 1995). However, the interpretation of the results is not straightforward, because of the various problems described above of weak or silenced or inappropriate expression of the genes in the array, which may be mistaken for a real loss of the DNA in a given lineage. In addition, perdurance of an overexpressed product could also introduce a bias. To minimize the problems, low copy number extrachromosomal transformants containing a piece of DNA largely including the gene of interest may be preferable.

A radically different kind of mosaic analysis, in which the group of cells in an expressing clone would not necessarily be linked by lineage, should be made possible by using transformation of loss-of-function alleles. To get mosaic animals expressing a gene in only a subset of cells, one could express the wild type gene under a promoter with an overlapping but more restricted pattern than that of the original gene. Alternatively, the *lin-14/lin-4* hybrid system (Goodwin et al., 1993 and Lee et al., 1994) should in principle allow to block translation of the gene in some cells, by expressing the small translational repressor RNAs encoded by *lin-4* under a selected promoter and the gene of interest fused to the *lin-14* 3'UTR in a larger set of cells. The resulting mosaics will contain a set of non-expressing cells in an animal otherwise expressing the gene, and will relate with respect to the restricted promoter experiment as a negative photomicrograph relates to the positive.

Concluding Remarks

This review illustrates how genetic transformation in *Caenorhabditis elegans* has become in just ten years an indispensable complement and extension of the classical genetic and developmental studies that made the worm famous.

First, through mutant rescue, it provides a fast and efficient tool to connect the genetic map or map of alleles to the physical map or map of genomic clones. Naturally, it becomes increasingly easy to clone any gene defined by mutation as more and more markers are linking the two maps together. A plan to saturate the genome with maternal and zygotic lethals has started (in particular in the Rose, Baillie, Riddle, Schnabel labs), but to date only 13% of the genes are identified by mutation. The construction of libraries of transgenic strains containing series of sequenced cosmids is also in progress and will provide a fine grain connection between the maps. Second, transformation allows one to describe the phenotypes of new alleles made *in vitro*. Third, transformation provides a functional test to study evolution, through the complementation of nematode mutants by homologous genes from other species. Fourth, transformation enables one to characterize the developmental patterns of expression of defined genes, by expression studies using reporters, such as β-galactosidase or GFP. Fifth, by easing mosaic analysis, transformation allows one to define the set of cells in which a gene has to be expressed for the animal to be wild type. Finally, transformation allows one to select for genes acting *in trans* on the pattern of expression, or on the phenotype obtained in transgenics, thereby helping to define the role of previously known genes, as well as new genes involved in a pathway.

Like all other experimental methods, genetic transformation has its limitations, which we described in this review. Nonetheless, transformation of *C. elegans* has quickly become one of the most useful and versatile tools for the analysis of gene function in the worm.

The complete sequence of the whole genome of *C. elegans* is rapidly made available to the community, with termination planned for 1998 (Waterston and Sulston, 1995). To exploit it, we would like to be able to get mutations in the many interesting new genes defined only by their sequences. Mutator strains and libraries of worms containing transposons inserted anywhere in the genome are available and can be searched by PCR (Zwaal *et al.*, 1993; Andachi and Kohara, pers. comm.; see also Maricq *et al.*, 1995), but most of these insertions are silent by construction, and it is not so easy to generate a loss-of-function allele.

The simplest way to mutate any gene would be to efficiently perform gene replacement, through homologous recombination or *in situ* directed mutagenesis. No serious attempts in that direction have yet been made in the nematode. But at least four fortuitous instances of homologous recombination between the injected and the host DNA have been reported (see for example Broverman *et al.*, 1993, with 2 gene replacements in 63 integrated lines). Also, one possible case of gene conversion after introduction of an oligonucleotide was mentioned by Mello *et al.* (1991). We can only hope that, in the next few years, a method to increase or select these rare spontaneous events will emerge, providing us with a way to replace or knock out any gene selected on the basis of its sequence.

Acknowledgements

We and our readers owe a big "merci" to Marty Chalfie for his creative comments on the manuscript. We wish to thank many of our nematode colleagues, in particular Andy Fire, and also Jean Thierry-Mieg, Georges Lutfalla and Danielle Brioudes for their help. We acknowledge support from the Institut National de la Santé et de la Recherche Médicale and the Centre National de la Recherche Scientifique.

References

Ahringer, J.A. and Kimble, J.E. (1991) Control of the sperm-oocyte switch in Caenorhabditis elegans hermaphrodites by the *fem-3* 3' untranslated region. *Nature* 349, 346–348

Albertson, D.G. and Thomson, J.N. (1982) The kinetochores of C. elegans. *Chromosoma* 86, 409–428

Albertson, D.G. and Thomson, J.N. (1993) Segregation of holocentric chromosomes at meiosis in the nematode, *Caenorhabditis elegans. Chromosome Research* 1, 15–26

Aroian, R.V., Koga, M., Mendel, J.E., Ohshima, Y. and Sternberg, P.W. (1990) The *let-23* gene necessary for *Caenorhabditis elegans* vulval induction encodes a tyrosine kinase of the EGF receptor subfamily. *Nature* 348, 693–699

Austin, J.A. and Kimble, J.E. (1987) *glp-1* is required in the germ line for regulation of the decision between mitosis and meiosis in C. elegans. *Cell* 51, 589–599

Austin, J.A. and Kimble, J.E. (1989) Transcript analysis of *glp-1* and *lin-12*, homologous genes required for cell interactions during development of C. elegans. *Cell* 58, 565–571

Barnes, W.M. (1994) PCR amplification of up to 35-kb DNA with high fidelity and high yield from lambda bacteriophage templates. *Proc. Natl. Acad. Sci. USA* 91, 2216–2220

Birchall, P.S., Fishpool, R.M. and Albertson, D.G. (1995) Expression patterns of predicted genes from the C. elegans genome sequence visualized by FISH in whole organism. *Nature genetics* 11, 314–320

Bowerman, B., Draper, B.W., Mello, C.C. and Priess, J.R. (1993) The maternal gene *skn-1* encodes a protein that is distributed unequally in early C. elegans embryos. *Cell* 74, 443–452

Brenner, S. (1974) The genetics of Caenorhabditis elegans. *Genetics* 77, 71–94

Broverman, S.A., MacMorris, M.M. and Blumenthal, T.E. (1993) Alteration of Caenorhabditis elegans gene-expression by targeted transformation. *Proc. Natl. Acad. Sci. USA* 90, 4359–4363

Chalfie, M., Tu, Y., Euskirchen, G., Ward, W.W. and Prasher, D.C. (1994) Green fluorescent protein as a marker for gene expression. *Science* 263, 802–805

Clark, D.V., Suleman, D.S., Beckenbach, K.A., Gilchrist, E. and Baillie, D.L. (1995) Molecular cloning and characterization of the *dpy-20* gene of C. elegans. *Mol. Gen. Genetics* 247, 367–378

Coulson, A.R., Waterston, R.H., Kiff, J.E., Sulston, J.E. and Kohara, Y. (1988) Genome linking with yeast artificial chromosomes. *Nature* 335, 184–186

Cowing, D.W. and Kenyon, C.J. (1992) Expression of the homeotic gene *mab-5* during Caenorhabditis elegans embryogenesis. *Development* 116, 481–490

Crittenden, S.L., Troemel, E.R., Evans, T.C. and Kimble, J.E. (1994) GLP-1 is localized to the mitotic region of the C. elegans germ line. *Development* 120, 2901–2911

Driscoll, M. and Chalfie, M. (1991) The *mec-4* gene is a member of a family of Caenorhabditis elegans genes that can mutate to induce neuronal degeneration. *Nature* 349, 588–593

Evans, T.C., Crittenden, S.L., Kodoyianni, V. and Kimble, J.E. (1994) Translational control of maternal *glp-1* mRNA establishes an asymmetry in the C. elegans embryo. *Cell* **77**, 183–194

Field, J., Nikawa, J.I., Broek, D., MacDonald, B., Rodgers, L., Wilson, L., Lerner, R. and Wigler, M. (1988) Purification of a RAS-responsive adenylyl cyclase complex from S.cerevisiae by use of an epitope addition method. *Mol. Cell Biol.* **8**, 2159–2165

Fire, A. (1986) Integrative transformation of C. elegans. *EMBO Journal* **5**, 2673–2680

Fire, A. and Waterston, R.H. (1989) Proper expression of myosin genes in transgenic nematodes. *EMBO Jl.* **8**, 3419–3428

Fire, A., Harrison, S.W. and Dixon, D.K. (1990a) A modular set of lacZ fusion vectors for studying gene expression in Caenorhabditis elegans. *Gene* **93**, 189–198

Fire, A., Kondo, K. and Waterston, R.H. (1990b) Vectors for low copy transformation of C. elegans. *Nucleic Acids Research* **18**, 4269–4270

Fire, A., Albertson, D.G., Harrison, S.W. and Moerman, D.G. (1991) Production of antisense RNA leads to effective and specific inhibition of gene expression in C. elegans muscle. *Development* **113**, 503–514

Fire, A. (1992) Histochemical techniques for locating Escherichia coli beta-galactosidase activity in transgenic organisms. *Genetic Analysis Techniques and Applications* **9**, 5–6

Gao, D.L. and Kimble, J.E. (1995) APX-1 can substitute for its homolg LAG-2 to direct cell interactions throughout Caenorhabditis elegans development. *Proc. Natl. Acad. Sci. USA* **92**, 9839–9842

Goodwin, E.B., Okkema, P.G., Evans, T.C. and Kimble, J.E. (1993) Translational regulation of *tra-2* by its 3′ untranslated region controls sexual identity in C. elegans. *Cell* **75**, 329–339

Granato, M., Schnabel, H. and Schnabel, R. (1994) *pha-1*, a selectable marker for gene-transfer in C. elegans. *Nucleic Acids Research* **22**, 1762–1763

Guo, S. and Kemphues K.J. (1995) *par-1*, a gene required for establishing polarity in *C. elegans* embryos, encodes a putative Ser/Thr kinase that is asymmetrically distributed. *Cell* **81**, 611–620

Hamelin, M., Zhou, Y.W., Su, M.-W., Scott, I.M. and Culotti, J.G. (1993) Expression of the *unc-5* guidance receptor in the touch neurons of *C. elegans* steers their axons dorsally. *Nature* **364**, 327–330

Han, M. and Sternberg, P.W. (1990) *let-60*, a gene that specifies cell fates during C. elegans vulval induction, encodes a ras protein. *Cell* **63**, 921–931

Hedgecock, E.M. and Herman, R.K. (1995) The *ncl-1* gene and genetic mosaics of Caenorhabditis elegans. *Genetics* **141**, 989–1006

Heim, R., Prasher, D. and Tsien, R. (1994) Wavelength mutations and posttranslational autooxidation of green fluorescent protein. *Proc. Natl. Acad. Sci.* **91**, 12501–12504

Hengartner, M.O. and Horvitz, H.R. (1994). C. elegans cell survival gene *ced-9* encodes a functional homolog of the mammalian protooncogene *bcl-2*. *Cell* **76**, 665–676

Herman, M.A., Vassilieva, L.L., Horvitz, H.R., Shaw, J.E. and Herman, R.K. (1995) The C. elegans gene *lin-44*, which controls the polarity of certain asymmetric cell divisions, encodes a Wnt protein and acts cell nonautonomously. *Cell* **83**, 101–110

Herman, R.K. (1995) Mosaic analysis. pp. 123–146, in Epstein, H.F. and Shakes, D.C. (eds.), *Methods in Cell Biology* **48**. *Caenorhabditis elegans Modern Biological Analysis of an Organism*. Academic Press

Hill, R.J. and Sternberg, P.W. (1992) The gene *lin-3* encodes an inductive signal for vulval development in C. elegans. *Nature* **358**, 470–476

Hope, I.A. (1991) 'Promoter trapping' in Caenorhabditis elegans. *Development* **113**, 388–408

Huang, L.S., Tzou, P. and Sternberg, P.W. (1994) The *lin-15* locus encodes two negative regulators of Caenorhabditis elegans vulval development. *Molecular Biology of the Cell* **5**, 395–411

Huang, M.X., Gu, G.Q., Ferguson, E.L. and Chalfie, M. (1995) A stomatin-like protein necessary for mechanosensation in C. elegans. *Nature* **378**, 292–295

Hunter, C.P. and Wood, W.B. (1992) Evidence from mosaic analysis of the masculinizing gene *her-1* for cell interactions in C. elegans sex determination. *Nature* **355**, 551–555

Hunter, C.P. and Kenyon, C.J. (1995) Specification of anteroposterior cell fates in Caenorhabditis elegans by Drosophila Hox proteins. *Nature* **377**, 229–232

Jia, Y., Xie, G. and Aamodt, E. (1996) *pag-3*, a *Caenorhabditis elegans* gene involved in touch neuron gene expression and coordinated movement. *Genetics* **142**, 141–147

Jones, D., Stringham-Durovic, E.G., Graham, R.W. and Candido, E.P.M. (1995) A portable regulatory element directs specific expression of the Caenorhabditis elegans ubiquitin gene *ubq-2* in the somatic gonad. *Dev. Biol.* **171**, 60–72

Kim, S.K. and Horvitz, H.R. (1990) The *Caenorhabditis elegans* gene *lin-10* in broadly expressed while required specifically for the determination of vulval cell fates. *Genes and Development* **4**, 357–371

Kimble, J.E., Hodgkin, J.A., Smith, T. and Smith, J.D. (1982) Suppression of an amber mutation by microinjection of suppressor tRNA in C. elegans. *Nature* **299**, 456–458

Klein, R.D. and Meyer, B.J. (1993) Independent domains of the *sdc-3* protein control sex determination and dosage compensation in C. elegans. *Cell* **72**, 349–364

Kramer, J.M. and Johnson, J.J. (1993) Analysis of mutations in the *sqt-1* and *rol-6* collagen genes of *Caenorhabditis elegans*. *Genetics* **135**, 1035, 1045

Kramer, J.M., French, R.P., Park, E.C. and Johnson, J.J. (1990) The *Caenorhabditis elegans rol-6* gene, which interacts with the *sqt-1* collage gene to determine organismal morphology, encodes a collagen. *Mol. Cell. Biol.* **10**, 2081–2089

Krause, M., Harrison, S.W., Xu, S.Q. and Chen, L.S. (1994) Elements regulating cell and stage-specific expression of the C. elegans myoD family homolog *hlh-1*. *Dev. Biol.* **166**, 133–148

Lackner, M.R., Kornfeld, K., Miller, L.M., Horvitz, H.R. and Kim, S.K. (1994) A MAP kinase homolog, *mpk-1*, is involved in ras-mediated induction of vulval cell fates in Caenorhabditis elegans. *Genes and Development* **8**, 160–173

Lee, R.C., Feinbaum R.L., and Ambros, V. (1993) The C. elegans heterochronic gene lin-4 encodes small RNAs with antisense complementarity to lin-14. *Cell* **75**, 843–854

Leung-Hagesteijn, C.J., Spence, A.M., Stern, B.D., Zhou, Y.W., Su, M.-W., Hedgecock, E.M. and Culotti, J.G. (1992) UNC-5, a transmembrane protein with immunoglobulin and thrombospondin type-1 domains, guides cell and pioneer axon migrations in C. elegans. *Cell* **71**, 289–299

Lynch, A.S., Briggs, D.A. and Hope, I.A. (1995) Developmental expression pattern screen for genes predicted in the C. elegans genome sequencing project. *Nature Genetics* **11**, 309–313

MacMorris, M.M., Broverman, S.A., Greenspoon, S., Lea, K., Madej, C., Blumenthal, T.E. and Spieth, J. (1992) Regulation of vitellogenin gene expression in transgenic Caenorhabditis elegans – Short sequences required for activation of the vit-2 promoter. *Mol. Cell. Biol.* **12**, 1652–1662

Maricq, A.V., Peckol, E., Driscoll, M. and Bargmann, C.I. (1995) Mechanosensory signalling in C. elegans mediated by the GLR-1 glutamate receptor. *Nature* **378**, 78–81

McKim, K.S., Howell, A.M. and Rose, A.M. (1988) The effects of translocations on recombination frequency in Caenorhabditis elegans. *Genetics* **120**, 987–1001

McKim, K.S., Peters, K. and Rose, A.M. (1993) Two types of sites required for meiotic chromosome-pairing in Caenorhabditis elegans. *Genetics* **134**, 749–768

Mello, C.C., Kramer, J.M., Stinchcomb, D.T. and Ambros, V. (1991) Efficient gene transfer in C. elegans: Extrachromosomal maintenance and integration of transforming sequences. *EMBO J.* **10**, 3959–3970

Mello, C.C., Draper, B.W., Krause, M., Weintraub, H. and Priess, J.R. (1992) The pie-1 and mex-1 genes and maternal control of blastomere identity in early C. elegans embryos. *Cell* **70**, 163–176

Mello, C.C. and Fire, A. (1995) DNA transformation., pp. 451–482, in Epstein, H.F. and Shakes, D.C. (eds.), *Methods in Cell Biology*, **48**. *Caenorhabditis elegans Modern Biological Analysis of an Organism*. Academic Press.

Miller, D.M. and Niemeyer, C.J. (1995) Expression of the unc-4 homeoprotein in Caenorhabditis elegans motor neurons specifies presynaptic input. *Development* **121**, 2877–2886

Miller, L.M., Gallegos, M.E., Morisseau, B.A. and Kim, S.K. (1993) lin-31, a Caenorhabditis elegans HNF-3/fork head transcription factor homolog, specifies three alternative cell fates in vulval development. *Genes and Development* **7**, 933–947

Mitani, S. (1995) Genetic regulation of mec-3 gene expression implicated in the specification of the mechanosensory neuron cell types in Caenorhabditis elegans. *Development, growth and differentiation* **37**, 551–557

Okkema, P.G., Harrison, S.W., Plunger, V., Aryana, A. and Fire, A. (1993) Sequence requirements for myosin gene-expression and regulation in Caenorhabditis elegans. *Genetics* **135**, 385–404

Perry, M.D., Li, W.Q., Trent, C., Robertson, B., Fire, A. Hageman, J.M. and Wood, W.B. (1993) Molecular characterization of the her-1 gene suggests a direct role in cell signaling during Caenorhabditis elegans sex determination. *Genes and Development* **7**, 216–228.

Priess, J.R., Schnabel, H. and Schnabel, R. (1987) The glp-1 locus and cellular interactions in early C. elegans embryos. *Cell* **51**, 601–611

Rhind, N.E., Miller, L.M., Kopczynski, J.B. and Meyer, B.J. (1995) xol-1 acts as an early switch in the C. elegans male/hermaphrodite decision. *Cell* **80**, 71–82

Roehl, H. and Kimble, J.E. (1993) Control of cell fate in C. elegans by a glp-1 peptide consisting primarily of ankryin repeats. *Nature* **364**, 632–635

Salser, S.J. and Kenyon, C.J. (1992) Activation of a C. elegans Antennapedia homolog in migrating cells controls their direction of migration. *Nature* **355**, 255–258

Seydoux, G. and Fire, A. (1995) Whole-mount *in situ* hybridization for the detection of RNA in Caenorhabditis elegans embryos. pp. 323–337. In Epstein, H.F. and Shakes, D.C., (eds.), *Methods in Cell Biology*, **48**. *Caenorhabditis elegans Modern Biological Analysis of an Organism*. Academic Press

Simske, J.S. and Kim, S.K. (1995) Sequential signaling during Caenorhabditis elegans vulval induction. *Nature* **377**, 142–146

Spence, A.M., Coulson, A.R. and Hodgkin, J.A. (1990) The product of fem-1, a nematode sex-determining gene, contains a motif found in cell cycle control proteins and receptors for cell-cell interactions. *Cell* **60**, 981–990

Stern, M.J., Marengere, L.E.M., Daly, R.J., Lowenstein, E.J., Kokel, M.J., Batzer, A.G., Olivier, P., Pawson, T. and Schlessinger, J. (1993) The human GRB2 and Drosophila drk genes can functionally replace the Caenorhabditis elegans cell signalling gene sem-5. *Mol. Biol. Cell* **4**, 1175–1188

Stinchcomb, D.T., Shaw, J.E., Carr, S.H. and Hirsh, D.I. (1985) Extrachromosomal DNA transformation of Caenorhabditis elegans. *Mol. Cell. Biol.* **5**, 3484–3496

Stone, S. and Shaw, J.E. (1993) A Caenorhabditis elegans act-4-lacZ fusion – Use as a transformation marker and analysis of tissue-specific expression. *Gene* **131**, 167–173

Stringham-Durovic, E.G., Dixon, D.K., Jones, D. and Candido, E.P.M. (1992) Temporal and spatial expression patterns of the small heat shock (hsp-16) genes in transgenic Caenorhabditis elegans. *Mol. Biol. Cell* **3**, 21–233

Sulston, J.E. (1988) Cell Lineage, pp. 123–155. In *The nematode C. elegans*, "Wood, W. (ed.) Cold Spring Harbor monograph series, **17**

Sulston, J.E. and Hodgkin, J.A. (1988) Methods, pp. 587–606. in The nematode C.elegans, Wood, W (ed.) Cold Spring Harbor monograph series, **17**

Sulston, J.E. and White, J.G. (1988) Parts List, p. 415–431. In *The nematode C. elegans*, Wood, W. (ed.) Cold Spring Harbor monograph series, **17**

Tabara, H., Motohashi, T. and Kohara, Y. (1996) A multiwell version of *in situ* hybridization on whole mount embryos of C. elegans. *Nucleic Acids Research* **24**, 2119–2124

Troemel, E.R., Chou, J.H., Dwyer, N.D., Colbert, H.A. and Bargmann, C.I. (1995) Divergent seven transmembrane receptors are candidate chemosensory receptors in C. elegans. *Cell* **83**, 207–218

Vaux, D.L., Weissman, I.L. and Kim, S.K. (1992) Prevention of programmed cell death in Caenorhabditis elegans by human *bcl-2*. *Science* **258**, 1955–1957

Villeneuve, A.M. (1994) A cis-acting locus that promotes crossing over between X chromosomes in Caenorhabditis elegans. *Genetics* **136**, 887–902

Waterston, R.H. and Sulston, J.E. (1995) The genome of Caenorhabditis elegans. *Proc. Natl. Acad. Sci. USA* **92**, 10836–10840

Way, J.C. and Chalfie, M. (1988) *mec-3*, a homeobox-containing gene that specifies differentiation of the touch receptor neurons in C. elegans. *Cell* **54**, 5–16

Way, J.C. and Chalfie, M. (1989) The *mec-3* gene of Caenorhabditis elegans requires its own product for maintained expression and is expressed in three neuronal cell types. *Genes and Development* **3**, 1823–1833

White, J.G. (1988) The Anatomy, pp. 81–122, in *The nematode C. elegans*, Wood, W. (ed.) Cold Spring Harbor monograph series **17**

Wightman, B., Ha, I. and Ruvkun, G.B. (1993) Posttranscriptional regulation of the heterochronic gene *lin-14* by *lin-4* mediates temporal pattern formation in C. elegans. *Cell* **75**, 855–862

Williams, B.D., Schrank, B., Huynh, C., Shownkeen, R. and Waterston, R.H. (1992) A genetic-mapping system in Caenorhabditis elegans based on polymorphic sequence-tagged sites. *Genetics* **131**, 609–624

Wood, W.B. (1988) *The nematode C. elegans*, Cold Spring Harbor monograph series **17**

Wu, Y. and Han, M. (1994) Suppression of activated *let-60* ras protein defines a role of Caenorhabditis elegans *sur-1 map kinase* in vulval differentiation. *Genes and Development* **8**, 147–159

Zorio, D.A.R., Cheng, N.S.N., Blumenthal, T.E. and Spieth, J. (1994) Operons as common form of chromosomal organization in C. elegans. *Nature* **372**, 270–272

Zwaal, R.R., Broeks, A., vanMeurs, J., Groenen, J.T.M. and Plasterk, R.H.A.(1993) Target-selected gene inactivation in Caenorhabditis elegans by using frozen transposon insertion mutant bank. *Proc. Natl. Acad. Sci. USA* **90**, 7431–7435

PART II, SECTION E

25. Gene Transfer Technology in Marine Invertebrates

E. Mialhe[1,2], Viviane Boulo[2], J.-P. Cadoret[2], V. Cedeño[1], C. Rousseau[1], E. Motte[1], S. Gendreau[2] and E. Bachère[2]

[1]Centro Nacional de Acuacultura e Investigaciones Marinas (CENAIM), Escuela Politecnica (ESPOL), Campus Prosperina, P.O. Box 09 01 4519, Guayaquil, Ecuador
[2]Unité de Recherche "Défense et Résistance chez les Invertébrès Marins"(DRIM), Institut Français de Recherche pour l'Exploitation de la Mer (IFREMER)-Centre National de la Recherche Scientifique (CNRS), Université de Montpellier II, 34000 Montpellier, France

Introduction

Invertebrates represent the most numerous animals in terms of species and individuals. They are ubiquitously distributed on the planet and their importance is extreme with regard to human activities: pollinisators and pests in agriculture; parasites and vectors of infectious diseases in medicine; shrimps, oysters and mussels in fishery and aquaculture.

Because of this economical and medical importance and by referring to the numerous applications of genetic transformation in plants and vertebrates, the development of gene transfer technologies for invertebrates appears as a prioritary topic that is still in its infancy.

For terrestrial invertebrates, gene transfer technologies have been developed and are routinely used only for *Drosophila* fruit flies (see Kaiser) and for a nematod, *Caenorhabditis elegans* (see D. Tierry-Mieg). Largely because of the availability and the reliability of these technologies, these species have acquired the status of models for basic research on molecular genetics and biology of development. For all the other terrestrial invertebrates groups, gene transfer technology is still speculative until the identification of efficient transformation systems (Handler and O'Brochta, 1991).

Concerning marine invertebrates, research on genetic transformation has been initiated in the eighties for sea urchins as models to study embryogenesis at the molecular level, a huge amount of informations being otherwise available about classical embryology of these primitive deuterostomes. A research strategy based on gene transfer technology has been developed as described below.

In the nineties, genetic transformation of molluscs and shrimps began to be explored. By referring to transgenic plants (Grumet, 1990 and Scholthof et al., 1993) and vertebrates (see this book) and like for insects vector of diseases (Miller et al., 1987 and Eggleston, 1991), genetic transformation has been actually considered as a promising strategy for producing pathogen-resistant animals, that could permit to control the infectious diseases that dramatically affect aquaculture productions all around the world (Mialhe et al.). Although significant progresses have been already achieved, as described below, shrimp and bivalve gene transfer technology is still considered as an uncertain project, specially because it is developed by a very limited number of scientists.

Gene Transfer Technology in Sea Urchins

The first studies related to sea urchin gene transfer technology were based on cytoplasmic microinjection of *Strongylocentrus purpuratus* unfertilized eggs that are then fertilized *in situ* by addition of a drop of sperm solution (Flytzanis et al., 1985 and McMahon et al., 1985).

Gametes are shed from sexually mature sea urchins after an intracoelomic injection of a 0.5 M KCl solution. The egg jelly coat is removed by treatment with acid sea water. The eggs are about 80 µm in diameter and their surface is negatively charged, that permits to immobilize them by electrostatic attraction to the surface of the injection dishes previously coated with a 1% protamine sulfate solution. Without the need of a holding micropipet, eggs are injected by using a continuous flowing micropipet of about 0.5–1.0 µm tip diameter. About

2 pL of DNA solution is injected into the cytoplasm, that is to say around 1.5% of the egg volume. Up to 300 eggs can be injected per hour with 90–95% undergoing normal fertilization and 40–50% completing normal embryogenesis to the pluteus stage, 3–4 days later. Occasionally, a batch of microinjected eggs completely fails to develop. Moreover, embryonic survival depends on the amount of DNA microinjected. In practice, it is recommended to use DNA solutions at concentrations below 32 μg/mL (that corresponds at a maximum injection of 12000 copies per 2 pL for 5 Kb DNA molcules). Microinjection of DNA solutions at concentrations equal to or greater than 320 μg/mL is completely toxic.

A protocol has been also developed to directly microinject the zygote nucleus, the main purpose of this procedure being to increase the number of embryonic cells containing DNA (Franks et al., 1988). This microinjection procedure has been established for the Gulf Coast sea urchin *Lytechinus variegatus* that has almost transparent eggs whereas *S. pururatus* has opaque eggs. The collect and the treatment of gametes are similar for the two species. Microinjection into the unfertilized egg pronucleus is not possible because it moves inside the cytoplasm at the contact of the microneedle. After fertilization and fusion of pronuclei, the resulting zygote pronucleus is more stably anchored inside the newly assembled cytoskeleton, that permits to introduce the continuously flowing microneedle. DNA solution is thus delivered into the nucleus but also dispersed inside the cytoplasm during the needle penetration. For a same duration of operations, the number of eggs that can be microinjected in the zygote nucleus is almost the same than for microinjection inside the cytoplasm. The percentage of eggs that complete normal embryogenesis to the feeding pluteus stage is also lower, with an average of about 60–65%.

More recently, transfection of *Strongylocentrotus purpuratus* sperm has been investigated as a mass transfection procedure permitting to vehicle DNA directly into the egg nucleus (Arezzo, 1989). Based on previous observation related to permeability of sperm membrane, experiments have been performed to transfect sperm (2 μL) by incubation for 1 hour in a solution of supercoiled or linearized plasmids (50 μg) diluted in sterile artificial sea water (30 μL). The transfected sperm is then mixed with eggs for *in situ* fertilization. The use of the reporter gene CAT placed under the control of heterologous promoters led to transient expression in embryos. Consequently, it was assumed that sea urchin sperm can be loaded with DNA that is then vehicled to the egg nucleus where it can express.

It has been established that the fate of of transfected DNA depends essentially on its physical form (Flytzanis et al., 1985; McMahon et al., 1985 and Franks et al., 1988). DNA microinjected as supercoils persists in the embryos, mainly in supercoiled or relaxed circular form, but without DNA amplification. Linearized plasmids are transformed into high-molecular-weight DNA after injection into the egg. Random ligation of heads and tails are observed resulting in concatenate structures. The concatenation of linear molecules is independant of the presence of staggered ends since it has been shown that "blunt-ended" as well as "sticky-ended" DNA molecules are end-to-end ligated after injection. Moreover, the terminal restriction sites are reformed, indicating little if any modification of the ends by intracellular nucleases. Based on experimental data, it may be assumed that the ligations occur soon after injection of linearized DNA, high molecular-weight DNA being found as soon as 60 min after injection. During early embryogenesis there is an amplification of the concatenated DNA, this amplification occuring independently of the presence of sea urchin sequences. Replicated DNA is present in more than 50% of 5-week larvae. Two to three months after metamorphosis, the frequency of positive animals ranged from 4 to 16% when the injection was made in the egg cytoplasm and about 36% when the injection was made in the zygote nucleus. Plasmid DNA is chiefly present as extrachromosomal concatenated structures. Integrated plasmid DNA concatenates have been found and homologies have been observed near the breakpoint between the end plasmid sequences and repeated sequences of the adjoining sea urchin DNA; The latter consist of short repetitive AT-rich sequences. Integration may thus result from homologous recombination (Flytzanis et al., 1985). Integrated sequences have been also found in the sperm of adult sea urchin issued from microinjected eggs.

Expression of reporter genes have been reported following microinjection into unfertilized egg cytoplasm or zygote nucleus and following sperm transfection. On the basis of transient expression analyses, a lot of studies have been performed on sea urchin promoters and genes that show the present routine character of gene transfer for sea urchin.

Gene Transfer Technology in Bivalve Molluscs and Shrimps

Molluscs

Gametes of bivalve molluscs are easily obtained by injection of 0.5 M KCl into the adductor muscle and fertilization is simply achieved by mixing gametes.

By using Lucifer Yellow as a reporter molecule, microinjection has been developed for 1-cell and early embryo stages, but unfertilized ovocytes revealed too sensitive to survive. Microholder and microneedles made with a two-stroke puller are necessary. Embryo microinjections are performed with the aid of micromanipulators mounted on an inverted microscope. In practice, it is possible to microinject up to 100 embryos per hour with survival rates ranging between 10% and 55% (Cadoret, 1992a).

Electroporation has been applied on early embryos to induce polyploïdy (Cadoret, 1992b) and will have to be evaluated as a possible mass transfection technique.

Aqueous biolistics, initially developed to transfect mosquito eggs, has been recently applied to mass transfection of ovocytes, uni- and pluricellular embryos of mussel and oyster (Cadoret et al., submitted). The determination and the optimization of biolistic parameters have been made on the basis of luciferase reporter gene transient expression.

In parallel to these investigations on transfection that are directly oriented to the obtention of transformed animals, lipofection of heart cells in primary culture has been developed and been proved suitable to analyse the functionality of heterologous promoters (Boulo et al., submitted).

Analyses about the fate of transfected DNA were first aimed at determining the functionality of heterologous promoters on the basis of transient expression of luciferase reporter gene. It has been shown that the hsp 70 (heat shock protein) promoter of Drosophila and the early promoter of CMV (cytomegalovirus) are efficient both in embryos and in heart cells. In the latter, the promoter of SV (simian virus) is also efficient (Cadoret et al., submitted and Boulo et al., submitted).

Shrimps

In vitro fertilization has been announced for the shrimp *Penaeus atzecus* (Clark et al., 1973) but this result revealed to be practically non reproducible. Consequently, transfection procedures have been only explored on naturally fertilized eggs and early embryos.

Microinjection of embryos has been successfully developed for different *Penaeus* species. The efficiency of the procedure has been controlled by checking the introduction of Lucifer Yellow and the subsequent survival. A chief limitation of this technique is linked to the rapid division of blastomers and consequently the small number of embryos that can be injected at the 1-cell stage (Gendreau, 1992).

Classical biolistics, that is referred to as "dry biolistics", has been tested on decapsulated artemia cysts and shown not to be very efficient (Gendreau et al., 1995). The modified biolistic procedure, named "aqueous biolistics" because DNA-coated particles are kept in suspension to be shooted (Mialhe and Miller, 1994), has been investigated and optimized with embryos of *P. stylirostris*. This technique is efficient to transfect thousands of early embryos (from 32 blastomer stage to blastula), survival reaching 60% (Rousseau et al., in prep., Rousseau, 1995).

By using aqueous biolistics to transfect embryos and by lipofecting ovarian cell in primary cultures, it has been showed on the basis of luciferase transient expression that the promoters for the Drosophila hsp 70 and human CMV early genes are efficient in shrimps (Rousseau, 1995).

Prospects

Gene transfer technology is presently studied in a number of marine invertebrates with very different prospects according to the group.

For sea urchins, investigations concern essentially embryogenesis and molecular biology of development. Because of this specific fundamental orientation, current gene transfer technology appears sufficiently controlled. As a matter of fact, DNA concatenation, extrachromosomal amplification and persistence are compatible with experimentations that are based on transient expression of constructs designed to analyse developmentally regulated promoters and genes. A limitation is related to the mosaic distribution of transfected DNA. However this limitation can be reduced by performing microinjection into the zygote nucleus rather than the egg cytoplasm and mass transfection of sperm could lead to non mosaic distribution. Evidence of DNA integration, that seems to result from homologous recombination between small AT-rich sequences shared by the plasmid and the chromosome of DNA, is a very promising way to make easily transgenic animals. As a matter of fact, it would be possible to flank transfected construct with repetitive sequences present in the host chromosome, such as telomeres (Edman, 1992). A chief advantage of gene transfer technology, based on homologous recombination, consists in the easiness to identify repetitive sequences compared to the identification of specific and/or functional transformation system, such as transposon or integrative virus.

Concerning gene transfer technology for bivalves and shrimps, the main objective is to produce transgenic strains that resist to pathogen through the expression of immune genes, viral genes or viral antisense sequences. It is consequently primordial to control integration of transfected DNA in order to get stable and inheritable resistance. Experiments to evaluate potential transformation systems are complicated by the generation times (about 9 months) and by the lack of cell lines although one has been recently established from shrimp lymphoid organs through transformation (Tapay et al., 1995). In the short term, experiments will be advantageously performed with heterologous transformation systems that are available and have already proved efficient in other animal systems. Particularly promising systems correspond to pseudotyped retroviral vectors (Burns et al., 1993) that have been successfully used to transform fishes (Lin et al., 1994) and rare possibly efficient in molluscs (Cheng, pers. comm.). A suitable way to use these transformation vectors could be to infect sperm in order to limit mosaic effect. Other potential heterologous transformation vectors have to be analyzed, some derived from insect densoviruses that are able to replicate or to integrate and others derived from transposons. In parallel, investigations are necessary to identify and to characterize transposons, viruses, repetitive sequences, promotors and immune genes of molluscs and shrimps. Taking into account the extent of these researches on genetic transformation of molluscs and shrimps, international networks are currently organized to create the necessary interactions between scientists from these different topics (Mialhe et al., 1995). The organization of these networks is supported by cooperation agencies because control of diseases is the priority for world shrimp aquaculture.

References

Arezzo, F. (1989) Sea urchin sperm as a vector of foreign genetic information. *Cell Biology International Reports* **13**, 391–403

Boulo, V., Cadoret, J.P., Le Marrec, F., Dorange, G., Mialhe, E. (submitted) Transient expression of luciferase reporter gene after lipofection in oyster (*Crassostrea gigas*) primary cell cultures

Burns, J.C., Friedmann, T., Driever, W., Burrascano, M. (1993) Vesicular stomatitis virus G glycoprotein pseudo typed retroviral vectors: concentration to very high titer and efficient gene transfer into mammalian and nonmammalian cells. *Proc. Natl. Acad. Sci. USA* **90**, 8033–8037

Cadoret, J.P. (1992a) Mise au point de méthodes de manipulation embryonnaire de mollusques bivalves. Application en génétique et pathologie infectieuse. Thèse *EPHE*, pp. 1–100 (IFREMER libraries)

Cadoret, J.P. (1992b) Electric field-induced polyploidy in mollusc embryos. *Aquaculture* **106**, 127–139

Cadoret, J.P., Boulo, V., Gendreau, S., Mialhe, E. (submitted) Transient expression of luciferase reporter gene in bivalve embryos using particle bombardment

Clark, W.H., Talbot, P., Neal, R.A., Mock, C.R., Salser, B.R. (1973) *In vitro* fertilization with non-motile spermatozoa of the brown shrimp *Penaeus aztecus*. *Marine Biology* **22**, 353–354

Edman, J.C. (1992) Isolation of telomerelike sequences from *Cryptococcus neoformans* and their use in high-efficiency transformation. *Molecular and Cellular Biology* **12**, 2777–2783

Eggleston, P. (1991) The control of insect-borne diseases through recombinant DNA technology. *Heredity* **66**, 161–172

Flytzanis, C.N., McMahon, A.P., Hough-Evans B.R., Katula, K.S., Britten, R.J., Davidson, E.H. (1985) Persistence and integration of cloned DNA in postembryonic sea urchins. *Developmental Biology* **108**, 431–442

Franks, R.R., Hough-Evans B.R., Britten, R.J., Davidson, E.H. (1988) Direct introduction of cloned DNA into the sea urchin zygote nucleus, and fate of injected DNA. *Development* **102**, 287–299

Gendreau, S. (1992) Recherches sur la transformation génétique des crustacés. *Thèse Université de Bretagne occidentale*, pp. 1–116 (IFREMER libraries)

Gendreau, S., Lardans, V., Cadoret, J.P., Mialhe, E. (1995) Transient expression of luciferase reporter gene after biolistic introduction into Artemia franciscana (Crustacea) embryos. *Aquaculture* **133**, 199–205

Grumet, R. (1990) Genetically engineered plant virus resistance. *HortScience* **25**, 508–512

Handler, A.M., O'Brochta, D.A. (1991) Prospects for gene transformation in insects. *Annual Review of Entomology* **36**, 159–183

Lin, S., Gaiano, N., Culp, P., Burns, J.C., Friedman, T., Yee, Y.K., Hopkins, N. (1994) Integration and germ line transmission of a pseudotyped retroviral vector in Zebrafish. *Science* **265**, 666–669

McMahon, A.P., Flytzanis, C.N., Hough-Evans B.R., Katula, K.S., Britten, R.J., Davidson, E.H. (1985) Introduction of cloned DNA into sea urchin egg cytoplasm: replication and persistence during embryogenesis. *Developmental Biology* **108**, 420–430

Mialhe, E., Bachère, E., Boulo, V., Cadoret, J.P. (1995) Strategy for research and international cooperation in marine invertebrate pathology, immunology and genetics. *Aquaculture* **132**, 33–41

Mialhe, E., Bachére, E., Boulo, V., Cadoret, J.P., Saraiva, E., Carrera, L., Calderon, J., Colwell, R. Bio-technology-based control of disease in marine invertebrates: development of molecular probe diagnostics and disease-resistant transgenic shrimp andmolluscs. *Molecular marine Biology and biotechnology*

Mialhe, E., Miller, L.H. (1994) Development of biolistic techniques for transfection of mosquito embryos (*Anopheles gambiae*). *Biotechniques* **16**, 924–931

Miller, L.H., Sakai, R.K., Romans, P., Gwadz R.W., Kantoff, P., Cran, H.G. (1987) Stable integration and expression of a bacterial gene in the mosquito *Anopheles gambiae*. *Science* **237**, 779–781

Rousseau, C. (1995) Développement de technologies de transformation génétique d'embryons et de cellules de crevettes pénéides. Thèse EPHE, pp. 1–76. (IFREMER libraries)

Rousseau, C., Cedeno, V., Robalino, J., Cadoret, J.P., Boulo, V., Mialhe, E. (in prep.) Transient expression of luciferase reporter gene in embryos of White Shrimp *Penaeus stylirostris* and analysis of two promoters using aqueous biolistic

Scholthof, K.B.G., Scholthof, H.B., Jackson, A.O. (1993) Control of plant virus diseases by pathogen-derived resistance in transgenic plants. *Plant Physiology* **102**, 7–12

Tapay, L.M., Lu, Y., Brock, J.A., Nadala, E.C. Jr, Loh, P.C. (1995) Transformation of primary cultures of shrimp (*Penaeus stylirostris*) lymphoid (Oka) organ with simian virus 40 (T) antigen. *Proc. Soc. Exp. Biol. Med.* **209**, 73–77

PART II, SECTION F

26. What are ES cells?

J.-E. Fléchon

Laboratoire de Biologie Cellulaire et Moléculaire, INRA Institut National de la Recherche Agronomique, 78352 Jouy-en-Josas, Cedex, France

What are ES and EG cells? Behind this fundamental question, about very useful biological tools restricted up to now to the murine species, is another practical question: is there any hope to obtain such totipotent stem cells in other mammalian species? In the absence of a present answer to the last question, we will at least try to define as well as possible the murine totipotent cells.

How can Murine ES Cells be Obtained?

The first differentiation occuring in cleaving mammalian embryos is the separation at the time of morula compaction of the "inner cells", which form the inner cell mass (ICM), from the "outer cells" which will constitute the first polarized extra embryonic epithelium or trophectoderm (Tarkowski and Wrobleska, 1967). Cells of the ICM of a young blastocyst, the next embryonic step characterized by an internal cavity or blastocoele, are normally the stem cells of all embryonic and extra embryonic cell lines, except (normally) trophectoderm cells. However, a transient reversability of ICM cells to trophectoderm cells has been even shown experimentally (Rossant et al., 1979).

The ICM cells are completely totipotent embryonic stem cells, at least for the embryo proper, from which it was tempting to obtain stem cell lines. The first lines of embryonic stem cells or embryo-derived stem cells (ES cells) were obtained independently by Evans and Kaufman (1981) and Martin (1981). In the first case, experimentally delayed blastocysts were cultured *in vitro*; from outgrowths, the ICMs were recovered and the isolated cells were subcultured on feeder cells. In the second case, ICMs were isolated by immunosurgery (Solter and Knowles, 1975) and cultured on a feeder layer in a conditioned medium. Later on it was found that implantation delay or conditioned medium were not necessary in presence of feeders (Axelrod, 1984).

New ES cell lines can be obtained if some rules are respected. It is recommended to use some strains or substrains of mice, such as 129/SV or 129/Ola, in order to obtain completely totipotent cells (Robertson et al., 1986; Pease et al., 1990 and Nichols et al., 1990). Other combinations are possible such as the inbred strain C57 BL/6 (Ledermann and Bürki, 1991; and Kawase et al., 1994) and DBA/1 lac J (Roach et al., 1995). The optimum stage for starting is the preimplantation embryo although postimplantation blastocysts were used (Wells et al., 1991). Cell lines were easily obtained from disaggregated morulae (Eistetter, 1989), however germline transmission was not reported. A feeder layer is generally used to isolate ES cells and to support their successive passages (Suemori and Nakatsuji, 1987). The main role of feeder cells is probably to provide growth factors necessary for proliferation and correlative inhibition of differentiation. The principal working factor is the leukemia inhibiting factor (LIF) as LIF defective fibroblasts do not allow ES cell culture (Stewart et al., 1992); when added in culture medium it allows the obtention of ES cells without feeder layers (Nichols et al., 1990 and Pease et al., 1990). Secreted by the uterus, LIF is necessary for implantation (Stewart et al., 1992) and may contribute to regulate proteinase activity of the mouse preimplantation embryo (Harvey et al., 1995). Embryos with LIF receptor disruption similarly suffer of abnormal placentation and do not survive after birth from several disorders

(Ware et al., 1995). For the establishment of ES cell lines, an alternative is to use other feeder cells known for their production of LIF, e.g. BRL cells (Smith and Hooper, 1987) which produce also other factors such as stem cell factor (Zsebo et al., 1990) or cells secreting LIF as a result of transfection (McMahon and Bradley, 1990). ES cells ex-press the transducing receptor component gp 130 which is common to LIF and other cytokines: ciliary neurotrophic factor or CNTF, oncostatin M or OSM, interleukin-6 or IL-6, etc... (Rose and Bruce, 1991; Ihle and Kerr, 1995). In fact CNTF may be also used to maintain the pluripotentiality of ES cells (Conover et al., 1993), but less efficiently than OSM (Piquet-Pellorce et al., 1994).

How are ES Cell Lines Maintained?

ES cells must be maintained by repeated passages on feeder layers, usually non proliferative mitomycin-C treated or X-irradiated fibroblasts (either permanent lines such as STO fibroblasts, or primary mouse embryonic fibroblasts), or less effectively without feeder layer on gelatin or extracellu- or lar matrix coated substrate in conditioned medium or in LIF-supplemented medium, in order to prevent spontaneous differentiation (reviewed by Wiles, 1993 and Abbondanzo et al., 1993). LIF seems to inhibit specifically differentiation into primitive ectoderm, but not into primitive endoderm (Shen and Leder, 1992).

The effect of LIF may depend on its species-specificity, on its natural or recombinant origin (Williams et al., 1988), and on purity and concentration. It should be noted also that two different forms of LIF exist, either diffusible or matrix-associated, whose production may depend on the activation of alternative promoters (Rathjen et al., 1990b). The best compromise, in decreasing order, is to use primary mouse embryo fibroblasts as feeder cells which maintain a stable karyotype of the ES cells (Suemori and Nakatsuji, 1987) or, in the absence of feeders and presence of LIF, fibroblast extracellular matrix as substrate (Abbondanzo et al., 1993). Note that LIF expression may be activated during ES cell differentiation; this could be a mechanism for feedback regulation of stem cell renewal (Rathjen et al., 1990a).

The culture media used, such as DMEM, contain serum and so are not well characterized, being susceptible of variations from batch to batch. The presence of differentiation factors such as retinoids may be eliminated by the use of neutralizing antibodies (Tamura et al., 1990) or by charcoal treatment (Van den Eijnden-Van Raaij et al., 1991). The presence of such factors may explain the so-called "spontaneous" differentiation of ES cells.

What are the Characteristics of ES Cells?

ES cells may be characterized by their morphology, their profile of gene expression and protein synthesis and by the use of cell markers for discriminating whether cells are differentiated or not.

Morphology

General Characteristics. ES cells appear as small, aggregated and unpolarized cells formings islands on the feeder layers (Evans and Kaufman, 1981; and Martin, 1981). ES cells show a high nucleo-cytoplasmic ratio and large nucleoli (Abbondanzo et al., 1993) indicating ribosomal RNA synthesis and a correlative high protein synthetic activity, at least relevant to cell proliferation.

Microscopic Cell Markers. Different cell markers have been used to characterize undifferentiated versus differentiated ES cells. These markers generally correspond to (glyco)proteins of the cytoskeleton, cell membrane and extracellular matrix and are demonstrated by immunocytochemical techniques. Markers may also be revealed by their eventual enzymatic activity. There are several classical positive markers of undifferentiated ES cells. Among them alkaline phosphatase (Wobus et al., 1984) which is equivalent to the cell surface nonspecific alkaline phosphatase of the inner cell mass of the mouse blastocyst (Mulnard and Huygens, 1978).

The other initial markers used were defined with antibodies to surface glycolipids (ECMA-7, Kemler, 1980), or embryoglycans (SSEA-1, Solter and Knowles, 1978 or its equivalent TEC-1, Draber and Pokorna, 1984), to a cell adhesion molecule, uvomorulin or E-cadherin (Decma-1, Vestweber and Kemler, 1985) and to cytokeratin 8 (Troma-1, Kemler et al., 1981). The latter marker is negative for undifferentiated cells and becomes positive after differentiation into epithelial cells, as in the blastocyst (Duprey et al., 1985). Other generally negative markers correspond to glycoproteins of the extracellular matrix such as fibronectin, laminin, etc... (Mummery et al., 1990). The corresponding transcription activities of most of these markers were checked by these authors as shown by the Table 26.1.

We have confirmed some of these results and precised that ES cells are unpolarized cells showing peripheral distribution of microtubules, f-actin (to which non-muscle myosin is bound, according to Slager et al., 1992) and E-cadherin (Horak et al., 1993), the latter being responsible of the aggregation of ES cells as it does for compaction. ES cells are lamin A/C negative as undifferentiated cells and their aggregates probably constitute a kind of community as they are connected by gap junctions

Table 26.1. Transcription and translation of marker genes of undifferentiated/differentiated ES cells (from Mummery et al., 1990).

	SSEAI	Uvomorulin	Cytokeratins 8	Fibronectin	Collagen IV	Laminin	Nidogen	tPA
mRNA	nd	+/− or +*	±/+	+/+	±/+	±/+	+/+	−
Protein	+/−	+/+*	−/+	/	nd	−/+	nd	nd

Legend: tPA: tissue plasminogen activator.
+: positive, ± : doubtful, −: negative, nd: not determined.
* epithelial cells.

(personal observation). A decreased metabolic cooperation between ES cells alters their developmental capacity in vitro (Sheardown and Hooper, 1992).

Contrary to Mummery et al. (1990), we have obtained an immunofluorescent staining for laminin in undifferentiated D3 ES cells, but not for cellular fibronectin (Horak et al., 1993). It should be stressed at this occasion that positivity for protein markers may depend, e.g. on the ES cell line and on the number of passages. ES cells also produce receptors for laminin, that is the integrin heterodimer α6 β1; however the α6 subunit is different (α6A) from that (α6B) expressed by differentiated cells (Cooper et al., 1991).

Cell Cycle Characteristics

As the alternative for ES cells, if we except death, is to proliferate or to differentiate, it is important to know the characteristics of the cell cycle of these totipotent cells. ES cells have a short cell cycle (t1/2 = 8 h) and an especially brief G1 phase (Savatier et al., 1994). Proliferating ES cells express low levels of cyclin E/CDK2 complexes and of p21 and p27 CDK inhibitors; their CDK4 associated kinase activity is not detectable (Savatier et al., 1995). What about other factors regulating the cell cycle, such as the phosphoprotein of the retinoblastoma susceptibility gene RB-1, a transcriptional repressor known for its tumour suppressor (antioncogene) function (reviewed by Helin and Harlow, 1993)? In exponentially growing ES cells, there is a decrease in RB content and no RB hypophosphorylation, that is immediate phosphorylation by cyclin-dependent kinases in the (virtual) G1 phase, and consequently no inhibition of the progression to S phase (Savatier et al., 1994). Absence of hypophosphorylated RB might be correlated to the tumorigenicity of ES cells; it may explain the "EIA"-like activity of ES cells which allow them to activate transcription at promoters containing E2F binding sites, since hypophosphorylated RB interacts with and down regulates the E2F transcription factor (Savatier, personnal communication).

In turn, genes implied in cell cycle progression such as c-myc and cdc 2 are regulated by E2F binding sites and E2F is associated in complexes with RB mainly in G1 and S phases, with p107, cdk2 and cyclin A in S phase, and with p107, cdk2 and cyclin E in G1 phase (reviewed by Helin and Harlow, 1993).

Gene Expression

Proliferating ES cells probably express all the "house-keeping" genes involved in the machinery of cell cycling and also the genes of at least some receptors to factors which allow them to escape cycling and to differentiate. We will review briefly the information available on these and other genes.

Transcription Factors in ES Cells. Indication on the available transcription factors present in ES cells can be deduced from the successful reporter gene expression obtained with some promoters (see other Chapters).

"Germ-line" specific transcription factors have been identified in ES cells, eg. Oct-3/4, Oct-5 (probably encoded by the same gene) and Oct-6 (Schöler et al., 1989a and b; Meijer et al., 1990 and Suzuki et al., 1990). The Oct transcription factors belong to the POU family of proteins characterized by a sequence binding to DNA octamer motif (ATGCAAAT) and homeodomain (Schöler, 1991); they are involved in the control of cell proliferation (review by Schöler 1991, Verrijzer and Van de Vliet, 1993) and may induce either transcriptional activation (Schöler, 1989b) or repression (Lenardo et al., 1989). In fact, Oct-3/4 acts as a transcription factor for many genes expressed specifically by pluripotent cells (Saijoh et al., 1996). Oct-3/4 is already expressed in the maturing oocyte (Schöler et al., 1990a and Rosner et al., 1990). The factors expressed in ES cells such as Oct-3/4 and Oct-6 are down regulated upon differentiation and have correlatively a developmentally regulated expression in embryo (Rosner *et al.*, 1990; Schöler *et al.*, 1990 and Suzuki *et al.*, 1990).

Transcription factors belonging to other families, such as the zinc-finger Rex-1 are transcribed in ES cells (Rogers *et al.*, 1991).

Alternative Splicing. Like in the other cell types, alternative splicing may occur in ES cells, by means of proteins such as smN, a small nuclear ribonucleoprotein, whose mRNA level decreases after differentiation (Sharpe *et al.*, 1990).

Homeobox Gene Expression. An homeobox gene belonging to a new homeodomain class, HesX1 is transcribed in ES cells and down regulated upon differentiation (Thomas *et al.*, 1995), similarly as Oct-3/4, Oct-6 which contain an homeobox (see above).

Growth Factors and Receptors. ES cells synthesize mRNAs of several growth factors of the FGF family (Heath *et al.*, 1989) such as FGF3 (Wilkinson *et al.*, 1988), FGF4 (Rappolee *et al.*, 1994) and another heparin-binding factor, MK (Nurcombe *et al.*, 1992); there is no down regulation during differentiation in all cases. The FGF receptors genes FGF-R1, 2, 3 and 4 are transcribed in ES cells; only FGF-R2 and 3 are up-regulated after differentiation and alternative splicing may occur (McDonald and Heath, 1994). Undifferentiated ES cells transcribe TGFβ1 gene (Katsube and Shimizu, 1991). They secrete latent TGFβ1 which becomes bound to extracellular matrix after differentiation (Slager *et al.*, 1993). βB inhibin is synthesized by undifferentiated ES cells and activin receptor II gene is transcribed, however TGFβ receptors seem lacking (review by Mummery *et al.*, 1993). Albano *et al.* (1993) found also βB inhibin mRNA in ES cells and a sharp decrease after differentiation.

The GH receptor gene is transcribed and translated in ES cells; the mRNA level is increased by retinoic acid treatment (Ohlsson *et al.*, 1993), implying that ES cells also express retinoic acid-receptors. In fact undifferentiated ES cells synthesize retinoid X receptors mRNA (Bain and Gottlieb, 1994). Of course, LIFR is also expressed in ES cells (Ware *et al.*, 1995).

Sex, X-Inactivation and Gene Imprinting. ES cells are genetically male or female according to the blastocyst of origin. X chromosomes receive during gametogenesis a sex-specific imprinting for gene inactivation in XX cells which would be erased in primitive ectoderm (reviewed by Tada *et al.*, 1993). It seems that parental imprinting is maintained in undifferentiated ES cells after multiple passages. Androgenetic ES cells, although able to give chimeras and a few newborns, induce cartilage defects supposed to originate from lack of female imprinting (Mann *et al.*, 1990).

X-inactivation is low in undifferentiated female ES cells and increases after differentiation, but it is inexistent in both cases when the inactivation centre is lost in one of the X chromosomes (Rastan and Robertson, 1985).

If imprinting involves DNA methylation, it should be noted that methylation may repress gene expression only in differentiated cells (Niwa, 1985). According to Weng *et al.* (1995), methylation, although present in ES cells, is increased upon differentiation.

What are the Potentialities of ES Cells?

Differentiation *In Vitro*. ES cells share the potentialities of the ICM cells from which they are derived. They are able to differentiate *in vitro*, in all the conditions which do not allow maintenance of totipotency (see above), either in monolayer or in aggregates (embryoid bodies) into all the derivatives of the ICM cells.

Monolayers The conditions required to obtain differentiation may be a high number of passages (Nagy *et al.*, 1993), absence of LIF and/or feeder cells, ES cell density and addition of factors such as retinoic acid or DMSO; without additional factors, a so called spontaneous differentiation occurs in monolayers which give rise to endoderm and ill-defined mesoderm-like cells, whereas with retinoic acid parietal-type endoderm cells are obtained (Mummery *et al.*, 1990). Retinoic acid can also induce neuron-like cells in vitro (Bain *et al.*, 1995). Lymphohematopoietic cells can be obtain by coculture with stromal cells lacking functional M-CSF (Nakano *et al.*, 1994).

Embryoid Bodies. When ES cells are cultured at high cell density on a non-adhesive surface (bacterial culture plastic), they form round embryoid bodies showing many similarities with embryo development *in vivo* (Doetschman *et al.*, 1985). The embryoid bodies develop an outer layer of endoderm-like cells and eventually a central cavity. When allowed to attach again and form outgrowths, the embryoid bodies can give rise to differentiated tissues such as myocardium, blood islands and hematopoietic stem cells (Chen and Kosco, 1993). Hematopoietic cell lines (Schmitt *et al.*, 1991; and Wiles and Keller, 1991) and hematopoietic stem cells (Potocnik *et al.*, 1994 and Palacios *et al.*, 1995) were obtained this way.

In vitro differentiation of ES cells appears to be an interesting model to study the factors implied in the establishment of a cell lineage, e.g. the hematopoietic lineage (Keller *et al.*, 1993) or muscle cells (Rohwedel *et al.*, 1994).

Differentiation *In Vivo*

Grafts Tumorogenicity of ES cells injected subcutaneously was already demonstrated by Martin (1981) and Magnuson et al. (1982).

When ES cells or embryoid bodies are implanted in different locations of immunodeficient mice, highly differentiated tissues can be obtained depending on the implantation sites (review by Chen and Kosco, 1993).

Chimeras ES cells can be injected into a morula or in the cavity of 3.5 day blastocysts, giving rise to chimeric mice in which ES cells take part in the development of all kind of tissues including the germ line (Robertson et al., 1986; Doetschman et al., 1987 and Thomson et al., 1989). This is the very interesting and particular characteristic of ES cells versus another kind of stem cells, the embryonal carcinoma (EC) cells, with one exception (Stewart and Mintz, 1982). EC cells will not be dealt with at all in the present review (see Robertson, 1987). ES cells are also able to contribute to the complete generation of embryos using tetraploid blastocysts as hosts (Nagy et al., 1993), however the process is very inefficient and not routinely feasible. Problems could arise when constituting chimeras with ES cells of one sex and blastocysts of the other one. Fortunately, it seems that most of the available ES cell lines are male. If injected in XX blastocysts, XY ES cells will give rise only to male gametes (Iannaccone et al., 1985). Aneuploid ES cells are of course not adequate for germ-line transmission, except 39X0 cells giving female gametes (Stewart et al., 1992).

Nuclear Transfer Although nuclei of 8-cell stage mouse embryos can be used for successful nuclear transfer, nuclear transplantation with ES cells has not given live foetuses (Tsunoda and Kato, 1993).

Use of ES Cells for Transgenesis. The properties of ES cells make them a splendid cell vector for the modification of the mouse genome versus the complicated and inefficient technique of DNA injection into pronuclei. All the techniques of gene insertion, mutation, deletion, gene targeting, gene knock out, gene trap... are treated elsewhere in this book (chapters by S. Viville and A. Nagy). The interest of ES cell is that the gene modifications introduced can be checked *in vitro* before cell transfer in the embryo.

The applications of these techniques particularly in developmental biology are too widespread to be reviewed here. We are just giving a non limitative list of the questions which can be addressed (Table 26.2) and will not either engage here in a discussion of the ethical problems which may arise.

The results of some of these studies are very exciting but sometimes surprising. Survival of some knock-out mice shows that there are many redundancies in systems of living cells. The principle of the method was also criticized as the result observed may correspond not to the simple absence of a gene, but to the reaction of a complex organism to the mutation (Routtenberg, 1995).

Can ES Cells be Obtained in other Mammalian Species?

The question was recently reviewed (Galli et al., 1994) and is analysed elsewhere (chapter by P.J. Donovan et al.), however if we use the criterion of germ-line competent cell for ES cells, no non murine ES cells have been obtained to our knowledge. In order to limit the flow of publications on putative ES cells it was proposed to make a preliminary test using a few essential markers (Fléchon, 1995). Embryonic cell lines obtained during these studies sometimes allow mosaics to be obtained (chapter by N. Strelchenko) and may constitute tools for the study of early embryogenesis. They may be useful insofar they correspond, e.g. to (primitive) endoderm (Van Stekelenburg-Hamers et al., 1995) or to specialized stem cells (Talbot et al., 1994).

No satisfactory explanation was given to this lack of success except that the mouse is not a general model for mammals and shows several particularities (rapid expression of zygote genome after fertilization, inbreeding, high frequency of tumors, etc...). It should be remembered that the obtention of ES may be restricted to some strains and that it

Table 26.2. Uses of ES cells.

Analysis of the characteristics of totipotent cells

Cell cycle studies (proliferation versus differentiation)

Detection of unknown genes involved in developmental biology (enhancer/promoter traps)

Function of an unknown gene

Confirmation of the role of a known gene

Analysis of gene structure/function by mutation

Gene imprinting

Study of the effect of chromosome rearrangements

Gene therapy

Acquisition of resistance to diseases

Obtention of models for human pathologies

Production of animals secreting hormones or other useful proteins is not always easy to repeat. Germ line transmission in chimeras also depends on ES cell and embryo strain (Schwartzberg et al., 1989), and the number of cell passages should be limited (Wiles, 1993).

The hope is to obtain enough totipotent cells from the blastocyst or another source to be able to produce transgenic animals, even if no lines can be established. As only morula blastomeres and blastocyst ICM cells among very early embryonic cells can, after nuclear transfer, give rise to foetal and live born development (see Collas et al., 1994 and Keefer et al., 1994), it is encouraging to note that ICM cells of sheep blastocysts could be used, after a few passages in vitro, to obtain at least a few newborns by nuclear transfer (Campbell et al., 1996).

We have now tools to characterize the cell populations used for nuclear transfer and we can also check that the reprogrammation of nuclei used is effective, that is that all gene expression is transiently repressed after nuclear transfer (Kaňka et al., 1996).

Are There Alternatives to ES Cells?

Even if available in other species, tumorigenic EC cells would be hazardous for application in transgenesis. However, cultured murine primordial germ cells (PGCs) can give rise to hematopoietic stem cells (Rich, 1995). Moreover, totipotent cells have been obtained from murine (PGCs) which are called embryonic germ (EG) cells (chapter by P.J. Donovan et al.,). PGCs are not considered as undifferentiated cells (McLaren, 1992), although they are the ancestors of the gamete stem cells. However, they are potentially immortal while somatic cells may have become mortal (Denis and Lacroix, 1993). The obtention of EG cells may be considered as an artificial bypass of meiosis, giving directly embryonic stem cells. EG cells look similar to ES cells although their gene expression may be slightly different (Matsui, personal communication). Imprinting may also be different in ES and EG cells, although, as in ES cells, methylation is low (or absent) in murine PGCs at the time they can derive EG cells (Kafri et al., 1992), and imprinting seems to be erased, at least when they reach the genital ridges (Szabo and Mann, 1995). So the risk is that imprinting is reestablished if gonadal cells are used; however earlier cells may not be imprinted at all, and their developmental potential is questionable (Rossant, 1993). Like ES cells, most murine female PGCs have two active X chromosomes (Tam et al., 1994). As in ES cells, Oct 3/4 is expressed in mouse PGCs (Schöler et al., 1989 and Rosner et al., 1990), however retinoic acid induces proliferation of PGCs (Koshimizu et al., 1995) whereas it should be avoided to obtain ES cells.

Besides tissue nonspecific alkaline phosphatase (TNAP, Lauwson and Hage, 1994), murine PGCs are characterised by markers such as SSEA-1 (chapter by P.J. Donovan et al.,). Identical or similar embryoglycans have been localized in PGCs with different monoclonal antibodies: 4C9 (Yoshinaga et al., 1991), Le^x (Kanai et al., 1992), EMA-1 (Hahnel and Eddy, 1986) and less defined polyclonal antibodies such as PG-1 (Heath, 1978). PGCs of other species are also alkaline phosphatase positive (bovine: Lavoir et al., 1994, rabbit: Moens et al., 1996). Furthermore its seems that at least some of the antibodies to embryoglycans cross-react with germ cells of other species such as rabbit (Moens et al., 1996), which opens the possibility to apply, for their isolation, techniques used for murine PGCs (chapter by P.J. Donovan et al.,) and to use double staining for cell characterization, as for putative ES cells. No cytokeratins, vimentin, nor lamins A/C were found in rabbit PGCs (Moens and Fléchon, 1995 unpublished). Noteworthy, murine EG cells co-express TNAP and TEC-1 and are lamin A/C negative (Vignon and Matsui, 1995 unpublished). So the PGCs, as the cells of the ICM, show already the same markers as ES or EG cells. The recurrent question is how to make them proliferate and not differentiate in vitro; this has not yet been solved in nonmurine species.

Unfortunately PGCs of other species do not appear better than murine PGCs for the obtention of chimeras (Moens et al., 1996). Nuclear transplantation of PGCs is not yet a simple nor productive technique in mouse (Kato and Tsunoda, 1995), pig (Liu et al., 1995) or rabbit (Moens et al., 1996; unpublished). It remains to find the (species-) specific factors or feeder cells necessary to maintain proliferation of PGCs in vitro. The hope is again to obtain, if not cell lines, at least cultures for a few passages without either cell degeneration or apoptosis (Pesce and De Felici, 1994) on one hand or differentiation on the other.

References

Abbondanzo, S.J., Gadi, I., Stewart, C.L. (1993) Derivation of embryonic stem cell lines. *Methods in Enzymology* **225**, 803–823

Albano, R.M., Groome, N. and Smith J.C. (1993) Activins are expressed in preimplantation mouse embryos and in ES and EC cells and are regulated on their differentiation. *Dev.* **117**, 711–723

Axelrod, H.R. (1984) Embryonic stem cell lines derived from blastocysts by a simple technique. *Dev. Biol.* **101**, 225–228

Bain, G. and Gottlieb, D.I. (1994) Expression of retinoid X receptors in P19 embryonal carcinoma cells and embryonic stem cells. *Bio. Bioph. Res. Com.* **200**, 1252–1256

Bain, G., Kitchens, D., Yao, M., Huettner, J.E. and Gottlieb, D.I. (1995) Embryonic stem cells express neuronal properties *in vitro. Develop. Biol.* **168**, 342–357

Campbell, K.H.S., Mc Whir, J., Ritchie, W.A. and Wilmut, (1996) Sheep cloned by nuclear transfer from a cultured cell line, *Nature* **380**, 64–66

Chen, U. and Kosco, M. (1993) Differentiation of mouse embryonic stem cells *in vitro*: III. Morphological evaluation of tissues developed after implantation of differentiated mouse embryoid bodies. *Dev. Dyn.* **197**, 217–226

Collas, P. and Barnes, F.L. (1994) Nuclear transplantation by microinjection of inner cell mass and granulosa cell nuclei. *Mol. Reprod. Dev.* **38**, 264–267

Conover, J.C., Ip, N.Y., Poueymirou, W.T., Bates, B., Goldfarb, M.P., DeChiara, T.M. and Yancopoulos, G.D. (1993) Ciliary neurotrophic factor maintains the pluripotentially of embryonic stem cells. *Dev.* **119**, 559–565

Cooper, H.M., Tamura, R.N. and Quaranta, V. (1991) The major laminin receptor of mouse embryonic stem cells is a novel isoform of the $\alpha 6\beta 1$ integrin. *J. Cell Biol.* **115**, 843–850

Denis, H. and Lacroix, J.C. (1993) The dichotomy between germ line and somatic line, and the origin of cell mortality. *TIG* **9**, 7–11

Doetschman, T.C., Eistetter H., Katz, M., Schmidt, W. and Kemler, R. (1985) The *in vitro* development of blastocyst-derived embryonic stem cell lines: formation of visceral yolk sac, blood islands and myocardium. *J. Embry. and Exp. Morph.* **87**, 27–45

Doetschman, T.C., Gregg, R.G., Maeda, N., Hooper, M.L., Melton, D.W. and Thomson, S. (1987) Targeted correction of a mutant HPRT gene in mouse embryonic stem cells. *Nature* **330**, 576–578

Draber, P. and Pokorna, Z. (1984) Differentiation antigens of mouse teratocarcinoma stem cells defined by monoclonal antibodies. *Cell Differentiation* **15**, 109–113

Duprey, P., Morello, D., Vasseur, M., Babinet, C., Condamine, H., Brulet, P. and Jacob, F. (1985) Expression of the cytokerain endoA gene during early mouse embryogenesis. *Proc. Natl. Acad. Sci. USA* **82**, 8535–8539

Eistetter, H.R. (1989) Pluripotent embryonal stem cell lines can be established from disaggregated mouse morulae. *Dev. Growth and Dif.* **31**, 275–282

Evans, M.J. and Kaufman, M.H. (1981) Establishment in culture of pluripotential cells from mouse embryos. *Nature* **292**, 154–156

Fléchon J.E. (1995) Request for a consensus on the definition of putative embryonic stem cells. *Mol. Reprod. Dev.* **41**, 274

Galli, C., Lazzari, G., Fléchon J.E. and Moor, R.M. (1994) Embryonic stem cells in farm animals, *Zygote* **2**, 385–389

Hahnel, A.C. and Eddy, E.M. (1986) The distribution of cell surface determinants of mouse embryonal carcinoma on early embryonic cells. *J. Reprod. Immunol.* **10**, 89–110

Harvey, M.B., Leco, K.J., Arcellana-Panlilio, M.Y., Zhang, X., Edwards D.R. and Schultz, G.A. (1995) Proteinase expression in early embryos is regulated by leukaemia inhibitory factor and epidermal growth factor. *Dev.* **121**, 1005–1014

Heath, J.K. (1978) Characterization of a xenogeneic antiserum raised against the fetal germ cells of the mouse: cross-reactivity with embryonal carcinoma cells. *Cell* **15**, 299–306

Heath, J.K., Paterno, G.D., Lindon, A.C. and Edwards, D.R. (1989) Expression of multiple heparin-binding growth factor species by murine embryonal carcinoma and embryonic stem cells, *Development* **107**, 113–122

Helin, K. and Harlow, E. (1993) The retinoblastoma protein as a transcriptional repressor. *Cell Biol.* **3**, 43–46

Horak, V., Fléchon, J.E. and Moens, A. (1993) Immunocytochemical detection of protein expression in undifferentiated mouse embryonic stem cells. *Int. J. Dev. Biol.* **37**, 7S

Iannaccone, P.M., Evans E.P. and Burtenshaw, M.D. (1985) Chromosomal sex and distribution of functional germ cells in a series of chimeric mice. *Exp. Cell Res.* **156,** 471–477

Ihle, J.N. and Kerr, I.M. (1995) Jaks and Stats in signaling by the cytokine receptor superfamily. *Elsevier* **11**, 69–74

Kafri, T., Ariel, M., Brandeis, M., Shemer, R., Urwen, L., McCarrey, J., Cedar, H. and Razin, A. (1992) Developmental pattern of gene-specific DNA methylation in the mouse embryo and germ line. *Genes Dev.* **6**, 705–711

Kanai, Y., Kawakami, H., Kanai-Azuma, M., Kurohmaru, M., Hirano, H. and Hayashi, Y. (1992) Changes in intracellular and cell surface localization of Le^x epitope during germ cell differentiation in fetal mice. *J. Vet. Med. Sci.* **54**, 297–303

Kaňka, J., Hozak, P., Heyman, Y., Chesne, P., Degrolard, J., Renard, J.P. and Fléchon, J.E. (1996) Transcriptional activity and nucleolar ultrastructure of embryonic rabbit nuclei after transplantation to enucleated oocytes. *Mol. Reprod. Dev.* **43**, 135–146

Kato, Y. and Tsunoda, Y. (1995) Germ cell nuclei of male fetal mice can support development of chimeras to midgestation following serial transplantation. *Develop.* **121**, 779–783

Katsube, K. and Shimizu, N. (1991) Elevation of transforming growth factor $\beta 1$ mRNA level in ES-D3 mouse embryonic stem cells cocultured with Balb/c3T3 A31 fibroblasts. *Cell Struc. Funct.* **16**, 375–382

Kawase, E., Suemori, H., Takahashi, N., Okazaki, K., Hashimoto, K. and Nakatsuji, N. (1994) Strain differences in establishment of mouse embryonic stem (ES) cell lines. *Int. J. Dev. Biol.* **38**, 385–390

Keefer, C.L., Stice, S.L. and Matthews, D.L. (1994) Bovine inner cell mass cells as donor nuclei in the production of nuclear transfer embryo and calves. *Biol. Reprod.* **50**, 935–939

Keller, G., Kennedy, M., Papayannopoulou and Wiles, M. (1993) Hematopoietic commitment during embryonic stem cell differentiation in culture. *Mol. Cell Biol.* **13**, 473–486

Kemler, R. (1980) Analysis of mouse embryonic cell differentiation. In: "Fortschritte der Zoologie", M. Lindauer, ed., Würzburg (Symposium "Progress in Developmental Biology"), Vol. 26, Fischer, Stuttgart, H.W. Sayer, ed. 175pp

Kemler, R., Brûlet, P., Schnebelen, M., Gaillard, J. and Jacob, F. (1981) Reactivity of monoclonal antibodies against intermediate filament proteins during embryonic development. *J. Embryol. exp. Morph.* **64**, 45–60

Koshimizu, U., Watanabe, M. and Nakatsuji, N. (1995) Retinoic acid is a potent growth activator of mouse primordial germ cells *in vitro*. *Dev. Biol.* **168**, 683–685

Lauwson, K.A. and Hage, W.J. (1994) Clonal analysis of the origin of primordial germ cells in the mouse. *Ciba Foundation Symp.* **182**, 68–91

Lavoir, M.C., Basrur, P.K. and Betteridge, K.J. (1994) Isolation and identification of germ cells from fetal bovine ovaries. *Mol. Reprod. Dev.* **37**, 413–424

Ledermann, B. and Bürki, K. (1991) Establishment of a germ-line competent C57BL/6 embryonic stem cell line. *Exp. Cell Res.* **197**, 254–258

Lenardo, M.J., Standt, L., Robbins, P., Kuang, A., Mulligan, R.C. and Baltimore, D. (1989) Repression of the IgH enhancer in teratocarcinoma cells associated with a novel octamer factor. *Science* **243**, 544–546

Liu, L., Moor, R.M., Laurie, S. and Notarianni, E. (1995) Nuclear remodelling and early development in cryopreserved, porcine primordial germ cells following nuclear transfer into *in vitro*-matured oocytes *Int. J. Dev. Biol.* **39**, 639–644

MacDonald, F.J. and Heath, J.K. (1994) Developmentally regulated expression of fibroblast growth factor receptor genes and splice variants by murine embryonic stem and embryonal carcinoma cells. *Dev. Genet.* **15**, 148–154

MacLaren, A. (1992) The quest for immortality. *Nature* **359**, 482–483

MacMahon, A.P. and Bradley, A. (1990) The Wnt-1 (int-1) proto-oncogene is required for development of a large region of the mouse brain. *Cell* **62**, 1073–1085

Magnuson, T., Epstein, C.J., Silver, L.M. and Martin, G.R. (1982) Pluripotent embryonic stem cell lines can be derived from t^{w5}/t^{w5} blastocysts *Nature* **298**, 750–753

Mann, J.R., Gadi, I., Harbison, M.L., Abbondanzo, S.J. and Stewart C.L. (1990) Androgenetic mouse embryonic stem cells are pluripotent and cause skeletal defects in chimeras. *Cell* **62**, 251–260

Martin, G.R. (1981) Isolation of a pluripotent cell line from early mouse embryos cultured in medium conditioned by teratocarcinoma stem cells. *Proc. Nat. Acad. Sci., USA* **78**, 7634–7638

Meijer, D., Graus, A., Kraay, R., Langeveld, A. and Mulder, M.P. and Grosveld, G. (1990) The octamer binding factor Oct6: cDNA cloning and expression in early embryonic cells. *Nucl. Acids Res.* **18**, 7357–7365

Moens, A., Betteridge, K.J., Brunet, A., Renard, J.P. (1996) Low levels of chimerism in rabbit fetuses produced from preimplantation embryos microinjected with fetal gonadal cells. *Mol. Reprod. Dev.* **43**, 38–46

Mulnard, J. and Huygens, R. (1978) Ultrastructural localization of non-specific alkaline phosphatase during cleavage and blastocyst formation in the mouse. *J. Embryol. Exp. Morph.* **44**, 121–131

Mummery, C.L., Feyen, A., Freund, E. and Shen, S. (1990) Characteristics of embryonic stem cell differentiation: a comparison with two embryonal carcinoma cell lines. *Cell Diff. Dev.* **30**, 195–206

Mummery, C.L. and Van Den Eijnden-Van Raaij, A.J.M. (1993) Type β transforming growth factors and activins in differentiating embryonal carcinoma cells, embryonic stem cells and early embryonic development. *Int. J. Dev. Biol.* **37**, 169–182

Nagy, A., Rossant, J., Nagy, R., Abramow-Newerly, W. and Roder, J.C. (1993) Derivation of completely cell culture-derived mice from early-passage embryonic stem cells. *Proc. Natl. Acad. Sci. USA* **90**, 8424–8428

Nakano, T., Kodama, H. and Honjo, T. (1994) Generation of lymphohematopoietic cells from embryonic stem cells in culture. *Science* **265**, 1098–1101

Nichols, J., Evans, E.P. and Smith, A.G. (1990) Establishment of germline-competent embryonic stem (ES) cells using differentiation-inhibiting activity. *Dev.* **110**, 1341–1348

Niwa, O. (1985) Suppression of the hypomethylated moloney leukemia virus genome in undifferentiated teratocarcinoma cells and inefficiency of transformation by a bacterial gene under control of the long terminal repeat. *Mol. Cell Biol.* **5**, 2325–2331

Nurcombe, V., Fraser, N., Herlaar, E. and Heath, J.K. (1992) MK: a pluripotential embryonic stem-cell-derived neuroregulatory factor. *Dev.* **116**, 1175–1183

Ohlsson, C., Lövstedt, K., Holmes, P.V., Nilsson, A., Carlsson, L. and Törnell, J. and (1993) Embryonic stem cells express growth hormone receptors: regulation by retinoic acid. *Endocr.* **133**, 2897–2903

Palacios, R., Golunski, E. and Samaridis, J. (1995) *In vitro* generation of hematopoietic stem cells from an embryonic stem cell line. *Proc. Natl. Acad. Sci. USA* **92**, 7530–7534

Pease, S., Braghetta, P., Gearing, D., Grail, D. and Williams, R.L. (1990) Isolation of embryonic stem (ES) cells in media supplemented with recombinant leukemia inhibitory factor (LIF). *Dev. Biol.* **141**, 344–352

Pesce, M. and De Felici, M. (1994) Apoptosis in mouse primordial germ cells: a study by transmission and scanning electron microscope. *Anat. Embryol.* **189**, 435–440

Piquet-Pellorce, C., Mereau, L.G.A. and Heath, J.K. (1994) Are LIF and related cytokines functionally equivalent? *Exp. Cell Res.* **213**, 340–347

Potocnik, A., Nielsen P.J. and Eichmann, K. (1994) *In vitro* generation of lymphoid precursors from embryonic stem cells. *EMBO J.* **13**, 5274–5283

Rappolee, D.A., Basilico, C., Patel, Y. and Werb, Z. (1994) Expression and function of FGF-4 in peri-implantation development in mouse embryos. *Dev.* **120**, 2259–2269

Rastan, S. and Robertson, E.J. (1985) X-chromosome deletions in embryo-derived (EK) cell lines associated with lack of X-chromosome inactivation. *J. Embryol. Exp. Morph.* **90**, 379–388

Rathjen, P.D., Nichols, J., Toth, S., Edwards, D.R., Heath, J.K. and Smith, A.G. (1990a) Developmentally programmed induction of differentiation inhibiting activity and the control of stem cell populations. *Genes Dev.* **4**, 2308–2318

Rathjen, P.D., Toth, S., Willis, A., Heath, J.K. and Smith, A.G. (1990b) Differentiation inhibiting activity is produced in matrix-associated and diffusible forms that are generated by alternate promoter usage. *Cell* **62**, 1105–1114

Rich, I. N. (1995) Primordial germ cells are capable of providing cells of the hematopoietic system *in vitro*. *Blood* **86**, 463–472

Roach, M.L., Stock, J.L., Byrum, R., Koller, B.H. and McNeish, J.D. (1995) A new embryonic stem cell line from DBA/1 lac J mice allows genetic modification in a murine model of human inflammation. *Exp. Cell Res.* **221**, 520–525

Robertson, E.J. (1987) Teratocarcinomas and embryonic stem cells, a practical approach IRL. *Oxford press, Washington DC*, **254** pp

Robertson, E.J., Bradley, A., Kuehn, M. and Evans, M. (1986) Germline transmission of genes introduced into cultured puripotential cells by retroviral vector. *Nature* **323**, 445–448

Rogers, M.B., Hosler, B.A. and Gudas, L.J. (1991) Specific expression of a retinoic acid-regulated, zinc-finger gene, Rex-1, in preimplantation embryos, trophoblast and spermatocytes. *Dev.* **113**, 815–824

Rohwedel, J., Maltsev, V., Bober, E., Arnold, H.H., Hescheler, J. and Wobus, A.M. (1994) Muscle cell differentiation of embryonic stem cells reflects myogenesis in vivo: developmentally regulated expression of myogenic determination genes and functional expression of Ionic currents. *Dev. Biol.* **164**, 87–101

Rose, T.M. and Bruce, A.G. (1991) Onconstatin M is a member of a cytokine family which included leukemia inhibitory factor, granulocyte colory–stimulating factor and interleukin-6. *Proc. Natl. Acad. Sci. USA* **88**, 8641–8645

Rosner, M.H., Vigano, M.A., Ozato, K., Timmons, P.M., Poirier, F., Rigby P.W.J. and Staudt, L.M. (1990) A POU-domain transcription factor in early stem cells and germ cells of the mammalian embryo. *Nature* **345**, 686–692

Rossant, J. and Lis, W.T. (1979) Potential of isolated mouse inner cell masses to form trophectoderm derivatives *in vivo*. *Dev. Biol.* **70**, 255–261.

Rossant, J. (1993) Immortal germ cells? *Current Biology* **3**, 47–49

Routtenberg, A. (1995) Knockout mouse fault lines. *Nature* **374**, 314–315

Saijoh, J., Fujü, H., Meno, C., Sato, M., Hirota, Y., Nagamatsu, S., Ikeda, M. and Hamada, H., (1996) A large proportion of pluripotent cell-specific genes are downstream targets of Oct-3 transcription factor. *Genes Cells* **1** (in press)

Savatier, P., Huang, S., Szekely, L., Wiman, K.G. and Samarut, J. (1994) Contrasting patterns of retinoblastoma protein expression in mouse embryonic stem cells and embryonic fibroblasts. *Oncogene* **9**, 809–818

Savatier, P., Lapillonne, H., Van Grunsven, L.A., Rudkin, B.B. and Samarut, J. (1995) Withdrawal of differentiation inhibitory activity/leukemia inhibitory factor up-regulates D-type cyclins and cyclin-dependent kinase inhibitors in mouse embryonic stem cells. *Oncogene* **12**, 309–322

Schmitt, R.M., Bruyns, E. and Snodgrass, H.R. (1991) Hematopoietic development of embryonic stem cells *in vitro* cytokine and receptor gene expression, *Genes Develop.* **5**, 728–740

Schöler, H.R. (1991) Octamania: The POU factors in murine development. *TIG* **7**, 323–329

Schöler, H.R., Balling, R., Hatzopoulos, A.K., Suzuki, N. and Gruss, P. (1989a) Octamer binding proteins confer transcriptional activity in early mouse embryogenesis. *EMBO J.* **8**, 2551–2557

Schöler, H.R., Hatzopoulos, A.K., Balling, R., Suzuki, N. and Gruss, P. (1989b) A family of octamer-specific proteins present during mouse embryogenesis: Evidence for germline-specific expression of an Oct factor. *EMBO J.* **8**, 2543–2550

Schöler, H.R., Dressler, G.R., Balling, R., Rohdewohld, H. and Gruss, P. (1990a) Oct-4: a germline-specific transcription factor mapping to the mouse t-complex. *EMBO J.* **9**, 2185–2195

Schwartzberg, P.L., Goff, S.P. and Robertson, E.J. (1989) Germ-line transmission of a c-abl mutation produced by targeted gene disruption in ES cells. *Science* **246**, 799

Sharpe, N.G., Williams, D.G. and Latchman, D.S. (1990) Regulated expression of the small nuclear ribonucleoprotein particle protein SmN in embryonic stem cell differentiation. *Mol. Cell. Biol.* **10**, 6817–6820

Sheardown, S.A. and Hooper, M.L. (1992) A relationship between gap junction-mediated intercellular communication and the in vitro developmental capacity of murine embryonic stem cells. *Exp. Cell Res.* **198**, 276–282

Shen, M.M. and Leder, P. (1992) Leukemia inhibitory factor is expressed by the preimplantation uterus and selectively blocks primitive ectoderm formation *in vitro*. *Proc. Natl. Acad. Sci. USA* **89**, 8240–8244

Slager, H.G., Freund, E., Buiting, A.M., Feijen, A. and Mummery, C.L. (1993) Secretion of transforming growth factor-beta isoforms by embryonic stem cells: isoform and latency are dependent on direction of differentiation. *J. Cell Physiol.* **156**, 247–256

Slager, H.G., Good, M.J., Schaart, G., Groenewoud, J.S. and Mummery, C.L. (1992) Organization of nonmuscle myosin during early murine embryonic differentiation. *Differentiation* **50**, 47–56

Smith, A.G. and Hooper, M.L. (1987) Buffalo rat liver cells produce a diffusible activity which inhibits the differentiation of murine embryonal carcinoma and embryonic stem cells. *Dev. Biol.* **121**, 1–9

Solter, D. and Knowles, B.B. (1975) Immunosurgery of mouse blastocyst. *Proc. Nat. Acad. Sci. USA.*, **72**, 5099–5102

Solter, D. and Knowles, B.B. (1978) Monoclonal antibody defining a stage-specific mouse embryonic antigen (SSEA-1), *Proc. Nat. Acad. Sci.* **75**, 5565–5569

Stewart, C.L., Kaspar, P., Brunet, L.J., Bhatt, H., Gadi, I., Köntgen F. and Abbondanzo, S.J. (1992) Blastocyst implantation depends on maternal expression of leukaemia inhibitory factor. *Nature* **359**, 76–79

Stewart, T.A. and Mintz, B. (1982) Recurrent germ-line transmission of the teratocarcinoma genome from the METT-1 culture line to progeny *in vivo*. *J. Exp. Zool.* **224**, 465–469

Suemori, H. and Nakatsuji, N. (1987) Establishment of the embryo-derived stem (ES) cell lines from mouse blastocysts : effects of the feeder cell layer. *Dev. Growth and Diff.* **29**, 133–139

Suzuki, N., Rohdewohld, H., Neuman, T., Gruss, P. and Schöler, H.R. (1990) Oct-6: a POU transcription factor expressed in embryonal stem cells and in the developing brain. *EMBO J.* **9**, 3723–3732

Szabo, P.E. and Mann, J.R. (1995) Biallelic expression of imprinted genes in the mouse germ line: implications for erasure, establishment, and mechanisms of genomic imprinting. *Genes Dev.* **9**, 1857–1868

Tada, T., Tada, M. and Takagi N. (1993) X chromosome retains the memory of its parental origin in murine embryonic stem cells. *Dev.* **119**, 813–821

Tamura, K., Ohsugi, K. and Ide, H. (1990) Distribution of retinoids in the chick limb bud: analysis with monoclonal antibody. *Dev. Biol.* **140**, 20–26

Talbot, N.C., Rexroad, C.E., Powell, A.M., Pursel, V.G., Carpena, T.J., Ogg, S.L. and Nel, N.D. (1994) A continuous culture of pluripotent fetal hepatocytes derived from the 8-day epiblast of the pig *in vitro* Cell *Dev. Biol.* **30A**, 843–850

Tam, P.P.L., Zhou, S.X. and Tan, S. (1994) X-chromosome activity of the mouse primordial germ cells revealed by the expression of an X-linked lacZ transgene. *Dev.* **120**, 2925–2932

Tarkowski, A.K. and Wrobleska, J. (1967) Development of blastomeres of mouse eggs isolated at the 4- and 8-cell blastomeres. *J. Embryol. Exp. Morph.* **18**, 155

Thomas, P.Q., Johnson, B.V., Rathjen, J. and Rathjen, P.D. (1995) Sequence, genomic organization, and expression of the novel homeobox gene Hesx1. *J. Biol. Chem* **270**, 3869–3875

Thomson, S., Clarke, A.R., Pow, A.M., Hooper, M.L., Melton, D.W. (1989) Germline transmission and expression of a corrected HPRT gene, produced by gene targeting in embryonic stem cells. *Cell* **56**, 313–321

Tsunoda, Y. and Kato, Y. (1993) Nuclear transplantation of embryonic stem cells in mice. *J. Reprod. Fert.* **98**, 537–540

Van Den Eijnden-Van Raaij, A.J.M., Van Achterberg, T.A.E., Van Der Kruijssen, C.M.M., Piersma, A.H., Huylebroeck, D., De Laat, S.W. and Mummery, C.L. (1991) *Mech. Dev.* **33**, 157–166

Van Steckelenburg-Hamers, A.E.P., Van Achterberg, T.A.E., Rebel, H.G., Fléchon, J.E., Campbell, K.H.S., Wiema, S.M. and Mummery, C.L. (1995) Isolation and characterization of permanent cell lines from inner cell mass of bovine blastocysts. *Mol. Reprod. Dev.* **40**, 444–455

Verrijzer, C.P. and Van der Vliet, P.C. (1993) POU domain transcription factors. *Bioch. Bioph. Acta* **1173**, 1–21

Vestweber, D. and Kemler, R. (1985) Identification of a putative cell adhesion domain of uvomorulin. *EMBO J.* **4**, 3393–3398

Ware, C.B., Horowitz, M.C., Renshaw, B.R., Hunt, J.S., Liggitt, D., Koblar, S.A., Gliniak, B.C., McKenna, H.J., Papayannopoulou, T., Thoma, B., Cheng, L., Donovan, P.J., Peschon, J.J., Bartlett, P.F., Willis, C.R., Wright, B.D., Carpenter, M.K., Davison, B.L. and Gearing, D.P. (1995) Targeted disruption of the low-affinity leukemia inhibitory factor receptor gene causes placental, skeletal, neural and metabolic defects and results in perinatal death. *Dev.* **121**, 1283–1299

Wells, D.N., McWhir, J., Hooper, M.L. and Wilmut, I. (1991) Factors influencing the isolation of murine embryonic stem cells. *Theriogenology* **35**, 293

Weng, A., Magnuson, T. and Storb, U. (1995) Strain-specific transgene methylation occurs early in mouse development and can be recapitulated in embryonic stem cells. *Dev.* **121**, 2853–2859

Wiles, M.V. (1993) Embryonic stem cell differentiation *in vitro*. *Meth. Enzymology* **225**, 900–918

Wiles, M.V. and Keller, G. (1991) Multiple hematopoietic lineages develop from embryonic stem cells (ES) cells in culture. *Development* **111**, 259–267

Wilkinson, D.G., Peters, G., Dickson, C. and McMahon, A.P. (1988) Expression of the FGF-related proto-oncogene int-2 during gastrulation and neurulation in the mouse. *EMBO J.* **7**, 691–695

Williams, R.L., Hilton, D.J., Pease, S., Willson, T.A., Stewart, C.L., Gearing, D.P. *et al.* (1988) Myeloid leukaemia inhibitory factor maintains the developmental potential of embryonic stem cells. *Nature* **336**, 684–688

Wobus, A.M., Holrhausen, H., Jäkel, P. Schöneich, J. (1984) Characterization of a pluripotent stem cell line derived from a mouse embryo. *Exp. Cell Res.* **152**, 212–219

Yoshinaga, K., Muramatsu, H. and Muramatsu, T. (1991) Immunohistochemical localization of the carbohydrate antigen 4C9 in the mouse embryo: a reliable marker of mouse primordial germ cells. *Differentiation* **48**, 75–82

Zsebo, K.M., Wypych, J., McNiece, I.K., Lu H.S., Smith, K.A., Karkare, S.B., Sachdev, R.K., Yuschenkoff, V.N., Birkett, N.C., Williams L.R., Satyagal, V.N., Tung, W., Bosselman, R.A., Mendiaz, E.A. and Langlaey K.E. (1990) Identification, purification, and biological characterization of hematopoietic stem cell factor from buffalo rat liver-conditioned medium. *Cell* **63**, 195–201

27. Formation of Mouse Chimeric Embryos from ES Cells

Andras Nagy

*Mount Sinai Hospital, Samuel Lunenfeld Research Institute,
600 University Ave., Toronto, Ont. M5G 1X5, Canada*

Introduction

In the past few years, mouse embryonic stem (ES) cell-mediated transgenic technologies have revolutionized mammalian genetics. The technologies take advantage of the four unique characteristics of the ES cell system; ES cell lines (i) are derived from a single mouse embryo, (ii) can be maintained permanently in culture, (iii) consist of pluripotent cells and (iv) are germline compatible if they are introduced back into chimeras. Introduction of exogenous DNA by electroporation into the ES cell is efficient and simple. The DNA vector integrates into the genome at a frequency of $\sim 10^{-3}$. In addition, numerous positive and negative selectable markers, originally developed for bacterial systems, are operational in ES cells, therefore their placement in the DNA vectors allows efficient selection for genetically altered cells. The gene-modified cell clones are introduced back into preimplantation stage embryos, either by blastocyst injection or by morula aggregation, to produce chimeras. An ES cell line with good developmental potential efficiently colonizes the germline of the chimeras and these animals transmit the genetic alteration into the next generation.

ES cell-mediated transgenesis certainly provides an alternative to the classical method of DNA injection into the pronucleus of zygotes. However, the real advantage of ES cell-mediated transgenesis is apparent when identification of special, low frequency integration is required. There are two large categories among these special integrations: one is gene trap (Skarnes, 1990), and the second is gene targeting (Capecchi, 1989). Gene trap strategies are based on a specifically designed, randomly integrating vector. The vector usually carries a *lacZ* reporter gene and thus upon integration displays the expression pattern of the "trapped" gene. The expression can provide information about the possible function of the gene. This method can be extremely useful in identifying novel genes expressed within particular lineages of development (Gossler *et al.*, 1990) or with certain physiological or developmental functions (Forrester *et al.*, 1996). Gene targeting, the other category, utilizes homologous recombination to integrate into a specific site of the genome. This approach provided for the first time in mammalian genetics the possibility to remove ("knockout") particular genes from the genome and study the effect of these mutations in animals. In addition to the increasing popularity of this technology to create null mutations in genes, a second phase of genome alteration strategies is evolving. These strategies further develop on the homologous recombination-based gene targeting technology by combining it with Cre/loxP site-specific recombination (Rossant and Nagy, 1995). These combined strategies allow introduction of changes ranging from subtle mutations, such as point mutations or small deletions (Gu *et al.*, 1994) to large genome rearrangements, such as chromosomal translocations (Smith *et al.*, 1995 and van Deursen *et al.*, 1995) or duplications (Ramirez-Solis *et al.*, 1995). Thus, we are close to the stage of being able to create all imaginable modifications in the mouse genome. This opens up valuable possibilities to explore the genetic mechanisms underlying normal and disease processes, and to create proper animal models for human genetic diseases.

Introduction of Altered ES Cells into the Mouse

Targeted ES cells still have the developmental potential to contribute to all lineages of the developing embryo proper when they are introduced back to the *in vivo* environment (Chambers et al., 1994). There are two different methods for this introduction:

1. injection of ES cells into a mouse blastocyst or morula by micromanipulations;
2. aggregation of a clump of ES cells with an eight cell stage embryo.

ES cell injection into blastocyst has been utilized for many years and numerous detailed descriptions of the technique are available (for example, Hogan et al., 1994; Papaioannou and Johson, 1993). The aggregation of ES cells with a blastomere stage embryo is simple, novel and quickly becoming a popular technology (Wood et al., 1993). In the following we briefly describe some technical aspects of the two methods to enable a comparison of the two and describe the advantages and disadvantages of each. Furthermore, as an extension of the ES cell aggregation technique we will discuss the use and some important aspects of completely ES cell-derived embryos made by ES cell ↔ tetraploid embryo aggregation (Nagy and Rossant, 1993).

Injecting ES Cells into Blastocyst/Morulas

Requirements:
  Instruments:
    – pair of micromanipulators
    – high power (inverted) microscope
    – dissecting microscope
    – CO_2 incubator
    – sterile hood
  Subjects:
    – blastocyst or morula stage host embryo – preferentially inbred C57
    – ES cells in single cell suspension
    – 2.5 dpc pseudopregnant recipient on the day of injection

Step 1.
Blastocysts are flushed from the uterus of 3.5 dpc females shortly before injection and stored in M16 at 37 C in a CO_2 incubator, as described in Hogan et al. (1994).

Step 2.
A single cell suspension of ES cells is prepared by trypsinizing and washing the cells in cold PBS.

Step 3.
Cells and blastocysts are placed in the injection chamber and the injection is performed as detailed in Papaioannou and Johson (1993) in detail.

Step 4.
The injected blastocysts are placed back in the CO_2 incubator for 1–3 hours to recover before they are transplanted back into the uterus of pseudopregnant females (Hogan et al., 1994).

Aggregating ES Cells with 8-Cell stage Embryos

Requirements:
  Instruments:
    – dissecting microscope
    – CO_2 incubator
    – sterile hood
  Subjects:
    – eight cell stage host embryo – preferentially outbred albino
    – clumps of 10–15 loosely connected ES cells
    – 2.5 dpc pseudopregnant recipient one day following the aggregation

Step 1.
An aggregation plate is prepared prior to flushing the embryos and the ES cell preparation. Then the prepared plate is placed in the incubator for CO_2 equilibration. For details of the preparation please refer to Nagy and Rossant (1993).

Step 2.
Outbred albino (such as CD1) females are routinely superovulated and mated. Two days later (2.5 dpc), these females are sacrificed and embryos are flushed from the oviduct and the upper part of the uterus, as described in Hogan et al. (1994). The embryos are placed in the incubator in M16 medium until all the embryos are collected. They are then treated with Acid Tyrod's to remove the zona pellucida and placed individually into the depressions made with an aggregation needle in the aggregation plate (Nagy and Rossant, 1993).

Step 3.
Single cell suspension ES cells are plated two days prior to the aggregation at low density ($<10^4$/ml) on gelatinized or feeder plate. Right before aggregation, the cells are briefly trypsinized, and the trypsin is stopped by adding medium to the plate. Clumps of cells are then picked directly from the plate, washed through a large drop of M16 and transferred to the aggregation plate (Nagy and Rossant, 1993).

Step 4.
A clump of 5–10 loosely connected cells is placed beside the embryo sitting in the depression of the aggregation plate. When all the aggregates are arranged the plate is placed back in the incubator for overnight culture. By the next day, the aggregates form a single embryonic structure of

an early blastocyst, and are transferred to 2.5 dpc pseudopregnant females.

Comparison of ES Cell Injection and ES Cell Aggregation

Before establishing the use of either method for ES cell introduction into embryos, consideration should be made of the pros and cons of each approach. In the following table some advantages (*italics*) *versus* disadvantages (plain) are listed. Not all of them are absolute, since depending on the actual needs, one might not require the advantage if it conflicts with the design of the experiment.

Injection	Aggregation
a. Expensive instruments (micromanipulators, microscope) are needed	*No need for expensive instruments*
b. Need highly skilled personnel to carry on injection, require long training time, not everyone can perfect	*No need for highly skilled personnel, require short training time, any one can be taught*
c. Inbred host embryo (C57)	*Outbred host embryo*
d. *Less requirement for optimal culture conditions*	Need for optimal *in vitro* culture conditions
e. *Less requirement for ES cells with excellent developmental potential for germline transmission*	Stronger requirement for excellent developmental potential
f. Frequent mixed type germline transmission	*Frequent overall germline transmission*
g. Inbred C57 females are used to test germline transmission	*Outbred CD1 females can be used to test germline transmission*

These pros and cons of each method are elaborated below:

a. The injection of ES cells into blastocyst or morula requires an expensive pair of high quality micromanipulators and high power microscope. On the other hand ES cell aggregation can be achieved under an ordinary dissecting microscope that is also used for embryo handling. Therefore, any laboratory can establish aggregation chimera production with a small investment.
b. Learning to use the micromanipulators requires 3 to 6 months of intensive practice and even after the investment of time, not everyone can perform ES cell injection efficiently. In contrast, aggregation does not require the use of micromanipulators. No special skill is needed to become productive.
c. Most ES cell lines are derived from the 129 inbred. They are introduced into host embryos of a different genotype to visualize ES cell contribution to the chimera and identify germline transmission. It was found that the use of C57/B1 inbred host embryo provided the highest incidence of germline transmission, and was accepted as standard protocol. C57/B1, however, is not easy to work with, because of inconsistent embryo production and fertility problems. ES cell aggregation on the other hand yields germline transmission with outbred hosts as well. Outbred strains are less expensive than C57/B1 mice and are usually more productive. Interestingly, we found that using C57/B1 embryos for aggregation chimeras did not increase the germline transmission frequency. In fact, the developmental rate and the level of chimerism in the surviving chimeras was lower (Nagy, unpublished).
d. ES cell injected blastocysts are exposed to the *in vitro* environment only for a few hours before and after the injection. In contrast, the aggregates are kept under *in vitro* conditions for at least 24 h. Therefore, for aggregation the *in vitro* culture conditions of the embryos must be optimal. Before starting ES cell aggregation, the conditions should be checked by culturing normal embryos for 2–3 days and then transplanting them into recipient mice.
e. Not all the ES cell lines that give germline transmission after blastocyst injection can give germline transmission by aggregation with blastomere stage embryos. The reason is not completely understood, although there are some indications that it might be linked to the genotypes that are placed together in the chimeras. If cells are injected into blastocysts they generally contribute less to the somatic part of the chimera than they do when they are aggregated (Nagy and Rossant, unpublished). Some ES cell lines do not allow survival of the chimeras above a certain level of ES cell contribution. A C57/B1 blastocyst host with 129 ES cells requires less somatic contribution from the 129 compartment to contribute to the germline. Therefore if the resultant chimeras are expected to have a compromised developmental potential due to the source of ES cells or the mutation introduced, they will have a better chance of survival if the cells are injected than if they are aggregated. Some ES cell lines, however, show very good developmental potential, allowing the survival of chimeras with high somatic contribution from the ES cell compartment (Nagy *et al.*, 1993 and Gail Martin, personal

communication). They should be the lines of choice for ES cell aggregation chimeras to achieve efficient germline transmission.

f. Interestingly most of the germline transmitter chimeras produced by aggregation are overall transmitters. They have no host embryo-derived germ cells. This is certainly advantageous, since in such a situation the test for germline transmission is shorter. Routinely, we discontinue the test after two negative litters fathered by aggregation chimeras. In contrast, injection chimeras are often mixed transmitters, which can make the introduction of ES cell-mediated mutations *in vivo* more laborious. From the sex distortion among aggregation chimeras, which favours males (using a male ES cell line), and from the ratio of transmitter *versus* non-transmitter chimeras, it seems likely that the male transmitter chimeras are produced when the male ES cells are aggregated with a female host embryo.

g. The albino host used for aggregation allows outbred albino partners in testcrosses to detect germline transmission. The eye pigmentation of the newborns is a simple, early indication of transmission.

Trouble-Shooting ES Cell Aggregation

Aggregation of ES cells with blastomere stage embryos is a novel technique with a rapidly increasing popularity, as it has some important advantages over the ES cell injection method. We will discuss here some of the typical problems that different laboratories have encountered before getting the technique working smoothly.

Improper ES Cell Line

The choice of ES cells is very important for the aggregation method. We have tested several commonly used ES cell lines. Of them, R1 (Nagy *et al.*, 1993) gave very consistent germline transmission with this method. The other lines, such as D3 (Doetschman *et al.*, 1985) and AB1 (McMahon and Bradley, 1990), frequently gave chimeric embryos with strong ES cell contribution, but they died at birth. The survivors were weak chimeras and did not give germline transmission. It is important to know that R1 is not a magical ES cell line, and it is possible to generate new ES cell lines with sufficient developmental potential for germline transmission by aggregation (Nagy, unpublished and Gail Martin, personal communication).

Slow Embryo Handling

The recovery of blastomere stage embryos from the oviduct is technically more difficult than flushing blastocysts from the uterus. Sometimes laboratories with established ES cell injection techniques but lacking experience with blastomere stage embryos can have problems during the learning period of oviduct flushing. Speed is always an important factor in handling embryos.

Suboptimal *in vitro* Culture Conditions

In vitro culture conditions must be optimal for ES cell aggregation, since the aggregates are cultured for at least 24 hours before they are transferred back into recipients. The *in vitro* conditions should be good enough to give a 50–70% frequency of normal postimplantation development after embryos are cultured from 4-cell stage to blastocyst stage.

Improper Aggregation Well

The aggregation wells made by pressing the aggregation needle into the plastic should have the right diameter and shape in order to promote aggregation. The surface of this depression should be smooth to avoid sticking of aggregates when they develop into blastocysts. The quality of the aggregation needle is critical (Nagy and Rossant, 1993).

Too Many ES Cells in Aggregates

Adding too many ES cells to the aggregates is one of the more common problems. The preimplantation and early postimplantation stage embryo has a limited capacity to compensate for increased cell number. We found that the developmental rate dropped if more than 15 cells were aggregated with an 8-cell stage embryo.

Transfer of Too Much ES Cell Medium into Aggregation Drops

This is the other common problem and is related to the issue of ES cell number. In a normal 3.5 dpc blastocyst, the ICM contains only 10–12 cells. Therefore it is important to slow down the cleavage of ES cells during *in vitro* aggregation. This can be easily achieved if no ES cell medium is transferred into the aggregation M16 drops. Washing the ES clumps in M16 before they are transferred to the embryo-containing drop is essential.

Generation of Completely ES Cell-derived Embryos/Animals with ES Cell ↔ Tetraploid Embryo Aggregates

A properly applied and shaped electric impulse efficiently fuses the two blastomeres of the 2-cell stage embryo into one single cell (Nagy *et al.*, 1993). These embryos, with twice the normal chromosome number (tetraploid embryo), develop normally through preimplantation. If they are aggregated with ES cells (diploid cells), the resultant chimera is extremely polarized with respect to cell contribution from the tetraploid blastomere and diploid ES cells. The embryo proper, amnion, and the mesoderm layer of the yolk sac are completely ES cell-derived, while the yolk sac endoderm and the trophoblast lineage of the placenta are tetraploid-embryo derived (Nagy et al., 1990) Recently several applications have been developed in which the ES cell ↔ tetraploid embryo aggregation had a key role:

A. Quick Characterization of Developmental Potential of ES Cells

Experimental designs have been developed that require a series of genome manipulations in ES cells and germline compatibility at the end. For such strategies, checking the developmental potential after each step of genome alterations is highly recommended. ES cell ↔ tetraploid embryo aggregation provides a quick test for developmental potential, since the requirement to support completely ES cell-derived embryonic development to term is very stringent. Those cells that can generate a newborn animal have a good chance of giving germline transmission in chimeras (Nagy et al., unpublished).

B. Establishing Expression Patterns of Trapped Genes

Gene trap approaches require characterization of the expression pattern of lacZ during development. Previously, normal chimeras were made and stained for lacZ. Because of the chimerism it was always uncertain that the pattern observed truly represented the expression of the trapped gene. ES cell ↔ tetraploid embryo chimeras eliminate this problem because the embryo is completely ES cell-derived (Forrester et al., 1996).

C. Phenotype Analysis

This technology has begun to be applied to in vivo phenotype analysis of mutations precluding the requirement for germline transmission and establishing mutant strains. We have recently shown that this approach is feasible and at least two times faster than the standard analysis (Carmeliet et al., 1996). Furthermore, it is relatively inexpensive because there is no need for animal breeding.

D. Separating Embryonic and Extraembryonic Phenotypes

A number of gene knockouts have resulted in defects in the placenta or yolk sac. These extraembryonic defects can create secondary embryonic phenotypes. The ES cell ↔ tetraploid embryo aggregation is a valuable tool for dissecting the embryonic from extraembryonic phenotypes, since wild type tetraploid embryos can provide a normal extraembryonic environment to the mutant embryo proper (Guillemot et al., 1994).

Predicting the Near Future

With the recently developed ease of chimera production, this technology is increasingly being used as an efficient tool to address specific questions about genetic mutations created in the mouse. Chimeras can provide novel information if a deficiency is cell autonomous in nature, or creates developmental lineage restrictions or abnormal behaviour. The ES cell ↔ tetraploid embryo aggregation further opens the horizon by allowing complete in vivo phenotype analysis directly from the tissue culture dish without creating animals. In the future we can expect to see an increasing number of studies applying these new technologies, because they not only save on time and money but also allow mammalian genetic studies to be done at a much larger scale than is possible if only mutant mouse lines must be established for all genome alterations created in ES cells. The power of these new mammalian genetic tools will provide novel insights into normal and disease aspects of mammalian biology.

Acknowledgements

I thank all of my colleagues, especially Janet Rossant, who significantly contributed to the collective knowledge presented here, and Bristol-Myers Squibb for its financial support to our Transgenic and ES Cell Facility. I thank Corrinne Lobe and Lynn Mar for critical reading of the manuscript.

References

Capecchi, M.R. (1989) Altering the genome by homologous recombination. *Science* **244**, 1288–1292

Carmeliet, P., Ferreira, V., Breier, G., Pollefeyt, S., Kieckens, L., Gertsenstein, M., Fahrig, M., Vandenhoeck, A., Harpal, K., Eberhardt, C., Declerq, C., Pawling, J., Moons, L., Collen, D., Risau, W. and Nagy, A. (1996) Abnormal blood vessel development and lethality in embryos lacking a single vascular endothelial growth factor allele. *Nature* **380**, 435–439

Chambers, C.A., Kang, J., Pawling, J., Huber, B., Hozumi, N. and Nagy, A. (1994) Exogenous Mtv-7 superantigen transgene expression in major histocompatibility complex class II I-E mice reconstituted with embryonic stem cell-derived hematopoietic stem cells. *Proc. Natl. Acad. Sci. USA* **91**, 1138–1142

Doetschman, T.C., Eistetter, H., Katz, M., Schmidt, W. and Kemler, R. (1985) The in vitro development of blastocyst-derived embryonic stem cell lines: formation of visceral yolk sac, blood islands and myocardium. *J. Embryol. Exp. Morph.* **87,** 27–45

Forrester, L.M., Nagy, A., Sam, M., Watt, A., Stevenson, L., Bernstein, A., Joyner, A. L. and Wurst, W. (1996) An induction gene trap screen in embryonic stem cells: Identification of genes that respond to retinoic acid *in vitro. Proc. Natl. Acad. Sci. USA* **93,** 1677–1682

Gossler, A., Joyner, A.L., Rossant, J. and Skarnes, W.C. (1990) Mouse embryonic stem cells and reporter constructs to detect developmentally regualted genes. *Science* **244,** 463–465

Gu, H., Marth, J.D., Orban, P.C., Mossmann, H. and Rajewsky, K. (1994) Deletion of a DNA polymerase β-gene segment in T cells using cell type-specific gene targeting. *Science* **265,** 103–106

Guillemot, F., Nagy, A., Auerbach, A., Rossant, J. and Joyner, A.L. (1994) Essential role of *Mash-2* in extraembryonic development. *Nature* **371,** 333–336

McMahon, A.P. and Bradley, A. (1990) The *Wnt*-1 (*int*-1) proto-oncogene is required for development of a large region of the mouse brain. *Cell* **62,** 1073–1085

Nagy, A., Gocza, E., Diaz, E.M., Prideaux, V.R., Ivanyi, E., Markkula, M. and Rossant, J. (1990) Embryonic stem cells alone are able to support fetal development in the mouse. *Development* **110,** 815–821

Nagy, A. and Rossant, J. (1993) Production of completely ES cell-derived fetuses. In *Gene Targeting: A Practical Approach.* (Ed. A. Joyner), pp. 147–179. IRL Press at Oxford University Press

Nagy, A., Rossant, J., Nagy, R., Abramow-Newerly, W. and Roder, J. (1993) Derivation of completely cell culture-derived mice from early-passage embryonic stem cells. *Proc. Natl. Acad. Sci. USA* **90,** 8424–8428

Papaioannou, V. and Johson, R. (1993) Production of chimeras and genetically defined offspring from targeted ES cells. In *Gene Targeting: A Practical Approach.* (Ed. A. Joyner), pp. 107–146. IRL Press at Oxford University Press

Ramirez-Solis, R., Liu, P. and Bradley, A. (1995) Chromosome engineering in mice. *Nature* **378,** 720–724

Rossant, J. and Nagy, A. (1995) Genome engineering: the new mouse genetics. *Nature Genetics* **1,** 592–594

Skarnes, W. C. (1990) Entrapment vectors: A new tool for mammalian genetics. *Biotechnology* **8,** 827–831

Smith, A.J.H., Sousa, M.A.D., Kwabi-Addo, B., Heppel-Parton, A., Impey, H. and Rabbitts, P. (1995) A site-directed chromosomal translocation induced in embryonic stem cells by Cre-*lox*P recombination. *Nature Genetics* **9,** 376–385

Van Deursen, J., Fornerod, M., van Rees, B. and Grosveld, G. (1995) Cremediated site-specific translocation between nonhomologous mouse chromosomes. *Proc. Natl. Acad. Sci. USA* **92,** 7376–7380

Wood, S.A., Allen, N.D., Rossant, J., Auerbach, A. and Nagy, A. (1993) Non-injection methods for the production of embryonic stem cell-embryo chimeras. *Nature* **365,** 87–89

PART II, SECTION F

28. Bovine Pluripotent Stem Cells for Transgenic Vectors

N. Strelchenko

ABS Global, Inc., DeForest, WI 53532-0459, USA

Introduction

The further development of targeted insertion of foreign genes into a combination of pluripotent stem cells has opened up great possibilities for livestock genetics. Transgenic animals have a potential benefit for both the agriculture and pharmaceutical industries. Gene transfer can be achieved by random and targeted insertion of DNA fragments into genome of domestic animals. The price of producing transgenic cows by injection into the pronucleus is approximately $100,000–$150,000 because the efficiency of this method is low and ranges 0.04–1.7% (Rexroad, 1988). Compared to this method, the use of embryonic stem cells for gene transfer in domestic species would be more advantageous. Figure 28.1 shows a scheme for the production of transgenic cows by using pluripotent stem cells as vectors for the foreign gene. The use of embryonic stem cells has several positive advantages. First, ES-cells might be propagated *in vitro* for many generations serving as an unlimited source of pluripotent cells. Second, ES-cells could be genetically transformed, selected and screened for integrated DNA, *in vitro*. Isolated bovine pluripotent stem cell lines from an embryo or transformed primordial germ cells are an unlimited source of pluripotent cells.

Nuclear transfer from bovine pluripotent stem cells is expected to play an important role in the production of normal and transgenic offspring from domestic animals. Underdeveloped placenta has been found after nuclear transfer from bovine stem cells. However, this problem can be circumvented.

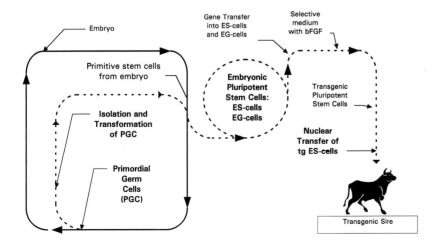

Figure 28.1. Production of transgenic animals by using embryonic stem cells and embryonic germ cells as vectors.

Bovine pluripotent stem cells will not become a really useful tool for transgenesis until the germ-line transmission into chimeric calves have been demonstrated.

Establishing Bovine Pluripotent Stem Cell Lines

The isolation of embryonic stem cells from embryos of large domestic animals is an important achievement in the livestock genetics. The first visual stage of any embryo differentiation is a separation of trophoblast cells from inner cell mass at the blastocyst stage. To prevent further differentiation, the cells should be disaggregated and seeded onto a feeder layer. The cells containing the smallest amount of cytoplasm and a high proliferation rate could be the embryonic stem cells. That strategy was introduced by Evans and Kufman (1981) and Martin (1981) for the isolation of murine embryonic stem cells. Later, their method, with slight modification, has been used for isolation of embryonic stem cells from other species: the Syrian golden hamster (Doetchman et al., 1988); pigs (Notariani et al., 1989); rabbit (Giles et al., 1991); cattle (Strelchenko et al., 1991); mink (Sukoyan et al., 1992); rat (Iannaccone et al., 1994).

These experiments have reported that isolated cell lines have fast proliferation in cell population and a low ratio of cytoplasm/nuclei. For disaggregation of bovine embryos, trypsin, pronase, or any suitable proteolytic enzyme could be used. As an alternative method, ICM might be isolated by immunosurgery (Van Stekelenburg-Haers et al., 1995). Establishing embryonic stem cell lines from cattle does not require a certain stage of development, breed, and strain, unlike murine ES-cells (Robertson, 1987). To establish bovine pluripotent stem cells, the embryos of any stage within 16-cells or hatched blastocyst (10–11 days) would be suitable. Mouse ES-cells have also been established from morula stage embryos (Eistetter, 1989). The early stage of embryos from other species is always preferable for the isolation of stem cells, (Sukoyan et al., 1993 and Strelchenko et al., 1994), but in our experiments we could not get constant results for establishing bovine embryonic stem cell lines from 2-, 4- and 8-cell stage bovine embryos. The study carried out at ABS Laboratory shows that disaggregation of bovine embryo is not necessary at the morula stage and that the entire morula without *zona pellucida* can be used. The most critical issue is then to obtain a tight contact between the feeder layers and blastomeres. In order to do this, bovine morulae or disaggregated blastomeres were placed under the blocked murine feeder layer. The blastomeres underwent morphological changes within 28–36 hrs and primary colonies of bovine pluripotent stem cells appeared.

Establishing bovine pluripotent stem cell lines from morula stage embryo has a success rate between 60–70% depending on the quality of embryos. After embryo disaggregation, blastomeres or cells (depending on embryo stage) were able to grow in colonies. Some predetermined blastomeres turned into giant and endoderm-like cells. The proliferation rate of isolated bovine pluripotent stem cells was slower compared to murine ES-cells, but this type of cell consistently has the fastest rate among the cell population. The method of cell trypsinisation and seeding affected cell appearance, but the morphology of embryonic stem cells have two major traits: the smallest amount of cytoplasm and the fastest proliferation rate in the population.

ES-cells need to be passaged relatively often due to their fast proliferation. Trypsin has been successfully used to passaging murine ES-cells (Evans, Martin, 1981). Attempts to use trypsin to transfer ES-cells have led in some cases to cell differentiation in cattle (Saito et al., 1992) and rabbit (Niemann et al., 1994). Although some researchers succeeded in passing ES-cells using trypsin with bovine (Van Stekelenburg-Haers et al., 1995) and rabbit (Graves et al., 1993) pluripotent embryonic stem cells. In our experiments we have observed that bovine and rabbit stem cells were loosing pluripotency and differentiated when trypsin was applied for passage. Maintenance of bovine embryonic stem cell lines in culture was carried out by the mechanical transfer of bovine pluripotent stem cells, having suitable morphology, onto fresh feeder layers (Strelchenko et al., 1994). Mechanical passage of pluripotent cells allows to make a preliminary selection of cells with suitable morphology and keep lower ratio of transferred cells between 1:30 and 1:60. In favor of the use of mechanical passage it is worth mentioning that coat chimeric rats were obtained in this way (Iannaccone et al., 1994). To initiate the growth of bovine pluripotent stem cells, a feeder layer is required. The primary feeder layer has been obtained from primary culture of murine embryonic fibroblasts. It is also possible to use it a primary culture of bovine embryonic fibroblasts, but with lower efficiency. The proliferation of the feeder layer should be blocked by mitomycin C or gamma-irradiation treatment before bovine pluripotent stem cells are seeded on it.

Subclonal pluripotent stem cell lines have been re-established from bovine nuclear transfer of pluripotent stem cells embryo (equal to second generation of nuclear transfer). Re-established cell lines have the same morphology as original cell lines and they are available for nuclear transfer. This way third and fourth generations of pluripotent stem cells through nuclear transfer were produced (Stice et al., 1995, in press).

The other source of bovine pluripotent stem cells is the transformation of primordial germ cells (PGC) into permanent cell line. The discovery of the presence of the basic fibroblast growth factor (bFGF) in the cultural medium has helped to transform murine PG cells into permanent line. PGC-ES cell lines (Embryonic Germ cells or EG-cells) have been established and chimeric mice have been produced (Matsui et al., 1992 and Resnick et al., 1992). Description of bovine primordial germ cells has been published recently (Lavoir et al., 1994).

At ABS Laboratory several attempts for isolating bovine PG cells has been done. The general strategy of the experiments was almost the same as in original publications Matsui et al. (1994) and Resnick et al. (1994), except that murine bFGF was replaced by bovine bFGF due to their different species affinity to cells. Bovine fetuses from 4 gm to 14 gm were taken from the slaughterhouse. The ages of those fetuses varied within 30–55 days Melton et al. (1951). Genital ridges of fetuses were isolated and disaggregated in a pronase solution. Cell suspension was seeded onto a mitotically inactivated feeder layer. At the same time a hormone "cocktail" was added to the supplemented medium, but after cell suspension. The mixture of hormones consisted of bovine basic fibroblast growth factor (bFGF), human stem cell factor (SF) and recombinant leukemia inhibitor factor (LIF). The concentrations of these hormones in several experiments was variable. The best results were obtained at concentration of 40 ng/ml; 20 ng/ml; 20 ng/ml for bFGF, SF and LIF respectively. Supplemented medium (Alpfa MEM) contained 10% FCS and 0.1 mM 2-mercaptoethanol. After two weeks of proliferation in vitro the "hormone cocktail" was removed from the medium.

The cells having a small ratio of cytoplasm/nuclei and a relatively fast proliferation were isolated and mechanically transferred onto fresh feeder layer. From 3 fetuses 3 cell lines were isolated. The morphology of transformed bovine PGC was similar to pluripotent stem cells having been isolated from bovine embryos. Some islands of transformed PG cells contained lipid bubbles similar to pluripotent stem cells derived from embryos.

It is difficult to overestimate the importance of bFGF in making a permanent proliferation of transformed PG cells. Basic FGF belongs to a family of growth factors which stimulate the proliferation of the almost all types of cells derived from the three germ layers. bFGF like platelet-derived growth factor (PDGF) prevents oligodendrocytes type 2 from differentiating (Bogler et al., 1990). In view of all the known facts it seemed reasonable to investigate the effect of bovine bFGF on bovine pluripotent stem cells. After adding bFGF, the population of cells derived from bovine embryos became more homogenous and their proliferation was accelerated. The optimal final concentration bFGF was 10–25 ng/ml to optimize the culture condition of the bovine pluripotent stem cells. The cloning efficiency of cells in the presence of bFGF was increased over two or three times after disaggregation of cells by pronase (Strelchenko, 1995, unpublished results).

Taking this to account, it will be important to add bovine of bFGF into selection medium to maintain their pluripotent status after genetic modification of the stem cells.

Pluripotency of the Embryonic Stem Cells

Differentiation of the Embryonic Pluripotent Stem Cells *In Vitro*

Bovine pluripotent stem cells in the suspension culture formed embryonic bodies (EB). The small EB contained several cells (5–10) and older EB resembled an inverted embryo and could reach 8–12 mm in diameter. Spontaneous formation of embryonal body started when free floating cell conglomerates or ICM-like cells appeared in culture. Within the next 2–4 days, conglomerates grow and form an internal cavity. In the early stage of its formation, the embryonic body wall consisted of one-cell layer. On the external surface of older bovine embryonic body cells, one or several groups looking like ICM of regular embryo were formed. Histological analysis of the transverse section of the older EB wall confirmed the location of ICM-like cells outside the EB and a two-layer wall consisting of ectoderm-like cells was covered by a thin layer of endoderm-like cells on the exterior. Detailed studies of histological sections of the old embryonal bodies (about 10 mm) show that different kinds of tissues could be found there, such as muscle or blood islands (red blood cells). Murine ES-cells produce two-layers EB with a similar structure: the internal ectoderm was covered by endoderm and Reinhart's membrane delineated them (Robertson et al., 1987). Some ES-cell lines from mink have produced multilayered EB (Sukoyan et al., 1992), but as with bovine pluripotent stem cells, no Reinhart's membrane was found (Stice et al., 1995, in press).

In the confluent cultures of bovine pluripotent stem cell lines spontaneous differentiation has been observed (Saito, 1992). The major types of differentiated cells were mesoderm-like cells and endoderm-like cells. Some lines of bovine stem cells differentiated, forming neurons and muscle.

Differentiation of murine ES-cells was induced by retinoic acid plus cAMP or DMSO (Robertson, 1987). Addition of these ingredients into the medium of bovine pluripotent stem cells induced differentiation of cells. The cytoplasm of cells increased within two days after inducer of differentiation were added. Bovine pluripotent stem cells were completely differentiated within 8–10 days of culture. No permanent cell line with differentiated traits (fibroblasts-like, ectoderm-like cells) was isolated. Most cells differentiated into fibroblasts, neurons, mesoderm. The addition of DMSO to the culture of bovine pluripotent stem cells induced synthesis of hemoglobin (Strelchenko, unpublished result).

The Ability Bovine Pluripotent Stem Cells to Differentiate *In Vivo*

Nuclear Transfer. The first attempt to use bovine pluripotent stem cells as a source of nuclei for nuclear transfer (NT) was not very successful (Saito et al., 1992). After the improvement of the technique for nuclear transfer of bovine blastomeres, progress has been made in the nuclear transfer of stem cells. To study the pluripotent properties of bovine pluripotent stem cells, nuclear transfer was carried out (Stice et al., 1994). This procedure includes the following steps. The matured and previously enucleated oocyte was fused with a single bovine pluripotent stem cell. A single cell suspension for nuclear transfer was achieved by temporary incubation in a pronase solution. The cell construction obtained after nuclear transfer is termed cybrid (Bunn et al., 1974), as it contains cytoplasm of oocyte (enucleated cell) and a nucleus with a relatively small amount of cytoplasm from pluripotent stem cells (intact cell). Developed embryos obtained after the nuclear transfer procedure were transferred into foster mothers which become pregnant.

On the 30th day of pregnancy the heartbeat of the fetus was registered by ultrasonic diagnosis. Within 45 to 55 days of pregnancy heartbeats were lost and the foetus died. All tissues belonging to three germ layers were present in NT-derived fetuses and were derived solely from an embryonic cell nucleus transferred into an enucleated oocyte. Histological sections of these fetuses showed correct and normal development. Histologists from University of Wisconsin-Madison did not find any abnormal development of the foetus. The reason why the fetus died is therefore unknown. For a fetus 45–55 days of pregnancy is a very critical stage. Because it is at this point that the establishment of the relationship between the fetus and the mother's organisms by attachment of the placenta to the uterus through cotyledons should occur (Melton et al., 1951).

The pregnancies initiated with embryos from nuclear transfer of ovine pluripotent stem cells were observed by another group of researchers (Moor et al., 1992). A real breakthrough was achieved when four lambs were obtained after the nuclear transfer of a short term (3 passages) culture of embryonal disk (Campbell et al., 1995). Cybrids were obtained after the nuclear transfer of murine ES-cell into oocyte did not develop into fetus, and formed only blastocyst (Tsunoda et al., 1993).

A general explanation for the lack of full-term pregnancies is that the bovine pluripotent stem cells may be more differentiated than ICM cells and were not completely reprogrammed after the NT-procedure and therefore could not produce normal placenta. When we compared the placenta of the nuclear transferred pluripotent stem cell embryo to a normal placenta (derived IFM-IVF embryo) we noticed that the number of cotyledons in the placenta of embryos resulting from nuclear transfer had decreased (10–15 units). Histological sections of NTES cotyledons confirmed that they were underdeveloped in both maternal and embryo sides.

Generation of Chimera. In order to get chimera, we used the traditional method which consists in injecting bovine pluripotent stem cells into the host blastocyst. Bovine pluripotent stem cells have the tendency to be incorporated into the ICM of the host blastocyst. Disaggregated pluripotent stem cells were labeled with a fluorochrome cell marker and were injected into the blastocyst. After a short period of culture, the host embryo with injected fluorochrome labeled pluripotent stem cells located in the ICM area. Unfortunately, no chimera was obtained due to the too small number of experiments.

Some progress in making chimeras has been achieved by aggregating blastomeres from IVF embryos with embryos obtained after nuclear transfer from bovine pluripotent stem cells. This fetus aborted after 85 days of pregnancy. The placenta of a normal fetus of this stage had about 200 cotyledons, whereas the placenta of chimeric fetus obtained after aggregation of NT pluripotent stem cells embryo with blastomeres of IVM-IVF embryo had only a few cotyledons (18–20). This is due to a high contribution of pluripotent stem cells in the formation of the placenta. Chimerism of this fetus was confirmed by PCR amplification of microsatellite markers from four different tissues (skin, heart, muscle and placenta) of the chimera. The chimeric fetus shares the same alleles (50%) as the pluripotent stem cell line and the sire of the *in vitro* produced embryo.

The study of the pluripotent properties of bovine PG cell lines is not finished, but some experiments have demonstrated their pluripotency by NT procedure (Strelchenko, unpublished results). Several developed embryos were obtained after the nuclear transfer of reprogrammed bovine PG cells. Two pregnancies were initiated after they were transferred into recipients. The presence of fetuses were diagnosed by ultrasonic screener, but the fetuses were not recovered. They aborted within 38–60 days (Strelchenko, 1995, unpublished results). We can only suggest that abortion occurred due to the lack of placenta formation.

The reason why placenta development was affected is not known. One possible explanation is that the beginning of erasure of imprinting could have ill-programmed placenta development. Still, it is not clear how long this process takes and in which stage of development it occurs (Rossant, 1993). The experiments on the mouse model suggest that the ES-cells have kept their original pattern of imprinting (Allen et al., 1994). It is not clear however, how it is kept intact. The study carried out on the inactivation of X chromosome (Strelchenko et al., 1995) in female bovine pluripotent cell lines does not contradict this hypothesis. The difference between murine embryonic stem cells and reprogrammed PG cells is the ability of the former to maintain the methylation pattern of the imprinted Igf2r gene *in vitro*. However, it should be noted that no correlation has been found between methylation and germline competence (Labosky et al., 1994). This could indicate that some changes are not reversible in the imprinting status in the ES-cells.

Conclusions and Practical Applications

The cell lines obtained from bovine embryos are pluripotent. Their pluripotent properties has been demonstrated *in vitro* and *in vivo*.

Primordial germ cells isolated from genital ridges of bovine fetuses are able to proliferate *in vitro* and generate permanent cell lines.

The cells obtained after the transfer of nucleus from the two kinds of stem cells into enucleated oocytes were pluripotent and were able to differentiate *in vivo*.

The animals resulting from high contribution of stem cells and showing a germ line transmission can be used for production of transgenic animals through stem cells. The transfer of nucleus from *in vitro* transfected stem cells is a very attractive and good way to generate transgenic cows. To reach this goal, several approaches could be used:

1. The aggregation of blastomeres of embryos obtained from nuclear transfer of bovine pluripotent stem cells or reprogrammed primordial germ cells, with regular or tetraploid embryos.

2. The transfer of primordial germ cells obtained from genital ridges of fetus derived from nuclear transfer of transfected bovine pluripotent stem cells or reprogrammed primordial germ cells, into abdominal cavity of the normal bovine fetus at the 35–45 days for to get germ line chimera.

Acknowledgments

I am deeply appreciative of the people without whose help this work would not have been done: J. Betthauser, P. Golueke, G. Jurgella, C. Keefer, L. Metthews, A.M. Paprocki, B. Scott and S. Stice.

References

Allen, N.D., Barton, S.C., Hilton, K., Norris, M.L. and Surani, M.A. (1994) A functional analysis of imprinting in parthenogenetic embryonic stem cells. *Development* **120**, 1473–1482

Bunn, C.L., Wallace, D.C., Eisenstadt and J.M. (1974) Cytoplasmic inheritance of chloramphenicol resistance in mouse tissue culture cells. *Proc. Natl. Acad. Sci. USA* **71**, 1618–1685

Campbell, K., McWhir, J., Ritchie, B. and Wilmut, I. (1995) Production of live lambs following nuclear transfer of cultured embryonic disc cells. *Theriogenology* **43**, 181

Cheng, L., Gearing, D.P., White, L.S., Compton, D.L., Schooley, K. and Donovan, P.J. (1994) Role of leukemia inhibitory factor and its receptor in mouse primordial germ cell growth. *Development* **120**, 3145–3153

Doetchman, T., Williams, P. and Maeda, N. (1988) Establishment of hamster blastocyst-derived embryonic stem cells. *Dev. Biol.* **127**, 224–227

Eistetter, H.R. (1989) Pluripotent embryonal stem cell lines can be established from disaggreggated mouse morulae. *Dev. Growth Diff.* **31**, 275–282

Giles, J.R., Yang, X., Mark, W. and Foote, R.H. (1993) Pluripotency of cultured rabbit inner cell mass cells detected by isozyme analysis and eye pigmentation of fetuses following injection into blastocysts or morulae. *Mol. Reprod. Dev.* **36**, 130–138

Graves, K.H. and Moreadith, R.W. (1993) Derivation and characterization of putative pluripotential embryonic stem cells from preimplantation rabbit embryos. *Mol. Reprod. Dev.*, **36**, 424–433

Giles, J.R., Yang, X., Mark, W. and Foote, R.H. (1992) Production of chimeric rabbit fetuses using cultured rabbit ICM cells. *Biol. Reprod.* **46**, 161

Iannaccone, P.M., Taborn, G.U., Garton, R.L., Caplice, M.D. and Brenin, D.R. (1994) Pluripotent embryonic stem cells from the rat are capable of producing chimeras. *Dev. Biol.* **163**, 288–292

Labosky, P.A., Barlow, D.P. and Hogan, B.L.M. (1994) Mouse embryonic germ (EG) cell lines, Transmission through the germline and differences in the methylation imprint of insulin-like growth factor 2 receptor (Igf2r) gene compared with embryonic stem (ES) cell lines. *Development* **120**, 3197–3204

Lavoir, M.-C., Basrur, P.K. and Betteridge, K.J. (1994) Isolation and identification of germ cells from fetal bovine ovaries. *Mol. Reprod. Dev.* **37**, 413–424

Matsui, Y., Toksoz, D., Nishikawa, S., Nishikawa, S.-I., Williams, D., Zsebo, K. and Hogan, B.L.M. (1991) Effect of Steel factor and leukemia inhibitory factor on murine primordial germ cells in culture. *Nature* **353**, 750–752

Matsui, Y., Zsebo, K. and Hogan, B.M.L. (1992) Derivation of pluripotential embryonic stem cells from murine primordial germ cells in culture. *Cell* **70**, 841–847

Melton, A.A., Berry R.O. and Butler, O.D. (1951) The interval between the time of ovulation and attachment of the embryo. *J. Anim. Sci.* **10**, 993–1005

Moor, R., Sun, F.Z. and Galli, C. (1992) Reconstruction of ungulate embryos by nuclear transplantation. *Anim. Reprod. Sci.* **28**, 423–431

Mueller, A.M. and Dzierzak, E.A. (1993) ES cells have only a limited lymphopoietic potential after adoptive transfer into mouse recipients. *Development.* **118**, 1343–1351

Notarianni, E., Galli, C., Laurie, S., Moor, R.M. and Evans, M.J. (1991) Derivation of pluripotent, embryonic cell lines from pig and sheep. Journal of the *Reproduction and Fertility* Suppl. **43**, 255–260

Notarianni, E., Laurie, S., Moor, R.M. and Evans, M.J. (1990) Maintenance and differentiation in culture of pluripotential embryonic cell lines from pig blastocysts. *J. Reprod. Fertil.* **41**, 51–56

Oshima, R.G., Howe, W.E., Tabor, J.M. and Trevor, K. (1983) Cytoskeletal proteins as markers of embryonal carcinoma differentiation. In: Silver, L.M., Martin, G.R. and Strickland, S. (eds.) *Teratocarcinoma Stem Cells* **10**, Cold Spring Harbor Laboratory, Cold Spring Harbor, NY, 1983, 51–61

Piedrahita, J.A., Anderson, G.B. and BonDurant, R.H. (1990) On the isolation of embryonic stem cells: comparative behavior of murine, porcine and ovine embryos. *Ther.* **34**, 879–891

Polejaeva, I.A., While, K., Ellis, L.C. and Reed, W.A. (1995) Isolation and long-term culture of mink and bovine embryonic stem-like cells. *Theriogenology* **43**, 300

Resnick, J.L., Bixler, L.S., Cheng, L. and Donovan, P.J. (1992), Long-term proliferation of mouse primordial germ cells in culture. *Nature* **359**, 550–552

Robertson, E.J. (1987) Embryo-derived stem cell lines. In: Robertson EJ (ed.) *Teratocarcinomas and Embryonic Stem Cells: A Practical Approach.*, IRL Press, Oxford, 71–112

Rossant, J. (1993) Immortal germ cells? *Curr. Biol.* **3**, 47–49

Saito, S., Strelchenko, N. and Niemann, H. (1992) Bovine embryonic stem cell-like cell lines cultured over several passages. *Roux's Arch. Dev. Biol.* **201**, 134–141

Sims, M. and First, N.L. (1991) Production of calves by transfer of nuclei from cultured inner cell mass cells. *Proc. Natl. Acad. Sci. USA.* **91**, 6143–6147

Stice, S.L., Strelchenko, N.S., Keefer, C.L. and Matthew, L. (1996) Pluripotent bovine embryonic cell lines direct embryonic development following nuclear transfer. *Biol. Reprod.* **54**, 100–110

Stice, S.L. and Strelchenko, N.S. (1996) Domestic animal embryonic stem cells: progress towards germ-line contribution. *J. Anim. Sci.* (accepted)

Sukoyan, M.A., Vatolin, S.Y., Golubitsa, A.N., Zhelezova, A.I., Semenova, L.A. and Serov, O.L. (1993) Embryonic stem cells derived from morulae, inner cell mass, and blastocysts of mink: Comparisons of their pluripotencies. *Mol. Reprod. Dev.* **36**, 148–158

Szabo, P. and Mann, J.R. (1994) Expression and methylation of imprinted genes during *in vitro* differentiation of mouse parthenogenetic and androgenetic embryonic stem cell lines. *Dev.* **120**, 1651–1660

Talbot, N.C., Rexroad, C.E., Pursel, V.G. and Powell, A.M. (1993) Alkaline phosphatase staining of pig and sheep epiblast cells in culture. *Mol. Reprod. Dev.* **36**, 139–147

Tsunoda, Y. and Kato, Y. (1993) Nuclear transplantation of embryonic stem cells in mice. *J. Reprod. Fertil.* **98**, 537–540

Van Stekelenburg-Haers, A.E.P., Van Achterberg, T.A.E., Rebel, H.G., Flechon, J.E., Campbell, K.H.S., Weima, S.M. and Mummery, C.H. (1995) Isolation and characteri-zation of permanent cell lines from inner cell mass cells of bovine blastocyst. *Mol. Reprod. Dev.* **40**, 444–454

PART II, SECTION F

29. Towards the Use of Primordial Germ Cells for Germline Modification

Peter J. Donovan*, James L. Resnick[†], Linzhao Cheng[‡] and Leslie F. Lock

ABL-Basic Research Program, Bldg 539, NCI-FCRDC, PO Box B, Frederick, MD 21702-1201, USA

Introduction

Introduction of new genetic material into the germline of an organism via transgenesis has become a widely used technique in developmental biology where expression or overexpression of normal or mutant proteins can have a profound effect on embryonic development as well as adult physiology. Similarly, deletion, disruption or replacement of genes *via* homologous recombination has yielded much new information on the roles of specific genes in development, growth, homeostasis and disease. Modification of the mouse germline traditionally has been achieved in two ways, introduction of DNA into a fertilized egg or into embryonic stem (ES) cells that can contribute to the germline of a chimeric mouse produced by injection of the altered ES cells into a host embryo (see, for example, Hogan *et al.*, 1996). The derivation of mouse ES cells has truly revolutionized the biology of the mouse as an experimental organism. The use of homologous recombination to carry out gene targeting in ES cells has allowed the null phenotype to be determined for (almost) any cloned gene. Mice carrying null mutations in specific genes have given important new insights into the role of many genes in development and disease and have provided a number of models of human disease. In addition, ES cells have been used in many different ways, by developmental biologists in particular, to dissect the biology of the mouse.

Although DNA microinjection is widely used for making transgenic flies, worms and to some extent domestic animals, in many species, the ability to collect zygotes of suitable quality and sufficient quantity is limited, compromising the usefulness of this approach. Similarly, the use of ES cells to manipulate the genome has been limited, thus far, to only a few strains of mice from which ES cells that contribute to the germline have been established. The ability of ES cells to enter the germline remains a poorly understood phenomenon (see Robertson, 1987). The effect of strain background on the penetrance of mutant alleles has been known for some time (see, for example, Silvers, 1979) but the effect on the phenotype of "knockout" mice is just becoming apparent (Donehower *et al.*, 1992; Ramirez-Solis *et al.*, 1993; Sibilia and Wagner, 1995; and Threadgill *et al.*, 1995). Also, in some instances, the ability to derive ES cell lines from any strain at will would be desirable. For instance, in some cases a mutation is caused by a null allele and "knocking in" a replacement gene would be the desired approach. This may require establishment of an ES cell line of a specific strain background. Derivation of ES cell lines from a variety of strains may also ease problems associated with using non-isogenic DNA for homologous recombination (te Riele *et al.*, 1992). Development of pluripotent stem (Es-like) cell lines from other organisms would, without doubt, have many uses. However, establishment of such lines has been limited to rat (Iannaccone *et al.*, 1994), hamster (Doetschman *et al.*, 1988), mink (Sukoyan *et al.*, 1992), pig and sheep (Notarianni *et al.*, 1991) and these cell lines do not contribute to the germline. The mouse is still the only organism from which pluripotent cells have been developed

---

*Author for Correspondence.
Present address: [†]Department of Molecular Genetics and Microbiology, University of Florida, 1600 Southwest Archer Road, Gainesville, FL 32610, USA.
[‡]Systemix Inc., 3155 Porter Drive, Palo Alto, CA 94304, USA.

which contribute to the germline (Bradley et al., 1984).

Alteration of the genome of the germ cells directly could potentially overcome some of these limitations. Recent developments in germ cell culture and germ cell transplantation techniques (Matsui et al., 1992; Resnick et al., 1992; Brinster and Avarbock, 1994; Brinster and Zimmermann, 1994; Hashimoto et al., 1992; and our own unpub. obs.) raise the possibility of alternative ways of introducing genetic material directly into germ cells of many different stages of development, including primordial germ cells (PGCs), spermatogonia and oogonia. Advances in PGC isolation and culture techniques may provide the opportunity to manipulate these PGCs *in vitro* while retaining their ability to populate the germline. PGCs and oogonia from female embryos can be transferred into the ovarian sac of an adult animal and mature into fertilizable oocytes that give rise to offspring (Hashimoto et al., 1992 and our own unpub. obs.). Recently, pluripotent stem cells called embryonic germ (EG) cells have been derived from PGCs (Matsui et al., 1992 and Resnick et al., 1992). EG cells can make germline chimeras when injected into host blastocysts (Stewart et al., 1994; Labosky et al., 1994a and Labosky et al., 1994b). Therefore, EG cells potentially provide an opportunity for genomic manipulation in a wide variety of mouse strains and other experimental organisms.

This review will deal with the potential for manipulating the germline using PGCs and their derivatives, including EG cells. Following a brief description of germ cell development in the mouse we will describe techniques for isolating and culturing PGCs. We will then describe results of studies on introducing DNA into PGCs and producing germline chimeras from PGCs. We will also describe techniques for deriving EG cell lines from PGCs and results of studies on introducing DNA into EG cells and making chimeras from such manipulated cells. These studies will also be contrasted with what is currently known about ES cell biology. Finally, we will discuss technological problems that need to be addressed if PGCs and their derivatives are to be used for modification of the germline and the future directions that this work is likely to take.

Germ Cell Development

The fusion of two germ cells, an egg and a sperm, at fertilization gives rise to a zygote that is totipotent and capable of giving rise to all the embryonic and extra-embryonic lineages (reviewed in Hogan et al., 1986). In the early embryo, PGCs are indistinguishable from the somatic cells. PGCs first become distinguishable from the somatic cells at 7.0 days post coitus (d.p.c.) because they begin to express an alkaline phosphatase izozyme, tissue non-specific alkaline phosphatase (TNAP). Using TNAP as a marker, PGCs can be traced in the embryo and their numbers determined. They are first detected outside of the embryonic gonad and migrate towards it over the next five days of embryonic development. PGC numbers increase from the 50–100 found in the 8.5 d.p.c. mouse embryo to the 25–30,000 found at 12.5 d.p.c. in the fetal gonad. At this time, the somatic lineages of the gonad undergo sexual differentiation which is first observed as the formation of testis cords in the male. In 12.5 d.p.c. male embryos, the PGCs enter a mitotic arrest. After birth, male germ cells resume mitosis forming the mitotic stem cell of the adult testis, the spermatogonium. It is these cells that give rise to meiotic derivatives that will undergo spermiogenesis to form mature sperm. In 12.5 d.p.c. female embryos, the PGCs enter directly into meiosis and form oogonia which arrest at meiotic prophase. In the ovary after birth, the oogonia resume meiosis (in waves) giving rise to fully mature oocytes (see McLaren, 1981 for review). The number of PGCs, oogonia or spermatogonia that can be isolated from embryonic or adult gonads is relatively large providing an abundant source of germline material for manipulation.

Mutations causing sterility in the mouse have identified some of the genes regulating germ cell development. Four distinct mutations have been identified, *Dominant White Spotting* (*W*), *Steel* (*Sl*), *Hertwigs Anemia* (*an*) and *germ cell deficient* (*gcd*). The *W* and *Sl* loci have been characterized at the molecular level and found to encode a receptor tyrosine kinase (c-Kit) and its cognate ligand, stem cell factor (SCF). The *an* and *gcd* loci have not yet been characterized at the molecular level. Characterization of such loci affecting PGC development is likely to lead to a better understanding of this important lineage.

PGC Isolation and Culture

Techniques have been developed for isolating PGCs from surrounding somatic cells of the embryo based on differences in cell size and cell surface antigen expression. A simple procedure for isolating PGCs involves dissociation of the gonad with EDTA and use of Percoll gradients for size separation of PGCs and somatic cells (De Felici and McLaren, 1982). An alternative method uses immunological reagents to isolate PGCs. During development, PGCs express a number of cell surface, differentiation antigens recognized by polyclonal antisera and monoclonal antibodies (Donovan et al., 1986). For example, PGCs from 8.5–13.5 d.p.c.

embryos, but not the surrounding somatic cells, express a cell surface carbohydrate differentiation antigen, the stage-specific embryonic antigen-1 (SSEA-1; Solter and Knowles, 1978; Fox et al., 1981 and Donovan et al., 1986). Monoclonal antibodies to the SSEA-1 antigen can be used to distinguish, and isolate, PGCs from the surrounding somatic cells by fluorescence-activated cell sorting (FACS; McCarrey et al., 1987 and Cheng et al., 1994). This methodology allows a high degree of PGC purification.

The ability of purified PGCs to survive in culture depends on their embryonic age. PGCs that have reached the embryonic gonad (11.5 d.p.c. onwards) survive for a limited period of time in culture in the absence of a feeder cell layer, whereas PGCs isolated before they reach the embryonic gonad require a feeder cell layer for survival. Over the past ten years, *in vitro* culture systems have been developed that allow PGCs to survive and proliferate for up to one week (reviewed by Buehr and McLaren, 1993; Cooke et al., 1993 and Donovan, 1994). PGC survival is best when they are cultured on confluent monolayers of mouse feeder cells, either the embryo fibroblast-derived cell line STO, the bone marrow-derived cell line Sl220 or the Sertoli cell-derived cell line TM4. The proliferative capacity of PGCs correlates with the age of embryo from which they are derived. For example, PGCs isolated from 8.5 d.p.c. embryos proliferate for 7–10 days in culture. In contrast, PGCs isolated from 11.5 d.p.c. embryos proliferate for only 1–2 days in culture. PGCs isolated from 12.5 d.p.c. embryos survive for a few days but do not proliferate. In general, PGCs proliferate as long in culture as they would in the embryo. PGC survival and proliferation is supported by the feeder cells probably because the feeder cells produce membrane-associated SCF and Leukemia Inhibitory Factor (LIF) as well as other factors that have not yet been characterized (see for example, Dolci et al., 1991; Matsui et al., 1991 and Godin et al., 1991).

A number of reagents can stimulate PGC proliferation in culture. Certain polypeptide growth factors stimulate PGC proliferation in culture without extending their longevity. For example, when PGCs are cultured on STO (or Sl220) cells, addition of exogenous LIF, stimulates their proliferation without extending the timecourse of their survival (DeFelici and Dolci, 1991; Matsui et al., 1991; Matsui et al., 1992 and Resnick et al., 1992). This is true of a number of other polypeptide growth factors as well as factors that stimulate cAMP levels in PGCs such as forskolin, follicle stimulating hormone (FSH) and cholera toxin (Dolci et al., 1993; and our own un-published observations). Thus these culture conditions may be used to introduce new genetic material directly into PGCs with the limitation that the cells will not survive for very long. Studies aimed at characterization of factors regulating PGC proliferation and differentiation will be important. In particular it will be necessary to develop conditions that allow PGC proliferation without concomitant differentiation.

Introduction of Cloned DNA into PGCs

Introduction and expression of cloned DNA in PGCs has not been optimized due to experimental difficulties such as dealing with limited numbers of cells which survive only a short period of time in culture. Standard transfection and electroporation techniques can be used to introduce DNA into PGCs but such techniques do not work well on primary cells and many of the parameters have not been worked out. If drug selection protocols are to be used to identify cells into which DNA has integrated, an efficient promoter is required. To date the number of genes known to be expressed in PGCs is relatively small and their regulatory elements, such as promoters, have not, in many cases, been well characterized. The use of ubiquitous promoters, which are found to be functional in many cell types, and viral promoters is currently being tested. The immediate early promoter of the human cytomegalovirus (CMV) can drive expression of the *lacZ* reporter gene in cultured PGCs, whereas the SV40 early promoter/enhancer or the long-terminal repeats (LTRs) of the Moloney murine leukemia virus (MuLV) and Moloney murine sarcoma virus (MuSV) cannot (Cheng, Resnick and Donovan, in Preparation). It is possible that PGCs, like ES cells, have mechanisms that regulate LTR activity. In support of this idea, mutant MuLVs, which can drive gene expression in ES cells, can also drive gene expression in PGCs (Cheng, Resnick and Donovan, in Preparation). An alternative approach to test the ability of promoters to drive expression in PGCs (or germ cells of any stage) is to examine expression in transgenic mice. For example, regulatory elements of the Oct-4 gene are sufficient to drive expression of a *lacZ* reporter gene in PGCs of transgenic mice (Yeom et al., 1996). However, the regulatory elements of the TNAP gene, while able to control expression of a *lacZ* reporter gene in PGCs, can only do so in the context of the whole genomic locus (MacGregor et al., 1995). Some work is clearly required in order to better understand the factors regulating gene expression in germ cells. An important obstacle to be overcome here is how to select for PGCs into which DNA has been introduced since the PGCs will only survive for short periods of time in these culture conditions. In the absence of conditions in which drug selection could be carried out, a selection protocol based on the use of green fluorescent protein (gfp) could be developed.

Chimera Formation by PGCs

Once DNA is introduced into PGCs they must be reintroduced into the germline of a mouse. PGCs are not efficient at reentering the germline when injected into the mouse blastocyst or 7.5–8.5 d.p.c. embryos. PGCs isolated from 8.5–11.5 d.p.c. embryos, when introduced into mouse blastocysts, did not contribute to the somatic tissues or to the germline (Donovan and Robertson, unpub. obs.; Papaioannou, pers. comm.; Stewart, pers. comm.). Similarly, PGCs, introduced into 7.5–8.5 d.p.c. embryos by a novel injection protocol developed for neural crest cells (Jaenisch, 1985), did not contribute to the somatic or germ lineages (Donovan, Snow and McLaren, unpublished observations). These data suggest that either PGCs, paradoxically, are restricted in developmental potency or that they are unable to survive long enough to contribute to any lineage. An alternative approach is to introduce PGCs into the ovarian sac of an adult female and allow them to mature as oocytes. When 11.5 d.p.c. genital ridges or 12.5 d.p.c. gonads are transplanted into the ovarian sac of a histocompatible host, the PGCs from oogonia that mature and can be fertilized to give rise to offspring (Hashimoto et al., 1992; and our own unpub. obs.). Indeed the gonad can be completely dissociated and reaggregated and the PGCs still give rise to offspring (Hashimoto et al., 1992). Using this approach, it may be possible, therefore, to isolate PGCs into culture, manipulate them *in vitro*, and re-introduce them into the germline. Once again, techniques for identifying PGCs into which DNA has successfully been introduced will need to be developed. Such technology may also require a better understanding of the factors that regulate PGC differentiation and entry into meiosis.

Derivation and Use of EG Cells

In culture PGC survival and proliferation is limited. However, PGCs can be effectively immortalized in the presence of a cocktail of growth factors, SCF, LIF and basic Fibroblast Growth Factor (bFGF) a confluent feeder layer of STO (or Sl220) cells (Matsui et al., 1992 and Resnick et al., 1992). In these conditions, PGC numbers increase dramatically during the first few days of culture and continue to do so until well after the time they would normally differentiate or die *in vitro*. Initially, the PGCs form small colonies of approximately 8–10 cells that expand as a monolayer on top of the feeder layer. By about 7–8 days of culture, these colonies begin to form multilayered clumps. By 9 days, these clumps are clearly visible to the naked eye. When the clumps of cells are dispersed by trypsinization and re-plated onto fresh feeder layers, they form new colonies that can be passaged indefinitely (Matsui et al., 1992 and Resnick et al., 1992). The resulting cells, termed embryonic germ (EG) cells are similar pluripotent ES cells in many respects. EG cells form embryoid bodies in culture, give rise to teratomas when introduced into histocompatible animals, and form germline chimeras when introduced into a host blastocyst (Matsui et al., 1992; Stewart et al., 1994; Labosky et al., 1994a; Labosky et al., 1994b; L. Jackson-Grusby and R. Jaenisch, pers. comm.; and U. Klemm and R. Mulligan, pers. comm. and our own unpub. obs.). Initial studies suggest that introduction and expression of cloned DNA into EG cell lines is likely to be more straightforward than into PGCs. EG cell lines can be treated in the same way as ES cells and can be manipulated *in vitro* by electroporation, transfection and drug selection (our own unpub. obs.). Indeed, Balb/c-derived EG cell lines have been used to create IL-4-deficient mice by homologous recombination using existing vectors and selection procedures (U. Klemm and R. Mulligan, pers. comm.). EG cells, therefore, represent perhaps the most viable mechanism by which PGCs or their derivatives can be used to generate transgenic animals.

EG cells may prove useful in overcoming some of the limitations encountered using ES cells. As discussed in the Introduction, in some instances, it would be advantageous to derive stem cell lines from any mouse strain, for example, to overcome potential problems with using non-isogenic DNA for homologous recombination or where a replacement ("knocking in") strategy is required. Most commonly used ES cell lines are derived from the 129/Sv strain. Recently, ES cells from a large number of mouse strains and F_1 hybrids have been established, but not all of the ES cell lines have been tested for their ability to transmit through the germline. EG cell lines have also been derived from a wide variety of mouse strains including inbred and outbred, as well as F_1 hybrid, strains (see Table 29.1). Once again, however, not all the cell lines have been tested for germline transmission. Nevertheless, if ES cell lines cannot be derived from a particular strain, then it may be possible to derive EG cells from that strain.

A second use of EG cell technology may be in the area of stem cell derivation from other organisms, particularly other mammals. The number of PGCs (EG cell progenitors) that could be derived from a single embryo greatly exceeds the number of inner cell mass (ICM) cells (ES cell progenitors) that could easily be obtained from blastocyst-stage embryos.

Table 29.1. List of mouse EG cell lines.

C57/Bl6J [1,4]

B6C3F$_1$ [2]

129/Sv [3]

ICR (Random bred)   (C57BL/6 x DBA)F$_1$ [1]

BALB/c [4,5]

FVB/N [4,5]

B6 Robertsonian (balanced) [6]

A/J [6]

---

[1] Matsui et al., 1992 and Labosky et al., 1994a and b.
[2] Resnick et al., 1992;   [3] Stewart et al., 1993.
[4] M. Klemm and R. Jaenisch, pers. comm.
[5] U. Klemm and R. Mulligan, pers. comm.
[6] K. Reese and J. Gearhart, pers. comm.

The percentages of PGCs that go on to make EG cells, or the percentage of ICM cells that go on to make ES cells, have not been accurately determined. Nevertheless, it seems obvious that starting with a larger pool of potential founder cells would be advantageous. Cows will carry a relatively large number of embryos at least until the period from which PGCs can be derived. We have successfully obtained large numbers of bovine embryos by laparotomy from superovulated cows and isolated PGCs from them. Depending on the age of the embryo, each genital ridge or gonad can yield between 5,000 and 35,000 PGCs. Furthermore, studies on the process of EG cell derivation are likely to increase our understanding of the mechanism of stem cell derivation and growth. This will aid attempts to develop truly pluripotent stem cell lines from other organisms. While understanding the biology and physiology of mouse ES cells will also undoubtedly aid in these efforts, at the moment our knowledge of the mechanism of ES cell derivation is limited. LIF is still the only factor known to be required for the establishment of ES cell lines (Williams et al., 1988 and Smith et al., 1988). In fact it is surprising that so little is known about the growth factor requirements and physiology of mouse ES cells! Thus it is difficult to establish rational conditions for deriving ES cells from, for example, bovine embryos since there are so few known parameters which can be modified. On the other hand the derivation of EG cells is known to require three polypeptide growth factors, SCF, LIF and bFGF (Matsui et al., 1992 and Resnick et al., 1992). To the best of our knowledge, however, attempts to derive EG cells from other mammals have been unsuccessful. In order for EG cell lines to be derived from other animals, therefore, a number of important questions need to be addressed and will be discussed below.

An important question is whether the specificity of the growth factors affects the ability to derive EG cell lines. For example, while murine SCF can act on human cells (albeit ten times less effectively than on murine cells), the human factor is one thousandfold less efficient at activating the murine c-Kit receptor. For this reason it may be important, not only to know exactly which factor within in growth factor family to use, but also to derive the growth factor from the species from which the EG cells are to be generated. In this regard it may also be necessary to use feeder cell lines from the species or strain of interest. While genetic evidence unequivocally demonstrates the role of SCF in PGC development, the role of bFGF and LIF remain unclear. In terms of the role of FGFs it will be important to know which FGF receptor(s) PGCs express and which member(s) of the FGF family are involved in PGC development. If EG cells are to be derived from other species it may be important to know exactly which type of FGF to use. At this time, nine members of the FGF family have been identified. Of these factors, FGF4 (K-FGF) seems the best candidate for the physiological ligand for PGCs since it is expressed in the primitive streak at around the time when PGCs are moving through the embryo (Niswander and Martin, 1992 and Drucker and Goldfarb, 1993). Recently the *Ter locus*, the major determinant of teratoma and teratocarcinoma formation in mice has been localized on mouse chromosome 18 close to the acidic FGF gene (aFGF or FGF1). This data suggests that aFGF may be a regulator of PGC proliferation and differentiation (Asada et al., 1994). Analysis of the FGF receptor subtype expressed by PGCs may help define which member of the FGF family is optimal for stimulating PGC proliferation and for deriving EG cells.

Another important question is whether LIF is the natural physiological ligand for PGCs. Again the answer to this question may have relevance to attempts to derive EG cells from other species. Although LIF acts on PGCs *in vitro*, we have been unable to identify LIF transcripts in the gonad (Cheng and Donovan, unpublished observations) and, moreover, mice lacking LIF produce viable gametes (Stewart et al., 1992). Two explanations for these data seem reasonable. First, that in the absence of LIF another factor can substitute for it. Indeed we have shown that other members of the hemopoietin family related to LIF, including ciliary neurotrophic factor (CNTF) and Oncostatin M (OSM) can stimulate PGC survival and proliferation in a manner similar to LIF (Cheng et al., 1992).

In fact, OSM is equivalent to LIF in its effects on PGCs (Cheng et al., 1992). These data are consistent with the ability of these ligands to bind to the LIFR/gp130 receptor complex. A second possibility is that LIF really is not the physiological ligand and that another factor fulfills that role. CNTF or OSM or an as yet undiscovered member of the hemopoietin family may fulfill this role. Intriguingly, a mutation causing sterility in the mouse (*germ cell deficient or gcd*) (Pellas et al., 1991) has recently been mapped to mouse chromosome 11 close to the LIF and OSM genes (Duncan et al., 1995). While it seems unlikely that *gcd* represents a mutation in the LIF gene (given the phenotype of the LIF knockout mouse) it is possible that gcd might be a mutation in the OSM gene or in another, possibly genetically linked, member of the hemopoietin family. Irrespective of whether the *gcd* mutation involves a member of the hemopoietin cytokine family of growth factors, understanding the role of these proteins in PGC development and EG cell derivation is an important issue.

Future Directions

In this chapter we have described recent advances in PGC culture and transplantation techniques. These advances may, in the near future, allow direct manipulation of germ cells *in vitro* which in turn can be re-introduced into the germline. Such techniques may allow gene targeting technology, currently an invaluable tool for manipulation of the mouse genome, to be transferred to other organisms, particularly other mammals. Female PGCs or oogonia transplanted into the adult ovary will mature into oocytes and can re-enter the germline, but it is still unclear whether these cells could easily be manipulated in culture. Further studies aimed at understanding PGC survival and differentiation *in vivo* and *in vitro* are clearly required. Characterization of factors produced by cell lines able to support PGC survival and proliferation will likely provide valuable information on growth factors involved in PGC development. Further studies on factors activating oocyte meiosis, such as sterols (Byskov et al., 1995), are also likely to yield valuable information of practical importance.

The area of greatest potential for using PGCs for germline manipulation is likely to be the use of EG cell technology. Development of pluripotent stem (ES-like) cell lines from other organisms would, without doubt, have many uses. A number of important questions about the mechanism of EG cell derivation remain. EG cell derivation requires three polypeptide growth factors, SCF, LIF and bFGF. The receptors for each of these factors is relatively well understood and the signaling pathways linking each of the receptors to the nucleus are being unraveled. SCF and bFGF bind to distinct receptor tyrosine kinases which, in response to ligand binding, activate a known number of substrates including cytoplasmic effectors such as Raf and Ras. To the extent to which they have been characterized, the signaling pathways downstream of the c-Kit and FGF receptors appear very similar. However, while SCF binding to c-Kit seems to be important for PGC survival (Dolci et al., 1991; Godin et al., 1991 and Matsui et al., 1991), bFGF seems to act as a mitogen (Matsui et al., 1992; Resnick et al., 1992). An important question, therefore, is how such similar signal transduction pathways give rise to such different effects on the germ cells.

Unlike SCF and bFGF, LIF binds to heterodimeric complex comprised of a low-affinity, ligand-binding receptor (LIFR) and a signaling subunit, gp130, which does not bind LIF directly. LIFR and gp130 are members of the hemopoietin cytokine receptor family that are not themselves active kinases. Together the complex of LIFR and gp130 forms a receptor for LIF that binds the factor with high affinity and which can transduce signals in response to ligand binding. Binding of LIF to its receptor causes receptor activation and in turn activation of members of the Jak family of cytoplasmic kinases. Ultimately activation of the LIF receptor complex leads to activation of members of the STAT family of transcription factors. Evidence from a number of laboratories suggest that LIF, like SCF, is a survival factor for germ cells and that signaling via gp130 is required for germ cell survival (DeFelici and Dolci, 1991; Pesce et al., 1993; Cheng et al., 1994; and Taga and Kishimoto, pers. comm.) but LIF also acts to suppress differentiation of ES and EG cells (Williams et al., 1988; Smith et al., 1988). Again it will be important to understand how the signaling pathway downstream of gp130 acts to effect PGC survival or to inhibit EG cell differentiation.

It seems likely that SCF, LIF and bFGF are not the only factors involved in PGC or EG cell growth. The wealth of information gathered about growth factor production and requirements of that other well known stem cell, the hemopoietic stem cell, is staggering and has allowed the successful isolation, culture and manipulation of hemopoietic stem cells from different species. Such studies on ES and EG cells would be immensely valuable. Given the number of growth factors and receptors that have been molecularly cloned, such an undertaking would be much more feasible today than it was for hematologists! These studies may help define which growth factors are required for derivation of ES and EG cells, what growth factors are produced by the feeder layer cells, and which ones are important for stem cell derivation. They will also provide information on which growth factors are produced by ES and EG cells themselves.

The utility of EG and ES cell lines for manipulation of the genome is absolutely dependent on the ability of these cells to transmit through the germline. Germline transmission remains an intriguing problem even though some of the parameters, such as sex of the stem cell and embryo and karyotype of the stem cell, are well known (see Robertson, 1987, for review). One area that may be relevant to germline transmission of EG cells is the role of imprinting. The skeletal defects observed in chimeras produced from some EG cell lines (Matsui et al., 1992) are reminiscent of defects seen in chimeras produced using androgenetic ES cells (Mann et al., 1990). This raises the question of the role that imprinting may play in the ability of stem cells to contribute to the somatic lineages and to enter the germline. Analysis of the methylation status of the insulin-like growth factor type 2 receptor gene (Igf2r) in EG cells and ES cells demonstrate that EG cells are different from ES cells in the methylation status of this gene. Moreover, there is no correlation between the ability of karyotypically-normal EG cell lines to transmit through the germline and the state of methylation the Igf2r gene. The potential role of methylation status and imprinting in germline transmission needs to be addressed further. Since the number of imprinted genes is growing steadily, it should be possible to analyze the imprinting status of these genes in many ES and EG cell lines and to correlate these data with the ability of each cell line to enter the germline.

Future studies are also likely to determine whether there are genetic factors that control the ability of PGCs or ICM cells to give rise to stem cells. Although 129/Sv animals seem to be the animal of choice for derivation of ES cells, such cells have been derived from a variety of mouse strains. Whether all such cell lines can transmit through the germline is an important question that remains to be addressed. Whether 129/Sv animals have some genetic predisposition to allow development of stem cells is also an important question that is under investigation (Kawase et al., 1994; and M. Klemm and R. Jaenisch, pers. comm.). Modern techniques of gene mapping that allow polygenic (complex) traits to be followed quickly from generation to generation could allow such questions to be answered. Genetic factors affecting formation of germ cell-derived tumors (teratomas and teratocarcinomas) in mice and humans (see Asada et al., 1994, for brief review) are also likely to be relevant to development of stem cells, particularly EG cells. An increasing number of mice carrying mutations created by targeted disruption of genes in ES cells display phenotypes in the germline and are likely to provide a wealth of information of PGC development and differentiation. Overall, more basic research on the development of the germline may ultimately allow modification of germ cells of many different stages. If such techniques can be developed they will have a variety of applications in animal transgenesis. Generation of genetically-modified animals will provide additional models of human disease and organs and tissues for transplantation.

Acknowledgements

We are grateful to Roger Pedersen, Tetsuya Taga, Tadamitsu kishimoto, Terry Magnuson, Trish Labosky, Brigid Hogan, Uwe Klemm, Richard Mulligan, Martina Klemm, Rudolph Jaenisch, Colin Stewart, Virginia Papaioannou, Kimberly Reese and John Gearhart for communicating results prior to publication and for helpful discussions on the subject of stem cell derivation. This research was supported in part by the National Cancer Institute, DHHS, under contract with ABL. The contents of this publication do not necessarily reflect the view or policies of the Department of Health and Human Services, nor does mention of trade names, commercial products, or organizations imply endorsement by the US Government.

References

Asada, Y., Varnum, D.S., Frankel, W.N. and Nadeau, J.H. (1994) A mutation in the Ter gene causing increased susceptibility to testicular teratomas maps to mouse chromosome 18. Nature Genetics 6, 363–368

Bradley, A., Evans, M., Kaufman, M.H. and Robertson, E. (1984) Formation of germ-line chimeras from embryo-derived teratocarcinoma cell lines. Nature 309, 255–256

Brinster, R.L. and Avarbock, M.R. (1994) Germline transmission of donor haplotype following spermatogonial transplantation. Proc. Natl. Acad. Sci. USA 91, 11303–11307

Brinster, R.L. and Zimmerman, J.W. (1994) Spermatogenesis following male germ-cell transplantation. Proc. Natl. Acad. Sci. USA 91, 11298–11302

Buehr, M. and McLaren, A. (1993) Isolation and culture of primordial germ cells. In P.M. Wassarman and M.L. DePamphilis, (eds.), Methods in Enzymology, Vol. 225, Academic Press, San Diego, pp. 58–77

Byskov, A.G., Anderson, C.Y., Nordholm, L., Thogersen, H., Xia, G., Wassmann, O., Andersen, J.V., Guddal, E. and Roed, T. (1995) Chemical structure of sterols that activate oocyte meiosis. Nature 374, 559–562.

Cheng, L., Gearing, D.P., White, L.S., Compton, D.L., Schooley, K. and Donovan, P.J. (1994) Role of leukemia inhibitory factor and its receptor in mouse primordial germ cell growth. Development 120, 3145–3153

Cooke, J.E., Godin, I., Ffrench-Constant, C., Heasman, J. and Wylie, C.C. (1993) Culture and manipulation of primordial germ cells. In P.M. Wassarman and M.L. DePamphlis (eds.), *Methods in Enzymology* Vol. **225**, Academic Press, San Diego, pp. 37–58

De Felici, M. and Dolci, S. (1991) Leukemia inhibitory factor sustains the survival of mouse primordial germ cells cultured on TM4 feeder layers. *Dev. Biol.* **147**, 281–284

De Felici, M. and McLaren, A. (1982) Isolation of mouse primordial germ cells. *Exp. Cell Res.* **142**, 476–482

Doetschman, T., Williams, P. and Maeda, N. (1988) Establishment of hamster blastocyst-derived embryonic stem (ES) cells. *Dev. Biol.* **127**, 224–227

Dolci, S., Pesce, M. and De Felici, M. (1993) Combined action of stem cell factor, leukemia inhibitory factor and cAMP on *in vitro* proliferation of mouse primordial germ cells. *Mol. Reprod. Dev.* **2**, 134–139

Dolci, S., Williams, D.E., Ernst, M.K. Resnick, J.L., Brannan, C.I., Lock, L.F., Lyman, S.D., Boswell, H.S. and Donovan, P.J. (1991) Requirements for mast cell growth factor for primordial germ cell survival in culture. *Nature* **353**, 809–811

Donehower, L.A., Harvey, M., Slagle, B.L., McArthur, M.J., Montgomery C.A., Jr., Butel, J.S. and Bradley, A. (1992) Mice deficient for p53 are developmentally normal but susceptible in spontaneous tumours. *Nature* **356**, 215–221

Donovan, P.J., (1993) Growth factor regulation of mouse primordial germ cell development. *Curr. Top. Dev. Biol.* **29**, 189–225

Donovan, P.J. Stott, D., Cairns, L.A., Heasman, J., and Wylie, C.C. (1986) Migratory and postmigratory muse primordial germ cells behave differently in culture. *Cell* **44**, 831–838

Drucker, B.J. and Goldfarb, M. (1993) Murine FGF-4 gene expression is spatially restricted within embryonic skeletal muscle and other tissues. *Mech. Dev.* **40**, 155–163

Duncan, M.K., Lieman, J. and Chada, K.K. (1995) The germ cell deficient locus maps to mouse Chromosome 11A2-3. *Mammalian Genome* **6**, 697–699

Escary, J.L., Perreau, J., Dumenil, D., Ezine, S. and Brulet, P. (1993) Leukaemia inhibitory factor is necessary for maintenance of haematopoietic stem cells and thymocyte stimulation. *Nature* **363**, 361–364

Fox, N.D.I., Martinez-Hernadez, A., Knowles, B.B. and Solter, D. (1981) Immunohistochemical localization of the early embryonic antigen (SSEA-1) in postimplantation mouse embryos and fetal and adult tissues. *Dev. Biol.* **83**, 391–398

Godin, I., Deed, R., Cooke, J., Zsebo, K., Dexter, M. and Wylie, C.C. (1991) Effects of the steel gene product on mouse primordial germ cells in culture. *Nature* **352**, 807–809

Hashimoto, K., Noguchi, M. and Nakatsuji, N. (1992) Mouse offspring derived from fetal ovaries or reaggregates which were cultured and transplanted into adult females. *Development Growth and Differentiation* **34**, 233–238

Hogan, M., Costantini, F. and Lacy, E. (eds.) (1986) *Manipulating the Mouse Embryo: A Laboratory Manual* Cold Spring Harbor Laboratory, New York, 113–114

Iannaccone, P.M., Taborn, G.U., Garton, R.L., Caplice, M.D. and Brenin, D.R. (1994) Pluripotent embryonic stem cells from the rat are capable of producing chimeras. *Dev. Biol.* **163**, 288–292

Jaenisch, R. (1985) Mammalian neural crest cells participate in normal embryonic development on microinjection into post-implantation mouse embryos. *Nature* **318** 181–183

Kawase, E., Suemori, H., Takahashi, N., Okazaki, K., Hashimoto, K. and Natatsuji, N. (1994) Strain difference in establishment of mouse embryonic stem (ES) cell lines. *Int. J. Dev. Biol.* **38**, 385–390

Labosky, P.A., Barlow, D.P. and Hogan, B.L.M. (1994) Embryonic germ cell lines and their derivation from mouse primordial germ cell. In *Germline Development* Wiley, Chichester, pp. 157–178

Kola, I., Davey, A. and Gough, N.M. (1990) Localization of the murine leukemia inhibitory factor gene near the centromere on chromosome 11. *Growth Factors* **2**, 235–240

MacGregor, G.R., Zambrowicz, B.P. and Soriano, P. (1995) Tissue non-specific alkaline phosphatase is expressed in both embryonic and extraembryonic lineages during mouse embryogenes but is not required for migration of primordial germ cells. *Development* **5**, 1487–1496

Mann, J.R., Gadi, I., Harbison, M.L., Abbondanzo, S.J. and Stewart, C.L. (1990) Androgenetic mouse embryonic stem cells are pluripotent and cause skeletal defects in chimeras; implications for genetic imprinting. *Cell* **62**, 251–260

Matsui, Y., Toksoz, D., Nishikawa, S., Nishikawa, S., Williams, D., Zsebo, K. and Hogan, B.L. (1991) Effect of Steel factor and leukaemia inhibitory factor on murine primordial germ cells in culture. *Nature* **353**, 750–752

Matsui, Y., Zsebo, K. and Hogan, B.L. (1992). Derivation of pluripotential embryonic stem cells from murine primordial germ cells in culture. *Cell* **70**, 841–847

McCarrey, J.R., Hsu, K.C., Eddy, E.M., Klevecz, R.R. and Bolen, J.L. (1987) Isolation of viable mouse primordial germ cells by antibody-directed flow sorting. *J. Exp. Zool.* **242**, 107–111

McLaren, A. (1981) *Germ Cells and Soma: A New Look at an Old Problem.* Yale University Press, New Haven, CT

Niswander, L. and Martin, G.R. (1992) FGF-4 expression during gastrulation, myogenesis, limb and tooth development in the mouse. *Development* **114**, 755–768

Notarianni, E., Galli, C., Laurie, S., Moor, R.M. and Evans, M.J. (1991) Derivation of pluriotent, embryonic cell lines from the pig and sheep. *J. Reprod. Fertil. Suppl.* **43**, pp. 255–260

Pellas, T., Ramachandran, B., Duncan, M., Pan, S., Marone, M. and Chada, K. (1991) Germ cell deficient (*gcd*), a novel insertional mutation manifested as infertility in transgenic mice. *Proc. Natl. Acad. Sci. USA* **88**, 8787–8791

Pesce, M., Farrace, M.G., Piacentini, M., Dolci, S. and De Felici, M. (1993) Stem cell factor and leukemia inhibitory factor promote primordial germ cell survival by suppressing programmed cell death (apoptosis). *Development* **4**, 1089–1094

Ramires-Solis, R., Zheng, H., Whiting, J., Krumlauf, R. and Bradley, A. (1990) Hoxb-4 (*Hox*-2.6) mutant mice show homeotic transformation of a cervical vertebra and defects in the closure of the sternal rediments. *Cell* **73**, 279–294

Resnick, J.L., Bixler, L.S., Cheng, L. and Donovan, P.J. (1992) Long-term proliferation of mouse primordial germ cells in culture. *Nature* **359**, 550–551

Robertson, E.J. (ed.) (1987) *Teratocarcinomas and Embryonic Stem Cells* IRL Press, New York

Sakurai, T., Katoh, H., Moriwaki, K., Noguchi, T., Noguchi, M. (1994) The *ter* primordial germ cell deficiency mutation maps near *Grl-1* on mouse chromosome 18. *Mamm. Genome* **5**, 333–336

Sibilia, M. and Wagner, E.F. (1995) Strain-dependent epithelial defects in mice lacking the EGF receptor. *Science* **269**, 234–238

Silvers, W.K. (1979) *The Coat Colors of Mice: A Model for Mammalian Gene Action and Interaction.* Springer-Verlag, N Y

Smith, A.G., Heath, J.K., Donaldson, D.D., Wong, G.G., Moreau, J., Stahl, M. and Rogers, D. (1988) Inhibition of pluripotential embryonic stem cell differentiation by purified polypeptides. *Nature* **336**, 688–690

Solter, D. and Knowles, B.B. (1978) Monoclonal antibody defining a stage-specific embryonic antigen (SSEA-1). *Proc. Natl. Acad. Sci. USA* **75**, 5565–5569

Stewart, C.L., Gadi, I. and Bhatt, H. (1994) Stem cells from primordial germ cells can reenter the germ line. *Dev. Biol.* **161**, 626–628

Stewart, C.L., Kaspar, P., Brunet, L.J., Bhatt, H., Gadi, I., Kontgen, F. and Abbondanzo, S. (1992) Blastocyst implantation depends on maternal expression of leukaemia inhibitory factor. *Nature* **359**, 76–79

Sukoyan, M.A., Golubitsa, A.N., Zhelezova, A.L., Shilov, A.G., Vatolin, S.Y., Maximovsky, L.P., Andreeva, L.E., McWhir, J., Pack, S.D., Bayborodin, S.I., et al. (1992) Isolation and cultivation of blastocyst-derived stem cell lines from American mink (*Mustela vision*). *Mol. Reprod. Dev.* **33**, 418–431

te Riele, H., Maandag, E.R. and Berns, A. (1992) Highly efficient gene targeting in embryonic stem cells through homologous recombination with isogenic DNA constructs. *Proc. Natl. Acad. Sci. USA* **89**, 5128–5132

Threadgill, D.W., Dlugosz, A.A., Hansen, L.A., Tennenbaum, T., Lichti, U., Yee, D., Lamantia, C., Mourton, T., Herrup, K., Harris, R.C. and Barnard, J.A. et. al. (1995) Targeted disruption of mouse EGF receptor: effect of genetic background on mutant phenotype. *Science* **269**, 230–234

Yuspa, S.H., Coffey, R.J. and Magnuson, T. (1995) Targeted disruption of mouse EGF receptor-effect of genetic background on mutant phenotype. *Science* **269**, 230–234

Williams, R.L. Hilton, D.J., Pease, S., Willson, T.A., Stewart, C.L., Gearing, D.P., Wagner, C.F., Metcalf, D., Nicola, N.A. and Gough, N.M. (1988) Myeloid leukaemia inhibitory factor maintains the developmental potential of embryonic stem cells. *Nature* **336**, 684–687

Yeom, Y.I., Fuhrmann, G., Ovitt, C.E., Brehm, A., Ohobo, K., Cross, M., Hubner, K. and Scholer, H.R. (1996) Germline regulatory element of Oct-4 specific for the totipotent cycle of embryonal cells. *Development* **122**, 881–894

PART II, SECTION G

30. The Adenovirus Vectors for Gene Delivery

J.F. Dedieu, E. Vigne, C. Orsini, M. Perricaudet and P. Yeh

CNRS URA 1301/Rhône-Poulenc Rorer Gencell, Laboratoire de Génétique des Virus Oncogènes, Institut Gustave Roussy, PR II, 39 rue Camille Desmoulins, 94805 Villejuif Cedex, France

Since their isolation in cultures from adenoïd tissues in 1953, human adenoviruses have received considerable attention. Initially studied from a clinical point of view, they rapidly became one of the most popular experimental models to unravel basic eukaryotic molecular biology (Berget *et al.*, 1977; Challberg and Kelly, 1979 and Whyte *et al.*, 1988). In recent years, adenoviruses have been extensively evaluated for somatic gene transfer in a number of preclinical studies and exhibited a tremendous potential for the treatment of a variety of severe human disorders (Rosenfeld *et al.*, 1992; Ragot *et al.*, 1993; Chen *et al.*, 1994; Horellou *et al.*, 1994 and Ohno *et al.*, 1994).

This manuscript will briefly review the basic features of adenovirus molecular biology in relation with its comprehensive utilization as a vector for gene delivery. Its main limitations will be pointed out, together with some of the approaches we, and others, are following to circumvent them. References will be limited to the most significant reviews and publications in the field.

Adenoviruses: Overview

Adenoviruses (Ad) are widespread in nature and nearly 100 Ads have been identified from man to bird (Horwitz, 1990). This includes 49 serotypes of human origin which are classified into 6 subgroups (A to F) based on their genome composition (G/C content), organisation (restriction profile) and homology (intra and inter genus), as well as the polypeptide profile of the viral particle, their haemaglutination pattern and their oncogenicity in newborn hamsters. Human serotypes 2 (Ad2) and 5 (Ad5) have been mostly studied and this knowledge is particularly useful for the rational design of recombinant adenoviruses (RAd). Both serotypes belong to subgroup C: they can transform non-permissive rodent cells *in vitro* but they are non-tumorigenic in newborn hamsters. In humans, Ad2 and Ad5 cause only mild diseases, namely mild upper and lower respiratory tract infections in infants and young children.

Adenoviruses are small, but complex, non enveloped DNA viruses with an icosahedric symmetry (Horwitz, 1990). The tridimensional structure of Ad2 has been obtained recently by image reconstruction from cryoelectron micrographs (Stewart *et al.*, 1991). The outer shell of the viral particle (60 to 90 nm in diameter depending on the serotype) is composed of 252 sub-units or capsomers: 240 hexons make up the 20 triangular faces and the edges of the icosahedron, while 12 pentons are located at the apexes of the icosahedron. Each penton is formed by a penton base with an attached fiber with a typical protruding tail-head shape. The fiber head (knob) mediates the initial binding of the virus to the cell surface through a yet unidentified receptor. A second receptor, a member of the integrin class of cell adhesion molecules, mediates the internalisation of the particle following its interaction with the RGD motif of the penton base. Ads can infect a large variety of adherent cells from many organs. Furthermore, and by contrast to retroviruses, infection is productive in both dividing and quiescent cells. Insertion of part of the adenovirus into the host genome by non homologous recombination after infection is unusual (Doerfler, 1968; and Wang and Taylor, 1993).

The virion contains a linear, double stranded, DNA molecule of approximately 36 kb in length which associates with a set of virally encoded proteins to form the viral core. The viral genome exhibits inverted terminal repeats of ca. 100 bp at either end of the molecule. A 55 Kd protein (the terminal protein) is also covalently linked to the 5′ end of each DNA strand and these unique terminal structures are crucial for viral DNA replication. Only DNA molecules with a size below 1.05 times that of the wild type genome can be accomodated by the viral capsid (Bett et al., 1993).

The adenovirus genome can be functionally divided into early and late transcription units (Figure 30.1), depending on their timely expression with regard to viral DNA replication (i.e., before and after, respectively). The pIX and IVa2 genes do not belong to either class as their transcription is initiated during viral DNA replication. The switch from early to late occurs approximately 8 hours post-infection, with the productive cycle lasting ca. 36 hours in established permissive cell lines.

There are five early regions: E1A, E1B, E2 (A and B), E3 and E4, and one major late transcription unit (MLTU). Both strands of the viral genome are coding, making its *in vitro* manipulation quite tedious and restricted to certain regions.

E1A, the so-called immediate early region, is the first region to be transcribed upon infection, very likely because binding sites for numerous endogenous cellular transcription factors, including a set of two binding sites for E2F, are present immediately upstream of its promoter. Through its major products, two proteins of 289 and 243 residues (known as R289 and R243), the E1A region elicit a variety of effects in the infected cells through a complex pattern of transcriptional gene activation and repression (Bayley and Mymryk, 1994). As far as the virus infectious cycle is concerned, the primary function of E1A is to activate the transcription of all other viral units *in trans* during the early phase of infection. E1A also induces quiescent cells to enter the S phase of the cell cycle, a prerequesite for efficient viral DNA replication but also a conflictual signal leading to upregulation of p53 and apoptosis.

The E1B region mainly encodes two proteins of 19 Kd and 55 Kd which independently suppress apoptosis (White, 1993). For example, the E1B 55 K protein binds and sequesters p53, while the E1B 19 K protein has the ability to block apoptosis induced by tumor necrosis factor (TNF), enabling many infected cells to escape from immune surveillance. In that regard, the combined action of the E1B proteins could be viewed as a strategy to sustain cell viability and virus production during infection. In late infected cells, the E1B 55 K protein is also complexed to one of the products of E4 region, the ORF6 protein (Sarnow, et al., 1984). This complex is believed to facilitate transport of late viral mRNAs

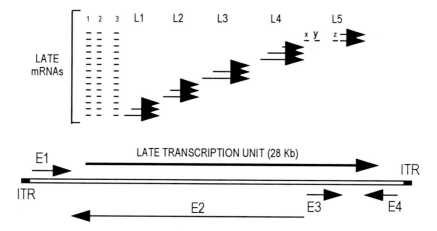

Figure 30.1. Schematic representation of the Ad5 genome and its major transcription units.
The Ad5 genome is compacted within a linear 36 kb chromosome. The main transcription units include 4 distinct early regions (E1 to E4) and a late transcription unit (LTU) which is alternatively polyadenylated and spliced to generate five (L1–L5) distinct families of late mRNAs. The switch from early to late gene expression takes place approximately 8 hours after infection. E1 proteins activate most other viral transcription units either directly or indirectly. E2 proteins include the DNA binding protein (DBP), the viral DNA polymerase and the terminal binding protein which are all directly involved in viral DNA replication. E3 proteins modulate the host immune response towards virally infected cells. E4 proteins are involved in the control of viral transcription, viral DNA replication, and viral late protein synthesis (see text). 1, 2 and 3 refer to the components of the tripartite leader present in all late mRNAs; x, y, and z refer to the components of the auxiliary leader present in some of the fiber-encoding mRNAs (L5). The pIX- and IVa2-encoding genes, and the VA RNAs have been omitted for clarity. The E1 region is deleted in most vectors and is replaced by the transgene expression cassette.

from the nucleus to the cytoplasm and/or to maintain their stability upon reaching the cytoplasm. This process is quite selective as it does not apply for most cellular mRNAs (Bridge and Ketner, 1990 and Bridge et al., 1991).

The E2A and E2B regions code for proteins which are required for viral DNA replication: a 72 Kd single stranded DNA binding protein (DBP), as well as a 140 Kd DNA polymerase and a 80 Kd precursor to the terminal protein (pTP), respectively. This region is also involved in regulating viral gene expression, at least in vitro. For example, the 72K DBP has been shown to upregulate its own expression as well as late gene expression, while repressing that of the E4 region (Chang and Shenk, 1990).

The functions of only four of the nine potential gene products of the E3 region have been characterized. These polypeptides somehow interfere with the immune recognition of the infected cells (Wold and Gooding, 1991). For example, a 19 Kd glycoprotein, gp19K, binds and sequesters certain alleles of MHC-I molecules within the endoplasmic reticulum, thereby impairing the CTL-mediated lysis of the infected cells. In addition, a 14.7 Kd protein and a 10.4 Kd/ 14.5 Kd complex prevent, by different mechanisms, the TNF-induced lysis of the infected cells. The E3 region has also been implicated in the downregulation of several adenoviral genes, including E1A, presumably at the level of translation (Zhang et al., 1994). Remarkably, the E3 region is dispensable for viral growth in vitro and is usually deleted to provide more cloning space.

The E4 region is a very complex transcription unit (Figure 30.2). It encodes for a 2.8 kb primary transcript which is heavily spliced into at least 24 mRNAs (Virtanen et al., 1984 and Dix and Leppard, 1993). At least 6 different E4 ORFs gene products have been identified, some of which being of unknown functions (Cutt et al., 1987; Bridge et al., 1993; Kleinberger and Shenk 1993; and Dix and Leppard, 1995). ORF3 and ORF6 encode essential, but redundant, activities required in the post-transcriptional processing of the major late transcription unit (MLTU) from which most of structural proteins are synthesized (Nordqvist and Akusjarvi, 1990; Ohman et al., 1993 and Nordqvist et al., 1994). The ORF6/7 protein functionally interacts with endogenous E2F transcription factors and upregulates the E2A promoter (Obert et al., 1994). This particular E4 protein is thus involved in the synthesis of the 72 K DBP, and as such indirectly interfere with viral DNA replication and MLTU expression (see above E2A functions).

The MLTU is a very large primary transcript of ca. 28 kb. Its processing generates approximately 20 mRNAs that can be grouped into five families (L1 to L5), each of which consisting of a set of alternatively spliced molecules with identical 3' ends. Most late mRNAs exhibit in their 5' end a common 201 bp leader sequence which is generated by the junction of three short exons. Sometimes, an internal exon is also included which encodes a 13.6 kD protein of unknown function (Soloway and Shenk, 1990). The late mRNAs encode most of the capsid proteins (e.g., the fiber, penton and hexon proteins), as well as structural proteins of the viral core, together with non structural proteins required for the assembly and maturation of the virions.

From Adenovirus to Adenovectors

The first recombinant adenovirus was recovered in 1981: it was replication competent and contained the gene coding for SV40 T antigen inserted in the E3 region (Solnick, 1981). The first replication-defective RAd was constructed in 1985 and was shown to express the hepatitis B virus surface antigen (Ballay et al., 1985). Since then, little has changed in the design of RAds. These so-called first generation

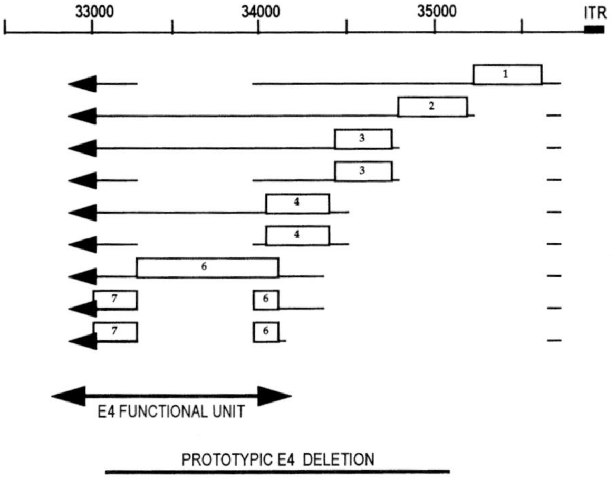

Figure 30.2. Genetic organization of the Ad5 E4 region. The main mRNAs encoding E4 ORF1, ORF2, ORF3, ORF4, ORF6 and ORF6/7 gene products have been described previously (Dix and Leppard 1993). The extent of the E4 deletion harbored by H2dl808 (a prototypic defective Ad2 E4 deletant) is indicated by a solid bar. The viral sequence encompassing ORF1, ORF2, ORF3, and ORF4 can be deleted without dramatic consequences for viral propagation on Hela cells (Hemstrom et al., 1988). Therefore, the distal part of the E4 region including ORF6 and ORF7 constitutes the minimal E4 functional unit which has been integrated in the IGRP2 cell line.

recombinant adenoviruses (FGRAds) are replication defective because they lack a 3 kb fragment encompassing the entire E1 region. The expression cassette for the transgene is mostly inserted in this region, although insertion within E3 is also possible. FGRAds are constructed and propagated on the 293 cell line, an artificial packaging cell line that provides *in trans* the missing E1 proteins. These immortalized cells have been derived from human embryonic kidney cells after integration of the Ad5 E1 genes, together with E1 flanking sequences, into the cellular genome (Graham *et al.*, 1977).

Deletions of additional regions which are not necessary for viral propagation *in vitro*, namely the entire E3 region and the proximal part of the E4 region, extend the cloning capacity of FGRAds to more than 8 Kb (Hemstrom *et al.*, 1988; Bett *et al.*, 1993 and Armentano *et al.*, 1995). FGRAds thus retain at least 80% of the viral genome. This represents a coding potential for more than 20 proteins, most of them being generated following extensive, E4 dependent, post-transcriptional processing of the MLTU.

FGRAds are usually constructed by homologous recombination in 293 cells between a restricted (non-infectious) viral genome and a plasmid containing the left end of the Ad genome, the transgene expression cassette, followed by adenoviral sequences overlapping with the restricted genome. Detailed protocols for the construction of RAds can be found in Graham and Prevec, 1991. Recently, a yeast artificial chromosome containing a complete adenoviral genome was constructed, from which infectious viral DNA can be recovered (Ketner *et al.*, 1994). The use of established yeast genetic tools will certainly make the construction and cloning of the recombinants easier and faster in a near future.

After plaque purification on 293 cells, the RAd is progressively amplified on these cells and the virus can be purified and concentrated by CsCl gradient centrifugations and/or chromatography. Viral stocks with titers as high as 10^{11} plaque forming units per ml (pfu/ml), i.e. 10^{12} to 10^{13} viral particles per ml, can routinely be prepared from one hundred, 100 mm, Petri dishes. This remarkable productivity allows the direct *in vivo* administration of highly purified recombinants.

Listing the theoretical advantages of adenoviruses as vectors for gene transfer, and possibly gene therapy, is easy. First, Ad2 and Ad5 are harmless in immunocompetent adults. Furthermore, RAds can be propagated and purified to very high titers, and they can accomodate a significant amount of foreign DNA. Also, they readily infect almost all adherent cell types *in vitro* as well as *in vivo*, and they efficiently express their transgene in both quiescent or dividing cells. Finally, they very rarely integrate into the host genome and yet their genome can persist in infected cells as a non replicating extrachromosomal entity for a sustained, but limited, period of time (see below).

Adenovirus Mediated Gene Transfer: From Excitement to Reality

Many preclinical studies have demonstrated the potential of FGRAds for gene transfer in various animal models. Therefore, a growing number of clinical trials have been initiated which target acquired (cancer), as well as hereditary diseases, including cystic fibrosis (for a recent compilation see Hum. Gene Ther. (1994) 5, pp. 1537–1551). Because Ad2- or Ad5-derived vectors can efficiently infect many cell types from many organs in many animal species (mice, rabbits, dogs, cotton rats, non-human primates ...), a valuable opportunity of adenovirus-mediated gene transfer (AMGT) is the generation of relevant animal models for human pathologies. This is particularly true for disorders affecting organs (e.g., the brain) for which other vectors either showed no or very little efficacy, or induced a non acceptable toxicity (Le Gal La Salle *et al.*, 1993 and Wood *et al.*, 1994). The possibility that adenoviral vectors could be engineered so that gene transfer will be efficient and long-lasting clearly opens new therapeutic perspectives for the treatment of human diseases for which animal models are still missing (e.g., Alzheimer disease).

However, intrinsic limitations of E1-deleted vectors have emerged during the course of these studies and have somehow tempered the initial enthusiasm of the scientific community with regard to AMGT.

One problem is associated with the use of the 293 cell line for the production of viral stocks, which frequently leads to their contamination by replication-competent adenovirus (RCA) (Lochmüller *et al.*, 1994). A second problem is the duration of expression of the transgene and the pathology associated with AMGT, which have been found in almost all cases respectively disappointingly low and high (Yang *et al.*, 1994a). Pre-existing or induced anti-adenovirus neutralizing antibodies are another major concern for AMGT, as they have been shown to severely reduce the efficiency of gene transfer following primary or secondary administration of RAds (Daï *et al.*, 1995 and Yang *et al.*, 1995a).

The transient nature of transgene expression following AMGT has been mainly documented in the lung and the liver. In these organs, AMGT is remarkably efficient (e.g., > 80% hepatocytes express the *lacZ* reporter gene a few days after systemic administration of the virus in mice). However, AMGT is also associated with a dose-dependent inflammation in immunocompetent animals so transgene expression usually does not exceed a few weeks.

These problems are not encountered when the virus is administered to newborn animals (Stratford-Perricaudet et al., 1990 and Ragot et al., 1993), to immunopriviledged organs (Bennett et al., 1994), or to human xenografts grown in immunodeficient mice (Engelhardt et al., 1993). Taken together, these observations outline the crucial importance of the immune system in the transient nature of transgene expression and the pathology concurrently associated with FGRAds.

From studies on AMGT in mouse liver, the hypothesis has emerged that cells harboring the FGRAd genome express, in addition to the desired transgene product, some adenoviral proteins despite the absence of the regulatory E1A gene (Yang et al., 1994a). This could occur through basal expression of viral promoters, or their upregulation by endogenous cellular factors displaying E1A-like transactivating properties. Presentation of viral antigens at the surface of the transduced cells in the context of major histocompatibility molecules is likely to lead to the specific clearance of the infected cells, and thus to the loss of transgene expression. Supports for this hypothesis have been provided in a wide variety of experimental models (Ginsberg et al., 1991; Engelhardt et al., 1994a; Engelhardt et al., 1994b; Yang et al., 1994b; Yang et al., 1994c; Daï et al., 1995; Kay et al., 1995 and Yang et al., 1995a).

Improvement of Adenovectors: Doubly Deleted Vectors

How can we improve AMGT as far as transient expression and pathology are concerned?

One approach is to handle/manipulate the host immune system to abrogate its reactivity against the vectors. This could be achieved by performing a neonatal administration of the FGRAd with the view of tolerizing the host to viral antigens, or by subjecting immune-competent recipients to immunosuppression. Both approaches have received experimental support (Ragot et al., 1993; Engelhardt et al., 1994b; Da, et al., 1995; Kay et al., 1995; Vilquin et al., 1995 and Yang et al., 1995b) but are far from being devoid of significant hurdles.

We, and others, are following a complementary approach, aimed at abolishing viral gene expression from the vector backbone by introducing an additional functional block within the backbone of E1-deleted vectors. Given the crucial role of E2A and E4 regions in the regulation of viral gene expression, these regions have been mostly targeted with the view of designing vectors exhibiting more silent features in infected cells.

A lacZ-expressing E1-deleted recombinant displaying the ts125 mutation, a thermosensitive mutation within E2A which inactivates the DBP at 39°C, has thus been engineered. Mice injected with this so-called second generation recombinant adenovirus (SGRAd) expressed the reporter gene in the liver for a longer period of time than animals injected with a non-mutant FGRAd. However the duration of lacZ expression was still below than that observed in nude mice (Engelhardt et al., 1994a). Importantly, the inflammatory response was both delayed and reduced, but not completely abolished, as compared with animals administred with an E1-deleted recombinant. Although a definite improvement over the first-generation of vectors, this approach is not entirely satisfactory, in particular because the mutated enzyme is partially functional at 37°C (Engelhardt et al., 1994a), and because the ts125 mutation is a single base pair mutation with a high probability of reversion (Kruijer et al., 1983).

Another approach in creating new, more crippled, adenoviral vectors relies on the deletion of the E4 region in addition to that of E1 (Krougliak and Graham, 1995; Wang et al., 1995 and Yeh et al., 1996). As a prerequesite, we have engineered a 293-derived cell line (IGRP2) capable of simultaneously transcomplementing both the E1 and the E4 functions. In addition to the E1 region, IGRP2 contains a minimal E4 functional unit (Figure 30.2) under the transcriptional control of the dexa methasone-inducible murine mammary tumor virus long terminal repeat (MMTV LTR). Functional analysis demonstrated that IGRP2 can propagate all tested E4-defective viruses, including H5dl1004 and H2dl808 (ORF1+), H5dl1007 and H5dl1011 (disruption of all E4 ORFs), and H5dl1014 (ORF4+) (Weinberg and Ketner, 1986; Bridge and Ketner, 1989 and Bridge et al., 1993). This E1E4-transcomplementing cell line has proven useful for the helper-free construction, propagation and plaque-purification of a set of four E1E4 doubly deleted lacZ-expressing recombinants (Yeh et al., 1996). These viruses can be grown on IGRP2 to generate purified viral stocks with a titer in the range of that of FGRAds (as determined by their lacZ transduction efficiency on non-permissive cells). They are currently being tested in vitro for their expected reduced expression of late proteins, as well as in vivo in a variety of animal models to assess their behavior in terms of duration of transgene expression and the host immune response to them.

It is envisioned that the availability of E1E4 doubly-defective adenoviral vectors will have important implications in terms of efficacy and safety following gene transfer. Indeed, deletion of the E4 region increases the cloning capacity of the recombinants to more than 10 kb. This improvement will make possible, or facilitate the manipulation of very large transgenes. More cloning capacity will also certainly be welcomed for the introduction of regulatory sequences aimed at facilitating a strong and/ or long-lasting transgene expression, or at restricting its expression, e.g., at a given moment or in a given cell type or organ. Furthermore, propagation of E1E4-doubly-defective recombinants within IGRP2 cells should significantly reduce the emergence of RCA during viral amplification. Indeed, the E4 functional unit within the IGRP2 cell line has been designed in such a way that no double homologous recombination can occur between the vector backbones and the chromosomal E4 unit, an event which precludes the reintroduction of a functional E4 region back into the viral backbones, and *a fortiori* the emergence of any fully replicative $E1^+E4^+$ virus.

Acknowledgements

We thank Alice Gillardeaux, Marjolaine Lepeut, Irène Mahfouz and Carole Jullien for excellent technical assistance. Many thanks to Carole for editorial assitance. The authors wish to acknowledge Rhône-Poulenc and the French Ministry of Research and Industry for financial support (BioAvenir).

References

Armentano, Armentano, D., Sookdeo, C.C., Hehir, K.M., Gregory, R.J., St. George, J.A., Prince, G.A., Wadsworth, S.C. and Smith, A.E. (1995) Characterization of an adenovirus gene transfer vector containing an E4 deletion. *Hum. Gene. Ther.* **6**, 1343–1353

Ballay, A., Levrero, M., Buendia, M.A., Tiollais, P. and Perricaudet, M. (1985) *In vitro* and *in vivo* synthesis of the hepatitis B virus surface antigen and of the receptor for polymerized human serum albumin from recombinant human adenoviruses. *EMBO J.* **4**, 3861–3865

Bayley, S.T. and Mymryk, J.S. (1994) Adenovirus E1A proteins and transformation. *Int. J. Oncol.* **5**, 425–444

Bennett, J., Wilson, J., Sun, D., Forbes, B. and Maguire, A. (1994) Adenovirus vector-mediated *in vivo* gene transfer into adult murine retina. *Invest. Ophtalmol. Visual Sci.* **35**, 2535–2542

Berget, S.M., Moore, C. and Sharp, P.A. (1977) Sliced segments at the 5' terminus of adenovirus 2 late mRNA. *Proc. Natl. Acad. Sci. USA* **74**, 3171–3173

Bett, A., Prevec, L. and Graham, F.L. (1993) Packaging capacity and stability of human adenovirus type 5 vectors. *J. Virol.* **67**, 5911–5921

Bridge, E. and Ketner, G. (1989) Redundant control of adenovirus late gene expression by early region 4. *J. Virol.* **63**, 631–638

Bridge, E. and Ketner, G. (1990) Interaction of adenoviral E4 and E1b products in late gene expression. *Virology* **174**, 345–353

Bridge, E., Hemstrom, C. and Petterson, U. (1991) Differential regulation of adenovirus late transcriptional units by the products of early region 4. *Virology* **183**, 260–266

Bridge, E., Medghalchi, S., Ubol, S., Leesong, M. and Ketner, G. (1993) Adenovirus early region 4 and viral DNA synthesis. *Virology* **193**, 794–801

Challberg, M. and Kelly, T.J. (1979) Adenovirus DNA replication *in vitro*. *Proc. Natl. Acad. Sci. USA* **76**, 655–658

Chang, L.-S. and Shenk, T. (1990) The adenovirus DNA-binding protein stimulates the rate of transcription directed by adenovirus and adeno-associated virus promoters. *J. Virol.* **64**, 2103–2109

Chen, S.H., Shine, H.D., J.C., G., Grossman, R. G. and Woo, S. (1994) Gene therapy for brain tumors: regression of experimental gliomas by adenovirus-mediated gene transfer *in vivo*. *Proc. Natl. Acad. Sci. USA* **91**, 3054–3057

Cutt, J.R., Shenk, T. and Hearing, P. (1987) Analysis of adenovirus early region 4-encoded polypeptides synthesized in productively infected cells. *J. Virol.* **61**, 543–552

Daï, Y., Schwarz, H.M., Gu, D., Zhang, W.-W., Sarvetnick, N. and Verma, I.M. (1995) Cellular and humoral immune responses to adenoviral vectors containing factor IX gene: tolerization of factor IX and vector antigens allows for long-term expression. *Proc. Natl. Acad. Sci. USA* **92**, 1401–1405

Dix, I. and Leppard, K.N. (1993) Regulated splicing of adenovirus type 5 E4 transcripts and regulated cytoplasmic accumulation of E4 mRNA. *J. Virol.* **67**, 3226–3231

Dix, I. and Leppard, K.N. (1995) Expression of adenovirus type 5 E4 Orf2 protein during lytic infection. *J. Gen. Virol.* **76**, 1051–1055

Doerfler, W. (1968) The fate of the DNA of adenovirus type 12 in baby hamster kidney cells. *Proc. Natl. Acad. Sci. USA* **60**, 636–643

Engelhardt, J.F., Yang, Y., Stratford-Perricaudet, L.D., Allen, E.D., Kozarsky, K., Perricaudet, M., Yankaskas, J.R. and Wilson, J.M. (1993) Direct gene transfer of human CFTR into human bronchial epithelia of xenografts with E1-deleted adenoviruses. *Nature Genet.* **4**, 27–34

Engelhardt, J.F., Ye, X., Doranz, B. and Wilson, J.M. (1994a) Ablation of E2A in recombinant adenoviruses improves transgene persistence and decreases inflammatory response in mouse liver. *Proc. Natl. Acad. Sci. USA* **91**, 6196–6200

Engelhardt, J.F., Litzky, L. and Wilson, J.M. (1994b) Prolonged transgene expression in cotton rat lung with recombinant adenoviruses defective in E2a. *Hum. Gene. Ther.* **5**, 1217–1229

Ginsberg, H.S., Moldawer, L.L., Sehgal, P.B., Redington, M., Kilian, P.L., Chanock, R.M. and Prince, G.A. (1991) A mouse model for investigating the molecular pathogenesis of adenovirus pneumonia. *Proc. Natl. Acad. Sci. USA* **88**, 1651–1655

Graham, F.L., Smiley, J., Russel, W.C. and Nairn, R. (1977) Characteristics of a human cell line transformed by DNA from human adenovirus type 5. *J. Gen. Virol.* **36**, 59–72

Graham, F.L. and Prevec, L. (1991) Manipulation of adenovirus vectors. In: *Gene transfer and expression protocols*. E.J. Murray. Clifton (Ed.), The Humana Press Inc. pp. 109–128.

Hemstrom, C., Nordqvist, K., Pettersson, U. and Virtanen, A. (1988) Gene product of region E4 of adenovirus type 5 modulates accumulation of certain viral polypeptides. *J. Virol.* **62**, 3258–3264

Horellou, P., Vigne, E., Castel, M.-N., Barnoud, P., Colin, P., Perricaudet, M., Dealère, P. and Mallet, J. (1994) Direct intracerebral gene transfer of an adenoviral vector expressing tyrosine hydroxylase in a rat model of Parkinson's disease. *Neuroreport* **6**, 49–53

Horwitz, M.S. (1990) Adenoviridae and their replication. In: *Virology*, second edition. Fields, B.N., Knipe, D.M., et al. (Eds.), pp. 1679–1721

Kay, M.A., Holterman, A.X., Meuse, L., Gown, A., Ochs, H.D., Linsley, P.S. and Wilson, C.B. (1995) Long-term hepatic adenovirus-mediated gene expression in mice following CTLA4Ig administration. *Nature genet.* **2**, 191–197

Ketner, G., Spencer, F., Tugendreich, S., Connelly, C. and Hieter, P. (1994) Efficient manipulation of the human adenovirus genome as an infectious yeast artificial chromosome clone. *Proc. Natl. Acad. Sci. USA* **91**, 6186–6190

Kleinberger, T. and Shenk, T. (1993) Adenovirus E4 ORF4 protein binds to protein phosphatase 2A, and the complex down regulates E1A-enhanced *junB* transcription. *J. Virol.* **67**, 7556–7560

Krougliak, V. and Graham, F.L. (1995) Development of cell lines capable of complementing E1, E4, and protein IX defective adenovirus type 5 mutants. *Hum. Gene Ther.* **6**, 1575–1586

Kruijer, W., Nicolas, J.C., van Schaik, F.M. and Sussenbach, J.S. (1983) Structure and function of DNA-binding proteins from revertants of adenovirus type 5 with a temperature sensitive DNA replication. *Virology* **124**, 425–433

Le Gal La Salle, G., Robert, J.J., Berrard, S., Ridoux, V., Stratford-Perricaudet, L.D., Perricaudet, M. and Mallet, J. (1993) An adenovirus vector for gene transfer into neurons and glia in the brain. *Science* **259**, 986–988

Lochmüller, H., Jani, A., Huard, J., Prescott, S., Simoneau, M., Massie, B., Karpati, G. and Acsadi, G. (1994) Emergence of early region 1-containing replication-competent adenovirus in stocks of replication-defective adenovirus recombinants (ΔE1 + ΔE3) during multiple passages in 293 cells. *Hum. Gene Ther.* **5**, 1485–1491

Nordqvist, K. and Akusjarvi, G. (1990) Adenovirus early region 4 stimulates mRNA accumulation via 5' introns. *Proc. Natl. Acad. Sci. USA* **87**, 9543–9547

Nordqvist, K., Ohman, K. and Akusjarvi, G. (1994) Human adenovirus encodes two proteins which have opposite effects on accumulation of alternatively spliced mRNAs. *Mol. Cell. Biol.* **14**, 437–445

Obert, S., O'Connor, R.J., Schmid, S. and Hearing, P. (1994) The adenovirus E4–6/7 protein transactivates the E2 promoter by inducing dimerization of a heteromeric E2F complex. *Mol. Cell. Biol.* **14**, 1333–1346

Ohman, K., Nordqvist, K. and Akusjarvi, G. (1993) Two adenovirus proteins with redundant activities in virus growth facilitates tripartite leader mRNA accumulation. *Virology* **194**, 50–58

Ohno, T., Gordon, D., San, H., Pompoli, V.J., Imperiale, M.J., Nabel, G.J. and Nabel, E.G. (1994) Gene therapy for vascular smooth muscle cell proliferation after arterial injury. *Science* **265**, 781–784

Ragot, T., Vincent, N., Chafey, P., Vigne, E., Gilgenkrantz, H., Couton, D., Cartaud, J., Briand, P., Kaplan, J.-C., Perricaudet, M. and Kahn, A. (1993) Efficient adenovirus mediated transfer of a human minidystrophin gene to skleletal muscle of mdx mice. *Nature* **361**, 647–650.

Rosenfeld, M.A., Yoshimura, K., Trapnell, B.C., Yoneyama, K., Rosenthal, E.R., Dalemans, W., Fukayama, M., Bargon, J., Stier, L.E., Startford-Perricaudet, L.D., Perricaudet, M., Guggino, W.B., Pavirani, A., Lecocq, J.P. and Crystal, R.G. (1992) *In vivo* transfer of the human cystic fibrosis transmembrane conductance regulator gene to the airway epithelium. *Cell* **68**, 143–155

Sarnow, P., Hearing, P., Anderson, C.W., Halbert, D.N., Shenk, T. and Levine, A.J. (1984) Adenovirus early region 1B 58,000-dalton tumor antigen is physically associated with an early region 4 25,000-dalton protein in productively infected cells. *J. Virol.* **49**, 692–700

Solnick, D. (1981) Construction of an adenovirus-SV40 recombinant producing SV40 T antigen from an adenoviral late promotor. *Cell* **24**, 135–143

Soloway, P.D. and Shenk, T. (1990) The adenovirus type 5 i-leader open reading frame functions *in cis* to reduce the half-life of L1 mRNAs. *J. Virol.* **64**, 551–558

Stewart, P.L., Burnett, R.M., Cyrklaff, M. and Fuller, S.D. (1991) Image reconstruction reveals the complex molecular organization of adenovirus. *Cell* **67**, 145–154

Stratford-Perricaudet, L.D., Levrero, M., Chasse, J.F., Perricaudet, M. and Briand, P. (1990) Evaluation of the transfer and expression in mice of an enzyme-encoding gene using a human adenovirus vector. *Hum. Gene Ther.* **1**, 241–256

Vilquin, J-T., Guérette, B., Kinoshita, I., Roy, B., Goulet, M., Gravel, C., Roy, R. and Tremblay, J.P. (1995) FK506 immunosuppression to control the immune response triggered by first-generation adenovirus-mediated gene transfer. *Hum. Gene Ther.* **6**, 1391–1401

Virtanen, A., Gilardi, P., Naslund, A., LeMoullec, J.M., Pettersson, U. and Perricaudet, M. (1984) mRNAs from human adenovirus early region 4. *J. Virol.* **51**, 822–831

Wang, Q. and Taylor, M.W. (1993) Correction of a deletion mutant by gene targeting with an adenovirus vector. *Mol. Cell. Biol.* **13**, 918–927

Wang, Q., Jia, X-C. and Finer, M.H. (1995) A packaging cell line for propagation of recombinant adenovirus vectors containing two lethal gene-region deletions. *Gene Ther.* **2**, 775–783

Weinberg, D.H. and Ketner, G. (1986) Adenoviral early region 4 is required for efficient viral DNA replication and for late gene expression. *J. Virol.* **57**, 833–838

White, E. (1993) Regulation of apoptosis by the transforming genes of the DNA tumor virus adenovirus. *Proc. Soc. Exp. Biol. Med.* **204**, 30–39

Whyte, P., Buchkovich, K., Horowitz, J.M., Friend, S.H., Raybuck, M., Weinbert, R.A. and Harlow, E. (1988) Association between an oncogene and an anti-oncogene: the adenovirus E1A proteins bind to the retinoblastoma gene product. *Nature* **334**, 124–129

Wold, W.S.M. and Gooding, L.R. (1991) Region E3 of adenovirus: a cassette of genes involved in host immunosurveillance and virus-cell interactions. *Virology* **184**, 1–8

Wood, M.J.A., Byrnes, A.P., Pfaff, D.W., Rabkin, S.D. and Charlton, H.M. (1994) Inflammatory effects of gene transfer into the CNS with defective HSV-1 vectors. *Gene Ther.* **1**, 283–291

Yang, Y., Nunes, F.A., Berencsi, K., Furth, E.E., Gonczol, E. and Wilson, J.M. (1994a) Cellular immunity to viral antigens limits E1-deleted adenoviruses for gene therapy. *Proc. Natl. Acad. Sci. U.S.A* **91**, 4407–4411

Yang, Y., Nunes, F.A., Berencsi, K., Gonczol, E., Engelhardt, J.F. and Wilson, J.M. (1994b) Inactivation of E2a in recombinant adenoviruses improves the prospect for gene therapy in cystic fibrosis. *Nature Genet* **7**, 362–369

Yang, Y., Ertl, H.C. and Wilson, J.M. (1994c) MHC class I-restricted cytotoxic T lymphocytes to viral antigens destroy hepatocytes in mice infected with E1-deleted recombinant adenoviruses. *Immunity* **1**, 433–442

Yang, Y., Li, Q., Ertl, H.C.J. and Wilson, J.M. (1995a) Cellular and humoral immune responses to viral antigens create barriers to lung-directed gene therapy with recombinant adenoviruses. *J. Virol.* **69**, 2004–2015

Yang, Y., Trinchieri, G. and Wilson, J.M. (1995b) Recombinant IL-12 prevents formation of blocking IgA antibodies to recombinant adenovirus and allows repeated gene therapy to mouse lung. *Nature Genet.* **1**, 890–893

Yeh, P., Dedieu, J.F., Orsini, C., Vigne, E., Denfle, P. and Perricaudet, M. (1996) Efficient dual transcomplementation of adenovirus E1 and E4 regions from a 293-derived cell line expressing a minimal E4 functional unit. *J. Virol.* **70**, 559–565

Zhang, X., Bellet, A.J.D., Tha Hla, R., Voss, T., Mullbacher, A. and Braithwaite, A.W. (1994) Down-regulation of human adenovirus E1a by E3 gene products: evidence for translational control of E1a by E3 14.5 K and/or E3 10.4 K products. *J. Gen. Virol.* **75**, 1943–1951

PART II, SECTION G

31. The Development of Adeno-Associated Virus as a Vector for *In Vivo* Gene Therapy

R. Jude Samulski

Gene Therapy Centre and Department of Pharmacology, University of North Carolina at Chapel Hill, 7119 Thurston-Bowles Building, CB 7352, Chapel Hill NC 27599-7352, USA

Significant developments in recombinant DNA technology and the increase in our understanding of the molecular basis of infectious agents such as viruses, have given rise to new opportunities for the development of novel delivery system for the correction of genetic diseases. The ability to identify, isolate, and sequence functional genes together with the recent development of these high efficient transducing viral vectors has made the treatment of human disease by gene transfer technologies a realistic possibility.

Several gene therapy strategies dealing with various viral vectors are now available. This report will focus on the use of the non-pathogenic human parvovirus Adeno-associated virus (AAV).

Adeno-Associated Virus

The development of viral vector using the human parvovirus, adeno-associated virus (AAV), has provided a novel delivery system for genes transfer into mammalian cells. This non-pathogenic virus possesses several unique properties which distinguish it from other viral vectors. Its advantages include stable integration of viral DNA into the host genome (Berns *et al.*, 1975; Cheung *et al.*, 1980; Hoggan *et al.*, 1972; Laughlin *et al.*, 1986 and McLaughlin *et al.*, 1988), lack of any associated human disease (Berns *et al.*, 1982), broad host range (Buller *et al.*, 1979 and Casto *et al.*, 1967), the ability to infect post miotic cells (Wong *et al.*, 1993), and the capacity to carry non-viral regulatory sequences without interference from the viral genome (Miller *et al.*, 1993a and Walsh *et al.*, 1992). In addition, superinfection immunity typically associated with other latent virus is not associated with AAV vectors (Lebkowski *et al.*, 1988 and McLaughlin, *et al.*, 1988).

Adeno-associated virus is a defective member of the parvovirus family. AAV can be propagated as a lytic virus or maintained as a parvovirus integrated into the host cell genome (Atchison *et al.*, 1965, Hoggan *et al.*, 1966 and Hoggan *et al.*, 1972). In a lytic infection, replication requires coinfection with either adenovirus (Atchison *et al.*, 1965; Hoggan *et al.*, 1966 and Melnick *et al.*, 1965), or herpes simplex virus (Buller *et al.*, 1981 and McPherson *et al.*, 1985); hence the classification of AAV as a "defective" virus. Vaccinia virus and maybe still other viruses can also provide at least partial helper function (Schlehofer *et al.*, 1986). The requirement of a *helper* virus for a productive infection has made understanding the wild type AAV life cycle difficult. However, from the point of view of a vector, its dependence has added a level of safety when generating non-replicative virions. In practice, this means that virus can be propagated under controlled conditions (only in the presence of Ad) thereby reducing unwanted vector spread, and providing a critical margin of safety when utilizing as a vector. One of the most interesting aspects of the AAV life cycle is the virus' ability to integrate into the host genome in the absence of helper virus. When AAV infects tissue culture cells in the absence of helper virus, it establishes latency by persisting in the host cell genome as an integrated provirus (Berns *et al.*, 1975; Cheung *et al.*, 1980; Handa *et al.*, 1977 and Hoggan *et al.*, 1972). Although AAV establishes a latent infection, it can be rescued from the chromosome and re-enter the lytic cycle if these cells are super infected with wild type helper virus. Analysis of various Ad mutants as helpers has determined that the lytic phase, either upon primary infection or upon rescue, requires the expression of the adenovirus early gene products (Richardson et al., 1981) E1a (Chang, 1989 and Richardson et al., 1984), E1b (Richardson et al., 1984 and Samulski et al., 1988), E2a (Jay et al., 1979), E4 (Carter et al., 1983; Laughlin et al., 1982; Richardson et al., 1981 and Richardson et al., 1984), and VA RNA (Janik et al., 1989 and West et al., 1987).

AAV Structure and Genetics

The AAV genome is encapsidated as a single-stranded DNA molecule of plus or minus polarity. Strands of both polarities are packaged, but it separate virus particles (Berns and Adler, 1972; Berns and Rose, 1970; Mayor et al., 1969 and Rose et al., 1969) and both strands are infectious (Samulski et al., 1987). Five serotypes of AAV have been identified, but the most extensively characterized in AAV-2. The non-enveloped virion is icosohedral in shape and one of the smallest that has been described, about 20–24 nm in diameter (Hoggan, 1970 and Tsao et al., 1991) with a density of $1.41\,g/cm^3$ (de la Maza and Carter, 1980a and de la Maza and Carter, 1980b). The relatively high density of AAV particles allows them to be easily separated by CsCl density centrifugation from adenovirus helper virus that has a density of approximately $1.35\,g/cm^3$ (de la Maza et al., 1980b). In addition, the AAV virion is resistant to a number of physical treatments that inactivate other viruses, such as heat treatment (56°C for 1 h), low pH, detergents and proteases (Bachmann et al., 1979) thereby further permitting the virions to be purified or concentrated.

The complete nucleotide sequence of AAV-2 consists of 4680 nucleotides (Srivastava et al., 1983). This single-stranded, linear DNA and contains two terminal inverted repeats that are 145 bp long (Gerry et al., 1973; Koczot et al., 1973 and Lusby et al., 1980). The first 125 bp of each repeat can form a T shaped hairpin structure which is composed of two small internal palindromes (B and C) flanked by a larger palindrome (A). These are the only cis sequences required for AAV replication. The AAV virion is composed of three structural proteins, VP1, VP2, and VP3 (Johnson et al., 1971, Johnson et al., 1977, Johnson et al., 1975 and Rose et al., 1971) There are at least four non-structural AAV proteins collectively termed the Rep proteins. These are named according to their molecular weights, Rep 78, Rep 68, Rep 52 and Rep 40 (Mendelson et al., 1986, Srivastava et al., 1983 and Trempe et al., 1987). In addition, two smaller Rep proteins have also been observed in AAV infected cells (Trempe et al., 1987), presumably resulting from translational initiation at internal ATG codons within the mRNA that produces the Rep 52 and Rep 40 proteins.

Production of Recombinant AAV

The present method for producing stocks of recombinant AAV utilizes a two component plasmid system divided in terms of the *cis* and *trans* functions necessary for replication, expression, and encapsidation of the recombinant virus. As mentioned above, the viral terminal repeats are the only elements required in *cis* and in the current packaging system flank the transgene on one of two plasmids. The second plasmid supplies the necessary *Rep* and *Cap* gene products in *trans*. An important consideration is that the two plasmid DNAs must be sufficiently non-homologous so as to preclude homologous recombination events which could generate wild-type AAV (Samulski et al., 1989). Although there are many variations on this theme, a widely used method for recombinant AAV production involves co-transfection of the human cell line 293 with a plasmid containing the internal region of the AAV genome, with adenovirus type 5 terminal repeats in place of the normal AAV terminal, and a second plasmid, the vector consisting of a heterologous gene (containing appropriate regulatory elements) flanked by the AAV inverted terminal repeats (Samulski et al., 1989). When these transfected cells are infected with adenovirus, rescue, replication and packaging of the foreign gene linked to the AAV terminal repeats into AAV particles occurs. The adenovirus genome activates the helper plasmid which enhances expression of the AAV genes. The *Rep* gene products recognize the AAV *cis*-acting terminal repeats on the recombinant vector containing the foreign gene, rescue the recombinant segments out of the plasmid, and begin to replicate them. The AAV capsids begin to accumulate, recognize the AAV *cis*-acting packaging signals located in the AAV terminal repeats and encapsidate the recombinant viral DNA into an AAV virion. The result of such a packaging scheme is an adenovirus helper virus and AAV particles carrying the recombinant heterologous genes. Adenovirus can then be removed by any of a number of physical separation strategies. In this packaging system one can generate helper-free stocks of recombinant AAV at titers of 10^5 to 10^6 transducting virus per millilitre. Although the number of genes built into AAV vectors and tested *in vitro* remains relatively small, there are few limitations to this step of the procedure. Recombinant genomes must be between 50% and 110% of wild type AAV size to be efficiently packaged into AAV particles (R.J. Samukski, unpublished and de la Maza et al., 1980b). This means that most recombinant AAV vectors can accommodate an insertion of up to about 4.5 kb in length.

Transduction of Non-Dividing Cells

Little is known about the relationship between viral infection and the cell cycle. In fact the mode of viral uptake has not been established and no cellular receptor has been identified to date. In tissue culture AAV has been shown to infect all established human cell lines so far examined (Laughlin et al., 1986, Lebkowski et al., 1988, McLaughlin et al., 1988, Samulski et al., 1989 and Tratschin et al., 1985). AAV also shows a similarly broad cellular host range in vivo, although this has not been characterized as extensively (R.J. Samulski et al., unpublished observations). However, the ability of AAV to infect truly terminally differentiated or nondividing cells has not been resolved satisfactorily. Retroviral vectors based on the murine leukemia virus require cell division for efficient transduction (Miller et al., 1990). For this reason there has been intense interest in the ability of recombinant AAV vectors to transduce nondividing cells. Recently, several reports have begun to shed light on this possibility. Wong et al., 1993 reported that recombinant AAV was able to transduce growth arrested human fibroblasts or 293 cells at the same, or better, efficiency as actively proliferating cells. However, since the cells were allowed to resume growing prior to being scored for transduction, there remains the possibility that the virus simply remained episomal and integrated after the cells entered S phase. A second report seems to confirm this possibility. Wherein, it was determined that the vector genomes can persist in stationary phase cells, but transduction preferentially occurs in cells after they have entered S phase (Russell et al., 1994). In this case transduction was assayed without subsequent stimulation and cell division. Although proliferating cells may be preferentially transduced by AAV vectors, proliferation may not be absolutely required for transduction. In fact, in vivo experiments have shown that recombinant AAV vectors are able to transduce cell populations that are thought to be largely quiescent at remarkably high efficiencies. These include human bone marrow progenitors (Goodman et al., 1994, Miller et al., 1994 and Zhou et al., 1993) in culture, and rat brain in vivo (Kaplitt et al., 1994) (McCown et al., in press). Although, the human bone marrow progenitor (CD34$^+$) cells were maintained in media containing growth factors, interleukin (IL)-3. IL-6, and stem-cell factor (SCF), analysis of colonies derived from these progenitors should allow the determination of the integration frequency in the primary hematopoeitic cells that gave rise to each colony. Histological analysis of rat brains injected with recombinant virus encoding β-galactosidase has demonstrated expression in many regions of the rat brain, including striatum, hippocampus and substantia nigra (McCown et al., in press, and During et al., unpublished). In addition to demonstrating the potential for gene therapy approaches into neuronal cells using AAV, this finding represents the most conclusive demonstration that recombinant AAV vectors can transduce non-dividing cells. It should be noted that AAV transduction measured by gene expression, as described by the majority of the above experiments, does not reflect viral integration, since gene expression can take place from episomal or integrated templates.

Recombinant AAV Experiments In Vitro

Analysis of in vitro studies presented to date suggest that several different kinds of transcriptional elements will be active in AAV vectors (see Table 31.1). These range from heterologous high-level viral promoter and enhancer elements (Chatterjee et al., 1992, Hermonant and Muzyczka, 1984, Lebkwski et al., 1988 and Muro-Cacho et al., 1990), various cellular enhancers linked to viral promoters (Ponnazhagan et al., 1993 and Zhou et al., 1993), tissue specific cellular control regions (Miller et al., 1993a, Miller et al., 1993b, Walsh et al., 1992 and Zhou et al., 1993), and inducible cellular control elements (Walsh et al., 1992). In addition to the traditional pol II promoters, polymerase III control elements and several snRNA, pol III and pol II transcriptional elements have been used in conjuction with recombinant AAV vectors (Rossi et al., 1994) (J.S. Bartlett et al., unpublished). The important conclusion from these studies was that in each case, these elements have been shown to function correctly and independent of their location within an AAV vector. Such preliminary results in tissue culture are encouraging, and await parallel studies in vivo.

Primary hematopoietic progenitor cells are a relevant cell type that has been used to determine the potential of recombinant AAV to transduce and express genes (Goodman et al., 1994, LaFace et al., 1988, Miller et al., 1994 and Zhou et al., 1993). Highly purified progenitor cell subject to rAAV infection demonstrated gene transfer and expression of the β-galactosidase reporter gene (Goodman et al., 1994). Successful transfer of the human FAC (Fanconi anemia C complementing) gene and a mutationally marked human γ-globin gene have been demonstrated after infecting human hematopoietic progenitor cells with AAV vectors (Miller et al., 1994 and Walsh et al., 1994). Molecular analysis demonstrated that AAV-transduced genes were present and

Table 31.1. Summary of gene transfer mediated by recombinant AAV vectors.

Gene	Promoter/Enhancer	Therapeutic Application	Cell Type/Tissue	Investigator
neo	SV40 early		Detroit 6, KB, L tk⁻	Hermonat et al. (1984)
CAT	MSV LTR		293, NC37, K562	Lebkowski et al. (1988)
neo			HL60, KG1a, U937	
neo	SV40 early		Murine progenitors	LaFace et al. (1988)
neo	SV40 early		HeLa, Detroit 6, mouse 3T3	Samulski et al. (1989)
neo	CMV 1E		293	Vincent et al. (1990)
neo	SV40 early		JY,V1, T cell	Muro-Cacho et al. (1992)
Anti-sense HIV	RWV LTR	HIV infection	293	Catterjee et al. (1992)
β-Galactosidase	CMV-IE		K562, Detroit 6, human CD34+	Goodman et al. (1992)
γ-Globin, neo	β-Globin LCRII	Hemoglobin disorders	K562	Walsh et al. (1992)
γ-Globin, neo	β-Globin LCRII	Hemolobin disorders	K562	Miller et al. (1993)
			Rhesus CD34⁺	Miller et al. (1993b)
			Human CD34⁺	Miller et al. (1994)
CAT	AAV ITR		293	Flotte et al. (1993)
CFTR	AAV p5	Cystic Fibrosis	IB3	Flotte et al. (1993b)
	AAV ITR		Nasal epithelium of rabbit	Flotte et al. (1993b)
α-Globin antisense/neo	a-Globin/SV40	β-Thallassemia	K562	Ponnazhagan et al. (1993)
GMCSF	SV40	Cancer	COS,MO7e	Luo et al. (1993)
neo	HSV-TK			
FACC	RSV-LTR	Fanconi anemia	EBV transformed human lymphoblasts Human CD34⁻	Walsh et al. (1994)
Anti-HIV ribozymes	Human tRNA^Met Human U6 Human U1 RSV LTR	HIV infection		Rossi et al. (1993)
neo	HSV-TK B19 HS-1/HSV-TK HS-2/B19		Human cord blood Mononuclear cells and Murine Progenitors	Zhou et al. (1993)
β-Globin	HS-2/β-globin	Hemoglobin disorders	K562, KB	Zhou et al. (1993)
β-Galactosidase	Murine U1 CMV-IF		CFT-1, A549, HeLa, 293	Bartlett et al. (1994)
Glucocerebrosidase	SV40 early	Lysomal storage disordeers	Murine and Patient fibroblasts	Wei et al. (1994)
Arylsulfatase A				
Tyrosine hydroxylzse	CMV-IE	Parkinson's disease	RAT brain	Kaplitt et al. (1994)
β-Galactosidase	DMV-IE		Detroit 6, K562, human CD34⁺, rhesus CD34⁻	Goodman et al. (1994)

mRNA was expressed at high levels in a majority of colonies. Biological analysis confirmed correction of the inherent cellular defects, supporting the potential of recombinant AAV to transduce human hematopoietic progenitors in a clinically relevant situation. Although these data support transduction of hematopoietic progenitors the efficiency of stable integration of the vector genome has not been determined. However, wild-type AAV infection of these progenitor cells has documented integration into the wtAAV DNA integration site on chromosome 19 suggesting that all steps required for AAV latency are functional in these cells (Goodman et al., 1994).

Primary cells transduced with recombinant AAV have included human liver hepatocytes, human fibroblasts, and explanted human glial cells (R.J. Samulski et al., unpublished observation) (Wei et al., 1994). So far the versatility of AAV gene transfer in tissue culture has met with paralleled success in primary cells.

AAV *In Vivo*

Detailed *in vivo* studies have been hampered by the technical difficulties in producing the quantity of recombinant virus needed for these experiments. In light of this limitation, initial efforts testing rAAV for transduction of rat colon and brain contained the β-galactosidase marker gene (Xiao X., Thesis, University of Pittsburgh, 1991). Gene expression was documented in both cases by intense blue staining of the transduced tissue. Other studies have demonstrated that a portion of the human cystic fibrosis transmembrane regulator (CFTR) gene can be delivered intrabronchially to rabbits (Flotte *et al.*, 1993). Transcripts from the introduced gene, and CFTR protein has been detected up to six months post-transfer. Although it is unclear whether or not sufficient protein is being produced to be of therapeutic value, this is an important experiment because it presented the first reported application of a potentially therapeutic recombinant AAV vector in an animal. Recent animal studies also include the introduction of recombinant AAV containing the human tyrosine hydroxylase (TH) gene into rat brain striatum (Kaplitt *et al.*, 1994). Evidence of TH protein has been demonstrated for up to two months. This vector-therapeutic gene combination may have potential for the genetic therapy of Parkinson's Disease and other neuronal disorders. It is important to note that no evidence for toxicity in the animals treated with these vectors was observed. This experiment extends early studies of AAV gene delivery in neuronal cells (Xiao and Samulski unpublished observation) and demonstrates for the first time, long term expression of a therapeutic gene.

Since recombinant AAV vectors are relatively new delivery systems, many questions remain to be answered regarding the biology of AAV and its applicability to gene therapy. Along with detailed *in vitro* studies, further characterization *in vivo* will be needed to ascertain the true value of this vector. A number of primary studies have been reported, and several more are underway. Many of the various experiments performed to date and a list of the genes transferred and cell type transduced have been compiled in Table 31.1. Some important questions that remain to be addressed by future studies include: (1) the efficiency of vector transduction *in vivo* (2) correct expression from these control elements after vector tansduction *in vivo*, and (3) the persistence of transgene expression *in vivo*.

References

Atchison, R.W., Casto, B.C., and Hammond, W.M. (1965) Adenovirus-associated defective virus particles. *Science* **149**, 754–756

Bachmann, P.A., Hoggan, M.D., Kurstak, E., Melnick, J.L., Pereira, H.G., Tattersall, P. and Vago, C. (1979) Parvoviridae: Second report. *Intervirology* **11**, 248–254

Berns, K.I. and Adler, S. (1972) Separation of two types of adeno-associated virus particles containing complementary polynucleotide chains. *J. Virol.* **5**, 693–699

Berns, K.I. Cheung, A., Ostrove, J., and Lewis, M. (1982) Adeno-associated Virus Latent Infection. In B. W. J. Mahy, A.C. Minson and G.K. Darby (eds.), *Virus Persistance* (pp. 249). Cambridge: Cambridge University Press.

Berns, K.I. Pinkerton, T.C., Thomas, G.F. and Hoggan, M.D. (1975) Detection of adeno-associated virus (AAV)-specific nucleotide sequences in DNA isolated form latently infected Detroit 6 cells. *Virology* **68**, 556–560

Berns, K.I. and Rose, J.A. (1970) Evidence for a single-stranded adeno-associated virus genome: isolation and separation of complementary single strands. *J. Virol.* **5**, 693–699

Buller, R.M., Janik, J.E., Sebring, E.D. and Rose, J.A. (1981) Herpes simplex virus types 1 and 2 completely help adenovirus-associated virus replication. *J. Virol.* **40**, 241–247

Buller, R.M., Straus, S.E. and Rose, I.A. (1979) Mechanism of host restriction of adenovirus-associated virus replication in african green monkey kidney cells. *J. Gen. Virol.* **43**, 663–672

Carter, B.J., Marcus-Sekura, C.J., Laughlin, C.A. and Ketner, G. (1983) Properties of an adenovirus type 2 mutant, Add1807, having a deletion near the right-hand genome terminus: failure to help AAV replication, *Virology* **126**, 505–516

Casto, B.C., Armstrong, J.A., Atchison, R.W. and Hammon, W.M. (1967) Studies on the relationship between adeno-associated virus type I (AAV-I) and adenoviruses. II. Inhibition of adenovirus plaques by AAV, its nature and specificity. *Virology* **33**, 452–458

Chang, L.-S., Shi, Y. and Shenk, T. (1989) Adeno-associated virus p5 promoter contains an adenovirus EIA inducible element and a binding site for the major late transcription factor. *J. Virol.* **63**, 3479–3488

Chatterjee, S., Johnson, P.R., and Wong, K.J. (1992) Dual-target inhibition of HIV-1 *in vitro* by means of an adeno-associated virus antisense vector. *Science* **258**, 1485–1488

Cheung, A.K., Hoggan, M.D., Hauswirth, W.W. and Berns, K.I. (1980) Integration of the adeno-associated virus genome into cellular DNA in latently infected human Detroit 6 cells. *J. Virol.* **33**, 739–748

de la Maza, L.M. and Carter, B.J. (1980a) Heavy and light particles of adeno-associated virus. *J. Virol.* **33**, 1129–1137

de la Maza, L.M. and Carter, B.J. (1980b) Molecular structure of adeno-associated virus variant DNA. *J. Biol. Chem.* **255**, 3194–3203

Flotte, T.R., Afione, S.A., Conrad, C., McGrath, S.A., Solow, R., Oka, H., Zeitlin, P.L., Guggino, W.B. and Carter, B.J. (1993) Stable *in vivo* expression of cystic fibrosis transmembrane regulator with an adeno-associated virus vector. *Proc. Natl. Acad. Sci. USA* **90**, 10613–10617

Gerry, H.W., Kelly, T.J.J. and Berns, K.I. (1973) Arrangement of nucleotide sequences in adeno-associated virus DNA. *J. Mol. Biol.* **79**, 207–225

Goodman, S., Xiao, X., Donahue, R.E., Moulton, A., Miller, J., Walsh, C., Young, N.S., Samulski, R.J. and Nienhuis, A.W. (1994) Recombinant adeno-associated virus-mediated gene transfer into hematopoietic progenitor cells. *Blood* **84**(5), 1492–1500

Handa, H., Shiroki, K. and Shimojo, H. (1977) Establishment and characterization of KB cell lines latently infected with adeno-associated virus type 1. *Virology* **82**, 84–92

Hermonat, P.L. and Muzyczka, N. (1984) Use of adeno-associated virus as a mammalian DNA cloning vector: transduction of neomycin resistance into mammalian tissue culture cells. *Proc. Natl. Acad. Sci. USA* **81**, 6466–6470

Hoggan, M.D. (1970) Adeno-associated viruses. *Prog. Med. Virol.* **12**, 211–239

Hoggan, M.D., Blacklow, N.R. and Rowe, W.P. (1966) Studies of small DNA viruses found in various adenovirus preparations: physical, biological, and immunological characteristics. *Proc. Natl. Acad. Sci. USA* **55**, 1457–1471

Hoggan, M.D., Thomas, G.F., Thomas, F.B. and Johnson, F.B. (1972) Continuous carriage of adenovirusassociated virus genome in cell culture in the absence of helper adenovirus. In *Proceedings of the Fourth Lepetite Colloquium* (pp. 243–249). Cocoya, Mexico:

Janik, J.E., Huston, M.M., Cho, K. and Rose, J.A. (1989) Efficient synthesis of adeno-associated virus structural proteins requires both adenovirus DNA binding protein and VA I RNA. *Virology* **168**, 320–329

Jay, F.T., De La Maza, L.M. and Carter, B.J. (1979) Parvovirus RNA transcripts containing sequences not present in mature mRNA: A method for isolation of putative mRNA precursor sequences. *Proc. Natl. Acad. Sci. USA* **76**, 625–629

Johnson, F.B., Ozer, H.L., and Hoggan, M.D. (1971) Structural proteins of adenovirus-associated virus type 3. *J. Virol.* **8**, 860–863

Johnson, F.B., Thomson, T.A., Taylor, P.A. and Vlazny, D.A. (1977) Molecular similarities among the adenovirus-associated virus polypeptides and evidence for a precursor protein. *Virology* **82**, 1–13

Johnson, F.B., Whitaker, C.W. and Hoggan, M.D. (1975) Structural polypeptides of adenovirus-associated virus top component. *Virology* **65**, 196–203

Kaplitt, M.G., Leone, P., Samulski, R.J., Xiao, X., Pfaff, D.W., O'Malley, K.L. and During, M.J. (1994) Adeno-associated virus vectors yield safe delivery and long term expression of potentially therapeutic genes into the mammalian brain. *Nature Genet.* **7** (October)

Koczot, F.J., Carter, B.J., Garon, C.F. and Rose, J.A. (1973) Self-complementarity of terminal sequences within plus or minus strands of adenovirus-associated virus DNA. *Proc. Natl. Acad. Sci. USA* **70**, 215–219

LaFace, D., Hermonat, P., Wakeland, E. and Peck, A. (1988) Gene transfer into hematopoietic progenitor cells mediated by an adeno-associated virus vector. *Virology* **162**(2), 483–486

Laughlin, C.A., Cardellichio, C.B. and Coon, H.C. (1986) Latent infection of KB cells with adeno-associated virus type 2. *J. Virol.* **60**, 515–524

Laughlin, C.A., Jones, N. and Carter, B.J. (1982) Effects of deletions in adenovirus region I genes upon replication of adeno-associated virus. *J. Virol.* **41**, 868–876

Lebkowski, J.S., McNally, M.M., Okarma, T.B. and Lerch, L.B. (1988) Adeno-associated virus: a vector system for efficient introduction of DNA into a variety of mammalian cell types. *Mol. Cell. Biol.* **8**, 3988–3996

Lusby, E., Fife, K.H. and Berns, K.I. (1980) Nucleotide sequences of the inverted terminal repetition in adeno-associated virus DNA. *J. Virol.* **34**, 402–409

Mayor, H.D., Torikai, K., Melnick, J. and Mandel, M. (1969) Plus and minus single-stranded DNA separately encapsidated in adeno-associated satellite virions. *Science* **166**, 1280–1282

McLaughlin, S.K., Collis, P., Hermonat, P.L. and Muzyczka, N. (1988) Adeno-associated virus general transduction vectors: analysis of proviral structures. *J. Virol.* **62**, 1963–1973

McPherson, R.A., Rosenthal, L.J. and Rose, J.A. (1985) Human cytomegalovirus completely helps adeno-associated virus replication. *Virology* **147**, 217–222

Melnick, J.L., Mayor, H.D., Smith, K.O. and Rapp, F. (1965) Association of 20 millimicron particles with adenoviruses. *J. Bacteriol.* **90**, 271–274

Mendelson, E., Trempe, J.P. and Carter, B.J. (1986) Identification of the trans-active rep proteins of adeno-associated virus by antibodies to a synthetic oligopeptide. *J. Virol.* **60**, 823–832

Milller, D.G., Adam, M.A. and Miller, A.D. (1990) *Mol. Cell. Biol.* **10**, 4239–4242

Miller, J.L., Donahue, R.E., Sellers, S.E., Samulski, R.J., Young, N.S. and Nienhuis, A.W. (1994) Recombinant adeno-associated virus (rAAV) mediated expression of a human g-globin gene in human progenitor derived erythroid cells. *Proc. Natl. Acad. Sci. USA*, in press.

Miller, J.L., Walsh, C.E., Ney, P.A., Samulski, R.J. and Nienhuis, A.W. (1993a) Single-copy transduction and expression of human gamma-globin in K562 erythroleukemia cells using recombinant adeno-associated virus vectors: The effect of mutations in NF-E2 and GATA-1 binding motifs within the hypersensitivity site 2 enhancer. *Blood* **82**, 1900–1906

Miller, J.L., Walsh, C.E., Samulski, R.J., Young, N.S. and Nienhuis, A.W. (1993b) Transfer and expression of the human g-globin gene in purified hematopoeitic progenitor cell from Rhesus bone marrow using a recombinant adeno-associated viral (rAAV) vector (abstract). In *Fifth Parvovirus Workshop*, Crystal River, FL.

Muro-Cacho, C., Samulski, R.J. and Kaplan, D. (1992) Gene transfer in human lymphocytes using a vector based on adeno-associated virus. *J. Immunother.* **11**, 231–237

Ponnazhagan, S., Nallari, M.L. and Srivastava, A. (1993) Suppression of human alpha-globin gene expression mediated by the recombinant adeno-associated virus 2-based antisense vectors. *J. Exp. Med.* **179**(2), 733–738

Richardson, W.D. and Westphal, W.D. (1981) A cascade of adenovirus early functions is required for expression of adeno-associated virus. *Cell* **27**, 133–141

Richardson, W.D. and Westphal, W.D. (1984) Requirement for eigher early region la or early rigion lb adenovirus gene products in the helper effect for adeno-associated virus. *J. Virol.* **51**, 404–410

Rose, J.A., Berns, K.I., Hoggan, M.D. and Koczot, F.J. (1969) Evidence for a single-stranded adenovirus-associated virus genome: Formation of a DNA density hybrid on release of viral DNA, *Proc. Natl. Acad. Sci. USA* **64**, 863–869

Rose, J.A., Maizel, J.K., and Shatkin, A.J. (1971) Structural proteins of adenovirus-associated viruses. *J. Virol.* **8**, 766–770

Rossi, J.J., Carbonnelle, C., Li, S., Chatterjee, S., Zaia, J.A., Larson, G. and Bertrand, E. (1994) Promoter for expression of transduced anti-HIV and SIV ribozymes. In *Gene Therapy, Keystone Symposia*, Copper Mountain, CO

Russell, D.W., Miller, A.D. and Alexander, I.E. (1994) Adeno-associated virus vectors perferentially transduce cells is S phase. *Proc. Natl. Acad. Sci. USA*. **91**, 8915–8919

Samulski, R.J., Chang, L.-S. and Shenk, T. (1987) A recombinant plasmid from which an infectious adeno-associated virus genome can be excised *in vitro* and its use to study viral replication. *J. Virol.* **61**, 3096–3101

Samulski, R.J. Chang, L.-S. and Shenk, T. (1989) Helper-free stocks of recombinant adeno-associated viruses: normal integration does not require viral gene expression. *J. Virol.* **63**, 3822–3828

Samulski, R.J., and Shenk, T. (1988) Adenovirus EIB 55-M, polypeptide facilitates timely cytoplasmic accumulation of adeno-associated virus mRNAs. *J. Virol.* **62**, 206–210

Schlehofer, J.R., Ehrbar, M., and zur Hausen, H. (1986) Vaccinia virus, herpes simplex virus, and carcinogens induce DNA amplification in a human cell line and support replication of a helpervirus dependent parvovirus. *Virology* **152**, 110–117

Srivastava, A., Lusby, E.W. and Berns, K.I. (1983) Nucleotide sequence and organization of the adeno-associated virus 2 genome. *J. Virol.* **45**, 555–564

Tratschin, J.-D., Miller, I.L., Smith, M.G. and Carter, B.J. (1985) Adeno-associated virus vector for high-frequency integration, expression, and rescue of genes in mammalian cells. *Mol. Cell. Biol.* **5**, 3251–3260

Trempe, J.P., Mendelson, E., and Carter, B.J. (1987) Characterization of adeno-associated virus rep proteins in human cells by antibodies raised against rep expressed in Escherechia coli. *Virology* **161**, 18–28

Tsao, J., Chapman, M.S., Agbandjo, M., Keller, W., Smith, K., Wu, H., Luo, M., Smith, T.J., Rossman, M.G., Compans, R.W. and Parrish, C.R. (1991) The three-dimensional structure of canine parvovirus and its functional implications. *Science* **25**, 1456–1464

Vincent, K.A., Moore, G.K. and Haigwood, N.L. (1990) Replication and packaging of HIV envelope genes in a novel adeno-associated virus vector system. *Vaccine* **90**, 353–359

Walsh, C.E., Liu, J.M., Xiao, X., Young, N.S., Nienhuis, A.W. and Samulski, R.J. (1992) Regulated high level expression of a human g-globin gene introduced into erythroid cells by an adeno-associated virus vector. *Proc. Natl. Acad. Sci. USA* **89**, 7257–7261

Walsh, C.E., Nienhuis, A.W., Samulski, R.J., Brown, M.G., Miller, J.L., Young, N.S. and Liu, J.M. (1994) Phenotypic correction of Fanconi anemia in human hematopoietic cells with a recombinant adeno-associated virus vector. *J. Clin. Invest.* **94**, 1440–1448

Wei, J.-F., Wei, F.-S., Samulski, R.J. and Barranger, J.A. (1994) Expression of the human glucocerebrosidase and arylsulfatase A genes in murine and patient primary fibroblasts transduced by an adeno-associated virus vector. *Gene. Ther.* **1**, 261–268

West, M.H.P., Trempe, J.P., Tratschin, J.-D. and Carter, B.J. (1987) Gene expression in adeno-associated virus vectors: the effects of chimeric mRNA structure, helper virus, and adenovirus VAI RNA. *Virology* **160**, 38–47

Wong, K.K., Podsakoff, G., Lu, D. and Chatterjee, S. (1993) High efficiency gene transfer into growth arrested cells utilizing an adeno-associated virus (AAV)-based vector (abstract). In *American Society of Hematology* St. Louis, MO.

Zhou, S.Z., Broxmeyer, H.E., Cooper, S., Harrington, M.A. and Srivastava, A. (1993) Adeno-associated virus 2-mediated gene transfer in murine hematopoietic progenitor cells. *Exp. Hamatol.* **21**(7), 928–933

PART II, SECTION G

32. Advances Toward Synthetic Retro-Vectoring

Clague P. Hodgson, Mary Ann Zink, Fauzia Solaiman, and Guoping Xu
Creighton University School of Medicine, Omaha, NE 68178

Introduction

Murine leukemia retrovirus (MLV) was used 20 years ago to produce the first transgenic mice (Jaenisch, 1976), illustrating the natural process by which RNA sequences are sometimes reverse-copied into DNA and integrated into the chromosomes of the germ line. Non-replication competent retro-vectors capable of introducing foreign DNA were developed in the 1980s (Wei et al., 1981 and Shimotohno and Temin, 1981), leading to their use in the first human gene therapy trials. In contrast to nonviral transfection methods and non-integrating viruses, retro-vectoring is capable of providing permanent integration and long-term expression. Retro-vectors have been used in 3/4 of gene therapy clinical trials to date, despite some problems. The biggest concern is the potential for recombination between *cis*- and *trans*-acting portions of vector systems during vector production, resulting in outbreaks of replication-competent retrovirus (RCR). Vector testing costs approximately $100,000 for each preparation of virus. Thus far, testing has detected RCR in 17% of the clinical lots RCR causes fatal neoplasms in primates (J. Ostrove, personal communication). Fortunately, humans have apparently not been exposed to RCR and no adverse effects have been observed in man. Other problems of retroviral vectors include: transcriptional inactivation of the long terminal repeat (LTR) promoter (by methylation), less than desired titer (usually 10^4–10^7 infectious units [IFU]/ml), vector size limited to 8 kb of inserted genes, insertional mutagenesis (and oncogene activation), inability to infect non-dividing cells, and inactivation of viral particles by human serum proteins (preventing *in vivo* delivery).

Fortunately, work in several labs is leading to rapid improvements in retro-vectorology. For example, the use of human (rather than mouse) vector producer cells eliminates recombinogenic or contaminant murine retroelements, improves titer, and reduces or eliminates complement inactivation (DePolo et al., 1995), making *in vivo* delivery more feasible. Studies of the HIV retrovirus have led to elucidation of its pathway for integration into non-mitotic cells (Bukrinski et al., 1992). This mechanism can potentially be incorporated into vector producer systems, overcoming the restriction to mitotic cell populations. Likewise, some cellular retroelements (such as Ty3 of yeast) undergo site-specific integration into desirable loci within the genome (Kirchner et al., 1995). If incorporated into vectoring systems, this type of integrase activity could help to control insertional mutagenesis and the related oncogene activation by retroviral vectors. Although these improvements are promising, retroviruses remain problematic.

Many gene therapists would prefer to use transfection methods that are safe and efficient, but these provide transient expression as opposed to permanent retroviral integration. Until a synthetic vector can be devised that combines these desirable characteristics, an alternative approach is to improve retroviral delivery by combining it with synthetic, non-recombinogenic vectors and synthetic particle delivery *via* liposomes.

Toward Synthetic Retroviral Particles

Ideally, it would be desirable to use enzymatic integration of DNA in the absence of virus. Some progress has been made toward this goal:

recA-coated DNA improves the efficiency of homologous recombination (Zarling, 1993), whereas integrase enzyme activity has been functionally separated from the reverse transcriptase molecule and used to integrate DNA sequences *in vitro* (Craigie *et al.*, 1990). Another innovation is the modification of tropism (infectivity of specific cells) through alterations of the viral particles themselves. For example, different viral envelope glycoprotein (env) tropisms have been used for years to target either mouse (ecotropic env), human (amphotropic env), or other host cell range variations (polytropic, xenotropic). Other proteins (such as asialoglycoprotein [Wu and Wu, 1988] or erythropoietin [Kasahara *et al.*, 1994]) that act as ligands for cellular receptors can be covalently or genetically added to the envelope in addition to env, to mimic env proteins and lead to selective attachment of vectors to targetable cells. However, it appears that the env protein (capable of internalizing the virus) must also be present in order for these targeting systems to work, and hence efforts are also underway to modify the existing env genes to suit different cellular receptors (Jolly and Barber, 1992). The vesicular stomatitis virus G protein (VSVG) can substitute for env completely, when added to viral particles (viral pseudotyping) (Yee *et al.*, 1994). This protein is toxic to cells, however, and it must be expressed transiently in order not to kill the vector producer cells (which form syncytia in its presence, due to its fusogenic properties). An advantage of this system is that the psuedotyped retrovirus can be concentrated and stored more effectively than pure retrovirus particles, and it will infect virtually any cell type.

Virosomes

Another useful viral modification was the discovery in our laboratory that a polycationic lipid, such as Lipofectamine (DOSPA: DOPE, GIBCO-BRL, Bethesda, MD), enhances the efficiency of retroviral gene transfer by 50–100 fold when added to viral preparations prior to transduction (Hodgson and Solaiman, 1996). These cationic materials are also known as liposomes, cytofectins, and cationic amphiphiles. However, it appears that the mechanism of their enhancement of viral infectivity is through negative charge neutralization rather than by masking or enclosure via lamellar liposomes, because the treatment was equally effective when added to the target cells rather than the virus. The proper viral env protein is required for infectivity, which is not improved by adding the transfection-enhancing chemical chloroquine or the transduction-enhancing material hexadimethrine bromide (polybrene). Although lipid treatment of virus is very much like the use of other polycations during transduction, the inclusion of the fatty acid moiety appears to be important for the added effect, beyond the 6–12 fold enhancement usually obtained by using polycations alone. Lipid enhancement worked equally well with retroviral or VL30-derived vectors with amphotropic or ecotropic virus and in human or mouse cells.

VL30 Vectors, Synthetic Vectors and Combinations

The genomes of mice and other vertebrates contain a variety of retroelements that might contribute to the production of safe and efficient vectors. It would be particularly useful to discover a vector that has little or no sequence similarity to the MLV used in vector producer cells, thus discouraging or eliminating RCR. The VL30 retrotransposons of mice are particularly interesting to us because some of the 200 or so genomic members are expressed in mouse cells as a parasitic RNA form, capable of packaging and transmission via retroviral virions, yet the sequences have little similarity to MLV. VL30 structure resembles that of a retrovirus, having long terminal repeats (LTRs) and other *cis*-acting elements but lacking viral or cellular structural genes. Interestingly, endogenous VL30 elements are efficiently transmitted by murine vector producer cells (Chakraborty *et al.*, 1994). Yet, there are no known consequences of VL30 transmission, such as evidence of RCR or of oncogene activation. Examination of VL30 elements reveals that they are highly conserved nucleic acid sequences, but that the U3 region of the LTR (the transcriptional promoter) is variable, depending upon the type of cell in which the VL30 is expressed. The differences in LTR structure between different elements can be correlated with their transcriptional characteristics, tissue-specificity, and hormone responsiveness. Thus, they are an evolving family of transcriptional controls that can potentially be used for gene therapy, transgenesis, etc.

A vector system was developed from the VL30 element NVL-3. Although it can be used with an internal promoter (Cook *et al.*, 1991), it was observed that the LTR transcriptional promoter is very powerful, permitting the expression of one or more genes. NVL-3 vectors are abundantly expressed as RNA in several human tissues and cell types that are targets for gene therapy, including fibroblasts, respiratory epithelium, skeletal muscle, mammary epithelium, and hematopoietic cells (B cells), in addition to a number of cancer cell types (Chakraborty *et al.*, 1995).

Synthetic vectors were made from the VL30 backbone by means of oligonucleotide-directed gene amplification of the essential *cis*-acting sequences required to make a vector (Chakraborty *et al.*, 1993). LTR-containing PCR fragments were repeatedly gel-isolated, digested, reisolated, ligated, reisolated, and inserted into a bacterial plasmid. The fragments made biologically active vectors with a high frequency. The smallest synthetic vector produces an RNA of only 1 kb in length, leaving 9 kb of "baggage space" for insert material. However, the addition of several hundred bases of packaging signals increases vector titers several fold, to $\sim 2 \times 10^5$ IFU/ml (but lower than the estimated 10^7 IFU/ml of endogenous VL30).

VL30 elements have a significant secondary structural potential. Due to the large number of false translational start (ATG) and stop signals, protein expression from VL30s was initially poor but has been improved by the addition of internal ribosome entry sites (Hodgson *et al.*, 1996a) (this permits ribosomes to translate two or more proteins from the same messenger RNA or to improve translational efficiency by internal entry). A variety of reporter genes and dominant, selectable marker constructs are available for investigations aimed at understanding transcription and translation *in vivo* and *in vitro*.

The VL30 vectors described above have some advantages over conventional MLV vectors. First, they have almost no sequence homology to MLV, discouraging homologous recombination and RCR (no RCR have been generated to date, even after a stringent, two-month long ping-pong test using classical complementation cell lines with ecotropic and amphotropic host range [Chakraborty *et al.*, 1995]). Second, the NVL-3 vectors and other VL30s have few promoter methylation sites (compared to MLV) and are expressed as abundant RNA *in vivo*. Thus, they are not likely to be transcriptionally inactivated by methylation. Recently, combination vectors have been introduced, using rat VL30 packaging signals together with MLV LTRs (Homann *et al.*, 1995) or mouse VL30 LTRs together with MLV backbones (French and Norton, 1995). Neither type produced RCR, and both varieties of chimeric vectors are reported to have titers comparable to good MLV vectors.

Transcriptional Targeting

Although NVL-3-derived vectors are faithfully expressed in human cells, this VL30 promoter is the only one yet characterized in gene transfer. Our present work is aimed at transcriptional targeting via the use of alternate LTRs. Comparison of a dozen known VL30 LTR sequences led to the identification of conserved and nonconserved domains of the U3 (promoter) region of the LTR (Hodgson *et al.*, 1996b). The conserved boundaries of these promoters provides a place for using the polymerase chain reaction to amplify the variable regions of U3. The promoters can be recovered either from genomic DNA, cDNA, or from RNA expressed in various cell types (G.P. Xu *et al.*, unpublished). For example, T-helper cell-specific VL30 promoters were isolated from a mouse cDNA library. Altogether, VL30 RNA constituted 0.3% of the mRNA sequences in T cells. Twenty clones were isolated, resulting in the recovery of two, closely related LTR types. One of these had three copies of a 35 bp enhancer element, whereas the other (apparently less abundant) type had only two enhancer copies. Throughout the promoters, small mutations, additions, and deletions differentiated the sequences of the T-cell-derived VL30 elements from those known to be expressed in other tissues.

Another method for isolating the U3 sequences was to use U3 primers in reverse transcriptase PCR reactions, copying cellular RNAs from different tissues (U3 promoters are expressed in mRNA). This approach resulted in rescue of U3 sequences from every tissue tested (a dozen tissues and cell types from the mouse). The U3s are polymorphic in size (200–400 bp), the specific size being characteristic for each cell type. In order to obtain the expressed U3 sequences without having to clone LTRs from each tissue, the amplification oligonucleotides have built-in *Bpm* 1 sites, enabling digestion of the DNA 14–16 bp away from the *Bpm* 1 site in a conserved region devoid of any other convenient restriction site. This feature permits the U3 sequences to be inserted precisely into a similarly digested, single LTR vector. The goal of this ongoing work in our laboratory is to facilitate transcriptional targeting without the introduction of competing transcriptional controls, such as internal promoters. Work from many other laboratories has added to our existing knowledge of tissue-, developmental-, and factor-specific gene regulation by mouse VL30 U3 sequences (reviewed in Hodgson *et al.*, 1996b). For example, various VL30 elements are regulated by oncogenes (*ras*, *fos-jun*), and by serum factors (epidermal growth factor, fibroblast growth factor and insulin), as well as by glucocorticoids, regulators of cyclic 3'–5' AMP, retinoids, regulators of intracellular calcium levels, and phorbol esters, to name a few. Comparison of the U3 sequences reveals a pattern of underlying signal transduction mechanisms that changes slightly from one LTR to another, providing for regulated expression. In addition to adding to our knowledge of how genes are expressed in different tissues, retrotransposon promoters offer additional choices for transcriptional targeting *via* retro-vectors.

References

Bukrinsky, M.I., Sharova, M.P., Dempsey, M.P., Sharova, N., Adzhubel, A., Spitz, L., Lewis, P., Goldfarb, D., Emerman, M. and Stevenson, M. (1992) Active nuclear import of human immunodeficiency virus type 1 preintegration complexes. *Proc. Natl. Acad. Sci. USA* **89**, 6580–6584

Chakraborty, A.K., Zink, M.A. and Hodgson, C.P. (1994) Transmission of endogenous VL30 retrotransposons by helper cells used in gene therapy. *Cancer Gene Ther.* **1**, 111–118

Chakraborty, A.K., Zink, M.A. and Hodgson, C.P. (1995) Expression of VL30 vectors in human cells that are targets for gene therapy. *Biochem. Biophys. Res. Comm.* **209**, 677–683

Chakraborty, A.K., Zink, M.A. and Hodgson, C.P. (1993) Synthetic retrotransposon vectors for gene therapy. *FASEB J.* **7**, 971–977

Cook, R.F., Cook, S.J. and Hodgson, C.P. (1991) Retrotransposon gene engineering. *Bio/Technology* **9**, 748–751

Craigie, R., Fujiwara, T., Bushman, F. (1990) The IN protein of Moloney murine leukemia virus processes the viral DNA ends and accomplishes their integration. *in vitro. Cell* **62**, 829–837

DePolo, N., DeJesus, C., Respess, J., Chang, S., Jolly, D., Mento, S. Differential sensitivity of retrovectors to human serum inactivation. *J. Cell. Biochem.* 1995, **21A**, pp. 404

French, N.S. and Norton, J.D. (1995) Construction of a retroviral vector incorporating mouse VL30 retrotransposon-derived, transcriptional regulatory signals. *Anal. Biochem.* **228**, 354–355

Hodgson, C.P., Zink, M.A. and Xu, G.P. (1996b) Structure and function of mouse VL30 sequences. In: Hodgson, C.P. (ed.), Retro-vectors for human gene therapy, Medical Intelligence Unit; R.G. Landes Company, Austin, TX, USA, pp. 73–102

Hodgson, C.P. and Solaiman, F. (1996) Virosomes: cationic liposomes enhance retroviral transduction. *Bio/Technology* (in press)

Hodgson, C.P., Chakraborty, A.K., Zink, M.A. and Xu, G.P. (1996a) Construction, transmission and expression of synthetic VL30 vectors. In: Hodgson, C.P. (ed.), *Retro-vectors for human gene therapy*, Medical Intelligence Unit; R.G. Landes Company, Austin, TX, USA, pp. 103–130

Homann, H.E., Poiteevin, Y., Torrent, C., Yu, Q., Darlix, J.-L. and Mehtali, M. (1995) New generation of safe, efficient retroviral vectors and packaging cell lines for application in gene therapy trials. *J. Cell. Biochem.* 1995, **21A**, p. 408

Jaenisch, R. (1976) Germ line integration and Mendelian transmission of the exogenous Moloney murine leukemia virus. *Proc. Natl. Acad. Sci. USA* **73**, 1260–1264

Jolly, D. and Barber, J. (1992) World Patent Publication WO9205266

Kasahara, N., Dozy, A.M. and Kan, Y.W. (1994) Tissue-specific targeting of retroviral vectors through ligand-receptor interactions. *Science* **266**, 1373–1376

Kirchner, J., Connolly, C.M. and Sandmeyer, S.B. (1995) Requirement of RNA polymerase III transcription factors for *in vitro*, position-specific integration of a retroviruslike element. *Science* **267**, 1488–1491

Wei, C.-M., Gibson, M., Spear, P.G. and Scolnick, E.M. (1981) Construction and isolation of a transmissible retrovirus containing the src gene of Harvey murine sarcoma virus and the thymidine kinase gene of herpes simplex virus type *J. Virol.* **39**, 935–944

Wu, G.Y., and Wu, C.H. (1988) Receptor-mediated gene delivery and expression *in vivo. J. Biol. Chem.* **263**, 14621–14624

Yee, J.-K., Mkyanohara, A., LaPorte, P., Bouic, K., Burns, J.C. and Friedman, T. (1994). A general method for the generation of high-titer, pantropic retroviral vectors: Highly efficient infection of primary hepatocytes. *Proc. Natl. Acad. Sci. (USA)* **91**, 9564–9568

Zarling, D.A. (1993) World Patent Publication WO9322443

PART II, SECTION G

33. The Biolistic

Pierre Couble

Centre de Génétique Moléculaire et Cellulaire, Université Claude Bernard Lyon 1 CNRS – UMR 5534
43, Boulevard du 11 Novembre 1918, 69622 Villeurbanne, Cedex, France

The method referred to as "biolistic" (a contraction of biological and ballistic) was developed by the group of Sanford (Sanford *et al.*, 1987 and Klein *et al.*, 1987), as an alternative to currently available methods for introducing foreign DNA in plants cells. It aims at both transient gene expression and stable transformation. The process consists in bombarding, at high velocity, DNA-coated microprojectiles into the nuclear genome of living cells.

Biolistic has been widely applied among plants geneticists and proved a very appropriate means for transforming various species especially those resistant to infection by *Agrobacterium* (see reviews by Johnston, 1990 and Stanford *et al.*, 1993 and references therein). Its practice spread out also to animal systems and is currently a source of new methodological concepts in applied and basic biological sciences.

From Gene Expression to Genetic Immunization

The use of biolistic to transform animal cells was delayed compared to plant cells by the necessity to reset more adapted conditions after the original protocol. This raised up a sustained interest and an increasing number of animal cells and tissues were proven amenable to transformation by biolistic, especially when conventional methods of introduction of DNA into cells had failed (see Johnston and Tang, 1993). Biolistic is effective in a wide range of cell types and cell sizes and in a wide range of phyla and species. Successful transformation has been reported in bacteria (Smith *et al.*, 1992) fungi (Armaleo *et al.*, 1990), protozoan (Vainstein *et al.*, 1994) and in many higher eucaryotes (see a survey in Table 33.1).

Recently, procedures were developed by which intact animals, instead of cells lines or dissected tissues, were used as targets. This procedure has shown to be remarkably efficient in eliciting specific antibody responses against the antigen encoded by the bombarded expression vector. Genetic immunization which bypasses the time-consuming and often costly purification of the antigen, was demonstrated by inducing anti-human growth hormone antibodies in mice whose ear or abdominal skin were bombarded with projectiles carrying a hGH expression vector (Tang *et al.*, 1992 and Eisenbraun *et al.*, 1993) and by the vaccination against *Mycoplasma pulmonis* infection of mice that received *Mycoplasma* DNA by biolistic onto their skin (Lai *et al.*, 1995).

The success of biolistic rests upon the use of specific apparatus, or accelerator gun, either static or hand-held, and upon the optimization of physical, as well as biological parameters. The following chapters are meant to describe the principal features of the most commonly used apparatus and the points that deserve attention for making the process more effective.

The Principles of Biolistic

The DNA Microcarrier

The method aims to propel at high speed, high mass density microprojectiles coated with the desired DNA, either into living cells or dissected tissues or *in situ*, into eggs or organs of intact organisms. The high mass density is confered by

Table 33.1. Applications of biolistic-mediated gene transfer in higher eucaryotes.

Organisms	Tissues and cell types	References
Insects		
Anopheles gambiae	embryo	Miahle and Miller, 1994
Bombyx mori	silkgland	Horard *et al.*, 1994
Drosophila melanogaster	embryo	Baldarelli and Lengyel, 1990
Fishes		
Loach, trout, zebrafish	fertilized eggs	Zelenin *et al.*, 1991
Amphibian		
Newt	regenerating limb	Pecorino *et al.*, 1994
Mammals		
Mouse	embryo	Zelenin *et al.*, 1993
Mouse	spleen cells	Hui *et al.*, 1994
Mouse	skin, liver (whole animal)	Williams *et al.*, 1991
Mouse	ear (whole animal)	Tang *et al.*, 1992
Mouse	abdominal skin (whole animal)	Eisenbraun *et al.*, 1993
Mouse	NIT 3T3 cells	Zelenin *et al.*, 1989
Mouse, hamster	T lymphocyte line, CHO	Fitzpatrick-McElligott, 1992
Rat	liver, kidney, mammary gland	Zelenin *et al.*, 1991
Rat	mammary epithelial cells	Thompson *et al.*, 1993
Rat, mouse	liver, skin, muscle	Yang *et al.*, 1990
Rat, rabbit, mice, monkey	various tissues (whole animals)	Cheng *et al.*, 1993

the use of heavy metals such as tungsten or gold which have a close specific density of respectively 19.1 and 19.3 g/cm^3. Tungsten is very inexpensive and available as graded powders of calibrated ships of a few micrometers of mean-diameter. The particles are heterogeneous in shape and represent often a broad range of sizes, but they are efficient in many situations. It was reported that tungsten could be toxic to certain cell types but this could be due to extensive oxidation of the metal prior to use. The reader is referred to the recommendations of Sanford *et al.* (1993) to handle tungsten projectiles.

Powders of round shaped, uniform gold particles of respectively 1 and 1.6 μm diameter are also available at Bio Rad, but the cost is higher.

The DNA is affixed to the surface of the metallic particles by precipitation in presence of calcium ions and spermidine, according to a classical procedure (Sanford *et al.*, 1993). The aqueous slurry of salted DNA and particles is usually deposited at the centre of a so-called macrocarrier (a disc membrane), let dry and introduced into the instrument. More recently, it was shown that keeping the DNA precipitate hydrated boosts considerably the yield of transformation, but this requires a specific modification of the archetypic instrument (see below).

The amount of metal particles employed for one shot averages 1.5 mg. With spheric, 1 μm large particles, such an amount launched over a circular surface of 2.5 cm of diameter would result in an even distribution of 1 particle per 3 μm^2.

The Gun Accelerator

Accelerating tungsten or gold microparticles to supersonic speeds can be achieved by a variety of methods that generate a gaz-shock, the driving force of the DNA-coated metallic ships. Gun powder, electric shock explosions and other procedures were tested, but the most safer and reproducible method consists in provoking a sudden blast of compressed helium. The only one commercially available instrument, PDS-1000/He from Dupont (distributed by Bio Rad, Richmond, USA) is built on this principle. It is described in Figure 33.1A.

The Gaz-shock Inducer. The superior part of the device consists in an aluminium chamber with a gas admission tube at the top and a ~1 cm circular opening at the bottom. This is sealed by an adapted kapton disc (Dupont), the property of which is to resist up to a given pressure of the compressed gas in the chamber. Its rupture induces the explosive release of gas that blows the microprojectiles underneath. The reproducibility of this step is inherent to the use of helium, a light and inert gas with the capacity to expand fast when blown into a container. It is therefore preferred to all other bottled neutral gases.

A large spectrum of rupture discs is available, which differ by their thickness. The commercially available discs are calibrated to break at graded pressures from 450 to 2200 psi. Many experiments are conducted below 1500 psi of helium.

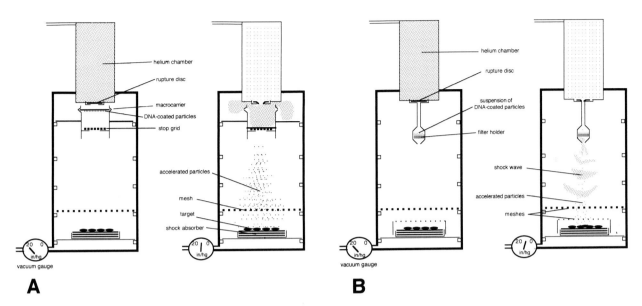

Figure 33.1. Schematic representation of instruments currently use for "dry" (A) and "aqueous" (B) biolistic.
A is a schema of the commercialized PDS-1000/He instrument that functions with a blast of helium. On the left, the apparatus is shown all set up, prior to reaching the pressure of rupture of the disc sealing the helium chamber. On the right, the disc has broken and the particles are propelled toward the target. Note that the shock wave energy is essentially absorbed in the upper part of the apparatus, which helps in reducing its deleterious effects onto the biological sample. The drawing in B is inspired Miahle and Miller (1994) who substituted the launching assembly of the PDS-1000/He instrument by a syringe filter holder to perform "aqueous" biolistic. The filter and its cartridge can be either made of stainless steel, plastic or teflon. The suspension of particles is retained by capillarity into the filter prior to the blast of gas. The shock wave "diffuses" throughout the upper piece and expands into the sample chamber, which requires that meshes and shock absorbers are used to protect the sample. Note that the diffusion of the particle projectiles is less than in the setting used for "dry" biolistic.

The Acceleration Barrel. The intermediary part is a stainless steel, cylinder-shaped piece that functions as an acceleration barrel. At the top, a so-called macrocarrier (a thin kapton membrane disc of 2.5 cm diameter) holds the dry particles with DNA, on its inferior face. The lower part of the barrel comprises a circular stop grid. Under the explosion of helium, the microcarrier and the particles are submitted to a violent acceleration and come up suddenly to speed. The course of the macrocarrier is abruptly interrupted by its collision with the grid placed on its path, whereas the accelerated microprojectiles are freed from the surface of the kapton disc and enter the bottom chamber. This is accompanied by a certain dispersion of the particles, making the surface of their impact wider than the surface of the grid.

The Sample Chamber. The bottom chamber receives the targeted sample. It is isolated from the exterior by a jointed plexiglass door and connected to a vacuum pump equipped with a gauge, to drive a depression along the path of the beam. This allows to reduce the deacceleration of particles in the chamber. The strength of vacuum supported by the biological material should be empirically set up for each tissue or cell tested; it varies in a range of ~ 5 to ~ 25 inches of Hg.

The sudden discharge of gas induces a shock wave whose effect is more or less pronounced according to the design of the apparatus, to the initial helium rupture pressure and to the level of residual air pressure in the sample chamber. In the PDS-1000/He instrument, part of the shock wave energy is absorbed by the arrest of the kapton macrocarrier onto the grid, preventing that too much helium enters the sample chamber. Displacement of residual air generated by the blast can cause severe cell damage, especially in the center of the impact. This can be partly circumvented by placing one or several wire meshes in between the launching assembly part and the target. The shock of the blast can be also attenuated by depositing the sample on a retractible basis made of a stack of wet filter papers or of agarose.

Hand-held Particle Delivery Instrument

A hand-held version of the static instrument described above has been developed by the group of Sanford (Williams *et al.*, 1991) to fire cloned DNA constructs directly into tissues of intact animals.

This type of apparatus was used to develop the concept of genetic immunization, as reported above. The apparatus functions with a similar launching assembly as PDS-1000/He, the design of which minimizes the deleterious effect of the shock wave, and is used therefore without air depression at the target interface. The sample chamber comprises a screen at its bottom and its design facilitates direction of the particle beam to internal organs.

"Aqueous Biolistic" or Hydrated better than Dried DNA

In the original description of the method, out of which PDS-1000/He like instruments were conceived, the DNA-coated particles are launched after being dried up. Dehydration allows an even distribution of the metallic ships onto the surface of the microcarrier, and, by then, the spreading of the projectiles onto the samples is uniform. The technique originally set up for plants was applied as such to animal cells and tissues before it appears that dehydration of the salted precipitate of nucleic acid around the particles prevents the mobilization of DNA in the host cell nuclei, and that higher yields of transfection could be reached with hydrated DNA, as reported by Miahle and Miller (1994) and by Horard et al. (1994).

The projection of DNA-coated particles in aqueous medium using the PDS-1000/He accelerator gun was carried out in our group on the silkgland of *Bombyx mori* (Horard et al., 1994), by simply spreading the suspension of projectiles in calcium and spermidine onto the microcarrier and proceeding at the shoot as usual. The efficacy of transfection was increased by at least two orders of magnitude when comparing the signal response with the same amount of DNA fired dry or hydrated. However, the sliding of the slurry in the process of acceleration renders it difficult to target the beam of particles to a precise area.

With the aim of better adapting the PDS-1000/He device to aqueous biolistics, Miahle and Miller (1994) proposed to substitute the original particle launch assembly part by a piece made of a narrow tube carrying a syringe filter holder at its inferior tip (Figure 33.1B). The piece is now interdependent with the helium chamber. The suspension of particles deposited onto the grid of the filter holder is retained by capillarity.

By this procedure, the narrowness of the tubing above the filter allows to reduce the pressure in the helium chamber. The configuration of the small chamber just underneath the filter also helps in homogenizing the spray of particles forced to beam out of the small 2.5 mm wide opening at its base. The shock wave is partly absorbed in the vortex process, but eventually hits the sample which should be protected by disperser wire meshes and by being placed on a retractile basis.

By experience, the modification described by Miahle and Miller greatly improves the easiness of biolistic at the use of animal cells and tissues.

Future Prospects

Biolistic applied to animal systems has allowed substantial progresses to transform cells and tissues that were reluctant to conventional methods such as lipofection, electroporation, injection, viral vector infection, etc. The range of the models where gene expression can be studied and exploited has thus broadened, allowing fundamental and applied concepts to develop. The dissection of individual retinoic acid receptors by using bombardment is an example of the potential of biolistic in fundamental research. The application of eliciting a specific immune response by simply bombarding the skin of a living animal is also very promising to induce vaccination against pathogens, and may substitute conventional strategies that use costly purification of antigens. More is expected of the development of genetic immunization by biolistic in medical and veterinarian gene therapy.

Biolistic can also be viewed as a means to construct transgenic animals, by bombarding either germ line cell precursors or mature gametes. However, the attempts made in this direction, in insects, fish or mammals, have not yet been successful. Reasons for this may be as diverse as the low probability that a particle hits the nucleus of one of the few gamete precursors when bombarding eggs or early developed embryos, the non-availability of appropriate gene transfer vectors for improving the efficiency or integration of the foreign gene into the host chromosomes, or the lack, in many systems, of appropriate selectable genetic markers for screening transgenics. Yet, even if adjustments may be required to adapt the physical parameters of the bombardment process to a variety of biological situations, biolistic remains an attractive method to treat large populations of cells or eggs at the same time.

References

Armaleo, D., Ye, G.N., Klein, T.M., Shark K.B., Sanford J.C. and Johnston, S.A. (1990) Biolistic nuclear transformation of *Saccharomyces cerevisiae* and other fungi. *Curr. Genet.* **17**, 97–103

Baldarelli, R.M. and Lengyel, J.A. (1990) Transient expression of DNA after ballistic introduction into *Drosophila* embryos. *Nucleic Acids Res.* **18**, 5903–5904

Cheng, L., Ziegelhoffer, P.R. and Yang, N.-S. (1993) *In vivo* promoter activity and transgene expression in mammalian somatic tissues evaluated by using particle bombardment. *Proc. Nat. Acad. Sci. USA* **90**, 4455–4459

Eisenbraun, W.D., Fuller, D.H. and Haynes, J.R. (1993) Examination of parameters affecting the elicitation of humoral responses by particle bombardment-mediated genetic immunization. *DNA Cell. Biol.* **12**, 791–797

Fitzpatrick-McElligott, S. (1992) Gene transfer to tumor-infiltrating lymphocytes and other mamalian somatic cells by microprojectile bombardment, *Biotechnology* **10**, 1036–1040

Hui, K.M, Sabapathy, T.K, Oei, A.A. and Ghia, T.F. (1994) Generation of allo-reactive cytotoxic T lymphocytes by particle bombardment-mediated gene transfer. *J. Immunol Methods* **171**, 147–155

Horard, B., Mangé, A., Pelissier, B. and Couble, P. (1994) *Bombyx* gene promoter analysis in transplanted silk-gland transformed by particle delivery system. *Insect Mol. Biol.* **3**, 261–265

Johnston, S.A. (1990) Biolistic transformation: microbes to mice. *Nature* **346**, 776–777

Johnston, S.A. and Tang, D.-C. (1993) The use of microparticle injection to introduce genes into animal cells *in vitro* and *in vivo*. *Genet. Eng.* (NY) **15**, 225–236

Klein, T.M., Wolf, E.D., Wu, R. and Sanford, J.C. (1987) High-velocity microprojectiles for delivering nucleic acids into cells. *Nature* **327**, 70–73

Lai, W.C., Bennett, M., Johnston, S.A., Barry, M.A. and Pakes, S.P. (1995) Protection against *Mycoplasma pulmonis* infection by genetic vaccination. *DNA Cell. Biol.* **14**, 643–651

Miahle, E. and Miller, L.H. (1994) Biolistic techniques for transfection of mosquito embryos (*Anopheles gambiae*). *Biotechniques* **16**, 924–931

Pecorino, L.T., Lo, D.C. and Brockes, J.P. (1994) Isoform-specific induction of a retinoid-responsive antigen after biolistic transfection of chimaeric retinoic acid/thyroid hormone receptors into a regenerating limb. *Development* **120**, 325–333

Sanford, J.C., Smith, F.D. and Russell, J.A. (1993) Optimizing the biolistic process for different biological applications. *Methods in Enzymology* **217**, 483–509

Smith, F.D., Harpending, P.R. and Sanford, J.C. (1992) Biolistic transformation of prokaryotes: factors that affect biolistic transformation of very small cells. *J. Gen. Microbiol.* **138**, 239–248

Tang, D.-C., De Vit, M. and Johnston, S.A. (1992) Genetic immunization is a simple method for eliciting an immune response. *Nature* **356**, 152–154

Thompson, T.A., Gould, M.N., Burkholder, J. and Yang, N.-S. (1993) Transient promoter activity in primary rat mammary epithelial cells evaluated using particle bombardment gene transfer. *In Vitro Cell. Dev. Biol.* **29**, 165–167

Vainstein, M.H., Alves, S.A., De Lima, B.D., Aragao, F.H. and Rech, E.L. (1994) Stable transformation in a flagellate trypanosomatid by microparticle bombardment. *Nucleic Acids Res.* **22**, 3263–3264

Williams, R.S., Johnston, S.A., Riedy, M., De Vit, M.J., McElligott, S.G. and Sanford, J.C. (1991) Introduction of foreign genes into tissues of living mice by DNA-coated microprojectiles. *Proc. Nat. Acad. Sci. USA* **88**, 2726–2730

Yang, N.-S., Burkholder, J., Roberts, B., Martinell, B. and McCabe, D. (1990) *In vivo* and *in vitro* gene transfer to mammalian somatic cells by particle bombardment. *Proc. Nat. Acad. Sci. USA* **87**, 9568–9572

Zelenin, A.V., Alimov, A.A., Zelenina, I.A., Semenova, M.L., Rodova, M.A., Chernov, B.K. and Kolesnikov, V.A. (1993) Transfer of foreign DNA into the cells of developing mouse embryos by microprojectile bombardment. *FEBS Letters* **315**, 29–32

Zelenin, A.V., Alimov, A.A., Barmintzev, V.A., Beniumov, A.O., Zelenina, I., Krasnov, A.M. and Kolesnikov, V.A. (1991) The delivery of foreign genes into fertilized fish eggs using high-velocity microprojectiles. *FEBS Letters* **287**, 118–120

Zelenin, A.V., Alimov, A.A., Titomirov, A.V., Karansky, A.V., Gorodetsky, S.J. and Kolesnikov, V.A. (1991) High-velocity mechanical transfer of the chloramphenical-acetyl transferase gene into rodent liver, kidney and mammary gland cells in organ explants and *in vivo*. *FEBS Letters* **280**, 94–96

Zelenin, AV., Titomirov, A.V. and Kolesnikov, V.A. (1991) Genetic transformation of mouse cultured cells with the help of high velocity mechanical DNA injection. *FEBS Letters* **244**, 65–67

PART III, SECTION A

34. The Fate of Microinjected Genes in Preimplantation Embryos

Robert J. Wall and Thomas G. Burdon

Gene Evaluation and Mapping Laboratory, Agricultural Research Service, USDA, Beltsville, MD, USA and Centre for Genome Research, University of Edinburgh, Edinburgh, Scotland, UK

Introduction

Transgenic animal technology has been employed in well over 6,000 published reports in the last 15 years. The proliferation of such studies attests to the power of this tool to address a wide variety of basic biological questions and further suggests that production of transgenic animals has become common practice at most research institutions. It is therefore somewhat surprising that almost nothing is known about the mechanism by which transgenes become integrated into an animals genome. The lack of attention to this aspect of transgenic technology may in part be attributed to the difficulty of addressing the question and the relative ease with which transgenic mice can be produced. Though the same techniques are employed to generate transgenic livestock, the efficiency of producing transgenic farm animals is about an order of magnitude lower than for transgenic mice. This inefficiency makes producing transgenic livestock costly, ranging from $ 25,000 for an expressing founder pig to over $ 300,000 for an expressing founder cow or bull. Understanding the mechanism of transgene integration could provide the insight needed to develop strategies for improving efficiency and lowering costs.

Most transgenic animals produced by microinjection contain an array of transgene copies integrated into the genome at a single locus. Therefore at a minimum, three distinct processes must occur: formation of the array; integration into the genome; and degradation of unincorporated DNA. These processes have been difficult to study in part because of the small amount of starting material (individual embryos) and the lack of appropriate molecular biological techniques.

When do Concatemers Form

In an attempt to determine when microinjected DNA concatemerises we conducted a series of experiments, the results of which are presented in Table 34.1 (Burdon and Wall, 1992). PCR analysis with primers designed to detect genes linked in a head-to-tail manner demonstrated that virtually all embryos contain ligated transgenes within 5–10 minutes of injection. Rapid ligation was not due to non-covalent association of the transgenes prior to injection since heating DNA to 65 C before injection or using genes with blunt or incompatible ends did not significantly impair the ligation of DNA ends. Furthermore, to test whether the rapid association of genes was a product of intramolecular ligation (circularisation of microinjected genes), 5' and 3' fragments were either coinjected or injected individually 2 hrs apart. Only in asynchronously injected embryos was ligation between genes significantly reduced, indicating that intermolecular ligation is an efficient process in pronuclei and likely to occur before transgene integration. The observation that transgenes are almost invariably organized as direct repeats (tandem head to tail arrangements), in transgenic animals, indicates that recombination also plays an key role in the generation of transgene arrays. Furthermore, a positive role for recombination in builiding the transgene array has been elegantly demonstrated by the reconstitution of an intact functional gene from coinjection of three overlapping DNA fragments, containing portions of the human serum albumin gene at an efficiency of 74% (Pieper *et al.*, 1992). Clearly both ligation and recombination play a central role in generating the transgene array.

Table 34.1. Analysis of transgene concatemer formation in 1-cell mouse*

Treatment	Embryos containing head-to-tail concatemers
Embryos incubated for 5–10 min., after transgene injection	100 (20/20)
Injection of transgenes into cytoplasm	30 (6/20)
Transgene heated before injection	95 (19/20)
Injection of blunt ended transgenes	100 (20/20)
Injection of transgenes with incompatible 5' and 3' ends	95 (19/20)
Coinjection of 5' amd 3' fragments	95 (19/20)
Asynchronous injection of 5' and 3' fragments	22 (4/18)

*Percentages (number with characteristic/number analyzed). Taken from Burdon and Wall, 1992.

When does Transgene Integration Occur

It has been commonly thought that transgenes integrate into the genome at the 1-cell stage of development, and as a consequence copies of transgenes are contained in all cells of the resulting animal. However, Pursel and colleagues (Pursel *et al.*, 1988) demonstrated that integration during the 1-cell stage is not an absolute requirement, producing transgenic pigs by microinjecting nuclei of 2-cell embryos.

Two other lines of evidence suggests that transgenes, introduced at the pronuclear stage of development, do not usually integrate into the genome prior to the first embryonic cleavage division. Transgene germline mosaicism is thought to be a result of late integration. An early study of over 200 transgenic mouse pedigrees suggested that about 30% of founder mice were germline mosaics (Wilkie *et al.*, 1986). A more recent study concluded that most transgenic founder mice are mosaic by comparing results of transmission frequency with PCR, and Southern blot analysis results of DNA extracted from transgenic mice (Whitelaw *et al.*, 1993). The authors of both of these studies concluded that mosaicism was a consequence of transgenes integrating after zygotic DNA replication. However, integration of transgenes after the 1-cell stage does not necessarily dictate that the subsequent animal will be mosaic. The fate of blastomeres is not determined until the 8-cell stage, at least in mice, where one-half or fewer blastomeres contribute to the embryo proper. It is therefore feasible for transgene integration to occur after the first cleavage division without generating mosaic founder animals (Wall and Seidel, Jr., 1992).

A somewhat more direct demonstration of post 1-cell transgene integration is based on PCR analysis of individual blastomeres (Burdon and Wall, 1992). In that study we examined 19 embryos at the 8-cell stage and found none which contained transgenes in more than four blastomeres. That result suggests integration did not occur in any of those embryos before the first mitotic S-phase and injected transgenes segregate as embryo development proceeds. Our data and that from a very similar study (Cousens *et al.*, 1994) are pooled and presented in Figure 34.1. In the combined data set, three embryos contained transgenes in all eight blastomers (6% of total) of 8-cell embryos. Estimates of transgenic efficiency, based on proportion of transgenic mice born, range from 15 to 30% and, therefore, it is unlikely that the three embryos with transgenes in all blastomeres were the only transgenic embryos generated in the two studies.

The disparity in transgene integration efficiency between laboratory and livestock species (one transgenic mouse per 40 eggs injected vs. one transgenic calf per 1,600 eggs injected) may offer some insight into the integration event. The timing of transgene introduction into zygotes, with respect to S-phase, differs between mice and livestock species (see Wall, 1996). Mouse eggs are microinjected about 8 hours after mating, resulting in DNA being introduced into zygotes during the beginning of S-phase. Sow and cow eggs are injected towards the end of the S-phase, possibly reducing the probability of an integration event before completion of DNA replication. Taken at face value, these

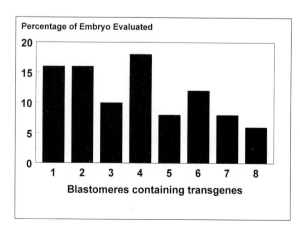

Figure 34.1. Distribution of blastomeres containing microinjected genes in 8-cell mouse embryos (n=50) (Combined data from Burdon and Wall, 1992 and Cousens *et al.*, 1994).

observations suggest integration efficiency may be enhanced by introducing transgenes early in an S-phase of a cell cycle. It can not be determined from these observations if introducing transgenes before the first mitotic S-phase is particularly advantageous.

Unintegrated Transgenes Persist

Transgenic production efficiency could be increased and costs could be dramatically reduced if a means were available to identify transgenic preimplantation embryos prior to transfer into recipients. The savings would be realized by reducing the number of embryo recipients required (e.g. in cattle the number of recipient cows could be reduced from hundreds to just a few). To date, attempts to identify transgenic embryos have failed, apparently because of the confounding persistence of unintegrated DNA. All of the studies towards this goal have been based on PCR analysis, with two notable exceptions (Tada et al., 1995; and Bondioli and Wall, 1996), which are based on the use of selectable genetic markers.

Most studies have clearly demonstrated that DNA injected into either pronuclei of zygotes or nuclei of 2-cell embryos is detectable at least until the blastocyst stage of development (Behboodi et al., 1993; Burdon and Wall, 1992; Cousens et al., 1994; Bowen et al., 1994; Page et al., 1995; Ninomiya et al., 1989 and Sparks et al., 1994). Because the studies were PCR based, it was not possible to distinguish between integrated and unintegrated transgenes. Additionally all of these studies detect transgenes in a higher proportion of morulae and blastocysts than is commonly detected in live born animals. The high frequency of PCR positive embryos observed suggests that either transgenic embryos are selectively lost at later stages of development or this approach is prone to generating false positive results. The later hypothesis is supported by two studies that showed no increase in efficiency by transferring only PCR positive embryos (Bowen et al., 1994 and Ninomiya et al., 1989). Many of those false positives presumably result from the detection of persisting unintegrated DNA.

Summary

Little direct evidence is available to determine the fate of DNA microinjected into preimplantation embryos. From indirect measurements and other observations it would appear that individual copies of transgenes become ligated into concatemers almost immediately; and those structures persist for an extended period of time (at least a week in developing bovine embryos). Furthermore, if transgenes do become integrated into the genome, the event does not necessarily occur prior to the first cleavage division.

References

Behboodi, E., Anderson, G.B., Horvat, S., Medrano, J.F., Murray, J.D. and Rowe, J.E. (1993) Microinjection of bovine embryos with a foreign gene and its detection at the blastocyst stage. *J. Dairy. Sci.* **76**, 3392–3399

Bondioli, K.R. and Wall, R.J. (1996) Positive selection of transgenic bovine embryos in culture. *Theriogenology* **45**, 345, abstract

Bowen, R.A., Reed, M.L., Schnieke, A., Seidel, G.E., Jr., Stacey, A., Thomas, W.K. and Kajikawa, O. (1994) Transgenic cattle resulting from biopsied embryos: expression of c-ski in a transgenic calf. *Biol. Reprod.* **50**, 664–668

Burdon, T.G. and Wall, R.J. (1992) Fate of microinjected genes in preimplantation mouse embryos. *Mol. Reprod. Devel.* **33**, 436–442

Cousens, C., Carver, A.S., Wilmut, I., Colman, A., Garner, I. and O'Neill, G.T. (1994) Use of PCR–based methods for selection of integrated transgenes in preimplantation embryos. *Mol. Reprod. Devel.* **39**, 384–391

Ninomiya, T., Hoshi, S., Mizuno, A., Nagao, M. and Yuki, A. (1989) Selection of mouse preimplantation embryos carrying exogenous DNA by polymerase chain reaction. *Mol. Reprod. Devel.* **1**, 242–248

Page, R.L., Canseco, R.S., Russell, C.G., Johnson, W.H., Velander, W.H. and Gwazdauskas, F.C. (1995) Transgene detection during early murine embryonic development after pronuclear microinjection. *Transgenic Res.* **4**, 12–17

Pieper, F.R., de Wit, I.C., Pronk, A.C., Kooiman, P.M., Strijker, R., Krimpenfort, P.J., Nuyens, J.H. and de Boer, H.A. (1992) Efficient generation of functional transgenes by homologous recombination in murine zygotes. *Nucleic Acids Res.* **20**, 1259–1264

Pursel, V.G., Miller, K.F., Pinkert, C.A., Palmiter, R.D., and Brinster, R.L. Effect of ovum cleavage stage at microinjection on embryonic survival and gene integration in pigs. *11th Internat. Congr. Anim. Reprod. Artif. Insem.* **4**, p. 480 Dublin: 1988

Sparks, A.E.T., Canseco, R.S., Russel, C.G., Johnson, J.L., Moll, H.D., Velander, W.H. and Gwazdauskas, F.C. (1994) Effects of time of deoxyribonucleic acid microinjection on gene detection and *in vitro* development of bovine embryos. *J. Dairy Sci.* **77**, 718–724

Tada, N., Sato, M., Hayashi, K., Kasai, K. and Ogawa, S. (1995) *In vitro* selection of transgenic mouse embryos in the presence of G-418. *Transgenics* **1**, 535–540

Wall, R.J. (1966) Modification of milk composition in transgenic animals. In: Miller, R.H. (Ed.) Proc. Beltsville Symosium XX: Biotechlogy's Role in the Genetic Improvement of Farm Animals (in press)

Wall, R.J. and Seidal, G., Jr. (1992) Transgenic farm animals–A critical analysis. *Theriogenology* **38**, 337–357

Whitelaw, C.B., Springbett, A.J., Webster, J. and Clark, J. (1993) The majority of G_0 transgenic mice are derived from mosaic embryos. *Transgenic Res.* **2**, 29–32

Wilkie, T.M., Brinster, R.L. and Palmiter, R.D. (1986) Germline and somatic mosaicism in transgenic mice. *Dev. Biol.* **118**, 9–18

PART III, SECTION A

35. Chromosomal Insertion of Foreign DNA

John O. Bishop
Centre for Genome Research, University of Edinburgh, King's Buildings, Edinburgh, EH9 3JQ, UK

Introduction

Presently, the 'normal' way to bring about random insertion of a foreign DNA into the mammalian chromosome is to microinject perhaps 50–100 identical molecules directly into one of the post-fertilisation pronuclei. In the mouse the same objective could be achieved, probably more economically, by incorporating a selective marker into the construct and introducing it into ES cells by electroporation. If necessary, the foreign DNA could be designed so as to allow the subsequent removal of the selective marker by site-specific recombination. However, this approach has not yet found favour.

Because of the difficulty of analysing the DNA of a single cell, and especially of distinguishing chromosomally integrated from residual extra-chromosomal foreign DNA (Cousens *et al.*, 1994), the available information bearing on the mechanism of integration is indirect, and comes from two sources, neither of which is completely satisfactory. The first is the structure of integrated foreign DNA molecules, which is generally determined long after the event from biopsies of weanlings, while the second is by analogy with events that occur when DNA is introduced into cells in culture (reviewed previously (Bishop and Smith, 1989)).

Structure of Transgenes

Typically a transgene becomes integrated in the form of a multicopy *array* containing anything up to hundreds of copies. However, on occasion the integration of only one foreign DNA molecule in whole or in part also occurs. As a rule, the genome of a transgenic founder animal carries a single array, but founders that carry two or even three arrays are not infrequent. Presumably, both the size and number of the arrays relate to the number of copies of the foreign DNA that were introduced, although other features of the DNA may also affect these parameters.

Predominantly, neighbouring copies within arrays are arranged in the same (*direct tandem*) orientation, although much more rarely they may be in opposite (*inverted*) orientation. We (Bishop and Smith, 1989) have presented arguments that support the explanation that arrays are assembled extrachromosomally prior to integration. We adopted the generally held view that the small number of inverted neighbours and presumably an equal number of direct neighbours are generated by the ligation of ends of two linear foreign DNA molecules and by illegitimate recombination between pairs of molecules. However, we also proposed that most of the vastly greater number of directly repeated neighbours are generated by homologous recombination, a proposal based on the well-documented process of extrachromosomal non-conservative recombination between foreign DNA molecules in transfected cultured mammalian cells (Lin *et al.*, 1984 and Wake *et al.*, 1985). This process is thought to be begun by exposure of one strand at each end of linear DNA molecules by a 5′ exonuclease, followed by base-pairing between exposed single-strands with the possibility of branch migration, nucleolytic removal of unpaired 3′ sequence and/or infilling of gaps, and finally ligation of the resulting single-strand nicks (Lin *et al.*, 1987). These processes have generally been studied in cultured cells by co-transfecting with dissimilar molecules (e.g. the same plasmid molecule linearised at different sites, or different linearised plasmid molecules) designed to allow particular types of recombinant molecule to be selected for. Without further assumptions they do not explain the generation of direct tandem arrays from a population of identical linear molecules.

Given that end-joining of linear molecules has been ruled out as an exclusive mechanism, there are two further ways in which direct arrays can be generated, namely by rolling circle replication and by recombination between *circular or circularly permuted* molecules. In mammalian cells only a special class of plasmids can replicate (e.g. replication in COS cells of plasmids that contain the SV40 replication origin) and rolling circle replication is unknown. Furthermore, rolling circle replication would not explain the fact that the junctions between neighbouring direct repeats are frequently damaged in a variety of different ways (Hamada et al., 1993 and Rohan et al., 1990). This last observation is however readily explained by recombination between molecules that have been circularised independently by end-joining within the cell.

Recombination between circularly permuted linear molecules was our favoured explanation for two reasons. First, the non-conservative recombination process known to occur in cultured cells requires linear molecules. Secondly, this mechanism would assemble long arrays of copies that are perfect in all respects except around the junctions between them, even from highly fragmented molecules, and is the only known mechanism that could do so. More recently it has been shown that a long molecule can be reconstituted intracellularly from shorter overlapping segments co-injected into nuclei of one-cell embryos (Shimoda et al., 1991 and Pieper et al., 1992). This is entirely consistent with the proposed non-conservative recombination mechanism.

To generate direct tandem arrays from a homogeneous population of linear molecules the requirements are twofold: (1) Some of the incoming linear molecules must be circularised by ligation or illegitimate recombination of their open ends. This reaction, being unimolecular, would be favoured over intermolecular joining. Subsequently there has been a report of end-joining (which did not distinguish between intra- and inter-molecular events) within 10 minutes of microinjection into nuclei of mouse one-cell embryos (Burdon and Wall, 1992); (2) Some at least of these circular molecules must be linearised by (more or less) random cleavages in order to generate circularly permuted molecules. Both circularisation of transfected linear molecules and random cleavage of transfected circular molecules occur in transfected animal cells (Bishop and Smith 1989).

Chromosomal Sites of Transgene Insertion

To hope to understand the mechanism of DNA insertion we need at the least to know the sequences of the transgene, the chromosome-transgene junctions (*CT junctions*) and the corresponding region of the chromosome prior to transgene insertion. Unfortunately, not many complete studies have been reported. In those that we have, a significant number show a limited homology between chromosome and foreign DNA at the junction. However, this is by no means universal. In this respect, like so many others, insertion of foreign DNA into the embryo genome strongly resembles the process that occurs in cultured cells (Konopka, 1988).

The commonest and most readily comprehensible sites of transgene integration are simple deletions of chromosomal DNA between the CT junctions (Covarrubias et al., 1987; Xiang et al., 1990; Karls et al., 1992; Hamada et al., 1993 and Hughes et al., 1993), which are consistent with double crossover events initiated, for example, by a 'double strand break repair' type of mechanism (Szostak et al., 1983) in which the ends of the foreign DNA molecule simulate a break or a gap in a homologue, or alternatively by the independent invasion of the chromosomal DNA duplex by the two ends of a linear molecule (or array).

Other events include duplications of flanking chromosomal DNA (Wilkie and Palmiter, 1987) and translocations (Francke et al., 1992). Like simple insertions, duplications of chromosomal DNA can be explained by double-strand break repair. However, both insertions and duplications can also be explained by supposing that the transgene insertions in question occurred during DNA replication and involved both daughter duplexes. The outcome would depend on the points at which CT junctions were generated. Relative to an intact duplex, lagging strand replication could possibly offer a more accessible target for invasion by a very poorly matched strand of foreign DNA, together with a local environment rich in proteins involved in DNA synthesis, replication and repair. If this is the case, some of the deletions and all duplications would involve the loss of the distal part of one of the daughter chromosomes. Presumably this would often be lethal to the developing embryo, but in some cases the death of the deficient cell might be tolerated, or daughter cells carrying the deficient chromosome might be viable. Cytological examination of non-transgenic offspring of transgenic founders might be rewarding in this respect.

Foriegn DNA is mainly Integrated into Chromosomal DNA that has Replicated at least Once since Fertilisation

The possibility that integration may involve replication loops focuses attention on the time at which integration occurs relative to DNA synthesis in the fertilised egg. In principle, by the use of PCR and restriction enzymes (such as *Dpn*I) that cleave foreign DNA at replication-sensitive sites, it may eventually be possible to determine directly the time at which microinjected foreign DNA becomes integrated into the chromosome (Cousens *et al.*, 1994). However, this has yet to be achieved. At present, the best estimate of integration time is indirect, based on the frequency of mosaicism amongst transgenic founders. The main weakness of this approach is, of course, that mosaicism may arise not only by an integration event that follows post-fertilisation replication of the DNA, but also by the loss of a 'daughter' transgene that has already replicated along with the chromosome.

Estimates of the frequency of mosaicism have varied from 15% to 62% (Brinster *et al.*, 1985 and Whitelaw *et al.*, 1993). Another estimate can now be made based on the male sterility associated with ectopic expression of herpes simplex virus (HSV) thymidine kinase (*tk*) reporter genes in the spermatids (Al-Shawi *et al.*, 1988, 1991). When under the control of the bovine thyroglobulin promoter the Type 1 HSV *tk* gene renders all transgenic carrier males sterile (Wallace *et al.*, 1994). However, males that carry a construct identical but for two single nucleotide alterations that convert the Met^{46} and Met^{60} codons to Leu codons, or another in which the HSV Type 2 *tk* gene replaces the Type 1 gene, are fertile (Ellison *et al.*, 1995). In these cases the sons and later male descendants of transgenic founders transmit the transgenes to 50% of their progeny, but in contrast the transgenic male founders, although generally fertile, mainly failed to transmit their transgenes (Ellison *et al.*, 1995). Recently, we were able to show by means of AI experiments that this is due to a failure of TK+ sperm to fertilise ova *in vivo* when in the presence of a similar number of normal (TK−) sperm (J.D. West, A.R. Ellison and J.O. Bishop, unpublished). Because of the pseudo-syncytial development of sperm, all sperm of a mouse heterozygous for a *tk* reporter gene are TK+ (Dym and Fawcett 1971; Erickson 1973 and Braun *et al.*, 1989) and consequently the *tk*+ and *tk*− sperm are not in competition. In a mosaic individual the situation is different, because *tk*+/− and *tk*−/− spermatogonia generate separate syncytia which develop independently of each other. The fertile founder males that do not transmit the *tk* transgenes can therefore be considered to be germ-line mosaics. By this criterion, we found that 65% (11/17) of male founders were germ-line mosaics (Ellison *et al.*, 1995). If, for the sake of argument, we include G$_0$ males that predominantly generated non-transgenic progeny, we have a figure of 82% (14/17) germ-line mosaics.

To estimate roughly the implications of this level of mosaicism, we can make the simplifying assumption that at the compaction of the morula 3 of 8 cells are chosen at random to form the inner cell mass, and that later determinative processes involve larger numbers of cells drawn from larger cell populations. Then, to explain the observed level of mosaicism, 90% of integration events would involve chromosomal DNA that has replicated at least once or is in the process of replication. This does not necessarily equate to integration into fully duplicated chromosomes since, as pointed out above, integration into a replication loop can most easily explain some of the attendant chromosomal aberrations.

Unexplained Properties of Integrated Foreign DNA

Some arrays of integrated foreign genes show two further peculiarities, which serve to link integration following injection into one-cell embryos yet more firmly with integration into the chromosomes of transfected cells: (1) Short 'filler' sequences of unknown origin are sometimes found at transgene-transgene junctions and CT junctions (Wilkie and Palmiter 1987 and Chen *et al.*, 1995); (2) On occasion, a substantial length of DNA from a remote region of the genome is found incorporated into the transgene array (Wilkie and Palmiter, 1987). Insertions that are apparently closely clustered, with intervening chromosomal DNA sequences (Covarrubias *et al.*, 1986 and Michalova *et al.*, 1988), may reflect a similar origin.

The filler sequences formally resemble the additions of sequence that occur at the V-D and D-J junctions of rearranged immunoglobulin genes. Similar sequences have been recorded at junction points in transfected cells (Woodworth-Gutai 1981; Wilson *et al.*, 1982 and Anderson *et al.*, 1984) and at chromosomal translocation junctions (Botchan *et al.*, 1980 and Hayday *et al.*, 1982). Whether they are synthesised *in situ*, or whether the nucleus contains a population of oligonucleotides, possibly single-stranded, that adventitiously close gaps between single-strand DNA extensions, is not known. Whatever the origin of filler sequences, their occurrence in transgene structures suggest that they are a manifestation of a common cellular activity that normally goes undetected.

The mechanism by which DNA sequences from remote chromosomal locations become incorporated into transgene arrays is equally obscure. They could result from the integration of a long extrachromosomal concatemer into two different chromosomal sites, followed by resolution of the complex in a way that does not produce a detrimental effect. An alternative, but perhaps less likely, mechanism would involve transgene integration at more than one site followed by a recombinatorial event, again without detrimental consequences for the cell. Either mechanism could explain an observed incidence of transgene-induced reciprocal translocation (Francke et al., 1992). The injection of DNA has a severe negative effect on embryo viability (Page et al., 1995). It is therefore quite possible that series of events like those proposed occur quite frequently, but usually cause embryo death and go undetected. Thus in order to entertain the possibility that such mechanisms occur, it is not necessary to suppose that their outcome is invariably benign.

Conclusions

Foreign DNA integrated into the genome following pronuclear microinjection usually, but not invariably, takes the form of a multimeric array. Extensive homology between the transgene and its integration site has not been observed. Both the arrays themselves and the chromosomes in the vicinity of integration sites take a wide variety of forms. The available evidence indicates that the arrays are generated extrachromosomally, partly by end-joining but mainly by a homologous recombination process which is likely to be non-conservative. Arrays are usually integrated into chromosomal DNA that has replicated or is in the process of replicating subsequent to fertilisation. The range of different configurations around the site of integration can all be explained quite readily, but the explanations are without mechanistic support and will remain so pending the emergence of a body of experimental data.

References

Al-Shawi, R., Burke, J., Jones, C.T., Simons, J.P. and Bishop, J.O. (1988) A Mup promoter-thymidine kinase reporter gene shows relaxed tissue-specific expression and confers male sterility upon transgenic mice. *Mol. Cell. Biol.* **8**, 4821–4828

Al-Shawi, R., Burke, J., Wallace, H., Jones, C., Harrison, S., Buxton, D., Maley, S., Chandley, A. and Bishop, J.O. (1991) The herpes simplex type 1 virus thymidine kinase is expressed in the testis of transgenic mice under the control of an internal promoter. *Mol. Cell. Biol.* **11**, 4207–4216

Anderson, R.A., Kato, S. and Camerini Otero, R.D. (1984) A pattern of partially homologous recombination in mouse L cells. *Proc. Natl. Acad. Sci. USA* **81**, 206–210

Bishop, J.O. and Smith, P. (1989) Mechanism of chromosome integration of microinjected DNA. *Mol. Biol. Med.* **6**, 283–298

Botchan, M., Stringer, J., Mitchison, T. and Sambrook, J. (1980) Integration and excision of SV40 DNA from the chromosome of a transformed cell. *Cell.* **20**, 143–152

Braun, R.E., Behringer, R.R., Peschon, J.J., Brinster, R.L. and Palmiter, R.D. (1989) Genetically haploid spermatids are genetically diploid. *Nature* **337**, 373–376

Brinster, R.L., Chen, H.Y., Trumbauer, M.E., Yagle, M.K. and Palmiter, R.D. (1985) Factors affecting the efficiency of introducing foreign DNA into mice by microinjecting eggs. *Proc. Natl. Acad. Sci. USA* **82**, 4438–4442

Burdon, T.G. and Wall, R.J. (1992) Fate of microinjected genes in preimplantation mouse embryos. *Mol. Reprod. Dev.* **33**, 436–442

Chen, C.M., Choo, K.B. and Cheng, W.T. (1995) Frequent deletions and sequence aberrations at the transgene junctions of transgenic mice carrying the papillomavirus regulatory and the SV40 TAg gene sequences. *Transgenic Res.* **4**, 52–59

Cousens, C., Carver, A.S., Wilmut, I., Colman, A., Garner, I. and O'Neill, G.T. (1994) Use of PCR-based methods for selection of integrated transgenes in preimplantation embryos. *Mol. Reprod. and Dev.* **39** 384–391

Covarrubias, L., Nishida, Y. and Mintz, B. (1986) Early postimplantation embryo lethality due to DNA rearrangements in a transgenic mouse strain. *Proc. Natl. Acad. Sci. USA* **83**, 6020–6024

Covarrubias, L., Nishida, Y. and Terao, M., D'Eustachio, P. and Mintz, B. (1987) Cellular DNA rearrangements and early developmental arrest caused by DNA insertion in transgenic mouse embryos. *Mol. Cell. Biol.* **7**, 2243–2247

Dym, M. and Fawcett, D.W. (1971) Further observations on the numbers of spermatogonia, spermatocytes and spermatids connected by intercellular bridges in the mammalian testis. *Biol. Reprod.* **4**, 195–215

Ellison, A.R., Wallace, H., Al-Shawi, R. and Bishop, J.O. (1995) Different transmission rates of herpesvirus thymidine kinase reporter transgenes from founder male parents and male parents of subsequent generations. *Mol. Reprod. Dev.* **41**, 425–434

Erickson, R.P. (1973) Haploid gene expression versus meiotic drive: the relevance of intercellular bridges during spermatogenesis. *Nature New Biol.* **243**, 210–212

Francke, U., Hsieh, C.L., Kelly, D., Lai, E. and Popko, B. (1992) Induced reciprocal translocation in transgenic mice near sites of transgene integration. *Mammalian Genome* **3**, 209–216

Hamada, T., Sasaki, H., Seki, R. and Sakaki, Y. (1993) Mechanism of chromosomal integration of transgenes in microinjected mouse eggs: sequence analysis of genome-transgene and transgene-transgene junctions at two loci. *Gene.* **128**, 197–202

Hayday, A., Ruley, H.E. and Fried, M. (1982) Structural and biological analysis of integrated polyoma virus DNA and its adjacent host sequences cloned from transformed rat cells. *J. Virol.* **44**, 67–77

Hughes, M.J., Lingrel, J.B., Krakowsky, J.M. and Anderson, K.P. (1993) A helix-loop-helix transcription factor-like gene is located at the mi locus. *J. Biol. Chem.* **268**, 20687–20690

Karls, U., Müller, U., Gilbert, D.J., Copeland, N.G., Jenkins, N.A. and Harbers, K. (1992) Structure, expression, and chromosome location of the gene for the beta subunit of brain-specific Ca_2^+/calmodulin-dependent protein kinase II identified by transgene integration in an embryonic lethal mouse mutant. *Mol. Cell. Biol.* **12**, 3644–3652

Konopka, A.K. (1988) Compilation of DNA strand exchange sites for non-homologous recombination in somatic cells. *Nucleic Acids Res.* **16**, 1739–1758

Lin, F.L., Sperle, K. and Sternberg, N. (1984) Model for homologous recombination during transfer of DNA into mouse L cells: role for DNA ends in the recombination process. *Mol. Cell. Biol.* **4**, 1020–1034

Lin, F.L., Sperle, K.M. and Sternberg, N.L. (1987) Extrachromosomal recombination in mammalian cells as studied with single- and double-stranded DNA substrates. *Mol. Cell. Biol.* **7**, 129–140

Michalova, K., Bucchini, D., Ripoche, M.A., Pictet, R. and Jami, J. (1988) Chromosome localization of the human insulin gene in transgenic mouse lines. *Hum. Genet.* **80**, 247–252

Page, R.L., Canseco, R.S., Russell, C.G., Johnson, J.L., Velander, W.H. and Gwazdauskas, F.C. (1995) Transgene detection during early murine embryonic development after pronuclear microinjection. *Transgenic Res.* **4**, 12–17

Pieper, F.R., de Wit, I.C., Pronk, A.C., Kooiman, P.M., Strijker, R., Krimpenfort, P.J., Nuyens, J.H. and de Boer, H.A. (1992) Efficient generation of functional transgenes by homologous recombination in murine zygotes. *Nucleic Acids Res.* **20**, 1259–1264

Rohan, R.M., King, D. and Frels, W.I. (1990) Direct sequencing of PCR-amplified junction fragments from tandemly repeated transgenes. *Nucleic Acids Res.* **18**, 6089–6095

Shimoda, K., Cai, X., Kuhara, T. and Maejima, K. (1991) Reconstruction of a large DNA fragment from coinjected small fragments by homologous recombination in fertilized mouse eggs. *Nucleic Acids Res.* **19**, 6654

Szostak, J.W., Orr-Weaver, T.L., Rothstein, R.J. and Stahl, F.W. (1983) The double-strand-break repair model for recombination (Review). *Cell* **33**, 25–35

Wake, C.T., Vernaleone, F. and Wilson, J.H. (1985) Topological requirements for homologous recombination among DNA molecules transfected into mammalian cells. *Mol. Cell Biol.* **5**, 2080–2089

Wallace, H., McLaren, K., Al-Shawi, R. and Bishop, J.O. (1994) Consequences of thyroid hormone deficiency induced by the specific ablation of thyroid follicle cells in adult transgenic mice. *J. Endocrinol.* **143**, 107–120

Whitelaw, C.B., Springbett, A.J., Webster, J. and Clark, J. (1993) The majority of G_0 transgenic mice are derived from mosaic embryos. *Transgenic Res.* **2**, 29–32

Wilkie, T.M. and Palmiter, R.D. (1987) Analysis of the integrant in MyK-103 transgenic mice in which males fail to transmit the integrant. *Mol. Cell Biol.* **7**, 1646–1655

Wilson, J.H., Berget, P.B. and Pipas, J.M. (1982) Somatic cells efficiently join unrelated DNA segments end-to-end. *Mol. Cell Biol.* **2**, 1258–1269

Woodworth-Gutai, M. (1981) Recombination in SV40-infected cells: viral DNA sequences at sites of circularization of transfecting linear DNA. *Virol.* **109**, 353–365

Xiang, X., Benson, K.F. and Chada, K. (1990) Mini-mouse: disruption of the pygmy locus in a transgenic insertional mutant. *Science* **247**, 967–969

PART III, SECTION A

36. Detection of Transgenes in the Preimplantation Stage Embryo

G.T. O'Neill

BBSRC/MRC Neuropathogenesis Unit, Institute of Animal Health, Ogston Building, King's Buildings, West Mains Road, Edinburgh EH9 3JF, Scotland

Introduction

The transgenic modification of animals has enabled significant advances to be made in our understanding of gene structure, function and expression. Introducing novel genetic constructs into the genome of numerous mammalian species has enabled modified spatio-temporal patterns of gene expression to be exploited. This has now lead to the development of new areas of biotechnology, notably that termed "genepharming", in which the traditional livestock species are genetically modified to produce novel proteins of both commercial and therapeutic value. Transgenic biology has also been examined as a means to radically alter quantitative characteristics (Clark et al., 1990; and Pursel and Rexroad, 1993).

The microinjection of DNA into the pronuclei of 1-cell embryos still remains the most efficient means to produce transgenic animals. In mouse, the production of transgenic offspring is now an established laboratory procedure, in which ~15% of first generation offspring (representing about 5-6% of injected 1-cell embryos) will carry integrated transgenes (Brinster et al., 1985). The production of transgenic offspring from other species, particularly pig and ungulates, has consistently been found to occur at a much lower incidence and has made the transgenic modification of these species, both in terms of animal and labour costs, an expensive undertaking. Technical and physical problems associated with the retrieval and microinjection of pronuclear stage embryos may be two of the factors that reduce the efficiency with which transgenic offspring occur. Apart from the strains of mouse most frequently used for microinjection, i.e. (C57Bl CBA)F1 and FVB, the visibility of the pronuclei of most species is obscured by the opaque and granular nature of the 1-cell cytoplasm. The use of *in vitro* maturation and *in vitro* fertilisation (IVM/IVF) techniques to produce larger numbers of "staged" cattle embryos for embryo manipulation and microinjection has not significantly improved the overall incidence of transgenic live births (Krimpenfort et al., 1991 and Eyestone, 1994). However it has provided an alternative route to address some of the problems currently associated with experimental and applied transgenesis in large animals, notably the visualisation and injection of pronuclei and the optimisation of *in vitro* conditions to maintain manipulated embryos in culture. It should also provide a large number of embryos to conduct studies into the detection and selection of putative first generation transgenic embryos. Investigation of the factors that control random heterologous chromosomal integration, DNA repair and replication (Rohan et al., 1990; and Burdon and Wall, 1992) may identify mechanisms to modify the actual incidence of chromosomal integration. The microinjection of specific promoter/enhancer constructs into the male- and female-derived pronuclei of murine embryos has also begun to identify the molecular mechanisms that activate and control gene expression in the early embryo (Majunder et al., 1993 and Henery et al., 1995). It has been proposed that there may be a link between the onset of genomic activation and the stage at which integration of exogenous DNA may be initiated (Whitelaw et al., 1993). In *Mus musculus*, zygotic genomic activation occurs during the 2-cell stage and the incidence of transgenesis in this species is relatively high. In the ungulate species, genomic activation occurs during the third cell cycle and the incidence of transgenesis at birth is less than 1% of injected embryos. As will be dicussed in a later section, chromosomal integration of DNA into a blastomere later than the 8-/16-cell stage will limit the contribution of the subsequent daughter cells to embryonic structures.

The selection of putative first generation (G_0) transgenic embryos during the preimplantation stages would reduce, specifically in relation to the domestic species, experimental costs as a smaller number of animals would be required to act as recipient females. With these species, DNA microinjection remains the only viable method for introducing transgenic modifications into the genome. Embryonic stem cell (ES) lines (Evans and Kaufman, 1981) and embryonic germ (EG) cell lines (Matsui *et al.*, 1992), provide a second route for genetic modification in rodents (Thomas and Capecchi, 1987; and Capecchi, 1989). Unfortunately, stable pluripotent cell lines with similar properties have yet to be isolated from the domestic species despite extensive effort expended upon this area of research over the past 10–15 years. The aim of this chapter is to review the strategies that have been used to investigate the detection of transgenes in early embryos and to discuss these finding in relation to future progress in transgenic embryo selection and animal biotechnology.

Detection of Exogenous DNA in Preimplantation Stage Embryos

The PCR reaction (Saiki *et al.*, 1988) enables a defined fragment of DNA to be exponentially amplified by thermostable DNA polymerases and is now one of the most widely used procedures to identify and analyse specific DNA sequences. It is particularly suited to the analysis of trace amounts of DNA, e.g. from single diploid cells or individual sperm cells (Li *et al.*, 1988) or the detection of a small number of transgenes in preimplantation embryos derived from established transgenic lines (King and Wall, 1988). It has also been used to analyse the persistance of DNA microinjected into 1-cell embryos and modifications to standard protocols, such as inverse-PCR (Ochman *et al.*, 1988) and anchored-PCR (Roux and Dhanarajan, 1990) have been proposed as a means to investigate transgene integration in early embryos (Ninomiya and Yuki, 1989). The use of fluorescent *in situ* hybridisation techniques to detect genes in preimplantation embryos will be discussed in the final section of this chapter.

The analysis of individual DNA microinjected embryos by the PCR reaction at several stages of preimplantation has repeatedly found that the DNA construct persists in the majority (80–100%) of pre- morulae stage embryos analysed (Ninomiya *et al.*, 1989, 1990; Burdon and Wall, 1992; Behboodi *et al.*, 1993, Cousens *et al.*, 1993; 1994 and Krisher *et al.*, 1994). In morulae and blastocyst stage embryos, the proportion in which detectable levels of microinjected DNA persisted was consistantly observed to drop to 25–35% of the surviving embryos. This value is severalfold higher than the incidence of transgenesis, as determined from "tail biopsy" analysis, observed in weaned offspring. These observations indicate that an unmodified PCR reaction will detect both integrated and nonintegrated forms of microinjected DNA even after several days of *in vivo* or *in vitro* culture. The stability of pronuclear injected DNA may result from the rapid concatomerization of the microinjected construct. Concatomerization of the pronuclear microinjected DNA has been reported to occur within 5–10 min-utes of microinjection (Burdon and Wall, 1992).

When the mouse blastocyst "hatches" from the zona pellucida, it rapidly invades the uterine lumen. The recovery of mouse embryos at days 5.5–6.5 *post coitum* is particularly difficult. In the domestic species, implantation of the embryo into the uterine wall does not occur until much later when extensive trophoblast expansion has altered the embryo into a structure that can be several centimeters in length. The developmental mechanisms that regulate the early development of ungulate embryos therefore offer an opportunity where recovery and analysis of the transgenic status of the microinjected embryo can be conducted through a larger developmental window. Such an approach may not only define the time period through which exogenous DNA can persist in a nonintegrated state but may indicate the basal incidence of transgenesis in pre- and peri-implantation stage embryos. This would be predictive of the incidence at birth as analyses of the developmental capacity of preimplantation embryos, as presented below, has found that embryolethality due to microinjection is largely confined to the preimplantational period. Krisher *et al.* (1994) have attemped to determine the limits of DNA persistance in microinjected IVM/IVF-derived bovine embryos. A significant number of embryos retained in culture for up to 14 days were still found by PCR analysis to carry exogenous DNA. Embryo explant cultures from 12 embryos that survived in culture for a further week were found not to carry exogenous DNA. No murine studies have conducted PCR analyses of embryos allowed to progress *in vitro* to form embryo explants or encouraged to undergo delayed implantation *in vivo* by transfer into ovarectomised recipients. These modifications could act as a model to the embryological events associated with pig and ungulate embryogenesis and provide further information about the basal incidence of DNA integration and concatomeric persistance at these stages of embryogenesis.

Numerous studies have examined the detrimental effects, both physical and developmental, of microinjection upon both murine and ungulate preimplantation embryo survival in vitro and in vivo (Walton et al., 1987; Rexroad and Wall, 1987; Rexroad et al., 1990; Behboodi et al 1993 and Kubisch, et al., 1995). The piercing of the cell and nuclear membranes followed by the introduction of many hundreds of copies of DNA in several femto litres of tris-EDTA based buffer results in the rapid degeneration of a small number of manipulated embryos. Those that cleave to the 2-cell stage and beyond have consistently been found to exhibit a significantly reduced ability to progress to the blastocyst stage. In our studies of microinjected embryo survival, (O'Neill and Wilmut, unpubl. obs.) only 25–30% of microinjected mouse embryos, in contrast 80–90% of non-injected "control" embryos, were found to progress from the 1-cell to blastocyst stage in vitro. The greatest developmental hurdle for a significant number of microinjected embryos was the transition from the compacted 8-cell stage/morulae to blastocyst. This may indicate that a reduction to the cell number, as a combined effect of in vitro culture and microinjection, in compacting embryos adversely affect their ability to complete blastocoel expansion and the differentiation of the inner cell mass (ICM) and trophectodermal lineages of the blastocyst. However, one study (Walker et al., 1990) has reported that no significant deleterious effects were observed to be associated with DNA microinjection techniques.

The detection of those embryos in which transgene integration has occurred will require not only the use of biopsy techniques that do not interfere significantly with the development of embryos that exhibit developmental retardation, but the development of a PCR-based protocol that can distinguish, e.g. by means of a biochemical difference, between integrated and non-integrated sequences. A small number of blastomeres can be removed from morulae and premorulae stage embryos using modifications to standard embryo manipulation techniques as described by Hogan et al. (1986). This microbiopsy material can be subjected to analysis and leaves the remainder embryos available for transfer to a recipient animal. It is also feasible to produce larger biopsy fragments by "embryo splitting" techniques (Szell and Hudson, 1991). Several studies with mouse (Bradbury et al., 1990), cattle (Kirkpatrick and Manson, 1993 and Machaty et al., 1993) and human embryos (Handyside et al., 1990) have shown that it is feasible to identify X and Y chromosome specific sequences in a blastomere biopsy and generate selective pregnancies from the remainder embryo. However, a reduction in the total cell number of the embryo is acknowledged to reduce the developmental ability of manipulated embryos. Biopsy of DNA microinjected embryos may further reduce the chances of successful pregnancies being established. PCR analysis of blastomere biopsy material removed from pronuclear microinjected embryos was initially considered by many as the most promising means to identify and subsequently transfer putative transgenic founder embryos. Even though the observed frequency of "transgenesis" during preimplantation was acknowledged to be higher than that observed at birth/weaning, the selection of those embryos that were found to be carrying exogenous DNA was believed to be a step in selecting against "negative" embryos. However, preliminary studies by Ninomiya et al. (1989) indicated that the PCR result derived from the analysis of demi-embryos did not correspond with the production of litters with an increased number of transgenic offspring. Analysis of 84 demi-embryos indicated that 30 carried exogenous DNA. The transfer of the 12 surviving embryos resulted in the birth of two offspring, one of which was transgenic. Transfer of 29 "negative" remainder embryos did not produce transgenic offspring. The results of this study has also indicated that the physical trauma of bisection or biopsy limits the expected postimplantation developmental success of embryos that have survived DNA microinjection.

Identification of Integrated Transgenes in Preimplantation Stage Embryos

Unmodified PCR analyses of individual embryos has clearly demonstrated that nonintegrated DNA persists within the embryos for several days. To identify first generation (G_0) transgenic embryos requires the development of analytical methods that can distinguish between recently integrated and episomal forms of the microinjected DNA construct. A suitable PCR-based transgene detection protocol, used in conjunction with embryo splitting or blastomere biopsy, could identify putative positive transgenic founder embryos suitable for transfer to recipient animals. One strategy has been to use dam methylase modified DNA for pronuclear microinjection. Dam methylation is a property of specific strains of E. coli that does not occur in mammalian systems and modifies adenine to N6 methyladenine (Hattman et al., 1978). The microinjected dam modified DNA will carry this biochemical "marker" until methyladenine is replaced by adenine by DNA integration/replication or repair. The *dam* methylase dependant endonuclease DpnI digests DNA at the recognition motif GA*TC (where A* represents methyladenine). In contrast to its isochizomer NdeI, DpnI does not digest *dam* negative or eukaryotic DNA. The incubation of *dam* positive DNA with DpnI, prior to being incorporated into a PCR reaction, will interfere with DNA amplification if PCR primers have been selected that flank one or more GA*TC motifs. Similarly, the incubation of an individual *dam* positive DNA microinjected embryo lysate with DpnI prior to PCR analysis should degrade the DNA that has not undergone modification through chromosomal integration/replication. DNA constructs that had been modified to a *dam* negative status would be available for PCR amplification as DpnI activity prior to PCR would not restrict the DNA at GATC motifs. This approach has been examined by Burdon and Wall (1992) and Cousens *et al.* (1994). However, the protocol is dependant on the premise that the non-integrated DNA does not undergo repair, replication or demethylation in the embryo during the time period between microinjection and PCR analysis. Analysis of microinjected embryos by DpnI/PCR-based methods (Burdon and Wall, 1994) at several stages of development indicated that one or more of these events did occur. Furthermore, in the subsequent study by Cousens *et al.* (1994), it was also observed that exposure of either *dam* positive or negative DNA to DpnI, even when the sequences examined did not have GATC motifs, inhibited the activity of the thermostabile DNA polymerase. This effect was abrogated by protease degradation of DpnI (a heat stable enzyme) prior to PCR. However, this introduced an increase the incidence of PCR "false positive" results. In a microreaction, DpnI was unable to degrade every copy of the exogenous *dam* modified DNA and the remainder molecules were available for PCR.

A modification of this approach, in which DNA used for microinjection was prepared with uridine residues, enabling the enzyme uracil DNA glycosylase (UDG) to degrade non-integrated DNA prior to PCR amplification (Longo *et al.*, 1990), has also been examined by Cousens *et al.* (1994). The results of this second approach also indicated that the non-integrated DNA underwent replication and/or repair during the early cell cycles of the preimplantation embryo. In 2- and 4-cell embryos, UDG treatment of uracil-DNA microinjected embryos reduced the proportion of PCR "positive" embryos from the expected 80–100% to 33% and 23% respectively. These values do not indicate the incidence of chromosomal integration but the ability of UDG to degrade uracil-DNA in the majority of embryos. The incidence of PCR "positive" embryos significantly increased to 67–70% in 8-cell and morula stage embryos. These latter results indicated that repair/replication of the uracil-DNA had occurred during the third or fourth cell cycle. The modification of microinjected DNA, either by replication or repair, prior to integration into host chromosomes infers that an approach based on the "biochemical tagging" of DNA is inappropriate to the identification of recently integrated transgenes by PCR-based methods. The use of PCR analysis of blastomere biopsy material may be of commercial value in occasions where the founder male transgenic animal derived from one of the domestic species exhibited a low transgene transmission frequency. Embryos, possibly produced by IVM/IVF procedures, could be screened for the presence of both the transgene and either X or Y chromosome specific sequences prior to their transfer to recipient hosts.

Distribution of Pronuclear Injected DNA in Preimplantation Stage Embryos

The preliminary results of Ninomiya *et al.* (1989) in which PCR analysis of embryo biopsy material was used to divide embryos into putative transgenic and "negative" embryos failed to promote the generation of litters with an increased number of transgenic offspring. The material provided for analysis did not accurately represent the status of the remainder embryo. The PCR analysis of individual blastomeres from microinjected embryos at the 4- and 8-cell stages has found that only a proportion carried copies of the exogenous construct (Burdon and Wall, 1992 and Cousens *et al.*, 1994). These results indicate that the concatomerised exogenous DNA may associate itself with nuclear or chromosomal elements that transfer the DNA to a restricted group of daughter cells. The results of both of these papers portray conditions where a microinjected embryo at the 8-cell stage may consist of blastomeres that carry exogenous DNA in several forms. If chromosomal integration occurred during or after the S phase in a 4-cell embryo, the 8-cell stage would exhibit 1 transgenic blastomere, possibly three with non-integated DNA and four that did not carry exogenous DNA. This mosaic distribution bears serious consequences in relation to the use of a single (or two) blastomere to analyse the transgenic status of a microinjected embryo, i.e. the biopsy may not be representative of the status of the remainder embryo. The incorporation of transgenic cells into the inner cell mass of the blastocyst also does not guarantee that they will give rise to embryonic structures. The cells of the ICM differentiate into parietal endoderm and primitive ectoderm and the contribution of transgenic cells to the former lineage may reduce the proportion of transgenic cells that contribute to the neonate. Furthermore, the daughter cells of a transgenic blastomere may contribute to an early tissue lineage that is wholly extraembryonic. The inability of some founder mice to transmit their transgenes to the next generation may not be due to factors such as meiotic disorder as a result of transgene insertion, but that the timing of the integration event precluded the daughter cells from being incorporated into the small number of (ICM) cells that give rise to the embryo and the germline (Merkett and Petters, 1978; Ginsberg et al., 1990). Biopsy of 8- and 16-cell stage embryos will in most cases involve the removal of the outer cells of the embryo. These are primarily destined to form the trophectodermal lineage of the blastocyst (Johnson and Ziomek, 1981). A combined molecular and statistical analysis of transgenic founder animals has revealed that the majority of Go founder mice exhibited mosiacism with respect to the transgene (Whitelaw et al., 1993). In this paper, a conservative estimation of the incidence of mosaicism in live born mice was proposed to be in the range of 60% and was believed to arise from integration events occurring during the second and subsequent cell cycles of the preimplantation period. The pattern of SV40-LacZ transgene expression in early mouse and bovine embryos has also identified that microinjected constructs are not uniformly distributed through the developing embryo (Takeda and Toyoda, 1991; and Kubisch et al., 1995). Although the location and fate of the LacZ constructs was analysed by a less sensitive assay, the similarity between their results and that produced by PCR analysis of individual blastomeres further confirms that microinjected DNA is not uniformly distributed within the preimplantation embryo.

Future Prospects for Transgene Analysis in Preimplantation Stage Embryos

The integration of exogenous DNA into the genome of the preimplantation embryo is an apparently random event. Integration is probably also dependent on a degree of damage being inflicted upon the integrity of the pronuclear DNA, the repair of which results in the incorporation of a concatomeric complex of the DNA construct. The ability to increase the incidence of transgenesis by the injection of both pronuclei may only increase the proportion of embryos that fail to progress to the later stages of preimplantation. In the absence of a working transgene selection/detection procedure for microinjected embryos, it may be more appropriate to transfer the embryos that have survived to the blastocyst stage into a suitable recipient animal. This does not increase the incidence of transgeneis, but would enable costs to be redirected to the purchase of animals as embryo donors and reduce the number of animals maintained as embryo recipients. Methods to increase the incidence of transgenesis may ultimately be promoted by either DNA/enzyme co-injection or the use of "binary" transgenic systems. In the latter, expression of specific recombinases can be induced to promote transgene insertion or excision into specific sites, as observed in model *Cre-lox* recombination systems (Baubonis and Saurer, 1993). Modifications to the approach of Schiestl and Petes (1991) can be used to catalyse chromosomal integration at specific sites within the genome when the exogenous DNA is co-injected with specific endonucleases. Fluorescent *in situ* hybridisation (FISH) techniques have been developed over recent years to give rapid and reliable diagnoses of the chromosome complement of human preimplantation stage embryos and blastomere biopsies. These techniques have enabled selective pregnancies to be established (Gimanez, et al., 1994 and Harper et al., 1994). The application of these techniques to detect transgenes in microinjected embryos may prove to be equally as problematic as the PCR-based procedures presented in this review. FISH analysis of these embryos may be predisposed to the detection of a nonspecific signal from both interphase and metaphase stage embryos, due to the presence of nonintegrated sequences associated with the nucleus. Moreover, the primary drawback to any form of preimplantation selection system is the established pattern of exogenous DNA mosaicism in microinjected embryos.

References

Baubonis, W. and Sauer, B. (1993) Genomic targeting with purified *cre* recombinase. *Nucl. Acid Res.* **21**, 2025–2029

Behboodi, E., Anderson, G.B., Horvat, S., Medrano, J.F., Murray, J.D. and Rowe, J.E. (1993) Microinjection of bovine embryos with a foreign gene and its detection at the blastocyst stage. *J. Dairy Sci.* **76**, 3392–3399

Bradbury, M.W., Isola, L.M. and Gordon (1990) Enzymatic amplification of a Y chromosome repeat in a single blastomere allows identification of the sex of preimplantation mouse embryos. *Proc. Natl. Acad. Sci. USA* **87**, 4053–4057

Brinster, R.L., Chen, H.Y., Trumbauer, M.E., Yagle, M.K. and Palmiter, R.D. (1985) Factors affecting the efficiency of introducing foreign DNA into mice by microinjecting eggs. *Proc. Natl. Acad. Sci. USA* **87**, 4438–4442

Burdon, T. and Wall, D. (1992) Fate of microinjected genes in preimplantation mouse embryos. *Mol. Reprod. Dev.* **33**, 436–442

Capecchi, M.R. (1989) Altering the genome by homologous recombination. *Science* **244**, 1288–1292

Clark, A.J., Archibald, A.L., McClenaghan, M., Simons, J.P., Whitelaw, C.B.A. and Wilmut, I. (1990) The germ-line manipulation of livestock: Progress during the past five years. *Proc. NZ Soc. Anim. Prod.* **50**, 167–178

Cousens, C., O'Neill, G.T., Wilmut, I. and Carver, A.S. (1993). Selection of preimplantation G_0 transgenic embryos. *J. Reprod. Fertil.* **11** 47 (abst)

Cousens, C., Carver, A.S., Wilmut, I., Colman, A., Garner, I. and O'Neill, G.T. (1994) Use of PCR-based methods for selection of integrated transgenes in preimplantation embryos. *Mol. Reprod. Dev.* **39**, 384–391

Evans, M.J. and Kaufman, M.H. (1981) Establishment in culture of pluripotential cells from mouse embryos. *Nature* **292**, 154–156

Eyestone, W.H. (1994) Challenges and progress in the production of transgenic cattle. *Reprod. Fertil Devel.* **6**, 647–652

Gimenez, C., Egozcue, J. and Vidal, F. (1994) Sexing sibling mouse blastomeres by polymerase chain-reaction and fluorescent *in situ* hybridization. *Hum. Reprod.* **9**, 2145–2149

Ginsberg, M., Snow, M.H.L. and McLaren, A. (1990) Primordial germ cells in the mouse embryo during gastrulation. *Development* **110**, 521–528

Handyside, A.H., Kontogianni, E.H., Hardy, K. and Winston, R.M.L. (1990). Pregnancies from biopsied human preimplantation embryos sexed by Y-specific DNA amplification. *Nature* **344**, 768–770

Harper, J.C., Coonen, E., Ramaekers, F.C.S., Delhanty, J.D.A., Handyside, A.H., Winston, R.M.L. and Hopman, A.H.N. (1994) Identification of the sex of human preimplantation stage embryos in 2 hours using an improved spreading method and fluorescent *in situ* hybriduszation. *Hum. Reprod.* **9**, 721–724

Hattman, S., Brooks, J.E. and Masurekar, M. (1978) Sequence specificity of the P1 modification methylase (mEcoP1) and the DNA methylase (m. Eco. *dam*) controlled by *Escherichia coli dam* gene. *J. Mol. Biol.* **126**, 367

Henery, C.C., Miranda, M., Wiekowski, M., Wilmut, I., and DePamplilis, M.L. (1995) Repression of gene-expression at the beginning of mouse development. *Devel. Biol.* **169**, 448–460

Hogan, B., Constantini, F. and Lacey, E. (1986) In *Manipulating the mouse embryo: a laboratory manual*. Cold Spring Harbor, NY. Cold Spring Harbor Press

Johnson, M.H. and Ziomek, C.A. (1981) The foundation of two distinct cell lineages within the mouse morula. *Cell* **24**, 71–80

King, D. and Wall, R.J. (1988) Identification of specific gene sequences in preimplantation embryos by genomic amplification: Detection of a transgene. *Mol. Reprod. Dev.* **1**, 57–62

Kirkpatrick, B.W. and Monson, R.L. (1993) Sensitive sex determination assay applicable to bovine embryos derived from IVM and IVF. *J. Reprod. Fertil.* **98**, 335–340

Krimpenfort, P., Rademakers, A., Eyestone, W., van der Schans, A., van den Broek, S., Kooiman, P., Kootwijk, E., Platenburg, G., Pieper, F., Strijker, R. and de Boer, H. (1991) Generation of transgenic dairy cattle using "*in vitro*" embryo production. *Bio/Technology* **9**, 845–847

Krisher, R.L., Gibbons, J.R., Gwazdauskas, F.C. and Eyestone, W.H. (1994) DNA detection frequency in microinjected bovine embryos following extended culture *in vitro*. *Theriogenology* **41**, 229

Kubisch, H.M., Hernandez-Ledezma, J.J., Larson, M.A., Sikes, J.D. and Roberts, R.M. (1995) Expression of two transgenes in *in vitro* matured and fertilized bovine zygotes after DNA microinjection. *J. Reprod. Fertil.* **104**, 133–139

Longo, M.C., Berninger, M.S. and Hartley, J.L. (1990) Use of uracil DNA glycosylase to control carry-over contamination in polymerase chain reactions. *Gene* **93**, 125–128

Machaty, Z., Paldi, A., Csaki, T., Varga, Z., Kiss, I., Barandi, Z. and Vajta, G. (1993) Biopsy and sex determination by PCR of IVF bovine embryos. *J. Reprod. Fertil.* **98**, 467–470

Majunder, S., Miranda, M. and DePamphilis, M.L. (1993) Analyses of gene expression in mouse preimplantation embryos demonstrates that the primary role of enhancers is to relieve repression of promoters. *EMBO* **12**, 1131–1140

Matsui, Y., Zsebo, K. and Hogan, B.L.M. (1992) Derivation of pluripotential embryonic stem cells from murine primordial germ cells in culture. *Cell* **70**, 841–847

Merkert, C.L. and Petters, R.M. (1978) Manufactured hexaparental mice show that adults are derived from three embryonic cells. *Science* **202**, 56–58

Ninomiya, T. and Yuki, A. (1989) Application of the polymerase chain reaction and inverted PCR to analyse transgenes in transgenic mice. *Agric. Biol. Chem.* **53**, 1729–1731

Ninomiya, T., Hoshi, M., Mizuno, A., Nagao, M. and Yuki, A. (1989) Selection of mouse preimplantation mouse embryos carrying exogenous DNA by polymerase chain reaction. *Mol. Rep. Dev.* **1**, 242–248

Ninomiya, T., Hoshi, M., Mizuno, A., Nagao, M. and Yuki, A. (1990) Selection of transgenic preimplantation embryos by PCR. *J. Reprod. Fertil.* **41** (suppl.), 222

Ochman, H., Gerber, A.S. and Hartl, D.L. (1988) Genetic applications of an inverse polymerase chain reaction. *Genetics* **120**, 621–623

Pursel, V.G. and Rexroad, C.E. (1993) Status of research with transgenic farm animals. *J. Anim. Sci.* **71** (suppl.3), 10–19

Rexroad, C.E. and Wall, R.J. (1987) Development of one-cell fertilized sheep ova following microinjection into pronuclei. *Theriogenology* **27**, 611–619

Rexroad, C.E., Powell, A.M., Rohan, R. and Wall, R.J. (1990) Evaluation of co-culture as a method for selecting viable microinjected sheep embryos for transfer. *Animal Biotech.* **1**, 1–10

Rohan, R., King, D. and Frels, W.I. (1990) Direct sequencing of PCR-amplified junction fragments from tandemly repeated transgenes. *Nucl. Acid Res.* **18**, 6089–6095

Roux, R.H. and Dhanarajan, P. (1990) A strategy for single site PCR amplification of dsDNA: Priming digested cloned or genomic DNA from an anchor-modified restriction site and a short internal sequence. *BioTecniques* **8**, 48–57

Saiki, R.K., Gelfand, D.H., Stoffel, S., Scharf, S.J., Higuchi, R., Horm, G.T., Mullis, K.B. and Erlich, H.A. (1988) Primer-directed enzymatic amplification of DNA with a thermolabile DNA polymerase. *Science* **239**, 487–491

Schiestl, R.H. and Petes, T.D. (1991) Integration of DNA fragments by illegitimate recombination in *Saccharomyces cerevisiae*. *Proc. Natl. Acad. Sci. USA* **88**, 7585–7589

Szell, A. and Hudson, R.H.H. (1991) Factors affecting the survival of bisected sheep embryos *in vivo*. *Theriogenolgy* **36**, 379–387

Takeda, S. and Toyoda, Y. (1991) Expression of SV40-LacZ gene in mouse preimplantation embryos after pronuclear microinjection. *Mol. Reprod. Dev.* **30**, 90–94

Thomas, K.R. and Capecchi, M.R. (1987) Site-directed mutagenesis by gene targetting in mouse embryo-derived stem cells. *Cell* **51**, 502–512

Walker, S.K., Heard, T.M., Verma, P.J., Rogers, G.E., Baeden, C.S., Sivaprasad, A.V., McLaughlin, K.J. and Seamark, R.F. (1990) *In vitro* assessment of the viability of sheep zygotes after pronuclear microinjection. *Reprod. Fertil. Dev.* **2**, 633–640

Walton, J.R., Murray, J.D., Marshall, J.T and Nancarrow, C.D. (1987) Zygote viability in gene transfer experiments. *Biol. Reprod.* **37**, 957–967

Whitelaw, C.B.A., Springbett, A.J., Webster, J. and Clark J. (1993) The majority of Go transgenic mice are derived from mosaic embryos. *Transgenic Res.* **2**, 29–32

PART III, SECTION A

37. Genetic Mosaicism in the Generation of Transgenic Mice

Yann Echelard

Genzyme Transgenics Corporation,
One Mountain Road, Framingham, Massachusetts, 01701-9322, USA

Genetic mosaicism is defined as the coexistence of two or more cell lines originating from a single zygote, within an individual. Genetic mosaicism often arises spontaneously and has been widely documented and reviewed (Stern, 1968; and Favor and Neuhäuser-Klaus, 1994).

The generation of transgenic animals seems particularly liable to favour the apparition of genetic mosaics at high frequencies. In some non-mammalian systems: Drosophila, sea urchin, Xenopus, fishes, chicken..., it has been shown that the microinjected DNA persists, and even replicates, for very long times in the dividing nuclei; and that most, if not all, transgenic founders are mosaic (Anders *et al.*, 1984; Flytzanis *et al.*, 1985; Hoffman and Goodman, 1987; Stuart *et al.*, 1988; and Bosselman *et al.*, 1989 reviewed in Houdebine and Chourrout, 1991). However in these cases, the generation of genetic mosaics is also a function of the techniques used to generate transgenic animals (cytoplasmic microinjection, utilization of circular plasmid, germ-cells infection consecutive to blastoderm microinjection).

In the case of transgenic mice and other mammalian systems, most commonly a linear transgene is microinjected into the pronuclei of fertilized eggs. The emergence of mosaics can then be explained by two types of events: persistence, and integration of the foreign DNA subsequent to the first DNA replication in the microinjected pronuclei (Wilkie *et al.*, 1986; and Burdon and Wall 1992), or, the probably less likely loss of an integrated transgene during early development (Ellison *et al.*, 1995).

A number of studies using transgenic mice have aimed to evaluate the frequency of mosaicism among founders. Originally, a study by Wilkie *et al.* (1986) concluded that approximately 30% of the transgenic founder mice are germline mosaic. This result was obtained by evaluating germline transmission data of 262 founder mice. However, it was acknowledged by the authors that 30% germline mosaicism was likely to be an underestimate of the true frequency of mosaicism. This, due to the facts that founders were screened by dot hybridization of tail DNA (a relatively insensitive technique) and that many of the transgenic lines included in the study had not produced enough offspring to reveal a significant deviation from normal hemizygous transmission.

Later, two sets of studies (Whitelaw *et al.*, 1990, Burdon and Wall, 1992 and Whitelaw *et al.*, 1993) have shown that the frequency of mosaicism following microinjection is probably even higher than 30%. Whitelaw *et al.* employed a combination of PCR and Southern blot analysis to detect and characterize transgenic animals. This study showed that out of 77 positive transgenic mice, at least 62% were either germline or somatic mosaic. The proportion of germline mosaic among the transgenic founders carrying at least one copy of the transgene (as detected by Southern blot analysis) was 23%, in line with the previous study of Wilkie *et al.* The difference could be traced to the large proportion (28.5%) of transgenic founders for which the integrated DNA could only be detected by the more sensitive PCR. Even the higher proportion of transgenic founders revealed by Whitelaw *et al.* are likely to be an underestimate, since there was no investigation of multi-integration events that could mask mosaic animals, and no investigation of extra-embryonic lineages.

Following a different approach, Burdon and Wall used the polymerase chain reaction (PCR) to track the transgene from the one-cell to the blastocyst stage. It was shown that 100% of the microinjected eight-cell embryos had a mosaic distribution of the microinjected DNA. In 19 out of 19 microinjected eight-cell embryos, the transgene was detected in four or less blastomeres. This analysis could not differentiate integrated from unintegrated microinjected DNA, and it could not be strictly predictive of the future status of the born founder. It is conceivable (although highly unlikely) that the 3 or 4 transgenic blastomeres from a mosaic eight-cell embryo could be the progenitors of most of the inner cell mass and the transgenic founder obtained could be identified as non-mosaic. However this study indicated that most microinjected integration must take place after the replication of pronuclear DNA.

To obtain a picture of transgene integration process in microinjected mouse embryos, we have used fluorescence *in situ* hybridization (FISH) to examine the status of a transgene in the developing microinjected egg (Lewis-Williams *et al.*, 1996, and manuscript in preparation). We microinjected a large Bovine α-*casein-lacZ* transgene (32.4 kb), for ease of FISH detection, into the pronuclei of 1-cell CD1 embryos. Microinjected embryos were cultured for 67 hours (8-cell to 16-cell stage), each embryo's zona pellucida was removed and blastomeres were dissociated (leaving individual embryos intact). Each embryo was immobilized and submitted to FISH using both an endogenous marker and a probe for the α-*casein-lacZ* transgene. All individual blastomere were scored for transgene-specific and endogenous signal (data summarized to Table 37.1). FISH allowed us to differentiate integrated from unintegrated DNA, permitting to deduct the probable moment of transgene integration. This experiment showed that transgene integration occurs mostly after the first DNA replication, in agreement with previous results described above. Two-thirds of embryos examined had at least one integration site; and 14% of all embryos (21% of embryos with integrations) had patterns consistent with multiple integrations. Most integrations occured at the two- and four-cell stage, but did also occur, with decreasing frequency up to the sixteen-cell stage (we did not examine later).

Mosaicism is an issue during basic studies involving small transgenic animals. Problems, as low level of transgene germline transmission, or masking of transgene-induced phenotypes by extensive mosaicism of the founders, can be overcome by generating large numbers of transgenic founder lines, as well as studying transgene-induced phenotypes in subsequent generations. Moreover, mosaicism can even be an advantage, allowing to generate viable animals with transgenes that would otherwise be lethal.

Table 37.1. Summary of FISH analysis of eight- and sixteen-cell embryos obtained by *in vitro* culture following microinjection with the Bovine α-*casein-lacZ* transgene. All embryos were analyzed with an endogenous mouse as well as an α-*casein-lacZ* (transgene) specific probes. * means that for these embryos there was presence of variable intensity FISH signals, which would indicate more than one integration with different copy number, some of the blastomeres containing more than one integration sites.

Percentage of blastomeres (per embryo) with transgene specific signal	Frequency	Probable integration stage
100	2/42 (4.8%)	one-cell
50	7/42 (16.7%)	two-cell
25	8/42 (19%)	four-cell
12.5	3/42 (7.1%)	eight-cell
6.25	2/42 (4.8%)	sixteen-cell
37.25	2/42 (4.8%)	four-cell + eight-cell*
62.5	1/42 (2.4%)	two-cell + eight-cell*
75	2/42 (4.8%)	two-cell + four-cell*
87.5	1/42 (2.4%)	two-cell + four-cell + eight-cell*
0	14/42 (33.3%)	no integration

However, the generation of genetic mosaic founders could be an obstacle in applied uses of transgenic animals for agricultural or pharmaceutical purposes, as for example, animal production improvement, the production of recombinant proteins in milk, transgenic animals for xenografts. Information obtained on the founder could be misleading, for example mosaicism could mask the effects of a deleterious integration, or , in the case of the milk production of recombinant proteins lead to the under evaluation of the production potential of a transgenic line. Low germline contribution could preclude the timely expansion of an interesting transgenic line.

Genetic mosaicism has been encountered in pigs (Pursel *et al.*, 1989), sheep (Simons *et al.*, 1988 and Rexroad *et al.*, 1991) and goats (Karl Ebert, Paul DiTullio and Harry Meade, personal communication). Although there are no systematic studies available mosaicism, is potentially as prevalent in large animals as in mice. With the amount of time necessary after the birth of the founder to obtain second generation animals (12–13 months for pigs, sheep and goats, 24 months for cattle), all information obtained from the founder animals has to be carefully interpreted, considering the possibility of mosaicism, in order to take judicious strategic decisions in the early development of a commercial transgenic herd.

References

Andres, A.C., Muellener, D.B. and Ryffel, G.U. (1984) Persistence, methylation and expression of vitellogenin gene derivatives after injection into fertilized eggs of *Xenopus laevis*. *Nucleic Acids Res.* **12**, 2283–2302

Bosselman, R.A., Hsu, R.Y., Boggs, T., Hu, S., Bruszewski, J., Ou, S., Kozar, L., Martin, F., Green, C., Jacobsen, F., Nicolson, M., Schultz, J.A., Semon, K.M., Rishell, W. and Stewart, R.G. (1989) Germline transmission of exogenous genes in the chicken. *Science* **243**, 533–535

Burdon, T.G. and Wall, R.J. (1992) Fate of microinjected genes in preimplantation mouse embryos. *Mol. Reprod. Dev.* **33**, 436–442

Ellison, A.R., Wallace, H., Al-Shawi, R. and Bishop, J.O. (1995) Different transmission rates of Herpesvirus thymidine kinase reporter transgenes from founder male parents and male parents of subsequent generations. *Mol. Reprod. Dev.* **41**, 425–434

Hoffman, F.M. and Goodman, W. (1987) Identification in transgenic animals of the Drosophila decapentaplegic sequences required for embryonic dorsal pattern formation. *Genes Dev.* **1**, 615–625

Houdebine, L.M. and Chourrout, D. (1991) Transgenesis in fish. *Experientia* **47**, 891–897

Favour, J. and Neuhäuser-Klaus, A. (1994) Genetic mosaicism in the mouse house. *Annu. Rev. Genet.* **28**, 27–47

Flytzanis, C.N., McMahon, A.P., Hough-Evans, B.R., Britten, R.J. and Davidson, E.H. (1985) Persistence and integration of cloned DNA in postembryonic sea urchins. *Devl. Biol.* **108**, 431–442

Lewis-Williams, J., Harvey, M., Wilburn, B., Destrempes, M. and Echelard Y. (1996) Analysis of transgenic mosaicism in microinjected mouse embryos using fluorescence *in situ* hybridization at various developmental time points. Abstract presented at the IETS meeting in Salt Lake City, 7–9 January 1996. Published in *Theriogenology*, January 1996

Pursel, V.G., Miller, K.F., Bolt, D.J., Pinkert, C.A., Hammer, R.E., Palmiter, R.D. and Brinster, R.L. (1989) Insertion of growth hormone genes ito pig embryos. In R.B. Heap, C.G. Prosser and G.E. Lamming (eds.), *Biotechnology in Growth Regulation*, Butterworths, London, pp. 181–188

Rexroad, C.E., Mayo, K., Bolt, D.J., Elsasser, T.H., Miller, K.F., Behringer, R.R., Palmitter, R.D. and Brinster, R.L. (1991) Transferrin- and albumin-directed expression of growth-related peptides in transgenic sheep. *J. Animal Sci.* **69**, 2995–3004

Simons, J.P., Wilmut, L., Clark, A.J., Archibald, A.L., Bishop, J.O. and Lathe, R. (1988) Gene transfer into sheep. *BioTechnology* **6**, 179–183

Stern, C. (1968) *Genetic Mosaics and Other Essays*, Harvard Univ. press, Cambridge

Stuart, G.W., McMurray, J.V. and Westerfield, M. (1988) Replication, integration and stable germline transmission of foreign sequences injected into early zebrafish embryos. *Development* **103**, 403–412

Wilkie, T.M., Brinster, R.L. and Palmiter, R.D. (1986) Germline mosaicism and somatic in transgenic mice. *Dev. Biol.* **118**, 9–18

Whitelaw, C.B.A., Archibald, A.L., Harris, S., McClenaghan, M., Simons, J.P., Springbett, A.J., Wallace, R. and Clark, A.J. (1990) Frequency of germline mosaicism in G_0 transgenic mice. *Mouse Genome* **88**, 114

Whitelaw, C.B.A., Springbett, A.J., Webster, J. and Clark, A.J. (1993) The majority of G_0 transgenic mice are derived from mosaic embryos. *Transgenic Res.* **2**, 29–32

PART III, SECTION A

38. Transgene Methylation in Transgenic Animals

Kenneth J. Snibson* and Timothy E. Adams

Centre for Animal Biotechnology, School of Verterinary Science, The University of Melbourne, Parkville, Victoria 3052, Australia

Introduction

It is now clear that DNA methylation has an important role in silencing genes in many biological contexts (Cedar, 1988 and Bird, 1993). Unfortunately, for workers who aspire to generate lines of transgenic animals which express introduced genes at consistent physiological levels, one context includes a repressive effect on transgene expression. Sequences introduced into the genome, whether by retroviral infection, microinjection or any other procedure commonly undergo de novo methylation (i.e. the addition of a methyl moiety to cytosine) (Gunthert et al., 1976; and Jaenisch and Jahner, 1984). One of the undesired consequences of de novo methylation in many cases is that it can repress the transcriptional activity of a transgene to well below physiologically effective levels. Aside from these detrimental effects, methylation plays a very important part in regulating expression of some transgenes in transgenic animals. In particular, the differential methylation of transgenes has been directly correlated with their correct tissue-specific (Shemer et al., 1990 and Salehi-Ashtiani et al., 1993), developmental (Shemer et al., 1991) and spatial expression (Donoghue et al., 1992 and Grieshammer et al., 1995).

One of the more interesting phenomena regarding transgene methylation is that the extent to which the transgene is methylated is sometimes found dependent on the sex and/or strain of the parent which donated the transgene (Hadchouel et al., 1987; Reik et al., 1987; Sapienza et al., 1987; Swain et al., 1987; Allen et al., 1990; Engler et al., 1991 and Sasaki et al., 1991). Parental- and strain-specific transgene methylation in some transgenic pedigrees has been implicated in the silencing of transcription of the introduced sequences (Hadchouel et al., 1987; Swain et al., 1987 and Allen et al., 1990). Typically, transgenes that are derived from the maternal transgenic parent exhibit high levels of methylation and reduced transcriptional activity. Little is known of why some transgenes are differentially modified in this manner with methylated cytosines. Much interest lies in the fact that parent-of-origin methylation and expression patterns seen in transgenic mice may resemble the phenomenon observed in a number of endogenous genes which is termed 'genomic imprinting'. Genomic imprinting, which is responsible for the distinct and complementary contribution of the two parental genomes to the developing embryo, refers to the epigenetic means by which a parent can determine whether its donated gene is repressed or expressed in offspring (McGrath and Solter, 1984). It is now clear that methylation plays an important role in the molecular mechanism of imprinting (Li et al., 1993). Why do 'methylation imprinted' transgenes sometimes behave like genomically imprinted endogenous genes? One possible explanation is that certain transgene sequences at particular chromosomal sites resemble 'imprinting boxes' of some endogenous genes which are subject to genomic imprinting (Barlow, 1993). 'Imprinting box' sequences are thought to direct an imprinted modification to the associated gene, where its imprinted expression pattern is part of normal mammalian development.

*Author for Correspondence.

DNA Methylation and Position Effects in Mammals

Most transgenes can be designed to include sufficient regulatory sequence information to confer, in the transgenic animal, appropriate developmental and tissue-specific expression. The major concern for *in vivo* gene transfer programs is, however, that the level of transgene expression is often inconsistent, varying between animals within a single lineage, and sometimes is completely extinguished after passage of the transgene to succeeding generations (Palmiter and Brinster, 1986). While some of these variable effects on expression can be associated with differential methylation of the transgene in single lines of transgenic animals, many of the effects are known to be due to the specific chromosomal environment at the site of transgene integration and are known as position effects. Some progress has been made recently in overcoming some of these obstacles to consistent transgene expression. Gene-transfer scientists, armed with a rapidly expanding knowledge of chromosomal organization, structure and functional units, are now able to in some cases characterize 'chromosomal organizing' sequences which, when linked to a transgene of interest, alleviate position effects on transgene transcription. These sequences, known as locus control regions (LCR) (reviewed by Orkin, 1990) and matrix or scaffold attachment regions (MARs or SARs) (reviewed by van-Driel *et al.*, 1991) allow a level of transgene expression that is essentially consistent with the endogenous gene, independent of the site of integration and dependent on transgene copy number.

Is there relationship between DNA methylation and position effects in mammals? We can gain intuitive insights into this issue by examining the broader context and discussing position effects in other organisms. Position effects are now recognized to be a universal phenomenon amongst eukaryotes. In yeast and *Drosophila* the phenomenon is referred as position-effect variegation (PEV) and is well characterized. Position-effect variegation is thought to be due to the effects of chromatin structure at the site of integration which influences the expression of an introduced gene. Moreover, the transcriptional activity of a gene, when placed in close proximity to transcriptionally inactive heterochromatin, is significantly repressed by the spreading of the surrounding heterochromatin. (Karpen, 1994 and Wolffe, 1994). What is of interest here is that the DNA of these two organisms is not modified by cytosine methylation (Bird, 1993). DNA methylation plays no part in the structural organization of chromatin in yeast and *Drosophila* and thus, for these two examples at least, is not involved in position-induced variation in the expression of an introduced gene. Conversely, in the mammalian genome the vast majority of chromatin is associated with methylated DNA and further, there is strong evidence that differential methylation determines higher order chromatin structure (Lewis and Bird, 1991; Eden and Cedar, 1994). Given the close relationship between chromatin assembly and PEV (Karpen, 1994) any influence DNA methylation has on position-induced variation in transgenic animals is likely to be an indirect one, through its intimate association with chromatin structure and function (Wilkins, 1990).

The Relationship Between Transgene Methylation and the Normal Function of Methylation in Eukaryotes

It is now thought that transgenes are methylated in mammalian cells because they inadvertently trigger a molecular host-defence mechanism which utilises DNA methylation to counter the effects of invading sequences (Doerfler, 1991, 1992 and Barlow, 1993). The host-defence mechanism can be best examined in organisms apart from vertebrate animals. This is because, unlike in all other organisms, in backboned animals DNA methylation appears to have evolved from serving a host-defence function to providing long-term (and reversible) global inhibition of endogenous gene expression (Bird, 1993). As such, the host-defence function of methylation in vertebrates cannot be inextricably divorced from its apparently essential role in global gene regulation (Li *et al.*, 1992). Where methylation exists in non-vertebrate animals, plants and fungi it is essentially confined to non-gene compartments of the genome and is apparently not required for repression of endogenous genes. The function of DNA methylation in these organisms as put forward by Bird (1993), is to distinguish between self and nonself DNA and then, more importantly, to repress and inactivate the exotic DNA, i.e. essentially a host-defence function. This phenomenon has been best illustrated in the filamentous fungi *Neurospora crassa* (Selker, 1993). The overall level of methylation in this organism is quite low, however when endogenous repeat sequences are transfected into *Neurospora* they are subject to considerable *de novo* methylation whether the sequences are inserted at the native chromosome site or at other sites (Selker *et al.*, 1993). Furthermore, methylated repeat sequences undergo a mutation process known as repeat-induced point mutation (RIP) in which the duplicate sequences are inactivated by a high rate of G.C to A.T point mutations. Other fungi and higher plants also have evolved silencing mechanisms directed against repeat sequences which appear to involve a role for methylation (Goyon and Faugeron, 1989; Freedman and Pukkila, 1993 and Flavell, 1994). Quite surprisingly, the proposed mechanisms in some cases involves regulating RNA processing as well as the more commonly observed effects at the level of transcription (Flavell, 1994).

Selfish sequences, transposable elements, proviruses or any other potentially damaging sequences are all threats to be countered in the complex genome. Accordingly, animals have retained methylation in their armoury against invading DNA. One of the consequences of this defence mechanism is that the host cell does not distinguish between hostile sequences and the integrated transgene. The host organism recognises the inserted transgene as foreign and sets in motion its molecular machinery, such as DNA methylation, to neutralise the 'exotic' DNA.

De Novo Methylation of Transgenes in Transgenic Animals

As discussed above, *de novo* methylation of transgene DNA can be considered part of the host-defence mechanism against invading DNA. A key question for biologists concerned with gene transfer is how do mammalian cells distinguish foreign DNA from their own sequences before setting in motion their molecular machinery to neutralise the threat? This issue must be addressed and strategically countered so that transgenes can be designed to avoid host recognition and silencing.

The molecular processes and/or complexes which initiate *de novo* methylation are only just beginning to be understood. The key enzyme responsible for methylating DNA in eukaryotes is DNA(cytosine-5)-methyltransferase (DNA Mtase). A single form of Mtase exists in adult mammalian tissues and is responsible for both *de novo* and maintenance methylation. This enzyme specifically adds a methyl group to the carbon-5 position on the cytosine ring but only where the cytosine molecule is followed by a guanosine residue (d-MCpG) (Bestor and Verdine, 1994). Cytosine is the only naturally modified base in mammalian tissues and essentially all methylation is restricted to the C in the CpG dinucleotide. Mtase methylates DNA in a highly organ- or tissue-specific manner and furthermore, the methylation patterns are faithfully inherited in somatic tissues. Specific *de novo* methylation as well as demethylation of previously methylated sequences are almost certainly involved in the formation of specific methylation patterns (Razin and Cedar, 1991; and Eden and Cedar, 1994). The mystery of precisely how a single form of Mtase establishes and maintains diverse methylation patterns seen in higher animals is still to be resolved. Addressing this question requires the elucidation of the particular details on what molecular properties/structures earmark transgene DNA (or any other DNA for that matter) for methylation. Other than the requirement for the dinucleotide CpG it is unclear to what extent specific sequence information residing in the transgene construct plays in directing *de novo* methylation (Orend *et al.*, 1995). Of course, bacterial plasmid sequences remaining in the construct after preparation for microinjection may provide a focal site for Mtase recognition and methylation as part of the host-defence response (Barlow, 1993). There are isolated reports that specific DNA fragments/elements may be responsible for the initiation of *de novo* methylation in flanking reporter gene sequences (Szyf *et al.*, 1990, Mummaneni *et al.*, 1993; and Hasse and Schulz, 1994). Other sequences have been characterised which direct demethylation of an associated transgene. In particular, an intronic immunoglobulin κ chain enhancer sequence, when placed in a downstream position relative to an associated κ chain, induces demethylation of the κ chain, promoting its tissue-specific transcription and thereby playing a role in B cell differentiation (Lichtenstein *et al.*, 1994). Aside from these sequence parameters much interest lies in the ability of Mtase to recognise and methylate 'unusual DNA' structures. Data from *in vitro* studies based on the ability of a purified mammalian Mtase to methylate various oligodeoxyribonucleotide substrates reveal that single-strand foldback secondary structures may provide the biological signal to either block or direct *de novo* methylation (Smith *et al.*, 1991; Smith *et al.*, 1992 and Christman *et al.*, 1995).

One intriguing possibility is that the very tandem repeat nature of many integrated transgenes form molecular structures at or near the replication fork which serve as targets for Mtase. Evidence is accumulating suggesting that mammalian cells may recognise and methylate repeat sequences in a manner similar to that described for fungi (Goyon and Faugeron, 1989; Freedman and Pukkila, 1993; and Selker *et al.*, 1993) and plants (Flavell, 1994). In transgenic plants, where research on inactivation of introduced sequences is more advanced than that for transgenic animals, it has been hypothesised that duplicated sequences may interact to form hybrid sequences (quadruplexes) which form a substrate for Mtase (Flavell, 1994). Indirect evidence in support for such a mechanism for methylation of repeat structures in mammals is that high copy repeat sequences in vertebrate genomes are normally highly methylated (Lewis and Bird, 1991). Stronger evidence has come from an unexpected source, the molecular dissection of the fragile X syndrome, a common cause of mental retardation in human males. The molecular basis of the fragile X syndrome centres on the methylation of an abnormally expanded triplet repeat sequence, (CGG)n, located in the 5' untranslated region of the FMR-1 gene. The triplet repeat is normally unmethylated and polymorphic in the human population, ranging from 5 to 50 repeat units. Once the repeat length increases to over 60 repeats the triplet repeat, upon maternal transmission, becomes unstable and prone to hyper-expansion. It is the abnormal methylation of the hyper-expanded repeat sequences and concomitant abolition of FMR-1 gene expression which manifests in the clinical signs of the syndrome (Heitz et al., 1991 and Oberle et al., 1991). Why should repeat sequences only become methylated in this way when the repeat size goes beyond the threshold of 60 repeat units? It turns out that unimolecular foldback structures formed from the C-rich strand of the FMR-1 triplet repeat are exceptional substrates for the human methyltransferase in vitro (Smith et al., 1994). Moreover, such secondary structure formed by the C-rich strand of the FMR-1 triplet repeat during triplet repeat expansion is sufficient to induce de novo methylation at the FMR-1 locus (Smith et al., 1994). Is secondary structure of repeat transgene sequences at the DNA replication complex the necessary catalyst for their de novo methylation, as it appears to be for the FMR-1 triplet repeat? In this regard it is interesting that where it has been specifically studied, the degree of methylation and ultimate silencing of a transgene repeat sequence correlates approximately with the copy number (Mehtali et al., 1990). As this has an obvious parallel with the methylation of expanded FMR-1 repeat sequences, it is tempting to speculate that, at high copy number, repeat transgene sequences are able to form molecular complexes or structures that can be recognised by Mtase for methylation. One significant outcome derived from these studies is that an effective way to reduce aberrant methylation and associated silencing of transgenes is to develop a strategy to limit the number of transgenes inserted at a particular genomic site to a single copy.

Once de novo methylation has been established methylation has been shown to gradually spread across the integrated sequences. This has been extensively studied in a number of mammalian cell lines containing integrated Adenovirus type 12 (Ad12) DNA (Doerfler, 1993a, b). Integrated Ad12 genomes become methylated in distinct patterns which subsequently spread across the viral genome progressively with time of subcultivation (Toth et al., 1990). Such methylation occurs predominantly in CpG dinucleotides and is faithfully maintained in somatic cell populations (Toth et al., 1990 and Doerfler, 1993b). Some light has been shed on the molecular basis for the spread of methylation from a primary site to flanking regions. In vitro studies conducted with purified human Mtase and partially methylated oligodeoxyribonucleotides reveal that methylated single-stranded foldback structures are able to direct methylation to a second site on the same strand (Smith et al., 1992 and Christman et al., 1995).

In some mammalian cells, the methylation spreads from the site of primary de novo methylation such that adjacent cytosines become methylated at non-CpG as well as CpG sites (Toth et al., 1990). This phenomenon is not solely restricted to cells in culture, as the spreading of methylation from CpG to non-CpG sites has been reported in transgenic mice which express sheep growth hormone (Snibson et al., 1995). Southern blot analysis with methylation-sensitive restriction endonucleases revealed the presence of methylated cytosines at CpC and CpA/T sites within the MToGH1 transgene of some F_1 and F_2 generation mice. More recently, mammalian methylation machinery was shown by the highly sensitive bisulphite sequencing protocol to be capable of de novo and maintenance methylation at CpNpG sites (Clark et al., 1995). The mechanism by which cytosines are de novo methylated and maintained at non-CpG sites is open to speculation at this stage. The maintenance methylation at CpG sites proceeds along a well established course based on the symmetry of the CpG dinucleotide in both DNA strands. As DNA replication proceeds hemimethylated CpG on the parent strand can serve as a template to direct methylation to the unmethylated CpG on the daughter strand. Conversely, non-CpG sites of methylation lack strand symmetry and therefore potential methylation sites in the opposite strand. Quite possibly the mechanism for methylation at non-CpG sites could be a random and non-templated process analogous to the widespread non-specific methylation of duplicated sequences in certain fungi (Goyon and Faugeron, 1989 and Selker et al., 1993). Alternatively, non-CpG methylation may be site-specific and directed by an unknown mechanism.

How does Methylation Inhibit Transgene Transcription?

Evidence derived from many experimental approaches over the past ten to fifteen years suggests that there are a variety of mechanisms by which DNA methylation represses transcriptional activity. The relative contribution of each mechanism to gene silencing in mammals is subject to much discussion. The simplest mechanism is that in some cases the methylation of key transcription factor binding sites directly interferes with the binding of transcription factors to these sites. This has been shown experimentally for a number of cellular factors which recognise sequences containing the CpG dinucleotide (Kovesdi *et al.*, 1987; Comb and Goodman, 1990; Bednarik *et al.*, 1991; Prendergast *et al.*, 1991; and Eden and Cedar, 1994). The result is reduced transcriptional activity of the associated gene (Tate and Bird, 1993). Although the binding of transcription factors is not always affected when their binding site is methylated at CpG sites (Holler *et al.*, 1988 and Ben Hattar *et al.*, 1989) it is likely that abnormal methylation at these sites within transgenes will reduce their transcription in a significant number of cases.

Methylation is also thought to repress transcription indirectly through an interaction with a methyl binding protein, MeCP1 (Meehan *et al.*, 1989, Boyes and Bird, 1991 and Boyes and Bird 1992). This protein binds DNA that is methylated at multiple CpG (at least 12) sites. A model for the role of MeCP1 in the regulation of transcription has been proposed by Bird (1992) that is based on relative promoter strength and density of methylated CpGs. Basically, an interaction of a weak promoter that is not highly methylated with the MeCP1 protein is sufficient to stall the transcriptional machinery. A low level of promoter methylation is unable to repress transcription of genes with strong promoters. According to the model, strong promoters are only inactivated by MeCP tightly bound to a promoter region exhibiting high density methylation. The attractiveness of the model is that the MeCP1 protein can combine with methylated CpG dinucleotides in any sequence context, and provides a mechanism whereby transcription of a host of genes can be globally controlled without the need for specific cognate sequences. Although the model deals specifically with the interrelationships of promoter strength and density of methylated CpGs in endogenous genes, there is no reason why MeCPs cannot interact with methylated transgenes in a similar manner to that described by Bird (1992). Indeed, MeCPs have been implicated in the transcriptional inactivation of introduced genes (Levine *et al.*, 1991 and Weng *et al.*, 1995). Results derived from the last-named study show that a murine transgene (designated HRD), which is subject to strain-specific methylation, is only transcribed in competent tissues when under-methylated. Interestingly, the bulk chromatin structure surrounding the HRD transgene does not vary with its transcriptional or DNA methylation status. The chromatin state is apparently not altered by transgene methylation, as DNase I enzyme sites surrounding the transgene are equally accessible whether methylation is present or not. The inability of the MspI enzyme to cut the methylated transgene (as opposed to the unmethylated transgene) is suggested to be due to a bound MeCP-like protein restricting MspI access to methylated CpG dinucleotides in the transgene (Weng *et al.*, 1995). To what extent MeCPs play a role in determining the level of transcriptional competence in the majority of methylated transgenes is still to be determined.

Other lines of evidence suggest that repressive effects of methylation are mediated through its influence on chromatin structure. Transcriptionally active genes are typically under-methylated and reside in open or accessible chromatin structure. Conversely, transcriptionally silent genes are surrounded by closed or inaccessible chromatin and are usually associated with a high degree of methylation (Lewis and Bird, 1991; and Eden and Cedar, 1994). Closed chromatin is predicted to restrict the access of various cellular factors required for gene transcription (Cedar, 1988; and Cedar and Razin, 1990). Closed chromatin structure associated with methylated DNA is likely to act as global repressor of transcription in a manner similar to that described above for MeCP proteins.

Conclusion

Since a report in the early 1980s first suggested that the methylation of transgene sequences in some cases correlated with their reduced expression in transgenic animals (Palmiter *et al.*, 1982) scientists are only now beginning to understand the nature and complexity of DNA methylation as it applies to introduced genes. The uncertainty of whether some transgenes will have altered methylation status in successive generations associated with the extinguishing of expression remains a major obstacle to the successful utilization of transgenic livestock for economic purposes. Clearly, if the goal is to have sufficient and stable long-term expression of introduced genes in transgenic animals then the 'methylation barrier' must be overcome.

References

Allen, N.D., Norris, M.L. and Surani, M.A. (1990) Epigenetic control of transgene expression and imprinting by genotype-specific modifiers. *Cell* **61**, 853–861

Barlow, D.P. (1993) Methylation and imprinting: From host defence to gene regulation? *Science* **260**, 309–311

Bednarik, D.P., Duckett, C., Kim, S.U., Perez, V.L., Griffis, K., Guenther, P.C. and Folks, T.M. (1991) DNA CpG methylation inhibits binding of NF-kappa B proteins to the HIV-1 long terminal repeat cognate DNA motifs. *New Biol.* **3**, 969–976

Ben Hattar, J., Beard P. and Jiricny J. (1989) Cytosine methylation in CTF and SP recognition sites of an HSV tk Promoter: Effects on transcription *in vivo* and on factor binding *in vitro*. *Nucleic Acids Res.* **17,** 10179–10190

Bestor, T.H. and Verdine, G.L. (1994) DNA methyltransferases. *Curr. Opin. Cell. Biol.* **6,** 380–9

Bird, A. (1992) The essentials of DNA methylation. *Cell* **70,** 5–8

Bird, A.P. (1993) Functions for DNA methylation in vertebrates. *Cold Spring Harb. Symp. Quant. Biol.* **58,** 281–5

Boyes, J. and Bird A. (1991) DNA methylation inhibits transcription indirectly via a methyl-CpG binding protein. *Cell* **64,** 1123–34

Boyes, J. and Bird A. (1992) Repression of genes by DNA methylation depends on CpG density and promoter strength: evidence for involvement of a methyl-CpG binding protein. *Embo. J.* **11,** 327–33

Cedar, H. (1988) DNA methylation and gene activity. *Cell* **53,** 3–4

Cedar, H. and Razin, A. (1990) DNA methylation and development. *Biochimica et Biophysica Acta* **1049,** 1–8

Christman, J.K., Sheikhnejad, G., Marasco, C.J. and Sufrin, J.R. (1995) 5-Methyl-2′-deoxycytidine in single-stranded DNA can act in cis to signal *de novo* DNA methylation. *Proc. Natl. Acad. Sci. USA* **92,** 7347–7351

Clark, S.J., Harrison, J. and Frommer, M. (1995) CpNpG methylation in mammalian cells. *Nature Genetics* **10,** 20–27

Comb, M. and Goodman, H.W. (1990) CpG Methylation inhibits proenkephalin gene expression and binding of the transcription factor AP-2. *Nucleic Acids Res.* **18,** 3975–3982

Doerfler, W. (1991) Patterns of DNA methylation – evolutionary vestiges of foreign DNA inactivation as a host defense mechanism. A proposal. *Biol. Chem. Hoppe Seyler* **372,** 557–564

Doerfler, W. (1992) DNA methylation: eukaryotic defense against the transcription of foreign genes? *Microb. Pathog.* **12,** 1–8

Doerfler, W. (1993a) Patterns of *de novo* DNA methylation and promoter inhibition: studies on the adenovirus and the human genomes. *Exs.* **64,** 262–299

Doerfler, W. (1993b) Adenoviral DNA integration and changes in DNA methylation patterns: a different view of insertional mutagenesis. *Prog. Nucleic Acid. Res. Mol. Biol.* **46,** 1–36

Donoghue, M., Patton, B.L., Sanes, J.R. and Merlie, J.P. (1992) An axial gradient of transgene methylation in murine skeletal muscle: genomic imprint of rostrocaudal position. *Development* **116,** 1101–1112

Eden, S. and Cedar, H. (1994) Role of DNA methylation in the regulation of transcription. *Curr. Opin. Genet. Dev.* **4,** 255–9

Engler, P., Haasch, D., Pinkert, C.A., Doglio, L., Glymour, M., Brinster, R. and Storb, U. (1991) A strain-specific modifier on mouse chromosome 4 controls the methylation of independent transgene loci. *Cell* **65,** 939–947

Flavell, R.B. (1994) Inactivation of gene expression in plants as a consequence of specific sequence duplication. *Proc. Natl. Acad. Sci. USA* **91,** 3490–3496

Freedman T. and Pukkila P.J. (1993) *De novo* methylation of repeated sequences in Coprinus cinereus. *Genetics* **135,** 357–66

Goyon, C. and Faugeron, G. (1989) Targeted transformation of Ascobolus immersus and *de novo* methylation of the resulting duplicated DNA sequences. *Molecular Cellular Biology* **9,** 2818

Grieshammer, U., McGrew, M.J. and Rosenthal, N. (1995) Role of methylation in maintenance of positionally restricted transgene expression in developing muscle. *Development* **121,** 2245–2253

Gunthert, U., Schweiger, M., Stupp, M. and Doerfler, W. (1976) DNA methylation in adenovirus, adenovirus transformed cells and host cells. *Proc. Natl. Acad. Sci. USA* **73,** 3923–3927

Hadchouel, M., Farza, H., Simon, D., Tiollais, P. and Pourcel, C. (1987) Maternal inhibition of hepatitis B surface antigen gene expression in transgenic mice correlates with *de novo* methylation. *Nature* **329,** 454–456

Hasse, A. and Schulz, W.A. (1994) Enhancment of reporter gene *de novo* methylation by DNA fragments from the α-fetoprotein control region. *J. Biol. Chem.* **269,** 1821–1826

Heitz, D., Rousseau, F., Devys, D., Saccone, S., Abderrahim, H., Le Paslier, D., Cohen, D., Vincent, A., Toniolo, D., Della Valle, G., Johnson, S., Schlessinger, D., Oberle, I. and Mandel, J.L. (1991) Isolation of sequences that span the fragile X and identification of a fragile X-related CpG island. *Science* **251,** 1236–1239

Holler M., Westin G., Jiricny J. and Schaffner W. (1988) Sp1 transcription factor binds DNA and activates transcription even when the binding site is CpG methylated. *Genes Dev.* **2,** 1127–1135

Jaenisch, R. and Jahner, D. (1984) Methylation, expression and chromosomal position of genes in mammals. *Biochimica et Biophysica Acta* **782,** 1–9

Karpen, G.H. (1994) Position-effect variegation and the new biology of heterochromatin. *Curr. Opin. Genet. Dev.* **4,** 281–291

Kovesdi, I., Reichel, R. and Nevins, J.R. (1987) Role of an adenovirus E2 promoter binding factor in EIA-mediated coordinate gene control. *Proc Natl Acad Sci USA* **84,** 2180–2184

Levine, A., Cantoni, G.L. and Razin, A. (1991) Inhibition of promoter activity by methylation: possible involvement of protein mediators. *Proc. Natl. Acad. Sci. USA* **88,** 6515–6518

Lewis, J. and Bird, A. (1991) DNA methylation and chromatin structure. *Febs. Lett.* **285,** 155–159

Li, E., Beard, C. and Jaenisch, R. (1993) Role for DNA methylation in genomic imprinting. *Nature* **366,** 362–365

Li E., Bestor, T.H. and Jaenisch, R. (1992) Targeted mutation of the DNA methyltransferase gene results in embryonic lethality. *Cell* **69,** 915–926

Lichtenstein, M., Keini, G., Cedar, H. and Bergman, Y. (1994) B cell-specific demethylation: A novel role for the intronic κ chain enhancer sequence. *Cell* **76,** 913–923

McGrath, J. and Solter, D. (1984) Completion of mouse embryogenesis requires both the maternal and paternal genomes. *Cell* **37,** 179–183

Meehan, R.R., Lewis, J.D., McKay, S., Kleiner, E.L. and Bird, A.P. (1989) Identification of a mammalian protein that binds specifically to DNA containing methylated CpGs. *Cell* **58,** 499–507

Mehtali, M., LeMeur, M. and Lathe, R. (1990) The methylation-free status of a housekeeping transgene is lost at high copy number. *Gene* **91**, 179–184

Mummaneni, P., Bishop, P.L. and Turker, M.S. (1993) A cis-acting element accounts for a conserved methylation pattern upstream of the mouse adenine phosphoribosyltransferase gene. *J. Biol. Chem.* **268**, 552–558

Oberle, I., Rousseau, F., Heitz, D., Kretz, C., Devys, D. Hanauer A., Boue J., Bertheas M.F. and Mandel J.L. (1991) Instability of a 550-base pair DNA segment and abnormal methylation in Fragile X syndrome. *Science* **252**, 1097–1102

Orend, G., Knoblauch, M., Kammer, C., Tjia, S.T., Schmitz, B., Linkwitz, A., Meyer, G., Maas J. and Doerfler, W. (1995) The initiation of *de novo* methylation of foreign DNA integrated into a mammalian genome is not exclusively targeted by nucleotide sequence. *J. Virol.* **69**, 1226–1242

Orkin, S.H. (1990) Globin gene regulation and switching, Circa 1990. *Cell* **63**, 665–672

Palmiter, R.D. and Brinster, R.L. (1986) Germ-line transformation of mice. *Ann. Rev. Genet.* **20**, 465–499

Palmiter, R.D. Chen, H.Y. and Brinster, R.L. (1982) Differential regulation of metallothionein-thymidine kinase fusion genes in transgenic mice and their offspring. *Cell* **29**, 701–710

Prendergast, G.C., Lawe, D. and Ziff, E.B. (1991) Association of *Myn*, the murine homolog of *Max* with *c-myc* stimulates methylation-sensitive DNA binding and *ras* cotransformation. *Cell* **65**, 395–407

Razin, A. and Cedar, H. (1991) DNA methylation and gene expression. *Microbiol. Rev.* **55**, 451–8

Reik, W., Collick, A., Norris, M.L., Barton, S.C. and Surani, M.A. (1987) Genomic imprinting determines methylation of parental alleles in transgenic mice. *Nature* **328**, 248–251

Salehi-Ashtiani, K., Widrow, R.J., Markert, C.L. and Goldberg, E. (1993) Testis-specific expression of a metallothionein I-driven transgene correlates with undermethylation of the locus in testicular DNA. *Proc. Natl. Acad. Sci. USA* **90**, 8886–8890

Sapienza, C., Peterson, A.C., Rossant, J. and Balling, R. (1987) Degree of methylation of transgenes is dependent on gamete of origin. *Nature* **328**, 251–254

Sasaki, H., Hamada,T., Ueda, T., Seki, R., Higashinakagawa, T. and Sakaki, Y. (1991) Inherited type of allelic methylation variations in a mouse chromosome region where an integrated transgene shows methylation imprinting. *Development* **111**, 573–81

Selker, E.U. (1993) Control of DNA methylation in fungi. *Exs* **64**, 212–217

Selker, E.U., Fritz, D.Y. and Singer, M.J. (1993) Dense nonsymmetrical DNA methylation resulting from repeat-induced point mutation in Neurospora. *Science* **262**, 1724–1748

Shemer, R., Kafri, T., O'Connell, A., Eisenberg, S., Breslow, J.L. and Razin, A. (1991) Methylation changes in the apolipoprotein AI gene during embryonic development of the mouse. *Proc. Natl. Acad. Sci. USA* **88**, 11300–11304

Shemer, R., Walsh, A., Eisenberg, S., Breslow, J.L. and Razin, A. (1990) Tissue-specific methylation patterns and expression of the human apolipoprotein AI gene. *J. Biol. Chem.* **265**, 1010–1015

Smith, S.S., Kan, J.L., Baker, D.J., Kaplan, B.E. and Dembek, P. (1991) Recognition of unusual DNA structures by human DNA (cytosine-5)methyltransferase. *Journal of Molecular Bioliogy* **217**, 39–51

Smith, S.S., Laayoun, A., Lingeman, R.G., Baker, D.J. and Riley, J. (1994) Hypermethylation of telomere-like foldbacks at codon 12 of the human c-Ha-ras gene and the trinucleotide repeat of the FMR-1 gene of fragile X. *J. Mol. Biol.* **243**, 143–51

Smith, S.S., Lingeman, R.G. and Kaplan, B.E. (1992) Recognition of foldback DNA by the human DNA (cytosine-5)-methyltransferase. *Biochemistry* **31**, 850–854

Snibson, K.J., Woodcock, D., Orian, J.M., Brandon, M.R. and Adams, T.E. (1995) Methylation and expression of a metallothionein promoter ovine growth hormone fusion gene (MToGH1) in transgenic mice. *Transgenic Res.* **4**, 114–122

Swain, J.L., Stewart, T.A. and Leder, P. (1987) Parental legacy determines methylation and expression of an autosomal transgene: A molecular mechanism for parental imprinting. *Cell* **50**, 719–727

Szyf, M., Tanigawa, G. and McCarthy, Jr., P.L. (1990) A DNA signal from the Thy-1 gene defines *de novo* methylation patterns in embryonic stem cells. *Mol. Cell Biol.* **10**, 4396–4400

Tate, P.H. and Bird, A. (1993) Effects of DNA methylation on DNA-binding proteins and gene expression. *Curr. Opin. Genet. Dev. Biol.* **3**, 226

Toth, M., Muller, U. and Doerfler, W. (1990) Establishment of *de Novo* DNA methylation patterns. transcription factor binding and deoxycytidine methylation at CpG and non-CpG sequences in an integrated adenovirus promoter. *Journal of Molecular Biology* **214**, 673–683

van-Driel, R., Humbel, B. and de-Jong, L. (1991) The nucleus: a black box being opened. *J. Cell Biochem.* **47**, 311–316

Weng, A., Engler, P. and Storb, U. (1995) The bulk chromatin structure of a murine transgene does not vary with its transcriptional or DNA methylation status. *Mol. Cell Biol.* **15**, 572–579

Wilkins, A.S. (1990) Position effects, methylation and inherited epigenetic states. *Bioessays* **12**, 385–386

Wolffe, A.P. (1994) Gene regulation. Insulating chromatin. *Curr. Biol.* **4**, 85–87

… PART III, SECTION A

39. Influence of Genomic Imprinting on Transgene Expression

J. Richard Chaillet

Department of Biological Sciences, University of Pittsburgh, Pittsburgh, PA 15260, USA

Introduction

Transgenesis in mammalian organisms is accomplished by the introduction and integration of DNA into the genome of the early embryo. Any of a number of techniques can be employed to facilitate the introduction of genetic material, although pronuclear injection of DNA into zygotes or the incorporation of genetically altered embryonic stem (ES) cells into early embryos are preferred. Regardless of the means by which transgenes are introduced, they must integrate into the genome for long-term propagation and vertical transmission. Once covalently included in the mouse genome, transgenes are subject to epigenetic modifications such as DNA methylation. The extent and the effect of methylation depend on many factors, including the sequence composition of the transgenes, and the genomic loci into which they integrate. Transgene methylation patterns may be associated with a reduction in the level of gene activity in the founder animals, and in a recurrent fashion in transgenic carriers of all future generations.

Even in instances in which a stably integrated and methylated transgene is transcriptionally active in the founder, it may be modified to a different extent in subsequent generations. An interesting and reproducible cause of this phenomenon is the process of genomic imprinting, which epigenetically modifies one or both of the parental alleles in the maternal and/or paternal germline and transmits the modification to the offspring. For the imprinted genes studied, the transmitted modification is DNA methylation which can be associated with transcriptional inactivation of the transgene. Thus, a situation may develop in which one of the parental alleles transmits a germline methylation pattern, and the opposite parental allele does not transmit its germline methylation. The former allele may be silent in the offspring due to retention of its methylation. As expected, the unmethylated allele acquires methylation in the early embryo, thereby becoming indistinguishable from the founder transgene allele. This interesting and important epigenetic modification of transgene behavior will be discussed in the context of a thoroughly investigated murine transgene, which undergoes a maternal germline modification, leading to transcriptional silencing of only the maternal allele in the offspring.

Developmental History and DNA Methylation

The fate of DNA introduced into the early embryo critically depends on the embryonic stage into which the DNA is introduced. Early observations in the area of epigenetic modification of the mammalian genome were made possible by the use of infectious Moloney murine leukemia virus (M-MuLV), which can be introduced into mouse cells at a wide variety of pre- and post-implantation embryonic stages (Jahner *et al.*, 1982). M-MuLV retroviral genomes introduced into mouse zygotes by microinjection of cloned (unmethylated) DNA, or into pre-implantation embryos by infection with essentially unmethylated M-MuLV, became *de novo* methylated and did not express. In contrast, when post-implantation mouse embryos were infected with the virus, no inhibition of virus expression and no *de novo* methylation were observed. An analogous differential effect between a pre- and post-implantation introduction of the retroviral genome was found in studies of embryonic carcinoma (EC) cells. In a similar fashion to the embryo studies, M-MuLV genomes introduced into undifferentiated EC cells became methylated and transcriptionally silent, whereas retroviruses introduced into differentiated cell lines remained unmethylated and active as demonstrated by productive retroviral infections (Stewart et al., 1982).

The aforementioned innovative studies indicate that the embryonic history of the integrated retrovirus can dramatically influence the extent of its DNA methylation and expression. *De novo* methylation activity in the early mammalian embryo targets unmethylated proviral genomes and results in their methylation and associated transcriptional silencing. This activity is absent from later embryonic stages. These observations made on retroviral genomes have been extended in two important ways. First, virtually all integrated transgene sequences (comprised of foreign or mouse sequences) and endogenous mouse sequences undergo *de novo* methylation during early embryogenesis. This methylation is subject to genetic (strain-specific) regulation (Engler et al., 1991; Chaillet et al., 1995 and Weng et al., 1995). The methylation acquired in this manner is inherited in somatic cell lineages, and is often related to the extent of transcriptional activity. Secondly, following *de novo* methylation in germ cells, methylation patterns can be transmitted to the embryo through the process of genomic imprinting. As will be evident from the following discussion, these gametic and embryonic processes have been elucidated in the mouse.

Since the important work performed on M-MuLV by Jaenisch and his colleagues, it has become very apparent that the mouse genome (and presumably all mammalian genomes) undergoes dramatic fluctuations in the extent of methylation. Mammalian DNA methylation is exclusively in the form of methylation at the 5 position of cytosine bases, and almost exclusively in the sequence context of a CpG dinucleotide (Bestor, 1993). The majority of CpG dinucleotides are subject to methylation, with the notable exception of constitutively unmethylated CpG islands (Bird, 1986). If we examine the overall (global) extent of genomic cytosine methylation, we see that genomic methylation is markedly reduced in fetal primordial germ cells, and in pre-implantation mouse embryos (Monk et al., 1987; Sanford et al., 1987 and Chaillet et al., 1991). In contrast, in differentiated cell types, such as gametes and cells from embryonic and adult somatic lineages, the degree of genomic methylation is very high. This is of the order of 60–80% of cytosine bases within CpG dinucleotides (Bestor, 1993).

The large difference in the level of genomic methylation between undifferentiated and differentiated cells arises because of changes in the genome associated with the reproductive cycle. Starting from the virtually unmethylated ground state of undifferentiated embryonic stem cells, the mouse genome becomes highly methylated in a manner coincident with cellular differentiation (Figure 39.1). Virtually all embryonic *de novo* methylation occurs during a relatively small window of time between the pre-implantation blastocyst stage and the implanted gastrulating embryo (Monk et al., 1987). As mouse transgenes are generally introduced into the

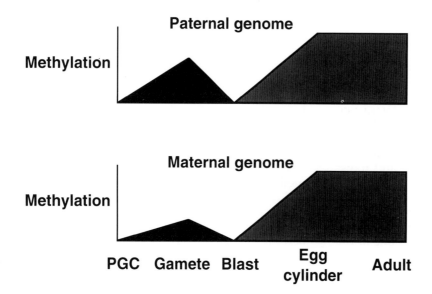

Figure 39.1. Global changes in DNA methylation during the reproductive cycle. Stages of the reproductive cycle are indicated on the horizontal axis. Gametogenesis is represented by PGC (primordial germ cells) and mature gametes (egg or sperm). Embryogenesis is represented by blast (blastocyst) and egg cylinder (early post-implantation embryo). The relative degree of CpG dinucleotide methylation within the genome is represented on the vertical axis.

early embryo or embryonic stem (ES) cells as unmethylated constructs, the transgene methylation patterns seen in differentiated somatic cells have likely been acquired during this wave of *de novo* methylation in early embryogenesis. The acquired methylation is then stably inherited in the somatic lineages of the later embryo and the adult. For many DNA constructs, this methylation will be associated with transcriptional inactivity of the transgene (Palmiter and Brinster, 1986; and Chaillet *et al.*, 1995). Except for alleles of imprinted genes (see below), methylation associated with gametes (sperm or unfertilized eggs) is lost soon after fertilization and thus does not directly contribute to embryonic methylation (Monk *et al.*, 1987 and Kafri *et al.*, 1992).

Genomic Imprinting

Besides the embryonic *de novo* methylation of transgene sequences, genomic imprinting is another potential impediment to reproducible transgene expression. To understand how genomic imprinting may influence transgene methylation and expression, we must first understand the relationship among the inheritance of epigenetic information, genomic imprinting, and normal mouse development. Afterwards, we will be able to discuss the effects of genomic imprinting on transgene methylation and expression.

Genomic imprinting can be defined as a regulatory process which distinguishes the parental origin of alleles such that one parental allele is transcriptionally active and the other silent. The importance of such a process for complete and normal mouse development arose from a series of embryological experiments in which the pronuclear composition of zygotes was changed. Normal mouse development did not occur in androgenetic embryos, in which the maternal pronucleus was removed and an additional paternal pronucleus from another zygote transplanted into the partially enucleated egg (McGrath and Solter, 1984; and Surani *et al.*, 1986). Likewise, when the paternal pronucleus of a fertilized egg was exchanged for a maternal one (gynogenetic egg), the mouse also failed to develop. Normal development only occurred if the developing zygote contained an intact maternal and an intact paternal pronucleus. The explanation for this developmental requirement of both a maternal and a paternal pronucleus is that the maternal and paternal haploid genetic contributions to the zygote encode two separate but complementary functions required for the developing embryo.

The notion arose from the nuclear transplantation experiments that the absolute biparental requirement for embryonic development is due to genomic imprinting. The maternal and paternal haploid genomes acquire different epigenetic imprints during gametogenesis. Afterwards, the imprints are transmitted to the embryo, creating a locus in which the two parental alleles are transcriptionally different (Solter, 1988). The consequence of genomic imprinting is that for some loci only the paternal allele is active, whereas for other loci only the maternal allele is active. The chromosomal location of imprinted loci has been estimated in a series of elegant experiments investigating the developmental outcome of uniparental disomies (Searle and Beechey, 1978; and Cattanach and Kirk, 1985). Individual imprinted genes have subsequently been identified which fall within these chromosomal domains (Barlow *et al.*, 1991; Bartolomei *et al.*, 1991; DeChiara *et al.*, 1991; Leff *et al.*, 1992; Hatada *et al.*, 1993; Giddings *et al.*, 1994; Hayashizaki *et al.*, 1994; Villar and Pedersen 1994; Guillemot *et al.*, 1995; and Hatada and Mukai, 1995).

Overall, these biological experiments demonstrate that a number of events must occur to create a locus in which the two parental alleles of an imprinted locus are functionally (transcriptionally) different (Table 39.1). First, one or both of the parental alleles must be uniquely marked (imprinted) during gametogenesis. Secondly, the imprints must be faithfully transmitted to the maturing embryo following fertilization and early embryonic development. Thirdly, inherited imprints must inhibit or promote gene expression in order to have a function consequence in the embryo. Lastly, parent-specific imprints must be reversible, because any allele at an imprinted locus must be capable of adopting the opposite parental imprint in a subsequent generation.

DNA methylation in mammals is the prime candidate for the genomic imprinting process. The reason for this is that the molecular process which modifies DNA double helices by methylating them possesses all of the aforementioned features which are required for the genomic imprinting process (Sasaki *et al.*, 1993 and Chaillet, 1992). These features are enumerated in parallel with the biological

Table 39.1. Genomic imprinting features.

Imprint	Methylation
1. Gametic epigenetic modification	Epigenetic modification
2. Transmitted to embryo	Heritable
3. Affects gene expression	Affects gene expression
4. Reversed in germ line	Removable

requirements of the imprinting process (Table 39.1). First, DNA methylation is a process which is known to epigenetically modify DNA sequences without affecting (mutating) the underlying DNA sequence. Secondly, DNA methylation patterns can be faithfully inherited following DNA replication (Bestor, 1990 and 1993). Thirdly, there is abundant evidence linking DNA methylation to the regulation of gene transcription (Cedar, 1988). Lastly, DNA methylation can be removed by active demethylation or by the absence of maintenance methylation normally associated with DNA replication. This removal also has no affect on the underlying genotype. Thus, the features of the DNA methylation process are precisely the desired features of a candidate genomic imprinting process.

A general role for DNA methylation in the process of genomic imprinting is supported by genetic experiments which examine the effect of decreases in the level of DNA cytosine methyltransferase activity on the expression of imprinted genes. Alleles of the methyltransferase gene with markedly reduced levels of activity were created with gene targeting techniques (Li et al., 1992). Mice heterozygous for these alleles are phenotypically normal. However, mice homozygous for these mutant alleles die in early embryogenesis. Interestingly, three genes which are normally imprinted in the early embryo have lost their imprinting in the homozygous mutant mice (Li et al., 1993). Whereas only one allele of each of the imprinted insulin-like growth factor II (*Igf2*), *H19*, and insulin-like growth factor II receptor (*Igf2/mpr*) genes is normally expressed, in the mutant mice either both alleles (*H19*) or no alleles (*Igf2* and *Igf2/mpr*) are expressed. The conclusion from these experiments is that methylation is important for at least one aspect of the imprinting process.

Transgene Imprinting

There is a rather small number of endogenous mouse genes (less than twenty) which are known to be genomically imprinted (Neumann et al., 1995). Because this represents a small percentage of the total number of known mouse genes, most genes in mammals are probably not imprinted. Curiously, there are only two murine transgenes which are known to be genomically imprinted, although there have been thousands of transgenes made, many of which have been examined for imprinting characteristics (Sasaki et al., 1993). Previously it was felt that a very large percentage of mouse transgenes are imprinted. This misconception arose because of the variety of mixed strain backgrounds on which the transgenes were evaluated. It has now been resolved that the apparent parent-specific imprinting of this latter category of transgenes is due to unlinked strain-specific modifiers which affect transgene methylation and/or expression in a parent-specific manner (Allen and Mooslehner, 1992 and Sasaki et al., 1993). When these "imprinted" transgenes are carefully studied on inbred strains, the alleles are indistinguishable. The maternal and paternal methylation patterns are invariably equivalent, and therefore the transgene loci are not imprinted. It is worth noting that even though the transgenes in these instances are not imprinted, the behavior of transgenes (methylation and expression) may vary depending on the strain background they are tested on (Sasaki et al., 1993). Thus, if the strain background of subsequent generations is different from that of the founder animal, the transgene behavior may be different even in the absence of *bona fide* transgene imprinting.

Two previously described transgenes, MPA434 and RSVIgmyc, exhibit *bona fide* transgene imprinting in inbred mice (Sasaki et al., 1993). The MPA434 transgenic line is imprinted at only a single integration site. Two other transgenic lines created with the same DNA construct as in MPA434, which have inserted at other genomic integration sites, are not imprinted. In contrast, the RSVIgmyc transgene is imprinted at virtually all sites of genomic integration (Chaillet et al., 1995). Therefore, the *cis*-acting signals for the imprinting of RSVIgmyc are within the transgene construct. Because an extensive series of experiments have been performed on the RSVIgmyc transgene to elucidate aspects of the imprinting process, this transgene will be discussed in more detail.

In an analogous fashion to the studies of developmental changes in DNA methylation which occur globally on genomic DNA sequences during mouse development, the methylation pattern of the RSVIgmyc transgene also undergoes large swings during its transit through the reproductive cycle of the mouse (Chaillet et al., 1991). These fluctuations are shown in a schematic fashion in Figure 39.2. Here, the relative degree of transgene methylation (deduced from the transgene methylation pattern) is followed over time. For the paternal transgene allele, the reproductive cycle changes in DNA methylation are not dissimilar from the changes which occur in the bulk DNA of the mouse genome (compare Figure 39.1 with Figure 39.2). However, a striking exception to the general trend in methylation is apparent when we examine the methylation of the maternal allele. In contrast to most DNA sequences and the paternal transgene allele, the gametic methylation of the maternal allele is transmitted to the embryo. Ultimately, because of the lower degree of *de novo* methylation placed on the paternal allele during embryogenesis, the two parental

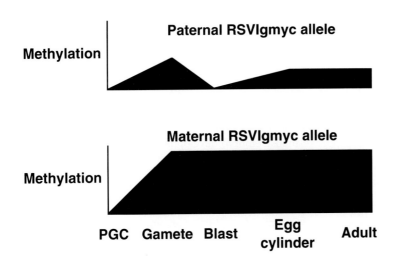

Figure 39.2. Changes in allele-specific DNA methylation of the imprinted RSVIgmyc transgene during the reproductive cycle. Gametogenesis and embryogenesis are represented on the horizontal axis in the same fashion as in Figure 39.1. The vertical axis represents the relative degree of CpG dinucleotide methylation on the transgene sequences.

alleles have different methylation patterns in the adult. This difference correlates with a difference in tissue-specific expression in the heart. We are led to the conclusion that there are two *de novo* methylation events during the reproductive cycle which can have an ultimate impact on transgene methylation and expression. For virtually all transgenes, *de novo* methylation during early embryogenesis occurs, and may have an impact on transgene expression. For the unusual transgene, the embryonic *de novo* methylation is superseded by a gamete-derived methylation which originates as *de novo* methylation in the germ line.

In a fundamental way, the process of mammalian DNA methylation provides an explanation for the requirements of establishing an epigenetic mark on an allele, and then providing a mechanism of propagating that mark. This holds true for methylation of the genome during either gametogenesis, embryogenesis, or postnatally. For the vast majority of genomic sequences, methylation levels are reduced to zero soon after fertilization, thereby isolating the embryo from the consequences of gametic *de novo* methylation. However, in the case of alleles of imprinted loci, gametic methylation is inherited by the embryo. The transcriptional consequences of this can supersede the consequences of embryonic *de novo* methylation seen in founder individuals and their transgenic descendants.

Conclusion

This chapter has focused on the influence of genomic imprinting on transgene expression in mammals. Because of the effects of genomic imprinting, the parental origin of both transgenes and endogenous genes can have significant consequences on their expression in embryonic and adult somatic tissue. Although this is a well-recognized phenomenon, only a few transgenes and endogenous genes are affected. A much more widespread and generally relevant process governing transgene expression is embryonic *de novo* methylation and its subsequent somatic-cell inheritance. In particular, embryonic *de novo* methylation is subject to the significant influences of strain-specific modifiers of transgene methylation and expression. Therefore, genetic background effects need to be considered when studying the behavior of mammalian transgenes.

References

Allen, N.D. and Mooslehner, K.A. (1992) Imprinting, transgene methylation, and genotype-specific modification. *Semin. Dev. Biol.* **3**, 87–98

Barlow, D.P., Stoger, R., Herrman, B.G., Saito, K. and Schweifer, N. (1991) The mouse insulin-like growth factor type-2 receptor is imprinted and closely linked to the *Tme* locus. *Nature* **349**, 84–87

Bartolomei, M.S., Zemel, S. and Tilghman, S.M. (1991) Parental imprinting of the mouse H19 gene. *Nature* **351**, 153–155

Bestor, T.H. (1990) DNA methylation: evolution of a bacterial immune function into a regulator of gene expression and genome structure in higher eukaryotes. *Philos. Trans. R. Soc. Lond.* (*Bio*) **326**, 179

Bestor, T.H. (1993) Methylation patterns in the vertebrate genome. *The Journal of NIH Research* **5**, 57–60

Bird, A.P. (1986) CpG islands and the function of DNA methylation. *Nature* **321**, 209–213

Cattanach, B.M. and Kirk, M. (1985) Differential activity of maternally and paternally derived chromosomal regions in mice. *Nature* **315**, 496–498

Cedar, H. (1988) DNA methylation and gene activity. *Cell* **53**, 3–4

Chaillet, J.R., Vogt, T.F., Beier, D.R. and Leder, P. (1991) Parental-specific methylation of an imprinted transgene is established during gametogenesis and progressively changes during embryogenesis. *Cell* **66**, 77–83

Chaillet, J.R. (1992) DNA methylation and genomic imprinting in the mouse. *Semin. Dev. Biol.* **3**, 99–105

Chaillet, J.R., Bader, D.S. and Leder, P. (1995) Regulation of genomic imprinting by gametic and embryonic processes. *Genes and Development* **9**, 1177–1187

DeChiara, T.M., Robertson, E.J. and Efstratiadis, A. (1991) Parental imprinting of the mouse insulin-like growth factor II gene. *Cell* **64**, 849–859

Engler, P., Haasch, D., Pinkert, C.A., Doglio, L., Glymour, M., Brinster, R. and Storb, U. (1991) A strain-specific modifier on mouse chromosome 4 controls the methylation of independent transgene loci. *Cell* **65**, 939–947

Giddings, S.J., King, C.D., Harman, K.W., Flood, J.F. and Carnaghi, L.R. (1994) Allele-specific inactivation of insulin 1 and 2 in the mouse yolk sac indicates imprinting. *Nature Genetics* **6**, 310–313

Guillemot, F., Caspary, T., Tilghman, S.M., Copeland, N.G., Gilbert, D.J., Jenkins, N.A., Anderson, D.J., Joyner, A.L., Rossant, J. and Nagy, A. (1995) Genomic imprinting of *Mash2*, a mouse gene required for trophoblast development. *Nature Genetics* **9**, 235–242

Hadchouel, M., Farza, H., Simon, D., Tiollais, P. and Pourcel, C. (1987) Maternal inhibition of hepatitis B surface antigen gene expression in transgenic mice correlates with *de novo* methylation. *Nature* **329**, 454–456

Hatada, I., Sugama, T. and Mukai, T. (1993) A new imprinted gene cloned by a methylation-sensitive genome scanning method. *Nucleic Acids Research* **21**, 5577–5582

Hatada, I. and Mukai, T. (1995) Genomic imprinting of p57KIP2, a cyclin-dependent kinase inhibitor, in mouse. *Nature Genetics* **11**, 204–206

Hayashizaki, Y., Shibata, H., Hirotsune, S., Sugino, H., Okazaki, Y., Sasaki, N., Hirose, K., Inoto, H., Okuizumi, H., Muramatsu, M., Komatsubara, H., Shiroishi, T., Moriwaki, K., Katsuki, M., Hatano, N., Sasaki, H., Ueda, T., Mise, N., Takagi, N., Plass, C. and Chapman, V.M. (1994) Identification of an imprinted U2af binding protein related sequence on mouse chromosome 11 using the RLGS method. *Nature Genetics* **6**, 33–40

Jahner, D., Stuhlmann, H., Stewart, C.L., Harbers, K., Lohler, J., Simon, I. and Jaenisch, R. (1982) De novo methylation and expression of retroviral genomes during mouse embryogenesis. *Nature* **298**, 623–628

Kafri, T., Ariel, M., Brandeis, M., Shemer, R., Urven, L., McCarrey, J., Cedar, H. and Razin, A. (1992) Developmental pattern of gene-specific DNA methylation in the mouse embryo and germ line. *Genes and Development* **6**, 705–714

Leff, S.E., Brannan, C.I., Reed, M.L., Ozcelik, T., Franke, U., Copeland, N.G. and Jenkins, N.A. (1992) Maternal imprinting of the mouse *Snrpn* gene and conserved linkage homology with the human Prader-Willi syndrome region. *Nature Genetics* **2**, 259–264

Li, E., Bestor, T.H. and Jaenisch, R. (1992) Targeted mutation of the DNA methyltransferase gene results in embryonic lethality. *Cell* **69**, 915–926

Li, E., Beard, C. and Jaenisch, R. (1993) Role for DNA methylation in genomic imprinting. *Nature* **366**, 362–365

McGrath, J. and Solter, D. (1984) Completion of mouse embryogenesis requires both the maternal and paternal genomes. *Cell* **37**, 179–183

Monk, M., Boubelik, M. and Lehnert, S. (1987) Temporal and regional changes in DNA methylation in the embryonic, extraembryonic and germ cell lineages during mouse embryo development. *Development* **99**, 371–382

Neumann, B., Kubicka, P. and Barlow, D.P. (1995) Characteristics of imprinted genes. *Nature Genetics* **9**, 12–13

Palmiter, R.D. and Brinster, R.L. (1986) Germ-line transformation of mice. *Annu. Rev. Genet.* **20**, 465–499

Reik, W., Collick, A., Norris, M.L., Barton, S.C. and Surani, M.A. (1987) Genomic imprinting determines methylation of parental alleles in transgenic mice. *Nature* **328**, 248–251

Sanford, J.P., Clark, H.J., Chapman, V.M. and Rossant, J. (1987) Differences in DNA methylation during oogenesis and spermatogenesis and their persistence during early embryogenesis in the mouse. *Genes and Development* **1**, 1039–1046

Sapienza, C., Peterson, A.C., Rossant, J. and Balling, R. (1987) Degree of methylation of transgenes is dependent on gamete of origin. *Nature* **328**, 251–154

Sapienza, C., Paquette, J., Tran, T.H. and Peterson, A. (1989) Epigenetic and genetic factors affect transgene methylation imprinting. *Development* **107**, 165–168

Sasaki, H., Allen, N.D. and Surani, M.A. (1993) DNA methylation and genomic imprinting in mammals. In J.P. Jost and H.P. Saluz, (eds.), *DNA methylation: Molecular Biology and Biological Significance*, Birkhauser Verlag, Basel, Switzerland pp. 469–486

Searle, A.G. and Beechey, C.V. (1978) Complementation studies with mouse translocations. *Cytogenet. Cell Genet.* **20**, 282–303

Solter, D. (1988) Differential imprinting and expression of maternal and paternal genomes. *Annu. Rev. Genet.* **22**, 127–146

Stewart, C.L., Stuhlmann, H., Jahner, D. and Jaenisch, R. (1982) De novo methylation, expression, and infectivity of retroviral genomes introduced into embryonal carcinoma cells. *Proc. Natl. Acad. Sci. USA* **79**, 4098–4102

Surani, M.A.H., Barton, S.C. and Norris, M.L. (1986) Nuclear transplantation in the mouse: Heritable differences between parental genomes after activation of the embryonic genome. *Cell* **45**, 127–136

Swain, J.L., Stewart, T.A. and Leder, P. (1987) Parental legacy determines methylation and expression of an autosomal transgene: A molecular mechanism for parental imprinting. *Cell* **50**, 719–727

Villar, A.J. and Pedersen, R.A. (1994) Parental imprinting of the *Mas* protooncogene in mouse. *Nature Genetics* **8**, 373–379

Weng, A., Magnuson, T. and Storb, U. (1995) Strain-specific transgene methylation occurs early in mouse development and can be recapitulated in embryonic stem cells. *Development* **121**, 2853–2859

PART III, SECTION B

40. The Use of Episomal Vectors for Transgenesis

Joé Attal[1], Marie Georges Stinnakre[2], Marie Claire Théron[1], Michel Terqui[3] and Louis Marie Houdebine[1]

[1] *Unité de Différenciation Cellulaire*
[2] *Laboratoire de Génétique Biochimique*
[3] *Laboratoire de Physiologie de la Reproduction des Mammifères Domestiques*
  Institut National de la Recherche Agronomique, 78352 Jouy-en-Josas Cedex, France

Circular stable vectors capable of autoreplicating and being transmitted to daughter cells are used for various purposes. Their advantages are known. They are present in cells in multicopies, they express their genes independently of the structure of the host cell genome, they are shuttle vectors and they can be rescued easily from their mammalian and bacterial hosts.

The Episomal Vectors Derived from Viral Genomes

The available vectors essentially derive from viral genomes which are naturally maintained in an episomal state. This is the case for SV40 genome which can autoreplicate in primate cells and is currently used in the monkey COS7 cells which express constitutively the viral T antigen (Gluzman, 1981). Similarly, vectors containing BK virus sequences are able to autoreplicate stably and to express efficiently the genes they harbour in human cells (Milanesi *et al.*, 1984). A fragment of the bovine papilloma virus is currently used as cloning and expression vector in mouse C127 cells (Stephens and Hentschel, 1987).

A part of the Epstein Barr Virus (EBV) is also able to autoreplicate, to be transmitted to daughter cells and to express efficiently the genes inserted in the vector. These vectors show good stability and capacity to harbour relatively long foreign DNA fragments. They are used to raise cDNA expression libraries (Margolskee *et al.*, 1988) to produce recombinant proteins (Kioussis *et al.*, 1987; Jalanko *et al.*, 1988; Young *et al.*, 1988 and Jalanko *et al.*, 1989), to analyse the mutation mechanisms (Du Bridge *et al.*, 1987) and to express antisense RNA (Groger *et al.*, 1989). They contain a well-identified origin of replication which is active only in the presence of, EBNAl, a specific EBV protein (Wysokenski and Yates, 1989). Optimal use of these vectors have been described recently (Teshigawara and Katsura, 1992 and Shen *et al.*, 1995). They have a major drawback: they are stable essentially in primate cells (Yates *et al.*, 1985 and Vidal *et al.*, 1990). Factors present only in these cells are thus necessary and this considerably limits the use of these vectors. Autoreplication of the vectors seems to take place in most cell type. It is rather the mechanism which allows the transmission of the replicated vector to daughter cells which is specifically active in primate cells. The natural mechanisms which allow episomal viral vectors to be transmitted in daughter cells are not well-known. Specific EBV proteins capable of binding to the viral genome and to the cell chromatin have been suspected to participate in transmission of the virus (Grogan *et al.*, 1983; Harris *et al.*, 1985 and Jiang *et al.*, 1991). The specific matrix attached region which has been found in EBV genome close to the origin of replication might also contribute to the transmission of the virus during cell division (Yankelevich *et al.*, 1992). Interestingly however, Trojan *et al.* (1993) reported that EBV genome based vector can be maintained spontaneously at an undetectable level as episomal vector in cultured rat cells and reamplified after a selection by the addition of hygromycin to the medium. Interestingly also, the reamplified vector which expressed anti- IGF-1 antisense RNA induced an activation of the immunosystem and an involution of glioblastoma.

Other episomal vectors based on the utilisation of other viruses such as vaccine or adenovirus are also used for specific purposes.

The Episomal Vectors Derived from Mammalian Genomes

Numerous sequences which are origins of replication are spread all along the genome of higher eucaryotes (De Pamphilis, 1993). The mechanism of action of these sequences is not well-known. They require transcription factors to work (Heintz, 1992). Replication and gene expression seem to have some antagonist effects (Haase *et al.*, 1994). Replication of DNA in eucaryotic cells seems to require attachment of chromatin to nuclear matrix (Amati *et al.*, 1990 and Gasser, 1991).

In Saccharomyces cerevisiae, the origin of replication is a well-defined short DNA sequence. In another yeast, Saccharomyces pombe, the origin of replication is spread over a relatively long stretch of DNA. The same is true in higher eucaryotes (De Pamphilis, 1993).

Several regions involved in mammalian DNA replication have been unambiguously identified. This is the case for sequences located in the vicinity of several genes such as those coding for dihydrofolate reductase (Burhans *et al.*, 1990), the CHO RPS14 locus (Tasheva and Roufa, 1994), CAD (Carroll *et al.*, 1987), adenosine deaminase (Carroll *et al.*, 1993), rRNA (Wegner *et al.*, 1989 and Little *et al.*, 1993) and laminin (Giacca *et al.*, 1994). Several of the plasmids which contain these origins of replication can be maintained to some extent in an episomal form. Interestingly, the vectors containing the origin of replication located in the intergenic region of rRNA genes are capable of autoreplicating as rolling circles and finally be integrated in a large number of copies. This allows high expression in the transfected cells (Hemann *et al.*, 1994).

In order to clone and identify sequences involved in mammalian DNA replication, the EBV origin of replication has been partly deleted from the classical vector. The resulting new vector was then used to clone functional replication sequence from human DNA libraries. A relatively large number of genomic sequences had the capacity to complement the defective EBV vector and to maintain it in a stable episomal form in human cells (Krysan *et al.*, 1989; Krysan *et al.*, 1991 and Heinzel *et al.*, 1991). Interestingly, some of these vectors were quite efficiently maintained in rodent cells (Krysan and Calos, 1993). Somewhat surprisingly, the authors of these studies concluded that any human genomic DNA fragment longer than 10 kb is able to generate an episomal vector when associated with the defective EBV vector. Although these results were very encouraging, it has not been reported so far that the EBV vectors containing mammalian genomic origin of replication are maintained in transgenic animals.

If origins of replications are really so numerous in mammalian genomes, the gene constructs used for transgenesis must sometimes contain such sequences. It is conceivable, although seemingly not proved, that some of these microinjected constructs give transgenics more efficiently than others because they autoreplicate and have thus more chance to be maintained during early embryo development and finally be integrated into the genome. Such an interpretation has been proposed to explain why in salmonids, microinjected DNA gives rise to the generation of many mosaic transgenic animals (Guyomard *et al.*, 1989).

Autonomous circular DNA has been found in several cell lines (Stanfield and Helinski, 1984; Rush and Misra, 1985; Kunisada and Yamagishi, 1987; and Wahl, 1989). These structures seem to participate to DNA amplification. Although they are candidate to be vectors for mammalian cells, their study seems very complex and they seem not to have been used so far.

The Episomal Vectors in Transgenic Animals

Several vectors have been reported to be maintained in an episomal state in transgenic mice. In some cases, the vectors were rescued from mouse tissues and were able to transform *E. coli*. They therefore shuttled between bacteria and animals.

A circular vector containing part of the bovine papilloma virus was used sucessfully to generate transgenic mice (Elbrecht *et al.*, 1987). This vector could be rescued and found in transformed bacteria. It was however unstable and practically not utilizable as a reliable vector for transgenesis.

Injection of the pPy LT1 plasmid containing the large T from polyoma virus into mouse embryos led to the generation of a set of new plasmids which derived from the injected pPy LT1. These plasmids were spontaneously maintained in an episomal state and could be transferred efficiently from mouse back to bacteria. When reinjected, these plasmids gave rise to the generation of a large proportion of transgenic mice harbouring the episomes (Leopold *et al.*, 1986). In our hands, the pPy LT1 microinjected into mouse embryo generated a rearranged plasmid but one of the episomal structure, p12B1, obtained by Leopold *et al.* (1986) never colonized mouse embryo and was never found as episome in adults. Potter and Lloyd (1987) observed the same event.

Several groups have shown independently that the upstream region of human *c-myc* gene contains a DNA sequence which can sustain autoreplication when added to a plasmid (Iguchi-Ariga *et al.*, 1988 and Mc Whinney *et al.*, 1995). Plasmids containing this *c-myc* sequence were reported to be transmitted to transgenic mice in an episomal state (Sudo *et al.*, 1990). In our hands, the human *c-myc* region did not provide to plasmids the capacity to remain episomal in transgenic mice.

A plasmid containing an association of Herpes simplex thymidine kinase gene and a fragment of *E. coli lacZ* gene was shown to be maintained in an episomal state in mouse Ltk$^-$ cells. The same plasmid was also found to remain in an episomal state in transgenic mice, to be rescued from mouse tissue and to replicate when transferred back into *E. coli* (Kiessing *et al.*, 1986). We were not able to obtain episomal forms of this plasmid (pSK1) in transgenic mice.

The reason why we and others were unable to obtain episomes from the above described plasmids is not clear. One possible explanation is that these plasmids can indeed autoreplicate but are too unstable to be observed easily. To tentatively improve the stability of the plasmids, we constructed a plasmid containing the three autoreplication sequences depicted above. This plasmid named pSKAC is described in Figure 40.1.

In order to evaluate its capacity to autoreplicate in an episomal form, the puromycin resistant gene under the control of SV40 promoter from the pBSpacΔp (De La Luna *et al.*, 1988) was introduced into the Ssp1 site of the plasmid pSKAC. This plasmid did not show a capacity higher than pBSpacΔp to confer resistance to puromycin in different cell types and no episome could be rescued from the resistant cells indicating that the plasmid was integrated.

pSKAC and the control plasmid devoid of the autoreplication sequences were transfected in parallel into cells. Plasmid DNA was extracted by the method of Hirt at different times after the transfection. The plasmids were visualized by Southern blotting. The plasmid pSKAC disappeared more slowly from the cells. It was still visible after one week whereas the control plasmid was disappeared.

The plasmid pSKAC in the circular form was microinjected into mouse embryo pronuclei. All the born mice harboured pSKAC sequences. This plasmid was therefore of very high efficiency to generate transgenic mice. The copy number was variable, ranging from 3 to 800. Examination of the DNA sequences isolated from the transgenic mice indicated that the original plasmid was rearranged in the same way in all the transgenic animals. The new rescued plasmid (4.2 Kb) was named pEpitrans. Its structure was not determined in detail. This plasmid was rescued from the tissues of the young transgenic mice. The plasmid was transferred into *E. coli* successfully and extracted in a non-modified form from the bacteria. The plasmid pEpitrans was injected into pig embryo pronuclei. It was found in tissues from almost all 30 days foetuses (12 out of 13). The plasmid could be rescued after an Hirt extraction and transferred to bacteria in an apparently unmodified form.

The pEpitrans is therefore a true shuttle vector. However, it was unstable and disappeared progressively from the tissues of the originally transgenic mice. Moreover, these mice were unable to transfer the episomal vector to their progeny.

Discussion

Although episomal vectors would be a precious tool for gene transfer into animals, no reliable material is presently available. Even not perfect,

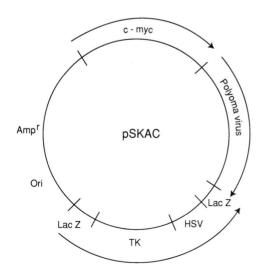

Figure 40.1. Schematic representation of pSKAC plasmid. The HindIII-KpnI upstream region of the human *c-myc* gene was inserted into the HindIII KpnI sites of ppolyIII plasmid (Lathe *et al.*, 1987). In the HindIII-BglII site of the resulting plasmid, the HindIII-BglII fragment from the pl2Bl plasmid (Leopold *et al.*, 1986) was introduced. The SspI-HaeII fragment of the resulting plasmid containing the *c-myc* and the pl2Bl fragments was blunted and inserted into the SspI site of the pSKI plasmid which contains two regions of LacZ gene, the thymidine kinase gene and a fragment of Herpes simplex virus (Kiesling *et al.*, 1986). The resulting plasmid (7.5 Kb) containing three origins of replication was named pSKAC. The arrows indicate the direction 5'P→3'OH for each DNA fragment.

episomal vectors would be very useful. The efficiency of transgenesis may be so high with such vectors that animals, even unable to transfer them to their progeny, would be of interest. Such vectors might even have the advantage of being not transmitted. From a biosafety point of view, the fact that they are shuttle structures capable of being transmitted from mammals to bacteria and back may also create serious biorisks for the environment. Deletion of the bacterial origin of replication before microinjection into embryos should eliminate the risk.

The experiments depicted above suggest that more or less potent autoreplication sequences can be found easily. The most potent can maintain vectors in an episomal state as long as the host cells are dividing rapidly. Quiescent or slowly growing cells seem unable to keep this kind of gene construct. Most likely, centromeric sequences or other DNA sequences having equivalent properties should be added to the plasmid already containing origins of replication. Centromeric sequences are only partly known in higher eucaryotes (Willard, 1990) and presently not utilizable for this purpose (Greider, 1991).

References

Amati, B., Pick L., Laroche, T. and Gasser, S.M. (1990) *EMBO J.* **9**, 4007–4016

Burhans, W.C., Vassilev, L.T., Caddle, M.S., Heintz, N.H. and DePamphilis, M.L. (1990) Identification of an origin of bidirectional DNA replication in mammalian chromosomes. *Cell* **62**, 955–965

Carroll, S.M., Gaudray, P., De Rose, M.L., Emery, J.F., Meinkoth, J.L., Nakkim, E., Subler, M., Von Hoff, D.D. and Wahl, G.M. (1987) Characterization of an episome produced in hamster cells that amplify a transfected CAD gene at high frequency: functional evidence for a mammalian replication origin. *Mol. Cell. Biol.* **7**, 1740–1750

Carroll, S.M., De Rose, M.L., Kolman, J.L., Nonet, G.H., Kelly, R.E. and Wahl, G.M. (1993) Localization of a bidirectional DNA replication origin in the native locus and in episomally amplified murine adenosine deaminase loci. *Mol. Cell. Biol.* **13**, 2971–2981

De La Luna, S., Soria, I., Pulido, D., Ortin, J. and Jiménez A. (1988) Efficient transformation of mammalian cells with constructs containing a puromycin-resistance marker. *Gene* **62**, 121–126

DePamphilis, M.L. (1993) Eukaryotic DNA replication: anatomy of an origin. *Annu. Rev. Biochem.* **62**, 29–63

DuBridge, R.B., Tang, P., Hsia, H.C., Leong, P.M., Miller, J.H. and Calos, M.P. (1987) Analysis of mutation in human cells by using an Epstein-Barr virus shuttle system. *Mol. Cell. Biol.* **7**, 379–387

Elbrecht, A., Demayo, F., Tsai, M.J. and O'Malley, B.W. (1987) Episomal maintenance of a bovine papilloma virus vector in transgenic mice. *Mol. Cell. Biol.* **7**, 1276–1279

Gasser, S.M. (1991) Replication origins, factors and attachment sites. *Cell Biol.* **3**, 407–413

Giacca, M., Zentilin, L., Norio, P., Diviacco, S., Dimitrova, D., Contreas, G., Biamonti, G., Perini, G., Weighardt, F., Riva, S. and Falaschi, A. (1994) Fine mapping of a replication origin of human DNA. *Proc. Natl. Acad. Sci. USA* **91**, 7119–7123

Gluzman, Y. (1981) SV40-transformed simian cells support the replication of early SV40 mutants. *Cell* **23**, 175–182

Greider, C.W. (1991) Telomeres. *Cell Biol.* **3**, 444–451

Grogan, E.A., Summers, W.P., Dowling, S., Shedd, D., Gradoville, L. and Miller, G. (1983) Two Epstein-Barr viral nuclear neoantigens distinguished by gene transfer, serology, and chromosome binding. *Proc. Natl. Acad. Sci. USA* **80**, 7650–7653

Groger, R.K., Morrow, D.M. and Tykocinski, M.L. (1989) Directional antisense and sense cDNA cloning using Epstein-Barr virus episomal expression vectors. *Gene* **81**, 285–294

Guyomard, R., Chourrout, D., Leroux, C., Houdebine, L.M. and Pourrain, F. (1989) Integration and germ line transmission of foreign genes microinjected into fertilized trout eggs. *Biochim.* **71**, 857–863

Haase, S.B., Heinzel, S.S. and Calos M.P. (1994) Transcription inhibits the replication of autonomously replicating plasmids in human cells. *Mol. Cell. Biol.* **14**, 2516–2524

Harris, A., Young, B.D. and Griffin, B.E. (1985) Random association of Epstein-Barr virus genomes with host cell metaphase chromosomes in Burkitt's lymphoma-derived cell lines. *J. Virol.* **56**, 328–332

Heintz, N.H. (1992) Transcription factors and the control of DNA replication. *Cell Biol.* **4**, 459–467

Heinzel, S.S., Krysan, P.J., Tran, C.T. and Calos, M.P. (1991) Autonomous DNA replication in human cells is affected by the size and the source of the DNA. *Mol. Cell. Biol.* **11**, 2263–2272

Hemann, C., Gärtner, E., Weidle, U.H. and Grummt, F. (1994) High-copy expression vector based on amplification-promoting sequences. *DNA and Cell Biol.* **13**, 437–445

Iguchi-Ariga, S.M.M., Okazaki, T., Itani, T., Ogata, M., Sato, Y. and Ariga, H., (1988) An initation site of DNA replication with transcriptional enhancer activity present upstream of the c-myc gene. *EMBO J.* **7**, 3135–3142

Jalanko, A., Kallio, A., Ruohonen-Lehto, M., Söderlund, H. and Ulmanen, I. (1988) An EBV-based mammalian cell expression vector for efficient expression of cloned coding sequences. *Biochem. Biophys. Acta* **949**, 206–212

Jalanko, A., Kallio, A., Salminen, M. and Ulmanen, I. (1989) Efficient synthesis of influenza virus hemagglutinin in mammalian cells with an extrachromosomal Epstein-Barr virus vector. *Gene* **78**, 287–296

Jiang, W., Wendel-Hansen, V., Lundkvist, A., Ringertz, N., Klein G. and Rosen A. (1991) Intranuclear distribution of Epstein-Barr virus-encoded nuclear antigens EBNA-1, -2, -3 and -5. *J. Cell Sci.* **99**, 497–502

Kiessling, U., Becker, K., Strauss, M., Schoeneich, J. and Geissler E. (1986) Rescue of a tk-plasmid from transgenic mice reveals its episomal transmission. *Mol. Gen. Genet.* **204**, 328–333

Kioussis, D., Wilson, F., Daniels, C., Leveton, C., Taverne, J. and Playfair, J.H.L. (1987) Expression and rescuing of a cloned human tumour necrosis factor gene using an EBV-based shuttle cosmid vector. *EMBO J.* **6**, 355–361

Krysan, P.J., Haase, S.B. and Calos, M.P. (1989) Isolation of human sequences that replicate autonomously in human cells. *Mol. Cell. Biol.* **9**, 1026–1033

Krysan, P.J. and Calos, M.P. (1991) Replication initiates at multiple locations on an autonomously replicating plasmid in human cells. *Mol. Cell. Biol.* **11**, 1464–1472

Krysan, P.J. and Calos, M.P. (1993) Epstein-Barr virus-based vectors that replicate in rodent cells. *Gene* **136**, 137–143

Kunisada, T. and Yamagishi H. (1987) Sequence organization of repetitive sequences enriched in small polydisperse circular DNAs from HeLa cells. *J. Mol. Biol.* **198**, 557–565

Lathe, R., Vilotte, J.L. and Clark, A.J. (1987) Plasmid and bacteriophage vectors for excision of intact inserts. *Gene* **57**, 193–201

Léopold, P., Vailly, J., Cuzin, F. and Rassoulzadegan, M. (1986) Maintenance of autonomous genetic elements in twelve transgenic mouse strains established after transfer of pPyLT1 DNA. *Biochem. Biophys. Res. Comm.* **141**, 1162–1169

Little, R.D., Platt, T.H.K. and Schildkraut, C.L. (1993) Initiation and termination of DNA replication in human rRNA genes. *Mol. Cell. Biol.* **13**, 6600–6613

Margolskee, R.F., Kavathas, P. and Berg, P. (1988) Epstein-barr virus shuttle vector for stable episomal replication of cDNA expression libraries in human cells. *Mol. Cell. Biol.* **8**, 2837–2847

McWhinney, C., Waltz, S.E. and Leffak, M. (1995) Cis-acting effects of sequences within 2.4-kb upstream of the human *c-myc* gene on autonomous plasmid replication in HeLa cells. *DNA Cell Biol.* **14**, 565–579

Milanesi, G., Barbanti-Brodano, G., Negrini, M., Lee, D. Corallini, A., Caputo, A., Grossi, M.P. and Ricciardi, R.P. (1984) BK virus-plasmid expression vector that persists episomally in human cells and shuttles into *Escherichia coli*. *Mol. Cell. Biol.* pp. 1551–1560

Potter, S.S. and Lloyd, J.A. (1987) pPyLTl does not always dictate the formation of autonomously replicating elements in transgenic mice. *Nucl. Acids Res.* **15**, 5482

Rush, M.G. and Misra, R. (1985) Extrachromosomal DNA in eucaryotes. *Plasmid* **14**, 177–191

Shen, E.S., Cooke, G.M. and Horlick, R.A. (1995) Improved expression cloning using reporter genes and Epstein-Barr virus *ori*-containing vectors. *Gene* **156**, 235–239

Stanfield, S.W. and Helinski, D.R. (1984) Cloning and characterization of small circular DNA from chinese hamster ovary cells. *Mol. Cell. Biol.* **4**, 173–180

Stephens, P.E. and Hentschel, C.C.G. (1987) The bovine papillomavirus genome and its uses as a eukaryotic vector. *Biochem.* **248**, 1–11

Sudo, K., Ogata, M., Sato, Y., Iguchi-Ariga, S.M.M. and Ariga, H. (1990) Cloned origin of DNA replication in *c-myc* gene can function and be transmitted in transgenic mice in an episomal state. *Nucl. Acids Res.* **18**, 5425–5432

Tasheva, E.S. and Roufa, D.J. (1994) A mammalian origin of bidirectional DNA replication within the chinese hamster ovary RPS14 locus. *Mol. Cell. Biol.* **14**, 5628–5635

Teshigawara, K. and Katsura, Y. (1992) A simple and efficient mammalian gene expression system using an EBV-based vector transfected by electroporation in G2/M phase. *Nucl. Acids Res.* **20**, 2607

Trojan, J., Johnson, T.R., Rudin, S.D., Tykocinski, M.L. and Ilan, J. (1993) Treatment and prevention of rat glioblastoma by immunogenic C6 cells expressing antisense insulin-like growth factor I RNA. *Science* **259**, 94–97

Vidal, M., Wrighton, C., Eccles, S., Burke, J. and Grosveld, F. (1990) Differences in human cell lines to support stable replication of Epstein-Barr virus-based vectors. *Biochem. Biophys. Acta* 172–177

Wahl, G.M. (1989) The importance of circular DNA in mammalian gene amplification. *Cancer Res.* **49**, 1333–1340

Wegner, M., Zastrow, G., Klavinius, A., Schwender, S., Muller, F., Luksza, H., Hoppe, J., Wienberg, J. and Grummt, F. (1989) *Cis*-acting sequences from mouse rDNA promote amplification and persistence in mouse L cells: implication of HMG-I in their function. *Nucl. Acids Res.* **17**, 9909–9932

Willard, H.F. (1990) Centromeres of mammalian chromosomes. *TIG* **6**, 410–416

Wysokenski, D.A. and Yates, J.L. (1989) Multiple EBNA1-binding sites are required to form an EBNA1-dependent enhancer and to activate a minimal replicative origin within *oriP* of Epstein-Barr virus. *J. Virol.* **63**, 2657–2666

Yankelevich, S., Kolman, J.L., Bodnar, J.W. and Miller, G. (1992) A nuclear matrix attachment region organizes the Epstein-Barr viral plasmid in Raji cells into a single DNA domain. *EMBO J.* **11**, 1165–1176

Yates, J.L., Warren, N. and Sugden, B. (1985) Stable replication of plasmids derived from Epstein-Barr virus in various mammalian cells. *Nature* **313**, 812

Young, J.M., Cheadle, C., Foulke, Jr., J.S., Drohen, W.N. and Sarver N. (1988) Utilization of an Epstein-Barr virus replicon as a eukaryotic expression vector. *Gene* **62**, 171–185

PART III, SECTION B

41. Insulation of Transgenes from Chromosomal Position Effects

Albrecht E. Sippel*, Harald Saueressig[†], Matthias C. Huber, Nicole Faust and Constanze Bonifer

Institut für Biologie III der Albert-Ludwigs-Universität, Schänzlestr. 1, D-79104 Freiburg/Br., Germany

Chromosomal Position Effects and the Transgene Problem

In recent years tremendous progress has been made in understanding the mechanisms of regulated transcription on the molecular level (McKnight, 1996). This applies in particular to the dissection of the components of the eukaryotic basic machinery for transcription and the molecular structure and function of numerous individual regulatory transcription factors. However, if viewed as a problem in its entirety, our knowledge about gene regulation must be evaluated with respect to our ability to transplant genes with predictable regulatory features into the genome of cells and organisms. The transfer of new genes into the germ line of farm animals is an attractive genetic route for new livestock production. The prospect that has most gripped the public imagination is gene therapy, that is somatic gene transplantation for medical purposes. It is safe to say that both fields are currently hampered considerably by our poor understanding of the mechanisms of regulation of transgenes when they are inserted stably into new host genomes.

Genes randomly inserted into host genomes are most often incorrectly regulated. They are not expressed in the desired cell type, or are expressed only transiently or at too low levels. Independently derived transgenic organisms or cell clones, respectively, show variable transcriptional efficiencies and variable tissue specific expression patterns of the same DNA construct depending on the chromosomal site of DNA integration. Chromosomal position effect (CPE) is a term used to describe phenomena in which the activity of a gene is affected by its location in the genome (Lima-de-Faria, 1983). Genomic position effects of genes were originally observed in *Drosophila* as position effect variegation (PEV) due to its conspicuous phenotype (for review see Henikoff, 1990). Today PEV is seen as a specific case of position effect, whereas CPE includes all phenomena by which the specific structural and functional features of the neighbouring chromatin influence the transcriptional performance of randomly inserted transgenes.

Difficulties caused by CPE can be circumvented by altering gene sequences at their natural genomic location by homologous recombination (Capecchi, 1989). Gene targeting by homologous recombination is, however, not applicable in cases when foreign genes with their own regulatory specificities are newly introduced into the host genome (i.e. transpecies transgenes) or when DNA transfer protocols have to be used which prevent cell-selection. For these cases strategies need to be developed to overcome the non-predictable influence of position effects. By better understanding the long distance influence of chromatin on gene regulation and the molecular mechanisms which cause CPE, we expect ultimately to be able to develop vector systems which render transgenes insulated from their neighbouring host environment, thus acting as independent regulatory units of transcription.

We cloned the genes for the avian egg white proteins lysozyme and ovomucoid (Sippel *et al.*, 1978 and Lindenmaier *et al.*, 1980) and for the

---

*Author for Correspondence.

[†] Present Address: Molecular Neurobiology Laboratory, The Salk Institute, 10010 N Torrey Pines Road La Jolla, CA 92037, USA.

mammalian milk proteins whey acidic protein (WAP) and several caseins (Henninghausen and Sippel, 1982a,b) in order to develop transgene vector systems for highly efficient tissue specific expression of commercially interesting proteins in secretory glands of farm animals. After studying for many years chromatin structure and the mechanisms of transcriptional control of the chicken lysozyme gene as endogenous gene, and as transgene in chicken cells, in transgenic mice and in mouse embryonal stem (ES-)cells as well as their hematopoietic derivatives, we are now able to draw some more general conclusions about transgene regulation, which open ways to counteract the disturbing influence of CPE. We have shown that randomly integrated transgenes can be insulated by two mechanisms from deregulating forces of neighbouring chromosomal regions. CPE is eliminated by the concerted action of a full set of regulatory cis-acting elements and parallely is buffered by the insulator function of border regions of chromatin domains. DNA of entire (natural) regulatory domains including both of these functions should provide a new generation of vectors which express transgenes correctly and predictably in their new host genomes.

The Chromatin Domain of the Chicken Lysozyme Gene

Early chromatin studies of the lysozyme locus in chicken cells (Figure 41.1A) were highly indicative in respect to its functional mechanisms of gene regulation. In steroid hormone induced tubular gland cells of chicken oviducts in which the gene is maximally active, the 4 kb transcription unit is located in the centre of a roughly 21 kb chromatin domain of elevated general DNase sensitivity (Jantzen et al., 1986 and Strätling et al., 1986). Since the chromatin in cells which do never express lysozyme was in a condensed configuration, it was concluded that elevated DNaseI sensitivity is characteristic for the active gene locus. In myelomonocytic cells of the hematopoietic system, a second cell type in which the gene is active, the domain of active chromatin extended to the same coordinates in 5' and 3' direction (Sippel et al., 1993). Both transition regions from "open" to condensed chromatin colocated with independently mapped matrix or scaffold attachment regions (Phi-Van and Strätling, 1988). This collocalization and the fact that in two different cell types the active chromatin

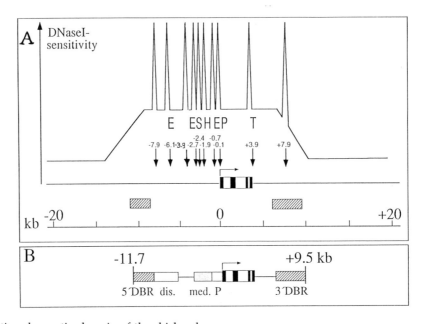

Figure 41.1. The active chromatin domain of the chicken lysozyme gene.
A. The relative DNaseI sensitivity of DNA in chromatin of isolated nuclei of laying hen oviduct cells and chicken promacrophages is shown over 40 kb of genomic DNA containing the lysozyme gene (exons black boxes, introns open boxes). Arrow heads point to the positions of regulatory elements, with numbers giving their location in kb in respect to the transcriptional start site (arrow). −6.1E early myeloid and oviduct cell enhancer (Grewal et al., 1992), −2.7 E late myeloid enhancer (Müller et al., 1990), −2.4S silencer (Baniahmad et al., 1990), −1.9H hormone responsive element in oviduct cells (Hecht et al., 1988) and −0.1P lysozyme promoter. Hatched boxes show the positions of MAR sequences collocating the chromatin domain border regions (5' and 3' DBR). B. The 21.2 kb wild type lysozyme gene domain DNA includes the two DBRs, the distal (dis.) and the medial (med.) enhancer region, the promoter (P) and the transcribed region.

terminated at the same DNA regions suggested a functional relationship between the S/MAR sequences and the domain border regions (DBRs).

A second feature of chromatin structure detected by DNase digestion is the presence of DNase hypersensitive sites. Altogether 10 DNaseI hypersensitive sites (DHS) were mapped flanking the lysozyme gene, all located within the domain of general DNase sensitivity (Tritton et al., 1984; Tritton et al., 1987 and Huber et al., 1996). No DHS was detected in the regions of the condensed chromatin extending 10 kb from the domain border regions outside the active chromatin domain.

Most indicative for gene regulation was the result that different sets of DHSs were characteristic for different states of transcriptional control of the gene locus. Since DHS in chromatin mark the position of multifactorial cis-regulatory elements, by analyzing chromatin structure of the gene locus in nuclei of various cell types, we literally could see which of these controlling elements are active and which are not (Rendelhuber, 1984). In this way and by using transient transfections of reporter gene constructs in tissue culture cells, we identified the regulatory specificity of 6 of the 8 cis-active elements (Figure 41.1A) in the 5′ flanking chromatin of the lysozyme gene (reviewed in Sippel et al., 1992). Beside the promoter region (P−0.1 kb), two enhancers were identified, which are appearing in myeloid precursor cells early in macrophage differentiation (E−6.1 kb; E−3.9 kb), a third one appears later in the monocytic state (E−2.7 kb) and a fourth one when monocytes or macrophages are activated by LPS or when oviduct cells are stimulated by steroid hormones (H−1.9 kb). Last, but not least, a transcriptional silencer element, present as DHS in all cells beside mature macrophages, was identified at −2.4 kb (S−2.4 kb). Each individual element appears to be responsible for a particular aspect of the global transcriptional control of the gene (Sippel et al., 1987 and Sippel et al., 1993).

Cis Requirements for Position-Independent Transgene Expression

It is conceptionally difficult to prove that the full complement of cis-active DNA sequences for the regulation of a gene is present on a distinct piece of genomic DNA. However, we reasoned, that the best way to approach this problem was to test, whether the entire 21.2 kb gene domain DNA, from the start of the 5′ DBR to the end of the 3′ DBR (Figure 41.1B), contains all the cis-active sequence information for the gene locus to function correctly throughout the entire ontogeny of the organism, that is to say, in every cell type of the body, at any developmental stage. Since no convenient transgenic chicken system was available, 7 independent transgenic mouse strains were made which carried variable copy-numbers of the complete wt transgene locus (Bonifer et al., 1990). The qualitative as well as quantitative analysis of transgene transcripts from different tissues of the mouse showed that the transpecies transgene was correctly regulated. It was expressed with the right cell-type specificity and its level of expression was independent of the random sites of genomic insertion (Table 41.1). The same wt transgene construct was stably transfected into mouse ES-cells and several independent clones were in vitro differentiated to macrophages. Also here copy-number dependent, correct cell-type specific expression of the transgene could be detected (Faust et al., 1994). Both results show that the 21.2 kb DNA of the chicken lysozyme locus is resistant to CPE and contains all sequences necessary for its correct regulation in its new host genome. We concluded that the entire domain DNA, as mapped in the chromatin of its natural host, constitutes the complete regulatory unit for transcription.

The stunning result of position independent expression of the wt locus offered the opportunity to map the cis requirements for the locus control function by deletion analysis. By again using the clonal position effect assay in transgenic mice (Bonifer et al., 1994a) and the ES-cell in vitro differentiation system for macrophages (Faust et al., in preparation), we could show (Table 41.1) that the transgene loses its copy number-dependent expression levels as soon as we delete one of the internal enhancer regions, (Table 41.1). Deletion of both DBRs alone does not deprive the regulatory domain of its function to resist position effects. However, deletion of DBRs plus any of the two enhancer regions leads frequently to ectopic expression of the transgene in non-typical tissues of transgenic mice. These results strongly suggest that the locus control function does not map to a particular regulatory LCR element in the gene domain, but rather is a feature of the cooperative action of most likely all regulatory elements of the gene locus.

The Concerted Action Model

The complete chicken gene locus acts in several respects in the mouse exactly as it does in its natural host, the chicken. Not only is this true for the cell-type specificity of transcription in hematopoietic cells and the LPS inducibility of gene activity in macrophages by LPS (Bonifer et al., 1994b), but also applies for differential levels of transcription in different macrophage types (Faust et al., 1996). We found that ES-cell derived embryonal type macrophages expressed a roughly 30-fold

Table 41.1. Expression of Chicken Lysozyme Constructs in Transgenic Mice and in Mouse ES Cell Derived Macrophages.

Transgenic system		DNA construct	Cell-specific expression (No. of strain/clones)		Expression in macrophages	
Type	No. of strains/clones	Deletion (Δ)	Wt[b] pattern	Ectopic[c]	Level	Copy No. dep.
tr.m.	7	−(wt)	7	0	high	+
tr.m.	5	Δd[a]	4	1	var	−
tr.m.	7	Δm	7	0	var[d]	−
tr.m.	5	ΔA	5	2	high	+
tr.m.	12	ΔA,d	0	6	var	−
tr.m.	4	ΔA,m	4	1	var	−
ES cell	5	−(wt)	5	0	low	+
ES cell	5	Δd	5	0	var	−
ES cell	5	Δm	4	1	var	−

[a] Deletion (Δ) from total wt domain (Figure 41.1B) of distal regulatory region (d), of medial regulatory region (m), of domain border regions (A).
[b] Wt expression pattern is defined by macrophage specific expression within the hematopoietic system.
[c] Ectopic expression is defined by expression detected in nontypical cell types like liver or lung cells.
[d] (var) Variable expression levels per gene copy from strain to strain, or clone to clone.

lower level of chicken lysozyme transcripts per transgene copy than it is found in adult-type macrophages of transgenic mice. This difference in transcription levels is also seen in chicken embryonic type versus adult-type macrophages.

The similarity of the mode of regulation of the gene in its natural host with its regulation as a transgene in another species can only be explained by usage of the same regulatory elements in both species. Hence, it did not come as a surprise, when the chromatin of the endogenous gene locus and of the transpecies transgene displayed the same pattern of DHS-elements (Huber et al., 1994 and Faust et al., in preparation), indicating that the same or very similar multifactorial protein-complexes form in both vertebrate species.

A careful qualitative and quantitative structural analysis of the chicken lysozyme chromatin in mouse macrophages by DNaseI and micrococcus nuclease digestion of chromatin shows that correct chromatin reassembly on transgene DNA is impeded if transgenes were subject to CPE due to deletion of one of the enhancer regions (Huber et al., 1996). The degree of suppression of enhancer or promoter protein complex formation (seen as DHS intensity) parallels the varying transcriptional levels of the transgene in transgenic macrophages of different transgenic mouse strains (Huber et al., 1994). According to higher or lower stress imposed by neighbouring chromatin, a higher or lower ratio of active versus inactive regulatory chromatin elements assemble on the various transgene repeats in their transgene clusters (Huber et al., 1996). This is most likely the result of a decreased level of regulatory complex stability in the chromatin of deletion constructs. An extreme case of inability to assemble enhancer protein complexes in transgene chromatin is observed when the promoter region of the lysozyme gene is deleted in the context of the entire domain (Huber and Saueressig, et al., in preparation).

Figure 41.2 shows a graphical outline of our model explaining the results from transcription and chromatin studies of the chicken lysozyme locus in transgenic mice or cells in culture. When the wt locus which contains all regulatory elements is transferred into the genome of a cell containing all necessary transcription factors, a highly stable multifactorial regulatory protein complex assembles in the chromatin by the concerted action of all elements. Its stability and its internal saturation of interacting protein surfaces may explain its resistance towards varying positional stress (CPE), resulting in position independent, copy number dependent expression levels (Figure 41.2a). As soon as regulatory cis-acting sequences are deleted, the in vivo

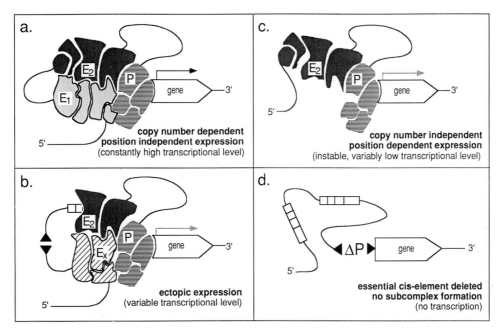

Figure 41.2. The "concerted action model" for eukaryotic gene regulation.
The schematic diagram shows multifactorial regulatory complex formation for four different transgene situations. a. Wt. situation; b. Ectopic expression of position dependent transcription; c. Low level position dependent expression and; d. inactive transgene due to promoter deletion, no regulatory complex assembly on enhancers possible. Same shading depict transcription factor proteins of the same regulatory element (E1, E2 lysozyme enhancer; Ex enhancer of neighbouring gene). Thin line shows loops of chromatin between regulatory elements, black triangles mark positions of DNA deletions in the transfected otherwise complete domain DNA-construct. Details of the "concerted action model" are described in the text.

assembly process appears to be impaired. Incomplete regulatory complexes, due to their decreased stability, may drop under their assembly threshold. Thus, different levels of stress imposed by neighbouring chromatin, cause the variable transcriptional efficiencies observed in different mouse lines or cell clones respectively (Figure 41.2c). The observed ectopic expression with incomplete regulatory transgene units is most likely explained by the deregulating activity of nearby regulatory elements with their different specificities as observed in enhancer trap experiments. These are able to interact because transcription factor protein surfaces are no longer saturated (Figure 41.2b). The deletion of an essential part of the regulatory complex (as for example the promoter and immediately upstream elements) can lead to a complete inability to assemble enhancer protein complexes (Figure 41.2d).

Our results indicate the necessity for cooperative transcription factor complex stabilization in the context of genomic chromatin. This "concerted action model" of gene regulation not only gives an explanation for the observed resistance of a complete locus towards position effects, but at the same time suggests a new mechanism for self-insulation. Due to their high stability and their lack of additional interacting protein surfaces the transcription factor complexes formed by complete natural regulatory units may be inert towards the deregulating influence of juxtaposed regulatory elements.

Domain Border Regions as Insulating Elements

We have seen that the complete regulatory part of the chicken lysozyme locus without its domain border regions was able to resist position effects in two transgenic systems (Table 41.1). The fact that a particular activity maps to one part of the domain, however, does not mean that there is no second part having the same activity. In cases in which one enhancer region is deleted the presence or absence of the domain border regions makes a significant difference for tissue specific expression. Lack of domain border regions increases deregulating influence by CPE.

A possible activity of domain border regions for increased resistance towards CPE of incomplete regulatory units was tested directly in clonal position effect assays in cells in culture (Stief et al., 1989). Reporter gene mini-loci containing the −6.1 kb enhancer and the promoter of the chicken lysozyme gene transfected stably into promacrophage cells were stimulated in their transcriptional efficiency and were significantly buffered towards position effects by the presence of the lysozyme 5′ domain border region (DBR, "A-element").

The experiment outlined in Figure 41.3A shows that this position effect buffering activity of the lysozyme domain border element is not restricted to chicken cells and to cooperation with the lysozyme enhancer. The presence of flanking chicken lysozyme 5′DBRs also stimulates and buffers position effects of a reporter gene driven by the BK-virus enhancer in mouse fibroblasts (Figure 41.3A). No such dual activity is seen when the reporter gene is framed by a non-domain border DNA-fragment of similar length.

When the clonal position effect assay is used in nondifferentiating, transformed cells in culture, it might be argued that differences in transcriptional activity of reporter genes might be due to intrinsic differences in cell clones rather than to the different sites of chromosomal location of the transgene construct. Figure 41.3B shows an experiment ruling this out. When the reporter gene construct is repeatedly inserted into the same chromosomal site via sequence specific recombination, transcriptional efficiencies do not vary. This is an experiment to definitely show CPE to be the cause of transcriptional variation of randomly integrated transgene constructs in a mammalian cell.

As to the mechanism by which the chicken lysozyme 5′-DBR functions, we had done an early experiment with transient transfections of various reporter gene constructs in promacrophage cells in culture (Stief et al., 1989). When we positioned the 5′-DBR at various sites in the reporter gene construct, we could prove that it interferred with promoter–enhancer interaction in a polar way. It counteracted enhancer stimulation of transcription only when it was placed between the enhancer and the promoter element and not when distal in respect to the enhancer. This polar interference with enhancer–promoter interaction was later taken to define what is now called an "insulator" function (Chung et al., 1993).

A few cases of insulator elements have since been found (for review see Wolffe, 1994 and Corces, 1995). They comprise a new class of genomic elements which suppress CPE when used as limits of transcriptionally active transgene domains. Their rather different source, DNA sequence and chromatin structure suggest different molecular mechanisms

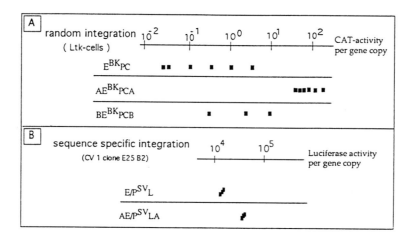

Figure 41.3. Position effects on reporter gene constructs in cells in culture.
A. Mouse fibroblasts (Ltk$^-$) were co-transfected with a neomycin resistance gene and the reporter gene construct containing the BK-virus enhancer (E^{BK}), the lysozyme promoter (P), and the bacterial CAT-gene (C) or the same construct either framed by the 5′ DBR (A for attachment element) of the chicken lysozyme gene (Stief et al., 1989) or a control DNA fragment (B) of the chicken lysozyme gene domain (Figure 41.1A) between the −6.1 kb enhancer and the −2.7 kb enhancer (2.9 kb aflII fragment). Individual neomycin resistant cell clones were grown, characterized for integrity and copy number of inserted reporter gene DNA and assayed for relative chloramphenicol acetyl transferase activity.
B. African green monkey kidney cells (CV-1 clone E25B2 from Stratagene) containing a FRT-lacz casette for site specific recombination with FLP yeast recombinase were cotransfected with the FLP expression plasmid (OG44) and plasmid DNA containing the SV40 enhancer promoter region (Theisen et al., 1986) and the luciferase gene (L) or the same reporter gene construct framed by the chicken lysozyme 5′-DBR (A). White lacZ minus cell clones were grown, characterized as described in A and assayed for relative luciferase activity.

for their respective insulating functions. It is obvious, that the scs-elements (Kellum and Schedl, 1991), the gypsy transposable element (Roseman et al., 1993), the 5' DHS-element of chicken β-globin gene (Chung et al., 1993) and the multiple S/MAR (but no DHS!) containing chicken lysozyme 5'-DBR are used to partition the genome into different types of functional and structural units, do this in very different ways and most likely on different hierarchical levels.

Two Mechanisms for Insulation: Perspectives for Better Transgene Vectors

We have got used to the fact that transcriptional units of genes may not be always uniquely definable. Genes proper may overlap, be contained in each other, or may have multiple alternative promoters, polyA-addition sites and terminators. If already transcription units of genes are that indistinct, what about their regulatory units? In eukaryotes intergenic DNA regions increase with biological complexity. What tells a particular enhancer which nearby promoter it is supposed to activate? What kinds of structural and functional partitioning prevents neighbouring genes and their regulatory elements to deregulate each other?

Our persistent work on the lysozyme gene locus has given us a few valuable insights into these questions. The results we got were partly due to certain advantages this locus offers for chromatin analysis. The relatively small size of the regulatory unit, the tissue specific expression in only a few clearly defined, easy to work with cell types, and the fact that transfer of a chicken gene into the mouse genome facilitates analysis, are only the most important to be named here. We found through rather time consuming clonal position effect assays in the transgenic mouse system, the transgenic ES-cell derived in vitro differentiation system and by stable transfections of cells in culture, that at least two parallel mechanisms exist which help to resist deregulation by neighbouring chromosomal regions. These two insulating functions for regulatory units of transcription are of a different and separable nature. Like it is inherent in the German word "Isolation", which stands for the meaning of both English words "insulation" and "isolation", genes possess a "transitive" and an "intransitive" mode of insulation function. Regulatory units of naturally evolved genes appear to assemble stable regulatory transcription factor complexes, which are inert towards positional stress and the influence of foreign regulatory elements. This feature of regulatory units self-isolates gene loci. In addition, at least some genes or gene clusters, as exemplified by the chicken lysozyme gene, do have insulator elements, blocking interference from outside by a different mechanism.

Why would genes have two different insulator functions? By deletion analysis of the wt chicken lysozyme domain DNA in transgenic mice we have seen that the full complement of regulatory elements in its natural configuration resists position effects without insulators. The reason might be found in the developmental situation of the gene locus. We have seen that during myeloid differentiation from early hematopoietic precursor cells to mature macrophages the 5'-flanking chromatin structure of the lysozyme gene locus dynamically changes with increasing transcriptional activity (Sippel et al., 1987; and Huber et al., 1995). With time more and more enhancer elements join the regulatory unit. Possibly cooperatively acting transcription factor complexes transiently pass through less stable and less inert modes on their way to the maximally stable conformation and might benefit considerably from a second, permanently active insulator function.

The results of basic research on mechanisms counteracting position effects has produced some valuable insights for gene-technology. Our results suggested that it is possible to design transgene vectors for the predictable and correct expression of randomly inserted transgenes. In fact, the chicken lysozyme 5'-DBR (5'-MAR, A-element) was used to suppress position effect caused variability of transgene constructs in plants (Mlynarova et al., 1994). Whereas the lysozyme 5'-DBR by itself does not appear to act cell type specificly, the entire gene domain DNA can be used as highly specific transgene vector for foreign genes as exemplified recently (Shen et al., 1995). The 15-lipoxygenase gene, when inserted into the chicken lysozyme regulatory domain DNA instead of the lysozyme gene proper, was not only specifically expressed in macrophages of transgenic rabbits but also exhibited high and permanent activity.

With time the complete set of regulatory elements or the extent of the regulatory unit of an increasing number of appropriate cellular gene loci will be defined. With these and their expected variety of regulatory specificity, better vectors will be available for medical somatic gene therapy as well as for transgenic organisms, may it be agricultural plants or animals.

Acknowledgements

Many former and present colleagues and collaborators, not appearing as authors, have contributed to our current knowledge of the chicken lysozyme gene domain and the mechanism of transgene regulation. We are greatly indebted to all of them for practical help, suggestions and scientific results.

We thank Claudia Tajeddine for secretarial work. Ongoing research from our group is supported by grants from the Deutsche Forschungsgemeinschaft, the Fonds der Chemischen Industrie, and the German Israeli Foundation for Scientific Research and Development.

References

Baniahmad, A., Steiner, C., Köhne, A.C. and Renkawitz, R. (1990) Modular structure of a chicken lysozyme silencer: involvement of an unusual thyroid hormone receptor binding site. Cell **61**, 505–514

Bonifer, C., Vidal, M., Grosveld, F. and Sippel, A.E. (1990) Tissue specific and position independent expression of the complete gene domain for chicken lysozyme in transgenic mice. EMBO J. **9**, 2843–2848

Bonifer, C., Yannoutsos, N., Krüger, G., Grosveld, F. and Sippel, A.E. (1994a) Dissection of the locus control function located on the chicken lysozyme gene domain in transgenic mice. Nucleic Acids Res. **22**, 4202–4210

Bonifer, C., Bosch, F.X., Faust, N., Schuhmann, A. and Sippel, A.E. (1994b) Evolution of gene regulation as revealed by differential regulation of the chicken lysozyme transgene and the endogenous mouse lysozyme gene in mouse macrophages. Europ. J. Biochem. **226**, 227–235

Capecchi, M.R. (1989) The new mouse genetics: altering the genome by gene targeting. Trends Genet. **5**, 70–76

Chung, J.H., Whiteley, M. and Felsenfeld, G. (1993) A 5′ element of the chicken beta-globin domain serves as an insulator in human erythroid cells and protects against position effect in Drosophila. Cell **74**, 505–514

Corces, V.G. (1995) Keeping enhancers under control. Nature. **376**, 462–463

Faust, N., Bonifer, C., Wiles, M.V. and Sippel, A.E. (1994) An in vitro differentiation system for the examination of transgene activation in mouse macrophages. DNA and Cell Biol. **13**, 901–907

Faust, N., Huber, M.C., Sippel, A.E. and Bonifer, C. (1996) Different macrophage populations develop from embryonic/fetal and adult hematopoietic tissues. Submitted

Fritton, H.P., Igo-Kemenes, T., Nowock, J., Strech-Jurk, U., Theissen, M. and Sippel, A.E. (1984) Alternative sets of DNaseI-hypersensitive sites characterize the various functional states of the chicken lysozyme gene. Nature **311**, 163–165

Fritton, H.P., Igo-Kemenes, T., Novock, J., Strech-Jurk, U., Theisen, M. and Sippel, A.E. (1987) DNaseI-hypersensitive sites in the chromatin structure of the lysozyme gene in steroid hormone target and non-target cells. Biol. Chem. Hoppe-Seyler **368**, 111–119

Grewal, T., Theisen, M., Borgmeyer, U., Grussenmeyer, T., Rupp, R.A.W., Stief, A., Quian, F., Hecht, A. and Sippel, A.E. (1992) The −6.1 kb chicken lysozyme enhancer is a multifactorial complex composed of several cell-specifically acting subelements. Mol. Cell Biol. **12**, 2339–2350

Hecht, A., Berkenstam, A., Strömstedt, P.-E., Gustafsson, J.A. and Sippel, A.E. (1988) A progesterone responsive element maps to the far upstream steroid dependent DNase hypersensitive site of chicken lysozyme chromatin. EMBO J. **7**, 2063–2073

Henikoff, S. (1990) Position-effect variegation after 60 years. Trends Genet. **6**, 422–426

Henninghausen, L.G. and Sippel, A.E. (1982a) Characterization and cloning of the mRNAs specific for the lactating mouse mammary gland. Europ. J. Biochem. **125**, 131–141

Henninghausen, L.G. and Sippel, A.E. (1982b) Mouse whey acidic protein is a novel member of the family of "four-disulfide core" proteins. Nucleic Acids Res. **10**, 2677–2684

Huber, M.C., Bosch, F.X., Sippel, A.E. and Bonifer, C. (1994) Chromosomal position effects in chicken lysozyme gene transgenic mice are correlated with suppression of DNaseI hypersensitive site formation. Nucleic Acids Res. **22**, 4195–4201

Huber, M.C., Graf, T., Sippel, A.E. and Bonifer, C. (1995) Dynamic changes in the chromatin of the chicken lysozyme gene during differentiation of multipotent progenitors to macrophages. DNA and Cell Biol. **14**, 397–402

Huber, M.C., Krüger, G. and Bonifer, C. (1996) Genomic position effects lead to an inefficient reorganization of nucleosomes in the 5′-regulatory region of the chicken lysozyme locus in transgenic mice. Nucleic Acids Res. **24**, 1443–1453

Jantzen, K., Fritton, H.P. and Igo-Kemenes, T. (1986) The DNaseI sensitive domain of chicken lysozyme gene spans 24 kb. Nucleic Acids Res. **14**, 6085–6099

Kellum, R. and Schedl, P. (1991) A position–effect assay for boundaries of higher-order chromosomal domains. Cell **64**, 941–950

Lima-de-Faria, A. (1983) in Molecular Evolution in organization of the chromosome, Elsevier, Amsterdam, p. 507–604

Lindenmaier, W., Nguyen-Huu, M.C., Lurz, R., Stratmann, M., Blin, N., Wurtz, T., Hauser, H.J., Giesecke, K., Land, H., Jeep, S., Grez, M., Sippel, A.E. and Schütz, G. (1980) The isolation and characterization of the chicken lysozyme and ovomucoid gene. J. Steroid Biochem. **12**, 211–218

McKnight, S.L. (1996) Transcription revisited: A commentary on the 1995 Cold Spring Harbor Laboratory meeting "Mechanisms of eukaryotic transcription", Genes a. Dev. **10**, 367–381

Mlynarova, L., Loonen, A., Heldens, J., Jansen, R.C., Keizer, P., Stiekema, W.J. and Nap, J.-P. (1994) Reduced position effect in mature transgenic plants conferred by the chicken lysozyme matrix-associated region. The Plant Cell **6**, 417–426

Müller, A., Grussenmeyer, T., Strech-Jurk, U., Theisen, M., Stief, A. and Sippel, A.E. (1990) Macrophage differentiation at the genome level: studies of lysozyme gene activation. Bone Marrow Transplantation **5**, 13–15

Phi-Van, L. and Strätling, W. (1988) The matrix attachment regions of the chicken lysozyme gene co-map with the boundaries of the chromatin domain. EMBO J. **7**, 655–664

Rendelhuber, T. (1984) Gene regulation – a step closer to the principles of eukaryotic transcriptional control. Nature **311**, 301

Roseman, R.R., Pirotta, V. and Geyer, P.K. (1993) The Su(Hw)protein insulates expression of the *Drosophila melanogaster white* gene chromosomal position effects. *EMBO J.* **12**, 435–442

Shen, J., Kühn, H., Petho-Schramm, A. and Chan, L. (1995) Transgenic rabbits with the integrated human 15-lipoxygenase gene driven by a lysozyme promoter: macrophage-specific expression and variable positional specificity of the transgenic enyzme. *The FASEB J.* **9**, 1623–1631

Sippel, A.E., Land, H., Lindenmaier, W., Nguyen-Huu, M.C., Wurtz, T., Timmis, K.N., Giesecke, K. and Schütz, G. (1978) Cloning of the chicken lysozyme structural gene sequences synthesized *in vitro*. *Nucleic Acids Res.* **5**, 3275–3294

Sippel, A.E., Borgmeyer, U., Püschel, A.W., Rupp, R.A.W., Stief, A., Strech-Jurk, U. and Theisen, M. (1987) In Hennig, W. (ed.) Results and Problems in Cell Differentiation. 14. *Structure and Function of Eukaryotic Chromosomes*. Springer-Verlag, Berlin, pp. 255–269

Sippel, A.E., Saueressig, H., Winter, D., Grewal, T., Faust, N., Hecht, A. and Bonifer, C. (1992) The regulatory domain organization of eukaryotic genomes–Implications for stable gene transfer. In Grosveld, F. and Kottias, G., (eds.), *Transgenic Animals*. Academic Press, London, pp. 1–26

Sippel, A.E., Schäfer, G., Faust, N., Saueressig, H., Hecht, A. and Bonifer, C. (1993) Chromatin domains constitute regulatory units for the control of eukaryotic genes. *Cold Spring Harbor Symp. on Quant. Biol.* **53**, 37–44

Stief, A., Winter, D.M., Strätling, W.H. and Sippel, A.E. (1989) A nuclear DNA attachment element mediates elevated and position-independent gene activity. *Nature* **341**, 343–345

Strätling, W.H., Dölle, A. and Sippel, A.E. (1986) Chromatin structure of the chicken lysozyme gene domain as determined by chromatin fractionation and micrococcal nuclease digestion. *Biochemistry* **25**, 495–502

Theisen, M., Stief, A. and Sippel, A.E. (1986) The lysozyme enhancer: Cell specific activation of the chicken lysozyme gene by a far upstream DNA element. *EMBO J.* **5**, 719–724

Wolffe, A.P. (1994) Insulating chromatin. *Curr. Biol.* **4**, 85–87

PART III, SECTION B

42. Transgene Rescue

A.J. Clark
Division of Molecular Biology, Roslin Institute, Roslin, Midlothian EH25 9PS, Scotland

Introduction

The expression of many transgenes is influenced strongly by their site of integration in the host genome. This includes, not only the nature of the host flanking sequences, but also the number and arrangement of transgene copies in the array. This is the main variable outwith the investigator's control and means that the frequency and level of transgene expression are not controllable. This phenomenon has been termed the position effect. In our attempts to express foreign proteins in the milk of transgenic animals we have encountered this effect constantly. We have employed sequences from the ovine milk protein gene β-lactoglobulin (BLG; Ali and Clark, 1988) to target expression of sequences encoding the human plasma proteins to the mammary gland. Transgenes comprising cDNA sequences encoding these proteins linked to the BLG promoter are expressed poorly and unpredictably (Whitelaw *et al.*, 1991) and indeed are very often completely silent. This is despite the fact that the same elements are capable of driving efficient expression of the unmodified BLG gene (Simons *et al.*, 1987 and Whitelaw *et al.*, 1991).

A number of strategies have been attempted to improve the expression of cDNA constructs in transgenic mice. The incorporation of homologous introns has been shown to enhance efficiency in a number of cases (Brinster *et al.*, 1988 and Whitelaw *et al.*, 1991). In some situations, however, the appropriate genomic clones may not be available or may be too large to manipulate readily. The inclusion of heterologous introns has been shown to improve transgene expression in some cases (Choi *et al.*, 1991 and Palmiter *et al.*, 1991) but there are no clear rules where they should be located in the construct. These considerations also apply to prokaryotic sequences which suffer from many of the same problems when it comes to achieving reliable expression.

We have developed a different strategy to improve the expression of cDNA constructs in transgenic mice. Since poorly expressed constructs appear to be highly influenced by chromosomal position effects we reasoned that selecting the site of integration rather than manipulating the construct *per se* might provide a strategy to "rescue" their expression. For targeting expression to the mammary gland the vicinity of an actively expressed milk protein gene could constitute a position which could enhance the expression of otherwise poorly expressed constructs. Targeting the site of integration by homologous recombination in ES cells would enable the introduction of transgenes in the vicinity of an endogenous milk protein gene. More simply, co-injection of two genes into pronuclei of mouse zygotes can result in their co-integration at a single site (Storb *et al.*, 1989 and Behringer *et al.*, 1989) and thus provides a direct method for integrating a poorly expressed transgene in the vicinity of one that is efficiently expressed. In this chapter I describe how this approach has been used to improve the expression of cDNA constructs encoding human α_1-antitrypsin (α_1AT), human factor IX (fIX) and other proteins in the mammary gland, as well as discussing the mechanism that may be involved.

Rescuing α_1AT cDNA Expression

To target expression to the mammary gland we have used sequences derived from the sheep milk protein gene β-lactoglobulin (BLG; Ali and Clark, 1988). The unmodified gene is expressed efficiently and specifically in the mammary gland of transgenic mice (Simons et al., 1987 and Harris et al., 1991). We have shown that essential regulatory elements are located within the proximal 406 bp of 5' flanking sequences (Whitelaw et al., 1992) and that this region contains the binding sites for a putative mammary gland-specific transcription factor (Watson et al., 1991, Burdon et al., 1994, Demmer et al., 1995). BLG transgenes comprising this region, the transcription unit and 3' flanking sequences are expressed in a position-independent manner; they are always expressed, and the levels of expression are related to copy number.

The BLG construct used in these studies comprised 4.0 kb of 5' and 1.9 kb of 3' flanking BLG sequences. To target expression of human α_1AT cDNA sequence to the mammary gland these same BLG elements were fused to a cDNA segment encoding human α_1AT to generate AATD (Figure 42.1).

BLG and AATD (Figure 42.1) were co-injected into mouse eggs in a 1:1 ratio; the constructs were not ligated. The co-integration frequency was high and more than 50% of the G_0 mice carried both transgenes. Breeding experiments indicated that the two transgenes were integrated at one locus since the BLG and AATD transgenes co-segregated into the G_1 generation (Clark et al., 1992). This was confirmed by more detailed Southern blotting experiments that identified restriction fragments containing both AATD and BLG specific sequences. Transgene copy number varied widely between the lines although, overall, the BLG and AATD copy numbers were similar, reflecting the molar ratio of the DNAs injected.

Milk was collected from mid-lactation females and assayed by ELISA. Human α_1AT was detected in the milk from seven of the nine BLG + AATD lines with the concentrations of the protein varying from 1–600 µg/ml (Table 42.1). Total RNA was prepared from the mammary gland and a selection of other tissues and probed with an α_1AT specific hybridisation probe. The expected AATD transcripts were detected in six of the seven lines and the steady-state levels correlated closely with concentration of the human protein measured in the milk, although we were unable to detect transcripts in the lowest expressing BAD line. No expression of AATD was detected in any of the other tissues analysed (for full details see Clark et al., 1992).

The efficiency of expression of AATD co-integrated with BLG contrasted markedly with the situation when it was integrated by itself. In this case only one line out of seven expressed AATD and this

Table 42.1. Rescue of α_1AT and fIX encoding transgenes with BLG.

Construct(s)	Expression frequency*	Expression levels (µg/ml)*
AATD	1/8	4.0
AATD + BLG	7/9	1.0–610
FIXD	0/9	0
FIXD + BLG	10/12	0.2–1.0
FIXDΔ3' + BLG	9/9	0.2–130

*The proportion of expressing lines and the range of expression levels of α_1AT or fIX in the milk are compared for singly integrated and co-integrated transgenes. Human α_1AT and fIX were estimated by ELISAs, which allowed their quantitation in mouse milk at levels above 0.2 µg/ml. Data from Clark et al., 1992 and Yull et al., 1995.

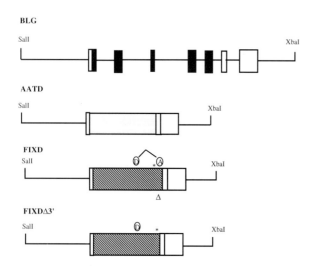

Figure 42.1. DNA constructs. The unmodified BLG gene comprises 4 kb of 5' flanking sequence and 1.9 kb of 3' flanking sequence. These same flanking sequences have been fused to cDNA sequences encoding α_1AT and fIX in AATD and FIXD. FIXDΔ3' contains a short deletion of fIX 3' non-coding sequences which removes the cryptic acceptor splice site present in FIXD. Line, 5',3' and intronic BLG sequences; open boxes, BLG non-translated exonic sequences; filled boxes, BLG coding exons; lightly shaded box, α_1AT cDNA sequences; hatched boxes, fIX cDNA sequences. The cryptic donor (D) and acceptor (A) splice sites in fIX are indicated, as are the fIX stop codon (*) and the 3' fIX sequences deleted in FIXDΔ3' (Δ).

showed only low levels of expression, contrasting with the AATD + BLG mice in which seven out of nine lines expressed the transgene, two producing human α_1AT in the milk at levels in excess of 100 µg/ml

Interestingly, the two lines which did not express the AATD transgene also failed to express the co-integrated BLG transgene. Thus activation of AATD appears to be correlated with the expression of the BLG transgene. In single transgenic mice the BLG transgene is invariably expressed as long as it comprises the 400 bp minimal promoter. The identification of these two non-expressing lines suggests that the AATD transgene can silence the BLG transgene in some transgene arrays.

Rescuing FIX cDNA Expression

FIXD comprises the 5′ and 3′ BLG sequences described above fused to a cDNA segment encoding human fIX. BLG and FIXD were co-injected into mouse eggs at a three to one ratio. Again the efficiency of co-injection was relatively high and twenty of the the G_0 animals obtained carried both the FIXD and BLG transgenes. A number of lines were established and they all co-segregated the two transgenes. On average, the BLG copy numbers were higher than FIX-D, reflecting the higher input ratio of BLG to FIXD in the injections.

Milk and RNA were collected from mid-lactation females. The levels of human fIX were measured by ELISA and this protein was detected in eight out of the twelve lines. These levels were, however, disappointingly low ranging from only 0.2–1.0 µg/ml. By contrast, Northern blotting experiments showed that a number of the expressing mouse lines exhibited relatively high levels of FIX-D transcripts in the mammary gland (in some lines more than 60 times the steady state level of human fIX mRNA in the human liver, the normal site of fIX synthesis; for full details see Clark et al., 1992). These results contrast with those obtained when the FIXD transgene was introduced into transgenic mice alone (Whitelaw et al., 1992). In this set of experiments none of the nine independently generated lines expressed the transgene. Thus, co-integration with BLG does activate the expression of FIXD, notwithstanding the low levels of protein secreted in the milk.

The discrepancy between the low level of fIX detected in the milk and the relatively abundant steady-state mRNA levels was investigated further. Careful Northern blotting showed that the transcripts were about 400 nt shorter than predicted from the structure of the FIXD transgene. RT-PCR was used to clone the transcripts from one of the BlX lines (Yull et al., 1995). Sequencing the products showed that these transcripts were shorter than predicted due to an internal deletion encompassing the sequences encoding the C-temminal 109 amino acids. Inspection of the sequences at the junction of this deletion showed that they corresponded to consensus donor and acceptor splice and, so, the shortened FIXD mRNA is almost certainly due to aberrant splicing of FIXD transcripts in the mammary gland. Additional RT-PCR experiments showed that other fIX cDNA constructs carry out the same aberrant splicing in the mammary gland again resulting in low levels of fIX protein production in the gland (Yull et al., 1996).

To correct the aberrant splicing a modified FIXD transgene FIXDΔ3′ was constructed in which the 3′ cryptic acceptor splice site was removed (Figure 42.1). This was co-injected with the BLG gene as before. Nine lines were established and all were shown to express human fIX in their milk. Analysis of the FIXDΔ3′ transcripts confirmed that the cryptic splicing had been abolished and levels of fIX in excess of 100 µg/ml were measured in the highest expressing line (Yull et al., 1995). The protein had a similar mobility to plasma derived fIX (Figure 42.2) and the purified material was fully active.

Other Gene Rescue Experiments

Rescue can provide a simple strategy for improving transgene expression and we have used this approach to facilitate the expression of a variety of cDNA constructs whose expression has been targeted to the mammary gland (e.g. Binas et al., 1995). Other workers have used BLG gene sequences to rescue expression of a BLG/protein-C cDNA construct. Like FIXD the BLG/protein-C construct was completely silent when integrated alone in the eleven lines tested, whereas six of the nine co-integrated lines expressed it (I. Garner and H. Lubon, unpublished results). This approach has also been used in transgenic sheep and rabbits and relatively high levels of protein C produced in the milk. Recently McKnight and co-workers (McKnight et al., 1995) have shown that co-integration of a WAP genomic transgene with a WAP-cDNA transgene activated expression of the latter. Ali and coworkers (S. Ali et al., unpublished observations) have used the rescue approach to try and enhance the expression of a bacterial cellulase, celE in either the pancreas or the gut using regulatory sequences derived from the elastase and intestinal FABP genes respectively, and observed a small increase in the frequency at which the bacterial gene was expressed.

In our experiments with BLG and BLG-derived transgenes the expression of the rescued gene is invariably lower than the unmodified BLG gene

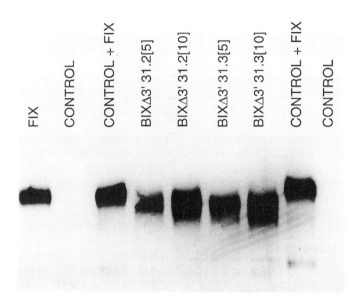

Figure 42.2. Expression of fIX in transgenic mouse milk. Milk samples from transgenic mouse line BIXΔ3'31 carrying FIXDΔ3' + BLG were electrophoresed under reducing conditions and analysed by Western blotting. The samples were diluted 1:200 and either 5 μl or 10 μl was loaded. fIX, 10 ng of human plasma derived fIX; CM, control mouse milk: CM + fIX, control milk + 10 ng of fIX. Data from Yull et al., 1995.

(Clark et al., 1992 and Binas et al., 1995). Similarly, McKnight et al. (1995) reported that the level of expression of the WAP cDNA construct did not exceed 1% of the co-integrated genomic construct. In the experiments targeting celE to the gut there was an increase in the frequency of expression of the co-integrated constructs but there was no indication that the overall levels of expression had improved (Ali et al., unpublished observations). Rescue, thus, appears to involve the activation of transgenes that are otherwise prone to silencing. For example, we have only ever seen the expression of FIXD when it has been co-integrated with a functional BLG gene. Nevertheless, the levels of expression of rescued cDNA constructs are generally quite low when compared to the rescuing gene or, indeed, to constructs comprising genomic sequences. Thus, the maximal level of human α_1AT expression in the milk obtained by rescue was 610 μg/ml, whereas levels in excess of 7 mg/ml have been achieved by targeting an α_1AT minigene to the mammary gland with the BLG promoter (Archibald et al., 1990). The reasons for this are unclear; so far gene rescue experiments have been restricted to cDNA or prokaryotic sequences and the lower levels of expression could reflect a requirement of introns for optimal chromatin structure (Liu et al., 1995) or mRNA processing.

Mechanism of Gene Rescue

The experiments described above show that the expression of cDNA constructs such as AATD and FIXD can be significantly enhanced by co-integration with the BLG gene. While we cannot exclude a mechanism whereby the BLG gene in some way targets the integration of these constructs to some type of permissive locus, there is little evidence that such a process occurs during integration. Therefore, mechanisms which are dependent upon interactions at the transgenic locus are considered to underlie rescue effects. In Figure 42.3, three possible models are outlined using BLG and BLG-derived transgene as schematic examples. In the first model the BLG gene provide positive effects on the adjacent gene through enhancer like sequences. The sequences responsible for this effect must be present within the transcription unit of the BLG gene since both AATD and FIXD contain the same 5' and 3' flanking sequences as the BLG transgene (Figure 42.1). In the second model an open chromatin configuration associated with the expression of the BLG gene spreads to encompass the adjacent rescued gene. In this model the expression of the BLG gene creates a permissive domain, for example, by allowing access of soluble transcription factors. It is known that transcription in the mammary gland is

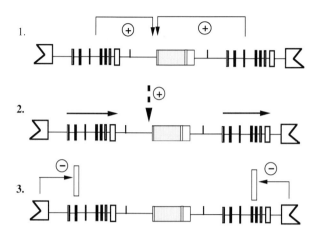

Figure 42.3. Models of transgene rescue. A site of integration in the chromosome is shown schematically in which a transgene such as AATD or FIXD is flanked by BLG transgenes. In 1, enhancer-like sequences present in the transcription unit of BLG activate the promoter elements of the adjacent transgene. In 2, the transcription of BLG creates a permissive domain that encompasses the co-integrated transgene. In 3, sequences within the BLG transcription unit insulate the rescued gene from negative effects of chromosomal sequences at the site of integration. Taken from Clark et al., 1992.

associated with changes in chromatin structure of the BLG gene and this is reflected in the appearance of discrete hypersensitive sites in the promoter region and with a general increase in DNAse sensitivity over this region of the gene (Whitelaw et al., 1992, 1995). In model 3 the BLG gene contains insulating elements such as the scs elements described in *Drosophila* (Kellum and Schedl, 1992) which serve to shield the transgene from the negative influences of the surrounding chromosomal sequences.

To discriminate between these models we have produced transgenic mice in which a promoterless BLG gene was co-integrated with FIXD (Yull et al., unpublished data). In these experiments co-integration with the promoterless BLG gene failed to rescue the expression of the FIXD transgene. This rules out models 1 and 3 since they both predict that the rescue effect is mediated by sequences present within the BLG transcription unit. Expression of the BLG gene is required and this is consistent with our own and others' work (McKnight et al., 1995) showing that rescue occurs only in lines of mice that express the rescuing gene.

Concluding Remarks

Transgene rescue can provide a simple method for enhancing the expression of constructs that are, otherwise, predominantly silent in transgenic mice. We have used this strategy to improve the expression of transgenes comprising cDNA segments in the mammary gland. Co-integration in the vicinity of a second, abundantly expressed, transgene encoding BLG enables the activation of predominantly silent cDNA gene constructs. The effect requires the expression of the rescuing transgene, but is not mediated by discrete enhancer-like elements within it. Rather, proximity to the efficiently expressed transgene may prevent silencing mechanisms that would otherwise suppress expression. Indeed, gene rescue and gene silencing may be related. Thus, in the experiments in which AATD was co-integrated with BLG, the two lines in which AATD was not expressed also failed to express BLG. It seems likely that this represents a silencing of BLG by AATD, since when BLG is introduced into transgenic mice alone it is always expressed (Whitelaw et al., 1992). In these rescue experiments the two genes were co-injected and the structures of the transgene arrays varied (Clark et al., 1992) and, presumably, this determined the balance between rescue and silencing.

Transgene rescue results in the activation of otherwise silent transgenes but the levels of expression obtained are considerably below those of the rescuing gene. For many applications this will not matter. This would include targeting the expression of growth factors, receptors and reporter genes where very high levels of expression are not required for efficacy. For protein production in milk, transgene rescue may not be able to deliver the very high levels of expression required for some products. For other products it is not clear if such high levels of protein production are desirable. For example, fIX requires complex post-translational modifications such as γ-carboxylation for biological activity, and it is not clear that the mammary gland would be able to carry these out fully at the very high levels of expression that the BLG gene is capable.

Acknowledgements

I would like to thank all my colleagues who contributed to the work described in this Chapter, most notably Fiona Yull, Bruce Whitelaw and Roberta

Wallace. This research was supported by the Biotechnology and Biological Sciences Research Council and PPL Therapeutics Ltd.

References

Ali, S. and Clark, A.J. (1988) Characterisation of the gene encoding ovine β-lactoglobulin. *J. Mol. Biol.* **199**, 415–426

Archibald, A.L., McClenaghan, M., Hornsey, V., Simons, J.P. and Clark, A.J. (1990) High level expression of biologically active human α-antitrypsin in transgenic mice. *Proc. Natl. Acad. Sci. (USA)* **87**, 5178–5182

Behringer, R.R., Ryan, T.M., Reilly, M.P., Asakura, T., Palmiter, R.D. and Brinster, R.L. (1989) Synthesis of functional hemoglobin in transgenic mice. *Science* **245**, 971–973

Binas, B., Gusterson, B., Wallace, R. and Clark, A.J. (1995) Epithelial proliferation and differentiation in the mammary gland do not correlate with cFABP expression during early pregnancy. *Dev. Gen.* **17**, 167–175

Brinster, R.L., Allen, J.M., Behringer, R.R. Gelinas, R.E., and Palmiter, R.D. (1988) Introns increase transcriptional efficiency in transgenic mice. *Proc. Natl. Acad Sci. (USA)* **85**, 836–840

Burdon, T., Maitland, A., Demmer, J., Clark, A.J., Wallace, R. and Watson, C.J. (1994) Regulation of the sheep β-lactoglobulin gene by lactogenic hormones is mediated by a transcription factor that binds an interferon-γ activation (GAS)-related element. *Mol. Endo.* **8**, 1528–1536

Choi, T., Huang, M., Gorman, C. and Jaenisch, R. (1991) A generic intron increases gene expression in transgenic mice. *Mol. Cell Biol.* **11**, 3070–3074

Clark, A.J., Cowper, A., Wallace, R., Wright, G. and Simons, J.P. (1992) Rescuing transgene expression by co-integration. *Bio/technology* **10**, 1450–1454

Demmer, J., Burdon, T., Djiane, J., Watson, C.J. and Clark, A.J. (1995) The proximal MPBF(MGF) binding site mediates the prolactin responsiveness of the BLG promoter. *Mol. Cell Endo.* **107**, 113–121

Harris, S., McClenaghan, M., Simons, J.P., Ali, S. and Clark, A.J. (1991) Developmental regulation of the sheep β-lactoglobulin gene in transgenic mice. *Dev. Gen.* **12**, 299–307

Kellum, R. and Schedl, P. (1991) A position effect assay for boundaries of higher order chromosomal domains. *Cell* **64**, 941–950

Liu, K., Sandgren, P., Palmiter, R.D. and Stein, A. (1995) Rat growth hormone gene introns stimulate nucleosomal alignment *in vitro* and in transgenic mice. *Proc. Natl. Acad. Sci. (USA)* **92**, 7724–7728

McKnight, R. A., Wall, R.J. and Hennighausen, L. (1995) Expression of genomic and cDNA transgenes after cointegration in transgenic mice. *Transgenic Res.* **4**, 39–43

Palmiter, R.D. Sandgren, E. P., Avarbock, M., Allen, D.D. and Brinster. R.L. (1991) Heterologous introns can enhance expression of transgenes in mice. *Proc. Natl. Acad. Sci (USA)* **88**, 478–482

Simons, J.P., McClenaghan, M. and Clark, A.J. (1987) Alteration of the quality of milk by expression of sheep β–lactoglobulin in transgenic mice. *Nature* **328**, 530–532

Storb, U., Pinkert, C. and Arp, B. (1986) Transgenic mice with mu and kappa genes encoding antiphosphorylcholine. *J. Exp. Med.* **164**, 627–641

Watson, C.J., Gordon, K.E., Robertson, M. and Clark, A.J. (1991) Interaction of DNA binding proteins with a milk protein gene promoter *in vitro*; identification of a mammary gland-specific factor. *Nuc. Acids Res.* **19**, 6603–6610

Whitelaw, C.B.A., Archibald, A.L., Harris, S., McClenaghan, M., Simons, J.P. and Clark, A.J. (1991) Targeting expression to the mammary gland; intronic sequences can enhance the efficiency of gene expression in transgenic mice. *Transgenic Res.* **1**, 3–13

Whitelaw, C.B.A., Harris, S., McClenaghan, M., Simons, J.P. and Clark, A.J. (1992) Position-independent expression of the ovine β-lactoglobulin gene in transgenic mice. *Biochem. J.* **286**, 31–39

Whitelaw, C.B.A. (1995) Regulation of β-lactoglobulin gene expression during the first stage of lactogenesis. *Biochem. Biophys. Res. Comm.* **209**, 1089–1083

Yull, F., Harold G., Cowper, A., Percy, J., Cottingham, I. and Clark, A.J. (1995) Fixing human factor IX (fIX): correction of a cryptic RNA splice enables the production of biologically active human factor IX in the mammary gland. *Proc. Natl. Acad. Sci. (USA)* **92**, 10899–10903

PART III, SECTION B

43. The Use of P1 Bacteriophage Clones to Generate Transgenic Animals

Sally P.A. McCormick[1,2,*], Charles M. Allan[1,2], John M. Taylor[1,2,3] and Stephen G. Young[1,2,4]

[1]The Gladstone Institute of Cardiovascular Disease, [2]Cardiovascular Research Institute, [3]Department of Physiology, [4]Department of Medicine, University of California, San Francisco, CA, USA

Introduction

The expression of mammalian genes in transgenic mice is routinely undertaken to study the functional consequences of overexpressing specific proteins and to understand the DNA sequences controlling gene expression. However, until relatively recently, expression of intact copies of large genes (>40 kb) and/or clusters of genes in transgenic animals has been difficult because of the limitations in the cloning capacity of commonly used vectors, such as plasmids, cosmids, and bacteriophage λ. For many years, investigators using transgenic animals have learned to live with the inherent limitations of these cloning vectors by constructing more compact "minigene" constructs from cDNA and genomic clones. However, minigene constructs invariably force the investigator to make assumptions about the regions of a gene that are functionally important for gene expression. The recent development of the P1 bacteriophage and yeast artificial chromosome (YAC) cloning systems has allowed investigators to generate transgenic animals with intact copies of larger genes. P1 bacteriophages (Sternberg, 1992) contain inserts of 80 to 90 kb, whereas the YACs can accommodate inserts ranging from 100 kb to more than 1 Mb. We have used P1 clones to generate transgenic mice for the purpose of studying important issues in mammalian lipoprotein metabolism.

Our initial studies using P1 clones for transgenic expression involved the human apo-B gene. The human apo-B gene is ~45 kb in length and contains 29 exons and 28 introns (Blackhart et al., 1986). The apo-B mRNA, which is 14 kb in length, codes for a large protein, apo-B100, containing 4326 amino acids. Apo-B100 plays a major structural role in the formation of several classes of atherogenic lipoproteins, including low density lipoproteins (LDL) and lipoprotein (a) [Lp(a)] (Young, 1990). Many attempts to generate human apo-B transgenic mice using minigene vectors yielded only one founder with barely detectable amounts of human apo-B expression (Chiesa et al., 1993 and Young et al., 1994), presumably because the DNA sequence elements that were important for appropriate levels of transgene expression were not included in the minigene construct. In contrast to the poor results with minigene vectors, an insert from a P1 clone spanning the entire human apo-B gene yielded high levels of apo-B expression in transgenic mice (Linton et al., 1993 and Callow et al., 1994) and transgenic rabbits (Fan et al.,). More recently, we have used modified P1 clones to study apo-B structure/function (McCormick et al., 1994a; 1995).

We have also used P1 clones to study the expression of the apo-E, apo-CI, apo-CII, and apo-CIV genes. These genes are located within a 45-kb cluster on human chromosome 19 (Myklebost and Rogne, 1988; Smit et al., 1988 and Allan et al., 1995). Apolipoproteins E, CI, and CII, which are synthesized in abundant amounts by the liver, are components of the plasma lipoproteins, and they have distinct and important roles in lipoprotein metabolism. Apo-CIV, which has only recently been described (Allan et al., 1995) and whose physiologic role is not fully understood, is also expressed in the liver. Several studies have used transgenic mice to examine the in vivo expression of this human apolipoprotein gene locus (Taylor

*Author for Correspondence.

et al., 1991; Shachter et al., 1993 and Dang et al.). Simonet et al. (1993) examined gene expression within this locus in transgenic mice and determined that the hepatic expression of the human apo-E and apo-CI genes was controlled by a distal hepatic control region (HCR-1), located 15 kb and 5 kb downstream from the apo-E and apo-CI genes, respectively. The apo-CII and apo-CIV genes may also be controlled by HCR-1 or by other distal elements, such as the recently described hepatic enhancer (HCR-2) located between the apo-CI' and apo-CIV genes (Allan et al.). While most of the early transgenic studies were performed with DNA from cosmid clones, none of these clones spanned the entire apoE/CI/CIV/CII gene locus. In particular, the most extensively used cosmid clone lacked most of the intergenic region between the apo-CI' pseudogene and the apo-CIV/apo-CII genes which contains the HCR-2 sequence (Lauer et al., 1988). Our recent studies of the regulation of this multigene locus have used P1 clones having inserts that span the entire locus.

This chapter reviews the techniques for preparation of P1 DNA for transgenic expression studies, using the human apo-B gene and the human apoE/CI/CIV/CII gene cluster as examples.

Human P1 Bacteriophage Clones

Human Apo-B P1 Clones

A P1 clone, p158, spanning the entire human apo-B gene was isolated from a human genomic P1 library using PCR screening (a service of Genome Systems, St. Louis, MO, USA) (Linton et al., 1993). A restriction map of p158 is depicted in Figure 43.1. The clone was 79.5 kb in length and contained 19 kb of 5' flanking sequences and 17.5 kb of 3' flanking sequences. A modified version of p158, designated p158/neo8, was generated by the insertion of a Tn10-based transposon (from the plasmid pZTneo5) into the coding sequences of the apo-B gene, using the techniques developed by Sternberg (1994). DNA sequencing of p158/neo8 revealed that the trans-

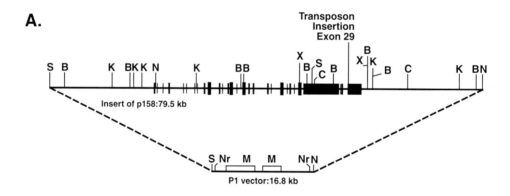

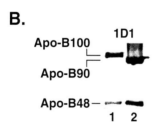

Figure 43.1. (A) Restriction map of p158, a P1 bacteriophage clone spanning the human apo-B gene. The insert from p158 contains all 29 exons of the apo-B gene and extends 19 kb 5' to the transcriptional start and 17.5 kb 3' of the polyadenlyation signal. S, SalI; B, BamHI; K, KpnI; X, XhoI; C, ClaI; N,NotI; Nr, NruI; and M, MluI. A modified form of p158, p158/neo8, was produced by the insertion of a transposon into exon 29 of the apo-B gene: p158/neo8 was predicted to yield a truncated form of apo-B100, apo-B90. Reproduced from McCormick et al. (1994b), with permission. (B) Western blot demonstrating human apo-B100 and human apo-B90 in the plasma of transgenic mice. Plasma (1μl) from an apoB-100 mouse (lane 1) and an apo-B90 (lane 2) mouse was subjected to electrophoresis on a 4% polyacrylamide-SDS gel and immunoblot analysis performed with the human apo-B-specific monoclonal antibody, 1D1 (Pease et al., 1990). The apo-B48 in both animals is produced as a result of hepatic editing of the apo-B mRNA (Davidson, 1994). Reproduced, with permission, from McCormick et al. (1994a).

poson was inserted into exon 29 of the apo-B gene (at apo-B cDNA nucleotide 12461) and that p158/neo8 would be predicted to code for a truncated form of apo-B, apo-B90, containing 4084 amino acids.

Human ApoE/CI/CIV/CII P1 Clone

Five P1 bacteriophage clones containing the human apo-CI and/or apo-CII gene were isolated from a human P1 genomic library (Allan et al., 1995). One clone, p1.198, contained the entire human apoE/CI/CIV/CII gene locus and lacked deletions or rearrangements, as judged by Southern blot analysis (Allan et al., 1995). The presence of HCR-1, located 15 kb downstream from the apo-E gene, was verified by Southern blot analysis and nucleotide sequencing. The presence of the HCR-2 element, located 25' kb downstream of the apo-E gene and between the apoC-I' and apoC-IV genes, was identified and characterized using subclones of p1.198, as described (Allan et al.).

Preparation of P1 DNA

To generate P1 DNA for transgenic experiments, we use the protocols described by McCormick et al. (1994b). E. coli (strain NS3529) harboring the P1 plasmid are grown in Terrific Broth containing 25 µg/ml of kanamycin. To make a large preparation of P1 DNA, a 500 ml volume of this culture medium is inoculated with 5 ml of an overnight culture and placed in a shaking incubator at 37 C. Isopropyl-β-D-thiogalactopyranoside (final concentration, 1 mM) is added to the culture after one hour of growth; the bacteria are then grown until an OD_{600} of 0.7–0.8 is reached. The P1 DNA is prepared from the E. coli by alkaline lysis using the Qiagen Plasmid Maxi kit (Qiagen, Chatsworth, CA, USA), according to the manufacturer's instructions. In general, we have noticed that incubation of the resuspended E. coli cells for five minutes with a 0.1 volume of a 10 mg/ml solution of lysozyme (Boehringer-Mannheim, Indianapolis, IN, USA) improves the recovery of P1 DNA. A longer incubation with lysozyme is undesirable because it leads to increased contamination with bacterial genomic DNA. The P1 DNA is purified from the bacterial lysate on two Qiagen-tip 500 columns, using the washing and elution buffers contained in the kit. At this point, we frequently perform a phenol/chloroform extraction before precipitating the DNA with 70% ethanol. After solubilizing the DNA in TE (10 mM Tris-HCl, pH 7.4, 1 mM EDTA), the concentration of the DNA is assessed by spectrophotometry. From a 500-ml culture, the typical yield of P1 DNA ranges from 25 to 100 µg.

Purifying the DNA Insert from the P1 vector

When the goal is to express a P1 clone in transgenic mice, it is desirable to remove vector sequences from the DNA fragment. This can frequently be accomplished by cleaving the P1 DNA at rare-cutting sites within the P1 polylinker (SfiI, NotI, or SalI). If these sites are present within the insert, the P1 vector can be cut at other rare-cutting sites, such as MluI or NruI. The complete P1 vector sequence can be found in the Genbank database. Because the apo-B gene contains an internal NotI and SalI site, p158 was cleaved with NruI to yield a 80-kb insert containing less than 600 bp of vector sequences. In the case of p158/neo8, the DNA was cleaved with MluI (because the transposon contained an NruI site). Cleavage of p158/neo8 with MluI generates a 91-kb insert containing approximately 7 kb of vector sequences (McCormick et al., 1994b). After cleaving the P1 preparation with an appropriate endonuclease, the P1 insert is purified from vector sequences on a pulsed-field agarose gel, using methods similar to those originally reported for the isolation of DNA from YACs (Schedl et al., 1993 and Peterson et al., 1993). Figure 43.2. illustrates the purification of the MluI fragment from p158/neo8. A total of 20 µg of the MluI-cleaved DNA was loaded onto a preparative 1% agarose pulsed-field gel (Sea Plaque GTG low melt agarose; FMC Bioproducts, Rockland, ME, USA), and electrophoresis was performed on a Beckman pulsed-field gel apparatus (GeneLine II System; Beckman, Fullerton, CA, USA) for 24 h at 350 V using a 6-sec pulse. Following pulsed-field gel electrophoresis, strips containing the molecular weight marker lanes and a thin (0.5–1 cm) strip of the preparative lane were stained with ethidium bromide to visualize the 91-kb MluI fragment. The stained agarose strips were notched to mark the location of the P1 insert DNA and then realigned with the main body of the gel. The unstained region of the preparative gel containing the P1 insert band was then excised from the gel. Figure 43.2A. shows a pulsed-field gel of MluI-digested p158/neo8 where the *entire* gel has been stained with ethidium bromide. The agarose strip containing the MluI fragment was equilibrated in gelase buffer (Epicentre Technologies, Madison, WI, USA) containing 100 mM NaCl, 30 µM spermine, 70 µM spermidine for 16 hrs at 4 C. The agarose was first melted at 68 C then equilibrated to 40 C before being digested with gelase (2U per ml of melted agarose). After spinning the DNA solution at 13,000 rpm for 10 min to pellet undigested agarose, the DNA solution was transferred to a fresh tube. At this stage, the DNA can be concentrated, if necessary, on a Millipore Ultrafree-MC

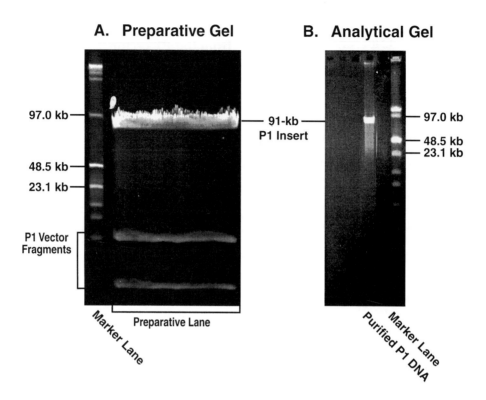

Figure 43.2. (A) Purification of P1 DNA by pulsed-field gel electrophoresis. A *Mlu*I digestion of p158/*neo*8 was electrophoresed on a 1% low-melt agarose pulsed-field gel. For this figure, the entire gel was stained with ethidium bromide, although normally, for the purpose of DNA purification, only flanking marker lanes would be stained. After marking the location of the high molecular weight DNA band, the band is excised from the gel. Low range pulsed-field gel markers (from New England Biolabs) are also loaded on the gel. Reproduced, with permission from McCormick *et al.*, (1994a) (B) An analytical 1% agarose pulsed-field gel showing the purified P1 DNA.

Filter Unit (Millipore, Bedford, MA, USA) (30,000 molecular weight limit). After concentration, the DNA solution should be left on the filter overnight at 4 C then removed the following morning using a wide-bore pipette. The DNA solution was dialyzed against microinjection buffer (10 mM Tris-HCl, pH 7.4, 250 μM EDTA) containing 100 mM NaCl, 30 μM spermine, 70 μM spermidine on a microdialysis membrane (type VS, 0.025 μM from Millipore). The concentration of purified P1 insert is assessed by spectrophotometry and by comparison to DNA markers of known concentration after electrophoresis on 1% agarose gels. The intactness of the purified P1 DNA insert is assessed by electrophoresis on a 1% agarose (Sea Kem GTG; FMC Bio-products) pulsed-field gel and staining with ethidium bromide as shown in Figure 43.2B. A typical P1 insert DNA preparation should yield a single band of the appropriate size with very little evidence of shearing (usually seen as a smear of DNA trailing the single intact band).

Generation of Transgenic Animals with P1 DNA

Transgenic Mice and Rabbits Expressing Human Apo-B

For generating transgenic mice with p158 and p158/*neo*8 (Figure 43.1A), the purified P1 insert DNA was adjusted to a concentration of 3 ng/μl in 10 mM Tris-HCl, pH 7.4, 250 μM EDTA containing 100 mM NaCl, 30 μM spermine, and 70 μM spermidine prior to microinjection into murine zygotes. In general, standard transgenic procedures were followed, except that 100 mM NaCl must be added to the microinjection buffer in order to prevent shearing of the DNA. The yield of transgenic founders among mouse pups for both constructs was ~15%. Transgenic founders expressing either apo-B100 and apo-B90 were identified by Southern analysis of tail DNA and by analyzing mouse plasma for the presence of human apo-B with a specific radioimmunoassay (Linton *et al.*, 1993 and

McCormick et al., 1994a). Figure 43.1B shows a western blot demonstrating the expression of human apo-B100 and human apo-B90 in transgenic mice. The levels of human apo-B in these mice (~50 mg/dl in high expressors) greatly exceeded the normal levels of mouse apo-B (~5 mg/dl) (Lusis et al., 1987).

The purified P1 insert DNA from p158 was also microinjected into rabbit zygotes, using standard procedures (Fan et al., 1994). The concentration of DNA in the microinjection buffer was 6 ng/μl. Transgenic rabbit founders were identified by Southern analysis of ear biopsy DNA and western blots of rabbit plasma. Four transgenic founders were obtained from 24 pups; the expression levels of human apo-B100 in these rabbits was similar to that obtained in the human apo-B100 transgenic mice (Fan et al., 1994).

Transgenic Mice Expressing the Human apo-E/C-I/C-IV/C-II Gene Locus

A 70-kb *Kpn*I genomic insert of p1.198, 198.KK (Figure 43.3A), extending ~5 kb upstream of the apoE gene to ~18 kb downstream of the apoC-II gene, was used to generate transgenic FVB/N mice. Using DNA purification and transgenic procedures identical to those described above, five transgenic founder mice were generated; 25% of the newborn pups were founders. Prolonged storage did not appear to adversely affect the linearized 198.KK fragment; two of the five lines were generated from DNA that had been stored at 4 C for more than 4 months. Analysis of two independent transgenic lines by RNase protection showed that all four apolipoprotein genes were expressed in the liver, as shown in Figure 43.3B. In addition, the relative levels of human apo-CII and apo-CIV as well as apo-E and apo-CI expression were similar to those observed in human liver (Allan et al., 1995), with apoC-IV being expressed at considerably lower levels. Thus, 198.KK appeared to contain all of the sequences required to direct appropriate hepatic expression of the entire locus. Site-directed mutagenesis of 198.KK will help to clarify the precise DNA sequences controlling the hepatic expression of each gene within this locus.

Transgenic Mice with Modified P1 Clones

For studies of protein structure/function and for studies of gene regulation, it would be desirable to generate transgenic mice with genetically modified P1 clones. Sternberg (1992) reported methods for interrupting P1 clones with Tn10-based transposons and suggested that this technique would be useful for generating truncated proteins for structure/function studies. The Tn10-based transposons integrate randomly into P1 DNA and contain a chloramphenicol resistance gene so it is possible to

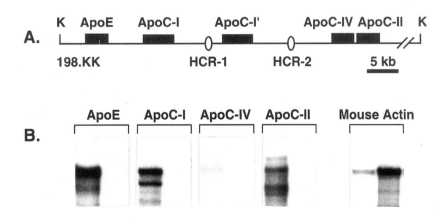

Figure 43.3. (A) A map of the 70-kb fragment of pl.198, 198.KK, containing the human apoE/C-I/C-II/C-IV gene locus. The locations of the human apo-E, CI, CIV and CII genes, and the apoC-I' pseudogene are indicated by the *closed boxes*. The positions of the two hepatic control regions, HCR-1 and HCR-2, are shown by the *open ovals*. Reproduced, with permission, from Allan et al., 1995; (B) Expression of human apoE, CI, CIV, and CII mRNAs in 198.KK transgenic mice, as determined by RNase protection analysis. The left and right lanes of each panel contain total RNA from the liver and intestine, respectively. The bands correspond to the expected protected fragments for human apoE, CI, C-II, or C-IV, and mouse actin control mRNAs. The human probes do not "cross-react" with the endogenous mouse apolipoprotein mRNAs (Simonet et al., 1993).

select for P1 clones containing transposon insertions. Also, the site of integration can be rapidly localized to within several hundred nucleotides by restriction analysis because the transposon contains several rare restriction sites, such as *Not*I, *Xho*I, and *Nru*I. DNA sequencing, using an oligonucleotide primer located within the transposon, can pinpoint the exact site of integration. We have found that high-quality DNA sequencing can be performed on P1 DNA prepared as described above. Transposon mutagenesis with the transposon-containing plasmid pZT*neo*5 has been used to modify the p158 apo-B clone producing a clone coding for a truncated apo-B, apo-B90 (McCormick *et al.*, 1994a). Apo-B90 was critically important in delineating the region of the apo-B molecule that covalently binds to apo(a).

More subtle genetic modifications of P1 clones, such as short insertions, deletions, or point mutations, can be achieved by one of two methods. One approach is to clone the insert from P1 into a YAC vector, then use gene-targeting techniques to modify the insert of the YAC (McCormick *et al.*, 1995 and McCormick *et al.*, 1996). Gene-targeting in yeast is relatively simple because the machinery for homologous recombination within yeast is extremely efficient and because there are multiple genetic markers that can be utilized to select for gene-targeting events. This YAC gene-targeting strategy was used by McCormick *et al.* (1995) to change a single amino acid residue within apo-B100. A single cysteine residue, cysteine-4326, was changed to glycine to test the hypothesis that apo-B100 cysteine-4326 participated in the disulfide bond between apo-B100 and apo(a) in the formation of Lp(a). The strategy of McCormick *et al.*, 1995 is illustrated in Figure 43.4. Briefly, an 87-kb *Mlu*I fragment of the human apo-B P1 clone, p158, was ligated into the *Mlu*I sites of YAC arms derived from the YAC vector, pYACRC (Smith, 1994). The high-molecular weight ligation product was then transformed into AB1380 yeast spheroplasts. The yeast were initially grown on plates lacking uracil and ultimately on plates lacking both uracil and tryptophan. Analysis of multiple TRP1/URA3-positive clones revealed that they contained a 99-kb YAC with the entire 87-kb *Mlu*I fragment. The URA3 marker in the right YAC arm was then replaced with a LYS2 gene so that the URA3 marker could be utilized in the "pop-in, pop-out" gene-targeting strategy (Scherer and Davis, 1979 and Rothstein, 1991).

For the "pop-in, pop-out" gene-targeting strategy, McCormick *et al.* (1995) generated a sequence insertion gene-targeting vector, using a URA3 containing yeast integrating plasmid and a short segment of the apo-B gene containing the Cys4326Gly point mutation. The Cys4326Gly point mutation was introduced into the apo-B YAC in a two-step gene-targeting process, as illustrated in Figure 43.4B. In the first step, the gene-targeting vector recombined with the cognate DNA sequences within the YAC, resulting in the integration of the entire gene-targeting plasmid into the YAC. In the second step, intrachromosomal recombination removed the plasmid sequences and the reduplicated apo-B gene sequences, leaving behind the targeted point mutation. Both PCR and Southern blots were used to identify yeast colonies that contained the correct gene-targeting events. The modified YAC DNA was then purified from yeast as previously described (McCormick *et al.*, 1995) and used to generate transgenic mice expressing a mutant form of human apo-B. Although the yield of YAC DNA from yeast cultures is substantially lower than that obtained from bacterial cultures harboring P1 clones, it is nevertheless possible to obtain pure YAC DNA, free of sheared fragments. We have found that the yield of transgenic animals using YAC DNA is equivalent to that obtained with P1 DNA. Analysis of the transgenic mice made with the modified YAC DNA revealed that the cysteine at apo-B100 residue 4326 was essential for the formation of the disulfide bond with apo(a) (McCormick *et al.*, 1995).

Another useful approach to modify P1-cloned inserts is to use RecA-Assisted Restriction Endonuclease (RARE) cleavage, a technique developed by Ferrin and Camerini-Otero (1991). This method can be used to remove a specific *Eco*RI fragment from a P1 clone, which typically could contain 10 to 20 internal *Eco*RI sites. The specific *Eco*RI fragment is mutated, then ligated back into the P1 clone. In the RARE cleavage procedure, two oligonucleotides and the bacterial protein, RecA, are used to protect two *Eco*RI sites from methylation by the enzyme *Eco*RI methylase. Following methylation and the removal of both oligonucleotides and the RecA, the non-methylated *Eco*RI sites can be cleaved with *Eco*RI to remove a short *Eco*RI fragment from the P1 clone. This *Eco*RI fragment is mutated by standard site-directed mutagenesis techniques, then ligated back into the P1 insert. This procedure was used to introduce specific point mutations into the apo-B P1 clone, p158, to generate transgenic mice expressing mutant apo-Bs (Callow and Rubin, 1995). Those studies confirmed that the elimination of apo-B100 cysteine-4326 prevents the covalent association of apo(a) and apo-B100. Other methylases are also available for use with the RARE cleavage system, so the strategy need not be confined to the use of *Eco*RI fragments (Callow and Rubin, 1995). A disadvantage of the RARE cleavage technique, which does not apply to the YAC

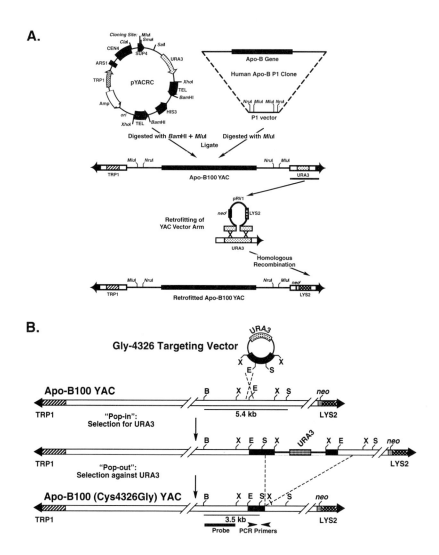

Figure 43.4. (A) Cloning of the P1 clone, p158, into the YAC vector, pYACRC. The apo-B P1 clone, p158, was cleaved with MluI and ligated into the dephosphorylated YAC vector arms generated by cleavage of pYAcRC with MluI and BamHI. The 99-kb ligation product was purified by pulsed-field gel electrophoresis and used to transform yeast spheroplasts made from the yeast strain AB1380. Transformants were initially selected on plates lacking uracil and then transferred onto plates lacking uracil and tryptophan to select for the TRP1 and URA3 genes on the YAC arms. All of the TRP1/URA3 transformants contained a 99-kb YAC. Southern blot analysis of a few of the YAC clones using several probes from different regions of the human apo-B gene revealed no obvious deletions or rearrangements. The URA3 gene in the right YAC arm was replaced with a LYS2 gene by retrofitting with a 9-kb insert from pRV1, which also contains a *neo* gene; this modification extended the length of the YAC to 108 kb. (B) A gene-targeting vector for mutating the YAC was constructed by cloning a 2.8-kb *Xba*I fragment spanning intron 28 to exon 29 into the yeast integrating vector, pRS406. The codon for cysteine-4326 was then changed to a glycine by site-directed mutagenesis; this mutation also created a new *Stu*I site. The targeting vector was linearized at the *Eco*RI site in exon 29 and introduced into spheroplasts containing the 108-kb YAC. Transformants were initially selected on plates lacking uracil and later grown on plates lacking uracil, tryptophan, and lysine. To identify targeted clones ("pop-in" step), yeast colonies were analyzed by pulsed-field-gel electrophoresis and Southern blot analysis. For the "pop-out" step, yeast harboring a targeted YAC were grown in medium lacking lysine overnight and plated onto 5-FOA plates (Rothstein, 1991). Colonies were subsequently transferred onto plates lacking tryptophan and lysine and analyzed by pulsed-field-gel electrophoresis and Southern blot analysis. The percentages of yeast colonies that were correctly targeted in the "pop-in" and "pop-out" steps were ~40% and ~25%, respectively. Reproduced, with permission, from McCormick *et al.* (1995).

gene-targeting approach, is that the appropriate restriction sites must be conveniently located with respect to the region-of-interest before this technique can be utilized. Moreover, it is necessary to have rather extensive DNA sequence information around the sites of interest prior to applying this technique.

Summary

P1 bacteriophage clones spanning large fragments of genomic DNA can be used to generate transgenic animals with high efficiency. In this chapter, we reviewed the techniques for purifying the DNA from P1 clones and showed that fragments as large as 91 kb can be used to achieve high levels of expression in both transgenic mice and transgenic rabbits. We also briefly reviewed current methods for modifying P1 clones. The ability to modify P1 clones will be important for understanding gene regulation and protein structure/function relationships.

References

Allan, C.M., Walker, D., Segrest, J.P. and Taylor, J.M. (1995) Identification and characterization of a new human gene (APOC4) in the apolipoprotein E, C-I, and C-II gene locus. *Genomics* **28**, 291–300

Allan, C.M., Walker, D. and Taylor, J.M. (1995) Evolutionary duplication of a hepatic control region in the human apolipoprotein E gene locus. Identification of a second region that confers high level and liver-specific expression of the human apolipoprotein E gene in transgenic mice. *J. Biol. Chem.* **270**, 26278–26281

Blackhart, B.D., Ludwig, E.M., Pierotti, V.R., Caiati, L., Onasch, M.A., Wallis, S.C., Powell, L., Pease, R., Knott, T.J., Chu, M.-L., Mahley, R.W., Scott, J., McCarthy, B.J. and Levy-Wilson, B. (1986) Structure of the human apolipoprotein B gene. *J. Biol. Chem.* **261**, 15364–15367

Callow, M.J. and Rubin, E.M. (1995) Site-specific mutagenesis demonstrates that cysteine 4326 of apolipoprotein B is required for covalent linkage with apolipoprotein(a) in vivo. *J. Biol. Chem.* **270**, 23914–23917

Callow, M.J., Stoltzfus, L.J., Lawn, R.M. and Rubin, E.M. (1994) Expression of human apolipoprotein B and assembly of lipoprotein(a) in transgenic mice. *Proc. Natl. Acad. Sci. USA* **91**, 2130–2134

Chiesa, G., Johnson, D.F., Yao, Z., Innerarity, T.L., Mahley, R.W., Young, S.G., Hammer, R.H. and Hobbs, H.H. (1993) Expression of human apolipoprotein B100 in transgenic mice. Editing of human apolipoprotein B100 mRNA. *J. Biol. Chem.* **268**, 23747–23750

Dang, Q., Walker, D., Taylor, S., Allan, S., Chin, P., Fan, J. and Taylor, J. (1995) Structure of the hepatic control region of the human apolipoprotein E/C-I gene locus. *J. Biol. Chem.* **270**, 22577–22585

Davidson, N.O. (1994) RNA editing of the apolipoprotein B gene. A mechanism to regulate the atherogenic potential of intestinal lipoproteins? *Trends Cardiovasc. Med.* **4**, 231–235

Fan, J., McCormick, S.P.A., Krauss, R.M., Taylor, S., Quan, R., Taylor, J.M. and Young, S.G. (1995) Overexpression of human apolipoprotein B100 transgenic rabbits results in increased levels of low density of lipoproteins and decreased levels of high density lipoproteins. *Arterioscler. Thromb. Vasc. Biol.* **15**, 1889–1899

Fan, J., Wang, J., Bensadoun, A., Lauer, S.J., Dang, Q., Mahley, R.W. and Taylor, J.M. (1994) Overexpression of hepatic lipase in transgenic rabbits leads to a marked reduction of plasma high density lipoproteins and intermediate density lipoproteins. *Proc. Natl. Acad. Sci. USA* **91**, 8724–8728

Ferrin, L.J. and Camerini-Otero, R.D. (1991) Selective cleavage of human DNA: RecA-assisted restriction endonuclease (RARE) cleavage. *Science* **254**, 1494–1497

Lauer, S.J., Walker, D., Elshourbagy, N.A., Reardon, C.A., Levy-Wilson, B. and Taylor, J.M. (1988) Two copies of the human apolipoprotein C-I gene are linked closely to the apolipoprotein E gene. *J. Biol. Chem.* **263**, 7277–7286

Linton, M.F., Farese Jr., R.V., Chiesa, G., Grass, D.S., Chin, P., Hammer, R.E., Hobbs, H.H. and Young, S.G. (1993) Transgenic mice expressing high plasma concentrations of human apolipoprotein B100 and lipoprotein(a). *J. Clin. Invest.* **92**, 3029–3037

Lusis, A.J., Taylor, B.A., Quon, D., Zollman, S. and LeBoeuf, R.C. (1987) Genetic factors controlling structure and expression of apolipoproteins B and E in mice. *J. Biol. Chem.* **262**, 7594–7604

McCormick, S.P.A., Linton, M.F., Hobbs, H.H., Taylor, S., Curtiss, L.K. and Young, S.G. (1994a) Expression of human apolipoprotein B90 in transgenic mice. Demonstration that apolipoprotein B90 lacks the structural requirements to form lipoprotein(a). *J. Biol. Chem.* **269**, 24284–24289

McCormick, S.P.A., Linton, M.F. and Young, S.G. (1994b) Expression of P1 DNA in mammalian cells and transgenic mice. *Genet. Anal. Tech. Appl.* **11**, 158–164

McCormick, S.P.A., Ng, J.K., Taylor, S., Flynn, L.M., Hammer, R.E. and Young, S.G. (1995) Mutagenesis of the human apolipoprotein B gene in a yeast artificial chromosome reveals the site of attachment for apolipoprotein(a). *Proc. Natl. Acad. Sci. USA* **92**, 10147–10151

McCormick, S.P.A., Peterson, K.R., Hammer, R.E., Clegg, C.H. and Young, S.G. (1996) Generation of transgenic mice from yeast artificial chromosome DNA that has been modified by gene targeting. *Trends Cardiovasc. Med.* **6**, 16–24

Myklebost, O. and Rogne, S. (1988) A physical map of the apolipoprotein gene cluster on human chromosome 19. *Hum. Genet.* **78**, 244–247

Pease, R.J., Milne, R.W., Jessup, W.K., Law, A., Provost, P., Fruchart, J.-C., Dean, R.T., Marcel, Y.L. and Scott, J. (1990) Use of bacterial expression cloning to localize the epitopes for a series of monoclonal antibodies against apolipoprotein B100. *J. Biol. Chem.* **265**, 553–568

Peterson, K.R., Clegg, C.H., Huxley, C., Josephson, B.M., Haugen, H.S., Furukawa, T. and Stamatoyannopoulos, T. (1993) Transgenic mice containing a 248-kb yeast artificial chromosome carrying the human β-globin locus display proper developmental control of human globin genes. *Proc. Natl. Acad. Sci. USA* **90**, 7593–7597

Rothstein, R. (1991) Targeting, disruption, replacement, and allele rescue: Integrative DNA transformation in yeast. *Methods Enzymol.* **194**, 281–301

Schedl, A., Montoliu, L., Kelsey, G. and Schütz, G. (1993) A yeast artificial chromosome covering the tyrosinase gene confers copy number-dependent expression in transgenic mice. *Nature* **362**, 258–261

Scherer, S. and Davis, R.W. (1979) Replacement of chromosome segments with altered DNA sequences constructed *in vitro*. *Proc. Natl. Acad. Sci. USA* **76**, 4951–4955

Shachter, N.S., Zhu, Y., Walsh, A., Breslow, J.L. and Smith, J.D. (1993) Localization of a liver-specific enhancer in the apolipoprotein E/C-I/C-II gene locus. *J. Lipid Res.* **34**, 1699–1707

Simonet, W.S., Bucay, N., Lauer, S.J. and Taylor, J.M. (1993) A far-downstream hepatocyte-specific control region directs expression of the linked human apolipoprotein E and C-I genes in transgenic mice. *J. Biol. Chem.* **268**, 8221–8229

Smit, M., van der Kooij-Meijs, E., Frants, R.R., Havekes, L. and Klasen, E.C. (1988) Apolipoprotein gene cluster on chromosome 19. Definite localization of the APOC2 gene and the polymorphic HpaI site associated with type III hyperlipoproteinemia. *Hum. Genet.* **78**, 90–93

Smith, D.R. (1994) Vectors and host strains for cloning and modification of yeast artificial chromosomes. In Nelson, D.L. and B.H. Brownstein (eds.), *YAC Libraries. A User's Guide* New York: W.H. Freeman and Company, 1–31

Sternberg, N. (1994) The P1 cloning system: Past and future. *Mamm. Genome* **5**, 397–404

Sternberg, N.L. (1992) Cloning high molecular weight DNA fragments by the bacteriophage P1 system. *Trends Genet.* **8**, 11–16

Taylor, J. M., Simonet, W.S., Bucay, N., Lauer, S.J. and de Silva, H.V. (1991) Expression of the human apolipoprotein E/apolipoprotein C-I gene locus in transgenic mice. *Curr. Opin. Lipidol.* **2**, 73–80

Young, S.G. (1990) Recent progress in understanding apolipoprotein B. *Circulation* **82**, 1574–1594

Young, S.G., Farese Jr., R.V., Pierotti, V.R., Taylor, S., Grass, D.S. and Linton, M.F. (1994) Transgenic mice expressing human apoB$_{100}$ and apoB$_{48}$. *Curr. Opin. Lipidol.* **5**, 94–101

PART III, SECTION B

44. Bacterial Artificial Chromosomes and Animal Transgenesis

Ken Dewar[1], Bruce W. Birren[2],* and Hadi Abderrahim[3]

[1] *Department of Biology, University of Pennsylvania, Philadelphia, PA, USA*
[2] *Whitehead Inst./MIT Center for Genome Research, 1 Kendall Square, Bldg 300 Cambridge MA 02139, USA*
[3] *Centre d'Etudes du Polymorphisme Humain, Paris, France*

Introduction

The ability to transfer large fragments of DNA into animals offers several important advantages for positional cloning, studies of gene function, and the production of transgenic animals. It allows the integration and expression of large and complex transcription units, including their neighboring regulatory regions. The integration of large inserts reduces position effects, resulting in a more position independent expression (Forget, 1993). It also allows the transfer of multiple coding units into an animal, where different genes could be put under the control of separate promoters. Cloning systems and techniques for working with large regions of cloned DNA are relatively recent developments in molecular biology. Chief among the concerns for using large insert clones as sources for generating transgenic animals are that the cloned segments should faithfully represent the original genomic sequences (that is, they must not have undergone any deletions or rearrangements) and that the DNA can be purified and manipulated easily.

The bacterial artificial chromosome (BAC) cloning system (Shizuya *et al.*, 1992) has been developed to stably maintain large fragments of genomic DNA (100–300 kb) in *E. coli*. While smaller than YACs (500 kb–2 Mb), the stability and ease of manipulation of BAC clones makes them attractive candidates for animal transgenesis.

The BAC Cloning System

The BAC system is based on the much studied F factor of *E. coli*. The maintenance of BAC clones at a single copy per cell appears to be critical to their ability to propagate mammalian DNA fragments with greater fidelity than multi-copy (i.e., cosmid) cloning vectors (Kim *et al.*, 1992). The high efficiency of BAC cloning results from the use of electroporation to introduce ligated molecules into *E. coli* host cells. Genomic libraries, containing clones ranging from 40 kb to over 300 kb, and offering many fold redundancy of coverage, have been constructed from a variety of mammalian species (Wang *et al.*, 1994; Cai *et al.*, 1995 and Kim *et al.*, 1996). The analysis of BAC clones and libraries employs standard molecular techniques. For example, BAC libraries can be screened by PCR or hybridization based approaches, and conventional alkaline lysis methods for plasmid DNA isolation allow the purification of BAC DNA from the *E. coli* host DNA. The quantity of BAC DNA recovered from 1–2 ml of culture is sufficient for restriction digestion and analytical or preparative agarose gel electrophoresis using ethidium bromide staining (Birren *et al.*, in press).

The BAC Vector and Construction of BAC Clones

The pBeloBAC11 vector (Kim *et al.*, 1996) shown in Figure 44.1 is a modification of the original BAC vector of Shizuya *et al.* (1992). The BAC vector contains the F-factor genes required for autonomous replication, copy number control, and partitioning as well as a gene encoding resistance to chloramphenicol for selection in bacteria. The pBeloBAC11 cloning cassette allows blue/white

*Author for Correspondence.

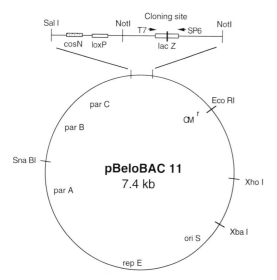

Figure 44.1. The pBeloBAC11 Vector. The vector pBeloBAC11 is capable of accepting cloned segments of DNA to over 300 kb. Mammalian DNA is cloned into either the *Bam*H1 or *Hind*III sites, which are immediately flanked by the T7 and SP6 promoters. Two *Not*I sites lie outside of these promoters, and digestion with *Not*I will usually excise the insert from the vector, usually as a single fragment. The *cos*N and *lox*P sites provide additional means for linearization of the clones. The genes encoded by *par* A, *par* B, *par* C, *rep* E ad *ori* S are those necessary for control of replication, copy number, and partitioning. The gene for chloramphenicol resistance (CM^r) provides a selectable marker for bacteria.

color selection to distinguish non-recombinant from recombinant clones. BAC libraries are prepared with this vector using size-selected genomic DNA that has been partially digested using enzymes that permit ligation into either the unique *Bam*HI or *Hind*III sites in the vector. Flanking these cloning sites are T7 and SP6 RNA polymerase transcription initiation sites. These can be used to generate end probes by either RNA transcription or PCR methods.

Linearizing BAC Clones

BAC DNA is purified from the host cell as a supercoiled circle. Converting these circular molecules into a linear form normally precedes both size determination and introduction of the BACs into recipient cells. Several alternative mechanisms for linearization are possible. The cloning site is flanked by two *Not* I sites, permitting cloned segments to be excised from the vector by *Not* I digestion. *Not* I sites occur infrequently in most mammalian genomes, and therefore *Not* I digestion of a BAC DNA molecule often will produce the entire cloned segment as a single linear DNA fragment. pBeloBAC11 also contains two other sites that can be used to linearize BAC clones. Treatment of BACs with the commercially available enzyme lambda terminase leads to cleavage at the unique *cos*N site. The *cos*N site does not naturally occur within mammalian DNA genomes, so terminase treatment will generate a full length BAC clone with vector sequences at each end. In addition, both linearization and introduction of new sequences into existing clones can be performed by *Cre*-mediated recombination at the *loxP* site (Sternberg, in press).

Obtaining BAC Clones with Genes and Regions of Interest

The large insert capacity of BACs allows most mammalian genes to be contained within a single BAC clone. Because BAC libraries consist of clones of a range of sizes (from 40 to over 300 kb), a library with redundant genomic coverage will usually contain many different sized clones for a particular genomic region. For transgenesis experiments, this allows flexibility and choice. The largest sized clones can be used when the location of the gene is not certain, or in other cases, the smallest clone known to contain the intact gene and its regulatory segments can be used. When complementation can be readily scored, the preparation of transgenic animals can be used to assist gene identification during positional cloning. In all of these cases, a group of overlapping clones covering the region of interest may be used to select candidate BACs for transgenic experiments.

Once BAC clones have been isolated by hybridization or PCR-based screening of a library, determining the extent of overlap between the related clones requires further analysis. Most frequently this involves restriction digestion and a comparison of the fragments produced from the different clones. These restriction "fingerprints" can be as simple as *Eco* RI or *Hind* III digestions followed by agarose gel electrophoresis, or can involve the more complex fingerprinting performed for generating cosmid contigs (Coulson et al., 1986). Given the amount of effort involved in generating transgenic animals, some time should be invested in confirming the fidelity of the cloned region before selecting the clones to be used for the production of transgenic animals. BAC clone verification can involve direct comparison of cloned sequences to genomic regions using PCR or by Southern hybridization. In addition, when multiple BAC clones are obtained that cover the same region, comparison of the clones to each other will likely indicate the presence of deletions or rearrangements, since identical independent

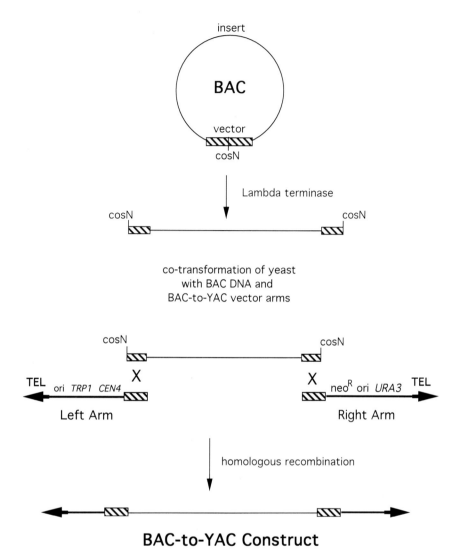

Figure 44.2. BAC to YAC conversion by co-transformation and homologous recombination in *Saccharomyces cerevisiae*. A two step process is used to convert segments cloned as BACs into YACs for subsequent transfer into mammalian cells. BAC DNA is first linearized *in vitro* with lambda terminase which cleaves at the *cos*N site within the vector. The linearized BAC molecules and the BAC-to-YAC vector arms are co-transformed into yeast. Each of the BAC-to-YAC arms carries a region of homology with a portion of the BAC vector, along with a yeast selectable marker. During transformation, homologous recombination takes place between the BAC vector sequences and the BAC-to-YAC arms, generating a YAC.

deletions or rearrangements have not been observed among the many BAC clones we have analyzed.

Methods for the Introduction of BACS into Recipient Animals

Microinjection of BAC DNA into Embryos

Microinjection of DNA into the male pronucleus of the fertilized egg (Gordon *et al.*, 1980) permits the direct introduction of DNA into the nucleus of the recipient cells. Microinjection of small insert DNA has been used to produce a variety of transgenic animals, including mice, rats, rabbits, sheep, goats, pigs, and cattle (Lathe and Mullins, 1993).

Generally, an amount of DNA representing 1–5 copies is injected at low concentration (1 ng/ml) (Smith *et al.*, 1995). As linear DNA appears to integrate at a higher frequency than circular supercoiled DNA, candidate BAC clones could be linearized as described above prior to microinjection.

To avoid mechanical breakage of the BAC DNA during the microinjection process, polyamines and high salt concentrations can be used. P1 clones and YACs (up to 250 kb) have been successfully microinjected using these methods (Schedl et al., 1993).

Introduction of BAC DNA into Embryonic Stem Cells

Large DNA inserts are very susceptible to mechanical breakage and possible rearrangements during the transfer process. The ability to introduce DNA fragments into embryonic stem (ES) cells permits the structural integrity and level of expression of the transferred sequences to be examined before blastocyst injection and the production of chimeric animals. However, this technique is limited to those animals from which ES cells have been derived.

The most efficient method of large DNA transfer into ES cells is lipofection. Linearized DNA is mixed with synthetic cationic lipid forming vesicles, which are then fused to the plasma membrane of the recipient ES cells (Strauss et al., 1993). In comparison with other ES cell transfection procedures (electroporation, calcium phosphate precipitation), lipofection gives higher efficiency and also minimizes mechanical breakage.

To enable the selection of cells transfected with BAC DNA, several approaches are possible. The BAC DNA could be co-lipofected (Choi et al., 1993) with a plasmid carrying a selectable marker (e.g., neomycin or hygromycin resistance). Alternatively, the BAC clone could be retrofitted to carry a drug resistance marker. Using a scheme similar to that employed for the modification of P1 clones (Sternberg, in press), a neomycin resistance gene could be inserted into the loxP site of existing clones by Cre-mediated recombination. The neo-BAC construct could then be linearized and used for lipofection.

To take advantage of the protocols that do not require purification of DNA for transferring YACs into embryos or ES cells (Forget, 1993), a strategy to convert BACs into YACs could be employed. YAC vector arms could be ligated directly to linearized BAC DNA, or added by homologous recombination as portrayed in Figure 44.2. After recovery of the newly formed YAC, a test of its integrity would be needed prior to transfection of the new construct into animal cells.

Conclusion

Large insert bacterial based genomic libraries are becoming increasingly available, with BAC and PAC (Ioannou et al., 1994) libraries now finding commercial distribution from a number of companies. The observations that these clones are appropriately sized to contain entire mammalian genes, that they can propagate mammalian genomic regions with high fidelity, and that conventional methods can be used for their analysis explains the increasing reliance on these systems for physical mapping, gene isolation and gene identification. For these same reasons, it is likely that their use as a reagent for construction of transgenic animals will soon become standard.

References

Birren, B., Mancino, V. and Shizuya, H. (in press) Bacterial artificial chromosomes. In Birren, Green, Hieter, Klapholz, and Myers, (eds.) *Genome Analysis: A Laboratory Manual*. Cold Spring Harbor Laboratory Press

Cai, L. Taylor, J.F., Wing, R.A., Gallager, D.S., Woo, S.S. and Davis, S.K. (1995) Construction and characterization of a bovine bacterial artificial chromosome library. *Genomics* **29**, 413–425

Choi, T.K., Hollenbach, P.W., Pearson, B.E., Ueda, R.M., Weddell, G.N., Kurahara, C.G., Woodhouse, C.S., Kay, R.M. and Loring, J.F. (1993) Transgenic mice containing a human heavy chain immunoglobulin gene fragment cloned in a yeast artificial chromosome [published erratum appears in Nat. Genet. 1993 Jul; 4(3):320]. *Nat. Genet.* **4**, 117–123

Coulson, A., Sulston, J., Brenner, S. and Karn, J. (1986) Toward a physical map of the genome of the nematode *Caenorhabditis elegans*. *Proc. Natl. Acad. Sci. USA* **83**, 7821–7825

Forget, B.G. (1993) YAC transgenes: bigger is probably better. *Proc. Natl. Acad. Sci. USA* **90**, 7909–7911

Gordon, J.W., Scangos, G.A., Plotkin, D.J., Barbosa, J.A. and Ruddle, F.H. (1980) *Proc. Natl. Acad. Sci. USA* **77**, 7380

Ioannou, P.A., Amemiya, C.T., Garnes, J., Kroisel, P.M., Shizuya, H., Chen, C., Batzer, M.A. and de Jong, P.J. (1994) A new bacteriophage P1-derived vector for the propagation of large human DNA fragments. *Nature Genetics* **6**, 84–89

Kim, U.-j., Birren, B.W., Slepak, T., Mancino, V., Boysen, C., Kang, H.-L., Simon, M.I. and Shizuya, H. (1996) Construction and characterization of a human bacterial artificial choromosome library. *Genomics* **34**, 213–218

Kim, U.-J., Shizuya, H., de Jong, P.J., Birren, B. and Simon, M. (1992) Stable propagation of cosmid sized human DNA inserts in an F factor based vector. *Nucleic Acids Res.* **20**, 1083–1085

Lathe, R. and Mullins, J.J. (1993) Transgenic animals as models for human disease – report of an EC Study Group. *Transgenic. Res.* **2**, 286–299

Schedl, A., Larin, Z., Montoliu, L., Thies, E., Kelsey, G., Lehrach, H. and Schutz, G. (1993) A method for the generation of YAC transgenic mice by pronuclear microinjection. *Nucleic Acids Res.* **21**, 4783–4787

Shizuya, H., Birren, B., Kim, U.-J., Mancino, V., Slepak, T., Tachiiri, Y. and Simon, M. (1992) Cloning and stable maintenance of 300-kilobase-pair fragments of human DNA in Escherichia coli using an F-factor-based vector. *Proc. Natl. Acad. Sci. USA* **89**, 8794–8797

Smith, D.J., Zhu, Y., Zhang, J., Cheng, J.F. and Rubin, E.M. (1995) Construction of a panel of transgenic mice containing a contiguous 2-Mb set of YAC/P1 clones from human chromosome 21q22.2. *Genomics* **27**, 425–434

Sternberg, N. (in press) The P1 Cloning System. In Birren, Green, Hieter, Klapholz, and Myers, (eds.) *Genome Analysis: A Laboratory Manual.* Cold Spring Harbor Laboratory Press

Strauss, W.M., Dausman, J., Beard, C., Johnson, C., Lawrence, J.B. and Jaenisch, R. (1993) Germ line transmission of a yeast artificial chromosome spanning the murine alpha 1(I) collagen locus. *Science* **259**, 1904–1907

Wang, M., Chen, X.-N., Shouse, S., Manson, J., Wu, Q., Li, Rumei, Wrestler, J., Noya, D., Sun, G.-Z., Korenberg, J., Lai, E. (1994) Construction and characterization of a human chromosome 2-specific BAC library. *Genomics* **24**, 527–533

PART III, SECTION B

45. The Use of Yeast Artificial Chromosomes for Transgenesis

Thorsten Umland, Lluis Montoliu[†] and Günther Schütz[*]

Division Molecular Biology of the Cell I, German Cancer Research Center, Im Neuenheimer Feld 280, D-69120 Heidelberg, Germany

The ability to stably integrate and express foreign genes in the mouse germline has brought about considerable advances in biomedical research. However, standard transgenes in mice often fail to reproduce the exact expression pattern of the endogenous copy (Jaenisch, 1988). Not only are transgenes often regulated inappropriately, their levels of expression can also vary widely between different founder lines (Palmiter and Brinster, 1986). These setbacks are due to the limited cloning capacity of standard plasmid/cosmid vectors. Frequently, cDNA-based minigenes have to be used which lack introns and possibly important regulatory elements. As the construct integrates randomly into the host genome, neighbouring sequences might silence, enhance or alter expression of the transgene. This phenomenon is known as "position effects" (Elgin, 1990; Wilson et al., 1990; Dillon and Grosveld, 1993 and 1994). Only a small number of transgenes have been shown to be insensitive to their chromosomal environment, their expression being position-independent and copy number dependent. This correct gene activity was shown to be due to "locus control regions" (Grosveld et al., 1987; Wang et al., 1992 and Palmiter et al., 1993) or "insulators", which shield the transgene from surrounding chromatin (Bonifer et al., 1990; Kellum and Schedl, 1991; McKnight et al., 1992; Chung et al., 1993; Kalos and Fournier, 1995; Choudhary et al., 1995; and Cai and Levine, 1995). However, these sequence elements do not occur in every gene, and if they do, they are usually located far from the coding region.

Again, the investigator finds himself confronted with the limited cloning capacity of customary vectors.

Yeast artificial chromosomes (YACs) are cloning vectors capable of stably maintaining heterologous DNA inserts exceeding 1 Mb in size (Burke et al., 1987). Thus, complex gene loci can be cloned and studied in conditions closely approaching their natural genomic context, including far-upstream regulatory sequences and elements crucial for the integrity of a chromosomal domain. Furthermore, due to the highly active homologous recombination system in yeast, specific sequences within the YACs can be modified at will (Schlessinger, 1990; Pavan et al., 1990; Hieter et al., 1990; Davies et al., 1992 and Spencer et al., 1993).

In the past two years, a number of groups have reported the use of YACs for gene transfer into the mouse germline (reviewed by Montoliu et al., 1994; Lamb and Gearhart, 1995). Table 45.1. shows a summary of the YAC transgenic mice generated to date. Three different techniques of gene transfer have been applied: spheroplast fusion, lipofection and microinjection. The first two involve the transfection of a YAC into embryonic stem (ES) cells, characterisation of stably transformed colonies and injection of these into mouse blastocysts to generate chimaeric mice. The third approach involves direct microinjection of the YAC DNA into pronuclei of fertilised mouse oocytes, with identification of the integrated transgene at the founder level. With all three techniques, nearly all of the transgenic mouse lines generated exhibited endogenous expression levels of the YAC-derived transgene, suggesting that the presence of large genomic regions including flanking and intronic sequences might be sufficient for a gene's faithful expression.

---

[*] Author for Correspondence.
[†] Present address: Department of Biochemistry and Molecular Biology, Faculty of Veterinary Medicine, University of Barcelona, 08193 Bellaterra (Barcelona), Spain.

Table 45.1. List of reported uses of YACs for transgenic experiments in mice.

Methodology	Gene	Size of YAC (kb)	Reference
Spheroplast fusion	human HPRT	670	Jakobovits et al. (1993)
	human Igκ light chain	300	Davies et al. (1993)
	human Ig (heavy and κ light chain)	195+240	Green et al. (1994)
Lipofection	mouse collagen (col1α1)	150	Strauss et al. (1993)
	human Ig heavy chain	100	Choi et al. (1993)
	human β-amyloid precursor protein (APP)	650	Lamb et al. (1993)
	human β-amyloid precursor protein (APP)	650	Pearson et al. (1993)
Microinjection	mouse tyrosinase	250	Schedl et al. (1993a)
	human β-globin locus	248	Peterson et al. (1993)
	human β-globin locus	150	Gaensler et al. (1993)
	human Apo(a)	270	Frazer et al. (1995)
	mouse tyrosinase	850	Montoliu et al. (unpublished)

Spheroplast Fusion

A prerequisite for this method of transfer is that the YAC must be modified prior to transfection: a selectable marker cassette has to be incorporated by homologous recombination within the yeast host strain. Then, the yeast cell wall is digested enzymatically to produce spheroplasts, which are subsequently fused to mouse ES cells in the presence of polyethylene glycol; varying portions of the yeast genome are transferred in the process. The marker gene in the YAC allows for selection of those ES cell clones which have stably integrated the YAC into their genome. After characterisation of positive colonies, chosen ES cells are injected into host blastocysts to generate chimaeric mice, a certain proportion of which will transmit the YAC through the germline.

Jakobovits et al. (1993) used this method with a 670 kb YAC carrying the human hypoxanthine phosphoribosyltransferase (HPRT) gene. 8 out of 20 ES cell clones which had survived selection in HAT (hypoxanthine–aminopterin–thymidine) medium turned out to have the complete, unrearranged YAC integrated into the genome. Transgenic mice generated with these ES cells expressed the human HPRT transgene at levels comparable to their murine counterpart. Varying amounts of yeast genomic DNA had co-integrated into the genome of these mice, without obvious impairment of their proper development.

Green et al. (1994) used this technique to produce mice expressing human antibodies. They isolated two YACs from a human YAC library, one 240 kb immunoglobulin (Ig) heavy chain YAC, one 195 kb κ light chain YAC, and retrofitted both of them with a human HPRT selectable marker. The YACs were transferred separately into HPRT-deficient mouse ES cells by spheroplast fusion and transformants selected in HAT-medium. Injection of ES cells containing an intact insert into blastocysts led to the generation of chimaeric mice. 50% of their agouti offspring had the unaltered YAC in their germline. Mice expressing human heavy chains were mated with mice expressing human κ light chains to produce double transgenics. These showed only weak levels of expression, though, and only a small proportion of B cells expressed both the heavy and κ genes simultaneously. In the cells they were expressed, they were shown to be rearranged to a human-like repertoire. Breeding these animals with mice deficient in mouse immunoglobulin production resulted in a strain with a high-level, antigen-specific human antibody response.

Among the advantages in using spheroplast fusion for the transfer of YACs is the fact that isolation of YAC DNA from the yeast host cells prior to transfection is not required. Furthermore, it has been observed that a relatively high percentage of the YACs remain intact in the course of transfer to ES cells. Also, positive ES cell clones can be characterised first and only those containing complete, unrearranged inserts are then used for the generation of transgenic mice.

One disadvantage is the comparatively long time it takes for the YAC to be stably transmitted through the mouse germline. Another drawback is that often varying amounts of yeast DNA are co-transferred and randomly integrate into the mouse genome, which may have undesirable effects on the transgenic lines generated. Sometimes, a number of ES cell clones can be selected which only have the YAC integrated (Davies et al., 1993).

Lipofection

Lipofection requires the YAC to be purified from the yeast cells via preparative pulsed-field gel electrophoresis (PFGE). The YAC DNA is isolated in the presence of spermine, followed by the addition of poly-L-lysine and finally the addition of a cationic lipid which leads to the formation of the transfection complex. Exposing the ES cells to this complex in suspension yields higher transfection efficiencies than when carried out on ES cells growing as a monolayer (Strauss et al., 1993). There are two options for the selection of positive ES cell clones: either the YAC is retrofitted with a selectable marker gene prior to transfection, or an unlinked marker plasmid is co-lipofected with the YAC. The latter leads to very low proportions of ES cells containing an intact YAC, in the order of 1% of the positive transformants (Choi et al., 1993; and Pearson and Choi, 1993)

Strauss et al. (1993) lipofected a 150 kb YAC carrying the mouse collagen gene Colα1 and a neomycin resistance gene in the left vector arm. They obtained 30 G418 resistant clones per 10^8 transfected ES cells, only 3 of which harboured unrearranged YAC DNA and 2 of these resulting in germline transmission. The transgenic offspring were found to express the exogenous Colα1 gene at levels comparable to the endogenous allele. These mice can be used for molecular complementation experiments with the Mov13 mouse strain, in which the Colα1 gene has been inactivated by a proviral insertion, leading to a recessive embryonic lethal phenotype.

The option of co-lipofecting a drug-selectable marker on a separate plasmid was chosen by Choi et al. (1993) with a 100 kb YAC covering an 85 kb human heavy chain immunoglobulin genomic fragment. This resulted in a very low efficiency of transfection, most of the resistant ES cell clones carrying fragmented YAC pieces. Of 3 unrearranged clones identified in a total of 1221 surviving colonies, one gave germline transmission. In this transgenic line, expression of the YAC-borne human μ chain was found to be very low compared to serum levels of endogenous mouse IgM. This could be partly attributed to competition from the mouse heavy chain locus, since human transgene expression was 10 fold higher in a background with the endogenous heavy chain alleles inactivated.

In an attempt to create a mouse model for Alzheimer's disease, the human β-amyloid precursor protein (APP) gene was introduced into the mouse by two independent groups, both using lipofection-mediated YAC transfer. Lamb et al. (1993) retrofitted a 650 kb YAC carrying the 400 kb human APP gene with a neomycin resistance gene and from 23 G418 resistant ES cell clones obtained 2 which contained the complete YAC. Both of these clones resulted in germline transmission. Expression analysis by quantitative RT-PCR using species-specific primers revealed expression levels of the multiple human transcripts comparable to endogenous mouse APP.

Similar results were obtained by Pearson and Choi (1993), the difference being that a PGKneo plasmid was co-lipofected with the unaltered YAC, resulting in a much lower proportion of primary resistant ES cell clones containing the entire APP gene (3 out of 240 analysed).

In general, the lipofection method shares with spheroplast-mediated fusion the advantage of allowing characterisation of transfected ES cells prior to the generation of mice. Since only the YAC is transferred to the mouse germline, the risk of detrimental effects from co-transfected yeast DNA is removed. However, a higher proportion of the drug-resistant ES cell clones obtained from lipofection carry fragmented YACs. Typically, about 10% of the primary resistant colonies contain the intact, unrearranged YAC (Strauss et al., 1993 and Lamb et al., 1993). This figure is reduced to 1% or less when co-lipofection of a separate marker plasmid with the YAC is applied (Pearson and Choi, 1993; and Choi et al., 1993). This compares to about 40% of positive transformants retaining a complete YAC after spheroplast transfection (Jakobovits et al., 1993).

Microinjection

YAC DNA can also be directly injected into the pronuclei of fertilised mouse oocytes. To achieve this, the YAC must be purified via preparative pulsed-field gel electrophoresis and maintained in a special buffer containing polyamines and/or high salt concentration to prevent breakage in solution and shearing upon passage through the microinjection needle. To achieve a sufficiently high copy number, the DNA must be concentrated, e.g. by two-dimensional gel electrophoresis, dialysis or ultrafiltration. In addition, the copy number of the YAC within the yeast cell can be increased up to 25-fold beforehand, by retrofitting the YAC with a conditional centromere; growing the yeast cells in galactose-containing medium will inactivate the centromere, leading to impaired segregation during cell division and enrichment of the YAC within the yeast cell (Smith et al., 1990 and Schedl et al., 1992; 1993b).

In our laboratory, transgenic mice were generated with a 250 kb YAC covering the 80 kb mouse tyrosinase locus, including 155 kb of 5' upstream sequence (Schedl et al., 1993a). Special features in the preparation of the YAC for microinjection included: an increase of the YAC copy number per yeast cell by inclusion of a conditional centromere; separation from the endogenous yeast chromosomes by PFGE and concentration via a second dimension gel electrophoresis; and membrane dialysis against a special microinjection buffer, which contained 100 mM NaCl and 100 μM polyamines (spermine and spermidine) to prevent shearing.

The YAC thus prepared was microinjected at a concentration of $4\,ng\,\mu L^{-1}$ into fertilised albino mouse oocytes. Albino mice have a point mutation in the first exon of the tyrosinase gene, leading to the production of non-functional protein (Jackson et al., 1990; Yokoyama et al., 1990 and Shibahara et al., 1990). From 24 newborns, 5 transgenics could be identified by the presence of tyrosinase activity and hence pigmentation in eyes and skin. In all of these cases the albino phenotype was fully rescued, the pigmentation level being indistinguishable from wild type. Expression analysis by quantitative RT-PCR, carried out on the three transmitting lines, demonstrated copy number-dependent, position-independent expression of the YAC-borne transgene, the expression level of one copy being comparable to the endogenous tyrosinase locus. In all lines, the integration of a complete YAC into the genome was shown by Southern analysis and fluorescence in situ hybridisation (FISH).

A YAC containing the human β-globin locus was microinjected by two independent groups. Peterson et al. (1993) used a 248 kb YAC spanning the 82 kb β-globin locus. After the first gel-purification, they concentrated the DNA by ultrafiltration in a series of centrifugation steps. Their microinjection buffer contained 100 mM NaCl and the DNA was diluted to $1\,ng\,\mu L^{-1}$ prior to injection. From 148 offspring generated, 17 transgenic mice were obtained, 10 of which carried an unrearranged YAC. One transgenic individual was bred further, and proper developmental control of the β-globin genes encompassed in the locus could be demonstrated in the F1 and F2 progeny of that line.

Similarly, a 150 kb YAC used by Gaensler et al. (1993) resulted in a level and pattern of expression of the β-globin genes that closely resembled that of the endogenous locus. Here, concentration of the DNA was achieved by dialysis against sucrose, followed by dialysis against standard microinjection buffer (10 mM Tris pH 7.4/0.1 mM EDTA). The YAC DNA was microinjected at a concentration of $2\text{–}4\,ng\,\mu L^{-1}$ and resulted in the generation of two transgenic founders with an intact insert.

Frazer et al. (1995) microinjected a 270 kb YAC encompassing the 110 kb human apolipoprotein(a) gene, which had been shown to represent an atherosclerosis risk factor in humans (Utermann 1989). The YAC was purified essentially using the method of Schedl et al. (1993b), with the exception that the injection buffer did not contain polyamines. From 190 offspring generated, 10 (5.3%) were transgenic founders, 4 of which contained the entire apo(a) locus. All of the transgenic lines expanded from these 4 founders produced high plasma levels of apo(a). Expression was confined to the liver which is also thought to be the major site of synthesis in humans.

In summary, pronuclear microinjection represents a very attractive method for the delivery of YACs directly into the germline of mice. Transgenic lines can be generated very rapidly, without the need for a detour via ES cells. Efficiencies for the generation of transgenics vary between 5 and 21% of total offspring, which is slightly below the range observed for standard transgenesis (10–30%; Brinster et al., 1985). Manipulation of the YAC to introduce a selectable marker cassette is not required. Only the YAC carrying the transgene is transferred, without any contaminating yeast DNA. Furthermore, we have shown that even large YACs, up to 1 Mb, can be delivered intact to the mouse germline. Electron microscopic studies carried out in our lab (Montoliu et al., 1995) enabled the visualisation of YACs in microinjection buffer containing high salt and polyamines (see Figure 45.1). Under these conditions, YACs up to 1 Mb in size appear as highly condensed spherical units with a diameter of less than 1 μm, which corresponds to the diameter of needles commonly used for pronuclear microinjection. We were able to generate transgenic mice by microinjection of an 850 kb YAC, which integrated intact into the mouse genome (Montoliu et al., unpublished). It has thus been shown that even the transfer of very large YACs is feasible by using microinjection and a specific buffer.

A disadvantage of this method is that a number of mice are generated with rearranged or fragmented

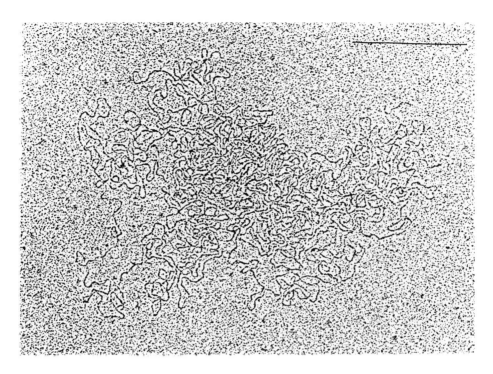

Figure 45.1. Electron micrograph of Tyr4 DNA (850 kb YAC covering the mouse tyrosinase gene; Montoliu et al., 1995). YAC DNA in microinjection buffer (see text) was adsorbed to a glow-discharged carbon grid. The bar represents 0.5 μm.

copies of the YAC, since there is no ES cell stage where clones can be characterised beforehand.

Procedure for the Purification and Preparation of YACs for Pronuclear Microinjection

Using the following protocol, YAC transgenic mice have been generated in our laboratory with efficiencies of 10–20% of mice born, which is comparable to the 10–30% observed in standard transgenesis via microinjection (Brinster et al., 1985). Figure 45.2 shows a schematic summary of the method described here.

1. Prepare high density agarose blocks from YAC containing yeast cells (see Schedl et al. (1995) for a more detailed description). Equilibrate these blocks in 0.5 × TAE two times for 30 minutes.
2. Load 6 to 7 of these blocks into a large preparative slot of a 1% Seaplaque GTG LMP agarose gel for pulsed field gel electrophoresis (PFGE), with 0.5 × TAE as running buffer. To the left and right of this slot, load blocks for marker lanes and a size standard. Seal the blocks by covering with 1% LMP agarose.
3. Run a preparative PFGE at 14 C using the conditions appropriate for the resolution of the YAC from the other yeast chromosomes.
4. After the electrophoresis, cut off the marker lanes and part of the preparative lane. Stain these by gentle shaking for 30 min in 0.5 × TAE containing 0.5 μg/ml ethidium bromide. Under a UV light, use a scalpel blade to mark the positions of the band containing the YAC and of one additional band containing an endogenous yeast chromosome (to be used as a marker lane for the second electrophoresis).
5. Reassemble the stained side parts with the unstained middle part of the gel. At the marked positions, cut out gel stripes from the middle part where the YAC and the marker chromosome are assumed to be. The remainder of the gel can be stained to confirm that the DNA has been precisely removed.
6. Equilibrate the gel slices two times for 15 min in 1 × TAE.
7. Align them in a minigel tray so that the direction of electrophoresis will be at a 90 angle to the previous PFGE. Embed the gel stripes in a 4% NuSieve GTG LMP agarose gel. Run for approximately 8 hours at 4 V/cm in circulating 1 × TAE buffer.
8. Again, cut the marker stripe away from the YAC stripe and stain it in 1 × TAE, 0.5 μg/ml ethidium bromide. The chromosomal DNA will

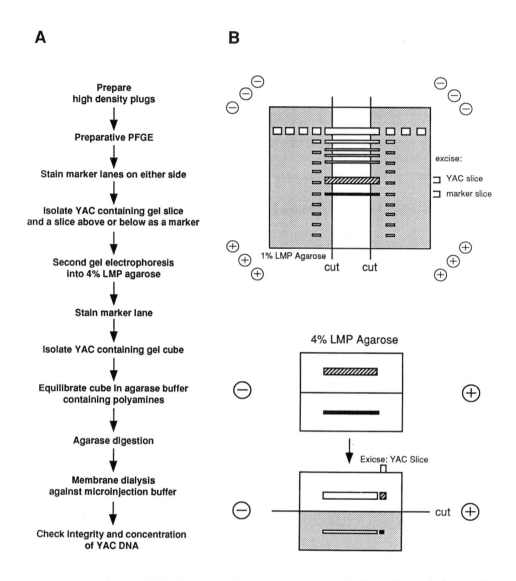

Figure 45.2. Purification of YAC DNA for pronuclear microinjection. A: Flow chart of the purification protocol. B: Schematic representation of the preparative PFGE and the second dimension gel electrophoresis for the isolation and concentration of YAC DNA.

now have run out of the gel stripe and into the high percentage NuSieve agarose, being concentrated in a small volume. Under UV light, mark the position where the DNA is now visible. Re-align the parts of the original gel and cut out the corresponding region from the unstained part containing the YAC.

9. Equilibrate this gel cube two times for 15 min. in 1 × TAE, 100 mM NaCl, 30 µM spermine, 70 µM spermidine.
10. Remove all liquid and weigh the agarose cube inside a pre-weighed Eppendorf tube.
11. Melt the agarose for 10 min at 65 C.
12. Keep for 10 min at 42 C before adding 4U of Gelase per 100 mg of gel.
13. Incubate the digestion at 42 C for 2 hours, mixing every 1/2 hour using a cut-off pipette tip.
14. Place the tube on ice for 5 min and check for complete digestion by pipetting several times with a cut-off pipette tip. If digestion is incomplete, melt for 10 min at 65 C and digest again at 42 C for at least one hour using the same amount of Gelase. Make sure the agarose is completely dissolved before proceeding with the next step.

15. Fill about 40 ml of injection buffer (10 mM Tris-HCl pH 7.5, 0.1 mM EDTA pH 8.0, 100 mM NaCl, 30 µM spermine, 70 µM spermidine) into a petri dish. Using a cut-off pipette tip, transfer the YAC DNA solution onto a dialysis filter (pore size 0.05 µm) floating on the surface of the injection buffer.
16. Dialyse for 3-4 hours.
17. Transfer this YAC DNA preparation to an Eppendorf tube and electrophorese 2 µl in an 0.8% agarose gel, together with DNA concentration standards. The DNA should be at a concentration of at least 2 ng/µl, to allow for an at least 1:1 dilution with injection buffer prior to microinjection. In our hands, diluting the DNA results in higher yields of transgenesis.
18. The YAC DNA can be stored at 4 C for at least half a year. Microinjection into pronuclei of fertilised mouse oocytes is carried out according to standard methods (Hogan et al., 1986).

Applications of YAC Transgenesis

The feasibility of introducing YACs into the germline of mice opens up a range of possible experiments which would be difficult to perform using standard vectors. The size of DNA that can be cloned in a YAC allows the analysis of complex transcription units; large gene clusters with far-upstream regulatory sequences can be investigated. By making use of the homologous recombination system in yeast, these genomic regions can be modified and the effect of the modifications studied *in vivo*. This will aid in the identification of important regulatory elements and mechanisms, as well as in the analysis of the developmental regulation of genes. One example for this is provided by work of Peterson et al. (1995), who introduced two different types of developmental mutations into YACs containing the β-globin locus. Analysis of the phenotypes of the corresponding transgenic mice gave new insights into the mechanisms of turning off γ-globin gene expression and switching to β-globin gene expression during fetal development.

Another example for the analysis of complex transcription units was carried out in our laboratory on a YAC containing the mouse tyrosinase gene (Montoliu et al., in preparation). In order to identify the element(s) conferring position-independence and copy number-dependence on transgene expression, a deletion series of the original YRT2 YAC (Schedl et al., 1993a; see above) was constructed, facilitated by the extensive homologous recombination system in yeast. By analysing the expression levels in the transgenic lines generated with the different deletion constructs, a locus control region could be identified at 12 kb upstream of the mouse tyrosinase gene harboured by the YACs.

Other genetic loci whose analysis could be facilitated by the use of YAC transgenesis would be, for example, the Hox gene clusters, the MHC locus, or the immunoglobulin gene loci, all of which encompass large regions of the genome, preventing their accessibility using other vector systems.

Even the investigation of higher order chromosomal structure and function becomes feasible when employing YACs. This includes the analysis of expression domains and domain boundaries, as well as questions concerning centromere and telomere function, origins of replication, imprinting, and X-inactivation. One of the genes known to play a role in the transcriptional silencing of one of the X-chromosomes in mammalian XX females is *Xist* (Rastan, 1994). However, neither the mechanism by which this is achieved, nor the exact sequences necessary for X-inactivation are known. Heard et al. (1994) created a deletion series of YACs containing the genomic region around the *Xist* gene. Using this deletion series in a biological assay will enable the dissection of this chromosomal region and the more precise determination of sequences indispensable for X-inactivation.

One feature of YACs in yeast that has not been achieved so far in mammalian cells is episomal stability. The development of MACs (mammalian artificial chromosomes; Brown, 1992) would bring about great advances in all kinds of biological investigations, as well as in gene therapy. It seems reasonable to use YACs as starting points for the introduction of components/sequences that confer episomal stability in mammals. One approach in this direction has been implemented by Taylor et al. (1994), who used retrofitting vectors to replace YAC telomeres with human telomeric DNA. Spheroplasting of these modified YACs into mammalian cells resulted in a low proportion of clones carrying a YAC with functional human telomeres.

YACs can also aid in the identification of genes by complementation experiments, in which a recessive mouse mutation is rescued by the introduction of a YAC carrying the gene(s) thought to be affected in the mutants. For instance, the tyrosinase-containing YRT2 YAC (Schedl et al., 1993a) not only rescued the albino phenotype as manifested by skin and eye pigmentation; it also corrected the abnormal retinal pathways associated with albinism, thereby proving that the underlying defect stems directly from the tyrosinase gene and the way in which melanin shapes the development of a part of the mammalian visual system (Jeffery et al., 1994).

Similarly, a YAC carrying the human PAX-6 gene was shown to rescue the *small eye* phenotype in transgenic mice (A. Schedl, pers. comm.). Surprisingly, overexpression of the PAX-6 gene in mice with a high copy number of the transgene resulted in severe abnormalities in the development of the eye, emphasising the importance of the strict regulation of this gene during development.

Another important role for YACs lies in the creation of mouse models for human diseases. Genes that are thought to be implicated in the aetiology of disease in humans can be tested in transgenic mice. Phenotypic effects of dosage imbalances as well as of specific mutations can be investigated. The creation of the APP-mice (Lamb et al., 1993; and Pearson and Choi, 1993) falls into this category of approach. Several missense mutations in the human APP gene (on chromosome 21) are known to lead to altered proteolytic processing of the protein, giving rise to extracellular deposits of the amyloid β-peptide (Aβ; Citron et al., 1992), an invariant feature of Alzheimer's disease (AD). Patients suffering from trisomy 21, or Down syndrome (DS), exhibit some neuropathologic characteristics reminiscent of AD, amongst numerous other phenotypic features. In order to test the hypothesis of a dosage imbalance of human APP leading to AD type features in DS individuals, YACs containing the APP genomic region were transferred to transgenic mice (Lamb et al., 1993 and Pearson and Choi, 1993). The mice indeed produced high levels of human APP mRNA and protein as well as levels of human Aβ comparable to mouse Aβ. However, they failed to show amyloid deposition or other neuropathologic features characteristic of AD or DS. This indicates that additional factors play a role in the aetiology of these complex diseases.

Atherosclerosis is another human disease where creation of YAC transgenic mice has contributed significantly towards establishing an animal model (Rubin and Smith, 1994). By introducing a YAC spanning the human apo(a) gene into transgenic mice, Frazer et al. (1995) were able to examine *in vivo* the regulation of this important atherosclerosis risk factor gene. These mice can now be used to test therapeutic reagents designed to lower the levels of apo(a).

YACs also facilitate the biotechnological production of foreign proteins in mice or other farm animals. Many proteins of interest are encoded by complex genetic loci too large for standard transgenesis. The genes encoding the human immunoglobulin heavy and light chains, for example, extend over a region of over 2 Mb each. Green et al. (1994) succeeded in the transfer of the major portions of both the heavy and κ light chain genes into YAC transgenic mice. By breeding these into a background deficient in mouse Ig production, they obtained a strain with an antigen-specific human antibody response, the Ig genes being rearranged to generate a human-like antibody repertoire. Mice like these will be of great benefit for the production of immunoglobulins for human therapeutic purposes.

Other genetic loci can be envisioned for which YAC transgenesis would provide a useful tool in mediating biotechnological production in animals, such as the gene for blood clotting factor VIII with 26 exons extending over 190 kb.

Moreover, not only mice can serve as hosts for YAC transgenes. Brem et al. (1995) managed to rescue the albino phenotype in rabbits by pronuclear microinjection of the YRT2 YAC containing the mouse tyrosinase gene (Schedl et al., 1993a; see above). The stable introduction of a YAC into the germline of a species other than the mouse will not only have great implications for the use of farm animals as "bioreactors"; it might also aid in providing transgenic donors for the xenotransplantation of organs from other species into humans (Cozzi and White, 1995; and Yannoutsos et al., 1995).

The advantage that large segments of DNA can easily be manipulated and transferred in order to guarantee faithful expression of genes carried by a YAC outweighs any difficulties encountered in the handling of the YACs. In summary, YAC transgenesis seems an ideal method for approaching a number of biological questions involving large and complex genetic loci.

Acknowledgements

The authors would like to thank Dr. A. Schedl for sharing unpublished data, Dr. H. Zentgraf for providing the electron micrograph of YAC Tyr4, and Drs. A.P. Monaghan, T. Mantamadiotis and H. Hiemisch for critical reading of the manuscript. We are also grateful to W. Fleischer for photographic work and M. Bock for secretarial assistance.

References

Bonifer, C., Vidal, M., Grosveld, F. and Sippel, A.E. (1990) Tissue specific and position independent expression of the complete gene domain for chicken lysozyme in transgenic mice. *EMBO* **9**, 2843–2848

Brem, G. et al. (1995) YAC transgenesis in farm animals: rescue of albinism in rabbits. *Molecular Rep. Dev.* (submitted)

Brinster, R.L., Chen, H.Y., Trumbauer, M.E., Yagle, M.K. and Palmiter, R.D. (1985) Factors affecting the efficiency of introducing foreign DNA into mice by microinjecting eggs. *Proc. Natl. Acad. Sci. USA* **82**, 4438–4442

Brown, W.R.A. (1992) Mammalian artificial chromosomes. *Curr. Opin. Genet. Dev.* **2**, 479–486

Burke, D.T., Carle, G.F. and Olson, M.V. (1987) Cloning of large segments of exogenous DNA into yeast by means of artificial chromosome vectors. *Science* **236**, 806–812

Cai, H. and Levine, M. (1995) Modulation of enhancer-promoter interactions by insulators in the *Drosophila* embryo. *Nature* **376**, 533–536

Choi, T.K., Hollenbach, P.W., Pearson, B.E., Ueda, R.M., Weddell, G.N., Kurahara, C.G., Woodhouse, C.S., Kay, R.M. and Loring, J.F. (1993) Transgenic mice containing a human heavy chain immunoglobulin gene fragment cloned in a yeast artificial chromosome. *Nature Genet.* **4**, 117–123

Choudary, S.K., Wykes, S.M., Kramer, J.A., Mohamed, A.N., Koppitch, F., Nelson, J.E., Krawetz, S.A. (1995) A haploid expressed gene cluster exists as a single chromatin domain in human sperm. *J. Biol. Chem.* **270**, 8755–8762

Chung, J.H., Whiteley, M. and Felsenfeld, G. (1993) A 5' element of the chicken β-globin domain serves as an insulator in human erythroid cells and protects against position effect in drosophila. *Cell* **74**, 505–514

Citron, M., Oltersdorf, T., Haass, C., McConlogue, L., Hung, A.Y., Seubert, P., Vigo-Pelfrey, C., Lieberburg, I. and Selkoe, D.J. (1992) Mutation of the β-amyloid precursor protein in familial Alzheimer's disease increases β-protein production. *Nature* **360**, 672–674

Cozzi, E. and White, D.J.G. (1995) The generation of transgenic pigs as potential organ donors for humans. *Nature Med.* **1**, 964–966

Davies, N.P., Rosewell, I.R. and Brüggemann, M. (1992) Targeted alterations in yeast artificial chromosomes for inter-species gene transfer. *Nucl. Acids Res.* **20**, 2693–2698.

Davies, N.P., Rosewell, I.R., Richardson, J.C., Cook, G.P., Neuberger, M.S., Brownstein, B.H., Norris, M.L. and Brüggemann, M. (1993) Creation of mice expressing human antibody light chains by introduction of a yeast artificial chromosome containing the core region of the human immunoglobulin κ locus. *Bio/Technol.* **11**, 911–914.

Dillon, N. and Grosveld, F. (1993) Transcriptional regulation of multigene loci: multilevel control. *Trends Genet.* **9**, 134–137

Dillon, N. and Grosveld, F. (1994) Chromatin domains as potential units of eukaryotic gene function. *Curr. Opin. Genet. Dev.* **4**, 260–264

Elgin, S.C.R. (1990) Chromatin structure and gene activity. *Curr. Opin. Cell Biol.* **2**, 437–445

Frazer, K.A., Narla, G., Zhang, J.L., Rubin, E.M. (1995) The apolipoprotein(a) gene is regulated by sex hormones and acute-phase inducers in YAC transgenic mice. *Nature Genet.* **9**, 424–431

Gaensler, K.M.L., Kitamura, M. and Kan, Y.W. (1993) Germ-line transmission and developmental regulation of a 150-kb yeast artificial chromosome containing the human β-globin locus in transgenic mice. *Proc. Natl. Acad. Sci. USA* **90**, 11381–11385

Green, L.L., Hardy, M.C., Maynard-Currie, C.E., Tsuda, H., Louie, D.M., Mendez, M.J., Abderrahim, H., Noguchi, M., Smith, D.H., Zeng, Y., David, N.E., Sasai, H., Garza, D., Brenner, D.G., Hales, J.F., McGuinness, R.P., Capon, D.J., Klapholz, S. and Jakobovits, A. (1994) Antigen-specific human monoclonal antibodies from mice engineered with human Ig heavy and light chain YACs. *Nature Genet.* **7**, 13–21

Grosveld, F., van Assendelft, G.B., Greaves, D.R. and Kollias, G. (1987) Position-independent, high-level expression of the human β-globin gene in transgenic mice. *Cell* **51**, 975–985

Heard, E., Avner, P. and Rothstein, R. (1994) Creation of a deletion series of mouse YACs covering a 500 kb region around Xist. *Nucl. Acids Res.* **22**, 1830–1837

Hieter, P., Connelly, C., Shero, J., McCormick, M.K., Antonarakis, S., Pavan, W. and Reeves, R. (1990) Yeast artificial chromosomes: promises kept and pending. In *Genome Analysis Volume 1: Genetic and Physical Mapping*, Cold Spring Harbor Laboratory Press, New York, pp. 83–120

Hogan, B., Constantini, F. and Lacy, E. (1986) In *Manipulating the Mouse Embryo*, Cold Spring Harbor Laboratory Press, New York

Jackson, I.J. and Bennet, D.C. (1990) Identification of the albino mutation of mouse tyrosinase by analysis of an *in vitro* revertant. *Proc. Natl. Acad. Sci. USA* **87**, 7010–7014

Jaenisch, R. (1988) Transgenic animals. *Science* **240**, 1468–1474

Jakobovits, A., Moore, A.L., Green, L.L., Vergara, G.J., Maynard-Currie, C.E., Austin, H.A. and Klapholz, S. (1993) Germ-line transmission and expression of a human-derived yeast artificial chromosome. *Nature* **362**, 255–258.

Jeffery, G., Schütz, G. and Montoliu, L. (1994) Correction of abnormal retinal pathways found with albinism by introduction of a functional tyrosinase gene in transgenic mice. *Dev. Biol.* **166**, 460–464

Kalos, M. and Fournier, R.E.K. (1995) Position-independent transgene expression mediated by boundary elements from the apolipoprotein B chromatin domain. *Mol. Cell. Biol.* **15**, 198–207

Kellum, R. and Schedl, P. (1991) A position-effect assay for boundaries of higher order chromosomal domains. *Cell* **64**, 941–950

Lamb, B.T., Sisodia, S.S., Lawler, A.M., Slunt, H.H., Kitt, C.A., Kearns, W.G., Pearson, P.L., Price, D.L. and Gearhart, J.D. (1993) Introduction and expression of the 400 kilobase amyloid precursor protein gene in transgenic mice. *Nature Genet.* **5**, 22–30

Lamb, B.T. and Gearhart, J.D. (1995) YAC transgenics and the study of genetics and human disease. *Curr. Opin. Genet. Dev.* **5**, 342–348

McKnight, R.A., Shamay, A., Sankaran, L., Wall, R.J. and Hennighausen, L. (1992) Matrix-attachment regions can impart position-independent regulation of a tissue-specific gene in transgenic mice. *Proc. Natl. Acad. Sci. USA* **89**, 6943–6947

Montoliu, L., Schedl, A., Kelsey, G. Zentgraf, H., Lichter, P., Schütz, G. (1994) Germ line transmission of yeast artificial chromosomes in transgenic mice. *Reprod. Fertil. Dev.* **6**, 577–584

Montoliu, L., Bock, C.T., Schtz, G. and Zentgraf, H. (1995) Visualization of large DNA molecules by electron microscopy with polyamines: application to the analysis of yeast endogenous and artificial chromosomes. *J. Mol. Biol.* **246**, 486–492

Palmiter, R.D. and Brinster, R.L. (1986) Germ-line transformation of mice. *Ann. Rev. Genet.* **20**, 465–499

Palmiter, R.D., Sandgren, E.P., Koeller, D.M. and Brinster, R.L. (1993) Distal regulatory elements from the mouse metallothionein locus stimulate gene expression in transgenic mice. *Mol. Cell. Biol.* **13**, 5266–5275

Pavan, W.J. and Hieter, P. and Reeves, R.H. (1990) Generation of deletion derivatives by targeted transformation of human-derived yeast artificial chromosomes. *Proc. Natl. Acad. Sci. USA* **87**, 1300–1304

Pearson, B.E. and Choi, T.K. (1993) Expression of the human β-amyloid precursor protein gene from a yeast artificial chromosome in transgenic mice. *Proc. Natl. Acad. Sci.* USA **90**, 10578–10582

Peterson, K.R., Clegg, C.H., Huxley, C., Josephson, B.M., Haugen, H.S., Furukawa, T. and Stamatoyannopoulos, G. (1993) Transgenic mice containing a 248-kb yeast artificial chromosome carrying the human β-globin locus display proper developmental control of human globin genes. *Proc. Natl. Acad. Sci. USA* **90**, 7593–7597

Peterson, K.R., Li, Q.L., Clegg, C.H., Furukawa, T., Navas, P.A., Norton, E.J., Kimbrough, T.G. and Stamatoyannopoulos, G. (1995) Use of yeast artificial chromosomes (YACs) in studies of mammalian development: Production of β-globin locus YAC mice carrying human globin developmental mutants. *Proc. Natl. Acad. Sci. USA* **92**, 5655–5659

Rastan, S. (1994) X chromosome inactivation and the XIST gene. *Curr. Opin. Genet. Dev.* **4**, 292–297

Rubin, E.M. and Smith, D.J. (1994) Atherosclerosis in mice: getting to the heart of a polygenic disorder. *Trends Genet.* **10**, 199–203

Schedl, A., Beermann, F., Thies, E., Montoliu, L., Kelsey, G. and Schütz, G. (1992) Transgenic mice generated by pronuclear injection of a yeast artificial chromosome. *Nucl. Acids Res.* **20**, 3073–3077

Schedl, A., Montoliu, L., Kelsey, G. and Schütz, G. (1993a) A yeast artificial chromosome covering the tyrosinase gene confers copy number-dependent expression in transgenic mice. *Nature* **362**, 258–260

Schedl, A., Larin, Z., Montoliu, L., Thies, E., Kelsey, G., Lehrach, H. and Schütz, G. (1993b) A method for the generation of YAC transgenic mice by pronuclear microinjection. *Nucl. Acids Res.* **21**, 4783–4787

Schedl, A., Grimes, B. and Montoliu, L. (1995) In *Methods in Molecular Biology: Yeast Artificial Chromosome Protocols.* Humana Press

Schlessinger, D. (1990) Yeast artificial chromosomes: tools for mapping and analysis of complex genomes. *Trends Genet.* **6**, 248–258

Shibahara, S., Okinaga, S., Tomita, Y., Takeda, A., Yamamoto, H., Sato, M. and Takeuchi, T. (1990) A point mutation in the tyrosinase gene of BALB/c albino mouse causing the cysteine-serine substitution at position 85. *Eur. J. Biochem.* **189**, 455–461

Smith, D.R., Smyth, A.P. and Moir, D.T. (1990) Amplification of large artificial chromosomes. *Proc. Natl. Acad. Sci. USA* **87**, 8242–8246

Spencer, F., Ketner, G., Connelly, C. and Hieter, P. (1993) Targeted recombination-based cloning and manipulation of large DNA segments in yeast. In *Methods: A Companion to Methods in Enzymology* **5**, 161–175

Strauss, W.M., Dausman, J., Beard, C., Johnson, C., Lawrence, J.B. and Jaenisch, R. (1993) Germ line transmission of a yeast artificial chromosome spanning the murine α_1(I) collagen locus. *Science* **259**, 1904–1907

Taylor, S.S., Larin, Z. and Smith, C.T. (1994) Addition of functional human telomeres to YACs. *Human Molec. Genet.* **3**, 1383–1386

Utermann, G. (1989) The mysteries of lipoprotein(a). *Science* **246**, 904–910

Wang, Y., Macke, J.P., Merbs, S.L., Zack, D.J., Klaunberg, B., Bennett, J., Gearhart, J. and Nathans, J. (1992) A locus control region adjacent to the human red and green visual pigment genes. *Neuron* **9**, 429–440

Wilson, C., Bellen, H.J. and Gehring, W.J. (1990) Position effects on eukaryotic gene expression. *Ann. Rev. Cell Biol.* **6**, 679–714

Yannoutsos, N., Langford, G.A., Cozzi, E., Lancaster, R., Elsome, K., Chen, P. and White, D.J.G. (1995) Production of pigs transgenic for human regulators of complement activation. *Transplantn. Proc.* **27**, 324–325

Yokoyama, T., Silversides, D.W., Waymire, K.G., Kwon, B.S., Takeuchi, T. and Overbeek, P.A. (1990) Conserved cysteine to serine mutation in tyrosinase is responsible for the classical albino mutation in laboratory mice. *Nucl. Acids Res.* **18**, 7293–7298

PART III, SECTION B

46. Gene Transfer Using Microdissected Chromosome Fragments

Jean Richa* and Cecilia Lo
Department of Biology, University of Pennsylvania, Philadelphia, PA 19104-6017, USA

An important application of genetic engineering entails the isolation or construction of genes and their introduction into cells or animals. The advent of recombinant DNA methodologies has made cloning the method of choice for retrieving and manipulating gene sequences for such gene transfer experiments. However, where the gene of interest is yet to be cloned or is comprised of a very large segment of DNA, metaphase chromosomes may be used as the source of donor genetic material (Kaiser *et al.*, 1987). In the latter instance, it is necessary to have a method to either select or target the gene or chromosomal region to be transferred. Thus transfection methods have been successfully used to transfer selectable marker genes from metaphase chromosomes into mammalian tissue culture cells (McBride and Peterson, 1980; and Pritchard and Goodfellow, 1986). The present technique of chromosome fragment dissection and injection combines the micromanipulation and retrieval of a selected chromosome fragment with its introduction into a mouse egg so as to make it part of a living mouse, that is the making of transgenic or "transomic" mice. This method of gene transfer may have unique applications which should complement other methods of transgenesis. The development of this technique may be of particular interest for pursuing mouse models of human diseases for which the genes involved are yet to be identified. Thus transferring into the mouse genome the human chromosome region spanning a disease locus may allow the debilitating phenotype to be recapitulated. Below we summarize the methodologies and discuss its application.

Chromosome Preparation

Tissue culture cells or primary cells may be used as the source of chromosomal material. We have utilized two different methods of chromosome preparation for such chromosome transfer experiments. Our studies suggest that the method of chromosome preparation may be critical to maintaining DNA integrity, with the microdrop method far superior in this regard (Ewart *et al.*, 1993).

Conventional Method of Metaphase Chromosome Preparation. To prepare for chromosome harvest, the cells were blocked at metaphase with colcemid (10 mg/ml) for 2 hrs, then collected by a brief trypsin/EDTA (0.05%/0.5%) treatment, spun down and resuspended in a hypotonic solution (0.075 M KCl) for 15–20 min at 37 C. Following this hypotonic shock, the cells were spun down and resuspended in ice cold methanol:acetic acid (3:1). After 5 min, the cells were taken through another change of fixative and incubated for 30 min on ice, then resuspended in a small volume of fresh fixative and splashed onto clean glass slides. This involves dropping small cell aliquots onto the slides and then vaporizing the fixative over a beaker of boiling water. Care must be taken not to extend the total fixation time beyond 45 min. This affects the spreading and overall configuration of the metaphase chromosomes.

Microdrop Method of Metaphase Chromosome Preparation. A flame pulled micropipette is used to aspirate mitotic cells from the culture vessel.

---

*Present address: Department of Genetics, University of Pennsylvania, 415 Curie Blvd., Philadelphia, PA 19104-6145.

These cells are then transferred into a hypotonic solution comprised of 3:1 water/Dulbecco's MEM and incubated for 2–3 min. After this brief hypotonic shock, the cells are transferred into ice cold 10:1 methanol/acetic acid for 20–30 s, then immediately deposited onto a glass slide. This method was far superior to many other methods of preparing metaphase chromosome preparation (Ewart et al., 1993). Thus using an in situ nick-translation assay to detect DNA strand scission, we found that chromosomes prepared by the conventional method, whether G-banded or not, exhibited extensive DNA breaks. In contrast, no detectable DNA damage was detected in chromosome spreads prepared by the microdrop method. We also observed that "aging" of chromosomes prepared by either method over several days may further compromise DNA integrity (Ewart et al., 1993). These observations would suggest that it is important to use fresh chromosome spreads prepared by the microdrop method to obtain chromosome fragments for the making of transomic mice.

Preparation of Dissection and Injection Pipettes

Dissection pipettes with 1–2 µm tip sizes were made from Kwikfil glass capillaries (1.0 mm o.d., 0.58 mm i.d., W.P. Instruments, Inc.), pulled on a vertical Narishige puller. For injection pipettes, glass capillaries (1.0 mm o.d., 0.6 mm, i.d., V.W.R) were cleaned with chromic acid, then baked in a 160 C oven for 10 min. Once cooled to room temperature, the capillaries were treated with NP-40, rinsed thoroughly and air dried. These capillaries were pulled on a horizontal puller (Industrial Science Associates, Inc.) to obtain a tip size of 0.1 µm. Prior to injection, pipette tips were enlarged to achieve a final size of 0.5–1.0 µm. For holding pipettes (80–100 mm o.d.), glass capillaries were hand pulled over a microburner, placed on a microforge and polished.

Dissection and Injection of Chromosome Fragments

Chromosome fragments were dissected from stained metaphase spreads. Staining was carried out for 1 h with basic fuchsin (0.1%), followed by a rinse with water and then the chromosome spreads are air dried. Fragments of 0.3–1.0 µm in size were cut in the dehydrated state using a dissection pipette mounted on a Leitz micromanipulator at a 45 angle to the glass slide. By lowering the tip onto the slide, a fragment is displaced by a sweep motion across the appropriate position of the chromosome (Figure 46.1). The displaced chromosome fragments were then covered with a droplet of injection buffer (1 mM Tris, pH 7.4/0.1 mM EDTA/0.05% basic fuchsin). An injection pipette connected to an oil-filled micrometer syringe was dipped into the droplet and then manipulated to dislodge the dissected fragment. The latter was then aspirated into the pipette. Subsequently, the pipette was inserted into the pronucleus of a mouse egg immobilized by the holding pipette in a glass depression slide filled with Hepes-buffered Whitten's medium. A volume of 3–4 picoliters containing the dissected chromosome fragment is then released and the pipette is withdrawn. Injected eggs that survive the injection process were cultured overnight to monitor their developmental progress.

Embryo Viability and the Making of Transomic Mice

We observed on average, 63% of the injected eggs survive and half of those develop to the blastocyst stage or beyond (Richa and Lo, 1989; and Ewart et al., 1993). This efficiency is comparable to that obtained in DNA microinjection experiments (Brinster et al., 1985). Moreover, the comparison becomes more important when the tip size of the injection pipette is taken into consideration. As

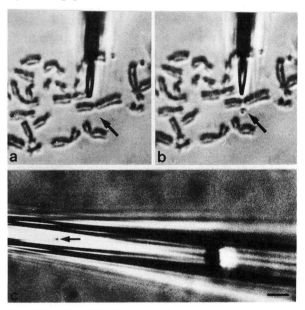

Figure 46.1. Dissection and retrieval of chromosome fragments. A human metaphase spread before (a) and after (b) the dissection. The arrows indicate the chromosome of choice (a) and the dissected fragment (b). In (c) the chromosome fragment is in the injection pipette (arrow). "Reprinted with permission from J. Richa and C.W. Lo, Science 245, 175–177. Copyright 1989 American Association for the Advancement of Science."

described earlier, the chromosome injection pipette carries a tip size 5–10 times larger than that used for the microinjection of DNA. Nevertheless, the viability of the injected eggs makes the injection of chromosome fragment a practical alternative.

In order to assess the integration of chromosomal DNA into the mouse genome, selected areas of various human chromosomes were used as a source of the injected fragments, and appropriate DNA probes were used in our detection schemes that included *in situ* hybridization, PCR, and Southern analysis (Richa and Lo, 1989; and Ewart *et al.*, 1993). In 50% of the blastocysts, hybridization signals were detected over more than one cell indicating that the injected fragments were integrated into individual mouse chromosomes. The analysis of 12.5 day embryos revealed that 50–90% contained DNA from the injected human chromosome fragment, with each embryo showing varying levels of mosaic (15–20%). When the injected eggs were allowed to develop to term, several mice were born and reached adulthood. Tail DNA analysis by Southern blot hybridization of one of these mice revealed the presence of human DNA sequences. This would suggest that at least a portion of the injected human chromosome fragment had been retained in the mouse genome (Richa and Lo, 1989, also Richa and Lo, unpublished observations).

In summary, it is possible to microdissect and retrieve a chromosome fragment and inject it into a mouse egg for the purpose of making a transgenic or "transomic" mouse. Analysis by several methods indicate that this method has comparable levels of transformation efficiency to that of DNA injection.

Perspectives

This "transomic" technique is advantageous in that it may make it possible to manipulate and transfer a large segment of chromosomal DNA. This is of particular importance when the transfer of a large gene or a gene cluster is sought. Although we have worked with chromosome fragments encompassing 15–20 megabases of DNA, it is likely that larger chromosome fragments can be transferred with this method. Moreover, this transomic approach may make it possible to make an animal model expressing a desired genetic trait without having to first clone the gene of interest. Lastly, as the results indicate, this method of gene transfer provides DNA integration efficiency very similar to that observed with the injection of cloned DNA, thereby further indicating that the transomic method may be a useful alternative method for gene transfer.

Nevertheless, as with any technique, the transomic method has some potential disadvantages. One major concern is that of DNA integrity in the metaphase chromosome prepration. In the future, it may be possible to harvest unfixed metaphase chromosomes with the aid of flow sorting methods for chromosome isolation (Bartholdi *et al.*, 1987). This in conjunction with chromosome dissection by laser technology (Hadano *et al.*, 1991 and Rajan-Separovic *et al.*, 1995) may help to improve, simplify, as well as add further precision to the transomic method for transferring genetic material.

It should be noted that an alternative method now exists for making transgenic mice containing large DNA inserts. This entails the use of yeast artificial chromosome (YAC) technology. Thus YAC vectors (Burke *et al.*, 1987) containing the desired DNA inserts can be directly microinjected into mouse eggs (Peterson *et al.*, 1993 and Schedl *et al.*, 1993), or such YACs can be transfected into ES cells from which transgenic mice can then be generated (Schedl *et al.*, 1993; Jakobovits *et al.*, 1993; Strauss *et al.*, 1993; and Lamb and Gearhart, 1995). This method, however, requires the prior cloning of the gene or gene(s) of interest in a YAC vector. Even within this context, the transomic technique may be the method of choice for manipulating genetic traits that have yet to be cloned or have been refractory to cloning, and also where very large DNA segments are to be transferred. A method which may complement the transomic technique is the engineering of chromosomal rearrangements into ES cells (Ramirez-Solis *et al.*, 1995), an approach which may allow the modeling of human diseases associated with specific chromosomal anomalies.

References

Bartholdi, M., Meyne, J., Albright, K., Luemann, M., Campbell, E., Chritton, D., Deaven, L.L. and Cram, L.S. (1987) Chromosome sorting by flow cytometry. *Method. Enzy.* **151**, 252–267

Brinster, R.L., Chen, H.Y., Trumbauer, M.E., Yagle, M. K. and Palmiter, R.D. (1985) Factors affecting the efficiency of introducing foreign DNA into mice by micoinjecting eggs. *Proc. Natl. Acad. Sci. USA* **82**, 4438–4442

Burke, D., Carle, G. and Olson, M. (1987) Cloning of large segments of exogenous DNA into yeast artificial chromosomes. *Science* **236**, 806–812

Ewart, J.E., Richa, J. and Lo, C.W. (1993) Widespread distribution of cells containing human DNA in embryos derived from mouse eggs injected with human chromosome fragments. *Hum. Gene Ther.* **4**, 597–607

Hadano, S., Watanabe, M., Yokoi, H., Kondo, I., Tsuchiya, H., Kanzawa, I., Wasaka, K. and Ikeda, J.-E. (1991). Laser microdissection and single unique primer PCR allow generation of regional chromosome DNA clones from a single human chromosome. *Genomics* **11**, 364–373

Jakobovits, A., Moore, A.L., Green, L.L., Vergara, G.J., Maynard-Curie, C.E., Austin, H.A. and Klapholz, S. (1993) Germ-line transmission and expression of a human-derived yeast artificial chromosome. *Nature* **362**, 255–258

Kaiser, R., Weber, J., Greschik, K.H., Edstrom, J.E., Driesel, A., Zengerling, A., Buchwald, M., Tsui, L.C. and Olek, K. (1987) Microdissection and microcloning of the long arm of human chromosome 7. *Mol. Biol. Reports* **12**, 3–6

Lamb, B.T. and Gearhart, J.D. (1995) YAC transgenics and the study of genetics and human disease. *Current Opinion in Genet. and Develop.* **5**, 342–348

McBride, O.W. and Peterson, J.L. (1980). Chromosome-mediated gene transfer in mammalian cells. *Ann. Rev. Genet.* **14**, 321–345

Peterson, K.R., Clegg, C.H., Huxley, C., Josephson, B.M., Haugen, H.S., Furukawa, T., Stamatoyannopoulos, G. (1993) Transgenic mice containing a 248-kb yeast artificial chromosome carrying the human β-globin locus display proper developmental control of human globin genes. *Proc. Natl. Acad. Sci. USA* **90**, 7593–7597

Pritchard, C. and Goodfellow, P.N. (1986). Development of new methods in human gene mapping: selection for fragments of the human Y chromosome after chromosome-mediated gene transfer. *EMBO J.* **5**, 979–985

Rajcan-Separovic, E., Wang, H.-S., Speevak, M.D., Janes, L., Korneluk, R.G., Wakasa, K. and Ikeda, J.-E. (1995). Identification of the region of double minutes in normal human cells by laser-based chromosome microdissection approach. *Hum. Genet.* **96**, 39–43

Ramirez-Solis, R., Liu, P. and Bradley, A. (1995) Chromosome engineering in mice. *Nature* **378**, 720–724

Richa, J., and Lo, C.W. (1989) Introduction of human DNA into mouse eggs by injection of dissected chromosome fragments. *Science* **245**, 175–177

Schedl, A., Montoliu, L., Kelsey, G. and Schutz, G. (1993) A yeast artificial chromosome covering the tyrosinase gene confers copy number-dependent expression in transgenic mice. *Nature* **362**, 258–261

Strauss, W.M., Dausman, J., Beard, C., Johnson, C., Lawrence, J.B. and Jaenisch, R. (1993). Germ-line transmission of a yeast artificial chromosome spanning the murine $\alpha 1(1)$ Collagen locus. *Science* **259**, 1904–1907

PART III, SECTION B

47. Prospects for Directing Temporal and Spatial Gene Expression in Transgenic Animals

Lothar Hennighausen[1],* and Priscilla A. Furth[2]

[1]*Laboratory of Biochemistry and Metabolism, National Institute of Diabetes, Digestive and Kidney Diseases, National Institutes of Health, Bldg. 10, Rm. 9N113, Bethesda, Maryland 20892-1812, USA*
[2]*Division of Infectious Diseases, Department of Medicine, University of Maryland Medical School and Baltimore Veterans Affairs Medical Center, Baltimore, Maryland 21201, USA*

Summary

Molecular mechanisms of development and disease can be studied in transgenic animals. Controlling the spatial and temporal expression patterns of transgenes, however, is a prerequisite for the elucidation of gene function in the whole organism. From the several binary systems which have been established to permit conditional activation of transgenes, the *tetracycline* responsive expression system has been the most successful one in animals. Transgenic mice carrying a *tet*R/VP16 hybrid gene (tTA) under different promoters have been used to temporally activate several-thousand fold the expression of genes under the control of a promoter containing *tetop* sequences. The importance of *tetracycline* regulatable systems goes beyond their ability to direct temporal gene expression in the context of the whole animal. In combination with Cre/lox recombination tools it will facilitate the deletion of genes from the genome in specific cells and at specific timepoints. This approach is required to test the function of genes whose presence is critical for embryonic development.

Why do We Need Inducible Gene Expression Systems?

Gain of function and loss of function experiments in animals have been used to elucidate the role of mammalian gene products in organ development and oncogenesis. Although it is possible to direct the expression of transgenes to defined tissues by employing specific genetic regulatory elements, conditional control of gene expression has been a challenge. For example, the promoter of the whey acidic protein (WAP) gene can target gene expression to mammary alveolar cells [1], but its temporal activation during puberty and pregnancy cannot be controlled by experimental conditions *in vivo* [2,3]. The inability to directly control the temporal expression of transgenes has important consequences for transgenic studies designed to elucidate the role of proteins involved in development. The phenotype observed will frequently reflect the stage of-development at which the transgene encoding the regulatory protein was first activated. For example, if expression of the transgene encoded protein causes lethality at a defined stage of development, it will not be possible to study its effect on later developmental stages. This is particularly important in cases in which a protein can play different roles which depend on the stage of development and/or the cell type. For example, depending on the time of expression during pregnancy, TGF-β can interfere with either ductal or alveolar development of the mammary gland [4–6].

Another application for inducible gene expression lies in the field of gene deletions. A substantial number of genes are required for mouse development, and their deletion from the genome in ES cells can result in early embryonic lethality. Through the combination of an inducible system with the Cre/lox technology it will be possible to delete genes at specific timepoints during development.

Tetracycline Responsive System

The development of regulatory circuits based on the tetracycline resistance operon *tet* from

*Author for Correspondence.
E-mail: henninghausen@nih.gov.

E.coli transposon Tn10 opened new approaches to controlling gene expression in eukaryotic cells [7–9]. In these system, transactivator proteins (tTA and rtTA) composed of either the wild-type or a mutant *tet* repressor, respectively, and the activating domain of viral protein VP16 of herpes simplex virus activate transcription from a minimal promoter fused to seven *tet* operator sequences from Tn10 (*tet*op) (Figure 47.1). The HCMVIE1 enhancer/promoter and the MMTV-LTR have been used to direct tTA expression in transgenic mice. These experiments demonstrated that *tet*op controlled target gene expression can be activated in different cell types [9,10]. Transcriptional activation in this constellation was repressed by administration of tetracycline to the animal and activated upon withdrawal of the antibiotic.

The classical *tetracycline* responsive system requires the continuous presence of *tetracyline* to keep the target gene silent (see Figure 47.1), a scenario which may be difficult to maintain in certain experimental settings. The identification and isolation of a mutant *tetracycline* repressor with altered DNA binding properties by Hillen and colleagues provided the tools to build an expression system that is silent in the absence of *tetracycline* and induced in its presence [7] (see Figure 47.1). The *tet* repressor in this reverse system only binds to the *tet*op in the presence of *tetracycline*. This system has been successfully tested in tissue culture cells, and experiments in transgenic mice are well underway.

Since expression levels of tTA transgenes can be low, Shockett and colleagues [11] modified the *tet* responsive system to permit an autoregulatory loop. *Tet*op sequences were ligated to the tTA transgene which resulted in higher levels of gene activation. However, basal gene activity in the presence of tetracycline was also higher, which may be undesirable in some instances.

Another approach to controlling gene activity uses the *tet*R to inhibit gene transcription. In this case *tet*R molecules bind to *tet*op sequences located at the transcriptional start site and block gene transcription in the absence of tetracycline. In the presence of tetracycline binding of the *tet*R to the *tet*op is greatly reduced and transcription is activated [12,13]. Although appealing, this system suffers from the drawback that a large number of *tet*R molecules are needed to silence transcription.

Other Systems

Other systems which permit conditional control of transgene expression have been explored in the past and are currently being developed (for references see [9,10]). O'Malley and colleagues have described a C-terminal deletion mutant of the progesterone receptor which fails to bind progesterone, but can bind its antagonist RU486 and can be used to activate genes containing progesterone response elements [14]. With RU486 as an agonist, a chimeric protein consisting of the steroid binding domain, the DNA binding domain of GAL4 and the VP16 activation domain can activate in cell lines a minimal promoter linked to GAL4 binding sites.

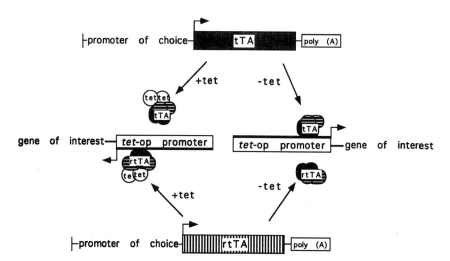

Figure 47.1. Comparison of the positive and negative *tet* responsive gene expression system.

Establishment of the utility of this system in transgenic animals is still pending. Another system being explored is based on the ecdysone receptor from Drosophila.

Inducible Gene Deletion and Targeting

Genes can be deleted from the genome of ES cells using homologous recombination and mice can be derived from such cells. The deletion of many genes from the genome of ES cells, however, is incompatible with the formation of a mouse, which demonstrates the importance of these molecules for developmental processes. Methodologies are being developed that permit the deletion of genes from the mouse genome at a specified time during ontogeny. Kuhn and colleagues [15] generated a transgene encoding the Cre enzyme which was under the control of the mouse Mx1 gene. This promoter is fairly inactive in healthy mice and can be induced by interferon α and β. These mice were crossed with mice carrying a DNA polymerase β gene flanked by loxP sites. Although the activation of the Cre enzyme was partially dependent on the presence of exogenous interferons, deletion of the polymerase β gene was seen in only a small percentage of cells. This observation highlights a current problem in transgenesis that not every cell within a given tissue expresses the transgene.

Currently several groups combine the *tet* responsive system with the CRE/lox recombination system. This combination has advantages over the Mx1 system described by Kuhn and colleagues, in that no or only little basal activity of Cre should be expected.

Hurdles

General problems linked to transgenesis together with those specific to the tetracycline system need to be overcome in order to successfully control gene expression. General hurdles of transgenesis include position dependent variations in the level and homogeneity of gene expression, even within a line. Heterogenous expression of transgenes within defined cell populations has been observed frequently. Extensive mosaic transgene expression occurs in mammary tissue of mice carrying hybrid genes under the control of the MMTV-LTR [16,17] the WAP [18] and the β lactoglobulin gene promoter. The degree of heterogeneity is not only reflected in the promoter elements used, which are also subject to a heterogenous expression within the endogenous loci [19] but also upon the site of integration. Mosaic gene expression may not be a problem in experiments studying the consequence of oncogene expression or that of a secreted protein. However, it poses serious problems in cases, such as the temporal excission of genes from the genome using the Cre/lox system, in which each cell within a given lineage or organ has to express the transgene.

Traditionally, binary systems are based on two transgenes which are independently integrated into the genome. This results in the excacerbation of any problems associated with transgenesis. Most importantly, the activation of the reporter gene by the transactivator will only be seen in cells in which both genomic integration sites are actually recognized by the transcription machinery. This conclusion is consistent with our findings that tTA induced expression of a *tet*op-lacZ reporter gene is extremely mosaic in certain tissues, including muscle [9] and mammary gland [10]. However, in other tissues, such as seminal vesicle and salivary gland, almost homogenous expression has been observed [10]. This suggests that an important determinant of the degree of mosaicism is the expression pattern of the transcriptional regulatory region used to activate the tTA gene. At the same time, markedly heterogenous β galactosidase staining was found in mammary tissue of double transgenic β lactating mice. This indicates that the degree of heterogenity from an individual promoter can vary between tissues.

Homogenous transgene expression may be achieved by choosing other promoter elements to control tTA expression. In contrast to the MMTV-LTR, the promoter/upstream region of the mouse WAP gene generally conveys a homogenous expression in lactating mammary tissue [19]. The introduction of sequences which can mediate more reliable transgene activity, such as MAR elements, could provide more homogeneous expression of tTA [3]. Another route of achieving reproducible expression pattern may be the introduction of the tTA into an endogenous gene *via* homologous recombination in ES cells.

In general, tetracycline delivery is performed through the implantation of slow release pellets and does not appear to be a major problem. However, complete abrogation of gene expression in some tissues, for example skin, may require higher levels of tetracycline administration or the use of different tetracycline analogues. While tetracycline is broadly distributed in tissues, the pharmacokinetics of tetracycline delivery for the *tet*-responsive system in individual tissues may vary. This could be relevant for studies related to embryonic development where rapid gene induction and repression is be required.

References

1. Pittius, C.W., Hennighausen, L., Lee, E., Westphal, H., Nichols, E., Vitale, J. and Gordon, K. (1988) *Proc. Natl. Acad. Sci. USA*, pp. 5874–5878

2. Burdon, T., Sankaran, L., Wall, R.J., Spencer, M. and Hennighausen, L.(1991) *J. Biol. Chem.* **266,** 6909–6914
3. McKnight, R.A., Shamay, A., Sankaran, L., Wall, R.J. and Hennighausen, L. (1992) *Proc. Natl. Acad. Sci. USA*, pp. 6943–6947
4. Jhappan, C., Geiser, A.G., Kordon, E.C., Bagheri, D., Hennighausen, L., Roberts, A.B., Smith, G.H. and Merlino, G. (1993) *EMBO J.* **12,** 1835–1845
5. Kordon, E., McKnight, R.A., Jhappan, C., Merlino, G., Hennighausen, L. and Smith, G.H. (1995) *Dev. Biol.* **167,** 47–61
6. Pierce, D.F., Jr., Johnson, M.D., Matsui, Y., Robinson, S.D., Gold, L.I., Purchio, A.F., Daniel, C.W., Hogan, B.L.M. and Moses, H.L. (1993) *Genes Dev.* **7,** 2308–2317
7. Gossen, M., Freundlieb, S., Bender, G., Muller, G., Hillen, W. and Bujard, H. (1995) *Science* **268,** 1766–1769
8. Gossen, M. and Bujard, H. (1992) *Proc. Natl. Acad. Sci. USA* **89,** 5547–5551
9. Furth, P.A., St. Onge, L., Boger, H., Gruss, P., Gossen, M., Kistner, A., Bujard, H. and Hennighausen, L. (1994) *Proc. Acad. Natl. Sci. USA* **91,** 9302–9306
10. Hennighausen, L., Wall, R., Tillmann, U., Li, M. and Furth, P.A. (1995) *J. Cell. Biochem.* **59,** 463–472
11. Shockett, P., Diflippantonio, M., Hellman, N. and Schatz, D.G. (1995) *Proc. Natl. Acad. Sci. USA* **92,** 6522–6526
12. Gatz, C., Frohberg, C. and Wendenburg, R. (1992) *Plant J.* **2,** 397–404
13. Gatz, C., Kaiser, A. and Wendenburg, R. (1991) *Mol. Gen. Genet.* **227,** 229–237
14. Wang, Y., O'Malley, B.W., Jr., Tsai, S.Y. and O'Malley, B.W. (1994) *Proc. Natl. Acad. Sci. USA* **91,** 8180–8184
15. Kuhn, R., Schwenk, F., Auget, M. and Rajewski, K. (1995) *Science* **269,** 1427–1429
16. Hennighausen, L., McKnight, R.A., Burdon, T., Baik, M., Wall, R.J. and Smith, G.H. (1994) *Cell Growth Diff.* **5,** 607–613
17. Muller, W.J., Sinn, E., Pattengale, P.K., Wallace, R. and Leder, P. (1988) *Cell* **54,** 105–115
18. Li, B., Greenberg, N., Stephens, L.C., Meyn, R., Medina, D. and Rosen, J.M. (1994) *Cell Growth and Differentiation* **5,** 711–721
19. Robinson, G.W., McKnight, R.A., Smith, G.H. and Hennighausen, L. (1995) *Development* **121,** 2079–2090

PART III, SECTION B

48. Mouse Genetic Manipulation via Homologous Recombination

Stéphane Viville

Institut de Génétique et de Biologie Moléculaire et Cellulaire, CNRS-INSERM-Université Louis Pasteur, Collège de France Parc d'innovation, 1, rue Laurent Fries BP 163, 67404 Illkirch, C.U. de Strasbourg, France

Introduction

For almost ten years now it has been possible to create, via homologous recombination in embryonic stem (ES) cells, a mouse line carrying a mutated gene. This technique, the so called "gene knock out", represents the most powerful genetic method to study the function of a particular gene product in the mouse because it allows the generation of mouse lines in which the protein is completely absent. Unfortunately, this technique suffers from two major disadvantages. First, it leaves at least one selectable gene in the studied locus which may influence the phenotype or the expression of neighbouring genes. Second, the mutation has an immediate and general effect, which, if the mutation is embryonic lethal, will prevent further study of the gene in the adult and will not allow tissue specific functional studies.

The use of new targeting vectors based on the Cre-*loxP* or FLP/FRT recombinase systems of bacteriophage P1 or of Saccharomyces cerevisiae, respectively, has greatly improved gene knock out technology and the ability to manipulate the mouse genome. Using such systems, it is now possible, not only to knock out a gene but also to manipulate the entire mouse genome almost at will. In this chapter, I will present a variety of the different approaches used in the genetic manipulations of the mouse via homologous recombination. The initial section will focus on the firsts generations of gene targeting vectors, subsequently the new generation of vectors and their applications will be discussed.

Gene Targeting, Already an "Old" Technique

The gene targeting strategy is based on two fundamental technical advances; the ability to: (i) obtain a mutated chromosomal locus, with a defined sequence alteration, *via* homologous recombination; and (ii) the existence of a embryonic stem (ES) cell line capable, after manipulation in culture, of contributing to the mouse germ line (see Figure 48.1).

Smithies *et al.* (1985) were one of the first groups to show that it is possible to mutate precisely a specific gene by homologous recombination in a mammalian cell line. They were able to mutate the β-globin gene in a human carcinoma cell line. The second major step was the establishment of murine cell lines that retain their pluripotentiality (Evans and Kaufman, 1981; and Martin, 1981). ES cell lines are derived from the inner cell mass (ICM) of 3.5 days postcoitum (dpc) blastocysts. Such cells can be grown in culture over long periods of time whilst preserving their capacity to colonise the ICM after injection into a 3.5 dpc blastocyst and to form a chimeric mouse. Obtaining chimeric animals from which the germ cells are derived from the injected ES cells makes it possible to establish mouse lines entirely derived from ES cells. If ES cells were first manipulated by homologous recombination it is possible to create a new mouse line deficient for the studied gene (see Figure 48.1) (Capecchi, 1989a,b and Pascoe, *et al.*, 1992). The use of the homologous recombination (HR) in ES cells makes it possible to create mice carrying a chosen gene mutation (Brandon *et al.*, 1995a,b,c).

Unfortunately the technique is not as easy as it looks. For example, it is very long and there are many steps that can fail:

The construction of the homologous recombination vectors can be tricky because genomic DNA is on the whole more difficult to manipulate than cDNA. Improvements in the Polymerase Chain

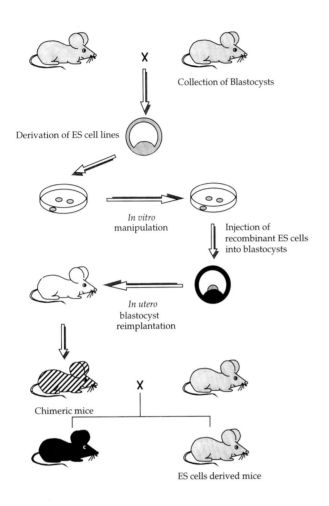

Figure 48.1. Mouse genetic manipulation *via* embryonic stem cells. Embryonic stem (ES) cells are derived from the inner cell mass (ICM) of 3.5 days post-coitum (dpc) blastocysts. They can be kept and manipulated in culture in a pluripotent state for long periods of time. The re-introduction of manipulated ES cells into 3.5 dpc blastocyst produces chimeric mice after reimplantation of the resultant embryos into pseudopregnant females. If these ES cells contribute to the germ line, mating of the chimeric mice will allow the derivation of a mouse line completely derived from the ES cell line. Chimeric mice are identified by coat colour.

Reaction (PCR) now makes it possible to amplify very big fragments (up to 35 kb (Barnes, 1994)), facilitating the construction of the recombinant vectors.

ES cells, despite the existence of excellent lines (line D3 from Gossler *et al.* (1986), line RI from Nagy *et al.* (1993) for example), are very sensitive to culture conditions. The ES cells have a natural tendency to differentiate and an unstable genome giving rise to chromosomal abnormalities which seem to confer a growth advantage over the normal cells but which make the cells unusable because they inhibit germinal transmission. Ideally, each laboratory using this technique should isolate its own ES cell lines in order to have a very early passage, thus ensuring the quality of a cell line.

Injection of ES cells in the blastocysts, although more accessible than one cell-stage embryo injection used for transgenesis, is delicate and demands sophisticated equipment. However, three innovations in ES cells transfer should be noted: First, morula injections (as practised in Brulet's laboratory (Lallemand and Brulet, 1990) are easier and give better results in terms of germ line transmission; unfortunately, this technique is accompanied by a very high embryonic mortality. Secondly, by co-culturing ES cells with eight cell stage embryos, it is possible to obtain aggregates which will form blastocysts *in vitro* the inner cell mass of which is colonised by the ES cells. After transfer of such *in vitro* derived blastocysts into the uterus of a pseudopregnant mouse, it is possible to obtain chimeric mice (Wood *et al.*, 1993). This is the most promising technique because of its simplicity and its effectiveness. Thirdly, it is possible to make aggregates between tetraploid morula cells, obtained by electrofusion, and ES cells (Nagy *et al.*, 1990). Such aggregates are likely to produce a blastocyst *in vitro* which, after reimplantation, gives rise to a viable mouse derived entirely from ES cells, the tetraploid cells taking part only in the formation of the extra-embryonic tissues; until now only a very early passage of the RI line has been described that allows this result (Nagy *et al.*, 1993).

Homologous recombination remains a rare event. The use of vectors allowing enrichment of recombined clones, that I will describe later, and the use of isogenic DNA (te Riele et al., 1992) facilitate the obtention and detection of recombinant clones.

The last step, the transfer of the blastocysts to the uterus of a pseudopregnant mouse is, thankfully, the least likely to cause problems.

The First Generation of Homologous Recombination Vectors

Two types of homologous recombination vectors are classically used.

Replacement Vectors

Replacement vectors are the traditional vectors, and are used to replace a targeted gene by a construction comprising homologous sequences disrupted by a selectable marker. Generally the neomycin resistance (*neo*) gene is used and it has a double role, (i) to mutate the studied gene and (ii) confer a property allowing the selection of the transfected cells (see Figure 48.2A).

Various methods are used to select cells that have undergone a homologous recombination event, including the use of a negative selection marker, such as the thymidine kinase (*tk*) gene which in the presence of Gancyclovir (Ganc) kill the cells in addition to the *neo* gene. The *tk* gene is placed at one end of the vector and will be eliminated at the time of a homologous recombination event; elimination of the *tk* gene confers to the transfected cells a resistant phenotype to Gancyclovir. Random integrations preserve the *tk* gene, cells are therefore sensitive and die when treated with Ganc. This is the principle of the double selection, positive and negative (Thomas et al., 1986).

On the other hand, it is possible to use a selectable gene lacking a poly-A signal which makes its transcript very unstable, the so called "poly A-less" technique. The vector is made in such a way that the integration by homologous recombination will provide a poly-A signal making the transcript of the selectable gene more stable and, in theory, selecting for cells having undergone a homologous recombination event will be selected (Donehower et al., 1992).

The "promoter less" technique consists of using a construction in which the selectable gene, lacking a promoter, is placed in the reading frame of the targeted gene. The expression of the selectable gene is only allowed in the case of a homologous recombination event. This technique seems to give excellent results (Schwartzberg et al., 1989). The studied gene, however, must be expressed in ES cells. Finally, it is possible to use a selectable gene corresponding to the fusion between the *β-galactosidase* and *neo* genes (the *β-geo* gene) which facilitates the establishment of the expression pattern of the studied gene (Friedrich and Soriano, 1991). The use of an Internal Ribosomal Entry Site (IRES), which allows, by creating a polycistronic messenger RNA, a re-initiation of the translation within an mRNA, and simplifies this type of construction making it no longer necessary to place the selectable gene in the reading frame of the targeted gene (Kim

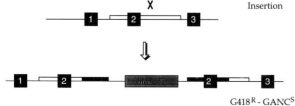

Figure 48.2. Two types of targeting vectors.
A. Replacement vectors are designed such that the regions of homology, separated by the selectable marker, are collinear with the endogenous sequences that they will replace. The marker has two functions, one is to give a selectable phenotype to the transfected cells and the second is to disrupt the targeted gene.
B. Insertion vectors are linearised such that the regions of homology remain adjacent. Such a vector is inserted into the endogenous gene provoking a duplication of the homologous sequences.

et al., 1992; Mountford et al., 1994; and Mountford and Smith, 1995).

Insertion Vectors

Insertion vectors are seldom used as following integration by homologous recombination, there is duplication of the homologous sequences making the event very unstable. A second homologous recombination, intra-chromosomal this time, may occur eliminating the newly introduced sequence and the selectable gene (see Figure 48.2B).

The two types of vectors described exhibit a major disadvantage in disturbing the studied locus because they introduce a new gene. The introduction of the selectable gene can distort the phenotype of the mutation, by disturbing the expression of adjacent genes or in the analysis of complex regulatory elements such as in the β-globin locus or the *Hox* clusters (Fiering et al., 1995 and Rijli et al., 1994; 1995). Moreover, these vectors do not allow the creation of fine changes. In order to circumvent these problems, new approaches were needed.

The Vectors of the Second Generation

The Knock-Out

In order to create changes of only a few base pairs and to leave the studied locus free of any selectable marker, various techniques were tested. The first, which seems the simplest, consists of the co-electroporation of the recombination vector carrying the desired mutation and a second vector carrying a selectable marker (Reid *et al.*, 1991). Indeed, if there is a homologous recombination event, the selectable vector should, theoretically, integrate at a another site. That would allow, by successive breeding of the chimeric mice, the segregation of both events and the elimination of the selectable gene. Such a method would make it possible to obtain a mouse line carrying a very fine mutation. Unfortunately, it seems that both vectors co-integrate at the same locus almost exclusively.

The second technique, consists of the microinjection of the mutated gene into ES cells without any selectable marker, and seeking the homologous recombination event by PCR (Zimmer and Gruss, 1989). Indeed, selection is not necessary because most microinjected cells will have integrated the DNA. The principal obstacles with this technique are the actual injection of the ES cells, which are very small, and the significant number of cells that it is necessary to inject. Moreover, it seems that they bear this treatment badly as so far, no line of mice has been obtained by this method

The third technique is based on an elegant idea that takes advantage of the second recombination event which occurs when using an insertion vector. Indeed, theoretically at least, it is possible to eliminate the selectable vector by a double recombination event while leaving a mutation in the studied gene; such mutation could theoretically be limited to a single base pair substitution. This involves the technique of the "hit and run" (Hasty *et al.*, 1991) or "in-out" (Valancius and Smithies, 1991). The principle of this technique, which is essentially an adaptation of the "pop-in/pop-out" approach developed to integrate mutation in the yeast genome (Scherer and Davis, 1979) is described in Figure 48.3. In the first step, an insertion vector carrying the desired change and two selectable genes, one for a positive selection and the other for a negative selection, is introduced by homologous recombination into the targeted gene. The second step is based on the intra-chromosomal recombination, which will eliminate either the sequences newly introduced, or the endogenous sequences, leaving in its place a mutated gene but not the selectable markers (Hasty *et al.*, 1991; and Valancius and Smithies 1991). So far, only one line of mice has been derived using this technique (Ramirez *et al.*, 1993) and it seems that the second step is difficult to achieve mainly because of its rarety and of the background of the selection.

The last approach, called the "tag and exchange" technique, is based on two successive recombination events, using two different replacement vectors. The first recombination, the tag, allows the introduction of the *tk* and *neo* genes into the targeted gene, and is selected positively through the *neo* gene. The second recombination, the exchange, is carried out in such a way as to introduce the desired change and to eliminate the selectable genes, cells are then resistant to the Gancyclovir (Figure 48.4) (Askew *et al.*, 1993 and Wu *et al.*, 1994). A similar approach has been successful using *HPRT* negative ES cells and through such an innovation a line of mutant mice has been obtained in which the exchange of a murine gene with its human homologue has been accomplished (Stacey *et al.*, 1994). The advantage of this technique is that even if the second step is not possible the first step produces a mutated cell line. Moreover, once the first step is carried out, it is possible to use these mutated cells to create various mutations in the studied gene. In light of the first results this approach seems rather promising, even if it involves longer periods of ES cell culture.

Unfortunately, the various techniques described in this section are difficult to carry out, the major problem stemming from the rarity of the homologous recombination event and the background of the selection, making the event difficult to isolate.

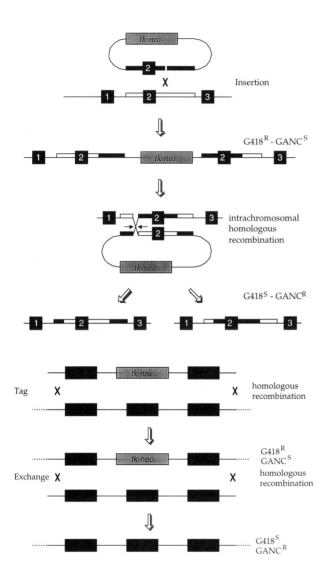

Figure 48.3. Introduction of a specific change into a given gene. The "hit and run" or "in-out" technique allows a specific mutation to be made which is free of any selectable marker but requires two homologous recombination steps. The first step is done with an insertion homologous recombinant vector comprising the desired mutation (represented by a black bar in exon two) and the *tk* and *neo* genes. The insertion (by homologous recombination) is selected by following the G418 resistant (G418^R) phenotype and leads to the duplication of the homologous regions. This duplication can cause a second intra-chromosomal homologous recombination event, resulting in either the loss of the newly introduced sequences, or of the endogenous sequences. The latter case results in the specific change desired. In both cases there is loss of the *tk* gene making the cells Gancyclovir resistant (GANC^R).

Figure 48.4. The "tag and exchange" technique. This technique is based on two homologous recombination events. One, the tag with a replacement vector, will mutate the targeted gene by introducing the *tk* and *neo* genes. The selection of the recombinant cells is done in the presence of G418. The second, the exchange, always by a replacement vector, will reintroduce the desired mutation (represented by a black bar in exon 2) into the studied gene while eliminating the *tk* and *neo* genes; the recombinant cells are then GANC^R.

The Knock In

Recently a similar technique has been used not just to knock out a gene but to knock one in. This has shown the functional redundancy of two genes, En1 and En2. In this experiment the En1 gene was replaced, *via* a homologous recombination event, by the En2 gene (Hanks *et al.*, 1995). This experiment clearly shows that the two genes are homologous enough to functionally replace each other and that their functional specificity is due only to their pattern of expression during development.

The New Generation of Recombinant Vectors

The limited results obtained with the vectors of second generation and always with the aim of making "clean" mutations has pushed investigators to invent a new type of vector. This involves vectors allowing mutagenesis without leaving a selectable gene and opens the possibility for spatio-temporally controlling the appearance of the mutation in the mouse. The latter point is important for the study of genes which cause, when they are mutated, early death of the animals thus preventing the study of later functions of the gene or the study of those genes that are expressed at different stages of development and/or in different tissues (Copp, 1995).

The Cre-*lox*P and the FLP/FRT Systems

These new vectors make use of the site specific recombination system of the P1 phage or of the yeast Saccharomyces cerevisiae.

The Cre-*lox*P System. The P1 phage possesses a recombinase called Cre (Abremski and Hoess, 1984) which interacts specifically with a 34 base pairs

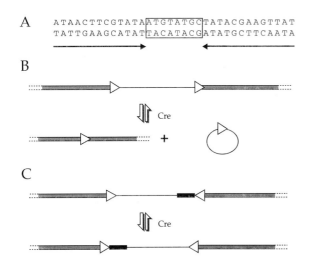

Figure 48.5. The *lox*P site and the Cre enzyme activity. A. Sequence of the *lox*P site recognised by the Cre enzyme. The *lox*P site is composed of two 13 bp palindromic sequences, underlined by arrows, separated by a conserved 8 bp sequence (framed). B. If two *lox*P sites (represented by open triangles) are placed in the same orientation, treatment with Cre enzyme leads to the deletion of the DNA fragment located between both sites in a circular form which will be quickly degraded. C. If the *lox*P sites are in opposite orientation, treatment with Cre enzyme leads to the inversion of the DNA fragment located between both sites. In both cases the reaction is reversible.

(bp)*lox*P site (Hoess and Abremski, 1984; and Hoess et al., 1986). The *lox*P site is composed of two palindromic sequences of 13 bp separated by a 8 bp conserved sequence. The recombination by the Cre enzyme between two *lox*P sites having an opposite orientations leads to the inversion of the DNA fragment lying between the two *lox*P sites. On the other hand, recombination between two sites having an identical orientation leads to the deletion of the DNA fragment. This reaction is reversible and a molecule containing a *lox*P site can integrate into another DNA molecule carrying an identical *lox*P site (Abremski et al., 1983 and Kilby et al., 1993) and (Figure 48.5).

The FLP/FRT System. The FLT/FRT system, which involves the recombinase FLP, recognising the FRT specific sites, is encoded by the 2 μ plasmid of the yeast Saccharomyces cerevisiae, can be used interchangebly with the Cre-*lox*P system (Kilby et al., 1993). O'Gorman et al. (1991) showed that it is possible to manipulate a DNA fragment via the FLP enzyme into mammalian cells having a FRT site in their genome. In order to study the regulation of the gene expression of the β-globin, Fiering et al. (1993) built a homologous recombination vector which allows them to eliminate the selectable marker after treatment by the FLP enzyme. Using a similar approach others have used the FLP/FRT system to eliminate the selectable marker after a homologous recombination event (Fiering et al., 1995; Jung et al., 1993 and Rickert et al., 1995). Just as with a Cre-*lox*P system (Medberry et al., 1995; Odell et al., 1990 and Qin et al., 1994), the FLP/FRT system can be used in plants (Lloyd and Davis, 1994).

These systems are now being used to create new homologous recombination vectors. I will concentrate mainly on examples based on the Cre-*lox*P system because it is the most common system in use in mouse at the moment.

The Cre-*lox*P System: Multiple Uses in Homologous Recombination

The Cre-*lox*P system was first used in combination with a homologous recombination technique in the study of the immunoglobin switch (Gu et al., 1993) and of the DNA polymerase β gene (Gu et al., 1994). There is now sufficient data proving the powerfulness of the system. The success obtained in these first studies suggested a large range of applications for the Cre-*lox*P system, the only limiting factor being the imagination...

Introduction of a Mutation and Elimination of the Selectable Marker. If the vector is constructed in such a way that selectable markers are flanked by *lox*P sites of the same orientation, it is possible, by treatment with the Cre enzyme to eliminate the selectable markers while leaving the gene carrying the selected mutation (see Figure 48.6). Once more, there must be two selectable markers: one to select for the recombination event, the other to select positively for the elimination of the selectable genes (Zou et al., 1994). However, it should be noted that if the selectable markers are eliminated a *lox*P site remains in the locus. This approach makes it possible to create precise mutations such as the introduction of a stop codon or to destroy a specific protein motif. Moreover, by this technique it is possible to eliminate a promoter, if the gene has several, an exon, if there is alternative splicing or an initiation site for translation, if the mRNA uses several.

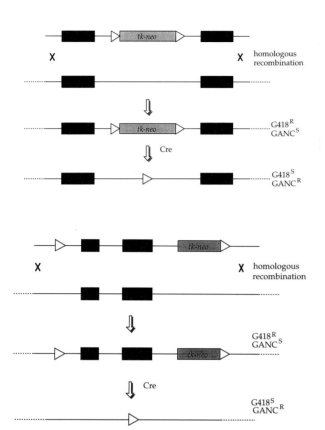

Figure 48.6. Introduction of a specific mutation using the Cre-*lox*P system. A replacement type vector is built in order to insert a selectable cassette, composed of the *tk* and *neo* genes, flanked by two *lox*P sites (open triangles), in a noncoding region of the studied gene, as well as the desired mutation (indicated by a black bar). The homologous recombination event is selected for on the basis of a G418^R phenotype of the ES cells. Subsequent treatment of the recombinant cells with Cre enzyme will eliminate the selectable markers, leaving in place a *lox*P site and the introduced mutation. This event is selected on the basis of the GANC^R phenotype of the Cre induced recombinant cells.

Figure 48.7. Spatio-temporal control of the appearance of a mutation. In order to control the appearance of a mutation, a replacement recombination vector is built in order to insert a *lox*P site (open triangle) into the studied gene along with a selectable cassette (*tk* and *neo* genes) with a *lox*P site at its end, placed so that the treatment by Cre eliminates not only the studied gene but also the selectable cassette. The homologous recombination event is selected for by the G418^R phenotype. The recombinant clones are either treated with Cre enzyme, resulting in the deletion of the studied gene, or they can be used to create mouse lines which will be used as described in Figure 48.8. Unfortunately, some forms of the *tk* gene render the male mouse sterile. It is possible however to replace the cassette *tk-neo* by the *hprt* gene and to use HPRT negative ES cells.

Spatio-Temporal Control of the Appearance of a Mutation. Using the Cre-*lox*P system it is possible to create mice for which the appearance of the mutation is controlled in a spatio-temporal way. For this purpose the recombination vector is designed so that the studied gene, as well as the selectable genes, (placed in an order that they, hopefully, do not affect the expression of the targeted gene) are surrounded by *lox*P sites (Figure 48.7). After introduction of the construct by homologous recombination, which should not disturb the expression of the targeted gene, two possibilities exist to produce the mutation: First treatment of the ES cells by the Cre enzyme, prior to introduction into a recipient blastocyst, will eliminate the targeted gene and selectable markers (a situation corresponding approximately to the type of mutation obtained using replacement vectors, except that it gives rise to a mutation which is free of selectable markers). The recombination can be carried out either by transfection of the ES cells with a Cre expressing vector (Gu et al., 1993; and Sauer and Henderson, 1988) or by lipofection of the enzyme itself into the cells (Baubonis and Sauer, 1993). Secondly, creation of mouse lines in which the gene of interest is flanked by two *lox*P sites followed by crossing with transgenic mice expressing the Cre enzyme at the desired time and place will make stage specific recombination possible. If the expression of the Cre gene is under the control of a tissue specific or inducible promoter or, better still, both at the same time, it is possible to control the appearance of the mutation in a spatio-temporal way. Indeed, by successive breeding of mice having one mutated allele and one allele surrounded by *lox*P sites over transgenic mice for the Cre gene, it is possible to control the appearance of the mutation (see Figure 48.8B). In theory, the expression pattern of the Cre transgene determines the appearance of the mutation. Such a system has been used successfully for the study of the DNA polymerase-β gene which I will describe later (Gu et al., 1994).

The major incoveniences of such an approach are the significant number of crossings necessary to establish the desired mice and the risk of causing a chromosomal translocation at the time of the Cre treatment in animals carrying *lox*P sites on two different chromosomes. Indeed, this property has been used to create such translocations in plants as well as in ES cells (Medberry et al., 1995; Qin et al., 1994; Smith et al., 1995 and van Deursen et al., 1995). However, the first results published suggest that

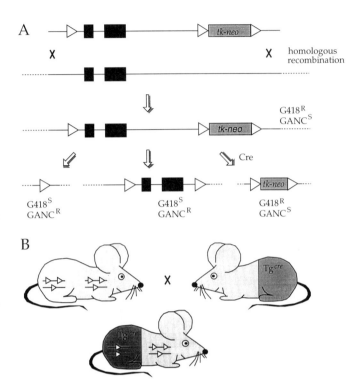

Figure 48.8. Use of a partial activity of the Cre enzyme. A. A recombination vector similar to that described in Figure 48.7 but with an extra loxP site flanking the selectable cassette is electroporated in ES cells. Recombinant ES cells are then treated by the Cre enzyme. Partial activity of the enzyme leads to the production of three genomic rearrangements, of which only two will be selectable by the GANC^R phenotype. This involves: (i) the complete deletion of the gene as well as the selectable markers; and (ii) the elimination of the selectable cassette. In the case of the elimination of the studied gene only, Gancyclovir selection cannot be used because cells retain the tk gene. The first event corresponds to the mutation of the studied gene; the second is used to create mice which can be bred with Cre transgenic mice. B. The breeding of mice obtained by injection of recombinant ES cells with the studied gene surrounded by two loxP sites, with Cre expressing transgenic mice (Tg-Cre) controls the appearance of the mutation in a spatio-temporal manner.

such rearrangements do not occur at a significant frequency (Gu et al., 1994 and Kuhn et al., 1995). The last potential drawback is the lack of documentation on the toxicity of the Cre enzyme in the mouse, although using Cre to control the expression of an oncogene in a specific tissue suggests an absence of toxicity (Lakso et al., 1992 and Orban et al., 1992).

Such a system allows the study of genes for which the mutation provokes a lethal phenotype during embryogenesis and also the study of the role of genes that present a complex pattern of expression. For this purpose the mutation is induced at the desired time and/or place.

The first described attempt to use a Cre-loxP based system (Gu et al., 1993) used the tk and neo genes as selectable genes. Unfortunately, with such an approach it is not possible to establish lines of mice since when the tk gene (at least some form of it) is expressed in the testes it renders male mice sterile (Al-Shawi et al., 1988; 1991; Braun et al., 1990 and Huttner et al., 1993). On the other hand, it is possible to use this technique, with two loxP sites, by using the hprt gene as a selectable marker and HPRT negative ES cells. The tk/sterility problem is probably one of the reasons that pushed Pr. Rajewsky's laboratory to develop a new and extremely elegant approach to achieve the spatio-temporal control of the appearance of a mutation (Gu et al., 1994).

Spatio-temporal control of the Cre enzyme expression can also be achieved with an adenovirus based vector that contains the Cre gene thus allowing infection of cells for delivery of the Cre enzyme (Anton and Graham, 1995 and Kanegae et al., 1995). This could also be used in vivo to infect organs. Finally, one can inject the Cre enzyme directly into one cell-stage embryo (Araki et al., 1995).

The Possibilities of Using Partial Activity of the Cre Enzyme. Gu et al. (1994) have provided an alternative to the previous use of the Cre-loxP system. They introduced not two but three loxP sites into the recombinant vector (Figure 48.8) and modulated the activity of the Cre enzyme in order to have a partial recombination. This partial reaction was obtained by using an enzyme lacking a nuclear translocation signal. In the vector two of the loxP sites flank the selectable cassette (composed of the tk and neo gene) while the third is placed in the studied gene without creating a change at the protein level. The treatment of the ES cells containing such a construction with the Cre enzyme, if the activity of Cre is partial, yields three types of deletion. Two of the deletions produce Gancyclovir resistant cells, the third gives rise to Gancyclovir sensitive cells and is not therefore selectable. This approach has been used to study the role of the DNA polymerase-β as knocking-out of this gene leads to an embryonic lethal phenotype precluding the study of this protein in T cell ontogeny. The first experiments were set up to prove the feasibility of the strategy. Unfortunately only a partial deletion of the gene was obtained in the T cell population.

Subsequently, experiments have proven that it is possible to get a 100% deletion in specific tissues in adults stage using an inducible promoter controlling Cre expression. In these experiments the mouse Mx1 promoter was used to drive Cre expression; Mx1 is activated upon application of interferon α or β or by polyinosinic-poly cytidylic acid (pI-pC). The selection of the correct transgenic line (from over 42 produced) followed by breeding with the above mentioned DNA β-polymerase mutated strain have shown that it is possible to get a 100% deletion of the DNA β-polymerase in the liver after induction of Cre expression with pI-pC. Furthermore, no deletion is detected before induction. Unfortunately, after induction of the recombination, deletion of the gene was also detected at a significant level in other organs such as the spleen, duodenum, heart, lung, uterus, thymus and kidney (Gu et al., 1994).

This approach still has several advantages over those described previously. It eliminates the *tk* gene in the ES cells thus resolving the problem of male sterility. It also leads to the establishment of a line of cells carrying the mutated gene in the absence of selectable markers and a line of cells where the gene is surrounded by *lox*P sites, again without selectable markers. The latter line is used to establish mice which, by crossing with mice transgenic for the Cre enzyme, will allow the spatio-temporal control of the mutation.

The major disadvantages of this approach come from the difficulty in controlling the activity of the Cre enzyme, the possible preferential accessibility of the *lox*P sites and the difficulty in creations of transgenic mice that express the Cre transgene at the right time and place and at a sufficient level in order to get a 100% deletion of the targeted gene. In these cases certain deletions may turn out to be impossible to obtain. It is worth noting that the frequency of the Cre induced recombination seems so high in the transfected ES cells that the counter selection with Gancyclovir is unnecessary. Therefore, there is no need to introduce the *tk* gene in the recombinant vector (Di Santo et al., 1995).

In order to create mice expressing the Cre enzyme with a very restricted profile Rickert et al. (1995) have introduced the Cre gene *via* homologous recombination in the reading frame of the CD19 gene. Furthermore, they eliminated the selectable *neo* gene using the FLP-FRT system, leaving the locus free of any heterologous promoter which could disturb the expression pattern of the Cre enzyme. Indeed a previous report has shown the possibility of using the FLP-FRT instead of the Cre-*lox*P system to eliminate sequences in the mouse genome after a homologous recombination event (Fiering et al., 1995; 1993 and Jung et al., 1993). With such an approach they simultaneously produced a CD19 deficient mice and a mice expressing the Cre gene only in B cells (Rickert et al., 1995).

In the near future there will be a plethora of transgenic mice expressing Cre in a wide variety of patterns. These will be obtained either by a classical transgenic approach or by the technique just described. A network for the distribution of such animals is already taking shape, for more information the reader can contact Dr. A. Nagy at the Samuel Lunenfeld Research Institute, Mont Sinaï Hospital Toronto, Canada.

Cre Mediated, Site-Directed Translocation between Mouse Chromosomes. The first experiments describing the use of the Cre-*lox*P system to induce translocations between nonhomologous chromosomes (Medberry et al., 1995 and Qin et al., 1994) was accomplished in the plant (Nicotiana tabacum). Indeed, this was the first time that manipulation of the genome at the chromosomal level was achieved in higher eukaryotes. Soon after, similar results were obtained in ES cells.

Two independent laboratories have shown that it is possible to induce a translocation between two mouse chromosomes using the Cre-*lox*P system. Smith et al. (1995) induced a translocation between chromosomes 15 and 12 a translocation normally associated with mouse plasmocytomas more precisely the translocation was made between the c-myc gene and the immunoglobulin heavy chain gene. Similarly, van Deursen et al. (1995) wanted to create a mouse model for a human acute myeloid leukaemia resulting from the translocation t(6;9) which fuses the DEK and the CAN genes. Both cases, clearly established the feasibility of the technique but neither of them describe founding a mouse line carrying the translocation. The approach chosen by Smith et al. (1995) seems easier because they are able to isolate directly and with a high frequency a recombinant clone (for each electroporation of the Cre expression vector they performed they were able to isolate a couple of translocated clones). To accomplish this two recombinant vectors were made, one for each gene involved in the translocation, each including part of a selectable marker (see Figure 48.9). The translocation event joined the two parts of the selectable marker and gave rise to a functional marker which was used for the selection of the translocation. Deursen et al. (1995) isolated their recombinant clones by PCR. The disadvantage of this approach is that the ES cells need to be kept in culture for a longer period of time and if the translocation is unstable the absence of selectable marker does not allow a counter selection. Such a technique is possible only if the orientation of the two genes involved in the translocation relative to the

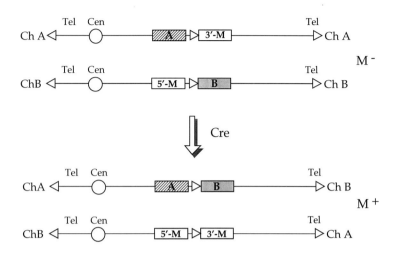

Figure 48.9. Cre induced chromosomal translocation. The introduction of a *lox*P site and complementary parts of a selectable marker onto different chromosomes will allow the induction of a chromosomal translocation. The A chromosome (ChA), containing a *lox*P site and the 3′ part of a selectable marker (3′-M), will translocate with the B chromosome (ChB) containing a *lox*P site and the 5′ end of the same selectable marker (5′-M). The Cre induced translocation will be selectable because it brings together the 5′-M and the 3′-M producing a functional marker (M). The orientation of the chromosomes are represented with respect to their centromere (Cen), the telomeres (Tel) are also indicated.

centromere is known because ortherwise one may obtain a chromosome with two centromeres and one without any centromere!

The possibility of inducing a desired chromosomal translocation into the mouse genome combined with the potential to do it *in vivo*, by using transgenic Cre animals, opens a new insight into the creation of murine models for some human cancers which result from chromosomal translocation. One can also imagine using such a system to set up a gene trap strategy to detected gene fusions involved in abnormal embryonic development or tumourogenesis.

The Still Hypothetical Use of the Cre-*lox*P System

Induction of Cre Activity Using Chemicals. Metzger *et al.* (1995) have created a fusion protein between the ligand-binding domain (LDB) of the human estrogen receptor and the Cre enzyme. This fusion protein renders the Cre activity completely estradiol-dependent. This type of experiment, which has been done in F9 cells, should also facilitate the conditional knock out of genes *in vivo* if a mutated LDB is used which responds only to a synthetic ligand. The combination of tissue specific expression of the chimeric receptor/recombinase protein and the inducibility of its activity should allow a very fine control of the spatio-temporal appearance of a mutation.

Logie and Stewart (1995) have set up a similar system using the FLP recombinase. In this case they also used the estrogen receptor LDB to create a fusion protein with the FLP recombinase. Such a system seems to work as well as the previous system. Another possibility is to use this system in combination with *in vitro* culture of mouse embryos. In such a case a diffusible ligand can be added to the culture medium in order to induce the mutation.

Creation of Large Deletions in the Mouse Genome. Since the induction of chromosomal translocation in feasible it should also be possible to create large deletions within a given chromosome. The introduction of *lox*P sites by homologous recombination separated by a large distance should allow the elimination of the DNA fragment between the two *lox*P sites after Cre treatment. According to the results obtained with the chromosomal translocation there should be no limitation on the size of the inducible deletion, other than viability.

Reintroduction of a Gene into a Given Locus. The recombination reaction induced by the Cre enzyme between two *lox*P sites is reversible (Figure 48.5), suggesting that the design of recombination vectors could allow not only the elimination of a gene but also the reintegration of a mutated form of the studied gene into a lone *lox*P site. Indeed, such experiments have been successfully carried out in mammalian cells (Fukushige and Sauer, 1992; and Sauer and Henderson, 1990). Initially, a promoterless selectable marker preceded by a *lox*P site was introduced into the cells. Subsequently, a promoter was introduced into the *lox*P site of the selectable marker by Cre mediated recombination (Fukushige and Sauer, 1992).

A similar approach could probably also be used in ES cells to reintroduce a DNA fragment into a given locus. For this purpose, one of the *lox*P sites would be placed between the selectable gene and its promoter. Treatment with Cre eliminates the targeted gene as well as the promoter of the selectable marker, making it silent. A second integration vector carrying a *lox*P site and a promoter placed

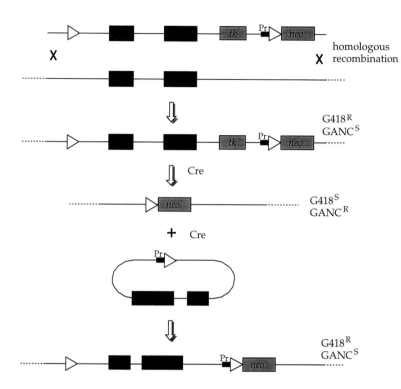

Figure 48.10. Attempting to reintroduce a DNA sequence in a given locus. Since the Cre mediated reaction is reversible, the modification of the vector described in Figure 48.7 opens the possibility of using the remaining *lox*P site (open triangle) to reintroduce a DNA fragment into the studied locus. For this purposes, a *lox*P site is placed between the *neo* gene and its promoter. After introduction of the construct by a homologous recombination event, treatment with the Cre enzyme eliminates the studied gene, the *tk* gene and the promoter of the *neo* gene, leaving in place a *lox*P site and a *neo* gene with its promoter (Pr) truncated. Cells can be selected according to their GANC^R phenotype. In a second step, a reintroduction vector is created so that a Cre induced recombination introduces a DNA fragment including the *neo* gene promoter and the desired mutation in the studied gene. Such cells should be G418^R. The recombinant ES cells obtained during the first step should be reusable for the introduction of different mutations.

in order to allow the re-expression of the selectable marker is then co-introduced with the Cre enzyme. A correct integration should be selected because the selectable marker recovers a promoter, thus allowing its expression (Figure 48.10).

The sophistication of the vectors used in such an approach makes it possible to imagine, after reintroduction of the DNA fragment, the elimination of the selectable genes. Indeed, it is possible to mutate the *lox*P site in such a way as to create a new site which will react only with identical sites (Hoess *et al.*, 1986). Thus, if the two types of *lox*P sites are used for the construction of the vectors at the time of the various recombination events, homologous or mediated by Cre, it is not only possible to reintroduce a DNA fragment into the studied locus but also to eliminate the markers used to detect such an event (Figure 48.11). Alternatively, the FLP/FRT system can be combined with the Cre-*lox*P system in order to achieve such a goal. This approach still needs to be rigorously tested.

Use of the Cre-*lox*P System in Trangenesis

With this system it is possible not only to conceive of a large number of experiments using homologous recombination, but it also gives multiple applications in transgenesis. Here are two examples. It makes it possible to reduce the number of copies of the transgene to only one. If a *lox*P site is placed in the transgene, integration of a tandem repeat of the transgene can be reduced to one copy of the transgene by treatment with the Cre enzyme (Lakso *et al.*, 1992 and Orban *et al.*, 1992).

Such a system also allows better study of genes for which overexpression is lethal and ensures a fine control of tissue specific expression. In such cases the transgene would be separated from its promoter by a DNA sequence which stops transcription and the stop sequence would be flanked by two *lox*P sites (Lakso *et al.*, 1992 and Orban *et al.*, 1992). This sequence could be eliminated by crossing these mice with transgenic mice expressing the Cre enzyme. The introduced DNA sequence can itself be a reporter gene allowing visualisation of the studied promoter (Orban *et al.*, 1992).

Starting with one transgene, two questions can be asked: what are the sequences which control the expression of the transgene and what is the function of the transgene? Usable marker genes are classically the β-galactosidase (*Lac Z*) or of the Chloramphenicol Acetyl Transferase (CAT) or, which may be of interest for *in vivo* studies, the recently described gene encoding the green fluorescent protein (GFP) (Chalfie *et al.*, 1994); GFP still needs to be tested in mammalian cells. In this type of experiment, there is also a reduction of the number of copies of the transgene (Lakso *et al.*, 1992 and Orban *et al.*, 1992).

Such types of experiment will certainly be used to address question about the fate map of the mouse embryo. For such purpose a transgene is made in

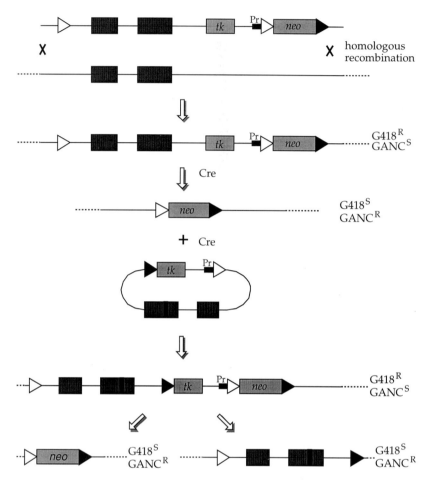

Figure 48.11. Reintroduction of a DNA sequence in a given locus and elimination of the selectable genes.
The selectable genes used in the experiment described in Figure 48.10 can interfere with the expression of the gene of interest, so it would be advantageous to try to eliminate them. For this purpose, one can imagine the use of two mutant *lox*P sites which will react only with themselves in the presence of the Cre enzyme. A recombinant vector similar to that described in Figure 48.7 but including a mutated *lox*P site (black triangle) at the 3' end of the *neo* gene is electroporated in ES cells. Treatment of the recombinant ES cells with Cre will eliminate the desired region of the gene leaving in place a promoterless *neo* gene surrounded by a wild type *lox*P site (open triangle-5' end) and a mutated *lox*P site (black triangle-3' end). A reintegration vector – including a wild type *lox*P site, in front of the *neo* gene promoter, a *tk* gene, a mutant *lox*P site upstream of the *tk* gene and the desired mutation, is introduced simultaneously with the Cre enzyme into the recombinant ES cells. The recombination induced by Cre should lead to the integration of this vector either in the mutant *lox*P site or in the wild type *lox*P site . Only the second event, however, confers a $G418^R$ phenotype to the cells because it brings a promoter to the *neo* gene. A second treatment with Cre should lead to recombination between either the wild type *lox*P sites or the mutated *lox*P site. The former recombination eliminates the reintroduced mutation while the latter eliminates the selectable cassette and leaves in place a mutated gene. This gene is then flanked by a wild type *lox*P site and a mutant *lox*P site. As for the experiment described in Figure 48.7, the *tk-neo* cassette can be replaced by the *hprt* gene, on the condition that HPRT negative ES cells are used.

such a way that the *LacZ* gene or the *GFP* gene is separated from an ubiquitous promoter by a stop sequence flanking by *lox*P sites. Crossing these mice with a Cre transgene that expresses the Cre enzyme only in the desired cell type will allow one to follow these cells during embryogenesis as the Cre enzyme will turn on the expression of the marker by eliminating the stop sequence. The marker gene thus becomes an inherited genetic marker for those cells that were treated with Cre, even if the Cre gene is no longer expressed. Such transgenic mice can be used in different studies simply by crossing them with different Cre transgenic mice.

Conclusion

In this article I have tried to review and suggest applications for the Cre-*lox*P system of the P1 phage, concentrating especially on experiences of homologous recombination. Indeed, the new generation of homologous recombination vectors should allow a finer study of mutated genes. Initially, they make it possible to create "clean" mutations where the selectable genes are eliminated from the studied locus. This makes it possible to envisage mutations such as:

- introducing a stop codon;
- suppressing a specific section of a protein;
- specifically eliminating a promoter or an exon;
- reducing the number of initiation sites of the translation if there are several.

Two realistic approaches to create specific clean mutations in a gene are now possible, using either the Cre-*lox*P system, which will always leave a *lox*P site in the studied locus, or the tag and exchange technique. The advantage of the Cre-*lox*P system is that it needs only one homologous recombination event and treatment with the Cre enzyme. Moreover, some of the constructions used with the Cre-*lox*P system also make it possible to carry out an exchange *via* homologous recombination.

The use of the Cre-*lox*P system allows the spatio-temporal control of the appearance of a mutation. The latter property is of fundamental importance in the study of genes critically involved in embryonic development. Indeed, the mutation of such genes often gives rise to an embryonic lethality which prohibits the study of these genes in the adult animal or in certain embryonic or adult tissues. Another considerable advantage of such an approach is the possibility of re-using the same selectable markers to create another mutation or mutations in the second allele in order to obtain mutant ES cells homozygote for the studied gene. Furthermore, most of mouse genetic manipulations based on the Cre-*lox*P system will involve only one homologous recombination event. Homologous recombination is a rare event while Cre induced recombination seems to be a very efficient process.

The three *lox*P sites cloned into the homologous recombination vector is an approach worth considering, as it allows the elimination of the selectable genes and the creation of two types of mutant mice. It is currently the most elegant approach.

The use of such a system should allow, in addition to the experiments described here, the creation of chromosomal translocations modelling human pathologies or the creation of very broad deletions in a given locus. Moreover, the combination of both systems, Cre-*lox*P and FLP/FRT or the use of mutated *lox*P sites makes it possible to imagine even more sophisticated manipulations of the murine genome.

I am convinced that it is now possible to realise any imaginable genetic manipulation in the mouse. No longer is the question "Can it be done?" but "Is it worth doing?" Indeed, some of the manipulations may be very difficult to achieve, time consuming and, finally very expensive.

In our day, I think that the rational approach to study the function of a gene is to knock it out. For this two constructions have to be made: (i) a replacement vector which will introduce a *LacZ* gene in addition to the *neo* gene (preferentially using a promoterless strategy if possible); and (ii) use of a Cre-3*lox*P site based construct. The second construction should be made at the same time as the first one, but used only if the phenotype is lethal. Since the technology is now available, only these two types of mutations will soon be acceptable.

In addition to the applications described in this chapter, numerous other experiments are realisable. Although only the application of these systems to the mouse has been described in this chapter, there is certainly opportunity to exploit these systems in other animal, parasitic or vegetable models.

References

Abremski, K. and Hoess, R. (1984) Bacteriophage P1 site-specific recombination, purification and properties of the Cre recombinase protein. *J. Biol. Chem.* **259**, 1509–1514

Abremski, K., Hoess, R. and Sternberg, N. (1983) Studies on the properties of P1 site-specific recombination: evidence for topologically unlinked products following recombination. *Cell* **32**, 1301–1311

Al-Shawi, R., Burke, J., Jones, C., Simons, J.P. and Bishop, J.O. (1988) A mup promoter-thymidine kinase reporter gene shows relaxed tissue-specific expression and confers male sterility upon transgenic mice. *Mol. Cell. Biol.* **8**, 4821–4828

Al-Shawi, R., Burke, J., Wallace, H., Jones, C., Harrison, S., Buxton, D., Maley, S., Chandley, A. and Bishop, J.O. (1991) The herpes simplex virus type 1 thymidine kinase is expressed in the testes of transgenic mice under the control of a cryptic promoter. *Mol. Cell. Biol.* **11**, 4207–4216

Anton, M. and Graham, F.L. (1995) Site-specific recombination mediated by an adenovirus vector expressing the Cre recombinase protein: a molecular switch for control of gene expression. *J. Virol.* **69**, 4600–4606

Araki, K., Araki, M., Miyazaki, J. and Vassalli, P. (1995) Site-specific recombination of a transgene in fertilized eggs by transient expression of Cre recombinase. *Proc. Natl. Acad. Sci. USA* **92**, 160–164

Askew, G.R., Doetschman, T. and Lingrel, J.B. (1993) Site-directed point mutations in embryonic stem cells: a gene-targeting tag-and-exchange strategy. *Mol. Cell. Biol.* **13**, 4115–4124

Barnes, W.M. (1994) PCR amplification of up to 35-kb DNA with high fidelity and high yield from λ bacteriophage templates. *Proc. Natl. Acad. Sci. USA* **91**, 2216–2220

Baubonis, W. and Sauer, B. (1993) Genomic targeting with purified Cre recombinase. *Nucleic Acids Res.* **21**, 2025–2029.

Brandon, E.P., Idzerda, R.L. and McKnight, G.S. (1995a) Targeting the mouse genome: a compendium of knockouts (part I). *Current Biology* **5**, 625–634

Brandon, E.P., Idzerda, R.L. and McKnight, G.S. (1995b) Targeting the mouse genome: a compendium of knockouts (part II). *Current Biology* **5**, 758–765

Brandon, E.P., Idzerda, R.L. and McKnight, G.S. (1995c) Targeting the mouse genome: a compendium of knockouts (part III). *Current Biology* **5**, 873–881

Braun, R.E., Lo, D., Pinkert, C.A., Widera, G., Flavell, R.A., Palmiter, R.D. and Brinster, R.L. (1990) Infertility in male transgenic mice: disruption of sperm development by HSV-*tk* expression in postmeiotic germ cells. *Biol. Reprod.* **43**, 684–693

Capecchi, M.R. (1989a) Altering the genome by homologous recombination. *Science* **244**, 1288–1292

Capecchi, M.R. (1989b) The new mouse genetics: altering the genome by gene targeting. *Trends Genet.* **5**, 70–76

Chalfie, M., Tu, Y., Euskirchen, G., Ward, W.W. and Prasher, D.C. (1994) Green fluorescent protein as a marker for gene expression. *Science* **263**, 802–805

Copp, A.J. (1995) Death before birth: clues from gene knockouts and mutations. *Trends Genet.* **11**, 87–93

Di Santo, J.P., Muller, W., Guy-Grand, D., Fischer, A. and Rajewsky, K. (1995) Lymphoid development in mice with a targeted deletion of the interleukin 2 receptor gamma chain. *Proc. Natl. Acad. Sci. USA* **92**, 377–381

Donehower, L.A., Harvey, M., Slagle, B.L., McArthur, M.J., Montgomery, C.J., Butel, J.S. and Bradley, A. (1992) Mice deficient for p53 are developmentally normal but susceptible to spontaneous tumours. *Nature* **356**, 215–221

Evans, M.J. and Kaufman, M.H. (1981) Establishment in culture of pluripotential cells from mouse embryos. *Nature* **292**, 154–156

Fiering, S., Epner, E., Robinson, K., Zhuang, Y., Telling, A., Hu, M., Martin, D.I.K., Enver, T., Ley, T.J. and Groudine, M. (1995) Targeted deletion of 5'HS2 of the murine β-globine LCR reveals that it is not essential for proper regulation of the β-globin locus. *Genes Dev.* **9**, 2203–2213

Fiering, S., Kim, C.G., Epner, E.M. and Groudine, M. (1993) An "in-out" strategy using gene targeting and FLP recombinase for the functional dissection of complex DNA regulatory elements: analysis of the β-globin locus region. *Proc. Natl. Acad. Sci. USA* **90**, 8469–8473

Friedrich, G. and Soriano, P. (1991) Promoter traps in embryonic stem cells: a genetic screen to identify and mutate developmental genes in mice. *Genes Dev.* **5**, 1513–1523

Fukushige, S. and Sauer, B. (1992) Genomic targeting with a positive-selection *lox* integration vector allows highly reproducible gene expression in mammalian cells. *Proc. Natl. Acad. Sci. USA* **89**, 7905–7909

Gossler, A., Doetschman, T., Korn, R., Serfling, E. and Kemler, R. (1986) Transgenesis by means of blastocyst-derived embryonic stem cell lines. *Proc. Natl. Acad. Sci. USA* **83**, 9065–9069

Gu, H., Marth, J.D., Orban, P.C., Mossmann, H., Rajewsky and K. (1994) Deletion of a DNA polymerase β gene segment in T cells using cell type-specific gene targeting. *Science* **265**, 103–106

Gu, H., Zou, Y.-R. and Rajewsky, K. (1993) Independent control of immunoglobulin switch recombination at individual switch regions evidenced through Cre-LoxP mediated gene targeting. *Cell* **73**, 1155–1164

Hanks, M., Wurst, W., Anson-Cartwright, L., Auerbach, A.B. and Joyner, A.L. (1995) Rescue of the En-1 mutant phenotype by replacement of En-1 with En-2. *Science* **269**, 679–682

Hasty, P., Ramirez, S.R., Krumlauf, R. and Bradley, A. (1991) Introduction of a subtle mutation into the Hox-2.6 locus in embryonic stem cells [published erratum appears in Nature 1991 Sep, 5;353(6339):94]. *Nature* **350**, 243–246

Hoess, R.H. and Abremski, K. (1984) Interaction of the bacteriophage P1 recombinase Cre with the recombining site *lox*P. *Proc. Natl. Acad. Sci. USA* **81**, 1026–1029

Hoess, R.H., Wierzbicki, A. and Abremski, K. (1986) The role of the *lox*P spacer region in P1 site-specific recombination. *Nucleic Acids Res.* **14**, 2287–2300

Huttner, K.M., Pudney, J., Milstone, D.S., Ladd, D. and Seidman, J.G. (1993) Flagellar and acrosomal abnormalities associated with testicular HSV-*tk* expression in the mouse. *Biol. Reprod.* **49**, 251–261

Jung, S., Rajewsky, K. and Radbruch, A. (1993) Shutdown of class switch recombination by deletion of a switch region control element. *Science* **259**, 984–987

Kanegae, Y., Lee, G., Sato, Y., Tanaka, M., Nakai, M., Sakaki, T., Sugano, S. and Saito, I. (1995) Efficient gene activation in mammalian cells by using recombinant adenovirus expressing site-specific Cre recombinase. *Nucl. Acids Res.* **23**, 3816–3821

Kilby, N.J., Snaith, M.R. and Murray, J.A.H. (1993) Site-specific recombinases: tools for genome engineering. *Trends Genet.* **9**, 413–420

Kim, D.G., Kang, H.M., Jang, S.K. and Shin, H.S. (1992) Construction of a bifunctional mRNA in the mouse by using the internal ribosomal emtry site of the encephalomyocarditis virus. *Mol. Cell. Biol.* **12**, 3636–3643

Kuhn, R., Schwenk, F., Aguet, M. and Rajewsky, K. (1995) Inducible gene targeting in mice. *Science* **269**, 1427–1429

Lakso, M., Sauer, B., Mosinger, B., Lee, E.J., Manning, R.W., Yu, S.-H., Mulder, K.L. and Westphal, H. (1992) Targeted oncongene activation by site-specific recombination in transgenic mice. *Proc. Natl. Acad. Sci. USA* **89**, 6232–6236

Lallemand, Y. and Brulet, P. (1990) An *in situ* assessment of the routes and extents of colonisation of the mouse embryo by embryonic stem cells and their descendants. *Development* **110**, 1241–1248

Lloyd, A.M. and Davis, R.W. (1994) Functional expression of the yeast FLP/FRT site-specific recombination system in Nicotiana tabacum. *Mol. Gen. Genet.* **242**, 653–657

Logie, C. and Stewart, A.F. (1995) Ligand-regulated site-specific recombination. *Proc. Natl. Acad. Sci. USA* **92**, 5940–5944

Martin, G.R. (1981) Isolation of a pluripotent cell line from early mouse embryos cultured in medium conditioned by teratocarcinoma stem cells. *Proc. Natl. Acad. Sci. USA* **78**, 7634–7638

Medberry, S.L., Dale, E., Qin, M. and Ow D.W. (1995) Intra-chromosomal rearrangements generated by Cre-lox site-specific recombination. *Nucleic Acids Res.* **23**, 485–490

Metzger, D., Clifford, J., Chiba, H. and Chambon, P. (1995) Conditional site-specific recombination in mammalian cells using a ligand-dependnet chimeric Cre recombinase. *Proc. Natl. Acad. Sci. USA* **92**, 6991–6995

Mountford, P., Zevnik, B., Düwel, A., Nichols, J., Li, M., Dani, C., Robertson, M., Chambers, I. and Smith, A. (1994) Dicistronic targeting constructs: reporters and modifiers of mammalian gene expression. *Proc. Natl. Acad. Sci. USA* **91**, 4303–4307

Mountford, P.S. and Smith, A.G. (1995) Internal ribosome entry sites and dicistronic RNAs in mammalian transgenesis. *Trends Genet.* **11**, 179–184

Nagy, A., Gocza, E., Diaz, E.M., Prideaux, V.R., Ivanyi, E., Markkula, M. and Rossant, J. (1990) Embryonic stem cells alone are able to support fetal development in the mouse. *Development* **110**, 815–821

Nagy, A., Rossant, J., Nagy, R., Abramow-newerly, W. and Roder, J.C. (1993) Derivation of completely cell culture-derived mice from early-passage embryonic stem cells. *Proc. Natl. Acad. Sci. USA* **90**, 8424–8428

O'Gorman, S., Fox, D.T. and Wahl, G.M. (1991) Recombinase-mediated gene activation and site-specific integration in mammalian cells. *Science* **251**, 1351–1355

Odell, J., Caimi, P., Sauer, B. and Russell, S. (1990) Site-directed recombination in the genome of transgenic tobacco. *Mol. Gen. Genet.* **223**, 369–378

Orban, P.C., Chui, D. and Marth, J.D. (1992) Tissue- and site-specific DNA recombination in transgenic mice. *Proc. Natl. Acad. Sci. USA* **89**, 6861–6865

Pascoe, W.S., Kemler, R. and Wooa, S.A. (1992) Genes and functions: trapping and targeting in embryonic stem cells. *Biochim. Biophys. Acta* **1114**, 209–221

Qin, M., Bayley, C., Stockton, T. and Ow, D.W. (1994) Cre recombinase-mediated site-specific recombination between plant chromosomes. *Proc. Natl. Acad. Sci. USA* **91**, 1706–1710

Ramirez, S.R., Zheng, H., Whiting, J., Krumlauf, R. and Bradley, A. (1993) *Hoxb*-4 (*Hox*-2.6) mutant mice show homeotic transformation of a cervical vertebra and defects in the closure of the sternal rudiments. *Cell* **73**, 279–294

Reid, L.H., Shesely, E.G., Kim, H.S. and Smithies, O. (1991) Cotransformation and gene targeting in mouse embryonic stem cells. *Mol. Cell. Biol.* **11**, 2769–2777

Rickert, R.C., Rajewsky, K. and Roes, J. (1995) Impairment of T-cell-dependnet B-cell responses and B-1 ceel development in CD19-deficient mice. *Nature* **376**, 352–355

Rijli, F.M., Dolle, P., Fraulob, V., Le Meur, M. and Chambon, P. (1994) Insertion of a targeting construct in a Hoxd-10 allele can influence the control of *Hoxd*-9 expression. *Dev. Dyn.* **201**, 366–377

Rijli, F.M., Matyas, R., Pellegrini, M., Dierich, A., Gruss, P., Dolle, P. and Chambon, P. (1995) Cryptorchidism and homeotic transformations of spinal nerves and vertebrae in *Hoxa*-10 mutant mice. *Proc. Natl. Acad. Sci. USA* **92**, 8185–8189

Sauer, B. and Henderson, N. (1988) Site-specific DNA recombination in mammalian cells by Cre recombinase of bacteriophage P1. *Proc. Natl. Acad. Sci. USA* **85**, 5166–5170

Sauer, B. and Henderson, N. (1990) Targeted insertion of exogenous DNA into the eukaryotic genome by the Cre recombinase. *New Biol.* **2**, 441–449

Scherer, S. and Davis, R.W. (1979) Replacement of chromosome segments with altered DNA sequences constructed *in vitro*. *Proc. Natl. Acad. Sci. USA* **76**, 4951–4955

Schwartzberg, P.L., Goff, S.P. and Robertson, E.J. (1989) Germ-line transmission of a c-abl mutation produced by targeted gene disruption in ES cells. *Science* **246**, 799–803

Smith, A.J.H., De Sousa, M.A., Kwabi-Addo, B., Heppell-Parton, A., Impey, H. and Rabbits, P. (1995) A site-directed chromosomal translocation induced in embryonic stem cells by Cre-*loxP* recombination. *Nature Genetics* **9**, 376–384

Smithies, O., Gregg, R.G., Boggs, S.S., Koralewski, M.A. and Kucherlapati, R.S. (1985) Insertion of DNA sequences into the human chromosomal β-globin locus by homologous recombinaison. *Nature* **317**, 230–234

Stacey, A., Schnieke, A., McWhir, J., Cooper, J., Colman, A. and Melton, D.W. (1994) Use of double-replacement gene targeting to replace the murine alpha-lactalbumin gene with its human counterpart in embryonic stem cells and mice. *Mol. Cell. Biol.* **14**, 1009–1016

Te Riele, H., Maandag, E.R. and Berns, A. (1992) Highly efficient gene targeting in embryonic stem cells through homologous recombination with isogenic DNA constructs. *Proc. Natl. Acad. Sci. USA* **89**, 5128–5132

Thomas, K.R., Folger, K.R. and Capecchi, M.R. (1986) High frequency targeting of genes to specific sites in the mammalian genome. *Cell* **44**, 419–428

Valancius, V. and Smithies, O. (1991) Testing an "in-out" targeting procedure for making subtle genomic modifications in mouse embryonic stem cells. *Mol. Cell. Biol.* **11**, 1402–1408

Van Deursen, J., Fornerod, M., van Rees, B. and Grosveld, G. (1995) Cre-mediated site-specific translocation between nonhomologous mouse chromosomes. *Proc. Natl. Acad. Sci. USA* **92**, 7376–7380

Wood, S.A., Pascoe, W.S., Schmidt, C., Kemler, R., Evans, M.J. and Allen, N.D. (1993) Simple and efficient production of embrionic stem cell-embryo chimeras by coculture. *Proc. Natl. Acad. Sci. USA* **90**, 4582–4585

Wu, H., Liu, X. and Jaenisch, R. (1994) Double replacement: strategy for efficient introduction of subtle mutations into the murine Col1a-1 gene by homologous recombination in embryonic stem cells. *Proc. Natl. Acad. Sci. USA* **91**, 2819–2823

Zimmer, A. and Gruss, P. (1989) Production of chimaeric mice containig embryonic stem (ES) cells carrying a homeoobox Hox 1.1 allele mutated by homologous recombination. *Nature* **338**, 150–153

Zou, Y.R., Muller, W., Gu, H. and Rajewsky, K. (1994) Cre-*loxP*-mediated gene replacement: a mouse strain producing humanized antibodies. *Curr. Biol.* **4**, 1099–1103

PART III, SECTION B

49. Improvements in Transgene Expression by Gene Targeting

Alexander J. Kind and Alan Colman
PPL Therapeutics, Roslin, Edinburgh, EH25 9PP, Scotland

It has long been observed that identical transgenes randomly integrated at different chromosomal locations can vary widely in the quantity and pattern of their expression. Where the intention is to achieve maximal expression of a transgene, this requires that large numbers of animals from independent founders are screened for expression. In species such as mouse where DNA microinjection is an efficient and cheap method of transgenesis this is not a major problem. However in domestic animals, the low efficiency of DNA microinjection and high cost of animal husbandry provide a strong incentive to reduce variation.

There are two approaches to overcoming the position effect. The first is to identify sequences which promote the desired pattern and level of gene expression and which can be incorporated into transgene constructs. Features in the chromosomal environment likely to affect transgene expression include the proximity of other transcribed regions, the presence of regulatory elements, their capacity for interaction with the transgene, and the local chromatin structure. Although major advances have been made in identifying elements which confer position independence on particular genes, most notably the locus control region of the β-globin gene cluster (Grosveld et al., 1987), a discrete element capable of conferring position independent expression onto a heterologous transgene in a wide variety of tissues has yet to be identified. The use of large cloned genomic fragments, e.g. YACs, can overcome position dependence by introducing a transgene within its normal genomic context, without the necessity for fine genetic analysis. However, this tends to reproduce the native pattern of gene expression, which is not always the intention. YACS are less suitable for expression in heterologous tissue because the large size of DNA fragments hinders the production of recombinant DNA constructs. Furthermore, it is frequently undesirable to introduce large regions of exogenous DNA into the host genome.

This account will concentrate on the second approach, which is to identify a chromosomal locus capable of supporting high expression then place the transgene there by gene targeting. This has the advantage that no chromosomal influences are excluded by a cloning step. It should however be pointed out that, although placement of a single transgene at an optimal site may yield the maximum expression per gene copy, this may be actually less than the maximum obtainable by random integration which could result from multiple integrations at a favourable site.

Gene targeting has proved an extraordinarily fruitful technique in mouse, but application to other species still awaits the demonstration of embryonic stem cells, or some functional equivalent, which can undergo genetic manipulation *in vitro* and then contribute to the germ line of an animal. Here we describe experimental strategies suitable for transgene placement in mouse.

Choice of Chromosomal Locus

A chromosomal location likely to support high expression in a chosen tissue may be chosen by experiment, or on the basis of endogenous gene expression.

One possible approach would be to screen a population of transgenic animals containing randomly integrated copies of a reporter construct, which should ideally resemble the transgene as closely as possible. Those integration sites shown to support high reporter gene expression are identified, cloned and used to target transgenes to the same position. However, it is practical to screen only a relatively tiny proportion of the total number of possible integration sites in transgenic animals. The use of stably transfected cultured cells provides a more comprehensive screen of integration sites, but cannot of course provide information on gene expression in other tissues. Cell culture systems which reliably model their *in vivo* counterparts are becoming increasingly available through improvements in cell derivation and culture techniques (e.g. Barcellos-Hoff *et al.*, 1989 and Noble *et al.*, 1992.). Cells should be transfected by a method likely to result in single copy number integrations, (e.g. electroporation) so that gene expression by an individual clone can be attributed to a particular integration site.

Alternatively, endogenous genes highly expressed in the tissue of choice represent loci likely to support high transgene expression. The precise placement of the transgene relative to the endogenous gene must be determined by experiment. For example, the transgene could be placed either closely adjacent to the endogenous gene or could replace the endogenous gene altogether, in which case the effect of the knockout should be taken into consideration.

Techniques of Transgene Placement

Numerous gene targeting techniques have been devised over recent years, two of these are practically suited to transgene placement.

The simplest method is the cotransfer of DNA linked to a positive selectable marker. This achieves placement in a single step, but deposits the marker gene at the target locus, which may be undesirable for some applications. This can be achieved using either an integration, or a replacement targeting vector (Deng *et al.*, 1993). The use of a replacement vector provides the option of removing host chromosome sequences in the targeting step, but carries the risk that DNA regions distant from the selectable marker will fail to cotransfer.

Double replacement gene targeting was first proposed by Evans (1989) as a means of introducing modifications into a gene of interest. It is also very suitable for the precise placement of transgenes. Figure 49.1. illustrates a double replacement scheme we have used to place a human α-lactalbumin transgene at the mouse α-lactalbumin locus (Stacey *et al.*, 1994). Two consecutive rounds of gene targeting are performed. In the first, a marker cassette capable of both positive and negative

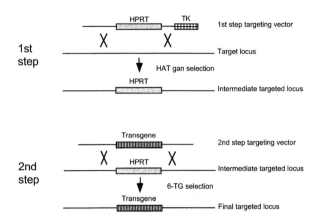

Figure 49.1. Double replacement gene targeting.

selection, in this case an HPRT minigene, is introduced at the target locus using a replacement targeting vector and positive selection. The use of the TK counter-selectable gene in the first step confers partial enrichment for targeted clones but is not obligatory. In the second round, the marker cassette is replaced at the target locus by the transgene using a replacement vector and negative selection.

The choice of marker cassettes for which both positive and negative selection is available is currently restricted to either HPRT or a combination of *neo* and *tk* genes (Askew *et al.*, 1993 and Wu *et al.*, 1994). Use of HPRT requires a HPRT deficient cell line, preferably capable of growth independent of feeder cells. Such a line is currently available in mice (Magin *et al.*, 1992), but would require derivation from ES cells of other species. The alternative *neo TK* can be used in wild type cells, but has the drawback that *TK* is a poor counter-selectable marker, being prone to extinction of expression without physical loss of the gene (Askew *et al.*, 1993). It is notable in this respect that the HSV TK coding sequence is rich in CpG pairs (121 CpG in 1128 bp of coding sequence) compared to human HPRT (9 CpG in 657 bp of coding sequence) and that 5-methyl cytosine in CpG pairs is known to mutate to thymine at high frequency (Laird *et al.*, 1994).

Double replacement gene targeting offers several advantages.

1. Transgene placement leaves the host genome otherwise undisturbed and free of exogenous marker genes.
2. Once a suitable locus has been identified and 1st step cells are available, other transgenes can be placed at the same site by retargeting with different second step vectors.
3. The 1st step construct can be designed to remove any amount of host sequence at the target

locus within the normal constraints of a replacement vector, i.e. 0–20 kb.

References

Askew, G.R., Doetschman, T. and Lingrel, J.B. (1993) Site directed point mutations in embryonic stem cells: a gene targeting tag-and-exchange strategy. *Mol. Cell. Biol.* **13**, 4115–4124

Barcellos-Hoff, M.H., Aggeler, J., Ram, T.G. and Bissell, M.J. (1989) Functional differentiation and alveoloar morphogenesis of primary mammary cultures on reconstituted basement membrane. *Development* **105**, 223–235

Deng, C., Thomas, K.R. and Capecchi, M.R. (1993) Location of crossovers during gene targeting with insertion and replacement vectors. *Mol. Cell. Biol.* **13**, 2134–2140.

Evans, M.J. (1989) Potential for genetic manipulation of mammals. *Mol. Biol. and Med.* **6**, 557–565

Grosveld, F., Blom van Assendelft, G., Greaves, D.R. and Kollias, G. (1987) Position-independent, high level expression of the human β-globin gene in transgenic mice. *Cell* **51**, 975–985

Laird, P.W. and Jaenisch, R. (1994) DNA methylation and cancer. *Hum. Mol. Gen.* **3**, 1487–1495

Magin, T.M., McWhir, J. and Melton, D.W. (1992) A new mouse embryonic stem cell line with good germ line contribution and gene targeting frequency. *Nucleic Acids Res.* **14**, 3795

Noble, M., Groves, A.K. Ataliotis, P. and Jat, P. (1992) From chance to choice in the generation of neural cell lines. *Brain Pathology* **2**, 39–46

Stacey, A.J., Schnieke, A., McWhir, J., Cooper, J., Colman, A. and Melton, D.W. (1994) Use of double replacement gene targeting to replace the murine α-lactalbumin gene with its human counterparet in embryonic stem cells and mice. *Mol. Cell Biol.* **4**, 1009–1016

Wu, H., Liu, X. and Jaenisch, R. (1994) Double replacement: strategy for effcient introduction of subtle mutations into the murine Col1 α-1 gene by homologous recombination in embryonic stem cells. *Proc. Natl. Acad. Sci. USA* **91**, 2819–2823

PART III, SECTION B

50. Nuclear Antisense RNA as a Tool to Inhibit Gene Expression

Gordon G. Carmichael
Department of Microbiology, University of Connecticut Health Center, Farmington CT, USA

Introduction

Current approaches to manipulate gene expression using antisense RNA involve the production within cells of RNA molecules that are complementary to a small region or the full length sequence of the gene to be targeted. To accomplish this, the DNA of interest is generally inserted in the antisense orientation immediately downstream of a constitutive or inducible promoter (Izant and Weintraub, 1985) in vectors that are capable of integrating into the host cell genome (i.e., retroviral vectors), or merely of existing in an episomal state within the host. Vector delivery can be via a number of techniques, including the recent use of liposomes to achieve long-term transgene expression *in vivo* (Thierry *et al.*, 1995).

A major problem with existing antisense RNA applications, however, is the fact that most have met with marginal or no success, either *in vitro* or *in vivo*. This might reflect an inherent inefficiency of antisense inhibition, or, alternatively, might indicate that alternative methods of antisense expression are needed. Most commonly used expression vectors contain RNA splicing signals for stable mRNA expression, and a polyadenylation signal for proper 3'-end formation. These elements allow the primary transcripts to be processed and exported to the cytoplasm just as the bulk of cellular messages. Efficient recognition of these RNA processing signals can lead to the accumulation of large amounts of antisense RNA in the cytoplasm. With such constructs, however, a 100–1000-fold molar excess of antisense RNA is often needed for effective inhibition of target gene expression. A common assumption, not supported by experimental data, is that effective antisense inhibition occurs at the level of translation.

Although antisense RNA has not proved particularly effective, there have been several reports of its successful application in transgenic systems. For example, Stockhaus *et al.* (1990), Hamilton *et al.* (1990) and more recently Höfgen *et al.* (1994) reported significant, dose-dependent inhibition of the expression of several target proteins in transgenic plants. A recent, and possibly more effective approach to antisense inhibition of gene expression (Liu *et al.*, 1994), and described here appears to mimic the natural process of antisense regulation that occurs in the nuclei of mammalian cells.

Natural Mammalian Antisense RNA Usually Acts in the Nucleus, and is More Effective There

Naturally-occurring antisense RNA has been found both in prokaryotes and in eukaryotes. In prokaryotes, where it was first discovered, many examples of antisense-mediated regulation of gene expression have been reported. The mechanism of action of this regulation is well understood for prokaryotes, and is usually at the level of translation (Simons, 1988). In eukaryotes, however, relatively few examples of antisense RNA-mediated gene regulation have been reported (Adelman *et al.*, 1987; Hildebrandt and Nellen, 1992; Lankenau *et al.*, 1994; Liu *et al.*, 1994; Nellen and Lichtenstein, 1993; Spencer *et al.*, 1986; Tosic *et al.*, 1990; Volk *et al.*, 1989; and Williams and Fried, 1986). Further, with the exception of the *C. elegans* lin-4 system (Lee *et al.*, 1993), this regulation is thought to occur primarily or exclusively within the nucleus (Cornelissen, 1989; and Murray and Crockett, 1992) and not at the translational level (Cornelissen, 1989; and Höfgen et al., 1994). In fact, the presence of double stranded RNA (dsRNA) in the cytoplasm may not be associated with natural antisense regulation, but rather with a more unnatural situation, such as that associated with some viral infections. In some cells, cytoplasmic dsRNA may also induce the expression of interferon. Finally, since there may be little degradation of RNA:RNA duplexes in the cytoplasm (Murray and Crockett, 1992), it is likely that inhibition would be more efficient if antisense RNAs could be retained in the nucleus.

Mechanism of Action of Nuclear Antisense RNA

The murine polyoma virus model system has provided a convenient system in which to study the mechanism of action of nuclear antisense RNA. At late times in the polyoma virus life cycle, nuclear antisense RNA is used to downregulate the expression of viral "early" genes (Liu et al., 1994). Recent work using this model system has revealed that target RNA molecules are not rapidly degraded, but rather are extensively modified (and consequently rendered inert for gene expression) within the nucleus. Almost half of the adenosines in the regions examined appear to have been converted to inosines (M. Kumar and G. Carmichael, submitted), most likely by the action of the enzyme double strand RNA specific adenosine deaminase (Bass and Weintraub, 1988 and Wagner et al., 1989). This enzyme is ubiquitous in eukaryotic cells, and is thought to reside primarily or exclusively within the nucleus (Bass and Weintraub, 1988 and Wagner et al., 1989). It is highly specific for dsRNA and before now its importance was inferred only from a few isolated cases of specific base modifications in several genes (Bass et al., 1989; Cattaneo, 1994; Cattaneo and Billeter, 1992 and Luo et al., 1990). The above results strongly suggest that the primary role of this enzyme in the cell may in fact be to effectively detect and eliminate naturally-occurring sense-antisense RNA hybrids in the nucleus. Finally, antisense-modified transcripts appear not to leave the nucleus but rather be degraded within that compartment (M. Kumar and G. Carmichael, submitted).

Nuclear Antisense Vectors

A new approach to antisense inhibition of targeted gene expression was recently reported using the polyoma virus model system (Liu et al., 1994). If the polyadenylation signal for 3'-end formation in an antisense expression vector was replaced with a self-cleaving ribozyme sequence then RNA polymerase II transcripts were produced without poly(A) at their 3'-ends. These molecules were incapable of export from the nucleus (Liu et al., 1994). In this work, a 2.4 kb fragment of the polyoma genome was expressed in the antisense orientation from the strong cytomegalovirus immediate early promoter, and 3'-end formation was by the hammerhead ribozyme sequence shown in Figure 50.1. The ribozyme cassette included a histone stem-loop structure to possibly stabilize cleaved transcripts against 3'–5' exonucleolytic degradation (Eckner et al., 1991). Using such ribozyme technology to retain antisense transcripts in the nucleus, it was demonstrated that only a modest excess of antisense RNA expression over target RNA expression was sufficient for significant gene regulation (Liu et al., 1994). Further, nuclear antisense

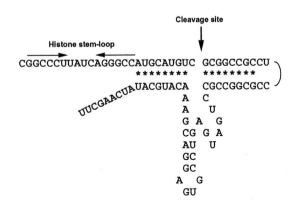

Figure 50.1. The sequence of the cis-cleaving hammerhead ribozyme used for 3'-end formation. The sequence is presented as a folded structure with the site of self cleavage indicated by the vertical arrow. A histone stem-loop structure has been included as a possible stabilizing element against 3' to 5' exonuclease degradation of ribozyme-cleaved RNAs. The sequence shown was inserted into the plasmid pBlueScript in order to provide numerous flanking restriction enzyme cutting sites for convenient excision and recloning into antisense vectors.

RNAs appeared to be more effective inhibitors of target gene expression than comparable polyadenylated antisense molecules, which were transported to the cytoplasm (Liu and Carmichael, 1994). More recent work has demonstrated that nuclear antisense RNA can also act in co-transfection studies to efficiently reduce mRNA levels from an HIV-1 subgenomic expression construct (Y. Huang and G. Carmichael, unpublished).

There are several important considerations for the design of an effective nuclear antisense construct. First, antisense expression should be driven from a promoter appropriate for the level of expression desired and the cells to be targeted. There are of course a very large number of available promoters that can be chosen, including both constitutive and inducible ones.

Second, and very importantly, the antisense RNA that is expressed must be relatively stable within the nucleus. While polyoma nuclear antisense molecules tend to be relatively stable, others do not, and the reason for this is obscure (Z. Liu, X. Li, Y. Huang and G. Carmichael, unpublished results). It is clear, however, that *cis*-elements can contribute greatly to the stability of RNA molecules in the nucleus. It has been observed that splicing signals (Barrett *et al.*, 1995), as well as other elements such as the HIV-1 Rev-responsive element (and Rev protein) and a long stretch of A residues (Y. Huang and G. Carmichael, submitted), can enhance the stability of nuclear RNA molecules. Therefore, it might be wise to include such an element (for example, an intron) into a nuclear antisense vector.

Third, the antisense should be specifically targeted to the nuclear compartment. This would allow not only more effective antisense regulation, but would also prevent the induction of interferon by cytoplasmic dsRNA. Nuclear retention can be achieved using a ribozyme for 3'-end formation, since polyadenylation appears to be a prerequisite for the nuclear export of non-histone mRNAs (Eckner *et al.*, 1991; Y. Huang and G. Carmichael, submitted). It should be stressed, however, that nuclear retention can also be achieved without resorting to ribozyme technology. A large body of evidence accumulated from studies of viral systems including retroviruses and polyoma virus has pointed out the importance of "weak" or inefficient RNA processing signals for nuclear retention of mRNAs. The nuclear retention of the naturally-occurring polyoma virus antisense transcripts results from inefficient polyadenylation, and the nuclear retention of HIV-1 unspliced, genomic transcripts at early times in the HIV-1 life cycle results from a combination of suboptimal splicing signals and the absence of the viral Rev protein (Chang and Sharp, 1989).

Finally, the antisense insert should be of sufficient length to assure effective action and should be targeted against accessible RNA sequences. The minimal length may be on the order of 100 nucleotides, since this is the minimum length of double stranded RNA that serves as an efficient substrate for dsRNA adenosine deaminase (Nishikura *et al.*, 1991). Although it is not yet known whether preferred targets include introns, exons, splicing signals or other processing signals, one should be aware that some regions of target RNAs may be inaccessible for dsRNA hybrid formation due either to secondary structure or to being complexed with specific hnRNP proteins or components of the splicing machinery. For example, antisense that covers splice sites might be less effective because these sites are normally bound by splicing factors, or might be more effective because splicing or RNA transport might be more efficiently inhibited. It is therefore suggested to first attempt antisense experiments using sequences as long as can be reasonably expressed.

References

Adelman, J.P., Bond, C.T., Douglass, J. and Herbert, E. (1987) Two mammalian genes transcribed from opposite strands of the same DNA locus. *Science* **235**, 1514–1517

Barrett, N.L., Li, X. and Carmichael, G.G. (1995) The sequence and context of the 5' splice site govern the nuclear stability of polyyoma virus late RNAs. *Nucleic Acids Res.*, in press

Bass, B.L. and Weintraub, H. (1988) An unwinding activity that covalently modifies its double-stranded RNA substrate. *Cell* **55**, 1089–1098

Bass, B.L. Weintraub, H., Cattaneo, R. and Billeter, M.A. (1989) Biased hypermutation of viral RNA genomes could be due to unwinding/modification of double-stranded RNA. *Cell* **56**, 331

Cattaneo, R. (1994) Biased (A→I) hypermutation of animal RNA virus genomes. *Curr. opin. Genet. Dev.* **4**, 895–900

Cattaneo, R. and Billeter, M.A. (1992) Mutations and A/I hypermutations in measles virus persistent infections. *Curr. Top. Microbiol. Immunol.* 63–74

Chang, D.D. and Sharp, P.A. (1989) Regulation by HIV rev depends upon recognition of splice sites. *Cell* **59**, 789–795

Cornelissen, M. (1989) Nuclear and cytoplasmic sites for anti-sense control. *Nucleic Acids Res.* **17**, 7203–7209

Eckner, R., Ellmeier, W. and Birnstiel, M.L. (1991) Mature messenger RNA 3' end formation stimulates RNA export from the nucleus. *EMBO J.* **10**, 3513–3522

Hamilton, A.J., Lycett, G.W. and Grierson, D. (1990) Antisense gene that inhibits synthesis of the hormone ethylene in transgenic plants. *Nature* **346**, 284–287

Hildebrandt, M. and and Nellen, W. (1992) Differential antisense transcription from the dictyostelium EB4 gene locus – implications on antisense-mediated regulation of messenger RNA stability. *Cell* **69**, 197–204

Hofgen, R., Axelsen, K.B., Kannangara, C.G., Schuttke, I., Pohlenz, H.D., Willmitzer, L., Grimm, B., Vonwettstein, D. (1994) A visible marker for antisense mRNA expression in plants – inhibition of chlorophyll synthesis with a Glutamate-1-semialdehyde aminotransferase antisense gene. *Proc. Natl. Acad. Sci. USA* **91**, 1726–1730

Izant, J.G. and Weintraub, H. (1985) Constitutive and conditional suppression of exogenous and endogenous genes by anti-sense RNA. *Science* **229**, 345–352

Lankenau, S., Corces, V.G. and Lankenau, D.-H. (1994) The *Drosophila micropia* retrotransposon encodes a testis-specific antisense RNA complementary to reverse transcriptase. *Mol. Cell. Biol.* **14**, 1764–1775

Lee, R.C., Feinbaum, R.L. and Ambros, V. (1993) The *C. elegans* heterochronic gene lin-4 encodes small RNAs with antisense complementarity to lin-14. *Cell* **75**, 843–854

Liu, Z. Batt, D.B. and Carmichael, G.G. (1994) Targeted nuclear antisense RNA mimics natural antisense-induced degradation of polyoma virus early RNA. *Proc. Natl. Acad. Sci. USA* **91**, 4258–4262

Liu, Z. and Carmichael, G.G. (1994) Nuclear antisense RNA: An efficient new method to inhibit gene expression. *Mol. Biotechnol.* **2**, 107–118

Luo, G.X., Chao, M., Hsieh, S.Y., Sureau, C., Nishikura, K. and Taylor, J. (1990) A specific base transition occurs on replicating hepatitis delta virus RNA. *J. Virol.* **64**, 1021–1027

Murray, J.A.H. and Crockett, N. (1992) Antisense techniques: an overview. *In Modern Cell Biology*, J.A.H. Murray, ed. (New York: Wiley-Liss, Inc.), pp. 1–49

Nellen, W. and Lichltenstein, C. (1993) What makes an messenger RNA anti-sensitive? *Trends Biochem. Sci.* **18**, 419–423

Nishikura, K., Yoo, C., Kim, U., Murray, J.M., Estes, P.A., Cash, F.E. and Liebhaber, S.A. (1991) Substrate specificity of the dsRNA unwinding/modifying activity. *EMBO J.* **10**, 3523–3532

Simons, R.W. (1988) Naturally occurring antisense RNA control – a brief review. *Gene* **72**, 35–44

Spencer, C.A., Gietz, R.D. and Hodgetts, R.B. (1986) overlapping transcription units in the Dopa decarboxylase region of Drosophila. *Nature* **322**, 279–281

Stockhaus, J., Hofer, M., Renger, G., Westhoff, P., Wydrznski, T. and Willmitzer, L. (1990) Anti-sense RNA efficiently inhibits formation of the 10 kD polypeptide of photosystem II in transgenic potato plants: Analysis of the role of the 10 kD protein. *EMBO J.* **9**, 3013–3021

Thierry, A.R., Lunardi-Iskandar, Y., Bryant, J.L., Rabinovich, P. and Gallo, R.C. (1995) Systemic gene therapy: Biodistribution and long-term expression of a transgene in mice. *Proc. Natl. Acad. Sci. USA* **92**, 9742–9746

Tosic, M., Roach, A., Rivaz, J.-C.D., Dolivo, M., Matthieu, J.-M. (1990) Post-transcriptional events are responsible for low expression of myelin basic protein in myelin deficient mice: role of natural antisense RNA. *EMBO J.* **9**, 401–406

Volk, R., Köster, M., Pöting, A., Hartmann, L. and Knöchel, W. (1989) An antisense transcript from the Xenopus laevis bFGF gene coding for an evolutionarily conserved 24 kd protein. *EMBO J.* **8**, 2983–2988

Wagner, R.W., Smith, J.E., Cooperman, B.S. and Nishikura, K. (1989) A double-stranded RNA unwinding activity introduces structural alterations by means of adenosine to inosine conversions in mammalian cells and Xenopus eggs. *Proc. Natl. Acad. Sci. USA* **86**, 2647–2651

Williams, T. and Fried, M. (1986) A mouse locus at which transcription from both DNA strands produces mRNAs complementary at their 3' ends. *Nature* **322**, 275–279

PART III, SECTION B

51. Controlled Genetic Ablation by Immunotoxin-MediatedCell Targeting

Kazuto Kobayashi[1,2], Ira Pastan[3] and Toshiharu Nagatsu[1,]*

[1] Institute for Comprehensive Medical Science, School of Medicine, Fujita Health University, Toyoake 470-11, Japan
[2] Research and Education Center for Genetic Information, Nara Institute of Science and Technology, Ikoma 630-01, Japan
[3] Laboratory of Molecular Biology, Division Cancer Biology, Diagnosis and Centers, National Cancer Institute, National Institutes of Health, Besthesda, MD 20892, USA

Introduction

Cell ablation techniques, that eliminate a specified cell population with particular identities from a complicated biological system, provide a powerful approach for an understanding of the molecular and cellular mechanisms involved in the development and physiology of a wide range of organisms. For example, ablation of different cell types by laser irradiation in the nematode has been an excellent tool to identify cell lineage relationships during development (Sulston and White, 1980 and Kimble, 1981). In the developing nervous system of the grasshopper and zebrafish, laser ablation of identified neurons and axon bundles has been used to elucidate the role of specific cell–cell recognition during axonal pathway finding (Raper et al., 1984; and Pike and Eisen, 1990). Also, chemical ablation with drugs that are toxic to specific neuronal types has been used to determine the physiological and behavioral roles of a specific neuronal pathway in mammalian brain function (Kostrzewa, 1989).

There is an alternative approach using transgenic animals that ablates a defined set of cell types based on the specificity of gene expression (Evans, 1989). Generally, genetic ablation of specific cell types during development has been induced by the expression of several toxic gene products, including diphtheria toxin and lectin ricin, under the control of a tissue-specific enhancer/promoter (Palmiter et al., 1987; Breitman et al., 1987; Behringer et al., 1988; Landel et al., 1988; Messing et al., 1992; Lowell et al., 1993 and Kalb et al., 1993). Furthermore, several modified methods for controlling toxin expression at desired periods have been developed in flies by generating temperature-sensitive mutations in the toxin gene (Bellen et al., 1992 and Moffat et al., 1992) and by inducing expression of the toxin gene with a flp-mediated site-specific recombination that removes a repressive DNA segment flanked by the target signals to express the downstream gene (Lin et al., 1995). An inducible method for obliterating dividing cells in mice has been to use the herpes simplex virus thymidine kinase (HSV-TK) gene (Borreli et al., 1988; 1989 and Heyman et al., 1989). This ablation is induced by treatment of nucleoside analogs that are converted to toxic compounds by HSV-TK expressed in transgenic mice. However, this method is not applicable for the elimination of postmitotic cells including neurons due to the inhibitory effect of the toxic compounds on DNA replication. Therefore it would be useful to establish a general strategy for controlled genetic ablation with transgenic mice.

Recently, we developed a novel technology termed immunotoxin-mediated cell targeting (IMCT) that can be applied to conditionally disrupt both mitotic and postmitotic cells in transgenic mice (Kobayashi et al., 1995). In this chapter we describe the use of IMCT to eliminate specific neurons from the brain neuronal network, and show that IMCT provides an approach to investigate the physiological and behavioral functions of specific neuronal types and to create experimental models for human neurodegerative disorders.

Experimental Strategy of IMCT

Immunotoxins are conjugates of monoclonal antibodies and toxins that kill animal cells bearing

*Author for Correspondence.

appropriate antigens, and are potential chemotherapeutic agents for the treatment of cancer and autoimmune disorders (Pastan et al., 1986; 1992 and Vitetta et al., 1987). Indeed, administration of these drugs to animals causes a selective loss of the cells expressing the target molecules (e.g. Case et al., 1989 and Lorberboum-Galski et al., 1989). To develop a novel system for controlled genetic ablation, our strategy of IMCT (Figure 51.1) is designed on the basis of the species-specific action of a recombinant immunotoxin, anti-Tac(Fv)-PE40 (Chaudhary et al., 1989 and Batra et al., 1990). Anti-Tac(Fv)-PE40 is composed of the variable heavy and light chains of the anti-Tac antibody, a monoclonal antibody against human interleukin-2 receptor α-subunit (IL-2Rα) fused to PE40, a truncated form of *Pseudomonas* exotoxin (see Figure 51.2). This immunotoxin selectively recognizes human IL-2R but does not cross-react with murine IL-2R. In our strategy, transgenic mice are generated that express human IL-2Rα under the control of a tissue-specific promoter functioning in the target neurons. Subsequently, these animals are treated with an appropriate dose of anti-Tac(Fv)-PE40. The immunotoxin is internalized by the neurons bearing human IL-2Rα, thereby leading to elimination of these neurons through inhibition of protein synthesis due to the exotoxin. Because human IL-2R does not respond to murine IL-2, the expression of the transgene products shows no biological effects on the transgenic animals. Also, the treatment with anti-Tac(Fv)-PE40 is nontoxic to normal animals because of the specificity of this immunotoxin. Therefore, some phenotypic effects can be induced as a consequence of the selective cell loss only in the transgenic mice treated with the immunotoxin.

Application of IMCT for Disrupting Specific Neuronal Types in Brain

Generation of Transgenic Mice Expressing IL-2Rα

In this study we used IMCT to conditionally disrupt the neurons containing dopamine β-hydroxylase (DBH) from the brain neuronal network. DBH is a monooxygenase that catalyzes the production of the neurotransmitter and hormone noradrenaline in the third step of catecholamine biosynthesis pathway, and is localized in noradrenergic and adrenergic neurons in the brain as well as in sympathetic ganglia and adrenomedullary chromaffin cells. To express human IL-2Rα in these cell types, the 4 kb 5′-flanking region of the human DBH gene was used as a tissue-specific promoter (Kobayashi et al., 1992 and Morita et al., 1993), and was fused upstream of the human IL-2Rα gene cassette (Figure 51.3A). Microinjection of the construct into fertilized mouse oocytes generated four independent founder mice that carry 50–100 copies of transgene per mouse genome. From these lines one transgenic line, designated DIL5-1, was selected on the basis of the expression level and pattern of the transgene. In the DIL5-1 line the human IL-2Rα transgene was tissue-specifically expressed in the DBH-containing neurons as well as adrenomedullary chromaffin cells (Figure 51.3B).

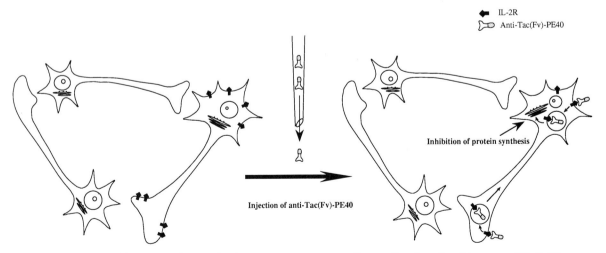

Figure 51.1. Strategy of IMCT. Human IL-2Rα is expressed in specific neuronal types of transgenic mice generated with an appropriate tissue-specific enhancer/promoter. Subsequently, adult transgenic mice are treated with anti-Tac (Fv)-PE40, and the neurons bearing human IL-2Rα are ablated by the cytotoxic activity of the immunotoxin.

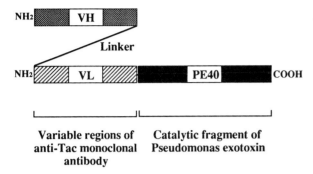

Figure 51.2. Schematic illustration of the structure of anti-Tac(Fv)-PE40. The variable domains of the heavy chain (VH) and light chain (VL) of anti-Tac antibody are connected with an insertion of 15-amino-acid linker, and these domains are fused to PE40, a truncated form of *Pseudomonas* exotoxin (Chaudhary *et al.*, 1989).

Disruption of IL-2Rα -Expressing Cells by Immunotoxin Treatment

To ablate the neurons expressing human IL-2Rα in the brain of the DIL5-1 transgenic mice, intracerebroventricular (i.c.v.) injection of anti-Tac(Fv)-PE40 was carried out with stereotactic technique as illustrated in Figure 51.4. A single i.c.v. injection of 0.2 μg of anti-Tac(Fv)-PE40 per mouse was nontoxic to nontransgenic control mice based on the behavioral and histological observations. In contrast, the immunotoxin-injected transgenic mice developed the behavioral abnormalities characterized by a decrease in locomotion and by ataxic behavior, and they finally died by day 5 after injection. The disruption of specific neurons in the transgenics by immunotoxin injection was ascertained by histological and biochemical studies. Figure 51.5A shows the results of the hematoxylin-eosin staining and immunostaining in the brain sections containing locus coeruleus (A6 neurons) as a representative tissue. Immunotoxin administration to transgenic mice led to a decrease in the number of cell bodies and axon fibers in A6 neurons as well as in the immunoreactivity against tyrosine hydroxylase (TH), the first enzyme of catecholamine biosynthetic pathway. This was accompanied by a significant decrease in DBH activity and noradrenaline level in various brain regions (Figure 51.5B,C). These results indicate that the i.c.v. injection of anti-Tac(Fv)-PE40 into transgenic animals causes a dramatic loss of DBH-containing neurons.

Advantages and Attentions

Our results demonstrate that it is possible to destroy specific neuronal types in transgenic mice engineered to express human IL-2Rα with a recombinant immunotoxin. Ablation of specific cell types can be induced by treatment with anti-Tac(Fv)-PE40 at desired periods. Also, the extent of cell ablation can be controlled by changing the dose of the immunotoxin and the number injections. Because the cytotoxicity of immunotoxins is based on their inhibitory effect on protein synthesis, IMCT will be applicable to other cell types in addition to neurons. This procedure should provide a general approach for controlled genetic ablation of both mitotic and postmitotic cells.

On the other hand, there are several attentions to apply IMCT as a general strategy. First, it is very important to select the transgenic lines reflecting the tissue-specific expression of the cell type of interest. Occasionally, transgene expression is affected by the information around the transgene integration site and by some other factors derived from the promoter and the reporter gene, which leads to expression of the transgene in ectopic sites, where the promoter used does not normally function (Palmiter and Brinster, 1986; and Russo *et al.*, 1988). Ablation of ectopic sites in addition to specific cell types by the immunotoxin treatment may lead to more complex change in the animal phenotype. To optimize the human IL-2Rα expression in the transgenic animals, the knock-in strategy with gene targeting might well be useful that introduces the transgene cassette into the targeted locus of the gene of interest, as used in other studies (Mansour *et al.*, 1990 and Pagano *et al.*, 1995). Second, selection of the injection site and dose of anti-Tac(Fv)-PE40 is a key factor in determining the extent of ablation of the cells and/or the localization of cell loss in a specified area. In our experiment described here the i.c.v. injection of the immunotoxin was carried out into cerebellomedullary cistern. Alternatively, the immunotoxin can be injected into the left or right lateral ventricle. These molecules injected into the ventricles possibly permeate various brain regions through the cerebrospinal fluid, but the cells localized near the injection site seem to be ablated more efficiently (unpublished data). In addition, anti-Tac(Fv)-PE40 can be injected into the specified area in the brain by stereotactic technique. Using this approach it would be possible to ablate specific nuclei or axon bundles expressing the target molecules localized around the injection site. In any case, the dose of the immunotoxin for proper ablation should be determined.

Future Aspects of IMCT

In this chapter we described an application of IMCT for conditional ablation of DBH-containing neurons in the brain. This technique provides an experimental

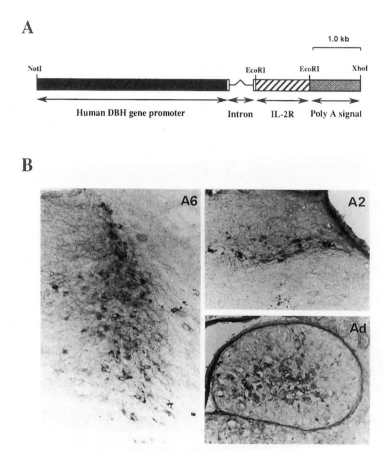

Figure 51.3. Generation of transgenic mice. (A) Structure of the transgene construct. The construct contains the 4 kb human DBH gene promoter, rabbit β-globin second intron, human IL-2Rα cDNA, rabbit β-globin polyadenylation signal, and SV40 early gene polyadenylation signal. (B) Tissue-specific expression of human IL-2Rα in the DIL5-1 transgenic mice. Immunohistochemical staining of tissue sections was carried out with anti-Tac antibody. Light microscopic images of locus coeruleus (A6) and nucleus tractus solitarius (A2) in the adult brain as well as adrenal gland (Ad) of the E16.5 embryos are shown.

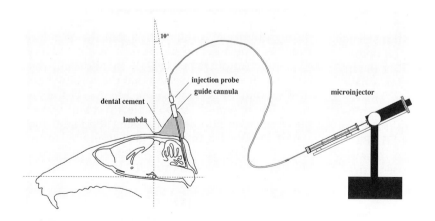

Figure 51.4. Schematic representation of the i.c.v. injection method. Adult mice were anesthetized with sodium pentobarbital, and were stereotaxically implanted with a 22-gauge guide cannula into cerebellomedullary cistern. Coordinates were 3.5 mm posterior to lambda and 3.7 mm ventral to the dorsal surface with an angle of 10 from the frontal plane. The guide cannula was fixed to the skull with dental cement and glue. At day 5–7 after the surgery, anti-Tac(Fv)-PE40 was administered through an injection probe placed inside the cannula at a constant velocity of 0.5 ml/min with a microinjection pump.

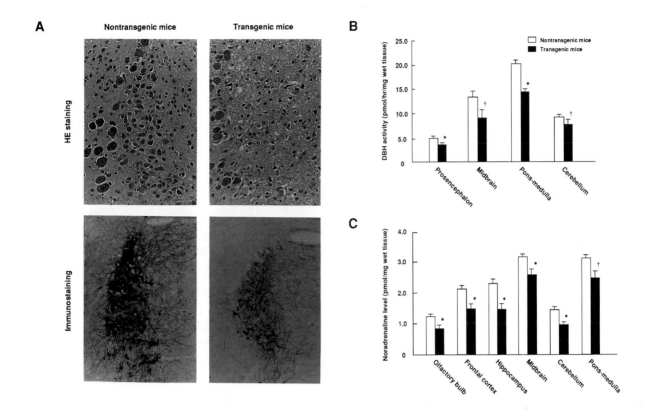

Figure 51.5. Inducible disruption of specific neuronal types by immunotoxin treatment. Nontransgenic and transgenic mice were treated i.c.v. with anti-Tac(Fv)-PE40 and they were sacrificed at day 5 after injection. (A) Histological detection of neural degeneration in the injected transgenic mice. Brain sections were stained with hematoxylin-eosin and were immunostained with anti-TH antibody. Light microscopic images of the sections containing A6 neurons are shown. (B and C) Decrease in DBH activity and noradrenaline level in the injected transgenic mice. Tissues were homogenized and supplied to DBH assay and catecholamine analysis. The values indicate mean SEM of data obtained from 9–11 independent animals. Significant differences from the treated nontransgenics according to Student's t test: *$P < 0.01$, † $P < 0.05$.

tool to elucidate the physiological and behavioral roles of specific neuronal types in the mammalian central nervous system. In contrast, several human brain disorders are known to be caused by degeneration of selective neurons, such as Parkinson, Huntington, and Alzheimer diseases. IMCT makes it possible to create animal models for these diseases by introducing a selective loss of the neurons involved in the pathogenesis. In addition, by devising the method of the immunotoxin treatment, such as intravenous and intrauterine injection, this technique will be applicable to eliminate specific cell types in the peripheral tissues and in the embryos at various developmental stages. IMCT has the potential to investigate a variety of cellular functions in animal physiology and to approach the mechanisms of cell-cell interactions during animal development.

Acknowledgement

We are grateful to H. Sawada for his technical support on histological examination.

References

Batra, J.K., FitzGerald, D., Gately, M., Chaudhary, V.K. and Pastan, I. (1990) Anti-Tac(Fv)-PE40, a single chain antibody *Pseudomonas* fusion protein directed at interleukin 2 receptor bearing cells. *J. Biol. Chem.* **265**, 15198–15202

Behringer, R.R., Mathews, L.S., Palmiter, R.D. and Brinster, R.L. (1988) Dwarf mice produced by genetic ablation of growth hormone-expressing cells. *Genes Dev.* **2**, 453–461

Bellen, H.J., D'Evelyn, D., Harvey, M. and Elledge, S.J. (1992) Isolation of temperature-sensitive diphtheria toxins in yeast and their effects on *Drosophila* cells *Development* **114**, 787–796

Borrelli, E., Heyman, R., Hsi, M. and Evans, R.M. (1988) Targeting of an inducible toxic phenotype in animal cells. *Proc. Natl. Acad. Sci. USA* **85**, 7572–7576

Borrelli, E., Heyman, R.A., Arias, C., Sawchenko, P.E. and Evans, R.M. (1988) Transgenic mice with inducible dwarfism. *Nature* **339**, 538–541

Breitman, M.L., Clapoff, S., Rossant, J., Tsui, L.-C., Glode, L.M., Maxwell, I.H. and Bernstein, A. (1987) Genetic ablation: targeted expression of a toxin gene causes microphthalmia in transgenic mice. *Science* **238**, 1563–1565

Case, J.P., Lorberboum-Galski, H., Lafyatis, R., Fitz Gerald, D., Wilder, R.L. and Pastan, I. (1989) Chimeric cytotoxin IL2-PE40 delays and mitigates adjuvant-induced arthritis in rats. *Proc. Natl. Acad. Sci. USA* **86**, 287–291

Chaudhary, V.K., Queen, C., Junghans, R.P., Waldmann, T.A., FitzGerald, D.J. and Pastan, I. (1989) A recombinant immunotoxin consisting of two antibody variable domains fused to *Pseudomonas* exotoxin. *Nature* **339**, 394–397

Evans, G.A. (1989) Dissecting mouse development with toxigenics. *Genes Dev.* **3**, 259–263

Hanks, M., Wurst, W., Anson-Cartwright, L., Auerbach, A.B. and Joyner, A.L. (1995) Rescue of the *En-1* mutant phenotype by replacement of *En-1* with *En-2*. *Science* **269**, 679–682

Heyman, R.A., Borrelli, E., Lesley, J., Anderson, D., Richman, D.D., Baird, S.M., Hyman, R. and Evans, R.M. (1989) Thymidine kinase obliteration: creation of transgenic mice with controlled immune deficiency. *Proc. Natl. Acad. Sci. USA* **86**, 2698–2702

Kalb, J.M., DiBenedetto, A.J. and Wolfner, M.F. (1993) Probing the function of *Drosophila melanogaster* accessory glands by directed cell ablation. *Proc. Natl. Acad. Sci. USA* **90**, 8093–8097

Kimble, J. (1981) Alterations in cell lineage following laser ablation of cells in the somatic gonad of *Caenorhabditis elegans*. *Dev. Biol.* **87**, 286–300

Kobayashi, K., Sasaoka, T., Morita, S., Nagatsu, I., Iguchi, A., Kurosawa, Y., Fujita, K., Nomura, T., Kimura, M., Katsuki, M. and Nagatsu, T. (1992) Genetic alteration of catecholamine specificity in transgenic mice. *Proc. Natl. Acad. Sci. USA* **89**, 1631–1635

Kobayashi, K., Morita, S., Sawada, H., Mizuguchi, T., Yamada, K., Nagatsu, I., Fujita, K., Kreitman, R.J., Pastan, I. and Nagatsu, T. (1995) Immunotoxin-mediated conditional disruption of specific neurons in transgenic mice. *Proc. Natl. Acad. Sci. USA* **92**, 1132–1136

Kostrzewa, R.M. (1989) Neurotoxins that affect central and peripheral catecholamine neurons. In A.A. Boulton, G.B. Baker and A.V. Juorio, (eds.), *Neuromethods 12. Drugs as Tools in Neurotransmitter Research*, Human Press, Clifton, New Jersey, pp. 1–48

Kuwada, J.Y. (1986) Cell recognition by neuronal growth cones in a simple vertebrate embryo. *Science* **233**, 740–746

Landel, C.P., Zhao, J., Bok, D. and Evans, G.A. (1988) Lens-specific expression of recombinant ricin induces developmental defects in the eyes of transgenic mice. *Genes Dev.* **2**, 1168–1178

Lin, D.M., Auld, V.J. and Goodman, C.S. (1995) Targeted neuronal cell ablation in the *Drosophila* embryo: pathfinding by follower growth cones in the absence of pioneers. *Neuron* **14**, 707–715

Lorberboum-Galski, H., Barrett, L.V., Kirkman, R.L., Ogata, M., Willingham, M.C., FitzGerald, D.J. and Pastan, I. (1989) Cardiac allograft survival in mice treated with IL-2-PE40. *Proc. Natl. Acad. Sci. USA* **86**, 1008–1012

Lowell. B.B., S-Susulic, V., Hamann, A., Lawitts, J.A., Himms-Hagan, J., Boyer, B.B., Kozak, L.P. and Flier, J.S. (1993) Development of obesity in transgenic mice after genetic ablation of brown adipose tissue. *Nature* **366**, 740–742

Mansour, S.L., Thomas, K.R., Deng, C. and Capecchi, M.R. (1990) Introduction of a *lacZ* reporter gene into the mouse *int-2* locus by homologous recombination. *Proc. Natl. Acad. Sci. USA* **87**, 7688–7692

Messing, A., Behringer, R.R., Hammang, J.P., Palmiter, R.D., Brinster, R.L. and Lemke, G. (1992) P_0 promoter directs expression of reporter and toxin genes to Schwann cells of transgenic mice. *Neuron* **8**, 507–520

Moffat, K.G., Gould, J.H., Smith, H.K. and O'Kane, C.J. (1992) Inducible cell ablation in *Drosophila* by cold-sensitive ricin A chain. *Development* **114**, 681–687

Morita, S., Kobayashi, K., Mizuguchi, T., Yamada, K., Nagatsu, I., Titani, K., Fujita, K., Hidaka, H. and Nagatsu, T. (1993) The 5'-flanking region of the human dopamine β-hydroxylase gene promotes neuron subtype-specific gene expression in the central nervous system of transgenic mice. *Mol. Brain Res.* **17**, 239–244

Palmiter, R.D. and Brinster, R.L. (1986) Germ-line transformation of mice. *Annu. Rev. Genet.* **20**, 465–499

Palmiter, R.D., Behringer, R.R., Quaife, C.J., Maxwell, F., Maxwell, I.H. and Brinster, R.L. (1987) Cell lineage ablation in transgenic mice by cell-specific expression of a toxin gene. *Cell* **50**, 435–443

Pastan, I., Willingham, M.C. and FitzGerald, D.J.P. (1986) Immunotoxins. *Cell* **47**, 641–648

Pastan, I., Chaudhary, V. and FitzGerald, D.J. (1992) Recombinant toxins as novel therapeutic agents. *Annu. Rev. Biochem.* **61**, 331–354

Pike, S.H. and Eisen, J.S. (1990) Identified primary motoneurons in embryonic zebrafish select appropriate pathways in the absence of other primary motoneurons. *J. Neurosci.* **10**, 44–49

Vitetta, E.S., Fulton, R.J., May, R.D., Till, M. and Uhr, J.W. (1987) Redesigning nature's poisons to create anti-tumor reagents. *Science* **238**, 1098–1104

Raper, J.A., Bastiani, M.J. and Goodman, C.S. (1984) Pathfinding by neuronal growth cones in grasshopper embryos. IV. The effects of ablating the A and P axons upon the behavior of the G growth cone. *J. Neurosci.* **4**, 2329–2345

Russo, A.F., Crenshaw III, E.B., Lira, S.A., Simmons, D.M., Swanson, L.W. and Rosenfeld, M.G. (1988) Neuronal expression of chimeric genes in transgenic mice. *Neuron* **1**, 311–320

Sulston, J.E. and White, J.G. (1980) Regulation and cell autonomy during postembryonic development of *Caenorhabditis elegans*. *Dev. Biol.* **78**, 577–597

PART III, SECTION B

52. Reporter Genes and Detection of Mutational Activity in Mice

Peter J. Stambrook[1] and Jay A. Tischfield[2]

[1] Department of Cell Biology, Neurobiology and Anatomy, University of Cincinnati, College of Medicine P.O. Box 670521, Cincinnati, OH 45267-0521
[2] Department of Medical Genetics, Indiana University School of Medicine, 975 West Walnut Street, Room IB 130, Indianapolis, IN 46202-5251

Reporter genes have been used as markers of biological activity for almost one-and-a-half centuries. Gregor Mendel took advantage of differences in flower color and seed morphology to describe the segregation of genetic traits in the green pea (Mendel, 1886). Almost 50 years later Thomas Hunt Morgan utilized changes in eye color in Drosophila to mark genetic changes (Morgan, 1910), and subsequently, Barbara McClintock so elegantly deduced the existence of mobile genetic elements by the analysis of differentially colored kernels of Maize (McClintock, 1950; 1965). Most endogenous reporter genes have been used for examination of mutational events and in vivo mutagenesis testing. The long history of coat color genetics served as the foundation for the development of the mouse spot test. In this test, mouse embryos that are mutant at several coat color loci are exposed to a chemical agent in utero and analyzed for the appearance of a colored patch of fur on an otherwise black background (Russell, et al. 1981). Reversion at any one of the loci will result in a colored spot. While the mouse spot test detects mutations within the melanocyte population, the Dlb-1 locus has been used as a marker for mutagenesis in the intestinal epithelium (Winton, et al. 1988 and Tao and Heddle, 1994). Dlb-1 encodes a glycoprotein for the Dolichos biflorus lectin receptor in the small intestine. Mice heterozygous at Dlb-1 can be analyzed by reaction with lectin coupled with horseradish peroxidase. Any stem cell that incurs a mutation in the functional Dlb-1 allele will not bind lectin and will be unstained. Thus, mutation in the functional allele will result in an unstained white ribbon of cells in otherwise stained intestinal villi (Winton et al., 1988; and Tao and Heddle, 1994). Conversely, reverse mutation in mice homozygous for the allele encoding a non-lectin binding form of the receptor produces intestinal villi with a stained strip of cells in an otherwise unstained background (Cosentino and Heddle, 1995). Mice treated with mutagens such as ethylnitrosourea (ENU), produce a predicted mutational dose-response relationship. Clearly, endogenous reporters have been valuable as indicators of mutational activity; however, their utility to date has been limited by the restricted number of tissues in which they serve as useful markers.

The development of technologies allowing genetic manipulation of multicellular organisms has expanded the potential roles of reporter genes. While new strategies for modifying eukaryotic genomes are continually being pursued (Brinster and Zimmermann, 1994; and Brinster and Avarbock, 1994), three approaches have formed the staple of this technology. All have their respective advantages and shortcomings. One of the earliest and most widely used is direct microinjection of foreign DNA into the pronucleus of the fertilized egg (Gordon et al., 1980). The DNA may integrate as a single copy, but most frequently integrates as multiple copies in tandem head-to-tail arrays. Integration appears random, but stable, and the transgenes are transmitted to progeny in Mendelian fashion. However, one complication is that chromosomal sites of integration may cause extensive rearrangement of flanking sequences. (Mark et al., 1985; Covarrubias et al., 1986; Mahon et al., 1988; and Wilkie and Palmiter, 1987). A second approach entails retroviral infection of preimplantation embryos (Jaenisch, 1976; Jahnerand Jaenisch, 1980; Huszar et al., 1985 and Rubenstein et al., 1986).

Although there is efficient retroviral expression in later embryonic stages, neither preimplantation embryos (Huszar et al., 1985; Jaenisch et al., 1975 and Sleigh, 1985) nor embryonal carcinoma (EC) cells (Stewart et al., 1982; Gautsch and Wilson, 1983 and Linney et al., 1984) appear able to support retroviral expression or replication. A characteristic of retroviral infection is that most infected cells sustain a single integration event (Jaenisch, 1988). The third approach, involves the modification of endogenous genes or the introduction of foreign genes into embryonal stem (ES) cells. These cells are derived from preimplantation blastocysts which are surgically removed from the uterus and placed into culture dishes with feeder layers of mitomycin C or UV-inactivated primary embryo fibroblasts (Robertson, 1987). Cells of the blastocyst inner cell mass emerge and proliferate, giving rise to clumps of aggregated ES cells. These cells, which can be disaggregated and genetically manipulated, can participate in the formation of all differentiated cell types following injection into a host blastocyst. The animals born from such composite embryos are chimeras comprised of cells derived from the injected ES cell population and from the parental blastocyst. If the ES cells populate the germ line, genetic traits encoded by the genome of these cells will be transmitted to progeny in Mendelian fashion (Gossler et al., 1986).

The reporter genes in predominant, but not exclusive, use are of prokaryotic origin. Bacterial chloramphenicol acetyltransferase (CAT), a bacterial enzyme absent in higher eukaryotes (Gorman et al., 1982) has been used extensively to localize tissue specific and ectopic expression when the CAT gene is placed under control of a eukaryotic promoter. Enzyme activity in tissues or cells of transgenic animals can be detected and quantitated by the degree to which tissue lysates will acetylate radiolabeled chloramphenicol (Severynse et al., 1995; Rindt et al., 1995 and Knotts et al., 1994). Although a histochemical approach has been developed for *in situ* detection of CAT (Donoghue et al., 1991), this approach has not been extensively exploited. Significantly, dissection of promoter elements that dictate levels and tissue specificity of expression are more accurately defined in transgenic mice than in cultured mammalian cells when using CAT as the reporter (Rindt et al., 1995; and Knotts et al., 1994). Bacterial β-galactosidase, encoded by *Lac Z*, has been used primarily as an *in situ* visual marker in transgenic organisms and provides a measure of transgene expression at the tissue and single cell level. The temporal and spacial distribution of promoter expression has been extensively studied in mouse embryos using fusion genes between a promoter of choice and *LacZ*, and assessing the distribution and intensity of β-galactosidase activity (see Cui et al., 1991 for review); however expression of *Lac Z* in post-natal mice is unreliable and is often extinguished after the mice are born (see Donoghue et al., 1991 for review). Although β-galactosidase activity can be measured in tissue extracts spectrophotometrically (Miller, 1972), its sensitivity is considerably lower than that of CAT. Thus, its utility has been confined primarily to *in situ* visualization of activity.

Several eukaryotic genes have found utility as reporter transgenes in transgenic mice. Expression of firefly luciferase transgenes have been utilized to monitor tissue specific expression directed by a variety of promoters (Crenshaw et al., 1989; and Morrey et al., 1992). The luciferase product can be detected and quantified enzymatically by the capacity of tissue lysates, in the presence of added luciferin and ATP, to produce light that can be measured sensitively with a luminometer. *In situ* localization of expression at the cellular level has been described using immunohistochemistry with anti-luciferase antisera (Harats et al., 1995). Other transgenes, such as mammalian growth hormone genes (Swift et al., 1989 and McGrane et al., 1990), Palmiter et al., 1991 and Ting et al., 1992) and the viral HSV*tk* gene (Palmiter et al., 1982; Al-Shawi et al., 1988 and Al-Shawi et al., 1991) also have provided a measure of transgene expression. More recently, the human placental alkaline phosphatase (PAP) gene has been described as a useful histochemical marker (Berger et al., 1988 and Lin and Culp, 1991). This transgene has several advantages. It is a small gene with 9 exons and 8 introns extending about 4.1 kb in length and encoding a heat-stable enzyme. Thus, tissue sections can be heated to 65°C to inactivate endogenous enzyme activity, but allow histochemical staining of the transgene product. Unlike *LacZ* the human PAP transgene appears to express reliably in all tissues and cell types examined in adult mice. The human PAP structural gene has been placed under control of the same human β-actin promoter as was previously employed for directing expression of *LacZ* (Cui et al., 1991). This construct was subsequently used to produce transgenic mice by pronuclear injection. Tissues from four independent lines were examined by histochemical staining, and in all tissues examined from all four lines expression was ubiquitous (DePrimo et al., in press).

A significant application of reporter genes in transgenic mice has been as monitors of mutagenic activity. There are currently 3 or 4 versions of transgenic mice (Gossen et al., 1989; Kohler et al., 1990; Burkhart et al., 1993 and Gossen et al., 1993), all produced by pronuclear injection, that are designed to detect mutagenic activity *in vivo*. While these models represent the state-of-the-art in current whole animal mutagen testing, they have numerous shortcomings, many of which have been summarized (Mirsalis et al., 1994). All of the current transgenic animal models use bacterial genes as the mutagenesis target (Mirsalis et al., 1994). They include the bacterial lac I and Lac Z genes in bacterio phage, or as a retrievable plasmid (Gossen et al., 1989; Kohler et al., 1990; Rogers et al., 1995 and Gossen et al., 1994), and a Sup F gene in bacteriophage ϕX174 (Burkhart et al., 1993). The former two involve forward mutation (i.e. loss of gene function), whereas the ϕX174 mutation assay involves reverse mutation (i.e. reversion of a mutant gene) and has a lower background mutant frequency (Burkhart et al., 1993 and Mirsalis et al., 1994). In each of the above cases, the prokaryotic target gene is recovered from the DNA of the exposed animal, reintroduced into bacteria, and the mutation frequency is quantitated by the number of plaques or colonies produced or by the number of blue plaques that arise following staining for β-galactosidase activity. The current process of testing is costly, tedious and labor intensive.

The use of bacterial genes as mutagenicity targets and use of multiple targets per genome are also issues of concern. The commercially available transgenic animals have as many as 40 copies of the target gene, each target gene being embedded within a bacterial lambda prophage that is necessary for reporter gene recovery, packaging into λ-phage heads, and infection of E. coli. Bacterial codon usage has a high CpG content, which is a substrate for methylation in a eukaryotic host (Douglas et al., 1994). The bacterial transgenes become highly methylated (Kohler et al., 1990) and are particularly susceptible to deamination of ^{me}C to T. The assay requires multiple steps, rendering reproducibility between laboratories an issue. The bacterial reporter genes lack a eukaryotic promoter and, therefore, are not expressed in the animal. This raises a potential for bias since repair of damaged or adducted DNA that is transcriptionally active is more efficient than that which is not (Lommel et al., 1995). Lastly, DNA adducts that are formed in the host and that normally would be resolved can be converted to mutations when replicated in E. coli, thereby producing an artificially elevated signal. Similarly, clonal proliferation of a cell that had incurred a mutation in a reporter gene would give rise to multiple plaques when, in fact, a single mutagenic event had occurred. That a transgene may respond differently to mutagen exposure than an endogenous reporter is highlighted by a recent report in which mice transgenic for lacI were subjected to daily treatment with ethylnitrosourea (ENU). In addition to being transgenic for lac I embedded within a λ prophage, the mice were heterozygous at Dlb-1^a which encodes a glyco protein receptor for the Dolichos biflorus lectin receptor in the small intestine (Shaver-Walker et al., 1995). Mutation in the functional allele results in an unstained white ribbon of cells in intestinal villi. Following chronic exposure to ENU, there was an accelerated frequency of mutations at Dlb-1^a as a function of time in the small intestine, indicating that daily exposure was producing more mutations at later times than at early times. In contrast, mutation frequency at the lacI transgene remained constant as a function of time.

To circumvent some of these issues and to address the above concerns regarding the use of bacterial transgenes as in vivo reporters of mutagenic activity, we have chosen to use a ubiquitously active endogenous gene at its resident location within the genome as a receptor. The target is the mouse Aprt gene which encodes the enzyme adenine phosphoribosyltransferase, APRT. APRT is a purine salvage enzyme that catalyzes the following reaction:

$$\text{adenine} + \text{PRPP} \xrightarrow{Mg^{++}} \text{AMP} + \text{PP}_i$$

where PRPP is 5 phosphoribosyl-1-pyrophosphate. This strategy entails producing a series of APRT deficient mice that have mutant Aprt alleles with point mutations that produce an inactive product. These genes can be rendered fully functional by reverse mutation (Schaff et al., 1990). The model is the equivalent of a whole animal Ames test in which every cell in a tester animal will be APRT deficient but revertible. In principle, every cell is a data point, reducing the number of animals required for analysis. As a first step towards this end, we have produced APRT deficient mice by targeted homologous recombination in ES cells, and have demonstrated that these mice are viable and fertile (Engle et al., 1996 and Engle et al., submitted).

The premise upon which the detection strategy is based derives from the observation that cells with APRT activity are able to sequester and accumulate exogenously administered tagged adenine and incorporate it into nucleic acids, whereas APRT negative cells cannot. The basis for this differential activity is that adenine administered to APRT$^-$ cells can get into and out of cells without impediment. In contrast, cells with APRT activity will ribophosphorylate adenine to AMP and thereby trap this charged molecule within the cell. The AMP is then converted to its di- and triphosphate derivatives, which are incorporated into RNA and DNA. The difference in behavior of APRT$^-$ and APRT$^+$ cells in response to exogenously applied adenine is presented schematically in Figure 52.1.

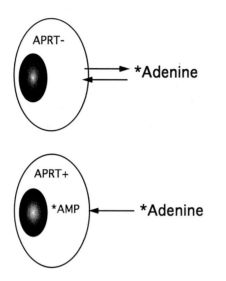

Figure 52.1. Schematic depiction showing that APRT deficient cells (upper) do not accumulate exogenously administered tagged adenine whereas cells with APRT activity (bottom) do (courtesy, Serinus Biotechnology, Inc.)

The animal model exploits the above salient characteristic of APRT$^+$ cells. If one were to inject tagged adenine into a mouse that carries mutant but revertible *Aprt* alleles, only those cells which will have become *Aprt*$^+$ by virtue of reverse mutation, and the descendants of these cells, will become marked by the accumulation of the tagged adenine. Thus, if one were to inject tagged adenine into such an APRT deficient mouse, any cell that undergoes reverse mutation would become marked by accumulation of tagged AMP and its incorporation into nucleic acids. The reversion events can be followed in several ways, which are depicted schematically in Figure 52.2. Animals injected with ^{14}C-adenine are allowed 48 hours, or more, in which to clear unincorporated adenine. At that time, individual organs are removed, fixed, sectioned and prepared for autoradiography. Any cell that has undergone reverse mutation at the *Aprt* target will become labeled. One advantage of this approach is that the exact type of cell that has incurred the mutation can be identified, information that cannot be obtained by current methodologies. We have tested the

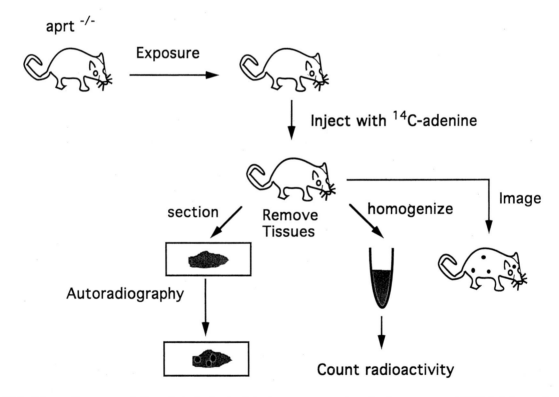

Figure 52.2. Schematic representation of strategies to detect reverse mutations in the *aprt gene*. APRT deficient mice with revertible *aprt* genes are exposed to a compound or an environment. Following varying exposure times, the animals are injected with ^{14}C-adenine. After 48 hours to allow for clearance of unincorporated labeled adenine, one of three strategies can, in principle, be followed. (1) remove tissues, section, coat slides with photographic emulsion, identify revertant cells by autoradiography; (2) remove tissues, homogenize and solubilize and count aliquots in scintillation counter; (3) inject with tagged adenine (e.g. fluorinated adenine), wait 48 hours and image (courtesy, Serinus Biotechnology, Inc.).

principle of this approach and shown that the methodology is sufficiently sensitive for detecting APRT⁺ cells within a background of APRT⁻ cells. A 1:50 mixture of *Aprt*⁺ and *Aprt*⁻ ES cells were injected subcutaneously into syngeneic 129Sv Aprt⁻ host mice, and tumors were allowed to form. Once tumors reached a size greater than 1 cm, the mice were injected with ^{14}C-adenine, and after 48 hours the tumors were removed, fixed, sectioned and subjected to autoradiography. Only a fraction of tumor cells were labeled with silver grains, whereas none of the host tissues became labeled. A more rapid analysis than that above, but less informative with respect to affected cell type, entails removal of individual organs, solubilization and scintillation counting to determine the relative increase in incorporated radioactivity due to reverse mutation. Lastly, a desirable alternative will be to use a modified adenine analog (e.g., a fluorinated analog) that will serve as an imaging contrast agent to mark the mutated cells in such a way that they can be detected non-invasively by imaging techniques. In that way it should be possible to follow the fate of marked cells or cell populations as a function of time.

References

Al-Shawi, R., Burke, J., Jones, C.T., Simons, J.P. and Bishop, J.O. (1988) A *mup* promoter-thymidine kinase reporter gene shows relaxed tissue-specific expression and confers male sterility upon transgenic mice. *Mol. Cell. Biol.* **8**, 4821–4828

Al-Shawi, R., Burke, J., Wallace, H., Jones, C., Harrison, S., Buxton, D., Maley, S., Chandley, A. and Bishop, J.O. (1991) The herpes simplex virus type 1 thymidine kinase is expressed in the testes of transgenic mice under the control of a cryptic promoter. *Mol. Cell. Biol.* **11**, 4207–4216

Berger, J., Hauber, J., Hauber, R., Geiger, R. and Cullen, B.R. (1988) Secreted placental alkaline phosphatase: a powerful new quantitative indicator of gene expression in eukaryotic cells. *Gene* **66**, 1–10

Brinster, R.L. and Zimmermann, J.W. (1994) Spermatogenesis following male germ-cell transplantation. *Proc. Natl. Acad. Sci. USA* **91** (24), 11298–112302

Brinster, R.L. and Avarbock, M.R. (1994) Germline transmission of donor haplotype following spermatogonial transplantation. *Proc. Natl. Acad. Sci. USA* **91** (24), 11303–11307

Burkhart, J.G., Burkhart, B.A. and Sampson, K.S. (1993) ENU-induced mutagenesis at a single A:T base pair in transgenic mice containing φX174. *Mutation Research* **292**, 69–81

Cosentino, L. and Heddle, J.A. (1995) The induction of dominant somatic mutations at the *Dlb*-1 locus. *Mutation Research* **346**(2), 115–119

Covarrubias, L., Nishida, Y. and Mintz, B. (1986) Early postimplantation embryo lethality due to DNA rearrangements in a transgenic mouse strain. *Proc. Natl. Acad. Sci. USA* **83**, 6020–6024

Crenshaw, E.B., Kalla, K., Simmons, D.M., Swanson, L.W. and Rosenfeld, M.G. (1989) Cell-specific expression of the prolactin gene in transgenic mice is controlled by synergistic interactions between promoter and enhancer elements. *Genes Dev.* **3**, 959–972

Cui, C., Wani, M., Wight, D., Kopchick, J. and Stambrook, P.J. (1994) Reporter genes in transgenic mice. *Transgenic Res.* **3**, 182–194

Deprimo, S.E., Stambrook, P.J. and Stringer, J.R., Human placental alkaline phosphatase as a histochemical marker of gene expression in transgenic mice. *Transgenic Res.* (in press)

Donoghue, M.J., Alvarez, J.D., Merlie, J.P. and Sanes, J.R (1991) Fiber type- and position-dependent expression of a myosin light chain-CAT transgene detected with a novel histochemical stain for CAT. *J. Cell Biol.* **115**, 423–434

Douglas, G.R., Gingerich, J.D., Gossen, J.A. and Bartlett, S.A. (1994) Sequence spectra of spontaneous *Lac Z* gene mutations in transgenic mouse somatic and germline tissues. *Mutagenesis* **9**, 451–458

Engle, S.J., Stockelman, M.G., Chen, J., Boivin, G., Moo-Nahm, Y., Davies, P.M., Ying, M.Y., Sahota, A., Simmonds, H.A., Stambrook, P.J. and Tischfield, J.A. (1996) A mouse model of heriditary nephrolithiasis. *Proc. Natl. Acad. Sci. USA* **93**, 5307–5312

Engle, S.J., Womer, D.E., Davies, P.M., Boivin, G., Sahota, A., Simmonds, H. Anne, Stambrook, P.J. and Tischfield, J.A. HPRT-APRT deficient mice are not as model for Lesch-Nyhan syndrome. (in press)

Gautsch, J.W. and Wilson, M.C. (1983) Delayed *de novo* methylation in teratocarcinoma suggests additional tissue-specific mechanisms for controlling gene expression. *Nature* **301**, 32–36

Gordon, J.W., Scangos, G.A., Plotkin, D.J., Barbosa, J.A. and Ruddle, F.H. (1980) Genetic transformation of mouse embryos by microinjection of purified DNA. *Proc. Natl. Acad. Sci. USA* **77**, 7380–7384

Gorman, C.M., Moffat, L.F. and Howard, B.H. (1982) Recombinant genomes which express chloramphenicol acetyltransferase in mammalian cells. *Mol. Cell. Biol.* **2**, 1944–1951

Gossen, J.A., DeLeeuw, W.J.F., Tan, C.H.T., Zwarthoff, E.C., Berends, F., Lohman, P.H.M., Knook, D.L. and Vijg, J. (1989)Efficient rescueof integrated shuttle vectors from transgenic mice: A model for studying mutations *in vivo*. *Proc. Natl. Acad. Sci. USA* **86**, 7971–7975

Gossen, J.A., deLeeuw, W.J.F., Molijn, A.C. and Vijg, J. (1993) Plasmid rescue from transgenic mouse DNA using LacI repressor protein conjugated to magnetic beads. *BioTechniques* **14**, 624–629

Gossen, J.A., de Leeuw, W.J. and Vijg, J. (1994) LacZ transgenic mouse models: their application in genetic toxicology. *Mutation Research* **307**, 451–459

Gossler, A., Doetschman, T., Korn, R. Serfling, E. and Kemler, R. (1986) Transgenesis by means of blastocyst-derived embryonic stem cell lines. *Proc. Natl. Acad. Sci. USA* **83**, 9065–9069

Harats, D., Kurihara, H., Belloni, P., Oakley, H., Ziober, A., Ackley, D., Cain, G., Kurihara, Y., Lawn, R. and

Sigal, E. (1995) Targeting gene expression to the vascular wall in transgenic mice using the murine preproendothelin-1 promoter. *J. Clin. Invest.* **95**, 1335–1344

Huszar, D., Balling, R., Kothary, R., Magli, M.C., Hozumi, N., Rossant, J. and Bernstein, A. (1985) Insertion of a bacterial gene into the mouse germ line using an infectious retrovirus vector. *Proc. Natl. Acad. Sci. USA* **82**, 8587–8591

Jaenisch, R. (1976) Germ line integration and mendelian transmission of the exogenous Moloney leukemia virus. *Proc. Natl. Acad. Sci. USA* **73**, 1260–1264

Jaenisch, R., Fan, F. and Croker, B. (1975) Infection of preimplantation mouse embryos and of newborn mice with leukemia virus: tissue distribution of viral DNA and RNA and leukemogenesis in the adult animal. *Proc. Natl. Acad. Sci. USA* **72**, 4008–4012

Jaenisch, R. (1988) Transgenic animals. *Science* **240**, 1468–1474

Jahner, D. and Jaenisch, R. (1980) Integration of Moloney leukaemia virus into the germ line of mice: correlation between site of integration and virus activation. *Nature* **287**, 456–458

Knotts, S., Rindt, H., Neumann, J., Robbins, J. (1994) *In vivo* regulation of the mouse beta myosin heavy chain gene. *J. Biol. Chem.* **269**, 31275–31282

Kohler, S., Provost, S., Kretz, P., Dycaico, M., Sorge, J. and Short, J.M. (1990) Development of a short-term, *in vivo*, mutagenesis assay: the effects of methylation on the recovery of a lambda phage shuttle vector from transgenic mice. *Nucleic Acids Res.* **18**, 3007

Lin, W.C. and Culp, L.A. (1991) Selectable plasmid vectors with alternative and ultrasensitive histochemical marker genes. *BioTechniques* **11**, 344–351

Linney, E., Davis, B., Overhauser, J., Chao, E. and Fan, H. (1984) Non-function of a Maloney murine leukaemia virus regulatory sequence in F9 embryonal carcinoma cells. *Nature* **308**, 470–472

Lommel, L., Carswell-Crumpton, C. and Hanawalt, P.C. (1995) Preferential repair of the transcribed DNA strand in the dihydrofolate reductase gene throughout the cell cycle in UV-irradiated human cells. *Mutation Research* **336**, 181–192

Mahon, K.A., Overbeek, P.A. and Westphal, H. (1988) Prenatal lethality in a transgenic mouse line is the result of a chromosomal translocation. *Proc. Natl. Acad. Sci. USA* **85**, 1165–1170

Mark, W., Signorelli, K. and Lacy, E. (1985) An insertional mutation in a transgenic mouse line results in developmental arrest at day 5 of gestation. *Cold Spring Harbor Symp. Quant. Biol.* **50**, 453–458

McClintock, B. (1950) The origin and behavior of mutable loci in maize. *Proc. Natl Acad. Sci. USA* **36**, 344–355

McClintock, B. (1965) The control of gene action in maize. *Brookhaven Symp. Biol.* **18**, 162–184

McGrane, M.M., Yun, J.S., Moorman, A.F., Lamers, W.H., Hendrick, G.K., Arafah, B.M., Park, E.A., Wagner, T.E. and Hanson, R.W. (1990) Metabolic effects of developmental, tissue, and cell specific expression of a chimeric phosphoenolpyruvate carboxykinase (GTP)/bovine growth hormone gene in transgenic mice. *J. Biol. Chem.* **265**, 22371–22379

Mendel, G. (1866) Versuche über pflangen-hybriden. *Verh. Naturf. Verein Brünn* **4**, 3–47. Translation printed in: Peters, J.A. ed. (1959) *Classic Papers in Genetics.* Prentice-Hall, Englewood Cliffs, NJ

Miller, J.H. (1972) Assay of β-galactosidase. In Miller, J.H. ed., *Experiments in Molecular Genetics*, pp. 352–355. Cold Coldspring Harbor Laboratory Press Spring Harbor, NY

Mirsalis, J.C., Monforte, J.A., and Winegar, R.A. (1994) Transgenic animal models for measuring mutations *in vivo. Critical Reviews in Toxicology* **24**, 255–280

Morgan, T.H. (1910) Sex linked inheritance in *Drosphila. Science* **32**, 120–122

Morrey, J.D., Bourn, S.M., Bunch, T.D., Sidwell, R.W. and Rosen, C.A. (1992) HIV-1 LTR activation model: evaluation of various agents in skin of transgenic mice. *J. Acquir. Immune Defic. Syndr.* **5**, 1195–1203

Palmiter, R.D., Sandgren, E.P., Avarbock, M.R., Allen, D.D. and Brinster, R.L. (1991) Heterologous introns can enhance expression of transgenes in mice. *Proc. Natl. Acad. Sci. USA* **88**, 478–482

Palmiter, R.D., Chen, H.Y. and Brinster, R.L. (1982) Differential regulation of metallothionein–thymidine kinase fusion genes in transgenic mice and their offspring. *Cell* **29**, 701–710

Rindt, H., Knotts, S. and Robbins, J. (1995) Segregation of cardiac and skeletal muscle-specific regulatory elements of the beta-myosin heavy chain gene. *Proc. Natl. Acad. Sci. USA* **92**, 1540–1544

Robertson, E.J. (1987) in: *Teratocarcinomas and Embryonic Stem Cells, A Practical Approach,* ed. E.J. Robertson, IRL Press, Oxford, Washington, DC, pp. 77–112

Rogers, B.J., Provost, G.S., Young, R.R., Putman, S.L. and Short, J.M. (1995) Intralaboratory optimization of mutant screening conditions used for a lambda/*lac I* transgenic mouse assay (I). *Mutation Research* **327**, 57–66

Rubenstein, J.L.R., Nicolas, J.F. and Jacob, F. (1986) Introduction of genes into preimplantation mouse embryos by use of a defective recombinant retrovirus. *Proc. Natl. Acad. Sci. USA* **83**, 366–368

Russell, L.B., Selby, P.B., von Halle, E., Sheridan, W. and Valcovic, L. (1981) Use of the mouse spot test in chemical mutagenesis: interpretation of past data and recommendations for future work. *Mutational Res.* **86**, 355–379

Schaff, D.A., Jarrett, R., Dlouhy, S., Ponniah, S., Stockelman, M., Stambrook, P.J., and Tischfield, J.A. (1990) Mouse transgenomes in human cells detect specific base substitutions. *Proc. Natl. Acad. Sci. USA* **87**, 8675–8679

Severynse, D.M., Colapietro, A.M., Box, T.L. and Caron, M.G. (1995) The human D1A dopamine receptor gene promoter directs expression of a reporter gene to the central nervous system in transgenic mice. *Molecular Brain Research* **30**, 336–346

Shaver-Walker, P.M., Urlando, C., Tao, K.S., Zhang, X.B. and Heddle, J.A., (1995) Enhanced somatic mutation rates induced in stem cells of mice by low chronic exposures to ethylnitrosourea. *Proc. Natl. Acad. Sci. USA* **92**, 11470–11474

Sleigh, M.J. (1985) Virus expression as a probe of regulatory events in early mouse embryogenesis. *Trends in Genet.* **1**, 17–21

Stewart, C.L., Stuhlmann, H., Jahner, D., and Jaenisch, R. (1982) *De novo* methylation, expression, and infectivity of retroviral genomes introduced into embryonal carcinoma cells. *Proc. Natl. Acad. Sci. USA* **79**, 4098–4102

Swift, G.H., Kruse, F., MacDonald, R.J. and Hammer, R.E. (1989) Differential requirements for cell-specific elastase I enhancer domains in transfected cells and transgenic mice. *Genes Dev.* **3**, 687–696

Tao, K.S. and Heddle, J.A. (1994) The accumulation and persistence of somatic mutations *in vivo*. *Mutagenesis* **9** (3), 187–191

Ting, C.N., Rosenberg, M.P., Snow, C.M., Samuelson, L.C. and Meisler, M.H. (1992) Endogenous retroviral sequences are required for tissue specific expression of a human salivary amylase gene. *Genes Dev.* **6**, 1457–1465

Wilkie, T.M. and Palmiter, R.D. (1987) Analysis of the integrant in MyK-103 transgenic mice in which males fail to transmit the integrant. *Molec. Cell Biol.* **7**, 1646–1655

Winton, D.J., Blount, M.A. and Ponder, B.A.J. (1988) A clonal marker induced by mutation in mouse intestinal epithelium. *Nature* **333**, 443–446

PART IV, SECTION A

53. Gene Activity in the Preimplantation Mouse Embryo

Sylvie Forlani and Jean-François Nicolas*

Unité de Biologie moléculaire du Développement, URA 1947 du Centre National de la Recherche Scientifique, Institut Pasteur 25, rue du Dr Roux, 75724 Paris Cedex 15, France

In this review, we describe the functioning of the nuclear material in the mouse from the stage of oocyte maturation to the 2-cell stage. During the past few years, it has been possible to define, using transgenes as a tool, modifications of gene activity in relation to other biochemical events. The part due to cytoplasmic events and the part due to nuclear events have been partially defined. It turns out that the transmission to the zygotic nuclei of information of maternal origin stored in the oocyte and activated after fertilization is crucial at these early stages while nuclear activity is stopped. However, numerous aspects of this period of transition from maternal to embryonic control remain to be described and understood. For instance, the time when the information of maternal origin is relayed to the zygotic nuclei remains to be determined as does the basis of the major change in the pattern of protein synthesis observed in the 2-cell embryo. Future studies based on a more intensive use of transgenes should allow a more comprehensive description of the dynamics of gene activity on which the transition from a gametic to an embryonic state is dependent.

The Gametes

The fertilized egg results from the fusion of the two germ cells, the oocyte and the spermatozoa, respectively the female and male gametes. These differentiated haploid cells, the products of several steps of division and maturation, have their own pattern of gene expression and their own epigenetic state. The oocyte is, during its growth, transcriptionally very active (see below). It stores RNA and maternal proteins on which the egg will develop, at least until it can produce its own proteins. In contrast, the transcriptionally inactive sperm cell is currently considered to provide no cytoplasmic contribution to the egg but mainly proteins involved in sperm entry and, as the oocyte, its genome.

The Oocyte

The ovaries of a new born female are composed of several thousand tetraploid oocytes, which are arrested in prophase of the first meiotic division (cells named oocytes I). They are the products of numerous cell divisions during embryogenesis. The growth of these cells initiates three days after birth and takes 17 days to be achieved. As a consequence, there is a cytoplasmic increase of RNA, proteins and organelles (Schultz, 1986). RNA molecules are mainly ribosomal (70% of the total synthesis) and polyadenylated RNA represents 10% of the RNA stock. The stock of proteins in the oocyte is mainly composed of structural proteins such as actin and tubulin, housekeeping enzymes used for metabolic pathways and oocyte-specific components such as the glycoproteins of the *zona pellucida* (ZP1, ZP2, ZP3). Histones represent 10% of the nuclear proteins of the oocyte (Schultz, 1986).

Growing oocytes display a transcriptional specificity, as shown by two major reports using transient expression of DNA constructs microinjected into the germinal vesicle (Dooley et al., 1989 and Bonnerot et al., 1991).

(1) LacZ reporter genes controlled by different promoters have been microinjected (1000 to 3000 copies of religated inserts) into oocytes of 13–14 day

*Author for Correspondence.

old females and analysed by X-gal staining 20 hours after injection (Bonnerot et al., 1991). There are no particular restrictions for the utilization of promoter sequences in these cells: promoters constituted of GC boxes alone (HPRT) or combined with a TATA box (SV40) are recognized as are promoters composed of a TATA box, a CAAT box and enhancer sequences like the promoter of the H2-K^b gene (which codes for a transplantation antigen of the K locus of the H2 complex). In contrast, specialized promoters are inactive: for instance, the promoter from the acetylcholine α-subunit receptor (AchR) gene is silent, demonstrating no ectopic expression of tissue-specific genes in the oocyte. The growing oocyte is not permissive for the Moloney murine leukemia virus (M-MuLV), a virus which is tightly negatively controlled in multipotential cells in culture. In oocytes, repression is likely to be mediated by a silencer element near the LTR (Bonnerot et al., 1991). (2) Plasmids combining the luciferase reporter gene with the viral promoter of the thymidine kinase gene (tk) have been microinjected (50 000 to 100 000 copies) into oocytes of 16–18 day old females (Dooley et al., 1989). The quantification of luminescence was carried out 24 hours after injection. As above, it was shown that a promoter composed of a TATA and a CAAT box and two binding sites for the Sp1 activator (the *tk* promoter) is active, although weakly (Table 53.1). In addition, the action of enhancers was studied. F101 is a polyoma virus mutant enhancer, permissive for the activation of the viral genome in F9 embryonal carcinoma (EC) cells (Linney and Donerly, 1983). Its action on the activity of the *tk* promoter in oocytes was tested. There is no detectable activation effect (Table 53.1) (Dooley et al., 1989 and Majumder et al., 1993). This result can be compared to the activity of the SV40 early promoter which consists of a TATA box, six tandemly arranged GC boxes for Sp1 binding and an enhancer region of 200 bp (located at −101 to −290). In order to make this promoter active in oocytes, three GC boxes are needed but the enhancer, which is necessary for expression in differentiated cells, is not required (Chalifour et al., 1987 and Bonnerot et al., 1991). Therefore, promoters seem unable to be activated at distance by viral enhancers in oocytes. This result suggests either that, for an unknown reason (the absence of a coactivator, for example) the enhancers cannot act on the promoter activity in the gamete or that the transactivators are missing at that stage. (3) Finally, the adenovirus early promoter EIIA, a promoter which was also assayed for activity in cells in culture, is active only when transactivated by the E1A viral protein, a putative indirect and sequence-independent activator (Murthy et al., 1985). In oocytes, EIIA is 9 fold more active than the *tk* promoter. It suggests that a E1A-like activity is present in these cells (Dooley et al., 1989). A similar result is obtained when EC cells are tested (Imperiale et al., 1984).

The results of these studies characterized a pattern of promoter utilization in the female gamete (Tables 53.1 and 53.2). It is interesting to note that exactly the same pattern was found for the fertilized egg (see below and Table 53.1), and EC cells (Table 53.2). Possible implications will be discussed later.

The transcriptional activity of the oocyte decreases as it reaches the end of the growth period. It ceases totally when the gamete enters the maturation phase, indicative of the resumption of meiosis (Wassarman and Letourneau, 1976 and De Leon et al., 1983). It has been shown recently that the activity of the SV40 Sp1 dependent promoter ceased in the fully grown oocyte (Worrad et al., 1994). It is not clearly established whether this is related to the concomitant decrease of the nuclear concentration of the transcriptional factors Sp1 and TBP (TATA box binding protein) at that stage (Worrad et al., 1994) or due to a more general inhibitory effect on transcription.

The maturation of the fully grown oocyte, which corresponds to the first meiotic reductive division, is under the control of two pituitary hormones (FSH and LH) and the maturation promoting factor (MPF) and involves some phosphorylation/dephosphorylation modification of proteins (Bornslaeger et al., 1986). Maturation, which occurs during a 16 hour period, causes the germinal vesicle breakdown (GVBD), the completion of the first meiotic division with the expulsion of the first polar body and a second arrest of the diploid oocyte (oocyte II) in metaphase of the second meiotic division (Schultz, 1986). The completion of meiosis in oocytes will be caused only by fertilization or by artificial activation of the gamete, leading to parthenogenesis.

The transcriptional machinery of the oocyte has no activity during maturation. Its reactivation will occur at a certain point in the development of the fertilized egg (see below). Translation of stored mRNA is still active at that stage, as shown by biochemical studies (Petzoldt et al., 1980 and Schultz, 1986). The mature oocyte can translate synthetic mRNA. With the maturation is also initiated a degradation of maternal mRNA synthetized during the growth of the oocyte (Bachvarova et al., 1985 and Bachvarova et al., 1989). This degradation will continue after fertilization.

The Sperm Cell

Different from what has been described for the oocyte, male germ cells multiply during the whole sexual life of the animal. These cells enter meiosis after birth and the mature male gamete has

Table 53.1. Utilization of Promoter and Enhancer for Expression of Microinjected DNA in Oocyte, 1-cell and 2-cell Embryos.

	Oocyte	Pronuclei of 1-cell embryos		2-cell Aphi	developping 2-cell
		♂ Aphi	♀ Aphi		
1- [TATA]	nd	++	nd	+	nd
2- [Sp1]-[CAAT]-[Sp1]-[TATA]	+/−	+++	+	+	+
3- [(Sp1)6 or (Gal4)5]-30 bp-[TATA]	nd	activated	nd	activated	nd
4- [(MyoD)2]-100bp-[CAAT]-[Sp1]-[TATA]	nd	activated	nd	nd	activated
5- [F101]-600 bp-[TATA]	nd	not activated	nd	activated	nd
6- [F101]-600 bp-[Sp1]-[CAAT]-[Sp1]-[TATA]	not activated	not activated	not activated	activated	activated

Experiments were conducted to evaluate the activity of simple promoters (1,2), their capacity to be transactivated when in close association with target sequences for transactivators (3,4) and the effect on their activity of the association at distance with a viral enhancer (5,6). (1) The adenovirus major late promoter's TATA box (2) The herpes virus thymidine kinase promoter (tk) (3) The TATA box of (1) associated to 6 Sp1 or 5 Gal4 binding sites (4) The acetylcholine α-subunit receptor (AchR) which contains 2 MyoD binding sites (5) The TATA box of (1) associated to the polyoma virus mutant enhancer F101 and (6) The tk promoter (2) associated to F101. These sequences combined to a reporter gene were microinjected in the germinal vesicle of growing oocytes, in the male or female pronucleus of aphidicolin (Aphi) treated 1-cell embryos or in one nucleus of aphidicolin treated (2-cell aphi) or developping 2-cell embryos (see text for detailed experimental procedures). The distance in bp is between the sequence located on the left and the transcription start site. The + signs (from +/− to +++) are qualitative indication of the promoter's activity (see text for details). The mention "activated" or "not activated" shows the effect on the promoter's activity of the sequences located upstream. To evaluate this effect, the reference is the activity observed for the promoters without F101 or (Sp1)6 (for 5–6 and 3, respectively) or in the absence of the exogenous transactivators Gal4-VP16 or MyoD (for 3 and 4, respectively) (TATA: TATA box; CAAT: CAAT box; Sp1: binding site for the transcription factor Sp1; nd: not determined) (from Majumder et al., 1993 and Bonnerot et al., 1991)

completed meiosis. Gene activity in the male germ line is poorly described. At the beginning of spermiogenesis, there is a maximal synthesis of mRNA and proteins as actin, tubulin and structural constituants of the head and the tail (Hecht, 1986). No study of transcriptional specificity has been described for the male gamete. The most striking aspect of this cell is that it is composed throughout its differentiation, of an overdimensioned nucleus containing several nucleoli, and few cytoplasm. This situation is extreme in the sperm cell, as there is almost no cytoplasm left. No RNA or protein synthesis is detectable at that stage.

It seems that the male genome is almost the only contribution of the spermatozoa to the fertilized egg. Two main aspects of the male DNA should be noted: first, it is overmethylated as compared to the female genome in the oocyte. This was verified for repeated sequences (Sanford et al., 1987) as well as for specific endogenous genes (Kafri et al., 1992; but see also Del Mazo et al., 1994). Second, sperm DNA is associated with specific basic nuclear proteins named P1 and P2 protamines. They are related to histones and their association with DNA generates a high condensed chromatin structure (Balhorn, 1982; and Nonchev and Tsanev, 1990). This structure could be involved in transcriptional silencing of the male gamete.

The Fertilized Egg

Fertilization causes the completion of the second meiotic division in the oocyte and creates the situation where the two parental genomes, isolated in two pronuclei, are in the same cytoplasm of maternal origin. This situation persists throughout the first cell cycle. In the mouse, this cycle is exclusively controlled by maternal information (mRNA, translation machinery and proteins). This conclusion was drawn from experiments where fertilized eggs were treated with α-amanitin, an inhibitor of the RNA polymerase II which blocked

Table 53.2. Transcriptional Specificity in Oocyte, 2-cell Embryo and EC cells.

Promoter	Oocyte	2-cell embryo	EC cells
HPRT	+	+	+
H2K^b	+	+	−
tk	+	+	+
SV40	+	+	+
EIIA	+	+	+
F101 py T Ag	+	+	+
M-MuLV	−	−	−
AchR	−	−	−

The promoters are described in the text. They were combined to a reporter gene and microinjected in growing oocytes or developing 2-cell embryos or transfected in EC cells. (+) and (−) indicate whether the promoter is expressed or not. Results obtained in EC cells are described in Dooley et al. (1989); Bonnerot et al. (1991) and Mélin et al. (1993).

all transcriptional activity. Neither the first cell cycle nor the pattern of protein synthesis are affected by this treatment (Flach et al., 1982 and Howlett and Bolton, 1985). Nevertheless, there are changes in the pattern of proteins in 1-cell embryos. Those changes are due to (1) post-translational modifications of maternal proteins (Howlett and Bolton, 1985); (2) translation of maternal messengers (Petzoldt et al., 1980); (3) degradation of maternal RNA, initiated at the time of the maturation in the oocyte, and still in progression in the 1-cell stage embryos (Bachvarova and DeLeon, 1980); (4) decrease in synthesis of 50 % of the proteins studied in the 1-cell stage (Latham et al., 1991).

The Male Pronucleus

Chromatin Modifications and DNA Replication. Soon after its entry in the cytoplasm of the oocyte, the male genome undergoes important chromatin changes. The chronology of these events has been established in *in vitro* fertilized eggs, as this treatment reduces the developmental asynchrony observed among *in vivo* fertilized eggs (Bolton et al., 1984). Protamines disappear from the male genome between 4 to 8 hours post-fertilization, as described by immuno-fluorescence analysis using antibodies against protamines and histones. They are replaced by maternal histones between 4 and 12 hours post-fertilization (Nonchev and Tsanev, 1990). Concomitantly with the disappearance of protamines, the male genome is surrounded by a nuclear membrane to form the male pronucleus (Howlett and Bolton, 1985). The chromatin structure is modified from a highly condensed one to a more relaxed nucleosomal structure.

This replacement of protamines by histones corresponds to a period of maximal decondensation for the male genome. It was suggested that this is necessary for the DNA to replicate (Nonchev and Tsanev, 1990). In fact, DNA replication in the male pronucleus begins between 10 and 12 hours after fertilization, after the replacement is completed (Figure 53.1). A densitometric quantification of DNA in *in vitro* fertilized eggs has established that DNA replication takes 7 hours (Howlett and Bolton, 1985). A slight accumulation of histones H4 acetylated on the lysine residues 5–8 and/or 12 at the periphery of the male pronucleus has been described in S/G2 and G2/M of the 1-cell stage (Worrad et al., 1995). Moreover, an immuno-fluorescence analysis has shown an increase of the nuclear concentration of maternal transcription factors Sp1 and TBP in the male pronucleus 6 hours (G1 phase) to 12 hours (G2 phase) after it was formed (Worrad et al., 1995).

Transcriptional Specificity in Arrested 1-cell Embryos. The study was conducted in S phase arrested embryos from *in vivo* fertilized eggs. In such embryos, the pronuclei remain physically separated. This can be achieved by cultivating embryos in medium supplemented with aphidicolin, an inhibitor of DNA polymerase α which prevents DNA replication and consequently cell cleavage (Ikegami et al., 1978). DePamphilis and colleagues use the following experimental procedure: plasmids composed of the luciferase reporter gene driven by different promoters are microinjected in the male pronucleus of arrested 1-cell embryos at 20–22 hphCG [(hphCG: hours post injection of human chorionic gonadotrophin. This hormone, which causes superovulation, is an analog of luteinizing hormone (LH). The time of injection serves as reference for the timing of the embryonic development (Figure 53.1)]. The plasmids are used at a concentration (200 ng/μl, which corresponds to 50 000–100 000 copies per injection) chosen to generate maximal luciferase activity. Luminescence is quantified in individual embryos 44 hours after injection. This corresponds to the time which is necessary for a developing embryo to reach the morula stage.

Several viral promoters composed of different elements were tested (Table 53.1): the adenovirus major late promoter's TATA box alone (Majumder et al., 1993), the polyoma virus T antigen promoter (Py TAg) composed of a TATA box and a CAAT box (Wiekowski et al., 1991 and Majumder et al., 1993), the herpes virus thymidine kinase promoter

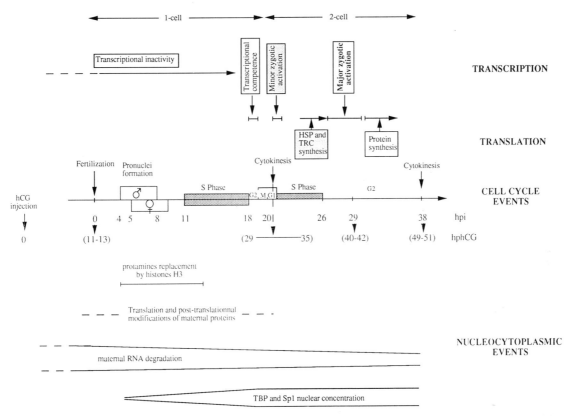

Figure 53.1. Timing of major Events during the first and second Cell Cycles following Fertilization in the Mouse. Schematic representation of some transcriptional, translational and nucleocytoplasmic events during the first and second cell cycles after fertilization in the mouse embryo. The phases of the cell cycle (G1, DNA replication (S phase), G2, cytokinesis) are also indicated. The time is in hours post insemination (hpi) for *in vitro* fertilized eggs. Below is given the correspondence in hours post injection of the human chorionic goadotrophin hormone (hphCG), for *in vivo* fertilized eggs. For *in vivo* fertilized eggs, the developmental asynchrony is more extended than for *in vitro* fertilized eggs, as shown for instance by the delay for the first cell cleavage (29–35 hphCG).

(*tk*) which, in addition to these two boxes, contains two Sp1 binding sites (Martinez-Salas *et al.*, 1989; Wiekowski *et al.*, 1991 and Majumder *et al.*, 1993). In aphidicolin treated 1-cell embryos, all these promoters were active in the male pronucleus. Each promoter displays a different but generally high level of activity. The activity of *tk* in the male pronucleus is 10-fold higher than that seen in oocytes. An analysis based on linker-scanning mutations which selectively inactivate each of the *tk* transcriptional elements, indicated that the two Sp1 binding sites and the CAAT box, to a lesser extent, were essential for the maximal activity of the *tk* promoter in the male pronucleus (Majumder *et al.*, 1993). Therefore, there is no special sequence requirements for gene expression in the male pronucleus, as a simple TATA box or a more complex promoter like *tk* show a similar level of activity. These results are consistent with the increasing pronuclear concentration of Sp1 and TBP factors at this stage (see above).

The activity of the EIIA adenovirus early promoter was also tested. The EIIA promoter displays a strong activity, 20-fold higher than that seen in oocytes, suggesting the presence in the male pronucleus of an E1A-like activity corresponding to an indirect and sequence-independent transactivation (Dooley *et al.*, 1989).

The capacity of promoters to be transactivated in the male pronucleus was studied by analyzing the activity generated by microinjected plasmids composed of a weak promoter associated with a reporter gene and target sequences for a DNA binding activator. The activity of the adenovirus major late promoter's TATA box associated to six Sp1 binding sites is 100 fold higher than the activity of the TATA box alone (Table 53.1). The same TATA box associated with five binding sites for the yeast transcription factor GAL4 and to the CAT reporter gene is activated only when co-injected with an expression vector for the fusion protein GAL4-VP16 (Majumder *et al.*, 1993). Another example is the acetylcholine α-subunit receptor (AchR) promoter which, when associated with the LacZ reporter gene and microinjected into the male pronucleus (1000 to 2000 copies of inserts), displays no detectable activity (Table 53.2) (Bonnerot et al., 1991). This promoter is activated only when provided with the MyoD factor, which binds two sites at 85 and 100 bp from the initiation start site (Table 53.1). These data demonstrate that transactivation of weak promoters by transactivating factors can occur in the male pronucleus of S phase arrested 1-cell embryos.

In contrast, no change of activity was reported for promoters associated at distance (600 bp upstream) with an enhancer sequence. This was verified for the tk promoter preceeded by 9 GAL4 binding sites when the GAL4-VP16 activator was present (but, unfortunately, the effect of these GAL4 sites in close association with the promoter was not assayed; therefore, it is not known whether this promoter is not already at its maximal activity) (Majumder et al., 1993). This was also verified for the F101 enhancer (described above) which has no effect on the activity of the tk or py T-Ag promoter (Table 53.1) (Wiekowski et al., 1991) nor on the activity of the adenovirus TATA box alone or closely associated with six Sp1 binding sites. A similar result was previously described for the tk promoter in the oocyte (see above). The presence and functionality of the F101 transactivator at these stages remain to be tested.

The Female Pronucleus

The female pronucleus is formed later than the male pronucleus, between 5 and 9.5 hours post *in vitro* fertilization (Howlett and Bolton, 1985) (Figure 53.1). Immunostaining against H3 histones strongly increases between 4 and 16 hours post fertilization (Nonchev and Tsanev, 1990). DNA replication occurs concomitantly with that in the male pronucleus. It begins between 10 and 12 hours post fertilization and lasts for 7 hours (Howlett and Bolton, 1985). The accumulation of acetylated H4, histones at the nuclear periphery described before in the male pronucleus is more pronounced in the female pronucleus at the transitions S/G2 and G2/M (Worrad et al., 1995). Sp1 and TBP transcription factors are less concentrated (50% and 30% less respectively) in the female pronucleus than in the male pronucleus (Worrad et al., 1994).

The transcriptional specificity of the female pronucleus has been poorly analysed. In aphidicolin treated 1-cell embryos (see above for experimental conditions), the viral tk promoter is activated but the activity is 5-fold decreased, as compared to that in the male pronucleus (Wiekowski et al., 1993). Moreover, the association of tk with the F101 enhancer has no stimulatory effect, as previously described for the oocytes and the male pronucleus (Table 53.1).

Transcriptional Competence before the Formation of the Zygotic Nucleus

Timing. The pronuclei in the 1-cell embryo derive from the two gametic nuclei which were transcriptionally inactive. Most of the experiments described above show that the two haploid pronuclei are competent for transcription. However, the embryos used in these experiments were in general artificially S-phase arrested at the 1-cell stage. They are morphological 1-cell embryos containing two separated pronuclei, but are assayed for expression at a time when a non-treated embryo reaches the 8-cell stage.

To time precisely the emergence of transcriptional capacity in the fertilized egg, the male pronucleus of developing 1-cell embryos was microinjected and screened for expression after variable delays following injection. Following the pronuclear injection of 1000 to 2000 copies of HPRT-LacZ inserts, Vernet et al. (1992) have demonstrated that activity begins at 18–19 hpi (hpi: hours post *in vitro* insemination), in G_2 of the first cell cycle (Figure 53.1). In addition, injection at different times followed by variable delays between injection and analysis of expression showed that the promoter's activity always initiated at the same time (18–19 hpi) (Vernet et al., 1992). Before that time, all the promoters assayed for expression (HPRT, SV40 and promoters of ubiquitously expressed genes β-actin and HMG) were inactive. Therefore, competence for transcription occurs at a specific time after fertilization (Vernet et al., 1992). It is probably a transcriptional regulation rather than a translational regulation since microinjected LacZ mRNA is translated in the 1-cell embryo as early as 12 hpi (Vernet et al., 1992). The mRNA maturation and translation machinery, of maternal origin, is effective at that stage.

Similar results were obtained with *in vivo* fertilized eggs after microinjection of 50 000 to 100 000 copies of SV40-luciferase plasmids in the male pronucleus (Ram and Schultz, 1993). In these experiments, the difficulty due to the developmental asynchrony inherent to *in vivo* fertilized eggs was diminished by injecting embryos synchronized according to their pronuclei formation and timed in relation to the development of non-injected control embryos. The SV40 early promoter was found to be active in injected embryos when the control embryos were about to cleave. On the contrary, no activity of the promoter was observed when embryos were microinjected following the same protocol in the female pronucleus (Ram and Schultz, 1993).

The emergence of transcriptional competence was also studied by transplanting early 2-cell stage diploid nuclei (normally competent for transcription but transcriptionally inactivated after an α-amanitine treatment) into 1-cell embryos. In this case, the expression of the endogenous gene of the TRC (transcription requiring complex) was followed (see below) (Latham et al., 1992). The expression of the TRC was detected in late 1-cell (30 hphCG) transplanted embryos but not in early 1-cell (24 hphCG) embryos.

These results can be compared with more direct biochemical analysis. Synthesis of poly (A)+ and poly (A) RNA during the 1-cell stage were previouly described (Clegg and Piko, 1982; 1983). Recently, expression of the Mhc class I genes (Major histocompatibility complex class I) of both paternal and maternal origin was observed by RT-PCR in the l-cell embryos (Sprinks et al., 1993). Moreover, the incorporation of microinjected BrUTP radionucleotides in nascent RNA transcripts of unknown nature was reported from 26 hphCG (Bouniol et al., l995). This incorporation was first observed in the male pronucleus then in both pronuclei. This timing is earlier than the one observed for the acquisition of the transcriptional competence (Vernet et al., 1992). It could be explained by the fact that the incorporation of BrUTP is the direct visualisation of transcription whereas a transgene expression is an indirect detection of transcription.

Taken together these data indicate that: (1) transcriptional competence (for plasmids and inserts) is acquired in the male pronucleus at the end of the G_2 phase of the first cell cycle (2) a 1-cell embryo of the same age is competent for the transcription of endogenous genes provided by a 2-cell nucleus; (3) endogenous transcriptional activity is already observed in late 1-cell embryos. The nature and function of this activity remains to be determined. These data suggest also that the acquisition of the transcriptional competence is directed by the 1-cell cytoplasm and, therefore, is under maternal control.

Regulation. Studies described above have shown that S phase arrested 1-cell embryos are competent for transcription of microinjected DNA sequences. These results indicate that the acquisition of the transcriptional competence in the 1-cell stage is independent of DNA replication and, therefore, also of a two-fold increase in DNA quantity, that is of the nucleo-cytoplasmic ratio. However, there is a shift in the time of acquisition of transcriptional competence in aphidicolin treated embryos. This latter is reported to be acquired at 40 hphCG which corresponds to a mid 2-cell embryos (Figure 53.1) (Martinez-Salas et al., 1989 and Wiekowski et al., 1991), rather than during G_2 of the first cell cycle (Vernet et al., 1992 and Ram and Schultz, 1993). For an unknown reason, the treatment by aphidicolin slows the acquisition of transcriptional competence. If this view is correct, it shows that this essential developmental event can be modulated.

It has also been observed that the female pronucleus was competent for transcription from the tk promoter in arrested 1-cell embryos (Wiekowski et al., 1993) but that the activity of the SV40 early promoter was undetectable in the same pronucleus of a developing l-cell embryo (Ram and Schultz, 1993). Two interpretations are possible. First, the delay between injection and analysis of expression could be too short for the developing embryo (10 hours) as compared to the arrested embryo (44 hours). It should be remembered that the activity in the female pronucleus of an S phase arrested 1-cell embryo is already five fold decreased as compared with the male pronucleus. Second, transcriptional competence could be acquired later in the female pronucleus than in the male pronucleus, an hypothesis which contrasts with the results of Sprinks et al. (1993).

The 2-Cell Embryo

In the mouse, the zygotic nucleus is formed when the two parental genomes are grouped in a unique nucleus after the first cell division. The G_1 phase of the 2-cell stage is very short (1 to 1.5 hours) and the S phase lasts approximately 5–6 hours (Figure 53.1) (Bolton et al., 1984). During the G_2 phase, a strong increase of the perinuclear localization of acetylated H4 histones is observed in the zygotic nuclei (Worrad et al., 1995). The nuclear concentration of Sp1 and TBP basal transcription factors is high, nearly the same as in the late 1-cell stage embryos (Figure 53.1) (Worrad et al., 1994). As observed for the acetylated H4 histones, a perinuclear accumulation of the RNA polymerase II is observed during the G_2 phase of 2-cell embryo (Worrad et al., 1995).

The Activation of the Zygotic Genome (ZGA)

The timing of ZGA was established in relation to sensitivity to α-amanitin (an inhibitor of RNA polymerase II) of the pattern of embryonic protein synthesis. This pattern can be characterized by autoradiography after labelling of the embryo with ^{35}S methionine and migration of the protein content of 20 to 50 embryos on mono or bidimensional gels. Such analysis showed that the first embryonic cell cycle is exclusively dependent on maternal products (see above). The timing of the ZGA was precisely described in in vitro fertilized eggs synchronized by pick-off after the first cell cleveage (Bolton et al., 1984). Two α-amanitin sensitive periods were localized in the 2-cell stage: the first was observed during the short G_1 phase (18–21 hpi), the second followed immediately the end of the DNA replication (26–29 hpi) (Figure 53.1) (Flach et al., 1982 and Bolton et al., 1984). These periods of zygotic transcription were immediately coupled to the translation of newly synthetized transcripts as the corresponding changes in the pattern of protein synthesis were detected at 23–26 hpi and 29–32 hpi, respectively (Figure 53.1) (Flach et al., 1982). The first synthesis of paternal proteins was also detected in 2-cell embryos (Sawicki et al., 1981). This was also considered as a landmark of the ZGA, as the sperm cell does not provide the oocyte with a cytoplasmic contribution.

The mouse is one of the mammals in which the earliest zygotic genome activation has been described. It occurs later at the 4–8 or 8–16-cell stages in numerous other mammals (rabbit, sheep, bovin, human) (reviewed in Telford et al., 1990). Such an early zygotic genome activation in the mouse may be related to the early degradation of maternal RNAs (Bachvarova et al., 1985 and Paynton et al., 1988).

The Minor Activation. The first phase of the zygotic genome activation corresponds to the synthesis of 73, 70 and 68 Kd proteins (Flach et al., 1982 and Conover et al., 1991). This phase is sometimes called the "minor activation" (Figure 53.1) (Vernet et al., 1992 and Schultz 1993). Only two proteins among this group have been clearly identified, the heat shock proteins HSP68 and HSP70 (Bensaude et al., 1983). The others, some nuclear associated proteins, are known as the TRC (for transcription requiring complex) (Conover et al., 1991). They are present in a high amount compared to the HSP70 (respectively 4 to 10% and 0.1 – 0.15% of the total proteins at the 2-cell stage (Conover et al., 1991 and Chastant et al., 1995). These proteins are specific to the 2-cell stage and are undetectable in the 4-cell embryo (Conover et al., 1991).

The heat shock proteins are produced in cells in response to different cellular stresses (Ananthan et al., 1986 and references therein). Therefore, the question is whether the synthesis of these proteins is induced in the 2-cell stage by the experimental procedure or occurs also in vivo. In fact, α-amanitin treament followed by analysis of the pattern of protein synthesis requires the in vitro culture of the embryos. These conditions may favour the synthesis of stress-induced proteins (Vernet et al., 1993). This question has been approached with transgenic mice containing an integrated reporter gene driven by the hsp70.1 (Christians et al., 1995) or hsp68 (Bevilacqua et al., 1995) promoter. In the hsp70.1-luciferase mice, the expression of the transgene was studied in non-cultured embryos. In these conditions, a weak activity of the hsp70.1 promoter is detected in transgenic 2-cell embryos (Christians et al., 1995). The expression was repressed at the end of the 2-cell stage. Therefore, there is probably an endogenous and non-induced zygotic expression of the hsp70.1 promoter at the beginning of the 2-cell stage. This activity was increased 5-fold in cultured embryos. The question remains to be solved for the hsp68 promoter since in the hsp68-LacZ mice, the expression was only analysed in embryos cultured from the early 1-cell stage. In these in vitro conditions, a weak activity of the hsp68-LacZ gene was observed (Bevilacqua et al., 1995).

The Major Activation. The second phase of the activation of the zygotic genome is followed by a major change in the pattern of transcriptional and protein synthesis (Flach et al., 1982 and Bolton et al., 1984). This phase called the "major activation" appears to be necessary for the embryo to pass the first cleavage and develop beyond the 2-cell stage.

The mRNAs contained in the G2 phase of the 2-cell stage differ markedly from those present in the oocyte: approximatively 50% of these 2-cell stage transcripts are absent from the maternal stock (Taylor and Piko, 1987 and Rothstein et al., 1992). Latham et al. (1991) have reported a quantitative analysis of 18 bidimensional protein gels obtained from embryos recovered at intervals of three hours over a period of 57 hours, from the early 1-cell to the end of the 4-cell stage. They have established that most of the changes in the pattern of protein synthesis can be attributed to (1) the repression of synthesis of proteins produced in the 1-cell stage, which reflects the degradation of the corresponding maternal mRNAs. This is the case for 40% of the studied proteins; (2) the induction of a transient synthesis at the 2-cell stage (for 6% of the proteins) or a continued synthesis from the 2-cell to the 4-cell stage (for 37%) – these proteins are most likely the products of newly transcribed genes; (3) the transient repression of synthesis of some proteins during the 2-cell stage which could result from the degradation of the maternal messengers followed by their replacement by the newly synthesized zygotic messengers (Latham et al., 1991).

Therefore, the major activation of the zygotic genome causes an extensive change in the pattern of protein synthesis as soon as the zygotic nucleus is generated. However, there are huge blanks in the identification of these proteins and of the corresponding genes. The presence of the tropomyosin 5 (TM5) and of the heat shock protein HSP73 among the newly synthesized proteins at the 2-cell stage was reported (Latham et al., 1991). The RT-PCR method gives a direct access to the genes expressed even at a low level and has been successfully applied to early embryos. A zygotic expression of the genes which code for the growth factor IGFII and its receptor IGFIIr has been described by RT-PCR in the 2-cell embryo (Rappolee et al., 1992). However, the zygotic nature of this expression is to be confirmed as the expression of the Igf2 and Igf2r genes has also been detected by the same method in oocytes (Latham et al., 1994). The zygotic expression of some genes may occur after the 2-cell stage as shown by the increase of the amount of certain PCR products between the 2-cell and the 8-cell stages (Zimmermann and Schultz, 1994). For example, a recent study demonstrates the production of the CB1-R mRNA, the brain form of the cannabinoid receptor, in the embryo from the 4-cell stage (Paria et al., 1995).

Control of Zygotic Genome Activation. Regulations which control the minor activation have been studied extensively (reviewed in Schultz, 1993). Latham et al. (1992) have shown that an early 2-cell stage nucleus transcriptionally inhibited by α-amanitin treatment can express the TRC genes, landmarks of the minor activation, after transfer into an enucleated late 1-cell embryo.

Therefore, as this expression in a normally developing embryo is detected only after the first cell cleavage (Conover et al., 1991), at least two steps should be required for the occurence of the minor activation: the acquisition of the transcriptional competence at the 1-cell stage and an additionnal delay involving events which are still to be characterized. Such an additionnal delay may be not required for the expression of certain endogenous genes at the late 1-cell stage (Sprinks et al., 1993 and Bouniol et al., 1995). The biological signification of this early expression remains to be determined as does the extent to which it can be considered to be an integral part of the minor activation.

The treatment of early 2-cell embryos by inhibitors of the cAMP-dependent protein kinase, like N-(2-(methylamino)ethyl)-5-isoquinoline-sulfonamide (H8), which has a reversible action, prevents TRC synthesis (Poueymirou and Schultz, 1989). Therefore the expression of the TRC genes depends on a phosphorylation activity at the beginning of the 2-cell stage. This is a specific requirement as the inhibition of another protein kinase, PKC, has no effect on TRC expression (Poueymirou and Schultz, 1989). This regulation is likely to be at the transcriptional level since the synthesis of the TRC is inhibited in H8-treated 1-cell embryos transferred into a medium lacking H8 and containing α-amanitin. It is not known whether this transcriptional requirement is specific for the TRC genes or also concerns other genes, that is, is more general. It may be distinct from the acquisition of transcriptional competence which is believed to occur in G2/M of the first cell cycle (Ram and Schultz, 1993). The fact that H8 treatment also prevents the expression of a plasmid containing the hsp70 promoter (Schwartz and Schultz, 1992) does not definitively answer the question of the general versus specific requirement as this control could still be specific to the genes of the minor activation. The test of various promoters by microinjection would clarify this point.

The production of HSP68 and HSP70 proteins is independent of DNA replication and of cell cleavage during the first cell cycle (Howlett, 1986 and Christians et al., 1995). Consequently, the formation of a zygotic nucleus is not a prerequesite to either the progress of the minor activation or the acquisition of the transcriptional competence. This has also been demonstrated by following the expression of HSP68 in uniparental embryos with two maternal (parthenogenotes) or paternal (androgenotes) genomes (Barra and Renard, 1988).

Similarly, the synthesis of proteins resulting from the major activation at the 2-cell stage is independent of the formation of a zygotic nucleus (Petzoldt et al., 1981). This synthesis occurs in an uniparental 1-cell stage embryo which does not cleave as a result of cytochalasin B treatment. Cytochalasin B inhibits cell division without affecting DNA replication or the nuclear development (Petzoldt et al., 1981 and Bolton et al., 1984). Therefore, as previously described for the minor activation, the major activation is independent of the cell cleavage. However, and in contrast, the major activation necessitates the completion of DNA replication during the first cell cycle (Howlett, 1986). The DNA replication which immediately precedes the time of the major activation at the 2-cell stage is not required (Bolton et al., 1984). The origin of this requirement is still to be elucidated. It has been proposed that DNA replication could be necessary for changes of chromatin structure required for gene expression (Wolffe, 1991).

A perinuclear accumulation of acetylated H4 histones and of the RNA polymerase II at the 2-cell stage has been recently described (Worrad et al., 1995). Acetylated histones are the sign of a relaxed chromatin structure and their presence is often correlated to a transcriptional activity (Turner, 1991 and Lee et al., 1993). Moreover, this particular location of the acetylated H4 histones in the zygotic nuclei is specifically observed at the 2-cell stage and is no longer observed after the inhibition of the first DNA replication (Worrad et al., 1995). These data suggest that changes of chromatin structure could be involved in the major activation of the zygotic genome in the mouse. Such implications have been demonstrated before in Xenopus (Prioleau et al., 1994 and Almouzni and Wolffe, 1995).

Transcriptional Specificity of the Zygotic Nucleus

The 2-cell embryo is the developmental stage of the mouse at which expression of the embryonic genome is detected. This could have allowed the study of zygotic transcriptional specificity on the basis of the expression of endogenous genes. However, this approach is not possible due to the major lack of information about the genes involved. This study was therefore conducted as described previously in the oocyte and the fertilized egg, using microinjection of exogenous DNA constructs in the nucleus of 2-cell embryos, either treated or not with aphidicolin.

S Phase Arrested 2-Cell Embryo. In a first series of experiments, the nucleus of one blastomere was microinjected with 50 000 to 100 000 copies of plasmids containing the luciferase gene driven by different promoters. Expression was analysed 44 hours after injection. The promoters previously tested in the pronuclei of the 1-cell stage (TATA box alone, py T-Ag, tk) (see above and Table 53.1) were all active in the nucleus of 2-cell embryos but the level of activity was decreased by 10 to 100 fold. The relative difference of activity between the promoters were exactly the same as those observed in the male pronucleus (Majumder et al., 1993). The adenovirus early promoter EIIA was also tested in 2-cell embryos exposed to aphidicolin during their G2 phase. The embryos reached therefore the 4-cell stage and were S phase arrested at that stage. The activity of the EIIA promoter was 2-fold less than in the male pronucleus (Wiekowski et al., 1991).

As observed in the male pronucleus, weak promoters (for instance, a TATA box) in close association with a target sequence for a transactivator were transactivated by endogenous (Sp1) or provided (GAL4-VP16) activators at the 2-cell stage (Table 53.1). The level of activation was similar in both stages (Majumder et al., 1993). The transactivation machinery appeared to act in a similar way in 1-cell and 2-cell embryos.

In contrast, a particular property of the zygotic nucleus was revealed by the fact that enhancers associated at a distance with the promoters increased their low activity to the level of activity observed for the same promoters without enhancers in the male pronucleus of arrested 1-cell embryos (Majumder et al., 1993). This was verified for every tested promoter, whether the enhancer was the target of an endogenous (for F101) (Table 53.1) or exogenous (for GAL4) transactivator.

The effect of several polyoma virus mutant enhancers selected for their ability to infect different lines of embryonic stem cells (ES) (enhancers py(ES)) was tested. In these experiments, they are placed 130 bp downstream of the polyoma T-Ag promoter. 50 000 to 100 000 copies of luciferase plasmids were injected in the nucleus of aphidicolin arrested 2-cell embryos (Mélin et al., 1993). These polyoma mutant enhancer sequences increased the activity of the py T-Ag promoter by a factor which was similar to that observed with the F101 enhancer. Similarly to F101, they showed no detectable effect on the activity of promoters in arrested 1-cell embryos. The most efficient mutant enhancers in 2-cell embryos were those selected in F9 EC cells, as was F101, and in ES cells. The sequences selected in PCC4 or LT1 EC cells showed a very low activation effect. The most efficient enhancers shared a similar configuration characterized by a tandem duplication of a sequence containing a point mutation (Fujimura et al., 1981 and Mélin et al., 1993). This mutation creates a site different by only a single base pair from the GT-IIC site in the SV40 enhancer which binds the human transcription enhancer factor 1 (TEF-1) (Davidson et al., 1988). A murine TEF-1, 95% homologous to the human factor, has recently been cloned (Blatt and DePamphilis, 1993). The question is now to determine whether this factor is a transactivator in the 2-cell embryo.

Developping 2-Cell Embryo. In the case of 2-cell embryos which developed during 44 hours after the microinjection (without aphidicolin treatment), the characteristics of expression were similar to those described in S phase arrested 2-cell embryos: the activity of promoters was low (for tk, py T-Ag, EIIA) and the association at a distance with an enhancer (F101 for tk and py (ES) sequences for py T-Ag) increased this activity (Table 53.1) (Wiekowski et al., 1991 and Mélin et al., 1993). Expression exclusively during the 2-cell stage of developping embryos was not examined.

Another study was conducted in 2-cell embryos after microinjection of 1000 to 3000 copies of inserts containing the LacZ reporter gene driven by different promoters. Injected 2-cell embryos were cultured 20 hours until they reached the 4-cell stage before X-gal staining. The pattern of utilization of promoters was the same as the one previously reported for the oocyte (Bonnerot et al., 1991) (Table 53.2). It was characterized by the utilization of GC boxes promoters (HPRT), containing a TATA box (SV40) or a TATA and a CAAT box with two additionnal enhancer sequences (H-2K[b]), the repression of the viral MoMu-LV sequences and the non utilization of the tissue-specific AchR promoter.

To summarize, two characteristics distinguish the expression of transgenes microinjected in the pronuclei of 1-cell embryos and the zygotic nuclei of 2-cell embryos: (1) the levels of activity of promoters alone (2) the ability of promoters to use enhancer sequences associated at distance to increase levels of expression. However, it should be noted, once again, that the origin of the inability of the 1-cell embryo to use these enhancer sequences is still unknown. It remains to be determined if this inability results from the absence of the corresponding transactivators rather than of a more general inhibition. In contrast, weak promoters are stimulated by closely associated enhancers both in the pronuclei and the zygotic nuclei (Table 53.1).

In an attempt to clarify the origin of those differences, transplantation experiments of microinjected nuclei between aphidicolin treated 1-cell and 2-cell stages were conducted. A male pronucleus was microinjected with plasmids in which the luciferase reporter gene was controlled by the py T-Ag promotor alone or preceded by the F101 enhancer and then transferred into a non-enucleated 2-cell embryo. In these conditions, the py promoter in the transferred pronucleus presented the characteristics of activity of the promoters microinjected in the nucleus of a 2-cell stage: the activity indicated a twofold decrease and was increased by the association, 130 bp upstream, with the F101 enhancer (Henery et al., 1995). In contrast, a plasmid microinjected in a 2-cell stage nucleus transferred into a 1-cell embryo did not follow the characteristics of activity observed in the male pronucleus of a 1-cell embryo. For instance, the level of activity did not reach that observed in a male pronucleus (Henery et al., 1995). All these results were confirmed with 2-cell recipients in G_2 phase. Therefore, it seems that the DNA injected in a 2-cell stage nucleus is subjected to modifications which are not reversed by the exposition to a 1-cell stage cytoplasm. However, as the association with the F101 enhancer does not influence the promoter activity in this transfered 2-cell nucleus, it appears that the arrested 1-cell embryo is unable to use the F101 enhancer, whether the DNA is localized either in a transferred 2-cell stage nucleus or in the male pronucleus. This result suggests that the ability or not to use the F101 enhancer is related to the cytoplasm into which the injected nucleus is placed.

From Maternal to Zygotic Control of Development: The Transition Period

This short review illustrates that in the mouse, the 1-cell stage and perhaps part of the 2-cell stage correspond to a transition period between maternal control and nuclear zygotic control of embryogenesis. The oocyte period is characterized by the formation of a highly specialized cell. It is followed by the acquisition of an embryonic state which will be radically different. The analysis of the pattern of protein synthesis reveals that the nuclear zygotic activity in the late 2-cell embryo starts with the synthesis of new mRNAs and proteins absent from the oocyte (Latham et al., 1991 and Rothstein et al., 1992). Therefore, zygotic genome activity is not simply a mere general reactivation of nuclear activity showing the previous oocytic specificity. It is characterized by a qualitative change in the utilization of the genome.

More generally, the data summarized in this review indicate that this transition has several aspects. The chronology of the biological and biochemical events (Figure 53.1) suggests the following hypothetical scheme. First, there is the need for an active cleaning of the cell, maybe to erase its oocytic characteristics. The final maturation of the oocyte initiates a degradation process of maternal RNA (Bachvarova et al., 1985 and Bachvarova et al., 1989). This degradation will continue after fertilization. Similarly, a significant decrease in the synthesis of numerous proteins is specific of 1-cell and 2-cell embryos (Latham et al., 1991). Second, there is the need of an arrest of transcription starting in the oocyte (Wassarman and Letourneau, 1976) and extending after fertilization until the late 1-cell stage (Ram and Schultz, 1993). This transcriptional arrest is perhaps necessary to facilitate the change of specificity of gene activity (see below) occuring during the transition period.

Fertilization induces changes of a different sort. Certain proteins are modified (for instance, some are phosphorylated (Howlett and Bolton, 1985)) and others are newly synthesized from the pool of maternal RNA (Latham et al., 1991). This indicates a translational activity perhaps crucial at this stage. Several of this newly synthesized proteins are 1-cell stage specific.

As the pronuclei are transcriptionally silent, all these changes are intrinsic to the cytoplasm of the oocyte and then of the fertilized egg. They occur independently of new nuclear information. It may be unique to this period of development to perform major cellular and biochemical changes by simply modifying material stored in advance and mobilised by fertilization.

These intrinsic cytoplasmic changes have another role. The embryo will modify the properties of the nuclei. The expression pattern of the zygotic nuclei of 2-cell embryos is distinct from that of the oocyte nucleus: some genes are repressed and other genes are activated. As androgenotes and gynogenotes appear to undergo correctly the first steps of development (Barra and Renard, 1988), this indicates that the nuclear material is neutral during this period. Indeed, the same pattern of expression is induced in a male pronucleus in the absence of a female pronucleus (and reciprocally) (Petzoldt et al., 1981). Therefore, it is clearly the cytoplasm which is informational at this stage. This is also illustrated by the ability of the egg to reinitiate even in older nuclei the expression of the genes first expressed in the 2-cell embryo (that is the hsp genes and genes of the TRC) (Latham et al., 1994).

What are the signals inducing the modifications of the maternal material? Is there a first nuclear relay before the major change of the pattern of expression at the 2-cell stage? The answer to this question depends on the real importance of the so called "minor zygotic activation" of the transition period. It is likely that at the end of the 1-cell stage, transcriptional competence is re-established: microinjected plasmids can be expressed (Vernet et al., 1992 and Ram and Schultz, 1993), as can endogenous genes from transplanted nuclei (Latham et al., 1992). It is even possible that certain pronuclear genes are already active (Sprinks et al., 1993 and Bouniol et al., 1995). This re-establishment of transcriptional competence is followed by zygotic nuclear activity during G_1 phase in nuclei of 2-cell embryos (Bensaude et al., 1983; Bolton et al., 1984 and Conover et al., 1991). This activity includes the synthesis of the TRC, proteins associated to the nuclei, and of the HSP. However, it is still unknown whether or not this minor zygotic activation is the nuclear zygotic interpretation of the maternal information contained in the cytoplasm, that is whether it is the maternal to zygotic relay. If indeed it is the case, the minor zygotic activation would constitute the primary event of development and presumably would be required for the initiation of the major zygotic activation.

However, it is still equally possible that the first link between maternal control elements and embryonic nuclear activity corresponds to the major zygotic activation in late 2-cell embryos. This doubt concerning the time when the link occurs also underlines the fact that it remains to be established whether there is or not a more simple step before the major zygotic activation in 2-cell embryo. One way to examine this question is to study the chronology of all these events in other mammals, the main point being to determine the existence of a minor phase of nuclear (or perhaps pronuclear) embryonic activity before the major zygotic activity. This information is still to be generated. A second way would be to study in the mouse the consequences of perturbing the minor phase on the major zygotic activation. It has been reported that the minor phase requires PKA-dependent phosphorylation of one or several substrates (Poueymirou and Schultz, 1989). It has also been shown that the appearance of the minor activation is dissociated from the acquisition of transcriptional competence and requires at least one more step. However, the importance of each of these steps for the major activation is still to be established.

Another question concerns what constitutes the step which follows the acquisition of the transcriptional competence and is necessary for the synthesis of the TRC. Could it involve changes in the nucleus or in the chromatin? The relative neutrality of the pronuclei in 1-cell embryo has already been discussed at least as long as it concerns the induction of the major activation (Petzoldt et al., 1981). This nuclear passivity is also illustrated by the discovery that plasmids microinjected in a male pronucleus transferred into 2-cell embryo behave like plasmids injected into the nucleus of 2-cell embryos (Henery et al., 1995). However, this does not mean that the pronuclei are not subjected to modifications. Quantitative analysis of plasmid activity show a difference between the nuclei of the 1-cell and those of the 2-cell embryos. For instance, a plasmid injected into a nuclei of a 2-cell embryo which is transferred into a 1-cell embryo gives a different level of activity that the same plasmid injected into a pronucleus (Henery et al., 1995). In addition, 1-cell embryos arrested in their development transcribe genes under the control of basic promoters at a surprisingly high level (Wiekowski et al., 1991 and Majumder et al., 1993). Finally, it has been hypothetized that in pronuclei promoters are insensitive to enhancers when they are associated at a distance (Wiekowski et al., 1991; Majumder et al., 1993 and Mélin et al., 1993). Although this needs to be confirmed as these experiments can still be interpreted as being due to the absence or to the inactivity of the corresponding transactivator factor, it remains that pronuclei have their specific way to transcribe plasmids. What could be the biological significance of this observation? It has been proposed that this could be related to the remodeling of chromatin associated with the exchange of protamines for histones in the male pronucleus (Wiekowski et al., 1993). It is also possible that in addition to a requirement to clean the cytoplasm, there is also a requirement to clean the two parental genomes. This cleaning may be required to prepare the appropriate initiation of nuclear events involved later in embryogenesis in the diversification of transcription patterns from equipotent cells.

Finally, it remains to establish whether modification of the pattern of expression in the 2-cell embryo is due to the generation of a new transcriptional specificity. Up to now, the emphasis has been on the striking similarity of the transcriptional specificity shared by the oocyte, 1-cell, 2-cell embryo and multipotential cells in culture (Bonnerot et al., 1991). This demonstrates the presence of common elements specific for all these cells. This similarity is maybe not surprising for the oocyte, 1-cell and 2-cell embryos as it illustrates the preparation by the oocyte for the transcriptional specificity of the 2-cell embryo. It is more surprising for multipotential cells in culture (Bonnerot et al., 1991 and Mélin et al., 1993), as they are usually considered to reflect the properties of cells of the morula and of the embryonic ectoderm (Graham, 1977). The maintenance of the same transcriptional specificity in all multipotential cells from the preimplantation period may be indicative of a stem cell mode of cellular differentiation for the first steps of embryogenesis, that is, a mode of diversification in which multipotential cells segregate various types of differentiated cells without changing themselves.

Acknowledgements

We thank L. Monfort and R. Kelly for their helpful advice and F. Kamel for typing the manuscript. This work was supported by the Centre National de la Recherche Scientifique. JFN is from the Institut National de la Santé et de la Recherche Médicale.

References

Almouzni, G. and Wolffe, A.P. (1995) Constraints on transcriptional activator function contribute to transcriptional quiescence during early Xenopus embryogenesis. *EMBO J.* **14**, 1752–1765

Ananthan, J., Goldberg, A.L. and Voellmy, R. (1986) Abnormal proteins serve as eukaryotic stress signals and trigger the activation of heat shock genes. *Science* **232**, 522–524

Bachvarova, R., Cohen, E.M., DeLeon, V., Tokunaga, K., Sakiyama, S. and Paynton, B.V. (1989) Amounts and modulation of actin mRNAs in mouse oocytes and embryos. *Development* **106**, 561–565

Bachvarova, R., DeLeon, V., Johnson, A., Kaplan, G. and Paynton, B.V. (1985) Changes in total RNA, polyadenylated RNA, and actin mRNA during meiotic maturation of mouse oocytes. *Dev. Biol.* **108**, 325–331

Bachvarova, R. and DeLeon, V. (1980) Polyadenylated RNA of mouse ova and loss of maternal RNA in early development. *Dev. Biol.* **74**, 1–8

Balhorn, R. (1982) A model for the structure of chromatin in mammalian sperm. *J. Cell Biol.* **93**, 298–305

Barra, J. and Renard, J.P. (1988) Diploid mouse embryos constructed at the late 2-cell stage from haploid parthenotes and androgenotes can develop to term. *Development* **102**, 773–779

Bensaude, O., Babinet, C., Morange, M. and Jacob, F. (1983) Heat shock proteins, first major products of zygotic gene activity in mouse embryo. *Nature* **305**, 331–333

Bevilacqua, A., Kinnunen, L.H., Bevilacqua, S. and Mangia, F. (1995) Stage-specific regulation of murine Hsp68 gene promoter in preimplantation mouse embryos. *Dev. Biol.* **170**, 467–478

Blatt, C. and DePamphilis, M.L. (1993) Striking homology between mouse and human transcription enhancer factor-1 (TEF-1). *Nucl. Acids Res.* **21**, 747–748

Bolton, V.N., Oades, P.J. and Johnson, M.H. (1984) The relationship between cleavage. DNA replication and gene expression in the mouse 2-cell embryo. *J. Embryol. Exp. Morph.* **79**, 139–163

Bonnerot, C., Vernet, M., Grimber, G., Briand, P. and Nicolas, J. F. (1991) Transcriptional selectivity in early mouse embryos–A qualitative study. *Nucl. Acids Res.* **19**, 7251–7257

Bornslaeger, E.A., Mattei, P. and Schultz, R.M. (1986) Involvment of cAMP-dependent protein kinase and protein phosphorylation in regulation of mouse oocyte maturation. *Dev. Biol.* **114**, 453–462

Bouniol, C., Nguyen, E. and Debey, P. (1995) Endogenous transcription occurs at the 1-cell stage in the mouse embryo. *Exp. Cell Res.* **218**, 57–62

Chalifour, L.E., Wirak, D.O., Hansen, U., Wassarman, P. and DePamphilis, M.L. (1987) Cis- and trans-acting sequences required for expression of simian virus 40 genes in mouse oocytes. *Genes Dev.* **1**, 1096–1106

Chastant, S., Christians, E., Campion, E. and Renard, J. P. (1995) Quantitative control of gene expression by nucleocytoplasmic interactions in early mouse embryos: consequence for reprogrammation by nuclear transfer. *Mol. Reprod. Dev.* in press

Christians, E., Campion, E., Thompson, E.M. and Renard, J.P. (1995) Expression of the HSP 70.1 gene, a landmark of early zygotic activity in the mouse embryo, is restricted to the first burst of transcription. *Development* **121**, 113–122

Clegg, K.B. and Piko, L. (1983) Quantitative aspects of RNA synthesis and polyadenylation in 1-cell and 2-cell mouse embryos. *J. Embryol. Exp. Morph.* **74**, 169–182

Clegg, K.B. and Piko, L. (1982) RNA synthesis and cytoplasmic polyadenylation in the one-cell mouse embryo. *Nature* **295**, 342–344

Conover, J.C., Temeles, G.L., Zimmermann, J.W., Burke, B. and Schultz, R.M. (1991) Stage-specific expression of a family of proteins that are major products of zygotic gene activation in the mouse embryo. *Dev. Biol.* **144**, 392–404

Davidson, I., Xiao, J.H., Rosales, R., Staub, A. and Chambon, P. (1988) The HeLa cell protein TEF-1 binds specifically and cooperatively to two SV40 enhancer motifs of unrelated sequences. *Cell* **54**, 931–942

De Leon, V., Johnson, A. and Bachvarova, R. (1983) Half-lives and relative amounts of stored and polysomal ribosomes and poly(A)+ RNA in mouse oocytes. *Dev. Biol.* **98**, 400–408

Del Mazo, J., Prantera, G., Torres, M. and Ferraro, M. (1994) DNA methylation changes during mouse spermatogenesis. *Chromosome Research* **2**, 147–152

Dooley, T.P., Miranda, M., Jones, N.C. and DePamphilis, M.L. (1989) Transactivation of the adenovirus EIIa promoter in the absence of adenovirus E1A protein is restricted to mouse oocytes and preimplantation embryos. *Development* **107**, 945–956

Flach, G., Johnson, M.H., Braude, P.R., Taylor, R.A.S. and Bolton, V.N. (1982) The transition form maternal to embryonic control in the 2-cell mouse embryo. *EMBO J.* **1**, 681–686

Fujimura, F.K., Deininger, P.L., Friedman, T. and E., L. (1981) Mutation near the polyoma DNA replication origin permits productive infection of F9 embryonal carcinoma cells. *Cell* **23**, 809–814

Graham, C.F. (1977) Teratocarcinoma cells and normal mouse embryogenesis. In *"Concepts in Mammalian Embryogenesis"*. Cambridge, Massachusets: MIT Press, 315–394,

Hecht, N. (1986) Regulation of gene expression during mammalian spermatogenesis, in: J. Rossant and R.A Pedersen, (eds.), *Experimental approaches to mammalian embryonic development*, Cambrige University Press, pp.151–181

Henery, C.A., Miranda, M., Wiekowski, M., Wilmut, I. and DePamphilis, M.L. (1995) Repression of gene expression at the beginning of mouse development. *Dev. Biol.* **169**, 448–460

Howlett, S.K. (1986) The effect of inhibiting DNA replication in the one-cell mouse embryo. *Roux's Arch. Dev. Biol.* **195**, 499–505

Howlett, S.K. and Bolton, V.N. (1985) Sequence and regulation of morphological and molecular events during the first cell cycle of mouse embryogenesis. *J. Embryol. Exp. Morph.* **87**, 175–206

Ikegami, S., Taguchi, T., Ohashi, M., Oguro, M., Nagano, H. and Mano, Y. (1978) Aphidicolin prevents mitotic cell division by interfering with the activity of DNA polynerase. *Nature* **275**, 458–460

Imperiale, M.J., Hung-Teh, K., Feldman, L.T., Nevins, J.R. and Strickland, S. (1984) Common control of the heat shock gene and early adenovirus gene: evidence for a cellular E1A-like activity. *Molec. Cell. Biol.* **4**, 867–874

Kafri, T., Ariel, M., Brandeis, M., Shemer, R., Urven, L. and McCarrey, J. (1992) Developmental pattern of gene-specific DNA methylation in the mouse embryo and germline. *Genes Dev.* **6**, 705–714

Latham, K.E., Doherty, A.S., Scott, C.D. and Schultz, R.M. (1994) *Igf2r* and *Igf2* gene expression in androgenetic, gynogenetic and parthenogenetic preimplantation mouse embryos: absence of regulation by genomic imprinting. *Genes Dev.* **8**, 290–299

Latham, K.E., Garrels, J.I., Chang, C. and Solter, D. (1991) Quantitative analysis of protein synthesis in mouse embryos. I. extensive reprogramming at the one- and two-cell stages. *Development* **112**, 921–932

Latham, K.E., Garrels, J.I. and Solter, D. (1994) Alterations in protein synthesis following transplantation of mouse 8-cell stage nuclei to enucleated 1-cell embryos. *Dev. Biol.* **163**, 341–350

Latham, K.E., Solter, D. and Schultz, R.M. (1992) Acquisition of a transcriptionally permissive state during the 1-cell stage of mouse embryogenesis. *Dev. Biol.* **149**, 457–462

Lee, D.Y., Hayes, J.J., Pruss, D. and Wolffe, A.P. (1993) A positive role for histone acetylation in transcription factor access to nucleosomal DNA. *Cell* **72**, 73–84

Linney, E. and Donerly, S. (1983) DNA fragments from F9 PyEC mutants increase expression of heterologous genes in transfected F9 cells. *Cell* **35**, 693–699

Majumder, S., Miranda, M. and DePamphilis, M.L. (1993) Analysis of gene expression in mouse preimplantation embryos demonstrates that the primary role of enhancers is to relieve repression of promoters. *EMBO J.* **12**, 1131–1140

Martinez-Salas, E., Linney, E., Hassell, J. and DePamphilis, M.L. (1989) The need for enhancers in gene expression first appears during mouse development with formation of the zygotic nucleus. *Genes Dev.* **3**, 1493–1506

Mélin, F., Miranda, M., Montreau, N., DePamphilis, M.L. and Blangy, D. (1993) Transcription enhancer factor-1 (TEF-1) DNA binding sites can specifically enhance gene expression at the beginning of mouse development. *EMBO J.* **12**, 4657–4666

Murthy, S.C.S., Bhat, G.P. and Thimmappaya, B. (1985) Adenovirus EIIA early promoter : transcriptional control elements and induction by the viral pre-early EIA gene, which appears to be sequence independent. *Proc. Natl. Acad. Sci. USA* **82**, 2230–2234

Nonchev, S. and Tsanev, R. (1990) Protamine-histone replacement and DNA replication in the male mouse pronucleus. *Mol. Reprod. Dev.* **25**, 72–76

Paria, B.C., Das, S.K. and Dey, K. (1995) The preimplantation mouse embryo is a target for cannabinoid ligand-receptor signaling. *Proc. Natl. Acad. Sci. USA* **92**, 9460–9464

Paynton, B., Rempel, R. and Bachvarova, R. (1988) Changes in state of adenylation and time course of degradation of maternal mRNAs during oocyte maturation and early development in the mouse. *Dev. Biol.* **129**, 304–314

Petzoldt, U., Hoppe, P.C. and Illmensee, K. (1980) Protein synthesis in enucleated fertilized and unfertilized mouse eggs. *Roux's Arch. Dev. Biol.* **189**, 215–219

Petzoldt, U., Illmensee, G.R., Bűrki, K., Hoppe, P.C. and Illmensee, K. (1981) Protein synthesis in microsurgically produced androgenetic and gynogenetic mouse embryos. *Mol. Gen. Genet.* **184**, 11–16

Poueymirou, W.T. and Schultz, R.M. (1989) Regulation of mouse preimplantation development: inhibition of synthesis of proteins in the two-cell embryo that require transcription by inhibitors of cAMP-dependent protein kinase. *Dev. Biol.* **133**, 588–599

Prioleau, M.-N., Huet, J., Sentennac, A. and Méchali, M. (1994) Competition between chromatin and transcription complex assembly regulates gene expression during early development. *Cell* **77**, 439–449

Ram, P.T. and Schultz, R.M. (1993) Reporter gene expression in G2 of the 1-cell mouse embryo. *Dev. Biol.* **156**, 552–556

Rappolee, D.A., Sturm, K.S., Behrendtsen, O., Schultz, G.A., Pedersen, R.A. and Werb, Z. (1992) Insulin-like growth factor II acts through an endogenous growth pathway regulated by imprinting in early mouse embryos. *Genes Dev.* **6**, 939–952

Rothstein, J.L., Johnson, D., DeLoia, J.A., Skowronski, J., Solter, D. and Knowles, B. (1992) Gene expression during preimplantation mouse development. *Genes Dev.* **6**, 1190–1201

Sanford, J.P., Clark, H.J., Chapman, V.M. and Rossant, J. (1987) Differences in DNA methylation during oogenesis and spermatogenesis and their persistence during early embryogenesis in the mouse. *Genes Dev.* **1**, 1039–1046

Sawicki, J.A., Magnuson, T. and Epstein, C.J. (1981) Evidence for expression of the paternal genome in the two-cell mouse embryo. *Nature* **294**, 450–451

Schultz, R.M. (1986) in: J. Rossant and R.A Pedersen, (eds.), *Experimental approaches to mammalian embryonic development*, Cambrige University Press, pp. 195–227

Schultz, R.M. (1993) Regulation of zygotic gene activation in the mouse. *BioEssays* **15**, 531–538

Schwartz, D.A. and Schultz, R.M. (1992) Zygotic gene activation in the mouse embryo: involvement of cyclic adenosine monophosphate-dependent protein kinase and appearance of an AP-1-like activity. *Mol. Reprod. Dev.* **32**, 209–216

Sprinks, M.T., Sellens, M.H., Dealtry, G.B. and Fernandez, N. (1993) Preimplantation mouse embryos express *Mhc* class I genes before the first cleavage division. *Immunogenetics* **38**, 35–40

Taylor, K.D. and Piko, L. (1987) Patterns of mRNA prevalence and expression of B1 and B2 transcripts in early mouse embryos. *Development* **101**, 877–892

Telford, N.A., Watson, A.J. and Schultz, G.A. (1990) Transition from maternal to embryonic control in early mammalian development: a comparison of several species. *Mol. Reprod. Dev.* **26**, 90–100

Turner, B. M. (1991) Histone acetylation and control of gene expression. *J. Cell Sci.* **99**, 13–20

Vernet, M., Bonnerot, C., Briand, P. and Nicolas, J. F. (1992) Changes in permissiveness for the expression of microinjected DNA during the first cleavages of mouse embryos. *Mech. of Dev.* **36**, 129–139

Vernet, M., Cavard, C., Zider, A., Fergelot, P., Grimber, G. and Briand, P. (1993) In vitro manipulation of early mouse embryos induces HIV1-LTR*lacZ* transgene expression. *Development* **119**, 1293–1300

Wassarman, P.M. and Letourneau, G.E. (1976) RNA synthesis in fully-grown mouse oocytes. *Nature* **261**, 73–74

Wiekowski, M., Miranda, M. and DePanphilis, M.L. (1993) Requirements for promoter activity in mouse oocytes and embryos distinguish paternal pronuclei from maternal and zygotic nuclei. *Dev. Biol.* **159**, 366–378

Wiekowski, M., Miranda, M. and DePhamphilis, M.L. (1991) Regulation of gene expression in preimplantation mouse embryos: effects of the zygotic clock and the first mitosis on promoter and enhancer activities. *Dev. Biol.* **147**, 403–414

Wolffe, A.P. (1991) Implications of DNA replication for eukaryotic gene expression. *J. Cell Sci.* **99**, 201–206

Worrad, D.M., Ram, P.T. and Schultz, R.M. (1994) Regulation of gene expression in the mouse oocyte and early preimplantation embryo: developmental changes in Sp1 and TATA box-binding protein, TBP. *Development* **120**, 2347–2357

Worrad, D.M., Turner, B.M. and Schultz, R.M. (1995) Temporally restricted spatial localization of acetylated isoforms of histone H4 and RNA polymerase II in the 2-cell mouse embryo. *Development* **121**, 2949–2959

Zimmermann, J.W. and Schultz, R.M. (1994) Analysis of gene expression in the preimplantation mouse embryo: use of mRNA diffenrential display. *Proc. Natl. Acad. Sci. USA* **91**, 5456–5460

PART IV, SECTION A

54. Insertional Mutagenesis in Transgenic Mice

Ulrich Rüther

Institut für Molekularbiologie, Medizinische Hochschule Hannover, 30625 Hannover, Germany

Aim

Analyzing the function of genes is the most promising approach to unravel the genetic program of development. This, however, is only possible when such genes are defined at both the genetic and at the molecular level. Although several hundred developmentally important genes are known cross species it is believed that this is only a minor fraction of the genes essential for proper development, in particular in mammals. Therefore, identification of novel genes is a fundamental prerequisite for understanding the molecular biology of development. Insertional mutagenesis is a powerful way to increase the number of known genes.

Problems and Progress

General

Different methods have been used to generate transgenic mice, each of which have resulted in comparable numbers of insertional mutants. Those mutants generated by retroviral infection or DNA microinjection have been investigated in great detail over the last 15 years. Although the number of publications is increasing every year, a recent review, which coveres through the end of 1993 and summarizes most of the studies, indicates an overall mutation frequency of about 5% in transgenic mice (Rijkers *et al.*, 1994). In this review I will not address frequencies and numbers since they have barely changed in the past two years. But first I will explain, with the help of two examples, the problems inherent in going from a mutant phenotype to the molecular analysis of a gene. Second, I will speculate on the future development of the field.

Specific

Insertional Mutagenesis by Microinjection of DNA. Most publications describing insertional mutagenesis are based on transgenic mice generated by microinjection of DNA. A representative example is the very successful and complete analysis from Kiran Chada's laboratory. The first publication about the mouse mutant $pg^{TgN40ACha}$ appeared some years ago (Xiang *et al.*, 1990). It was shown that this mutant is allelic to the *pygmy* mutation (*pg*), which is characterized by dwarfism independent of the growth hormone pathway. The integration of the transgene in this new mutant facilitated the cloning of the integration site and thus served as the entry point toward a molecular analysis of pygmy mice. Analysis of the transgene-induced mutation revealed a large deletion at the pygmy locus. The finding of a deletion at transgene integrations sites is not unusual and makes the identification of the mutated gene sometimes complicated. In this case, however, the comparison of different *pg* mutations deliniated a common deletion of about 56 kb (Zhou *et al.*, 1995). Unfortunately subsequent analysis revealed that more than 90% of this deletion was upstream of the gene. Nevertheless, intensive investigation of this 56 kb region finally identified a homologue of the *Hmgi-c* gene (Zhou *et al.*, 1995). Using this candidate gene the authors documented the absence of expression of *Hmgi-c* gene in different *pg* mutants mice including the one derived by insertion of the transgene. The pygmy phenotype was confirmed by the generation of *Hmgi-c* knock-out mice (Zhou *et al.*, 1995).

This example is characteristic for such an analysis. First, it shows the time frame required to perform a convincing molecular investigation of this type of mouse mutation, i.e. about five years. But it also shows that this approach can reveal a known gene, which was not previously thought to be involved in the pygmy phenotype, as an essential component for proper growth regulation. Finally, fifty years after the detection of *pygmy* this analysis has led to a molecular explanation of the mutation.

Insertional Mutagenesis by Gene Trapping. An alternative approach to investigate the mutagenic effect of integrated DNA is gene trapping. This technique has several important differences to the DNA injection methods. First, the DNA construct used for integration is specially designed for the detection of genes. In principle, the construct contains a *lacZ* reporter gene and, as a selectable marker, the neomycin resistance gene (*neo*). The *lacZ* gene lacks a promoter and a start codon but carries instead a splice acceptor site at its 5' end. Upon integration of the construct into mouse genomic DNA, a *lacZ* gene activity can only appear when the integration has occured in the intron of a gene in the correct orientation and when splicing fuses the endogenous ATG in frame with the *lacZ* reporter. Second, the endogenous gene is disrupted as a consequence of integration but its expression can be visualized by X-gal staining for *lacZ* activity. Third, the gene trap construct is electroporated into mouse embryonal stem (ES) cells, which are selected for neo resistance. Clones that stain blue with X-gal (about 5% in total) are candidates for a trapped gene relevant for development. This relevance can be monitored by the generation of chimeric mice and whole-mount X-gal staining. Like the DNA-injection approach germ-line transmission of the trapped allele and breeding of homozygous mice can result in mutant phenotypes. In contrast to the DNA injection approach the mutated gene can be cloned via a cDNA analysis of the sequences fused 5' to the *lacZ* reporter.

The initial description of this approach was published by Gossler et al. (1989). Since that time several different constructs and a few novel mouse mutants and genes have been described using this method. A representative example is a recently published mutation from the group of Toru Higashinakagawa. They have used a novel type of gene trap vector where the neomycin resistance gene must be fused in frame with an endogenous gene. The neo stop codon is followed by an internal ribosome entry site (IRES), a start codon and the *lacZ* gene. The advantage of this construct is that every ES cell clone which is growing after neo selection should express the *lacZ* reporter from this dicistronic mRNA.

This idea was confirmed experimentally. 41 of 44 neo-resistant colonies showed *lacZ* activity. Eighteen of the X-gal-positive clones were used for the generation of chimeric mice which were then stained with X-gal at day 10.5 of development. One showed a pattern of high expression restricted to midbrain-hindbrain boundary (Takeuchi et al., 1995). After germ-line transmission of the trapped gene, its expression was studied in more detail during development. This revealed strong expression first in the mid-hindbrain region and then in the cerebellum from day 14.5 onwards. Although faint expression was also detected in other structures, this pattern suggested a function of the trapped gene in the formation of defined regions of the neural tube. This interpretation was supported by the phenotype of homozygous mutated mice. These mice died about day 15.5 in development displaying an abnormal groove in a region anterior of the mid-hindbrain region and a defect in neural tube closure in the same region (Takeuchi et al., 1995). According to this phenotype the mutant was named *jumonji* (*jmj*).

The fusion mRNA was cloned from the ES cell clone and the novel sequence obtained was used as a probe to derive the full-length cDNA of the *jmj* gene. Subsequent molecular analysis confirmed the relationship of the novel gene sequence to the mutant phenotype.

This is a representative example for gene trap publications. When germ-line transmission of the ES cell clone is obtained it is easy to compare the expression pattern of the trapped gene and the mutant phenotype and thus to decide whether a further molecular investigation is of interest. Without laborious analysis the affected gene can be cloned and analyzed functionally. In the case cited above the gene trap approach has led to the identification of a new mouse mutation for which no mutant allele has been described. Furthermore, the gene is novel and shows homology to the human retinoblastoma-binding protein RBP-2 (Takeuchi et al., 1995). The overall time required for this investigation is a matter of speculation but is certainly much shorter in comparison to the analysis of the $pg^{TgN40ACha}$.

Prospects

What have we learned from these two examples? In the past insertional mutagenesis was only a byproduct of most transgenic mouse studies. As a consequence cloning of the transgene integration site was sometimes very complicated, since the construct was designed for the expression of the transgene and not for recloning. Second, the mutagenic effect of the integrated DNA only became obvious after mating of the transgene integration site to homozygozity (except for dominant mutations). Thus, a frequency of 5% mutants among all the transgenic lines meant that twenty times more lines had to be bred to identify one mutant. These problems do not exist in the gene trap system. Here, the expression pattern of the trapped gene in chimeric animals can already be used to discriminate individual ES clones. Furthermore, different prescreening approaches are under investigated to allow an initial test for trapping at the ES cell level. For example, ES cell clones are treated with different factors to direct differentiation into certain cell types and *lacZ* gene activity is monitored in parallel. Alternatively, the trapped gene is partially analyzed by using a combination of PCR and sequencing. The sequence obtained is then screened for known motifs before the corresponding ES cell clones are used for the generation of chimeric mice. Several groups have been seduced by the gene trap approach and have started specific programs to search for novel genes. It is easy to forsee that the majority of the insertional mutagenesis studies will arise from the gene trap in]vestigations. This system will speed up the collection of interesting genes important for mammalian development.

Acknowledgements

I would like to thank Bob Hipskind for critical reading of the manuscript. The work of my group in the frame of insertional mutagenesis is supported by the Volkswagen-Stiftung, the Human Mobility and Capital Program (Gene-trap Network) and the SFB 271.

References

Gossler, A., Joyner, A.L., Rossant, J. and Skarnes, W.C. (1988) Mouse embryonic stem cells and reporter constructs to detect developmentally regulated genes. *Science* **244**, 463–465

Rijkers, A., Peetz, A. and Rüther, U. (1994) Insertional mutagenesis in transgenic mice. *Transgenic Res.* **3**, 203–215

Takeuchi, T., Yamazaki, Y., Katoh-Fukui, Y., Tsuchiya, R., Kondo, S., Motoyama, J. and Higashinakagawa, T. (1995) Gene trap capture of a novel gene, *jumonji*, required for neural tube formation. *Genes and Dev.* **9**, 1211–1222

Xiang, X., Benson, K.F. and Chada, K. (1990) Mini-Mouse: Disruption of the Pygmy Locus in a transgenic Insertional Mutant. *Science* **247**, 967–969

Zhou, X., Benson, K.F., Ashar, H.R. and Chada, K. (1995) Mutation responsible for the mouse pygmy phenotype in the developmentally regulated factor HMGI-C. *Nature* **376**, 771–774

PART IV, SECTION A

55. Enhancer-Trap Studies of the *Drosophila* Brain

J. Douglas Armstrong and Kim Kaiser

Division of Molecular Genetics, IBLS, University of Glasgow, Glasgow G11 6NU, UK

Drosophila Development

The fruit fly, *Drosophila melanogaster*, is amongst the best studied of multicellular eukaryotes. It has a short life-span, a small size, and simple culture conditions that facilitate study in the most primitive of laboratories (Ashburner, 1989). Despite its small size and apparent simplicity, *D. melanogaster* is nevertheless a valuable model for a wide range of biological processes, including development, physiology (at both cellular and organismal levels) and more recently the structure and function of the brain (Davis and Han, 1996) .

D. melanogaster goes through a series of developmental stages (Bate and Arias, 1993). Major elements of the body plan are established in the embryo, from which hatches a small larva. This simple organism does little other than feed for a period of several days, moulting twice in the process. In the wild, the food source would be rotting fruit, and the larva is thus adapted for life in a semi-liquid environment. Even at this stage, the organism exhibits a range of behaviours appropriate to its environment. Eventually, the full-grown larva crawls away from the food, ceases movement, and attaches itself to a stable anchor point. A hard outer case is secreted, and the larva becomes a pupa. Within the protective case, a number of major developmental changes take place. Large elements of larval structures, including the nervous system, degenerate or are otherwise re-modelled, and are replaced by adult structures. The adult fly hatches a few days later. In summary, there are two major phases of *D. melanogaster* development, the first embryonic and the second pupal.

Enhancer-Trap Elements as Cell-Markers

The use in *Drosophila* of P-element transgenesis (described in Chapter 23; Kaiser, 1996) has enabled a new wave of scientific discovery. It has found application in the study of almost all biological processes. Here we focus on recent developments in enhancer-trap approaches, and in particular the value of cell-type specific markers for unravelling cellular heterogeneity in complex organs, of which the brain is a prime example.

Enhancer-trap elements are effectively markers of gene expression. They can be used to trace cell lineages, though one must always bear in mind that a gene expressed in one cell may not necessarily be expressed in its daughter cells. One of the first uses of enhancer-trapping for lineage marking was in a study of neurogenesis in the embryo (O'Kane and Gehring, 1987). At the other end of development, enhancer-trapping has revealed subtle maturation of gene expression during adult ageing (Rogina and Helfand, 1995).

Of particular interest has been an attempt at a comprehensive enhancer-trap screen for embryonic anatomical markers (Hartenstein and Jan, 1992). 171 P{lacZ} lines were described as having expression patterns corresponding to known structures. In most cases, the expression patterns were evident before the final stage of development of any particular structure. By following changes in expression patterns through embryogenesis, several novel aspects of embryonic development were observed. In addition, the staining patterns revealed a number of previously unreported embryonic structures. Many lines revealed elements of the nervous system, few of which could be easily identified.

One important feature of the P{*lacZ*} element is that it reveals only cell nuclei, as a consequence of *lacZ* being fused to a nuclear localisation determinant (Bier *et al.*, 1989). This can be a significant drawback in cases where cell-body position alone tells one little about the nature of the cells revealed. This problem can be particularly acute in the context of the nervous system, and for which markers that reveal cell morphology would be far preferable.

Using P{GAL4} to Map Neuronal Morphology

For most cells, nuclear expression is a reasonable marker for the position and identity of the cell itself. The *D. melanogaster* CNS is organised somewhat differently from other tissues, however. Neural cell bodies lie in a thin cortex surrounding the bulk of neural tissue (Power, 1943). Each cell body sends a single process or neurite inwards, which then gives rise to separate axonal and dendrititic projections, collectively forming tracts and large regions of synapsis known as neuropil. Only a few CNS neurons can be characterised by cell body position alone. A reporter capable of revealing neuronal projection patterns would clearly be preferable to a nuclear tag (Smith and O'Kane, 1990). The P{GAL4} enhancer-trap element (Brand and Perrimon, 1993 and Kaiser, 1993), when used in conjunction with a cytoplasmically-localised secondary reporter (e.g. *lacZ* lacking a nuclear targeting signal) provides such a marker. A number of other secondary reporters have been specifically developed to study different aspects of neuronal anatomy (Giniger *et al.*, 1993; Brand, 1995 and Yeh *et al.*, 1995).

One of the first reports of use of the P{GAL4} system for anatomical mapping of the CNS described lines having reporter expression within the mushroom bodies (Yang *et al.*, 1995). Mushroom bodies (MBs) are large phylogenetically conserved insect brain elements that have been implicated in associative learning and memory, and in a variety of complex functions including courtship, motor control, and spatial recognition and target location (Huber, 1960; Erber *et al.*, 1987 and Mizunami *et al.*, 1993). In *D. melanogaster*, single gene mutations that cause defective mushroom body anatomies (e.g. *mushroom body miniature* and *mushroom bodies deranged*) have been shown to interfere significantly with olfactory associative learning (Heisenberg *et al.*, 1985 and Heisenberg, 1989). Olfactory learning is even more profoundly affected by ablation of mushroom body neuroblasts at an early stage of development, depleting the adult brain of mushroom body intrinsic neurons (deBelle and Heisenberg, 1994). Additional support for a role of *Drosophila* mushroom bodies in olfactory learning derives from studies of 'biochemical' learning mutants. Expression of the learning genes *dunce* and *rutabaga*, for example, occurs predominantly, though not exclusively, in the mushroom bodies (Nighorn *et al.*, 1991; Han *et al.*, 1992 and Davis, 1993). Finally, gynandromorph analysis implicates *Drosophila* mushroom bodies, or adjacent neuropils, in control of the male courtship repertoire (Hall, 1979), a behaviour that relies heavily on olfaction. Taken together, the picture that emerges is of a specialised neuropil involved in associating and storing multimodal sensory information, thereby providing the organism with memory, and predictive behaviour.

Both structural (Strausfeld, 1976 and Mobbs, 1982) and functional (Mauelshagen, 1993; Mizunami *et al.*, 1993; and Laurent and Davidowitz, 1994) studies suggest that the fundamental computational properties of mushroom bodies are provided by the intrinsic neurons. Known as Kenyon cells, these arise from dense clusters of cell bodies dorsal and posterior to each brain hemisphere. Their dendrites form the mushroom body calyces, large regions of input from the olfactory lobes, while beneath each calyx the Kenyon cell axons converge to form a stalk-like structure, the pedunculus. The pedunculus extends almost to the front of the brain, at which point it gives rise to a number of lobes: a dorsally-projecting α lobe, and a β/γ lobe complex projecting towards the mid-line (see Figure 55.1).

Using the P{GAL4} enhancer-trap approach, Yang *et al.* (1995) obtained several lines in which axonal processes corresponding to *D. melanogaster* Kenyon cells were revealed (of which Figure 55.2d is an example). Different lines exhibited subtle differences in expression pattern, however. Rather than being homogenous, the mushroom bodies were thus shown to be compound neuropils, in which parallel sub-components exhibit discrete patterns of gene expression (summarised in Figure 55.1). Different patterns corresponded to hitherto unobserved differences in Kenyon cell trajectory, and placement. On the basis of this unexpected complexity, Yang *et al.* (1995) proposed a model for mushroom body function in which parallel channels of information flow, perhaps with different computational properties, subserve different behavioural roles.

Structure/Function Relationships within the MBS

An important feature of the P{GAL4} system and indeed the motivating factor in its design, was to allow facile targeting of gene expression to particular cell types. Though still in its infancy, the potential power of the system is already evident (Brand and Dormand, 1995).

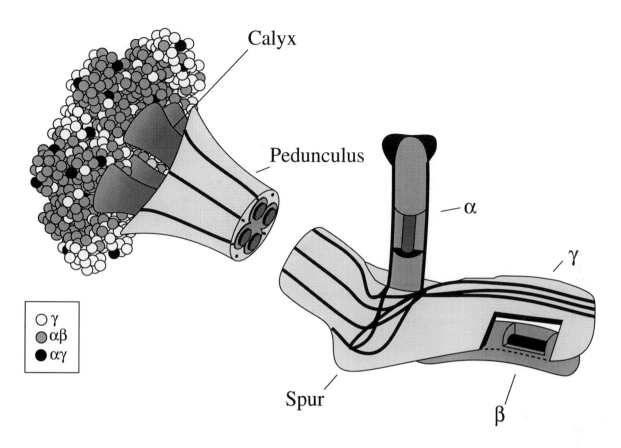

Figure 55.1. Schematic view of mushroom body substructure.
Major MB structural features revealed by analysis of P{GAL4} expression patterns (Armstrong, 1995 and Yang et al., 1995). Continuous substructures are similarly shaded. The MB is drawn severed at the pedunculus for clarity. Regions of the lobes and calyx have been cut away to reveal deeper structure. The calyx, pedunculus and spur (at the bifurcation of the lobes) are indicated. The α, β and γ lobes are indicated.

In the context of the adult brain, GAL4 directed cell manipulation has been used to study the neural circuitry underlying sex-specific behaviours. In two related studies (Ferveur et al., 1995 and O'Dell et al., 1995) the feminising gene *transformer* was expressed under GAL4 control within different brain regions. Expression in lines that marked particular subsets of mushroom body intrinsic neurons caused males to exhibit non-discriminatory sexual behaviour; they courted mature males in addition to females. Expression of *transformer* in other mushroom body domains had no such effect. These data support the view that genetically-defined subsets of mushroom body intrinsic neurons perform different functional roles.

Mushroom Body Development

Unlike many other elements of the CNS, the mushroom bodies are present throughout development, in the embryo, larva and pupa, as well as the adult brain (Ito and Hotta, 1992). During metamorphosis however, there appear to be changes in the number of Kenyon cell processes without a corresponding change in cell body number. In explanation, it has been suggested that during pupation some Kenyon cell processes degenerate and that new ones then arise from the old cell bodies (Technau and Heisenberg, 1982). In an ongoing study, we are using lines that exhibit GAL4 expression within the mushroom bodies in a new approach to such phenomena. Figure 55.2 illustrates the MB staining pattern at four developmental stages within the brain of the same P{GAL4} line, 201Y. Only a specific subset of Kenyon cells is revealed in this line (Yang et al., 1995)

The larval pattern (Figure 55.2a) is clearly different, particularly with respect to lobe shape, from the pattern seen in late pupae and adults (Figure 55.2d). Careful analysis of the intervening stages, many more than the selection shown here, reveals the nature of the transition. In particular, staining of the

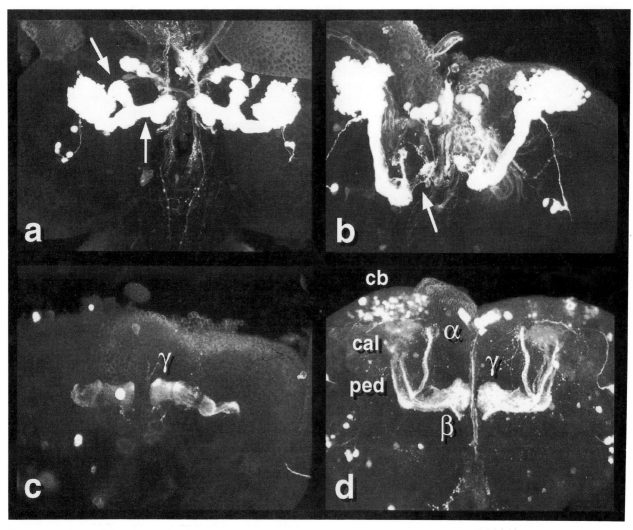

Figure 55.2. Pupal development of the mushroom bodies.
Frontal confocal reconstruction of the Kenyon cell staining pattern in the brain of P{GAL4} line 201Y. Each reconstruction shows the MBs embedded within the rest of the central brain, which exhibits bilateral symmetry about the midline. For details of the staining method see Yang *et al.* (1995).
(a) Expression pattern in late 3rd instar larva. (b) 12 hour pupa; the arrow points to the degenerating larval lobes. (c) 36 hour pupa; the developing γ-lobe is labelled (the rest of the mushroom bodies are omitted from this reconstruction). (d) 56 hour pupa; the pattern is essentially that of the adult. The Kenyon cell body region (cb), calyx (cal), pedunculus (ped), the α-lobe, β-lobe and γ-lobes are labelled.

larval lobes virtually disappears during the first 12 hours of pupal development (Figure 55.2b arrows). Some Kenyon fibres within the pedunculus remain stained, though it is not possible to judge the proportion. After 36 hours (Figure 55.2c), a staining pattern corresponding to the adult γ-lobe has become readily apparent. The α and β lobes are not yet visible, however. Only 24 hours later do they eventually appear, by which time the pattern is effectively that of the adult (Figure 55.2d). Expression patterns in several other P{GAL4} lines reveal similar, though not identical, changes in MB anatomy taking place over roughly the same time scale (not shown).

These results confirm that considerable reorganisation of the mushroom bodies takes place during metamorphosis. Some elements of the larval lobes may remain through pupation, but most would appear to undergo partial or complete degeneration. Given that the mushroom bodies are good candidates for the fly memory centres (Erber *et al.*, 1980; Balling *et al.*, 1987 and Davis, 1993) and that there is some evidence for the retention of memory through metamorphosis (Cambiazo et al., 1994), the exact details of mushroom body re-organisation may eventually shed some light on which components, if any, embody such long-lasting memories.

Limitations of Enhancer-Based Developmental Studies

The P{GAL4} system has a limitation in common with all other enhancer-trap approaches, at least with respect to descriptions of anatomical development. This is that temporal changes in the expression characteristics of the enhancer driving the reporter can never be ruled out. Embryogenesis and metamorphosis, whilst being generally accepted as the most important developmental stages, are the most problematic in terms of lineage tracing, since many genes are likely to exhibit changes in their expression pattern at these times. We have observed several cases in which changes of a staining pattern occur far too rapidly to be accounted for by anatomical effects alone. In two of the lines described by Yang et al. (1995) for example, the Kenyon cell expression pattern appears simultaneously in several hundred cells at a time close to the end of pupal development, and when the mushroom bodies are virtually complete. These lines would thus appear to be clear cases of temporal changes in enhancer activity.

The above limitation has been turned to advantage, in studies of ageing processes in the adult. The fly appears to become fully developed within a day or so of hatching from the pupal case, and little further changes might be envisaged. However, a survey of enhancer-trap lines involving quantitative analysis of reporter activity reveals marked changes, possibly senescence related, in expression levels during the adult lifetime (Helfand et al., 1995; and Rogina and Helfand, 1995).

Until recently, another limitation was the requirement for a fixation step in reporter visualisation. This makes it impossible to follow kinetic aspects of development in the living organism. The advent of Green Fluorescent Protein (GFP) as a marker, expressed in *D. melanogaster* under UAS_G control, provides a partial solution to this problem (Brand, 1995; and Yeh et al., 1995). GFP is a naturally fluorescent protein which allows cells to be visualised by fluorescent microscopy without the need for fixation or any other traumatic treatment (with the exception of the fluorescent microscopy itself). Expression of GFP in tissues should allow developmental processes to be followed *in situ* and in real time.

Summary

The P{GAL4} enhancer-trap system is still in its own development phase, yet its potential as both a cell specific marker for revealing developmental processes, and also as a targeted gene expression system for *in vivo* manipulation of the same processes, is clearly evident. Of course, the P{GAL4} system is far from the only strategy for following *D. melanogaster* development. It is just one of a whole suite of tools and techniques, many involving transgenesis in one form or another. The idea behind the P{GAL4} system seems unique to *D. melanogaster* at present, however. It will be interesting to see it translated to other organisms.

References

Armstrong, J.D. (1995) Structural Characterisation of the *Drosophila* Mushroom Bodies. Ph.D. Thesis, University of Glasgow

Ashburner, M. (1989) *Drosophila: a laboratory manual*. Cold Spring Harbour Laboratory Press

Balling, A., Technau, G.M. and Heisenberg, M. (1987) Are the structural changes in adult *Drosophila* mushroom bodies memory traces? Studies on biochemical learning mutants. *Journal of Neuroscience* **4**, 65–73

Bate, M. and Arias, A.M. (1993) *The Development of Drosophila melanogaster*. Cold Spring Harbour Laboratory Press

Bier, E., Vaessin, H., Shepherd, S., Lee, K., McCall, K., Barbel, S., Ackerman, L., Carretto, R., Uremura, T., Grell, E., Jan, L.Y. and Jan, Y.N. (1989) Searching for pattern and mutation in the *Drosophila* genome with a P-lacZ vector. *Genes and Development* **3**, 1273–1287

Brand, A. and Perrimon, N. (1993) Targeted gene expression as a means of altering cell fates and generating dominant phenotypes. *Development* **118**, 401–415

Brand, A.H. (1995) GFP in *Drosophila*. *Trends in Genetics*, **11**, 324–325

Brand, A.H. and Dormand, E.L. (1995) The GAL4 system as a tool for unraveling the mysteries of the *Drosophila* nervous system. *Current Opinion in Neurobiology* **5**, 572–578

Cambiazo, V., Tully, T. and Krase, L. (1994) Memory through metamorphosis in normal and mutant *Drosophila*. *J. Neurosci.* **14**, 68–74

Davis, R.L. (1993) Mushroom bodies and *Drosophila* learning. *Neuron* **11**, 1–14

Davis, R.L. and Han, K.-H. (1996) Mushrooming mushroom bodies. *Current Biology* **6**, 146–148

DeBelle, J.S. and Heisenberg, M. (1994) Associative odor learning in *Drosophila* is abolished by chemical ablation of mushroom bodies. *Science* **236**, 692–695

Erber, J., Homberg, U. and Gronenberg, W. (1987) Functional roles of the mushroom bodies in insects. In A. P. Gupta (Eds.), *Arthropod Brain* (pp. 485–511). New York: Wiley-Interscience

Erber, J., Mashur, T. and Menzel, R. (1980) Localisation of short-term memory in the brain of the bee, *Apis mellifera*. *Physiol. Entomol.* **5**, 343–358

Ferveur, J.-F., Störtkuhl, K.F., Stocker, R.F. and Greenspan, R.J. (1995) Genetic feminisation of brain structures and changed sexual orientation in male *Drosophila melanogaster*. *Science* **2647**, 902–905

Giniger, E., Wells, W., Jan, L.Y. and Jan, Y.N. (1993) Tracing neurons with a kinesin-β-galactosidase fusion protein. *Roux's Arch. Dev. Biol.* **202**, 112–122

Hall, J.C. (1979) Control of Male Reproductive Behavior by the Central Nervous System of *Drosophila*: Dissection of a Courtship Pathway by Genetic Mosaics. *Genetics* **92**, 437–457

Han, P.-L., Levin, L.R., Reed, R.R. and Davis, R.L. (1992) Preferential expression of the *Drosophila rutabaga* gene in mushroom bodies, neural centers for learning in insects. *Neuron* **9**, 619–627

Hartenstein, V. and Jan, Y.N. (1992) Studying *Drosophila* embryogenesis with P-*lacZ* enhancer trap lines. *Rous Arch. Dev. Biol.* **201**, 194–220

Heisenberg, M. (1989) Genetic approach to learning and memory (mnemogenetics) in *Drosophila melanogaster*. In Rahmann (Eds.), *Fundamentals of Meomory Formation: Neuronal Plasticity and Brain Function Stuttgart*, New York: Gustav Fischer Verlag

Heisenberg, M., Borst, A., Wagner, S. and Byers, D. (1985) *Drosophila* mushroom body mutants are deficient in olfactory learning. *J. Neurogentics* **2**, 1–30

Helfand, S.L., Blake, K.J., Rogina, B., Stracks, M.D., Centurion, A. and Naprta, B. (1995) Temporal Patterns of Gene Expression in the Antenna of the Adult *Drosophila melanogaster*. *Genetics* **140**, 549–555

Huber, F. (1960) Untersuchungenüber die Function des Zentralnervensytems und insbesondere des Gehirns bei der Fortbewegung und der Lautäußerung der Grillen. *Z. vgl. Physiol.* **44**, 60–132

Ito, K. and Hotta, Y. (1992) Proliferation pattern of postembyonic neuroblasts in the brain of *Drosophila melanogaster*. *Dev. Biol.* **149**, 134–148

Kaiser, K. (1993) A second generation of enhancer-traps. *Current Biol.* **3**, 560–562

Kaiser, K. (1996) Gene transfer in *Drosophila melanogaster*. This volume

Laurent, G. and Davidowitz, H. (1994) Encoding of olfactory information wth oscillating neural assemblies. *Science* **265**, 1872–1874

Mauelshagen, J. (1993) Neural correlates of olfactory learning pardigms in an identified neruon in the honeybee brain. *J. Neurophys.* **69**, 609–625

Mizunami, M., Weibrecht, J.M. and Strausfeld, N.J. (1993) A new role for the insect mushroom bodies: place memory and motor control. In R. D. Beer, R. Ritzman and T. McKenna (Eds.), *Biological Neural Networks in Invertebrate Neuroethology and Robotics* (pp. 199–225). Cambridge: Academic Press

Mobbs, P.G. (1982) The Brain of the honeybee *Apis mellifera* 1. The connections and spatial organisation of the mushroom body. *Phil. Trans. R. Soc. London* **2948**, 309–354

Nighorn, A., Healy, M.J. and Davis, R.L. (1991) The cyclic AMP phosphodiesterase encoded by the *Drosophila* dunce gene is concentrated in the mushroom body neuropil. *Neuron* **6**, 455–467

O'Dell, K.M.C., Armstrong, J.D., Yang, M.Y. and Kaiser, K. (1995) Functional dissection of the *Drosophila* mushroom bodies by selective feminisation of genetically defined sub-compartments. *Neuron* **15**, 55–61

O'Kane, C.J. and Gehring, W.J. (1987) Detection *in situ* of genomic regulatory elements in *Drosophila*. *Proc. Natl. Acad. Sci. USA* **84**, 9123–9127

Power, M.E. (1943) The brain of *Drosophila* melanogster. *J. Morphol.* **72**, 517–559

Rogina, B. and Helfand, S.L. (1995) Regulation of Gene Expression Is Linked to Life Span in Adult *Drosophila*. *Genetics* **141**, 1043–1048

Smith, H.K. and O'Kane, C.J. (1990) Use of a cytoplasmically located P-*lacZ* fusion to identify cell shape by enhancer trapping in *Drosophila*. *Roux's Arch. Dev. Biol.* **200**, 306–311

Strausfeld, N.J (1976) *Atlas of an Insect Brain*. Heidelberg, New York: Springer

Technau, G.M. and Heisenberg, M. (1982) Neural reorganisation during metamorphosis of the corpora pedunculata in *Drosophila melanogaster*. *Nature* **295**, 405–407

Yang, M.-Y., Armstrong, J.D., Vilinsky, I., Strausfeld, N.J. and Kaiser, K. (1995) Subdivision of the *Drosophila* mushroom bodies by enhancer-trap expression patterns. *Neuron* **15**, 45–54

Yeh, E., Gustafson, K. and Boulianne, G.L. (1995) Green Fluorescent Protein as a vital marker and reporter of gene expression in *Drosophila*. *P.N.A.S.* **92**, 7036–7040

PART IV, SECTION A

56. Transgenic Strategies for the Study of Mouse Development: An Overview

Charles Babinet

Institut Pasteur – Unité du Biologie du Développement, URA CNRS 1960, 25, rue du Docteur Roux, 75724 Paris, Cedex 15, France

Introduction

The last decade has witnessed tremendous progress in our insights into the way complex organisms are built from a single cell, the zygote (Davidson, 1994). It had long been recognized that all the cells of a developing multicellular embryo possess the same genetic information and therefore that its orderly development into an organism made of many different tissues and organs is based on the differential expression, tightly controlled in time and space of genes or groups of genes in the cells generated by the division of the zygote (Davidson, 1986). However, a very important notion has emerged which no one would have predicted only ten years ago, namely that despite noticeable variations from one animal species to another, the general molecular circuits used to mediate the building of an organism are astonishingly conserved (Davidson, 1994). Two examples will illustrate this point of pivotal importance. The first is concerned with the discovery in *Drosophila* of a group of genes, the *HOM-C* complex which was demonstrated to be involved in the determination of segmental identity along the anteroposterior axis (Lawrence and Morata, 1994). The surprise came when it was realized that these genes, which code for transcription factors containing a specific DNA binding homeodomain (Gehring, 1987 and Gehring *et al.*, 1994) have been remarkably conserved during evolution, with *Drosophila* homologs being found in all vertebrates, including man and mouse (where they are called Hox-Complex (Hox-C). Furthermore and most importantly all the available evidence indicated that *HOM-C* and *Hox-C* genes share a common function also conserved during evolution namely to give positional information to populations of cells along different axes, in particular the anteroposterior axis (Krumlauf, 1994). A second striking example of the fact that some fundamental mechanisms are at work in the making of any multicellular organism pertains to the process of embryonic induction in which cell type A instructs an uncommitted cell type X to adopt cell fate B (Lemaire and Gurdon, 1994). Embryonic induction has been recognized a long time ago. However only in the recent years was it realized that all metazoans may share homologous systems of signalling ligands, receptors and transcription factors at work in this process (Davidson, 1994; Egan and Weinberg, 1993; and Lemaire and Gurdon, 1994). Furthermore, a same signalling system may be used in various types of induction. This is perhaps best exemplified by the recent studies on *hedgehog* (*hh*) genes. Originally isolated in *Drosophila*, *hh* was shown to play various inductive roles in this organism (Perrimon, 1995). Subsequently a vertebrate homolog, *Sonic hedgehog*, cloned by homology with the *Drosophila hh* has been isolated in several species (see Ingham, 1995 and references therein) shown to play a signalling role in such embryologic processes as the patterning of the neural tube and of the limb bud as well as the differentiation of somites (Bumcrot and McMahon, 1995; Ingham, 1995; Johnson and Tabin, 1995; and Placzek, 1995). This progress towards a unified view of the basic processes of development and their underlying molecular mechanisms has been made possible by the experimental exploration of different organisms and also by the advent of many new and powerful technologies (Davidson, 1994). Among them, the ability to generate genetically modified – transgenic – organisms at will has become, in recent years, of primary importance, particularly in the mouse. Indeed, since the first descriptions of mouse germline modification by exogenous DNA more than fifteen years ago (reviewed in Babinet *et al.*, 1989; Jaenisch, 1988; and Palmiter and Brinster, 1986), thousands of transgenic lines of mice have been generated to address various aspects of mouse development. Due to space limitations it is impossible to review all these studies. Rather, we will try and illustrate, on the basis of selected examples, the specific contribution of transgenesis to the study of mouse development.

General Aspects

It is important to briefly recall the procedures for making transgenic mice (for more information on this topic see other chapters in this book). Apart from infection of embryos with retroviruses which has been used relatively unfrequently (for review see Jaenisch, 1988; and Palmiter and Brinster, 1986), there are two prevalent routes to generate transgenic mice: (1) Direct microinjection of DNA into the pronuclei of fertilized eggs. This procedure gives rise to mice in which the transgene has integrated randomly in the mouse genome generally as tandem head to tail arrays of variable length. Thus neither the number of copies integrated nor the site of integration are controlled. However, as we will see below, this route of transgenesis is instrumental in delineating the cis-acting regulatory sequences implicated in the correct temporal and spatial gene expression necessary for normal development and in offering a means of targeting expression of a given gene product into ectopic sites to try and find clues to its function; (2) A second route is the use of embryonic stem cells (ES). These cells, which are derived directly from the culture of blastocysts (Evans and Kaufman, 1981; Martin, 1981 reviewed in Robertson, 1987) have the remarkable ability to colonize a host embryo including its germline; thus previous introduction of genes into the cultured ES cells allows one to generate transgenic mice *via* the production of germline chimeras (Gossler *et al.*, 1986 and Robertson *et al.*, 1986). However the most important virtue of the use of ES cells to make transgenic mice is that desirable and low-frequency genetic alterations may be selected and verified in cultured cells and then reintroduced in the animal. Such a scenario has opened enormous possibilities in the study of development; indeed, methods have been developed which permit the replacement of a given endogenous allele by a mutated one *via* homologous recombination, thus paving the way for the elucidation of gene function *via* targeted mutagenesis (Bronson and Smithies, 1994; Capecchi, 1989; and Rossant and Joyner, 1989).

Transgenesis by Addition of Cloned DNA Sequences

Gene Regulation and Development

As already mentioned, it is essential for an organism to develop normally, that genes be expressed at the proper time and in the proper cells. Therefore understanding the mechanisms of gene regulation during development is one essential aspect of its study. The use of transgenic mice is invaluable to fulfil this goal; indeed, transgene expression can be studied in all cell types and at any desired time, in the context of the developing organism where all the possible regulatory circuits are at work. Finally, different versions of a given gene may be used to generate transgenic mice, thereby permitting to map the functional regions of the gene implicated in its correct regulation. However, one difficulty of this strategy has to be taken into account, which lies in the fact that transgene integration is apparently random; therefore neighbouring sequences might interfere, both qualitatively and quantitatively, with the proper regulation of the transgene under study. Such an effect is referred to as "position effect" (Wilson *et al.*, 1990) and might obscure the analysis of transgene regulation. Interestingly, specific DNA sequences called Locus Control Region have been discovered in the vicinity of some genes which render transgenes copy-number dependent and position independent (reviewed in Dillon and Grosveld, 1993).

Innumerable individual reports addressing the question of gene regulation have been published during the past years. In particular, since the first demonstration of the activity of an enhancer in controlling the expression of a gene in an *in vivo* context (Hammer *et al.*, 1987), many experiments of this type have been conducted to study the regulation of genes specific of a tissue or cell type, i.e. genes expressed in terminal differentiation (for review see Kollias and Grosveld, 1992). Here, we will consider more specifically genes involved in developmental processes and take the case of the *Hox-C* genes as an example of this kind of approach. As already mentioned these genes have been shown to be evolutionary homologs of *Drosophila HOM-C* genes. The *HOM-C* complex comprises 8 adjacent genes, the mutations of which results in homeotic transformations, i.e. the change of one body part into the likeness of another. The 38 murine *Hox* genes are organized in four complexes, each on a different chromosome and containing up to thirteen paralogous genes distributed over approximately 120 Kb. Furthermore and intriguingly, there is a correlation between the physical order of the genes along the chromosome and their spatially restricted expression along the anteroposterior axis of the embryo in different tissues or organs. This property, referred to as colinearity holds true also for the time at which expression of the genes initiates during development and in the induction of *Hox* genes by retinoic acid (Figure 56.1).

Thus, there is a distinct combination of expressed *Hox* genes in cells along the anteroposterior axis and at a given time of development, referred to as the *Hox* code (Hunt *et al.*, 1991; Kessel and Gruss, 1991; Krumlauf, 1994; and McGinnis and Krumlauf, 1992). That the *Hox* code has functional implications was amply demonstrated by various experimental approaches including ectopic expression or knockout of *Hox* genes (see below). Thus, identification of regulatory elements concurring to the proper spatial and temporal expression of *Hox* genes is one of the key questions to understand the genetic mechanisms at work in determining positional information and establishment of body plan (Hunt *et al.*, 1991; Kessel and Gruss, 1991 reviewed in McGinnis and Krumlauf, 1992). To try and address this issue, transgenic technology has proven particularly adapted; the general scenario is fairly simple in principle. A reporter gene, the *E. coli* β-galactosidase whose activity is easily detectable by a simple staining procedure, is coupled with various putative regulatory regions of a given *Hox* gene and the expression of the transgenes is monitored in the resulting transgenic mice: comparison with the normal endogenous expression permits to draw conclusions as to the regulatory elements acting to recapitulate this expression. Examples of this type of studies are given in Table 56.1. The results may be summarized as follows: (1) In a number of cases, it has been possible, using given regions of an *Hox* gene, to recapitulate its pattern of expression, thus suggesting that regulated expression could take place outside of its normal chromosomal context (see for example Behringer *et al.*, 1993 and Gerard *et al.*, 1993). This came as a surprise in view of the colinearity and the extraordinary conservation of chromosomal organization between Drosophila HOM-C and vertebrate *Hox*-C genes, suggesting that this physical organization might be essential for proper regulation and expression. However, it should be noted that in several cases the complete endogenous pattern of expression could not be reproduced, even when using large genomic regions around the gene (see for example Eid *et al.*, 1993; Kress *et al.*, 1990 and Vogels *et al.*, 1993); furthermore, although the monitoring of gene expression by β-galactosidase activity is efficient, it has proven difficult to make a precise comparison with endogenous activity; in addition this approach does not address such aspects as level of expression, differential transcripts, post-transcriptional regulation, etc. Thus it remains quite possible that differences in the precise pattern of expression between transgene and endogenous expression have escaped analysis and that the truely correct expression of *Hox* genes might depend on a clustered organization. (2) It has been clearly demonstrated that regulation of *Hox* gene expression depended on a complex combinatorial action of discrete regulatory elements, acting either positively or negatively and responsible for various aspects of expression pattern, including

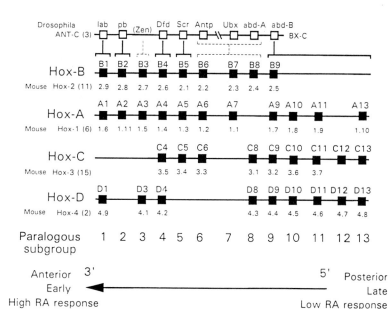

Figure 56.1. Chromosomal organization of the *Drosophila* HOM-C homeotic complex and the four vertebrae *Hox* complexes. There are thirteen paralogous murine groups corresponding to the different genes of the *Drosophila* HOM-C complex. Not all complexes have members representative of each group. Above and below the boxes are the new (Scott, 1992) and former nomenclature respectively. The large arrow at the bottom indicates the sense of transcription of the genes in the different clusters as well as their colinearity with respect to temporal and spatial expression and response to retinoic acid (RA) (courtesy of R. Krumlauf).

Table 56.1. Examples of the use of transgenic mice to identify the regulatory elements contributing to the pattern of expression of *Hox* genes during embryonic development. In all these studies, the E. *coli* β-galactosidase was used as a reporter gene fused to different regions of a given Hox gene.

GENE	Main Conclusions
Hoxa-4	4 kb of 5′ flanking sequences plus 1 kb of structural gene sufficient for correct expression. A positive element necessary for expression present in a 2 kb 5′ flanking region.[1]
Hoxa-5	An enhancer found in a 912 bp region in the 5′ end of the gene sufficient for expression in the brachial region of the embryonic spinal cord.[2]
Hoxa-7	3.6 kb of 5′ plus 1.7 kb of 3′ flanking sequences sufficient for correct initial expression, but insufficient for later expression.[3]
	Three different elements (A: 3.6 kb 5′ sequences; B: 130 bp 5′ nontranslated sequences; C: intragenic sequences encompassing 1st and 2nd exon), shown to contribute to spatial restriction of expression.[4]
	Definition of a minimal enhancer element, 470 bp long located 1.6 kb 5′ of promoter specifying anterior boundary of expression during embryonal development.[5]
Hoxb-1	Two enhancers found in 3′ flanking region of the gene cooperating in the proper early expression of the gene. One of these contains a retinoic acid response element essential for early expression.[6]
	Identification of a repressor element, containing a retinoic acid response element and responsible for the normal restriction of expression to rhombomere 4 in the hind brain.[7]
Hoxb-4	Two regions, one 3 kb long 3′ of the gene, another 1.2 kb containing the intron and part of the 2 flanking exons, identified as a spatially specific enhancer.[8]
Hoxb-7	A 1.3 kb long sequence 5′ of the promoter sufficient for expression in intermediate mesoderm derivatives but not in other parts of mesoderm or neurectoderm.[9]
	Identification of three cis-acting elements 5′ of the gene cooperating for assigning boundaries of expression along the enteroposterior axis.[10]
Hoxc-8	Identification of a 138 bp DNA fragment (3 kb 5′ to the gene) sufficient for posterior neural tube expression. Combinatorial interactions suggested to take place within this fragment between several elements to achieve expression in this region.[11,12]

[1] Behringer *et al.* (1993); [2] Tuggle *et al.* (1990); [3] Püschel *et al.* (1990); [4] Püschel *et al.* (1991); [5] Knittel *et al.* (1995); [6] Marshall *et al.* (1994); [7] Studer *et al.* (1994); [8] Whiting *et al.* (1991); [9] Kress *et al.* (1990); [10] Vogels *et al.* (1993); [11] Bieberich *et al.* (1990); [12] Shashikant *et al.* (1995).

onset, maintenance or anteroposterior boundaries of expression as well as tissue specific expression. (3) In several instances, it was shown that discrete regulatory elements located in noncoding sequences were conserved between vertebrate species not only structurally but functionally: for example, an enhancer element of the mouse *Hoxa-7* gene could be replaced by the human homologue of *HOX A7* and reproduce a correct anterior boundary of expression in transgenic mice (Knittel *et al.*, 1995). These results come as a further suggestion of conserved regulatory elements in evolution. In summary, the use of transgenic mice has permitted to uncover numerous regulatory elements which act in concert to control the proper expression of *Hox* genes during mouse embryonic development. These observations in turn raise the important question of the transacting factors interacting with these ele- ments and therefore of the upstream regulatory genes controlling *Hox* gene expression. A few studies have begun to address this question using transgenic methodology. For example, it was shown that *Krox*-20, a zinc-finger protein was involved in rhombomere specific activation of *Hoxb*-2 during hindbrain segmentation. An initial evidence came from the demonstration that ectopic expression of *Krox*-20 could induce the expression of a fusion transgene containing an *Hoxb*-2 enhancer linked to β-galactosidase coding sequences as a reporter (Sham *et al.*, 1993). The involvement of *Krox*-20 in *Hoxb*-2 transcriptional activation was further demonstrated by the analysis of *Hoxb*-2 expression in *Krox*-20 knock-out mice (see below): indeed, in these mice, in the rhombomere 5 region, which normally coexpresses *Krox*-20 and *Hoxb*-2, the latter was extincted (Schneider-Maunoury *et al.*, 1993). Interestingly, a similar type of control by *Krox*-20 was demonstrated for the expression of

Hoxa-2, the only other paralog of Hoxb-2, in rhombomeres 3 and 5 (Nonchev et al., 1996). In another series of experiments Marshall et al., 1994 and Studer et al., 1994 demonstrated that retinoids and their nuclear receptors are involved in rhombomere 4 restricted expression of Hoxb-1. Furthermore, it was recently shown, using a reporter transgene scenario, that this restricted expression was mediated by an autoregulatory loop involving Hoxb-1 itself (Pöpperl et al., 1995). Collectively, these studies illustrate the usefulness of transgenic methodology to try and delineate the regulatory pathways, including the role of cis-acting DNA sequences and the corresponding trans-acting factors, which are at work in complex developmental processes. Finally, it is important to note that the potential of this approach has been extended, in recent years, by the generation of transgenic mice bearing very large sequences (up to several hundred kilobases) of exogenous DNA sequences cloned in Yeast artificial chromosomes (YAC). This will make it possible to tackle the study of large genes or gene clusters and to take into account factors like putative remote control elements or the long-range effects of regulatory elements which are at work in phenomenons like X-chromosome inactivation or imprinting (reviewed in Forget, 1993 and Montoliu et al., 1993, see also the chapter on YAC transgenic mice in this book).

Gain-of-Function Genetic Modifications by Ectopic Gene Expression in Transgenic Mice

Using appropriate transgene constructs, it is possible to deregulate the normal expression of a given gene. Depending on the promoter used and the coding sequences fused to it, enhanced or ectopic expression of either the normal or of a modified gene product may be obtained. Thus, the transgenic approach permits to create gain-of-function genetic modifications, the effect of which may give insights into the biological role of the gene product under study. Such a scenario has been extensively used for the study of putative developmental genes. In the case of Hox genes, whose main features we have described above, the first demonstration in mouse of an homeotic transformation analogous to the one observed in Drosophila HOM-C gene, was given by this type of experiment: in effect, ectopic expression of Hoxa-7 under the regulation of a chicken β-actin promoter, resulted in posterior transformations of vertebrae, thus giving the first hint that Hoxa-7 is a developmental control gene (Balling et al., 1989 and Kessel et al., 1990). Since then, several other gain-of-function experiments have been performed for other Hox genes, contributing to shed light on various aspects of their developmental function (Hoxa-1: Zhang et al., 1994; Hoxb-8: Charite et al., 1994; Hoxd-4: Lufkin et al., 1992; Hoxb-8 and Hoxc-8: Pollock et al., 1995; and Hoxc-6: Jegalian and De, 1992). Another example concerns Wnt-1, a member of the Wnt gene family, which consists of at least fifteen members. Wnt genes are expressed in precise antero-posterior domains along the axis of the central nervous system (CNS), and code for secreted glycoproteins. Together these properties are consistent with a role in establishing regional differentiation of the CNS, via a system of intercellular signalling. To investigate the putative role of Wnt-1 in this signalling system, Dickinson et al., 1994 generated transgenic mice expressing Wnt-1 under the control of Hoxb-4 regulatory sequences which resulted in expanding the normal dorsal domain of Wnt-1 expression to include ventricular cells all along the dorsoventral axis of the spinal cord. A dramatic increase in the number of dividing cells followed in the ventricular region and a concomittant ventricular expansion. This experiment provided strong evidence that Wnt-1 is implicated in the control of cell growth which is an important aspect of the orderly development of the complex structure of the neural tube. This was an interesting addition, as regards the developmental role of Wnt-1, to the study of embryos bearing a Wnt-1 null mutation which resulted in the deletion of part of the dorsal rostral hindbrain (McMahon and Bradley, 1990; and Thomas and Capecchi, 1990; see also McMahon et al., 1992 and Shimamura et al., 1994).

Transgenes as Insertional Mutagens

This section deals with the potential mutagenic effect of transgene insertion and the opportunity it offers to identify new genes of developmental importance.

(a) Random Transgenic Mutations. When any exogenous DNA is introduced into the mouse germline, it may integrate into the genome in such a way that it perturbs the expression of an endogenous gene(s), therefore causing a mutation, which may result in a phenotype when transgene insertion is rendered homozygous. Thus systematic intercrosses between transgenic mice become a way of screening for insertional mutations. The main virtue of this approach is that the transgene can serve as a molecular tag to recover the affected gene. There is however an important drawback to this strategy, namely the fact that the recovery of a phenotype following transgene insertion is entirely serendipitous, and therefore a mere by-product of transgenic studies initiated for other purposes (for review see Gridley et al., 1987; Meisler, 1992; Rijkers et al., 1994; and U. Rütter, this book). The overall frequency of insertional mutations in transgenic mice has been estimated to be between 5% and 10%, whether they were obtained by retroviral infection or pronuclear injection (Palmiter and Brinster, 1986). However an important difference between these two routes for generating insertional mutations should be mentioned. Indeed, whereas proviral integration is "clean" with only one copy integrated and a typical 4–6 bp duplication of cellular DNA at the site of integration (Grandgenett and Mumm, 1990), normal exogenous DNA integrates as large head-to-tail tandem arrays and complex structural changes (particularly deletions) frequently take place at the site of integration (see for example Cheng and Costantini, 1993; Palmiter and Brinster, 1986; and Woychick et al., 1985 reviewed in Rijkers et al., 1994). Furthermore it appears that, in the case of retroviral insertions, transcribed regions are generally found close to the insertion site and are likely to represent the gene putatively affected by the integration event (see Gridley et al., 1987). One drawback, though, of the use of retroviral vectors for insertional mutagenesis is that the integration site does not seem random; thus there might be an important bias in the type of genes which may be tagged in this way (Gridley et al., 1987). Although several dozens of insertional mutations have been documented, only for a minority of them has the primary molecular defect – i.e., the affected gene – been identified. However among the latter, several appear particularly interesting from a developmental point of view and have permitted to isolate and characterize for example the *limb deformity* gene which codes for a family of proteins, the formins, implicated in limb and kidney development (Chan et al., 1995; Maas et al., 1990; Woychick et al., 1990 and Woychick et al., 1985), the *microphthalmia* gene which codes for a basic-helix-loop-helix leucine zipper transcription factor and is implicated in multiple developmental processes (Hodgkinson et al., 1993 and Hughes et al., 1993, see also Moore, 1995) the *reelin* gene which codes for an extracellular matrix protein involved in cell adhesion and is important for the control of neuronal migration in the developing brain (D'Arcangelo et al., 1995 and Miao et al., 1994) or the *nodal* gene which codes for a novel member of the transforming growth factor-β family which is essential for the process of gastrulation (Conlon et al., 1991 and Zhou et al., 1993). Thus notwithstanding the difficulties bound to it, transgene insertional mutagenesis has permitted to uncover a number of genes of developmental importance and in view of the interesting phenotypes described for other insertional mutants (see Meisler, 1992; Rijkers et al., 1994 and U. Rütter, this book), it is expected that the list of these genes should grow in the near future.

(b) A More Directed Approach to the Identification of Developmental Genes: The Enhancer or Gene Trap Strategy. In contrast to the possible mutagenic effect of any transgene which relies on the chance insertion of the transgene in a gene of interest, new approaches have been developed based on the use of "entrapment" vectors which are specifically constructed to reveal the activity of endogenous genes or their cis-acting regulatory sequences (reviewed in Gridley et al., 1987; Hill and Wurst, 1993; Rossant and Joyner, 1989; and Skarnes, 1993). A generic entrapment vector typically contains a reporter gene generally *E. coli* β-galactosidase and a selector gene. It may be constructed either as plasmid or as retroviral vector. After transfection into or retroviral infection of ES cells and the use of an appropriate selection, the activity of the reporter gene may be monitored in ES cells or during the development of chimeras derived from blastocysts injected with the modified ES cells. Typically, two broad types of entrapment vectors may be used: in the first type (enhancer trap vector) the reporter gene coding sequences are linked to a minimal promoter, and the activity of the reporter gene will depend on the presence of cis-acting regulatory sequences of an endogenous gene in the vicinity of the integration site. In the second type (gene trap vector) the reporter gene does not comprise a promoter and is placed in the vector with or without an acceptor splice in its 5' part. In the former case, the reporter gene may be expressed even if it is integrated into an intron whereas in the latter it must be integrated in frame into an exon of an endogenous gene (reviewed in Skarnes, 1993). The reporter gene may be the *E. coli* β-galactosidase in which case an independent selection cassette must be added in the vector, or a selectable gene (von Melchner et al., 1992) or a fusion between β-galactosidase and the selectable gene (Camus et al., 1996; and Friedrich and Soriano, 1991). In the last two cases, only the insertions into genes active in ES cells will be selected. An alternative scenario has also been described in which, in addition to the 5' splice acceptor fused to the β-galactosidase gene, an independent selection cassette is placed 3' of the construct which contains a *neo* gene with a splice donor but without a polyadenylation signal. Thus, efficient expression of the cassette should depend on the integration into a transcription unit and the use of its polyadenylation signal (Niwa et al., 1993 and Yoshida et al., 1995).

The potential advantages of gene trap vectors are three-fold: (1) the reporter gene activity may be monitored in ES cells as well as in chimeras or their progeny; it is anticipated that it should mimic the activity of the endogenous trapped gene and permit the analysis of its expression pattern during development and therefore give clues to the possible involvement of the trapped gene in a developmental process. (2) The insertion of the vector is likely to interrupt the proper functioning of the trapped gene and therefore be mutagenic; thus phenotypic abnormalities can be screened in mice homozygous or heterozygous for the insertion. (3) The occurrence of a fusion transcript should facilitate the cloning of the interrupted genes in particular by the use of rapid amplification of cDNA ends by PCR (Frohman et al., 1988). Several large screens of gene (Friedrich and Soriano, 1991 and Wurst et al., 1995) or enhancer trap vector (Korn et al., 1992) insertions have been performed and demonstrated that indeed entrapment vectors can uncover genes whose expression was either temporally or spatially restricted, therefore developmentally regulated; furthermore, in two cases where the trapped gene has been cloned it was shown that the reporter gene reflected endogenous gene activity (Skarnes et al., 1992). In addition, examination of several dozens of insertions transmitted to the germline indicates that around 40% of them cause embryonic lethal phenotypes (Friedrich and Soriano, 1991; Skarnes et al., 1992 and von Melchner et al., 1992). Several genes have already been identified using the entrapment vector strategy and shown to be important in development (Chen et al., 1994; DeGregori et al., 1994; Neuhaus et al., 1994; Skarnes et al., 1992; Skarnes et al., 1995; Soininen et al., 1992 and Takeuchi et al., 1995) and many more insertions are awaiting cloning of the interrupted genes. Taken together, these results give strong support to the premise of the gene trap strategy that it permits to isolate new genes important in development and is therefore an important tool for the molecular dissection of embryonic development.

From Gene Addition to Gene Replacement: Targeted Mutagenesis *via* Homologous Recombination in Embryonic Stem Cells

Mutational analysis is one of the key approaches to the full understanding of developmental processes: indeed, the analysis of the consequences of a mutation in a gene is a classical and privileged way to gain access to its function. In mouse, in spite of numerous spontaneous and experimentally induced mutations exhibiting various phenotypes, the mutational analysis has been hindered, in particular by the physical size and complexity of the genome, which made it difficult to go from the phenotype to the genotype, in other words to identify the gene, the mutation of which induced a given phenotype. Due to these limitations, alternative strategies have been developed to identify murine developmental control genes. One of these,

as already mentioned, was to use insertional mutagenesis in transgenic mice (see above) which has the virtue to tag a gene and therefore to give a molecular means to identify it and to induce a mutation and hopefully a phenotype. Another powerful way to identify developmentally important genes stemmed from the realization that functional domains of these genes are very often conserved. Thus, starting from the key developmental genes cloned in organisms relatively more easily amenable to genetic analysis like *Drosophila*, it became possible to isolate their mammalian homologs; indeed, this strategy was very successful and numerous genes or gene families putatively involved in the control of murine development were identified in this way, including, for example, the *POU*, *Pax* and *Hox* (reviewed in Kessel and Gruss, 1990) or the *hedgehog* (reviewed in Ingham, 1995) families of murine genes; considerable efforts were then made to describe their spatial and temporal pattern of expression.

Here again, the case of *Hox* genes is paradigmatic: the deployment of their spatial and temporal pattern of expression was reminiscent of that found in *Drosophila HOM-C* genes in several respects, such as their partially overlapping expression domains along the anteroposterior axis and in different organs and their precisely defined anterior expression boundaries. This was a strong indication that they might share some functional role with that of *Drosophila HOM-C* genes. However, the lack of mutants allowing to demonstrate the effects of impaired gene expression rendered difficult to draw a definitive conclusion.

The advent of targeted mutagenesis has radically changed the situation in the late eighties. Indeed the use of ES cells combined with homologous recombination of an incoming modified version of a gene opened the way to generate mice with mutations in any gene, as long as it had been cloned (reviewed in Bradley et al., 1992; Bronson and Smithies, 1994; and Capecchi, 1989). Thus, mice bearing loss-of-function mutations in many of the 38 murine *Hox* genes have been generated and the effect of these mutations, either in the heterozygous or the homozygous state has been reported. Overall and despite the complexity of the various phenotypes observed including homeotic transformations and various morphological anomalies, these studies have revealed the essential role of *Hox* genes in organizing the mouse body plan, particularly the segmental patterning of the embryo, including hind brain and neural crest, pharyngeal arches and its derivatives or vertebrate spine (for reviews, see Krumlauf, 1993; 1994; Mark et al., 1995 and Wilkinson, 1993) and the formation of a limb (Duboule, 1994). In addition, the availability of these knock-out mice permits to create by appropriate crosses compound mutants bearing more than one mutated *Hox* gene and to examine the resulting phenotype in comparison to the one obtained in the single mutant (Condie and Capecchi, 1994; Horan *et al.*, 1995 and Rancourt *et al.*, 1995). Such an approach, in its beginning, should help to define more precisely the role of individual *Hox* genes and their interactions in the formation of the mammalian body plan. Finally, it should be mentioned that targeted gene disruption may also shed light on the regulatory cascade controlling *Hox* gene expression *in vivo*. We have already seen that the disruption of *Krox*-20 had permitted to demonstrate a direct and positive regulation of *Hoxb*-2 by *Krox*-20 (Schneider-Maunoury *et al.*, 1993 and Sham *et al.*, 1993). Two recent studies further illustrate this point. In the first, inactivation of *Cdx1*, a murine homologue of the *Drosophila caudal* gene coding for a transcription factor, resulted in mice showing anterior homeotic transformations of vertebrae. These changes were accompanied by a shift in *Hox* gene expression towards more caudal domains, in the somitic mesoderm. This was taken as evidence of the *Cdx1* gene controlling directly *Hox* gene expression. The evidence was further strengthened by the discovery of *Cdx1* binding motives in *Hox* gene regulatory sequences and the demonstration of the transactivation of the *Hoxa*-7 promoter by *Cdx1* in *ex vivo* experiments (Subramanian *et al.*, 1995). In the second study, the introduction into the germ line of a null mutation in *Mll* gene revealed yet another factor acting to control *Hox* gene expression; the product of *Mll* gene, MLL, is a protein which possesses a highly conserved SET domain also found in *Drosophila* trithorax and Polycomb group genes which have been shown to regulate *HOM-C* genes (Simon, 1995). While homeotic transformations were observed in *Mll* +/− heterozygous mice with concomitant shift in anterior boundaries of expression of *Hoxa*-7 and *Hoxc*-9, the expression of these two genes was completely abolished in *Mll* −/− embryos, thus demonstrating that *Mll* positively regulates *Hox* gene expression (Yu *et al.*, 1995).

Due to the extremely powerful character of targeted mutagenesis in assessing gene function it is no wonder that despite its relatively demanding conditions, it has been increasingly used since the first demonstration of targeted mutagenesis in ES cells (Doetschman *et al.*, 1987; Thomas and Capecchi, 1987; reviewed in Capecchi, 1989; and Bronson and Smithies, 1994). Thus, several hundred genes have already been targeted, using homologous recombination in ES cells and the list is growing steadily (see Brandon *et al.*, 1995a,b,c). Among those, a good proportion has been shown to be involved in numerous and various basic mechanisms of embryogenesis therefore giving insights into their function in development. A detailed review of each of these studies is not within the scope of this chapter. Rather, we will discuss some general aspects of the results of targeted mutagenesis as they have emerged from these studies, to end with a description of the current improvements in its use.

Patterns of Gene Expression and Biological Function

The prominent expression of a protein in one or another cell type is generally taken as an evidence of a functional role in this cell type. However, examination of many gene knock-out mice has shown that this contention is rarely verified. Rather, phenotypes are generally observed only in a subset of the expression domains where the genes are expressed. Table 56.2 shows the example, among many others, of the gene coding for lymphoid enhancer factor 1 (*LEF*-1), a sequence-specific DNA-binding protein. The knock-out of *LEF*-1 results in a remarkable phenotype, demonstrating its implication in the development of several organs that require inductive epithelial mesenchymal interactions (van Genderen *et al.*, 1994). However, several other organs expressing *LEF*-1 appear to develop normally. This situation, encountered in *LEF*-1 knock-out as well as in many other cases, could simply be due to the general requirements of gene control mechanisms. Thus it has been argued that, whereas a given gene should of course be expressed in a cell where its product is needed, it would not have necessarily to be turned off in other cells where it is useless, as long as it is not toxic for these cells.

Table 56.2. An example of divergence between expression domains and phenotype: the case of *LEF*-1 gene knock-out mice.

Expression sites	*Phenotype*
T-cells	
Pituitary gland	
Inner ear	
Kidney	
Genital eminence	
Tooth germs	+
Mammary glands	+
Hair follicules	+
Trigeminal nerve	+

In other words, private negative controls for any of the thousands of proteins made by an organism would be too expensive and even unnecessary (Erickson, 1993). Although difficult to prove experimentally, this hypothesis cannot be eliminated.

Functional Redundancy

Yet another reason why the lack of a gene product in a given cell or tissue would not result in a phenotype could be coexpression of other gene products sustaining a similar function. This is the concept of functional redundancy (discussed in Thomas, 1993), which is likely to operate when disrupting genes belonging to a family of related genes. Whereas true genetic redundancy is most probably impossible from an evolutionary point of view (see Brookfield, 1992 and Thomas, 1993), minimal phenotypes of gene knock-outs could be explained in this way. Two interesting examples will illustrate this point. (1) In vertebrates, the myogenic regulatory factors (MRFs) constitute a family of four basic helix-loop-helix transcription factors including MyoD, Myf-5, myogenin and MRF4, which have the ability when expressed ectopically in non-muscle cells to induce the skeletal muscle differentiation program (Weintraub et al., 1991). Neither Myo-D (Rudnicki et al., 1992) nor Myf-5 (Braun et al., 1992) knock-out mice revealed any abnormalities in skeletal muscle. However in MyoD knock-out mice, Myf-5 mRNA was up regulated, whereas in mice lacking Myf-5, skeletal muscle development, though normal, was delayed until Myo-D was expressed. Taken together, these results suggested that Myf-5 and Myo-D can support skeletal muscle development each on its own, and therefore can be considered as functionally redundant. This hypothesis was further substantiated by the generation of Myo-D–Myf-5 double knock-out mice, which exhibited a total absence of skeletal myoblasts and muscle (Rudnicki et al., 1993; reviewed in Olson and Klein, 1994; Weintraub, 1993; and Rudnicki and Jaenisch, 1995). (2) In mice, there are two Engrailed (En) genes, En-1 and En-2 which are homologues of the Drosophila segmentation gene engrailed. They both encode homeodomain-containing transcription factors, which share small highly conserved domains but which overall exhibit only 55% identity with each other. En-1 and En-2 have essentially overlapping patterns of expression in a presumptive midhindbrain region but the former is expressed earlier than the latter. Disruption of En-1 and En-2 results in contrasting phenotypes: En-1 mutants die around birth with various defects (Wurst et al., 1994) while En-2 mutants are viable and fertile, only exhibiting an embryonic reduction in cerebellar size and postnatal alteration of foliation (Joyner et al., 1991 and Millen et al., 1994). This could be due either to differences in their biochemical function or to the difference in the temporal expression of the two genes. To address this question, En-2 coding sequences were inserted into the En-1 locus by homologous recombination which altogether resulted in disrupting En-1, and in rendering En-2 expression dependent on En-1 regulatory sequences. Dramatically, En-1 mutant defects were completely rescued. This indicated that En-1 and En-2 gene products share the same biological function and that the only difference between En-1 and En-2 gene function resides in their divergent pattern of expression. The evolutionary implication of these experiments should be underlined: they suggest that genes could acquire new developmental functions through modifications of not only their coding sequences but also of their regulatory sequences, resulting in modified patterns of expression. Interestingly, an analogous situation has been observed in Drosophila: indeed, it was shown that the *paired* (*prd*), *gooseberry* (*gsb*) and *gooseberry neuro* (*gsbn*) genes, have distinct patterns of expression and distinct developmental function; yet their products share the same biochemical function, and therefore can be considered as functionally equivalent (Li and Noll, 1994). Thus, as in the case of En-1–En-2 murine genes, *prd*, *gsb* and *gsbn* genes appear to have gained new developmental functions by acquisition of different cis-regulatory regions.

Absence of Phenotype: Superfluous Genes?

Even more extreme cases of absence of phenotypes after gene disruption have been documented, for example in tenascin (Saga et al., 1992) and vimentin (Colucci-Guyon et al., 1994) genes which encode an extracellular matrix protein and an intermediate filament (IF) protein respectively. However and in contrast to the cases of Myf-5–Myo-D and En-1–En-2 genes there was no obvious candidate gene to take over the putative function of tenascin (Saga et al., 1992) and in vimentin knock-out mice it appeared that no other member of the large family of IF was synthesized in place of vimentin. These genes are expressed in specific patterns during embryonic development; furthermore, both genes are highly conserved in vertebrates. Yet and most surprisingly mice lacking vimentin or tenascin are apparently normal and fertile. At their face value, these results indicate that these two proteins are not essential in normal development. Should one conclude however that these genes are superfluous? This is highly improbable, if not impossible from an evolutionary point of view, as already mentioned in the case of apparently redundant gene function; indeed, if a gene were totally superfluous, and therefore not subjected to selective constraints, the constant rate of mutations would inevitably result in its inactivation. Other explanations must therefore be found. These probably lie in the difficulty of finding the right test for detecting a phenotype. Thus it could be that the phenotype has escaped the analysis, either because it is slight, or because it would be manifested only in unusual, e.g. pathological situations. Finally one possibility cannot be dismissed which pertains to the fact that search for a phenotype is made in the conditions of a laboratory animal colony. Therefore, possible changes in genetic fitness bound to a given mutation cannot be taken into account. For example, a 1% fitness reduction, significant in an evolutionary sense, would necessitate a sample size of 600,000 to be uncovered (Brookfield, 1992).

Recent Progress in Targeted Mutagenesis

Though powerful, the strategy of introducing null mutations in a given gene to gain insights into its biological role has two main limitations: firstly, when it results in a lethal phenotype, it precludes analysis of the possible function of the knock-out gene later in development; secondly, for obvious reasons, it would be desirable to introduce other types of gene modifications like point mutations, deletions or mutations in regulatory sequences, etc. To circumvent these limitations, new scenarios and new types of targeting vectors are being developed.

(a) New Types of Mutations. When introducing a point mutation into a gene, one would like the final modified allele be free of foreign sequences, i.e. the selectable marker gene normally used for generating null mutations, which might interfere with the regulation of the targeted gene. Several strategies have been designed to fulfil this goal (see the chapter by S. Viville, this book). We will briefly describe two of them which are probably the most promising and both comprise two steps. The first one uses selection for or against expression of hypoxanthine phosphoribosyl transferase (HPRT). $Hprt^-$ ES cells (bearing a mutation in the endogenous hprt gene, on chromosome X) are used and targeted with a construct containing an *hprt* selectable marker (see Figure 56.2). $Hprt^+$ homologous recombinants, are in turn targeted with a construct made of sequences homologous to the gene of interest and bearing a mutation. Selection for $hprt^-$ clones finally permits the isolation of cell lines containing the modified allele, free of any foreign sequences. It should be noted that the clones selected in the first step are a base to generate ES lines with various mutations in the gene of interest (Stacey *et al.*, 1994; reviewed in Bronson and Smithies, 1994). The second approach uses site

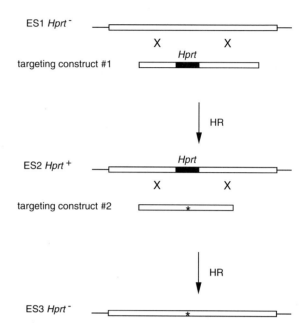

Figure 56.2. Schematic view of the use of HPRT selection in homologous recombinants to create "clean" mutations. ES1 cells (HPRT$^-$) are transfected with a first targeting contruct containing sequences homologous to the gene of interest and an HPRT expression cassette. After selection for HPRT$^+$, homologous recombinants (ES2) are screened. ES2 cells (HPRT$^+$) are then transfected with a second targeting construct made of homologous sequences to the gene of interest and bearing a mutation (*). Selection for HPRT$^-$ should give rise only to ES cells (ES3) bearing the mutation and completely devoid of foreign sequences (for details see Stacey *et al.*, 1994; and Bronson and Smithies, 1994) (HR = homologous recombination).

specific recombinases, several of which are known (reviewed in Kilby *et al.*, 1993). The use of Cre recombinase has been particularly investigated in targeted mutagenesis in ES cells. This enzyme, isolated from bacteriophage P1, can induce excision of a DNA sequence flanked by two copies of a specific 34 bp-long sequence, the *Lox P* site. After excision, the only remaining foreign sequence is one *Lox P* site. Thus, after introduction into ES cells of a construct containing a mutation in a given gene and the selectable marker flanked by *Lox P* sites, homologous recombinant clones are isolated; subsequently the transfection of a Cre recombinase expression vector into these clones promotes the excision of the selectable marker sequences (Figure 56.3) (see for example Disanto *et al.*, 1995; Zou *et al.*, 1994; reviewed in Bronson and Smithies, 1994). Finally, it should be noted that a potentially very interesting use of Cre-*Lox P* system lies in the derivation of ES cells and/or mice bearing either

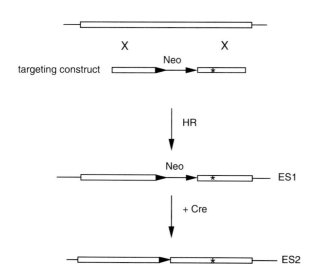

Figure 56.3. Simplified view of the use of the Cre-*LoxP* system to generate subtle "clean" mutations.
The targeting construct is made of sequences homologous to the endogenous gene of interest, and containing a particular mutation (∗). A selectable marker (Neo), flanked by *LoxP* sequences (▶) is placed in an intron of the gene of interest. The ES1 homologous recombinant cells are then transfected by a Cre recombinase expression vector, which results in the excision of the selectable marker. This gives rise to ES2 cells, bearing a mutation in the gene of interest and devoid of the selectable marker sequences; the only remaining foreign sequence is the 34 bp *LoxP* element (for details see Disanto *et al.*, 1995; and Bronson and Smithies, 1994) (HR = homologous recombination).

translocations (Smith *et al.*, 1995) or long-range deletions (3–4 centiMorgans) (Ramirez-Solis *et al.*, 1995). These methods greatly extend the range of designed modifications that can be made to the mouse genome. The availability of mice with chromosomal rearrangements should help in the study of mouse development for two main reasons; first, they will facilitate the genetic analysis of particular chromosome regions; second they should serve as models of developmental defects, as it has been shown that chromosomal rearrangements are an important cause of fetal loss in humans (Epstein, 1986).

(b) Conditional Mutations. One further prospect of the Cre-*Lox P* system of great interest for developmental studies is the possibility of creating conditional mutations. In this scenario, two transgenic lines are used (Figure 56.4). By gene targeting in ES cells, a first line A is generated which contains the gene of interest flanked by two *Lox-P* sites ("floxed" gene) in such a configuration that it does not perturb gene functioning; the second line, B, is a conventional transgenic line containing a fusion transgene made of a promoter and regulatory sequences fused to the coding sequences of the Cre recombinase. When crossing lines A and B, and depending on the promoter used, the excision of the gene of interest will ensue, thus inducing the mutation in only a restricted cell lineage or tissue. Such a strategy should be extremely useful in that it will permit in principle to assess the effect of a mutation at any given time or in any cell type of the developing embryo. The feasibility of this strategy has been recently demonstrated, using DNA polymerase β as the "floxed" gene and a fusion transgene targeting the expression of Cre in T cells. This resulted in the deletion of the DNA polymerase β, specifically in

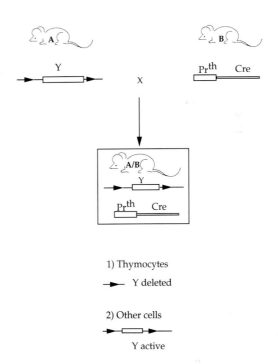

Figure 56.4. The use of the Cre-*LoxP* system to create conditional mutations: a schematic view. Transgenic mouse A, obtained via homologous recombination in ES cells contains the gene of interest (Y) flanked by two *LoxP* (▶), which do not prevent normal functioning of Y. Transgenic mouse B, obtained by pronuclear injection contains a fusion transgene made of a promoter specifically active in thymocytes (P_r^{th}) linked to the coding sequences of Cre recombinase. Crossing mice A and B, results in a mouse in which Cre recombinase will be expressed only in thymocytes resulting in excision of Y gene. In other cells, the "floxed" Y gene functions normally (for details see Gu *et al.*, 1994).

T cells (Gu et al., 1994). In a variation of this scenario, Kuhn et al., 1995 generated double transgenic mice containing the "floxed" DNA polymerase β gene and a fusion transgene made of the interferon inducible promoter of the mouse *Mx*1 gene linked to Cre recombinase coding sequences; treatment of these mice with interferon induced the deletion of the DNA polymerase β in various organs of the transgenic mice.

A very interesting alternative means of creating transgenic mice in which the Cre recombinase synthesis could be induced at a given time and in specific cells, is suggested by the results of Gossen et al., 1995. Indeed, these authors engineered a fusion protein made of a mutated version of the tetracycline repressor sequences linked to the Herpes VP16 activating sequences (called rtTA). They demonstrated that, only in the presence of tetracycline, the fusion protein could bind to tetracycline operator (*tet O*) DNA sequences, thereby promoting the activation of a reporter gene (either luciferase or β-galactosidase) linked to them. These experiments were performed in cell cultures and in transgenic mice. It should be possible to extend this scenario to the control of Cre recombinase synthesis. Thus, transgenic mice could be generated, containing two transgenes, one expressing rtTA and the other made of *tet O* sequences linked to the Cre recombinase coding sequences. Thus depending on the specificity of the promoter used to express the fusion protein, such transgenic mice should express, upon tetracycline treatment, the recombinase in a given type of cells, at the time of treatment.

Concluding Remarks

The transgenic methodology has been instrumental in such important aspects of the study of development as gene regulation *in vivo*, search of new genes of developmental importance and the analysis of their function in the various and complex processes which underlie the development of the mammalian embryo. In this regard, perhaps the most promising aspect of mouse germline modification lies in the ability to replace a gene by a mutated version, thereby opening a way to illuminate its putative function in development. Up until very recently, almost all the mutations generated by targeted mutagenesis were null mutations. However, new methods are being developed which will permit to create more subtle mutations or conditional ones and therefore to significantly broaden the potential of targeted mutagenesis. Thus, the ability to introduce exogenous genetic information into the germline of the mouse has given those studying mouse development a new and remarkably powerful tool, which was not available only ten to fifteen years ago. Of course this tool cannot tackle all the questions raised by mouse development and other approaches like classical genetics, experimental embryology or the study of other organisms should not be underestimated. Nevertheless, it is clear that transgenesis will continue in the coming years to contribute to our increasing knowledge of mammalian development.

Acknowledgements

I wish to thank my colleagues Patricia Baldacci, Dominique Morello and Michel Cohen-Tannoudji for their helpful comments and critical reading of the manuscript, Christophe Poirier for help in preparing the drawings and Ms. Isabelle Fleurance for her inexhaustible patience during the preparation of the manuscript.

References

Babinet, C., Morello, D. and Renard, J.-P. (1989) Transgenic mice. *Genome* **31**, 938–949

Balling, R., Mutter, G., Gruss, P. and Kessel, M. (1989) Craniofacial abnormalities induced by ectopic expression of the homeobox gene *Hox*-1.1 in transgenic mice. *Cell* **58**, 337–347

Behringer, R.R., Crotty, D.A., Tennyson, V.M., Brinster, R.L., Palmiter, R.D. and Wolgemuth, D.J. (1993) Sequences 5' of the homeobox of the *Hox*-1.4 gene direct tissue-specific expression of *lacZ* during mouse development. *Development* **117**, 823–833

Bieberich, C.J., Utset, M.F., Awgulewitsch, A. and Ruddle, F.H. (1990) Evidence for positive and negative regulation of the *Hox*-3.1 gene. *Proc. Natl. Acad. Sci. USA* **87**, 8462–8466

Bradley, A., Hasty, P., Davis, A. and Ramirez-Solis, R. (1992) Modifying the mouse: design and desire. *Biotechnology* **10**, 534–539

Brandon, E.P., Idzerda, R.L. and McKnight, G.S. (1995a) Targeting the mouse genome: a compendium of knockouts (part I). *Curr. Biol.* **5**, 625–634

Brandon, E.P., Idzerda, R.L. and McKnight, G.S. (1995b) Targeting the mouse genome: a compendium of knockouts (part II). *Curr. Biol.* **5**, 758–765

Brandon, E.P., Idzerda, R.L. and McKnight, G.S. (1995c) Targeting the mouse genome: a compendium of knockouts (part III). *Curr. Biol.* **5**, 873–881

Braun, T., Rudnicki, M.A., Arnold, H.H. and Jaenisch, R. (1992) Targeted inactivation of the muscle regulatory gene *Myf-5* results in abnormal rib development and perinatal death. *Cell* **71**, 369–382

Bronson, S.K. and Smithies, O. (1994) Altering mice by homologous recombination using embryonic stem cells. *J. Biol. Chem.* **269**, 27155–27158

Brookfield, J. (1992) Can genes be truly redundant? *Curr. Biol.* **10**, 553–554

Bumcrot, D.A. and McMahon, A.P. (1995) Sonic signals somites. *Curr. Biol.* **5**, 612–614

Camus, A., Kress, C., Babinet, C. and Barra, J. (1996) Unexpected behavior of a gene trap vector comprising a fusion between the Sh *ble* and the *lacZ* genes. *Mol. Reprod. Dev.* (in press)

Capecchi, M.R. (1989) Altering the genome by homologous recombination. *Science* **244**, 1288–1292

Chan, D.C., Wynshaw-Boris, A. and Leder, P. (1995) Formin isoforms are differentially expressed in the mouse embryo and are required for normal expression of *fgf*-4 and *shh* in the limb bud. *Development* **121**, 3151–3162

Charite, J., De, G.W., Shen, S. and Deschamps, J. (1994) Ectopic expression of *Hoxb*-8 causes duplication of the ZPA in the forelimb and homeotic transformation of axial structures. *Cell* **78**, 589–601

Chen, Z., Friedrich, G.A. and Soriano, P. (1994) Transcriptional enhancer factor 1 disruption by a retroviral gene trap leads to heart defects and embryonic lethality in mice. *Genes Dev.* **8**, 2293–2301

Cheng, S.S. and Costantini, F. (1993) Morula decompaction (*mdn*), a preimplantation recessive lethal defect in a transgenic mouse line. *Dev. Biol.* **156**, 265–277

Colucci-Guyon, E., Portier, M.-M., Dunia, I., Paulin, D., Pournin, S. and Babinet, C. (1994) Mice lacking vimentin develop and reproduce without an obvious phenotype. *Cell* **79**, 679–694

Condie, B.G. and Capecchi, M.R. (1994) Mice with targeted disruptions in the paralogous genes *hoxa*-3 and *hoxd*-3 reveal synergistic interactions. *Nature* **370**, 304–307

Conlon, F.L., Barth, K.S. and Robertson, E.J. (1991) A novel retrovirally induced embryonic lethal mutation in the mouse: assessment of the developmental fate of embryonic stem cells homozygous for the 413.d proviral integration. *Development* **111**, 969–981

D'Arcangelo, G., Miao, G.G., Chen, S.-C., Soares, H.D., Morgan, J.I. and Curran, T. (1995) A protein related to extracellular matrix proteins deleted in the mouse mutant *reeler*. *Nature* **374**, 719–723

Davidson, E.H. (1986) *Gene activity in early development*, Academic Press Inc., London

Davidson, E.H. (1994) Molecular biology of embryonic development: how far have we come in the last ten years? *BioEssays* **16**, 603–615

DeGregori, J., Russ, A., von Melchner, H., Rayburn, H., Priyaranjan, P., Jenkins, N.A., Copeland, N.G. and Ruley, H.E. (1994) A murine homolog of the yeast RNA1 gene is required for post-implantation development. *Genes Dev.* **8**, 265–276

Dickinson, M.E., Krumlauf, R. and McMahon, A.P. (1994) Evidence for a mitogenic effect of *Wnt*-1 in the developing mammalian central nervous system. *Development* **120**, 1453–1471

Dillon, N. and Grosveld, F. (1993) Transcriptional regulation of multigene loci: multilevel control. *Trends Genet.* **9**, 134–137

DiSanto, J., Müller, W., Guy-Grand, D., Fischer, A. and Rajewsky, K. (1995) Lymphoid development in mice with a targeted deletion of the interleukin 2 receptor γ chain. *Proc. Natl. Acad. Sci. USA* **92**, 377–381

Doetschman, T., Gregg, R.G., Maeda, N., Hooper, M.L., Melton, D.W., Thompson, S. and Smithies, O. (1987) Targetted correction of a mutant *HPRT* gene in mouse embryonic stem cells. *Nature* **330**, 576–578

Duboule, D. (1994) How to make a limb? *Science* **266**, 575–576

Egan, S.E. and Weinberg, R.A. (1993) The pathway to signal achievement. *Nature* **365**, 781–782

Eid, R., Koseki, H. and Schughart, K. (1993) Analysis of *LacZ* reporter genes in transgenic embryos suggests the presence of several cis-acting regulatory elements in the murine *Hoxb*-6 gene. *Dev. Dyn.* **196**, 205–216

Epstein, C.J. (1986) *The consequences of chromosome imbalance. Principles, mechanisms and models*, Cambridge University Press, Cambridge

Erickson, H.P. (1993) Gene knockouts of c-src, transforming growth factor beta 1, and tenascin suggest superfluous, nonfunctional expression of proteins. *J. Cell Biol.* **120**, 1079–1081

Evans, M.J. and Kaufman, M.H. (1981) Establishment in culture of pluripotential cells from mouse embryos. *Nature* **292**, 154–156

Forget, B.G. (1993) YAC transgenes: bigger is probably better. *Proc. Natl. Acad. Sci. USA* **90**, 7909–7911

Friedrich, G. and Soriano, P. (1991) Promoter traps in embryonic stem cells: a genetic screen to identify and mutate developmental genes in mice. *Genes and Dev.* **5**, 1513–1523

Frohman, M.A., Dush, M.K. and Martin, G.R. (1988) Rapid production of full-length cDNAs from rare transcripts: amplification using a single gene-specific oligonucleotide primer. *Proc. Natl. Acad. Sci. USA* **85**, 8998–9002

Gehring, W.J. (1987) Homeo boxes in the study of development. *Science* **236**, 1245–1252

Gehring, W.J., Qian, Y.Q., Billeter, M., Furukubo-Tokunaga, K., Schier, A.F., Resendez-Perez, D., Affolter, M., Otting, G. and Wüthrich, K. (1994) Homeodomain-DNA recognition. *Cell* **78**, 211–223

Gerard, M., Duboule, D. and Zakany, J. (1993) Structure and activity of regulatory elements involved in the activation of the *Hoxd*-11 gene during late gastrulation. *EMBO J.* **12**, 3539–3550

Gossen, M., Freundlieb, S., Bender, G., Müller, G., Hillen, W. and Bujard, H. (1995) Transcriptional activation by tetracyclines in mammalian cells. *Science* **268**, 1766–1769

Gossler, A., Doetschman, T., Korn, R., Serfling, E. and Kemler, R. (1986) Transgenesis by means of blastocyst-derived embryonic stem cell lines. *Proc. Natl. Acad. Sci. USA* **83**, 9065–9069

Grandgenett, D.P. and Mumm, S.R. (1990) Unravelling retrovirus integration. *Cell* **60**, 3–4

Gridley, T., Soriano, P. and Jaenisch, R. (1987) Insertional mutagenesis in mice. *Trends Genet.* **3**, 162–166

Gu, H., Marth, J.D., Orban, P.C., Mossmann, H. and Rajewsky, K. (1994) Deletion of a DNA polymerase beta gene segment in T cells using cell type-specific gene targeting. *Science* **265**, 103–106

Hammer, R.E., Swift, G.H., Ornitz, D.M., Quaife, C.J., Palmiter, R.D., Brinster, R.L. and McDonald, R.J. (1987) The rat elastase 1 regulatory element is an enhacer that directs correct cell specificity and developmental onset of expression in transgenic mice. *Mol. Cell. Biol.* **7**, 2956–2967

Hill, D.P. and Wurst, W. (1993) Gene and enhancer trapping: mutagenic strategies for developmental studies. *Curr. Top. Dev. Biol.* **28**, 181–206

Hodgkinson, C.A., Moore, K.J., Nakayama, A., Steingrimsson, E., Copeland, N.G., Jenkins, N.A. and Arnheiter, H. (1993) Mutation at the mouse microphthalmia locus are associated with defects in a gene encoding a novel basic-helix-loop-helix-zipper protein. *Cell* **74**, 395–404

Horan, G.S., Ramirez, S.R., Featherstone, M.S., Wolgemuth, D.J., Bradley, A. and Behringer, R.R. (1995) Compound mutants for the paralogous *hoxa*-4, *hoxb*-4, and *hoxd*-4 genes show more complete homeotic transformations and a dose-dependent increase in the number of vertebrae transformed. *Genes Dev.* **9**, 1667–1677

Hughes, M.J., Lingrel, J.B., Krakowsky, J.M. and Anderson, K.P. (1993) A helix-loop-helix transcription factor-like gene is located at the *mi* locus. *J. Biol. Chem.* **268**, 20687–20690

Hunt, P., Whiting, J., Nonchev, S., Sham, M.-H., Marshall, H., Graham, A., Cook, M., Allemann, R., Rigby, P.W., Gulisano, M., Faiella, A., Boncinelli, E. and Krumlauf, R. (1991) The branchial *Hox* code and its implications for gene regulation, patterning of the nervous system and head evolution. *Development Suppe.* **2**, 63–77

Ingham, P.W. (1995) Signalling by hedgehog family proteins in *Drosophila* and vertebrate development. *Curr. Op. Genet. Dev.* **5**, 492–498

Jaenisch, R. (1988) Transgenic animals. *Science* **240**, 1468–1474

Jegalian, B.G. and De Robertis, E. M. (1992) Homeotic transformations in the mouse induced by overexpression of a human *Hox*3.3 transgene. *Cell* **71**, 901–910

Johnson, R.L. and Tabin, C. (1995) The long and short of *hedgehog* signaling. *Cell* **81**, 313–316

Joyner, A.L., Herrup, K., Auerbach, B.A., Davis, C.A. and Rossant, J. (1991) Subtle cerebellar phenotype in mice homozygous for a targeted deletion of the *En-2* homeobox. *Science* **251**, 1239–1243

Kessel, M., Balling, R. and Gruss, P. (1990) Variations of cervical vertebrae after expression of a *Hox*-1.1 transgene in mice. *Cell* **61**, 301–308

Kessel, M. and Gruss, P. (1990) Murine developmental control genes. *Science* **249**, 374–379

Kessel, M. and Gruss, P. (1991) Homeotic transformations of murine vertebrae and concomitant alteration of *Hox* codes induced by retinoic acid. *Cell* **67**, 89–104

Kilby, N.J., Snaith, M.R. and Murray, J.A. (1993) Site-specific recombinases: tools for genome engineering. *Trends Genet.* **9**, 413–421

Knittel, T., Kessel, M., Kim, M.H. and Gruss, P. (1995) A conserved enhancer of the human and murine *Hoxa*-7 gene specifies the anterior boundary of expression during embryonal development. *Development* **121**, 1077–1088

Kollias, G. and Grosveld, F. (1992) The study of gene regulation in transgenic mice. In F. Grosveld and G. Kollias, (eds.), *Transgenic animals*, Academic Press, London, 79–98

Korn, R., Schoor, M., Neuhaus, H., Henseling, U., Soininen, R., Zachgo, J. and Gossler, A. (1992) Enhancer trap integrations in mouse embryonic stem cells give rise to staining patterns in chimaeric embryos with a high frequency and detect endogenous genes. *Mech. Dev.* **39**, 95–109

Kress, C., Vogels, R., De, G.W., Bonnerot, C., Meijlink, F., Nicolas, J.F. and Deschamps, J. (1990) *Hox*-2.3 upstream sequences mediate *lacZ* expression in intermediate mesoderm derivatives of transgenic mice. *Development* **109**, 775–786

Krumlauf, R. (1993) *Hox* genes and pattern formation in the branchial region of the vertebrate head. *Trends Genet.* **9**, 106–112

Krumlauf, R. (1994) *Hox* genes in vertebrate development. *Cell* **78**, 191–201

Kuhn, R., Schwenk, F., Aguet, M. and Rajewsky, K. (1995) Inducible gene targeting in mice. *Science* **269**, 1427–1429

Lawrence, P.A. and Morata, G. (1994) Homeobox genes: Their function in drosophila segmentation and pattern formation. *Cell* **78**, 181–189

Lemaire, P. and Gurdon, J.B. (1994) Vertebrate embryonic inductions. *BioEssays* **16**, 617–620

Li, X. and Noll, M. (1994) Evolution of distinct developmental functions of three *Drosophila* genes by acquisition of different *cis*-regulatory regions. *Nature* **367**, 83–87

Lufkin, T., Mark, M., Hart, C.P., Dolle, P., Le, M.M. and Chambon, P. (1992) Homeotic transformation of the occipital bones of the skull by ectopic expression of a homeobox gene. *Nature* **359**, 835–841

Maas, R.L., Zeller, R., Woychik, R.P., Vogt, T.F. and Leder, P. (1990) Disruption of formin-encoding transcripts in two mutant *limb deformity* alleles. *Nature* **346**, 853–855

Mark, M., Rijli, F.M. and Chambon, P. (1995) Alteration of *Hox* gene expression in the branchial region of the head causes homeotic transformations, hindbrain segmentation defects and atavistic changes. *Seminars Dev. Biol.* **6**, 275–284

Marshall, H., Studer, M., Popperl, H., Aparicio, S., Kuroiwa, A., Brenner, S. and Krumlauf, R. (1994) A conserved retinoic acid response element required for early expression of the homeobox gene *Hoxb*-1. *Nature* **370**, 567–571

Martin, G.R. (1981) Isolation of a pluripotent cell line from early mouse embryos cultured in medium conditioned by teratocarcinoma stem cells. *Proc. Natl. Acad. Sci. USA* **78**, 7634–7638

McGinnis, W. and Krumlauf, R. (1992) Homeobox genes and axial patterning. *Cell* **68**, 283–302

McMahon, A.P. and Bradley, A. (1990) The *Wnt*-1 (*int*-1) proto-oncogene is required for development of a large region of the mouse brain. *Cell* **62**, 1073–1085

McMahon, A.P., Joyner, A.L., Bradley, A. and McMahon, J.A. (1992) The midbrain-hindbrain phenotype of *Wnt*-1$^-$/*Wnt*-1$^-$ mice results from stepwise deletion of engrailed-expressing cells by 9.5 days postcoitum. *Cell* **69**, 581–595

Meisler, M.H. (1992) Insertional mutation of 'classical' and novel genes in transgenic mice. *Trends Genet.* **8**, 341–344

Miao, G.G., Smeyne, R.J., D'Arcangelo, G., Copeland, N.G., Jenkins, N.A., Morgan, J.I. and Curran, T. (1994) Isolation of an allele of *reeler* by insertional mutagenesis. *Proc. Natl. Acad. Sci. USA* **91**, 11050–11054

Millen, K.J., Wurst, W., Herrup, K. and Joyner, A.L. (1994) Abnormal embryonic cerebellar development and patterning of postnatal foliation in two mouse *Engrailed*-2 mutants. *Development* **120**, 695–706

Montoliu, L., Schedl, A., Kelsey, G., Lichter, P., Larin, Z., Lehrach, H. and Schutz, G. (1993) Generation of transgenic mice with yeast artificial chromosomes. *Cold Spring Harb. Symp. Quant. Biol.* **58**, 55–62

Moore, K.J. (1995) Insight into the *microphthalmia* gene. *Trends Genet.* **11**, 442–448

Neuhaus, H., Bettenhausen, B., Bilinski, P., Simon-Chazottes, D., Guénet, J.-L. and Gossler, A. (1994) *Etl2*, a novel putative type-I cytokine receptor expressed during mouse embryogenesis at high levels in skin and cells with skeletogenic potential. *Dev. Biol.* **166**, 531–542

Niwa, H., Araki, K., Kimura, S., Taniguchi, S., Wakasugi, S. and Yamamura, K. (1993) An efficient gene-trap method using Poly A trap vectors and characterization of gene trap events. *J. Biochem.* **113**, 343–349

Nonchev, S., Vesque, C., Maconochie, M., Seitanidou, T., Ariza-McNaughton, L., Frain, M., Marshall, H., Sham, M.H., Krumlaub, R. and Charnay, P. (1996) Segmental expression of *Hoxa*-2 in the hindbrain is directly regulated by *Krox*-20. *Development* **122**, 543–554

Olson, E.N. and Klein, W.H. (1994) bHLH factors in muscle development: dead lines and commitments, what to leave in and what to leave out. *Genes Dev.* **8**, 1–8

Palmiter, R.D. and Brinster, R.L. (1986) Germ-line transformation of mice. *Ann. Rev. Genet.* **20**, 465–499

Perrimon, N. (1995) Hedgehog and beyond. *Cell* **80**, 517–520

Placzek, M. (1995) The role of the notochord and floor plate in inductive interactions. *Curr. Op. Genet. Dev.* **5**, 499–506

Pollock, R.A., Sreenath, T., Ngo, L. and Bieberich, C.J. (1995) Gain of function mutations for paralogous *Hox* genes: implications for the evolution of *Hox* gene function. *Proc. Natl. Acad. Sci. USA* **92**, 4492–4496

Pöpperl, H., Bienz, M., Studer, M., Chan, S. K., Aparicio, S., Brenner, S., Mann, R.S. and Krumlauf, R. (1995) Segmental expression of *Hoxb*-1 is controlled by a highly conserved autoregulatory loop dependent upon *exd/pbx*. *Cell* **81**, 1031–1042

Puschel, A.W., Balling, R. and Gruss, P. (1990) Position-specific activity of the *Hox*1.1 promoter in transgenic mice. *Development* **108**, 435–442

Puschel, A.W., Balling, R. and Gruss, P. (1991) Separate elements cause lineage restriction and specify boundaries of *Hox*-1.1 expression. *Development* **112**, 279–287

Ramirez-Solis, R., Liu, P. and Bradley, A. (1995) Chromosome engineering in mice. *Nature* **378**, 720–724

Rancourt, D.E., Tsuzuki, T. and Capecchi, M.R. (1995) Genetic interaction between *hoxb*-5 and *hoxb*-6 is revealed by nonallelic noncomplementation. *Genes Dev.* **9**, 108–122

Rijkers, T., Peetz, A. and Rütter, U. (1994) Insertional mutagenesis in transgenic mice. *Transgenic Res.* **3**, 203–215

Robertson, E., Bradley, A., Kuehn, M. and Evans, M. (1986) Germ-line transmission of genes introduced into cultured pluripotential cells by retroviral vector. *Nature* **323**, 445–448

Robertson, E.J. (1987) *Teratocarcinomas and embryonic stem cells: a practical approach*, IRL Press, Oxford–Washington DC

Rossant, J. and Joyner, A.L. (1989) Towards a molecular-genetic analysis of mammalian development. *TIG* **5**, 277–283

Rudnicki, M.A., Braun, T., Hinuma, S. and Jaenisch, R. (1992) Inactivation of *MyoD* in mice leads to up-regulation of the myogenic HLH gene *Myf*-5 and results in apparently normal muscle development. *Cell* **71**, 383–390

Rudnicki, M.A. and Jaenisch, R. (1995) The *MyoD* family of transcription factors and skeletal myogenesis. *Bioessays* **17**, 203–209

Rudnicki, M.A., Schnegelsberg, P.N., Stead, R.H., Braun, T., Arnold, H.H. and Jaenisch, R. (1993) MyoD or Myf-5 is required for the formation of skeletal muscle. *Cell* **75**, 1351–1359

Saga, Y., Yagi, T., Ikawa, Y., Sakakura, T. and Aizawa, S. (1992) Mice develop normally without tenascin. *Genes Dev.* **6**, 1821–1831

Schneider-Maunoury, S., Topilko, P., Seitandou, T., Levi, G., Cohen, T.M., Pournin, S., Babinet, C. and Charnay, P. (1993) Disruption of *Krox*-20 results in alteration of rhombomeres 3 and 5 in the developing hindbrain. *Cell* **75**, 1199–1214

Scott, M.P. (1992) Vertebrate homeobox gene nomenclature. *Cell* **71**, 551–553

Sham, M.H., Vesque, C., Nonchev, S., Marshall, H., Frain, M., Gupta, R.D., Whiting, J., Wilkinson, D., Charnay, P. and Krumlauf, R. (1993) The zinc finger gene *Krox*-20 regulates *HoxB2* (*Hox*2.8) during hindbrain segmentation. *Cell* **72**, 183–196

Shashikant, C.S., Bieberich, C.J., Belting, H.-G., Wang, J.C.H., Borbély, M.A. and Ruddle, F.H. (1995) Regulation of *Hoxc*-8 during mouse embryonic development: identification and characterization of critical elements involved in early neural tube expression. *Development* **121**, 4339–4347

Shimamura, K., Hirano, S., McMahon, A.P. and Takeichi, M. (1994) *Wnt*-1-dependent regulation of local E-cadherin and alpha N-catenin expression in the embryonic mouse brain. *Development* **120**, 2225–2234

Simon, J. (1995) Locking in stable states of gene expression: transcriptional control during *Drosophila* development. *Curr. Opin. Cell. Biol.* **7**, 376–385

Skarnes, W.C. (1993) The identification of new genes: gene trapping in transgenic mice. *Curr. Op. Biotech.* **4**, 684–689

Skarnes, W.C., Auerbach, B.A. and Joyner, A.L. (1992) A gene trap approach in mouse embryonic stem cells: the *LacZ* reporter is activated by splicing, reflects endogenous gene expression, and is mutagenic in mice. *Genes Dev.* **6**, 903–918

Skarnes, W.C., Moss, J.E., Hurtley, S.M. and Beddington, R.S.P. (1995) Capturing genes encoding membrane and secreted proteins important for mouse development. *Proc. Natl. Acad. Sci. USA* **92**, 6592–6596

Smith, A.J., De, S.M., Kwabi, A.B., Heppell, P.A., Impey, H. and Rabbitts, P. (1995) A site-directed chromosomal translocation induced in embryonic stem cells by Cre-loxP recombination. *Nat. Genet.* **9**, 376–385

Soininen, R., Schoor, M., Henseling, U., Tepe, C., Kisters-Woike, B., Rossant, J. and Gossler, A. (1992) The mouse *Enhancer trap locus* 1 (*Etl*-1): a novel mammalian gene related to *Drosophila* and yeast transcriptional regulator genes. *Mech. Dev.* **39**, 111–123

Stacey, A., Schnieke, A., McWhir, J., Cooper, J., Colman, A. and Melton, D.W. (1994) Use of double-replacement gene targeting to replace the murine alpha-lactalbumin gene with its human counterpart in embryonic stem cells and mice. *Mol. Cell. Biol.* **14**, 1009–1016

Studer, M., Pöpperl, H., Marshall, H., Kuroiwa, A. and Krumlauf, R. (1994) Role of a conserved retinoic acid response element in rhombomere restriction of *Hoxb*-1. *Science* **265**, 1728–1732

Subramanian, V., Meyer, B.I. and Gruss, P. (1995) Disruption of the murine homeobox gene *Cdx*1 affects axial skeletal identities by altering the mesodermal expression domains of *Hox* genes. *Cell* **83**, 641–653

Takeuchi, T., Yamazaki, Y., Katoh-Fukui, Y., Tsuchiya, R., Kondo, S., Motoyama, J. and Higashinakagawa, T. (1995) Gene trap capture of a novel mouse gene, *jumonji*, required for neural tube formation. *Genes Dev.* **9**, 1211–1222

Thomas, J.H. (1993) Thinking about genetic redundancy. *Trends Genet.* **9**, 395–399

Thomas, K.R. and Capecchi, M.R. (1987) Site-directed mutagenesis by gene targeting in mouse embryo-derived stem cells. *Cell* **51**, 503–512

Thomas, K.R. and Capecchi, M.R. (1990) Targeted disruption of the murine *int*-1 proto-oncogene resulting in severe abnormalities in midbrain and cerebellar development. *Nature* **346**, 847–850

Tuggle, C.K., Zakany, J., Cianetti, L., Peschle, C. and Nguyen, H.M. (1990) Region-specific enhancers near two mammalian homeo box genes define adjacent rostrocaudal domains in the central nervous system. *Genes Dev.* **4**, 180–189

van Genderen, C., Okamura, R.M., Farinas, I., Quo, R.G., Parslow, T.G., Bruhn, L. and Grosschedl, R. (1994) Development of several organs that require inductive epithelial-mesenchymal interactions is impaired in *LEF*-1-deficient mice. *Genes Dev.* **8**, 2691–26703

Vogels, R., Charite, J., De, G.W. and Deschamps, J. (1993) Proximal cis-acting elements cooperate to set *Hoxb*-7 (*Hox*-2.3) expression boundaries in transgenic mice. *Development* **118**, 71–82

von Melchner, H., De Gregori, J., Rayburn, H., Reddy, S., Friedel, C. and Ruley, H.E. (1992) Selective disruption of genes expressed in totipotent embryonal stem cells. *Genes Dev.* **6**, 919–927

Weintraub, H. (1993) The *MyoD* family and myogenesis: redundancy, networks, and thresholds. *Cell* **75**, 1241–1244

Weintraub, H., Davis, R., Tapscott, S., Thayer, M., Krause, M., Benezra, R., Blackwell, T.K., Turner, D., Rupp, R., Hollenberg, S., Zhuang, Y. and Lassar, A. (1991) The *myoD* gene family: nodal point during specification of the muscle cell lineage. *Science* **251**, 761–766

Whiting, J., Marshall, H., Cook, M., Krumlauf, R., Rigby, P.W., Stott, D. and Allemann, R.K. (1991) Multiple spatially specific enhancers are required to reconstruct the pattern of *Hox*-2.6 gene expression. *Genes Dev.* **5**, 2048–2059

Wilkinson, D.G. (1993) Molecular mechanisms of segmental patterning in the vertebrate hindbrain and neural crest. *Bioessays* **15**, 499–505

Wilson, C., Bellen, H.J. and Gehring, W.J. (1990) Position effects on eukaryotic gene expression. *Annu. Rev. Cell. Biol.* **6**, 679–714

Woychick, R.P., Maas, R.L., Zeller, R., Vogt, T.F. and Leder, P. (1990) Formins: proteins deduced from the alternative transcripts of the *limb deformity* gene. *Nature* **346**, 850–853

Woychick, R.P., Stewart, T.A., Davis, L.G., D'Eustachio, P. and Leder, P. (1985) An inherited limb deformity created by insertional mutagenesis in a transgenic mouse. *Nature* **318**, 36–40

Wurst, W., Auerbach, A.B. and Joyner, A.L. (1994) Multiple developmental defects in *Engrailed*-1 mutant mice: an early mid-hindbrain deletion and patterning defects in forelimbs and sternum. *Development* **120**, 2065–2075

Wurst, W., Rossant, J., Prideaux, V., Kownacka, M., Joyner, A., Hill, D.P., Guillemot, F., Gasca, S., Cado, D., Auerbach, A. and Ang, S.-L. (1995) A large-scale gene-trap screen for insertional mutations in developmentally regulated genes in mice. *Genetics* **139**, 889–899

Yoshida, M., Yagi, T., Furuta, Y., Takayanagi, K., Kominami, R., Takeda, N., Tokunaga, T., Chiba, J., Ikawa, Y. and Aizawa, S. (1995) A new strategy of gene trapping in ES cells using 3' RACE. *Transgenic Res.* **4**, 277–287

Yu, B.D., Hess, J.L., Horning, S.E., Brown, G.A. and Korsmeyer, S.J. (1995) Altered *Hox* expression and segmental identity in *Mll*-mutant mice. *Nature* **378**, 505–508

Zhang, M., Kim, H.J., Marshall, H., Gendron, M.M., Lucas, D.A., Baron, A., Gudas, L.J., Gridley, T., Krumlauf, R. and Grippo, J.F. (1994) Ectopic *Hoxa*-1 induces rhombomere transformation in mouse hindbrain. *Development* **120**, 2431–2442

Zhou, X., Sasaki, H., Lowe, L., Hogan, B.L.M. and Kuehn, M.R. (1993) *Nodal* is a novel *TGF-β*-like gene expressed in the mouse node during gastrulation. *Nature* **361**, 543–547

Zou, Y.R., Muller, W., Gu, H. and Rajewsky, K. (1994) Cre-*Lox* P-mediated gene replacement: a mouse strain producing humanized antibodies. *Curr. Biol.* **4**, 1099–1103

PART IV, SECTION A

57. Gene Transfer in Laboratory Fish: Model Organisms for the Analysis of Gene Function

Christoph Winkler and Manfred Schartl*

Physiological Chemistry I, Biocenter, University of Wuerzburg, Am Hubland, 97074 Wuerzburg, Germany

Small Aquarium Fish Species as Laboratory Models

Small aquarium fish species have attracted increasing attention during the last years as valuable experimental tools for developmental-, neuro-, and tumorbiology (reviewed by Nüsslein-Volhard, 1994; Driever et al., 1994 and Schartl, 1995). Over the last 10 years the zebrafish, *Danio rerio*, a small egglaying teleost from the Ganges river system in India has become an accepted laboratory model especially for studies in developmental biology. Its suitability for large scale mutagenesis screens will allow the genetic analysis of vertebrate embryogenesis. Another freshwater teleost, the Japanese medaka, *Oryzias latipes*, that inhabits the rice-fields of South-East Asia is a well studied experimental system for tumor biology since many decades. In addition, a variety of spontaneous mutants affecting sex determination, body shape and pigmentation have been collected and described over the years (for review see Ozato and Wakamatsu, 1994).

The unique characteristics of fish, namely easy breeding, large brood size, transparency of embryos and relatively small genome size make them ideal models for the analysis of gene function in vertebrates. One important tool for such studies is the availability of efficient gene transfer methods allowing the ectopic expression of genes of interest in the whole animal rather than in a cell culture system.

Several fish species show characteristics that make them particularly suited for gene transfer like short generation time and the daily availability of large numbers of eggs. Egg production is induced by a specific photoperiod, so that under artificial light cycles embryos can be produced throughout the year in the laboratory. The embryos, compared to that of mice, can be easily obtained and do not have to be reimplanted into pseudopregnant females, but can be reared in a balanced salt solution after microinjection. Teleost embryos generally are of relatively large size thereby allowing easy micromanipulation and microinjection. Thus, large numbers of eggs and embryos can be processed in a relatively short time. Their tough chorion protects them against environmental stress, but in most species it is soft enough to be penetrated by injection needles at least for the first hours of development. Several methods of gene transfer have been developed in different species. Recently, techniques utilizing pseudotyped retroviral vectors have been established yielding efficient rates of transgene integration and transmission (Burns et al., 1993 and Lin et al., 1994a). Also, high rates of DNA uptake in fish embryos were reported using gene transfer by electroporation (Buono and Linser, 1992 and Powers et al., 1992). The introduced DNA was transiently expressed, however, transmission of the transgene to the progeny was obtained only in a single case (Inoue et al., 1990). The most commonly used technique to date is injection of plasmid DNA into the cytoplasm of early embryos (i.e. at the one or two cell stage) (Table 57.1). This method is most elaborated in both laboratory fish, the zebrafish (Stuart et al., 1988, 1990) and the Japanese medaka (Chong and Vielkind, 1989 and Winkler et al., 1991), as well as some other species with importance for aquaculture such as trout (Guyomard et al., 1989),

---

*Author for Correspondence.

Table 57.1. Gene transfer into zebrafish and medaka.

	Gene	Promoter	Expression (F_0)	Transmission	Expression (F_1)	Reference
Zebrafish						
	hygromycin	SV40	nd	+		Stuart et al., 1988
	CAT	RSV, SV40	+	+	+	Stuart et al., 1990
	lacZ	RSV, SV40	+	+		Culp et al., 1991
	lacZ	RSV, mhs	+	+	+	Bayer et al., 1992
	lacZ	ependymin	+	nd	nd	Rinder et al., 1992
	lacZ	mHox 1.1, hHOX 3.3	+	nd	nd	Westerfield et al., 1992
	lacZ	XEF-1	+	+	+	Lin et al., 1994b
	CAT, Neo, hygro	hsp70	+	+		Gibbs et al., 1994
	lacZ	rGAP-43	+	nd	nd	Reinhard et al., 1994
	GFP	XEF-1	+	+	+	Amsterdam et al., 1995
	CAT	β-actin	+	+	+	Caldovic and Hackett, 1995
Medaka						
Cytoplasmic	hGH	mMT, β-actin, rCCK, Tk	+	+	+	Lu et al., 1992
	CAT	RSV, SV40	+	nd	nd	Chong and Vielkind, 1989
	CAT	AFP	+	nd	nd	Gong et al., 1991
	CAT, lacZ	hMT, CMVTk, RSV, SV40, X47	+	nd	nd	Winkler et al., 1991
	Xmrk, HER/Xmrk	CMVTk	+	nd	nd	Winkler et al., 1994
Nuclear	δ-Crystallin	δ-Crystallin	+	nd	nd	Ozato et al., 1986
	δ-Crystallin	δ-Crystallin	+	nd	nd	Inoue et al., 1989
	LUC	RSV	+	nd	nd	Tamiya et al., 1990
	LUC	ChMT, hsp70	+	+		Sato et al., 1992
	lacZ	β-actin	+	+	+	Takagi et al., 1994
	lacZ	RSV, CMV, MMLV	+	nd	nd	Tsai et al., 1995

Abbreviations used:
AFP   anti freeze protein
CAT   chloramphenicol acetyl transferase
ChMT   chinese hamster metallothionein
CMV   Cytomegalovirus enhancer + promoter
CMVTk CMV enhancer + Thymidine kinase promoter
GFP   green fluorescent protein
HER/Xmrk human epidermal growth factor receptor/Xmrk chimera
hGH   human growth hormone
hMT,mMt human, mouse metallothionein
hsp70 *Drosophila* heatshock promoter 70
lacZ   β-galactosidase
LUC   luciferase
mhs   mouse heatshock promoter
MMLV   Moloney murine leukemia virus
Neo   neomycin
rCCK   rat cholecystokinin
rGAP   ratgrowth associated protein
RSV   Rous sarcoma virus long terminal repeat
SV40   Simian Virus 40
Tk   Thymidine kinase
X47   *Xiphophorus* metallothionein-like
XEF   *Xenopus* elongation factor
Xmrk   *Xiphophorus* melanoma receptor kinase

salmon (Shears et al., 1991), tilapia (Maclean et al., 1992), catfish (Dunham et al., 1992 and Volckaert et al., 1994) and carp (Chen et al., 1993). After cytoplasmic injection no transmission of the transgene was reported in medaka, whereas transmission occurred in zebrafish (4–5% Stuart et al., 1988 and 16% Culp et al., 1991). The reported rates, however, were low compared to transmission rates in transgenic trout that have been obtained using a similar technique (50% Guyomard et al., 1989 and Chourrout et al., 1989). In every case of transmission the number of positive F1 progeny was well below 50% indicating a high degree of mosaicism in the germline of the founder fish (trout 10–30%; zebrafish 18–30%). This is apparently due to the integration of cyto-plasmically injected DNA only at advanced stages of development. In addition to cytoplasmic injection, so far only the medaka offers the unique possibility to inject DNA directly into the germinal vesicle of unfertilized oocytes (Ozato et al., 1986). Premature oocytes can be isolated from female fish 9 hours before the anticipated time of ovulation, which coincides with the onset of light phase. The isolated oocytes show a prominent, large nucleus that can easily be microinjected through the soft surrounding follicular layer. The oocytes are subsequently matured *in vitro* in a standard medium without expensive additives. After spontaneous ovulation or experimental removal of the follicular layer, the mature oocytes can be fertilized *in vitro* by incubating them with sperm suspension. With this method integration and expression of injected plasmid DNA, as well as transmission to the next generation and expression in the progeny has been described (Ozato et al., 1989; Takagi et al., 1994 and Winkler, unpublished).

Analysis of Promoter Activity by Gene Transfer

Transcriptional regulation *in vivo* has been characterized so far mainly by employing transient expression assays after injection of plasmid DNA into the cytoplasm of early embryos (1–4 cell stage). The foreign DNA persists throughout embryonic and larval development. It is replicated extrachromosomally, becomes methylated and is expressed transiently (Stuart et al., 1988 and Winkler et al., 1991). Expression of injected plasmid DNA starts at the midblastula stage (i.e. midblastula transition) and is strictly dependent on the promoter elements used (Winkler et al., 1992).

In medaka, as well as in zebrafish, analyses of cis-acting regulatory sequences have been described using different reporter genes. As reporters β-galactosidase (lacZ), chloramphenicol acetyl transferase (CAT), as well as firefly luciferase (LUC) are being used because of their effectiveness in tissue culture experiments. Recently, expression vectors containing the *Aequorea victoria* green fluorescent protein (GFP) as a reporter have been developed and successfully applied in zebrafish (Amsterdam et al., 1995). In addition, the expression of GFP fusion proteins in zebrafish embryos *in vivo* could be demonstrated (Peters et al., 1995). This offers the possibility of efficient *in vivo* screening for putative transgenic animals with appropriate expression of the introduced gene. If weak promoters or small quantitative differences are to be analysed *in vitro* and *in vivo*, the luciferase gene has been proposed as reporter of choice for studies on fish gene expression because of the extremely sensitive detection methods (Sekkali et al., 1994).

Many heterologous as well as homologous promoters and enhancers have been tested for their spatial and temporal activity in medaka and zebrafish embryos. Most of the viral regulatory elements analysed so far show no restriction of expression to any tissue. On the other hand some tissue-specific promoters have been identified from vertebrates which drive expression of a transgene only in a specific subset of cell types. For example, expression of a lacZ reporter under control of the rat GAP-43 promoter was preferentially detected in neural tissues of zebrafish at stages and sites where many neurons become postmitotic and express the endogenous GAP-43 gene (Reinhard et al., 1994). Also, tissue- and stage specific activation of the zebrafish ependymin promoter exclusively in meningeal cells of microinjected embryos could be shown to correspond exactly to the activity of the endogenous promoter (Rinder et al., 1992). So far, no promoter has been described to be inducible by exogenous stimuli under the conditions of transient expression from extrachromosomal DNA. Promoters, known to be inducible in other systems, e.g. the metallothionein or heatshock promoters, already show a high basal activity during early developmental stages in injected embryos and can not be superinduced above this level by treatment of the embryos with heavy metals or temperature shift (Winkler et al., 1991; Hong et al., 1993; and Winkler, unpublished). Taken together, to date a panel of different promoter and enhancer sequences is available that drive transient expression of a transgene of interest at different levels and in different tissues of a developing fish embryo (reviewed in Hackett, 1993). In medaka, the LTRs of Rous sarcoma virus (RSV) and the mouse mammary tumor virus (MMTV) as well as the cytomegalovirus (CMV) promoter/enhancer have also been tested after nuclear injection into oocytes yielding different expression profiles in injected embryos (Tsai et al., 1995). In zebrafish, the promoters tested so far in stable transgenic fish, such as the promoters of RSV, Simian virus 40 (SV40) and the promoter for the Xenopus elongation factor gene XEF-1, showed overall expression, rather than a tissue specific activity pattern (Stuart et al., 1988; 1990; Culp et al., 1991 and Lin et al., 1994b). In one transgenic zebrafish line carrying a lacZ reporter under the control of a mouse heat-shock promoter, spatially and temporally restricted expression in primary sensory neurons was described. This specific expression pattern was discussed to be due to the integration of the transgene in the vicinity of a tissue specific enhancer thereby indicating that an "enhancer trap" approach for cloning of differentially expressed genes might be possible in fish (Bayer and Campos-Ortega 1992).

Analysis of Gene Function by Injection of RNA

So far, most studies on gene function in fish have been performed by injection of synthetic RNA into the cytoplasm of early embryos in order to ensure rapid and equal distribution of expressed proteins throughout the developing embryo. This strategy was used to circumvent some limitations of DNA transfer into early embryos, namely the high degree of mosaic expression and the onset of expression as late as the blastula stage. Many developmentally important genes have been analysed by the effect of their ectopic expression on normal development (e.g. sonic hedgehog, Krauss et al., 1993; activin βB, Wittbrodt and Rosa, 1994; wnt 4, Ungar et al., 1995; wnt 8, 8b, Kelly et al., 1995a and embryonic fibroblast growth factor eFGF, Griffin et al., 1995). The specific effects observed as gain-of-function phenotypes gave indications on the normal functions of the tested genes, in particular during the process of mesoderm induction and the patterning of the embryonic axes (Table 57.2).

Interference with normal gene function was also obtained by the expression of dominant negative mutant proteins. This strategy can be used if a protein–protein interaction is a prerequisite for expression of the biochemical phenotype. Interference with the functional interaction abolishes the biochemical activity, thus leading to a loss-of-function phenotype. Components of signalling cascades that are instrumental in embryonic development, such as the growth and differentiation factor activin βB (Wittbrodt and Rosa, 1994) or the FGF receptor (Griffin et al., 1995) have been analysed for their function in vertebrate axis formation by expression of mutated forms in zebrafish and medaka. On top of this, injection of an antibody into the yolk of zebrafish embryos has been successfully utilized to study the function of pax(b) during midbrain-hindbrain development (Krauss et al., 1992).

Analysis of Gene Function by Injection of DNA: Oncogene Activity after Transient Gene Transfer in Medaka

Cytoplasmic injection of foreign DNA leads to a transient expression of the injected genes throughout embryonic and larval development. One important

Table 57.2. Functional analysis by injection of synthetic RNA into early fish embryos.

Injected RNA	Species	Observed phenotype	Reference
Gain of function			
nodal	ZF	induction of secondary axis	Toyama et al., 1995
activin βB	MF	induction of multiple axes	Wittbrodt and Rosa, 1994
eFGF	ZF	retarded gastrulation, induction of posterior markers	Griffin et al., 1995
wnt 4	ZF	defects in anterior structures	Ungar et al., 1995
wnt 8, 8b	ZF	dorsalization	Kelly et al., 1995a
sonic hedgehog	ZF	forebrain defects, induction of floorplate markers	Barth and Wilson, 1995; Krauss et al., 1993
β-catenin	ZF	induction of secondary axis	Kelly et al., 1995b
N-cadherin	ZF	microaggregation of deep cells	Bitzur et al., 1994
Xhox 3	ZF	disruption of anterior-posterior axis	Barro et al., 1994
Loss of function			
FGF receptor	ZF	suppression of trunk and tail formation	Griffin et al., 1995
activin βB	MF	suppression of mesoderm formation	Wittbrodt and Rosa, 1994

feature is the high degree of mosaicism leading to expression in single cells or small clusters of cells of different tissues. Whereas the high degree of mosaicism sometimes is a major drawback for the functional analysis of genes instrumental in early pattern formation of the developing embryo, this phenomenon provides an important advantage for studies of effects of activated oncogenes or their cellular counterparts during tumor induction. A cytoplasmatically injected medaka embryo ectopically expressing a certain oncogene more likely mimics the situation in a developing tumor *in situ* than a stable transgenic animal, because in the initial stage of a tumor single transformed cells or small clusters of transformed cells exist and interact with non-transformed cells in their normal environment. In addition, microinjection of DNA encoding oncogenes into early embryos rather than RNA can be advantageous, as overall expression of a gene with high tumor inducing potential would probably have severe effects on the survival rate during early development. Indication for this comes from injections of activated cytoplasmic kinases into early medaka embryos (Winkler, in preparation). We have utilized the medaka system to functionally characterize a novel oncogene designated *Xmrk*. This gene encodes a receptor tyrosine kinase of the EGF-receptor family whose transcriptional activation appears to be the underlying mechanism for spontaneous melanoma formation in hybrids of platyfish and swordtails of the genus *Xiphophorus* (Wittbrodt *et al.*, 1989 and Adam *et al.*, 1993). *Xiphophorus* is a livebearing fish, which so far is not accessible to gene transfer, because early embryos cannot be cultivated *in vitro*. Its close phylogenetic relationship to the medaka prompted us to use both species for a heterologous gene transfer experiment. Ectopic overexpression of the *Xmrk* oncogene after microinjection into early medaka embryos led to the induction of embryonal tumors with high incidence (approx. 10%) and short latency period (6–7 days) demonstrating the high tumorigenic potential of this gene *in vivo* (Winkler *et al.*, 1994). However, tumor formation was observed only in some embryonic tissues although the transgene was expressed ubiquitously throughout the whole embryo under the control of a non cell type specific promoter. Transformation was only observed in those tissues that express the corresponding *Xmrk* proto-oncogene indicating the importance of the cellular context for tumor induction (Winkler *et al.*, 1994). One possible explanation is that the activating ligand is restricted to certain tissues or organs. To analyse the ligand dependency of *Xmrk* tumor formation *in vivo* a chimeric receptor was used. This receptor consisted of the extracellular domain of the human epithelial growth factor receptor (hEGFR) and the intracellular kinase domain of *Xmrk*. Expression of this receptor alone in medaka embryos did not lead to the induction of tumors although strong ubiquitous expression could be demonstrated in microinjected embryos by immunohistochemistry. However, after coexpression of the corresponding activating ligand, human transforming growth factor alpha (hTGFα), tumors appeared again with high frequency and short latency period. Utilizing a whole embryo system, this demonstrated the ligand dependent activation of a receptor tyrosine kinase resulting in tumor formation *in vivo*. Although both, the ligand and the chimeric receptor, were expressed ubiquitously in this set of experiments, again tumors occurred only in those tissues that express the corresponding proto-oncogene. This shows that an *Xmrk* specific intracellular signal transduction machinery is important for the process of tumor induction. Only those tissues that contain the appropriate transduction components are competent for tumor formation *in vivo*. Using this system for transient ectopic expression of candidate genes during fish embryogenesis such molecules can now be identified and characterized that interact with the *Xmrk* receptor *in vivo* and represent components of the *Xmrk* specific signal transduction cascade.

Analysis of Gene Function in Stable Transgenic Medaka Fish

Stable transgenic medaka fish obtained by nuclear injection into oocytes also have been used for functional gene analysis. In an attempt to rescue a mutant pigmentation phenotype, stable transgenic medaka fish were produced using a mouse tyrosinase cDNA under the control of its own regulatory region. As recipient an orange red variety was employed that lacks melanin-containing pigment cells in the skin. Fish were obtained that exhibited the full wildtype pigmentation phenotype (Matsumoto *et al.*, 1992). Surprisingly, the transgenic founder female was not mosaic in soma and germline as seen in all other stable transgenic experiments with medaka performed so far (Ozato *et al.*, 1989; Takagi *et al.*, 1994 and Winkler, unpublished). The phenotypic result was also unexpected as the orange–red variety is due to a mutation that does not affect tyrosinase on the mRNA (Inagaki *et al.*, 1994) or protein/enzyme level (Hishida *et al.*, 1961). In our own experiments using a similar transgene for injection into the tyrosinase-negative albino medaka (Tomita, 1975), mosaic expression, F_0 germline mosaicism and an absolute lethality at larval stages of the expressed transgene obviously due to high levels of toxic melanin metabolites were observed. However, at the onset of transgene expression during late embryonic stages and hatching a rescue of the albino phenotype was observed (Hyodo-Taguchi, Winkler, and Schartl unpublished). The incorrect expression of the transgene leading to lethality in late larval stages was most probably caused by the heterologous promoter used in this assay. With the availability of the medaka tyrosinase gene (Inagaki et al., 1994) all advantages of homologous gene transfer are now at hand. Besides the functional analysis of melanin synthesis *in vivo*, homologous tyrosinase expression in transgenic medaka of albino genetic background could then also be used as an efficient marker to screen for transgenic fish. This is especially important, as rates of integration and expression in transgenic progeny is relatively low as compared to mice. Tyrosinase as a visible dominant marker for successful transgene integration would also open the field for an insertional mutagenesis in fish.

Towards Transgenic "Knock-out" Fish: Establishment of Embryonic Stem Cells

Although the approach of dominant negative mutants offers first insights into the function of certain genes it has many drawbacks. Firstly, dominant negative mutants are limited to studies of processes where protein–protein interactions are involved. Secondly, instead of inhibiting a specific receptor signalling pathway, ectopically expressed receptor mutants may block whole families of related receptors. Finally, the interaction of a receptor mutant with an activating ligand at ectopic sites and developmental stages in the embryo can prevent the ligand's function on other receptor systems. The observed phenotype then would reflect an impairment of unknown receptor systems rather than a potential function of the receptor of interest. Alternatively, the functional analysis of a given gene is possible by the creation of mutant animals that have lost the capacity to produce the gene product. This can be done in individuals with mutations in the gene in question that partially impair or totally abolish its function. Such mutations may occur spontaneously after mutagenesis but will be extremely rare or time consuming to obtain. The experimental production of such mutations at will is desired. In mice, the technique of gene targeting using homologous recombination in embryonic stem cells has provided a powerful system for this type of analysis of gene function (reviewed by Joyner, 1991). As outlined above, when compared to mice, fish offer the advantages of easy handling and continuous observation of all embryonic stages. The establishment of gene targeting in fish would therefore facilitate the analysis of gene function in vertebrates.

To establish gene targeting via homologous recombination in a fish system two important techniques have to be developed. First, it has to be possible to transplant cells into embryos at early stages in a way that the transplanted cells take part in embryonic development to produce a genetically mosaic chimera. The successful production of chimera in medaka (Wakamatsu et al., 1993; Hong and Winkler, unpublished) and zebrafish (Lin et al., 1992) has been reported. Secondly, the technique of culturing embryonic stem cells has to be established in fish. Currently, there are many attempts to propagate early embryonic cells from zebrafish and medaka *in vitro*. In zebrafish, cultivation of blastula stage embryonic cells was possible for up to 14 days without loss of pluripotency (Sun et al., 1995). In medaka, cultivation of early embryonal cells was possible for more than 1 year either in the presence (Wakamatsu et al., 1994) or absence of feeder layers (Hong and Schartl, 1996; Hong et al., submitted). The cultivated cells retained their potency to differentiate into many cell types after induction *in vitro*. The cells that show a normal chromosomal set, can be transfected by conventional methods. Expression of reporter genes driven by a variety of promoters has been demonstrated (Hong et al., in preparation). Future studies will show whether these cells have retained their full developmental potential and thus are able to survive after transplantation, as well as to participate in the germ line of a recipient embryo.

Perspectives

Small aquarium fish represent easy to handle yet powerful experimental systems to study gene function in vertebrates. Easy breeding in the laboratory, large brood size and the transparency of embryos allow a detailed study of developmental processes. Small genome size, the possibility of ploidy manipulation and the availability of genomic maps for zebrafish (Postlethwait et al., 1994), medaka (Wada et al., 1995) and swordtails (Morizot et al., 1991) are the basis for a genetic approach to analyse gene function. Established gene transfer methods make ectopic expression of genes of interest possible in either a transient or stable transgenic situation. Although stable integration of foreign DNA has been demonstrated for both, the medaka and zebrafish, factors affecting the efficiency of gene transfer are still to be determined. So far, integration of introduced DNA in fish is usually determined by successful transmission of the DNA to the progeny instead of characterizing "junction fragments" of introduced DNA with endogenous sequences. The high degree of germ line mosaicism in transgenic fish, however, makes the identification of transgenic progenies difficult and time consuming.

Therefore, a detailed analysis that elucidates whether the size of the introduced DNA, its conformation (linear versus circular) or the presence of bacterial vector sequences has an influence on the rate of integration or expression is still lacking. The availability of visible dominant markers of integration like tyrosinase or the green fluorescent protein, the expression of which can be detected in living embryos, will allow the rapid identification of stable transgenic progenies. On this basis a detailed study of the factors affecting the efficiency of integration, transmission and expression of an introduced transgene will be possible. Efficient methods for gene transfer will allow ectopic expression of developmental regulators or their dominant negative mutant forms. This together with an easily accessible fish embryo is expected to facilitate the functional analysis of developmentally important genes. After establishment of embryonic stem cells in fish a much more sophisticated analysis of gene function by homologous recombination may be expected in a powerful, small vertebrate system.

Acknowledgements

We are grateful to Dr. Yunhan Hong for sharing unpublished results, Drs. J. Altschmied and I. Schlupp for critically reading the manuscript. The work of the authors was supported by the European Community through BRIDGE and BIOTECH programmes.

References

Adam, D., Dimitrijevic, N. and Schartl, M. (1993) Tumor suppression in *Xiphophorus* by an accidentally acquired promoter. *Science* **259**, 816–819

Amsterdam, A., Lin, S. and Hopkins, N. (1995) The *Aequoria victoria* green fluorescent protein can be used as a reporter in live zebrafish embryos. *Dev. Biol.* **171**, 123–129

Barro, O., Joly, C., Condamine, H. and Bouklebache, H. (1994) Widespread expression of the Xenopus homeobox gene Xhox3 in zebrafish eggs causes a disruption of the anterior–posterior axis. *Int. J. Dev. Biol.* **38**, 613–622

Barth, K.A. and Wilson, S.W. (1995) Expression of zebrafish nk2.2 is influenced by sonic hedgehog/vertebrate hedgehog-1 and demarcates a zone of neuronal differentiation in the embryonic forebrain. *Development* **121**, 1755–1768

Bayer, T.A. and Campos-Ortega, J.A. (1992) A transgene containing lacZ is expressed in primary sensory neurons in zebrafish. *Development* **115**, 421–426

Bitzur, S., Kam, Z. and Geiger, B. (1994) Structure and distribution of N-cadherin in developing zebrafish embryos: morphogenetic effects of ectopic over-expression. *Dev. Dyn.* **201**, 121–136

Buono, R.J. and Linser, P.J. (1992) Transient expression of RSVCAT in transgenic zebrafish made by electroporation. *Mol. Mar. Biol. Biotech.* **1**, 271–275

Burns, J.C., Friedmann, T., Driever, W., Burrascano, M. and Yee, J.K. (1993) Vesicular stomatitis virus G glycoprotein pseudotyped retroviral vectors: concentration to very high titer and efficient gene transfer into mammalian and nonmammalian cells. *Proc. Natl. Acad. Sci. USA* **90**, 8033–8037

Caldovic, L. and Hackett, P.B., Jr. (1995) Development of position-independent expression vectors and their transfer into transgenic fish. *Mol. Mar. Biol. Biotech.* **4**, 51–61

Chen, T.T., Kight, K., Lin, C.M., Powers, D.A., Hayat, M., Chatakondi, N., Ramboux, A.C., Duncan, P.L. and Dunham, R.A. (1993) Expression and inheritance of RSVLTR-rtGH1 complementary DNA in the transgenic common carp, Cyprinus carpio. *Mol. Mar. Biol. Biotech.* **2**, 88–95

Chong, S.S.C. and Vielkind, J.R. (1989) Expression and fate of CAT reporter gene microinjected into fertilized medaka (Oryzias latipes) eggs in the form of plasmid DNA, recombinant phage particles and its DNA. *Theor. Appl. Genet.* **78**, 369–380

Chourrout, D., Guyomard, R. and Houdebine, L.M. (1989) Techniques for the development of transgenic fish: a review. *J. Cell. Biochem. Suppl.* **13**, 167–180

Culp, P., Nüesslein-Volhard, C. and Hopkins, N. (1991) High frequency germ-line transmission of plasmid DNA sequences injected into fertilized zebrafish eggs. *Proc. Natl. Acad. Sci. USA* **88**; 7953–7957

Driever, W., Stemple, D., Schier, A. and Solnica-Krezel, L. (1994) Zebrafish: genetic tools for studying vertebrate development. *Trends Genet.* **10**, 152–159

Dunham, R.A., Ramboux, A.C., Duncan, P.L., Hayat, M., Chen, T.T., Lin, C.M., Kight, K., Gonzales-Villasenor, L.I. and Powers, D.A. (1992) Transfer, expression and inheritance of salmonid growth hormone gene in channel catfish, Ictalurus punctatus, and effect on performance traits. *Mol. Mar. Biol. Biotech.* **1**, 380–389

Gibbs, P.D., Gray, A. and Thorgaard, G. (1994) Inheritance of P element and reporter gene sequences in zebrafish. *Mol. Mar. Biol. Biotech.* **3**, 317–326

Griffin, K., Patient, R. and Holder, N. (1995) Analysis of FGF function in normal and no tail zebrafish embryos reveals separate mechanisms for formation of the trunk and the tail. *Development* **121**, 2983–2994

Gong, Z., Hew, C.L. and Vielkind, J.R. (1991) Functional analysis and temporal expression of promoter regions from fish antifreeze protein genes in transgenic Japanese medaka embryos. *Mol. Mar. Biol. Biotech.* **1**, 64–9

Guyomard, R., Chourrout, D., Leroux, C., Houdebine, L.M. and Pourrain, F. (1989). Integration and germ line transmission of foreign genes microinjected into fertilized trout eggs. *Biochimie* **71**, 857–863

Hackett, P.B. (1993) The molecular biology of transgenic fish. In P. Hochachka and T. Mommsen, (eds.), *The Biochemistry and Molecular Biology of Fishes* **2**, 207–240

Hishida, T., Tomita, H. and Yamamoto, T. (1961) Melanin formation in color varieties of the medaka (*Oryzias latipes*). *Embryologia* **5**, 335–346

Hong, Y., Winkler, C., Brem, G. and Schartl, M. (1993) Development of a heavy metal-inducible fish specific expression vector for gene transfer in vitro and in vivo. *Aquaculture* **111**, 215–226

Hong, Y. and Schartl, M. (1996) Estabilshment and growth responses of early medakafish (*Oryzias latipes*) embryonic cells in feeder layer-free cultures. *Mol. Mar. Biol. Biotech.* **5**, 93–104

Hong, Y., Winkler, C. and Schartl, M. Pluripotency and differentiation of embryonic stem cell lines from the medakafish (*Oryzias latipes*). Submitted

Inagaki, H., Bessho, Y., Koga, A. and Hori, H. (1994) Expression of the tyrosinase-encoding gene in a colorless melanophore mutant of the medaka fish, *Oryzias latipes*. *Gene* **150**, 319–324

Inoue, K., Ozato, K., Kondoh, H., Iwamatsu, T., Wakamatsu, Y., Fujita, T. and Okada, T.S. (1989) Stage-dependent expression of the chicken δ-crystallin gene in transgenic fish embryos. *Cell Diff. Dev.* **27**, 57–68

Inoue, K., Yamashita, S., Hata, J., Kabeno, S., Asada, S., Nagahisa, E. and Fujita, T. (1990) Electroporation as a new technique for producing transgenic fish. *Cell Diff. Dev.* **29**, 123–128

Joyner, A.L. (1991) Gene targeting and gene trap screens using embryonic stem cells: New approaches to mammalian development. *Bioessays* **13**, 649–656

Kelly, G.M., Greenstein, P., Erezyilmaz, D.F. and Moon, R.T. (1995a) Zebrafish *wnt8* and *wnt8b* share a common activity but are involved in distinct developmental pathways. *Development* **121**, 1787–1799

Kelly, G.M., Erezyilmaz, D.F. and Moon, R.T. (1995b) Induction of a secondary embryonic axis in zebrafish occurs following the overexpression of *β-catenin*. *Mech. Dev.* **53**, 1–13

Krauss, S., Maden, M., Holder, N. and Wilson, S.W. (1992) Zebrafish pax(b) is involved in the formation of the midbrain-hindbrain boundary. *Nature* **360**, 87–89

Krauss, S., Concordet, J.P. and Ingham, P.W. (1993) A functionally conserved homolog of the Drosophila segment polarity gene hh is expressed in tissues with polarizing activity in zebrafish embryos. *Cell* **75**, 1431–1444

Lin, S., Long, W., Chen, J. and Hopkins, N. (1992) Production of germ-line chimeras in zebrafish by cell transplants from genetically pigmented to albino embryos. *Proc. Natl. Acad. Sci. USA* **89**, 4519–4523

Lin, S., Gaiano, N., Culp, P., Burns, J.C., Friedmann, T., Yee, J.K. and Hopkins, N. (1994a) Integration and germ-line transmission of a pseudotyped retroviral vector in zebrafish. *Science* **265**, 666–669

Lin, S., Yang, S. and Hopkins, N. (1994b) lacZ expression in germline transgenic zebrafish can be detected in living embryos. *Dev. Biol.* **161**, 77–83

Lu, J.K., Chen, T.T., Chrisman, C.L., Andrisani, O.M. and Dixon, J.E. (1992) Integration, expression and germ-line transmission of foreign growth hormone genes in medaka (*Oryzias latipes*). *Mol. Mar. Biol. Biotech.* **1**, 366–375

Maclean, N., Iyengar, A., Rahman, A., Sulaiman, Z. and Penman, D. (1992) Transgene transmission and expression in rainbow trout and tilapia. *Mol. Mar. Biol. Biotech.* **1**, 355–365

Matsumoto, J., Akiyama, T., Hirose, E., Nakamura, M., Yamamoto, H. and Takeuchi, T. (1992) Expression and transmission of wild-type pigmentation in the skin of transgenic orange-colored variants of medaka (Oryzias latipes) bearing the gene for mouse tyrosinase. *Pigment Cell Res.* **5**, 322–327

Morizot, D.C., Slaugenhaupt, S.A., Kallman, K.D. and Chakravarti, A. (1991) Genetic linkage map of fishes of the genus *Xiphophorus* (Teleostei: Poeciliidae) *Genetics* **127**, 399–410

Nüsslein-Volhard, C. (1994) Of flies and fishes. *Science* **266**, 572–574

Ozato, K., Kondoh, H., Inohara, H., Iwamatsu, T., Wakamatsu, Y. and Okada, T.S. (1986) Production of transgenic fish: Introduction and expression of chicken δ-crystallin gene in medaka embryos. *Cell Differ.* **19**, 237–244

Ozato, K., Inoue, K. and Wakamatsu, Y. (1989) Transgenic fish: Biological and technical problems. *Zool. Sci.* **6**, 445–457

Ozato, K. and Wakamatsu, Y. (1994) Developmental genetics of medaka. *Develop. Growth and Differ.* **36**, 437–443

Peters, K.G., Rao, P.S., Bell, B.S. and Kindman, L.A. (1995) Green fluorescent fusion proteins: powerful tools for monitoring protein expression in live zebrafish embryos. *Dev. Biol.* **171**, 252–257

Postlethwait, J.H., Johnson, S.L., Midson, C.N., Talbot, W.S., Gates, M., Ballinger, E.W., Africa, D., Andrews, R., Carl, T., Eisen, J.S., Horne, S., Kimmel, C.B., Hutchinson, M., Johnson, M. and Rodriguez, A. (1994) A genetic linkage map for the zebrafish. *Science* **264**, 699–703

Powers, D.A., Hereford, L., Cole, T., Chen, T.T., Lin, C.M., Kight, K., Creech, K. and Dunham, R. (1992) Electroporation: a method for transferring genes into the gamestes of zebrafish (Brachydanio rerio), channel catfish (Ictalurus punctatus), and common carp (Cyprinus carpio). *Mol. Mar. Biol. Biotech.* **1**, 301–308

Reinhard, E., Nedivi, E., Wegner, J., Pate Skene, J.H. and Westerfield, M. (1994) Neural selective activation and temporal regulation of a mammalian GAP-43 promoter in zebrafish. *Development* **120**, 1767–1775

Rinder, H., Bayer, T.A., Gertzen, E.M. and Hoffmann, W. (1992) Molecular analysis of the ependymin gene and functional test of its promoter region by transient expression in *Brachydanio rerio*. *DNA Cell Biol.* **11**, 425–432

Sato, A., Komura, J., Masahito, P., Matsukuma, S., Aoki, K. and Ishikawa, T. (1992) Firefly luciferase gene transmission and expression in transgenic medaka (*Oryzias latipes*). *Mol. Mar. Biol. Biotech.* **1**, 318–325

Schartl, M. (1995) Platyfish and swordtails: a genetic system for the analysis of molecular mechanisms in tumor formation. *Trends Genet.* **11**, 185–189

Sekkali, B., Belayew, A., Martial, J.A., Hellemans, B.A., Ollevier, F. and Volckaert, F.A. (1994) A comparative study of reporter gene activities in fish cells and embryos. *Mol. Mar. Biol. Biotech.* **3**, 30–34.

Shears, M.A., Fletcher, F.L., Hew, C.L., Gauthier, S. and Davies, P.L. (1991) Transfer, expression and stable inheritance of antifreeze protein genes in Atlantic salmon (Salmo salmar). *Mol. Mar. Biol. Biotech.* **1**, 58–63

Stuart, G.W., McMurray, J.V. and Westerfield, M. (1988) Replication, integration and stable germ line transmission of foreign sequences injected into early zebrafish embryos. *Development* **103**, 403–412

Stuart, G., Vielkind, J., McMurray, J. and Westerfield, M. (1990) Stable lines of transgenic zebrafish exhibit reproducible patterns of transgene expression. *Development* **109**, 577–584

Sun, L., Bradford, C.S., Ghosh, C., Collodi, P., Barnes, D.W. (1995) ES-like cell cultures derived from early zebrafish embryos. *Mol. Mar. Biol. Biotech.* **4**, 193–199

Takagi, S., Sasado, T., Tamiya, G., Ozato, K., Wakamatsu, Y., Takeshita, A. and Kimura, M. (1994) An efficient expression vector for transgenic medaka construction. *Mol. Mar. Biol. Biotech.* **3**, 192–199

Tamiya, E., Sugiyama, T., Masaki, K., Hirose, A., Okoshi, T. and Karube, I. (1990) Spatial imaging of luciferase gene expression in transgenic fish. *Nucleic Acids Res.* **18**, 1072

Tomita, H. (1975) Mutant genes in the medaka. In: Yamamoto, T. (ed.), *Medaka (Killifish), Biology and Strains*. Yugakusha, Tokyo, pp. 251–272

Toyama, R., O'Connell, M.L., Wright, C.V., Kuehn, M.R. and Dawid, I.B. (1995) Nodal induces ectopic goosecoid and lim1 expression and axis duplication in zebrafish. *Development* **121**, 383–391

Tsai, H.J., Wang, S.H., Inoue, K., Takagi, S., Kimura, M., Wakamatsu, Y. and Ozato, K. (1995) Initiation of transgenic lacZ gene expression in medaka (*Oryzias latipes*) embryos. *Mol. Mar. Biol. Biotech.* **4**, 1–9

Ungar, A.R., Kelly, G.M. and Moon, R.T. (1995) Wnt4 affects morphogenesis when misexpressed in the zebrafish embryo. *Mech. Dev.* **52**, 153–164

Volckaert, F.A., Hellemans, B.A., Galbusera, P. Ollevier, F., Sekkali, B. and Belayew, A. (1994) Replication, expression, and fate of foreign DNA during embryonic and larval development of the African catfish (*Clarias gariepinus*). *Mol. Mar. Biol. Biotech.* **3**, 57–69

Wada, H., Naruse, K., Shimada, A. and Shima, A. (1995) Genetic linkage map of a fish, the Japanese medaka *Oryzias latipes*. *Mol. Mar. Biol. Biotech.* **4**, 269–274

Wakamatsu, Y., Ozato, K., Hashimoto, H., Kinoshita, M., Sakaguchi, M., Hyodo-Taguchi, Y. and Tomita, H. (1993) Generation of germ-line chimeras in medaka (*Oryzias latipes*). *Mol. Mar. Biol. Biotech.* **2**, 325–332

Wakamatsu, Y., Ozato, K. and Sasado, T. (1994) Establishment of a pluripotent cell line derived from a medaka (*Oryzias latipes*) blastula embryo. *Mol. Mar. Biol. Biotech.* **3**, 185–191

Westerfield, M., Wegner, J., Jegalian, B.G., DeRobertis, E.M. and Püeschel, A.W. (1992) Specific activation of mammalian Hox promoters in mosaic transgenic zebrafish. *Genes and Develop.* **6**, 591–598

Winkler, C., Vielkind, J.R. and Schartl, M. (1991) Transient expression of foreign DNA during embryonic and larval development of the medaka fish (*Oryzias latipes*). *Mol. Gen. Genet.* **226**, 129–140

Winkler, C., Hong, Y., Wittbrodt, J. and Schartl, M. (1992) Analysis of heterologous and homologous promoters and enhancers *in vitro* and *in vivo* by gene transfer into Japanese medaka (*Oryzias latipes*) and *Xiphophorus*. *Mol. Mar. Biol. Biotech.* **1**, 326–337

Winkler C., Wittbrodt, J., Lammers, R., Ullrich, A. and Schartl, M. (1994) Ligand-dependent tumor induction in medakafish embryos by a *Xmrk* receptor tyrosine kinase transgene. *Oncogene* **9**, 1517–1525

Wittbrodt, J., Adam, D., Malitschek, B., Mäueler, W., Raulf, F., Telling, A., Robertson, S.M. and Schartl, M. (1989) Novel putative receptor tyrosine kinase encoded by the melanoma-inducing *Tu* locus in *Xiphophorus*. *Nature* **341**, 415–421

Wittbrodt, J. and Rosa, F.M. (1994) Disruption of mesoderm and axis formation in fish by ectopic expression of activin variants: the role of maternal activin. *Genes and Develop.* **8**, 1448–1462

PART IV, SECTION A

58. The Preparation of Human Antibodies from Mice Harbouring Human Immunoglobulin Loci

Marianne Brüggemann
Developmental Immunology Laboratory, The Babraham Institute, Babraham, Cambridge CB2 4AT, UK

Introduction

Transgenic mice have been generated which carry human antibody heavy (H) and light (L) chain genes in germline configuration on minigene constructs or regions similar in size to the endogenous loci. The gene constructs contained variable (V) region genes, diversity (D) segments, joining (J) segments and one or two constant (C) regions for the H chain, and Vs, Js and one C gene for the L chain. The experiments showed rearrangements of the transgenes, leading to a human antibody repertoire, which demonstrated the possibility of eliciting specific human responses after immunisation. Evaluation of human immunoglobulin (Ig) expression in these mice necessitates consideration of the following: The complexity of the Ig loci regarding the gene content and the large size of the H, and κ and λL chain region; the efficient use of the human genes and regulatory elements in the mouse; and, the methodologies to assemble and manipulate Ig gene constructs such as the recent use of yeast artificial chromosomes (YACs) and gene silencing approaches.

Gene Constructs and Transfer

Ig Gene Segments

Lymphoid specific expression, following DNA rearrangement, is directed by the V–(D)–J–C unit driven by a V-gene promoter and Ig enhancers (Pettersson *et al.*, 1990). For the κ locus 2 enhancers, one between J–C and the second some distance 3' of C_κ, play a major role in high expression, isotype exclusion and somatic mutation (Neuberger and Milstein, 1995). In the initial experiments, using germline configured V_H–D–J_H–C_μ or V_κ–J_κ–C_κ constructs, research groups expected problems with efficient DNA rearrangements of the human genes transferred into mice. In anticipation of these problems their gene constructs contained not only germline gene segments for either H or L chain genes placed into proximity to each other, but also mouse regulatory sequences such as the enhancer to drive expression (Brüggemann *et al.*, 1989 and Taylor *et al.*, 1992) (see Figure 58.1). However, in more recent experiments it became clear that additional mouse sequences were not required as long as equivalent human segments were included (Brüggemann *et al.*, 1991). Construction of a useful heavy chain region is more challenging than that of a light chain region as there are many more V, D, J and C gene segments which are spaced a considerable distance apart. The other problem with heavy chain constructs is that the regulation of high expression levels and allelic exclusion has not been achieved (Brüggemann *et al.*, 1989 and Taylor *et al.*, 1992); in fact promoter use with the J–C intron enhancer as well as addition of the enhancer 3' of C_α does not secure high expression. However, an important question was how many gene segments are needed to obtain a diverse set of antibodies? For the human heavy chain the minimum number of segments which has been used to obtain a limited human antibody repertoire includes 1 V, 4 Ds, the J region with the 6 Js, and the μ constant region, including the switch region and membrane μ (Brüggemann *et al.*, 1989). For human κ light chain expression a region has shown to be efficient when it contains 1 V, the region from J_κ to C_κ and the 3' enhancer (Taylor *et al.*, 1992). In this way rather

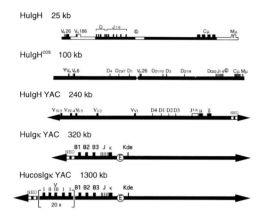

Figure 58.1. Human Ig H and L chain regions transferred and expressed in transgenic mice. From top to bottom: Expression was achieved first with a 25 kb HuIgH minigene construct containing mouse (thick unfilled lines) and human (thick black lines) H chain segments in germline configuration (Brüggemann et al., 1989). A 100 kb fragment, HuIgH^{cos}, containing an authentic human region was expressed from 2 co-integrated cosmids (Brüggemann et al., 1991). This was followed by the use of ES cells for the transfer of YACs. For the H chain, a 240 kb region, HuIgH YAC, was successfully transferred and expressed (Choi et al., 1993; Green et al., 1994; M. Neuberger and MB unpublished). For the κ light chain, a 320 kb YAC, HuIg$_\kappa$ YAC (Davies et al., 1993 and Green et al., 1994), and a V gene extended 1.3 Mb HucosIg$_\kappa$ YAC were successfully transferred and expressed (Zou et al., 1996). The extended YAC was obtained by site-specific integration of about 20 copies of a cosmid containing 5 different V$_\kappa$ genes from various subgroups.

small gene constructs, sometimes less than 30 kb, have been assembled from cloned and sequenced human Ig gene segments. It is clear that D and J recombinations significantly diversify the last hypervariable region of an antibody which makes the need for many V genes rather speculative. In addition, the results from transgenic mice showed that expression levels, rather than the number of V genes were crucial. It may therefore be possible to obtain a human antibody repertoire from a few V genes, ideally one from each of the seven different V$_H$ families, as long as controlled expression is secured.

Transgene Introduction

Transgenic mice carrying relatively small H and L chain constructs have been obtained by pronuclear microinjection procedures. The largest regions integrated and successfully expressed were human IgH loci of 60 kb and 100 kb (Brüggemann et al., 1991; Taylor et al., 1992 and Schedl et al., 1993). For the introduction of larger regions, microinjection and DNA purification can be avoided using embryonic stem (ES) cells and YAC transfer by fusion. ES cells can then be used for blastocyst injection to obtain chimaeric mice which have been shown to give germline transmission of the transgene (Green et al., 1994 and Zou et al., 1996). The advantages of using YACs are that cloned authentic regions can be well over 1 mega base (Mb) and DNA modifications in yeast are easily achievable by site-specific integration (Guthrie and Fink, 1991; Nelson and Brownstein, 1994). The core region of the IgH and IgL loci has been cloned onto YACs which are up to 300 kb in size (Davies et al., 1993; Choi et al., 1993 and Green et al., 1994) (see Figure 58.1), however, two disadvantages were immediately apparent from the clones isolated: a lack of V genes and C genes. To remedy this, overlapping homologous regions were successfully used to add V genes and recently also C genes (Davies and Brüggemann 1993; Zou et al., 1996 and Popov et al., 1996).

The transfer of large authentic Ig loci on YACs into cells and animals has been achieved by lipofection and protoplast fusion of ES cells (Choi et al., 1993; Davies et al., 1993; Jakobovits, 1994; and Lamb and Gearhart, 1995). Both methods were dependent on a marker gene to allow for selection of ES cell clones containing exogenous DNA. The most frequently used selective marker was the gene containing neomycin resistance which was added to the YAC by homologous integration (Davies et al., 1992; 1993). The advantages of the protoplast fusion procedure over lipofection treatment, which uses purified YAC DNA carried in lipid micelles, is that it does not involve DNA handling and thus, there is no limit to the size of the transferred YAC (Davies et al., 1996). The experiments showed that one complete YAC can be introduced without a substantial amount of yeast DNA (Davies et al., 1993 and Green et al., 1994). So far the largest YAC that has been transferred and successfully expressed is 1.3 Mb and contains a large part of the human κ light chain region (see Figure 58.1) with about 100 V genes (Zou et al., 1996).

Human Antibody Expression

Rearrangement and Ig Levels in Transgenic Mice

In order to assess whether DNA rearrangements of the human gene segments would lead to antibodies consisting of human μ or γ heavy chains or human κ light chains, transgenic mouse serum was titrated in ELISA assays. It was apparent that human antibodies were expressed (Brüggemann et al., 1989; Brüggemann et al., 1991; Taylor et al., 1992; Tuaillon *et al.*, 1993; Choi *et al.*, 1993 and Green *et al.*, 1994), however, the levels were disappointingly low. This was confirmed when mouse splenic B-cells were used in immunofluorescence analysis where only a low percentage stained for human μ (Brüggemann *et al.*, 1989). In addition, class switching, from μ to γ expression, within a V–D–J–μ–γ minilocus was a rare event and subsequently the level of human IgG was very low (Tuaillon *et al.*, 1993).

The maximum amount of human IgM found in the serum of some transgenic mice was about 100 μg/ml and the levels in hybridoma supernatants were up to $10\,\mu$g/ml – similar to those produced by mouse hybridoma lines (Wagner *et al.*, 1994a, 1994b). In transgenic mice carrying a very large Ig$_\kappa$ YAC we found up to $50\,\mu$g/ml for human Ig$_\kappa$, a very reasonable expression level not found in κ minigene transgenic mice (Zou *et al.*, 1996). However, given that the transferred region and the number of V genes had been dramatically increased, these amounts are still relatively low. A comparison of transgenic mouse serum levels are shown in figure 58.2 with several interesting observations: At best, human antibody levels were 10 times lower than those found in human serum. Indeed, in most transgenic lines studied concentrations were considerably lower (Taylor *et al.*, 1992; Tuaillon *et al.*, 1993; Choi *et al.*, 1993; Green *et al.*, 1994 and Wagner *et al.*, 1994a). The exception was that the antibody concentrations found in cultures of mouse B-cell hybridomas expressing human Ig could be similar to those found for normal mouse hybridomas. This implies that reasonable expression rates can be achieved for human Ig, especially when no mouse antibodies are expressed simultaneously, but that the number of cells which achieve this is very low.

Transgene Expression in Knock-out Mice

Human antibody expression has been drastically improved in transgenic IgH or IgL mice crossed with animals in which the endogenous H and L loci were silenced by gene targeting (Melton, 1994). Such knock-out experiments involved the use of embryonic stem cells modified by integration of a gene replacement construct into the mouse antibody locus. Mice generated in that way have part of C_μ disrupted or the J_H region deleted to silence H chain production (Kitamura *et al.*, 1991 and Chen *et al.*, 1993a). For silencing of the L chain, C_κ and/or J_κ were disrupted or deleted (Chen *et al.*, 1993b; Zou *et al.*, 1993a; Sanchez *et al.*, 1994 and Zou *et al.*, 1995). In such mice there were no H or κ light chains expressed, in fact H chain disruption led to the lack of B-cell generation. However, a most surprising and pleasing result was that, when crossing mice homozygous for the silenced endogenous locus with human Ig transgenic mice, the levels of human antibody increased dramatically (Choi *et al.*, 1993; Lonberg *et al.*, 1994; Green *et al.*, 1994; Wagner *et al.*, 1994a; 1994b and Zou *et al.*, 1996). Figure 58.2 shows the comparison of human IgM and Ig$_\kappa$ production in mice with active or silenced endogenous loci. In transgenic mice carrying human H or L chain genes crossed to homozygosity into the mouse H^- or κ^- background, human IgM levels could reach about $1/2$ (0.5–1 μg/ml) of those found in human serum, and for human κ antibodies about $1/8$ (1 μg/ml) of those found in human serum. It is likely that the rather reduced human Ig$_\kappa$ light chain concentrations are a result of competition with mouse Ig$_\lambda$ light chain expression which has yet to be silenced.

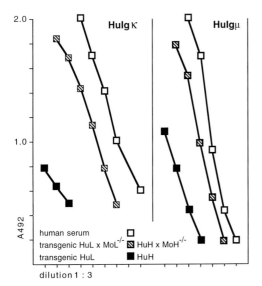

Figure 58.2. Ig serum concentrations for human κ and human μ. In ELISA titrations human serum was compared with serum from transgenic mice carrying a human L or H chain region. In addition, a comparison is shown of mice containing a human L or H chain transgene bred to homozygosity with mice in which the endogenous κL (HuL × MoL$^{-/-}$) or H (HuH × MoH$^{-/-}$) chain locus was silenced. Serum levels are the following: For human Ig$_\kappa$ 8 mg/ml were detected in human serum, 1 mg/ml in transgenic human κL mice crossed into the mouse L$^-$ background and only 5–50 μg/ml in transgenic human κL mice (Zou *et al.*, 1996). For human IgM 1–1.5 mg/ml were present in human serum, 0.5–1 mg/ml in transgenic human μ H chain mice crossed into the mouse H$^-$ background and only 10–100 μg/ml in transgenic human μ H mice (Wagner *et al.*, 1994a; 1994b).

Further improvements of the human antibody production in transgenic mice could be achieved by replacing mouse antibody genes, or indeed the mouse Ig loci, with human genes. Successful attempts have been made to replace mouse V and C region genes with the aim of using mouse control elements to drive human antibody expression. Exchanging mouse C_κ for human C_κ produced normal antibody concentrations with chimaeric mouse V_κ and human C_κ light chains (Zou et al., 1993b). After immunisation these mice expressed high affinity chimaeric antibodies using their hypermutation and selection mechanism. A similar approach has been carried out for V_H region replacement where a rearranged V–D–J replaced the mouse J_H locus (Taki et al., 1993). Since expression and allelic exclusion were driven by mouse regulatory elements, the experiments allowed analysis of hypermutation and class switching. In addition, a repertoire of high affinity chimaeric antibodies could be obtained after immunisation. The use of endogenous regulatory sequences in combination with transgenic exons appears to have the advantage of securing controlled expression rates. The exchange of small gene segments is presently feasible but the use of large YACs to replace genomic regions may become possible in the near future.

Repertoire Formation and Antibody Response

Diverse Ig Gene Rearrangements

Transgenic antibody diversity has been analysed by concentrating on the generation of hybridomas and PCR reactions to assess hypermutation in the B-cell repertoire. In minigene transgenic mice, where low levels of human Ig were expressed, the introduced V_H and V_L genes appeared to be predominantly in germline configuration even after immunisation (Brüggemann et al., 1991; Taylor et al., 1992 and Tuaillon et al., 1993). However, even though extensive junctional diversity in the rearranged transgenes has been found, specific human antibodies after immunisation were rare (Brüggemann and Neuberger, 1991). For the μH chain, human D segments have been shown to be expressed with extensive additions of non-encoded sequences, the N-regions at the V–D–J joins (Brüggemann et al., 1991; Taylor et al., 1992 and Tuaillon et al., 1993). There was a bias towards the use of the most proximal D segment, D_{HQ52}. In addition $J_H 4$ was most frequently found to be rearranged (Brüggemann et al., 1991; Tuaillon et al., 1993 and Wagner et al., 1994a). In human κ light chains, J1 was most frequently rearranged, but, as for the heavy chain, all J segments were being utilised (Taylor et al., 1992). In H chain transgenic mice the CDR3 diversity of the human antibody genes was quite similar to that found in human fetal liver development but has not been reported in the adult human repertoire (Tuaillon et al., 1993). However, when larger loci regions on YACs were introduced a more adult human-like repertoire was obtained and the preferential use of D_{HQ52} and $J_\kappa 1$ was not found (Green et al., 1994).

Hypermutation and Diversity

In the initial experiments with transgenic mice carrying human immunoglobulin miniloci it was shown that the human genes rearrange and express a diverse repertoire based on (D)–J rearrangements only, but not on V gene mutation. Immunisation of these animals yielded a response, but the specific antibodies were almost always of mouse origin. This changed dramatically when human IgH and IgL mice were bred with endogenous knockout animals in which the H chain or the κ L chain locus was silenced. Here, antibody responses after immunisation with a variety of immunogens, including human cells, resulted in specific Igs consisting of human H and L chains (Lonberg et al., 1994; Green et al., 1994 and Wagner et al., 1994b). From the experiments it was clear that human anti-human antibodies could easily be obtained, either polyclonal in serum or monoclonal after fusion of the transgenic mouse spleen cells with myeloma cells. Somatic hypermutation is expected in mouse B lymphocytes which have encountered antigen challenges and indeed specific human hybridoma antibodies and c-DNA sequences obtained from B-cell pools showed germline differences due to somatic mutation (Wagner et al., 1994b and Lonberg et al., 1994). It was also shown that, because of the antigen-selection advantage, antibody affinities were quite reasonable and this will allow the generation of therapeutically useful authentic human antibodies. Although the human genes can utilise expression, selection and mutation mechanisms provided by the mouse B-cells, the efficiency with which they do so is clearly reduced. It will be interesting to find out whether the provision of V or C region genes or as yet unidentified regulatory sequences – in short, driving the generation of antibody repertoires from large almost locus size regions – provides a transgenic construct with authentic expression capabilities.

Conclusions

It is clear that human Ig genes introduced in germline configuration can be rearranged, effectively expressed and mutated in the mouse background. A comparison of the transgenic lines shows that larger, authentic regions lead to better expression which may mean that the addition of more segments or control elements is essential. Human antibody production is most efficient in a mouse knock-out background and this lack of competition resulted in specific human antibody repertoires after immunisation.

Acknowledgements

I thank Lucy Bowden for helpful comments on the manuscript and the Babraham Institute for support of the work.

References

Brüggemann, M., Caskey, H.M., Teale, C., Waldmann, H., Williams, G.T., Surani, M.A. and Neuberger, M.S. (1989) A repertoire of monoclonal antibodies with human heavy-chains from transgenic mice. *Proc. Natl. Acad. Sci. USA* **86**, 6709–6713

Brüggemann, M., Spicer, C., Buluwela, L., Rosewell, I., Barton, S., Surani, M.A. and Rabbitts, T.H. (1991) Human antibody production in transgenic mice: Expression from 100 kb of the human IgH locus. *Eur. J. Immunol.* **21**, 1323–1326

Brüggemann, M. and Neuberger, M.S. (1991) Generation of antibody repertoires in transgenic mice. *Methods* **2**, 159–165

Chen, J., Trounstine, M., Alt, F.W., Young, F., Kurahara, C., Loring, J.F. and Huszar, D. (1993a) Immunoglobulin gene rearrangement in B cell deficient mice generated by targeted deletion of the J_H locus. *Int. Immunol.* **5**, 647–656

Chen, J., Trounstine, M., Kurahara, C., Young, F., Kuo, C.-C., Xu, Y., Loring, J.F., Alt, F.W. and Huszar, D. (1993b) B cell development in mice that lack one or both immunoglobulin κ light chain genes. *EMBO J.* **12**, 821–830

Choi, T.K., Hollenbach, P.W., Pearson, B.E., Ueda, R.M., Weddell, G.N., Kurahara, C.G., Woodhouse, C.S., Kay, R.M. and Loring, J.F. (1993) Transgenic mice containing a human heavy chain immunoglobulin gene fragment cloned in a yeast artificial chromosome. *Nature Genetics* **4**, 117–123

Davies, N.P., Rosewell, I.R. and Brüggemann, M. (1992) Targeted alterations in yeast artificial chromosomes for inter-species gene transfer. *Nucl. Acids Res.* **20**, 2693–2698

Davies, N.P. and Brüggemann, M. (1993) Extension of yeast artificial chromosomes by cosmid multimers. *Nucl. Acids Res.* **21**, 767–768

Davies, N.P., Rosewell, I.R., Richardson, J.C., Cook, G.P., Neuberger, M.S., Brownstein, B.H., Norris, M.L. and Brüggemann, M. (1993) Creation of mice expressing human antibody light chains by introduction of a yeast artificial chromosome containing the core region of the human immunoglobulin κ locus. *Bio/Technology* **11**, 911–914

Davies, N.P., Popov, A.V., Zou, X. and Brüggemann, M. (1996) Human antibody repertoires in transgenic mice: Manipulation and transfer of YACs. *Antibody Engineering: A Practical Approach* IRL, 59–76

Green, L.L., Hardy, M.C., Maynard-Currie, C.E., Tsuda, H., Louie, D.M., Mendez, M.J., Abderrahim, H., Noguchi, M., Smith, D.H., Zeng, Y., David, N.E., Sasai, H., Garza, D., Brenner, D.G., Hales, J.F., McGuinness, R.P., Capon, D.J., Klapholz, S. and Jakobovits, A. (1994) Antigen-specific human monoclonal antibodies from mice engineered with human Ig heavy and light chain YACs. *Nature Genetics* **7**, 13–21

Guthrie, C. and Fink, G.R. (1991) Guide to yeast genetics and molecular biology. *Meth. Enzymol.* **194**, Academic Press

Jakobovits, A. (1994) Humanizing the mouse genome. *Current Biol.* **4**, 761–763

Kitamura, D., Roes, J., Kühn, R. and Rajewsky, K. (1991) A B cell-deficient mouse by targeted disruption of the immunoglobulin μ chain gene. *Nature* **350**, 423–426

Lamb, B.T. and Gearhart, J.D. (1995) YAC transgenesis and the study of genetics and human disease. *Current Opinion in Genetics and Development* **5**, 342–348

Lonberg, N., Taylor, L.D., Harding, F.A., Trounstine, M., Higgins, K.M., Schramm, S.R., Kuo, C.-C., Mashayekh, R., Wymore, K., McCabe, J.G., Munoz-O'Regan, D., O'Donnell, S.L., Lapachet, E.S.G., Bengoechea, T., Fishwild, D.M., Carmack, C.E., Kay, R.M. and Huszar, D. (1994) Antigen-specific human antibodies from mice comprising four distinct genetic modifications. *Nature* **368**, 856–859

Melton, D. (1994) Gene targeting in the mouse. *BioEssays* **16**, 633–638

Nelson, D.L. and Brownstein, B.H. (1994) YAC libraries: A user's guide. *UWBC Biotechnical Resource Series* Freeman and Co., New York

Neuberger, M.S. and Milstein, C. (1995) Somatic hypermutation. *Curr. Opin. Immunol.* **7**, 248–254

Pettersson, S., Cook, G., Brüggemann, M., Williams, G.T. and Neuberger, M.S. (1990) A second B-cell specific enhancer is 3' of the immunoglobulin heavy chain locus. *Nature* **344**, 165–168

Popov, A.V., Bützler, C., Frippiat, J.-P., Lefranc, M.-P. and Brüggemann, M. (1995) Assembly and extension of yeast artificial chromosomes to build up a large locus, *Gene*, in press

Sanchez, P., Drapier, A.-M., Cohen-Tannoudji, M., Collucci, E., Babimet, C. and Cazenave, P.-A. (1994) Compartmentalization of λ subtype expression in the B cell repertoire of mice with a disrupted or normal C_κ gene segment. *Int. Immunol.* **6**, 711–719

Schedl, A., Larin, Z., Montoliu, L., Thies, E., Kelsey, G., Lehrach, H. and Schütz, G. (1993) A method for the generation of YAC transgenic mice by pronuclear microinjection. *Nucl. Acids. Res.* **21**, 4783–4787

Taki, S., Meiering, M. and Rajewsky, K. (1993). Targeted insertion of a variable region gene into the immunoglobulin heavy chain locus. *Science* **262**, 1268–1271

Taylor, L.D., Carmack, C.E., Schramm, S.R., Mashayekh, R., Higgins, K.M., Kuo, C.-C., Woodhouse, C., Kay, R.M. and Lonberg, N. (1992) A transgenic mouse that expresses a diversity of human sequence heavy and light chain immunoglobulins. *Nucl. Acids Res.* **20**, 6287–6295

Tuaillon, N., Taylor, L.D., Lonberg, N., Tucker, P.W. and Capra, D. (1993) Human immunoglobulin heavy-chain minilocus recombination in transgenic mice: Gene-segment use in μ and γ transcripts. *Proc. Natl. Acad. Sci. USA* **90**, 3720–3724

Wagner, S.D., Williams, G.T., Larson, T., Neuberger, M.S., Kitamura, D., Rajewsky, K., Xian, J. and Brüggemann, M. (1994a) Antibodies generated from human immunoglobulin miniloci in transgenic mice. *Nucl. Acids Res.* **22**, 1389–1393

Wagner, S.D., Popov, A.V., Davies, S.L., Xian, J., Neuberger, M.S. and Brüggemann, M. (1994b) The diversity of antigen-specific monoclonal antibodies from transgenic mice bearing human immunoglobulin gene miniloci. *Eur. J. Immunol.* **24**, 2672–2681

Zou, Y.-R., Takeda, S. and Rajewsky, K. (1993a) Gene targeting in the Ig_κ locus: efficient generation of λ chain-expressing B-cells, independent of gene rearrangements in Ig_κ. *EMBO J.* **12**, 811–820

Zou, Y.-R., Gu, H. and Rajewsky, K. (1993b) Generation of a mouse strain that produces immunoglobulin κ chains with human constant regions. *Science* **262**, 1271–1274

Zou, X., Xian, J., Popov, A.V., Rosewell, I.R., Müller, M. and Brüggemann, M. (1995) Subtle differences in antibody responses and hypermutation of λ light chains in mice with a disrupted κ constant region. *Eur. J. Immunol.* **25**, 2154–2162

Zou, X., Xian, J., Davies, N.P., Popov, A.V. and Brüggemann, M. (1996) Dominant expression of a 1.3 Mb human Ig_κ locus YAC replacing mouse light chain production. *FASEB J.*, in press

PART IV, SECTION B

59. Standardization of Transgenic Lines: from Founder to an Established Animal Model

Dorothée Carvallo[1,*], Georges Canard[2] and David Tucker[3]

[1]*Château de Villandry, 37510 Villandry, France*
[2]*Transgenic Alliance, Domaine des Oncins, 69210, Saint Germain sur l' Arbresle, France*
[3]*Zeneca Pharmaceuticals, Mereside, Alderley Park, Macclesfield, Cheshire SK10 4TG, UK*

Since the development, in the 1980s, of routine procedures for gene addition or modification to generate transgenic animals, principally mice, it may be estimated that several thousands of transgenic lines have been produced worldwide. For the most part these have been designed to answer specific questions concerning the role of genes in biological processes such as development, immune function, endocrinology and oncogenesis. Nevertheless, there has been increasing realization that animals harboring new gene combinations have the potential to provide more reliable animal models in a range of applications, notably in the testing of pharmaceuticals for their efficacy against specific disease processes. For instance, mice expressing oncogenes are often predisposed to tumorigenesis and furnish a testing ground for the evaluation of antitumor drugs (see, for instance, Dexter *et al.*, 1993). In another example, mice lacking gene products such as the cystic fibrosis transmembrane receptor (CFTR) can simulate the pathophysiology of cystic fibrosis (Dorin *et al.*, 1992) so permitting the *in vivo* assessment of novel treatments such as gene therapy. Ultimately it may be envisaged that transgenic animals may become the model of choice, not only for testing the efficacy of pharmaceuticals, but also for routine toxicological testing of a variety of products destined for human use.

The production of transgenic models for use in these applications differs considerably from the use of transgenic animals in fundamental research. Large numbers of animals are required but also inter-animal variation (principally due to genetic background and health status) must be kept to an absolute minimum. For this reason the standardization of transgenic lines has become a prerequisite for their exploitation by the healthcare industries. In this article we discuss our experience with standardization of transgenic animals; we make recommendations to be considered by workers faced with the problem of converting a transgenic line produced by an academic research laboratory into a 'reagent-grade' model acceptable to industry.

The Requirement for Pathogen-Free Animal Models

Health Profile Provides a Major Source of Inter-animal Variation

Few rodent pathogens generate overt clinical signs in infected individuals; although temporary disease symptoms can be observed at the time of infection, the animal often recovers. However, more often than not the animal remains pathogen positive and can transmit to other animals in the same colony leading, in the worst case scenario, to seroconversion of all animals in that colony. In view of the difficulties of eradication some animal facilities are obliged to maintain entire colonies with pathogen-positive status. However, there are considerable disadvantages to this. First and foremost, infections persisting in the colony often interfere with reproductive function and can lead to a dramatic drop in breeding efficiency. Second, animals of different genetic backgrounds differ considerably in their susceptibility and even different individuals of a given line can differ in the extent to which they are

* Author for Correspondence.

affected. These factors lead to increased inter-individual variation. Third, certain pathogens can interfere with selected organs to which transgene expression may be targeted. As an illustration, some strains of mouse hepatitis virus (MHV) cause brain lesions: animals harboring such virus strains would be unacceptable for the study of neurodegenerative processes in transgenic animals (reviewed by the National Research Council, 1991). In addition, the emergence of transgenic technology has dramatically increased the number of transfers of lines from one facility to the next and the concomitant spread of common rodent specific infectious agents; these (e.g. MHV, MVM, *Pasteurella pneumotropica*) are frequently found in transgenic mouse colonies.

A high-quality health profile is the only sustainable strategy. In a research facility, the elimination of all animals following an outbreak of infection is the method of choice for decontamination. Because transgenic lines are often unique, this does not afford a practical solution. Instead it is necessary to rederive the transgenic line into a clean facility by embryo transfer or aseptic hysterectomy.

Rederivation Procedure

Embryo transfer is likely to become the preferred method for the rederivation of transgenic lines because it appears to be microbiologically safer than aseptic hysterectomy. In addition, the techniques and equipment required are often available in laboratories undertaking transgenic research (Reetz *et al.*, 1988 and Rouleau *et al.*, 1993). Preimplantation mouse embryos are essentially impervious to the entry of microorganisms (Carthew *et al.*, 1983; 1985) due to the impermeability of the zona pellucida. Exogenous contaminants are removed by rinsing the embryos in sterile media. An exception is for the elimination of mycoplasmas (Hill and Stalley, 1991). Here embryos from mycoplasma-infected colonies require to be cultured in M16 medium for 24 hours prior to cryopreservation or embryo transfer in order to reduce the contamination levels to below the infectious dose. These routine procedures, when combined with cryopreservation, generate a stock of frozen pathogen-free transgenic embryos ready for rederivation by embryo transfer. By mating transgenic males with superovulated F1 hybrid females, sufficient embryos for rederivation by embryo transfer can be collected from a single mouse or from a small colony. This method provides significant savings of cost and time when compared to aseptic hysterectomy. In addition, for only marginally increased effort, further embryos can be collected and kept frozen as security.

Maintenance of Clean Rodent Colonies

The key to both rederivation techniques is the ability to maintain, without pathogen spread, contaminated colonies in the close vicinity of pathogen-free colonies of foster or surrogate mothers. This entails an absolute physical separation between the contaminated and clean colonies, notably including the use of separate personnel for routine maintenance of the two colonies. To this end, several approaches may be used. We have opted for the following: all contaminated colonies are kept to an absolute minimum in terms of the number of animals, and for a minimum time, confined in a quarantine facility with housing in negative-pressure isolators. All lines originating from outside the facility are considered to be contaminated – it is inadvisable to rely on health monitoring reports accompanying incoming animals as these are often inaccurate and seldom comprehensive. Even the 'complete' health monitoring screens carried out on isolator-quarantined colonies are not fully predictive, particularly when one has to rely on a one-time sampling and a small number of animals (for instance, agents like *Pasteurella pneumotropica* are not easily diagnosed and specific sampling and determination is used; MHV is not always detected by serology). A further procedure employed to minimize the spread of contamination is the disinfection of waste from contaminated stock (disinfection before exit from isolators followed by rapid incineration).

Once rederived, the transgenic animal lines are housed in a semi-barriered system comprising positive pressure, HEPA-filtered ventilation, barrier autoclave for cages and equipment, irradiated feed and bedding, sterile filtration of water and the use of filter top cages. All animal manipulations are performed under aseptic conditions in a horizontal laminar flow hood. Non-transgenic animals required for matings are either bred in the same area (nucleus colonies of the most commonly-used strains are easily maintained), or are bred in separate isolators or filter-top cage systems. This point is of particular importance for groups generating their own transgenic lines. If founder stock is to be generated and maintained free of infectious disease, it is inadvisable to rely on surrogate mothers that have been bred in barriered facilities (which is the case for most commercially available mice). These are potential vectors of contamination, as there is always a lag time between outbreak of an infection and the diagnosis/detection of the infection. The use of non-transgenic stock bred in the end-user's facility is to be recommended. Figure 59.1 summarizes recommended procedures for health management of transgenic colonies.

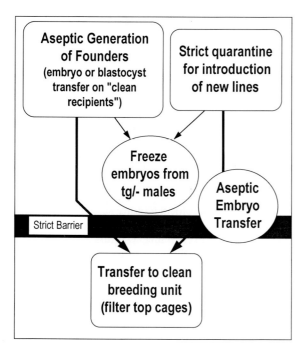

Figure 59.1. Health profile management of transgenic colonies.

Standardization of Genetic Background to Reduce Variation between Individuals

The phenotypic expression of a transgenic modification, either by gene addition or replacement, clearly depends on the extent to which the genetic background can compensate for the change introduced by the experimenter. An obvious example is the development of tumors in mice overexpressing oncogenes. While overexpression of the oncogene is the primary event that ultimately leads to uncontrolled cell proliferation, other events are necessary to lead to tumor development. The nature and frequency of such secondary events will depend at least partially on other genes, for instance the mutational activation of cooperating oncogenes or the inactivation of tumor-suppressor genes (potentially via endogenous retroviruses integrated into and inseparable from the rodent genome). Each such gene may be present in different allelic versions in different rodent strains and thus may require different levels of activation or down-regulation so as to cooperate in oncogenesis. A second example is the study of the effect of transgene expression on brain function. Different inbred lines clearly show quantitative and qualitative differences in a number of behavioral tests (Upchurch and Wehner, 1988; and Lhotellier and Cohen-Salmon, 1991).

Most Transgenic Founders have an Undefined Genetic Background

Hybrids between different inbred strains offer many advantages to the experimenter because they provide more robust and numerous one-cell embryos than embryos of a single inbred background. For this reason F1 hybrids have been routinely used to generate transgenic animals by pronuclear microinjection. C57BL/6 × SJL and C57BL/6 × CBA F1's are among the most commonly used hybrids. The founder animals generated using these lines are most often F2 hybrids. For lines obtained by gene targeting in ES cells, the vast majority derive from various 129 sublines (e.g., 129/Ola, 129/Sv). To generate a congenic inbred strain, chimeras should ideally be mated to the 129 background. Genetically, however, the 129 line is poorly characterized, is known to have poor breeding performance, and is impaired in a number of other respects. In most cases, chimeras have therefore been mated to C57BL/6 animals (or CBA), resulting in founders which are effectively F1's between 129 and C57BL/6. These founders are then inter-crossed to homozygosity with respect to the transgene. As above, the resulting progeny have the heterogenous background of F2 hybrids.

Drawbacks of the Most Commonly Used Mating Schemes

These approaches, while their implementation has offered a short-term pragmatic solution to the investigator, have a number of disadvantages that should be taken into consideration.

Maintenance of Colonies Heterozygous for the Transgene. When animals heterozygous for the transgene are used, the most common breeding scheme is to mate transgenic heterozygous animals to an F1 hybrid produced by intercrossing two inbred strains. The resulting progeny are thus an undefined mixture of the parental genetic backgrounds. The distribution of variable genes is random in each individual, thus all individuals are genetically different. However, this heterogeneity between individuals is stable with time if animals of the identical F1 hybrid are used for crossing at each successive generation. In this case, non-transgenic littermates may then be used as controls. However, the high genetic variability between individuals will translate into scattered responses in testing, impacting on the size of the population needed for each experiment.

One alternative comprises systematic backcrossing of to one of the parental inbred strains. Batches of test animals may be produced at each generation by mating transgenic males in harem with 5 inbred females. The first F(n + 1) males produced are used to replace their F(n) male parent as soon as they mature. In this scheme, nontransgenic littermates afford control animals though it must be accepted that each generation is different in terms of allele diversity. The genetic background may be considered stabilized at generation F6.

In the two mating schemes, testing of all the progeny is necessary to identify transgenic animals. If a cheap test is not available and polymerase chain reaction (PCR) has to be used, the cost of such breeding is, in our hands, approximately 4 to 5 times that of breeding a colony homozygous for the transgene. For this reason, and when the insertion mutation is not homozygous lethal, homozygous colonies are often derived by intercrossing transgenic animals.

Maintenance of Colonies Homozygous for the Transgene. For knock-out lines, in order to analyse the effect of a gene deletion, heterozygous progeny of the founders are inter-crossed. If animals homozygous for the deletion are both viable and reproductively competent, the deletion mutation may be maintained as a homozygous colony. When homozygous breeding pairs are first obtained there is a choice to be made between two mating schemes for maintenance of the colony:

(a) The colony is maintained using strict brother–sister matings. The end result will be the generation of an inbred recombinant strain between the two parent lines, which will be considered fully inbred only after 20 generations of brother–sister mating (98.6% homozygosity at all loci). This technique has two drawbacks. First, there is no insurance that this new inbred line will be viable and progressive loss of viability/reproductive performance can often be encountered; indeed, the partially inbred line may be lost after 4 to 6 generations of brother–sister matings. This is why, if using this scheme, it is wise to perform brother–sister matings with several lines in parallel. Second, there is no adequate control line, and it is impossible to generate one (allele fixation within the inbred line takes place essentially at random).

(b) The alternative is to maintain the colony as if it were outbred, by using a systematic rotation of breeders to minimize inbreeding. We employ Robertson's mating scheme as we feel it is the easier to establish (for a comparison of various minimum inbreeding rotation scheme see Rapp, 1972; and Rapp and Burrow, 1979). However, while such techniques are fairly efficient at large primary colony size (for 120 pairs, the increase in inbreeding will be around 0.2% per generation), this is far from true for small transgenic colonies (our biggest transgenic colonies are commonly 40–50 pairs). The increase in inbreeding will then be of a few percent per generation resulting in a definite genetic shift towards an inbred line. If the line is intended to be used in drug testing, such a shift may be unacceptable – experiments downstream will be performed on animals significantly different from the ones used during the initial stages of model validation. We have experienced this drift with one knock-out colony (original background 129/Ola × swiss), where, in 3 generations, the most interesting phenotypic feature of the line appeared to be lost despite the presence of the original gene targeting event. Another example, now with gene-addition transgenic animals, further illustrates this point: of 4 homozygous colonies, all derived from the same mouse line and maintained with the above rotational mating protocol (colony size between 20 and 40 breeding pairs), we have observed over 2 years (5 generations), and in each line, a significant decrease (30 to 50%) in the productivity (as measured by the number of pups weaned/female/week), thus illustrating a significant shift in the characteristics of these lines. Also, using the above mating scheme, no adequate control line exists.

Suggestions for the Genetic Management of Transgenic colonies. The principal drawbacks highlighted above were (i) inter-individual variations (likely to induce a low signal-to-noise ratio in test results); (ii) lack of a good control line; and (iii) progressive shift of the genetic background. We therefore argue that end-users may wish to consider backcrossing their transgenic lines to a defined inbred background, unless there are good reasons not to do so (e.g. a short-term experiment; a very precise yes-or-no test system). Backcrossing a transgene or gene deletion to an inbred background. The inbred line to be used for systematic backcrossing must be carefully chosen, depending on the area of research. An obvious consideration is to select a line that is well characterized vis a vis drug-testing (for example C57BL/6 is the most commonly used model in atherosclerosis research). Ideally, transgenic founders should be backcrossed onto multiple inbred lines as the effect of the transgene or gene modification may be significantly different when placed on different genetic backgrounds. Ultimately, this offers the option of intercrossing different congenic lines so as to obtain information on other genes that act as 'modifiers' for the transgene-induced phenotype under study.

Six generations of backcrossing will take 98.4% of the genes to homozygosity, which for most applications should be sufficient. Ten generations is sometimes necessary to reduce residual heterozygosity (99.9% of genes should then be homozygous). Backcrossing is an investment in time: six generations of backcrossing takes 18 months. It is therefore highly desirable to generate transgenic lines directly onto an inbred background whenever possible. Many over expressors are today generated directly in inbred strains (FVB, C57BL/6). However, to our knowledge, there is only one group that has successfully generated knock-out mouse lines using a C57BL/6 ES cell line though other groups have employed ES cells derived from a C57BL/6 × 129 hybrid background, shortening the backcross time (to C57BL/6 mice) by one generation but possibly complicating matters for subsequent crossing to other inbred strains.

Rapid production of animals while backcrossing takes place. Most users would wish to avoid waiting 1.5 to 2 years before commencing work on their transgenic line. For this reason, we breed initially from the background provided: either crossing to non-transgenic F1 hybrids (when heterozygotous animals can be used), or by managing the colony with minimum inbreeding protocols (if homozygous). In the latter case, particularly if dealing with a knock-out line where heterozygotes afford appropriate control animals, our suggestions for maintaining the line are as follows:

(a) Maintain a small colony of homozygous animals by a minimum inbreeding protocol,
(b) Produce heterozygous animals by mating the homozygotes to an inbred line (most preferably one already present in the founder background),
(c) Cross the above homozygous animals to the heterozygous animals. The resulting progeny should be 50% homozygous, 50% control (heterozygous). Though all animals will have different genetic backgrounds, the allele distribution should be similar in the homozygous and the control.

Simultaneously the transgenic lines should be crossed to one or both of the founder inbred strains so as to generate the inbred line(s) that will become the standardized model in the longer term.

Generating Congenic F1 Hybrids, Homozygous for the Transgene/Gene Targeting Event. A transgenic line congenic to a well characterized inbred line has substantial advantages over lines whose background is undefined. However, congenic lines suffer from the drawbacks familiar in inbred strains of mice. First, breeding performance is reduced over that exhibited by F1 hybrids (0.5 pups weaned per female per week on average compared to 1.5 to 2 in an F1 × F1 cross). Also, inbred lines are known to be extremely sensitive to changes in the environment. This sensitivity results in variability of test data obtained within a group and/or with time. To reduce this variability, while maintaining the advantages of having well defined genetics, we suggest the following breeding protocol:

Commence by backcrossing the transgene to two different inbred backgrounds. At F6, intercross to derive 2 congenic homozygous transgenic lines. The two homozygous lines are then intercrossed to generate F1 hybrid animals homozygous for the transgene/gene deletion. The non-transgenic F1 hybrid can be used as a suitable control. This strategy is summarized in Figure 59.2. F1 hybrids are (in principle) genetically identical, but unlike their inbred parents are heterozygous at all loci where there are allelic differences between the two strains. They are less sensitive to environmental changes and therefore produce more uniform test results and more reproducible responses over time. We are adopting this approach with the $Apoe^{m1Unc}$ line (Piedrahita et al., 1992), with both the C57BL/6 and a DBA/2 background. The objective is to obtain a line with a more reproducible development of arterosclerotic plaques. If successful, this approach will provide a reduction in the number of animals necessary for testing this type of drug therapy. The second advantage of the above strategy is that the breeding performance of the F1 hybrid is generally twice that of the parent line. This should achieve a double reduction in cost – a reduced cost of production and a reduction in the number of animals used.

Genetic Stability of the Transgenic and its Expression. Finally, and especially for transgenic animals generated by gene addition, it is essential to verify the transgene for integrity and stability of expression. Indeed, genetic standardization comprises rapid generation turn-over, thus increasing the risk of transgene re-arrangement or change in expression levels. Although transgene loss or rearrangement is thought to be rare, there have been instances where transgene expression levels are progressively lost over successive generations. However, we believe it is generally safe to verify the level of transgene expression and transcription (mRNA and corresponding protein expression level) in a small number of animals at each generation.

The standardization of a transgenic line is a long-term and costly commitment, and often more so than the initial generation of the founders. Clearly there are trade-offs to be made between the extent of the standardization and the final use of the line. Going through the above procedure of creating several congenic inbred and F1 hybrid transgenic lines from a single founder can be only justified for those models that will ultimately be used in large quantities for new drug development. However, we feel that an early analysis of the issues raised above may result in real savings in terms of time and cost, as well as generating more effective research tools, so ultimately reducing the overall number of animals required for drug development and validation.

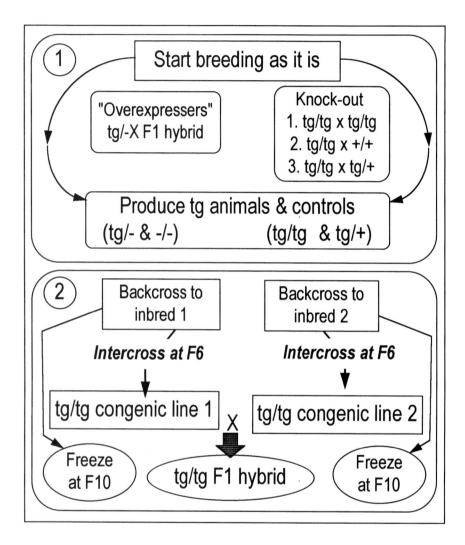

Figure 59.2. Genetic management of transgenic colonies.

Acknowledgements

We thank Pr. Richard Lathe (Center for Genome Research, Edinburgh, United Kingdom) for his useful comments and review of this chapter. We also thank Dr. Pierre Gallix (Rhone Poulenc Rorer Research Center, Vitry, France) for his close collaboration in the development of the above procedures.

References

Carthew, P., Wood, M.J. and Kirby, C. (1983) Elimination of Sendai (parainfluenza type 1) virus infection from mice by embryo transfer. *J. Reprod. Fert.* **69**, 253–257

Carthew, P., Wood, M.J. and Kirby, C. (1985) Pathogenicity of mouse hepatitis virus for preimplantation mouse embryos. *J. Reprod. Fert.* **73**, 207–213

Dorin, J. *et al.* (1992) Cystic fibrosis in the mouse by targeted insertional mutagenesis. *Nature* **359**, 211–215

Dexter, D.L., Diamond, M., Crevelling, J. and Chen, S. (1993) Chemotherapy of mammary carcinomas in ras transgenic mice. *Investigational New Drugs* **11**, 161–168

Hill, C.H. and Stalley, P.G. (1991) *Mycoplasma pulmonis* infection with regards to embryo freezing and hysterectomy rederivation. *Laboratory Animal Science* **41**(6), 563–566

Lhotellier, L. and Cohen-Salmon, C. (1991) Senescence of motor behavior in mice: contribution of genetic methods. *European Bulletin of Cognitive Psychology*, **11**(1), 27–53

National Research Concil (1991) *Infectious Diseases of Mice and Rats*, National Academy Press, Washington D.C., USA

Piedrahita, J.A., Zhang, S.H., Hagaman, J.R., Oliver, P.M. and Maeda, N. (1992) Generation of mice carrying a mutant apolipoprotein E gene inactivated by gene targetting in embryonic stem cells. *Proc. Natl. Acad. Sci. USA* **89**, 4471–4475

Reetz, I.C., Wullenweber-Schmit, M., Kraft, V. and Hedrich, H.J. (1988) Rederivation of inbred strains of mice by means of embryo transfer. *Laboratory Animal Science* 38(6), 696–701

Rapp, K.G. (1972) HAN-Rotation a new system for rigorous outbreeding. *Z. Versuchstierk. Bd.* 14, 133–142

Rapp, K.G. and Burrow, K. (1979), Genetic problems due to small numbers of outbred animals within an isolator. *Clin. and Experim. Gnotobiotics*, Zbl. Bakt. Suppl. 7, Gustav Fischer Verlag, Stutgart, New York

Rouleau, A.M.J., Kovacs, P.R., Kunz, H.W. and Amstrong, D.T. (1993) Decontamination of rat embryos and transfer to specific pathogen-free recipients for the production of a breeding colony. *Laboratory Animal Science* 43(6), 611–615

Upchurch, M. and Wehner, J.M. (1988) TI DBA/2Ibg mice are incapable of cholinergically-based learning in the Morris water task. *Pharmacology, Biochemistry and Behavior* 29(2), 325–329

Upchurch, M., Wehner, J.M. (1988) Differences between inbred strains of mice in Morris water maze performance. *Behavior Genetics* 18(1), 55–68

PART IV, SECTION B

60. The Use of Mouse Knockouts to Study Tumor Suppressor Genes

Christopher J. Kemp

Program in Cancer Biology, Division of Public Health Science, Fred Hutchinson Cancer Research Center
1124 Columbia St., Seattle, WA 98104, USA

This chapter will focus on the uses and interpretations of mouse knockout models of classes of genes whose function is to prevent or suppress the development of cancer. These include the classic tumor suppressor genes as well as genes involved in DNA repair and tumor progression. While these models have been extremely useful they have also produced unexpected phenotypes. It will be shown that the absence of an expected phenotype in a knockout model can, in some cases, be attributed to the lack of appropriate environmental co-factors or to an inappropriate genetic background. A case will also be made to expand the definition of "tumor suppressor gene" to include any gene whose loss of function results in an increased incidence or rate of tumor formation or degree of malignancy.

General Considerations on Interpretation of Knockout Phenotypes

One of the most powerful ways to study a gene's function is to examine the phenotype in the absence of gene product. The development of technology to create knockout mice by gene targeting in embryonic stem cells holds much promise in this regard. It is hard to imagine a conceptually simpler yet as realistic a system to determine the role of a mammalian gene product than to compare two groups of mice which are identical in virtually all aspects save the presence or absence of a single gene product. In many cases the promise has been fulfilled and much has been learned about the function of a variety of genes which would have been difficult or impossible without this technology.

However this power is limiting, and in many cases the phenotypes of the knockout mice have proven very difficult to interpret or even misleading. For example, many show an unexpected or even the absence of a phenotype. This is seen even for highly conserved genes. Does the lack of a phenotype imply the gene is not essential? The evolutionary conservation of the gene's structure indicates it must have been essential to the overall evolutionary fitness of at least the near ancestors to *Mus musculus*. To attribute the absence of a perceived phenotype to redundancy, where other gene products can compensate and perform the same function as the knocked out gene is unsatisfying due to the same evolutionary argument. Partial redundancy between related genes undoubtedly exists, but unless gene duplication occured very recently in evolutionary history, the gene products are unlikely to have completely redundant phenotypes. It seems clear then, that the absence of an observable phenotype is misleading, perhaps as a result of the unnaturally benign conditions of the laboratory. An (unfortunately) untestable prediction is that phenotypes effecting some aspect of reproductive fitness would emerge if the mice were living and breeding in the wild. A testable prediction is that if the knockouts are appropriately stressed or challenged in the laboratory, novel phenotypes will emerge which can be interpreted to imply the gene's function.

The types of genes knocked out may be roughly divided into two categories: those which are known to contribute to human disease by virtue of their inadvertent mutational inactivation and those not known to have a mutant counterpart but whose function is deemed important and worthy of study. In the latter case, the phenotype of the knockout is generally more difficult to predict as there exists no natural precedent. In the former case, the absence of a functional gene product is already known or suspected of producing a particular phenotype in humans and *a priori* it is predicted that the mouse knockout will approximate the human disease condition. In reality however, the phenotype of the mouse knockouts are rarely entirely as predicted.

Tumor Suppressor and Anti-Tumor genes

Tumor suppressor genes are in the category of genes which are known or suspected to contribute to human disease by virtue to their inactivation. The concept of tumor suppressor genes is rooted in early *in vitro* experiments demonstrating suppression of the malignant phenotype of tumor cells when fused to normal cells (Harris, 1988). This was taken to indicate that the tumor cells had lost the function of genes which were still intact in the normal cells and which could compensate in the fusion to revert the malignant phenotype. However this has led to a narrow definition of tumor suppressor genes as those which can actively *post hoc* suppress tumor formation. By this strict criteria, many genes which inhibit or suppress tumor development by a variety of mechanisms, for example DNA repair, paracrine or exocrine growth control, angiogenesis, or metastasis, are excluded. Thus discussions of cancer genes tend to focus on oncogenes and classical tumor suppressor genes and leave out an entire class of genes which can play a critical role in tumorigenesis. To address this, a broader definition of "tumor suppressor gene" may be in order to include any gene whose loss of function results in an increased incidence or rate of tumor formation or degree of malignancy. These would include the classically defined tumor suppressor genes as well as genes involved in DNA repair and segregation, metastasis, angiogenesis, inhibitory growth factors, etc.

Inactivation of the classical tumor suppressor genes may occur by at least two pathways; as an inherited mutant allele where all somatic cells carry the mutation; or by somatically acquired inactivation, either by mutation or viral inactivation. Examples of inherited defects in tumor suppressor genes include Retinoblastoma (mutant Rb gene), Li Fraumeni syndrome (mutant p53 gene), Neurofibromatosis (mutant NF-1 gene) and Wilm's tumor (mutant WT-1 gene in at least some cases). The former two proteins are involved in cell cycle regulation (Riley *et al.*, 1994; and Donehower and Bradley, 1993), NF-1 negatively regulates ras signaling (Bernards, 1995), and WT-1 has properties of a kidney specific transcription factor (Hastie, 1994). These conditions have all been recently reviewed (Hastie, 1994; Riley *et al.*, 1994; Bernards, 1995 and Malkin, 1994). In all these cases the disease is dominant or semi-dominant at the organismal level, as inheritance of a single defective allele is sufficient to cause disease.

The second and far more common contribution of the classical tumor suppressor genes is by somatically acquired mutation or viral inactivation. This can involve the exact same genes as in the inherited cases or other genes, many of which have yet to be characterized. In general both alleles need to be inactivated in the tumor; that is, they are recessive at the cellular level. A unique feature of this as a disease mechanism is that the mutation is presumed to occur in a single somatic cell, providing it with a selective advantage leading to clonal expansion and eventually growth into a detectable tumor. Such a mechanism has not been documented for the DNA repair class of tumor suppressor genes. Knockout models, where one allele is already inactivated, which accelerates the response, will be essential to fully understand the natural history of this complex and little understood process.

Another class of genes which have an antitumorigenic phenotype are those involved in DNA repair and/or maintenance of proper genetic segregation. The existence of these genes was first suggested by inherited cancer syndromes. Examples include xeroderma pigmentosum (mutant XP genes), Bloom's syndrome (gene to be identified), ataxia telangiectasia (mutant ATM gene), Fanconi's anemia (mutant FA gene), hereditary nonpolyposis colorectal cancer (HNPCC) (mutant mismatch repair genes) and several less well characterized syndromes (Friedberg *et al.*, 1995). These syndromes all result in an increased incidence of tumorigenesis at particular sites and are also characterized by decreased fidelity of DNA repair or segregation. Cloning and characterisation of some of the responsible genes has shown that, as predicted, they have suffered inactivating mutations. Thus one function of these gene products is to prevent cancer by insuring faithful segregation of genetic material. They differ from the inherited defects of classical tumor suppressor genes in that they are usually, but not always, recessive at the organismal level; mutations in both alleles need to be inherited. There is, however suggestive evidence that AT heterozygotes may have an increased susceptibility to tumor formation (Swift *et al.*, 1991). Also, reintroduction of these genes is not likely to suppress the tumorigenic phenotype. Despite these difference with the classical tumor suppressor genes, the phenotype is similar: increase rate of cancer in their absence, and so it is argued these should also be considered tumor suppressor genes.

In general these inherited syndromes display a fairly narrow range of susceptibility to tumor types (Riley et al., 1994; Bernards, 1995; Hastie, 1994 and Malkin, 1994) (Table 60.1). This tissue specificity of tumor formation is poorly understood. In a small number of cases it may be explained on the basis of restricted tissue expression of the gene product, for example the WT-1 gene product is expressed during nephrogenesis and probably contributes to differentiation of this tissue (Hastie, 1994). However, p53, Rb, and many DNA repair proteins are ubiquitously expressed and especially in the case of Rb it is difficult to explain why retinoblasts are uniquely susceptible in patients with a constitutive defect in Rb (Hamel et al., 1993). Also the tumor spectrum is rarely identical when comparing inherited vs. somatic mutations for reasons that are again not understood (Table 60.1). For example patients inheriting a mutant Rb allele develop predominantly retinoblastoma and to a lesser extent osteosarcomas. Yet many "spontaneous" tumor types including bladder, breast, prostate, sarcomas and small cell lung carcinoma suffer somatic mutations in Rb (Riley et al., 1994). There is a similar difference in tumors involving NF-1 (Table 60.1). Why are these spectra so different? It is hoped that knockout models will help explain some of these paradoxes.

Tumor Suppressor Gene Knockouts

In the handful of classical tumor suppressor knockouts so far created, the majority have shown that the gene is required for development, as the homozygous knockouts die *in utero*. In three cases, discussed more fully below, the heterozygotes are apparently normal until they develop tumors at particular sites. In each case, the remaining wild type allele is often lost in the tumors, creating null mosaics. This implies at least two functions of these genes, an essential one during development and an additional one to prevent or retard tumor development. This discussion will focus on the latter phenotypes. Several examples will be briefly presented followed by a general discussion of interpretation of phenotypes especially those which are unexpected.

Retinoblastoma Gene

One of the most unusual knockouts is that of the Rb deficient mouse (Lee et al., 1992; Jacks et al., 1992 and Clarke et al., 1992). The homozygous null is embryonic lethal which is perhaps not surprising given the ubiquitous expression of the Rb protein and its role in cell cycle regulation. However, the Rb heterozygotes, rather than developing retinoblastoma, instead develop pituitary tumors at 100% incidence Hu et al. (1994). This tissue is very small in the mouse, consisting of fewer than 200,000 cells indicating an extreme sensitivity of this cell type. These tumors showed loss of the remaining wild type allele, as assessed by Southern blot analysis, indicating somatic mutation and clonal selection had occurred. The tissue discordance between mice and humans will be discussed below but it will be pointed out here that mice can indeed develop retinoblastoma as was shown in transgenic mice which express the SV40 T antigen in the retina (Windle et al., 1990).

Wilm's Tumor Gene

In the case of a knockout model for the Wilm's tumor gene, the homozygote null is again embryonic lethal. The mice fail to undergo kidney development in keeping with the role of WT-1 as an essential transcription factor (Kreidberg et al., 1993). The heterozygotes however, failed to develop tumors of the kidney or any other tissue.

Neurofibromatosis Type 1 Gene

The NF-1 homozygote knockout mice are embryonic lethals, showing abnormal cardiac development (Brannan et al., 1994). The heterozygotes survive but do not show the classical symptoms of human neurofibromatosis including neurofibromas, iris Lisch nodules or pigmentation defects (Jacks et al., 1994c). However, they are predisposed to phaeochromocytoma and myeloid leukaemia, both of which are seen in the human disease. Loss of the remaining wild type allele is detected in roughly 50% of the tumors.

p53 Gene

The "mother of all" knockouts, that of the p53 tumor suppressor gene, has by far yielded the greatest wealth of information, surprises, and practical benefit. Just prior to the creation of the first p53 knockout mouse, there was accumulating evidence indicating a critical role for p53 in cell cycle progression (Casey et al., 1991). This, plus the high levels of p53 expression during development (Schmid et al., 1991), led to predictions of embryonic lethality for p53 homozygous nulls. Again confounding the predictions, the p53 nulls were developmentally normal as well as fertile showing p53 is not indispensable for development (Donehower et al., 1992). However, the awe inspiring power of this model and indeed of this enigmatic protein soon emerged when the nulls began to develop highly malignant tumors at an early age. In the three independent p53 knockouts now generated, the great majority of mice spontaneously develop highly malignant, lethal tumors by 6 months of age, predominantly lymphomas and sarcomas, with a

Table 60.1. Tumor spectra involving tumor suppressor genes in human vs. mouse*.

Human Inherited	Human Somatic	Mouse Inherited
Retinoblastoma (Rb)		
Retinoblastoma	Bladder carcinoma	Pituitary tumor
Osteosarcoma	Small cell lung carcinoma	
	Breast carcinoma	
Neurofibromatosis type 1 (NF-1)		
Neurofibrosarcoma	Melanoma	Pheochromocytoma
Pheochromocytoma	Colon adenocarcinoma	Myeloid leukemia
Astrocytoma	Myelodysplastic syndrome	
Wilm's tumor (Wt-1)		
Nephroblastoma	Unknown	No tumors
Li Fraumeni syndrome (p53)		
Breast carcinoma	Most tumor types	Lymphoma
Sarcoma		Sarcoma
Brain tumor		Testicular tumor
Leukemia		Others infrequently
Adrenocortical carcinoma		
Xeroderma pigmentosum (XPA, XPC)		
Squamous cell carcinoma	Unknown	Squamous cell carcinoma
Hereditary nonpolyposis colorectal cancer (MSH2, PMS2)		
Colorectal carcinoma	Unknown	Lymphoma
Plus others		Sarcoma

*Characteristic tumor types observed in human inherited cancer conditions are listed, as well as primary tumor types observed in the mouse knockouts of the same genes, given in parentheses. Also indicated are "spontaneous" human tumor types where somatic mutations in the same genes have been detected. This list is not intended to be comprehensive. Reference sources are in text.

broad range of other types occasional represented (Donehower *et al.*, 1992; Purdie *et al.*, 1994 and Jacks *et al.*, 1994a) (Table 60.1). The heterozygotes also develop tumors with a very high frequency albeit with a longer latency; approximately 50% incidence at 1 year of age. The tumor spectrum is slightly different than in the nulls with fewer lymphomas and a higher percentage of sarcomas (Harvey *et al.*, 1993b). Just over 50% of the tumors from the heterozygotes suffer somatic loss of the remaining wild type p53 allele (Purdie *et al.*, 1994; Harvey *et al.*, 1993b; Jacks *et al.*, 1994b and Kemp *et al.*, 1994).

Alpha Inhibin Gene

Inhibins are growth inhibitory peptides of the TGFβ family and thus potentially have a tumor suppressing phenotype. Unlike the cases above, there is at present no evidence for a role of this gene in human cancer Homozygous knockout mice were generated to study the phenotype of this gene. These mice developed normally until all developed gonadal stromal tumors (Shikone *et al.*, 1994). This situation illustrates another use of the knockouts; that is to reveal tumor suppressing phenotypes of genes previously not demonstrated to have such activity. In this case, the flow of information would be the reverse and studies would next focus on whether human cancers mimic those seen in mice. This would provide the knockout mice with a predictive rather than confirming role in identifying tumor suppressor genes.

DNA Repair Genes

Several mouse knockouts of DNA repair genes have recently been generated as models of some of the human repair deficient syndromes. Two xeroderma pigmentosum complementation group genes have been knocked out, XPA and XPC (Sands *et al.*, 1995; Nakane *et al.*, 1995 and de Vries *et al.*, 1995). The nulls are viable and apparently have no overt phenotype. However, they are highly susceptible to

UV-induced squamous cell carcinoma, confirming the important role of these genes in preventing tumorigenic mutations caused by exposure to the sun. Several genes involved in mismatch repair including MSH2, PMS2, PMS1, and MLH1 have been shown to be mutated and responsible for the human cancer syndrome hereditary nonpolyposis colorectal carcinoma (HNPCC) (Lynch and Lynch, 1994). These patients are predisposed to the development of colorectal cancer as well as a range of other tumor types. Mouse knockout models have been made in MSH2 (Reitmair et al., 1995 and de Wind et al., 1995) and PMS2 (Baker et al., 1995). Early results indicate that the MSH2 null mice are viable and apparently normal. They are predisposed to lymphoma development but so far have not developed colorectal or other tumor types characteristic of human HNPCC (Lynch and Lynch, 1994) (Table 60.1). The PMS2 null mice are also viable and predisposed to lymphomas and sarcomas again unlike their human counterpart. These DNA repair knockouts will be extremely useful for determining the interaction of environmental carcinogens with particular pathways of DNA repair. Other very useful experiments will be to determine if the heterozygous carriers are in any way susceptible which would have implications for human health due to the much higher percentage of heterozygous carriers.

Interpretations of Tumor Phenotypes

Thus, the Rb, Nf1, p53, XP, and mismatch repair knockouts have fulfilled a primary obligation, that of proving that these genes are in fact tumor suppressing. The case with the WT1 knockout is still open as these mice have yet to show tumor predisposition. What else have we learned from these models? The first surprise is frequent lack of concordance between tumor types seen in some of the mouse models vs. the human equivalent (Table 60.1). Explanations for these observations fall into two categories. The first is that the equivalent tissue in the mouse is in fact predisposed, but due to the low sensitivity of the system, e.g. simply observing groups of mice for spontaneous tumors, this predisposition is not detected. The lack of detection could be because the target tissue may consist of too few cells or there may be insufficient time for them to accumulate additional genetic events required for tumor development. Alternatively there may be environmental co-factors such as carcinogens or tumor promoters which humans but not laboratory mice are exposed. Also, the genetic background of the mice may be unfavorable to the expression of the tumor phenotype in a given tissue. The other less satisfying category of "explanation" is that mice are fundamentally different to humans and the same molecules may have different roles in the two organisms, making many models inherently imperfect. Only the first category of explanation can be easily tested.

Environmental Factors

One of the first such experiments was done with the p53 deficient mice. p53 mutations are found in human tumors derived from most tissues. As most human cancers arise from epithelial tissue it was expected that the p53 deficient mice would be predisposed to carcinoma development, which are tumors derived from epithelial tissues. However carcinomas were only very rarely observed. Was this due to the lack of co-factors or to an inherent difference in the role of p53 between mice and humans? Treatment of the dorsal skin of these mice with the carcinogen 7, 12-dimethylbenzanthracene did not induce a greater number of skin tumors in the p53 deficient mice compared to wild type but those that did arise progressed much more readily to malignancy (Kemp et al., 1993). Thus, p53 deficiency has no apparent phenotype in untreated mouse skin. Only after carcinogen treatment and the emergence of benign tumors does the phenotype emerge.

In the human version of xeroderma pigmentosum, tumors frequently develop at sun exposed sites implicating UV radiation as a co-factor. In several recent elegant studies, knockout mice deficient for XP complementation group genes, XPA and XPC, had no obvious defects or tumor susceptibility until they were treated with UV radiation and thence showed a dramatic sensitivity to skin tumor induction (Sands et al., 1995; Nakane et al., 1995 and de Vries et al., 1995). This confirms this agent as a co-factor and provides a very useful and realistic model to study other agents which could enhance or block tumor development.

Genetic Factors

In another example, the p53 deficient mice did not develop tumors of the mammary gland despite the increased incidence of these tumors in Li Fraumeni syndrome. However, when crossed to transgenic mice which over-express the Wnt-1 oncogene in breast tissue, the p53 deficient mice showed dramatically enhanced rate of mammary gland tumorigenesis (Donehower et al., 1995). Here again the effect of p53 deficiency is only detectable in pre-existing tumors. There are other examples of genetic background effects which can either enhance or suppress the tumor phenotype in tumor suppressor gene deficient mice (Dietrich et al., 1993 and Harvey et al., 1993a). These illustrate that the genetic background can also strongly influence the expression of a knockout phenotype and it will be useful to identify and characterize the particular modifier genes involved.

In at least several cases, the lack of co-factors has not explained the lack of concordance in the mouse models. Again with regard to p53, mutations are frequently observed in human hepatocellular carcinoma, but the knockout mice did not develop this tumor type. In order to determine if p53 plays a role in the later stages of hepatocarcinogenesis as was seen in the skin, liver tumors were induced in wild type and p53 deficient mice by injection of the hepatocarcinogen diethylnitrosamine (Kemp, 1995). There was no detectable difference in any stage of tumorigenesis between groups of mice. Consistent with this, mouse liver tumors induced in wild type mice rarely show p53 mutations and both results indicate a lack of a role for p53 in mouse liver tumor development. In this instance then, there appears to be a genuine difference in the role of p53 in human vs. mouse hepatocellular carcinoma which is at present difficult to explain.

Thus, the apparent lack of tumor development in the mouse tumor suppressor gene knockouts can at least in some cases be due to inadequate co-carcinogenic events, be they environmental or genetic. The initial disappointment may actually prove to be a blessing in disguise as it allows study of agents, genes, or processes which specifically enhance the expression of the phenotype of a given gene. This information will contribute to an understand of the etiology of human cancers. For example, p53 plays a very prominent role in human colorectal carcinoma, a very common tumor type, which requires a good model for study. However, the lack of tumor development in the colon of the p53 deficient mice is at present a disappointment. The same argument applies to the lack of colorectal cancer in the mismatch repair deficient mice. Perhaps the appropriate co-carcinogen stimulus such as a high fat diet or different genetic background will soon be found to produce such tumors in these mice. These types of experiments will be very valuable to ferret out which suspected environmental agents contribute to murine and by extrapolation, human cancer.

On the other side of the coin, tumors which appear in knockout mice but are not part of the corresponding human syndrome are more difficult to explain. Perhaps it should be obvious that the same genes will have different functions in different organisims. The most notable example is the almost exclusive occurrence of pituitary tumors in Rb deficient mice (Hu et al., 1994). In this instance, the best "explanation" is that the Rb protein plays a somewhat unique and evidently powerful tumor suppressing role in cells of the pituitary gland of mice. The equivalent tissue in humans is rudimentary and this might explain the lack of such tumors in humans with Rb defects. It remains to be seen if co-factors such as UV radiation or a different genetic background(s) can be found which will reveal a propensity to retinoblastoma or if these mice can be shown to be predisposed to other tumor types such as bladder and lung, in which Rb mutations are frequently found.

Other Uses of the Models

In Vivo Studies

It is desirable to go beyond the simple goal of demonstrating that a particular gene is a *bonafide* tumor suppressor gene. Examples of follow up uses are: (1) Using the mice or cells derived from them to understand the basic function of the gene product. (2) Understanding the etiology of the human counterpart disease with the ultimate aim of prevention. (3) Using the model to improve cancer treatment. (4) As a highly sensitive model for carcinogen testing.

Again the p53 deficient mice have set the standard for how such models can be immensely beneficial. Rather than just yield descriptive studies these mice have been essential for generating entirely new ways of thinking about p53 function as well as providing reagents to test developing ideas. Below several examples will be discussed.

Dietary factors have long been suspected of contributing to human cancer. The influence of calorie restriction and other dietary factors on tumor development were tested on the p53 null mice. Both calorie restriction and the steroid hormone dihydroepiandrosterone, DHEA, significantly delayed tumor development (Hursting et al., 1994 and Hursting et al., 1995). It is not known how these agents achieve their effects, but these studies show that they are not mediated through p53. Ionizing radiation is a well known human carcinogen, however, little is known about its carcinogenic mechanism of action. For example, no consistent genetic targets have been identified. The p53 knockout mice represent an ideal model to test several ideas. (1) In the heterozygotes, is the single wild type p53 allele a radiation target? (2) As p53 plays a key role in radiation induced apoptosis, are the p53 null mice susceptible to radiation carcinogenesis? A single whole body dose of ionizing radiation as low as 1 Gy was shown to dramatically shorten the tumor latency in the p53 nulls with no effect in the wild types indicating a major role of p53 in protecting against radiation carcinogenesis (Kemp et al., 1994). The tumor spectrum was remarkably similar to that seen spontaneously in p53 nulls indicating that radiation was accelerating the development of these tumors and there may be additional genetic targets in addition to p53. The tumor latency was also dramatically shortened in the p53 heterozygotes exposed to a similar radiation treatment. This was unexpected and suggests that heterozygous carriers of radiation sensitivity disorders may well be at increased risk to radiation carcinogenesis. In the future many additional agents will no doubt be tested using this model.

In all three heterozygous knockout models, NF-1, Rb, and p53, spontaneous tumors show complete loss of the wild type allele as assayed by Southern blot analysis of tumor DNA. It is really quite fascinating that with three different genes on three different chromosomes and in different tissues, the same process is faithfully recapitulated in these models; that is, somatic loss of the wild type allele followed by clonal expansion into a malignant tumor. These mice represent a tremendous resource to study this complex, important and little understood process. It will be interesting to compare the mechanisms and rate of loss between genes, between mice and humans, and with different treatments. For example, the 50–70% rate of loss of the wild type p53 allele in spontaneous tumors in p53 heterozygous mice is consistent with studies of spontaneous reduction to homozygosity at other loci (Klinedinst and Drinkwater, 1991). In contrast, in the radiation-induced tumors in the p53 heterozygotes, the frequency of loss was 97%, much higher than spontaneous, indicating p53 itself was a target of radiation-induced gene loss (Kemp et al., 1994). Such a phenomenon would be impossible to detect without an animal model. The rate of loss or mutation of the remaining wild type allele could also vary substantially between tissues. This could be one explanation for the tissue specificity seen in heterozygous knockouts or even in human cancers.

In Vitro Studies

Derivation of primary cell cultures from knockout mice has proven to be an extremely useful resource for testing predictions about gene function at the cellular level. They also provide some of the best negative controls for many types of experiments. For example, p53 null cell lines have contributed seminally to the ideas about the role of p53 in DNA damage induced G1 arrest (Keurbitz et al., 1992), apoptosis (Lowe et al., 1993a; 1994b), gene amplification and genetic instability (Livingstone et al., 1992), spindle checkpoint (Cross et al., 1995) and chemosensitivity (Lowe et al., 1993a) among others. In several cases these results can be confirmed and extended *in vivo* which again has yielded benefits. For example, whole body irradiation was shown to induce apoptosis in thymocytes in wild type mice,

but not in p53 deficient mice confirming that p53-dependent apoptosis is not a tissue culture artifact (Lowe et al., 1993b and Clarke et al., 1993). A further use of *in vivo* studies is to demonstrate tissue or cell type specificity. Thus in similarly irradiated mice, p53 induction is localized to the proliferating compartments of both the GI tract (cells at the base of the crypt) (Merritt et al., 1994) and skin (hair follicle). These are precisely the same cells that undergo apoptosis and again this response is not seen in the p53 null mice. This indicates that p53 induction and apoptosis may be both cell type and cell cycle dependent. This is more than a curious observation as tumors frequently evade chemotherapy-induced apoptosis by little understood mechanisms and p53 may play a key role in this process (Lowe et al., 1994a and Kohn et al., 1994). It is clear that cell lines derived from other knockouts will also be useful for generating new hypothesis, many of which can then be confirmed *in vivo*.

Conclusion

Rather than providing an exhaustive review, this chapter was intended to illustrate examples of some promises and pitfalls in knockouts of genes whose function is to prevent or suppress cancer development. In common with other types of knockouts, the null mice or cells derived from them are useful for describing function. In addition they are useful for identifying co-factors which modulate the tumorigenic phenotype such as environmental and genetic factors. Further, some tumor suppressor genes are unique in that the heterozygotes can develop a phenotype through somatic inactivation of the remaining wild type allele leading to clonal expansion and genetic mosaicism. Understanding this complex process requires a complex system. The beauty and power of gene knockouts is they provide a window into this complexity through the eyes of a single gene product. If we learn where and how to look, we will more fully understand how these gene products prevent the malignant evolution of cells. This in turn will suggest strategies to prevent their loss, restore their function, or selectively target these cell types for destruction. Although a tall order, the days of the tumor suppressor gene knock outs are early and their full potential is yet to be realized.

References

Baker, S.M., Bronner, C.E., Zhang, L., Plug, A.W., Robatzek, M., Warren, G., Elliott, E.A., Yu, J., Ashley, T., Arnheim, N., Flavell, R.A. and Liskay, R.M. (1995) Male mice defective in the DNA mismatch repair gene

PMS2 exhibit abnormal chromosome synapsis in meiosis. *Cell* **82**, 309–319

Bernards, A. (1995) Neurofibromatosis type 1 and Ras-mediated signalling: filling in the GAPS. *Biochimica Et Biophysica Acta* **1242**, 43–59

Brannan, C.I., Perkins, A.S., Vogel, K.S., Ratner, N., Nordlund, M.L., Reid, S.W., Buchberg, A.M., Jenkins, N.A., Parada, L.F. and Copeland, N.G. (1994) Targeted disruption of the neurofibromatosis type-1 gene leads to developmental abnormalities in heart and various neural crest-derived tissues. *Genes and Development* **8**, 1019–1029

Casey, G., Lo-Hsueh, M., Lopez, M.E., Vogelstein, B. and Stanbridge, E.J. (1991) Growth suppression of human breast cancer cells by the introduction of a wild-type p53 gene. *Oncogene* **6**, 1791–1797.

Clarke, A.R., Maandag, E.R., van Roon, M., van der Lugt, N.M., van der Valk, M., Berns, A. and te Riele, H. (1992) Requirement for a functional Rb-1 gene in murine development. *Nature* **359**, 328–330

Clarke, A.R., Purdie, C.A., Harrison, D.J., Morris, R.G., Bird, C.C., Hooper, M.L. and Wyllie, A.H. (1993) Thymocyte apoptosis induced by p53-dependent and independent pathways. *Nature* **362**, 849–852

Cross, S.M., Sanchez, C.A., Morgan, C.A., Schimke, M.K., Ramel, S., Idzerda, R.L. and Reid, B.J. (1995) A p53-dependent mouse spindle checkpoint. *Science* **267**, 1353–1356

De Vries, A., van Oostrom, C.T.M., Hofhuis, F.M.A., Dortant, P.M., Berg, R.J.W., de Gruijl, F.R., Wester, P.W., Van Kreijl, C.F., Capel, P.J.A., van Steeg, H. and Verbeek, S.J. (1995) Increased susceptibility to ultraviolet-B and carcinogens of mice lacking the DNA excision repair gene XPA. *Nature* **377**, 169–173

De Wind, N., Dekker, M., Berns, A., Radman, M. and te Riele, H. (1995) Inactivation of the mouse Msh2 gene results in mismatch repair deficiency, methylation tolerance, hyperrecombination, and predispostion to cancer. *Cell* **82**, 321–330

Dietrich, W.F., Lander, E.S., Smith, J.S., Moser, A.R., Gould, K.A., Luongo, C., Borenstein, N. and Dove, W. (1993) Genetic identification of Mom-1, a major modifier locus affecting Min-induced intestinal neoplasia in the mouse. *Cell* **75**, 631–639

Donehower, L.A., Harvey, M., Slagle, B.L., McArthur, M.J., Montgomery, C.A., Jr., Butel, J.S. and Bradley, A. (1992) Mice deficient for p53 are developmentally normal but susceptible to spontaneous tumours. *Nature* **356**, 215–221

Donehower, L.A., Godley, L.A., Aldaz, C.M., Pyle, R., Shi, Y.P., Pinkel, D., Gray, J., Medina, D. and Varmus, H.E. (1995) Deficiency of p53 accelerates mammary tumorigenesis in Wnt-1 transgenic mice and promotes chromosomal instability. *Genes and Development* **9**, 882–895

Donehower, L.A. and Bradley, A. (1993) The tumor suppressor p53. [Review]. *Biochimica Et Biophysica Acta* **1155**, 181–205

Friedberg, E.C., Walker, G.C. and Siede, W. (1995) Human hereditary diseases with defective processing of DNA damage. In *DNA repair and mutagenesis* (Anonymous, ed.), ASM Press, Washington D.C., pp. 633–685

Hamel, P.A., Phillips, R.A., Muncaster, M. and Gallie, B.L. (1993) Speculations on the roles of RB1 in tissue-specific differentiation, tumor initiation, and tumor progression. *The FASEB Journal* **7**, 846–854

Harris, H. (1988) The analysis of malignancy by cell fusion: the position in 1988. *Cancer Res.* **48**, 3302–3306

Harvey, M., McArthur, M.J., Montgomery, C.A., Jr., Bradley, A. and Donehower, L.A. (1993a) Genetic background alters the spectrum of tumors that develop in p53-deficient mice. *Faseb Journal* **7**, 938–943

Harvey, M., McArthur, M.J., Montgomery, C.A., Jr., Butel, J.S., Bradley, A. and Donehower, L.A. (1993b) Spontaneous and carcinogen-induced tumorigenesis in p53-deficient mice. *Nature Genetics* **5**, 225–229

Hastie, N.D. (1994) The genetics of Wilm's tumor-a case of disrupted development. *Annual Review of Genetics* **28**, 523–558

Hu, N., Gutsmann, A., Herbert, D.C., Bradley, A., Lee, W.H. and Lee, E.Y. (1994) Heterozygous Rb-1 delta 20/+mice are predisposed to tumors of the pituitary gland with a nearly complete penetrance. *Oncogene* **9**, 1021–1027

Hursting, S.D., Perkins, S.N. and Phang, J.M. (1994) Calorie restriction delays spontaneous tumorigenesis in p53-knockout transgenic mice. *Proc. Natl. Acad. Sci. USA* **91**, 7036–7040

Hursting, S.D., Perkins, S.N., Haines, D.C., Ward, J.M. and Phang, J.M. (1995) Chemoprevention of spontaneous tumorigenesis in p53-knockout mice. *Cancer Research* **55**, 3949–3953

Jacks, T., Fazeli, A., Schmitt, E.M., Bronson, R.T., Goodell, M.A. and Weinberg, R.A. (1992) Effects of an Rb mutation in the mouse. *Nature* **359**, 295–300

Jacks, T., Remington, L., Williams, B.O., Schmitt, E.M., Halachmi, S., Bronson, R.T. and Weinberg, R.A. (1994a) Tumor spectrum analysis in p53-mutant mice. *Current Biol.* **4**, 1–7

Jacks, T., Remington, L., Williams, B.O., Schmitt, E.M., Halachmi, S., Bronson R.T. and Weinberg, R.A. (1994b) Tumor spectrum analysis in p53-mutant mice. *Current Biol.* **4**, 1–7

Jacks, T., Shih, T.S., Schmitt, E.M., Bronson, R.T., Bernards, A. and Weinberg, R.A. (1994c) Tumour predisposition in mice heterozygous for a targeted mutation in Nf1. *Nature Genetics* **7**, 353–361

Kemp, C.J., Donehower, L.A., Bradley, A. and Balmain, A. (1993) Reduction of p53 gene dosage does not increase initiation or promotion but enhances malignant progression of chemically induced skin tumors. *Cell* **74**, 813–822

Kemp, C.J., Wheldon, T. and Balmain, A. (1994) p53 deficient mice are extremely susceptible to radiation-induced tumorigenesis. *Nature Genetics* **8**, 66–69

Kemp, C.J. (1995) Hepatocarcinogenesis in p53 deficient mice. *Molecular Carcinogenesis* **12**, 132–136.

Keurbitz, S.J., Plunkett, B.S., Walsh, W.V. and Kastan, M.B. (1992) Wild-type p53 is a cell cycle checkpoint determinant following irradiation. *Proc. Natl. Acad. Sci. USA* **89**, 7491–7495.

Klinedinst, D.K. and Drinkwater, N.R. (1991) Reduction to homozygosity is the predominant spontaneous mutational event in cultured human lymphoblastoic cells. *Mutation Research* **250**, 365–374

Kohn, K.W., Jackman, J. and O'Connor, P.M. (1994) Cell cycle control and cancer chemotherapy. *Journal of Cellular Biochemistry* **54**, 440–452

Kreidberg, J.A., Sariola, H., Loring, J.M., Maeda, M., Pelletier, J. and Housman, D. (1993) WT-1 is required for early kidney development. *Cell* **74**, 679–691

Lee, E.Y., Chang, C.Y., Hu, N., Wang, Y.C., Lai, C.C., Herrup, K., Lee, W.H. and Bradley, A. (1992) Mice deficient for Rb are nonviable and show defects in neurogenesis and haematopoiesis. *Nature* **359**, 288–294

Livingstone, L.R., White, A., Sprouse, J., Livanos, E., Jacks, T. and Tlsty, T.D. (1992) Altered cell cycle arrest and gene amplification potential accompany loss of wild-type p53. *Cell* **70**, 923–935

Lowe, S.W., Ruley, H.E., Jacks, T. and Housman, D.E. (1993a). p53-dependent apoptosis modulates the cytotoxicity of anticancer agents. *Cell* **74**, 957–967

Lowe, S.W., Schmitt, E.M., Smith, S.W., Osborne, B.A. and Jacks, T. (1993b) p53 is required for radiation-induced apoptosis in mouse thymocytes. *Nature* **362**, 847–849

Lowe, S.W., Bodis, S., McClatchey, A., Remington, L., Ruley, H.E., Fisher, D.E. and Jacks, T. (1994a) p53 status and the efficacy of cancer therapy *in vivo*. *Science* **266**, 807–810

Lowe, S.W., Jacks, T., Housman, D.E. and Ruley, H.E. (1994b). Abrogation of oncogene-associated apoptosis allows transformation of p53-deficient cells. *Proc. Natl. Acad. Sci. USA* **91**, 2026–2030

Lynch, H.T. and Lynch, J.F. (1994) 25 years of HNPCC. *Anticancer Research* **14**(4B), 1617–1624

Malkin, D. (1994) p53 and the Li-Fraumeni syndrome. *Biochimica Et Biophysica Acta* **1198**, 197–213

Merritt, A.J., Potten, C.S., Kemp, C.J., Hickman, J.A., Lane, D.P. and Hall, P.A. (1994) The role of spontaneous and radiation-induced apoptosis in the gastrointestinal tract of normal and p53 deficient mice. *Cancer Research* **54**, 614–617

Nakane, H., Takeuchi, S., Yuba, S., Saijo, M., Nakatsu, Y., Murai, H., Nakatsuru, Y., Ishikawa, T., Hirota, S., Kitamura, Y., Kato, Y., Tsunoda, Y., Miyauchi, H., Horio, T., Tokunaga, T., Matsunaga, T., Nikaido, O., Nishimune, Y., Okada, Y. and Tanaka, K. (1995) High incidence of ultraviolet-B or chemical carcinogen-induced skin tumours in mice lacking the xeroderma pigmentosum group A gene. *Nature* **377**, 165–168

Purdie, C.A., Harrison, D.J., Peter, A., Dobbie, L., White, S., Howie, S.E.M., Salter, D.M., Bird, C.C., Wyllie, A.H., Hooper, M.L. and Clarke, A.R. (1994) Tumour incidence, spectrum and ploidy in mice with a large deletion in the p53 gene. *Oncogene* **9**, 603–609

Reitmair, A.H., Schmits, R., Ewel, A., Bapat, B., Redston, M., Mitri, A., Waterhouse, P., Mittrucker, H.-W., Wakeham, A., Liu, B., Thomason, A., Griesser, H., Gallinger, S., Ballhausen, W.G., Fishel, R. and Mak, T.W. (1995) MSH2 deficient mice are viable and susceptible to lymphoid tumours. *Nature Genetics* **11**, 64–70

Riley, D.J., Lee, E.Y.-H.P. and Lee, W. (1994) The retinoblastoma protein: more that a tumor suppressor. *Annual Review of Cell Biology* **10**, 1–29

Sands, A.T., Abuin, A., Sanchez, A., Conti, C.J. and Bradley, A. (1995) High susceptibility to ultraviolet-induced carcinogenesis in mice lacking XPC. *Nature* **377**, 162–165

Schmid, P., Lorenz, A., Hameister, H. and Montenarh, M. (1991) Expression of p53 during mouse embryogenesis. *Development* **113**, 857–865

Shikone, T., Matzuk, M.M., Perlas, E., Finegold, M.J., Lewis, K.A., Vale, W. and Hsueh, A.J. (1994) Characterization of gonadal sex cord-stromal tumor cell lines from inhibin-alpha and p53-deficient mice: the role of activin as an autocrine growth factor. *Molecular Endocrinology* **8**, 983–995

Swift, M., Morrell, D., Massey, R.B. and Chase, C.L. (1991) Incidence of cancer in 161 families affected by ataxia-telangiectasia. *New England Journal of Medicine* **325**, 1831–1836

Windle, J.J., Albert, D.M., O'Brien, J.M., Marcus, D.M., Disteche, C.M., Bernards, R. and Mellon, P.L. (1990) Retinoblastoma in transgenic mice. Nature **343**, 665–669

PART IV, SECTION B

61. The Generation of New Immortalized Cell Lines through Transgenesis

Manfred Theisen[†] and Andrea Pavirani[*]

Department of Molecular and Cellular Biology, Transgene S.A., 67082 Strasbourg, Cedex, France

Introduction

The ever growing need for new and defined cell lines for basic research or applied biotechnology has led to the development of several techniques to obtain permanent cell lines of a specific origin. The transfection of primary cell cultures with vectors expressing *onc*-genes such as SV40 large T antigen (TAg) has been described (Chou, 1989 and Shay *et al.*, 1991). *In vitro* transformed cell clones are expanded into cell lines that can be tested for the presence of specific differentiation markers and characteristics of the original primary cells. An alternative to such a technique is the use of transgenic mice to obtain novel specialized cell lines which can also be tailored for the expression, of complex recombinant proteins (Theisen *et al.*, 1995). Cell lines are derived by *onc*-gene-mediated *in vivo* immortalization (trans-immortalization). Tissue specific regulatory DNA elements are used to target the expression of an *onc*-gene to specific tissues and cell types of transgenic mice. This results in tumors which can then be cultured *in vitro* to establish permanent cell lines. Furthermore, co-expression of a gene coding for a human recombinant protein in the transgenic tumor cells results in the derivation of cell lines producing the recombinant protein.

[*] Author for Correspondence.
[†] Present address: Laboratoire de, Biologie Cellulaire et Moleculaire, Biocem S.A., 63170, Aubiere, France.

Choice of Promoters

The fusion of *onc*-genes to defined tissue specific regulatory DNA elements results in transgenic constructs which have been used to obtain programmed tumorigenesis in transgenic mice (Messing *et al.*, 1985 and Jaenisch, 1988). The expression of SV40 TAg under the control of the metallothionein promoter led to the development of coroid plexus tumors from which cell lines could be established (Brinster *et al.*, 1984). Transgenic mice containing the same promoter-*onc*-gene fusion have also been used to establish non-malignant hepatocyte cell lines from liver tissue (Paul *et al.*, 1988). Hanahan (1985) obtained pancreatic β cell lines by expressing the SV40 TAg *onc*-gene using the 5′ regulatory sequences of the insulin gene. Expression of either the c-*myc* or SV40 TAg *onc*-genes under the control of the immunoglobulin (Ig) heavy chain enhancer led to the development of lymphomas in 16 out of 19 independent founders (Pavirani *et al.*, 1989). No other tumors were detected. Human α1-antitrypsin (α1-AT) promoter driven expression of TAg, c-*myc* or a combination of both *onc*-genes led specifically to liver hyperplasia and hepatomas in 8 out of 8 independent founders (Dalemans *et al.*, 1990 and Perraud *et al.*, 1991). In some cases transformation by the transgenic *onc*-gene under the control of a tissue specific promoter can occur in other tissues than those, in which the endogenous promoter is expected to be functional. An example is the expression of SV40 TAg driven by the human CFTR promoter (Perraud *et al.*, 1992). CFTR, the cystic fibrosis transmembrane conductance regulator, whose defect is responsible for cystic fibrosis, is expressed in epithelial cells of several organs and tissues including lungs, pancreas and gastrointestinal tract, the major sites in which the clinical manifestations of the disease are evident (Jetten et al., 1989). The CFTR promoter, whose transcriptional activity in epithelial cells was demonstrated (Yoshimura et al., 1991), did not induce TAg-dependent tumor formation in all organs that normally express the CFTR gene but led to the formation of malignant ependymomas and choroid plexus carcinomas (Perraud et al., 1992). The fusion of SV40 TAg to the human villin gene promoter did not lead to the transformation of brush border epithelia of the gastrointestinal or urogenital tracts, in which the endogenous gene is expressed, but to the development of neuro-endocrinal tumors of the stomach (Theisen et al., unpublished results).

The use of trans-immortalisation techniques to create new cell lines has been of particular interest for neurobiological studies, in which immortalized neurons could be derived from SV40 TAg induced tumors in transgenic mice (Hammang et al., 1990). Mellon et al. (1990) immortalized hypothalamic GnRH neurons by trans-immortalization.

A valuable modification of the technique of onc-gene expression directed by a tissue specific promoter has been described by Noble et al. (1995), who have generated mice which were transgenic for a thermosensitive SV40 TAg transgene (tsA58) under the control of the ubiquitous interferon-inducible $H2K^b$ class I antigen promoter (see below).

Choice of onc-Genes

Trans-immortalization may lead eventually to the derivation of cell lines which are different in several aspects from the original cell type. The choice of the transforming onc-gene can thereby influence the degree of difference between the original cell in vivo and the cultured immortalized cell. The following example illustrates such a difference in the case of trans-immortalized cells that were obtained by in vivo transformation of hepatocytes by either c-myc or SV40 TAg.

Cells from liver tumor tissues of transgenic mice that expressed either the c-myc or SV40 TAg onc-gene under the control of the $\alpha 1$-AT promoter were cultured as described (Dalemans et al., 1990 and Perraud et al., 1991). During culture morphological differences of c-myc and TAg transformed cells were evident. c-myc cells began to form clusters and degenerating cells containing multiple nuclei were present. In contrast TAg-transformed cells grew as a homogenous monolayer of mononuclear cells that showed a hepatocyte-like morphology. Hepatocyte-like morphology of c-myc cells became visible only later, at a time when TAg cell cultures formed typical hepatic structures like bile canaliculi.

The differentiation status of cell lines was determined by RNA dot blot analysis for the hepatic marker genes albumin, mouse $\alpha 1$-AT, α-fetoprotein, transferrin and acute-phase protein. Expression was compared to normal mouse liver cells and Chinese hamster ovary (CHO) cells as a non-hepatic control. Hepatic marker genes such as albumin, $\alpha 1$-AT and transferrin were expressed at comparable levels in both c-myc and TAg-derived lines. In contrast to c-myc-derived cell lines, TAg transformed cells expressed high levels of the fetal marker α-fetoprotein and acute-phase protein, which is a marker of a stressed metabolic state. The co-expression of albumin and α-fetoprotein in the TAg-derived cell lines is in contrast with the natural state, in which expression of both genes is mutually exclusive (Woodworth and Isom, 1987).

To avoid transformation-dependent differences between the trans-immortalized cell line and the original cell type, Noble et al. (1995) have generated the $H2K^b$ tsA58 transgenic mouse. tsA58 is a thermolabile mutant of SV40 TAg (Tegtmeyer, 1975), that can transform and immortalize cells only at the permissive temperature of 33°C. The tsA58 transgene is expressed under the control of the interferon inducible $H2K^b$ class I antigen promoter. Primary cells from transgenic organs are grown at 33°C under permissive conditions for immortalization. A shift to 39.5°C allows the cells to undergo normal differentiation thus leading to cultured cells of the desired non-transformed phenotype. So far astrocyte, myoblast and osteoclast cell lines have been generated from $H2K^b$ tsA58 transgenic mice. Mutant cell lines can be obtained by mating $H2K^b$ tsA58 mice to mutant strains and isolating cells from heterozygous offspring.

Production of Recombinant Human Proteins in Trans-Immortalized Cells

The production of recombinant pharmacologically active protein by transfection of procaryotic hosts, yeast or standard mammalian cell lines such as CHO cells may be of limited use in particular cases, mainly due to the fact that these heterologous systems are incapable to perform post-translational modifications that are necessary for a full biological activity of the protein (Courtney et al., 1984; Kaufmann et al., 1986 and Balland et al., 1988). Trans-immortalization strategies can be applied for the production of human recombinant proteins in cell lines that were derived from transgenic mouse tumors originating from the natural expressing cell type (Dalemans et al., 1990 and Jallat et al., 1990). Tissue specific regulatory DNA elements are used to target expression of an onc-gene and a gene of interest to specific tissues and cell types of transgenic mice that are adapted for the synthesis of fully active recombinant proteins (example: transgenic mouse liver tumors for the production of liver specific proteins as α1-AT or coagulation factors). The co-expression of an *onc*-gene and the gene of interest result in tumors whose cells express the recombinant protein (see as a review Theisen *et al.*, 1995). Tumor cells are cultured *in vitro* to establish permanent cell lines that produce the recombinant protein. Human α1-AT was produced in trans-hepatoma derived cells (Dalemans *et al.*, 1990 and Perraud *et al.*, 1991). α1-AT is a serine protease inhibitor and it is synthesized primarily in the liver. Transgenic mice carrying the human α1-AT gene and the c-*myc* gene both under the control of the human α1-AT promoter expressed up to 2 mg of circulating human α1-AT per ml of serum. Mice developed c-*myc*-induced hepatomas at an average age of 1 year. Trans-hepatoma cell lines were established. Secretion of 4 to 5 μg/ml/24 h/10^6 cells α1-AT was stable for several generations of five independent cell lines. Trans-hepatomas that were derived from transgenic mice expressing human α1-AT, mouse c-*myc* and TAg under the control of the α1-AT promoter showed levels of 4 to 8 μg/ml/24 h/10^6 cells (Jallat *et al.*, 1990). α1-AT was correctly processed and glycosylated.

A variant form of human α1-AT, the Pittsburgh variant α1-AT [Arg] (Owen *et al.*, 1983; Courtney *et al.*, 1985) was produced in trans-lymphoma derived cells (Pavirani *et al.*, 1989). Co-expression of the α1-AT [Arg] gene and either c-*myc* or SV40 TAg in lymphocytes of transgenic mice was targeted by regulatory sequences of the Ig heavy chain enhancer, heavy chain promoter and kappa light chain promoter. Levels of human α1-AT [Arg] in serum were between 0.3 and 3 μg/ml for TAg mice and 0.3 and 20 μg/ml in c-*myc* mice. Trans-immortalized cell lines from transgenic mice secreted active α1-AT [Arg] into the culture medium at levels of 60 ng/10^6cells/24 h.

One of the most remarkable examples for the use of the trans-immortalization approach is the production of recombinant coagulation factor IX (FIX). FIX is a vitamin K-dependent protease involved in the intrinsic pathway of the coagulation cascade (Jackson and Nemerson, 1980; and Furie and Furie, 1988). FIX deficiency is responsible for hemophilia B, a chromosome X-linked recessive disorder that occurs in 1 in 30,000 males. The only treatment of this disease is so far the infusion of human FIX preparations that are purified from donor blood. FIX needs at least four post-translational modifications to obtain its full pharmacological activity, thus making its production in many established mammalian cell lines inefficient. Transgenic mice that express the recombinant FIX gene under the control of its own promoter were generated. Up to 36 μg/ml of FIX were detected in the plasma of these mice. The transgenic mouse derived protein showed full coagulation activity and correct post-translational processing. FIX transgenics were crossed with c-*myc* or TAg transgenic mice. Double transgenics developed hepatocellular carcinomas and cell lines were derived from these tumors. These cells secreted 0.15 to 0.7 μg/ml/24 hr/10^6 cells of fully active FIX into the culture medium (Jallat *et al.*, 1990).

Conclusions

Trans-immortalization (see as a review Theisen *et al.*, 1995), which utilises *in vivo* *onc*-gene-mediated transformation of specifically targeted cell types coupled to *in vitro* tissue culture, allows the production of specific new cell lines within a relatively short period of time. The technique has certain advantages as compared to classical *in vitro* transformation protocols. For *in vitro* immortalization large numbers of primary cells have to be generated which are often difficult to obtain and to culture. These cells have to be exposed to transforming agents or have to be tranfected with tranforming genes present on either viral or plasmid vectors (Chou, 1989 and Shay *et al.*, 1991). After co-transfection of the transforming gene and a selectable marker gene relatively long selection periods are necessary to obtain a limited number of cell clones that have integrated both the selectable marker and the transforming gene. Different clones with different copy numbers and integration sites of the transfected DNA may show a large variability in their phenotypes and differentiation states. In contrast trans-immortalized cells are issued from a tumor that develops in a transgenic animal. Once brought into culture these cells are already transformed and a tedious selection is not necessary. In addition all cells in culture are issued from one tumor and may represent a single (clonal) and unique transgene integration and immortalization event within the genome of the original transgenic animal.

The major advantage of trans-immortalization lies in the generation of specialized cells for the production of recombinant human proteins (Dalemans *et al.*, 1990 and Jallat *et al.*, 1990). The cells allow the production of pharmacologically active proteins, a task which is not always fulfilled using classical *in vitro* technologies (Courtney *et al.*, 1984; Kaufmann *et al.*, 1986 and Balland *et al.*, 1988). In addition the correct expression of the recombinant protein in an *in vivo* situation can be monitored prior to culturing the trans-immortalized cells. A loss of expression of the recombinant gene during lengthy amplification and selection protocols is thereby avoided.

Currently one drawback of the trans-immortalization technique is its limitation to generate mainly mouse-derived cell lines. Technological developments that allow the generation of other transgenic species (for example other laboratory and farm animals) will expand the number of available specialized cells from a whole variety of different species.

Acknowledgements

We thank Bruce Acres for critical reading of the manuscript; Mike Courtney for supporting our work; all colleagues at TRANSGENE for their research and technical help in the trans-immortalization projects. Work on factor IX was supported by Institut Mérieux, Lyon, France and work on CFTR was financed in part by the Association Française de Lutte contre la Mucoviscidose (AFLM).

References

Balland, A., Faure, T., Carvallo, D., Cordier, P., Ulrich, P., Fournet, B., De La Salle, H. and Lecocq, J.P. (1988) Characterization of two differently processed forms of human recombinant factor IX synthesized in CHO cells transformed with polycistronic vectors. *Eur. J. Biochem.* **172**, 565–572

Brinster, R.L., Chen, H.Y., Messing, A., van Dyke, T., Levine, A. J. and Palmiter, R. D. (1984) Transgenic mice harboring SV40 T antigen genes develop characteristic brain tumors. *Cell* **37**, 367–379

Chou, J.Y. (1989) Differentiated mammalian cell lines immortalized by temperature sensitive tumor viruses. *Mol. Endocrinol.* **3**, 1511–1514

Courtney, M., Buchwalder, A., Tessier, L.H., Jaye, M., Benavente, A., Balland, A., Kohli, V., Lathe, R., Tolstoshev, P. and Lecocq, J.P. (1984) High-level production of biologically active human α1-antitrypsin in *Escherichia coli*. *Proc. Natl. Acad. Sci. USA* **81**, 669–673

Courtney, M., Jallat, S., Tessier, L.H., Benavente, A., Crystal, R.G. and Lecocq, J.P. (1985) Synthesis in *E. coli* of α1-antitrypsin variants of therapeutic potential for emphysema and thrombosis. *Nature* (London) **313**, 149–151

Dalemans, W., Perraud, F., Le Meur, M., Gerlinger, P., Courtney, M. and Pavirani, A. (1990) Heterologous protein expression by trans-immortalized differentiated liver cell lines derived from transgenic mice. *Biologicals* **18**, 191–198

Furie, B. and Furie, B.C. (1988) The molecular basis of blood coagulation. *Cell* **53**, 505–518.

Hammang, J.P., Baetge, E.E., Behringer, R.R., Brinster, R.L., Palmiter, R. D. and Messing, A. (1990) Immortalized mouse neurons derived from SV40 TAg induced tumors in transgenic mice. *Neuron* **4**, 775–782

Hanahan, D. (1985) Heritable formation of pancreatic β-cell tumors in transgenic mice expressing recombinant insulin-simian virus 40 oncogenes. *Nature* (London) **315**, 115–122

Jackson, C.M. and Nemerson, Y. (1980) Blood coagulation. *Annu. Rev. Biochem.* **49**, 765–811

Jaenisch, R. (1988) Transgenic animals. *Science* **240**, 1468–1474

Jallat, S., Perraud, F., Dalemans, W., Balland, A., Dieterle, A., Faure, T., Meulien, P. and Pavirani, A. (1990) Characterization of recombinant human factor IX expressed in transgenic mice and in trans-immortalized hepatic cell lines. *EMBO J.* **9**, 3295–3301

Jetten, A.M., Yankaskas, J.R., Stutts, M.J., Williamsen, N.J. and Boucher, R.C. (1989) Persistence of abnormal chloride conductance regulation in transformed cystic fibrosis epithelia. *Science* **244**, 1472–1475

Kaufmann, R.J., Wasley, L.C., Furie, B.C., Furie, B. and Shoemaker, C.B. (1986) Expression, purification and characterization of recombinant γ-carboxylated Factor IX synthesized in chinese hamster ovary cells. *J. Biol. Chem.* **261**, 9622–9628

Mellon, P.L., Windle, J.J., Goldsmith, P.C., Padula, C.A., Roberts, J.L. and Weiner, R.I. (1990) Immortalization of hypothalamic GnRH neurons by genetically targeted tumorigenesis. *Neuron* **5**, 1–10

Messing, A., Chen, H.Y., Palmiter, R.D. and Brinster, R. L. (1985) Peripheral neuropathies, hepatocellular carcinomas and islet cell adenomas in transgenic mice. *Nature* (London) **316**, 461–463

Noble, M., Groves, A.K., Ataliotis, P., Ikram, Z. and Jat, P.S. (1995) The H-2KbtsA58 transgenic mouse: a new tool for the rapid generation of novel cell lines. *Transg. Res.* **4**, 215–225

Owen, M.C., Brennan, S.O., Lewis, J.H. and Carrell, R.W. (1983) Mutation of antitrypsin to antithrombin. α1-Antitrypsin Pittsburgh (358 Met → Arg), a fatal bleeding disorder. *N. Engl. J. Med.* **309**, 694–698

Paul, D., Höhne, M. and Hoffmann, B. (1988) Immortalized differentiated hepatocyte lines derived from transgenic mice harboring SV40 T-antigen genes. *Exp. Cell. Res.* **175**, 354–365

Pavirani, A., Skern, T., Le Meur, M., Lutz, Y., Lathe, R., Crystal, R.G., Fuchs, J.P., Gerlinger, P. and Courtney, M. (1989) Recombinant proteins of therapeutic interest expressed by lymphoid cell lines derived from transgenic mice. *Bio/Technology* **7**, 1049–1054

Perraud, F., Dalemans, W., Grenerault, J.L., Dreyer, D., Ali-Hadji, D., Faure, T. and Pavirani, A. (1991) Characterization of trans-immortalized hepatic cell lines established from transgenic mice. *Exp. Cell. Res.* **195**, 59–65

Perraud, F., Yoshimura, K., Louis, B., Dalemans, W., Ali-Hadji, D., Schultz, H., Claudepierre, M.C., Danel, C., Bellocq, J.P., Crystal, R.G., Lecocq, J.P. and Pavirani, A. (1992) The promoter of the human cystic fibrosis transmembrane conductance regulator gene directing SV40 T antigen expression induces malignant proliferation of ependymal cells in transgenic mice. *Oncogene* **7**, 993–997

Shay, J.W., Wright, W.E. and Werbin, H. (1991) Defining the molecular mechanisms of human cell immortalization. *Bioch. Bioph. Acta* **1072**, 1–7

Tegtmeyer, P. (1975) Function of simian virus 40 gene A in transforming infection. *J. Virol.* **15**, 613–618

Theisen, M., Perraud, F., Ali-Hadji, D. and Pavirani, A. (1995) Trans-immortalization strategies in transgenics. In G.M. Monastersky, and J.M. Robl, (eds.) *Strategies in Transgenic Animal Science* ASM Press, Washington DC

Woodworth, C.D. and Isom, H.C. (1987) Regulation of albumin gene expression in a series of rat hepatocyte cell lines immortalized by simian virus 40 and maintained in chemically defined medium. *Mol. Cell. Biol.* **7**, 3740–3748

Yoshimura, K., Nakamura, H., Trapnell, B.C., Dalemans, W., Pavirani, A., Lecocq, J.P. and Crystal, R.G. (1991) The cystic fibrosis gene has a housekeeping-type promoter and is expressed at low levels in cells of epithelial origin. *J. Biol. Chem.* **266**, 9140–9144

PART IV, SECTION B

62. The Use of Transgenic Mammals for AIDS Studies

Tania Sorg and Majid Mehtali
Department of Gene Therapy, Transgene, 11 rue de Molsheim, 67000 Strasbourg, France

Introduction

Animal models are essential in Acquired Immunodeficiency syndrome (AIDS) research for the evaluation of vaccine strategies or the screening of antiviral agents. Unfortunately, the only animals sensitive to the Human Immunodeficiency Virus (HIV) are the gibbon and chimpanzee; the former is not available for biomedical research and the latter is available only in limited numbers and in addition does not develop AIDS. Simian Immunodeficiency Virus (SIV) infection of macaques represents the best AIDS model available, as the infected animals suffer from opportunistic infections, wasting, diarrhea and other symptoms associated with AIDS in humans. However, such monkeys are expensive and in relatively short supply, and do not develop any disease following experimental infection with HIV-1. Since an ideal animal model satisfying all the biological and economical characteristics described in Table I does not naturally exist, the development of alternate models using available gene transfer technologies should be seriously considered. Indeed, transgenic mouse technology already proved a valuable approach to provide new animal models for the study of human viral diseases: transgenic mice susceptible to human viruses (Ren *et al.*, 1990) or with pathologies related to human viral diseases were described (Chisari *et al.*, 1989 and Green *et al.*, 1989). Transgenesis might therefore similarly provide novel models that circumvent some of the limitations associated with the currently available animal models for AIDS.

Table 62.1. Characteristics of an ideal animal model for HIV infection.

Host

    Readily and inexpensively obtained
    Easily handled and housed

    Well-characterized immunology and physiology

    Species-specific reagents available (monoclonal antibodies, immortalized cell lines)

Agent

    HIV-1

    Closely related virus with similar genomic structure and replication cycle

Pathogenesis

    Routes of infection: mucosal, intraveinous, perinatal

    AIDS induced in a time-frame compatible with the studies

    CD4+ target cells (lymphocytes, monocytes, macrophages)

    Disease progression characterized by: progressive loss of CD4+ cells, decline in antiviral antibodies rise in circulating virus levels

    Immunosuppressive disease with opportunistic infections

Transgenic Animals Sensitive to HIV-1 Infection

The characterization of the human CD4 (hCD4) cell surface receptor as the principal determinant of susceptibility of human cells to HIV-1 infection (Klatzman et al., 1984) led logically to the proposal that transgenic mice expressing the hCD4 molecule may constitute a novel animal model sensitive to HIV-1 infection. This hypothesis was further supported by the demonstration that murine cells transfected with the provirus genomic DNA can secrete infectious virions (Adachi et al., 1986 and Levy et al., 1986). Unfortunately, despite an efficient binding of HIV-1 to the cell surface, CD4-expressing mouse cell lines were found to remain refractory to HIV-1 infection (Maddon et al., 1986), making the construction of CD4$^+$ transgenic mice less promising that expected at first. The absence of susceptibility of CD4$^+$ transgenic mice to HIV-1 infection was later confirmed by Lores et al. (1992) and by ourself (Mehtali et al., 1991a), suggesting the existence in murine cells of a molecular barrier that hinders the penetration of the virus. Several reports suggest that HIV-1 infection of non-human cells might require the presence of a second molecule, but no such co-receptor has however been yet identified despite some controversial claims about the potential role of the CD26 molecule (Callebaut et al., 1993). The possibility of generating HIV-1 sensitive transgenic mice should therefore be reconsidered in the light of new findings regarding the nature of this putative co-receptor. hCD4$^+$ transgenic mice proved nonetheless useful models for the *in vivo* analysis of the interactions between the CD4 receptor and the virus envelope. Wang et al. (1994) have thus shown that administration of HIV envelope proteins can induce *in vivo* a massive deletion of CD4$^+$ T cells, suggesting that soluble envelope might be involved in some aspects of AIDS pathogenesis.

In contrast to the mouse, the rabbit has been described as being weakly susceptible to infection with high virus doses (Filice et al., 1988 and Kulaga et al., 1989). The demonstration that expression of the hCD4 molecule significantly increases the susceptibility of rabbit T cell lines to HIV-1 infection suggests that the rabbit might constitute a better model than the mouse for expression of hCD4 (Yamamura et al., 1991 and Hague et al., 1992). Transgenic rabbits expressing the *hCD4* gene specifically in T lymphocytes were thus generated and, as expected, were shown to have aquired an increased sensitivity to HIV-1 infection (Dunn et al., 1995). Successful infection could be demonstrated by detection of HIV DNA in peripheral blood lymphocytes, seroconversion to various HIV-1 proteins, and virus isolation from lymph nodes. These transgenic rabbits remain however much less sensitive to HIV-1 than humans or chimpanzees, and no disease has yet been observed. Further investigation of these transgenic rabbits is required to determine whether they might constitute a novel attractive animal model.

Transgenic Mice Carrying the Full Length HIV-1 Progenome

To circumvent the barriers associated with the cross-species virus infection, Leonard et al. (1988) generated transgenic mice carrying the full length proviral genome: 12 founder mice were produced, all healthy during the course of the study. However, the offspring of one founder developed symptoms characteristic of human AIDS, such as lymphadenopathy, pulmonary infiltrates, spleno-megaly and epidermal hyperplasia, followed by death at about 25 days of age. Moreover, virus could be recovered from several organs in all affected animals tested, with the epidermis containing the most substantial number of HIV-1 expressing cells. It remains however unclear the reason why only one transgenic line developed AIDS-like disorders, and what are the molecular mechanisms responsible for these pathologies. In contrast, mice carrying the HIV-1 genome inserted downstream the Mouse Mammary Tumor Virus (MMTV) promoter did not develop any particular abnormalities, despite high levels of viral proteins found in the transgenic mammary glands (Jolicoeur et al., 1992). This absence of pathology is most probably due to the strict specificity of expression of the MMTV promoter for the mammary tissue.

Transgenic Mice Carrying Incomplete HIV-1 Progenomes

Transgenic mice carrying incomplete proviral genomes have been produced to more precisely determine the role of defined viral genetic regions in AIDS pathogenesis. Transgenic mice with an HIV-1 provirus deleted from the *gag* and *pol* genes, but still encoding the envelope and the regulatory genes *tat, rev, nef, vpu, vpr* and *vif* were shown to develop a severe nephrotic syndrome characterized by a progressive glomerulosclerosis (Dickie et al., 1991 and Kopp et al., 1992) and cutaneous disorders manifested as diffuse epidermal hyperplasia (Kopp et al., 1993). Santoro et al. (1994) noticed that these same mice also exhibited marked cachexia, growth retardation, lymphoproliferation with a reduction in the percentage of CD4$^+$ cells but an increase in the absolute number of splenic CD4$^+$ and CD8$^+$ T cells. The viral gene products responsible for these

AIDS-like pathologies were not identified, but such results stimulated the production and analysis of mice transgenic for single viral genes.

Transgenic Mice Expressing HIV Regulatory Genes

Organization of the lentivirus genomes, and more particularly of HIV-1, is much more complex than the structure of other retroviruses (Figure 62.1). In addition to the classical *gag*, *pol* and *env* genes encoding structural and enzymatic proteins, the HIV genome also encodes at least six other regulatory proteins. While the function of some of these proteins (TAT, REV) has been precisely determined, the role of most of the viral regulatory proteins remains obscure (Cullen, 1992). Transgenic mice expressing individual viral regulatory genes might give insights into their respective biological functions.

***Tat* Transgenic Mice.** The major role of *Tat* is to activate the elongation of viral transcripts by interacting with the TAR (*Trans*-Activating Responsive Region) sequence localized on the nascent RNA strand. Other functions of *Tat* have been more recently characterized. In its extracellular form, *Tat*: (i) inhibits T-cell antigen-specific responses (Subramanyam *et al.*, 1993) (ii) increases the expression of IL-6 receptor on human B cell lines (Puri *et al.*, 1992) and the expression of TNF-β (Sastry *et al.*, 1990) (iii) stimulates the growth of spindle-like cells derived from Kaposi's sarcoma lesions from AIDS patients. Moreover, its ability to repress the expression of the class I Major Histocompatibility Complex (Howcroft *et al.*, 1993) suggests a role for *Tat* in viral escape to specific cytotoxic T lymphocytes. Other effects of *Tat* have also been reported such as the down-regulation of the interferon induced p68 kinase (Roy *et al.*, 1990), repression of manganese superoxide dismutase expression (Flores *et al.*, 1993), activation of extracellular matrix proteins (Taylor *et al.*, 1992), and interference with tumour angiogenesis in cell lines transplanted in *nude* mice (Huber *et al.*, 1992). These observations led to the hypothesis that *Tat* may act as a general transcriptional cellular factor.

Transgenic mice carrying the *tat* gene under the control of the HIV-LTR were shown to express Tat ubiquitously and to develop skin pathologies similar to Kaposi's sarcoma of AIDS patients (Vogel *et al.*, 1988). An increase of the liver cancer incidence was also found in tat males over 18 months of age (Vogel *et al.*, 1991). Our studies showed that transgenic mice expressing *Tat* specifically in T cells presented no particular anomalies (Mehtali *et al.*, 1991b and Mehtali *et al.*, 1992). In contrast, 30 percent of the mice expressing *Tat* ubiquitously showed a depletion of lymphocytes in thymus and spleen, a complete disorganization of the epidermis structure and a marked hyperkeratosis; no Kaposi's sarcoma-like disorder was however observed (Mehtali *et al.*, 1991a and Mehtali *et al.*, 1991b).

Tat-transgenic mice are not well adapted for the *in vivo* evaluation of anti-viral compounds. The pathologies observed are often found only in a small number of animals and cannot be used as quantitative markers to monitor the anti-viral activity of candidate drugs. In order to develop animals with a quantitative *Tat*- and LTR-dependent phenotype, transgenic mice expressing the *tat* gene specifically in T lymphocytes were crossed with mice carrying the human *alpha antitrypsine* cDNA (hαAT) under the control of the HIV-LTR promoter (Mehtali *et al.*, 1992). Double-transgenic mice were generated that strongly produce the Tat protein in T lymphocytes, which in turn secrete the *Tat*-inducible hαAT reporter protein in the blood circulation. Since circulating levels of hαAT can be precisely determined by ELISA assay, these double transgenic mice constitute a novel murine model for the evaluation of anti-LTR and/or anti-Tat drugs. Any variation in plasmatic hAT concentration after treatment might reflect an effect of the drug on either Tat or LTR activities, although an effect of such drugs on cellular factors interacting with *Tat* and/or LTR cannot be completely excluded.

Nef Transgenic Mice. *Nef* was originally shown in T cell lines to downregulate viral replication (Terwilliger *et al.*, 1986) and CD4 cell surface levels (Guy *et al.*, 1987), and to repress HIV-LTR transcription (Ahmad *et al.*, 1988). In contrast, Nef accelerates viral replication in primary lymphocytes (DeRonde *et al.*, 1992), and a strong selective pressure for a functional Nef protein was described *in vivo* in SIV-infected Rhesus macaques (Kestler *et al.*,

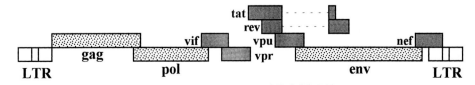

Figure 62.1. Genomic structure of HIV-1

1991). To further elucidate Nef functions, transgenic mice expressing *nef* in T cells were generated (Brady et al., 1993 and Skowronski et al., 1993). These mice showed a marked depletion of peripheral CD4 lymphocytes and a dramatic increase in the level of thymocytes activation and proliferation in response to polyclonal stimulation. This cellular activation may contribute *in vivo* to a more efficient production of infectious viruses. The down-regulation of CD4 is probably a consequence of a direct interaction between Nef and CD4 in the Golgi, blocking the progression of CD4 to the cell surface. Dickie et al. (1993) have investigated the pathophysiological consequences of Nef expression driven by the HIV- or the MMTV-LTR. Nef expression was detected in the epidermal basal layer of HIV-LTR/*nef* mice. These animals developed spontaneously discrete proliferative skin lesions resembling papillomas, often accompanied by a progressive ulceration of the epidermal layer. In MMTV-LTR/*nef* animals, Nef expression was directed to the mammary glands, but without pathological effects.

***Rev* Transgenic Mice.** Rev function is essential to HIV-1 replication. Rev binds specifically to its RRE (Rev Responsive Element) target-sequence localized on viral RNAs and stimulates the expression of late viral proteins by inducing the active transport of the viral transcripts from the nucleus to the cytoplasm (Cullen, 1992). Since early *in vitro* studies have shown the lack of Rev functions in non-human cells (Trono et al., 1990), very few groups attempted to produce *rev*-transgenic mice. Transgenic mice carrying a *gag-pol* deleted progenome and developing a severe nephrotic syndrome were reported to strongly express Rev in the sclerotic glomeruli (Kopp et al., 1992). But this phenomenological observation does not allow any conclusion regarding the real role of Rev in the described renal pathology. Moreover, mice ubiquitously expressing *rev* did not present such renal disorders but developed very aggressive lymphomas after one year of age (Mehtali et al., 1991b). The precise implication of Rev in the development of such tumors remains unclear.

Transgenic Mice Expressing the gp120 *env* Gene

It is well documented that HIV-1 induces encephalopathy in AIDS patients (Janssen, 1991). The demonstration that the gp120 viral envelope is neurotoxic *in vitro* (Kimes et al., 1991 and Glowa et al., 1992) is an indication of a possible role for this structural protein in the development of AIDS dementia complex. Transgenic mice specifically expressing the *env* gene in brain tissues were produced to evaluate *in vivo* the potential alterations associated with envelope production and elucidate the mechanisms of gp120 neurotoxicity (Thomas et al., 1994; Toggas et al., 1994 and Berrada et al., 1995). Neuronal expression of the Env protein was demonstrated and neuropathological evaluation of the transgenic mice showed an extensive Central Nervous System (CNS) damage, with vacuolizations of dendrites, large neuron loss, and widespread reactive astrocytosis. These observations support the notion that Env might be involved in the HIV related neuropathology. It remains however to be determined whether gp120 is directly responsible for these effects or if intermediate molecules are involved.

Transgenic Mice Carrying Reporter Genes Under the Control of HIV-1 LTR Promoter Sequences

In order to study the regulation of the HIV-LTR promoter sequences *in vivo* and in absence of virus replication, transgenic mice carrying reporter genes like the *E. coli* chloramphenicol acetyl transferase (CAT) gene (Khillan et al., 1988) or the human hAT cDNA (Mehtali et al., 1992) under the control of the HIV-LTR promoter were generated. Cross-breeding of the LTR-*cat* animals with transgenic mice expressing Tat specifically in the lens generated double transgenic animals with a 5 fold increase in CAT activity in eye tissues (Khillan et al., 1988). Similarly, double-transgenic mice carrying both the LTR-hαAT transgene and a T cell specific *tat* expression cassette synthesized the hαAT protein only in the thymus and presented a 20-fold increase in plasmatic hαAT levels compared to LTR-hαAT single transgenic animals (Mehtali et al., 1992). These observations confirm the ability of *Tat* to induce the LTR activity in murine tissues. It was also shown that CAT activity was 10 fold higher in placenta and uterus of LTR-*cat* mice during pregnancy (Furth et al., 1990) suggesting that hormonal induction of the LTR contributes to the perinatal transmission of HIV. Frucht et al. (1991) also showed that UV irradiation increases the CAT activity up to 30 fold in the skin. This was confirmed in LTR-*tat* transgenic mice expressing the *tat* gene selectively in epidermal structures (Vogel et al., 1992), where the steady-state level of *tat* RNA was 10 fold higher after UV exposure.

Transgenic mice carrying the β-*Galactosidase* (*lacZ*) reporter gene under the control of LTR sequences isolated from a CNS-derived HIV-1 strain have been produced to study the regulation of LTR expression in the brain (Corboy et al., 1992). *LacZ* RNA and protein were found in various neuronal regions of the brain and in the retina, in contrast to previous studies with LTR sequences from non-CNS-derived HIV strains (Leonard et al., 1988; Mehtali et al., 1992 and Vogel et al., 1992). The molecular mechanisms of this neuro-adaptation remain however obscure.

Conclusion

The transgenic animal technology proved very useful in confirming the direct role of some viral genes in the development of AIDS. However, several drawbacks severely limit the potential interest of these animal models. First, all reported observations are mostly phenomenological and no clear characterization of the mechanisms of action of the viral proteins was generated by these experiments. A more detailed, but also more difficult, molecular analysis of the affected tissues is required in order to elucidate these mechanisms. Moreover, the mouse model has intrinsic limitations. Both Tat and Rev proteins were shown to be less or not active in murine cells; other viral proteins might similarly be less or not functional in the mouse, and cellular factors essential for HIV-1 replication might be missing in murine cells. This might explain the reason why no HIV-sensitive transgenic animals allowing an efficient *in vivo* viral replication could be generated. In addition, laboratory mice are inbred animals and a higher or weaker sensitivity of some strains to individual gene products cannot be excluded; we indeed observed that an ubiquitous expression of the *tat* gene induced skin disorders only in outbred CD1 mice and not in inbred C57Bl/6 × SJL animals (Mehtali et al., 1991b).

Despite these limitations, transgenic animals offer the unique opportunity to correlate molecular mechanisms discovered in cell culture systems with responses observed at the organism level.

"Note added in proofs: the production of a small transgenic animal model sensitive to HIV1 infection was recently made more realistic by the identification of two cofactors required for HIV1 fusion and entry (Feng et al., Science 272,872; Alkhatib et al., Science 272, 1966, 1996)"

Acknowledgement

We thank A. Pavirani for critical reading of the manuscript.

References

Adachi, A., Gendelman, H.E., Koenig, S., Folks, T., Willey, R., Rabson, A. and Martin, M.A. (1986) Production of AIDS-associated retroviru in human and non-human cells transfected with an infectious molecular clone. *J. Virol.* **59**, 284–291

Ahmad, N. and Venkatesan, S. (1988) Nef protein of HIV-1 is a transcriptional repressor of HIV-1 LTR. *Science* **241**, 1481–1485

Berrada, F., Ma, D., Michaud, J., Doucet, G., Giroux L. and Elbaz, A.K. (1995) Neuronal expression of HIV-1 Env protein in transgenic mice: distribution in the central nervous system and pathological alterations. *J. Virol.* **69**, 6770–6778

Brady, H.J.M., Pennington, D.J., Miles C.G. and Dzierzak, E.A. (1993) CD4 cell surface downregulation in HIV-1 Nef transgenic mice is a consequence of intracellular sequestration. *EMBO Journal* **12**, 4923–4932

Callebaut, C., Krust, B., Jacotot, E. and Hovanessian, A.G. (1993) T cell activation antigen, CD26, as a cofactor for entry of HIV in $CD4^+$ cells. *Science* **262**, 2045–2050

Chisari F.V., Klopchin, K., Moryama, T., Pasquinelli, C., Dunsford, H.A., Sell, S., Pinkert, C.A., Brinster R.L., and Palmiter R.D. (1989) Molecular pathogenesis of hepatocellular carcinoma in Hepatitis B virus transgenic mice. *Cell* **59**, 1145–1156

Corboy, J.R., Buzy, J.M., Zink, M.C. and Clements, J.E. (1992) Expression directed from HIV long terminal repeats in the central nervous system of transgenic mice. *Science* **258**, 1804–1808

Cullen, B.R. (1992) Mechanism of action of regulatory proteins encoded by complex retroviruses. *Microbiological Reviews* **56**, 375–394

De Ronde, A., Klaver, B., Keulen, W., Smit, L. and Goodsmit, J. (1992) Natural HIV-1 Nef accelerates virus replication in primary human lymphocytes. *Virology* **188**, 391–395

Dickie, P., Felser, J., Eckhaus, M., Bryant, J., Silver, J., Marinos, N. and Notkins, A.L. (1991) HIV-associated nephropathy in transgenic mice expressing HIV-1 genes. *Virology* **185**, 109–119

Dickie, P., Ramsdell, F., Notkins, A.L. and Venkatesan, S. (1993) Spontaneous and inducible epidermal hyperplasia in transgenic mice expressing HIV-1 Nef. *Virology* **197**, 431–438

Dunn, C.S., Mehtali, M., Houdebine, L.M., Gut, J.-P., Kirn, A. and Aubertin, A.-M. (1995) Human immunodeficiency virus type 1 infection of human CD4-transgenic rabbits. *J. Gen. Virology* **76**, 1327–1336

Filice, G., Cereda, P.M. and Varnier, O.E. (1988) Infection of rabbits with human immunodeficiency virus. *Nature (London)* **335**, 366–369

Flores, S.C., Marecki, J.C., Harper, K.P., Bose, S.K., Nelson, S.K. and McCord, J.M. (1993) Tat protein of human immunodeficiency virus type I represses expression of manganese superoxide dismutase in Hela cells. *Proc. Natl. Acad. Sci. USA* **90**, 7632–7636

Frucht, D.M., Lamperth, L., Vicenzi, E., Belchon, J.H. and Martin, M.A. (1991) Ultraviolet radiation increases HIV-long terminal repeat-directed expression in transgenic mice. *AIDS Research and Human Retroviruses* **7**, 729–733

Furth, P.A., Westphal, H. and Hennighausen, L. (1990) Expression from the HIV-LTR is stimulated by glucocorticoids and pregnancy. *AIDS Research and Human Retroviruses* **6**, 553–560

Glowa, J.R., Parlilio, L.V., Brenneman, D.E., Gozes, I., Fridkin, M. and Hill, J.M. (1992) Learning impairment following intracerebral administration of the HIV envelope protein gp120 or a VIP antagonist. *Brain Research* **570**, 49–53

Green, J.E., Hinrichs, S.H., Vogel J., and Jay. G. (1989) Exocrinopathy resembling Sjögren's syndrome in HTLVI tat transgenic mice. *Nature (London)* **341**: 72–74

Guy, B., Kieny, M.-P., Rivire, Y., Le Peuch, C., Dott, K., Girard, M., Montagnier, L. and Lecocq, J.-P. (1987) HIV F/3' orf encodes a phosphorylated GTP-binding protein resembling an oncogene product. *Nature (London)* **330**, 266–269

Hague, B.F., Sawasdikosol, S., Brown, T.J., Lee, K., Recker, D.P. and Kindt, T.J. (1992) CD4 and its role in infection of rabbit cell lines by human immunodeficiency virus type 1. *Proc. Natl. Acad. Sci. USA* **89**, 7963–7967

Howcroft, T.K., Strebel, K., Martin, M.A. and Singer, D.S. (1993) Repression of MHC class I gene promoter activity by two-exon Tat of HIV. *Science* **260**, 1320–1322

Huber, B.E., Richards, C.A., Martin, J.L. and Wirth, P.J. (1992) Alterations in tumour angiogenesis associated with stable expression of the HIV Tat gene. *Molecular Carcinogenesis* **5**, 293–300

Janssen, R.S. Nomenclature and research case definitions for neurological manifestations of HIV-1 infection. *Neurology* **778**, 773–785

Jolicoeur, P., Laperrire A. and Beaulieu, N. (1992) Efficient production of human immunodeficiency virus proteins in transgenic mice. *J. Virol.* **66**, 3904–3908

Kestler, H.W., Ringler, D.J., Mori, K., Panicalli, D.L., Shehgal, P.K., Daniel, M.D. and Desrosiers, R.C. (1991) Importance of the Nef gene for maintenance of high virus loads and for development of AIDS. *Cell* **65**, 651–662

Khillan, J.S., Deen, K.C., Yu, S.H., Sweet, R.W., Rosenborg, M. and Westphal, H. (1988) Gene transactivation mediated by Tat gene of human immunodeficiency virus in transgenic mice. *Nucleic Acids Res.* **16**, 1423–1430

Kimes, A.S., London, E.D., Szabo, G., Raymon, L. and Tabakoff, B. (1991) Reduction of cerebral glucose utilization by the HIV envelope glycoprotein gp-120. *Experimental Neurology* **112**, 224–228

Klatzman, D., Champagne, E., Shamaret, S., Gruest, J., Guetard, D., Hercend, T., Gluckman, J.C. and Mantagnier, L. (1984) T-lymphocyte T4 molecule behaves as the receptor for human retrovirus LAV. *Nature* **312**, 767–768

Kopp, J.B., Klotman, M.E., Adler, S.H., Broggeman, L.A., Dickie, P., Marinos, N.J., Eckhaus, M., Bryant, J.L., Notkins, A.L. and Klotman, P.E. (1992) Progressive glomerulosclerosis and enhanced renal accumulation of basement membrane components in mice transgenic for human immunodeficiency type 1 genes. *Proc. Natl. Acad. Sci. USA* **89**, 1577–1581

Kopp, J.B., Rooney, J.F., Woikenberg C., Dorfmon, N., Marinos, N.J., Bryant, J.L., Katz, S.I., Notkins, A.L. and Klotman, P.E. (1993) Cutaneous disorders and viral gene expression in HIV-1 transgenic mice. *AIDS Research and Human Retroviruses* **9**, 267–275

Kulaga, H., Folks, T., Rutlegde, R., Truckenmiller, M.E., Gugel, E. and Kindt, T.J. (1989) Infection of rabbits with human immunodeficiency virus : a small animal model for AIDS. *J. Exp. Med.* **169**, 321–326

Leonard, J.M., Abramczuk, J.W., Pezen, D.S., Rutledge, R., Belcher, J.H., Hakim, F., Shearer, G., Lamperth, L., Travis, W., Fredrickson, T., Notkins, D.L. and Martin, M.A. (1988) Development of disease and virus recovery in transgenic mice containing HIV proviral DNA. *Science* **242**, 1665–1670

Levy, J.A., Cheng-Mayer, C., Dina, D. and Luciw, P.A. (1986) AIDS retrovirus replicates in transfected human and animal fibroblasts. *Science* **232**, 998–1001

Lores, P., Boucher, V., Mackay, C., Pla, M., Von Boehmer, H., Jami, J., Barré-Sinoussi, F. and Weill, J.-C. (1992) Expression of human CD4 in transgenic mice does not confer sensitivity to human immunodeficiency virus infection. *AIDS Research and Human Retroviruses* **8**, 2063–2071

Maddon, P.J., Dalgleish, A.G., McDougal, J.S., Clapman, P.R., Weiss, R.A. and Axel, R. (1986) The T4 gene encodes the AIDS virus receptor and is expressed in the immune system and in the brain. *Cell* **47**, 333–348

Mehtali, M., Acres, B, and Kieny, M.-P. (1991a) Transgenic mice expressing HIV genes for the *in vivo* evaluation for anti-HIV drugs. In *Viral quantitation in HIV infection*, 97–111. Edited by Andrieu, J.M. Paris: John Libbey Eurotext.

Mehtali, M., Munschy, M., Caillaud, J.M., and Kieny, M.-P. (1991b) HIV regulatory genes induce AIDS-like pathologies in transgenic mice. In *Colloque des Cents Gardes*, 25–30. Edited by Valette L. and Girard. M. Lyon: Fondation Merieux.

Mehtali, M., Munschy, M., Ali-Hadji, D. and Kieny, M.-P. (1992) A novel transgenic mouse model for the *in vivo* evaluation of anti-human immunodeficiency virus type 1 drugs. *AIDS Research and Human Retroviruses* **8**, 1959–1965

Puri, R.K. and Aggarwal, B.B. (1992) Human immunodeficiency virus type 1 Tat up-regulates interleukin 4 receptors on a human B lymphoblastoid cell line. *Cancer Research* **52**, 3787–3790

Ren R., Costantini, F., Gagacz, E.J., Lee, J.J. and Racaniello. U.R. (1990). Transgenic mice expressing the human poliovirus receptor: a new model for poliomyelitis. *Cell* **63**, 353–362

Roy, S., Katze, M.G., Parkin, N.T., Edery, I., Hovanessian, A.G. and Sonenberg, H. (1990) Control of the interferon-induced 68-kilodalton protein kinase by the HIV-1 Tat gene product. *Science* **247**, 1216–1219

Santoro, T.J., Bryant, J.L., Pellicoro, J., Klotman, M.E., Kopp, J.B., Bruggeman, L.A., Franks, R.R., Notkins, A.L. and Klotman, P.E. (1994) Growth failure and AIDS-like cachexia syndrome in HIV-1 transgenic mice. *Virology* **201**, 147–151

Sastry, K.J., Reddy, R.H.R., Pandita, R., Totpal, K. and Aggarwal, B.B. (1990) HIV-1 tat gene induces tumour necrosis factor-β (lymphotoxin) in a human B-lymphoblastoid cell line. *J Biol Chem* **265**, 20091–20093

Skowronski, J., Parks, D. and Mariani, R. (1993) Altered T cell activation and development in transgenic mice expressing the HIV-1 nef gene. *EMBO Journal* **12**, 703–713

Subramanyam, M., Gutueil, W.G., Bachovchin, W.W. and Huber, B.J. (1993) Mechanism of HIV-1 tat induced inhibition of antigen-specific T cell responsiveness. *J Immunol.* **150**, 2544–2553

Taylor, J.P., Cupp, C., Diaz, A., Choudoury, M., Khalilik, K., Jimenaz, S.A. and Amini, S. (1992) Activation of expression of genes coding for extracellular matrix proteins in Tat-producing glioblastoma cells. *Proc. Natl. Acad. Sci. USA* **89**, 9617–9621

Terwilliger, E., Sodorski, J.G., Rosen, C.A. and Haseltine, W.A. (1986) Effects of mutations within the 3' orf open reading frame region of human T-cell lymphotropic virus type III (HTLV-III/LAV) on replication and cytopathogenicity. *J. Virol.* **60**, 754–760

Thomas, F.P., Chalk, C., Lalonde R., Robitaille, Y. and Jolicoeur P. (1994) Expression of HIV-1 in the nervous system of transgenic mice leads to neurological disease. *J. Virol.* **68**, 7099–7107

Toggas, S.M., Masliah, E., Rockenstein, E.M., Rall, G.F., Abraham, C.R. and Mucke, L. (1994) Central nervous system damage produced by expression of the HIV-1 coat protein gp120 in transgenic mice. *Nature (London)* **367**, 188–193

Trono, D. and Baltimore, D. (1990) A human cell factor is essential for HIV-1 Rev action. *EMBO Journal* **9**, 4155–4160

Vogel, J., Hinrichs, S.H., Reynolds, R.K., Luciw, P.A. and Jay, G. (1988) The HIV tat gene induces dermal lesions resembling Kaposi's sarcoma in transgenic mice. *Nature (London)* **335**, 606–611

Vogel, J., Hinrichs, S.H., Napolitano, L.A., Ngo, L. and Jay, G. (1991) Liver cancer in transgenic mice carrying the human immunodeficiency virus tat gene. *Cancer Research* **51**, 6686–6690

Vogel, J., Cepsoan, M., Tschacher, E., Napolitano, L.A. and Jay, G. (1992) UV activation of human immunodeficiency virus gene expression in transgenic mice. *J. Virol.* **66**, 1–5

Wang, Z., Orlikowsky, T., Dudhane, A., Mittler, R., Blum, M., Lacy, E., Riethmüller, G. and Hoffmann, M.K. (1994) Deletion of T lymphocytes in human CD4 transgenic mice induced by HIV-gp120 and gp120-specific antibodies from AIDS patients. *Eur. J. Immunol.* **24**, 1553–1557

Yamamura, Y., Kotani, M., Chowdhury, Md I.,H., Yamamoto, N., Yamaguchi, K., Karasuyama, H., Katsura, Y. and Miyasaka, M. (1991) Infection of human $CD4^+$ rabbit cells with HIV-1: the possibility of the rabbit as a model for HIV-1 infection. *Intern. Immunol.* **3**, 1183–1187

PART IV, SECTION B

63. The Role of Mouse Models in the Development of New Therapies for Cystic Fibrosis

Gerry McLachlan* and David J. Porteous

Molecular Genetics Section, MRC Human Genetics Unit, Western General Hospital EH42XU, Edinburgh, UK

Cystic Fibrosis

Cystic fibrosis (CF) is the most common fatal autosomal recessive genetic disease in the Caucasian population. The disease is characterized by a defect in cyclic AMP (cAMP) dependent chloride ion transport which results in abnormalities in epithelial secretion. The gene responsible for the disease, the cystic fibrosis transmembrane conductance regulator (CFTR) gene, was cloned in 1989 (Riordan *et al.*) opening the door to treatment by somatic gene therapy. The progressive lung disease which dominates morbidity and mortality is not present at birth but develops through accumulation of mucus leading to opportunistic and persistent pulmonary infection. Given that it is a recessive disease and that the lung is easily accessible via the airways, CF is therefore ideal for developing somatic gene therapy protocols. *In vivo* studies, ideally in a genetic laboratory animal model are needed to allow evaluation of the safety and efficacy of this approach.

Generation of Mouse Models

The first step in the generation of a CF mouse model was the isolation of the murine homologue, *Cftr* (Tata *et al.*, 1991). It is highly similar to the human gene (78% overall identity at the amino acid level), and the majority of CF mutations lie at conserved residues, which argues for conservation of function across species. The probability that a genetic mouse model for CF would mimic the clinical phenotype was given weight by the demonstration that the ion transport properties of wild-type mice are similar to those of normal human subjects (Smith *et al.*, 1992).

'Knock-out' Models of Cystic Fibrosis

Four 'knockout' models have been described using slightly different gene targeting strategies in embryonal stem (ES) cells; replacement vectors ($cftr^{m1UNC}$, $cftr^{m1Cam}$), or insertional vectors ($cftr^{m1HGU}$, $cftr^{m1Bay}$) (Table 63.1). The first strategy was designed to replace an exon of the *Cftr* gene with an incomplete exon running into plasmid sequence containing stop codons in all three reading frames. These mutations result in absolute "nulls" since there is no mechanism for reversion to wild-type. The second strategy generates a duplication of exon sequences. This insertion into the genomic target occurs without loss of sequence therefore reversion to wild-type by intrachromosomal recombination through the region of duplication is a formal possibility. Aberrant splicing around the insertional mutation can in some instances result in wild-type mRNA. This does not apparently occur with the $cftr^{m1Bay}$ which results in complete exon duplication, but does with the $cftr^{m1HGU}$ insertion which results in the introduction of a disrupted exon. The $cftr^{m1HGU}$ allele is slightly 'leaky' and a low level of wild-type mRNA is expressed.

Clinically Relevant 'Mutation-specific' Models

The value of creating clinically relevant mutations in the mouse lies in being able to compare genotype and phenotype on a controlled genetic background and also to test novel therapeutic strategies that

*Author for Correspondence.

Table 63.1. Mouse models of cystic fibrosis.

	'Knock-out' Mouse models				'Mutation specific' Mouse models			
Mutant allele*	$cftr^{m1UNC}$	$cftr^{m1Cam}$	$cftr^{m1Bay}$	$cftr^{m1HGU}$	$cftr^{\Delta F508Cam}$	$cftr^{\Delta F508Rot}$	$cftr^{\Delta F508Uta}$	$cftr^{G551D}$
Type of mutation	replacement	replacement	insertion	insertion	replacement	insertion 'hit and run'	replacement	replacement
Genomic Site	exon 10	exon 10	exon 3	exon 10	exon 10	exon 10	exon 10	exon 11
Level of expression	none detectable	none detectable	none detectable	~10% of normal level	15% of normal level	normal level	normal except ↓intestine	53% of normal level
Reference	Snouwaert et al (1992)	Ratcliffe et al., (1993)	O'Neal et al., (1993)	Dorin et al., (1992)	Colledge et al., (1995)	van Doorninck et al., (1995)	Zeiher et al., (1995)	Delaney et al., (1995)

*The designation of the *cftr* mutant alleles for the 'knock-out' mice follows the nomenclature recommendations of the Committee on Standardised Genetic Nomenclature for Mice and uses abbreviations of the laboratories in which they were derived (UNC; University of North Carolina, Cam; Cambridge UK, HGU; Human Genetics Unit, Edinburgh, Bay; Baylor, Texas). For the purposes of this chapter the 'mutation-specific' models have been assigned similar nomenclature which includes the nature of the mutation. The three ΔF508 models have the abbreviations of where they were created to distinguish them from one another (Cam; Cambridge UK, Rot; Rotterdam, Uta; Utah).

may be specific for a particular mutant protein. In the majority of CF patients, somatic gene therapy strategies will introduce the wild type transgene into cells with endogenous mutant CFTR protein. These 'mutation-specific' models will allow investigators to mimic this situation *in vivo*. Of the four models created to date, three carry the major ΔF508 mutation and one carries the G551D mutation (Table 63.1). Two of the ΔF508 models ($cftr^{\Delta F508Cam}$, $cftr^{\Delta F508Uta}$) and the G551D model ($cftr^{G551D}$) were created by replacement gene-targeting. The other ΔF508 model ($cftr^{\Delta F508Rot}$) was created using an insertional gene-targeting vector adapted to create a precise mutation by the 'hit and run' strategy (Hasty et al., 1991).

Phenotype of the CF Mice

Electrophysiology

Despite the differences in the targeted gene disruptions, all the 'knock-out' lines of CF mice showed evidence for an abnormality in cAMP-mediated chloride ion conductance in the airways and intestine, which is characteristic of the disease (Ratcliff et al., 1993; Clarke et al., 1992; Smith et al., 1995; Colledge et al., 1995; van Doorninck et al., 1995; Delaney et al., 1995 and Zeiher et al., 1995). These abnormalities are summarised in Tables 63.2 and 63.3. Similar changes are reported for the mutation-specific mice although in the $cftr^{\Delta F508Rot}$ ileum there is a significant residual forskolin response. CF-associated abnormalities of the pancreas are not reported to the same degree as humans in any of the mice. The bioelectric properties of the pancreas in $cftr^{m1HGU}$ mice have been examined and the ratio of cAMP/calcium mediated chloride transport is very small compared to humans (Gray et al., 1994). It is speculated that this high level of Ca^{2+}-regulated chloride conductance compensates for the defect in cAMP-mediated transport.

Intestinal Disease

The different mutant mice all possess similarities in pathophysiological phenotypes including gut abnormalities (mucin accumulation with dilation of the crypts and engorgement and an increase in goblet cells) (Snouwaert et al., 1992; Dorin et al., 1992; Ratcliff et al., 1993; O'Neal et al., 1993; Colledge et al., 1995; van Doorninck et al., 1995; Delaney et al., 1995 and Zeiher et al., 1995). With the exception of the $cftr^{m1HGU}$, the severity of the intestinal phenotype is such that the majority of homozygous mutant offspring die as a result of intestinal blockage that is similar to meconium ileus (Table 63.4). In the mutation-specific lines this phenotype is less severe and in the case of the $cftr^{\Delta F508Rot}$ mice there is no evidence of the lethal obstruction. The difference between these ΔF508 lines, which essentially carry the same mutation, could be due to genetic background variation or a difference in the levels of transcription from the targeted alleles. It has been shown that in some cases transcription from alleles created by replacement targeting is reduced compared to the normal allele due to the presence of the selection cassette in the intron. This appears to be the case with the $cftr^{\Delta F508Cam}$ and the $cftr^{G551D}$ alleles (Table 63.1). It has previously been reported that residual intestinal chloride transport in a small proportion of ΔF/ΔF patients is correlated with mild disease (Veeze et al., 1994). Further evidence to support this comes from the $cftr^{G551D}$ mouse which shows similar characteristics to the null mice except that there is a better survival rate (Table 63.4). This reproduces the reduced incidence of meconium ileus in CF individuals carrying this mutation (Hamosh et al., 1993). It is hypothesised that this is due to residual forskolin response in the caecum (Delaney et al., 1995). Comparison of these different mouse models on the same genetic background will be valuable in determining the precise relationship between genotype and phenotype.

The high incidence of CF in the Caucasian population has led to speculation that heterozygotes may benefit from some selective advantage, perhaps in protection against dehydration in times of cholera epidemic (Quinton, 1994). The first evidence to support this theory in the CF mice came from a study by Cuthbert et al., (1994) who studied the response of the gut of wild type and $cftr^{m1Cam}$ mice to the 15aa peptide guanylin. This peptide was isolated from rat small intestine and exhibits structural homology with a heat stable enterotoxin (STa) produced by *E. coli* and is responsible for some forms of secretory diarrhoea. This peptide increased electrogenic chloride secretion in the wild-type intestine. However its action was abolished in CF mice. This suggests that the CFTR chloride channel is involved in mediating the effects of STa toxin and that a possible heterozygote advantage exists for CF mutations. Cholera causes death by inducing massive secretory diarrhoea as a result of the stimulation of fluid secretion in the gut by cholera toxin. Recent results in the $cftr^{m1UNC}$ mouse, which shows intermediate levels of intestinal fluid secretion in heterozygotes, compared to normal or mutant homozygotes, in response to cholera toxin lends credence to this hypothesis (Gabriel et al., 1994)

Lung Disease in CF Mice

The major cause of morbidity and mortality in patients with CF is lung disease, and the usefulness of mouse models in mimicking the human disease,

Table 63.2. Comparison of the bioelectric phenotype of cystic fibrosis respiratory tract in man and mouse.

Tissue	Measurement	Man	$cftr^{m1UNC}$ Mouse	$cftr^{m1Cam}$ Mouse	$cftr^{m1Bay}$ Mouse	$cftr^{m1HGU}$ Mouse	$cftr^{ΔF508Cam}$ Mouse	$cftr^{ΔF508Rot}$ Mouse	$cftr^{ΔF508Uta}$ Mouse	$cftr^{G551D}$ Mouse
Nasal Epithelium	Baseline PD	↑	↑	↑	Not Reported	↑	Not Reported	↑	↑	↑
	Response to amiloride (sodium absorption)	↑	Not Reported	Not Reported	Not Reported	↑	Not Reported	↑	↑	↑
	cAMP mediated chloride conductance	↓	↓	Not Reported	Not Reported	↓	Not Reported	Not Reported	↓	↓
	Calcium related chloride conductance	Preserved	Preserved	Not Reported	Not Reported	Not Reported	Not Reported	Not Reported	Not Reported	Preserved
Lower Repiratory Tract	Baseline PD	Preserved or ↑	Preserved	↓	Preserved (fetal)	Preserved	Reduced	Not Reported	Not Reported	Preserved
	Response to amiloride (sodium absorption)	↑	Not Reported	↓	Not Reported	↓	Preserved	Not Reported	Not Reported	Preserved
	cAMP mediated chloride conductance	↓	↓ by 100% in cultured tracheal cells. preserved in Ussing chamber	↓ by 75% in Ussing chamber	↓ by 70% in cultured fetal tracheal cells	↓ by 50% in Ussing chamber	↓ by 0–60% in Ussing chamber	Preserved in Ussing chamber	Not Reported	↓ in Ussing chamber
	Calcium related chloride conductance	Preserved	Preserved	Not Reported Preserved?	Not Reported	Preserved	↑	Not Reported	Not Reported	↑

Table 63.3. Comparison of the bioelectric phenotype of cystic fibrosis intestinal tract in man and mouse.

Measurement	Man	$cftr^{m1UNC}$ Mouse	$cftr^{m1Cam}$ Mouse	$cftr^{m1Bay}$ Mouse	$cftr^{m1HGU}$ Mouse	$cftr^{\Delta F508Cam}$ Mouse	$cftr^{\Delta F508Rot}$ Mouse	$cftr^{\Delta F508Uta}$ Mouse	$cftr^{G551D}$ Mouse
Baseline PD	↓ or preserved	→	→	Preserved	→	→	→	→	→
cAMP mediated chloride conducted	↓	↓ by 100%	↓ by 100%	↓ by 80%	↓ by 65%	↓ by 100%	↓ by ≤45%	↓ by 100%	↓ by 99% (95% in caecum)
Calcium related chloride conductance	↓	→	→	Not Reported	→	→	Preserved	Not Reported	Not Reported
Sodium-Glucose cotransport	↑	Preserved	Not Reported	Not Reported	Preserved	Not Reported	Preserved	Not Reported	Not Reported

Table 63.4. Comparison of cystic fibrosis phenotypes in man and mouse–intestinal disease.

Species	Allele	Perinatal	Post Weaning	Body weight at weaning
Man	R117H & DF508	~10–20% meconium ileus or meconium plug in large intestine in newborns	Distal intestinal obstruction syndrome in ~20% of children & adults	Failure to thrive and reduced growth rate
Mouse	$cftr^{m1UNC}$	~50% within 1 week of birth ~40% death at weaning	<5% survive to adulthood	10-50% reduced
	$cftr^{m1Cam}$	~80% within 1 week of birth ~10% death at weaning	<5% survive to adulthood	50% reduced
	$cftr^{m1Bay}$	~40% within 1 week	% <<50% survive to adulthood.	70% reduced
	$cftr^{m1HGU}$	~5% within 1 week ~2% death at weaning	<93% survive to adulthood (2% small/large intestine blockage which may be fatal)	No body weight reduction
	$cftr^{\Delta F508Cam}$	35% at 16 days vs 60% for null mice	death rate at thirty days similar to null mice	Not Reported
	$cftr^{\Delta F508Rot}$	No severe intestinal phenotype	Not reported	Reduction of body weight
	$cftr^{\Delta F508Uia}$	~10% within 1 week	40% survive to adulthood	50% reduced
	$cftr^{G551D}$	Not reported	67% survive at 35 days in SPF conditions 27% survive at 35 days instandard conditions	30–50% reduced

or their use in testing therapeutic strategies, is largely dependent on how well they model this aspect of the disease. In humans, submucosal glands of the proximal airways appear to be the major site of *CFTR* expression in the respiratory tract (Engelhardt et al., 1992) with only a low level of expression in the lung epithelia. By contrast, mice have very few submucosal glands which are located in the trachea, and this has raised the question of whether CF mutant mice would be capable of developing CF lung disease. Only minor pathological alterations have been observed in the lungs and upper airways of the $cftr^{m1UNC}$ and $cftr^{m1Bay}$ mice, even, in the case of the $cftr^{m1UNC}$ mice, after exposure to *S.aureus* (Snouwaert et al., 1995) or *P. aeruginosa* (Gosselin et al., 1995) infection. The $cftr^{m1HGU}$ mice demonstrate good long-term survival and show no significant difference in the incidence of pathological changes under standard conditions. Following repeated exposure to CF-associated pathogens, the $cftr^{m1HGU}$ mice have a reduced capacity to clear either *Staphylococcus aureus* or *Burkholderia cepacia*, and were shown to develop pathogen-specific histopathology that is characteristic of human CF patients (Davidson et al., 1995). It may be that infection with more than one pathogen in a particular order is important in onset and development of the severe lung disease seen in CF patients therefore these questions can now be addressed in the mouse model. There have been reports that CF patients show increased susceptibility to *Pseudomonas* through increased adherence to the airway epithelial cell surface (Imundo et al., 1995) and in particular to ΔF508/ΔF508 cells (Zar et al., 1995). It is of interest to determine whether the ΔF508 mice are likewise more susceptible to *P. aeruginosa* infection.

Prospects for Therapy

The generation of CF mice has allowed the possibility of *in vivo* testing of novel therapeutic strategies for CF, either by modifying the mutant protein or supplying functional protein by gene therapy.

Pharmacological

Mice that partially or completely lack functional *Cftr* can be used to test the safety and efficacy of pharmacological approaches such as drugs targeted to alternative mechanisms for regulating fluid transport. The same trafficking defect appears to be present in $cftr^{\Delta F508Cam}$ mice as in patients (Colledge et al., 1995). This model will enable novel drug therapies, perhaps based on interactions between CFTR and the specific chaperone molecule calnexin (Pind et al., 1994), to allow trafficking of the mutant protein to the cell surface to be tested. In contrast the protein produced by the G551D mutation is correctly localised but is incorrectly regulated. The $cftr^{m1G551D}$ mice will be of value in testing pharmacological agents which alter CFTR regulation. A recent report from Lloyd Mills et al. (1995) shows that the $cftr^{m1HGU}$ mouse mimics the CFTR-related glycoprotein secretion defect in the human submandibular salivary gland and can be partially corrected by increasing cAMP levels. They report that the submandibular gland is analogous to the sub-mucosal glands in regulation of exocytosis and therefore development of a pharmaceutical therapy may be relevant in preventing CF symptoms. In the lower respiratory tract, cAMP-dependent chloride ion conductance was found to be abolished in cultured tracheae from $cftr^{m1UNC}$ mice but Using chamber studies of tracheae from $cftr^{m1UNC}$ show forskolin responses equivalent to those of +/+ mice. It has been suggested that this is mediated through an increase in intracellular Ca^{2+} (Clarke et al., 1994) since Ca^{2+}-mediated chloride responses are known to be preserved in $cftr^{m1UNC}$, $cftr^{m1Cam}$ and $cftr^{m1HGU}$ mice and in man (Smith et al., 1995 and Hyde et al., 1993). Pharmacologic intervention directed at activating the alternative chloride conductance in the airways could slow the progression of CF lung disease.

Gene Therapy

The mouse models can also be of value in testing somatic gene-therapy approaches. Before these strategies are transferred to CF patients there are many questions, primarily concerning safety, which can and should be addressed in animals. Whitsett et al. (1992) reported that expression of *hCFTR* in airway epithelial cells of transgenic mice, under the control of the sufactant protein C promoter, was non-toxic. Subsequently, two gene-therapy protocols involving delivery of cDNA constructs complexed with cationic lipids to the respiratory tract of CF mice have shown correction of the bioelectric defect in these mice. Direct tracheal instillation of *CFTR* cDNA complexed with DOTMA/DOPE showed evidence for complete correction of the electrophysiological defect in tracheal explants of four out of six animals reported (Hyde et al., 1993). Aerosol nebulization of *CFTR* cDNA complexed to DC-chol/DOPE showed evidence for partial, to complete, correction of the electrophysiological defect in the nose (measured *in vivo*) and in the trachea (*in vitro*) (Alton et al., 1993). In both these studies there was no evidence of any toxic effect or histological damage. Topical application of DNA/liposomes was used in an attempt to correct the electrophysiological defect in the intestine of CF mice (Alton et al., 1993) but was without a marked effect. This may have been due to an unfavourable environment or simply the fact that endogenous CFTR expression is much higher in the intestinal crypts than in the airways and therefore would require a significantly higher efficiency of gene transfer. By contrast attempts to use CF mice to demonstrate correction of the CF defect in CF mouse airways by recombinant CFTR-expressing adenovirus have not proved as successful. It was shown that repeated high doses were required to achieve partial (50%) correction of the chloride transport defect. This has been ascribed to inefficient in vivo adenoviral transduction of columnar cells of the respiratory epithelium (Grubb et al., 1994).

Initial studies in patients indicate that current vectors for gene delivery are relatively inefficient and repeated treatment may be required for therapeutic benefit. In the case of adenoviruses, this can result in unacceptable immune responses which decrease the efficacy with every dose and produces inflammatory responses to adenovirus proteins. By contrast, up to five successive doses of DNA/DOTAP showed no evidence of histological damage despite an increasing promoter dependent dose reponse (McLachlan et al., 1995). This means that further refinement of vectors, CFTR expression constructs and methods of delivery are necessary and that a suitable model for testing any improvements will be important. In the CF mouse, it will be important to determine how safe, efficient, repeatable and long lasting any of these gene-therapy strategies prove to be, so that the treatments can be advanced to the clinic.

A further problem for CF gene therapy is the ability to measure efficacy in relation to clinical end points such as halting the decline in lung function or a reduction in the amount of infection since this can take many years. The ability to induce lung disease in CF mice may prove to be a valuable method of assessing what levels of correction are required to prevent this.

It is important to establish the relationship between CFTR activity, chloride ion transport and clinical phenotype to estimate what level of correction is necessary to ameliorate the disease. The variation in residual levels of CFTR activity and corresponding disease severity in the CF mouse models may be a good indication. The $cftr^{m1HGU}$ mice have approximately 10% residual activity which appears to be sufficient to prevent the severe intestinal phenotype. By intercrossing the different strains to create compound heterozygotes with intermediate levels of residual CFTR expression and chloride conductance, Dorin et al. (1995) have demonstrated that 5% CFTR is sufficient to have a significant effect on survival.

Transgenic Complementation of CF Mice

Modifying the phenotype of CF mice other than by alterations to the Cftr gene itself is possible using classical transgenic techniques. Transgenic mice in which the level and tissue specificity of CFTR expression is determined by the promoter used, the integration site, and the copy number of the transgene can be crossed with 'knock-out' CF mice to produce lines in which the only source of CFTR is derived from the introduced transgene. Lines of transgenic mice have been created that express normal human CFTR under the control of the rat intestinal fatty-acid binding protein (FABPi) (Zhou et al., 1994 and James et al., 1994) or villin (Auerbach et al., 1994) promoters, both of which should direct gut-specific expression. Increased survival was observed when these lines were crossed into $cftr^{tm1UNC}$ homozygote mice (Zhou et al., 1994). Importantly, functional correction of the intestinal defect was observed by expression of human CFTR even though the profile of FABPi promoter-driven expression was distinct from that of the endogenous murine Cftr mRNA in wild-type mice. However, in the case of mice generated from crosses involving villin-$CFTR^{+/-}$ $cftr^{tm1UNC}$ mice, offspring that were transgene negative also appeared to show increased survival (Auerbach et al., 1994), and it is possible that alterations to the genetic background generated during the breeding programme may have affected the phenotype. These mice provide further support for the use of gene-therapy treatments to correct the basic defect in CF. They will also be useful in determining the abundance and distribution of CFTR expression that are required to correct the physiological and histological abnormalities in the intestine and will allow a more thorough investigation of the respiratory phenotype in $cftr^{tm1UNC}$ mice than has so far been possible.

There are several disadvantages of these classical cDNA transgenic studies. CFTR is being expressed under the control of ubiquitous or tissue specific promoters which have different expression profiles from normal CFTR and transcription from these constructs is often affected by the genomic site of integration of the transgene (termed 'position effect'). It has recently been shown that large genomic transgene constructs, produced as Yeast Artificial Chromosomes (YAC's), are capable of overcoming 'position effects' imposed by the site of integration to allow normal expression of the transgene (Schedl et al., 1993). Attempts to examine human CFTR promoter function in transgenic mice have so far been unsuccessful however, it has recently been reported that a YAC containing a 310 kb fragment of the human genome including the *CFTR* gene is able to restore chloride secretory activity to at least normal levels in the epithelia lining the colon, caecum and jejunum of CF mutant mice (Cuthbert *et al.*, 1995). This vector has the advantage that it contains the *CFTR* promoter region with the endogenous regulatory elements in the control of *CFTR* gene expression. This type of experiment will facilitate the dissection of the *CFTR* promoter region and identify which elements are required to mimic expression of the endogenous gene leading to the design of better gene therapy vectors. Gene therapy in CF patients will be required for the patients lifetime thus the effects of long term transgene expression should be investigated in the mouse models. Ideally, vectors will be constructed which can replicate extrachromosomally and segregate in mammalian cells thus avoiding the potential risk of genomic mutation with an integrative vector. These would be particularly useful if 'stem-cells' could be identified and targeted in the airways and intestinal tract and may lead to life-long correction.

Concluding Remarks

It is clear that the CF mutant mice collectively model many of the hallmarks of the disease in remarkable detail. There can be little doubt that these mice are proving to be a valuable tool not only in increasing our understanding of the disease but in the development of improved therapies.

References

Alton, E.W.F.W., Middleton, P.G., Caplen, N.J., Smith, S.N., Steel, D.M., Munkonge, F.M., Jeffrey, P.K., Geddes, D.M., Hart, S.L., Williamson, R., Fasold, K.I., Miller, A.D., Dickinson, P., Stevenson, B.J., McLachlan, G., Dorin, J.R. and Porteous, D.J. (1993) Non-invasive liposome-mediated gene delivery can correct the ion transport defect in cystic fibrosis mutant mice. *Nature Genetics* **5**, 135–142

Auerbach, W., Robine, B., Chen, M., Caillot, E., Lu, Z., Pringault, E., Rochwerger, L., Naruszewicz, I., Louvard, D. and Buchwald, M. (1994) Transgenic correction of the intestinal defect of CF mice by expression of the human CFTR under the control of the villin promoter. *Pediatr. Pulmonol.* **10**, 197

Clarke, L.L., Grubb, B.R., Gabriel, S.E., Smithies, O., Koller, B.H. and Boucher, R.C. (1992) Defective epithelial chloride transport in a gene-targeted mouse model of cystic fibrosis. *Science* **257**, 1125–1128

Clarke, L.L., Grubb, B.R., Yankaskas, J.R., Cotton, C.U., McKenzie, A. and Boucher, R.C. (1994) Relationship of a non-cystic fibrosis transmembrane conductance regulator-mediated chloride conductance to organ-level disease in Cftr (−/−) mice. *Proc. Natl. Acad. Sci. USA* **91**, 479–483

Colledge, W.H., Abella, B.S., Southern, K.W., Ratcliff, R., Jiang, C., Cheng, S.H., MacVinish, L.J., Anderson, J.R., Cuthbert, A.W. and Evans, M.J. (1995) Generation and characterisation of a ΔF508 cystic fibrosis mouse model. *Nature Genetics* **10**, 445–452

Cuthbert, A.W., Hickman, M.E., MacVinish, L.J., Evans, M.J., Colledge, W.H., Ratcliff, R., Seale, P.W., Humphrey, P.P.A. (1994) Chloride secretion in response to guanylin in colonic epithelia from normal and transgenic cystic fibrosis mice. *Br. J. Pharmacol.* **112**, 31–36

Cuthbert, A., MacVinish, L.J., Evans, M.J., Ratcliff, R., Colledge, W.H. and Huxley, C. (1995) Epithelial function of the murine cystic fibrosis gut *20th European Cystic Fibrosis Conference, Brussels, Belgium* – 18–21 June, L47

Davidson, D.J., Dorin, J.R., McLachlan, G., Ranaldi, V., Lamb, D., Doherty, C., Govan, J. and Porteous, D.J. (1995) Lung disease in the cystic fibrosis mouse exposed to bacterial pathogens. *Nature Genet.* **9**, 351–357

Delaney, S.J., Alton, A.W.F.W., Smith, S.N., Lunn, D.P., Farley, R., Lovelock, P.K., Thomson, S.A., Hume, D.A., Lamb, D., Porteous, D.J., Dorin, J.R. and Wainright, B.J. (1995) A cystic fibrosis mouse model carrying the common missense mutation G551D. *EMBO J.* in press

Dorin, J.R., Dickinson, P., Alton, E.W.F.W., Smith, S.N., Geddes, D.M., Stevenson, B.J., Kimber, W.L., Fleming, S., Clarke, A.R., Hooper, M.L., Anderson, L., Beddington, R.S.P. and Porteous, D.J. (1992) Cystic fibrosis in the mouse by targeted insertional mutagenesis. *Nature* **359**, 211–215

Dorin, J.R., Webb, S., Farini, E., Delaney, S., Wainwright, B., Smith, S., Farley, R., Alton, E.W.F.W. and Porteous, D.J. (1995) Phenotypic consequences of CFTR modulation in mutant mice: Implications for somatic gene therapy. *Pediatr. Pulmonol.* **12**, 231

Engelhardt, J.F., Yankaskas, J.R., Ernst, S.A., Yang, Y., Marino, C.R., Boucher, R.C., Cohn, J.A. and Wilson, J.M. (1992) Submucosal glands are the predominant site of CFTR expression in human bronchus. *Nature Genet.* **2**, 240–248

Gabriel, S.E., Brigman, K.M., Koller, B.H., Boucher, R.C. and Stutts, M.J. (1994) Cystic fibrosis heterozygote resistance to cholera toxin in the cystic fibrosis mouse model. *Science* **266**, 107–109

Gosselin, D., Boulé, M., Eidelman, D.H., Griesenbach, U., Stevenson, M.M., Tsui, L.-C. and Radzioch, D. (1995) Effect of CFTR gene defect on the host response to acute *Pseudomonas aeruginosa* lung infection in mice. *Pediatr. Pulmonol.* **12**, 272

Gray, M.A., Winpenny, J.P., Porteous, D.J., Dorin, J.R. and Argent, B.E. (1994) CFTR and calcium-activated chloride currents in pancreatic duct cells of a transgenic CF mouse. *Am. J. Physiol.* **266**, C213–221

Grubb, B.R., Pickles, R.J., Ye, H., Yankaskas, J.R., Vick, R.N., Engelhardt, J.F., Wilson, J.M., Johnson, L.G. and Boucher, R.C. (1994) Inefficient gene transfer by adenovirus vector to cystic fibrosis airway epithelia of mice and humans. *Nature* **371**, 802–806

Hamosh, A. and Corey, M. (1993) Cystic fibrosis genotype/phenotype consortium: correlation between genotype and phenotype in cystic fibrosis. *New Engl. J. Med.* **329**, 1308–1313

Hasty, P., Ramirez-Solls, R., Krumlauf R. and Bradley, A. (1991) Introduction of a subtle mutation into the *Hox-2.6* locus in embryonic stem cells. *Nature* **350**, 243–246

Hyde, S.C., Gill, D.R., Higgins, C.F., Trezise, A.E.O., McVinish, L.J., Cuthbert, A.W., Ratcliff, R., Evans, M.J. and Colledge, W.H. (1993) Correction of the ion transport defect in cystic fibrosis transgenic mice by gene therapy. *Nature* **362**, 250–255

Imundo, L., Barasch, J., Prince, A. and Al-Awqati, Q. (1995) Cystic fibrosis epithelial cells have a receptor for pathogenic bacteria on their apical surface. *Proc. Natl. Acad. Sci. USA* **92**, 3019–3023

James, R.J., Dickinson, P., Dorin, J.R., Webb, S. and Porteous, D.J. (1994) Tissue specific complementation of cystic fibrosis mutant mice. *Pediatr. Pulmonol.* **10**, 195

Lloyd Mills, C., Dorin, J.R., Davidson, D.J., Porteous, D.J., Alton, E.W.F.W., Dormer, R.L. and McPherson, M.A. (1995) Decreased β-adrenergic stimulation of glycoprotein secretion in CF mice submandibular glands: reversal by the methylxanthine, IBMX. *Biochem. Biophys. Res. Comm.* **215**, 674–681

McLachlan, G., Davidson, D.J., Stevenson, B.J., Dickinson, P., Davidson-Smith, H., Dorin, J.R. and Porteous, D.J. (1995) Evaluation *in vitro* and *in vivo* of cationic liposome-expression construct complexes for cystic fibrosis gene therapy. *Gene Therapy* **2**, 614–622

O'Neal, W.K., Hasty, P., McCray, P.B., Casey, B., Rivera-Pérez, J., Welsh, M.J., Beaudet, A.L. and Bradley, A. (1993) A severe phenotype in mice with a duplication of exon 3 in the cystic fibrosis locus. *Hum. Mol. Genet.* **2**, 1561–1569

Pind, S., Riordan, J.R. and Williams, D.B. (1994) Participation of the ER chaperone calnexin (p88, IP90) in the biogenesis of CFTR. *J. Biol. Chem.* **269**, 12784–12788

Quinton, P.M. (1994) What is good about cystic fibrosis? *Curr. Biol.* **4**, 742–743

Ratcliff, R., Evans, M.J., Cuthbert, A.W., MacVinish, L.J., Foster, D., Anderson, J.R. and Colledge, W.H. (1993) Production of a severe cystic fibrosis mutation in mice by gene targeting. *Nature Genetics* **4**, 35–41

Riordan, J.R., Rommens J.M., Kerem, B.S., Alon, N., Rozmahel, R., Grzelczak, Z., Zielenski, J., Lok, S., Plavsic, N., Chou, J-L., Drumm, M.L., Iannuzzi, M.C., Collins, F.S., Tsui, L.-C. (1989) Identification of the cystic fibrosis gene: chromosome walking and jumping. *Science* **245**, 1066–1073

Schedl, A., Montoliu, L., Kelsey, G. and Schütz, G. (1993) A yeast artificial chromosome covering the tyrosinase gene confers copy number dependent expression in transgenic mice. *Nature* **362**, 258–261

Smith, S.N, Alton, E.W.F.W. and Geddes, D.M. (1992) Ion transport characteristics of the murine trachea and caecum. *Clin. Sci.* **82**, 667–672

Smith S.N., Steel D.M., Middleton P.G., Munkonge, F.M., Geddes, D.M., Caplen, N.J., Porteous, D.J., Dorin, J.R., Alton, E.W.F.W. (1995) Bioelectric characteristics of the exon 10 insertional mouse model of cystic fibrosis closely mimic those found in man. *Am. J. Physiol.* **268**, C297–C307

Snouwaert, J.N., Brigman, K.K., Latour, A.M., Malouf, N.N., Boucher, R.C., Smithies, O. and Koller, B.H. (1992) An animal model for cystic fibrosis made by gene trageting. *Science* **257**, 1083–1088

Snouwaert, J.N., Brigman, K.K., Latour, A.M., Iraj, E., Schwab, U., Gilmour, M.I. and Koller, B.H. (1995) A murine model of cystic fibrosis. *Am. J. Respir. Crit. Care Med.* **151**, S59–64

Tata, F., Stanier, P., Wicking, C., Halford, S., Kruyer, H., Lench, N.J., Scambler, P.J., Hansen, C., Braman, J.C., Williamson, R. and Wainwright, B.J. (1991) Cloning the mouse homologue of the human cystic fibrosis transmembrane conductance regulator gene. *Genomics* **10**, 301–307

van Doorninck, J.H., French, P.J., Verbeek, E., Peters, R.H.P.C., Morreau, H., Bijman, J. and Scholte, B.J. (1995) A mouse model for the cystic fibrosis ΔF508 mutation. *Emb. J.* **14**, 4403–4411

Veeze, H.J., Halley, D.J., Bijman, J., de Jongste, J.C., de Jong, H.R. and Sinaasappel, M. (1994) Detection of mild clinical symptoms in cystic fibrosis patients. Residual chloride secretion measured in rectal biopsies in relation to the genotype. *J. Clin. Invest.* **93**, 461–466

Whitsett, J.A., Dey, C.R., Stripp, B.R., Wikenheiser, K.A., Clark, J.C., Wert, S.E., Gregory, R.J., Smith, A.E., Cohn, J.A., Wilson, J.M. and Engelhardt, J. (1992) Human cystic fibrosis transmembrane conductance regulator directed to respiratory epithelial cells of transgenic mice. *Nature Genetics* **2**, 13–20

Zar, H., Saiman, L., Quittell, L. and Prince, A. (1995) Binding of *Pseudomonas aeruginosa* to respiratory epithelial cells from patients with various mutations in the cystic fibrosis transmembrane regulator. *J. Pediatr.* **126**, 230–233

Zeiher, B., Eichwald, E., Zabner, J., Smith, J., McCray, P., Puga, A. and Cappechi, M. (1995) Characterisation of a ΔF508 mouse model of cystic fibrosis. *J. Clin. Invest.* **96**, 2051–2064

Zhou, L., Dey, C.R., Wert, S.E., DuVall, M.D., Frizzell, R.A. and Whitsett, J.A. (1994) Correction of lethal intestinal defect in a mouse model of cystic fibrosis by human *CFTR*. *Science* **266**, 1705–1708

PART IV, SECTION B

64. Transgenic Animals in Atherosclerosis Research

Miles W. Miller and Edward M. Rubin
Human Genome Center, Berkeley National Laboratory, Berkeley, CA 94720, USA

Introduction

The most common cause of death in the industrialized nations is atherosclerosis (Ross 1986, WHO 1985). Although this condition has long been known to have a genetic predisposition, determining the roles of the multitude of genes believed to contribute to the disorder has been a daunting task. Abnormalities of the lipid metabolism have been established as major factors increasing an individual's susceptibility to coronary artery disease. At least 17 different genes encode proteins or enzymes that are involved in the assembly and processing of lipoprotein particles. Lipoprotein particles have been classically categorized by their protein content and density: either high, low, intermediate, or very low density lipoproteins (HDL, LDL, IDL, VLDL), or as chylomicrons and chylomicron remnants. Lipoprotein particles also vary in the type and the amount of associated protein, apolipoproteins, as well as in the percentage of their other ingredients, such as triglycerides, phospholipids, and cholesterol (Mackness and Durrington, 1992 and Groff et al., 1995).

The use of the mouse as a model to study atherosclerosis has lagged behind its use to study a variety of other human conditions. This relates to the relative resistance of mice to develop atherosclerosis and the availability of other mammals such as rabbits and pigs who are significantly more susceptible to atherosclerosis. In the last 8 years there has been marked increase in the use of mice in atherosclerosis studies. This relates to the development of atherogenic diets tolerated by this organism which result in reproducible lesions, and the power of transgenic and gene knockout technologies, available only in the mouse, to address the particular hypothesis relevant to human atherosclerosis (Paigen et al., 1994).

The atherogenic consequences of increasing or decreasing the level of expression in mice of a variety of genes involved in lipid transport are providing unique insights into this common human condition. These studies of genetically engineered animals have allowed the testing of a variety of hypotheses concerning atherosclerosis in a manner not feasible in humans. In this chapter, we have selected and summarized some of the research using transgenic mice to assess the impact of various genes involved in lipid transport on atherosclerosis susceptibility.

Genetic Engineering of Mice with Reduced Susceptibility to Atherosclerosis

Apolipoprotein AI and HDL

A high level of HDL in an individual has long been associated with reduced risk for developing atherosclerosis, while a low level of HDL is one of the strongest indicators of increased risk for developing coronary artery disease. Despite the strong association of HDL levels and susceptibility to coronary artery disease, little direct information has been available assessing whether HDL acts directly to prevent coronary artery disease, or is but an associated biochemical marker. To examine this issue directly, transgenic mice have been created that overexpress apolipoprotein AI (apoAI), a structural component of HDL that determines its level (Walsh et al., 1989 and Rubin et al., 1991a). As predicted, overexpression of a human apoAI transgene in mice resulted in an increased plasma level of HDL.

In several studies the apoAI transgenic mice with increased HDL plasma levels have been shown to have markedly reduced susceptibility to atherosclerosis (Rubin et al., 1991b; Schultz et al., 1993; Liu, et al., 1994; Paszty et al., 1994; and Plump et al., 1994). These consistent results demonstrating decreased atherosclerosis in animals who have been engineered to have increased levels of apoAI and HDL provide very strong support for a direct role of this lipoprotein in the prevention of coronary artery disease.

Genetically Engineered Mice with Increased Susceptibility to Atherogenesis

Low Density Lipoprotein Receptor Knockout Mice

Familial hypercholesterolemia (FH) is a common cause of markedly increased atherosclerosis susceptibility in humans. This condition is the result of an absence or an abnormal function of the low density lipoprotein (LDL) receptor. Due to the inability of FH patients to efficiently remove LDL from their plasma, these individuals invariably have marked increases in plasma cholesterol and develop coronary artery disease at a young age. Recently, an animal model for FH has been developed through gene targeting of the mouse LDL receptor in embryonic stem cells. These transgenic mice lacking the LDL receptor have plasma cholesterol levels several-fold higher then control animals. Consistent with the human condition, these animals also have markedly increased susceptibility to atherosclerosis. They develop atherosclerotic lesions throughout their vasculature which in many ways mimic lesions observed in humans lacking LDL receptors (Ishibashi et al., 1993).

Apolipoprotein E Knockout Mice

Two laboratories have used gene targeting of embryonic stem cells to create mice that lack apolipoprotein E (apoE) (Piedrahita et al. 1992; Plump et al. 1992; and Zhang et al., 1992). ApoE is found in nearly all classes of lipoproteins and serves as a ligand in the receptor-mediated clearance of these lipid rich macromolecules. Extensive analysis of the apoE knockout mice have demonstrated that these animals are extremely susceptible to atherosclerosis independent of dietary intake. Mice fed a mouse chow diet containing nearly 10-fold less fat and cholesterol than the typical human diet still develop extensive atherosclerosis throughout their arterial system. The apoE knockout mice have proven to be extremely useful substrates for studying a variety of issues related to atherosclerosis due to the severity and human-like nature of their vascular disease.

Apolipoprotein B Transgenics

LDL is the major non-HDL lipoprotein in the plasma of humans while in the mouse it is VLDL. Generation of transgenic mice overexpressing apolipoprotein B (apoB), the primary protein component of both LDL and VLDL, has been hindered by the extremely large size of the apoB protein (greater than 4500 amino acids) and genomic size (43 kb). Recently, these obstacles have been overcome through the creation of transgenic mice with a P1 phagemid containing a 75 kb insert including the entire human apoB gene plus significant flanking DNA (Callow et al., 1994 and Linton et al., 1993). By this approach two labs have independently generated mice with high level expression of human apoB and significant plasma levels of a human-like LDL particle. Like humans with marked elevations of LDL, the apoB transgenic mice are also susceptible to the development of atherosclerosis. Although the apoB transgenics develop a milder form of atherosclerosis than the apoE and the LDL receptor knockout mice, their lipoprotein profile more closely mimics that seen in humans with coronary artery disease (Callow et al., 1995 and Purcell-Huynh et al., 1995).

Apolipoprotein (a)

High plasma levels of apolipoprotein (a) (apo(a)) serve as a major independent risk factor for atherosclerosis in humans. Despite the importance of apo(a) as an atherogenic risk factor, relatively little is known about biological properties of this molecule. In part, this arises from the fact that apo(a) is present in the plasma of only a limited number of animals, including Old World monkeys, great apes, humans and the European hedgehog. The absence of apo(a) in the organisms utilized in most biomedical research (i.e., rodents, dogs, rabbits) has limited studies into the properties of this molecule. Transgenic mice have been engineered expressing either a human apo(a) cDNA associated with a mouse transferrin promoter (Chiesa et al., 1992), or an apo(a) gene contained within a 270 kb yeast artificial chromosome (Frazer et al., 1995). Studies of the animals transgenic for the apo(a) cDNA have shown that they develop significantly more diet-induced atherogenesis than non-transgenic control animals. A distinct aspect of the atherogenesis noted in apo(a) transgenics is that it does not appear to be lipid associated. The apo(a) in these mice is found free in the plasma, unassociated with murine lipoproteins and the transgenic and non-transgenic animals have similar lipoprotein profiles, despite differences in their atherogenesis susceptibility.

Summary

A basic question arising from these studies of atherosclerosis in transgenic mice has to do with the relevance of the murine data to the human condition.

As progressively more mouse atherogenesis studies are performed comparison of the murine results with information generated from indirect human studies is leading to the emergence of a valid assessment of the relevance of the mouse studies to the human condition. The data from atherogenesis studies carried out to date in mice, where levels of more than 12 different genes involved in lipid transport have been manipulated, have lead to conclusions consistent with our knowledge of the properties of these genes in humans (Table 64.1). These findings support the analysis of atherogenesis in the genetically engineered mouse as a valid means for deriving insights concerning the human condition.

Table 64.1. Results of genetic manipulations on risk of atherosclerosis.

Genetic manipulation	Predicted risk in humans	Tested risk in mice	References
↑ apoAI	↓	↓	1
↑ apoAII	↑	↑	2,3
↓ apoB	↓	↓	4
↑ apo(a)	↑	↑	5
↑ apo(a) + apoB	↑	↑	6
↑ apoE	↓	↓	7
↓ apoE	↑	↑	8,9
↑ LCAT	↓	↓	10,11,12
↑ LDLR	↓	↓	13,14
↓ LDLR	↑	↑	15
↑ LPL	↓	↓	16,17,18

Upward arrow (↑), increase; downward arrow (↓), decrease.
[1] Rubin et al., 1991.
[2] Schultz et al., 1993.
[3] Warden et al., 1993.
[4] Farese et al., 1995.
[5] Lawn et al., 1992.
[6] Callow et al., 1995.
[7] Stoltzfus and Rubin, 1993.
[8] Plump et al., 1992.
[9] Zhang et al., 1992.
[10] Francone et al., 1995.
[11] Vaisman et al., 1995.
[12] Mehlum et al., 1995.
[13] Hofmann et al., 1988.
[14] Brown and Goldstein, 1992.
[15] Ishibashi et al., 1993.
[16] Shimada et al., 1993.
[17] Liu, M.S. et al., 1994.
[18] Zsigmond et al., 1994.

Acknowledgements

This work was supported by a National Institutes of Health Grant to E.R., PPG HL18574, and a grant funded by the National Dairy Promotion and Research Board and administered in cooperation with the National Dairy Council. E.R. is an American Heart Association Established Investigator. Research was conducted at the Lawrence Berkeley National Laboratory (Dept. of Energy Contract DE-AC0376SF00098), University of California, Berkeley.

References

Brown, M.S. and Goldstein, J.L. (1992) Koch's postulates for cholesterol. *Cell* **71**, 187–188

Callow, M.J., Stoltzfus, L.J., Lawn, R.M. and Rubin, E.M. (1994) Expression of human apolipoprotein B and assembly of lipoprotein(a) in transgenic mice. *Proc. Nat'l. Acad. Sci. USA* **91**, 2130–2136

Callow, M.J., Verstuyft, J., Tangirala, R., Palinski, W. and Rubin, E.M. (1995) Atherogenesis in transgenic mice with human apolipoprotein B and lipoprotein(a). *J. Clin. Invest.* **96**, 1639–1646

Chiesa, G., Hobbs, H.H., Koschinsky, M.L., Lawn, R.M., Maika, S.D. and Hammer, R.E. (1992) Reconstitution of lipoprotein(a) by infusion of human low density lipoprotein into transgenic mice expressing human apolipoprotein(a). *J. Biol. Chem.* **267**, 24369–24374

Farese, R.V., Ruland, S.L., Flynn, L.M., Stokowski, R.P. and Young, S.G. (1995) Knockout of the mouse apolipoprotein B gene results in embryonic lethality in homozygotes and protection against diet-induced hypercholesterolemia in heterozygotes. *Proc. Nat'l. Acad. Sci. USA* **92**, 1774–1778

Francone, O.L., Gong, E.L., Ng, D.S., Fielding, C.J. and Rubin, E.M. (1995) Expression of human lecithin-cholesterol acyltransferase in transgenic mice. *J. Clin. Invest.* **96**, 1440–1448

Frazer, K.A., Narla, G., Zhang, J.L. and Rubin, E.M. (1995) The apolipoprotein(a) gene is regulated by sex hormones and acute-phase inducers in YAC transgenic mice. *Nature Genetics* **9**, 424–431

Groff, J.L., Gropper, S.S. and Hunt, S.M. (1995) *Advanced nutrition and human metabolism*, St. Paul, Minnesota p. 129

Hofmann, S.L., Russell, D.W., Brown, M.S., Goldstein, J.L. and Hammer, R.E. (1988) Overexpression of low density lipoprotein (LDL) receptor eliminates LDL from plasma in transgenic mice. *Science.* **239**, 1277–1281

Ishibashi, S., Brown, M.S., Goldstein, J.L., Gerard, R.D., Hammer, R.E. and Herz, J. (1993) Hypercholesterolemia in low density lipoprotein receptor knockout mice and its reversal by adenovirus-mediated gene delivery. *J. Clin. Invest.* **92**, 883–893

Lawn, R.M., Wade, D.P., Hammer, R.E., Chiesa, G., Verstuyft, J.G. and Rubin, E.M. (1992) Atherogenesis in transgenic mice expressing human apolipoprotein (a). *Nature* **360**, 670–672

Linton, M.F., Farese Jr., R.V., Chiesa, G., Grass, D.S., Chin, P., Hammer, R.E., Hobbs, H.H. and Young, S.G. (1993) Transgenic mice expressing high plasma concentrations of human apolipoprotein B100 and lipoprotein(a). *J. Clin. Invest.* **92**, 3029–3037

Liu, A.C., Lawn, R.M., Verstuyft, J.G. and Rubin, E.M. (1994) Human apolipoprotein A-I prevents atherosclerosis associated with apolipoprotein(a) in transgenic mice. *J. Lip. Res.* **35**, 2263–2267

Liu, M.S., Jirik, F.R., LeBoeuf, R.C., Henderson, H., Castellani, L.W., Lusis, A.J., Ma, Y., Forsythe, I.J., Zhang, H., Kirk, E., Brunzell, J.D. and Hayden, M.R. (1994) Alteration of lipid profiles in plasma of transgenic mice expressing human lipoprotein lipase. *J. Biol. Chem.* **269**, 11417–11424

Mackness, M.I. and Durrington, P.N. (1992) Lipoprotein separation and analysis for clinical studies. In C.A. Converse and E.R. Skinner, (eds.), *Lipoprotein Analysis*, Oxford, UK. p. 3

Mehlum, A., Staels, B., Duverger, N., Tailleux, A., Castro, G., Fievet, C., Luc, G., Fruchart, J.C., Olivecrona, G. Skretting, G. et al. (1995) Tissue-specific expression of the human gene for lecithin:cholesterol acyltransferase in transgenic mice alters blood lipids, lipoproteins and lipases toward a less atherogenic profile. *Eu. J. Biochem.* **230**, 567–575

Paigen, B., Plump, A.S. and Rubin, E.M. (1994) The mouse as a model for human cardiovascular disease and hyperlipidemia. *Curr. Opin. Lipid.* **5**, 258–264

Paszty, C., Maeda, N., Verstuyft, J. and Rubin, E.M. (1994) Apolipoprotein A-I Transgene Corrects Apolipoprotein E Deficiency-Induced Atherosclerosis in Mice. *J. Clin. Invest.* **94**, 899–903

Piedrahita, J.A., Zhang, S.H., Hagaman, J.R., Oliver, P.M. and Maeda, N. (1992) Generation of mice carrying a mutant apolipoprotein E gene inactivated by gene targeting in embryonic stem cells. *Proc. Nat'l. Acad. Sci., USA* **89**, 4471–4475

Plump, A.S., Scott, C.J. and Breslow, J.L. (1994) Human apolipoprotein AI gene expression increases high density lipoprotein and suppresses atherosclerosis in the apolipoprotein E deficient mouse. *Proc. Nat'l Acad. Sci. USA* **91**, 9607–9611

Plump, A.S., Smith, J.D., Hayek, T., Aalto–Setala, K., Walsh, A., Verstuyft, J.G., Rubin, E.M. and Breslow, J.L. (1992) Severe hypercholesterolemia and atherosclerosis in apolipoprotein E-deficient mice created by homologous recombination in ES cells. *Cell* **71**, 343–353

Purcell-Huynh, D.A., Farese Jr., R.V., Johnson, D.F., Flynn, L.M., Pierotti, V., Newland, D.L., Linton, M.F., Sanan, D.A. and Young, S.G. (1995) Transgenic mice expressing high levels of human apolipoprotein B develop severe atherosclerotic lesions in response to a high fat diet. *J. Clin. Invest.* **95**, 2246–2257

Ross, R. (1986) The pathogenesis of atherosclerosis – an update. *N. Engl. J. Med.* **314**, 488–500

Rubin, E.M., Ishida, B.Y., Clift, S.M. and Krauss, R.M. (1991a) Expression of human apolipoprotein A-I in transgenic mice results in reduced plasma levels of murine apolipoprotein A-I and the appearance of two new high density lipoprotein size subclasses. *Proc. Nat'l. Acad. Sci. USA* **88**, 434–438

Rubin, E.M., Krauss, R.M., Spangler, E.A., Verstuyft, J.G. and Clift, S.M. (1991b) Inhibition of early atherogenesis in transgenic mice by human apolipoprotein AI. *Nature* **353**, 265–267

Schultz, J.R., Verstuyft, J.G., Gong, E.L., Nichols, A.V. and Rubin, E.M. (1993) Protein composition determines the anti-atherogenic properties of HDL in transgenic mice. *Nature* **365**, 762–764

Shimada, M., Shimano, H., Gotoda, T., Yamamoto, K., Kawamura, M., Inaba, T., Yazaki, Y. and Yamada, N. (1993) Overexpression of human lipoprotein lipase in transgenic mice. Resistance to diet-induced hypertriglyceridemia and hypercholesterolemia. *J. Biol. Chem.* **268**, 17924–17929

Stoltzfus, L. and Rubin, E.M. (1993) Atherogenesis. Insights from the study of transgenic and gene-targeted mice. *Trends Cardiovasc. Med.* **3**, 130–134

Vaisman, B.L., Klein, H., Rouis, M., Berard, A.M., Kindt, M.R., Talley, G.D., Meyn, S.M., Hoyt, R.F., Marcovina, S.M., Albers, J.J., Hoeg, J.M., Brewer, H.B. and Santamarina-Fojo, S. (1995) Overexpression of human lecithin cholesterol acyltransferase leads to hyperalphalipoproteinemia in transgenic mice. *J. Biol. Chem.* **270**, 12269–12275

Walsh, A., Ito, Y. and Breslow, J. (1989) High levels of human apolipoprotein A-I in transgenic mice result in increased plasma levels of small high density lipoprotein (HDL) particles comparable to human HDL_3. *J. Biol. Chem.* **264**, 6488–6494

Warden, C.H., Hedrick, C.C., Qiao, J.H., Castellani, L.W. and Lusis, A.J. (1993) Atherosclerosis in transgenic mice overexpressing apolipoprotein A-II. *Science* **261**, 469–472

World Health Organization (1985) Classification of atherosclerotic lesions. *WHO Tech. Rep. Serv.* **143**, 1–20

Zhang, S.H., Reddick, R.L., Piedrahita, J.A. and Maeda, N. (1992) Spontaneous hypercholesterolemia and arterial lesions in mice lacking apolipoprotein E. *Science* **258**, 468–471

Zsigmond, E., Scheffler, E., Forte, T.M., Potenz, R., Wu, W. and Chan, L. (1994) Transgenic mice expressing human lipoprotein lipase driven by the mouse metallothionein promoter. A phenotype associated with increased perinatal morality and reduced plasma very low density lipoprotein of normal size. *J. Biol. Chem.* **269**, 18757–18766

PART IV, SECTION B

65. Genetically Engineered Mice in Obesity Research

Bradford B. Lowell

Division of Endocrinology, Department of Medicine, Beth Israel Hospital and Harvard Medical School, Boston, Massachusetts 02215, USA

Introduction

Obesity is a prevalent disorder that often leads to diabetes and cardiovascular disease. The development of obesity is thought to represent the dysregulation of various homeostatic mechanisms which normally function to maintain body weight. However, the nature of these protective mechanisms, and the events which lead to their dysregulation, remain poorly understood. Because obesity develops within the context of a living organism, and involves the collaboration of many different organs, tissues and glands, it has been refractory to *in vitro* tissue culture based approaches. Therefore, it has been difficult to develop and test molecular and cellular hypotheses which might account for its pathogenesis. Because the study of body fat regulation requires *in vivo* systems, forward (positional cloning, candidate gene approaches) and reverse (transgenics, gene targeting) mouse genetic techniques are significantly impacting on obesity research. Indeed, the potential value of positional cloning was recently realized with the identification of leptin, the gene responsible for extreme obesity in *ob/ob* mice (Zhang et al., 1994). This advance has provided tremendous insight into mechanisms which operate to control body fat stores. Transgenics and gene targeting are also contributing importantly to obesity research and examples of such studies are the subject of this chapter.

Obesity is the result of positive energy balance (i.e., energy intake greater than energy expenditure). Calories ingested in excess of calories expended are stored as triglyceride in white adipocytes. Energy expenditure is complex being the net sum of calories expended to maintain cellular functions (ion gradients, enzymatic reactions, etc.), calories expended to perform physical activity, and calories expended in order to modulate energy balance (sometimes referred to as facultative energy expenditure). This latter category is of great interest as it is highly regulated and can change dramatically depending upon the nutritional status of the organism. Despite general agreement regarding the existence of facultative energy expenditure, uncertainty persists regarding its molecular and cellular basis. Leading hypotheses include substrate cycling in liver and muscle (Newsholme, 1978), and uncoupled mitochondrial respiration in brown adipose tissue (Himms-Hagen, 1989). Brown adipocytes are unique in their expression of uncoupling protein (UCP), an inner membrane mitochondrial protein which uncouples oxidative phosphorylation (Nicholls and Locke, 1984). Importantly, tight control exists over uncoupled respiration in brown fat. As a result, energy expenditure in brown fat ranges over many orders of magnitude, and is controlled primarily by sympathetic innervation. In summary, the regulation of body fat is thought to involve a classic feedback loop (Figure 65.1) composed of the following elements: *afferent systems* including leptin which senses perturbed total body fat stores, *the brain* which analyzes the fat-derived signal and integrates it with other afferent signals, and *efferent systems* which restore fat stores to normal through changes in food intake and energy expenditure (brown fat representing a significant component). Transgenic and gene knockout mice should be useful in unraveling the details of this feedback loop.

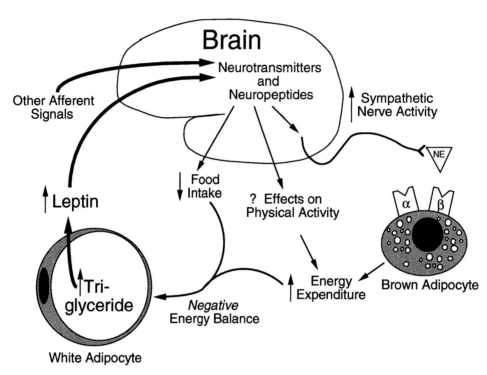

Figure 65.1. Feedback loop controlling total body fat stores. Details concerning the proposed feedback loop are described within the text. A primary events causes increased fat stores resulting in increased circulating leptin levels. Subsequent arrows shown on the diagram reflect effects of increased leptin concentrations (i.e. increased leptin produces decreased food intake and increased energy expenditure, resulting in negative energy balance). NE=norepinephrine, α, and β= α and β-adrenergic receptors).

Promoter/Enhancer Analyses of Adipocyte-Specific Genes

Transgenic studies have contributed importantly towards the identification of adipocyte-specific *cis*-regulatory elements. Such studies have had two significant outcomes for obesity research. First, identification of adipocyte-specific *cis*-regulatory elements has led to the isolation of relevant *trans*-acting factors, critically advancing knowledge of adipocyte differentiation. Second, adipocyte-specific *cis*-regulatory elements have been used to drive fat-specific expression of heterologous genes in transgenic mice. The genes which have been intensively studied to date include adipocyte fatty acid binding protein (aP2 or ALBP), which is expressed uniquely in white and brown adipocytes, and uncoupling protein (UCP), which is expressed exclusively in brown adipocytes.

Transgenic mice bearing ALBP promoter/enhancer – reporter gene constructs have been generated and a region required for fat-specific expression of ALBP *in vivo* has been identified (Ross *et al.*, 1990). As will be reviewed below, this element has been used extensively to drive expression of heterologous genes in white and brown fat of transgenic mice. Additional *in vitro* studies have led to identification of the responsible *cis*-regulatory element (direct repeat-1), and its corresponding *trans*-acting factor, peroxisome proliferator-activated receptor γ2 (PPARγ2) (Tontonoz *et al.*, 1994a and 1994b). PPARγ2, in synergy with CCAAT/enhancer-binding protein α (C/EBPα) (Christy *et al.*, 1989), transactivates transcription of many adipocyte specific genes, thereby promoting adipocyte differentiation.

Similar studies have been performed using the UCP promoter, and a region has been identified which confers brown fat-specific expression in transgenic mice (Boyer and Kozak, 1991 and Cassard-Doulcier *et al.*, 1993). As will be described later, this promoter element has been used to drive expression of heterologous genes in brown adipocytes of transgenic mice. Unfortunately, attempts to precisely define the brown adipocytes-specific *cis*-regulatory elements, and to identify relevant *trans*-acting factors, have been complicated by the presence of unexpected complexity, including interactions between regulatory sites. Nevertheless, it is anticipated that in the future these studies will provide key insights into the regulation of UCP gene expression, and ultimately increase understanding of brown adipocyte differentiation.

Leptin is another gene for which transgenic analyses of promoter/enhancer activities is likely to be informative. Leptin is expressed exclusively in adipocytes and is secreted into the blood stream (Maffei et al., 1995), where it functions as a hormone communicating the status of fat stores to the brain (Figure 65.1). Consequently, leptin gene expression is dramatically increased by obesity and decreased with fasting (Maffei et al., 1995; Funahashi et al., 1995; Frederich et al., 1995 and Ogawa et al., 1995). The mechanisms accounting for this "nutritional" regulation of leptin gene expression are unknown, and are of significant interest. Since it is difficult or impossible to recreate obesity and fasting conditions using *in vitro* cell culture systems, it is likely that transgenic analyses of leptin promoter/enhancer function will be required in order to identify pathways through which nutritional status regulates leptin gene expression.

Transgenic Perturbation of Energy Expenditure

As reviewed earlier, whole body energy expenditure is composed of many components. Two potentially important sources of regulated energy expenditure include substrate cycling (Newsholme 1978) and uncoupled mitochondrial respiration in brown adipose tissue (Himms-Hagen, 1989). Transgenic modulation of these two components, and the physiologic consequences of each, are described below.

One possible substrate cycle involves cytoplasmic glycerol 3-phosphate dehydrogenase (GPDH), which catalyzes the reduction of dihydroxyacetone phosphate to glycerol 3-phosphate. This reaction performs two separate functions: the generation of precursor for triglyceride and phospholipid synthesis, or the transfer of cytoplasmic NADH into the mitochondria *via* the glycerol phosphate shuttle. Usually, cytoplasmic NADH is transferred into mitochondria by way of the malate/aspartate shuttle leading to the conversion of 1 NADH molecule to 3 ATP molecules. In contrast, the glycerol phosphate shuttle performs the same function but ultimately produces only 2 ATP molecules. Thus, overexpression of cytoplasmic GPDH would be expected to increase activity of the glycerol phosphate shuttle leading to inefficient metabolism. Kozak et al. (1991) has generated transgenic mice which overexpress cytoplasmic GPDH by approximately 50–200 fold. These mice are characterized by increased energy expenditure and markedly reduced white fat stores. In additon, GPDH overexpression and the ensuing increase in energy expenditure prevents obesity in genetically predisposed *db/db* mice (Kozak et al., 1994). These studies demonstrate that modulation of a metabolic enzyme can significantly impact on energy expenditure, and support the hypothesis that substrate cycling can importantly effect body fat stores.

It has also been hypothesized that energy expenditure in brown adipose tissue functions to protect against the development of obesity (Rothwell and Stock, 1979 and Himms-Hagen, 1989). In order to evaluate this possibility, we recently generated transgenic mice which have decreased brown fat (Lowell et al., 1993). In this study, the promoter for UCP was used to drive expression of diptheria toxin-A chain selectively in brown fat. The resulting transgenic animals were characterized by reduced brown fat and marked obesity. Initially, obesity occurred in the absence of hyperphagia indicating a reduction in energy expenditure. However, as obesity advanced, hyperphagia developed raising the possibility that brown fat might also play an unexpected role in regulating food intake. More recently, we demonstrated that transgenic mice with decreased brown fat have markedly enhanced susceptibility to diet-induced obesity (Hamann et al., in press), thus supporting the hypothesis that brown fat protects against obesity caused by calorically dense diets. Overall, these studies strongly support the importance of brown fat in regulating body fat stores.

Transgenic Perturbation of Adipocyte Function

A number of studies have utilized the ALBP (aP2) promoter/enhancer to express heterologous genes in adipocytes. Transgenic mice overexpressing the insulin-sensitive glucose transporter (GLUT4) in adipocytes are characterized by fat cell hyperplasia and increased fat stores (Shepherd et al., 1993). This finding is of interest as gene knockout mice lacking GLUT4 have markedly decreased fat stores (Katz et al., 1995). Thus, alterations in the capacity to transport glucose into adipocytes significantly impacts on adipositiy. In another study, the ALBP promoter/enhancer was used to target expression of SV40 large T antigen (Tag) to adipocytes (Ross et al., 1992). As the ALBP promoter/enhancer is active in white and brown adipocytes, Tag was found to be expressed in both cell types. Remarkably, brown but not white fat tumors appeared raising the possibility of important cell-type differences in cell cycle control. Similarly, brown but not white fat tumors were noted when the α-amylase promoter was used to drive expression of Tag (Fox et al., 1989). Brown adipocyte cell lines have been created from Tag expressing brown fat tumors, facilitating investigations of uncoupling protien gene expression and brown fat cell metabolism (Ross et al., 1992 and Kozak et al., 1992). Finally, numerous additional studies are in progress where the ALBP promoter/enhancer is being used drive expression of other heterologous genes including UCP (Kozak et al., 1994), various adrenergic receptors (Soloveva et al., 1994), and leptin, to name a few. These studies, and others to be performed in the future, will continue to provide important insights into adipocyte function, as well as the effects of adipocyte dysfunction on systemic energy balance.

Transgenic Perturbation of Central Nervous System Function

Many neurotransmitters, neuropeptides and hormones have been speculated to play important roles in controlling body fat content. At present, transgenic and gene knockout technologies are only beginning to be used to investigate central regulation of body fat stores. In one relevant study, gene knockout mice lacking serotonin receptors (5-HT$_{2c}$ subtype) have been created (Tecott et al., 1995). Serotonin has been suspected of playing an important role in regulating food intake and drugs which enhance serotonin transmission via inhibition of serotonin re-uptake, such as dexfenfluramine, are being used as appetite suppressants. Of interest, mice lacking 5-HT$_{2c}$ receptors were found to have increased food intake, increased body fat and resistance to appetite suppressive effects of a serotonin agonist. These mice should be useful for investigating the mechanisms by which serotonin regulates energy balance. At present, other studies are underway in which gene targeting and transgenic approaches are being used to modulate levels of many additional potential central mediators of energy balance, such as neuropeptide Y. It is anticipated that these studies will significantly advance our understanding of central regulation of energy balance.

Using Transgenic Techniques to Determine Causes of Obesity in Genetically Obese Mice

The mutations responsible for causing a number of single gene obesity syndromes in rodents have recently been identified. In the case of *ob/ob* mice, the adipocyte secreted gene product leptin is absent, and as reviewed in Figure 65.1, this suggests a clear hypothesis relating leptin deficiency with development of obesity (Zhang et al., 1994 and Rink, 1994). However, for other obese rodents such as A^y mice (autosomal dominant mutation) and *fat/fat* mice (autosomal recessive mutation), it is less clear how the identified mutations might lead to obesity. For example, in A^y mice, a chromosomal rearrangement causes ubiquitous expression of the *agouti* gene product, which in wild-type mice is limited to hair follicles (Bultman et al., 1992 and Miller et al., 1993). Recently, it has been demonstrated that transgenic mice on a wild-type background bearing either a β-actin promoter-agoutic expression construct or a phosphoglycerate kinase promoter-agoutc expression construct, have wide spread expression of *agouti* and develop obesity (Klebig et al., 1995 and Perry et al., 1995). Thus, it is clear that ectopic expression of *agouti* in one or more sites causes obesity. However, the site or sites where ectopic *agouti* expression produces obesity is unknown, and is of considerable interest. Recently, *agouti* protein has been shown to antagonize melanocyte stimulating hormone (MSH) induced stimulation of the MSH receptor (expressed in hair follicles) and the related melanocortin-4 receptor (expressed in the brain) (Lu et al., 1994). As the brain plays an important role in controlling body fat stores, it is possible that obesity in A^y mice is due to antagonism of central melanocortin-4 receptors, or possibly antagonism of related melanocortin receptors which are also expressed in the brain. Alternatively, it has also been shown that *agouti* protein causes an increase in intracellular free calcium concentration in skeletal muscle, suggesting that some or all of its pathologic effects might be mediated by ectopic expression outside of the central nervous system (Zemel et al., 1995). In order to identify the site or sites where ectopic expression of *agouti* leads to the development of obesity, transgenic mice are presently being generated using a variety of tissue-specific promoters.

Mutation of the gene encoding carboxypeptidase E (CPE), a prohormone processing enzyme, is responsible for obesity in *fat/fat* mice (Naggert et al., 1995). It is assumed that abnormalities in the processing of one or more hormones leads to the development of obesity. However, analogous to the situation for *agouti*, it is unclear how this mutation produces obesity. By attempting to rescue *fat/fat* mice with tissue-specific transgenes expressing wild-type CPE, it should be possible to determine the site or sites where decreased CPE activity leads to obesity, thereby restricting the list of potential hormones which, when processed abnormally, cause significant obesity.

Genetically Engineered Mice in Drug Development

Transgenic and gene knockout mice are being used in pharmaceutical drug development as a means of learning more about potential durg targets, or to create mice which have human but not rodent drug targets ("humanized mice"). The β_3-adrenergic receptor (β_3-AR) is an example of a potentially important drug target in obesity research. β_3-ARs are expressed predominantly in white and brown adipose tissue, and β_3-selective agonist treatment of rodents produces a brown adipose tissue-mediated increase in energy expenditure and protection from obesity (Arch et al., 1984). Thus β_3-selective agonists are potential anti-obesity drugs. We have recently used gene targeting to investigate a number of issues relating to the physiology and pharmacology of β_3-ARs (Susulic et al., 1995). Mice that were homozygous for a disrupted β_3-AR allele had undetectable levels of intact β_3-AR mRNA, lacked functional β_3-ARs, and had modestly increased fat stores (females more than males), indicating that β_3-ARs play a role in regulating energy balance. Importantly, β_1 but not β_2-AR mRNA levels up-regulated in white and brown adipose tissue of β_3-AR deficient mice (brown more than white), strongly implying that β_3-ARs mediate physiologically relevant signalling under normal conditions, and that "cross-talk" exists between β_3-ARs and β_1-AR gene expression. Finally, acute treatment of wild-type mice with CL 316, 243, a β_3-selective agonist, increased serum levels of free fatty acids and insulin, increased energy expenditure, and reduced food intake. These effects were completely absent in β_3-AR deficient mice, proving that these actions of CL are mediated exclusively by β_3-ARs. β_3-AR deficient mice should be useful as a means to better understand the physiology and pharmacology of β_3-ARs.

Originally, β_3-selective agonists were identified by screening compounds for anti-obesity effects in rodents. However, significant differences between human and rodent β_3-AR amino acid sequences exist, and consequently most available β_3-selective agonists have low potency against human β_3-ARs. In order to create an improved test-system, and to obtain information regarding human β_3-AR promoter/enhancer activities, we have transgenically introduced 75 kb of human β_3-AR genomic DNA (P1 phagemid) into knockout mice lacking functional murine β_3-ARs. These "humanized" mice express human β_3-AR mRNA in brown adipose tissue, but not in other sites, and do not express murine β_3-ARs. Therefore, these animals should assist in the search for effective β_3-agonists, and should be useful as a means of locating human promoter/enhancer elements which confer tissue specificity and regulation on human β_3-AR gene expression.

Summary

Transgenic and gene knockout techniques are beginning to impact on obesity research. Because these approaches permit one to make molecular alterations *in vivo*, it is anticipated that they will lead to important contributions in the future. Also, as other gene mutations responsible for producing obesity are identified, genetic engineering in mice is likely to play an important part in unraveling the mechanisms by which these mutations produce obesity. Finally, further refinements in methodology such as inducible promoter systems (chapter by L. Hennighahsen et al.), tissue-specific gene knockout techniques, and controlled gene knockout methods (chapter by S. Viville) will significantly increase the value of these approaches in the future.

References

Arch, J.R.S., Ainsworth, A.T., Cawthorne, M.A., Piercy, V., Sennitt, M.V., Thody, V.E., Wilson, C. and Wilson, S. (1984) Atypical β-adrenoceptor on brown adipocytes as target for anti-obesity drugs. *Nature* **309**, 163–165

Boyer, B. and Kozak, L.P. (1991) The mitochondrial uncoupling protein gene in brown fat: correlation between Dnase I hypersensitivity and expression in transgenic mice. *Mol. Cell Biol.* **11**, 4147–4156

Bultman, S.J., Michaud, E.J. and Woychik, R.P. (1992) Molecular characterization of the mouse agouti locus. *Cell* **71**, 1195–1204

Cassard-Doulcier, A.M., Gelly, C., Fox, N., Schrementi, J., Raimbault, S., Klaus, S., Forest, C., Bouillaud, F. and Ricquier, D. (1993) Tissue-specific and β-adrenergic regulation of the mitochondrial uncoupling protein gene: control by *cis*-acting elements in the 5'-flanking region. *Mol Endocrinology* **7**, 497–506

Christy, R.J., Yang, V.W., Ntambi, J.M., Geiman, D.E., Lanschulz, W.M., Friedman, A.D., Nakabeppu, Y., Kelly, T.J. and Lane, M.D. (1989) Differentiation-induced gene expression in 3T3-L1 preadipocytes: CCAAT/enhancer binding protein interacts with and activates the promoters of two adipocyte-specific genes. *Genes Dev.* **3**, 1323–1335

Fox, N., Crooke, R., Hwang, L.S., Schibler, U., Knowles, B.B. and Solter, D. (1989) Metastatic hibernomas in transgenic mice expressing an α-amylase-SV40 T antigen hybrid gene. *Science* **244**, 460–463

Frederich, R.C., Löllmann, B., Hamann, A., Napolitano-Rosen, A., Kahn, B.B. and Flier, J.S. (1995) Expression of ob mRNA and its encoded protein in rodents: impact of nutrition and obesity. *J. Clin. Invest* **96**, 1658–1663

Funahashi, T., Shimomura, I., Hiraoka, H., Arai, T., Takahashi, M., Nakamura, T., Nozaki, S., Yamashita, S., Takemura, K., Tokunaga, K. and Matsuzawa, Y. (1995) Enhanced expression of rat obese (ob) gene in adipose tissue of ventromedial hypothalamus (VMH)-lesioned rats. *Biochem. Biophys. Res. Comm.* **211**, 469–475

Hamann, A., Flier, J.S. and Lowell, B.B. (in press) Decreased brown fat markedly enhances susceptibility to diet-induced obesity, diabetes and hyperlipidemia. *Endocrinology*

Himms-Hagen, J. (1989) Brown adipose tissue thermogenesis and obesity. *Prog. Lipid. Res.* **28**, 67–115

Katz, E.B., stenbit, A.E., Hatton, K., DePinho, R. and Charron, M.J. (1995) Cardiac and adipose tissue abnormalities but not diabetes in mice deficient in GLUT4. *Nature* **377**, 151–155

Klebig, M.L., Wilkinson, J.E., Geisler, J.G. and Woychik, R.P. (1995) Ectopic expression of the agouti gene in transgenic mice causes obesity, features of type II diabetes, and yellow fur. *Proc. Natl. Acad. Sci. USA* **92**, 4728–4732

Kozak, L.P., Kozak, U.C. and Clarke, G.T. (1991) Abnormal brown and white fat development in transgenic mice overexpressing glycerol 3-phosphate dehydrogenase. *Genes and Dev.* **5**, 2256–2264

Kozak, U.C., Hekd, W., Kreutter, D. and Kozak, L.P. (1992) Adrenergic regulation of the mitochondrial uncoupling protein gene in brown fat tumor cells. *Mol. Endocrinology* **6**, 763–772

Kozak, L.P., Kozak, U. and Kpecky, J. (1994) Genes of energy balance: modulation in transgenic mice. *J. Cellular Biochem.* Supplement **18A**, 155

Lowell, B.B., S-Susulic, V., Hamann, A., Lawitts, J.A., Himms-Hagen, J., Boyer, B.B., Kozak, L.P. and Flier, J.S. (1993) Development of obesity in transgenic mice after genetic ablation of brown adipose tissue. *Nature* **366**, 740–742

Lu, D., Willard, D., Patel, I.R., Kadwell, S., Overton, L., Kost, T., Luther, M., Chen, W., Woychik, R.P., Wilkison, W.O. and Cone, R.D. (1994) Agouti protein is an antagonist of the melanocyte-stimulating-hormone receptor. *Nature* **371**, 799–802

Maffei, M., Fei, H., Lee, G.-H., Dani, C., Leroy, P., Zhang, Y., Proenca, R., Negrel, R., Ailhaud, G. and Friedman, J.M. (1995) Increased expression in adipocytes of ob RNA in mice with lesions of the hypothalamus and with mutations at the *db* locus. *Proc. Natl. Acad. Sci. USA* **92**, 6957–6960

Miller, M.W., Duhl, D.M.J., Vrieling, H., Cordes, S.P., Ollmann, M.M., Winkes, B.M. and Barsh, G.S. (1993) Cloning of the mouse *agouti* gene predicts a secreted protein ubiquitously expressed in mice carrying the *lethal yellow* mutation. *Genes and Dev.* **7**, 454–467

Naggert, J.K., Fricker, L.D., Varlamov, O., Nishina, P.M., Rouille, Yl, Steiner, D.F., Carroll, R.J., Paigen, B.J. and Leiter, E.H. (1995) Hyperproinsulinaemia in obese fat/fat mice associated with a carboxypeptidase E mutation which reduces enzyme activity. *Nature Genetics* **10**, 135–142

Newsholme, E.A. (1978) Substrate cycles: their metabolic, energetic and thermic consequences in man. *Biochem. Soc. Symp.* **43**, 183–205

Nicholls, D.G. and Locke, R.M. (1984) Thermogeneic mechanisms in brown fat. *Physol. Rev.* **64**, 1–64

Ogawa, Y., Masuzaki, H., Isse, N., Okasaki, T., Mori, K., Shigemoto, M., Satoh, N., Tamura, N., Hosoda, K., Yoshimasa, Y., Jingami, H., Kawada, T. and Nakao, K. (1995) Molecular cloning of rat *obese* cDNA and augmented gene expression in genetically obese Zucker faty (*fa/fa*) rats. *J. Clin. Invest.* **96**, 1647–1652

Perry, W.L., Hustad, C.M., Swing, D.A., Jenkins, N.A. and Copeland, N.G. (1995) A transgenic mouse assay for agouti protein activity. *Genetics* **140**, 267–274

Rink, T.J. (1994) In search of a satiety factor. *Nature* **372**, 406–407

Ross, S.R., Graves, R.A., Greenstein, A., Platt, K.A., Shyu, H., Mellovitz, B. and Spiegelman, B.M. (1990) A fat-specific enhancer is the primary determinant of gene expression for adipocyte P2 *in vivo*. *Proc. Natl. Acad. Sci. USA* **87**, 9590–9594

Ross, S.R., Choy, L., Graves, R.A., Fox, N., Soleveva, V., Klaus, S., Ricquier, D. and Spiegelman, B.M. (1992) Hybrinoma formation in transgenic mice and isolation of a brown adipocyte cell line expressing the uncoupling protein gene. *Proc. Natl. Acad. Sci. USA* **89**, 7561–7565

Rothwell, N.J. and Stock, M.J. (1979) A role for brown adipose tissue in diet induced thermogenesis. *Nature* **281**, 31–35

Shepherd, P.R., Gnudi, L., Toozo, E., Yang, H., Leach, F. and Kahn, B.B. (1993) Adipose cell hyperplasia and enhanced glucose disposal in transgenic mice overexpressing GLUT4 selectively in adipose tissue. *J. Biol. Chem.* **268**, 22243–22246

Soloveva, V., Graves, R.A., Spiegelman, B.M. and Ross, S.R. (1994) Transgenic mice with altered adrenergic receptor levels in adipose tissue. *J. Cellular Biochem.* supplement **18A**, 172

Susulic, V.S., Frederich, R.C., Lawitts, J., Tozzo, E., Kahn, B.B., Harper, M.-E., Himms-Hagen, J., Flier, J.S. and Lowell, B.B. (1995) Targeted disruption of the β_3-adrenergnic receptor gene. *J. Biol. Chem.* **270**, 29483–29492

Tecott, L.H., Sun, L.M., Akana, S.F., Strack, A.M., Lowenstein, D.H., Dallman, M.F. and Julius, D. (1995) Eating disorder and epilepsy in mice lacking 5-HT$_{2c}$ serotonin receptors. *Nature* **374**, 542–546

Tontonoz, P., Hu, E., Graves, R.A., Budavari, A.I. and Spiegelman, B.M. (1994a) mPPARγ2: tissue-specific regulator of an andipocyte enhancer. *Genes Dev.* **8**, 1224–1234

Tontonoz, P., Hu, E., and Spiegelman, B.M. (1994b) Stimulation of adipogenesis in fibroblasts by PPARγ2, a lipid-activated transcription factor. *Cell* **79**, 1147–1156

Zemel, M.B., Kim, J.H., Woychik, R.P., Michaud, E.J., Kadwell, S.H., Patel, I.R. and Wilkison, W.O. (1995) Agouti regulation of intracellular calcium: role in the insulin resistance of vaible yellow mice. *Proc. Natl. Acad. Sci. USA* **92**, 4733–4737

Zhang, Y., Proenca, R., Maffel, M., Barone, M., Leopold, L. and Friedman, J.M. (1994) Positional cloning of the mouse *obese* gene and its human homologue. *Nature* **372**, 425–432

PART IV, SECTION B

66. The Generation and Use of Transgenic Animals for Xenotransplantation

Jeffrey L. Platt[1,*] and John S. Logan[2]

[1]Departments of Surgery, Pediatrics and Immunology, Box 2605 DUMC, Duke University, Durham, North Carolina 27710, USA
[2]Nextran, Princeton, New Jersey

Introduction

Transplantation is the preferred treatment for chronic failure of each of the major organs; yet clinical application is severely restricted, not by biological hurdles, but by a shortage of donors. The dimensions of this shortage are such that fewer than 50% of the procedures needed can ever be carried out (Evans et al., 1992). One potential solution to this problem is to use animal organs in lieu of human organs for transplantation, that is xenotransplantation. Xenotransplantation has been carried out at various times using non-human primates as donors. Although use of non-human primates has enjoyed some very limited success, the very large number organs needed impels consideration of non-primates, such as the pig as organ donors.

The major hurdle to transplanting organs of non-primates into humans is rejection. A porcine organ transplanted into an unmanipulated primate or into a human is rejected in minutes to a few hours by a process known as hyperacute rejection (Platt et al., 1991). Hyperacute rejection is mediated by the binding to the graft of xenoreactive natural antibodies of the recipient leading to activation of the complement system. Complement activation on blood vessels in the newly transplanted organ causes hemorrhage, edema and thrombosis leading to destruction of the organ. If hyperacute rejection is avoided, by depleting natural antibodies or inactivating complement, the graft is subject to rejection over a period of days to weeks by a process known as acute vascular xenograft rejeection (Leventhal et al., 1993). The cause of acute vascular xenograft rejection has not been determined with certainty, but it seems to involve the action xenoreactive antibodies and perhaps of small amounts of complement on the endothelial lining of blood vessels leading to procoagulant and pro-inflammatory changes (Saadi et al., 1995). Beyond acute vascular rejection there are likely to be other types of rejection, such as cellular rejection, which are seen in allografts. Although these more delayed types of rejection may prove limiting there is some evidence that currently available immunosuppressive therapies can deal effectively with them.

The Pathogenesis of Xengraft Rejection as a Basis for Therapy

During the past several years, the components of the host's immune system that would cause the rejection of a porcine organ by a human recipient have been elucidated (Figure 66.1). These components – xenoreactive natural antibodies and the complement system – constitute a potent defense against invasive microorganisms; in fact, individuals with deficiencies of natural antibodies or complement may be subject to overwhelming infection. Because of the critical role of natural immunity in host defense, there has been much interest in the use of genetic or biochemical techniques to treat the xenograft donor rather than focusing therapeutic strategies on the recipient. This chapter will review current knowledge and our experience in this area, focusing especially on how transgenic techniques might be applied to overcoming the hurdles to xenotransplantation posed by natural immunity.

*Author for Correspondence.

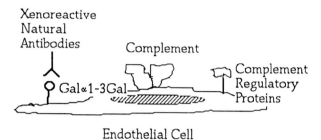

Figure 66.1. Components of the host's natural immunity involved in rejection of xenografts. The rejection of a xenograft is initiated by binding of xenoreactive antibodies of the recipient to endothelial cell antigens, predominantly Galα1-3Gal. Antibody binding activates the complement system of the recipient leading to deposition of complement in the graft. Ordinarily, complement-mediated injury is inhibited by complement regulatory proteins; however, these proteins do not effectively control foreign complement, thus the reaction can proceed to destroy the graft.

Use of Transgene to Decrease Antigen Expression

The rejection of a porcine xenograft by a primate is initiated by xenoreactive IgM antibodies directed predominantly against a disaccharide, Galα1-3Gal (Good et al., 1992 and Sandrin et al., 1993). The synthesis of Galα1-3Gal is catalyzed by α1,3-galactosyltransferase which is expressed in functional form by lower animals such as pigs (Galili et al., 1991). This enzyme adds galactose via an α-linkage to Galβ1-4GlcNAc-R to yield Galα1-3Galβ1-4GLcNAc-R (Figure 66.2). Humans, apes and Old World monkeys do not have a functional α1,3-galactosyltransferase gene but have two pseudo-genes; these species make natural antibodies specific for Galα1-3Gal, whereas lower animals do not.

Some investigators have suggested that the α1,3-galactosyltransferase gene might eventually be disrupted by gene targeting (Cooper et al., 1993). However, gene targeting by homologous recombination in embryonic stem cells has not been possible in pigs due to the lack of an appropriate pluripotent embryonic stem cell. Sandrin and Hutchinson have suggested an alternative approach, the expression as the product of transgenes of other glycosyltransferases which would compete with α1,3galactosyltransferase for synthesizing the terminus of oligosaccharide chains. For example, α1,2-fucosyltransferase (H transferase) utilizes Galβ1-4GlcNAc-R as does α1,3galactosyltransferase; its expression in a cell might divert synthesis away from Galα1-3Galβ1-4GlcNAc to Fucα1-2Galb1-4GlcNAc (H antigen) (Figure 66.2). Studies in cultured cells and in transgenic mice suggest that this strategy is, infact, effective-porcine and murine cells expressing human H transferase make substantially less Galα1-3Gal (Sandrin et al., 1995). A. Sharma with us applied the same concept to the development of transgenic pigs, expressing H transferase under control of H-2Kb and chick β-actin promoters (Sharma et al., 1995). Preliminary analysis of founder pigs reveals that a very significant decrease in expression of Galα1-3Gal has been achieved. The extent to which this approach actually decreases the binding of complement fixing xenoreactive antibodies to porcine cells and the implications for xenotransplantation remain to be tested; however, it seems not unlikely given the preliminary results that the method will be shown to decrease if not eliminate the humoral reactions initiated by natural antibodies.

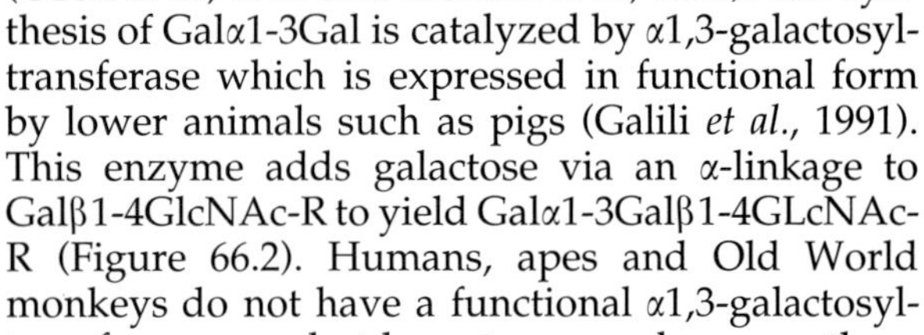

Complement Regulation in Xenotransplantation

Transgenic animals have also been developed to address the problem of complement in xenotransplantation. The complement system is a series of more than twenty proteins in the plasma which kill foreign organisms and facilitate their phagocytosis (Figure 66.3). Activation of the complement system leads to activation of the components C3 and C4 in plasma and deposition on the surfaces of cells. Although deposition of complement may be specifically targeted by antibodies to the foreign cell surface, the solution phase reactions can lead to deposition of complement on autologous cells. The ability of the complement system to selectively damage foreign organisms and spare host cells thus depends in part on the functioning of complement regulatory proteins, of which more than ten have been described. These proteins inhibit complement activation on the surface of autologous cells. Some complement regulatory proteins function far more effectively against homologous complement than

$$\text{Galβ1-4GalNAc-R + UDP-Gal} \xrightarrow{\alpha 1,3\ GT} \text{Galα1-3Galβ1-4GlcNAc-R}$$

$$\text{+ GDP-Fuc} \xrightarrow{\alpha 1,2\ FT} \text{Fucα1-2Galβ1-4GlcNAc-R}$$

Figure 66.2. Synthesis of Galα1-3Gal, the major antigen recognized by human xenoreactive antibodies. Synthesis of Galα1-3Gal by α1,3-galactosyltransferase may be inhibited by the presence of other glycosyltranferases, such as α1,2-fucosyltransferase, which diverts synthesis of oligosaccharide chains toward the addition of a different terminal sugar.

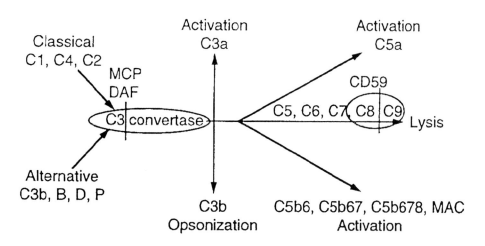

Figure 66.3. The complement system and its control. The complement system may be activated via the classical or alternative pathways to yield C3 convertase, the pivotal enzyme complex. C3 convertase is controlled by various regulators including the cell surface molecules membrane cofactor protein (MCP) and decay accelerating factor (DAF). Activation of the complement system generates various effector molecules including the anaphylotoxins (C3a and C5a), an opsonin (C3bi) and lytic complexes (the membrane attack complex or MAC). The formation of lytic complexes is controlled by a cell membrane protein, CD59.

against heterologous complement. As a result, the exposure of a cell to heterologous complement may lead to damage or lysis of the cell, while exposure to homologous complement under the same conditions has little effect. The importance of complement regulatory proteins in maintaining tissue homeostasis is suggested by the fact that absence of decay accelerating factor and CD59 on erythrocytes leads to the clinical syndrome of paroxysmal nocturnal hemoglobinuria (Nicholson-Weller et al., 1983; Pangburn et al., 1983). The importance of complement regulatory proteins in maintaining tissue integrity is suggested by the work of Matsuo et al., who found that administration of F(ab')2 or Fab fragments of antibodies specific for the rat homologue of decay accelerating factor causes a rapid deposition of C3 on blood vessel walls, alteration in vascular tone and permeability and loss of leukocytes and platelets from the circulation (Matsuo et al., 1994).

Expression of Human Complement Regulatory Proteins in Heterologous Cells

Dalmasso (Dalmasso et al., 1991 and Platt et al., 1990) and Miyagawa (Miyagawa et al., 1988) proposed that a xenograft would be more susceptible to complement-mediated injury than an allograft because complement regulatory proteins expressed in the xenograft would fail to control activation of the recipient's complement system. Based on this concept, Dalmasso isolated human decay accelerating factor and inserted it through its phosphatidylinositol anchor directly into the cell membranes of cultured porcine endothelial cells. Decay accelerating factor caused dose-dependent inhibition of complement-mediated cytotoxicity and inhibition of iC3b-mediated neutrophil adhesion.

Expression of Human Complement Regulatory Proteins in Transgenic Mice

To overcome the problem posed by the species-restricted functioning of complement regulatory proteins, various groups have undertaken the development of transgenic animals which would express complement regulatory proteins of recipient origin. As a preliminary step, Cary and White developed transgenic mice expressing human decay accelerating factor in various tissues under control of the endogenous promoter of decay accelerating factor (Cary et al., 1993). The cells of the transgenic mice were found to resist lysis by human complement. We developed transgenic mice expressing human decay accelerating factor, CD59 and membrane co-factor protein using a variety of promoters to control expression (Kooyman et al., 1995; Diamond et al., 1995b and Byrne et al., 1995a). Three modes of expression were tested – indirect expression using globin promoters, direct expression using heterologous promoters and direct expression using modified homologous promoters. As a first approach, Kooyman used the human α-globin promoter which yields erythroid-specific expression in pigs (Logan *et al.*, 1994), to express human CD59 and decay accelerating factor in transgenic mice (Kooyman *et al.*, 1994). As expected, the PI-linked proteins were expressed in murine erythrocytes. Surprisingly, the human proteins were also found on endothelial cells of murine heart, kidney and liver (Kooyman *et al.*, 1995). The mechanism mediating expression in parenchymal organs was not direct expression of the human proteins, for the murine tissues free of hematopoietic cells had no detectable RNA for these proteins. Rather, the PI-linked proteins were shown to transfer passively to endothelium (Kagan *et al.*, 1994). The functional properties of the transferred protein were tested by McCurry (McCurry *et al.*, 1995b). When organs containing human decay accelerating factor and CD59 expressed by the α-globin promoter were perfused with human plasma, the activation of complement at the level of C8/C9 was nearly completely inhibited; similar results were achieved when the organs were connected to baboons through an *ex vivo* circuit. The results provided the first evidence that expression of human complement regulatory proteins in a transgenic animal might help to control the activation of complement when the organs of such an animal are exposed to human complement. The observation that GPI-linked proteins could transfer spontaneously from erythrocytes to endothelial cells also offered a novel approach to modifying the composition of endothelial cell membranes.

Diamond *et al.* used another strategy to achieve expression of human complement regulatory proteins in transgenic animals. The CD59 gene was re-structured as recently described (Diamond *et al.*, 1994) to yield a minigene which was used for expression of human CD59 and membrane cofactor protein. This construct produced widespread, but low-level constitutive expression of the genes in various parenchymal organs. Although the human genes were expressed at significantly lower levels than in human tissues, the organs from the transgenic mice exhibited significantly improved ability to control activation of human and baboon complement (Diamond *et al.*, 1995a).

As a third approach, Byrne used the chick β-actin promoter to control expression of human CD59 and the H-2K promoter to control expression of human DAF in transgenic mice (Byrne *et al.*, 1995a). The organs from these animals expressed both of the human proteins and in *ex vivo* perfusion models exhibited resistance to the activation of human and baboon complement (McCurry *et al.*, 1995b). It was thus clear from these experiments that these proteins could control the activation of human and baboon complement; however, whether they might contribute to preventing hyperacute rejection remained to be analyzed by the development of transgenic pigs.

Expression of Human Complement Regulatory Proteins in Transgenic Pigs

Testing whether the homologous restriction of complement regulatory proteins would actually contribute to the susceptibility of an organ to hyperacute rejection and whether this problem could be overcome by "genetic engineering" would require the development of trangenic pigs whose organs could be used for xenografts. One effort to develop such animals was that of White and colleagues who developed transgenic pigs expressing human DAF under control of the DAF promoter (Rosengard *et al.*, 1995). Fodor *et al.* used the H-2K promoter to drive expression of CD59 in a trangenic pig (Fodor *et al.*, 1994). In both systems, lymphocytes isolated from the transgenic pigs showed resistance to lysis by human complement.

Our own efforts allowed direct evaluation of a xenograft model. Using the erythroid-specific α-globin expression system, Kooyman and Martin developed transgenic pigs (Kooyman *et al.*, 1995). The erythrocytes of the transgenic animals were shown to express the human proteins at a level comparable to the level in human erythrocytes. As in mice, the human proteins transferred spontaneously from the surface of erythrocytes to endothelium in various organs, although the level of expression was considerably below the level observed in human organs. McCurry transplanted the hearts from three of these transgenic pigs into baboons (McCurry *et al.*, 1995a). In two cases the transplanted organs did not undergo hyperacute rejection but functioned for up to 30 hours. This result was achieved despite the only transient and low level expression of the proteins which could be achieved with this system. Of greater import were the results of histologic studies which revealed that in all three cases that tissue injury was notably less than in xenografts of normal porcine hearts in unmodified baboons.

More recently, hearts from transgenic pigs expressing human CD59 and decay accelerating factor under control of the chick β-actin and H-Kb promoters were transplanted into baboons (Byrne *et al.*, 1995b). These organs, with the human proteins expressed constitutively, functioned for days and biopsies of the transplants revealed a notable decrease in the amoung of complement deposited in the grafts and striking protection from complement-mediated tissue damage.

The results of our initial studies involving the transplantation of organs from pigs expressing human complement regulatory proteins suggested that these proteins play a very important role in maintaining the integrity of a graft in the early post-transplant period and that the species specificity of the proteins is an important aspect of the susceptibility to xenograft rejection (Platt, 1995). A second conclusion concerns our finding that the transgenic organ xenografts exhibited such significant protection even though in some cases, terminal complement complexes were deposited in the organ. This result is consistent with the idea that under conditions in which complement is activated in a graft, as for example in ABO-incompatible organ transplants, hyperacute rejection is not the invariable outcome and that intrinsic resistance to complement-mediated injury is indeed an important factor in avoiding rejection. The third and perhaps most important conclusion is that these results underscore the value of focusing therapy on the donor rather than on the recipient. Such a focus is bound to reduce the complications of immunosuppression. Recent announcements about the very prolonged survival of xenografts expressing human decay accelerating factor in primates treated with drugs for suppression of antibody synthesis (Thompson, 1995) encourage the view that the combination of therapies aimed at antibodies and at complement may be successful.

Acknowledgement

Work described in this paper was supported in part by grants from the National Institutes of Health (HL52297, HL50985 and HI46810).

References

Byrne, G.W., McCurry, K.R., Kagan, D., McClellan, S., Quinn, C., Martin, M.J., Platt, J.L. and Logan, J.S. (1995a). Protection of xenogenic endothelium from human complement by expression of CD59 or DAF in transgenic mice. *Transplantation* **60**, 1149–1156

Byrne, G.W., McCurry, K.R., Martin, M.J., McClellan, S.M., Platt, J.L. and Logan J.S. (1995b). Expression of human CD59 and DAF in transgenic pigs: intrinsic regulation of complement activity. Submitted

Cary, N., Moody, J., Yannoutsos, N., Wallwork, J. and White, D. (1993). Tissue expression of human decay accelerating factor, a regulator of complement activation expressed in mice: a potential approach to inhibition of hyperacute xenograft rejection. *Transpl. Proc.* **25**, 400–401

Cooper, D.K.C., Good, A.H., Koren, E., Oriol, R., Malcom, A.J., Ippolito, R.M., Neethling, F.A., Ye, Y., Romano, E. and Zuhdi, N. (1993). Identification of α-galactosyl and other carbohydrate epitopes that are bound by human anti-pig antibodies: relevance to discordant xenografting in man. *Transplant Immunology* **1**, 198–205

Dalmasso, A.P., Vercellotti, G.M., Platt, J.L. and Bach, F.H. (1991). Inhibition of complement-mediated endothelial cell cytotoxicity by decay accelerating factor: Potential for prevention xenograft hyperacute rejection. *Transplantation* **52**, 530–533

Diamond, L.E., Oldham, E.R., Platt, J.L., Waldman, H., Tone, M., Walsh, L.A. and Logan, J.S. (1994). Cell and tissue specific expression of a human CD59 minigene in transgenic mice. *Transpl. Proc.* **26**, 1239

Diamond, L.E., McCurry, K.R., Oldham, E.R., McClellan, S.B., Martin, M.J., Platt, J.L. and Logan, J.S. (1996). Characterization of transgenic pigs expressing functionally active human CD59 on cardiac endothelium **61**, 1241–1249

Diamond, L.E., McCurry, K.R., Platt, J.L., Oldham, E.R., Tone, M., Waldmann, H. and Logan, J.S. (1995). A human CD59 minigene encodes a biologically functional complement regulatory protein in transgenic mice. *Transplant Immunology* **3**, 305–312

Evans, R.W., Orians, C.E. and Ascher, N.L. (1992). The potential supply of organ donors: an assessment of the efficiency of organ procurement efforts in the United States. *JAMA* **267**, 239–246

Fodor, W.L., Williams, B.L., Matis, L.A., Madri, J.A., Rollins, S.A., Knight, J.W., Velander, W. and Squinto, S.P. (1994). Expression of a functional human complement inhibitor in a transgenic pig as a model for the prevention of xenogeneic hyperacute organ rejection. *Proc. Natl. Acad. Sci. USA* **91**, 11153–11157

Galili, U. and Swanson, K. (1991). Gene sequences suggest inactivation of α-1, 3-galactosyltransferase in catarrhines after the divergence of apes from monkeys. *Proc. Natl. Acad. Sci. USA* **88**, 7401–7404

Good, A.H., Cooper, D.K.C., Malcolm, A.J., Ippolito, R.M., Koren, E., Neethling, F.A., Ye, Y., Zuhdi, N. and Lamontagne, L.R. (1992). Identification of carbohydrate structures that bind human antiporcine antibodies: implications for discordant xenografting in humans. *Transpl. Proc.* **24**, 559–562

Kagan, D.T., Platt, J.L., Logan, J.S. and Byrne, G.W. (1994). Expression of complement regulatory factors using heterologous promoters in transgenic mice. *Transpl. Proc.* **26**, 1242

Kooyman, D., Byrne, G.W., McClellan, S., Nielsen, D.L., Kagan, D.T., Coffman, T., Masahide, T., Waldmann, H., Platt, J.L. and Logan, J.S. (1994). Erythroid-specific expression of human CD59 and transfer to vascular endothelial cells. *Transpl. Proc.* **26**, 1241

Kooyman, D.L., Byrne, G.W., McClellan, S., Nielsen, D., Tone, M., McCurry, K.R., Coffman, T.M., Waldmann, H., Platt, J.L. and Logan, J.S. (1995). In vivo transfer of GPI-linked complement restriction factors from erythrocytes to the endothelium. *Science* **269**, 89–92

Leventhal, J.R., Matas, A.J., Sun, L.H., Reif, S., Bolman, R.M., III, Dalmasso, A.P. and Platt, J.L. (1993). The immunopathology of cardiac xenograft rejection in the guinea pig to rat model. *Transplantation* **56**, 1–8

Logan, J.S. and Martin, M.J. (1994). Transgenic swine as a recombinant production system for human hemoglobin. *Methods Enzymol.* **231**, 435–445

Matsuo, S., Ichida, S., Takizawa, H., Okada, N., Baranyi, L., Iguchi, A., Morgan, B.P. and Okada, H. (1994). In vivo effects of monoclonal antibodies that functionally inhibit complement regulatory proteins in rats. *J. Exp. Med.* **180**, 1619–1627

McCurry, K.R., Kooyman, D.L., Alvarado, C.G., Cotterell, A.H., Martin, M.J., Logan, J.S. and Platt, J.L. (1995a). Human complement regulatory proteins protect swine-to-primate cadriac xenografts from humoral injury. *Nature Medicine* **1**, 423–427

McCurry, K.R., Kooyman, D.L., Diamond, L., Byrne, G., Logan, J. and Platt, J.L. (1995b). Transgenic expression of human complement regulatory proteins in mice results in diminished complement deposition during organ xenoperfusion. *Transplantation* **59**, 1177–1182

Miyagawa, S., Hirose, H., Shirakura, R., Naka, Y., Nakata, S., Kawashima, Y., Seya, T., Matsumoto, M., Uenaka, A. and Kitamura, H. (1988). The mechanism of discordant xenograft rejection. *Transplantation* **46**, 825–830

Nicholson-Weller, A., March, J.P., Rosenfeld, S.I. and Austen, F. (1983). Affected erythrocytes of patients with paroxysmal nocturnal hemoglobinuria are deficient in the complement regulatory protein, decay accelerating factor. *Proc. Natl. Acad. Sci. USA* **80**, 5066–5070

Pangburn, M.K., Schreiber, R.D. and Muller-Eberhard, H.J. (1983). Deficiency of an erythrocyte membrane protein with complement regulatory activity in paroxysmal nocturnal hemoglobinuria. *Proc. Natl. Acad. Sci. USA* **80**, 5430–5434

Platt, J.L., Vercellotti, G.M., Dalmasso, A.P., Matas, A.J., Bolman, R.M., Najarian, J.S. and Bach, F.H. (1990). Transplantation of discordant xenografts: a review of progress. *Immunol. Today* **11**, 450–456

Platt, J.L., Fischel, R.J., Matas, A.J., Reif, S.A., Bolman, R.M. and Bach, F.H. (1991). Immunopathology of hyperacute xenograft rejection in a swine-to-primate model. *Transplantation* **52**, 214–220

Platt, J.L. *Hyperacute xenograft rejection,* Austin: R.G. Landes Company, 1995

Rosengard, A.M., Cary, N.R.B., Langford, G.A., Tucker, A.W., Wallwork, J. and White, D.J.G. (1995). Tissue expression of human complement inhibitor, decay accelerating factor, in transgenic pigs. *Transplantation* **59**, 1325–1333

Saadi, S., Ihrcke, N.S. and Platt, J.L. (1995) Pathophysiology of xenograft rejection. R. Lieberman and R. Morris, (eds.), *Principles of immunomodulatory drug development in transplantation and autoimmunitiy*, New York, NY: Raven Press.

Sandrin, M.S., Vaughan, H.A., Dabkowski, P.L. and McKenzie, I.F.C. (1993). Anti-pig IgM antibodies in human serum react predominantly with Galα(1,3)Gal epitopes. *Proc. Natl. Acad. Sci. USA* **90**, 11391–11395

Sandrin, M.S., Fodor, W.L., Mouhtouris, E., Osman, N., Cohney, S., Rollins, S.A. and Guilmette, E.R., Setter, E.,Squinto S.P., McKenzie, I.F.C. (1995). Enzymatic remodelling of the carbohydrate surface of a xenogenic cell substantially reduces human antibody binding and complement-mediated cytolysis. *Nature Medicine* **1**, 1261–1267

Sharma, A., Okabe, J.F., Birch, P., Platt, J.L. and Logan, J.S. (1996). Reduction in the level of Gal (α1,3) Gal in transgenic mice and pigs by the expression of an α1,2 fucosyltransferase. *Pro. Natl. Acad. Sci. USA*, in press

Thompson, C. (1995). Humanised pigs hearts boost xenotransplantation. *Lancet* **346**, 766

67. The Preparation of Recombinant Proteins from Mouse and Rabbit Milk for Biomedical and Pharmaceutical Studies

Marie Georges Stinnakre[1], Micheline Massoud[2], Céline Viglietta[3] and L.M. Houdebine[3]

[1] *Laboratoire de Génétique Biochimique,* [2] *Laboratoire de Physiologie Animale*
[3] *Unité de Différentiation Cellulaire, Institut National de la Recherche Agronomique, 78352 Jouy-en-Josas, Cedex, France*

Preparation of recombinant proteins for biochemical studies or for biomedical use is one of the successes of biotechnology. Using milk as a possible source of recombinant proteins has been envisaged soon after the generation of the first transgenic animals. The first demonstration that this was possible was given by Simons *et al.* (1987) who succeeded in generating transgenic mice secreting sheep β-lactoglobulin in their milk. The following chapters of this book show that this process is becoming an industrial practice. Ideally, farm animals should be used to produce recombinant proteins for pharmaceutical use. Smaller animals, essentially mouse and rabbit, remain useful to prepare small amounts of proteins.

Numerous studies have shown that the efficiency of a gene construct *in vivo* cannot be predicted from studies using cell transfection (Palmiter *et al.*, 1991 and Petitclerc *et al.*, 1995). Results obtained with mice can generally be extrapolated to farm animals and the use of this laboratory species remains a compulsory test to evaluate the efficiency of a gene construct to be transferred to farm animals.

Transgenic animals are used to produce recombinant proteins not only because milk may be the source of very large amounts of proteins but also because mammary cells are expected to proceed to post-translational modifications (cleavage, glycosylation, γ-carboxylation, etc...) with good efficiency. In this respect, mouse and rabbit milk may be of some predictive value, although a certain number of data indicate that mammary gland from rodents, pig and small ruminants may have different capacity to proceed to the post-translational modifications. The direct gene transfer into the mammary gland of adult ruminants using viral vectors may be more predictive for this purpose (Archer *et al.*, 1994 and Bremel *et al.*, this book, 199).

To study biochemical properties of a protein, including crystallization, a minimum amount of this molecule must often be obtained. Milk may be one interesting source of the recombinant protein since several milligrams of foreign proteins per milliliter of milk can be obtained (see the following papers). Using transgenic mice may be a relatively confortable approach, although the amount of a given recombinant protein in milk remains unpredictable.

In case the biochemical studies need more of the recombinant proteins or if preclinical studies of the protein are necessary, transgenic rabbit may be used. Lactating rabbits provide to their offspring daily more than 200 ml of milk which is very rich in proteins. These animals can thus be the source of relatively large amount of recombinant protein. In case a pharmaceutical protein has to be produced in quantity not larger than one kilogram per year the use of transgenic rabbits may be envisaged at an industrial scale.

The Direct Milk Collection from Isolated Mammary Gland

Mouse milking is relatively time consuming and of limited efficiency. An observation done in our laboratory many years ago showed us that milk stored in the mammary gland is spontaneously and almost quantitatively released when the isolated tissue is incubated on ice. Under the influence of low temperature, the mammary gland contracts and releases its milk spontaneously and quantitatively. The release is complete after an overnight incubation and it is almost achieved after only 4 hours. During incubation, mammary gland may be kept in a Petri dish. More efficiently, the tissue can be incubated in cold phosphate buffer standard to which anti-proteases can be added. The buffer helps the milk which is somewhat viscous, to be more efficiently released from the gland.

This protocol was used with mice harbouring a construct containing the rabbit whey acidic promoter and the human growth hormone gene. This construct proved to be quite efficient, leading a mean hGH concentration of 8 mg per ml milk in different mouse lines (Devinoy et al., 1994). To optimize milk collection, the transgenic lactating mice were separated from their offspring for 24 hours to obtain the maximum milk accumulation. The animals were then sacrificed and the mammary gland was collected and incubated in cold. About 1.5 ml of milk was obtained from each animal. Polyacrylamide gel electrophoresis showed that the hGH released with milk was by no means degraded in these conditions, even after an overnight incubation in cold. hGH is known to have a high prolactin-like activity. The milk released from the transgenic mice after incubation of the isolated mammary gland was used to induce milk synthesis in isolated rabbit mammary cells in culture. The test showed with no ambiguity that the hGH obtained in this way was not only undegraded but fully active (Stinnakre et al., 1992).

This method is simple to use and it can be applied to other species. More than 150 ml of milk was obtained in this way from the mammary gland of individual transgenic rabbits. The major drawback of this method is of course that it implies the breeding and the sacrifice of lactating females. Their offspring can be adopted by a foster mother. The transgenic females can be used later to obtain milk.

Milking Lactating Mice

Only ruminants have a cisternae in their mammary gland where a major part of the milk is stored between each milking. In these species, the only barrier to get milk is a sphincter. This barrier can be crossed easily just by using vacuum. The milk stored within the mammary ducts and alveoli can be obtained only after an ejection reflex which is triggered by oxytocin. Oxytocin is secreted from posterior pituitary when the nipples are stimulated by offspring or by a milking machine. A nervous pathway triggers oxytocin secretion from pituitary. Oxytocin induces a contraction of myoepithelial cells in the mammary gland. The resulting internal pressure provokes a spontaneous milk release from the mammary gland which requires only a moderate vacuum. The ejection reflex is quite easily blocked as soon as the female is stressed. In case of stress, part of the blockade takes place at the mammary gland level itself. Injection of oxytocin is then only weakly efficient.

In practice, lactating mice at any stage of lactation are separated from their offspring for one to 24 hours to favour milk accumulation in the mammary gland. Various amounts of oxytocin (0.3 to 5 IU) are injected intraperitoneally to mice according to the quantity of milk to be collected. Mice may then be mildly anaesthetized. Milk can be collected a few minutes later. When small amounts of milk are wanted, secretion can be provoked by a gentle massage of the mammary glands. Milk can be collected using a Pasteur pipette. When large amounts of milk are required, aspiration with a controlled vacuum can be done with a plastic tube applied to nipples. Up to 1–1.5 ml milk can be obtained in this way. These procedures can be repeated during a lactation period. The amount of milk and the frequency of collections must be adjusted in such a way as to leave enough milk for the pups.

Milking Lactating Rabbits

Milk collection is essentially similar in rabbit and mouse. Rabbit females naturally milk their offspring only once per day. The lactating females can thus be isolated from their offspring for one day without any damage for the mammary gland. The rabbit is very sensitive to stress and milk secretion is often blocked even after injection of oxytocin. A stressed female which failed to give milk delivers it soon after and with a quite good efficiency to its offspring. Manipulating the females the days before milking to accustom them to the protocol significantly reduces the stress. A female thus usually gives its milk more efficiently after a few days of milking. Anesthetizing the animals seems not very helpful. Injecting substances like chlorpromazine may contribute to reduce the stress and it may favour lactation by inducing prolactin secretion. Oxytocin 5 IU is injected intramuscularly. Injection of oxytocin intravenously is more difficult to do and of lower efficiency due to the very short half-life of the hormone. This observation was also done by Duby et al. (1993). A moderate vacuum must then be applied to nipples to collect milk. Both continuous and alternative vacuum can be used. Plastic tubes adapted to the diameter of the nipples are connected to the source of vacuum. Any exceeding vacuum does not allow a better milk collection. The opposite is even observed. An excess of vacuum causes some suffering to the animal and finally a stress and a significant damage to the nipples. Aspiration can be performed simultaneously to several nipples on condition the source of vacuum is sufficient. After oxytocin injection and until the end of milking, the animals are preferably kept in a containment device which prevents them from moving while leaving an easy access to nipples.

Duration of milking is of about 30 minutes. The mean amount of milk obtained per nipples is usually 10–20 ml.

Many observations done in numerous species have shown that lactation is rapidly and irreversibly impaired when milk is not quantitatively released from the mammary gland. Although in optimized conditions all the milk can be removed from the rabbit mammary gland (Lebas, 1970), it is preferable, in practice, to collect only part of the milk (no more than 100 ml). The remaining milk is sufficient to feed offspring which deliver to their mother a good physiological stimulus and succeed easily in emptying the mammary gland. This protocol also does not oblige technicians to collected milk on unworked days. The milk is then entirely collected by offspring.

Duration of lactation is of about 4–5 weeks in rabbit. Up to 1 liter of milk can be obtained in this way per lactation and per female. A rabbit female can lactate up to 8 fold per year. Transgenic rabbit can be obtained relatively easily (Viglietta et al., this book) and concentrations of recombinant proteins above 1 mg/ml have been obtained in rabbit milk. This was the case for human IGF-1 (Brem et al., 1994) and human superoxide dismutase (L. Hansson et al., unpublished data). Rabbit can thus be considered as amenable to the production of recombinant proteins at an industrial scale.

References

Archer, J.S., Kennan, W.S., Gould, M.N. and Bremel, R.D. (1994) Human growth hormone (hGH) secretion in milk of goats after direct transfer of the hGH gene into the mammary gland by using replication-defective retrovirus vectors. *Proc. Natl. Acad. Sci.* **91**, 6840–6844

Brem, G., Hartl, P., Besenfelder, U., Wolf, E., Zinovieva, N. and Pfaller, R. (1994) Expression of synthetic cDNA sequences encoding human insulin-like growth factor-1 (IGF-1) in the mammary gland of transgenic rabbits. *Gene* **149**, 351–355

Devinoy, E., Thépot, D., Pavirani, A., Stinnakre, M.G., Fontaine, M.L., Grabowski, H., Puissant, C. and Houdebine L.M. (1994) High level production of human growth hormone in the milk of transgenic mice: the upstream region of the rabbit whey acidic protein (WAP) gene targets transgene expression to the mammary gland. *Transgenic Research* **3**, 79–89

Duby, R.T., Cunniff, M.B., Belak, J.M., Balise, J.J. and Robl, J.M. (1993) Effect of milking frequency on collection of milk from nursing New Zealand white rabbits. *Animal Bio-technology* **4**, 31–42

Lebas, F. (1970) Description d'une machine à traire les lapines. *Ann. Zootech.* **19**, 223–228

Palmiter, R.D., Sandgren, E.P., Avarbock, M.R., Allen, D.D. and Brinster, R.L. (1991) Heterologous introns can enhance expression of transgenes in mice. *Proc. Natl. Acad. Sci. USA* **88**, 478–482

Petitclerc, D., Attal, J., Théron, M.C., Bearzotti, M., Bolifraud, P., Kann, G., Stinnakre, M.-G., Pointu, H., Puissant, C. and Houdebine, L.M. (1995) The effect of various introns and transcription terminators on the efficiency of expression vectors in various cultured cell lines and in the mammary gland of transgenic mice. *Journal of Biotechnology* **40**, 169–178

Simons, J.P., McClenaghan, M. and Clark, A.J. (1987) Alteration of the quality of milk by expression of sheep β-lactogobulin in transgenic mouse. *Nature* **328**, 530–532

Stinnakre, M.G., Devinoy, E., Thépot, D., Chêne, M., Bayat-Sarmadi, M., Grabowski, H. and Houdebine, L.M. (1992) Quantitative collection of milk and active recombinant proteins from the mammary glands of transgenic mice. *Animal Biotechnlogy* **3**, 245–255

Viglietta, C., Massoud, M. and Houdebine L.M. The generation of transgenic rabbits, (this book)

PART IV, SECTION C

68. Production of Complex Human Pharmaceuticals in the Milk of Transgenic Goats Using the Goats Beta Casein Promoter

P. DiTullio*, K.M. Ebert[1], J. Pollock[2], T. Edmunds[2] and H.M. Meade

[1]School of Medicine, Dental Medicine and Veterinary Medicine, Tufts University, 200 Westboro Rd, N. Grafton, MA 01536, USA
[2]Genzyme Transgenics Corp. and Genzyme Corp., 1 Mountain Rd, Framingham, MA 01701, USA

The mammary gland has been the focus of intense research over the past decade aimed at the efficient, cost effective production of complex human pharmaceuticals. Several groups have targeted the expression of heterologous proteins to the mammary glands of goats (Ebert et al., 1991), sheep (Wright et al., 1991), and cows (Krimpenfort et al., 1991) to assess the ability of the gland to synthesize complex, post-translationally modified proteins. In this section, we will discuss the production of a glycosylation variant of human tissue plasminogen activator (LAtPA) in the milk of transgenic goats.

In goats, the predominant milk protein is beta casein present at 10–20 g/L. We have previously described the cloning and characterization of the goat beta casein encoding gene (CSN2) (Roberts et al., 1992). The 18.5 kb clone contained 4.2 kb of promoter sequence and 5.3 kb of 3' flanking sequence. In transgenic mice, maximal expression of the CSN2 transgene was estimated to be 50% of the endogenous mouse beta casein mRNA. These results indicated that the CSN2 promoter isolated was sufficient for targeting high level expression to the mammary gland.

To test the ability of the goat beta casein promoter to direct high level expression of a heterologous protein to the mammary gland of transgenic goats, the LAtPA cDNA was fused between exons 2 and 7 of the goat beta casein gene (Ebert et al., 1994). The construct was designed in this fashion for several reasons. First, the fusion of the LAtPA cDNA into exon 2 instead of exon 1 allowed for the splicing of intron 1 and added the complete 5' UT region of the CSN2 gene to the LAtPA cDNA. Second, the addition of exons 7, 8 and 9 allowed for the splicing of two additional introns and the 3' UT region. The presence of introns has been shown to increase the transcriptional efficiency of a transgene (Brinster et al., 1988) and the 5' UT and 3' UT may effect mRNA stability (Blackburn et al., 1982).

The 15 kb goat beta casein-human LAtPA transgene was prepared for microinjection and microinjected into goat embryos as previously described (Selgrath et al., 1990). A total of 295 embryos were microinjected and transferred into 78 synchronized recipients out of which 42 (54%) became pregnant. A total of 60 kids were born and five were determined to be transgenic for the goat beta casein-human LAtPA transgene (3 females, 2 males). One female died at one month of age due to sudden infant death and was not analyzed further. Southern analysis of the remaining four transgenic founders showed copy number to vary from less than 1 to 16 copies (Table 68.1). In order to study genetic and expression stability, all four founder animals were bred to obtain milk for expression and progeny for germline analysis.

Genetic stability of a transgene is an important criteria in the maintenance and propagation of a herd for pharmaceutical production. The loss of transgene copies or rearrangements of the transgene can have deleterious effects on expression which will limit production capabilities. Southern analysis of progeny from all four transgenic lines showed copy number to remain stable except in the line 53–90 (Table 68.1). In goat 53–90, the copy number was found to be less than 1 copy while in goat 53–10 the copy number was estimated at 2–4.

---

*Author for Correspondence.

Table 68.1. Expression of the beta casein–human LAtPA transgene in goats.

					LAtPA Production (mg/ml)		
Founder	1st Gen.	2nd Gen.	Sex	Copy #	1st	2nd	Induced
30–89			F	1–2	2.5	2.0	1.5
	30–1		M	1–2			
		30–1–2	F	1–2	3.5		2.5
		30–1–4	F	1–2	3.0		1.5
		30–1–6	F	1–2	2.5*		1.5
53–90			M	<1			
	53–10		F	2–4	4.5		7.0
57–90			M	1			
	57–4		F	1	N.D.		2.0
	57–6		F	1	3.0		N.D.
81–90			M	12–16	2.0	1.5	N.D.
	81–1		F	12–16	1.5	1.5	1.2
		81–1–3	M	12–16			
		81–1–4	M	12–16			

N.D. = Not determined, * Pre-term abortion
Data for line 30–89 has been summarized from Ebert *et al.*, 1994 and 81–90 from Ebert and DiTullio, 1995.

The increase in copy number between the F_0 and F_1 generations is indicative of the founding animal being a mosaic. This is supported by the fact that 53–90 transmitted the transgene to only one out of ten progeny analyzed instead of the normal 50%. Germline stability of this construct will continue to be studied as additional generations and progeny become available for testing.

Following parturition, transgenic females from all four transgenic lines were milked twice daily and monitored for milk production and expression of human LAtPA. Milk production in the four lines varied greatly with regard to daily production and duration of lactation. Goats from the 81–90 line produced LAtPA at 1.5 g/L and showed a normal lactation with a peak daily production of 4 liters per day and a lactation of 80–100 days. The 53–90 line produced a higher level of LAtPA of 4.5 g/L and showed a peak daily production of 1 liter and a lactation of only 30 days. The truncated lactation appeared to correlate with expression of a high level of LAtPA. As previously described, goats expressing high levels of LAtPA displayed signs similar to mastitis which was attributed to a solubility problem of the LAtPA at concentrations above 1mg/ml and may be the cause of the shortened lactation profile (Ebert *et al.*, 1994). Goats expressing low levels of LAtPA or g/L levels of other heterologous proteins have shown normal lactations of greater than 250 days and daily production of 2–4 liters per day (unpublished data). In contrast to the variable milk production, the expression of LAtPA remained consistent throughout a single lactation as well as between lactations and generations (Table 68.1).

The evaluation of four transgenic goat lines for production is a long drawn out procedure. From the time of microinjection, a female line requires approximately 18 months before milk can be collected while a male line requires 24 months. Recently, a protocol for the induction of lactation in virgin animals has been described by Ebert *et al.* (1994). Using this protocol goats from the lines 53–90, 57–90 and 81–90 were induced to lactate. The results, which are presented in Table 68.1, support the previously published data that induction can accurately reflect the expression levels of a natural lactation. Only goat 53–10 showed a higher level of expression during induction than natural lactation which cannot presently be explained.

Characterization of the transgenically produced human LAtPA at the protein level has been limited. Western blot analysis of the milk showed the LAtPA to be produced predominantly in the two chain form (data not shown). Purification of the human LAtPA was performed as previously published (Denman *et al.*, 1991) and resulted in a similar yield and purity. Monosaccharide analysis of the transgenic LAtPA revealed less sialic acid and galactose than C127 derived material and the presence of N-acetylgalactosamine which is consistent with the results published for the low expressing

1–89 line (Denman *et al.*, 1991). The effect of these differences on protein function is not known and will require a more in depth analysis of carbohydrate structure and *in vivo* activity.

In summary, the data shows that the goat beta casein promoter is capable of directing high level expression of a heterologous protein to the mammary gland of a transgenic goat. The promoter appears unique in its ability to achieve g/L expression levels from a cDNA construct. Additional experiments demonstrate this is not specific to the LAtPA cDNA, levels of 3–4 g/L have been achieved for human anti-thrombin III (unpublished data) and 5 g/L for a human monoclonal antibody (DiTullio *et al.*, 1995) in transgenic goats utilizing cDNAs. This result, however, cannot be generalized to include all cDNAs and must be evaluated on an individual basis. Characterization of the transgenic LAtPA revealed the protein to be fully glycosylated with some differences in monosaccharide content whose effect on protein function *in vivo* is presently unknown.

References

Blackburn, D.E., Hobbs, A.A. and Rosen, J.M. (1982) Rat β-casein cDNA: sequence analysis and evolutionary comparisons. *Nucl. Acids Res.* **10**, 2295–2307

Brinster, R.L., Allen, J.M., Behringer, R.R., Gelinas, R.E. and Palmiter, R.D. (1988) Introns increase transcriptional efficiency in transgenic mice. *Proc. Natl. Acad. Sci. USA* **85**, 836–840

Denman, J., Hayes, M., O'Day, C., Edmunds, T., Bartlett, C., Hirani, S., Ebert, K.M., Gordon, K. and McPherson, J.M. (1991) Transgenic expression of a variant of human tissue-type plasminogen activator in goat milk: Purification and characterization of the recombinant enzyme. *Bio/Technology* **9**, 839–843

DiTullio, P., Ebert, K.M., Pollock, J., Pollock, D., Harvey, M., Williams, J., Wilburn, B., Friedman, B.A., Marshall, D., Barry, C., Ayer, S. and Meade, H. (1995) High level production of human monoclonal antibody in the milk of transgenic mice and a transgenic goat. *IBC Conf. on Monoclonal Antibody Production and Purification*, San Francisco, CA, April 1995

Ebert, K.M., Selgrath, J.P., DiTullio, P., Denman, J., Smith, T.E., Mushtaq, M.A., Schindler, J.E., Monastersky, G.M., Vitale, J.A. and Gordon, K. (1991) Transgenic production of a variant of human tissue-type plasminogen activator in goat milk: Generation of transgenic goats and analysis of expression. *Bio/Technology* **9**, 835–838

Ebert, K.M., DiTullio, P., Barry, C.A., Schindler, J.A., Ayres, S.L., Smith, T.E., Pellerin, L.J., Meade, H.M., Denman, J. and Roberts, B. (1994) Induction of human tissue plasminogen activator in the mammary gland of transgenic goats. *Bio/Technology* **12**, 699–702

Ebert, K.M. and DiTullio, P. (1995) The production of human pharmaceuticals in milk of transgenic animals. *The Natural Enviroment: Interdisciplinary Views: Proceedings*, Boston, MA, June 1995

Krimpenfort, P., Rademakers, A., Eyestone, W., van der Schans, A., van den Broek, S., Kooiman, P., Kootwijk, E., Platenburg, G., Pieper, F., Strijker, R. and de Boer, H. (1991) Generation of transgenic dairy cattle using 'in vitro' embryo production. *Bio/Technology* **9**, 844–847

Roberts, B., DiTullio, P., Vitale, J., Hehir, K. and Gordon, K. (1992) Cloning of the goat β-casein-encoding gene and expression in transgenic mice. *Gene* **121**, 255–262

Selgrath, J.P., Memon, M.A., Smith, T.E. and Ebert, K.M. (1990) Collection and transfer of microinjectable embryos from dairy goats. *Theriogenology* **34**, 1195–1205

Wright, G., Carver, A., Cottom, D., Reeves, D., Scott, A., Simons, P., Wilmut, I., Garner, I. and Colman, A. (1991) High level expression of active human alpha-1-antitrypsin in the milk of transgenic sheep. *Bio/Technology* **9**, 830–834

PART IV, SECTION C

69. Purification of Recombinant Proteins from Sheep's Milk

Gordon Wright and Alan Colman

PPL Therapeutics Ltd. Roslin, Edinburgh, EH25 9PP, Scotland, UK

Introduction

Proteins produced in milk through transgenic technology could be put to a number of uses, e.g., for therapeutics, foodstuffs or for research purposes. This chapter will focus on production and purification of recombinant proteins for therapeutic use, in particular those administered to humans by injection. How to obtain high standards of animal husbandary and milk collection have to be considered as well as how to obtain the product purity desired and assure it is safe and free to pathogens. This chapter will briefly attempt to address some to these issues.

Animal Husbandry and Milk Collection

Sheep like all animals, are susceptible to a number of diseases. Virus particles can be difficult to remove from protein products since, unlike bacteria, they cannot be completely removed by sub-micron filtration. Of particular importance are those virus which could potentially be infectious to humans if administred intravenously. Animals such as sheep, pigs or cows cannot be kept completely pathogen free so the highest standards of animal husbandry are essential in limiting the extent of the problem.

General control is achieved in the following ways:

1. Having a defined health status to which all new animals to be introduced must comply;
2. Maintaining producer animals under quarantine like conditions;
3. Visual inspection for ill health;
4. Laboratory and/or post mortem examination of animals suspected of ill health;
5. Collection of milk only from healthy animals.

Where there is particular concern about a specific virus measures can be put in place such as:

1. Routine serological testing of animals.
2. Viral screens on bulk milk collected.
3. Avoiding conditions and/or locations which may lead to specific disease problems.

The emotive issue of 'non-conventional virus' such as scrapie would fall into this category. Although scrapie has never been shown to be transferable to man and has not been detected in the milk of infected animals, the lack of understanding of the nature of the infective particles and the absence of a rapid detection system for their presence, has meant that there is a considerable focus on maintaing a flock which is free from scrapie. Monitoring of a flock could be by periodic culling of a certain number of animals followed by histopathologycal examination of tissues as well as regular and close veterinary examination of production animals. Care shoud be taken in the sourcing of feed and rendered animal material should certainly be avoided.

The Collection of raw product expressed by the mammary gland involves milking of animals. Prewashes of udders and special care of milking equipment can improve standards but sterile milk cannot be guaranteed on a routine basis as it is prone to environmental contamination by bacteria during the milking operation. Levels of bacteria can vary greatly from individually milked sheep from zero to 10,000 cfu/ml. However, pooled fractions obtained from well maintained milking parlours are normally much closer to the lower end of this scale.

Milk has a large somatic cell count (10^5–10^6 cells/ml) and freezing or harsh treatment of milk may result in rupturing of these cells. Milk also represents an excellent growth medium and storage of milk prior to processing is therefore an important issue.

Milk Constituents

Sheep milk is very similar to cows milk. It is high in minerals of which calcium is the most abundant at about 2 g/l. Lactose is the major carbohydrate (~ 40 g/l). These low molecular weight components of milk are relatively easy to remove and will generally be lost during processing steps aimed at separating one protein from another.

The lipid content of sheep milk is normally somewhat higher (50–70 g/l) than found is cows milk (~ 40 g/l). The protein content of sheep milk is about 50 g/l of which about 80% are caseins. The casein proteins are unusual in that they form large micellar structures which give milk its colloidal nature. The average diameter of casein micelles in milk is around 0.2 µm [1] which makes sterile filtration before casein removal particularly difficult.

The total number of different proteins in milk is not certain. However the number is extremely large. As well as proteins specifically produced by the mammary gland such as beta-lactoglobulin (~ 3 g/l), and alphalactalbumin (~ 1 g/l), there is a fairly large amount of immunoglobulin (~ 1 g/l) and almost every plasma protein appears to be present in milk although at a much lower concentration that found in blood. Of these, serum albumin is probably the major protein (~ 0.8 g/l).

Proteases are present in milk with the most prominent being plasmin [2]. However, the activity of these have, in our experience with a large number of different transgenically produced proteins, not as yet resulted in any problems of protein cleavage or loss of biological activity.

Purification

The fat content is present in milk as lipid micelles which will float on milk as cream when centrifuged. The lipid content of milk can therefore be radically reduced by using a skimming centrifuge. The liquid streams are produced (cream and skimmed milk) by these centrifuges and there is also the added advantage that a third solid phase of cells/debris remains within the skimmer. Although not sterile after this operation, the material is however more suitable for storage (e.g., after freezing) because of the greatly reduced cellular and bacterial content.

Skimming will remove 95–98% of milk lipid. Higher levels of removal can be obtained but this can result in greater product loss as the skimmed milk content of the cream increases. Levels of lipid after skimming can still be fairly high (1–4 g/l) and therefore consideration of how to remove the remaining lipid is important if the lifetime of expensive chromatography, used later in processing, is to be maximised. Filtration of milk before complete lipid removal can lead to rapid blocking of both micro and ultra filters.

Casein represents the next problem for purification. Its tendency to aggregate can lead to precipitation on chromatography columns and even without aggregation, casein micelles will readily block filters. The shear bulk of casein in milk (40 g/l) and its size and properties mean that filtration techniques for its removal can be problematic and often result in large losses of product. The dairly industry has however, shown an interest in casein removal by ultrafiltration (ceramic and organic membranes) and there is considerable literature on this available [3].

The traditional method of casein precipitation for cheese production is by addition of renin to milk. This is obtained from calves stomach. Care should however be taken before adding any material with a biological origin, since these should be from a controlled source which is equally pathogen free. Additives with a biological origin can also lead to further, perhaps variable, protein or other contamination.

An attractive method of removing casein, because of its simplicity, is to reduce the pH of milk. Caseins will precipitate by reducing the pH to 4.5. However, many protein products of interest will lose their activity at this pH and may even precipitate themselves. Another option is to precipitate by addition of an agent such as polyethylene glycol or a salt such as ammonium sulphate. If such techniques are used, a suitable concentration of additive has to be found which does not also cause the recombinant protein to come out of solution, and this will depend on the properties of the protein product of interest. It may not always be possible to find such a concentration. If a suitable partition can be found, it may be possible to convert it to a liquid/liquid two phase separation rather than liquid/solid [4].

Another option is to break up and solubilise the caseins e.g., by using EDTA [5]. This may allow early use of chromatography for casein removal.

Whatever the methods employed for lipid and casein removal, the completeness is critical, since small amounts of lipid and/or casein can lead to eventual chromatography failure. The whey fraction obtained may therefore need further clarified before going onto conventional chromatography columns. The use of fluidised or expanded bed chromatography [6] may be useful at this stage (or before) since it will tend to be less susceptible to a feed stock which is not completely clarified.

The speed of operation in getting from milk to a solution clear enough to microfilter is an important consideration since even after casein and lipid removal, the 'whey' will allow growth of any remaining bacteria.

Subsequent purification becomes more dependant on the properties of the protein product of interest (in relation of remaining whey proteins) than on the properties of milk itself. The major whey proteins have acidic iso-electric points and therefore ion-exchange chromatography of a transgenically produced protein with a basic pI could lead to a dramatic purification. However, as with the purification of recombinant proteins from other production systems such as fermentation or blood fractionation, several other forms of chromatography can give excellent results. The choice will depend on the properties of the recombinant protein and the cost of using such chromatography at a large scale.

Perhaps the most effective, because of its high specificity, is immunoaffinity chromatography. However, the antibodies used can be so expensive as to make the manufacture of the protein at commercial scale unrealistic. The antibodies may be labile, particularly if harsh chemicals must be used for cleaning between batch runs. The source of the antibody used should also be considered since it may directly or indirectly have been obtained from animal tissue, therefore bringing additional viral contamination concerns.

Once a purification process has been developed, its potential viral removal capability should be investigated. This is normally achieved by scaling down the process to laboratory size and measuring the removal of model virus and virus of specific concern which have been spiked into it.

Addition of viral killing steps to the process, such as dry heat, pasteurisation or solvent/detergent treatment should also be considered. Finding such a killing step for scrapie or BSE is extremely difficult since the type of conditions which have been demonstrated to destroy infectivity of these agents, such as incubation with 1 M-NaOH for 1 hour or autoclaving at 134–138 C for 1 hour, are seldom tolerated by proteins. Demonstration that the process is capable of removing scrapie infectivity, should it be present in milk, is therefore very important in assuring product safety.

Conclusion

Transgenic technology, although a relatively new method of production of therapeutically useful proteins, shares several features with other methods of production which have been around for many years. It is similar to other recombinant technologies in that a genetically engineered, single production line is utilised. It is similar to blood fractionation technology in that the bulk material is pooled from many individual donors. Finally, it inherits the history of safety of the use in humans of animal derived products, though like all therapeutic products of non human origin high purity is necessary to avoid immunogenic problems.

Purification of recombinant proteins from milk should be possible for almost any product, some of the purification steps will be generic to milk as a source and the rest dependant on the properties of the protein product itself.

References

1. Rose, D. and Colvin, J.R. (1966) Appearance and size of micelles from bovine milk. *J. Dairy Sci.* **49**, 1091
2. Schaar J. (1985). Plasmin activity and proteose-peptone content of individual milks. *J. Dairy Sci.* **52**, 369
3. International dairy Federation (1989) The use of ultrafiltration technology in cheesemaking. *Bulletin* **240**
4. Harris, P., Wright, G., Andrews, A.T., Pyle., D.L. and Asenjo, J.A. (1993). The application of aqueous two phase partitioning to the purification of proteins from transgenic sheep milk. *Proc. 6th European Conf. Biotech.* Ed. by L.Alberghina, Frontali L. and Sensi. P. Elsevier Press, Vol. 9, 381
5. Lin, S.H.C., Leong, S.L., Dewan., R.K., Bloomfield, V.A. and Morr, C.V. (1972) Casein: disruption of micelles by removal of calcium. *Biochemistry* **11**, 1818
6. Chase, H.A. (1994) Fluidised bed chromatography: purification of proteins by adsorption chromatography in expanded beds. *TIBTECH* **12**, 296

PART IV, SECTION D

70. The Modification of Milk Protein Composition through Transgenesis: Progress and Problems

Jean-Claude Mercier and Jean-Luc Vilotte

Laboratoire de Génétique Biochimique et de Cytogénétique, INRA-CRJ, 78352 Jouy-en-Josas, Cedex, France

Introduction

The mammary gland synthesizes and secretes large amounts of specific proteins, up to one kg of protein per lactation day in cow. Milk composition can differ widely between species. In ruminants the four phosphoproteins β-, α_{s1}-, α_{s2}-, and κ-casein, associated as micelles through calcium phosphate bridges, account for 80% of the total protein. Major specific whey proteins are β-lactoglobulin, and α-lactalbumin. In contrast, human milk (10 g/l protein as compared to 33 g/l in cow) contains 70% of whey proteins, mainly α-lactalbumin, lactoferrin, and lysozyme, and is devoid of β-lactoglobulin. Human micelles are smaller and contain β- and κ-casein and minute amounts of α_{s1}-casein. Whey acidic protein (WAP) is an abundant phosphoprotein only found in rodent, rabbit and camel milks.

Milk and derived products provide up to 30 per cent of dietary proteins in developed Western countries and are important worldwide in infant nutrition. Milk proteins are high quality nutriments due to their content in essential amino acids, calcium and inorganic phosphate, and the high digestibility of caseins. Whey proteins have also distinct physiological properties. α-lactalbumin induces lactose synthesis in modifying the substrate specificity of a Golgian galactosyltransferase, lactoferrin is an iron carrier and might have a bacteriostatic function when competing with iron-demanding bacteria, lysozyme destroys the bacterial cell walls, and β-lactoglobulin can bind small hydrophobic ligands such as retinol.

Until now, continuous improvement of milk production and slight differences in milk composition among breeds have been obtained through selection. Although quite effective for improving existing traits, there are limitations inherent in the conventional methods of breeding (Mercier, 1986). Hence the early interest for applying genetic engineering and gene transfer to farm animals (Palmiter et al., 1982; Hammer et al., 1985 and Lathe et al., 1986). In the 1980s, the rapid development of molecular and cellular biology has made possible the identification, isolation, characterization and modifications of milk protein-encoding cDNAs and genes (Figure 70.1), their chromosomal assignment, and expression studies in various host cells and transgenic animals (Mercier and Vilotte, 1993 and Vilotte and L'Huillier, 1995). The better knowledge of the structure and function of the major milk protein genes from several species including ruminants has already been used for (i) typing and selecting domestic animals with interesting dairy traits, e.g., quantitative and qualitative improvement of goat milk and cheese which is under progress through selection of sires and dams carrying peculiar α_{s1}- and β-casein alleles (Ng-Kwai-Hang and Grosclaude, 1992), (ii) creating transgenic dairy animals producing exogenous proteins in their milk, mainly pharmaceuticals (reviewed by Vilotte and L'Huillier, 1995, and discussed in the next chapter).

Potential changes in milk composition or in the primary structures of milk proteins and their presumed advantageous effects upon the physicochemical, nutritional and technological properties of milk and products derived from it, have already been discussed in several reviews (Mercier, 1986; Jimenez-Flores and Richardson, 1988; Creamer et al., 1988; Bremel et al., 1989; Wilmut et al., 1990; Hitchin et al., 1992; Clark, 1992; Richardson et al.,

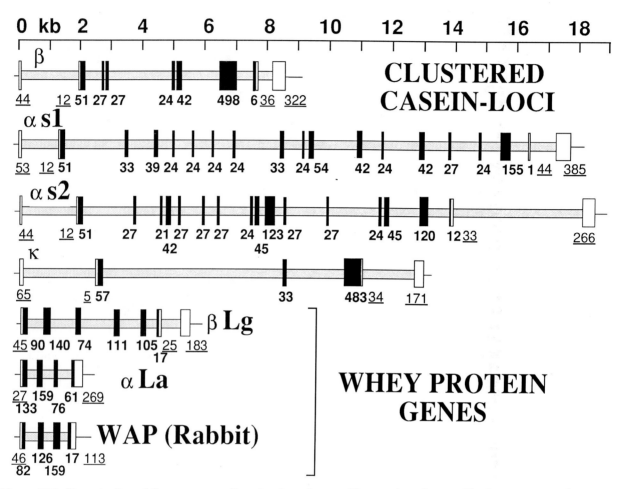

Figure 70.1. Organization of the genes encoding the six major specific proteins of cow milk, β-, α_{s1}-, α_{s2}-, and κ-casein, and both whey proteins α-lactalbumin (αLa) and β-lactoglobulin (βLg), and rabbit whey acidic protein (WAP). Adapted from Mercier and Vilotte (1993).
Only the exons, which are represented as high boxes, are not at scale. Their base pair sizes are indicated below each drawing. The coding frame is in black and underlined numbers refer to untranslated (white box) regions.

1992; Martin and Grosclaude, 1993; and Vilotte and L'Huillier, 1995). In this chapter, we will focus on some projects in progress to illustrate the actual potential of transgenesis but also the difficulties met in practice with respect to the present knowledge on gene expression and the gene transfer methodology currently available.

Genomic Integration of an Additional Wild-Type or Engineered Milk Protein-Encoding Gene

The transgene can be either a wild-type gene possibly from another species, or an engineered gene encoding a native or a modified milk protein. In the latter case and as discussed further, the main problem is to know which mutation(s) should be introduced into the coding frame for obtaining a novel protein with interesting economic features. Until now, suitable recipient cells for gene targeting are not yet available in domestic species, so that a modified gene cannot be substituted to its native counterpart through homologous recombination. Consequently, present transgenic farm animals can produce either a higher amount of a native milk protein or both the wild-type protein and its modified homologue.

Expression of Milk Protein-Encoding Transgenes

Significant modifications of milk composition have been achieved by the successful high expression of several milk protein-encoding transgenes in animals,

Table 70.1. Expression of native or hybrid milk protein-encoding transgenes. Adapted from Vilotte and L'Huillier (1995) with permission of CAB International.

Native gene or construct	Transgenic species	% of transgenic lines expressing*	Highest expression level (mg/ml of milk)	Average content of native protein in (mg/ml)	Reference
oβLG	mouse	8/8	23	Sheep 2.8; mouse 0	Simons et al., 1987
bαLA	mouse	3/6	0.45	cow 1.3; mouse 0.8	Vilotte et al., 1989
bαLA	mouse	3/3	1.5		Bleck and Bremel 1994
bαLA	rat	11/17	2.4	rat?	Hochi et al., 1992
hαLA[a]	mouse	1/1	14-fold the mouse allele	man ≥ 2	Stacey et al., 1995
cαLA	mouse	5/10	3.7		Soulier et al., 1992
gαLA	mouse	1/1	good level not quantified	goat ~2	Maschio et al., 1991
rWAP	mouse	8/9	95% of endogenous WAP?	guinea-pig?	Bayna et al., 1990
rWAP	mouse	16/18	5-fold endogenous WAP	rat?; mouse 2	Dale et al., 1992
mWAP	pig	3/3	1	pig 0	Wall et al., 1991
mWAP	pig	5/5	1.5		Shamay et al., 1991
rβCAS	mouse	3/5	1% of endogenous βCAS	rat?; mouse 25?	Lee et al., 1988
cβCAS	mouse	9/9	24	goat 12	Persuy et al. 1992
cβCAS	mouse	6/6	45% of endogenous βCAS		Roberts et al., 1992
bβCAS	mouse	12/12	20	cow 10	Rijnkels et al., 1995
bαs1CAS	mouse	8/8	20	cow 10; mouse 20?	Rijnkels, pers. commun.
bαs2CAS	mouse	0/3	undetected	cow 2.6; mouse?	Rijnkels et al., 1995
bκCAS	mouse	0/7	undetected	cow 3.3; mouse 2?	Rijnkels et al., 1995
cκCAS	mouse	5/5	20·10⁻³	goat 4	Persuy et al., 1995
bβCAS::hLZ	mouse	1/8	traces	cow 10 βCAS; man 0.4 LZ	Maga et al., 1994
bαlCAS::hLZ	mouse	2/5	116% of endogenous WAP	cow 10 αs1CAS	Maga et al., 1994
bαs1CAS::hLF	mouse	?	~1	man 1.7 LF	Platenburg et al., 1994
obLG::hSA	mouse	?	10	sheep 2.8 βLG; man 0.4SA	Barash et al., 1994
cβCAS::cκCAS	mouse	8/8	3	goat 12 βCAS, 4κCAS	Persuy et al., 1995
bαLA::bβCAS	mouse	3/4	10	cow 1.3 αLA, 10 βCAS	Bleck et al., 1995

*Number of lines expressing the transgene compared with the number of transgenic lines analysed for expression. [a] the human α-lactalbumin gene was targeted to the murine αLA locus through homologous recombination. αLA, α-lactalbumin; LF, lactoferrin; LZ, lysozyme; CAS, casein; SA, serum albumin (blood origin). WAP, whey acidic protein. b, bovine; c, caprine; g, guinea-pig; h, human; m, murine; o, ovine; r, rat. The highest expression level is that observed in hemizygous transgenic animals whereas the contents of endogenous milk proteins refer to the expression of pairs of alleles. The latter data were taken from many different sources which have not been referenced for lack of space. ? not known or possible content. ::, novel junction (fusion or insertion). For example, bβCAS::hLZ indicates a hybrid gene specifying human lysozyme under the control of 5' and possibly 3' regulatory sequences of the bovine β-casein gene.

mainly mice (Table 70.1). In the overall, it was shown that (i) expression was mammary-specific, integration-site dependent and copy number independent, (ii) short (1–3 kb) 5′ and 3′ flanking sequences were sufficient for high and specific expression of the transcription unit. This is in accordance with the localization of several important regulatory motifs in the proximal 5′ flanking region, using DNA footprinting and comparative functional analyses of 5′ truncated genes and hybrid genes in mammary cell lines and transgenic animals (reviewed by Mercier and Vilotte, 1993; and see Groner and Gouilleux, 1995, for the β-casein gene). However, native bovine α_{s2}-casein and κ-casein genes were not expressed in transgenic mice, despite the occurrence of long flanking sequences (Rijnkels et al., 1995). Similarly, a very low level of expression was observed with a caprine κ-casein construct only lacking a 0.5-kb LINE sequence in intron 2 (Persuy et al., 1995). Expression studies of yeast artificial chromosomes (YAC) containing milk protein genes, including the four clustered casein genes, are obviously the next step to identify other regulatory elements.

The wealth of structural and functional data now available for major milk protein-encoding genes are already useful for selecting presumed functional homologs in the genomic libraries of another mammal species.

Increasing the Whey Protein Content

A human-like milk enriched with human lysozyme and lactoferrin might enhance defence against gastrointestinal infections and promote iron transport in the digestive tract. As summarized in Table 70.1, a construct comprising human lysozyme cDNA driven by the bovine α_{s1}-casein promoter yielded a high level of lysozyme in mice, whereas that driven by the β-casein promoter was not expressed (Maga et al., 1994). Low levels of human lactoferrin were obtained with the relevant cDNA driven by the bovine α_{s1}-casein promoter (Platenburg et al., 1994). Obtention of a transgenic bull carrying a human lactoferrin construct was also reported (Krimpenfort et al., 1991).

Increasing the Individual Casein Content of Milk

Increasing the κ-Casein Content. In the mammary gland, calcium-induced precipitation of α_{s1}-, α_{s2}- and β-casein is prevented by their association with κ-casein as submicelles stably bound together through calcium phosphate bridges (Rollema, 1992). The average size of micelles appears to be correlated with the ratio κ-casein/calcium-sensitive caseins. A higher κ-casein content was found to be associated with a smaller micelle size (Rollema, 1992). Furthermore, heat stability of milk was markedly improved by addition of κ-casein (Singh and Creamer, 1992). Because of the key roles of κ-casein, the relevant gene is one of the most obvious candidate for transgenesis.

Unfortunately, present attempts to express native κ-casein genes (Table 70.1) or a hybrid gene driven by a κ-casein promoter (Ninomiya et al., 1994) have been unsuccessful. In contrast, transgenic mice carrying a caprine βCas::κCas fusion gene produced a milk containing up to 3 mg caprine κ-casein per ml (Persuy et al., 1995), to be compared with the 1 and 2 mg/ml yields of one endogenous κ-casein allele in mouse and goat. The recombinant caprine κ-casein with a Arg-Glu-Gly-Ala N-terminal extension occurred essentially in murine micelles and was able to protect caprine α_{s1}-casein against Ca^{2+}-induced precipitation, in forming micelles. Comparative analyses of the average micelle size in milks from wild-type mice and transgenic mice carrying one or two inherited transgene loci, are in progress. This hybrid gene is thus quite suitable for microinjection into eggs from a farm animal.

Increasing the β-Casein Content. β-casein is the most abundant casein in ruminant and human milks. Artificial enrichment of milk with β-casein modifies the cheese-making process. In addition to a slight reduction of the rennet clotting time and a slight increase of the syneresis rate, the firmness of the curd can be increased up to 50%. A reverse effect was observed by adding dephosphorylated β-casein (Richardson et al., 1992). In accordance with these results, goat milk lacking β-casein presented longer rennet coagulation times and the curd firmness was consistently poorer (Chianese et al., 1993). For industrials preparing hydrolysates of casein, a milk with a higher β-casein content would be inherently richer in peptides with opioid activity (Fox and Flynn, 1992) which are mostly derived from β-casein. The native caprine β-casein gene successfully tested in mice which produced up to 25 mg exogenous β-casein per ml of milk (Persuy et al., 1992) is being used as such for hopefully generating transgenic goats over producing β-casein.

Increasing the α_{s1}- or α_{s2}-Casein Content. Artificial enrichment of milk with the highly phosphorylated α_{s1}- or α_{s2}-casein has been much less documented. A higher content of α_{s2}-casein which contains 10–13 phosphoserine and two cyteine residues in the bovine species would obviously enhance the nutritional value of casein which is deficient in sulfur-containing amino acid, and would presumably increase micelle stability. The bovine α_{s1}-casein gene (M. Rijnkels, personal communication) and hybrid genes driven by the α_{s1}-casein promoter (Meade et al., 1990; Ninomiya et al., 1994 and Maga et al., 1994) have been well expressed in mice but attempts to express the bovine α_{s2}-casein gene have been unsuccessful (Rijnkels et al., 1995).

Production of Modified Milk Proteins with Novel Properties

Prerequisite for such Production in Transgenic Animals. Such a goal requires the precise knowledge of the amino acid or prosthetic group substitution(s), addition(s) or deletion(s) which might confer interesting properties to a given protein. At present, it is not easy to predict the conformation and physicochemical properties of a polypeptide on the basis of the primary structure. Fortunately, cDNAs can be easily modified by site-directed mutagenesis. The availability of expression vectors suitable for bacteria, yeast, fungus or animal cells, has made possible the production of recombinant proteins for structure-function studies in order to elaborate polypeptides with novel technological properties. However, most proteins are not simple amino acid chains as they undergo co- or/and post-translational modifications which often are essential to biological activity. Thus, in the case of calcium-sensitive caseins which are highly phosphorylated, the host cell should have a kinase(s) able to specifically recognize the tripeptide phosphorylation sites (Mercier, 1981) in order to produce a true casein. Although the lack of adequate phosphorylation may not be redhibitory for studying some properties of these three caseins, it would nevertheless be a serious disadvantage as phosphate groups play a key role in casein conformation, calcium bridges formation and resistance to proteolysis. Some genetic mutations proposed for milk proteins have been reported by Richardson et al. (1992).

Expression of Modified Milk Protein cDNAs in Different Host Cells. Several recombinant milk proteins have been produced using different systems. Caprine (Kumagai et al., 1990; 1991; 1992) and bovine (Wang et al., 1989 and Grobler et al., 1994) α-lactalbumin, bovine β-casein (Simons et al., 1993) and κ-casein (Oh and Richardson, 1991), caprine β-lactoglobulin (Persuy, personal communication) in *Escherichia coli*; bovine (Viaene et al., 1991) and caprine (Takeda et al., 1990) α-lactalbumin and bovine β-casein (Jimenez-Flores et al., 1990) in yeast; murine α-lactalbumin in cos cells and baculovirus-infected insect cells (Soulier et al., 1995). Some studies on α-lactalbumin were aimed at a better knowledge of the substructures interacting with and modifying the β-galactosyltransferase substrate specificity to promote lactose synthesis (Grobler et al., 1994 and Soulier et al., 1995). α-lactalbumin with lysozyme activity was also successfully engineered by exon 2 exchange between both evolutionary related genes (Kumagai et al., 1991; 1992). Experiments with recombinant caseins dealt with improvement of the sulfur content of κ-casein (Oh and Richardson, 1991) and of the chymosin-resistance of β-casein to prevent the release of bitter peptides (Simons et al., 1993). β-casein forms produced in *Saccharomyces cerevisiae* (Jimenez-Flores et al., 1990) differed in their phosphate groups and O-linked sugar contents, but their precise location was not determined.

Genomic Integration of a Transgene Specifying a Product Altering Milk Composition

The occurrence or high content of some milk components can be considered as disadvantageous for some dairy practices and products. Knockout of a gene or decrease of its promoter strength using site-directed mutagenesis can be routinely performed *in vitro* but gene exchange at the locus for generating transgenics has been so far restricted to the mouse species. However, there are possible indirect approaches for preventing the technologically disavantageous effect of a natural component of milk or lowering its content as illustrated with a few examples. Secretion of an inhibitor to prevent an awkward enzymatic activity or secretion of an enzyme acting on the component in excess are restricted to specific problems. Methodologies aimed at lowering the level of expression which are based on the inhibition of mRNA expression, are of more general use. Classical antisense RNA acts stoichiometrically without destroying the targeted RNA and this technique requires both an adequate rate of synthesis of antisense RNA and the cellular co-localization with the target RNA to be effective. In contrast, a ribozyme has the ability to repeat many cycles of selective binding, cleavage and dissociation (Sigurdsson and Eckstein, 1995). The latter methodology seems to be the most adapted for reducing the level of expression of genes with strong promoters such as those encoding major milk proteins, provided there is also co-localization of the ribozyme and target RNAs.

Preventing Plasmin Activity in Milk through Secretion of a Specific Inhibitor

Caseins are the most digestible milk proteins owing to their rather loose structure which promotes proteases activity. β- and α_{s2}-casein, and to a lesser extent α_{s1}-casein, are quite sensitive to plasmin, a serine protease occurring naturally in milk together with plasminogen. Thus, β-casein, the most abundant casein in ruminant milks, undergoes limited proteolysis by plasmin yielding the so-called γ_1-, γ_2- and γ_3-caseins and proteose-peptones 5 and 8. The 1.6 g/l average peptide content can be fivefold higher in milks richer in plasmin. This can be disadvantageous as casein proteolysis decreases the curd yield, and can induce organoleptic defects and gelation of UHT milk. A milk enriched with a specific inhibitor of either plasmin or plasminogen activator could therefore be attractive for the process industry.

Lowering the Lactose Content through Secretion of an Active Lactase or Ribozyme-Mediated Decrease of α-Lactalbumin mRNA Expression

Lactose, the most abundant milk sugar plays several key roles. It determines the water content of milk due to its osmotic pressure and it is a source of energy for the offspring and the microorganisms involved in yoghurt fermentation and cheese ripening. However, decreasing the lactose content might be advantageous for several reasons. It would counterbalance the milk components dilution resulting from selection of overproducing dairy animals, thus reducing the volume of milk to collect and the cost of transportation. Furthermore, this would promote milk consumption by lactose-intolerant people. Different approaches have been considered to lower the lactose content.

Secretion of an Active β-Galactosidase in Milk. A hybrid transgene driven by a strong milk protein promoter and encoding a lactase able to convert lactose into glucose and galactose under physiological conditions is one possibility yet to be tested. Such a milk could be used for consumption by lactose malabsorbers.

Lowering the α-Lactalbumin mRNA Level with a Specific Ribozyme. A MMTV-based construct expressing ribozymes targeted against the bovine α- lactalbumin mRNA was designed and tested in mice (L'Huillier et al., 1996). The level of bovine α-lactalbumin mRNA was half-reduced in double transgenic mice expressing both the hammerhead ribozyme and its bovine α-lactalbumin mRNA target whereas the level of murine α-lactalbumin mRNA remained unaffected. The bovine α-lactalbumin content of murine milk fell from 0.4 mg to 0.2 mg/ml, demonstrating the powerful possibilities of this methodology which could possibly be applied to larger animals and to other milk protein-encoding genes.

It remains that the relationship between the α-lactalbumin and lactose contents is not yet clearly determined. This can be achieved by crossbreeding transgenic mice homozygous for the α-lactalbumin null allele (Stinnakre et al., 1994) with other transgenic mice, each one carrying one heterologous α-lactalbumin transgene with a different level of expression (Vilotte et al., 1989 and Soulier et al., 1992). In the second generation some individuals will only carry the single functional heterologous locus, allowing analysis of lactose concentration in milks differing in their α-lactalbumin contents.

Producing a Milk Devoid of β-Lactoglobulin

β-lactoglobulin, the major heat-labile whey protein of ruminant milk, has a high nutritional value owing to its high content in essential amino acids including cysteine. It is able to bind small hydrophobic ligands such as retinol but its actual biological function, if any, is unknown. It is generally assumed that a milk depleted from β-lactoglobulin would be a better source of humanized milk. Both arguments – the nonoccurrence of the protein in human milk and its well-known allergenic properties – should be considered with care because other foreign milk proteins including caseins can also induce milk protein intolerance in neonates.

Presently, a reasonable genetic approach is the systematic search for individuals which might carry a null β-lactoglobulin allele, which proved successful for the α_{s1}- and β-casein loci in the goat species (Ng-Kwai-Hang and Grosclaude, 1992). The use of antisense or ribozyme anti-β-lactoglobulin mRNA is appealing as discussed above, but they should be very potent to prevent the synthesis of 3 g of β-lactoglobulin in the bovine species! The methodology of choice is obviously the knockout of the β-lactoglobulin gene which relies on the availability of true ES cells in dairy species.

Gene Exchange at the Locus of a Milk Protein-Encoding Gene

In the past few years, gene targeting in embryonic stem cells has been widely used for generating mice with a deficient gene in order to study its biological function (knockout experiments). More recently, homologous recombination was also successfully used to introduce subtle mutations at a given locus (gene replacement experiments). Unfortunately, these sophisticated techniques are so far restricted to the mice species since transmission to the germ line has not been demonstrated yet for ES-like cells isolated from domestic species. The transplantation of spermatogonial stem-cell engineered *in vitro* is a promising alternative under development (Brinster and Zimmermann, 1994 and Brinster and Avarbock, 1994).

Lactation and Milk Composition of Mice with Modified Endogenous Milk Protein Loci

β-casein–(Kumar et al., 1994) and α-lactalbumin–(Stinnakre et al., 1994 and Stacey et al., 1994; 1995) deficient mice have been created by gene targeting in ES cells in order to characterize the biological effect on mammary gland development, milk composition, lactation, and the offspring growth. Murine milk lacking β-casein contained smaller micelles and a slight higher content of other proteins. Growth of pups was slightly affected presumably because of the reduced total protein concentration. Milk from mice homozygous at the knockout α-lactalbumin locus was highly viscous preventing the suckling by offspring. Lactose was undetected (Stinnakre et al., 1994 and Stacey et al., 1995), thus demonstrating that only α-lactalbumin is able to induce lactose synthesis by β-galactosyltransferase. Lactation and lactose synthesis was restored by inserting a functional human α-lactalbumin gene at the deficient locus (Stacey et al., 1995). The human transgene expressed 15-fold greater mRNA than its murine counterpart in heterozygous mice, indicating that the former contained the cis-acting motifs responsible for the high rate of α-lactalbumin synthesis in the human species.

Prospects for Gene Targeting in Transgenic Farm Animals

Gene targeting has two major advantages. Integration of any functional mutated gene occurs at a precise locus where the suitable chromatin environment warrants the expression in contrast with aleatory integration which results in unpredictable levels of expression if any and could be detrimental to the animal. Furthermore, substitution of a gene with a modified reading frame would allow the production of solely the novel protein in animals homozygous at the modified locus. In this respect, site-directed mutagenesis applied to milk protein cDNA and comparative studies of the properties of recombinant proteins should be more developed.

This methodology provides for example another way to decrease the lactose content of milk through modification of the α-lactalbumin gene. Adequate modification of the promoter-enhancer region would result in a substantial lower transcription rate and a reduced α-lactalbumin content. As previously discussed in the ribozyme section, the main difficulty is to reach the critical level of α-lactalbumin affecting lactose synthesis. The more interesting alternative is a modification of the nucleotide sequence encoding the domain interacting with the β-galactosyltransferase. A mutant α-lactalbumin with lower affinity for the enzyme might result in the reduction of lactose synthesis. In this case, a prerequisite is the precise knowledge of the mutation(s) altering the function of α-lactalbumin in the lactose synthetase complex. As previously mentioned, studies of structure-function relationships are in progress using recombinant α-lactalbumins expressed from cDNA. The modification of the α-lactalbumin gene for reducing lactose synthesis is obviously an interesting but long range project now restricted to mouse to test the feasibility and because it is based on gene substitution at the α-lactalbumin locus.

Conclusion

The powerful possibilities of transgenesis applied to dairy animals are obvious and some remarkable results have already been obtained in using the mammary gland as a bioreactor for producing pharmaceuticals. But much progress remains to be done before routinely using transgenesis for generating farm animals producing milk for nontherapeutic use. In the present state of the art, it is difficult to predict that a construct will be functional because of insufficient knowledge on gene transcription, pre-mRNA processing, mRNA and protein stability. Integration of the microinjected transgene is aleatory resulting in highly variable levels of expression, and possible detrimental effects. Nevertheless, expression was less position-dependent in the case of constructs containing an insulator from the chicken globin locus control region (Y. Echelard, Genzyme, personal communication). Integration often occurs after the first cellular division(s) resulting in mosaic transgenic founders and consequently a low transmission of the transgene to the progeny.

The present constraints and the low efficiency of trangenesis can be minimized when a hundred million dollar pharmaceutical market can be covered with the production of a few transgenic animals. They cannot be ignored in the case of less money-rewarding projects aimed at improving the nutritional and/or technological properties of milk. Therefore, costly and time-consuming transgenesis should be applied to dairy animals for specific purposes that cannot be addressed by classical genetics. Such projects must be sound enough and must rely on convincing evidence that the modification will be beneficial to the farmer, the consumer or the industrial while being not detrimental to the animal and its transgenic progeny. In principle, simple concepts are easier to handle and minimize possible hazards for the health of the transgenic farm animal. In this respect, creation of animals overproducing a protein naturally occurring in milk such as a casein or a whey protein, or a foreign protein without known side effect seems to be a reasonable goal. In contrast, manipulation of genes encoding proteins with pleiotropic effects, such as hormones and enzymes acting in different biological pathways is more risky. For example, because of the consumer concern about fat consumption, a decrease of the milk fat content and a higher ratio unsaturated/saturated fat is highly desirable. It has been suggested to introduce a Δ9 desaturase-encoding gene functional in the mammary gland, but potential adverse side effects such as a change in the plasticity of mammary membranes cannot be excluded. Obviously, the feasibility of any project must be first thoroughly tested in the mouse species or possibly in a better milk producer such as the rabbit.

The second step is the creation of a dairy farm animal model, goat, sheep or cow, able to produce the novel milk in amounts sufficient for practical evaluation of its nutritional or/and industrial value. Rationale for this purpose is the use of both transgenic animals and individuals which naturally produce particular milks. Thus, transgenic caprine milks enriched with β- and κ-casein will be compared to individual goat milks differing in the α_{s1}-casein (0 to 7 g/l) and β-casein (0,6 and 12 g/l) contents (Ng-Kwai-Hang and Grosclaude, 1992). Such studies should provide useful information on the actual effect of the relative casein ratios on micelle structure and properties. So far, this has been mainly studied by making artificial micelles or by comparing micelles from different species which was more open to criticism. Any expected improvement of milk deduced from assays based on artificial enrichment must be confirmed in transgenic milks since the transgene might be expressed at the expense of endogenous milk protein genes. Thus, high expression of ovine β-lactoglobulin in mice resulted in a dramatic decrease of β-casein and WAP contents (McClenaghan et al., 1995). A priori, this competition might not occur in mammary glands not working at full capacity, such as those of goats naturally deficient in α_{s1}- or β-casein.

If the economical value of the novel milk is actually interesting, a herd of transgenic animals could be constituted provided a few precautions are taken. The aleatory integration of a transgene is a possible drawback as it might disrupt an important locus and affect the function of the endogenous gene with adverse consequence on the health or genetic value of the animal with respect to economic traits. Therefore, the use of any transgenic animal with interesting dairy properties for transgene introgression into a population must be considered with great care as slight detrimental effects might be detected only after thorough analysis of a large progeny.

So far the more sophisticated gene targeting to a precise locus in ES cells has been restricted to the mouse species. The availability of appropriate recipient cells in domestic species will be a major break-through. It will allow insertion of the transgene into a permissive chromatin environment, i.e. a more precise tuning of expression, and gene replacement.

In view of the present results mainly obtained in mice, modification of milk composition through transgenesis is a promising way for improving existing products and extending the uses of milk components. Farm animal models are being developed to study in detail the novel properties of milk. Milk for the para-medical industry, i.e. humanized milk, is likely to be produced first. In the foreseeable future, herds producing many different types of milk will presumably be available, allowing consumers and industrials to choose à la carte the milk best adapted to their needs.

References

Barash, I., Faerman, A., Ratovitsky, T., Puzis, R., Nathan, M., Hurwitz, R. and Shani, M. (1994) Ectopic expression of β-lactoglobulin/human serum albumin fusion genes in transgenic mice: hormonal regulation and *in situ* localization. *Transgenic Res.* **3**, 141–151

Bayna, E.M. and Rosen, J.M. (1990) Tissue-specific, high level expression of the rat whey acidic protein gene in transgenic mice. *Nucleic Acids Res.* **18**, 2977–2985

Bleck, G.T. and Bremel, R.D. (1994) Variation in expression of a bovine α-lactalbumin transgene in milk of transgenic mice. *J. Dairy Sci.* **77**, 1897–1904

Bleck, G.T., Jiménez-Flores, R. and Bremel, R.D. (1995) Abnormal properties of milk from transgenic mice expressing bovine β-casein under control of the bovine α-lactalbumin 5' flanking region. *Intern. Dairy J.* **5**, 619–632

Bremel, R.D., Yom, H.C. and Bleck, G.T. (1989) Alteration of milk composition using molecular genetics. *J. Dairy Sci.* **72**, 2826–2833

Brinster, R.L. and Avarbock, M.R. (1994) Germline transmission of donor haplotype following spermatogonial transplantation. *Proc. Natl. Acad. Sci. USA* **91**, 11303–11307

Brinster, R.L. and Zimmermann, J.W. (1994) Spermatogenesis following male germ-cell transplantation. *Proc. Natl. Acad. Sci. USA* **91**, 11298–11302

Burdon, T., Sankaran, L., Wall, R.J., Spencer, M. and Hennighausen, L. (1991) Expression of a whey acidic protein transgene during mammary development. *J. Biol. Chem.* **266**, 6909–6914

Chianese, L., Garro, G., Nicolai, M.A., Mauriello, R., Ferranti, P., Pizzano, R., Cappuccio, U., Laezza, P., Addeo, F., Ramunno, L., Rando, A. and Rubino, R. (1993) The nature of β-casein heterogeneity in caprine milk. *Le Lait* **73**, 533–547

Clark, A.J. (1992) Prospects for the genetic engineering of milk. *J. Cell. Biochem.* **49**, 121–127

Creamer, L.K., Jimenez-Flores, R. and Richardson, T. (1988) Genetic modification of food proteins. *TIBTECH* **6**, 163–169

Dale, T.C., Krnacik, M.J., Schmidhauser, C., Yang, C.L.Q., Bissell, M.J. and Rosen, J.M. (1992) High-level expression of the rat whey acidic protein gene is mediated by elements in the promoter and 3' untranslated region. *Mol. Cell. Biol.* **12**, 905–914

Fox, P.F. and Flynn, A. (1992) Biological properties of milk proteins. In P.F. Fox (Ed.), *Advanced Dairy Chemistry-I: Proteins*, Elsevier Applied Science, London, pp. 255–284

Grobler, J.A., Wang, M., Pike, A.C.W. and Brew, K. (1994) Study by mutagenesis of the roles of two aromatic clusters of α-lactalbumin in aspects of its action in the lactose synthase system. *J. Biol. Chem.* **269**, 5106–5114

Groner, B. and Gouilleux, F. (1995) Prolactin-mediated gene activation in mammary epithelial cells. *Current opinion in Genetics and Development* **5**, 587–594

Hammer, R.E., Parsel, V.G., Rexroad, C.E. Jr., Wall, R.J., Bolt, D.J., Ebert, K.M., Palmiter, R.D. and Brinster, R.L. (1985) Production of transgenic rabbits, sheep and pigs by microinjection. *Nature* **315**, 680–683

Harris, S., McClenaghan, M., Simons, J.P., Ali, S. and Clark, A.J. (1991) Developmental regulation of the sheep β-lactoglobulin gene in the mammary gland of transgenic mice. *Dev. Genetics* **12**, 299–307

Hitchin, E., Clark, A.J., Stevenson, E.M., Leaver, J. and Holt, C. (1992) Engineering of milk proteins. In P. Goodenough (Ed.), *Protein Engineering. Proceedings of an AFRC Conference on Protein Engineering in the Agricultural and Food Industry.* CPL Press, Newbury, pp. 147–154

Hochi, S.-I., Ninomiya, T., Waga-Homma, M., Sagara, J. and Yuki, A. (1992) Secretion of bovine α-lactalbumin into the milk of transgenic rats. *Mol. Reprod. Develop.* **33**, 160–164

Jimenez-Flores, R. and Richardson, T. (1988) Genetic engineering of the caseins to modify the behavior of milk during processing: a review. *J. Dairy Sci.* **71**, 2640–2654

Jimenez-Flores, R., Richardson, T. and Bisson, L.F. (1990) Expression of bovine β-casein in *Saccharomyces cerevisiae* and characterization of the protein produced in vivo. *J. Agric. Food Chem.* **38**, 1134–1141

Krimpenfort, P., Rademakers, A., Eyesqtone, W., Van der Schans, A.V., Van den Broek, S., Kooiman, P., Kootwijk, E., Platenburg, G., Pieper, F., Strijker, R. and De Boer, H. (1991) Generation of transgenic dairy cattle using *in vitro* embryo production. *BioTechnology* **9**, 844–847.

Kumagai, I., Takeda, S., Hibino, T. and Miura, K.I. (1990) Expression of goat α-lactalbumin in *Escherichia coli* and its refolding to biologically active protein. *Protein Engineering* **3**, 449–452

Kumagai, I., Takeda, S. and Miura, K.I. (1991) Introduction of enzymatic activity to α-lactalbumin by artificial exon shuffling. *Proc. Japan. Acad.* **67B**, 184–187

Kumagai, I., Takeda, S. and Miura, K.I. (1992) Functional conversion of the homologous proteins α-lactalbumin and lysozyme by exon exchange. *Proc. Natl. Acad. Sci. USA* **89**, 5887–5891

Kumar, S., Clarke, A.R., Hooper, M.L., Horne, D.S., Law, A.J.R., Leaver, J., Springbett, A., Stevenson, E. and Simons, J.P. (1994) Milk composition and lactation of β-casein-deficient mice. *Proc. Natl. Acad. Sci. USA* **91**, 6138–6142

L'Huillier, P., Soulier, S., Stinnakre, M.G., Lepourry, L., Davis, S.R., Mercier, J.C. and Vilotte, J.L. (1996) Efficient and specific ribozyme-mediated reduction of bovine α-lactalbumin expression in double transgenic mice. *Proc. Natl. Acad. Sci. USA*, in press

Lathe, R., Clark, A.J., Archibald, A.L., Bishop, J.O., Simons, P. and Wilmut, I. (1986) Novel products from livestock. In C. Smith, J.W.B. King and J.C. McKay, (Eds.), *Exploiting new technologies in animal breeding. Genetic development.* Oxford University Press, Oxford, pp. 91–102

Lee, K.F., Demayo, F.J., Atiee, S.H. and Rosen, J.M. (1988) Tissue-specific expression of the rat β-casein gene in transgenic mice. *Nucleic Acids Res.* **16**, 1027–1041

Maga, E.A., Anderson, G.B., Huang, M.C. and Murray, J.D. (1994) Expression of human lysozyme mRNA in the mammary gland of transgenic mice. *Transgenic Res.* **3**, 36–42

Martin, P. and Grosclaude, F. (1993) Improvement of milk protein quality by gene technology. *Livestock Production Science* **35**, 95–115

Maschio, A., Brickell, P.M., Kioussis, D., Mellor, A.L., Katz, D. and Craig, R.K. (1991) Transgenic mice carrying the guinea-pig α-lactalbumin gene transcribe milk protein genes in their sebaceous glands during lactation. *Biochem. J.* **275**, 459–467

McClenaghan, M., Springbett, A., Wallace, R.M., Wilde, C.J. and Clark, A.J. (1995) Secretory proteins compete for production in the mammary gland of transgenic mice. *Biochem. J.* **310**, 637–641

Meade, H., Gates, L., Lacy, E. and Lonberg, N. (1990) Bovine $α_{s1}$-casein gene sequences direct high level expression of active human urokinase in mouse milk. *Bio/Technology* **8**, 443–446

Mercier, J.C. (1981) Phosphorylation of caseins. Present evidence for an amino acid triplet code post-translationally recognized by specific kinases. *Biochimie* **63**, 1–17

Mercier, J.C. (1986) Genetic engineering applied to milk producing animals. Some expectations. In C. Smith, J.W.B. King and J.C. McKay (Eds.), *Exploiting New Technologies in Animal Breeding – Genetic Developments.*, Oxford Univ. Press, pp. 122–131

Mercier, J.C. and Vilotte, J.L. (1993) Structure and function of milk protein genes. *J. Dairy Sci.* **76**, 3079–3098

Ng-Kwai-Hang, K.G. and Grosclaude, F. (1992) Genetic polymorphism of milk protein. In P.F. Fox (Ed.), *Advanced Dairy Chemistry-1: Proteins.* Elsevier Applied Science, London, pp. 405–455

Ninomiya, T., Hirabayashi, M., Sagara, J. and Yuki, A. (1994) Functions of milk protein gene 5' flanking regions on human growth hormone gene. *Molec. Reprod. Develop.* **37**, 276–283

Oh, S. and Richardson, T. (1991) Genetic engineering of bovine κ-casein to improve its nutritional quality. *J. Agric. Food Chem.* **39**, 422–427

Palmiter, R.D., Brinster, R.L., Hammer, R.E., Trumbauer, M.E., Rosenfeld, M.G., Birnberg, N.C. and Evans, R.M. (1982) Dramatic growth of mice that develop from eggs microinjected with metallothionein-growth hormone fusion gene. *Nature* **300**, 611–615

Persuy, M.A., Stinnakre, M.G., Printz, C., Mahé, M.F. and Mercier, J.C. (1992) High expression of the caprine β-casein gene in transgenic mice. *Eur. J. Biochem.* **205**, 887–893

Persuy, M.A., Legrain, S., Printz, C., Stinnakre, M.G., Lepourry, L., Brignon, G. and Mercier, J.C (1995) High-level, stage- and mammary-tissue-specific expression of a caprine κ-casein-encoding minigene driven by a β-casein promoter in transgenic mice. *Gene* **165**, 291–296

Platenburg, G.J., Kootwijk, E.P.A., Kooiman, P.M., Woloshuk, S.L., Nuijens, J.H., Krimpenfort, P.J.A., Pieper, F.R., Boer de, H.A. and Strijker, R. (1994) Expression of human lactoferrin in milk of transgenic mice. *Transgenic Res.* **3**, 99–108

Richardson, T., Oh, S., Jiménez-Flores, R., Kumosinski, T.F., Brown, E.M. and Farrell, H.M. Jr. (1992) Molecular modeling and genetic engineering of milk proteins. In P.F. Fox (Ed.), *Advanced Dairy Chemistry-1: Proteins.* Elsevier Applied Science, London, pp. 545–577

Rijnkels, M., Kooiman, P.M., Krimpenfort, P.J.A., Boer de, H.A. and Pieper, F.R. (1995) Expression analysis of the individual bovine β-, α_{s2}- and κ-casein genes in transgenic mice. *Biochem. J.* **311**, 929–937

Roberts, B., DiTullio, P., Vitale, J., Hehir, K. and Gordon, K. (1992) Cloning of the goat β-casein-encoding gene and expression in transgenic mice. *Gene* **121**, 255–262

Rollema, H.S. (1992) Casein association and micelle formation. In P.F. Fox (Ed.), *Advanced dairy Chemistry-1: proteins.* Elsevier, London and New York, pp. 111–140

Shamay, A., Solinas, S., Pursel, V.G., McKnight, R.A., Alexander, L., Beattie, C., Hennighausen, L. and Wall, R.J. (1991) Production of the mouse whey acidic protein in transgenic pigs during lactation. *J. Anim. Sci.* **69**, 4552–4562

Sigurdsson, S.Th. and Eckstein, F. (1995) Structure-function relationships of hammerhead ribozymes: from understanding to applications. *TIBTECH* **13**, 286–289

Simons, G., Heuvel, W., van den., Reynen, T., Frijters, A., Rutten, G., Slangen, C.J., Groenen, M., Vos, W.M.de. and Siezen, R.J. (1993) overproduction of bovine β-casein in *Escherichia coli* and engineering of its main chymosin cleavage site. *Protein Engineering* **6**, 763–770

Simons, J.P., McClenaghan, M. and Clark, A.J. (1987) Alteration of the quality of milk by expression of sheep β-lactoglobulin in transgenic mice. *Nature* **328**, 530–532

Singh, H. and Creamer, L.K. (1992) Heat stability of milk. In P.F. Fox (Ed.), *Advanced Dairy Chemistry-l: Proteins.* Elsevier Applied Science, London, pp. 621–656

Soulier, S., Vilotte, J.L., Stinnakre, M.G. and Mercier, J.C. (1992) Expression analysis of ruminant α-lactalbumin in transgenic mice: developmental regulation and general location of important *cis*-regulatory elements. *Fed. Eur. Biol. Soc. Lett.* **297**, 13–18

Soulier, S., Vilotte, J.L. and Mercier, J.C. (1995) Expression of wild-type and mutant murine α-lactalbumin cDNAs in baculovirus-infected insect cells. *J. Agric. Food Chem.* **43**, 1392–1395

Stacey, A., Schnieke, A., McWhir, J., Cooper, J., Colman, A. and Melton, D.W. (1994) Use of double-replacement gene targeting to replace the murine α-lactabumin gene with its human counterpart in embryonic stem cells and mice. *Mol. Cell. Biol.* **14**, 1009–1016

Stacey, A., Schnieke, A., Kerr, M., Scott, A., McKee, C., Cottingham, I., Binas, B., Wilde, C. and Colman, A. (1995) Lactation is disrupted by α-lactalbumin deficiency and can be restored by human α-lactalbumin gene replacement in mice. *Proc. Natl. Acad. Sci. USA* **92**, 2835–2839

Stinnakre, M.G., Vilotte, J.L., Soulier, S. and Mercier, J.C. (1994) Creation and phenotyping analysis of α-lactalbumin-deficient mice. *Proc. Natl. Acad. Sci. USA* **91**, 6544–6548

Takeda, S., Tamaki, E., Miura, K.I. and Kumaga, I. (1990) Expression and secretion of goat α-lactalbumin as an active protein in *Saccharomyces cerevisiae*. *Biochem. Biophys. Res. Commun.* **173**, 741–747

Viaene, A., Volckaert, G., Joniau, M., de Baetselier, A. and van Cauwelaert, F. (1991) Efficient expression of bovine α-lactalbumin in *Saccharomyces cerevisiae*. *Eur. J. Biochem.* **202**, 471–477

Vilotte, J.L., Soulier, S., Stinnakre, M.G., Massoud, M. and Mercier, J.C. (1989) Efficient tissue-specific expression of bovine α-lactalbumin in transgenic mice. *Eur. J. Biochem.* **186**, 43–48

Vilotte, J.L. and L'Huillier, P. (1995) Modification of milk protein composition by gene transfer. In C.J.C. Phillips, (Ed.), *Progress in Dairy Science.* CAB International, Wallingford, pp. 281–309

Wall, R.J., Pursel, V.G., Shamay, A., McKnight, R.A., Pittius, C.W. and Hennighausen, L. (1991) High-level synthesis of a heterologous milk protein in the mammary glands of transgenic swine. *Proc. Natl. Acad. Sci. USA* **88**, 1696–1700

Wall, R.J., McKnight, R.A., Shamay, A. and Hennighausen, L. (1992) Matrix attachment sequences improve genetic control of a mammary gland transgene in mice. *Theriogenology* **37**, 319

Wang, M., Scott, W.A., Rao, K.R., Udey, J., Conner, G.E. and Brew, K. (1989) Recombinant bovine α-lactalbumin obtained by limited proteolysis of a fusion protein expressed at high levels in *Escherichia coli*. *J. Biol. Chem.* **264**, 21116–21121

Whitelaw, C.B.A., Harris, S., McClenaghan, M., Simons, J.P. and Clark, A.J. (1992) Position-independent expression of the ovine β-lactoglobulin gene in transgenic mice. *Biochem. J.* **286**, 31–39

Wilmut, I., Archibald, A.L., Harris, S., McClenaghan, M., Simons, J.P., Whitelaw, C.B.A. and Clark, A.J. (1990) Modification of milk composition. *J. Reprod. Fertil.* **41**, 135–146

PART IV, SECTION E

71. The Use of Host-Derived Antiviral Genes

Ellen Meier* and Heinz Arnheiter

Laboratory of Developmental Neurogenetics, National Institue of Neurological Disorders and Stroke, National Institutes of Health, Building 36, Room 5D04, 36 Convent Drive MSC 4160, Bethesda, MD 20892, USA

Successful viral replication requires complex interactions between viral and host components, which have to be present at the right times, intracellular locations, and concentrations. Any host- or virus-derived factor that modulates these interactions is likely to change a cell's susceptibility to infection. Virus-derived proteins have been studied extensively in vertebrate cells, and some of them have been expressed successfully from transgenes to create virus-resistant cells or animals (Friedman *et al.*, 1988; Smith and DeLuca, 1992). Host-derived intracellular resistance factors, however, are only beginning to be characterized at the molecular level, although genetic variation in susceptibility to various viral diseases independent of the immune system has been observed in many vertebrate species (Dveksler *et al.*, 1991; Fitzgerald and Shellam, 1991; Freund *et al.*, 1992; Lindenmann, 1962; Sangster *et al.*, 1994 and for review, see Müller and Brem, 1991). Several reasons account for the fact that, at present, the genes for only a small number of host-derived resistance factors have been mapped or cloned in vertebrates: First, the resistance phenotype is often specified by more than one genetic locus with each locus having only a relatively small contribution. Second, the resistance phenotype can be modified by the humoral and cellular immune responses. Third, the availability of genetic markers is very limited in most animal species.

A characteristic feature of most defense mechanisms is that individually they do not completely block viral infection. Even though, however, they usually assure that the virus titer stays low enough to limit damage before the T- and B-cell mediated immune system is activated to clear the virus. Thus, it is conceivable that resistance of vertebrate animals to specific viral diseases may be achieved through genetic engineering even if the transgene product does not block viral replication completely but only limits the spread of the virus. An example of a host factor that confers resistance by limiting the spread of a specific virus and that has been exploited to genetically engineer vertebrate animals resistant to specific viral diseases is the family of α/β interferon-induced Mx proteins.

The Mx Genes and Antiviral Resistance

In mice, resistance to infection with influenza A and B viruses (Lindenmann, 1962) and the related tick-borne Thogoto virus (Haller *et al.*, 1995) is controlled by a single, dominant autosomal locus, the Mx1 locus, located on chromosome 16 (for reviews, see Arnheiter and Meier, 1990 and Staeheli *et al.*, 1993). Resistant mice carry at least one copy of the Mx1$^+$ allele, which codes for a large, nuclear GTPase (Mr = 72,000), the Mx1 protein. Susceptible mice, i.e. all inbred strains with the exception of A2G and SL/NiA, carry one of two Mx1$^-$ alleles, which due to mutations have the coding capacity for only truncated Mx1 protein versions (Staeheli *et al.*, 1988). The Mx1 protein mediated resistance depends on the action of α/β interferon which is induced by the viral infection and in turn regulates the transcriptional activity of the Mx1 gene. Thus, in uninfected Mx1$^+$ mice, no α/β interferon is produced, and neither Mx1 mRNA nor Mx1 protein are detectable.

---

*Author for Correspondence:
  Center for Neurovirology Allegheny University of the Health Sciences Broad and Vine Streets Philadelphia, PA 19102-1192.

In virus infected Mx1$^+$ mice the infected cells synthesize and secrete α/β interferon which binds to the α/β interferon receptor and induces the transcription of the Mx1 gene resulting in high levels of Mx1 protein. The Mx1 protein very efficiently blocks an early step in the infectious cycle by inhibiting the accumulation of influenza viral mRNAs (Krug et al., 1985). It has been suggested that the Mx1 protein interferes with the PB2 subunit of the influenza virus-encoded RNA-dependent RNA polymerase (Huang et al., 1992), but the exact mechanism remains to be elucidated. In infected Mx1$^-$ mice, α/β interferon is also induced, but, because of the absence of a functional Mx1 protein, it is insufficient to protect against the infection.

Interferon-induced homologs of the murine Mx1 protein exist in vertebrate species ranging from fish to man (for review, see Arnheiter et al., 1996). In contrast to mice, which produce only one Mx protein that accumulates in the nucleus, most species synthesize at least two different, usually cytoplasmic Mx proteins. Besides the murine Mx1 cDNA, 10 other Mx cDNAs or DNAs originating from 7 different species (human (Aebi et al., 1989; Horisberger et al., 1990), rat (Meier et al., 1990), sheep (Charleston and Stewart, 1993), pig (Müller et al., 1992), duck (Bernasconi et al., 1995), chicken (Bazzigher et al., 1993), and fish (Staeheli et al., 1989)) have been cloned and sequenced. A comparison of these sequences reveals a high degree of amino acid identity over the entire lengths of the proteins. The most prominent structural features in all Mx proteins are a tripartite GTP-binding site, which is located in the amino-terminal half, and two leucine zipper regions, which are located close to the carboxy-terminus and which may be involved in oligomerization (Melen et al., 1992). Both, GTP-binding and GTPase activity have been demonstrated for the murine Mx1 (Nakayama et al., 1991), the human MxA (Horisberger, 1992), and the rat Mx2 and Mx3 proteins (Arnheiter et al., 1996), demonstrating that Mx proteins are indeed GTPases. Most interestingly, the amino terminal halves of Mx proteins share about 50% identity with the amino-terminal halves of dynamin (Obar et al., 1990), Vps1 (Rothman et al., 1990), and MGM1 (Jones and Fangman, 1992), three constitutively expressed GTPases, while the carboxy-terminal halves are completely divergent. The functions of these Mx-related proteins apparently are quite divergent: Dynamin, which is found in drosophila, sea urchin and vertebrates, plays a role in endocytosis of clathrin-coated vesicles (Takei et al., 1995). The yeast protein Vps1 is involved in exocytosis (Rothman et al., 1990) and the yeast protein MGM1 in maintenance of mitochondrial DNA (Jones and Fangman, 1992).

The question arises whether the non-murine Mx proteins protect their respective hosts against influenza or possibly other viral infections. So far, a genetic approach to this question has not been possible since Mx mutants have not yet been identified in species other than mice. Thus, in order to assess the antiviral activity of non-murine Mx proteins, several groups have resorted to an in vitro assay which is based on the constitutive expression of Mx cDNA constructs in cultured Mx1$^-$ mouse cells, followed by superinfection of these cells with virus. In this assay, the nuclear murine Mx1 protein has the same antiviral specificity as in vivo: it protects cells against influenza A and tick-borne Thogoto virus, but not against vesicular stomatits virus (VSV) (Haller et al., 1995 and Staeheli et al., 1986). When assayed in the same way, the human MxA protein which accumulates in the cytoplasm has a broader antiviral specificity than the murine Mx1 protein: it renders cells resistant not only to influenza A and Thogoto virus, but also to VSV, measles, Hantaan, La Crosse, and Rift Valley fever virus (Frese et al., 1995a; Frese et al., 1995b; Pavlovic et al., 1990 and Schnorr et al., 1993). In contrast, the rat Mx1 protein, a nuclear protein, protects cells against influenza virus and VSV, and the rat Mx2 protein, a cytoplasmic protein, confers resistance only to VSV (Meier et al., 1990) Interestingly, when the rMx2 protein is moved into the nucleus by way of appending a foreign nuclear transport signal to its amino terminus, the protein loses anti-VSV activity and gains anti-influenza virus activity, indicating that this protein's antiviral specificity is determined entirely by its subcellular localization (Johannes et al., 1993). Several Mx proteins such as the human MxB, rat Mx3, duck Mx, and chicken Mx protein have no antiviral activity against either influenza or VSV (Bazzigher et al., 1993; Bernasconi et al., 1995; Meier et al., 1990 and Pavlovic et al., 1990). In summary, when expressed in Mx1$^-$ mouse cells, the antiviral specificities of Mx proteins are quite heterogeneous, and some non-murine Mx proteins may have no antiviral activity at all or their target viruses have not yet been discovered.

Mx Transgenic Animals

Mx cDNAs which encode proteins with potent anti-viral activity appear to be perfect candidates for intracellular immunization (Baltimore, 1988) of susceptible species against specific viruses. For example, one could envision obtaining chickens and turkeys resistant to influenza viruses by introducing the murine Mx1 cDNA into their genomes. Both, chickens and turkeys are highly susceptible to avian influenza A viruses, which cause sporadic outbreaks, called fowl plague. As mentioned above, chickens have an Mx protein of their own, but this protein apparently is devoid of anti-influenza virus activity (Bernasconi et al., 1995); the situation in turkeys with respect to endogenous Mx proteins is unknown at present.

Transgenic Mice Expressing the Murine Mx1 Protein

To explore the feasibility of intracellular immunization with an Mx cDNA in a vertebrate species, Arnheiter et al. (1990) chose to use the murine Mx1 cDNA and Mx1$^-$ mice as recipient animals for the following reasons: (1) the Mx1 cDNA confers robust resistance to influenza and Thogoto virus when introduced into cultured cells, (2) Mx1$^-$ mice (i.e. all inbred mouse strains with the exception of A2G and SL/NiA) are highly susceptible to influenza virus and devoid of endogenous Mx1 protein, and (3) DNA can easily be introduced into the mouse genome using transgenic technology.

Initially, transgenic mice carrying the Mx1 cDNA under the control of a constitutive promoter (either the SV40 early promoter/enhancer or the human metallothionein IIA promoter) were generated. It was reasoned that the animals would have a head-start in fighting the intruding virus, if the Mx1 protein were already present before the virus enters the cells. However, all transgenic lines generated with these constitutive promoter constructs had undetectable levels of Mx1 protein and succumbed to influenza virus infections (Arnheiter et al., unpublished; Meier et al., unpublished). Recently, using the mouse 3-hydroxy-3-methyl-glutaryl coenzyme A reductase promoter (muHMG), transgenic mice were generated that constitutively expressed low level of Mx1 protein in few organs, particularly the brain (Kolb et al., 1992). These mice were resistant to the neurotropic influenza A/NWS virus strain, albeit not to the same degree as naturally resistant A2G mice. But most notably, they retained some degree of resistance when antibodies to α/β interferon were injected, while A2G mice became susceptible. Thus, in the absence of interferon, Mx1 protein can act in vivo, and mice can tolerate constitutive Mx1 expression, at least in the brain. However, the possibility exists that constitutive expression of the normally α/β interferon-inducible Mx1 protein may be deleterious to other organs such as the epithelium of the respiratory tract, the target organ of influenza viruses.

Therefore, transgenic animals were produced in which Mx1 protein expression was controlled by its own α/β interferon-inducible promoter (Arnheiter et al., 1990). Several lines were obtained which expressed the Mx1 protein only in the presence of α/β interferon or α/β interferon-inducers such as virus or poly-IC. Upon such treatment, the Mx1 protein levels differed depending on the line, but as in A2G mice, transgenic Mx1 protein was expressed in all organs tested. Were these transgenic animals protected against influenza and Thogoto virus? To answer this question, homozygous mice of the high-expressor line 979 were infected with influnza A/PR/8/34 virus, a pneumotropic mouse-adapted human influenza virus strain (Arnheiter et al., 1996). Heterozygous mice of this line synthesize 63% as much Mx1 protein as A2G mice, when measured in the liver. The virus was given in 10-fold serial dilutions and each virus dose was administered to 5 animals. All transgenic animals survived and sero-converted, even at the highest virus dose. In contrast, non-transgenic littermates died between 4 and 12 days postinfection. In fact, line 979 mice were about as resistant to influenza A/PR/8/34 virus as A2G mice in that they survived at least a 10,000 fold higher virus titer than non-transgenic Mx1$^-$ mice. A detailed analysis showed how the resistance is mediated by Mx1 protein: Upon infection of 979 mice with influenza virus, a few cells take up and amplify the virus, and α/β interferon is induced and released. This interferon acts locally by binding to the α/β interferon receptor on neighboring cells resulting in the induction of Mx1 protein. Consequently, the initially infected cells become surrounded by Mx1-positive cells which cannot amplify the virus and thereby stop the virus from spreading. This gives the immune system enough time to eliminate the virus and the animal survives. In non-transgenic animals, much as in transgenic mice only a few cells become infected initially, and α/β interferon is induced; but no functional Mx1 protein accumulates and the virus spreads unhindered. By the time the immune system is activated, viral infection has apparently inflicted too much damage for the animal to recover. Line 979, but not non-transgenic control mice, were also resistant to influenza A/NWS virus, a neurotropic influenza virus strain (Arnheiter et al., 1990), and Thogoto virus (Haller et al., 1995). These data demonstrate that the murine Mx1 cDNA is capable of transferring heritable specific antiviral resistance to an otherwise susceptible vertebrate host.

The analysis of additional transgenic lines demonstrated that the amount of Mx1 protein is a crucial factor in anti-viral resistance (Arnheiter et al., 1990). As mentioned above, line 979 mice, which synthesized 63% as much Mx1 protein as A2G mice, were approximately as resistant as A2G mice. In contrast, line 1009 mice, which synthesized only 3% as much Mx1 protein as A2G mice, were as susceptible to the neurotropic influenza virus A/NWS strain as non-transgenic Mx1-mice.

Surprisingly, mice of a third line (964), which synthesized 11% as much Mx1 protein as A2G mice, were resistant when given a high virus dose, but died when given a low dose of the same virus. These mice could be protected against a low virus dose, when α/β interferon was given to them along with the virus, while non-transgenic animals treated in the same way succumbed. This suggested that the susceptibility of 964 mice to a low virus dose was due to α/β interferon levels which were too low to induce sufficient amounts of Mx1 protein in order to stop the spread of the virus. Thus, in the murine Mx1 system the decision over death and survival depends on a subtle balance between the amounts of virus, α/β interferon, and Mx1 protein.

Transgenic Mice Expressing the Human MxA Protein

Besides the murine Mx1 protein, the human MxA cDNA has also been introduced into Mx1⁻ mice (Pavlovic et al., 1995). Transgenic mice that expressed the human MxA protein constitutively in the brain and other organs under the control of the muHMG promoter survived challenges with Thogoto virus; and although they did not survive infections with the neurotopic influenza A/NWS virus and VSV, the mean survival times with both viruses was increased. Thus, in mice the MxA protein displayed similar intrinsic antiviral specificities as in cultured cells, but since the expression levels were low, resistance was not very pronounced.

Transgenic mice made with the inducible murine Mx1 promoter linked to the human MxA cDNA were only marginally protected to neurotropic influenza A/NWS virus and Thogoto virus (Pavlovic et al., 1995). Again, this low level resistance was most likely due to low level expression of MxA protein in these lines.

Concluding Remarks

The generation of virus resistant animals is an attractive goal since it shoud not only improve the welfare of animals but also reduce breeding costs. Traditionally, resistance has been achieved through standard vaccinations. However, vaccines are not available for all viruses, difficult to administer in some cases, and usually ineffective in already infected animals. The successful generation of influenza virus-resistant Mx transgenic mice attests to the feasibility of obtaining virus-resistant mammals by an alternative approach, i.e. intracellular immunization via the transgenic route, and at the same time it alerts us to potential practical problems. The first problem is related to the choice of the promoter as the major control element for resistance gene expression in the viral target cells. Synthesis of a resistance protein in the wrong cell, at the wrong time, or at the wrong level may incapacitate or even kill the animal, while lack or low levels of expression in the viral target cells do not confer resistance. The α/β interferon-inducible murine Mx1 or any other virus-inducible promoter has the advantage of keeping the expression of a transgene to a minimum in the absence of a viral infection. However, as demonstrated by the susceptibility of the low Mx1 expressor line to a low virus dose, a disadvantage of a virus-inducible promoter may be that the invading virus replicates to too high titers before the promoter is sufficiently activated. This observation may have important implications for intracellular vaccination strategies, since in nature most infections occur at low virus concentration. The second problem has to do with insufficient expression levels of a resistance protein in the transgenic animal because either the transgene is expressed in a mosaic fashion, i.e. only in a subset of cells of the same kind, or the levels of resistance protein per cell are too low. To minimize deleterious effects of resistance gene expression on the host and at the same time achieve best anti-viral protection, expression of the resistance gene should be limited to the viral target cells. The best way to achieve this may be by placing the expression of a resistance gene under the control of an endogenous promoter by homologous recombination in embryonic stem cells. At present, intracellular immunization with a host-derived resistance gene still is a relatively new approach which hopefully can be developed into a broadly applicable procedure and extended to other resistance genes and non-murine species.

References

Aebi, M., Fäh, J., Hurt, N., Samuel, C.E., Thomis, D., Bazzigher, L., Pavlovic, J., Haller, O. Staeheli, P. (1989) cDNA structures and regulation of two interferon-induced human Mx proteins. *Mol. Cell. Biol.* **9**, 5062–5072

Arnheiter, H., Frese, M., Kambadur, R., Meier, E. Haller, O. (1996) Mx transgenic mice – animal models of health. *Curr – Top. Microbiol. Immunol.* **206**, 119–147

Arnheiter, H. Meier, E. (1990) Mx proteins: antiviral proteins by chance or necessity? *New Biol.* **2**, 815–857

Arnheiter, H., Skuntz, S., Noteborn, M., Chang, S. Meier, E. (1990) Transgenic mice with intracellular immunity to influenza virus. *Cell* **62**, 51–61

Baltimore, D. (1988) Intracellular immunization. *Nature* **335**, 395–396

Bazzigher, L., Schwartz, A. Staeheli, P. (1993) No enhanced resistance of murine and avian cells expressing cloned duck Mx protein. *Virology* **195**, 100–112

Bernasconi, D., Schultz, U. Staeheli, P. (1995) The interferon-induced Mx protein of chickens lacks antiviral activity. *J. Interferon Cytokine Res.* **15**, 47–53

Charleston, B. Stewart, H.J. (1993) An interferon-induced Mx protein: cDNA sequence and high-level expression in the endometrium of pregnant sheep. *Gene* **137** 327–331

Dveksler, G.S., Pensiero, M.N., Cardellicho, C.B., Williams, R.K., Jiang, G.-S., Holmes, K.V. Dieffenbach, C.V. (1991) Cloning of the mouse hepatitis receptor: expression in human and hamster cell lines confers susceptibility to MHV. *J. Virol.* **65**, 6881–6891

Fitzgerald, N.A. Shellam, G.R. (1991) Host genetic influences on fetal susceptibility to murine cytomegalovirus after maternal or fetal infection. *J. Infect. Dis.* **163**, 276–281

Frese, M., Kochs, G., Meier-Dieter, U., Siebler, J., Staeheli, P. Haller, O. (1995a) Human MxA protein inhibits tick-born Thogoto but not Dhori virus. *J. Virol.* **69**, 3904–3909

Frese, M., Kochs, G., Feldmann, H., Hertkorn, C. Haller, O. (1995b) Inhibion of Bunya-, and Hantaviruses by human MxA protein. *J. Virol.* **70**, in press

Friedman, A.D., Triezenberg, S.J. McKnight, S.L. (1988) Expression of a truncated viral trans-activator selectively impedes lytic infection by its cognate virus. *Nature* **335**, 452–454

Freund, R., Dubensky, T., Bronson, R., Sotniko, A., Carroll, J. Benjamin, T. (1992) Polyoma tumorigenesis in mice: evidence for dominant resistance and dominant susceptibility genes of the host. *Virology* **191**, 724–731

Haller, O., Frese, M., Rost, D., Nuttal, P.A. Kochs, G. (1995) Tick-born Thogoto virus infection in mice is inhibited by the orthomyxovirus resistance gene product Mx1. *J. Virol.* **69**, 2596–2607

Horisberger, M.A. (1992) Interferon-induced human protein MxA is a GTPase which binds transiently to cellular proteins. *J. Virol.* **66**, 4705–4709

Horisberger, M.A., McMaster, G.K., Zeller, H., Wathelet, M.G., Dellis, J. Content, J. (1990) Cloning and sequence analysis of cDNAs for interferon- and virus-induced human Mx proteins reveal that they contain putative guanine nucleotide-binding sites: functional study of the corresponding gene promoter. *J. Virol.* **64**, 1171–1181

Huang, T., Pavlovic, J., Staeheli, P. Krystal, M. (1992) Overexpression of the influenza virus polymerase can titrate out inhibition by the murine MxA protein. *J. Virol.* **66**, 4705–4709

Johannes, L., Arnheiter, H. Meier, E. (1993) Switch in antiviral specificity of a GTPase upon translocation from the cytoplasm to the nucleus. *J. Virol.* **67**, 1653–1657

Jones, B. Fangman, W. (1992) Mitochondrial DNA maintenance in yeast requires a protein containing a region related to the GTP binding domain of dynamin. *Genes. Dev.* **6**, 380–389

Kolb, E., Laine, E., Strehler, D. Staeheli, P. (1992) Resistance to influenza virus infection of Mx transgenic mice expressing Mx protein under the control of two constitutive promoters. *J. Virol.* **66**, 1709–1716

Krug, R.M., Shaw, M., Broni, B., Shapiro, G. Haller, O. (1985) Inhibition of influenza viral mRNA synthesis in cells expressing the interferon-induced Mx gene product. *J. Virol.* **56**, 201–206

Lindenmann, J. (1962) Resistance of mice to mouse adapted influenza A virus. *Virology* **16**, 203–204

Meier, E., Kunz, G., Haller, O. Arnheiter, H. (1990) Activity of rat Mx proteins against a rhabdovirus. *J. Virol.* **64**, 6263–6269

Melen, K., Ronni, T., Broni, B., Krug, R.M., von Bonsdorff, C.-H. Julkunen, I. (1992) Interferon-induced Mx proteins form oligomers and contain a putative leucin zipper. *J. Biol. Chem.* **267**, 25898–25907

Müller, M. Brem, G. (1991) Disease resistance in farm animals. *Experientia* **47**, 923–934

Müller, M., Winnaker, E.-L. Brem, G. (1992) Molecular cloning of porcine Mx cDNAs: new members of a family of interferon-inducible proteins with homology to GTP-binding proteins. *J. Interferon. Res.* **12**, 119–129

Nakayama, M., Nagato, K., Kato, A. Ishihama, A. (1991) Interferon-inducible mouse Mx1 protein that confers resistance to influenza virus is a GTPase. *J. Biol. Chem.* **266**, 21404–21408

Obar, R.A., Collins, C.A., Hammarback, J.A., Shpetner, H.S. Vallee, R.B. (1990) Molecular cloning of the microtubule-associated mechanochemical enzyme dynamin reveals homology with a new family of GTP-binding proteins. *Nature* **347**, 256–261

Pavlovic, J., Arzet, H.A., Hefti, H.P., Frese, M., Rost, D., Ernst, B., Kolb, E., Staeheli, P. Haller, O. (1995) Enhanced resistance of transgenic mice expressing the human MxA protein. *J. Virol.* **69**, 4506–4510

Pavlovic, J., Zürcher, T., Haller, O. Staeheli, P. (1990) Resistance to influenza virus and vesicular stomatitis virus conferred by expression of human MxA protein. *J. Virol.* **64**, 3370–3375

Rothman, J.H., Raymond, C.K., Gilbert, T., O'Hara, P.J. Stevens, T.H. (1990) A putative GTP binding protein homologous to interferon-inducible Mx proteins performs an essential function in yeast protein sorting. *Cell.* **61**, 1063–1074

Sangster, M.Y., Urosevic, N., Mansfield, J.P., Mackenzie, J.S. Shellam, G.R. (1994) Mapping the Flv locus controlling resistance to flaviviruses on mouse chromosome 5. *J. Virol.* **68**, 448–452

Schnorr, J.-J., Schneider-Schaulies, S., Simon-Jödicke, A., Pavlovic, J. Horisberger, M.A. ter Meulen, V. (1993) MxA-dependent inhibition of measles virus glycoprotein synthesis in a stably transfected human monocytic cell line. *J. Virol.* **67**, 4760–4768

Smith, C.A. DeLuca, N.A. (1992) Transdominant inhibition of herpes simplex virus growth in transgenic mice. *Virology* **191**, 581–588

Staeheli, P., Grob, R., Meier, E., Sutcliffe, J.G. Haller, O. (1988) Influenza virus-susceptible mice carry Mx genes with a large deletion or a nonsense mutation. *Mol. Cell. Biol.* **8**, 4518–4528

Staeheli, P., Haller, O., Boll, W., Lindenmann, J. Weissmann, C. (1986) Mx protein: constitutive expression in 3T3 cells transformed with cloned Mx cDNA confers selective resistance to influenza virus. *Cell.* **44**, 147–158

Staeheli, P., Pitossi, F. Pavlovic, J. (1993) Mx Proteins: GTPases with antiviral activity. *Trends Cell. Biol.* **3**, 268–272

Staeheli, P., Yu, Y.-X., Grob, R. Haller, O. (1989) A double-stranded RNA inducible fish gene homologous to the murine influenza virus resistance gene Mx. *Mol. Cell. Biol.* **9**, 3117–3121

Takei, K., McPherson, P.S. De Camilli, P. (1995) Tubular membrane invaginations coated by dynamin rings are induced by GTP-γS in nerve terminals. *Nature* **374**, 186–189

PART IV, SECTION E

72. The Use of Antisense RNA and Ribozyme Sequences to Prevent Infectious Diseases in Transgenic Animals

Lei Han and Thomas A. Wagner

Department of Laboratory Medicine and Pathology, Medical School Box 609, Room 6-159, Jackson Hall, UMHC, 321 Caurch Street S.E. Minneapolis MN 55455, USA

The discovery of the natural biological function of antisense RNA was made in 1981 by Jun-ichi Tomizawa (Tomizawa et al., 1982). In the past 5 years, this new technology has been extensively developed. It has been applied *in vitro* and *in vivo* studies in both transgenic animals and plants. The main biological forms of antisense molecules have been designated as antisense oligonucleotides or antisense RNA molecules. Among the antisense oligonucleotides, chemically modified molecules with uncharged sugar-phosphate "back-bones", such as phosphorothioate, P-ethoxy and O-methyl oligonucleotides have been developed to enhance the entrance of oligonucleotides through the cell membrane and resist degradation by nucleases inside the cells (Ts'o et al., 1991). Antisense studies which have involved transgenic animals are limited. In the past five years 2736 antisense studies and 5067 transgenic studies have been reported. However, only 101 of the studies are related to *in vivo* transgenic antisense research and only 39 of the 101 involve animal studies. There are only two reports where antisense technology was applied in transgenic animals to prevent infectious diseases (Han et al., 1991) and (Sasaki et al., 1993).

The precise mechanisms, whereby antisense RNA can inhibit the production of specific proteins, was the basic topic studied initially after the discovery of antisense function in 1981. The mechanism of hybridization of antisense molecules with the complementary messenger RNA to arrest protein translation has been proven (Levis et al., 1992). Other research has suggested that antisense RNA can act both in the cell nucleus and in the cytoplasm. It may also prevent messenger RNA from being spliced by stable binding to it in the nucleus (Weintraub, 1990). With regard to mRNA transport, antisense RNA can arrest protein translation by preventing the export of messenger, sense RNA, from the nucleus to ribosome's in the cytoplasm (Wold et al., 1990). Recent experiments involving antisense RNA in mouse embryos have shown that duplex RNA is a senstive target for a double-stranded RNA nuclease (Nishikura, 1992). The cellular nucleases cut messenger RNA at the positions where antisense RNA binds to them (Weintraub et al., 1990). Another interesting study showed that a cellular adenosine deaminase can modify adenosine residues in double-stranded RNAs from A-U to I (inosine)-U to change the normal coding, thereby preventing the messenger RNA from producing a biologically functional protein (Bass et al., 1988). This mechanism is a way to modify a functional sense messenger RNA to a nonsense dysfunctional mRNA at the post-transcriptional level. Interestingly, this process is activated by duplex RNA formation resulting from Antisense/sense hybridization.

By blocking the mRNA translation on ribosome; increasing the sensitivity of mRNA to cellular dsRNA ribonuclease; inhibiting the export of mRNA from the nucleus and base modification of mRNA, antisense molecules can inhibit the complementary mRNA at multiple levels in the replication cycle of a gene. These features suggests the multiple potential of antisense RNA molecules for inhibiting the production of detrimental gene products. In contrast to small antisense oligonucleotides, larger antisense RNA molecules have a higher efficiency and stronger inhibition of the target mRNA (Ao et al., 1991). Prior studies have also shown that the duplex of RNA:RNA is a particularly stable form of hybridization (Brantl *et al.*, 1994), and detectable in transfected cells (Wang *et al.*, 1993).

Our study as the first report of inhibition of virus replication with antisense RNA *in vivo* supports the above prior studies and the hypothesis that intracellular sense:antisense RNA duplexes could be stable and efficient enough to arrest virus replication. However, the region chosen as the target sequence for antisense RNA was different in our study than in these previous studies. In our study, the Moloney murine leukemia virus (M-MuLV) packaging (ψ) sequence was chosen as the antisense target sequence. The reason for that choice is twofold. First, the packaging sequence is an RNA sequence crucial in the retroviral life cycle, so that an RNA:RNA duplex inhibiting this packaging step would block virus replication. Second, and most significant, the viral RNA packaging process involves a specific viral RNA/viral core protein interaction. This RNA/protein interaction is of considerably lower affinity than RNA/RNA duplex formation. Therefore, antisense RNA inhibition of packaging should be very effective since the viral RNA genome will preferentially bind to the complementary anti-sense RNA over binding to the core viral protein. Such a mechanism of inhibition is quite different from other mechanisms which involve a more equal competition between the anti-sense RNA and another RNA target such as the ribosomal RNA. Gene constructs microinjected into one-cell mouse embryos were recombinant sequences expressing antisense RNA complementary to the packaging sequences. The M-MuLV packaging (ψ) sequences were positioned in reverse orientation under the transcriptional regulation of promoter/regulatory elements from the M-MuLV long terminal repeat (LTR) in the expression plasmid termed pLψas, or the cytomegalovirus (CMV) immediate-early region in the expression plasmid temed pCψas. Chromosomal integration, germ-line transmission, and expression of the antisense (ψ) RNA in the transgenic mice were confirmed by Southern and Northern blot hybridization. The transgenic male Lψas and Cψas mice (C57B6/SJL) were mated to nontransgenic females. Within several hours after birth (a period allowing maximal immunological tolerance of the virus), each offspring was injected i.p. with 0.1 ml containing 10^5 PFU of infectious M-MuLV Virions. At the age of 4 weeks, the mouse pups were determined to be transgenic or nontransgenic by slot blot hybridization of their tail DNA. At 12–14 weeks of age, transgenic and control mice were sacrificed and assayed for the presence of symptoms of leukemia. At the end of the assay period, none of the transgenic mice expressing the antisense RNA complementary to the retroviral packaging sequences were judged to be leukemic, while 31% of control mice showed severe leukemic symptoms (three of them had died) and others were at different degrees of sickness. This strong inhibition of M-MuLV replication-dependent leukemia initiation indicates the possibility of using antisense RNA molecules complementary to viral sequences as a potential technology to prevent infectious diseases in animals.

Since the antisense RNA produced in these transgenic mice contains sequences complementary only to M-MuLV packaging sequences and not to coding sequences (Korman *et al.*, 1989), the most obvious site of interference in the viral replication cycle would be in the packaging step. It may be that RNA complementary to ω sequences is highly effective at competing with ψ/capsid protein interaction. An interesting aspect of this study was shown by Northern hybridization analysis of cellular material with and without expression of the anti-sense RNA against the packaging sequence following infection with M-MuLV. While supernatant from non-antisense expressing cells showed virus particles containing viral RNA, supernatant from anti-sense expressing cells showed viral particles but Northern hybridization analysis showed these particles to be devoid of viral RNA. Indeed, the anti-sense expressing cells upon viral infection produced empty, non-infectious, viral particles.

One of the two other reports of antisense RNA inhibition of infectious diseases in transgenic animals is also related to the viral packaging process (Sasaki *et al.*, 1993). These researchers chose the nucleocapsid protein (N protein) gene of the mouse hepatitis virus (MHV) as the target sequence to inhibit with antisense RNA. The viral nucleocapsid protein plays an important role in the viral packaging step (Meric *et al.*, 1988). It can recognize and bind to the viral packaging sequence, then carry the viral genome into the packaging process to form virion particles. The nucleocapsid protein gene of mouse hepatitis virus shows a specific interaction with the packaging sequence in the viral leader RNA (Stohlman *et al.*, 1988). The N protein of mouse hepatitis virus is also important in viral transcription and replication (Baric *et al.*, 1988). In their study, the full length of the N protein sequence was placed under the control of Rouse sarcoma virus (RSV) long terminal repeat (LTR) in the antisense orientation. In the transgenic mice carrying the N protein antisense sequence, antisense RNA could be detected in various tissues including the liver and brain, the target organs of MHV infection. With the lethal challenge of MHV at a dosage of 5 to 10 × 10^2 PFU through the external nares, the transgenic mice expressing the antisense RNA were shown to be more resistant to infection leading to death than non-transgenic mice. The level of inhibitory effect is directly related to the expression level of the antisense RNA.

The above reported research has shown the ability of antisense RNA to protect animals against viral infections *in vivo*. Another feature of the work of these two groups is that both chose the packaging process as the target step in the viral replication cycle and demonstrate efficient protection of the animals. It is worth mentioning that the packaging process is an essential and common step in any viral replication cycle. Among some types of viruses, the packaging sequences contain an identical or highly conserved sequence, such as retroviruses (Lever et al., 1989) and Coronaviridae (Spaan et al., 1982). Therefore, inhibiting the packaging process with antisense RNA could supply broad-range protection to animals which could be infected by viral sub-groups or different strains.

There is another study which is involved in both *in vitro* and *in vitro* antisense studies for inhibition of infectious diseases (Ernst et al., 1991). Rabbits were chosen as the transgenic animals and introduced an antisense RNA gene against the E1a region of adenovirus h5 (Ad5) into the rabbit pronuclei. The efficacy of transgenesis was 36% and the antisense RNA expression was retained in the offspring. The viral challenge was preformed *in vitro*. The kidney tissue was obtained from newborn rabbits and primary kidney cell cultures were raised from transgenic and normal rabbit tissues. To evaluate the resistance to adenovirus driven by the antisense RNA gene, the adenoviral suspension was plated on the kidney cell cultures raised from the transgenic and normal rabbit kidney. Two of four cell lines possessing the antisense RNA transgene were estimated to be 90% to 98% more resistant to Ah5 than a normal kidney cell line. This inhibition was targeted to adenovirus early gene expression.

Ribozymes, a similar but more efficient molecule for RNA translation inhibition (at least in theory), were reported (Cech et al., 1986) following the discovery and study of antisense molecules. Ribozymes, like antisense RNA. are RNA molecules and can bind to their complementary sequences.

However, they have an additional function to that of antisense molecules. Ribozymes function to cleave the target RNA sequence (Prody et al., 1986). An essential secondary structure of all the ribozymes for that enzymatic activity is the hammerhead structure (Hutchins et al., 1986). The hammerhead consists of three stems and a catalytic center containing 13 conserved nucleotides (5' **GAAAC**N_n (the first stem) **GUN**(the site of cleavage N_n (the second stem) **CUGANGAN**$_n$ (the third stem) (Uhlenbeck et al., 1987). On the target RNA molecule, the cleavage occurs at 3' to the GUN triplet where N can be C, U, or A (Sheldon et al., 1989). Hammerhead ribozymes were first described for the avocado sun blotch viroid (Hutchins et al., 1986) and the satellite RNAs of the tobacco ringspot virus (Buzayan et al., 1986). Recently, more attention has been drawn to ribozymes as potential anti-virus therapeutic agents (Sarver et al., 1991). The first case of ribozyme suppression of gene expression in the reporter gene CAT (chloramphenicol acetyltransferase) in monkey cells (COS1) was reported in 1989 (Cameron et al.). In the past five years, 410 ribozyme studies have been reported; 105 of the studies were related to viral research. However, only 5 transgenic studies have involved ribozymes, and none of them were related to prevention of infectious diseases. Among the 5 studies, two were reported in Drosophila (Heinrich et al., 1993) (Zhao et al., 1993) and one in tobacco plants (Wegener et al., 1994). The other studies were performed in transgenic mice, but related to the beta 2-microglobulin gene (Larsson et al., 1994) and to the pancreatic beta-cell glucokinase gene (Efrat et al., 1994). Initial *in vitro* studies, (129 studies reported in the past 5 years), were carried out in cell-free systems and reported that, in most cases, the efficiency of cleavage had been much less than quantitative (Dropulic et al., 1991). The *in vivo* studies actually were tissue culture studies, or intracellular studies. 71 ribozyme tissue culture studies were reported in the past 5 years.

So far, there are no transgenic ribozyme animal models established to show resistance to or prevention of infectious diseases. Ribozymes have all of the attributes of an antisense RNA that functions catalytically (Haseloff et al., 1988). Actually, ribozymes are antisense RNAs with an added domain, or added function that cleaves the substrate RNA. In theory, it can only be a better and more efficient inhibitor to target gene replication. When appropriately introduced into cells, ribozymes theoretically, (since no such work has been done yet), can efficiently and specifically destroy (viral) RNA, thus inhibiting (virus) expression. In addition, because ribozymes function as enzymes during a given reaction, it is released from the substrate and remains unconsumed and therefore can be used repeatedly (Sarver et al., 1990). Two recent reports make the ribozyme transgenic animal study for prevention of disease more convincing. One of them is an *in vitro* study (Yu et al., 1995) carried out in CD34$^+$ cells from human fetal cord blood. A hairpin ribozyme was designed to target the human immunodeficiency virus type 1 (HIV-1) RNA in the leader sequence (at nucleotides +111/112 relative to the transcription initiation site). The CD34$^+$ cells, the hematopoietic progenitor cells, were transduced with the retroviral vectors carrying the ribozyme sequences and selected. The stable ribozyme CD34+ cells were then challenged with the macrophage tropic stain HIV-1/Bal. The infection was monitored by the antigen capture assay. The results showed that the progeny macrophage-like cells derived from the genetically altered stem/progenitor cells with an anti-HIV hairpin ribozyme can quite efficiently resist HIV-1 replication (up to 90%). The second report was an *in vivo* transgenic work (Larsson *et al.*, 1994). The researchers produced transgenic mice carrying the hammerhead ribozyme constructs (Rz-c) targeting at exon II of the mouse β-microglobulin mRNA. Seven Rz-c transgenic founder animals were identified and analyzed. Expression of the ribozyme transgene was tested for and detected in the lung, kidney and spleen. Expression was accompanied with reduction of the beta-microglobulin mRNA levels of heterozygous (Rz+/−) animals compared to nontransgenic litter mates. The effect was most pronounced in the lung with more than 90% beta-microglobulin mRNA reduction in individual mice.

In conclusion, antisense technology has a strong potential to be applied to the prevention of infectious diseases in animals. The studies which combine antisense and transgenic animal production have been limited. This may be the result of the time and expense required for transgenic studies. As mentioned above, ribozymes contain similar properties and additional RNA degradative function than antisense molecules. The principles which have been proven in antisense studies should be applicable to ribozyme research. The results reviewed here encourage the continued efforts to generate transgenic animals expressing anti-sense RNA and ribozymes directed against viral pathogens to provide protection against these pathogens.

References

Ao, A., Erickson, R.P., Bevilacqua, A. and Karolyi, J. (1991) Antisense inhibition of beta-glucuronidase expression in pre implantation mouse embryos, a comparison of transgenes and oligodeoxynucleotides. *Antisense Research and Development* **1**(1), 1–10

Barric, R.S., Nelson, G.W., Fleming, J.O., Deans, R.J., Keck, J.G., Casteel N. and Stohlman, S.A. (1988) Interactions between coronavirus nucleocapsid protein and viral RNAs, implications for viral transcription. *Journal of Virology* **62**, 4280–4287

Cech, T.R. and Bass, B.L. (1986) Biological catalysis by RNA. *Annual Review of Biochemistry* **55**, 599–629

Bass, B.L. and Weintraub, H. (1988) An unwinding activity that covalently modifies its double-stranded RNA substrate. *Cell* **55**(6), 1089–1098

Brantl, S. and Wagner, E.G. (1994) Antisense RNA-mediated transcriptional attenuation occurs faster than stable antisense/target RNA pairing, an *in vitro* study of plasmid pIP501. *EMBO Journal* **13**(15), 3599–3607

Buzayan, J.M., Gerlach, W.L. and Bruening, G. (1986) Satellite tobacco ringspot virus RNA, a subset of the RNA sequence is sufficient for automatic processing. *Proc. Natl. Acad. Sci. USA* **83**, 8859–8862

Cameron, F.H. and Jennings, P.A. (1989) Specific gene suppression by engineered ribozymes in monkey cells. *Proc. Natl. Acad. Sci. USA* **86**, 9139–9143

Dropulic, B., Lin, N.H., Martin, M.A. and K.T. Jeang. (1991) Functional characterization of a U5 ribozyme, intracellular suppression of human immunodeficiency virus type 1 expression. *Virology* **66**(3), 1432–1441

Efrat, S., Leiser, M., Wu, Y.J., Fusco-DeMane, D., Emran, O.A., Surana, M., Jetton, T.L., Magnuson, M.A., Weir, G. and Fleischer, N. (1994) Ribozyme-mediated attenuation of pancreatic beta-cell glucokinase expression in transgenic mice results in impaired glucose-induced insulin secretion. *Proc. Natl. Acad. Sci. USA* **91**(6), 2051–1055

Ernst, L.K., Zakcharchenko, V.I., Suraeva, N.M. Ponomareva, T.I., Miroshnichenko, O.I., Prokofev, M.I. and Tikchonenko, T.I. (1991) Transgenic rabbits with antisense RNA gene targeted at adenovirus H5. *Theriogenology* **35**, 1257–1271

Han, L., Yun, J.S. and T.E. Wagner. (1991) Inhibition of Moloney murine leukemia virus-induced leukemia in transgenic mice expressing antisense RNA complementary to the retroviral packaging sequences. *Proc. Natl. Acad. Sci. USA* **88**, 4313–4317

Hasloff, J. and Gerlach, W.L. (1988) Simple RNA enzymes with new and highly specific endoribonuclease activities. *Nature* (London) **334**, 585–591

Heinrich, J., Tabler, C., M. and Louis, C. (1993) Attenuation of white gene expression in transgenic *Drosophila melanogaster*, possible role of a catalytic antisense RNA. *Development Genetics* **14**(4), 258–265

Hutchins, C.J., Rathjen, P.D., Forster, A.C. and Symons, R.H. (1986) Self-cleavage of plus and minus RNA transcripts of avocado sunblotch viriod. *Nucleic Acids Res.* **14**, 3627–3640

Kim, S.K. and Wold, B.J. (1985) Stable reduction of thymidine kinase activity in cells expressing high levels of anti-sense RNA. *Cell* **42**(1), 129–138

Korman, A.J., Frantz, J.D., Strominger, J.L. and Mulligan, R.C. (1987) Expression of human class II major histocompatibility complex antigens using retrovirus vectors. *Proc. Natl. Acad. Sci. USA* **84**(8), 2150–2154

Larsson, S., Hotchkiss, G., Andabg, M., Nyholm, T., Inzunza, J., Jansson, I. and Ahrlund-Richter, L. (1994) Reduced beta 2-microglobulin mRNA levels in transgenic mice expressing a designed hammerhead ribozyme. *Nucleic Acids Research* **22**(12), 2242–2248

Lever, A., Gottlinger, H., Haseltine, W. and Sodroski, J. (1989) Identification of a sequence required for efficient packaging of human immunodeficiency virus type 1 RNA into virions. *Journal of Virology* **63**(9), 4085–4087

Levis, C., Tronchet, M., Meyer M. and Albouy, J. (1992) Effects of antisense oligodeoxynucleotide hybridization on *in vitro* translation of potato virus Y RNA. *Virus Genes* **6**(1), 33–46

Meric, C., Gouilloud, E. and Spahr, P.F. (1988) Mutations in Rous sarcoma virus nucleocapsid protein p12 (NC), deletions of cys-his boxes. *Journal of Virology* **62**(9), 3328–3333

Nishikura, K. (1992) Modulation of double-stranded RNAs *in vivo* by RNA duplex unwindase. *Annals of the New York Academy of Sciences* **660**, 240–250

Prody, G.A., Bakos, J.T., Buzayan, J.M., Schneider, I.R. and Bruening, G. (1986) Autolytic processing of dimeric plant virus satellite RNA. *Science* **231**, 1577–1580

Sarver N., Cantin, E.M., Chang, P.S., Zaia, J.A., Ladne, P.A., Stephens, D.A. and Rossi, J.J. (1990) Ribozymes as potential anti-HIV-1 therapeutic agents. *Science* **247**, 1222–1225

Sarver, N. (1991) Ribozymes, a new frontier in anti-HIV strategy. *Antisense Research and Development* **1**(4), 373–378

Sasaki, N., Hayashi, M., Aoyama, S., Yamashita, T., Miyoshi, I., Kasai, N. and Namioka, S. (1993) Transgenic mice with antisense RNA against the nucleocapsid protein mRNA of mouse hepatitis virus. *Journal of Veterinary Medical Science* **55**(4), 549–554

Sheldon, C.C. and Symons, R.H. (1989) Mutagenesis analysis of a self-cleavage RNA. *Nucleic Acids Research* **17**(14), 5679–5685

Spaan, W.J.M., Rottier, P.J.M., Horzinek, M.C. and van der Zeijst, B.A.M. (1982) Sequence relationships between the genome and the intracellular RNA species 1, 3, 6 and 7 of mouse hepatitis virus strain A59. *Journal of Virology* **42**, 432–439

Stohlman, S.A., Baric, R.S., Nelson, G.N., Soe, L.H., Welter, L.M. and Deans, R.J. (1988) Specific interaction between coronavirus leader RNA and nucleocapsid protein. *Journal of Virology* **62**, 4288–4295

Tomizawa, J.I. (1982) The importance of RNA secondary structure in CoIE1 primer formation. *Cell* **31**(3 pt 2), 575–583

Ts'o, P.O.P. (1991) Nonionic oligonucleotide analogues (Matagen) as anticodic agents in duplex and triplex formation. *Antisense Research and Development* **1**(3), 273–276

Uhlenbeck, O.C. (1987) A small catalytic oligoribonucleotide. *Nature (London)* **328**, 596–600

Wagener, D., Steinecke, P., Herget, T., Petereit, I., Philipp, C. and Schreier, P.H. (1994) Expression of a receptor gene is reduced by a ribozyme in transgenic plants. *Molecular and General Genetics* **245**(4), 465–470

Wang, S. and Dolnick, B.J. (1993) Quantitative evaluation of intracellular sense, antisense RNA hybrid duplexes. *Nucleic Acids Research* **21**(18), 4383–4391

Yu, M., Leavitt, M.C., Maruyama, M., Yamada, O., Young, D., Ho A.D. and Wong-Staal, F. (1995) Intracellular immunization of human fetal cord blood stem/progenitor cells with a ribowyme against human immunodeficiency virus type 1. *Proc. Natl. Acad. Sci. USA* **92**, 699–703

Weintraub, H.M. (1990) Antisense RNA and DNA. *Scientific American* **262**(1), 40–46

Zhao, J.J. and Pick, L. (1993) Generating loss-function phenotypes of the fushi tarazu gene with a targeted ribozyme in drosophila. *Nature* **365**(6445), 448–451

PART IV, SECTION E

73. Antibody Encoding Transgenes – Their Potential Use in Congenital and Intracellular Immunisation of Farm Animals

Mathias Müller, Ulrich H. Weidle and Gottfried Brem

Institute for Animal Breeding and Genetics, Veterinärmedizinische Universität Wien, Linke Bahngasse 11, A-1030 Vienna, Austria

The possibility of expressing foreign genes in mammals by gene transfer has opened new dimensions in animal breeding and husbandry (Brem et al., 1985; Hammer, et al., 1985 and reviewed by Brem and Müller, 1994). An important aspect is the improvement of animal health by transgenic means (Müller and Brem, 1991; 1994; 1995). Approaches to reduce disease susceptibility of livestock will be a benefit in terms of animal welfare and will also be of economic importance. The costs of disease have been estimated to account for 10–20% of total production costs. Historically, the control or elimination of infectious agents in farm animals has depended on the use of vaccines and drugs, quarantine safeguards and eradication. Attempts to select for improved disease resistance by conventional breeding programs is hampered by several problems (Müller and Brem, 1991 and refs. therein). Novel immunisation strategies based on nucleic acid technologies focus on two main issues: additive gene transfer and the development of nucleic acid vaccines. The aim is to stably or transiently express components known to provide or influence non-specific or specific host defence mechanisms against infectious pathogens. Referring to the site and mode of action and the source of the effective agent the strategies are termed 'intracellular', 'genetic' and 'congenital' immunisation (see Figure 73.1). Here we discuss the possibility of using antibody (Ab) encoding gene constructs in gene transfer experiments. Two approaches envisage the use of such genes in the improvement of health 'congenital' and 'intracellular' immunisation.

Physiologically, B lymphocytes are responsible for the humoral immune response by producing and secreting Abs. The immunoglobulins (Ig) consist of two light (L) and heavy (H) chain polypeptides. The genes encoding these chains are composed of several exons corresponding to the constant (C) region and the variable (V) region of the Ab. The V portions are not fixed genetic entities but are joined somatically from separated gene segments by recombination (Ig gene rearrangement) during B cell maturation. Although B cells are diploid and hence all gene segments involved in Ig rearrangement exist in two copies, only one Ig allele is rearranged and expressed (allelic exclusion). The hybridoma cell technique (Köhler and Milstein, 1975) has extremely simplified the cloning of functional rearranged Ab encoding genes. Transgenic mice expressing monoclonal Ab (MAb) genes have been instrumental in analysing the immune system (Storb, 1987 and Iglesias, 1991). Ig transgenes also cause allelic exclusion of the endogenous counterpart and this mechanism seems to operate even if the difference between the transgenic and the endogenous isotypes are quite pronounced (Rusconi and Köhler, 1985).

'Congenital' immunisation of animals utilises the transgenic expression and germ line transmission of a gene encoding an Ig specific for a pathogen and therefore providing congenital immunity without prior exposure to that pathogen. The feasibility of this approach was tested in mice and farm animals in two independent studies.

Since the production of Abs to various polysaccharides (e.g. phosphorylcholine) is potentially protective against pathogenic bacteria, Lo et al. (1991) generated transgenic mice, sheep and pigs harbouring genes encoding the murine H(α) and L(κ) chains for Abs against phosphorylcholine (PC). Of each species two transgenic founders with the

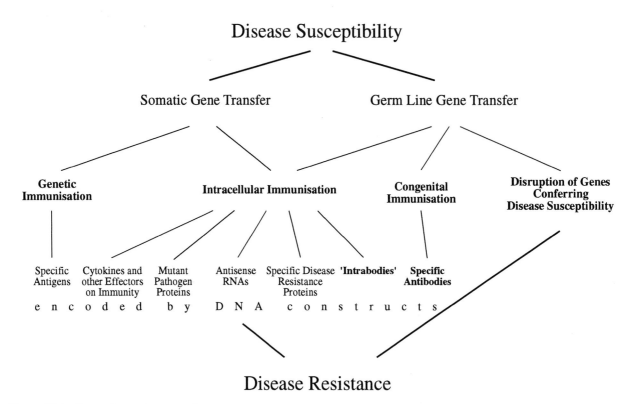

Figure 73.1. Strategies to improve disease resistance by transgenic means (adapted from Müller and Brem, 1995). The disruption of specific genes by homologous recombination ('gene knockout') requires totipotential embryonic cells, which remain to be established for livestock species. Somatic gene transfer approaches are reviewed elsewhere (Mulligan, 1993; Müller and Brem, 1995). The other concepts are described in the text, mainly focussing on antibody encoding gene constructs. A prerequisite for the practical use of the possible applicable methods in veterinary medicine and animal genetics is an evaluation of the benefits and costs and of the safety aspects.

integration of one or both intact transgenes was described. High serum levels (ranging from 300–1300 μg/mL) of transgene Ab (IgA) were detected in transgenic mice and pigs. In the transgenic sheep no serum expression of anti-PC Abs was found. Despite the absence of a functional mouse L chain transgene in the pigs, both animals secreted mouse transgene Abs into their serum. The secreted Abs presumably included endogenous pig L chains. Allelic exclusion, i.e. the suppression of endogenous immunoglobulin rearrangement and expression, was observed in only one transgenic mouse, the endogenous Ig production of the other transgenic animals was unaffected. Little if any of the transgene Ig in the sera examined showed binding specificity for the antigen. Thus a functionality in terms of the transgene Ab could not be demonstrated.

In the second gene transfer experiment (Weidle et al., 1991) the genes for the H (γ) and L chain (κ) genes of a MAb directed against 4-hydroxy-3-nitrophenylacetate (NP) were introduced into the germ line of mice, rabbits and pigs. The results of the gene transfer experiments are summarised in Table 73.1. The secretion of the anti-NP Ab into the milk of the transgenic founder was examined during the second lactation (U.H. Weidle and G. Brem, unpublished). MAb titers of > 100 μg/mL milk were found during the first three days of the lactation. In the remainder of the lactation the titer rapidly went down to 10–3 μg/mL on days 4–6 and was > 1 μg/mL from day 12 on. Transgenic Ab was purified from serum of transgenic rabbits and pigs and was shown to have two intact binding sites for the antigen when analysed in ELISA. However, in isoelectric focusing only a small fraction of the transgenic product matched to the mouse MAb. This findings could be due to heterologous Ab by association of endogenous L chains with the mouse transgene H chains (or vice versa). The electrophoretic differences could be also attributed to species – and cell-type-specific posttranslational modifications (e.g. glycosylation, deamidation) as has been observed for hybridoma cells cultured in differing media (for refs. see Weidle et al., 1991).

Table 73.1. Production of a monoclonal antibody (MAb) gene in transgenic animals (Weidle et al., 1991).

Species	Born animals	Transgenic founders	Integration frequency (%)	Copies integrated	Expressing transgenic lines	MAB titer (µg/mL)
Mouse	16	3	19	1–3	2	<40
Rabbit	43	3	7	40–60	3	<300
Pig	30	2*	7	≈50	1*	>1000

* Transgenic pig died.

Despite the unexpected findings, both experiments illustrate the potential of introducing beneficial traits such as germ-line-encoded immunity into farm animals (rabbits, sheep, pigs). Mouse Ig transgenes can be expressed in a tissue-specific manner and their expression does not necessarily interfere with the expression of the endogenous Igs. The finding of the milk secretion of the xenogenic Ab at the beginning of the lactation offers an interesting strategy to protect against infectious newborn diseases. It remains to be investigated, however, whether the efforts required for optimising the concept of 'congenital immunisation' are justified by its benefits in terms of increasing disease resistance in a certain species (for additional refs. see Kooyman and Pinkert, 1994). Following this route one has also to keep in mind that a given infectious pathogen will readily be able to escape the transgenic animal's immunity by changing its antigenic determinants.

The term 'intracellular immunisation' (Baltimore, 1988) was originally used for the overexpression in the host of an aberrant form (dominant-negative mutant) of a viral protein that is able to interfere strongly with the replication of the wild type virus. This definition was then extended to all approaches based on intracellular expression of transgene products which inhibit the replication of pathogens in host organisms (see Figure 73.1). Intracellular immunisation in mammalian cells can be also achieved by targeting the expression of recombinant antibodies to cell compartments where Abs are physiologically not directed (Biocca et al., 1990). This is achieved by mutation or deletion of the hydrophobic leader sequence for secretion of the recombinant Abs. Intracellularly the Abs can be directed to the desired subcellular compartment by adding the appropriate localisation signal (Biocca et al., 1995). Intracellular Abs, also termed 'intrabodies', represent a versatlile tool to modulate the function of selected intracellular gene products ('phenotypic knockout') but also promise to be powerful in the defence against infectious pathogens (for reviews see Chen et al., 1994; and Biocca and Cattaneo, 1995). The recombinant Abs most often utilised for this approach have been single chain Fv fragments, but other formats (whole Abs, antigen binding fragments) have been used as well (e.g. Mhashilkar et al., 1995). Single-chain Fv proteins are recombinant Abs derived from MAbs and are able to bind to their antigens with similar affinities as the parental Abs (Bird et al., 1988). They consist of only the variable L chain and the variable H chain domains covalently linked by an engineered polypeptide linker. Due to their small size, the lack of assembling requirements and the easy derivatisation (Griffiths et al., 1994) they are well suited for the design of new therapeutics and transgenic approaches. Currently, intrabodies are mainly investigated for their potential in human gene therapy. In agriculture the approach has been successfully tested in plants (Tavladoraki et al., 1993), but awaits its establishment in transgenic farm animals. As mentioned for 'congenital' immunisation the application of this strategy in animal breeding and production highly depends on costs/benefits considerations.

Another important aspect of using Ab construct transgenesis is the production of human Abs in animals. The concept is to generate an animal that, upon challenge with an antigen, will produce a human Ig response. Two objectives have to be achieved by transgenic technologies: (i) The endogenous Ig loci of the animal must be rendered inactive. This requires homologous recombination in ES cells. (ii) The human H and L chain loci have to be introduced into the germ line of the animal. Since these loci are very large (up to 5 megabases in their entirety) either mini loci containing a smaller number of variable regions and/or large gene constructs cloned in yeast artificial chromosome (YAC) vectors have to be generated. The development of fully humanised MAbs with therapeutic potential in transgenic mice has been reported by several groups (e.g. Green et al., 1994; Lonberg et al., 1994 and Zou et al., 1994). The transfer of this technology to a larger animal is not feasible because of lack of ES cell lines in other species, thus preventing gene targeting experiments.

In addition to the difficulties to be solved for successfully transferring Ab encoding genes, the attempts to improve health of livestock by the means of germ line gene transfer are facing several obstacles. (i) There is an obvious difference in scale and in reproductive biology between experiments in mice and livestock. Most gene transfers into livestock are carried out by microinjection of the DNA into zygotes. On average only about 1–3 % of embryos microinjected and transferred result in transgenic newborns. (ii) DNA microinjection results in random integration of the gene constructs in the host genome. Thus the transgene expression often underlies 'chromosomal position effects', which results in uncertainty about the expected tissue specificity and in variation of transgene expression levels in different transgenic lines carrying identical DNA constructs (reviewed in Brem, 1993). One approach to achieve strict spatio-temporal pattern of expression from genes of interest is the use of YAC gene constructs providing extensive sequences flanking the coding unit of the gene in order to avoid unwanted side effects of transgene expression. Recently, the first YAC DNA transgenic farm animals have been generated (Brem et al., 1996). (iii) Totipotential cell lines derived from farm animals are currently not available. Without doubt, the most exciting development will be the *in vitro* establishment of embryonic stem (ES) cells from these species. The availability of this technique will give new impetus to gene transfer in farm animals, because it will not only provide the possibility of additive gene transfer and homologous recombination but will also notably reduce problems such as low efficiency, non-expression of transgenes or insertional mutations.

References

Baltimore, D. (1988) Intracellular immunization. *Nature* **335**, 395–396

Biocca, S. and Cattaneo, A. (1995) Intracellular immunization: antibody targeting to subcellular compartments. *Trends Cell Biol.* **5**, 248–252

Biocca, S., Neuberger, M.S. and Cattaneo, A. (1990) Expression and targetting of intracellular antibodies in mammalian cells. *EMBO J.* **1**, 101–108

Biocca, S., Ruberti, F., Tafani, M., Pierandrei-Amaldi, P. and Catteneo, A. (1995) Redox state of single chain Fv fragments targeted to the endoplasmatic reticulum, cytosol and mitochondria. *Bio/Technology* **13**, 1110–1115

Bird, K.E., Hardman, K.D., Jacobson, J.W., Johnson, S., Kaufman, B., Lee, S.M., Pope, S.H., Riordan, G.S. and Whitlow, M. (1988) Single-chain antigen-binding proteins. *Science* **242**, 423–426

Brem, G. (1993) Transgenic animals. In Rehm, H.-J., Reed, G., Pühler, A. and Stadler, P. (eds.), *Biotechnology*, Weinheim, FRG, VCH, pp. 745–832

Brem, G., Besenfelder, U., Aigner, B., Müller, M., Liebl, I., Schütz, G. and Montoliu, L. (1996) YAC transgenesis in farm animals: rescue of albinism in rabbits. *Mol. Reprod. Dev.* **44**, 56–62

Brem, G., Brenig, B., Goodman, H.M., Selden, R.C., Graf, F., Kruff, B., Springmann, K., Hondele, J., Meyer, J. and Winnacker, E.-L., et al., (1985) Production of transgenic mice, rabbits and pigs by microinjection into pronuclei. *Zuchthygiene* **20**, 251–252

Brem, G. and Müller, M. (1994) Large transgenic animals. In N. Maclean, (ed.) *Animals with novel genes*, Cambridge, UK, Cambridge University Press, pp. 179–244

Chen, S.Y., Bagley, J. and Marasco, W.A. (1994) Intracellular antibodies a new class of therapeutic molecules for gene therapy. *Hum. Gene Ther.* **5**, 595–601

Green, L.L., Hardy, M.C., Maynard-Currie, C.E., Tsuda, H., Louie, D.M., Mendez, M.J., Abderrahim, H., Noguchi, M., Smith, D.H. and Zeng, Y., et al. (1994) Antigen-specific human monoclonal antibodies from mice engineered with human Ig heavy and light chain YACs. *Nat. Genet.* **7**, 13–21

Griffiths, A.D., Williams, S.C., Hartley, O., Tomlinson, I.M., Waterhouse, P., Crosby, W.L., Kontermann, R.E., Jones, P.T., Low, N.L. and John Allison, T., et al. (1994) Isolation of high affinity human antibodies directly from large synthetic repertoires. *EMBO J.* **13**, 3245–3260

Hammer, R.E., Pursel, V.G., Rexroad, C.E.J., Wall, R.J., Palmiter, R.D. and Brinster, R.L. (1985) Production of transgenic rabbits, sheep and pigs by microinjection. *Nature* **315**, 680–683

Iglesias, A. (1991) Analysis of the immune system with transgenic mice: B cell development and lymphokines. *Experientia* **47**, 878–884

Köhler, G. and Milstein, C. (1975) Continous culture of fused cells secreting antibody of predefined species. *Nature* **256**, 495–497

Kooyman, D.L. and Pinkert, C.A. (1994) Transgenic mice expression a chimaeric anti-E.coli immunoglobulin alpha heavy chain gene. *Transgenic Res.* **3**, 167–175

Lo, D., Pursel, V., Linto, P.J., Sandgren, E., Behringer, R., Rexroad, C., Palmiter, R.D. and Brinster, R.L. (1991) Expression of mouse IgA by transgenic mice, pigs and sheep. *Eur. J. Immunol.* **21**, 25–30

Lonberg, N., Taylor, L.D., Harding, F.A., Trounstine, M., Higgins, K.M., Schramm, S.R., Kuo, C.C., Mashayekh, R., Wymore, K. and McCabe, J.G., et al., (1994) Antigen-specific human antibodies from mice comprising four distinct genetic modifications. *Nature* **368**, 856–859

Mhashilkar, A.M., Bagley, J., Chen, S.Y., Szilvay, A.M., Helland, D.G. and Marasco, W.A. (1995) Inhibition of HIV-1 Tat-mediated LTR transactivation and HIV-1 infection by anti-Tat single chain intrabodies. *EMBO J.* **14**, 1542–1551

Mulligan, R.C. (1993) The basic science of gene therapy. *Science*, **260**, 926–932

Müller, M. and Brem, G. (1991) Disease resistance in farm animals. *Experientia* **47**, 923–934

Müller, M. and Brem, G. (1994) Transgenic strategies to increase disease resistance in livestock. *Reprod. Fert. Dev.* **6**, 605–613

Müller, M. and Brem, G. (1996) Intracellular, genetic or congenital immunisation – transgenic approaches to increase disease resistance of farm animals. *J. Biotechnol.* **44**, 233–242

Rusconi, S. and Köhler, G. (1985) Transmission and expression of a specific pair of rearrangend immunoglobulin μ snf κ genes in a transgenic mouse line. *Nature* **314**, 330–334

Storb, U. (1987) Immunoglobulin transgenic mice. *Annu. Rev. Immunol.* **5**, 151–174

Tavladoraki, P., Benvenuto, E., Trinca, S., De Martinis, D., Cattaneo, A. and Galeffi, P. (1993) Transgenic plants expressing a functional single-chain Fv antibody are specifically protected from virus attack. *Nature* **366**, 469–472

Weidle, U.H., Lenz, H. and Brem, G. (1991) Genes encoding a mouse monoclonal antibody are expressed in transgenic mice, rabbits and pigs. *Gene* **98**, 185–191

Zou, Y.-R., Müller, W., Gu, H. and Rajewsky, K. (1994) Cre-loxP-mediated gene replacement: a mouse strain producing humanized antibodies. *Curr. Biol.* **4**, 1099–1103

PART IV, SECTION E

74. Intracellular Antibodies (Intrabodies): Potential Applications in Transgenic Animal Research and Engineered Resistance to Pathogens

Susan Dana Jones[1] and Wayne A. Marasco[2]

[1]IntraImmune Therapies Inc., PO Box 15599, Boston, MA 02215, USA, [2]Division of Human Retrovirology, Dana Farber Cancer Institute, Harvard Medical School, 44 Binney Street, Boston, MA 02115, USA

Introduction

Recent advances in antibody engineering have allowed the genes encoding antibodies to be manipulated so that the antigen binding domain can be expressed intracellularly (Marasco et al., 1993). The use of intracellular antibodies, termed intrabodies, in transgenic animal research and eventual development of pathogen resistant transgenic farm animals is one of the many exciting applications. Recent successes in the use of intrabodies to inhibit cellular and viral functions have generally used a single chain antibody (scFv) in which the variable regions of the heavy and light chains are linked together via a flexible linker. By adding known intracellular trafficking signals, the scFvs can be localized to specific cellular compartments such as the nucleus, the endoplasmic reticulum, the inner surface of the plasma membrane, the cytoplasm, and the mitochondria (Marasco et al., 1993; Mhashilar et al., 1995 and Biocca et al., 1995). In the whole animal, the expression of an intrabody can be restricted temporally or spatially by using a promoter that is either tissue-specific, active at only a certain time in development, or both. Hence, inhibition of a gene product that is required for development can occur after the crucial stage is passed. It is also noteworthy that this technology is applicable to antigens that are not protein, such as specific sugar or lipid moieties. Therefore, the power of the immune system can be brought to the intracellular milieu and can be used to inhibit a variety of functions within the cell.

Recent Applications

Successful applications of the technology include inhibition of HIV-1 replication, functional inactivation of p21^{ras}, and phenotypic knockout of growth factor receptors (Marasco et al., 1994; Biocca et al., 1994; Richardson et al., 1995; Deshane et al., 1995 and Graus-Porta et al., 1995). It is this latter application, the phenotypic knockout of cellular function, that will be broadly applicable to research in transgenic animals. The current data on intrabodies has shown that cells can tolerate high levels of intrabody gene expression with no apparent effect on cell growth or phenotype, provided that the intrabody target is not necessary for cell survival (Richardson and Marasco, 1995). The ability to inhibit viral growth in a host cell by providing an intrabody against a virally-encoded protein allows the further application of engineering genetic resistance to known host pathogens without inhibiting development or cellular function.

Recent experiments to test the efficacy of intrabody inhibition of HIV-1 replication have been done using intrabodies directed against a highly conserved domain of the HIV-1 envelope protein gp120, against the activation domain of the transcriptional activator protein *tat*, and against the *rev* protein, which is essential for transport of full length viral RNA from the nucleus. Cultured cells expressing any of these intrabodies have shown a marked reduction in viral titer from cells that were infected or transfected with laboratory or primary isolates of HIV-1. The anti-gp120 intrabody, scFv105, was directed to the endoplasmic reticulum by inclusion of the heavy chain leader sequence. Although an ER retention signal is often used to retain proteins in the membrane, the scFv105 protein does not require such a signal. In fact, addition of the KDEL sequence to this intrabody causes the protein to be unstable and less effective in inhibition of HIV-1 replication (Marasco et al., 1993). The intrabodies against the nuclear proteins *tat* and *rev* were effective when localized to the cytoplasm, perhaps because they inhibit nuclear transport of the proteins (Mhashilkar et al., 1995). It should be noted that, of two anti-*tat* intrabodies that were evaluated, only one showed nearly complete inhibition of viral replication. This intrabody was known through mutagenesis experiments to be directed against the activation domain of the *tat* protein. Therefore, it may be necessary to synthesize and screen panels of intrabodies to generate some that are directed against critical epitopes and that will be effective at inhibition of the target molecule. The parameters that govern efficacy of specific intrabodies are not clear, although it is likely that the higher affinity intrabodies will be more effective. Therefore, new intrabodies should be fully evaluated in cell culture before they are introduced into animal models.

Single chain intrabodies are an effective tool for inhibition of transport of secreted or cell surface proteins to the plasma membrane, as shown by the work using the anti-gp120 intrabody. This concept has also been tested using intrabodies directed against growth factor receptors. The human interleukin 2 receptor (IL2-R) was downregulated in a T cell leukemia cell line that expresses an intrabody against the α subunit of this receptor (Richardson et al., 1995). The intrabody-containing cells showed no cell surface IL2Rα. Likewise, the cell surface expression of the ErbB2 receptor could be markedly decreased in a cell line that overexpresses this receptor when the line contains a gene encoding an intrabody against ErbB2 that is targeted to the endoplasmic reticulum (Beerli et al., 1994 and Graus-Porta et al., 1995). Since ErbB2 is activated in certain breast and ovarian carcinomas, it has been proposed to use this intrabody to treat patients with these cancers.

Proteins that function in the cytoplasm have also been inhibited by intrabodies. In addition to the anti-HIV-1 antibodies mentioned above, intrabodies have been designed against the guanine nucleotide binding protein p21 *ras*. *Ras* is involved in the control of cell growth and differentiation and is implicated in the development of several cancers. Injection of *Xenopus* oocytes with mRNA encoding an anti-*Ras* intrabody inhibited insulin-dependent meiotic maturation of the cell., a *ras* dependent process (Biocca et al., 1994 and Biocca et al., 1993). This inhibition reflects the activity of the intrabody in the cytoplasm. It has been found, however, that the half life of the intrabodies in the cytoplasm is shorter than in the ER, and perhaps additional modifications of the intrabody structure would allow a longer half life and therefore greater efficacy (Biocca et al., 1995).

Studies of Gene Function in Knockout Mice

Advances in transgenic technology in the last few years have allowed the creation of mice in which a single gene is deleted in the whole organism (Capecchi, 1989; and Frohman and Martin, 1989). This powerful molecular approach to the study of gene function has been used to determine the role of several genes in development, tumorigenesis, and metabolism. A limitation of this approach is that if the gene of interest is also required for complete development (i.e., until birth), then a knockout mouse cannot be made because the founder mice will die in utero. One solution has been the creation of regulable systems such that the gene of interest is not removed until a certain stage of development has been reached, or is only removed in a certain tissue. The *Cre* enzyme from the bacteriophage P1 is used in one such regulable system (Gu et al., 1993; Lakso et al., 1992 and Orban et al., 1992). *Cre* will delete DNA that is between two copies of its target sequence, *lox*P. The *Cre* gene can be placed under the control of a tissue-specific or temporal-specific promoter, and the gene of interest can be flanked by *lox*P sequences. Upon expression of the *Cre* enzyme, the gene will be removed and the function of the gene can be evaluated. This system has been successfully used to analyze the role of DNA polymerase β in cells of the immune system (Gu et al., 1994). In a unique twist of this approach, the deletion of a DNA segment by *Cre* can also be used to activate a gene at a certain time, if the DNA that is removed contains sequences that block gene activation.

Intrabodies for Studies of Gene Function

Another system that could be used for temporal and spatial inhibition of gene function is the regulable expression of a gene that directly provides an inhibitory effect on the target molecule. Because of their high specificity and stability, intracellular antibodies provide a powerful alternatives to other inhibitory molecules such as dominant negative mutants, antisense RNA, or ribozymes. There are few theoretical constraints to the use of intrabodies, and in fact these may be the only option for inhibition of non-protein components of the cellular machinery such as sugar, DNA, or a soluble metabolite.

Intracellular antibody genes could be introduced into transgenic animals using either the standard technique of microinjection of fertilized eggs, or could be introduced using transfection of embryonic stem cells. In either case, the gene could be placed under control of specific regulatory elements such that it is only expressed at certain times or places. By use of an inducible promoter, the time at which intrabody expression occurs can be controlled by the investigator. Thus, a gene thought to play a role in tumorigenesis could be inactivated by intrabody expression at various time points in the animal's life. It may be possible to combine these approaches to regulable gene expression, such that intrabody gene expression could be induced in a subset of tissues.

Several possible targets for intrabody-mediated inhibition of gene function can be envisioned. The many components of various signal transduction pathways could be inhibited either singly or together, to determine which components participate in the same or different pathways. Elegant studies in which tumor suppresser genes are knocked out have shown that some of these genes are involved in directing the cells toward apoptosis when the cell accumulates too many mutations in its DNA for effective repair (Williams et al., 1994 and Jacks et al., 1994). The suppression of these genes by intrabodies later in the life of the cell or the animal might help elucidate what the signals are that lead a cell to this pathway, and at what time in the growth and development of an animal these genes are no longer needed. It is also possible to use the intrabody technology to activate an enzyme or a transcription factor that is normally inhibited by another cellular factor. One example of this would be the transcription factor NFκB, which is functionally inactive while bound to the inhibition IκB (Nolan et al., 1991). If an intrabody against IκB were expressed, it could allow NFκB to become active at an aberrant time in the cell cycle and the effects on cell growth and neoplastic phenotype could be examined. This approach is applicable to any inhibitor molecule, or any enzyme that performs a modification of a cellular factor that renders the factor inactive.

Direct activation of a cellular component might be accomplished by an intrabody. For example, steroid receptors are normally found in the cytoplasm when the steroid hormone is not present. Upon stimulation of cells with the cognizant ligand, these receptors are translocated to the nucleus of the cell where they bind to DNA and effect changes in the transcriptional profile of the cell (Yamamoto, 1985). It is conceivable that an intrabody that is precisely directed to the hormone-binding domain might allow the receptor to be translocated to the nucleus and alter transcription of its responsive genes in the absence of systemic hormone. Hence the effect of overactive estrogen receptor on cancers of the ovaries, uterus, and breast could be examined without the confusion of the effects of systemic estrogen administration.

There are some applications in which the alternatives such as gene knockouts would be either technically difficult or impossible to achieve. For example, the growth factor receptor bound protein 2 (Grb2) links tyrosine-phosphorylated proteins such as growth factor receptors to the Ras signaling pathway (Chardin et al., 1995). Grb2 binds to phosphorylated tyrosines through a Src homology 2 (SH2) domain, and to a component of the Ras pathway through its Src homology 3 (SH3) domain. An isoform of grb2 which lacks the SH2 domain has been described (Fath et al., 1994). This protein, grb3-3, retains the ability to interact with the Ras signaling pathway, but cannot bind to phosphorylated tyrosines. Grb3-3 targets the cells for apoptosis by acting as a dominant negative mutant of growth factor responsiveness. As would be expected for an essential component of signal transduction, a knockout of grb2 is lethal. By designing an intrabody directed against the unique epitope of grb3-3, the role of this protein in the cell cycle can be evaluated. Since grb3-3 is expressed from the grb2 gene, it is not possible to address this question with a traditional gene knockout, since grb3-3 cannot be selectively deleted. It is likely that inhibition of grb3-3 function using an intrabody would not be lethal. However, high levels of grb3-3 are expressed in rat thymus during the stage of massive negative selection of thymocytes (Fath et al., 1994). Hence, it might be more effective to allow grb3-3 to be expressed until this developmental stage has been passed, and then inhibit its expression using an intrabody under control of a regulable promoter.

Another application that is unique to the intrabody approach is the possibility of directing an intrabody toward a conserved epitope that is shared by a family of proteins. This would generate a multilocus knockout, and the effects of this type of deletion of gene function could be examined. This is especially useful in systems where there is redundancy in gene function such that a knock out of an individual component of the gene family would not render an observable phenotype. Some possible targets of this approach include the steroid receptors, either systemically or in individual target organs, the rel/bcl family of transcription factors, or a set of components of the signal transcription factors, or a set of components of the signal transduction pathway, such as the SH2 domains or phosphorylated tyrosines. These multilocus knockouts would be lethal if engineered by standard means and would be technically difficult even using the Cre-loxP system. However, these could be easily engineered using the appropriate intrabody genes under control of regulable promoters.

Engineered Resistance to Pathogens

One commercial application that can be envisioned for the intrabody technology is the creation of animals that contains intrabody genes against common pathogens, as has been reported for plants (Tavladoraki, et al., 1993; and Benvenuto and Tavlodoraki, 1995). This would be particularly applicable in farm animals, where the relevant disease-causing organisms are well known. For example, poultry could be protected against infection by Salmonella if their cells contained an intrabody against a conserved epitope of an essential protein of the Salmonella bacterium. A more effective strategy would be to provide low levels of intrabodies against multiple target proteins in the disease causing organism, thereby reducing the chance that the pathogen could escape by mutagenesis. The effect of long term expression of intrabodies on the animals lifespan and phenotype would have to be evaluated before this application could be developed.

One experimental system that is being used to test the theory of engineering resistance to pathogens involves the use of the anti-HIV-1 antibodies discussed above. Rhesus monkeys are susceptible to an HIV-like virus termed Simian Immunodeficiency Virus (SIV). A hybrid has been constructed that contains the envelope protein of HIV-1, as well as the HIV-1 regulatory proteins *tat* and *rev*. The remainder of the virus is of SIV origin. This hybrid SHIV virus is able to infect monkeys, whereas HIV-1 is not (Li et al., 1992). It has been used to test vaccine strategies that aim to elicit an immune response against the HIV-1 envelope protein. The monkey model for HIV-1 treatment can also be used to test the intrabody strategy, both as a therapy in a monkey that is already infected, and for the purposes of this discussion, as a preventative measure that could be engineered into susceptible cells of monkeys. For example, the bone marrow from an adult monkey can be ablated and replaced with autologous hematopoietic stem cells derived from cord blood of the animal that was harvested at birth. The stem cells can be transduced with retroviruses containing the anti-HIV-1 intrabody genes scFv105 and anti-*tat*. These cells will replenish the simian immune system such that every cell of the immune system contains these resistance genes. The monkeys thus treated can then be infected with SHIV virus, and the viral burden in these animals can be compared to that of control animals over time. Because the SHIV virus contains the HIV-1 envelope and *tat* protein, it should be neutralized by this intrabody combination. In this model, the engineered protection would pertain only to the SHIV-susceptible cells of the immune system. In a fully transgenic animal, the intrabody against known pathogens could also be expressed only in the susceptible cells by using tissue specific promoter elements as discussed. It is also possible to put such intrabodies under the control of an inducible promoter, but this would be less effective since an exposure to the organism could occur when the gene was not activated. The intrabody gene expression could be activated by the infectious agent itself, thereby allowing expression only in the face of an infection. Experiments in which the anti-HIV-1 intrabody scFv105 was controlled by the HIV-1 LTR promoter have shown that this is an effective method of inhibiting viral replication (Chen et al., 1994). This approach is useful if the promoters of the infectious organism are well characterized.

Research Directions

Perhaps the most pressing question that must be answered before there is a widespread use of this technology in animal research is the fate and behavior of intrabody-expressing cells in the whole animal. If the genes are introduced by transducing a subset of the animals cells, the animal's immune system may recognize the cells as foreign and destroy them. This could be avoided by the use of autologous intrabody genes, but these may not always be available. It is possible that there will not be such a problem with intrabody genes. In human clinical trials, patients have had their cells transduced with retroviral constructs that contain the bacterial gene for neomycin resistance. These cells have persisted for up to 18 months without destruction by the immune system (Brenner et al., 1993a,b).

Animals that are transgenic for the intrabody gene will not recognize the intrabody protein as foreign. However, the phenotype of the intrabody expressing cells should be unaltered from that of the normal cells if this is to be a useful technology in farm animals. In a number of experiments in which cellular functions of intrabody gene transduced T cells lines have been examined, morphology, surface phenotype, replication rate, and response to mitogenic stimulation have been normal (Chen et al., 1994). In human peripheral blood mononuclear cells that were transduced with the anti HIV-1 intrabody gene scFv105, the growth rate and the relative levels of $CD4^+$ and $CD8^+$ cells was unchanged from that of control cells (Jones et al., 1995). In these experiments, the intrabody expression was under the control of a constitutive promoter, and was therefore continuous. These encouraging results suggest that intrabody gene expression will not affect cellular phenotype. This should be examined in other cell types and with additional intrabody genes.

Intrabodies will need to be expressed at reasonable levels for long periods of time if they are to be effective, especially in the transgenic applications that have been discussed. Transgenes can be silenced by a number of mechanisms including deletion and methylation of the DNA. This problem is a general problem of transgene expression and is not unique to intrabodies. Further improvements in transgene expression and gene transfer technology will be needed to overcome the silencing of genes. It is also imperative that sufficient levels of intrabodies be maintained in the cell. This can be a relatively low number if the goal is to inhibit a pathogenic organism upon entry into the cell, since initial numbers of pathogens are not high. However, if a regulable promoter is used, then the burst of intrabody expression needs to be sufficient to inhibit the organism after it has begun to replicate. The balance of transgene expression to that of the target proteins can be evaluated.

Summary

The emerging technology of antibody engineering has led to the development of intracellular antibody expression. These intrabodies provide a powerful new tool to selectively inhibit microbial or cellular functions. A broad scope of applications in research, clinical, and veterinary applications can be envisioned. A number of basic questions regarding intrabody design and expression will need to be answered before there is widespread clinical or veterinary use. The eventual use of this technology to create disease-resistant breeds of farm animals can be envisioned.

References

Beerli, R.R., Wels, W. and Hynes, N.E. (1994) Autocrine inhibition of the epidermal growth factor receptor by intracellular expression of a single-chain antibody. *Biochem. Biophys. Res. Commun.* **204**, 666–672

Benvenuto, E. and Tavladoraki, P. (1995) Immunotherapy of plant viral diseases. *Trends in Microbiology* **3** 272–275

Biocca, S., Ruberti, F., Tafani, M., Pieranderi-Amaldi, P. and Cattaneo, A. (1995) Redox state of single chain Fv fragments targeted to the endoplasmic reticulum, cytosol, and mitochondria. *Bio/Technology* **13**, 1110–1115

Biocca, S., Pierandrei-Amaldi, P., Campioni, N. and Cattaneo, A. (1994) Intracellular immunization with cytosolic recombinant antibodies. *Bio/Technology* **12**, 396–399

Biocca, S., Pieranderi-Amaldi, P. and Cattaneo, A. (1993) Intracellular expression of anti-p21*ras* single chain Fv fragments inhibits meiotic maturation of xenopus oocytes. *Biochem. Biophys. Res. Commun.* **197**, 422–427

Brenner, M.K., Rill, D.R., Moen, R.C., Krance, R.A., Mirro, J., Jr., Anderson, W.F. and Ihle, J.N. (1993a) Gene-marking to trace origin of relapse after autologous bone-marrow transplantation. *Lancet* **341**, 85–86

Brenner, M.K., Rill, D.R., Holladay, M.S., Heslop, H.E., Moen, R.C., Buschle, M., Krance, R.A., Santana, V.M., Anderson. W.F. and Ihle, J.N. (1993b) Gene marking to determine whether autologous marrow infusion restores long term haemopoiesis in cancer patients. *Lancet* **342**, 1134–1137

Capecchi, M.R. (1989) Altering the genome by homologous recombination. *Science* **244**, 1288–1292

Chardin, P., Cussac, D., Maignan, S. and Ducruix, A. (1995) The grb2 adaptor. *FEBS Lett.* **369**, 47–51

Chen, S.-Y., Bagley, J. and Marasco, W.A. (1994) Intracellular antibodies as a new class of therapeutic molecules for gene therapy. *Human Gene Therapy* **5**, 595–601

Deshane, J., Cabrera, G., Grim, J.E., Siegal, G.P., Pike, J., Alvarez, R.D. and Curiel D.T. (1995) Targeted eradication of ovarian cancer mediated by intracellular expression of anti-erbB-2 single-chain antibody. *Gynecologic Oncology* **59**, 8–14

Fath, I., Schweighoffer, F., Rey, I., Multon, M.C., Boiziau, J., Duchesne, M. and Tocque, B. (1994) Cloning of grb2 isoform with apoptotic properties. *Science* **264**, 971–974

Frohman, M.A. and Martin, G.R. (1989) Cut, paste, and save: new approaches to altering specific genes in mice. *Cell* **56**, 145–147

Graus-Porta, D., Beerli, R.R. and Hynes, N.E. (1995) Single-chain antibody-mediated intracellular retention of ErbB-2 impairs *neu* differentiation factor and epidermal growth factor signalling. *Molec. Cell Biol.* **15**, 1182–1191

Gu, H., Marth, J.D., Orban, P.C., Mossmann, H. and Rajewsky, K. (1994) Deletion of a DNA polumerase β gene segment in T cell using cells type-specific genes targeting. *Science* **265**, 103–106

Gu, H., Zou, Y.-R. and Rajewsky, K. (1993) Independent control of immunoglobulin switch recombination at individual switch regions evidenced through *Cre-loxP*-mediated gene targeting. *Cell* **73**, 1155–1164

Jacks, T., Remington, L., Williams, B.O., Schmitt, E.M., Halachmi, S., Bronson, R.T. and Weinberg, R.A. (1994) Tumor spectrum analysis in p53-mutant mice. *Current Biology* **4**, 1–7

Jones, S.D., Porter-Brooks, J., Eberhardt, B., Chen, S.-Y., Mhashilkar, A., Marasco, W.A. and Ramstedt, U. (1995) Gene Therapy for HIV Using Intracellular Antibodies. *J. Cell. Biol.* **21A**, 395

Lakso, M., Sauer, B., Mosinger, B. Jr., Lee, E.J., Manning, R.W., Yu, S.H., Mulder, K.L. and Westphal, H. (1992) Targeted oncogene activation by site-specific recombination in transgenic mice. *Proc. Natl. Acad. Sci.* **89**, 6232–6236

Li, J., Lord, C.I., Haseltine, W., Letvin, N.L. and Sodroski, J. (1992) Infection of cynomolgus monkeys with a chimeric HIV-1/SIV$_{mac}$ virus that expresses the HIV-1 envelope glycoproteins. *J. AIDS* **5**, 639–646

Marasco, W.A., Haseltine, W.A. and Chen, S.-Y. (1993) Design, intracellular expression, and activity of a human anti-HIV-1 gp120 single chain antibody. *Proc. Natl. Acad. Sci.* **90**, 7889–7893

Mhashilkar, A.M., Bagley, J., Chen, S.-Y., Szilvay, A.M., Helland, D.G. and Marasco, W.A. (1995) Inhibition of HIV-1 tat-mediated LTR transactivation and HIV-1 infection by anti-tat single chain intrabodies. *EMBO J.* **14**, 1542–1551

Nolan, G.P., Ghosh, S., Liou, H.C., Tempst, P., and Baltimore, D. (1991) DNA binding and I-κB inhibition of the cloned p65 subunit of NF-κ B, a related polypeptide. *Cell* **64**, 961–969

Orban, P.C., Chui, D. and Marth, J.D. (1992) Tissue-and site-specific DNA recombination in transgenic mice. *Proc. Natl. Acad. Sci.* **89**, 6861–6865

Richardson, J.H., Sodroski, J.G., Waldmann, T.A., and Marasco, W.A. (1995) Phenotypic knockout of the high-affinity human interleukin 2 receptor by intracellular single-chain antibodies against the alpha subunit of the receptor. *Proc. Natl. Acad. Sci.* **92**, 3137–3141

Richardson, J. and Marasco, W.A. (1995) Intracellular antibodies: development and therapeutic potential. *Trends Biotechnol.* **13**, 306–310

Tavladoraki, P., Benvenuto, E., Trinca, S., De Martinis, D., Cattaneo, A. and Galeffi, P. (1993) Transgenic plants expressing a functional single-chain Fv antibody are specifically protected from virus attack. *Nature* **366**, 469–472

Williams, B.O., Remington, L., Albert, D.M., Mukai, S., Bronson, R.T. and Jacks, T. (1994) Cooperative tumorigenic effects of germline mutations in Rb and p53. *Nature Genetics* **7**, 480–484

Yamamoto, K.R. (1985) Steroid receptor regulated transcription of specific genes and gene networks. *Ann. Rev. Genet.* **19**, 209–252

PART IV, SECTION F

75. Improved Wool Production from Insulin-Like Growth Factor 1 Targeted to the Wool Follicle in Transgenic Sheep

D.W. Bullock[1,2,*], S. Damak[1], N.P. Jay[2], H.-Y. Su[1,2] and G.K. Barrell[2]

[1]Centre for Molecular Biology and [2]Animal and Veterinary Sciences Group, Lincoln University, PO Box 84, Canterbury, New Zealand

Introduction

Despite success in producing transgenic sheep (Clark et al., 1990; Ward and Nancarrow, 1991; and Pursel and Rexroad, 1993), improvement of production traits by genetic engineering remains an elusive goal (Rexroad et al., 1991 and Powell et al., 1994). In 1989 we began a transgenic sheep programme by attempting to target heterologous gene expression to the wool follicle, using a mouse keratin (KER) promoter. Success in these experiments (Damak et al., 1996a) led us to link the KER promoter to a sheep insulin-like growth factor 1 (IGF1) cDNA, with the aim of affecting follicular metabolism and thus wool production or properties. The results demonstrated heritable, stable expression of the transgene. A progeny test in 1994 revealed increased fleece weight in transgenic animals, with little effect on fibre characteristics (Damak et al., 1996b).

Targeting Expression to the Wool Follicle

Transgenic Animals

To establish the technology and to demonstrate expression in the wool follicle, transgenic sheep were produced with the mouse KER promoter linked to the bacterial chloramphenicol acetyl transferase (CAT) gene. This construct had been used previously to target the hair follicle in transgenic mice (McNab et al., 1990). Thirty-one lambs were born after microinjection and transfer of 371 embryos, of which 4 were transgenic. Transgene copy number varied from less than one to more than 20 in one ram expressing CAT activity in the keratogenous zone of the cortical region in the wool fibre. This ram (Line one) was mated to random-bred ewes, resulting in 80 lambs born of which 36 were transgenic.

Gene Expression

Out of 26 G1 transgenic lambs from Line one, tested at 3-month intervals during the first year of age, 9 showed CAT expression in skin. Measurement of clean wool weight taken from mid-side patch samples and of CAT activity showed that expression of the transgene paralleled the expected seasonal changes in the rate of wool growth (Figure 75.1).

The results demonstrated a high rate of success in producing transgenic sheep. The mouse promoter is shown to function normally in transgenic sheep, directing gene expression to the wool follicle with a pattern that correlates with the physiological activity of the follicle in terms of wool growth. This keratin promoter appears to be useful for targeting genes that might affect wool production or properties.

Wool Production and Properties

Transgenic Animals

As a candidate gene for the promotion of wool growth, we selected IGF1 because of its well established properties as a somatomedin (Froesch et al., 1985) with both mitogenic and morphogenic activity (Sara and Hall, 1990). While this work was in progress, Philpott et al. (1994) reported that IGF1 stimulates the elongation of human hair in tissue

*Author for Correspondence.

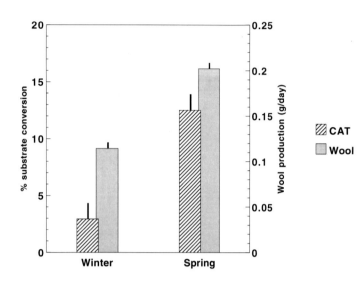

Figure 75.1. Relationship between CAT activity and seasonal changes in the rate of wool growth (Mean ± s.d.).

culture. We microinjected a DNA construct consisting of the mouse KER promoter linked to a sheep IGF1 cDNA (Damak et al., 1996b). From 591 embryos, 5 transgenic animals were produced among 103 lambs born. One transgenic ram (Line nine) was mated to 43 random-bred non-transgenic ewes, producing 85 lambs of which 43 were transgenic.

Gene Expression

Ribonuclease Protection Assay was used to distinguish transgene expression from endogenous IGF1 mRNA. Among the G1 transgenic progeny of Ram nine, 20 out of 25 lambs tested showed expression of the transgene in the skin. All lambs showed skin expression of endogenous IGF1, at about half the level of the transgene.

Progeny Testing

Wool production and properties were compared between transgenics and their non-transgenic half-sibs during the lamb year (Table 75.1). Using means adjusted for body weight as a covariate, wool production averaged 6% greater ($p=0.028$) in transgenics than non-transgenics. Transgenic status had no significant effect on body weight, fibre diameter, or medullation. There was a small but significant ($p=0.042$) increase in bulk. Average staple strength was not significantly affected by transgenic status, but transgenic males showed a reduction compared to transgenic females or non-transgenic animals.

Overexpression of IGF1 in the wool follicle resulted in a significant gain in clean fleece weight. The mechanism of this effect requires investigation, but the lack of a systemic effect on body weight suggests that the growth factor has a direct influence on the wool follicle. No attempt was made to select for high quality wool among the embryo donors. The increase in fleece weight and preservation of fibre characteristics is an encouraging demonstration of the potential of this technology for

Table 75.1. Wool production and fibre characteristics at yearling shearing (Mean ± s.d.).

	Transgenic		Non-transgenic	
	Male	Female	Male	Female
Number of animals	12	10	16	14
Clean fleece weight (Kg)	5.10 ±0.14	5.14 ±0.17	4.67 ±0.12	4.97 ±0.13
Percent yield	73.24 ±0.98	79.39 ±1.17	73.25 ±0.86	79.53 ±0.90
Rate of wool growth (g/day)	114.3 ±0.30	115.4 ±0.36	104.5 ±0.27	111.4 ±0.28
Fibre diameter (microns)	37.07 ±0.63	38.28 ±0.76	36.73 ±0.56	36.59 ±0.59
Medullation (units)	1.99 ±0.20	1.93 ±0.26	2.10 ±0.17	1.93 ±0.18
Bulk (cm³/g)	19.83 ±0.25	19.82 ±0.30	19.41 ±0.22	19.15 ±0.23
Staple Strength (Newtons/Ktex)	29.20 ±3.05	47.93 ±3.64	38.96 ±2.69	41.67 ±2.82
Body weight (Kg)	74.00 ±1.47	70.39 ±1.69	75.31 ±1.27	73.09 ±1.36

improving performance. To our knowledge, this is the first reported improvement of a production trait by genetic engineering of a farm animal without detrimental effects on health or reproduction. Progeny testing of these two-year old animals and of second-generation transgenics is continuing and results will be reported at a later date.

References

Clark, A.J., Archibald, A.L., McClenaghan, M., Simons, J.P., Whitelaw, C.B.A. and Wilmut, I. (1990) The germline manipulation of livestock: progress during the past five years. *Proc. NZ Soc. Anim. Prod.* **50**, 167–177

Damak, S., Jay, N.P., Barrell, G.K. and Bullock, D.W. (1996a) Targeting gene expression to the wool follicle in transgenic sheep. *Bio/Technology*, **14**, 181–184

Damak, S., Su, H.-Y., Jay, N.P. and Bullock, D.W. (1996b) Improved wool production in transgenic sheep expressing insulin-like growth factor 1. *Bio/Technology* **14**, 185–188

Froesch, E.R., Schmid, C., Schwander, J. and Zapf, J. (1985) Actions of insulin-like growth factors. *Annu. Rev. Physiol.* **47**, 443–467

McNab, A.R., Andrus, P., Wagner, T.E., Buhl, A.E., Waldon, D.J., Kawabe, T.T., Rea, T.J., Groppi, V. and Vogeli, G. (1990) Hair-specific expression of chloramphenicol acetyl transferase in transgenic mice under the control of an ultra-high-sulfur keratin promoter. *Proc. Natl. Acad. Sci. USA* **87**, 6848–6852

Philpott, M.P., Sanders D.A. and Kealy. T. (1994) Effects of insulin and insulin-like growth factors on cultured human hair follicles: IGF-1 at physiological concentrations is an important regulator of hair follicle growth in vitro. *J. Invest. Dermatol.* **102**, 857–861

Powell, B.C., Walker, S.K., Bowden, C.S., Sivaprasad, A.V. and Rogers, G.E. (1994) Transgenic sheep and wool growth: possibilities and current status. *Reprod. Fertil. Dev.* **6**, 615–623

Pursel, V.G. and Rexroad, C.E. (1993) Recent progress in the transgenic modification of swine and sheep. *Mol. Reprod. Dev.* **36**, 251–254

Rexroad, C.E., Jr., Mayo, K.M., Bolt, D.J., Elasser, T.H., Miller, K.F., Behringer, R.R., Palmiter, R.D. and Brinster, R.L. (1991) Transferrin- and albumin-directed expression of growth-related peptides in transgenic sheep. *J. Anim. Sci.* **69**, 2995–3004

Sara, V.R. and Hall, K. (1990) Insulin-like growth factors and their binding proteins. *Physiol. Rev.* **70**, 591–614

Ward, K.A. and Nancarrow, C.D. (1991) The genetic engineering of production traits in domestic animals. *Experientia* **47**, 913–922

PART IV SECTION F

76. The Transfer of Isolated Genes Having Known Functions

Kevin A. Ward

CSIRO, *Division of Animal Production, Blacktown, NSW 2148, Australia*

Introduction

Domesticated animals have been crucial to the development of modern affluent societies and the productivity of these animals is today an important factor in maintaining social stability. The increases in production that have been achieved with the application of modern technology are impressive. For example, even confining our attention to recent years where accurate records are available, beef and mutton from world rangelands have increased by an average of 6.5% per annum in the period from 1950 to 1990 (Brown, 1994) and similar dramatic gains are evident in the supply of other major animal products. Unfortunately, general environmental degradation and more specific degradation of pastures and rangelands is creating a situation which is increasingly hostile to high productivity. The inevitable effect of this negative pressure is a significant reduction in the rate of gain of animal productivity, with suggestions that total production may be near maximum level (Brown, 1995). This is a sobering thought in view of the predicted increase in world human population over the next 25 years to at least 8 billion people. Unless both crop and animal productivity can be increased throughout this period, a significant decrease in average global per capita consumption is inevitable.

Throughout history, animal productivity has been incrementally improved by the application of selective breeding, in which animals of superior phenotype are used as the breeding stock for the next generation. With the application of quantitative genetic theory, this approach has proven highly effective as evidenced by the contrast in performance of today's animals with their original progenitors. Annual improvement of many production characters can be increased by between 1% and 3% by using modern statistical tools to aid in selecting complex genetic traits. It would be useful, however, if this rate of progress could be increased.

One way to increase the current rate of productivity improvement is to introduce genes of "major effect" to existing highly productive animals in order to provide a new base for quantitative genetic selection. This may be achievable by making use of the novel technology of genetic engineering, in which single, characterised pieces of recombinant DNA can be stably integrated directly into mammalian genomes. The technology rests on the experiments demonstrating successful transfer of recombinant DNA to laboratory mice by microinjection of the pronuclei of single-cell embryos (for review, Palmiter and Brinster, 1986) and the pioneering experiments of Palmiter *et al.* (1982; 1983) demonstrating that suitably modified genes could be used to alter animal phenotype. The subsequent demonstration that the pronuclear microinjection technique was also applicable to domestic animal species (Hammer *et al.*, 1985) raised the possibility of using the approach for productivity improvement. Thus was initiated a global effort by a number of laboratories skilled in molecular techniques and also possessing the knowledge of domestic animal physiology necessary for the identification of areas where productivity might be increased and for proper evaluation of the transgenic animals that might be produced. As the various research programs progressed, however, it became increasingly apparent that identifying and using genes of major effect to manipulate some of the more obvious physiological processes of domestic animals, for example, circulating levels of growth hormone, was more likely to reduce rather than increase production efficiency (Pursel et al., 1989 and Nancarrow et al., 1991). This may be the result of the long period of natural selection applied to the lineage during evolution, compounded by the intensive selection process that has been applied to domestic animals for many hundreds of years. Together, these have resulted in highly effective gene combinations that may be readily disturbed by the introduction of a single gene of major effect. If this is so, obtaining substantial increases in productivity may require gene modifications significantly more subtle than those attempted to date.

The major areas of animal production that are amenable to manipulation by genetic engineering are the endocrine system, intermediary metabolism, structural proteins, the immune system and novel approaches to disease resistance. The genes for transfer may be selected on the basis of known function and predicted effect on phenotype, or by the isolation of genes by genome mapping. The latter approach will be covered separately. In this review, I will summarise the progress that is being made in the areas identified above, using the former approach in which candidate genes are selected, modified for appropriate function in the target species and then transferred by standard protocols.

Manipulation of the Endocrine System

The manipulation of endocrine status in domestic animals centres largely around the modification of genes encoding growth hormone and has the longest history in domestic animal genetic engineering, undoubtedly resulting from the original growth hormone research in laboratory mice pioneered by Palmiter et al. (1982; 1983). It has been thoroughly reviewed (Rexroad et al., 1989 and Pursel et al., 1989) and hence will be only briefly summarised here. The concept, based on the effects of systemically injected growth hormone on growth in domestic animals, was that by increasing the level of circulating growth hormone, domestic animal species would grow faster and more efficiently utilise available feed (Daughaday et al., 1975). Accordingly, various GH-encoding genomic sequences were spliced to a range of mammalian promoters and used to produce transgenic pigs, sheep, cattle and fish (fish will be considered elsewhere in this book). This work provided a clear example of the existence of limits to physiological manipulation of our productive animals. None of the early animals has proven to be of commercial significance, and many have died very early in life. In most cases, the levels of circulating growth hormone were very high, in sheep reaching as high as 3000 ng/ml in the blood (Nancarrow et al., 1991). Many physical abnormalities were noted in these animals, the most common being aberrations in bone growth, an impaired immune system and pronounced renal dysfunction. The symptoms were classically those observed in acromegaly in humans resulting from an excess of high circulating levels of growth hormone. Attempts were made to reduce the amounts of growth hormone produced in transgenic animals by modifying the promoter sequence ahead of the GH coding sequence (Wagner, 1989) or by attempting to effect slower GH release by utilising a gene encoding the releasing hormone for GH (Pursel et al., 1989). However, these attempts were at best of limited success in the production of animals that could be used commercially.

There has been very little reported on research into the regulation of animal growth by GH manipulation following that described above. A recent report describing pathological changes in pigs harbouring a bovine growth hormone gene regulated by a phospho enolpyruvate carboxykinase promoter (Pinkert et al., 1994) is consistent with earlier findings. Nevertheless, it is possible that if the concentration of circulating growth hormone could be maintained at a level only slightly higher than that found in a non-transgenic animal, faster growth and improved feed utilisation efficiency would follow as found in animals regularly given systemic injections of GH. It remains to be seen if this can be achieved.

Manipulation of Intermediary Metabolism

One of the more exciting areas of application of genetic engineering to domestic animals is the possibility of altering intermediary metabolism in order to increase the availability of substrates identified as rate-limiting for specific production traits. During evolution, mammals have lost many genes that encode enzymes critical for the function of various biochemical pathways. As a result, the molecules which these pathways previously produced must now be provided by the diet, and inadequate supplies of these essential nutrients inevitably restrict productivity. If the missing genes could be replaced, it is likely that the resulting functional pathways would be able to remove such dietary requirements. The appropriate coding sequences for the missing enzymes can frequently be found intact in bacteria and it is therefore possible to isolate these genes, modify them for appropriate eukaryotic expression and transfer them to domestic animals as appropriate. A good illustration of the potential of this approach is given by on going research in sheep to provide these animals with additional cysteine. The current status of this work is summarised as follows:

Introduction of a Cysteine Biosynthetic Pathway to Sheep

Merino sheep produce on average about 4–5 kg of wool per annum. This wool, like all hair fibres, is composed of a complex mixture of keratin proteins that are chiefly characterised by a high content of the amino acid cysteine. Thus, sheep have a high requirement for cysteine which, just for wool production alone, amounts to about 5 g per day (Reis, 1967 and Reis, 1979). Because the biochemical pathway for cysteine biosynthesis from inorganic sulphur is absent in all mammals, cysteine can only be synthesised from methionine or alternatively, must be provided in the diet. There is now substantial evidence to show that on pastures that are of poor quality, the amount of cysteine available limits wool growth in sheep. While it is possible to supplement the diet of sheep on poorer pastures, this is expensive because the cysteine supplement must be "protected" in order to prevent its rapid degradation in the rumen compartment of the sheep's digestive tract (Reis, 1979). An attractive long-term solution would be to enable sheep to synthesise cysteine *de novo* from the sulphide that is present in the digestive tract (Ward and Nancarrow, 1991). To achieve this, it is necessary to provide the animals with two enzymes, serine transacetylase and O-acetylserine sulfhydrylase, which together catalyze the conversion of serine to cysteine as shown in Figure 76.1. If these enzymes were present in the appropriate cells of the sheep's digestive tract, it should be possible for this tissue to synthesise cysteine, because the necessary substrates, serine, acetyl-CoA and sulphide, are all available.

In *Escherichia coli* and *Salmonella typhimurium*, the *cysE* gene encodes the enzyme serine transacetylase (Denk and Bock, 1987) and the *cysK* gene encodes O-acetylserine sulfhydrylase (Byrne *et al.*, 1988). Both of the genes have been isolated and characterised, thus providing the necessary functional genetic material to commence replacement of the pathway for cysteine biosynthesis in mammals. In order to provide transcription of the two genes, several different promoters have been used. These include the sheep metallothionein-Ia promoter (Peterson and Mercer, 1986), the Rous sarcoma virus long terminal repeat promoter (Bawden *et al.*, 1995) and the mouse phosphoglycerate kinase-1 promoter (Bawden *et al.*, 1995). The research associated with the sheep MT-Ia promoter is summarised below; the research using the other promoters differs only in detail. The sheep MT-Ia promoter is known to direct expression in cells of the small intestine and therefore the bacterial *cysE* and *cysK* coding sequences were each joined to separate copies of this promoter. To facilitate transfer of the modified gene sequences to mammalian embryos, a single piece of DNA was prepared that contains both bacterial genes each regulated by the MT-Ia promoter as shown in Figure 76.2.

The gene MTCEK1 was shown to be expressed in cells in tissue culture (Leish *et al.*, 1993) where it enabled cell extracts to carry out the biosynthesis of cysteine from serine and Na_2S *in vitro*. This demonstrated that both bacterial proteins could be stably synthesised by mammalian cells, a prerequisite to extending the work to animals. The gene was then used to generate transgenic mice. In these animals, expression of both bacterial genes was zinc-inducible and primarily found in the cells of the small intestine, confirmed by *in situ* hybridisation studies, Northern blots and enzyme assays. *In vitro* incubation of tissue slices of small intestine from the transgenic animals in a medium containing Na_2S demonstrated that the tissue was capable of *de novo* biosynthesis of cysteine.

To test whether the transgenic mice were able to synthesise sufficient cysteine to alter the phenotype, transgenic and non-transgenic mice were placed on an artificial diet containing only trace amounts of the two sulphur amino acids, cysteine and methionine, but supplemented with Na_2S. The non-transgenic control animals showed a predictable loss of body weight and a substantial loss of hair from actively growing follicles, while the transgenic mice continued to grow normally and to produce

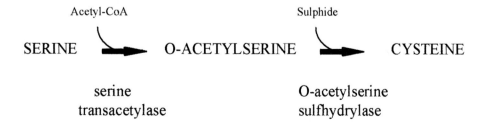

Figure 76.1. The pathway for cysteine biosynthesis.

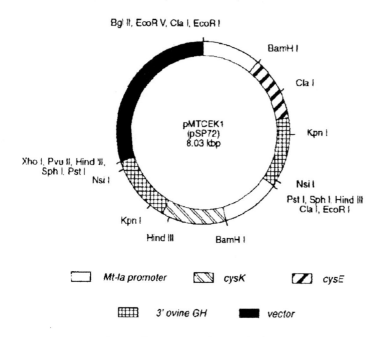

Figure 76.2. The plasmid pMTCEK1 encoding serine transacetylase and O-acetylserine sulfhydrylase.

hair indistinguishable from animals on a normal diet (Figure 76.3) (Ward et al., 1994).

These results support the proposal that the transgenic mice contain a functional cysteine biosynthetic pathway in the cells of the small intestine and that this pathway is able to utilise endogenous substrates and an external source of sulphide to carry out the *de novo* biosynthesis of cysteine in quantities sufficient to replace normal dietary requirements. The final goal of this research is to provide this functional pathway in sheep and the appropriate gene transfer experiments are currently in progress. Transgenic sheep have been produced containing the *cysE* and *cysK* genes from *E. coli* and *S. typhimurium*, but to date, expression levels have been very low and no change in animal phenotype has been observed (Bawden et al., 1995).

Introduction of Other Possible Pathways

The results reported above demonstrate that, in principle, it is possible to alter the intermediary metabolism of animals by transferring functional genes from diverse sources without apparently disturbing the homeostasis of the recipient animals. Clearly, limits to the degree of alteration would be expected, but at least in transgenic mice, these limits appear to be tolerant of significant changes. This is further supported by the results obtained in a second area of research aimed at modifying mammalian intermediary metabolism, where the goal is the introduction of a functional glyoxylate cycle (Ward and Nancarrow, 1991). The general concepts and gene modifications are essentially as described above for the introduction of a cysteine pathway, and the goal of the work is to provide ruminants with the ability to synthesise glucose from the abundant sources of acetate that are produced by fermentation in the ruminal compartment of the sheep's digestive tract. The enzymes that are required for a glyoxylate cycle are isocitrate lyase, encoded in *E. coli* by the *aceA* gene, and malate synthase, encoded by the *aceB* gene. In research that has largely paralleled the work on the cysteine biosynthetic pathway, a gene has been constructed as for pMTCEK1 except that the *cysE* and *cysK* sequences are replaced by the coding sequences of the *aceA* and *aceB* genes. This gene has been shown to produce functional enzymes in mammalian cells in culture and more recently, has been used to generate transgenic mice. The mice have not yet been subjected to detailed physiological examination, but it has been shown that appropriate transcripts are being produced in the intestinal tissue of these mice and that the relevant enzymes are readily detectable in tissue homogenates. Thus, it appears highly likely that these mice possess in the cytoplasm of at least some of the cells of the small intestine, the two enzymes necessary for a functional glyoxylate cycle. The animals are phenotypically indistinguishable from nontransgenic mice, suggesting that they are tolerant of the presence of

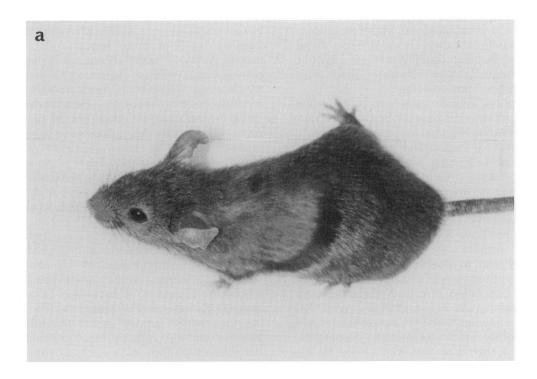

Figure 76.3 The effect of a cysteine-deficient diet on hair growth in (a) control mice and (b) transgenic mice containing a cysteine biosynthetic pathway.

the new enzymes. Since the glyoxylate cycle interacts significantly with substrates common both to glycolysis and the tricarboxylic acid cycle, these results suggest that significant alterations might be possible in intermediary metabolism without causing harm to animal homeostasis.

For monogastric animals such as pigs, the amino acids lysine and threonine are particularly important when the animal's feed is based on cereals. Supplementation of the feed in the order of 3.8 kg/tonne for lysine and 1.8 kg/tonne for threonine is required for optimal protein utilisation. A significant production advantage may result if the genes encoding the necessary biosynthetic pathways could be transferred into these animals and possible approaches have been discussed (Rees et al., 1990). These authors devised a computer-based modelling program to determine the compatibility of such new pathways with existing biochemistry in pigs and concluded that both threonine and lysine synthesis would be possible provided that some of the enzymes were produced in defined ratios one to other. From a technical viewpoint, the transfer of the threonine pathway would be the easier of the two because it involved the transfer of three genes compared with at least eight genes for a functional lysine biosynthetic pathway. However, as discussed below, a recent innovation in plants may now make it more feasible to consider the introduction of longer pathways to animals. In the light of the recent successful demonstration of a cysteine pathway transfer, it is to be hoped that the possibilities of threonine and lysine biosynthesis in monogastric animals will be given further attention.

Technical Limitations

There are technical limitations at present that restrict the application of the transfer of multiple genes encoding biochemical pathways. The most important of these is the size restriction that applies to the piece of DNA to be transferred, since this places an immediate restriction on the number of coding sequences in the transgene. With current techniques, DNA up to 15 kb can be transferred with acceptable efficiency to mice and even larger pieces of DNA have been transferred at low efficiency (Jakobovits et al., 1993 and Schedl et al., 1993). If the approach used to transfer the cysteine biosynthetic pathway is used, each coding sequence with its associated promoter and "spacer" DNA occupies about 3 kb, so that pathways of 3–5 enzymes would be all that could be realistically considered. Recently, however, a unique method of overcoming this limit has been described for the introduction of a biochemical pathway into the plant Nicotiana. This involved the construction of a single DNA sequence which encoded 3 enzymes but all encoded on a single RNA transcript regulated by one 5' promoter sequence. Engineered at precise positions in the encoded polypeptide were plant-specific protease cleavage sites which allowed cleavage of the translation product into three separate enzyme molecules (von Bodman et al., 1995). In the example cited, one of the three enzymes encoded by this unique gene was in fact the protease needed to cleave the fusion polypeptide into its three components. There is obvious potential for this approach to encode metabolic pathways that involve many enzyme steps and it will be interesting to apply similar concepts to mammalian systems.

The manipulation of intermediary metabolism in order to increase domestic animal productivity is still in its earliest stages of development, with the attempts to increase wool growth being the most advanced. However, it may also be possible to increase the efficiency of milk production by altering the energy supply in the mammary gland, increase muscle growth by supplementing the supplies of amino acids such as threonine and lysine and alter fat composition and amount by addition of additional enzymes involved in lipid biosynthesis and modification.

The Manipulation of Structural Proteins

The modification of economically-important structural proteins is an attractive application of genetic engineering for improving productivity. It is difficult to envisage how structural proteins such as collagen or elastin that have an important role *in vivo* can be significantly altered, but the proteins that comprise wool fibres offer significant opportunities and there is currently research attempting to modify the properties of wool by this approach. As described earlier, wool is composed of a complex mixture of keratin proteins for which many of the genes have been isolated and characterised (Powell et al., 1986). In addition, studies over a number of years have identified promoters which have the potential to direct the expression of these genes to different sites within the wool follicle (Rogers and Powell, 1993). This is an attractive target because manipulating the wool fibre is unlikely to have any deleterious effect on the animal itself. The major challenge of the work is understanding how modification of wool composition can predicably alter its properties and efficiency of production. Recent progress is most encouraging. It has been shown that a wool keratin protein can be synthesised in mouse hair follicles in amounts sufficient to cause phenotypic alterations to the mouse hair (Powell and Rogers, 1990) and more recently in sheep, that different wool proteins can be produced in cells of mature wool fibres in quantities that alter the wool phenotype (Rogers and Powell, 1993). It would seem only a matter of time before a combination of proteins results in a fibre with properties uniquely different to existing wool. It will require full testing of this new fibre from shearing to fabric production in order to determine if the new wool is of advantage to the industry, but as our knowledge of the structural relationships of the various wool proteins increases, wool designed-for-purpose is a definite possibility.

Conclusion

Genetic engineering has been promoted to agriculture for over a decade as the solution to many of its current problems. It is becoming obvious, however, that the time-frame for most genetic engineering applications applied to animals is at least 10 years and in most cases is likely to be much longer. In addition, as many of the experiments initiated some years ago advance from studies in laboratory animals to the stage of application in the target species, problems specific to the individual species are being identified and require further research. This is particularly apparent in those projects that involve modifying animal physiology because of the inherent danger of disturbing the homeostatic balance of the modified animal. It is nevertheless encouraging to see that significant changes can be made to the metabolic processes of laboratory mice without apparently harming their overall fitness for growth and survival. While the fine detail may be different for larger animals, it is unlikely that the major principles will be any different from those established in the smaller animal species. Thus, it would seem probable that we will soon see some of the research that has been proven in mice being tested to the full in domestic species.

References

Bawden, C.S., Sivaprasad, A.V., Verma, S.K., Walker, S.K., and Rogers, G.E. (1995) Expression of bacterial cysteine biosynthesis genes in transgenic mice and sheep: toward a new *in vivo* amino acid biosynthesis pathway and improved wool growth. *Transgenic Research* **4**, 87–104

Brown, L.R. (1994) Facing Food Insecurity. In Lester R. Brown, (ed.), *State of the World 1994*, Earthscan Publications, London, pp. 177–197

Brown, L.R. (1995) Nature's Limits. In Lester R. Brown, (ed.), *State of the World 1995*, Earthscan Publications, London, pp. 3–20

Byrne, C.R., Monroe, R.S., Ward, K.A. and Kredich, N.M. (1988) DNA sequences of the *csyK* regions of *S. typhimurium* and *E. coli* and linkage of the *cysK* regions to *pts H. J. Bacteriol.* **170**, 3150–3157

Daughaday, W.H., Herington, A.C. and Phillips, L.S. (1975) The regulation of growth by endocrines. *Annu. Rev. Physiol.* **37**, 211–244

Denk, D. and Bock, A. (1987) L-cysteine biosynthesis in *Escherichia coli*: nucleotide sequence and expression of the serine acetyltransferase (*cysE*) gene from the wild-type and a cysteine-excreting mutant. *J. Gen. Microbiol.* **133**, 515–525

Hammer, R.E., Pursel, V.G., Rexroad, C.E., Jr., Wall, R.J., Bolt, D.J., Ebert, K.M., Palmiter, R.D. and Brinster, R.L. (1985) Production of transgenic rabbits, sheep and pigs by microinjection. *Nature* **315**, 680–683

Jakobovits, A., Moore, A.L., Green, L.L., Verggara, G.J., Maynard-Currie, C.E., Austin, H.A. and Klapholz, S. (1993) Germ-line transmission and expression of a human-derived yeast artificial chromosome. *Nature* **362**, 255–258

Leish, Z., Byrne, C.R., Hunt, C.L. and Ward, K.A. (1993) Introduction and expression of the bacterial genes *cysE* and *cysK* in eukaryotic cells. *Appl. Environ. Microbiol.* **59**, 892–898

Nancarrow, C.D., Marshall, J.T.A., Clarkson, J.L., Murray, J.D., Millard, R.M., Shanahan, C.M., Wynn, P.C. and Ward, K.A. (1991) Expression and physiology of performance regulating genes in transgenic sheep. *J. Reprod. Fertil.* **43**(suppl), 277–291

Palmiter, R.D., Brinster, R.L., Hammer, R.E., Trumbauer, M.E., Rosenfeld, M.G., Birnberg, N.C. and Evans, R.M. (1982) Dramatic growth of mice that develop from eggs microinjected with metallothionein-growth hormone fusion genes stimulate growth in mice. *Nature* **300**, 611–615

Palmiter, R.D., Norstedt, R.E., Gelinas, R.E., Hammer, R.E. and Brinster, R.L. (1983) Metallothionein-human GH fusion genes stimulate growth in mice. *Science* **222**, 809–814

Palmiter, R.D. and Brinster, R.L. (1986) Germline transformation of mice. *Annu. Rev. Genet.* **20**, 465–500

Peterson, M.G. and Mercer, J.F.B. (1986) Structure and regulation of the sheep metallothionein-Ia gene. *Eur. J. Biochem.* **160**, 579–585

Pinkert, C.A., Galbreath, E.J., Yang, C.W. and Striker, L.J. (1994) Liver, renal and subcutaneous histopathology in PEPCK-BGH transgenic pigs. *Transgenic Research* **3**, 410–405

Powell, B.C., Cam, G.R., Feitz, M.J. and Rogers, G.E. (1986) Clustered arrangement of keratin intermediate filament genes. *Proc. Natl. Acad. Sci. USA* **83**, 5048–5052

Powell, B.C. and Rogers, G.E. (1990) Cyclic hair-loss and regrowth in transgenic mice overexpressing an intermediate filament gene. *Embo J.* **9**, 1485–1493

Pursel, V.G., Pinkert, C.A., Miller, K.F., Bolt, D.J., Campbell, R.G., Palmiter, R.D., Brinster, R.L. and Hammer, R.E. (1989) Genetic engineering of livestock. *Science* **244**, 1281–1288

Rees, W.D., Flint, H.J. and Fuller, M.F. (1990) A molecular biological approach to reducing dietary amino acid needs. *Biotechnology* **8**, 629–633

Reis, P.J. (1967) The growth and composition of wool. *Aust. J. Biol. Sci.* **20**, 809–825

Reis, P.J. (1979) Effect of amino acids on the growth and properties of wool. In Black, J.L. and Reis, P.J. (eds.),

Physiological and Environmental Limitations to Wool Growth. University of New England Publishing Unit, Armidale, Australia, pp. 223–242

Rexroad, C.E. Jr., Hammer, R.E., Bolt, D.J., Mayo, K.M., Frohman, L.A., Palmiter, R.D. and Brinster, R.L. (1989) Production of transgenic sheep with growth regulating genes. *Mol. Reprod. Develop.* **1**, 164–169

Rogers, G.E., and Powell, B.C. (1993) Organization and expression of hair follicle genes. *J. Invest. Dermatol.* **101**(suppl), 50S–55S

Schedl, A., Montoliu, L., Kelsey, G. and Schutz, G. (1993) A yeast artificial chromosome covering the tyrosinase gene confers copy number-independent expression in transgenic mice. *Nature* **362**, 258–261

Von Bodman, S.B., Domier, L.L. and Farrand, S.K. (1995) Expression of multiple eukaryotic genes from a single promoter in *Nicotiana*. *Biotechnology* **13**, 587–591

Wagner, T.E. (1989) Development of transgenic pigs. *J. Cell. Biochem. suppl.* **13B**, 164

Ward, K.A., Leish, Z., Bonsing, J., Nishimura, N., Cam, G.R., Brownlee, A.G. and Nancarrow, C.D. (1994) Preventing Hairloss in Mice. *Nature* **371**, 563–564

Ward, K.A. and Nancarrow, C.D. (1991) The genetic engineering of production traits in domestic animals. *Experientia* **47**, 913–922

PART IV, SECTION F

77. Recent Progress in Mammalian Genomics and its Implication for the Selection of Candidate Transgenes in Livestock Species

Michel Georges
Department of Genetics, Faculty of Veterinary Medicine, University of Liège, Belgium

Since the pioneering experiments of Palmiter and Brinster (Palmiter *et al.*, 1982) producing giant mice by injection into a fertilized mouse egg of a rat growth-hormone structural gene under control of a metallothionein promotor, the perspectives of transgenics as the ultimate tool for the genetic improvement of livestock have spurred the imagination of many scientists and laymen alike. While conventional breeding strategies are limited to the exploitation of the genetic variation pre-existing within the species, if not the breed of interest, transgenics on the contrary opened possibilities to exploit genetic variation across species boundaries, and even more, to exploit "artefactual" genetic variants created *in vitro*.

If transgenesis has become an integral part of the arsenal used by plant breeders, the equivalent methods have proven much more difficult to implement in animal genetics. This reflects the convergence of a number of complicating factors, including technical and economic hurdles associated with the production of transgenic livestock, as well as concerns about public perception. It is often contended, however, that the implementation of transgenic techniques in livestock has also been hampered considerably by the limited choice of suitable transgenes.

The remarkable progress that we have been witnessing in recent years in the analysis of complex mammalian genomes – a scientific discipline now commonly referred to as genomics – may soon alleviate the latter concern, i.e., the limited choice of transgenes. Most of the required information will likely be a *direct* result of the massive investments that are presently being allocated to the analysis of the human and – to a slightly lesser degree – mouse genomes, epitomized by the Human Genome Project. In addition these megaprojects are *indirectly* driving the field of livestock genomics, and helping us to unravel the molecular biology of economically important production traits. These advances will increasingly point towards genes ameanable to transgenic manipulation in livestock.

The Human and Mouse Genome Projects

Approximately five years ago, the human genetics community committed itself to the Human Genome project: a major undertaking expected to culminate in the obtention of the complete sequence of the human genome around the year 2005. The way towards this ultimate goal was beaconed with a number of milestones of which several are now completed. The first of these was the generation of a high density marker map spanning the entire genome. In 1994, the group of Jean Weissenbach published a linkage map comprising more than 2,000 highly polymorphic microsatellite markers (Gyapay *et al.*, 1994). This set of markers, as well as continuously improving methods for high throughput genotyping, have rendered the mapping of genes underlying single-locus traits virtually trivial provided segregating family material is at hand, and has even allowed to gain some preliminary insight in the genetic determinisims of complex traits. Moreover, this high density marker map has formed the scaffold for STS content mapping, that in combination with fingerprinting data and information on cross-hybridization of Alu-PCR products, has allowed for the generation of a nearly complete YAC-based physical map of the human genome: the second milestone (Chumakov *et al.*, 1995). This YAC-based primary physical map will progressively mature into a more refined sequence ready map, exploiting user-friendlier procaryotic cloning systems such as PACs and cosmids. In the meantime sequencing efforts have focused their attention on the transcribed portion of the genome, as obtained from a variety of cDNA libraries representing at least thirty distinct tissues. Recently, nearly 90,000 distinct partial complementary DNA sequences – referred to as Expressed Sequence Tags (ESTs) – were described totalling 83 million nucleotides (Adams et al., 1995). One third of these sequences resulted from the assembly of overlapping ESTs into Tentative Human Consensus (THC) sequences, yielding full-length transcripts in many cases. These THCs and ESTs are presently being positioned onto the physical map by assignment to specific YACs or through the use of radiation hybrids. These efforts should result in the obtainment of a nearly complete human transcript map in the next year.

In parallel, comparable efforts are being devoted to the study of the murine genome: generation of a high density microsatellite based marker map, construction of a complete YAC-based physical map and efforts towards transcript sequencing and mapping (f.i. Dietrich et al., 1994). While for a minority of this plethora of newly identified genes, some preliminary ideas about their role may be obtained from sequence similarities observed with known genes and gene-families, as a general rule their precise function remains elusive. Obviously, the daunting task of deciphering the human (and for that matter mammalian) genetic program will challenge the inventiveness of biologists decades if not centuries to come. Two strategies are likely to contribute in a decisive manner to the elucidation of gene function in the near future: positional candidate cloning and the production of knock-out mice.

Positional candidate cloning is the logic extension of positional cloning (Collins, 1995). This latter strategy consists in (1) the initial chromosome mapping or positioning of the gene or genes underlying a trait of interest using linkage analysis and related strategies, followed by (2) the construction of a physical map of the identified chromosome region, (3) the identification of the genes lying within this chromosome segment, and (4) the determination of the culprit gene through the identification of phenotype-specific sequence variation. By April 1995, Collins (1995) reported the succesful positional cloning of 42 human genes, all disease-causing. Similar progresses were reported in mice, the most publicised of which probably being the positional cloning of the *leptin* gene causing the obese phenotype of *ob* mice (Zhang et al., 1994).

These numbers are expected to increase dramatically in the near future as a result of the major genome analysis efforts. Indeed, the benefit of the Human and Mouse Genome Projects to positional cloning is to directly address issues 2 and 3 for the entire genome, which is expected to immensely accelerate the identification of genes underlying segregating pathological as well physiological phenotypes: the positional candidate cloning approach.

The elucidation of the role of an increasing number of genes cloned by the positional candidate approach, both in man and in the mouse, can confidently be expected in the near future, and will likely provide animal geneticists with an increasing number of candidate genes to manipulate by transgenesis. It is noteworthy that animal breeders have used mice to model selection for all economically relevant production traits for decades. Strains of mice are available that could be exploited to great advantage using the positional candidate approach, and that might provide a subset of prime candidate genes for further *in vitro* manipulation. This strategy is illustrated by the ongoing efforts to positionnaly clone the *hg* gene enhancing growth in mice (Horvat and Medrano, 1994).

Positional candidate cloning will only shed light on the function of those genes underlying identified segregating phenotypes. A more general strategy to obtain information about the function of a newly identified gene is to knock the gene out by homologous recombination in murine ES cells. The ES cells heterozygous for the loss-of-function mutation are then injected in mouse blastocysts, and the descendants of the resulting germ-line chimaeras intercrossed to produce homozygotes for the *de novo* mutation. The phenotypes exhibited by such "knock-out" mice provides some clues about the normal function of the corresponding gene. A compendium of phenotypes associated with 263 reported gene knock-outs was recently published (Brandon et al., 1995). It is reasonable to expect that the number of such experiments will dramatically increase in the near future. While most phenotypes associated with knock-out experiments reported so far are debilitating to say the least, it would seem very surprising if at least some of the obtained knock-out phenotypes would not point towards genes of potential interest to animal breeders and deserving further manipulation in livestock.

Livestock Genomics

It is noteworthy that the community of animal geneticists has been relatively reluctant to invest into the new discipline of genomics when compared to plant, human and mice genetics. This is likely due in part to the realisation that the majority of economically important traits in livestock are typical multifactorial traits, i.e. determined by a multitude of genes acting in concert with a myriad of environmental effects, and are therefore the most difficult ones to tackle using genomic strategies. Moreover, given the spectacular genetic progress achieved by means of conventional breeding programs, some scepticism prevailed as to the cost-effectiveness of biotechnology in livestock production.

Although both concerns remain valid to a great extent, we have nevertheless witnessed remarkable progress in the field of livestock genomics over the last five years, undoubtedly catalyzed by the Human Genome Initiative. Considerable efforts are now being devoted to elucidate the molecular basis of the observed variation for most economically important traits.

The so-called "candidate gene approach" (Collins, 1995) has led to the identification of a limited number of genes and sometimes mutations therein that influence traits of relevance to livestock production. As its name implies, this approach consists in studying the effect on the phenotype of interest of genetic variants uncovered in genes which are prime candidates given their known physiological function. It is important to realize that the thousand or so genes whose function is known to a sufficient extend to qualify as candidate genes only represent of the order of 1% of the total number of genes composing the genome of a typical mammal. This method therefore has an inherently low *a priori* chance of success. Genes causing recessive disorders in livestock identified using the candidate gene approach include the *ryanodine receptor* underlying malignant hyperthermia and the associated Porcine Stress Syndrome in pigs (Fuji *et al.*, 1991), *thyroglobulin* that when mutated causes goitre in goat (Veenboer *et al.*, 1993) and cattle (Ricketts *et al.*, 1987), the CD18 gene causing bovine leucocyte adhesion deficiency (BLAD) (Schuster *et al.*, 1992), *uridine monophosphate synthetase* causing DUMPS in cattle (Schwenger et al., 1993), *arginosuccinate synthetase* causing bovine citrullinemia (Dennis *et al.*, 1989), α-*glucosidase* causing Pompe's disease in cattle (Wisselaar *et al.*, 1993), the α *subunit of the skeletal muscle sodium channel* gene causing periodic paralysis in the horse (Rudolph *et al.*, 1992). Mutations in the α-*MSH receptor* have been shown to underly the black-red coat color polymorphism observed in cattle (Klungland *et al.*, 1995 and Charlier, personal communication). Growing evidence is being presented for a contribution of genetic variants of milk protein genes, both caseins and whey proteins, to the population variance of milk yield in cattle and goat (f.i. Grosclaude, 1988 and Mahé *et al.*, 1993). More recently, evidence has been reported supporting an effect of *oestrogen receptor* variants on litter size in pigs (Rotschild *et al.*, 1994).

As in human and rodent genetics however, the importance of the candidate gene approach is expected to be superseded very quickly by positional cloning approaches which are more generally applicable. Implementation of these strategies, however, requires the preliminary development of mapping tools suitable for livestock species. Several groups have devoted their activities towards that goal during the last five years. As a results of these efforts, comprehensive microsatellite-based marker maps are rapidly becoming available for the most important livestock species: cattle, pig, sheep and poultry in particular (Beattie, 1994). Large insert libraries (YACs and BACs) (e.g. Libert *et al.*, 1994 and Cai *et al.*, 1995) as well as radiation hybrid panels are being generated. Moreover, the animal genetics community, with its long-standing tradition in quantitative genetics, is making a significant contribution in the development of statistical mapping methods (f.i. Haley and Knott, 1992).

Succesful implementations of this arsenal of new tools for the mapping of economically important genes has recently begun. As expected the first successes dealt with monogenic traits, both recessive disorders as well as production traits. Genes causing recessive disorders in livestock recently localized by linkage strategies include bovine *progressive degenerative myeloencephalopathy* (or Weaver) (Georges *et al.*, 1993), bovine *syndactyly* (Charlier *et al.*, 1996), and the *roan* locus involved in the determination of *White Heifer Disease* (Charlier *et al.*, 1996). The *Polled* genes determining the presence or absence of horns in cattle (Georges *et al.*, 1994), the *Booroola* ovine fecundity gene (Montgomery *et al.*, 1994) the *callipyge* locus causing a muscular hypertrophy in sheep (Cockett *et al.*, 1994), the *mh* locus causing double-muscling in cattle (Charlier *et al.*, 1995), the *Rn* or *Napole* locus affecting carcass quality in pigs (Milan *et al.*, 1995 and Leif Andersson, personal communication), the *Dominant White* coat-color locus in pigs (Johansson *et al.*, 1992), the *dominant white* locus in poultry (Ruyter-Spira *et al.*, 1996) are amongst the economically relevant monogenic traits recently located on the respective marker maps using linkage approaches. As denser maps become available, genotyping becomes more efficient and statistical methods improve, the location of single-gene traits becomes trivial in livestock species as well, as long as appropriate pedigree material is available.

Moreover the first whole genome scans performed in livestock species have lead to the location of Quantitative Trait Loci, i.e. polygenes underlying multifactorial traits. A first study pinpointed a pair of loci accounting for part of the difference observed between wild-boar and domestic pig for growth and carcass characteristics (Andersson *et al.*, 1994).

The second study identified five autosomal loci influencing milk yield, still segregating within highly selected dairy populations and therefore representing the molecular substrate of ongoing selection programs (Georges et al., 1995). Both studies demonstrated the feasibility of QTL mapping in livestock, and are only a prelude of what will likely be a rich harvest of QTL in the different livestock species resulting from the multitude of ongoing projects.

At present, the actual identity of all these mapped genes remains unknown, awaiting their positional cloning. While the efforts required to positionally clone a gene following its location by linkage analysis should not be underestimated, animal geneticist will greatly benefit from the Human and Mouse Genome Initiatives in their efforts. Comparison of gene maps across mammals has indeed demonstrated a remarkable conservation of gene order. It has for instance been estimated that as few as 150 to 200 chromosomal breaks separate the mouse and human genome (Nadeau and Taylor, 1984), and this number seems even less when comparing bovine and human. Even if more refined studies can only uncover additional reaarangements, the conservation of synteny remains striking. Therefore, the efforts to produce complete transcript maps in the human and mouse, shifting cloning strategies in these species from positional towards positional candidate, will have similar repercussions in livestock species as long as strategies are developed that allow for high resolution comparative mapping, i.e. allowing to determine boundaries on the human map flanking the human homologue of one's favourite livestock gene. Considerable efforts are being devoted in order to facilitate such cross-species referencing. Zoo-fish, that is chromosome painting realized with complex heterologous probes, is one strategy that has recently been applied succesfully to define chromosomal homologies between the human, pig and cattle genomes (Rettenberger et al., 1995 and Solinas-Toldo et al., 1995). Methods with higher resolution are likely to emerge in the near future.

Prospects for the identification of genes underlying important traits for livestock production using positional cloning strategies might therefore seem bright. While it is indeed reasonable to predict the succesful outcome of such endeavours in the not too distant future for genes underlying monogenic traits, the outlook might be considerably darker when considering quantitative traits. Indeed, present QTL mapping efforts in livestock roughly delineate a chromosomal segment likely to contain one or more genes influencing the trait of interest. The size of such segment is of the order of 10–20 cM or 10 to 20 million base pairs which might contain as much as 300 to 600 genes. Devising strategies to pinpoint the gene or genes accounting for the observed effect within the defined interval remains a major, unsolved intellectual challenge.

It is important to realize that the major driving force behind these QTL mapping efforts is not the perspective to clone the underlying genes. What motivates QTL mappers in a first instance is the linkage disequilibrium that is inherently revealed between the QTL and the linked genetic markers as part of the mapping effort (whether within a pedigree or at the population level). Indeed, even without the actual knowledge of the causal genes and mutations, this linkage disequilibrium allows one to monitor and direct the segregation of the identified chromosomal segment, with the QTL it contains, in the population of interest. This so-called Marker Assisted Selection (MAS) is expected to increase genetic response by a combination of improved accuracy of selection, reduction of generation interval and increased selection differential. Because of its inherent relative simplicity, as it does not require a detailed molecular knowledge of the manipulated genes, this form of "marker assisted engineering" potentially associated with new reproductive technologies allowing for a reduction in generation interval (Georges, 1991), is likely to precede genetic engineering of livestock based on transgenic methods.

Conclusions

The flourishing of genomics as a new scientific discipline will lead to the identification of numerous genes affecting diverse biological functions of potential relevance to animal production. These genes will be manipulated initially by marker assisted selection allowing to exploit the genetic variation preexisting within animal populations of interest. This preexisting genetic variation will eventually be complemented with transgenic variation created *de novo*, also amenable to marker-aided manipulation as already witnessed in plant breeding. While a considerable number of technical hurdles remain to be addressed before seeing these methods implemented in livestock production, the major challenge faced by animal geneticist might be to convince the taxpayer to continue investing in research pursuing enhanced productivity in agriculture, while the major challenge towards feeding the world population increasingly becomes distribution rather than production.

References

Adams, M. and 84 co-authors (1995) Initial assessment of human gene diversity and expression patterns based upon 83 million nucleotides of cDNA sequence. *Nature* **377**, 3–174

Andersson *et al.* (1994) Genetic mapping of quantitatve trait loci for growth and fatness in pigs. *Science* **263**, 1771–1774

Beattie, C. (1994) Livestock genome maps. *Trends in Genetics* **10**(9), 334–338

Brandon, E.P., Idzerda, R.L., McKnight G.S. (1995) Targeting the mouse genome, a compendium of knockouts. *Current Biology* **5**(6–7–8), 625–634, 758–765, 873–881

Cai, L., Taylor, J.F., Wing, R.A., Gallagher, D.S., Woo, S.-S., Davis, S.K. (1995) *Genomics* **29**, 413–425

Charlier, C., Coppieters, W., Farnir, F., Grobet, L., Leroy, P., Michaux, C., Mni, M., Schwers, A., Vanmanshoven, P., Hanset, R. and Georges, M. (1995) The mh gene causing double-muscling in cattle maps to bovine chromosome 2. *Mammalian Genome* **6**, 788–792

Charlier, C., Denys, B., Belanche, J.I., Coppieters, W., Grobet, L., Mni, M., Womack, J., Hanset, R. and Georges, M. (1996) Microsatellite mapping of a major determinant of *White Heifer Disease*; the bovine *roan* locus. *Mammalian Genome* **7**, 138–142

Charlier, C. and Farnir, F., Berzi, P., Vanmanshoven, P., Brouwers, B., Georges, M. (1996) IBD mapping of recessive disorders in livestock: application to bovine syndactyly. Genome Research in press

Chumakov, I.M. and 61 co-authors (1995) A YAC contig map of the human genome. *Nature* **377**, 175–297

Cockett, N.E., Jackson, S.P., Shay, T.D., Nielsen, D., Green, R.D. and Georges, M. (1994) Chromosomal localization of the callipyge gene in sheep (Ovis aries) using bovine DNA markers. *Proceedings of the National Academy of Sciences USA* **91**, 3019–3023

Collins, F.S. (1995) Positional cloning moves from perditional to traditional. *Nature Genetics* **9**, 347–350

Dennis, J.A., Healy, P.J., Beaudet, A.L. and O'Brien, W.E. (1989) Molecular definition of bovine arginosuccinate synthetase deficiency. *Proceedings of the National Academy of Sciences USA* **86**, 7947–7951

Dietrich, W.F. and 14 co-authors (1994) A genetic map of the mouse with 4006 simple sequence length polymorphisms. *Nature Genetics* **7**, 220–245

Fuji, J., Otsu, K., Zorzato, F., Deleon, S., Khanna, V.K., Weiler, J.E. O'Brien, P.J. and MacLennan, D.H. (1991) Identification of a mutation in the porcine ryanodine receptor associated with malignant hyperthermia. *Science* **253**, 448-451

Georges, M. (1991) Perspectives for marker assisted selection and velogenetics in animal breeding. Current Communications in Cell and Molecular Biology 4, Animal applications of research in mammalian development. *Cold Spring Harbor Laboratory Press*, pp. 285–325

Georges, M., Lathrop, M., Dietz, A.B., Lefort, A., Libert, F., Mishra, A., Nielsen, D., Sargeant, L.S., Steele, M.R., Zhao, X., Leipold, H. and Womack, J.E. (1993) Microsatellite mapping of the gene causing weaver disease in cattle will allow the study of an associated QTL. *Proceedings of the National Academy of Sciences USA* **90**, 1058–1062

Georges, M., Drinkwater, R., Lefort, A., Libert, F., King, T., Mishra, A., Nielsen, D., Sargeant, L.S., Sorensen, A., Steele, M.R., Zhao, X., Womack, J.E. and Hetzel, J. (1993) Microsatellite mapping of a gene affecting horn development in Bos taurus. *Nature Genetics* **4**, 206–210

Georges, M., Nielsen, D., Mackinnon, M., Mishra, A., Okimoto, R., Pasquino, A.T., Sargeant, L.S., Sorensen, A., Steele, M.R., Zhao, X., Womack, J.E. and Hoeschele, I. (1995) Mapping quantitative trait loci controlling milk production by exploiting progeny testing. *Genetics* **139**, 907–920

Grosclaude, F. (1988) Le polymorphisme génétique des principales lactoprotéines bovines. *INRA Production Animale* **1**, 5–17

Gyapay, G., Morisette, J., Vignal, A., Dib, C., Fizames, C., Millasseau, P., Marc, S., Bernardi, G., Lathrop, M. and Weissenbach, J. (1994) The 1993–1994 Généthon human genetic linkage map. *Nature Genetics* **7**, 246–339

Haley, C.S. and Knott, S. (1992) A simple regression method for mapping quantitative trait loci in line crosses using flanking markers. *Heredity* **69**, 315–324

Horvat, S. and Medrano, J.F. (1994) Interval mapping of high growth (hg), a major locus that increases weight gain in mice. *Genetics* **139**, 1737–1748

Johansson, M., Ellegren, H., Marklund, L., Gustavsson, U., Ringmar-Cederberg, E., Andersson, K., Edfors-Lilja, I. and Andersson, L. (1992) The gene for dominant white color in the pig is closely linked to ALB and PDGFRA on chromosome 8. *Genomics* **14**, 965–969

Klungland, H. and Lien, S. (1995) *Mammalian Genome* **6**, 636–639

Libert, F., Lefort, A., Okimoto, R. and Georges, M. (1993) Construction of a bovine genomic library of large yeast artificial chromosome clones. *Genomics* **18**, 270–276

Mahé, M.F., Manfredi, E., Ricordeau, G., Piacere, A. and Grosclaude, F. (1993) Effects of the s1 casein polymorphism on goat dairy performances, a within-sire analysis of alpine bucks. *Génétique Selection Evolution* **26**, 151–158

Milan, D., Le Roy, P., Woloszyn, N., Caritez, J.C., Elsen, J.M., Sellier, P. and Gellin, J. (1995) *Génétique Slection Evolution* **27**, 195–199

Montgomery, G.W. *et al.* (1994) The ovine Booroola fecundity gene (FecB) is linked to markers from a region of human chromosome 4q. *Nature Genetics* **4**, 410–414

Nadeau and Taylor (1984) Length of chromosomal segments conserved since the divergence of man and mouse. *Proceedings of the National Academy of Sciences USA* **81**, 814–818

Palmiter, R.D., Brinster, R.L., Hammer, R.E., Trumbauer, M.E., Rosenfeld, M.G., Birnberg, N.C. and Evans, R.M. (1982) Dramatic growth of mice that develop from eggs microinjected with metallothionein-growth hormone fusion genes. *Nature* **300**, 611–615

Rettenberger, G., Klett, C., Zecher, U., Kunz, J., Vogel, W. and Hameister, H. (1995) Visualisation of the conservation of synteny between humans and pigs by heterologous chromosomal painting. *Genomics* **26**, 372–378

Ricketts, M.H., Simons, M.J., Parma, J., Mercken, L., Dong, O. and Vassart, G. (1987) A nonsense mutation causes hereditary goitre in the Afrikander cattle and unmasks alternative splicing of thryglobulin transcripts. *Proceedings of the National Academy of Sciences USA* **84**, 3181–3184

Rotschild, M.F., Jacobson, C., Vaske, D.A., Tuggle, C.K., Short, T.H., Sasaki, S., Eckhardt, G.R. and McLaren, D.G. (1994) A major gene for litter size in pigs. *Proceedings of the 5th World Congress on Genetics applied to Livestock Production*, Guelph August 7–12, 1994, **21**, 225–228

Rudolph, J.A., Spier, S.J., Byrns, G., Rojas, C.V., Bernoco, D. and Hoffman, E.P. (1992) Periodic paralysis in Quarter Horses: a sodium channel mutation disseminated by selective breeding. *Nature Genetics* **2**, 144–147

Ruyter-Spira, C.P., van der Poel, J.J. and Groenen, M.A.M. (1996) Bulked segregant analysis using microsatellites: *mapping of the dominant white locus in chicken*, submitted for publication.

Sabinas-Toldo, S., Lengauer, C. and Fries, R. (1995) Comparative genome map of man and cattle. *Genomics* **27**, 489–496

Schwenger, B., Schober, S. and Simon, D. (1993) DUMPS cattle carry a point mutation in the uridine monophosphate synthase gene. *Genomics* **16**, 241–244

Shuster, D.E., Kehrli, M.E., Ackermann, M.R. and Gilbert, R.O. (1992) Identification and prevalence of a genetic defect that causes leukocyte adhesion deficiency in Holstein cattle. *Proceedings of the National Academy of Sciences USA* **89**, 9225–9229

Veenboer, G.J.M. and Devijlder, J.J.M. (1993) Molecular basis of the thyroglobulin synthesis defect in dutch goats. *Endocrinology* **132**, 377–381

Wisselaar, H.A., Hermans, M.M.P., Visser, W.J., Kroos, M.A., Oostra, B.A., Aspden, W., Harrison, B., Hetzel, D.J.S., Reuser, A.J.J. and Drinkwater, R.D. (1993) Biochemical Genetics of Glycogenosis Type-II in Brahman cattle. *Biochemican and Biophysical Research Communications* **190**, 941–947

Zhang *et al.* (1994) Positional cloning of the mouse obese gene and its human homologue. *Nature* **372**, 425–432

PART V

78. To Save or Not to Save: The Role of Repositories in a Period of Rapidly Expanding Development of Genetically Engineered Strains of Mice

John J. Sharp and Larry E. Mobraaten
The Jackson Laboratory, Bar Harbor, Maine, USA

Introduction

The ability to selectively alter the mouse genome began 15 years ago with the first reports of transgenic mice (Gordon et al., 1980). More recently (Mansour et al., 1988 and Capecchi, 1989) targeted mutant mice have been created in which a selected gene is disrupted (or 'knocked out') through homologous recombination in embryonic stem cells (ES cells). New targeting technologies are being rapidly developed which allow tissue specific (Gu et al., 1994) and temporal control (Kuhn et al., 1995) of the gene targeting event. These methods of producing genetically engineered mice have produced hundreds of strains that provide powerful tools for biomedical research. The ability to add to and selectively alter the mouse genome opens an exciting new era of research into the genetic bases of human health and disease and also provide powerful tools for devising and testing novel therapeutic approaches.

We have included a single figure with this chapter because it dramatically illustrates the problem facing any repository, the large number of strains being produced and used. Figure 78.1 graphs current and projected publications where transgenic or targeted mutant mice were produced or used in the study. Because some individual strains may appear in several studies these data do not represent the actual number of strains being produced, but they do represent the trend in production of new strains. This is especially true for targeted mutant mice where the large majority of publications are reports of newly developed strains.

In 1992 a special meeting was held at the Mouse Molecular Genetics conference in Cold Spring Harbor where concerns were expressed regarding the health, preservation and free access to these strains. It was the consensus of those present at this meeting, that there was a critical need for a central repository to archive and distribute important transgenic, targeted and chemically mutagenized mice (Culliton, 1993). Similar sentiments were expressed at the *Workshop on Sharing Laboratory Resources in Biological Research* sponsored by The National Academy of Sciences National Research Council (Anderson, 1993). In response to these concerns and following requests from the scientific community, The Jackson Laboratory established the Induced Mutant Resource (IMR) in September 1992 whose function is to collect, preserve and distribute genetically engineered strains of mice (Sharp and Davisson, 1994). Since that time the IMR has been designated as the *National Repository for Transgenic and Targeted Mutant Mice* by the National Institutes of Health (NIH) (Curran, 1994).

Once the decision was made to create a new genetic resource, the essential components of an ideal repository were identified based on our experience with other genetic resources at the Laboratory (i.e. the Mouse Mutant Resource which maintains spontaneous mutants and the Frozen Embryo Repository which preserves them). Subsequently a plan was developed which incorporated many of these components into the basic functions of the IMR. In this article we will describe these functions, what we consider to be an ideal repository and a prospectus for the future.

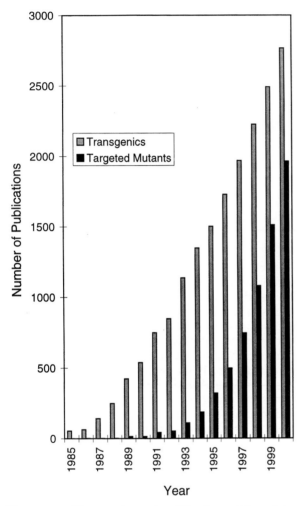

Figure 78.1. The number of publications citing either transgenic mice or targeted mice. Current (1985–1995) values were obtained using SilverPlatter to search MEDLINE. Values for 1996–2001 were projected utilizing the historical rates and trends for transgenic publications. In this projection the rate of increase publications citing targeted mutant mice is increasing 3.4 fold faster than publications citing transgenic mice. Such a trend would be expected.

Functions of the IMR

Identification and Selection of Strains

Ideally, a repository would collect and distribute every genetically engineered mouse strain that is produced. As illustrated in Figure 78.1, it is not possible, or perhaps not even desirable, to archive or distribute every strain being produced. It is therefore necessary to select from among the total, those of the greatest biomedical importance, realizing that predicting future, or even current, biomedical importance is fraught with uncertainty. What follows is a description of the process used to identify and select mutants for inclusion into the IMR.

The first step in the selection process is to identify the strains being produced. New strains are identified by the following routes: (1) strains are offered by the investigator who made them; (2) regular literature scans are carried out; (3) Jackson Laboratory scientists or scientists outside The Jackson Laboratory suggest possible mutants; and (4) suggestions are made by IMR Advisory Board or Associated Board members (The IMR Advisory Board is a seven member board that meets annually at The Laboratory and the Associated board is comprised of scientists who are asked to keep us informed about important mutants in their field of expertise).

Information gathered about mutants being considered for admission is presented to an internal Laboratory committee, the Genetic Resources Committee (GRC). The final decision on taking a strain rests with this committee, which meets monthly. Criteria for selecting mutants is based on existing guidelines for importing mice to the Laboratory's Genetic Resources. These criteria include: (1) the immediate need for use in biomedical research; (2) the numbers of requests for animals being received by the investigators who created them; (3) the potential for future research; (4) the time and effort needed to replace or recreate the mutant or strain; and (5) the uniqueness of the mutation or strain. Because of the subjective nature of the selection criteria, advice is frequently sought from Jackson Laboratory scientific staff, outside investigators and/or IMR board members. If an offered strain is accepted the investigator is contacted and importation is scheduled. If a strain identified by one of the other methods is accepted then the investigator is contacted to see if he or she would be willing to contribute it to the repository.

Standardizing the Health of Strains

A critical function of any repository is assuring and maintaining the health of stocks that it distributes, and in fact, the wide difference in health of strains being exchanged between research colonies was one of the concerns that led to the establishment of the IMR. Inadequate surveillance could potentially threaten the health of colonies worldwide. The Jackson Laboratory has an importation program designed to free incoming animals from viruses and other pathogens they carry. All strains arriving at The Laboratory are placed in isolators in the Importation Facility where they are monitored and hysterectomy or embryo rederived into existing colonies. We receive animals originating from multiple research colonies with diverse levels of animal health monitoring and rederivation is absolutely necessary to protect existing Laboratory colonies and insure the health of animals being distributed. This is illustrated by the fact that within the past 5 years we have detected evidence of current or prior infection with one or more viruses in as many as 75% of mice being imported, with nearly 50% of these infections being mouse hepatitis virus.

The importation process rids animals of the pathogens they carry but this process may also affect the phenotype of the strain. We have on occasion observed an alteration or delay of onset of the phenotype following rederivation. We attempt to characterize the phenotype of every strain received but financial constraints hinder a full characterization. We are creating a database that will provide any information we do collect. This information will be available via the Jackson Laboratory home page on the World Wide Web.

Cryopreservation

As soon as breeding mice are available after passing the importation process, cryopreservation of 8-cell embryos is initiated to provide back-up in case the strain is accidentally lost as a result of breeding failure, disease, genetic contamination, or other misfortune. Although many stocks could be re-imported, or even reconstructed, the investment in time and cost to the point of introduction into the IMR easily warrants the effort to cryopreserve a small number of embryos to insure the ability to readily reconstitute a stock if necessary. The cryopreservation of embryos at this initial entry point also serves to preserve the original genotype where it is subject to change as a result of further backcrossing onto genetically defined backgrounds.

Because many transgenic or knock-out stocks arrive on genetically heterogeneous backgrounds it is not necessary for these stocks to use sib or other close relative matings to produce embryos for cryopreservation. In fact, it is often expeditious to cross transgene or mutant carrier males to readily available C57BL/6J females, producing hetero- or hemizygous embryos for purposes of preservation. The stock can then be reconstituted by mating mice recovered from frozen embryos *inter se* or with mice from the maternal strain. Progeny are then genotyped for selection of breeders for the subsequent generation. In addition to availability there are two advantages in using a strain such as C57BL/6J for this. First, the superovulatory response of C57BL/6J females is well defined, yielding a predictably efficient cryopreservation of the transgene or mutant. Secondly, C57BL/6J is a commonly used genetic background for many mutants and transgenes, and the use of this strain for cryopreservation accomplishes the initial or, in some cases, additional backcrosses to the inbred strain.

For stocks in which the transgene or mutant is fixed in the homozygous state, it may be desirable to produce embryos by mating males with females from the same stock. The disadvantages are the difficulty of obtaining sufficient numbers of female donors and superovulatory responses may be less certain with these females. The primary advantage is that embryos will be homozygous and do not require typing to select appropriate breeders for the next generation.

A minimum of 200 embryos that are at least hetero- or hemizygous are preserved at the outset. In the case where it is necessary to use a hetero- or hemi-zygous male as the stud mated to wild-type females, it is necessary to preserve twice as many embryos since only half will be carrying the transgene or mutation. The minimum of 200 embryos provides reasonable expectation of being able to reconstitute the stock in case breeders are accidentally lost. Those stocks for which demand has diminished or for which "cryo only" preservation has been recommended, a minimum of 500 carrier embryos is preferred. The extra embryos not only confer added probability of recovery of the stock, but also allow multiple recoveries to be made over time without requiring the expansion of breeding stock with each recovery for the purpose of replacing thawed embryos. Then, when withdrawals reduce the account to the 200 embryo minimum, the next recovery should be used to produce replacement embryos.

Many successful cryopreservation protocols for mammalian embryos have now been described. They basically fall into two general groups, those based on equilibrium freezing where cooling rates are generally low requiring equipment or devices which can control the rate of cooling, and non-equilibrium freezing where rates are ultra-rapid, obviating the need for controlled rate freezers. Because of nearly 20 years of success with the cryopreservation of 8-cell embryos using a controlled rate freezing method, our laboratory continues to use this method. It is basically the method first described by Whittingham *et al.* (1972) when the technique was first reported in 1972. It is fully described in Mobraaten (1986).

Genetic Monitoring

Genetic monitoring (genetic typing) of strains constitutes one of the most expensive and necessary aspects of operating a repository for genetically engineered mice. Most of these strains have no visible phenotype and therefore must be genotyped to verify they are carrying the appropriate transgene or targeted mutation. Genetic typing is required to: (1) verify the genotype of the animals received by the repository; (2) identify heterozygotes for distribution for those strains where the homozygous mutants are embryonic lethals or are not viable; (3) annually verify the genotype of strains being distributed as homozygotes; (4) identify carrier animals (heterozygotes) for strains being backcrossed onto a defined genetic background; and (5) verify the genetic background for those strains on a defined genetic background. Identifying heterozygotes for distribution represents better than 80% of the genotyping activity because 35% of the strains currently held by the IMR must be distributed as hetero- or hemi-zygotes and we must type every animal produced from these strains.

The IMR will genotype about 40,000 samples in 1996 and this figure is expected to grow in subsequent years. Because of this large volume, the polymerase chain reaction (PCR) is utilized for the majority of genetic typing. PCR is rapid, the reaction conditions may be standardized, it does not require the use of radioisotopes, and it is adaptable to automation. Where possible, flow cytometry (FACS analysis) may be substituted for PCR and has the advantage that it also provides a verification of the mutant phenotype. Southern blotting is used to verify the copy number of transgenics on an annual basis and may be utilized for genotyping before a PCR protocol is developed. The development of a robust PCR protocol is necessary for reliable, high-volume genetic typing. Maintaining this robustness is highly dependent on the quality of the DNA template utilized in the PCR reaction. For this reason we place great emphasis on standardizing DNA extraction procedures. Blood is utilized as the major source of DNA although tail tips may also be used.

In addition to technicians in the typing laboratory, there are personnel dedicated to developing and/or improving PCR protocols. We also have a program to follow and develop new typing and PCR product-detection methodologies. A semi-automated hardware and software genotyping system from the Applied Biosystems division of Perkin Elmer is utilized for some routine typing and is also being used to develop quantitative PCR methods for copy number verification of transgenic strains.

Distribution

Providing new mutations to investigators as rapidly as possible is a major goal of the IMR program and should be the goal of any repository that supplies animals. All strains are normally distributed on a first-come-first-serve basis. We have found it necessary to categorize the availability status of strains because the demand for a strain may be greater than our ability to supply it. In order to distribute strains as quickly as possible we begin to distribute excess mice while establishing the colony. These strains are usually listed as *Minimal Distribution* in our strain list. Once a colony is established but demand is low, the strain is listed as *Limited Distribution* and it is possible to obtain limited numbers of mice. More popular strains are moved to our production facility and are listed as *Available* once the colony has been expanded to meet demand. If there is a single request for a large number of mice of a particular strain, we will ship in smaller groups interspersed with shipments to other investigators so that the resources are distributed equitably. Colonies are expanded or reduced according to demand.

Distribution of strains from the IMR in 1995 increased more than 300% from levels in 1994. Early 1996 figures indicate a trend towards continuing growth. These figures show that the scientific community is utilizing the IMR and documents the need for a central repository that distributes animals. Archiving embryos and ES cell lines must also be a component of a repository but researchers will experience a considerable delay in obtaining live mice maintained solely by archiving. If one goal of a repository is to speed the rate of scientific research then it should provide live animals, at least for more widely utilized strains. Our own rapid growth has begun to stretch our own resources and we are currently in the process of evaluating current research trends in order to anticipate future needs.

Strain Development and Model Development

Most currently used ES cell lines are derived from the 129 genetic background and researchers often mate a chimera obtained from a gene targeting event with a C57BL/6 (B6) animal. Subsequent intercrossing of these F_1 progeny results in offspring of mixed B6/129 genetic background and most targeted mutant strains arrive at the Laboratory on this segregating background. Some researchers also mate their chimera with a 129 animal, thus providing the mutation on a pure 129 background. In general we do not import mutants on the 129 background because of space considerations. The breeding characteristics of the 129 strain are so poor that the colony size must be very large to provide animals for distribution. Many transgenic strains also arrive on a segregating background because F_1 embryos are often used to inject the DNA construct. However, the types of experiments that can be performed with mutations on segregating genetic background is limited and such mutations will find their greatest utility if backcrossed onto a defined genetic background. For this strain development we backcross animals for 10 to 12 generations, a process that is expensive and takes about two years to complete. Although we initially planned to transfer most mutations on mixed backgrounds to the C57BL/6J background, we now find that space limitations force us to select mutants for strain development. Backgrounds other than C57BL/6J are also considered for backcrossing.

It is anticipated that the numbers of mutants arriving on segregating backgrounds will decrease as ES cell lines derived from inbred strain mice other than 129 become more generally available (for example Kagi et al., 1994).

The IMR also undertakes genetic development of new disease models by transferring mutant genes or transgenes to defined genetic backgrounds and by combining multiple mutant genes, or transgenes into a single animal. The majority of model development is being carried out with funds provided by the Department of the Army Breast Cancer Initiative and consequently involves the development of new models for breast cancer.

Information Resources

Information about the strains held by the IMR may be obtained via a direct mailing or via the World Wide Web (WWW). We are currently accepting 8 to 9 new strains each month and thus the mailed list is out of date rather quickly. The most current list may be obtained via the WWW. The address of the Jackson Laboratory home page is *http://www.jax.org* and a link to the IMR home page is provided from here. Information provided about each strain includes: reference number, genetic background, availability, gene symbol, chromosomal location, and reference(s). A link to MGD (Mouse Genome Database) is also provided for each mutant. Additional information available at the WWW site includes: genotyping protocols, informational data sheets for many of the mutants, access to technical support and an on-line form for submitting a mutant to the IMR. It is also possible to be added to a list and receive a monthly update of new IMR additions. Links to other databases such as TBASE (described elsewhere in this book) and GDB are also provided.

The IMR, as a genetic resource, manages a large number of strains and we are currently adding about 100 new strains per year. A database is maintained for the internal tracking of strains and to record information such as the observance of an altered phenotype. A database with references to targeted and transgenic mutants is also maintained in order to identify potential new IMR strains. A third database is maintained to track, record and report all genotyping results. We are currently in the process of constructing a single database that will integrate all of these functions and that will make much of this information available via the WWW.

Funding and Self-Sufficiency

To our knowledge there has never been a living culture collection, for any organism, that was totally self-sufficient and able to support itself solely from user fees (Matthews and Qualset, 1994). The responsibility of a genetic resource is to collect and maintain genetic variants of scientific importance, and to select them by criteria that include factors other than potential distribution levels. If distribution level were the only criteria used for selecting IMR mutants, the number available to the scientific community would be very small indeed.

Complete financial self-sufficiency requires that the cost of importation, and cryopreservation be recovered as well as costs for strain development, genetic typing, maintenance and distribution. If these costs were to be totally recovered, it would make the price of these animals extraordinarily high and a barrier to experimental design. The majority of IMR mutants will not be distributed in large enough numbers to permit full cost recovery although they are of considerable biomedical importance. The diversity of strains available is also an important value at a repository. We have therefore sought funding to support the importation, cryopreservation and strain development of these stains. The mission of the IMR to collect and distribute these biomedically important mutants cuts across the entire spectrum of human disease and it is therefore reasonable to expect support from a variety of government and voluntary health care agencies. The response to requests for funding has been very positive. Startup funding was provided by the March of Dimes, the American Cancer Society, the American Heart Association, the Cystic Fibrosis Foundation, the Multiple Sclerosis Society and the Amyotropic Lateral Sclerosis Foundation. Subsequent funding has been obtained from the Howard Hughes Medical Institute, from two branches of the NIH, NCRR and NIAID, and from the Department of the Army Breast Cancer Initiative.

The current plan for self-sufficiency of the IMR is based on the expectation that a majority of the biomedically important transgenic, chemically induced and targeted mice imported into The Jackson Laboratory will not be distributed in large enough numbers to permit full cost recovery for their importation, cryopreservation, strain development and distribution, and still be affordable to researchers. We expect then, that some supplementation will be a necessary component of the IMR and indeed for any repository operating as a genetic resource.

Legal and Technology Transfer Issues

The Jackson Laboratory has taken the position that genetically engineered strains of mice are research tools, developed to a large extent with government funds and as such, should be made freely available to the scientific community. The Laboratory encourages other institutions to provide these mice to the IMR without licensing in the interest of making their mice available to the scientific community as quickly as possible. This position has been maintained with respect to distributing mice to scientists at academic or other non-profit research institutions.

However, to obtain these mice we have sometimes been obliged to engage in distribution licenses, providing among other things for payment of royalties on mice sold or facilitating the obtaining of use licenses from for-profits. In the latter situation, the Laboratory may seek in turn to be reimbursed for the costs involved in importing and maintaining these mice. Whether subjected or not to licensing agreements, these mice are still distributed in the same manner as mice from the Laboratory's other colonies. About one third of the strains being distributed by the IMR require that commercial entities obtain a license from the originating institution in order to use the strain.

Current Status

As of this writing the IMR has accepted more than 250 mutant stocks with an active distribution of more than 170. The remaining stocks are either being expanded for distribution, are waiting importation or have been archived.

Components of an Ideal Repository

Overview

The ideal repository, of course, is one in which all scientifically valuable transgenic animals are preserved and can be made available at a standard high level of health quality in a timely fashion to investigators at a reasonable cost. As the literature based on the use of transgenic animals is exponentially growing, it is clear that the number of transgenic animals themselves is also growing exponentially. An ideal repository is one that can expand at the same pace. In a climate of limited resources for research, however, it may not be possible to maintain an ideal repository, so in order to maximize the proportion of transgenics that can be preserved, it is important to preserve and provide resources in the most efficient manner possible. This may call for new and more economical methods of preservation as well as facilities designed for maximum efficiency of function for this specialized effort. New methods may, however, have a price in terms of the timeliness of availability. For example, transgenic strains with infrequent use for a period of time are most economically preserved by cryopreservation of embryos or sperm as opposed to maintaining active breeding colonies, but several months may be required before mature, experimental animals of the appropriate genotype can be made available after being requested.

Selection of Mutants

If all transgenic strains developed cannot be preserved, priorities based on valid criteria or the collective wisdom of an appropriately selected board of advisors should be assigned before individual strains are eligible to be preserved. This is a difficult task since oftentimes the future scientific value of a particular transgenic model cannot readily be predicted. There is a certain sense of prophetic self fulfillment if a strain is not preserved–it will never become valuable.

Efficient preservation of transgenic strains may call for cooperation and possible sharing of preserved material between regionally dispersed centers. Total centralization, while possibly efficient, may not be the most desirable means because of necessary remoteness from some research laboratories. Total de-centralization, on the other hand, where each transgenic laboratory preserves its own products lacks standardization in terms of health status, availability, distribution ability, or even security of preservation.

Cryopreservation

Cryopreservation of embryos has been a proven method of preservation of mouse stocks for some 20 years. It is generally agreed that embryos or gametes stored at liquid nitrogen temperatures (196°C) should remain viable for time periods on the order of centuries providing the storage is carefully managed. Cryopreservation of embryos is the only practical and sure means of preserving those stocks for which homozygosity is necessary or desired. These would be stocks for which a unique genetic background must be preserved or for which progeny typing after recovery from frozen embryos would be impractical or undesirable.

On the other hand, many transgenic and targeted mutant strains could be cryopreserved by freezing and storing sperm from carrier males. Thawed sperm could then be used to fertilize oocytes from a commonly available inbred or hybrid female to produce hetero- or hemizygous embryos. Cryopreservation of sperm would be far more cost effective than producing and cryopreserving embryos. While methods for cryopreserving sperm from mice remain to be perfected to a level of success achieved for other mammalian species, an illustration of the potential efficacy of sperm cryopreservation can be made from reported results. A single male can yield sufficient sperm to fertilize about 32 *in vitro* fertilization cultures, each containing 30 eggs. With 50% fertilization and 17% live born recovery of transferred zygotes, as reported by Penfold *et al.* (1993), a single male, then, could yield an average of 81 live births. Furthermore, since only a very few males would be needed to cryopreserve a strain, the building up of a colony solely for the purpose of providing donors, as is the case in embryo cryopreservation, is not required. This is usually the most costly part of embryo cryopreservation. With embryo cryopreservation one can expect to obtain an average of 16 embryos from a superovulated female donor of a typical strain such as C57BL/6J (Mobraaten, 1981). At an overall live-born recovery of 20%–25% for inbred strains, a single female donor will then yield an average of only 4 live births, compared to the 81 expected form a single sperm donor.

A reliable method for mouse sperm cryopreservation, then, would provide an economic means to allow the preservation of a much larger proportion of all transgenic and targeted mutants being produced than is practical at the present time. The cost of ultra-cold storage on a per stock basis is relatively insignificant, minimizing the concern about the cost of accumulating too many transgenic stocks. The virtue of being able to preserve nearly everything that has been described in peer reviewed published literature is that it makes that material available for future research whether for verification of initial findings or for further study. Experience with the cryopreservation of spontaneous mutants and other genetically defined mouse strains over the past 18 years amply shows that many such strains are revisited after their initial discovery or development and subsequent periods of quiescence in frozen storage. At the present time, from a bank which contains approximately 700 stocks of genetically defined mice available only as frozen embryos, researchers throughout the world are requesting reconstitution of about 100 stocks per year. Many fruitful lines of research would not have been possible or as productive had such stocks been terminated after their initial use.

Upgrading Health Status

Collecting transgenic strains in a central repository offers an opportunity for re-derivation and the maintenance of all strains at the highest level of health status. Combining an importation procedure with a cryopreservation program confers a certain efficiency on the process. Since embryos, or eventually sperm, are taken for cryopreservation, they can be taken from animals of uncertain health status in an aseptic manner, washed, and used, either before or after freezing, to produce animals of defined health status. Transgenic strains acquired by a repository solely for cryopreservation need not be re-derived by conventional methods since taking embryos from such strains constitutes the essential step in a re-derivation process. As long as the embryos are transferred into recipients of defined health status, they are re-derived each time a reconstitution is made. It is recommended that animals derived from such reconstitutions be monitored as further security against a breech in the prevention of disease transmission.

Genetic Quality Control

Providing animals, embryos or ES cells from genetically engineered strains of mice necessitates a commitment to verifying the genetic integrity of the strains being supplied. Investigators receiving strains (whether as live animals, embryos or ES cells) that have no visible phenotype rely on the repository to supply material with the correct genotype. Consequently, the repository must dedicate considerable resources to verify the genetic integrity of the material it is supplying.

It is unreasonable to expect that usable, robust genotyping protocols will be provided by the investigator and therefore, the repository should also be capable of developing and implementing its own typing protocols. In addition, the ideal repository would dedicate some effort to developing and/or employing new methodologies that have the potential for reducing the costs associated with genotyping.

Distribution

An ideal repository should be in a position to distribute either live animals of defined health status or cryopreserved embryos (or sperm). Many laboratories using transgenic animals for research do not have facilities or expertise to derive developed animals from frozen embryos or gametes. The rapidly increasing demand for animals from the IMR as described earlier amply illustrates this need. Animals distributed to such laboratories should be of the highest health quality, not only to meet the requirements of the animal facility associated with the requesting laboratory (the number of animal facilities for research requiring barrier maintained animals is constantly increasing), but also to encourage the use of such high quality animals in research.

Laboratories possessing the facilities and expertise to produce developed animals from frozen embryos or gametes often request embryos or gametes for the purpose of being able to re-derive the strain to their particular health standards. Shipping stresses on animals are also avoided by transporting frozen embryos or gametes. For some countries, quarantine regulations governing the importation of embryos or gametes are much less strict than for young or adult animals.

Information Resources

Basic information on each stock preserved must be recorded and made available. Material preserved without information is virtually non-existent and it is questionable whether it should have been preserved in the first place.

In an era of information explosion, contributed to in large part by the remarkable surge of molecular and transgenic methodology, the management of large amounts of information must be carefully designed, verified, and secured. Many different and sophisticated database management applications are available and informatics has become an important part of any repository.

Information must not only be preserved but needs to be disseminated as well if the transgenic model is to be at all useful. The dissemination of information has taken a giant leap forward with the advent of the World Wide Web. Other means of distribution will be needed as well, but this one advance has made the provision and acquisition of information easier by orders of magnitude.

The Future

Genetically engineered strains of mice have become critical tools for biomedical research into the cause and treatment of human disease but the full scientific benefit of these strains will only be realized if they are made readily available to the scientific community. Clearly the current demand for archived strains and the rapidly growing demand for strains held by the IMR has established the need and usefulness of central repositories to maintain and distribute these strains. In our opinion bio-medical research will best be served by the formation of regional central repositories which cooperate towards a common goal. It will be impossible for a single repository to assume total responsibility for all genetically engineered strains and it is imperative that a cooperative effort be established in order to prevent potential competition and distribute the associated costs. One such repository being formed in Europe is EMMA (the European Mouse Mutant Archive). Current plans call for EMMA to act primairly as an archive for frozen embryos with little or no distribution of live animals. Scientists from the The Jackson Laboratory have been providing assistance in establishing EMMA. We are also aware of an effort in Japan to start a repository similar to the IMR.

As the Human Genome Project begins to identify the total spectrum of genes and other functional regions of the genome, genetically engineered murine models will continue to be produced, certainly at an even greater rate than they are being produced today. The challenge for the future will be to meet the increased demand for the preservation and distribution of these strains with, what has always been, limited resources.

References

Anderson, C. (1993) Researchers win decision on knockout mouse pricing. *Science* **260,** 23–24

Capecchi, M.R. (1989) Altering the genome by homologous recombination. *Science* **244,** 1288–1292

Culliton, B.J. (1993) A home for the mighty mouse. *Nature* **364,** 744

Curran, M. (1994) New Resource for Genetically Engineered Mice. *NCRR Reporter* **18,** 12–13

Gordon, J.W., Scangos, G.A., Plotkin, D.J., Barbosa, J.A. and Ruddle, F.H. (1980) Genetic transformation of mouse embryos by microinjection of purified DNA. *Proceedings of the National Academy of Sciences USA* **77,** 7380–7384

Gu, H, Marth, J.D., Orban, P.C., Mossmann, H. and Rajewsky, K. (1994) Deletion of a DNA polymerase beta gene segment in T cells using cell type-specific gene targeting. *Science* **265,** 103–106

Kagi, D., Ledermann, B., Burki, K., Seiler, P., Odermatt, B., Olsen, J.J., Podack, E.R., Zinkernagel, R.M. and Hengartner, H. (1994) cytotoxicity mediated by T cells and natural killer cells is greatly impaired in perforin-deficient mice. *Nature* **369,** 31–37

Kuhn, R., Schwenk, F., Aguet, M. and Rajewsky, K. (1995) Inducible gene targeting in mice. *Science* **269,** 1427–1429

Mansour, S.L., Thomas, K.R. and Capecchi, M.R. (1988) Disruption of the proto-oncogene int-2 in mouse embryo-derived stem cells: A general strategy for targeting mutations to non-selectable genes. *Nature* **336,** 348–352

Matthews, K. and Qualset, C.O. (1994) Living Culture collections a National Assett: Experts set recommendations for the future. *Diversity* **10,** 48–49

Mobraaten, L.E. (1981) The Jackson Laboratory Genetic Stocks Resource Repository in *Frozen Storage of Laboratory Animals.*, Gustav Fischer Verlag, Stuttgart. pp. 165–177

Mobraaten, L.E. (1986) Mouse embryo cryobanking. *J. In Vitro Fert. Emb. Transfer* **3,** 28–32

Noben-Trauth, N., Kohler, G., Burki, K., Ledermann, B. (1996) Efficient targeting of the IL-4 gene in a BALB/c embryonic stem cell line. Transgenic Res. 5: (In Press).

Penfold, L.M. and Moore, H.D. (1993) A new method for cryopreservation of mouse spermatozoa. *J. Reprod. Fertil.* **99,** 131–134

Sharp, J.J. and Davisson, M.T. (1994) The Jackson Laboratory Induced Mutant Resource. *Lab Animal* **23** 32–40

Whittingham, D., Leibo, S. and Mazur, P. (1972) Survival of mouse embryos frozen to 196°C and 269°C. *Science* **178,** 411–414

PART V

79. TBASE: The Relationalized Database of Transgenic Animals and Targeted Mutations

Anna V. Anagnostopoulos

Division of Biomedical Information Sciences, The Johns Hopkins University School of Medicine, 2024E. Monument Street, Baltimore, Maryland, 21205-2236, USA

Recent Investigative Approaches and their Impact on Transgenic Data Growth and Management

Remarkable accomplishments in embryo culture, gamete micromanipulation and recombinant DNA technology, originally applied as unrelated technical advances, have taken place during the last two decades, and these areas have now coalesced to create an entirely novel investigative approach to various aspects of genetics. Germline gene transfer into the pronuclei of fertilized ova *via* microinjection, has resulted in the generation and use of numerous transgenic animals as powerful probes into complex biological systems. Similarly, gene targeting, particularly in mouse embryonic stem (ES) cells, has been widely used to generate a variety of mutations in a number of distinct loci, so that the phenotypic consequences of specific genetic modifications may be assessed.

Technical Advances in Genetic Manipulations: Transgenesis and Gene Targeting

Insertion of exogenous DNA during the creation of transgenic animals may lead to the generation of very interesting mutants. Since the advent of transgenic mouse technology about ten years ago, numerous transgenic mice have been made carrying wild-type or mutant forms of many genes, under the transcriptional regulation of various promoters/enhancers. In addition, transgenic mice have provided a new source of mouse mutations, since the introduced DNA can act as a mutagen if it integrates randomly into loci. In most cases, however, it is laborious to identify the locus in which the insertion has occurred, as these sequences have not been previously identified. Thus, the major limitation inherent in the genetic analysis of the mammalian genome has been the production of defined mutations that are transmittable through the germline. These shortcomings have now been circumvented by the development of gene targeting. Unlike traditional transgenic technologies, in which exogenous DNA is inserted randomly in multiple copies into various chromosomal locations, gene targeting relies on homologous recombination between the exogenous DNA sequences and the target locus. Thus, while transgenesis normally leaves homologous sequences unaffected, gene targeting places sequences at one site in the genome. Depending on the nature and preparation of the construct, the endogenous nonselectable sequences can be either replaced and/or disrupted with replacement or insertion vectors containing a selectable marker. Smithies *et al.* (1985) provided the first example of "targeting" into an endogenous gene. They disrupted the beta-globin gene with the neomycin (neo^r) and bacterial suppressor (supF) gene and used a "phage rescue assay" to detect and clone targeted cells. Just a few years ago, data from null mutant mice produced by the two pioneering investigators, Capecchi and Smithies, entered the scientific literature in a trickle. Now the fields of immunology, developmental biology, neurobiology, cancer genetics and many others, are generating an increasing stream of data from these designer mutants (see Data Growth and Management).

To date, gene targeting has been primarily used to examine gene function by creating mice that carry null mutations at particular loci ("knockouts"). With time, and as more sophisticated methods for making subtle mutations become available, gene targeting will have other applications such as *in vivo* protein structure function studies, analysis of DNA regulatory elements, production of mouse models of human diseases and production of dominant mutants. Techniques to make subtle modifications at a defined site, such as within translated, untranslated, transcribed or nontranscribed regions, have been recently developed by a number of groups. These strategies include (a) coelectroporation, which generates recombinants in a single step, (b) the "hit-and-run" approach, which uses two rounds of single reciprocal recombination, and (c) the "double-hit" strategy, which uses two rounds of double reciprocal recombination. The most powerful, "double-hit" gene replacement, relies on the exchange of a negative marker at the target site with vector sequences. Virtually any genetic alteration (deletions, insertions, exchange of control regions, and exchange of coding sequences) could potentially be made and isolated by selection. Although these techniques have not been widely used to date, their feasibility has been clearly demonstrated (Bradley, 1993).

As the complexities of multifactorial human genetic diseases are unraveled, correlations are observed, genetic linkages revealed and candidate genes identified and studied at an ever-increasing rate (Smithies, 1993). Consequently, gene targeting allows a synthetic approach: defined mutations can be created, their individual effects studied (against more than one genetic background, if desired), and be combined with other genetic factors by breeding. In the case of the ApoE knockout for instance, analyses of the double heterozygotes and homozygotes should prove useful in elucidating the multifactorial basis of the genetic predisposition to atherosclerosis (Robbins, 1993). In addition, phenotypic analysis of double mutants may clarify whether redundancy is the main factor underlying the lack of correlation often observed between gene expression patterns and mutant phenotypes. If redundancy is the case, then screens based on mutant phenotypes will only uncover aspects of a gene's function that are not redundant. Considering the hundreds of known genes likely to affect a complex biological system, like the immune system, and which are now amenable to gene targeting, it would not be surprising to see an exponential growth in the number of mouse strains with mutations affecting immune functions. Thus, while the relative rarity of natural mutants was an important limitation to the study of *in vivo* immune responses in the past, it is conceivable that in the future one limitation to these studies may be the logistics of maintaining the large number of mouse strains that will be generated from targeted ES cells, from interbreeding targeted strains with one another, with existing transgenic mice, and with naturally occurring mutant strains (Koller, 1992). Indeed, the very success of these genetic manipulations, has raised important issues pertaining to the access and availability of popular knockout and transgenic animals (Anderson, 1993).

Evidently, the development of the technique of germline gene targeting in mice is a compelling example of how various technical advances are brought together to solve a long standing problem. The development of the polymerase chain reaction for detecting rare genetic alterations has greatly contributed to the solution. Moreover, the recent emergence of yeast artificial chromosome (YAC) clones, is expected to increase dramatically as large genome centers generate additional physical mapping data. YACs will, in turn, be used in the construction of *in vivo* libraries in "transpolygenic" mice designed to contain multiple transgene insertions. Transpolygenic mice are expected to represent yet another category of data populating TBASE in the imminent future.

Despite the value of targeted murine models, a mouse is not a human nor, in a number of cases, the most appropriate animal model. Transgenic technology has already been adapted to domestic livestock, poultry and fish, in view of its potential commercial exploitation in agriculture and biotechnology. Similarly, efforts are under way in a number of laboratories to develop ES cells capable of germline transmission for the hamster, rat, guinea pig and rabbit. Although no successes have yet been reported, it may be anticipated that the methodological and technological problems are solvable. In birds, for instance, pronuclear microinjection into large yolky eggs, has been proven inefficient, and thus replaced by DNA integration from a cytoplasmic site. Also, in the fowl, a novel methodology has been recently described referred to as ballistic transfection of avian primordial germ cell *in ovo*. This method employs direct firing of tungsten DNA vector-coated microprojectiles in this avian tissue, without the need of subsequent special culture techniques (Li *et al.*, 1995).

TBASE: Its Significance and Long-Term Objectives

TBASE is the computerized community database that stores and organizes the exploding literature describing transgenic animals and targeted mutants. It contains information pertaining to the experimental methodologies and techniques conducted for their production, characterization, attributes and current maintenance, as well as any applications that they may have in the medical, agricultural or industrial fields. As the central data repository, TBASE provides the scientific community with a methodical procedure for accessing and openly communicating findings generated worldwide. TBASE is designed to be a significant component of a larger information infrastructure for computational biology. Indeed, it is a unique attempt to systemize published and unpublished data, and allow retrieval of relevant material through sophisticated queries and transparent linking of related genomic databases. Its full adoption by the community will enable users to share data electronically, avoid duplication of experimentation, and exploit the collaborative possibilities it offers at an international level.

The long-term objectives of TBASE as a community database may be listed as follows:

(a) To enhance TBASE awareness and promote its adoption as the central animal community database that maintains comprehensive and current coverage of all transgenic animals and targeted mutants generated internationally.
(b) To implement a relational database schema that enhances the role of TBASE as a powerful research tool with considerable contribution to genome informatics, as opposed to a static archive of scientific publications or unreported findings.
(c) To ensure that such a schema emphasizes, at any given time, emerging scientific needs, controversial findings, and technological advancements.
(d) To develop enhanced applications for data browsing and editing over the Internet, and design software submission tools with error-checking capabilities for direct electronic submission.
(e) To enable and encourage electronic submission as the primary mode of data deposition, while continuing to process direct paper entry forms, and proceeding with literature scanning as the basic data submission strategies.
(f) To further an active cooperation between the funding agencies and journals and effectively promote and press direct data submission concurrent with, or prior to publication.
(g) To evolve TBASE as a semantically consistent component of a broader information infrastructure for computational biology, that enables direct query links among related genomic databases through transparent linking.
(h) To train and allocate additional TBASE personnel in direct data extraction, processing and entry, in order to maximize TBASE utility.
(i) To utilize a board of scientific advisors to serve as a filtering mechanism and provide guidance and curation in controversial issues, such as transgenic nomenclature, unpublished findings and others.
(j) To maintain a bulletin board where news, general comments, technical advances and future TBASE-related meetings will be announced electronically.

Data Growth and Management

The marked increase observed over the last few years in the number of transgenic animals and targeted germline mutations is illustrated in Figures 79.1 and 79.2. In particular, the gene targeting approach has gained tremendous popularity over the last few years (Figure 79.2), and been successfully used and recognized as a powerful genetic tool. Mutant phenotypic analyses reported through December 1991 included phenotypes of only two homeobox-containing genes En-2 (Joyner et al., 1991) and Hox1.5 (Chisaka and Capecchi, 1991), three proto-oncogenes wnt-1 (Thomas and Capecchi, 1990 and McMahon and Bradley, 1990), src (Soriano et al., 1991) and myb (Mucenski et al., 1991), four genes in addition to myb involved in the hematopoietic system (beta-2 microglobulin (Zilstra et al., 1990 and Koller et al., 1990), GATA-1 (Pevny et al., 1991), immunoglobulin mu chain (Kitamura et al., 1991) and Lyt-2 (Fung-Leung et al., 1991), and the growth factor IGF-II (DeChiara et al., 1990 and DeChiara et al., 1991). The count has increased dramatically since then, as reflected by the number of relevant Medline Citations retrievable during the specified time intervals (Figures 79.1 and 79.2). For instance, since May 1995 there have been at least seven reported transgenic mouse models attempting to reproduce the pathology and phenotypic characteristics of Alzheimer's disease alone (Lamb, 1995). Likewise, the total number of abstracts presented on transgenics and null mutants at the Cold Spring Harbor Mouse Molecular Genetics Meeting mounted to 98 and 88, respectively, reflecting 69.9% of all data presented in 1992, and to 73 and 145 abstracts, respectively, reflecting 80.7% of all data presented in 1994 (Figure 79.3). An even greater percentage was noted during the course of 1995 (data not shown), incorporating emerging data on "inducible" targeted mutants.

TBASE is strictly an animal database, which currently focuses on the mouse as the predominant mammalian model. Spontaneous and chemically induced mutants are presently excluded from the database. The expected feasibility to create targeted lines in other species (hamster, rat, rabbit, pig, poultry, etc.) and the emergence of transpolygenic

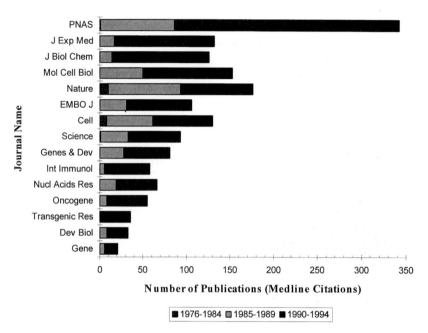

Figure 79.1. Total number of publications on transgenic animals presented as Medline citations from regularly scanned journals during the indicated time periods.

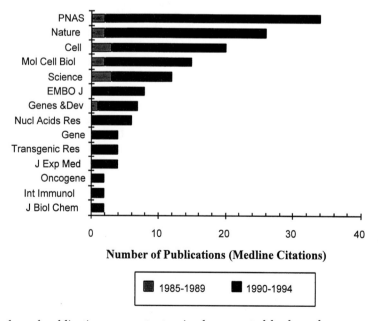

Figure 79.2. Total number of publications on mutant animals generated by homologous recombination in embryonic stem cells, and presented as Medline citations from regularly scanned journals during the indicated time periods.

Data Categories Presented as Abstracts at the CSH Mouse Molecular Genetics Meeting in 1992.

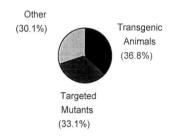

Data Categories Presented as Abstracts at the CSH Mouse Molecular Genetics Meeting in 1994.

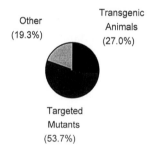

Figure 79.3. Data categories presented as abstracts of papers at the Cold Spring Harbor (CSH) Mouse Molecular Genetics Meetings in 1992 and 1994. Note that the percentage of data pertaining to targeted mutants has increased from 33.1% to 53.7%, whereas the percentage of data on transgenic animals has decreased from 36.8% to 27.0% during this two-year period. Clearly, the data on targeted mutations and transgenic animals combined represent the majority of data presented at recent CSH Meetings (69.9% in 1992 and 80.7% in 1994).

mice will have a considerable impact on the growth of the database. Furthermore, as the central repository of data generated by transgenesis and gene targeting, TBASE must also allow the data acquisition strategies to reflect advances in technological and investigational approaches. Presently, TBASE represents 792 laboratory entries (that is, address of institution where the contact person is located), 647 distinct DNA constructions, 804 transgenic, and 782 targeted mutants generated via homologous recombination in ES cells. Based on previous observations (Chipperfield et al., 1993), it is expected that use of the electronic data submission scheme will have a marked effect on the rate of data entry and, thus, growth of TBASE (see Data Acquisition and Submission Methods). Systematic promotion of TBASE awareness and input of *unpublished* data from the original investigators, will undoubtedly contribute to the expansion of the database. TBASE awareness is to be broadened through constructive interaction with principal investigators, frequent on-line demonstrations of the database at major related meetings, and an appealing, dynamic Internet exposure. These and other issues will be discussed in detail in the following sections of this chapter.

Promoting Tbase Awareness and Usability

Reciprocal Exposure and Constructive Feedback between TBASE and the Scientific Community

Originally funded by the National Institute of Environmental Health Sciences (NIEHS), TBASE was at first operated jointly by the Oak Ridge National Laboratory (ORNL) and the Genome Data Base (GDB) at Johns Hopkins University (JHU). TBASE is presently funded by the Department of Energy (DOE) and NIEHS, and is operated by the Division of Biomedical Information Sciences, The Johns Hopkins University School of Medicine.

Upon allocation of TBASE to the Division of Biomedical Information Sciences at JHU in 1993, a small "campaign" was initiated to promote the role of TBASE as a community animal database. As a first step, a large number of letters were selectively mailed to familiarize the community with the current administration and status of the project, and re-establish contacts that had been previously identified as potential TBASE users. More importantly, paper submission forms (see Data Acquisition and Submission Methods), were generated for direct data entry. In some cases, data and "contact" information were indeed modified or updated, as appropriate, and returned with remarks.

Encouraged and stimulated by the eager response, and voluntary recommendations of potential users, we concluded that TBASE should enter a phase of schematic reorganization and expansion (see Progressive Steps in TBASE Schema Revision and Data Collection). This was partly accomplished by actively pursuing constructive ideas, comments or criticisms from the community experts. An on-going "symbiotic" relationship between TBASE and the scientific community was decidedly established. The scientific community were to benefit from the systemized availability of the data; in turn, TBASE would benefit from user suggestions for schema revisions, and accommodate emerging needs and technical advances. Incoming contributions varied from simple assistance in creating controlled vocabulary lists, to full formulation of complex queries of interest. The process was facilitated by engaging and consulting persons of high expertise on genetic and informatics issues at the GDB in Baltimore, Maryland. TBASE was introduced as a powerful genome resource and informatics tool in several publications (Woychik et al., 1993; Takahashi et al., 1994; Fasman et al., 1994 and Anagnostopoulos et al., in preparation). Following the same rationale, a number of on-line demonstrations were presented at the Mammalian Developmental Mutants Workshop at NIH (April 1994), the International Mouse Molecular Genetics Meeting at Cold Spring Harbor (August 1994), the DOE Workshop on Transgenic Mice and Targeted Mutations in Washington, DC (October 1994), the International Mouse Molecular Genetics Meeting in Heidelberg, Germany (August 1995), and the Transgenic Technology Conference in San Diego (November 1995). The valuable observations that were voiced at these meetings were critically analyzed by TBASE personnel, and "transcribed" into successive schematic enhancements.

History – Initial TBASE Record Format

During most of the transitional period that incurred until December 1994, TBASE existed as a flat file database. The initial record format was composed of a single table of interrelated fields, based on an equally simplistic Data Dictionary. The initial record format did not incorporate any static reference values, such as controlled vocabulary lists, in any fields. It became increasingly obvious that most fields were amenable to further structural organization. For instance, a text field such as "Host background", initially of 150 characters in length, could be conveniently dissected into (a) "Type of ES cells", with an appended menu of commonly used embryonic stem cell lines, i.e. "AB1", "CCE", "D3", "J1", "E14TG2a", "R1", or "Not Entered" (if unknown, or non-applicable as in transgenesis), (b) "Recipient Strain", describing the host animal strain, (c) "Foster Mother Strain", describing the strain of pseudopregnant females used in transplantation of manipulated blastocysts, and (d) "Comments on Host Background", describing effects of individual mutations studied against more than one background. Moreover, the original flat file format did not accommodate "pointers" (links) to supporting databases, such as a Citation database, Contact Persons database, or related genomic databases. A single accession number (denoted as TG-###-##-###) was serially assigned per record (entry) of related fields. As a result, individual fields such as Contact, References, or DNA Construct were in fact "copied and pasted" in several records of the database – each record being retrievable by a unique accession number – when these fields were common to a number of distinct transgenic lines generated by random insertional mutagenesis. Retrieval of individual fields by accession number was clearly not feasible. Close examination of the remaining fields revealed that numerous modifications were possible. The existing format and Data Dictionary were to act as scaffolds for the construction of an elaborate, multi-table schema that, unlike its precedent, would allow linking of interrelated data within TBASE, and between TBASE and other databases. The final outcome of this reconstruction is presented in the following sections of this chapter.

Upgrading TBASE Accessibility and Searchability

At first, the underlying data were maintained in an interactive database management system and reports from that system were generated via a user-friendly Gopher Client at JHU (Woychik et al., 1993). The Gopher software was available without charge through the TBASE office at JHU. The flat file format allowed the performance of relatively unsophisticated searches. This type of query, entered in plain English, is particularly serviceable to the novice or less probing user. Boolean operators ("and", "or", "not") or double quotes (" ") and wildcards (*) could be used to restrict or broaden a given search. A more dynamic World Wide Web (WWW) Interface was subsequently released from the Division of Biomedical Information

Sciences in August 1994. TBASE became accessible on Mosaic, a multimedia WWW client, which displays embedded "hyperlinks" (underlined or colored words and phrases) and images within documents. The WWW interface was first presented and received with great enthusiasm at the Mouse Molecular Genetics Meeting at Cold Spring Harbor in 1994, with the Uniform Resource Locator (URL) *http://www.gdb.org/Dan/tbase/tbase.html*. This interface enabled the retrieval of hyperlinks across the Internet, with REBASE (Restriction Enzyme Database) and GDB being the first two related databases to be "rudimentally" linked to TBASE. Hyperlinks to REBASE, which were embedded in the DNA construct description, provided prompt retrieval of information related to restriction enzyme type, commercial availability, and recognition sequences; links to GDB were embedded as citation details (including abstracts) and contact information (including postal and e-mail addresses, phone and fax numbers). Links to OMIM (On-Line Mendelian Inheritance In Man) were also available. The TBASE introductory (Home) page was linked to data submission information and "paper version" entry forms accessible via URL:

http://www.gdb.org/Dan/tbase/submit.html. It outlined a brief history of the database, "forms support" specifications, examples, and brief searching instructions. As a WWW interface, it provided more flexible search options (text and check box searches), allowing the exploration of all, or individual fields in the database with the Boolean operators described above. Detailed representative queries and searching instructions are currently available to the occasional user via *http://www.gdb.org/Dan/tbase/docs/instr.html*.

Progressive Steps in TBASE Schema Revision and Data Collection

Management of the Literature and Contact Components into Separate Databases

Major schematic modifications were sequentially designed in order to direct the evolution of TBASE towards a relationalized (Sybase-based) schema. A number of meaningful enhancements were incorporated along the way, starting with the management of all TBASE-related citations into CitDB, a separate literature database initially devised to serve the Genome Database (GDB). CitDB, is ultimately intended to provide a single literature reference source for all public genomic databases, including OMIM, GSDB (Genome Sequence DataBase), EGAD, SST among others (Fasman *et al.*, 1994). CitDB is primarily built up from bibliographical material that is loaded directly from Medline as journal articles, using search strategies that retrieve TBASE-relevant material (see Data Acquisition and Submission Methods). At present, CitDB stores nearly 47,337 citation entries, 3,157 of which are TBASE-related. A complete citation record in CitDB is uniquely identified by a permanent GDBID number, includes an abstract, and matches a Medline record, where available. *Unpublished* data in CitDB are entered as "personal communication" with a specified contact person, and receive unique GDBID accession numbers. Similarly, contact investigators were progressively cataloged in the GDB Contact Manager. Again, each entry, specifying name, complete postal and e-mail address, phone and fax numbers of the contact person, was identified by a permanent GDBID number. Currently, 400 of the 12,210 contacts available in GDB have been cataloged in association with TBASE. During the preliminary stages of schematic revision, Citation and Contact GDBID numbers were merely assigned and cataloged in the appropriate fields. The significance of accession numbers in linking, and rapid retrieval of individual data items in a relational database will become apparent in the following sections.

Evolution of Data Acquisition and Submission Methods

One of the initial changes that took place at JHU was the derivation and distribution of paper entry forms that were specifically designed to reflect the evolving database schema. It soon became apparent that reliance on TBASE personnel to directly input data into the database cannot be effectively scaled with increasing data volume. Until then, the vast majority of data had been entered manually by TBASE personnel following direct extraction from the literature (see Data Acquisition and Submission Methods). These forms faithfully represented the major objects in the evolving schema, and mandated completion of specific information, such as contact information, references, and DNA construction details. They were to be completed by the initial investigators, who are, after all, the most authentic sources of data. A set of three direct paper entry forms (Construct, Line, and Laboratory) accompanied all introductory correspondence, and were distributed at major related events, in order to compile newly discussed data following a poster or platform presentation. In addition, the paper forms were made available by mail or fax upon request from the Division of Biomedical Information Sciences at JHU. They later became available on the Web, where they could be printed out, completed and returned by mail or fax, as indicated. Individual pages of each set were designed to be self-explanatory and independent of the rest, so as to be used separately in case of simple data modifications. The rationale employed was that, until alternative methods of data acquisition and submission became handy, processing of paper entry forms would be the primary method of reliable, direct data entry. Paper entry forms were expected to invigorate interest in TBASE, and secure timely submissions that reflected the momentum of data generation. Inevitably, they became subject to all successive enhancements that occurred during the period of schematic revision.

Emphasis on Murine Targeted Mutagenesis

Over 800 new entries were added by one person to the database during the schema revision process. Due to the wide use of homologous recombination in ES cells, and expressed community request, TBASE initially focused on maintaining comprehensive coverage of all targeted mutations reported to date. Moreover, TBASE concentrated on the mouse as the predominant mammalian model. Specific emphasis is still placed on cataloging novel "knockout" mice, which have defined genetic modifications, with an extensive, detailed description of their resulting mutant phenotypes. An "unbiased" distribution of data with respect to Genus, transgenesis and gene targeting will be then undertaken, additionally incorporating "knockout" animals which serve as recipients for transgenes that encode the respective disrupted gene.

Current TBASE Status

Schema Design and Data Dictionary

Implementation of the relational database schema allows rapid retrieval of individual data items, such as distinct transgenic animal/line entries, via unique, permanent, artificial identifiers, known as accession numbers. Accession numbers appear as "TBASE id" numbers in TBASE, and are assigned to all major data objects, such as LINE, LABORATORY and DNACON (Figure 79.4). Consider, for instance, the microinjection of one particular DNA construct that is identified by a unique DNACON TBASE id number: random DNA insertion may lead to the generation of many transgenic mutants that are phenotypically distinct or exhibit varying degrees of transgene expression. Each of these mutants is uniquely identified by an unchanging LINE TBASE id number, enabling one to retrieve all transgenics that are linked to that DNA construct by just submitting the DNACON TBASE id number. TBASE id numbers compensate for data items that are not guaranteed to be permanent and singular, such as transgenic nomenclature or locus symbolism; they are promptly expunged and never recycled when an object is deleted. Although they lack intrinsic informational value, they do display a format that characterizes them as TBASE identifiers. (Note that the original "TG-###-##-###" notation has been replaced by the "TBASE ID" format). This is especially advantageous when they are used as external pointers to TBASE, linking embedded data hyperlinks within related databases (GDB, OMIM, MGD, etc.) across an integrated federation of genomic databases.

Reorganization and restructuring of the schema permitted the execution of complex sophisticated queries which was not feasible in the past. For instance, it is now possible to retrieve all recorded targeted mutant mice, which are generated by homologous recombination in embryonic stem cells, and exhibit an abnormal phenotype in the *heterozygous* state (+ / −); or, all those targeted mutants which exhibit embryonic, perinatal or postnatal lethality when a gene of interest is disrupted; or, lastly, all the data generated by author C.M.W. Roberts on knockout mice with a disrupted homeobox gene (Hox*) and an abnormal phenotype of asplenia. Clearly, such structured queries could not be performed previously, with TBASE based on a flat file format, or by just searching Medline. (It is important to note that Medline does not exist in an on-line standard searchable form and fails to accommodate all journals, books, abstracts, conference and workshop proceedings, etc.) Use of controlled-value vocabulary in certain cases (Tables 79.1 and 79.3) was an additional enhancement aimed towards scientific data integrity.

In actuality, TBASE is not just a computerized archive of scientific publications (Tables 79.1–79.3). In its most recent Sybase-based schematic version (Figure 79.4), TBASE allows for external links with a number of related databases such as GDB (Genome Data Base), OMIM (On-Line Mendelian Inheritance In Man), MGD (Mouse Genome Database), REbase (Restriction Enzyme database), and others; it accommodates additional submitter's comments for each major object and includes editing history of data modifications/annotations; it provides direct reference to the designated contact investigator, who may furnish details about generating or acquiring a maintained line; it includes addresses of laboratories collaborating on a given project for easier contact access within the United States or abroad; it describes "related" mutant or transgenic lines (parental and derived) in the case of "double" transgenics/mutants; it incorporates an "application" field that briefly describes any relevance to disease models or industrial/agricultural exploitations; it attempts to provide ILAR laboratory codes, and unique and

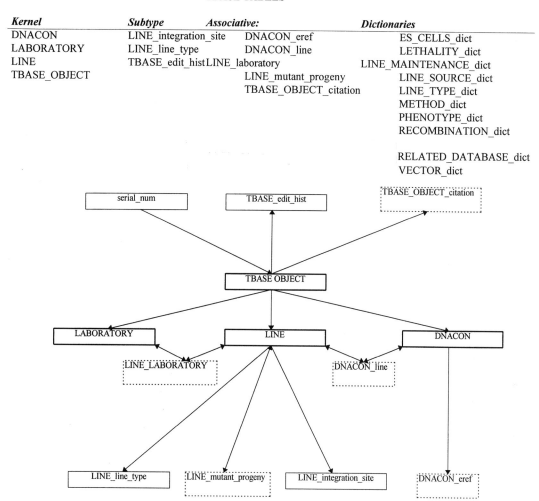

Figure 79.4. The TBASE schema depicts the relational components of TBASE, including the kernel, subtype, associative and dictionary tables. Kernel tables are functionally independent of other tables and describe the fundamental TBASE objects (DNA Construct, Laboratory and Line). Associative tables show links between objects (for instance, "DNACON-line" shows possible links between one DNA construct and many lines). Subtype tables contain additional data specific to TBASE Kernel tables (for instance, "TBASE-edit-hist" provides information on modifications made to a TBASE object).

descriptive names for individual lines, where ILAR nomenclature is not applied, or line naming by the original authors fails to be systematic or singular. In addition, TBASE accounts for both published and unpublished material, so that data that are unavailable in the literature may be accessed, and duplication of time-consuming and costly, yet unreported, efforts be avoided. Unpublished material is entered in the database as "personal communication" with a specified contact person; characterization of unpublished data, with respect to the validity or authenticity of the findings, naturally relies on the browser's personal evaluation and/or communication with the appropriate contact person. Inclusion of such information is considered a public disclosure, and TBASE can be cited as a source when referencing a mutant or transgenic animal not elsewhere described in the literature (Woychik et al., 1993).

TBASE Integration into a Federation of Genomic Databases

The strong urgency for creating a federation of genomic databases was fully realized and discussed by bioinformatics specialists at the DOE Informatics Summit which took place in Baltimore in April 1993. It was concluded that "achieving coordination and interoperability among genome databases and other informatics systems must be of the highest priority."

Table 79.1. TBASE data dictionary.*
TBASE TABLE: TBASE_OBJECT.
DESCRIPTION: Kernel table containing data on TBASE objects (LINE, DNACON, LABORATORY).

Column	Data type	Null?	Description
add_date	datetime	not null	Date entry added (month, day, year)
modify_date	datetime	not null	Date entry modified (month, day, year)
submitter	varchar(50)	not null	GDB accession number (pointer to Contact Database in GDB)
submitter_comment	text	null	Any comments that the submitter would like to make
tbase_id	integer	not null	Unique TBASE accession number

TBASE TABLE: DNACON.
DESCRIPTION: Kernel table containing data on DNA construct.

Column	Data type	Null?	Description
common_name	varchar(255)	not null	Common name of organism used (e.g. mouse, pig, rat, etc.)
description	text	not null	Description of the DNA construct that was introduced as a transgene, or a description of the vector that was used to target a specific gene by homologous recombination
genus	varchar(255)	null	Genus of organism used
locus	varchar(80)	not null	Locus symbol
species	varchar(255)	null	Species of organism used
tbase_id	integer	not null	Unique TBASE accession number
vector	varchar(50)	not null	See VECTOR_dict (Controlled vocabulary for type of vector used)

TBASE TABLE: LINE.
DESCRIPTION: Kernel table containing data on line.

Column	Data type	Null?	Description
application	text	null	Information relating to animal disease models, industrial, agricultural, pharmaceutical applications, etc.
common_name	varchar(255)	not null	Common name of organism used (e.g. mouse, pig, rat, etc.)
contact	varchar(50)	not null	GDB accession number (pointer to Contact Database in GDB, indicating person to contact to arrange acquisition of mutant; includes full address, phone and fax numbers, e-mail address)
comment_host_background	varchar(255)	null	Comment(s) on host background (includes information on effects of mutation studied against different genetic backgrounds; strain from which embryonic stem cell line is derived; embryonic background currently used to maintain the line, etc.)
copy_number	varchar(10)	null	Transgene copy number

TBASE TABLE: LINE (*continued*).
DESCRIPTION: Kernel table containing data on line.

Column	Data type	Null?	Description
crosses	text	null	Information relating to the effects of crossing the line (transgenic animal or targeted mutant) entered with other transgenic, targeted mutant, of other mutuant lines
duplicates	integer	null	Boolean values relating to presence of tandem, head-to-tail array of transgenes inserted at a single chromosomal locus; values can have one of only two states (-1 for "yes" and 0 for "no")
ES_cells	varchar(50)	not null	See ES_CELLS_dict (Controlled vocabulary table for embryonic stem cell line used)
foster_mother_strain	varchar(80)	null	Strain of pseudopregnant foster mother used
gene_expression	text	null	Temporal/spatial molecular expression of the transgene or targeted allele (includes methods on detection, restriction enzymes and probes used, comments on ectopic expression, etc.)
genus	varchar(255)	null	Genus of organism used
heterozygous_phenotype	varchar(50)	null	See PHENOTYPE_dict (Controlled vocabulary for phenotype)
homozygous_phenotype	varchar(50)	null	See PHENOTYPE_dict (Controlled vocabulary for phenotype)
comment_mutation	text	null	Comment(s) on random insertional mutation caused by the integration of the transgene into the host chromosome, or comment(s) on targeted mutation generated by homologous recombination
lethality	varchar(50)	not null	See LETHALITY_dict (Controlled vocabulary for lethality)
line_maintenance	varchar(50)	null	See LINE_MAINTENANCE_dict (Controlled vocabulary for line_maintenance)
line_source	varchar(50)	not null	See SOURCE_dict (Controlled vocabulary for line-source)
method	varchar(50)	not null	See METHOD_dict (Controlled vocabulary for method)
multiple_sites	int	null	Boolean values relating to occurrence of multiple integration sites; values can have one of only two states (-1 for "yes" and 0 for "no")
name	varchar(255)	not null	Name of the line as it appears in original publication; alternatively, the name assigned by the submitter to best represent construct and gene(s) involved.
Phenotype	text	null	Detailed description of mutant phenotype (includes methods of detection or examination, details on tissue/stage specificity, anatomy, histopathology, fertility, behavioral changes, neurological deficits, etc.)
recipient_strain	varchar(80)	null	Genetic background (strain) of host embryo
recombination	varchar(50)	not null	See RECOMBINATION_dict (Controlled vocabulary for recombination)

TBASE TABLE: LINE (*continued*).
DESCRIPTION: Kernel table containing data on line.

Column	Data type	Null?	Description
special_handling	text	null	Special handling instructions associated with using the line
species	varchar(255)	null	Species of organism used
tbase_id	integer	not null	Unique TBASE accession number

TBASE TABLE: LABORATORY.
DESCRIPTION: Kernel table containing data on Laboratory.

Column	Type	Null?	Description
comment	text	null	Default: "This entry was prepared by the TBASE staff located at Johns Hopkins University from the citation listed below. Verification and completion has been requested from the author(s). Collaborating laboratories include:"
description	varchar(255)	not null	Full address of the laboratory where the contact person is located
ilar_code	char(5)	null	ILAR code assigned for laboratory
tbase_id	integer	not null	Unique TBASE accession number

TBASE TABLE: LINE_integration_site.
DESCRIPTION: Subtype table containing data on integration site(s).

Column	Type	Null?	Description
comment	text	null	Mapping of the transgene integration site or the site of the targeted mutation on the host chromosome; includes effects of integration site on the expression of the introduced transgene (position effects), etc.
locus	varchar(80)	null	Locus symbol
locus_db	varchar(50)	null	Related (external) database where the locus information is stored
locus_id	varchar(80)	null	Related (external) database accession number
tbase_id	integer	not null	Unique TBASE accession number

TBASE TABLE: LINE_line_type.
DESCRIPTION: Subtype table on type of line.

Column	Type	Null?	Description
lt_description	text	null	Description of the line
tbase_id	integer	not null	Unique Tbase accession number
type	varchar(80)	not null	Type of the line (e.g. Transgenic, Targeted mutant, etc.)

TBASE TABLE: TBASE_edit_hist.
DESCRIPTION: Subtype table summarizing any modifications made to a database object.

Column	Type	Null?	Description
comment	varchar(255)	null	Modifications, corrections, updates made to an object; includes reasons for change
editor	integer	not null	Unique GDB accession number (pointer to Contact Database)
modify_date	datetime	not null	Date entry was modified
tbase_id	integer	not null	Unique TBASE accession number

TBASE TABLE: serial_num.
DESCRIPTION: Used to assign unique accession numbers.

Column	Type	Null?	Description
serial_num_id	integer	not null	Unique TBASE accession number
serial_num_incr_val	smallint	not null	Increment value for assigning the next accession number
serial_num_limit	integer	not null	Maximum accession number

*The Data Dictionary provides definitions and descriptions of the tables, columns and data types that are used in TBASE. The table and column names are designed to be descriptive of their contents. Column names are listed alphabetically, and each column entry contains its null characterization in the respective table.

Table 79.2. Associative TBASE tables.

TBASE TABLE: DNACON_eref.
DESCRIPTION: Relation between DNA constructs and related (external) databases.

Column	Type	Null?	Description
comment	varchar(255)	not null	Comment on a given DNA construct provided in external database
db	varchar(50)	not null	Name of related (external) database
id	varchar(80)	not null	Related (external) database accession number
tbase_id	integer	not null	Unique TBASE accession number

TBASE TABLE: DNACON_line.
DESCRIPTION: Relation between DNA constructs and lines.

Column	Type	Null?	Description
DNACON_tbase_id	integer	not null	Unique TBASE accession number (foreign key to DNACON table)
line_tbase_id	integer	not null	Unique TBASE accession number (foreign key to LINE table)

TBASE TABLE: LINE_laboratory.
DESCRIPTION: Relation between lines and laboratories.

Column	Type	Null?	Description
line_tbase_id	integer	not null	TBASE accession number (foreign key to LINE table)
laboratory_tbase_id	integer	not null	TBASE accession number (foreign key to LABORATORY table)

TBASE TABLE: LINE_mutant_progeny.
DESCRIPTION: Relation between parental lines and mutant progeny.

Column	Type	Null?	Description
mutant_progeny_tbase_id	integer	not null	Unique TBASE accession number (foreign key to LINE table)
parental_line_tbase_id	integer	not null	Unique TBASE accession number (foreign key to LINE table)

TBASE TABLE: TBASE_OBJECT_citation.
DESCRIPTION: Relation between CitDB and TBASE objects.

Column	Type	Null?	Description
citation_id	integer	not null	GDB accession number (pointer to CitDB)
tbase id	integer	not null	Unique TBASE accession number

Table 79.3. TBASE Dictionary Tables.*

TBASE TABLE: ES_CELLS_dict.
DESCRIPTION: Controlled vocabulary table for embryonic stem cell line used.

Column	Type	Values
es_cells	varchar(50)	"A3.1", "AB1", "AB2.1", "BK4", "C57BL/6-Thy1.1", "CC1.2", "CCE", "CJ7", "D3", "E-14", "E14TG2a", "EK-CP1", "ES-D3 C-12", "ESDeltaP-37/10", "J1", "J7", "Not Applicable", "Not Entered", "PJ5", "R1", "TG4 hprt-", "TT2", "Unavailable", "WW6", "52-25"

TBASE TABLE: LETHALITY_dict.
DESCRIPTION: Controlled vocabulary for lethality associated with transgenesis or mutagenesis.

Column	Type	Values
lethality	varchar(50)	"Embryonic", "NO", "Not Entered", "Perinatal", "Postnatal", "YES"

TBASE TABLE: LINE_MAINTENANCE_dict.
DESCRIPTION: Controlled vocabulary for maintenance of line.

Column	Type	Values
line_maintenance	varchar(50)	"Breeders", "Founders", "Frozen embryos", "No", "Not Entered", "Yes"

TBASE TABLE: LINE_SOURCE_dict.
DESCRIPTION: Controlled vocabulary for "source" of line.

Column	Type	Values
line_source	varchar(50)	"Not Entered", "cytoplasmic", "embryonic stem cells", "pronuclear"

TBASE TABLE: LINE_TYPE_dict.
DESCRIPTION: Controlled vocabulary for type of line.

Column	Type	Values
line_type	varchar(50)	"Not Entered", "Random insertional", "Targeted insertional", "Transgenic"

TBASE TABLE: METHOD_dict.
DESCRIPTION: Controlled vocabulary for method.

Column	Type	Values
method	varchar(50)	"Not Entered", "calcium phosphate precipitation", "electroporation", "lipid-mediated transfer", "microinjection", "retroviral infection", "transfection"

TBASE TABLE: PHENOTYPE_dict.
DESCRIPTION: Controlled vocabulary for phenotype of heterozygous and homozygous mutants.

Column	Type	Values
phenotype	varchar(50)	"Altered", "Not Entered", "Wild-type"

TBASE TABLE: RECOMBINATION_dict.
DESCRIPTION: Controlled vocabulary for type recombination involved.

Column	Type	Values
recombination	varchar(50)	"Not Entered", "homologous", "non-homologous"

TBASE TABLE: RELATED_DATABASE_dict.
DESCRIPTION: Controlled vocabulary for names of related (external) databases integrated with TBASE.

Column	Type	Values
related_databases	varchar(50)	"GDB", "MGD", "MLC", "OMIM", "Pigbase", "Not Entered"

TBASE TABLE: VECTOR_dict.
DESCRIPTION: Controlled vocabulary for type of vector used.

Column	Type	Values
vector	varchar(50)	"Cosmid", "Cytomegalovirus", "Herpes simplex virus", "Non-replicating retrovirus", "Not Entered", "Other", "Phage", "Plasmid", "Replicating retrovirus", "YAC"

* Tables that contain relatively static reference values (such as controlled vocabulary lists, definitions of numeric keys or letter codes) are considered as "Dictionary Tables". Dictionary tables names use the "_dict" suffix (e.g. ES_CELLS_dict).

Moreover, "the computational infrastructure genome research (indeed, of biological research)" was envisioned as "a federated information infrastructure of interlocking pieces," where data would be expected to "flow electronically over networks from producers to databases to users". Experts agreed that "the distributed nature of biological research will require the development of multiple software and database projects" and that "users must be able to retrieve related data from multiple databases such as GDB, PIR, Medline [...] without having to make separate queries to the databases, then integrate the results themselves" (Robbins et al., 1994).

As a community database, TBASE must acknowledge the biological interdependence of material in multiple databases, and support integrated queries involving multiple databases. The latest schematic enhancements were indeed devised in view of TBASE serving the entire scientific family. TBASE must be a semantically consistent component of a broader information infrastructure for computational biology. It presently accommodates a "universal" means of creating links to supporting databases, via stable external identifiers (accession numbers) and common data concepts. As already mentioned, TBASE has been linked to CitDB and the GDB Contact Manager (soon to become an expanded genome Registry Database) through GDBID accession numbers. Additional database links (connections between objects in distinct databases) are expected to connect TBASE to a number of related databases including GDB, OMIM, MGD, Pigbase, and others. A constructive dialog has already been initiated between the TBASE principal investigator at JHU and MGD and Induced Mutant Resources (IMR) at the Jackson Laboratory, Bar Harbor, Maine. OMIM, the continuously updated on-line version of Dr. Victor A. McKusick's book *Mendelian Inheritance In Man*, was until recently operated at the same performance site as GDB and TBASE; integration with OMIM is expected to tie TBASE to human genetic disorder entries.

Viewed as a long-term objective, database linking will necessitate the expansion of TBASE personnel to engage professionals in identifying shared objects and executing the links. A human genetic disorder that is mimicked by a given murine transgenic model is readily identified as a putative hyperlink between TBASE "Application" and "OMIM" entries. However, some links are expected to be more subjective to establish than others. For instance, it is evident that linking TBASE to MGD will enable the user to "couple" transgenic/targeted mutant data with murine mapping data, matrix data, probe and clone information, as well as homology information for human and many other mammalian species, via a multitude of hyperlinks embedded in MGD/MLC (Mouse Locus Catalog). At first glance, this type of link, is to be accomplished via "Locus Symbol", which represents a common data concept between TBASE and MLC. The creation of this hyperlink may seem simple and straightforward for mutants generated by gene targeting, where the disrupted allele and its symbol are known. Its implementation, however, reveals the following complexity: as of November 1993, a large number of gene symbols have been modified or recalled to reflect the new nomenclature guidelines, that have been approved by the International Committee for Standardized Genetic Nomenclature in Mice at the Seventh International Mouse Genome Conference, Hamamatsu, Japan. For instance, the vast majority of homeotic genes, which control cell fates during embryonic development, have had their previous names withdrawn, or new symbols assigned. Hoxa, the homeobox A cluster gene, was previously known as Hox-1; similarly, Hoxb8, the homeobox B8 gene, was previously known as Hox-2.4 etc. While most of the new symbols and nomenclature rules are currently observed, many investigators continue to refer to genes with their outdated symbols. In addition, all publications generated prior to November 1993, describe the genes with their original names and symbols. This sort of naming/symbol discrepancies may present problems when searching for a gene with such diverse symbols as the ones mentioned above. Clearly, both the old and the new symbols and names must be searchable, to completely satisfy a given query. A proposed mechanism to optimize locus queries is to establish a table of aliases for each gene that will include the official MGD locus symbol plus any locus symbol(s) used previous publications. More complexities are expected to be encountered for other shared objects; their solution will require close cooperation between developers to maintain semantic consistency, and interoperability among related databases.

Data Acquisition and Submission Methods

TBASE should be viewed as a dynamic database with a potentially unlimited number of entries, expected to grow considerably as more experimental methods and lines are generated. Data acquisition is primarily effected by Active Literature Scanning and Manual Data Entry, as well as Processing of Direct Submission Forms.

Active Literature Scanning and Manual Data Entry

Literature scanning refers to direct data extraction from the scientific literature through regular examination of over twenty journals (Figure 79.5).

These journals have been statistically identified as periodically stable sources of information on transgenesis and gene targeting. Active literature scanning has so far been the primary mode of data accumulation, and has ensured that the database faithfully reflects the general direction of related research. It is expected to remain a significant component of the data acquisition effort, even as other data acquisition methods become available (see below). In an attempt to avoid neglecting important data, *keyword*-related references from additional journals are also screened. Relevant references are loaded directly from Medline 2000+ into CitDB, which will eventually represent a convergent citation resource for a number of related genomic databases. "Current Contents" are also scanned for latest updates of TBASE-related citations (last six months), whereas inspection of the "Uncover Service" provides a current awareness and table-of-contents electronic delivery from selected journals on a weekly basis. Each reference in CitDB receives its own GDBID number, enabling retrieval of all data associated with a particular reference by entering a single accession number.

Journal Abbreviation

ADV IMMUNOL
ANN REV IMMUNOL
BIOESSAYS
BIOTECHNIQUES
CELL
CELL IMMUNOL
DEV BIOL
ENZYME
EUR J IMMUNOL
GENES DEV
IMMUNOL TODAY
INT IMMUNOL
J CELL BIOCHEM
J EXP MED
MOL CELL BIOL
NATURE
NATURE GENETICS
NUCL ACIDS RES
PROC NATL ACAD SCI
SEM DEV BIOL
SEM IMMUNOL
TRANSGENIC RESEARCH
TRENDS BIOCHEM SCI
TRENDS CELL BIOL

Figure 79.5. List of scientific journals frequently scanned for TBASE.

Following Medline searching, the data are subsequently processed and *manually* entered in the database by TBASE staff. Processing entails (a) physical retrieval, xeroxing and filing of the articles, and most importantly, (b) thorough understanding and cautious "dissection" of the relevant material at hand. The latter necessitates that TBASE staff be already familiar with the subject matter, or adequately trained in data extraction. Undoubtedly, the rate at which the material is scanned and selected is much faster than that with which it is entered in the database. This is due to the fact that TBASE is largely phenotype-oriented and involves the typing of lengthy texts in the "Phenotypic Analysis" field. Given the present situation of restricted staff, it is not surprising that keeping the database current remains extremely laborious. It is not unusual to have to confront a several-month backlog of selected but non-accessioned material, when emphasis is placed on issues other than data entry. Considering the dramatic growth in data generation, and ensuing demand to make the data available in TBASE, it becomes imperative that two things take place: (a) the utilization of direct submission by the scientific community; and, (b) the allocation of additional TBASE personnel for the purpose of data extraction and entry. It is important to attract and train more data entry personnel; once proficient in extracting selectable material, such personnel will prove invaluable in the utility and expansion of the database.

Processing of Direct Submission Forms

In the future, data which are not manually entered by TBASE staff through the literature scanning process, are expected to arrive as either direct paper or electronic submissions. Close cooperation with journal editors will ensure direct data deposition that is concurrent with, or prior to publication. *Nature* has quite recently taken an impressive step in reinforcing access to research data and material (Nature (editorial), p. 191, 1996). As of January 1996, no paper on transgenesis or gene targeting will be accepted without a reference to the database accession number, or to the Internet address at which the data can be accessed from the time of publication. Similarly, research materials, such as DNA constructs, must be made available to interested parties upon request, as a condition of publication. Although over 98% of the data have been entered manually so far, direct paper submissions are expected to increase as TBASE becomes fully adopted by the scientific community. In addition, as electronic submission tools become available, more research groups will be prepared to commit resources to the production of data deposition to

TBASE. The two proposed mechanisms of direct data submission are discussed below:

(a) *Paper Entry Forms.* The use of direct paper entry forms has already been discussed in this chapter (see Evolution of Data Acquisition and Submission Methods). Naturally, original forms became subject to all successive enhancements that occurred during the period of schematic revision. These forms are distributed to research centers and at major TBASE-related events, in order to accommodate newly generated or unreported data. They are also available on the Web (URL: *http://www.gdb.org/Dan/tbase/submit.html*, as before), and by mail or fax upon request. In the future all paper submission forms received by the TBASE office, will be filed according to the date of their receipt. They will be subsequently reviewed for possible omissions or errors in their contents. Erroneous, or incomplete forms will be sent back to the submitter, along with an appropriate message. Correctly completed submissions will be then filed by the date of receipt and processed by TBASE staff. Immediate priority will be given to processing completed paper forms over manual entry of selected material, as a gesture of appreciation towards data contributors. Once a submission is complete, TBASE staff will then notify the original submitters requesting confirmation of entry by fax or electronic mail.

(b) *Electronic Entry Forms.* Electronic submission tools are currently being developed based on prototypes for other databases such as Kidbase, the database created for studying genetic disorders of the kidney (URL: *http://www.gdb.org/kidbase.html*). The basic structure and contents of electronic submission forms will reflect the archetypal paper entry forms, modified accordingly (Dan Jacobson, personal communication). Also, special software tools will be designed to help eliminate problems of semantics. The original submitter will be requested to complete, edit and prepare his/her entry for submission; the submitter will send the final version to the TBASE operator, and each entry will be assigned a unique TBASE accession number. This accession number may be used to access the record at a later date. In addition, accession numbers will be used to crosslink new and earlier related entries. The TBASE operator will, in turn, review the record and contact the submitter by e-mail.

Use of robust electronic submission scheme is predicted to have a remarkable impact on the data of data entry and database growth, as previously reported for GDB, OMIM and other databases (Chipperfield *et al.*, 1993). Electronic submission tools and their error-checking capabilities will ensure efficient and timely deposition of the data.

One of the long-term objectives of this endeavor is to eventually enforce electronic submission as the primary mode of data deposition, while still processing direct paper entry forms, and scanning the literature as part of the basic extraction strategy. It is our hope that other journals will follow the lead of *Nature* and require accession numbers as a condition of publication. This will encourage direct submission and reduce the reliance on literature scanning.

TBASE Data Curation

Proposed Curation Process

Curation of data stored in TBASE will be effected primarily through feedback provided by the original investigators or submitters. At present, there are no full-time curators to maintain scientific integrity; besides, this task would soon become insurmountable due to the exploding data growth. As a result, original investigators and contact individuals will be requested to verify all data that have been entered manually through the literature scanning process. In parallel, the electronic submission scheme will permit direct submitters to edit their entries prior to submitting. Modifications or corrections with additional comments should eventually be sent by the original investigator as an e-mail message to *tbase@gdb.org*. Once received by the TBASE operators, a given modification will be reported as a comment in the "edit history" section, and receive a unique TBASE accession number, modification date and a GDBID number identifying the original editor. In addition, scientific integrity will be reinforced by a group of scientific advisors, who will be charged with the broad responsibility of supervising the general evolution of the database (see Proposed Role of TBASE Advisory Committee below). The curation operations themselves are therefore expected to undergo changes to accommodate future database revisions.

Besides maintaining scientific data integrity, TBASE personnel must ensure that citation integrity is well preserved. In this respect, incomplete records loaded as "Current Contents" in CitDB, will eventually become replaced with the full Medline source details and abstracts through Medline loading. Furthermore, TBASE must enjoy professional guarantee that no corruption of the database occurs during future transitions from one release of the database to the another. TBASE developers involved in the schema evolution, interface design, and data dissemination must produce scripts that run periodically within the production database to filter out any errors that may have been introduced in the process.

Proposed Role of the TBASE Advisory Committee

As any dependable community database, TBASE must be subject to periodic evaluation by an independent panel of experts. A successful panel should consist of a combination of individuals of high expertise on molecular genetics, transgenesis, gene targeting and various aspects of bioinformatics. Engaging experts of diverse areas of specialization will promote the credibility of the database, and provide optimal guidance and curation in TBASE-related issues. Thus, the committee selection process should be based on the fact that, collectively, these advisors must be able to assist in biological issues or controversies, as well as database development and operability. These individuals should be energetic, prolific scientists who are in a position to follow the momentum of scientific developments, and fully appreciate the role of TBASE as a vital component of a broader information infrastructure. In addition, they must be (a) willing to make the long-term commitment to attend future TBASE-organized meetings, (b) act as influential mediators in identifying users' needs and relaying constructive comments, and (c) exert reasonable pressure on the community and journals to promote direct data submission. At the time of writing this manuscript, efforts are being made to assemble a committee of five members who meet the above criteria. Principle investigators are being contacted at several research facilities in the United States in the hope that the first meeting will take place in early 1996.

Maintenance of an Information Server

A server has been recently created, providing links to TBASE-related announcements and general information. Announcements are categorized by subject to include information on (a) upcoming TBASE-related meetings, seminars, conferences and workshops, (b) courses and demonstrations on transgenesis and gene targeting, (c) books, reviews, and laboratory manuals, (d) technical novelties, and (e) video guides. General information categories include (a) transgenic facilities worldwide, (b) nomenclature guidelines, (c) animal welfare legislation and regulations, and (d) multiple links to additional sites of interest. Database users are able to send their constructive feedback or criticism through a "Drop us a Note" hyperlink, reinforcing the relationship between TBASE and the scientific community. Original investigators and submitters may initially use the *tbase@gdb.org* alias to curate data which they have generated *themselves*, but consider erroneously entered. Finally, a mailing list service may also be considered in the future, to facilitate the electronic flow of ideas and discussions between listed subscribers. Although viewed as a distant objective, this may be accomplished through programs which automate the management of Internet mailing lists, in a moderated fashion, such as Majordomo.

Data Distribution

TBASE is currently accessible via *Netscape* and *Mosaic*, both multimedia World Wide Web (WWW or W3) browsers, via the Uniform Resource Locator (URL): *http://www.gdb.org/Dan/tbase/tbase.html*. The WWW was originally developed in 1990 at the European Laboratory for Particle Physics (CERN), Geneva, Switzerland, to share manuscripts and other technical documentation among high energy physics laboratories throughout the world (Berners-Lee *et al.*, 1992). It is a "universe of network-accessible information", that manages and retrieves hyperlinked data across the Internet. Client software display WWW documents dynamically, invoking various external viewer programs for the numerous standard multimedia data formats that are currently available. The World Wide Web server will provide suitable forms-based query interfaces, and enable the user to formulate the specific, structured queries that are allowable through the current relational database schema. A general bioinformatics World Wide Web Server has been developed at the Genome Database at JHU to provide access to TBASE, in addition to the well established GDB, OMIM and several other biological databases (URL: *http://www.gdb.org/hopkins.html*).

TBASE Contact Information

Any questions concerning access or data submission to TBASE should be directed to the User Support Line by dialing (410) 955-9705 or sending an e-mail message to *help@gdb.org*. Comments or suggestions may be addressed to: Anna V. Anagnostopoulos, Division of Biomedical Information Sciences, The Johns Hopkins University School of Medicine, 2024 E. Monument Street, Baltimore, MD 21205-2236 (tel: (410) 614-3226, fax: (410) 614-0434, e-mail address: *anna@gdb.org*).

References

Anderson, C. (1993) Intellectual Property. Researchers win decision on knockout mouse pricing. *Science* (News and Comment), **260**, 23–24

Berners-Lee, T.J., Cailliau, R., Groff, J-F., Pollermann, B. (1992) World Wide Web: The Information Universe. In *Electronic Networking: Research, Applications and Policy*, Meckler Publishing, Westport, CT, pp. 52–58

Bradley, A. (1993) Site-directed mutagenesis in the mouse. *Recent Progress in the Hormone Research* **48**, 237–251

Chipperfield, M.A., Porter, C.J., Talbot, C.C. Jr., Anagnostopoulos, A.V., Campbell, J. and Cuticchia, A.J. (1993) Growth of data in the Genome Data Base since CCM92 and methods for access. In A.J. Cuttichia and P.L. Pearson (eds.), *Human Gene Mapping, 1993*, Baltimore, MD: The Johns Hopkins University Press, pp. 3–5

Chisaka, O. and Capecchi, M.R. (1991) Regionally restricted developmental defects resulting from targeted disruption of the mouse homeobox gene hox-1. *Nature* **350**, 473–479

DeChiara, T.M., Efstratiadis, A. and Robertson, E.J. (1990) A growth-deficiency phenotype in heterozygous mice carrying an insulin-like growth factor II gene disrupted by gene targeting. *Nature* **345**, 78–80

DeChiara, T.M., Robertson, E.J. and Efstratiadis, A. (1991) Parental imprinting of the mouse insulin-like growth factor II gene. *Cell* **64**, 849–860

Fasman, K. H., Cuticchia, J.A. and Kingsbury, D.T. (1994) The GDB™ Human Genome Data Base anno 1994. *Nucleic Acids Research* **22**, 3462–3469

Fung-Leung, W.-P., Schilham, M.W., Rahemtulla, A., Kundig, T.M., Vollenweider, M., Potter, J., van Ewijk, W. and Mak, T.W. (1991) CD8 is needed for development of cytotoxic T cells but not helper T cells. *Cell* **65**, 443–449

Joyner, A.L., Herrup, K., Auerbach, B.A., Davis, C.A. and Rossant, J. (1991) Subtle cerebellar phenotype in mice homozygous for a targeted mutation of the En-2 homeobox. *Science* **251**, 1239–1243

Kitamura, D., Roes, J., Kuhn, R. and Rajewsky, K. (1991) A B cell-deficient mouse by targeted disruption of the membrane exon of the immunoglobulin mu chain gene. *Nature* **350**, 423–426

Koller, B.H. (1992) Altering genes in animals by gene targeting. *Annual Review of Immunology* **10**, 705–730

Koller, B.H., Marrach, P., Kappler, J.W. and Smithies, O. (1990) Normal development of mice deficient in beta$_2$ M, MHC class I proteins, and CD8$^+$ T cells. *Science* **248**, 1227–1230

Lamb, B. (1995) Making models for Alzheimer's disease. *Nature Genetics* **9**, 4–6

Li, Y., Behnam, J. and Simkiss, K. (1995) Ballistic transfection of avian primordial germ cells in ovo. *Transgenic Research* **4**, 26–29

McMahon, A.P. and Bradley, A. (1990) The Wnt-1 (int-1) proto-oncogene is required for development of a large region of the mouse brain. *Cell* **62**, 1073–1085

Mucenski, M.L., McLain, K., Kier, A.B., Swerdlow, S.H., Schreiner, C.M., Miller, T.A., Pietryga, D.W., Scott, W.J. Jr. and Potter, S.S. (1991) A functional c-myb gene is required for normal murine fetal hepatic hematopoiesis. *Cell* **65**, 677–689

Pevny, L., Simon, M.C., Robertson, E., Klein, W.H., Tsai, S.-F., D'Agati, V., Orkin, S.H. and Costantini, F. (1991) Erythroid differentiation on chimeric mice blocked by a targeted mutation in the gene for transcription factor GATA-1. *Nature* **349**, 257–260

Reinforcing access to research data (Editorial) (1996) *Nature* **379**, 191

Robbins, J., et al. (1994) Report of the Invitational DOE Workshop on Genome Informatics, 26–27 April 1993, Baltimore, Maryland. Genome Informatics I: Community Databases. *Journal of Computational Biology* **1**, 173–190

Robbins, J. (1993) Gene targeting: The precise manipulation of the mammalian genome. *Circulation Research* **73**, 3–9

Smithies, O. (1993) Animal models of human genetic diseases. *Trends in Genetics* **9**, 112–116

Smithies, O., Gregg, R.J., Boggs, S.S., Koralewski, M.A. and Kucherlapati, R.S. (1985) Insertion of DNA sequences into the human chromosome beta-globin locus by homologous recombination. *Nature* **317**, 230–234

Soriano, P., Montgomery, C., Geske, R. and Bradley, A. (1991) Targeted disruption of the c-src proto-oncogene leads to osteoporosis in mice. *Cell* **64**, 693–702

Takahashi, J.S., Pinto, L.H. and Hotz Vitaterna, M. (1994) Forward and reverse genetic approaches to behavior in mice. *Science* **264**, 1724–1733

Thomas, K.R. and Capecchi, M.R. (1990) Targeted disruption of the murine int-1 proto-oncogene resulting in severe abnormalities in midbrain and cerebellar development. *Nature* **346**, 847–850

Woychik, R.P., Wassom, J.S., Jacobson, D.A. and Kingsbury, D.T. (1993) TBASE: a computerized database for transgenic animals and targeted mutants. *Nature* **363**, 375–376

Zilstra, M., Bix, M., Simister, N.E., Loring, J.M., Raulet, D.H. and Jaenisch, R. (1990) beta-2 microglobulin deficient mice lack CD4$^-$ 8$^+$ cytolytic T cells. *Nature* **344**, 742–746

PART V

80. The Patenting of Transgenic Animals

Jacques Warcoin

Cabinet Regimbeau, 26, Avenue Kléber, 75116 Paris, France

Demonstrators brandishing placards: "Einspruch! Kein Patent auf Leben!" (No patent on life) and other street demonstrations are hardly routine occurrences in the patent field, though they do give the flavour of the problems posed by patents relating to transgenic animals, even if, incidentally, the claims in question are equally applicable to other subject matter including, for example, transgenic plants and human genes.

In order to try to understand the problems at issue, we need to reexamine certain basic concepts relating to patents and relating to the history of patents associated with the phenomenon of life.

Patents in General

A patent is a property right of limited duration (generally 20 years) which is granted by a country or a group of countries (European patent) to the holder of the patent, for an invention.

This patent confers a monopoly right in the invention, the clearest consequence of which is the power to prohibit an unauthorized third party from working the said invention.

It must be clearly understood that a patent is not the right to do something, like the authorization to place a medicinal product on the market; it is a right to prohibit something. This feature is important in the field we are now examining.

Naturally, this monopoly right can be granted only:

For any inventions which are susceptible of industrial application, which are new and which involve an inventive step (Art. 52(1) EPC*).

In essence, these conditions mean that the invention must be capable of being "used in some kind of industry" (industrial application), must not have been described as such in the prior art (novelty) and must not be obvious to the person skilled in the art (inventive step).

There is no need to examine in detail the matter of interpretation of these criteria, but they can be applied to inventions in the field of biotechnology, and specifically to transgenic animals, without insurmountable difficulties.

Thus, for example, a model mouse for testing products for use in the treatment of cancer can obviously have an industrial use, especially in the pharmaceutical industry. It is possible to determine, by reference to the prior literature, whether or not this mouse has previously been described as such, and hence whether it is new. As for the matter of knowing whether the obtaining of this mouse was or was not obvious to those skilled in the art, this depends on the existing documents, but it is quite conceivable that such an analysis is feasible and that it is therefore possible to determine whether or not this mouse involves an inventive step.

Exceptions to Patentability

In addition to the above criteria, most laws – and, in particular, the European Convention – lay down exceptions in principle to patentability, which are listed in Article 53 EPC*:

European patents shall not be granted in respect of:

(a) inventions the publication or exploitation of which would be contrary to ordre public or morality, ... ;

(b) plant or animal varieties or essentially biological processes for the production of plants or animals; this provision does not apply to microbiological processes or the products thereof.

These concepts of "public order", "morality" and "animal varieties" constitute the essential basis for the arguments advanced by those opposed to the patentability of transgenic animals.

Historical Developments

It must nevertheless be noted that, in the early days of biotechnology, other arguments were advanced – in particular, the very fact that living beings could not, as a matter of principle, be appropriated by patent – but this form of argument, which is lacking in any legal basis, especially in Europe, has never been upheld.

This argument, which endeavoured to rely upon an overriding principle, has in any case been expressly rejected in the United States since the Chakrabarty case of 1980.

What happened in 1980 was that the US Supreme Court, in Diamond vs. Chakrabarty, recognized the patentability of a bacteria strain (namely *Pseudomonas*) in US Patents 3,813,316 and 4,259,444.

Although this decision created a considerable furore in Europe, with headlines such as "Life can be patented!", it deals with problems of which some are highly specific to the United States, and does not entirely have the force of precedent with which it has been credited.

Be that as it may, the decision did accept the principle that living matter could be patented provided that man had intervened in its "production".

A number of other cases in the United States, in particular Bergey, have strengthened and clarified this position.

In the decision "Ex parte Allen", the Board considered that treating an oyster under high pressure at a certain stage of development to get a polyploid oyster was a patentable subject matter. It was one of the very first decisions dealing with the patentability of higher organisms. Yet because the said polyploid oyster was similar to earlier polyploid oysters, the patent was denied on the basis of obviousness. Nevertheless, the patentability of higher organisms was confirmed.

At the same period, in Europe, the patentability of microorganisms was recognized, as was the patentability of plants (T49/83 and T320/87).

It was on 12 April 1988 that the US Patent Office granted the first patent for a transgenic animal (obtained by genetic recombination); (it was the first because previous patents had been content to circumvent the problem with cell-type claims).

This patent, US 4,736,822, was granted to Harvard University, the inventors being Leder *et al.*, and relates to a mouse referred to as "oncomouse". The corresponding European Patent, EP 85/304,490, comprises, in addition to claims for:

Method for producing a transgenic non-human mammalian animal having ...

Method of testing a material suspected of being a carcinogen ...

a claim which lit the fuse:

A transgenic non-human mammalian animal whose germ cells and somatic cells contain an activated oncogene sequence as a result of chromosomal incorporation into the animal genome or into the genome of an ancestor of said animal, said oncogene optionally being further defined according to any one of Claims 3 to 10.

The patent was refused by the Examination Division on the ground, in particular, that animal varieties are excluded from patentability (Article 53(b) EPC*).

The Board of Appeal, to which an appeal was lodged from this decision, delivered on 3 October 1990 its decision T19/90 which represents the essential basis for ideas relating to the patentability of transgenic animals at the European Patent Office (EPO).

The "Oncomouse" Decision – T19/90

The Board of Appeal, having noted that Article 53(b) EPC*, which provides that animal varieties are not patentable, was an exception and to be interpreted strictly, also noted that the terminology used for this exclusion was different in the three languages: "races animales", "animal varieties" and "Tierarten", and that it was therefore not the intent of the legislator to exclude "animals" in general from patentability, which would not have resulted in these problems of translation.

Since the Examination Division had not demonstrated that the claim related to "animal varieties", the Board of Appeal referred the decision back to the Examination Division on this point.

With regard to Article 53(a) EPC*, which relates to the non-patentability of inventions which would be contrary to *ordre public* and morality, the Board of Appeal saw fit to give a ruling although this point had not been raised in the notification of rejection.

First, it affirmed the principle whereby patent law is not the proper instrument for solving problems associated with the use of genetic manipulations, especially in terms of ethics.

It should be recalled at this point that a patent is not a right to do something but a right to prohibit others from doing something. Thus, the non-existence of a patent in no way restricts (far from it) the possibility for any person to work on transgenic animals, whether this be ethically correct or otherwise.

The Board of Appeal conceded, however, that it could be a matter for the EPO to strike a balance between the benefits offered to mankind by the invention in question and the suffering of the animals and any impact on the environment.

Finally, the last problem which arose related to the scope of the invention. The Examination Division believed that the description contained in the patent would not enable the invention to be applied to absolutely any mammal. The Board of Appeal recalled the principles of proof, indicating that it was a matter for the Examination Division to prove that serious doubts, supported by verifiable facts, existed with regard to the possibility of reproducing the invention, and that this was not so in the present case.

The file having been referred back to the Examination Division, the latter reached the following conclusion:

> Mammals occupy a place in the taxonomic classification which is higher than that of species, and furthermore a variety is a subgroup of a species, and that under these conditions a claim relating to mammals is patentable.

It will be seen that the same reasoning in connection with plants and plant varieties, in the Plant Genetic Systems case (T356/93), led to a diametrically opposite conclusion, to the effect that since plants include varieties, they cannot be patentable.

With regard to Article 53(a) EPC*, the Examination Division restated the position adopted by the Board of Appeal, to the effect that:

- the invention was of value to mankind,
- the environmental risks were limited, and, finally,
- the suffering of the animals was not increased (by relation to the number of non-transgenic animals needed for the same applications).

It may be wondered whether these considerations do indeed have any place in the analysis of the patentability of an invention, especially the assessment of the "balance" which, ultimately, threatens to lead the European Office into philosophical contradictions which are far from being within its remit.

The patent was therefore granted.

It should be noted that, in the meantime, the French National Institute of Industrial Property has granted a patent in respect of a transgenic animal (double transgenesis) on 27 September 1991 under number 2,637,613 to the company Transgène, this being just before the Examination Division reached its final decision.

However, in accordance with the European procedure, once the European "oncomouse" patent had been granted the so-called opposition period began. A great many oppositions were filed, and the opponents are to be heard during the month of November 1995. In essence, the grounds of opposition recapitulate the arguments set out earlier – that the invention is contrary to ordre public and morality, and that "animal varieties" are not patentable.

It seems probable that the Opposition Division will uphold, in part, the position adopted by the previous instances with regard to the problems relating to *ordre public* and morality, despite the political and legal pressures being exerted by the ecological and religious organizations, in particular, which are supporting the opposition.

However, with regard to the problem relating to the concept of "animal varieties", the recent decision T356/93 has been able to shed further light.

The Plant Genetic Systems Decision, T356/93

Curiously enough, it is in connection with plants that the most interesting developments have occurred with regard to the patentability of higher organisms.

In decision T356/93, the problem of *ordre public* and morality had to be reviewed by a Board of Appeal to which it had been explicitly referred following an opposition by Greenpeace to a Plant Genetic Systems (PGS) patent which related to recombinant plants (cf. in particular the definitions under headings 5 and 6).

In this decision, which could be applied to transgenic animals, the substance of what was said is as follows (Point 5):

> Under Article 53(a) EPC*, inventions the exploitation of which is likely seriously to prejudice the environment are to be excluded from patentability as being contrary to *ordre public* ...

In the present case, it was held that the plants at issue did not constitute an ecological threat.

The concept of being contrary to "morality" was set aside because, according to the Board, the invention, the cells and the plants "cannot be considered to be wrong as such in the light of conventionally accepted standards of conduct of European culture" (Point 18).

On the other hand, it was decided that claims relating to plants were not acceptable because the concept of "plant" included the concept of "plant varieties" and was thus not protectable:

> A product claim which embraces within its subject-matter plant varieties is not patentable.

As has been seen, applying identical logic in the case of animals, the other Board arrived at the opposite conclusion.

The question being a vitally important one for both plants and animals, the matter of principle was referred to the Enlarged Board of Appeal, where the proceedings are pending under number G 3/95. The question asked is as follows:

> Does a claim which relates to plants or animals but wherein specific plant or animal varieties are not individually claimed contravene the prohibition on patenting in Article 53(b) EPC* if it embraces plant or animal varieties?

At present, the controversy on these various points is still raging.

Nevertheless, it seems to be established that transgenic animals, as living entities, are not excluded from patentability. The concept of "variety" has yet to be defined (an undertaking upon which no one is willing to embark) and, since we know that a variety is excluded from patentability, must this result in the non-patentability of a higher classification such as transgenic mammals?

In this respect, the answer given by the Enlarged Board of Appeal with regard to plants will be decisive.

> As regard *ordre public* and morality, the concept of "balance" will probably tend to become generalized, even if in the majority of cases it has only an intellectual value (since, even if the fate of mankind is not at stake, in a great many cases animals, whether transgenic or otherwise, are destined to be sacrificed). What is at issue is much more the problem of animal experimentation as such than patent law.

The problems which remain to be solved comprise defining the specific criteria of novelty and inventive step as applicable to transgenic animals; these are more traditional and more easily comprehended matters which will be able to be clarified by subsequent case law.

The Alternative Solutions

In order to avoid the problems posed by patents in this field, mention has been made of creating an animal obtainment certificate, similar to the plant obtainment certificate, in other words specific protection relating to new animal varieties (for quite incomprehensible reasons this form of protection never seems to create any problem for the defenders of the environment). This certificate, which would be valid for transgenic varieties, would be equally valid for varieties obtained by traditional crossing procedures.

This solution, which seems an attractive one, does have a number of shortcomings – those which arose when an attempt was made to arrange for plant patents and plant obtainment certificates to coexist. Finding acceptable solutions to these problems took 10 years, and no one knows at present whether those solutions are satisfactory; in some circumstances, the creation of an animal variety certificate has to be approached with caution.

Finally, in certain cases and in order to avoid the problems associated with explicit claims relating to transgenic animals, alternative solutions can be adopted.

When the transgenic animal is intended to be used either in a "screening" process or to produce a subproduct, it is possible, rather than claiming the animal as such, to claim the method employing the animal, which eliminates the problem associated with the "animal variety". Thus, a claim relating to a process for the production of Factor IX, characterized in that milk is taken from a transgenic ewe having as a genetic feature a recombinant coding for factor IX, said Factor IX being extracted from the milk, does not come up against the above-mentioned problems. (See also the claims of the "Onco-mouse" patent.)

Finally, in some cases, it is possible to claim the method which makes it possible to obtain the transgenic animal.

The conditions governing the European patent provide that the product obtained by the process is likewise protected. In the present case, this amounts to indirect protection of the animal obtained.

However, this type of claim then comes up against two further problems:

> The specific provision of Article 53(b) EPC* lays down that protection cannot be granted to "essentially biological processes for the production... of animals..."; many commentators offer ways of interpreting this somewhat ambiguous provision, since after all everything which relates to life is biological.

The applicable rule appears in the European Directives:

> The question whether a process is "essentially biological" is one of degree, depending on the extent to which there is a technical intervention by man in the process; if such intervention plays a significant part in determining or controlling the result, the process would not be excluded

from patentability. A method of crossing, interbreeding or selectively breeding, say, a horse ... would be essentially biological and therefore unpatentable.

So, to summarize, this exclusion applies only to traditional processes of producing a new animal variety (see T320/87). But what about biotechnology or multi-step processes comprising a recombinant step and a biological step of regeneration and replication? The answer of the Board of Appeal in PGS T356/93 is clear: if one step uses a recombinant technology, it is enough to consider that the process as a whole is not essentially biological and so is not excluded from patentability.

But the second problem is: assuming that the product (animal or plant) of said process is protected, what about the second generation? No decisions have been handed down on this point.

The last solution, also adopted both in the PGS decision and in the Oncomouse decision, is the use of "cell" claims, which according to all decisions are patentable and may in the future be used to sue a transgenic animal containing said cell!

In the PGS decision, nevertheless, the Board has decided (point 40.7):

A claim is not allowable if the grant of a patent in respect of the invention defined in said claim is conducive to an evasion of a provision of the EPC* establishing an exception to patentability ...

Judges think of everything.

Conclusions

In the United States, patents for transgenic animals have been granted, together with patents for the use of said animal and the process for its preparation.

In Europe, the situation is unclear at present. We have to wait for the decision of the Enlarged Board of Appeal.

Anyway, the rejection of said patent as being contrary to "morality" and *ordre public* has been withdrawn with some reasonable limitations.

It seems for the time being that processes for producing transgenic animals, processes using said animals and cells of animals are patentable.

*EPC: European Patent Convention

References

Decision of Technical Board of Appeal T356/93, EPO *Official Journal* **8**, 1995

Decision of Technical Board of Appeal T19/90, EPO *Official Journal* **476**, 1990

Houdebine, L.M., "Quels animaux conviendrait-il de faire breveter?" In *Cahier Agricultures* **2**, 1993, 343–345

Gallochat, A., "Brevetabilité des Animaux Myc-Mouse et le Brevet Harvard: Conséquence", RDPI

Bergman, B., *La Protection des Innovations Biologiques*, Maison Larcier S.A., Brussels

Cooper, I.P., *Biotechnology and the Law*, eds., Clark, Boardman Callaghan

Wegner, H.C., *Patent Law in Biotechnology*, ed., Stockton

Note Added in Proof

The European Patent Office took a tougher stand in relation with claims relating to transgenic plants or animals inasmuch as it has indicated that because of the present case-law, it would no longer accept this type of claims. However, at the present time, about 700 patent applications are concerned and are waiting for a first decision.

PART V

81. The Biosafety Problems of Transgenic Animals

Louis Marie Houdebine

*Unité de Différenciation Cellulaire, Institut National de la Recherche Agronomique,
78352 Jouy-en-Josas, Cedex, France*

Transgenesis offers researchers an unprecedented possibility to induce mutations in living organisms and even to cross the species barrier. Indeed, natural gene transfer from one species to another is a rare event in nature. The mechanisms which control genome expression are so complex that some of the biological effects of a transgene cannot be predicted in most cases. This is of course one of the major reasons why experimentators use transgenesis which is expected to modify the natural physiology of living organisms. Creating new and more or less exotic physiological situations is obviously one of the best ways to decipher the natural mechanisms which control life.

Experimental gene transfer is not expected to create situations fundamentally different from those generated by evolution. Indeed, the natural evolutionary process has accomplished and is still accomplishing such profound mutations in living organisms that it has given and still gives rise to the generation of new species. We may therefore consider that gene transfer into living organisms does not generate *per se* a fundamentally new type of risk. Hence, evaluating the risks of animal genome manipulation and defining containments for transgenic animals consist essentially in determining when to follow already known biosafety rules rather than to imagine quite new protection devices and practice. However, transgenesis leads to the generation of a relatively large number of mutated living organisms in a short period of time and in a concentrated area. This may create real biorisks for human beings and for the environment. The uncontrolled release of genetically modified animals into nature might trigger quite significant changes in some wild animal species that the majority of human beings wish to keep essentially in the present state.

The biorisks generated by transgenesis in animals result from different and independent parameters: the animal species, the method used to transfer the foreign gene, the nature of the foreign gene and the fate of the transgenic animals.

The Biorisks Due to Animal Species

A transgenic animal has *a priori* some unknown biological properties. It must therefore not escape, live and multiply in the environment or transmit its transgene to the corresponding wild animals until the reasoned decision to do so has been taken. Appropriate containments in facilities must be adapted to each animal species.

Containments have no reason to be as strict for animals such as cow or sheep which have little chance to escape and no chance to cross with wild animals, as for mouse, rat, rabbit, fly, salmon, etc... For laboratory rodents, simple obstacles to their free moving (cages, step at the door of facilities) are sufficient. For birds, appropriate grids are necessary to prevent their flying away.

For aquatic animals, several devices can be used: filters or electric fields at the exit of tanks, grids over tanks to prevent birds from capturing the animals and transporting them away to wild ponds, to rivers or to sea, decontamination of water in special tanks containing sterilizing chemical agents before releasing it to the sewer. For insects, appropriate filters, cooled or U.V. irradiated areas with automatic devices to detect any flying object may separate cages from the exterior of the facilities. This list is not complete and specific containments have to be imagined case by case according to the species.

Facilities harbouring transgenic animals must be accessible only to persons involved in the experimentation. Thieves and ecoterrorists must be kept away using appropriate locks.

In practice, dissemination of transgenes and not only of transgenic animals must be controlled. Dissemination of transgenes through normal reproduction must absolutely not take place. Transgenic animals must therefore be bred in areas which are sufficiently isolated to prevent uncontrolled mating with non-transgenic animals. Tags indicating GMO (Genetically Modified Organism) must be present on each cage or tank.

The Biorisks Due to the Method Used for Gene Transfer

Foreign genes can be transferred to animals using different physicochemical methods: microinjection, electroporation, lipofection and biolistics. These methods can be applied to germ cells, early embryos and to ES cells. None of these methods have intrinsic biorisks except for biolistics which generates aerosols which may carry the foreign genes.

Retroviral and adenoviral vectors are sometimes used. These vectors are also those used for human gene therapy. Their utilization involves biorisks which are well identified. Special care must be taken to make sure that these vectors cannot propagate without the help of a wild virus of the same family or transcomplementing cells. Transgenic animals obtained with viral vectors must be kept isolated until it has been proven that they do not release recombinant infectious viral particles.

The Biorisks Due to the Nature of Foreign Genes

The major biorisks due to transgenes are for the animals themselves. The expression of a foreign gene or the inactivation of an endogenous gene frequently induces physiological changes in the animal which may make them suffer or even die. Abnormal situations are created on purpose in this way, to tentatively reveal the normal mechanisms which control life of animals.

Some transgenes may have no intrinsic deleterious effect but may contain known or unknown mobile elements and genes. This situation is likely to be encountered when long genomic DNA fragments included in P1 phages, BAC and YAC vectors are used. These putative mobile elements and unknown genes may have deleterious effects and generate some biorisks.

Some transgenes code for toxins. In these cases, animals are very likely to suffer from the toxins. In other cases, toxicity may be more cryptic. Prions are a good example. Their effect is slow and more or less species specific and not observable early after transmission. These animals must of course be manipulated and kept in conditions which prevent the dissemination of the toxins to experimentators or to other animals.

Some transgenes may have *per se* no intrinsic deleterious effect for animals, humans or environment, but they may modify the animals in such a way that they can become very dangerous in specific conditions. Genes coding for receptors of human or animal viruses are a typical case in point. The animals expressing these genes may become a particularly dangerous source of virus. These animals must be isolated in a proper way to avoid uncontrolled infection which may be caused by carriers such as insects. After experimental infections, these animals must be confined in a containment corresponding to the pathogenicity level of the virus.

Transgenic animals harbouring complete or partial but functional viral genomes must be kept in conditions corresponding to the pathogenicity of the virus.

A specific new type of problems is emerging with the transfer of organs form one animal species to another. This is particularly the case for the xenografting of pig organs to humans. In these situations not one or a few genes but a complete animal genome is transferred to other animals or to humans. The possibility to transfer prions or to generate new pathogens compatible with pigs and humans is far from being negligible (Chapman *et al.*, 1995). A specific study of these new biological situations must be done with a special care to evaluate and prevent biorisks which cannot be easily predicted.

The Biorisks Due to the Use of Transgenic Animals

Animals which receive a foreign gene at the embryo stage and which appear to be non-transgenic using PCR or any other sensitive test should not be used for uncontrolled reproduction. They may be highly mosaic and their transgene may have escaped detection. These animals may however be used as recipients for foreign embryos or for any other purpose not involving transmission of their genes. These animals may even be used for human consumption on condition that the microinjected gene does not have any expected deleterious effect. In this case, the decision to use these animals for human consumption must be taken after an evaluation of an *ad hoc* commission. These animals should then be in all cases sold dead to avoid any uncontrolled reproduction and transmission of the transgenes. This problem is of course significant essentially for farm animals which are a costly biological material.

Transgenic animals must be eliminated at the end of experiments. Normal procedures may be used when animals were not at risk. Procedures commonly used to eliminate viremic animals must be used when the transgenic animals harbour a viral genome.

The Biorisks Due to Transportation of Transgenic Animals

Problems are of course generated by the transportation of transgenic animals. These animals must be kept in conditions which do not allow any escape into the environment, even in case of accident. Additional containments determined case by case are of course required for animals harbouring pathogens. Transportation of transgenic animals therefore does not constitute a specific safety problem. The introduction of transgenic animals in new facilities may create unpredictable situations due to the presence of other animals. A quarantine or even a specific isolation of animals may be justified.

Alternatively, non-transgenic pregnant females harbouring transgenic foetuses may be used for transportation. Frozen embryos or sperm may also be used safely.

The Biorisks Due to the Dissemination of Transgenic Animals in Herds or the Environment

Some opponents to biotechnology claim that highly sophisticated breeding techniques including transgenesis may lead to a dangerous reduction of biodiversity. Paradoxically, the same persons are against the use of transgenesis because it creates new animals. Transgenesis intrinsicly increases biodiversity. Herds in the future may suffer from transgenesis if an insufficient number of founders are used to spread a transgene of interest. Geneticists have a long term expertise to control consanguinity in herds. It is therefore highly improbable that they will not be vigilant enough to solve such a well-identified problem.

Dissemination of transgenic animals in the environment has not yet occured. Animals of interest are in fact not available. Carps harbouring a foreign growth hormone gene and showing accelerated growth are presently probably the only candidates for a release in wild waters (Fox, 1992). Obviously, a long observation of transgenic animals will be necessary before deciding to disseminate them in nature. Their specific way of life and the presence or the absence of similar wild animals in the environment will have to be taken into consideration. The experience of transgenic plant release in the environment will be a good model to define conditions of dissemination for transgenic animals.

Interestingly, transgenic plants have been recently classified according to two criteria of biorisks: (i) the probability of the transgenes to be transmitted to wild relatives (ii) the potentially enhanced competitiveness of the transgenic plants towards wild relatives (Ahl Goy and Duesing, 1996). Three groups has been defined. In group I are plants with a minimal probability of gene flow to wild relatives. In group II are plants with a low probability of gene flow to wild relatives. In group III are plants with a high probability of gene flow to wild relatives. Independently, in class I are plants with a minimal advantage or a disadvantage towards wild relatives. In class IIa are plants with a low advantage under selective pressure (e.g. plants harbouring a gene for herbicide tolerance). In class IIa are plants with a low advantage in natural conditions (e.g. plants harbouring a resistance gene against pathogens). In class III are plants with a high advantage in all physiological situations. The evaluation of the potential risks takes into account these two independent and complementary parameters. These mode of classification might be adapted to transgenic animals prepared to be disseminated in the environment.

A Possible Classification of Transgenic Animals According to Biorisk

Standards have to be defined to help scientists to work in the best safety conditions and also to help the bioindustry to know what will be in conformity with government regulations. Such standards are being defined in the European Union (Aldridge, 1995).

A possible classification of biorisks and the corresponding containments may be the following. This classification is presently used by the french commission of biorisk evaluation (Commission de Génie Génétique).

Classification of Animals Harbouring a Foreign Gene (Gene Therapy and Transgenesis)

Class 1

- Animals releasing no mobile foreign DNA (example: mice obtained after microinjection of

non-viral DNA into embryo pronuclei, after ES cell transfection or after injection of naked non-viral DNA into muscle).
- Animals potentially releasing class 1 mobile foreign DNA.
- Animals harbouring a gene having no known harmful effect for humans or the environment.

Class 2
- Animals potentially releasing class 2 mobile foreign DNA (example: mice or chicken infected by viral vectors at the embryo stage or in adults)
- Animals harbouring a foreign gene, mobile or not, but having a potentially harmful effect for humans or the environment (animals harbouring a gene coding for a prion, for a viral receptor etc...)

Class 3
- Animals potentially releasing class 3 mobile foreign DNA.
- Animals harbouring foreign DNA, mobile or not, but having a potentially very harmful effect for humans or the environment.

Class 4
- Animals potentially releasing class 4 mobile foreign DNA.
- Animals harbouring foreign DNA, mobile or not, but having a potentially particularly harmful effect for humans or environment.

Containment for Animals Harbouring Foreign DNA

Facilities Level 1

They are conventional facilities with specific containments preventing dissemination of the animals in the environment. These specific containments shall be adapted to each animal species (examples: specific physical barriers for mammals and birds, filters and electric barriers for aquatic animals, chemical system to inactivate gametes in water for aquatic animals, filters, cooled atmosphere, space with U.V. light and automatic system to detect escape for insects, etc...).

The animals shall be identified individually or housed in identified spaces (cages, tanks, etc...)

The animals shall be eliminated at the end of experiments.

Facilities Level 2

Containments are those described for facilities level 1 with the following additional provisions.

Animals shall be housed in conditions preventing dissemination of the mobile foreign DNA (Filters over cages preventing dissemination of the mobile DNA by dust or aerosols, use of safety cabinets, etc...).

In case the animals harbour foreign DNA, mobile or not, but having a potentially harmful effect for humans or the environment, reinforced specific barriers shall be added

- to prevent any dissemination of experimental animals in the environment (whichever species is concerned)
- to prevent living foreign organisms (insects, parasites...) from carrying pathogens to transgenic animals which have become more sensitive to these pathogens because of the transgenes
- to prevent dissemination of pathogens which may have contaminated transgenic animals because of the transgenes.

Waste shall be inactivated by autoclaving or any other efficient procedure.
Animals shall be inactivated by autoclaving, incineration or any other efficient procedure.

Facilities Level 3

Conditions of animal housing are those of facilities level 2 with the containment described for the manipulation of genetically modified organisms of level 3.

Facilities Level 4

Conditions of animal housing are those of facilities levels 2 with the containment described for the manipulation of genetically modified organisms of level 4.

References

Aldridge, S. (1995) European biotechnology standards-update. *Elsevier Science* **13**, 239–242

Ahl Goy, P. and Duesing, J.H. (1996) Assessing the environmental impact of gene transfer to wild relatives. *Biotechnology* **14**, 39–40

Chapman, L.E., Folks, T.M., Salomon, D.R., Patterson, A.P., Eggerman, T.E. and Noguchi, P.D. (1995) Xenotransplantation and xenogeneic infections. *N. Engl. J. Med.* **433**, 1498–1501

Fox, J.L. (1992) Transplantation fish cast shadow. *Biotechnology* **10**, 1090–1091

PART V

82. Food Safety Evaluation of Transgenic Animals

Margaret Ann Miller and John C. Matheson III

US Food and Drug Administration, Center for Veterinary Medicine, 7500 Standish Place, Rockville, Maryland, 20855, USA

Recombinant DNA (rDNA) and related genetic modification techniques are being applied to species of animals from which food for humans is derived (food-producing animals). The objectives of these modifications are diverse, including production of biologics and pharmaceuticals, optimization of nutritional value of derived food products, improvement in animal growth rate and reproduction, prevention of animal disease, selection of superior strains, identification of individuals, and preservation of genetic stocks. Many of these applications fall outside the purview of the Federal Food Drug and Cosmetic (FFD&C) Act. In other cases, the product claims for the genetic modification and the animals containing them are very similar to those of products currently regulated by the US Food and Drug Administration (FDA). For example, when a growth hormone gene cassette is inserted into the genome of a zygote to improve growth and feed efficiency of the individual and/or its offspring, this gene cassette and the gene expression product should be regulated as a new animal drug like growth hormone itself. Thus, although the agency has not published a formal policy on the regulation of these "transgenic" animals, it is clear that some of these techniques and product claims may require regulation under the FFD&C Act.

Products regulated as new animal drugs in the United States are subject to a pre-market review process to determine efficacy and safety. Product safety includes target animal safety, safety to the environment and safety for consumers of food derived from treated animals. The new animal drug approval process permits experimental animals to enter the food supply after the FDA determines that the food products are safe for human consumption. Although future products may extend outside the jurisdiction of the FFD&C Act, our current regulations can accommodate the current generation of genetically modified animals for which food safety may be of concern.

This chapter describes and compares the approaches to food safety used by the FDA Center for Veterinary Medicine when evaluating the safety of food derived from animals treated with exogenous chemicals that are animal drugs, protein hormones that are animals drugs, and transgenic animals that have been modified in a manner that is equivalent to having been treated with an animal drug.

Introduction

The Food and Drug Administration (FDA) is the primary US federal agency responsible for the regulation of food and food additives, pharmaceuticals, biologics and medical devices for humans, and drugs, feed, food additives and medical devices for animals. FDA's authority as set out in the Federal Food, Drug and Cosmetic (FFD&C, 1993) Act gives the agency authority to determine the safety of food, direct and indirect food additives, and contaminants. The regulatory standard for demonstrating safety is different for these different categories of products, however. The food safety standard for direct and indirect food additives is much higher than for foods and contaminants. A food is considered safe unless it contains a poisonous substance which is injurious to health; while a food additive is considered safe only if there is a "reasonable certainty of no harm".

During the past few years, the FDA has published several policy statements and guidelines discussing the agency's position on the regulation of products produced by biotechnology. For example, in May 1992, FDA published a document in the FEDERAL REGISTER entitled "Statement of Policy: Foods Derived from New Plant Varieties" (FDA, 1992). This document outlines the decision-making process and safety assessment that food producers should follow prior to marketing "transgenic" plants as human food and animal feed. Also, FDA's Center for Biologics Evaluation and Research has published a guidance document "Points to Consider in the Manufacture and Testing of Therapeutic Products for Human Use Derived from Transgenic Animals" (FDA, 1995). This document discusses guidance for the manufacture of pharmaceuticals by "Biopharm" animals as well as recommendation that the FDA determine food safety before these animals enter the human food supply.

Other US federal agencies also regulate genetically modified animals. The regulatory pathway for agricultural biotechnology products is similar to the regulatory pathway for agricultural products in general. The United States Department of Agriculture (USDA), the Environmental Protection Agency (EPA) and FDA are the primary regulatory agencies responsible for products of agricultural biotechnology. In 1986, these agencies published the Coordinated Framework for the Regulation of Biotechnology (Office of Science and Technology Policy, 1986) describing the comprehensive Federal policy for ensuring the safety of biotechnology research and products. EPA regulates all pesticides under the Federal Insecticide, Fungicide and Rodenticide Act (FIFRA) and controls the use of genetically engineered microorganisms (GEMs) under Toxic Substances Control Act (TSCA). EPA is not principally involved in regulating transgenic animals.

USDA has authority to regulate viruses under the Virus, Serum and Toxin Act. Therefore, when viral vectors are used to introduce the new genetic material into animals, the viral component of the gene construct may be under the jurisdiction of USDA's Animal and Plant Health Inspection Service. USDA's Food Safety and Inspection Service (FSIS) is responsible for ensuring the safety, wholesomeness and accurate labeling of all meat, meat food products and poultry products under the Federal Meat Inspection Act and Poultry Products Inspection Act. FSIS has developed a Points to Consider document "Safety Evaluation of Transgenic Animals from Transgenic Animal Research" (FSIS, 1994). Also, there are many unsuccessful attempts to create stable germ line changes because the new genetic material either fails to integrate into the genome or the integrated gene is not expressed. These animals are called "No-Takes" by the FSIS, and these animals can enter the food supply if they are nontransgenic by DNA analysis, if they lack measurable gene product, if they do not have transgene-associated traits, and if they are healthy (FSIS, 1991).

FDA is the primary agency responsible for the regulation of food products intended for human consumption. FDA has responsibility for the safety of milk and dairy products, fish and shellfish and animal drug products. To date (August 1996), FDA has not published a formal policy statement on how the FFD & C Act applies to the regulation of new gene transfer technologies when applied to food-producing animals. In most of the transgenic animal experiments conducted to date, the new genetic material is introduced into the germ line or somatic cells. With the exception of the "Biopharm" animals, in many cases, the expression product and the product claims for the transgene are identical or very similar to products which are currently regulated by the agency as animal drugs. ["Biopharm" animals are defined here as animals that have been modified to manufacture a human or veterinary drug or biologic substance, food additive, pesticide or other product of commercial value that is harvested from the milk, blood or other tissues of the animal. Biopharm animals are, in their manner of use, analogous to the fermentation organisms that are used to manfacture antibiotics and enzymes.]

The FFD & C Act defines drugs based upon their functional claims rather than their chemical structure or manufacturing source. The term "drug" includes "articles intended for use in the diagnosis, cure, mitigation, treatment or prevention of disease in man or other animals; and articles (other than food) intended to affect the structure or any function of the body of man or other animals; . . ." Thus, some transgenic animals will be regulated under the animal drug provisions of the FFD & C Act. When the genetic material is introduced into somatic or germ cells to produce a phenotypic change in the animal or its offspring, it would be considered to be an animal drug. On the other hand, when a genetic procedure is used to map the genome and phenotypic change is achieved through traditional breeding, it would not be considered to be involving the use of an animal drug.

Animal Drug Approval Process

Animal drugs are subject to a pre-market review process. Before any new animal drug is legally marketed for use in food-producing animals, the sponsor must demonstrate to the FDA that the drug is safe and effective in the target animals, safe for humans consuming edible products from treated animals and safe for the environment. Furthermore, the firm must prove that they can reproducibly manufacture the product to specified standards of purity, strength and identity.

The animal drug approval process is primarily designed for drug sponsors seeking agency approval to market drugs in interstate commerce. However, this process can be used to conduct research on food-producing animals so that the experimental animals can enter the food supply. Researchers conducting experiments with animal drugs in food-producing animals can establish an Investigational New Animal Drug Application (INAD) with the FDA Center for Veterinary Medicine. The INAD is confidential. The Center cannot reveal the existence of an INAD or discuss any of the data in the application, without the sponsor's permission.

Under the INAD process, researchers have an opportunity to conduct all the studies required to support the commercial marketing approval. Once the sponsor demonstrates the safety of food products from the treated animals, the experimental animals can enter the food supply. When the sponsor is confident that they want to pursue an approval of their investigational product, they file a New Animal Drug Application (NADA). The agency permits the marketing of new animal drugs in interstate commerce by approving NADAs.

Human Food Safety Requirements for Animal Drugs

While the FDA Center for Food Safety and Applied Nutrition (CFSAN) regulates the vast majority of food for humans, i.e., food commodities consumed directly by humans, the Center for Veterinary Medicine (CVM) ensures the safety of animal drug residues which are regulated as indirect food additives. The safety determination for animal drug residues in food for humans ("human food safety") is not dictated by the drug claims but by the chemical nature and biological activity of the drug entity. The human food safety concern for animal drug residues focuses on the assessment of the effect of chronic low level exposure to drug residues in edible tissues. Generally, the concentration of drug residues in edible tissues are not high enough to produce acute toxicity.

For most animal drugs, the human safety evaluation involves the completion of the standard battery of toxicology test and residue and metabolism studies (FDA, 1994). The standard battery of toxicology tests includes a subchronic (90 day) feeding study in a non-rodent and rodent species, a battery of short-term genetic toxicology tests, a multi-generation reproduction study in rats and a teratology study in rats. These toxicology tests are designed to determine the dose at which the compound produces an adverse effect in test animals and a dose at which the drug produces no adverse effect, i.e., the no observed effect level (NOEL). The observed effect does not always represent a toxic effect. Pharmacological, hormonal, and microbiological effects are all considered adverse biological effects in the human food safety evaluation. For certain drug classes, additional specialized tests are required to establish a NOEL for these physiological effects.

In addition to the general food safety provisions of the FFD & C Act, the Delaney Clause prohibits the use of carcinogenic compounds as animal drugs when they leave residues in food. Therefore, if the animal drug or its metabolities are structurally related to a demonstrated carcinogen or if the compound tests positive in the gene toxicology studies, additional carcinogenicity studies may be required.

Following completion of all the toxicology studies, the NOEL of the most sensitive effect from the most appropriate toxicology study is divided by a safety factor to determine an acceptable daily intake (ADI). The ADI represents the highest amount of total residue of the compound that is allowed in edible tissues of the target animal. This total residue consists of parent compound, free metabolites, and metabolites that are covalently bound to endogenous molecules, and is very difficult to measure on a routine basis. Rather than measuring the total residues in all edible tissues, CVM establishes one value, a tolerance, in one tissue, the target tissue for monitoring drug residues. The tolerance is established so that when residues are below the tolerance in the target tissue, the whole carcass is safe. To establish the tolerance for the drug, the drug sponsor conducts a residue depletion study to determine when the concentration of total drug residues is below the ADI.

This food safety assessment process works very well for drugs that have toxicology endpoints which display a clearly observable dose-response relationship. This food safety assessment is not appropriate for determining the food safety of all animal drugs and is certainly not applicable to transgenic animals.

The FFD & C Act also requires a determination of the safety of compounds formed in food as a result of drug treatment. For "traditional" animal drug products this provision serves as a basis for an examination of metabolites. For the recombinant protein products, CVM has interpreted this provision to require examination of the food safety implications of secondary metabolic changes which occur as a result of drug treatment. For genetically modified animals this provision could require an assessment of intended or unintended pleiotrophic changes.

Recombinant Animal Drug Products

To date, the biotechnology products approved by CVM have been proteins produced by recombinant DNA technology using a bacterial fermentation system. The desired gene is isolated and fused with plasmid DNA. The recombinant plasmid is cloned or inserted into a gram negative bacterial host, usually *Escherichia coli*. Under fermentation conditions, these transformed microorganisms become factories which produce large quantities of the protein hormone product at relatively low cost. The protein product is isolated and purified from the bacteria. When treated under defined conditions, the product assumes a conformation which is biologically active. In many respects the production of recombinant protein hormones is not substantially different from the production of other new animal drugs made by fermentation processes.

The food safety evaluation for recombinant protein hormone products is similar to that performed for other protein products approved as animal drugs. On the other hand, the toxicology studies conducted to demonstrate the human food safety of protein products are considerably different from those just described for more traditional animal drug products. With the consent of the drug's sponsors, CVM has published a comprehensive review of the food safety evaluation for bovine somatotropin describing the studies required for this recombinant protein hormone product (Guyer and Juskevich, 1990).

The initial food safety evaluation for bovine somatotropin determined that the recombinant molecules was fundamentally equivalent to naturally occuring pituitary-derived somatotropin. Natural bovine somatotropins have been extensively characterized and these residues pose no toxicological risk for humans bacause: (1) bovine somatotropin is a protein hormone which is readily degraded in the gastrointestinal tract, and (2) bovine somatotropins are largely species specific and biologically inactive in humans even when injected in large quantities. Although the recombinant bovine somatotropins were structurally very similar to the endogenous compounds, sponsors were asked to perform studies demonstrating the lack of oral bioactivity in rats.

For protein hormones that are biologically active in humans and for recombinant proteins with altered amino acid sequences, the human food safety assessment compares the oral bioavailability of the protein to the concentration of the protein residues in tissues. For compounds with extremely low oral bioavailability , it is possible for humans to consume the entire dose administered to the food animal with no food safety concerns. In other cases, the sponsor must quantify the protein hormone concentration in edible tissue to assure an adequate margin of safety between the oral dose where no bioavailability is demonstrated and the level expected to be consumed by humans as residues.

Food Safety Determination of Transgenic Animals

The diversity of transgenic animal technologies that might result in food products being offered for human consumption demand a flexible scientific approach to meeting the food safety statutory standard. A brief survey of the types of transgenic animals being produced supports this contention.

Transgenic Animals Containing an Animal Drug. The statutory food safety requirements for animal drug residues in genetically modified animals are the same as those for other animal drugs. Basically, the food products produced from genetically modified animals must be as safe as those from nontransgenic animals; and the sponsor of the transgenic animals must demonstrate safety of the animal products before the animals can enter the food supply. The standard battery of toxicology studies used to establish the safety of "traditional" animal drugs are not appropriate for assessing the safety of a transgene in a genetically modified animals. Also, the "traditional" withdrawal period may not apply to transgenic animals. Although it may be possible to "turn-off" the expression of the transgene, and thereby limit exposure to the expression product; it will not eliminate the transgene from the animal. In cases where there are food safety concerns for the expression products and not the transgene, a tolerance approach could apply.

CVM does not plan to offer a standard set of guidelines on how the food safety determination for transgenic animals must be conducted. Sponsors seeking approval for transgenic animals with animal drug claims are, instead, encouraged to contact the CVM early in the development process so that a case-specific human food safety program can be designed. While a case-by-case approach may be frustrating for those seeking certainty and international harmony in food safety decisions for food products derived from transgenic animals, the diversity of research being conducted with transgenic food animal species needs to be recognized.

Many different transgenic animal products could be marketed for food, yet few examples have been presented for evaluation. Transgenic animals modified specifically to improve the economics of producing them for food or to improve their desirability as food are expected to ultimately be the common situation. Gene therapy may be used to combat disease in specific tissues, such as treatment of mastitis, or to create secondary sites for the release of growth hormones or other production enhancement gene products. Stable germ lines containing genetic modifications that relate to productivity or disease resistance are also likely, particularly in fish (Du *et al.*, 1992; Rahman and Maclean, 1992 and Xie *et al.*, 1993).

Generally, the human food safety assessment for transgenic animals will be more similar to that performed on recombinant protein products than that performed on the "traditional" drugs. The product's sponsor will need to demonstrate the safety of: (1) the transgene, including the promoter and other unexpressed regions; (2) the expression products and (3) in some cases, pleiotrophic effects, in edible animal products. Information on the biology of the genetic modification from the scientific literature, data on the biochemical characterization of the transgene and the expression products, information on the mode of action, data on the quantity of transgenes and expression products and studies investigating oral bioavailability of the expressed protein will be useful in performing the food safety assessment. Basically, the safety of the animal cannot be determined without a thorough understanding of the processes used to produce the animal.

Biopharm Animals. Biopharm animals have been genetically modified to manufacture a human or veterinary drug or biologic substance, a food additive or other product of commercial value. The substance is then harvested from milk, blood or other tissue of the animal. The genetic modification can be either a germ line, heritable modification (Ebert *et al.*, 1991; Krimpenfort *et al.*, 1991 and Swanson *et al.*, 1992) or a somatic cell or gene therapy involving the introduction of the modified genes into cells of a particular tissue of an individual (Wolff, 1991). The main emphasis of these efforts is on harnessing the metabolic capabilities of the animal to produce a product in lieu of using, for example, chemical synthesis, fermentation, or extraction from a dilute natural source.

Biopharm animals are usually of the same species that humans use for food, so it can be anticipated that sometimes it will be desired to salvage them for food, thereby avoiding other more costly means of carcass disposition. Safety evaluation of food derived from biopharm animals would include, in addition to the factors discussed above, an evaluation of effect of the management of the animals on their residue profile. Animal management would be examined for the potential for unsafe residues of drugs and other chemicals that were used during the utilization of the animal as a protein factory.

Not surprisingly, biopharm animals are sometimes managed with research compounds, unapproved and approved drugs and biologics in order to facilitate production of biopharm products. Because of later purification steps, these research compounds do not appear as contaminants in the biopharm product, however the residues may remain in portions of the animal that might be offered for food. If the safety of these residues in animal tissues cannot be demonstrated, the animals cannot be offered for food. It may be a challenge for biopharm animal managers to manage both for the prod-uction of the biopharm product and for the animal, its milk or eggs, to be suitable to be offered as food. Regulation of transgene expression and the tolerance-withdrawal time approach might well apply in reconditioning biopharm animals to be suitable for food for humans.

Biopharm animals may be the first transgenic animals to be offered as food for humans.

Transgenic Animals Containing Food Additives and Color Additives. A third grouping of transgenic animals are those modified in a manner to affect their quality as food for humans. Examples, might include cattle producing more nutritionally complete milk, fish that produce more omega-3 fatty acids, and farm raised trout whose flesh is pinker. It is anticipated that CFSAN, rather than CVM, will evaluate these types of modifications under the food additive, color additive or Generally Recognized As Safe (GRAS) provisions of the FFD&C Act. Although approaches to food safety evaluation are similar under both areas of the FFD&C Act, a discussion of this grouping of transgenic animals is outside the scope of this chapter.

Summary and Conclusions

Many of the product claims being anticipated for transgenic animals, for example, improved growth, improved feed efficacy, improved carcass characteristics, and improved disease resistance, are the same as animal drug claims. Any regulation of transgenic animals under the FFD&C Act will require a demonstration of human food safety before the animals enter the human food supply. The food safety evaluation under the animal drug provisions of the FFD&C Act is science-based and its inherent flexibility can accommodate the additional products and animals carrying or sold with animal drug claims. If the human food safety determination can be made on purely scientific grounds, it should not be a major barrier to approval. Other issues, like product labeling and international acceptance of transgenic animal food products in commerce, may be more difficult to resolve. The ultimate success of the transgenic animals in agriculture will depend upon consumer acceptance of the product. FDA review and approval should improve consumer confidence in the safety of these products.

References

Du, S.J., et al. (1992) Growth enhancement in transgenic Atlantic salmon by the use of an "all fish" chimeric growth hormone gene construct. *Bio/technology* **10**, 176–181

Ebert, K.E., et al. (1991) Transgenic production of a varient of human tissue-type plasminogen activator in goat milk: Generation of transgenic goats and analysis of expression. *Bio/Technology* **9**, 835–838

FFD & C Act. (1993) *Federal Food, Drug and Cosmetic Act as amended through February 1993*. ISBN 0-16-041900-X U.S. Government Printing Office, Superintendent of Documents, Mail Stop SSOP, Washington, DC 20402–9328, USA

Food and Drug Administration (1995) *Points to Consider in the Manufacture and Testing of Therapeutic Products for Human Use Derived from Transgenic Animals*. Docket Number 95D-0131. Congressional and Consumer Affairs Branch (HFM-12), Center for Biologics Evaluation and Research, 1401 Rockville Pike, Rockville, MD 20852-1448, USA

Food and Drug Administration (1994) *General Principles for evaluating the Safety of Compounds Used in Food-Producing Animals* Center for Veterinary Medicine. 7500 Standish Place, Rockville, MD, 20855, USA July

Food and Drug Administration (1992) Statement of Policy: Foods derived from new plant varieties. *Federal Register* **57**, 22894, May 29

Food Safety Inspection Service (1994) *Points to Consider for the Evaluation of Transgenic Animals from Transgenic Animal Research* Docket Number 93-019N. Technology Transfer and Coordination Staff, Food Safety and Inspection Service, US Department of Agriculture, Washington, DC 20250, USA (Availability announced in the *Federal Register* **59**, 12582, March 17)

Food Safety Inspection Service (1991) Livestock and poultry connected with biotechnology research. *Federal Register* **56**, 67054–67055, December 27

Guyer, C.G. and Juskevich, J.C. (1990) Bovine Growth Hormone: Human food safety evaluation. *Science* **249**, 875–884

Krimpenfort, P., et al. (1991) Generation of transgenic dairy cattle using 'in vitro' embryo production. *Bio/Technology* **9**, 844–847

Office of Science and Technology Policy. (1986) Coordinated framework for regulation of biotechnology. *Federal Register* **51**, 23302–23350

Rahman, M.A, and Maclean, N. (1992) Production of transgenic tilapia (*Oreochromis niloticus*) by one-cell-stage microinjection. *Aquaculture* **105**, 219–232

Swanson, M.E., et al. (1992) Production of functional human hemoglobin in transgenic swine. *Bio/Technology* **10**, 557–559

Wolff, J.A., et al. (1991) Conditions affecting direct gene transfer into rodent muscle *in vivo*. *BioTechniques* **11**, 474–485

Xie, Y., et al. (1993) Gene transfer via electroporation in fish. *Aquaculture* **111**, 207–213

PART V

83. From a Moral Point of View: Ethical Problems of Animal Transgenesis

Egbert Schroten
Center for Bioethics and Health Law, Utrecht University

Introduction

Homo sapiens sapiens is an *animal morale*. By implication, any human activity can be seen from a moral point of view. To put it in a simplistic way: we can always ask whether human actions are (morally) good or bad. To put it more technically: we can evaluate human actions in the light of principles, norms, values and virtues. That is what ethics is about. It is systematical or methodical reflection on human actions in the light of principles, norms, values and virtues which are present in our society.

Of course, in this contribution on ethical problems of animal transgenesis, it is not my intention to dive into ethical theory in general. Nevertheless, I want to make some preliminary remarks in order to show that, from a moral point of view, we are already confronted with (moral) problems, even before we concentrate, in applied ethics, on specific issues like animal transgenesis.

The Predicament of Ethics

The first problem is, so to say, ethics itself. Ethics is not (an empirical) science but a philosophical discipline. Let there be no misunderstanding: ethics, especially applied ethics, needs science in the sense that it needs reliable factual information, in order to be relevant in practice. But that is only one part of the story. The other part is philosophical – *in casu* normative – argumentation. And the point I want to make here is that there are various approaches in normative argumentation which may result in different moral positions. To give an example: a principle oriented or deontological approach, in which the sanctity of life principle is central, will probably result in an absolute rejection of euthanasia, whereas a strict utilitarian approach, in which pleasure and pain are balanced against each other, will not.

But there is not merely a difference in approaches in ethics. We also have to face the problem of pluralism in society. On a global scale we see different cultures with different value systems (e.g., concerning human rights), norms and virtues. In Europe, there is often talk of a cultural difference between north and south. But, as we all know, even in our own societies people have different priorities as to values and norms.

As to this phenomenon of pluralism, we could look for private, historical and cultural reasons, in the context of post-modern society. But it may, at least partly and superficially, also be explained by the existence, in our society, of a plurality of interests. There is an interplay of a wide variety of individual, industrial, scientific, social, political and many other interests, which are all value laden. For instance, values like health, profit, economic growth, knowledge, power and fame are presupposed in this clash of interests. On a personal level we may 'solve' this problem of pluralism (partly) by an attitude of tolerance or by choosing the safety of group morality, in public policy, however, we have to find to find compromises. But, from a moral point of view, compromises compromise...!

This is, in short, the predicament of ethics: pluralism in moral reasoning, pluralism in society. Nevertheless we go on doing ethics, for we still want to know whether specific human actions are (morally) good or bad. Does that make sense? I think it does: Even if ethics does not give 'objective', everlasting – let alone: divine – answers to moral questions, it can help people to understand these questions and to take their own (moral) responsibilities, by clarifying, analysing, arguing, discussing, and suggesting solutions from a moral point of view.

Animal Transgenesis

Now, over to animal transgenesis. It is not necessary to elaborate on the technical aspects of this form of biotechnology, as it is done elsewhere in this book. I shall not discuss definitions either (see for instance Galloux 1995, 3ff). In this contribution I shall take 'animal transgenesis' as changing the genetic make up of animals by introducing genes (or gene-constructs) or DNA from other organisms (including animals). In other words, it is a specific form of germ line genetic modification. It goes without saying, however, that much of what is written below will be relevant for other forms of animal biotechnology as well.

It is nearly a truism, nowadays, to state the biotechnology in general (cf. EC White paper 1993: chapter 5B) and genetic modification of animals in particular may be seen as a very promising and crucial technology, with far reaching consequences for society. It is of paramount importance for basic research in the biosciences (e.g., knock out mice for research on the functioning of genes), for biomedical research (e.g., research on cancer, xenografting and disease resistance), for the production of therapeutic agents (e.g., blood clotting factors) and for quantity and quality improvement of food producing animals. This could lead us to a *prima facie* conclusion that, from a moral point of view, animal transgenesis is good and should, therefore, be promoted, because it serves the development of science, the improvement of health care and of food supply in the world, to mention only these attractive consequences for society.

Public Concern

Nevertheless, it seems to be a truism too, that, in spite of this optimistic outlook, there is much public concern about genetic modification (of animals). In Eurobarometer 39.1 (1993), a survey on what Europeans think about biotechnology and genetic engineering, one of the key results is that – granted considerable differences among the member states of the EU – 48% of the interviewees believe that biotechnology will improve our way of life in the next 20 years whereas 15% thinks the opposite. The term 'genetic engineering', however, has a more negative connotation than the term 'biotechnology' in general and the "level of 'optimism' regarding genetic engineering has lessened considerably since the last survey" (in 1991). Public attitude towards research on farm animals is not supportive at all. Moreover, there is a 'massive' demand for governmental control and for 'clear ethical rules' guiding biotechnological research.

Public concern about animal biotechnology is a complex phenomenon (Report, 1990; and Hamstra and Feenstra, 1994). Paul Thompson (Davis, 1994, 38) rightly speaks of a hybridization of concerns that have been traditionally considered distinct: concerns about social and economic consequences as well as about consequences for environment, human life and health care become intermingled with concerns about animals and with religious and philosophical ideas about the human animal relationship and the place of man in nature.

In trying to clarify what are the ethical problems of animal transgenesis it is helpful to analyse public concern. Anxieties are important for ethics because they are, at least at first sight, 'markers' of values at risk, and thus of potential moral problems. One might reply that there is no moral problem here. Public concern should, rather, be explained by the fact that people lack information. This is certainly right. There is an information gap between science, technology and industry on one side, and the public on the other. Eurobarometer, too, points out that public attitude depends on what is known of the subject. This may result in too much optimism on the one hand and uneasy feelings on the other. Moreover, when this knowledge gap is filled with science fiction stories concerning the makeability of life, fear of risks and unexpected side effects, it will have a 'Jurassic Park effect', as I have labelled it elsewhere (Davis, 1994, 41).

However, how important this explanation of public concern may be, it would be a simplification to leave it at that. Analysis of public debate from a moral point of view shows that much more is going on. In an attempt to demonstrate this, I shall make use of a Dutch inventory and analysis of moral arguments in public debate on animal biotechnology (Brom 1995). But let it be clear: This is not a Dutch idiosyncrasy. Most of these arguments are present in international publications as well (e.g., Hastings Center Report, 1994; Davis, 1994; The Hague, 1994 and Galloux, 1995).

Arguments in Public Debate

Let me start with the argument that genetic modification may threaten or harm animal health and welfare. As to animal health it is possible to distinguish between two perspectives (Brom, 1995, with reference to Rollin and Lorz): (1) The owner's perspective, in which health is seen as 'aptitude': an animal is healthy when it is, physiologically, in a good condition to fulfil the function for which it is held. (2) The animal's perspective, in which health is not seen in terms of 'aptitude' even not merely in terms of absence of illness and disorder, but in terms of species specific behavior as well.

Public concern has to do with this second perspective, in which the concept of health comes close to the concept of welfare. That is why I take them together here. People are anxious and sometimes outraged about the degree of animal suffering in terms of disorder, pain, stress and lack of harmony with the environment. In the context of transgenesis, the 'oncomouse' is often used as an example. Rumours about 'knock out' mice and of manipulation of species specific behavior of farm animals in view of bioindustry add fresh fuel to this concern.

There is much discussion about the definition and measurement of animal welfare. A vital point in this discussion is the so called analogy postulate: On the basis of an analogy in anatomy, physiology and behavior between humans and animals it is reasonable to suppose that there is an analogy in experience of pain and stress as well. I am in support of Brom's suggestion (1995, 82) to constrain the meaning of animal welfare to actual 'wellbeing', in order to make a clear distinction between welfare and concepts like integrity and naturalness (see below).

An argument in connection with this moral concern about animal health and welfare is based on the anxiousness about animal exploitation. This – like the former – is not a new argument. It goes back to intensive breeding and farming, where animals are used as (production) machines, as mere instruments for human (economic) interests (e.g., Visser/Grommers, 1988). Genetic modification is seen as another step into this 'instrumentalization' or *Verdinglichung* of animals. They are manipulated in order to make them suitable for human interests, without very much taking into account their own interests and integrity. Animals are reduced to their functions, which we are trying to maximalize. Patenting of genetically modified organisms may be seen as proof for this tendency to treat animals as things. The rejection, in the beginning of 1995, by the European Parliament, of a draft directive on the legal protection of biotechnological inventions, can be seen as a sign of public concern – even reluctancy – about (certain aspects of) animal biotechnology.

In the previous paragraphs, the word 'integrity' was used. This leads us to another argument why animal transgenesis is taken to be morally problematic: it may harm animal integrity. The problem here is that 'integrity' is not a strictly descriptive word. A certain image of the animal is presupposed. Apart from health and welfare there is 'something more' in the animal that deserves our respect. This 'something more' may be phrased as wholeness, unaffectedness, normality. We should not, for instance, cut a dog's tail or a chicken's bill. In the context of animal transgenesis, the argument runs that bringing foreign genes or DNA into the genetic make up of animals, for instance a human gene construct coding for lactoferrine in a cow, would infringe animal integrity.

It could be replied that integrity, in a biological context, raises question marks. It looks like a static concept, whereas biologists deal with organisms, i.e., with dynamic development and processes. Moreover, genetic changes are 'normal' in nature. But, here, the answer may be that what happens in nature is not just a moral reason for man to do the same. If so, that would have absurd consequences! It is possible, then, to refine the argument by saying that it is not allowed for man to harm or to change intentionally or artificially the genetic make up of animals. In other words respect for animal integrity has to do with our responsibility in the way we treat animals.

However, now a difficult question is looming: granted that there is something like animal integrity, why should we respect it? This may look like a fundamental question, but, before dealing with it, I want to treat two arguments against animal transgenesis which are related to the one on animal integrity, namely that it is seen as an infringement of the intrinsic value of animals and of naturalness.

In the context of biotechnology, the term 'intrinsic value of animals' is frequently used, sometimes even as a basic concept in ethical theory or in public policy, for instance in The Netherlands (Brom/Schroten, 1993). It is not meant in the traditional sense, as a value to be pursued for its own sake, but rather in the sense that animals have an inherent worthiness, a worthiness of their own. It is a way to express that animals, as sentient beings, are having, so to say, a 'plus' apart from their instrumental or functional value. As in the case of animal integrity, it has an important implication: animals become morally relevant. They are proper objects of moral concern, which means that it is not allowed to do with them just whatever we would like to. Taken as a term to characterize the moral status of animals, it may be compared – *mutatis mutandis* of course – with a term like 'human dignity' which, too, is a basic concept in ethical theory, for instance in medical ethics.

Often, genetic modification is seen as morally problematic because it is unnatural. This, again, is a rather vague argument for it is not clear what is meant by naturalness. Does it presuppose a romantic view of a 'virgin' state of nature, untouched by human beings? Does it presuppose a (created) natural order, which forms the basis of a – highly contested – natural law ethics? Is it not possible to reply that biotechnology cannot be unnatural because it makes use of natural processes? However, how relevant these (and other) questions may be, we should not neglect this argument altogether. It may be interpreted, as Brom (1995, 111) does, as the expression of a general feeling of uneasiness about technocratic and bureaucratic tendencies in our culture, as an intuitive concern caused by an increasing (bio-)technological domination of our natural environment. This may end up in questions like: what kind of 'natural order' do we bequeath to our children? What kind of life do we want? What kind of society do we want? It is the everlasting tension between *ars* and *natura*, which always crops up in connection with a breakthrough in technology. But that does not mean that we should not take it seriously.

But here we could repeat the question: granted that nobody likes technocracy, granted that we should be careful and that we should care for animals, why should animal transgenesis not be permitted. What is so special about it? Animal transgenesis is, indeed, manipulation of nature, but as such it is simply a continuation of what mankind did from prehistoric times onwards. Domestication, selective breeding, transgenesis are virtually the same thing: regulation of nature for man's interest.

This way of arguing, however, begs the question (Schroten, 1992a, 164f). It cannot be denied that man has always tried to control nature, but the question is where the differences between past and present become apparent. And than, looking at the technology of animal transgenesis, some even speak of a 'break in trend' (e.g. Report, 1990) or more in general, of 'new biology' (Polkinghorne, 1986). Whether one would support this assessment or not, there are differences in (1) the nature of controlling, which may be characterized by the term 'externalization' because a vital part of the technology is done in the laboratory; (2) the level of controlling (from organisms to cells, DNA and genes); (3) the range (crossing the boundaries of species, even kingdoms); and (4) the (potential) efficiency of the technology.

Against this background one can hear the argument, in public debate, that genetic modification, especially germ line modification, is not allowed because, in using this technology, man is 'playing God'. This term does not necessarily have a religious meaning – although this is very well possible – but it may be taken as a moral metaphor. It can be interpreted as an attempt to articulate the opinion that, in interfering with the genetic make up of organisms (and thus with the fundamentals of life), science and technology (i.e., man) are transgressing boundaries which they are not allowed to transgress. It is the expression of a deep concern that man has forgotten human finiteness in acting as if (s)he were the almighty God moulding life to his/her will, as if (s)he had a 'providential' knowledge of all consequences and implications of what (s)he is doing, *quod non*! Genetic modification, especially germ line modification, is taken to be a sign of *hubris*, of pride and arrogance, which is dangerous for society and for environment.

Elsewhere (Schroten, 1992b) I have been trying to show that this philosophical argument may be interpreted in two ways: in a practical and in a more principled way. As to the last possibility I argued that, in the end, I do not agree, taking my arguments from (christian) theology. 'Playing God' may be seen as a bold metaphor for man's stewardship in creation. In other words, both man's leading position and responsibility are underlined. But that does not mean that I do not see the worries which are present in the use of this metaphor in the context of biotechnology. In practice, it can rightly be seen as a warning to be cautious in using this 'new' technology, keeping in mind the human predicament, including finiteness of man.

Meanwhile, this position could be taken as a form of anthropocentrism. Indeed it is. From a christian theological point of view – but the same approach can be met in Jewish, Muslim and Humanist thought – man is seen as the 'crown' of Creation, with a vocation of stewardship (which, it should be repeated, implies responsibility!). One could safely say that anthropocentrism is predominant in Western culture. There are, however, not only in Hinduism for instance, but in Western tradition and in modern thinking as well, other approaches, more or less rejecting human superiority over nature, specifically over animals. We could think of some presocratic philosophers (e.g., Pythagoras *cum suis*), of Jeremy Bentham, Albert Schweitzer and of 'Deep Ecology' in our days (Galloux, 1995). We could even say that, nowadays, we are witnessing a more general shift from anthropocentrism to a more 'eco-oriented' way of thinking (Davis, 1994, 41). A movement like Greenpeace, and the support it has, may be taken as a clear example thereof. There is no reason to neglect this ecological perspective. On the contrary, it is a valuable criticism of technocracy in Western society.

Public Policy

Looking back to the moral arguments about animal transgenesis in public debate, it may be clear that they often partly overlap each other. Moreover, behind them one can easily find other (more fundamental) questions, for instance: what kind of society do we want? What is the place of man in nature (Creation)? How should we look on the human-animal relationship? What is the meaning of science and technology, and what should be their place in our society? What about public policy? What direction should it take and who is to take the final decisions?

In this contribution it is not possible to elaborate on all these questions – let alone to solve them! What can be concluded, however, is that animal transgenesis, in spite of scientific, economic and social advantages, involves ethical problems and other, more fundamental, questions behind them. This – from a moral point of view – problematic situation, increased by plurality in society, creates considerable difficulties for public policy making, since it presupposes a certain degree of social support. But, although it will be very difficult to find a broad consensus about genetic modification of animals in our pluralistic societies, it might be worthwhile to look for a 'narrow' consensus, which could serve as a basis for public policy. I want to conclude my contribution by making some suggestions in that direction.

In the first place I would recommend to take the public concern seriously. As I said above, it would be a simplification to reduce it to a (though real!) lack of information. It should be seen as a 'marker' of values too, and thus of moral problems, which should be articulated and dealt with. In a democratic society this is a moral duty. But there is a pragmatic reason as well: policy needs support. As Van Vugt says (Davis, 1994, 5): "We are convinced that animal biotechnology can only become successful when society wants it". It is, therefore, not only important to get people informed, in order to bridge the gap between science/technology and public opinion, but also to promote public debate in order to let people participate in decision making.

My second remark concerns the human-animal relationship. In the light of what has been said about the shift from anthropocentrism to a more 'eco-oriented' way of thinking, I would recommend a moderate anthropocentrism. Most people would agree, I think, that animals will remain necessary for important human interests like food production and health care. As animal transgenesis can be shown to serve these important human interests, it should not be taken as *a priori* ethically unacceptable. However, as our fellow creatures, animals deserve our moral concern in terms of – at least – respect for their health and welfare. That implies a critical (moral) assessment of animal biotechnology in general, and thus of animal transgenesis.

One of the forms this moral concern for animals could take is to look for alternatives for genetic modification (Alternatives, 1993). The question of the availability of alternatives, however is not a simple one. There are, at least, three questions involved. In the first place a factual question: are there alternative technologies available which bring forward (approximately) the same results? For instance, in the case of the bull Herman, lactoferrine produced by fungi. But, even then, there is another, preliminary, question: what do we see as an alternative or a 'real' alternative? An alternative which can be developed (in theory)? Which is being developed? An alternative which is already there? In order to be able to answer these questions it is important that the aim (or output) of a specific biotechnological experiment is clearly defined, which, especially in fundamental research, will not always be the case. And, in the third place, even if there is a real alternative, there is the question of its assessment. Animal transgenesis may be (morally) problematic, but what about the alternative? We could make a comparison on several points like environmental impact, efficacy, costs, purity, social economic aspects. Some of these questions cannot be met by giving just more information. Sooner or later we have to cut a knot. On what arguments? In short, looking for alternatives is important but it is not an easy solution for ethical and political problems of animal transgenesis.

Next, with my 'moderate anthropocentrism' in mind, we should be aware of the 'Janus character' of science and technology. The Roman god Janus had two faces. So science and technology have. They are important for progress and, sometimes, for the improvement of the human predicament. There is, however, a price to be paid. Pollution could be mentioned as a general example. In the context of health care there are the 'victims of medical technology', for instance comatose patients and severely handicapped newborns. In animal biotechnology, for our wellbeing, a considerable price (of suffering) is paid by animals. That is an important reason why moral concern for animals should lead to an ethics and a policy of prudence, which could take the form of a 'no unless policy' (Report, 1990).

As to a 'no unless policy', there should be no misunderstanding. I would suggest to make a distinction between two interpretations: (1) in principle and (2) in practice. From a moral point of view, as I have indicated above, I am not fundamentally against animal biotechnology. In the light of a 'moderate anthropocentrism' it can be justified, under certain conditions, to change the genetic make up of animals. This position might be phrased as a 'conditional yes'. However, in view of public concern, in order to take moral respect for animals seriously and in order to create a basis for public policy, I would recommend a 'no unless policy'. It means, in practice, the formation of (independent) licensing bodies for the assessment of animal transgenesis, perhaps of animal biotechnology in general. The idea behind it is that (a specific proposal for) animal transgenesis is only morally acceptable when an independent group of people, in which public concern is represented, can be convinced by good arguments of its necessity. In this context it may be useful to stress that the logic of the market (economy) or scientific development alone, although they are morally relevant, do not run the show in ethics.

In view of these licensing bodies it is of paramount importance to develop international guidelines for animal transgenesis (biotechnology). Science and technology are not confined within national borders, but they are a global affair. So, when there are no international guidelines, people will be played off against one another. In the light of the (expected) impact of biotechnology it might even be useful to think of an international convention, in which various aspects of genetic modification (even patenting) could be dealt with. This, again will take the form of a 'narrow consensus' but nevertheless, it will serve as a necessary basis for international and national public policy.

Acknowledgement

I wish to thank Dr F.W.A. Brom for his valuable comments on a previous version of this chapter.

References

Alternatives (1993) A discussion paper. The assessment of possible alternatives in the context of the ethical evaluation of gene transfer in animals. Issued by The Provisional Committee on Ethical Evaluation of Genetic Modification of Animals, P.O. Box 8359, NL-3501 RJ, Utrecht

Brom, F.W.A. (1995) *Waarom is biotechnologie bij dieren moreel problematisch? Een inventarisatie en analyse van de argumenten uit het publieke debat* (Why is biotechnology on animals morally problematic? An inventory and analysis of the arguments in public debate). Utrecht University, Center for Bioethics and Health Law

Brom, F.W.A. and Schroten E. (1993), Ethical questions around animal biotechnology. The Dutch approach. *Livestock Production Science* **36**, pp. 99–107

Davis (1994). *Proceedings of the International Workshop on Animal Biotechnology Issues*, University of California, Davis

EC White Paper (1993) European Commission, *Growth, Competitiveness, Employment. White Paper*. Supplement 6/93 of Bulletin of the European Communities

Hastings Center Report 24/1 (1994), Special Supplement, 'The Brave New World of Animal Biotechnology'

Eurobarometer 39.1 (1993) *Biotechnology and Genetic Engeneering: What Europeans Think About It in 1993*. Report written for the European Commission by INRA, October 1993

Galloux, J.C. (1995) *Etude relative aux implications éthiques des animaux transgéniques (Aspects juridiques)*. Study for the Group of Advisors on the Ethical Implications of Biotechnology of the European Commission. Brussels

Hamstra, A.M. and Feenstra, M.H. (1994) *Publiek Debat: Genetische modificatie van dieren, mag dat? Projectverslag en evaluatie* (Public Debate: Genetic modification of animals, is that allowed? Project report and evaluation). SWOKA, The Hague, NL

Peacocke, A. (1986) *God and the New Biology*, London

Report (1990) *Ethics and Biotechnology in Animals*. Report by the Advisory Committee on Ethics and Biotechnology in Animals. Wageningen NL

Schroten, E. (1992a) Embryo production and manipulation: ethical aspects. *Animal Production Science* **28**, pp. 163–169

Schroten, E. (1992b) Playing God. Some Theological Comments on a Metaphor. Van Den Brink, G. *et al.*, eds., *Christian Faith and Philosophical Theology. Essays in Honour of Vincent Brümmer* Kampen, Kok/Pharos, 1992, pp. 186–196

Visser, M.B.H. and Grommers, F.J., eds. (1988) *Dier of Ding. Objectivering van dieren* (Animal or Thing. Animals as objects). Pudoc Wageningen, NL

Index

adenoassociated viral vectors: 197
adenoviral vectors: 189
AIDS: 427
animal production: 15, 28, 113, 473, 495, 507, 511, 519
animal repository: 525
antibody: 397, 495, 501
antisense RNA: 327, 495
atherosclerosis: 273, 445

bacterial artificial chromosome (BAC): 283, 289
biolistic: 101, 153, 209
biosafety: 559
bird: 61, 69, 75, 83, 95, 101,

Caenorhabditis elegans: 137
cancer: 411
chimaera: 65, 76, 169, 176, 182
chromosome: 299
concatemer: 109, 123, 152, 215, 219
cow: 27, 42, 51, 173
Cre/lox: 307
cystic fibrosis: 435

database (Tbase): 533
disease resistance: 15, 483, 490, 495, 501
development: 137, 371, 387
DNA integration efficiency: 12, 20, 29–33, 37, 43 72, 79, 90, 109, 121, 123, 134, 139, 215, 219
dominant negative: 3
double replacement gene targeting: 324
Drosophila: 133, 365

electroporation: 42, 129
embryo culture: 8, 45–48, 51, 61, 69, 120, 138, 345

embryo cryopreservation: 55
embryonic stem cells (ES cells): 9, 157, 167, 173, 179, 307, 377, 378
enhancer trap: 144, 171, 362, 365, 376
episomal DNA: 72, 109, 123, 139, 216, 251
ethics: 569
expression vectors: 109, 303, 307

fish: 105, 119, 129, 387
FLP/FRT: 307
food safety: 563

gene targeting: 307, 323
gene therapy: 189, 442
genetic ablation: 331
goat: 19, 45, 465
growth hormone (GH): 112

homologous recombination: 33, 142, 220, 278, 285, 289, 307, 377, 392

immunoglobulin: 397
imprinting: 241, 245
in vitro fertilization: 47, 52
insertional mutagenesis: 361, 375
insulator: 238, 257
intrabody: 501

knock out: 307, 392, 400, 411, 435, 446

LCR: 259
LIF: 157, 183
liposomes: 42, 65, 95, 291

MAR (SAR): 238, 263

marine invertebrates: 151
marker genes: 3, 71, 81, 91, 121, 140, 143, 153, 317, 337, 345
methylation: 237, 245
microinjection: 8, 12, 16, 20, 27, 29–31, 37–39, 62, 70, 105, 120, 123, 151, 289, 300
milk: 461, 465, 469, 473
model: 9, 11, 16, 273, 463, 411, 421, 427, 435, 445, 449, 455
mosaicism: 3, 12, 16, 72, 109, 123, 140, 145, 153, 167, 233

nuclear transfer: 2, 176

obesity: 449
oncogene: 411
oocyte maturation: 45, 51

P element: 133
P_1 bacteriophage vectors: 273
packaging cell lines: 87
patenting: 553
pharmaceuticals: 15, 461, 465, 469
pig: 15, 455
position effect: 141, 238, 257
primordial germ cells (PGC): 65, 75, 101, 162, 179, 249

rabbit: 11, 461,
rat: 7

retroviral vectors: 83, 205
ribozyme: 327, 495

salmonids: 105
sheep: 23, 45, 469, 507
somatic cells: 189, 197, 207
spermatozoa: 41–43, 95, 130
spheroplast fusion: 96
superovulation: 8, 12, 16, 19, 23

tetracycline: 303
transgene detection: 3, 24, 31, 43, 97, 225
transgene expression: 71, 91, 97, 109, 121, 124, 134, 139, 143, 153, 220, 240, 248, 267, 345
transgene rescue: 267
transgene transmission: 12, 16, 25, 102, 109, 125
transimmortalization: 11, 421
transnomy: 300
transposon: 135, 205

wool: 507

xenopus: 123
xenotransplantation: 455

yeast artificial chromosome (YAC): 279